Thomas M. Smith
University of Virginia

Robert Leo Smith
West Virginia University, Emeritus

PEARSON

Boston Columbus Indianapolis New York San Francisco Upper Saddle River
Amsterdam Cape Town Dubai London Madrid Milan Munich Paris Montréal Toronto
Delhi Mexico City São Paulo Sydney Hong Kong Seoul Singapore Taipei Tokyo

Senior Acquisitions Editor: Star MacKenzie Burruto
Project Manager: Margaret Young
Program Manager: Anna Amato
Editorial Assistant: Maja Sidzinska
Text Permissions Project Manager: William Opaluch
Executive Editorial Manager: Ginnie Simione-Jutson
Program Management Team Lead: Michael Early
Project Management Team Lead: David Zielonka
Production Management and Compositor: Integra
Design Manager: Derek Bacchus
Interior and Cover Designer: Tani Hasegawa
Illustrator: Imagineering
Photo Permissions Management: Lumina Datamatics
Photo Research: Steve Merland, Lumina Datamatics
Photo Lead: Donna Kalal
Manufacturing Buyer: Stacey Weinberger
Executive Marketing Manager: Lauren Harp
Cover Photo Credit: Paul Nicklen/National Geographic Creative

Credits and acknowledgments for materials borrowed from other sources and reproduced, with permission, in this textbook appear on the appropriate page within the text [**or starting on p. C-1**].

Library of Congress Cataloging-in-Publication Data

Smith, T. M. (Thomas Michael), 1955-
Elements of ecology / Thomas M. Smith.—9th ed.
pages cm
Summary: An introductory textbook for college students.
ISBN 978-0-321-93418-5
1. Ecology. 2. Ecology—Textbooks. 3. Ecology—Study and teaching (Higher) I. Title.
QH541.S624 2015
577—dc23

2014021187

ISBN 10: 0-321-93418-0; ISBN 13: 978-0-321-93418-5 (Student edition)
ISBN 10: 0-321-99491-4; ISBN 13: 978-0-321-99491-2 (a la Carte)

www.pearsonhighered.com

14 2019

CONTENTS

PART 5 COMMUNITY ECOLOGY

PART 6 ECOSYSTEM ECOLOGY

CHAPTER 20 Ecosystem Energetics 439

CHAPTER 21 Decomposition and Nutrient Cycling 464

CHAPTER 25 Coastal and Wetland Ecosystems 561

CHAPTER 26 Large-Scale Patterns of Biological Diversity 575

CHAPTER 27 The Ecology of Climate Change 592

PREFACE

The first edition of *Elements of Ecology* appeared in 1976 as a short version of *Ecology and Field Biology*. Since that time, *Elements of Ecology* has evolved into a textbook intended for use in a one-semester introduction to ecology course. Although the primary readership will be students majoring in the life sciences, in writing this text we were guided by our belief that ecology should be part of a liberal education. We believe that students who major in such diverse fields as economics, sociology, engineering, political science, law, history, English, languages, and the like should have some basic understanding of ecology for the simple reason that it has an impact on their lives.

New for the Ninth Edition

For those familiar with this text, you will notice a number of changes in this new edition of *Elements of Ecology*. In addition to dramatic improvements to the illustrations and updating many of the examples and topics to reflect the most recent research and results in the field of ecology, we have made a number of changes in the organization and content of the text. An important objective of the text is to use the concept of adaptation through natural selection as a framework for unifying the study of ecology, linking pattern and process across the hierarchical levels of ecological study: individual organisms, populations, communities, and ecosystems. Many of the changes made in previous editions have focused on this objective, and the changes to this edition continue to work toward this goal.

Treatment of Metapopulations

Beginning with the 7th Edition we included a separate chapter covering the topic of metapopulations (Chapter 12, 8th edition) for the first time. It was our opinion that the study of metapopulations had become a central focus in both landscape and conservation ecology and that it merited a more detailed treatment within the framework of introductory ecology. Although this chapter has consistently received high praise from reviewers, comments have suggested to us that the chapter functions more as a reference for the instructors rather than a chapter that is directly assigned in course readings. The reason for this is that most courses do not have the time to cover metapopulations as a separate subject, but rather incorporate an introduction to metapopulations in the broader context of the discussion of population structure. To address these concerns, in the 9th edition we have deleted the separate chapter on metapopulations and moved the discussion to Chapter 19: Landscape Dynamics.

Expanded Coverage of Landscape Ecology

The incorporation of metapopulation dynamics into Chapter 19 was a part of a larger, overall revision of Landscape Dynamics in the 9th edition. Chapter 19 has been reorganized and now includes a much broader coverage of topics and presentation of current research.

Reorganization of Materials Relating to Human Ecology

In the past three editions, the ecology of human-environment interactions has been presented in Part Eight–Human Ecology. This section of the text has been comprised of three chapters that address three of the leading environmental issues: environmental sustainability and natural resources; declining biodiversity; and climate change. The objective of these chapters was to illustrate how the science of ecology forms the foundation for understanding these important environmental issues. Based on current reviewer comments it appears that although instructors feel that the materials presented in Part Eight are important, most are not able to allocate the time to address these issues as separate topics within the constraints of a single-semester course. The question then becomes one of how to best introduce these topics within the text so that they can be better incorporated into the structure of courses that are currently being taught.

After much thought, in the 9th edition we have addressed issues of human ecology throughout the text, moving most of the topics and the materials covered in Part Eight to the various chapters where the basic ecological concepts that underlying these topics are first introduced. The topics and materials that we covered in Chapter 28 (*Population Growth, Resource Use and Environmental Sustainability*) and Chapter 29 (*Habitat Loss, Biodiversity, and Conservation*) of the 8th edition are now examined in the new feature, **Ecological Issues and Applications**, at the end of each chapter. This new feature covers a wide range of topics such as ocean acidification, plant response to elevated atmospheric carbon dioxide, the development of aquatic "dead zones" in coastal environments, sustainable resource management, genetic engineering, the consequences of habitat loss, and the conservation of threatened and endangered species.

New Coverage of the Ecology of Climate Change

Although topics addressed in Chapters 28 and 29 of the 8th edition are now covered throughout the text in the **Ecological Issues and Applications** sections, the topic of global climate change (Chapter 30, 8th edition) is addressed in a separate chapter – Chapter 27 (The Ecology of Climate Change) in the 9th edition. Given the growing body of ecological research relating to recent and future projected climate change, we feel that it is necessary to cover this critical topic in an organized fashion within the framework of a separate chapter. This new chapter, however, is quite different from the chapter covering this topic in the 8th edition, which examined an array of topics relating to the greenhouse effect, projections of future climate change, and the potential impacts on ecological systems, agriculture, coastal environments and human health. In the 9th edition we have focused on the ecology of climate change, presenting research that examines the response of ecological systems (from individuals to ecosystems) to recent climate change over the past century, and how ecologists are trying to understand the implications of future climate change resulting from human activities.

Updated References and Research Case Studies to Reflect Current Ecological Research

It is essential that any science textbook reflect the current advances in research. On the other hand, it is important that they to provide an historical context by presenting references to the classic studies that developed the basic concepts that form the foundation of their science. In our text we try to set a balance between these two objectives, presenting both the classic research studies that established the foundational concepts of ecology, and presenting the new advances in the field. In the 9th edition we have undertaken a systematic review of the research and references presented in each chapter to make sure that they reflect the recent literature. Those familiar with the 8th edition will notice significant changes in the research case studies presented in each chapter.

Updated *Field Studies*

The *Field Studies* features function to introduce students to actual scientists in the field of ecology, allowing the reader to identify with individuals that are conducting the research that is presented in text. The body of research presented also functions to complement the materials/subjects presented in the main body of the chapter. In the 9th edition we have updated references for the researchers who were profiled in the 8th edition. In addition, two new Field Studies features have been added to Chapter 5 (Adaptation and Natural Selection) and Chapter 8 (Properties of Populations). These two new features profile scientists whose research is in the new and growing fields of ecological genetics.

Redesign of Art Program

For the 9th edition, the entire art program was revised to bring a consistent and updated presentation style throughout the text, with the added benefit of using color to highlight and clarify important concepts.

Structure and Content

The structure and content of the text is guided by our basic belief that: (1) the fundamental unit in the study of ecology is the individual organism, and (2) the concept of adaptation through natural selection provides the framework for unifying the study of ecology at higher levels of organization: populations, communities, and ecosystems. A central theme of the text is the concept of trade-offs—that the set of adaptations (characteristics) that enable an organism to survive, grow, and reproduce under one set of environmental conditions inevitably impose constraints on its ability to function (survive, grow, and reproduce) equally well under different environmental conditions. These environmental conditions include both the physical environment as well as the variety of organisms (both the same and different species) that occupy the same habitat. This basic framework provides a basis for understanding the dynamics of populations at both an evolutionary and demographic scale.

The text begins with an introduction to the science of ecology in Chapter 1 (The Nature of Ecology). The remainder of the text is divided into eight parts. Part One examines the constraints imposed on living organisms by the physical environment, both aquatic and terrestrial. Part Two begins by examining how these constraints imposed by the environment function as agents of change through the process of natural selection, the process through which adaptations evolve. The remainder of Part Two explores specific adaptations of organisms to the physical environment, considering both organisms that derive their energy from the sun (autotrophs) and those that derive their energy from the consumption and break-down of plant and animal tissues (heterotrophs).

Part Three examines the properties of populations, with an emphasis on how characteristics expressed at the level of the individual organisms ultimately determine the collective dynamics of the population. As such, population **dynamics are viewed as a function of life history** characteristics that are a product of evolution by natural selection. Part Four extends our discussion from interactions among individuals of the same species to interactions among populations of different species (interspecific interactions). In these chapters we expand our view of adaptations to the environment from one dominated by the physical environment, to the role of species interactions in the process of natural selection and on the dynamics of populations.

Part Five explores the topic of ecological communities. This discussion draws upon topics covered in Parts Two through Four to examine the factors that influence the distribution and abundance of species across environmental gradients, both spatial and temporal.

Part Six combines the discussions of ecological communities (Part Five) and the physical environment (Part One) to develop the concept of the ecosystem. Here the focus is on the flow of energy and matter through natural systems. Part Seven continues the discussion of communities and ecosystems in the context of biogeography, examining the broad-scale distribution of terrestrial and aquatic ecosystems, as well as regional and global patterns of biological diversity. The book then finishes by examining the critical environmental issue of climate change, both in the recent past, as well as the potential for future climate change as a result of human activities.

Throughout the text, in the new feature, **Ecological Issues & Applications**, we examine the application of the science of ecology to understand current environmental issues related to human activities, addressing important current environmental issues relating to population growth, sustainable resource use, and the declining biological diversity of the planet. The objective of these discussions is to explore the role of the science of ecology in both understanding and addressing these critical environmental issues.

Throughout the text we explore the science of ecology by drawing upon current research, providing examples that enable the reader to develop an understanding of species natural history, the ecology of place (specific ecosystems), and the basic process of science.

Associated Materials

Personalize Learning with MasteringBiology®

www.masteringbiology.com

- **New! MasteringBiology** is an online homework, tutorial, and assessment product that improves results by helping students quickly master concepts. Students benefit from self-paced tutorials that feature immediate wrong-answer feedback and hints that emulate the office-hour experience to help keep students on track. With a wide range of interactive, engaging, and assignable activities, students are encouraged to actively learn and retain tough course concepts. Specific features include:
  - **MasteringBiology assignment options reinforce basic ecology concepts presented in each chapter for students to learn and practice outside of class.**
    - **A wide variety of assignable and automatically-graded Coaching Activities**, including **GraphIt, QuantifyIt**, and **InvestigateIt** activities, allow students to practice and review key concepts and essential skills.
    - **MapMaster™ Interactive map activities** act as a mini-GIS tool, allowing students to layer thematic maps for analyzing patterns and data at regional and global scales. Multiple-choice and short-answer assessment questions are organized around the themes of ecosystems, physical environments, and populations.
    - **Reading Questions** keep students on track and allow them to test their understanding of ecology concepts.

Instructor's Resource DVD for Elements of Ecology

0321977947 / 9780321977946

The Instructor Resource DVD puts all of your lecture resources in one easy-to-reach place:

- High-quality electronic versions of photos and illustrations form the book

- All of the illustrations and photos from the text presentation-ready JPEG files
- Customizable PowerPoint® lecture presentations
- Classroom Response System questions in PowerPoint
- Test Item File in Microsoft Word
- TestGen test generation and management software
- All resources are organized by chapter.

TestGen Test Bank (Download Only) for Elements of Ecology

0321977955 / 9780321977953

TestGen is a computerized test generator that lets instructors view and edit *Test Bank* questions, transfer questions to tests, and print the test in a variety of customized formats. This *Test Bank* includes over 2,000 multiple choice, true/false, and short answer/essay questions. Questions are correlated to the revised U.S. National Geography Standards, the book's Learning Outcomes, and Bloom's Taxonomy to help teachers better map the assessments against both broad and specific teaching and learning objectives. The *Test Bank* is also available in Microsoft Word®, and is importable into Blackboard. www.pearsonhighered.com/irc

Acknowledgments

No textbook is a product of the authors alone. The material this book covers represents the work of hundreds of ecological researchers who have spent lifetimes in the field and the laboratory. Their published experimental results, observations, and conceptual thinking provide the raw material out of which the textbook is fashioned. We particularly acknowledge and thank the thirteen ecologists that are featured in the Field Studies boxes. Their cooperation in providing artwork and photographs is greatly appreciated.

Revision of a textbook depends heavily on the input of users who point out mistakes and opportunities. We took these suggestions seriously and incorporated most of them. We are deeply grateful to the following reviewers for their helpful comments and suggestions on how to improve this edition:

Fernando Agudelo-Silva, *College of Marin*
Brad Basehore, *Harrisburg Area Community College*
James Biardi, *Fairfield University*
Steve Blumenshine, *California State University, Fresno*
Emily Boone, *University of Richmond*
William Brown, *State University of New York at Fredonia*
Brian Butterfield, *Freed-Hardeman University*
Liane Cochran-Stafira, *Saint Xavier University*
Francie Cuffney, *Meredith College*
Elizabeth Davis-Berg, *Columbia College Chicago*
Hazel Delcourt, *College of Coastal Georgia*
Bart Durham, *Lubbock Christian University*
Bob Ford, *Frederick Community College*
Patricia Grove, *College of Mount Saint Vincent*
Rick Hammer, *Hardin-Simmons University*
Tania Jogesh, *University of Illinois Urbana-Champaign*
Claudia Jolls, *East Carolina University*
Douglas Kane, *Defiance College*
Ned Knight, *Linfield College*
John Korstad, *Oral Roberts University*
Kate Lajtha, *Oregon State University*
Maureen Leupold, *Genesee Community College*
William McClain, *Davis & Elkins College*
Beth Pauley, *University of Charleston*
William Pearson, *University of Louisville*
Helene Peters, *Clearwater Christian College*
Carl Pratt, *Immaculata University*
Vanessa Quinn, *Purdue University North Central*
Tara Ramsey, *University of Rochester*
James Refenes, *Concordia University Ann Arbor*
Lee Rogers, *Washington State University, Tri-Cities*
Rachel Schultz, *State University of New York at Plattsburgh*
Cindy Shannon, *Mt. San Antonio College*
Walter Shriner, *Mt. Hood Community College*
Tim Tibbetts, *Monmouth College*
Randall Tracy, *Worcester State University*
Robert Wallace, *Ripon College*
Vicki Watson, *University of Montana*
John Williams, *South Carolina State University*

Reviewers of Previous Editions:

Peter Alpert, *University of Massachusetts*
John Anderson, *College of the Atlantic*
Morgan Barrows, *Saddleback College*
Paul Bartell, *Texas A&M University*
Christopher Beck, *Emory University*
Steve Blumenshine, *California State University, Fresno*
Judith Bramble, *DePaul University*
Nancy Broshot, *Linfield College*
Chris Brown, *Tennessee Tech University*
Evert Brown, *Casper College*
William Brown, *State University of New York at Fredonia*
David Bybee, *Brigham Young University, Hawaii*
Dan Capuano, *Hudson Valley Community College*
Brian Chabot, *Cornell University*
Mitchell Cruzan, *Portland State University*
Robert Curry, *Villanova University*
Richard Deslippe, *Texas Tech University*
Darren Divine, *Community College of Southern Nevada*
Curt Elderkin, *The College of New Jersey*
Mike Farabee, *Estrella Mountain Community College*

Lauchlan Fraser, *University of Akron*
Sandi Gardner, *Triton College*
E.O. Garton, *University of Idaho*
Frank Gilliam, *Marshall University*
Brett Goodwin, *University of North Dakota*
James Gould, *Princeton University*
Mark Grover, *Southern Utah University*
Mark Gustafson, *Texas Lutheran University*
Greg Haenel, *Elon University*
William Hallahan, *Nazareth College*
Douglas Hallett, *Northern Arizona University*
Gregg Hartvigsen, *State University of New York at Geneseo*
Floyd Hayes, *Pacific Union College*
Michael Heithaus, *Florida International University*
Jessica Hellman, *Notre Dame University*
Gerlinde Hoebel, *University of Wisconsin, Milwaukee*
Jason Hoeksema, *University of California at Santa Cruz Sue Hum-Musser*
Western Illinois *University*
John Jaenike, *University of Rochester*
John Jahoda, *Bridgewater State University*
Stephen Johnson, *William Penn University*
Doug Keran, *Central Lakes Community College*
Jacob Kerby, *University of South Dakota*
Jeff Klahn, *University of Iowa*
Jamie Kneitel, *California State University, Sacramento*
Ned Knight, *Linfield College*
Frank Kuserk, *Moravian College*
Kate Lajtha, *Oregon State University*
Vic Landrum, *Washburn University*
James Lewis, *Fordham University*
Richard Lutz, *Rutgers University*
Richard MacMillen, *University of California at Irvine*
Ken Marion, *University of Alabama, Birmingham*
Deborah Marr, *Indiana University at South Bend*
Chris Migliaccio, *Miami Dade Community College*
Don Miles, *Ohio University*
L. Maynard Moe, *California State University, Bakersfield*
Sherri Morris, *Bradley University*
Steve O'Kane, *University of Northern Iowa*
Matthew Parris, *University of Memphis*
David Pindel, *Corning Community College*
James Refenes, *Concordia University Ann Arbor*
Ryan Rehmeir, *Simpson College*
Seith Reice, *University of North Carolina*
Rich Relyea, *University of Pittsburgh*
Carl Rhodes, *College of San Mateo*
Eric Ribbens, *Western Illinois University*
Robin Richardson, *Winona State University*
B.K. Robertson, *Alabama State University*
Thomas Rosburg, *Drake University*
Irene Rossell, *University of North Carolina, Asheville*
Tatiana Roth, *Coppin State College*
Rowan Sage, *University of Toronto*
Nathan Sanders, *University of Tennessee*
Thomas Sarro, *Mount Saint Mary College*
Maynard Schaus, *Virginia Wesleyan College*
Erik Scully, *Towson University*
Wendy Sera, *University of Maryland*
Daniela Shebitz, *Kean University*
Barbara Shoplock, *Florida State University*
Mark Smith, *Chaffey College*
Paul Snelgrove, *Memorial University of Newfoundland*
Amy Sprinkle, *Jefferson Community College Southwest*
Alan Stam, *Capital University*
Christopher Swan, *University of Maryland*
Alessandro Tagliabue, *Stanford University*
Charles Trick, *University of Western Ontario*
Peter Turchin, *University of Connecticut*
Neal Voelz, St. *Cloud State University*
Joe von Fischer, *Colorado State University*
Mitch Wagener, *Western Connecticut State University*
David Webster, *University of North Carolina at Wilmington*
Jake Weltzin, *University of Tennessee*

The publication of a modern textbook requires the work of many editors to handle the specialized tasks of development, photography, graphic design, illustration, copy editing, and production, to name only a few. We'd like to thank the Editorial team for the dedication and support they gave this project throughout the publication process, especially acquisitions editor Star MacKenzie for her editorial guidance. Her ideas and efforts have helped to shape this edition. We'd also like to thank the rest of the team—Anna Amato, Margaret Young, Laura Murray, Jana Pratt, and Maja Sidzinska. We also appreciate the efforts of Angel Chavez at Integra-Chicago, for keeping the book on schedule.

Through it all our families, especially our spouses Nancy and Alice, had to endure the throes of book production. Their love, understanding, and support provide the balanced environment that makes our work possible.

Thomas M. Smith

Robert Leo Smith

CHAPTER 1 The Nature of Ecology

Scientists collect blood samples from a sedated lioness that has been fitted with a GPS tracking collar as part of an ongoing study of the ecology of lions inhabiting the Selous Game Reserve in Tanzania.

CHAPTER GUIDE

THE COLOR PHOTOGRAPH OF EARTHRISE, taken by *Apollo 8* astronaut William A. Anders on December 24, 1968, is a powerful and eloquent image (**Figure 1.1**). One leading environmentalist has rightfully described it as "the most influential environmental photograph ever taken." Inspired by the photograph, economist Kenneth E. Boulding summed up the finite nature of our planet as viewed in the context of the vast expanse of space in his metaphor "spaceship Earth." What had been perceived throughout human history as a limitless frontier had suddenly become a tiny sphere: limited in its resources, crowded by an ever-expanding human population, and threatened by our use of the atmosphere and the oceans as repositories for our consumptive wastes.

A little more than a year later, on April 22, 1970, as many as 20 million Americans participated in environmental rallies, demonstrations, and other activities as part of the first Earth Day. The *New York Times* commented on the astonishing rise in environmental awareness, stating that "Rising concern about the environmental crisis is sweeping the nation's campuses with an intensity that may be on its way to eclipsing student discontent over the war in Vietnam." Now, more than four decades later, the human population has nearly doubled (3.7 billion in 1970; 7.2 billion as of 2014). Ever-growing demand for basic resources such as food and fuel has created a new array of environmental concerns: resource use and environmental sustainability, the declining biological diversity of our planet, and the potential for human activity to significantly change Earth's climate. The environmental movement born in the 1970s continues today, and at its core is the belief in the need to redefine our relationship with nature. To do so requires an understanding of nature, and ecology is the particular field of study that provides that understanding.

1.1 Ecology Is the Study of the Relationship between Organisms and Their Environment

With the growing environmental movement of the late 1960s and early 1970s, ecology—until then familiar only to a relatively small number of academic and applied biologists—was suddenly thrust into the limelight (see this chapter, *Ecological Issues & Applications*). Hailed as a framework for understanding the relationship of humans to their environment, *ecology* became a household word that appeared in newspapers, magazines, and books—although the term was often misused. Even now, people confuse it with terms such as *environment* and *environmentalism*. Ecology is neither. Environmentalism is activism with a stated aim of protecting the natural environment, particularly from the negative impacts of human activities. This activism often takes the form of public education programs, advocacy, legislation, and treaties.

So what is ecology? Ecology is a science. According to one accepted definition, **ecology** *is the scientific study of the relationships between organisms and their environment.* That definition is satisfactory so long as one considers *relationships* and *environment* in their fullest meanings. Environment includes the physical and chemical conditions as well as the biological or living components of an organism's surroundings. Relationships include interactions with the physical world as well as with members of the same and other species.

The term *ecology* comes from the Greek words *oikos,* meaning "the family household," and *logy*, meaning "the study of." It has the same root word as *economics*, meaning "management of the household." In fact, the German zoologist Ernst Haeckel, who originally coined the term *ecology* in 1866, made explicit reference to this link when he wrote:

> By ecology we mean the body of knowledge concerning the economy of nature—the investigation of the total relations of the animal both to its inorganic and to its organic; including above all, its friendly and inimical relations with those animals and plants with which it comes directly or indirectly into contact—in a word, ecology is the study of all those complex interrelationships referred to by Darwin as the conditions of the struggle for existence.

Haeckel's emphasis on the relation of ecology to the new and revolutionary ideas put forth in Charles Darwin's *The Origin of Species* (1859) is important. Darwin's theory of natural selection (which Haeckel called "the struggle for existence") is a cornerstone of the science of ecology. It is a mechanism allowing the study of ecology to go beyond descriptions of natural history and examine the processes that control the distribution and abundance of organisms.

Figure 1.1 Photograph of Earthrise taken by *Apollo 8* astronaut William A. Anders on December 24, 1968.

1.2 Organisms Interact with the Environment in the Context of the Ecosystem

Organisms interact with their environment at many levels. The physical and chemical conditions surrounding an organism—such as ambient temperature, moisture, concentrations of oxygen and carbon dioxide, and light intensity—all influence basic physiological processes crucial to survival and growth. An organism must acquire essential resources from the surrounding environment, and in doing so, must protect itself from becoming food for other organisms. It must recognize friend from foe, differentiating between potential mates and possible predators. All of this effort is an attempt to succeed at the ultimate goal of all living organisms: to pass their genes on to successive generations.

The environment in which each organism carries out this struggle for existence is a place—a physical location in time and space. It can be as large and as stable as an ocean or as small and as transient as a puddle on the soil surface after a spring rain. This environment includes both the physical conditions and the array of organisms that coexist within its confines. This entity is what ecologists refer to as the ecosystem.

Organisms interact with the environment in the context of the **ecosystem**. The *eco–* part of the word relates to the environment. The *–system* part implies that the ecosystem functions as a collection of related parts that function as a unit. The automobile engine is an example of a system: components, such as the ignition and fuel pump, function together within the broader context of the engine. Likewise, the ecosystem consists of interacting components that function as a unit. Broadly, the ecosystem consists of two basic interacting components: the living, or **biotic**, and the nonliving (physical and chemical), or **abiotic**.

Consider a natural ecosystem, such as a forest (**Figure 1.2**). The physical (abiotic) component of the forest consists of the atmosphere, climate, soil, and water. The biotic component includes the many different organisms—plants, animals, and microbes—that inhabit the forest. Relationships are complex in that each organism not only responds to the abiotic environment but also modifies it and, in doing so, becomes part of the broader environment itself. The trees in the canopy of a forest intercept the sunlight and use this energy to fuel the process of photosynthesis. As a result, the trees modify the environment of the plants below them, reducing the sunlight and lowering air temperature. Birds foraging on insects in the litter layer of fallen leaves reduce insect numbers and modify the environment for other organisms that depend on this shared food resource. By reducing the populations of insects they feed on, the birds are also indirectly influencing the interactions among different insect species that inhabit the forest floor. We will explore these complex interactions between the living and the nonliving environment in greater detail in succeeding chapters.

1.3 Ecological Systems Form a Hierarchy

The various kinds of organisms that inhabit our forest make up populations. The term *population* has many uses and meanings in other fields of study. In ecology, a **population** is a group of individuals of the same species that occupy a given area. Populations of plants and animals in an ecosystem do not function independently of one another. Some populations compete with other populations for limited resources, such as food, water, or space. In other cases, one population is the food resource for another. Two populations may mutually benefit each other, each doing better in the presence of the other. All populations of different species living and interacting within an ecosystem are referred to collectively as a **community**.

We can now see that the ecosystem, consisting of the biotic community and the abiotic environment, has many levels (**Figure 1.3**). On one level, individual organisms both respond to and influence the abiotic environment. At the next level, individuals of the same species form populations, such as a population of white oak trees or gray squirrels within a forest. Further, individuals of these populations interact among themselves and with individuals of other species to form a community.

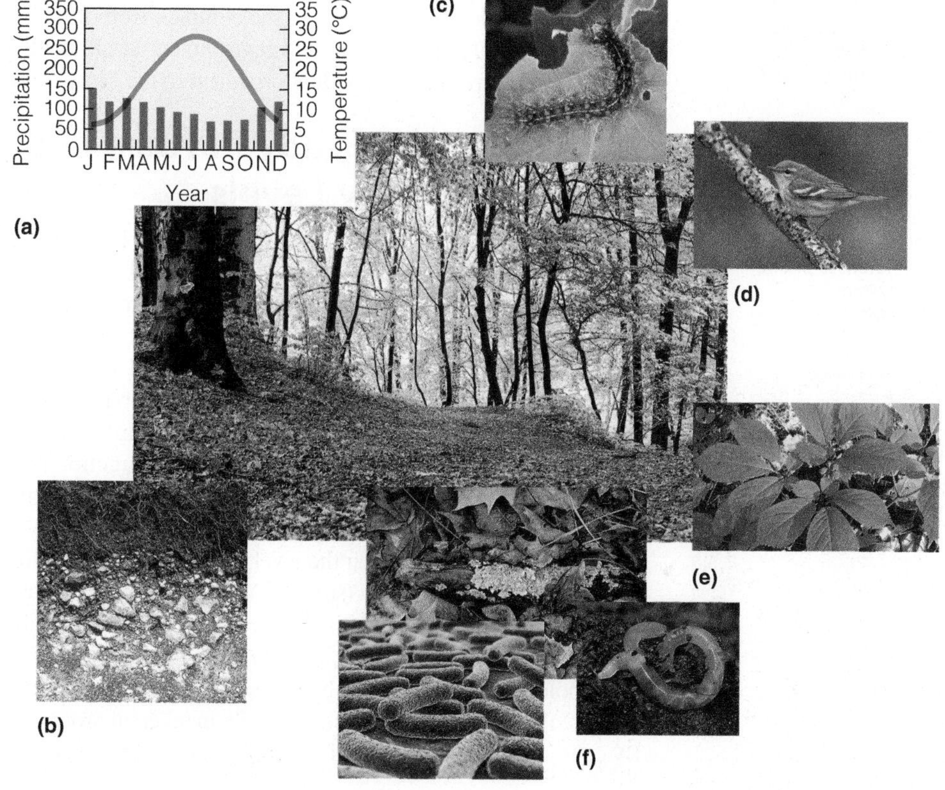

Figure 1.2 Example of the components and interactions that define a forest ecosystem. The abiotic components of the ecosystem, including the (a) climate and (b) soil, directly influence the forest trees. (c) Herbivores feed on the canopy, (d) while predators such as this warbler feed upon insects. (e) The forest canopy intercepts light, modifying its availability for understory plants. (f) A variety of decomposers, both large and small, feed on dead organic matter on the forest floor, and in doing so, release nutrients to the soil that provide for the growth of plants.

Individual

What characteristics allow the *Echinacea* to survive, grow, and reproduce in the environment of the prairie grasslands of central North America?

Population

Is the population of this species increasing, decreasing, or remaining relatively constant from year to year?

Community

How does this species interact with other species of plants and animals in the prairie community?

Ecosystem

How do yearly variations in rainfall influence the productivity of plants in this prairie grassland ecosystem?

Landscape

How do variations in topography and soils across the landscape influence patterns of species composition and diversity in the different prairie communities?

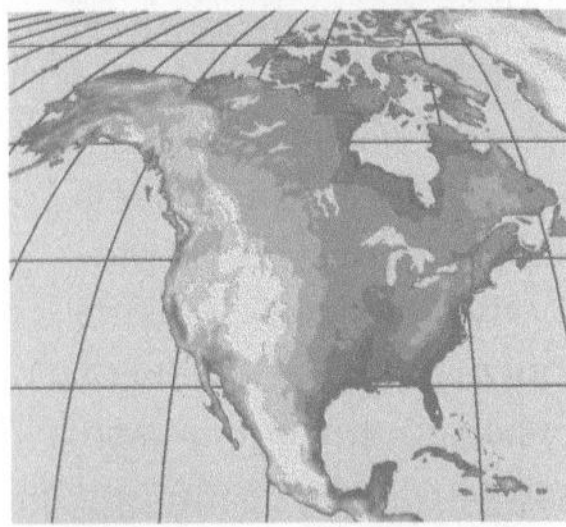

Biome

What features of geology and regional climate determine the transition from forest to prairie grassland ecosystems in North America?

Biosphere

What is the role of the grassland biome in the global carbon cycle?

Figure 1.3 The hierarchy of ecological systems.

Herbivores consume plants, predators eat prey, and individuals compete for limited resources. When individuals die, other organisms consume and break down their remains, recycling the nutrients contained in their dead tissues back into the soil.

Organisms interact with the environment in the context of the ecosystem, yet all communities and ecosystems exist in the broader spatial context of the **landscape**—an area of land (or water) composed of a patchwork of communities and ecosystems. At the spatial scale of the landscape, communities and ecosystems are linked through such processes as the dispersal of organisms and the exchange of materials and energy.

Although each ecosystem on the landscape is distinct in that it is composed of a unique combination of physical conditions (such as topography and soils) and associated sets of plant and animal populations (communities), the broad-scale patterns of climate and geology characterizing our planet give rise to regional patterns in the geographic distribution of ecosystems (see Chapter 2). Geographic regions having similar geological and climatic conditions (patterns of temperature, precipitation, and seasonality) support similar types of communities and ecosystems. For example, warm temperatures, high rates of precipitation, and a lack of seasonality characterize the world's equatorial regions. These warm, wet conditions year-round support vigorous plant growth and highly productive, evergreen forests known as tropical rain forests (see Chapter 23). The broad-scale regions dominated by similar types of ecosystems, such as tropical rain forests, grasslands, and deserts, are referred to as **biomes**.

The highest level of organization of ecological systems is the **biosphere**—the thin layer surrounding the Earth that supports all of life. In the context of the biosphere, all ecosystems, both on land and in the water, are linked through their interactions—exchanges of materials and energy—with the other components of the Earth system: atmosphere, hydrosphere, and geosphere. Ecology is the study of the complex web of interactions between organisms and their environment at all levels of organization—from the individual organism to the biosphere.

1.4 Ecologists Study Pattern and Process at Many Levels

As we shift our focus across the different levels in the hierarchy of ecological systems—from the individual organism to the biosphere—a different and unique set of patterns and processes emerges, and subsequently a different set of questions and approaches for studying these patterns and processes is required (see Figure 1.3). The result is that the broader science of ecology is composed of a range of subdisciplines—from physiological ecology, which focuses on the functioning of individual organisms, to the perspective of Earth's environment as an integrated system forming the basis of global ecology.

Ecologists who focus on the level of the individual examine how features of morphology (structure), physiology, and behavior influence that organism's ability to survive, grow, and reproduce in its environment. Conversely, how do these same characteristics (morphology, physiology, and behavior) function to constrain the organism's ability to function successfully in other environments? By contrasting the characteristics of different species that occupy different environments, these ecologists gain insights into the factors influencing the distribution of species.

At the individual level, birth and death are discrete events. Yet when we examine the collective of individuals that make up a population, these same processes are continuous as individuals are born and die. At the population level, birth and death are expressed as rates, and the focus of study shifts to examining the numbers of individuals in the population and how these numbers change through time. Populations also have a distribution in space, leading to such questions as how are individuals spatially distributed within an area, and how do the population's characteristics (numbers and rates of birth and death) change from location to location?

As we expand our view of nature to include the variety of plant and animal species that occupy an area, the ecological community, a new set of patterns and processes emerges. At this level of the hierarchy, the primary focus is on factors influencing the relative abundances of various species coexisting within the community. What is the nature of the interactions among the species, and how do these interactions influence the dynamics of the different species' populations?

The diversity of organisms comprising the community modify as well as respond to their surrounding physical environment, and so together the biotic and abiotic components of the environment interact to form an integrated system—the ecosystem. At the ecosystem level, the emphasis shifts from species to the collective properties characterizing the flow of energy and nutrients through the combined physical and biological system. At what rate are energy and nutrients converted into living tissues (termed *biomass*)? In turn, what processes govern the rate at which energy and nutrients in the form of organic matter (living and dead tissues) are broken down and converted into inorganic forms? What environmental factors limit these processes governing the flow of energy and nutrients through the ecosystem?

As we expand our perspective even further, the landscape may be viewed as a patchwork of ecosystems whose boundaries are defined by distinctive changes in the underlying physical environment or species composition. At the landscape level, questions focus on identifying factors that give rise to the spatial extent and arrangement of the various ecosystems that make up the landscape, and ecologists explore the consequences of these spatial patterns on such processes as the dispersal of organisms, the exchange of energy and nutrients between adjacent ecosystems, and the propagation of disturbances such as fire or disease.

At a continental to global scale, the questions focus on the broad-scale distribution of different ecosystem types or biomes. How do patterns of biological diversity (the number of different types of species inhabiting the ecosystem) vary geographically across the different biomes? Why do tropical rain forests support a greater diversity of species than do forest ecosystems in the temperate regions? What environmental factors determine the geographic distribution of the different biome types (e.g., forest, grassland, and desert)?

Finally, at the biosphere level, the emphasis is on the linkages between ecosystems and other components of the earth system, such as the atmosphere. For example, how does the exchange of energy and materials between terrestrial ecosystems and the atmosphere influence regional and global climate patterns? Certain processes, such as movement of the element carbon between ecosystems and the atmosphere, operate at a global scale and require ecologists to collaborate with oceanographers, geologists, and atmospheric scientists.

Throughout our discussion, we have used this hierarchical view of nature and the unique set of patterns and process associated with each level—the individual population, community, ecosystem, landscape, biome, and biosphere—as an organizing framework for studying the science of ecology. In fact, the science of ecology is functionally organized into subdisciplines based on these different levels of organization, each using an array of specialized approaches and methodologies to address the unique set of questions that emerge at these different levels of ecological organization. The patterns and processes at these different levels of organization are linked, however, and identifying these linkages is our objective. For example, at the individual organism level, characteristics such as size, longevity, age at reproduction, and degree of parental care will directly influence rates of birth and survival for the collective of individuals comprising the species' population. At the community level, the same population will be influenced both positively and negatively through its interactions with populations of other species. In turn, the relative mix of species that make up the community will influence the collective properties of energy and nutrient exchange at the ecosystem level. As we shall see, patterns and processes at each level—from individuals to ecosystems—are intrinsically linked in a web of cause and effect with the patterns and processes operating at the other levels of this organizational hierarchy.

1.5 Ecologists Investigate Nature Using the Scientific Method

Although each level in the hierarchy of ecological systems has a unique set of questions on which ecologists focus their research, all ecological studies have one thing in common: they include the process known as the scientific method (**Figure 1.4**). This method demonstrates the power and limitations of science, and taken individually, each step of the scientific method involves commonplace procedures. Yet taken together, these procedures form a powerful tool for understanding nature.

All science begins with observation. In fact, this first step in the process defines the domain of science: if something cannot be observed, it cannot be investigated by science. The observation need not be direct, however. For example, scientists cannot directly observe the nucleus of an atom, yet its structure can be explored indirectly through a variety of methods. Secondly, the observation must be repeatable—able to be made by multiple observers. This constraint helps to minimize unsuspected bias, when an individual might observe what they *want* or think they *ought* to observe.

The second step in the scientific method is defining a problem—forming a question regarding the observation that has been made. For example, an ecologist working in the prairie grasslands of North America might observe that the growth and productivity (the rate at which plant biomass is being produced

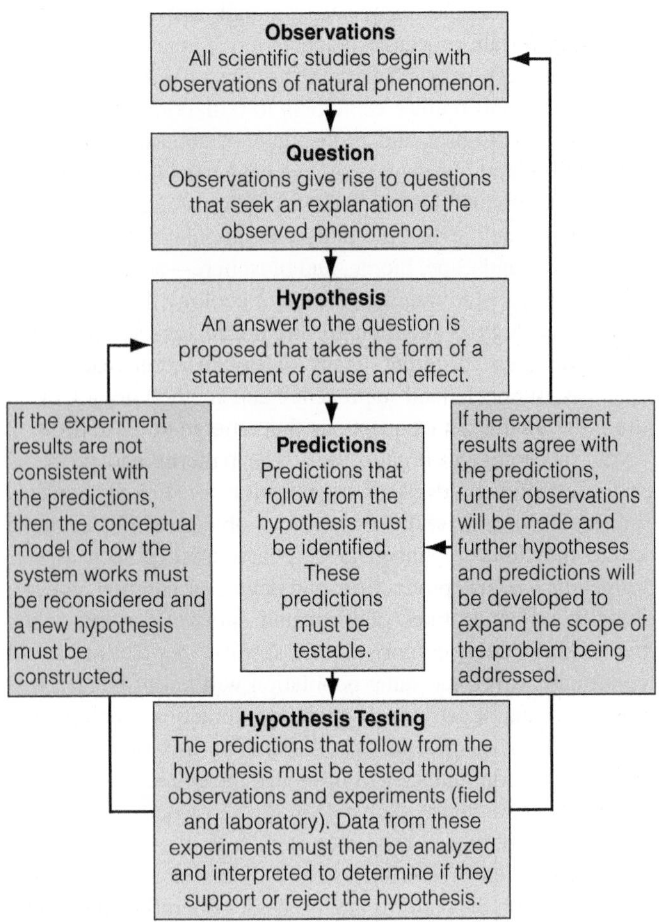

Figure 1.4 A simple representation of the scientific method.

per unit area per unit time: grams per meter squared per year [g/m²/yr]) of grasses varies across the landscape. From this observation the ecologist may formulate the question, what environmental factors result in the observed variations in grassland productivity across the landscape? The question typically focuses on seeking an explanation for the observed patterns.

Once a question (problem) has been established, the next step is to develop a hypothesis. A **hypothesis** is an educated guess about what the answer to the question may be. The process of developing a hypothesis is guided by experience and knowledge, and it should be a statement of cause and effect that can be tested. For example, based on her knowledge that nitrogen availability varies across the different soil types found in the region and that nitrogen is an important nutrient limiting plant growth, the ecologist might hypothesize that *the observed variations in the growth and productivity of grasses across the prairie landscape are a result of differences in the availability of soil nitrogen*. As a statement of cause and effect, certain predictions follow from the hypothesis. If soil nitrogen is the factor limiting the growth and productivity of plants in the prairie grasslands, then grass productivity should be greater in areas with higher levels of soil nitrogen than in areas with lower levels of soil nitrogen. The next step is testing the hypothesis to see if the predictions that follow from the hypothesis do indeed hold true. This step requires gathering data (see **Quantifying Ecology 1.1**).

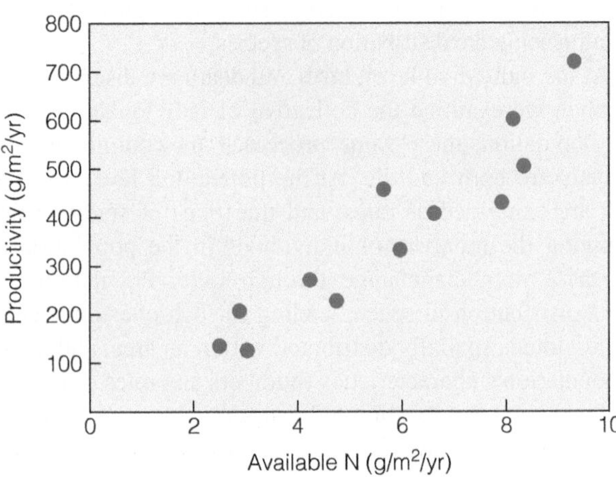

Figure 1.5 The response of grassland production to soil nitrogen availability. Nitrogen (N), the independent variable, is plotted on the *x*-axis; grassland productivity, the dependent variable, is plotted on the *y*-axis.

Interpreting Ecological Data

Q1. *In the above graph, which variable is the independent variable? Which is the dependent variable? Why?*

Q2. *Would you describe the relationship between available nitrogen and grassland productivity as positive or negative (inverse)?*

To test this hypothesis, the ecologist may gather data in several ways. The first approach might be a field study to examine how patterns of soil nitrogen and grass productivity covary (vary together) across the landscape. If nitrogen is controlling grassland productivity, productivity should increase with increasing soil nitrogen. The ecologist would measure nitrogen availability and grassland productivity at various sites across the landscape. Then, the relationship between these two variables, nitrogen and productivity, could be expressed graphically (see **Quantifying Ecology 1.2** on pages 8 and 9 to learn more about working with graphical data). Visit MasteringBiology at www.masteringbiology.com to work with histograms and scatter plots.

After you've become familiar with scatter plots, you'll see the graph of **Figure 1.5** shows nitrogen availability on the horizontal or x-axis and grassland productivity on the vertical or y-axis. This arrangement is important. The scientist is assuming that nitrogen is the cause and that grassland productivity is the effect. Because nitrogen (x) is the cause, we refer to it as the independent variable. Because it is hypothesized that grassland productivity (y) is influenced by the availability of nitrogen, we refer to it as the dependent variable. Visit MasteringBiology at www.masteringbiology.com for a tutorial on reading and interpreting graphs.

From the observations plotted in Figure 1.5, it is apparent that grassland productivity does, in fact, increase with increasing availability of nitrogen in the soil. Therefore, the data support the hypothesis. Had the data shown no relationship between grassland productivity and nitrogen, the ecologist would have rejected the hypothesis and sought a new explanation for the observed differences in grassland productivity across the landscape. However, although the data suggest that grassland

QUANTIFYING ECOLOGY 1.1 Classifying Ecological Data

All ecological studies involve collecting data that includes observations and measurements for testing hypotheses and drawing conclusions about a population. The term *population* in this context refers to a **statistical population**. An investigator is highly unlikely to gather observations on *all* members of a total population, so the part of the population actually observed is referred to as a **sample**. From this sample data, the investigator will draw her conclusions about the population as a whole. However, not all data are of the same type; and the type of data collected in a study directly influences the mode of presentation, types of analyses that can be performed, and interpretations that can be made.

At the broadest level, data can be classified as either categorical or numerical. **Categorical data** are *qualitative*, that is, observations that fall into separate and distinct categories. The resulting data are labels or categories, such as the color of hair or feathers, sex, or reproductive status (pre-reproductive, reproductive, post-reproductive). Categorical data can be further subdivided into two categories: nominal and ordinal. **Nominal data** are categorical data in which objects fall into unordered categories, such as the previous examples of hair color or sex. In contrast, **ordinal data** are categorical data in which order is important, such as the example of reproductive status. In the special case where only two categories exist, such as in the case of presence or absence of a trait, categorical data are referred to as **binary.** Both nominal and ordinal data can be binary.

With **numerical data**, objects are "measured" based on some *quantitative* trait. The resulting data are a set of numbers, such as height, length, or weight. Numerical data can be subdivided into two categories: discrete and continuous. For **discrete data**, only certain values are possible, such as with integer values or counts. Examples include the number of offspring, number of seeds produced by a plant, or number of times a hummingbird visits a flower during the course of a day. With **continuous data**, any value within an interval theoretically is possible, limited only by the ability of the measurement device. Examples of this type of data include height, weight, or concentration.

1. What type of data does the variable "available N" (the *x*-axis) represent in Figure 1.5?
2. How might you transform this variable (available nitrogen) into categorical data? Would it be considered ordinal or nominal?

production does increase with increasing soil nitrogen, they do not prove that nitrogen is the *only* factor controlling grass growth and production. Some other factor that varies with nitrogen availability, such as soil moisture or acidity, may actually be responsible for the observed relationship. To test the hypothesis another way, the ecologist may choose to do an experiment. An experiment is a test under controlled conditions performed to examine the validity of a hypothesis. In designing the experiment, the scientist will try to isolate the presumed causal agent—in this case, nitrogen availability.

The scientist may decide to do a field experiment (**Figure 1.6**), adding nitrogen to some field sites and not to others. The investigator controls the independent variable (levels of nitrogen) in a predetermined way, to reflect observed variations in soil nitrogen availability across the landscape, and monitors the response of the dependent variable (plant growth). By observing the differences in productivity between the grasslands fertilized with nitrogen and those that were not, the investigator tries to test whether nitrogen is the causal agent. However, in choosing the experimental sites, the ecologist must try to locate areas where other factors that may influence productivity, such as moisture and acidity, are similar. Otherwise, she cannot be sure which factor is responsible for the observed differences in productivity among the sites.

Finally, the ecologist might try a third approach—a series of laboratory experiments (**Figure 1.7**). Laboratory experiments give the investigator much more control over the environmental conditions. For example, she can grow the native grasses in the greenhouse under conditions of controlled temperature, soil acidity, and water availability. If the plants exhibit increased growth with higher nitrogen fertilization, the investigator has further evidence in support of the hypothesis. Nevertheless, she faces a limitation common to all laboratory experiments; that is, the results are not directly applicable in the field. The response of grass plants under controlled laboratory conditions may not be the same as their response under natural conditions in the field. There, the plants are part of the ecosystem and interact with other plants, animals, and the

Figure 1.6 Field experiment at the Cedar Creek Long Term Ecological Research (LTER) site in central Minnesota, operated by the University of Minnesota. Experimental plots such as these are used to examine the effects of elevated nitrogen deposition, increased concentrations of atmospheric carbon dioxide, and loss of biodiversity on ecosystem functioning.

QUANTIFYING ECOLOGY 1.2 Displaying Ecological Data: Histograms and Scatter Plots

Whichever type of data an observer collects (see Quantifying Ecology 1.1), the process of interpretation typically begins with a graphical display of observations. The most common method of displaying a single data set is constructing a **frequency distribution**. A frequency distribution is a count of the number of observations (frequency) having a given score or value. For example, consider this set of observations regarding flower color in a sample of 100 pea plants:

Flower color	Purple	Pink	White
Frequency	50	35	15

These data are categorical and nominal since the categories have no inherent order.

Frequency distributions are likewise used to display continuous data. This set of continuous data represents body lengths (in centimeters) of 20 sunfish sampled from a pond:

8.83, 9.25, 8.77, 10.38, 9.31, 8.92, 10.22, 7.95, 9.74, 9.51, 9.66, 10.42, 10.35, 8.82, 9.45, 7.84, 11.24, 11.06, 9.84, 10.75

With continuous data, the frequency of each value is often a single instance because multiple data points are unlikely to be exactly the same. Therefore, continuous data are normally grouped into discrete categories, with each category representing a defined range of values. Each category must not overlap; each observation must belong to only one category. For example, the body length data could be grouped into discrete categories:

Body length (intervals, cm)	Number of individuals
7.00–7.99	2
8.00–8.99	4
9.00–9.99	7
10.00–10.99	5
11.00–11.99	2

Once the observations have been grouped into categories, the resulting frequency distribution can then be displayed as a **histogram** (type of bar graph; **Figure 1a**). The *x*-axis represents the discrete intervals of body length, and the *y*-axis represents the number of individuals whose body length falls within each given interval.

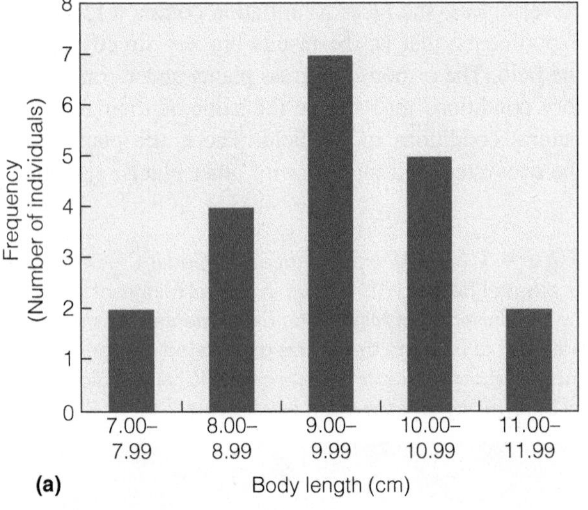

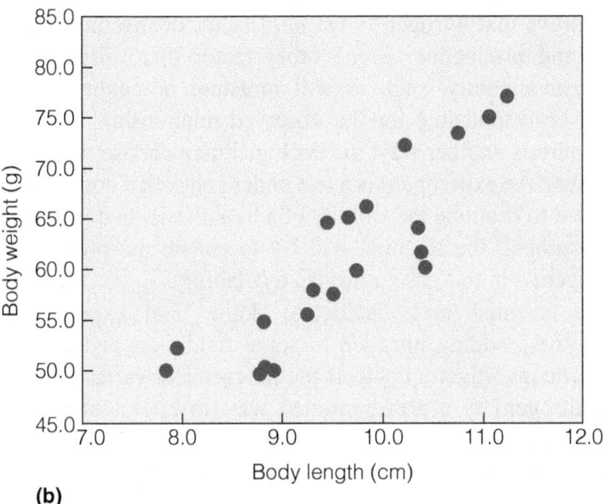

Figure 1 (a) An example of a histogram relating the number of individuals belonging to different categories of body length from a sample of the sunfish population. (b) Scatter plot relating body length (x-axis) and body weight (y-axis) for the sample of sunfish presented in (a).

physical environment. Despite this limitation, the ecologist has accumulated additional data describing the basic growth response of the plants to nitrogen availability.

Having conducted several experiments that confirm the link between patterns of grass productivity to nitrogen availability, the ecologist may now wish to explore this relationship further, to see how the relationship between productivity and nitrogen is influenced by other environmental factors that vary across the prairie landscape. For example, how do differences in rainfall and soil moisture across the region influence the relationship between grass production and soil nitrogen? Once again hypotheses are developed, predictions made, and experiments conducted. As the ecologist develops a more detailed understanding of how various environmental factors interact with

In effect, the continuous data are transformed into categorical data for the purposes of graphical display. Unless there are previous reasons for defining categories, defining intervals is part of the data interpretation process and the search for patterns. For example, how would the pattern represented by the histogram in Figure 1a differ if the intervals were in units of 1 but started with 7.50 (7.50–8.49, 8.50–9.49, etc.)?

Often, however, the researcher is examining the relationship between two variables or sets of observations. When both variables are numerical, the most common method of graphically displaying the data is by using a scatter plot. A **scatter plot** is constructed by defining two axes (*x* and *y*), each representing one of the two variables being examined. For example, suppose the researcher who collected the observations of body length for sunfish netted from the pond also measured their weight in grams. The investigator might be interested in whether there is a relationship between body length and weight in sunfish.

In this example, body length would be the *x*-axis, or independent variable (Section 1.5), and body weight would be the *y*-axis, or dependent variable. Once the two axes are defined, each individual (sunfish) can be plotted as a point on the graph, with the position of the point being defined by its respective values of body length and weight (**Figure 1b**).

Scatter plots can be described as belonging to one of three general patterns, as shown in **Figure 2**. In plot (a) there is a general trend for *y* to increase with increasing values of *x*. In this case the relationship between *x* and *y* is said to be positive (as with the example of body length and weight for sunfish). In plot (b) the pattern is reversed, and *y* decreases with increasing values of *x*. In this case the relationship between *x* and *y* is said to be negative, or inverse. In plot (c) there is no apparent relationship between *x* and *y*.

You will find many types of graphs throughout our discussion but most will be histograms and scatter plots. No matter which type of graph is presented, ask yourself the same set of questions—listed below—to help interpret the results. Review this set of questions by applying them to the graphs in Figure 1. What do you find out?

1. What type of data do the observations represent?
2. What variables do each of the axes represent, and what are their units (cm, g, color, etc.)?
3. How do values of *y* (the dependent variable) vary with values of *x* (the independent variable)?

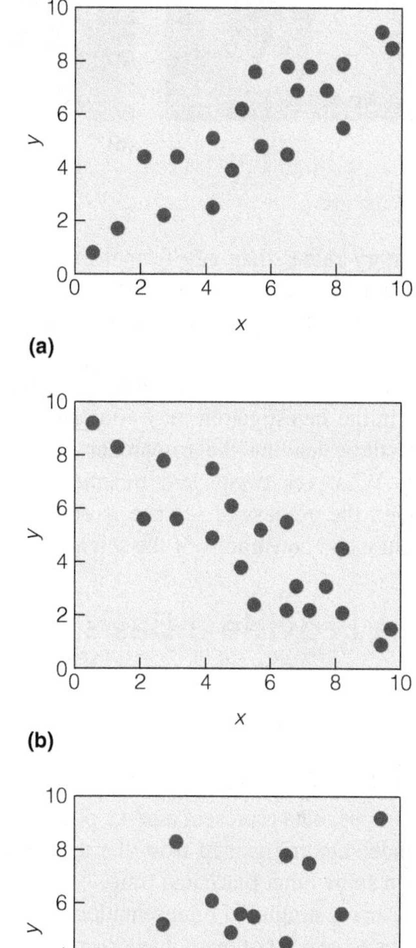

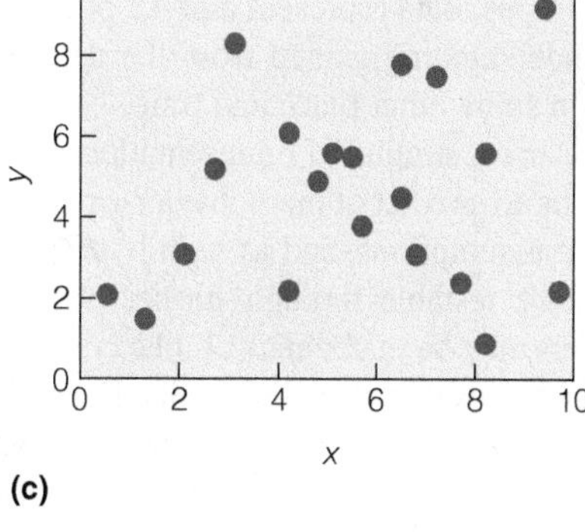

Figure 2 Three general patterns for scatter plots.

Go to Analyzing Ecological Data at www.masteringbiology.com to further explore how to display data graphically.

soil nitrogen to control grass production, a more general theory of the influence of environmental factors controlling grass production in the grassland prairies may emerge. A **theory** is an integrated set of hypotheses that together explain a broader set of observations than any single hypothesis—such as a general theory of environmental controls on productivity of the prairie grassland ecosystems of North America.

Although the diagram of the scientific method presented in Figure 1.4 represents the process of scientific investigation as a sequence of well-defined steps that proceeds in a linear fashion, in reality, the process of scientific research often proceeds in a nonlinear fashion. Scientists often begin an investigation based on readings of previously published studies, discussions with colleagues, or informal observations made in

(a)

(b)

Figure 1.7 (a) Undergraduate research students at Harvard Forest erect temporary greenhouses that were used to create different carbon dioxide (CO_2) treatments for a series of experiments directed at testing the response of ragweed (*Ambrosia artemisiifolia*) to elevated atmospheric CO_2. (b) Response to elevated CO_2 was determined by measuring the growth, morphology, and reproductive characteristics of individual plants from different populations.

the field or laboratory rather than any formal process. Often during hypothesis testing, observations may lead the researcher to modify the experimental design or redefine the original hypothesis. In reality, the practice of science involves unexpected twist and turns. In some cases, unexpected observations or results during the initial investigation may completely change the scope of the study, leading the researcher in directions never anticipated. Whatever twists and unanticipated turns may occur, however, the process of science is defined by the fundamental structure and constraints of the scientific method.

1.6 Models Provide a Basis for Predictions

Scientists use the understanding derived from observation and experiments to develop models. Data are limited to the special case of what happened when the measurements were made. Like photographs, data represent a given place and time. Models use the understanding gained from the data to predict what will happen in some other place and time.

Models are abstract, simplified representations of real systems. They allow us to predict some behavior or response using a set of explicit assumptions, and as with hypotheses, these predictions should be testable through further observation or experiments. Models may be mathematical, like computer simulations, or they may be verbally descriptive, like Darwin's theory of evolution by natural selection (see Chapter 5). Hypotheses are models, although the term *model* is typically reserved for circumstances in which the hypothesis has at least some limited support through observations and experimental results. For example, the hypothesis relating grass production to nitrogen availability is a model. It predicts that plant productivity will increase with increasing nitrogen availability. However, this prediction is qualitative—it does not predict *how much* plant productivity will increase. In contrast, mathematical models usually offer quantitative predictions. For example, from the data in Figure 1.5, we can develop a regression equation—a form of statistical model—to predict the amount of grassland productivity per unit of nitrogen in the soil (**Figure 1.8**). Visit MasteringBiology at www.masteringbiology.com to review regression analysis.

All of the approaches just discussed—observation, experimentation, hypothesis testing, and development of models—appear throughout our discussion to illustrate basic concepts and relationships. They are the basic tools of science. For every topic,

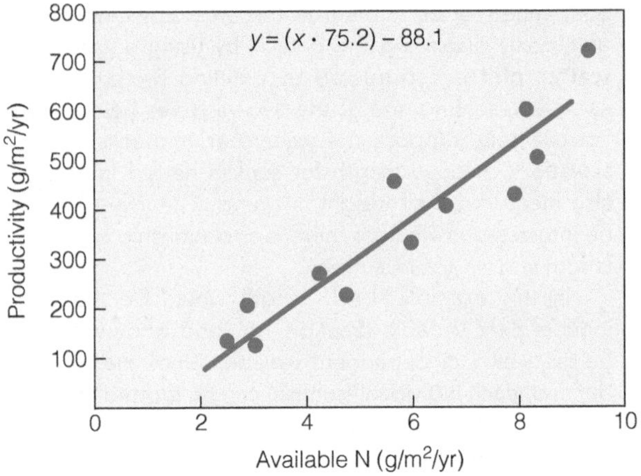

Figure 1.8 A simple linear regression model to predict grassland productivity (y-axis) from nitrogen availability (x-axis). The general form of the equation is $y = (x \times b) + a$, where b is the slope of the line (75.2) and a is the y-intercept (–88.1), or the value of y where the line intersects the y-axis (when $x = 0$).

Interpreting Ecological Data

Q1. *How could you use the simple linear regression model presented to predict productivity for a grassland site not included in the graph?*

Q2. *What is the predicted productivity for a site with available nitrogen of 5 g/m²/yr? (Use the linear regression equation.)*

an array of figures and tables present the observations, experimental data, and model predictions used to test specific hypotheses regarding pattern and process at the different levels of ecological organization. Being able to analyze and interpret the data presented in these figures and tables is essential to your understanding of the science of ecology. To help you develop these skills, we have annotated certain figures and tables to guide you in their interpretation. In other cases, we pose questions that ask you to interpret, analyze, and draw conclusions from the data presented. These figures and tables are labeled *Interpreting Ecological Data*. (See Figure 2.15 on page 23 for the first example.)

1.7 Uncertainty Is an Inherent Feature of Science

Collecting observations, developing and testing hypotheses, and constructing predictive models all form the backbone of the scientific method (see Figure 1.4). It is a continuous process of testing and correcting concepts to arrive at explanations for the variation we observe in the world around us, thus unifying observations that on first inspection seem unconnected. The difference between science and art is that, although both pursuits involve creation of concepts, in science, the exploration of concepts is limited to the facts. In science, the only valid means of judging a concept is by testing its empirical truth.

However, scientific concepts have no permanence because they are only our interpretations of natural phenomena. We are limited to inspecting only a part of nature because to understand, we have to simplify. As discussed in Section 1.5, in designing experiments, we control the pertinent factors and try to eliminate others that may confuse the results. Our intent is to focus on a subset of nature from which we can establish cause and effect. The trade-off is that whatever cause and effect we succeed in identifying represents only a partial connection to the nature we hope to understand. For that reason, when experiments and observations support our hypotheses, and when the predictions of the models are verified, our job is still not complete. We work to loosen the constraints imposed by the need to simplify so that we can understand. We expand our hypothesis to cover a broader range of conditions and once again begin testing its ability to explain our new observations.

It may sound odd at first, but science is a search for evidence that proves our concepts wrong. Rarely is there only one possible explanation for an observation. As a result, any number of hypotheses may be developed that might be consistent with an observation. The determination that experimental data are consistent with a hypothesis does not prove that the hypothesis is true. The real goal of hypothesis testing is to eliminate incorrect ideas. Thus, we must follow a process of elimination, searching for evidence that proves a hypothesis wrong. Science is essentially a self-correcting activity, dependent on the continuous process of debate. Dissent is the activity of science, fueled by free inquiry and independence of thought. To the outside observer, this essential process of debate may appear to be a shortcoming. After all, we depend on science for the development of technology and the ability to solve problems. For the world's current environmental issues, the solutions may well involve difficult ethical, social, and economic decisions. In this case, the uncertainty inherent in science is discomforting. However, we must not mistake uncertainty for confusion, nor should we allow disagreement among scientists to become an excuse for inaction. Instead, we need to understand the uncertainty so that we may balance it against the costs of inaction.

1.8 Ecology Has Strong Ties to Other Disciplines

The complex interactions taking place within ecological systems involve all kinds of physical, chemical, and biological processes. To study these interactions, ecologists must draw on other sciences. This dependence makes ecology an interdisciplinary science.

Although we explore topics that are typically the subject of disciplines such as biochemistry, physiology, and genetics, we do so only in the context of understanding the interplay of organisms with their environment. The study of how plants take up carbon dioxide and lose water, for example, belongs to plant physiology (see Chapter 6). Ecology looks at how these processes respond to variations in rainfall and temperature. This information is crucial to understanding the distribution and abundance of plant populations and the structure and function of ecosystems on land. Likewise, we must draw on many of the physical sciences, such as geology, hydrology, and meteorology. They help us chart other ways in which organisms and environments interact. For instance, as plants take up water, they influence soil moisture and the patterns of surface water flow. As they lose water to the atmosphere, they increase atmospheric water content and influence regional patterns of precipitation. The geology of an area influences the availability of nutrients and water for plant growth. In each example, other scientific disciplines are crucial to understanding how individual organisms both respond to and shape their environment.

In the 21st century, ecology is entering a new frontier, one that requires expanding our view of ecology to include the dominant role of humans in nature. Among the many environmental problems facing humanity, four broad and interrelated areas are crucial: human population growth, biological diversity, sustainability, and global climate change. As the human population increased from approximately 500 million to more than 7 billion in the past two centuries, dramatic changes in land use have altered Earth's surface. The clearing of forests for agriculture has destroyed many natural habitats, resulting in a rate of species extinction that is unprecedented in Earth's history. In addition, the expanding human population is exploiting natural resources at unsustainable levels. As a result of the growing demand for energy from fossil fuels that is needed to sustain economic growth, the chemistry of the atmosphere is changing in ways that are altering Earth's climate. These environmental problems are ecological in nature, and the science of ecology is essential to understanding their causes and identifying ways to mitigate their impacts. Addressing these issues, however, requires a broader interdisciplinary framework to better understand their historical, social, legal, political, and ethical dimensions. That broader framework is known as **environmental science**. Environmental science examines the impact of humans on the natural environment and as such covers a wide range of topics including agronomy, soils, demography, agriculture, energy, and hydrology, to name but a few.

Throughout the text, we use the *Ecological Issues & Applications* sections of each chapter to highlight topics relating to current environmental issues regarding human impacts on the environment and to illustrate the importance of the science of ecology to better understanding the human relationship with the environment.

1.9 The Individual Is the Basic Unit of Ecology

As we noted previously, ecology encompasses a broad area of investigation—from the individual organism to the biosphere. Our study of the science of ecology uses this hierarchical framework in the chapters that follow. We begin with the individual organism, examining the processes it uses and constraints it faces in maintaining life under varying environmental conditions. The individual organism forms the basic unit in ecology. The individual senses and responds to the prevailing physical environment. The collective properties of individual births and deaths drive the dynamics of populations, and individuals of different species interact with one another in the context of the community. But perhaps most importantly, the individual, through the process of reproduction, passes genetic information to successive individuals, defining the nature of individuals that will compose future populations, communities, and ecosystems. At the individual level we can begin to understand the mechanisms that give rise to the diversity of life and ecosystems on Earth—mechanisms that are governed by the process of natural selection. But before embarking on our study of ecological systems, we examine characteristics of the abiotic (physical and chemical) environment that function to sustain and constrain the patterns of life on our planet.

ECOLOGICAL Issues & Applications

Ecology Has a Rich History

The genealogy of most sciences is direct. Tracing the roots of chemistry and physics is relatively easy. The science of ecology is different. Its roots are complex and intertwined with a wide array of scientific advances that have occurred in other disciplines within the biological and physical sciences. Although the term *ecology* did not appear until the mid-19th century and took another century to enter the vernacular, the idea of ecology is much older.

Arguably, ecology goes back to the ancient Greek scholar Theophrastus, a friend of Aristotle, who wrote about the relations between organisms and the environment. On the other hand, ecology as we know it today has vital roots in plant geography and natural history.

In the 1800s, botanists began exploring and mapping the world's vegetation. One of the early plant geographers was Carl Ludwig Willdenow (1765–1812). He pointed out that similar climates supported vegetation similar in form, even though the species were different. Another was Friedrich Heinrich Alexander von Humboldt (1769–1859), for whom the Humboldt Current, flowing along the west coast of South America, is named. He spent five years exploring Latin America, including the Orinoco and Amazon rivers. Humboldt correlated vegetation with environmental characteristics and coined the term *plant association*. The recognition that the form and function of plants within a region reflects the constraints imposed by the physical environment led the way for a new generation of scientists that explored the relationship between plant biology and plant geography (see Chapter 23).

Among this new generation of plant geographers was Johannes Warming (1841–1924) at the University of Copenhagen, who studied the tropical vegetation of Brazil. He wrote the first text on plant ecology, *Plantesamfund*. Warming integrated plant morphology, physiology, taxonomy, and biogeography into a coherent whole. This book had a tremendous influence on the development of ecology.

Meanwhile, activities in other areas of natural history also assumed important roles. One was the voyage of Charles Darwin (1809–1882) on the *Beagle*. Working for years on notes and collections from this trip, Darwin compared similarities and dissimilarities among organisms within and among continents. He attributed differences to geological barriers. He noted how successive groups of plants and animals, distinct yet obviously related, replaced one another.

Developing his theory of evolution and the origin of species, Darwin came across the writings of Thomas Malthus (1766–1834). An economist, Malthus advanced the principle that populations grow in a geometric fashion, doubling at regular intervals until they outstrip the food supply. Ultimately, a "strong, constantly operating force such as sickness and premature death" would restrain the population. From this concept Darwin developed the idea of "natural selection" as the mechanism guiding the evolution of species (see Chapter 5).

Meanwhile, unbeknownst to Darwin, an Austrian monk, Gregor Mendel (1822–1884), was studying the transmission of characteristics from one generation of pea plants to another in his garden. Mendel's work on inheritance and Darwin's work on natural selection provided the foundation for the study of evolution and adaptation, the field of **population genetics**.

Darwin's theory of natural selection, combined with the new understanding of genetics (the means by which characteristics are transmitted from one generation to the next) provided the mechanism for understanding the link between organisms and their environment, which is the focus of ecology.

Early ecologists, particularly plant ecologists, were concerned with observing the patterns of organisms in nature, and attempting to understand how patterns were formed and maintained by interactions with the physical environment. Some, notably Frederic E. Clements (**Figure 1.9**), sought some system of organizing nature. He proposed that the plant community behaves as a complex organism or *superorganism*

Figure 1.9 The ecologist Frederic E. Clements in the field collecting data.

that grows and develops through stages to a mature or climax state (see Chapter 16). His idea was accepted and advanced by many ecologists. A few ecologists, however, notably Arthur G. Tansley, did not share this view. In its place Tansley advanced a holistic and integrated ecological concept that combined living organisms and their physical environment into a system, which he called the ecosystem (see Chapter 20).

Whereas the early plant ecologists were concerned mostly with terrestrial vegetation, another group of European biologists was interested in the relationship between aquatic plants and animals and their environment. They advanced the ideas of organic nutrient cycling and feeding levels, using the terms *producers* and *consumers*. Their work influenced a young limnologist at the University of Minnesota, R. A. Lindeman. He traced "energy-available" relationships within a lake community. His 1942 paper, "The Trophic-Dynamic Aspects of Ecology," marked the beginning of **ecosystem ecology**, the study of whole living systems.

Lindeman's theory stimulated further pioneering work in the area of energy flow and nutrient cycling by G. E. Hutchinson of Yale University (**Figure 1.10**) and E. P. and H. T. Odum of the University of Georgia. Their work became a foundation of ecosystem ecology. The use of radioactive tracers, a product of the atomic age, to measure the movements of energy and nutrients through ecosystems and the use of computers to analyze large amounts of data stimulated the development of **systems ecology**, the application of general system theory and methods to ecology.

Animal ecology initially developed largely independently of the early developments in plant ecology. The beginnings of animal ecology can be traced to two Europeans, R. Hesse of Germany and Charles Elton of England. Elton's *Animal Ecology* (1927) and Hesse's *Tiergeographie auf logischer grundlage* (1924), translated into English as *Ecological Animal Geography*, strongly influenced the development of animal ecology in the United States. Charles Adams and Victor Shelford were two pioneering U.S. animal ecologists. Adams published the first textbook on animal ecology, *A Guide to the Study of Animal Ecology* (1913). Shelford wrote *Animal Communities in Temperate America* (1913).

Figure 1.10 Ecologist G. Evelyn Hutchinson in his lab at Yale University.

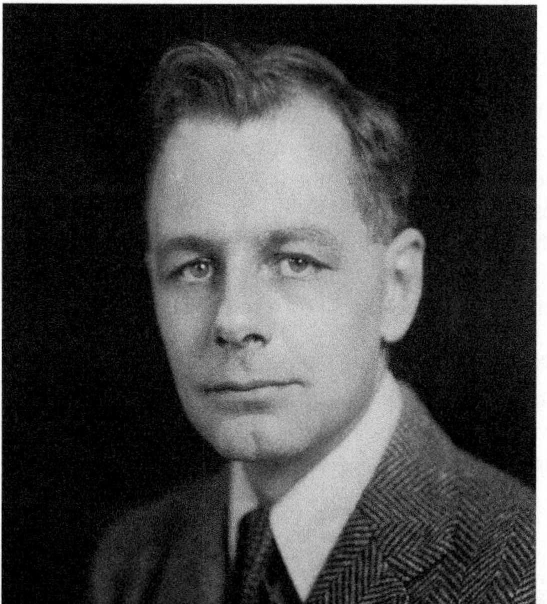

Shelford gave a new direction to ecology by stressing the interrelationship between plants and animals. Ecology became a science of communities. Some previous European ecologists, particularly the marine biologist Karl Mobius, had developed the general concept of the community. In his essay "An Oyster Bank is a Biocenose" (1877), Mobius explained that the oyster bank, although dominated by one animal, was really a complex community of many interdependent organisms. He proposed the word *biocenose* for such a community. The word comes from the Greek, meaning *life having something in common*.

The appearance in 1949 of the encyclopedic *Principles of Animal Ecology* by five second-generation ecologists from the University of Chicago (W. C. Allee, A. E. Emerson, Thomas Park, Orlando Park, and K. P. Schmidt) pointed to the direction that modern ecology would take. It emphasized feeding relationships and energy budgets, population dynamics, and natural selection and evolution.

During the period of development of the field of animal ecology, natural history observations also focused on the behavior of animals. This focus on animal behavior began with 19th-century behavioral studies including those of ants by William Wheeler and of South American monkeys by Charles Carpenter. Later, the pioneering studies of Konrad Lorenz and Niko Tinbergen on the role of imprinting and instinct in the social life of animals, particularly birds and fish, gave rise to **ethology**. It spawned an offshoot, **behavioral ecology**, exemplified by L. E. Howard's early study on territoriality in birds. Behavioral ecology is concerned with intraspecific and interspecific relationships such as mating, foraging, defense, and how behavior is influenced by natural selection.

The writings of the economist Malthus that were so influential in the development of Darwin's ideas regarding the origin of species also stimulated the study of natural populations. The study of populations in the early 20th century branched into two fields. One, **population ecology**, is concerned with population growth (including birthrates and death rates), regulation and intraspecific and interspecific competition, mutualism, and predation. The other, a combination of population genetics and population ecology is **evolutionary ecology**, which deals with the role of natural selection in physical and behavioral adaptations and speciation. Focusing on adaptations, **physiological ecology** is concerned with the responses of individual organisms to temperature, moisture, light, and other environmental conditions.

Closely associated with population and evolutionary ecology is **community ecology**, with its focus on species interactions. One of the major objectives of community ecology is to understand the origin, maintenance, and consequences of species diversity within ecological communities.

With advances in biology, physics, and chemistry throughout the latter part of the 20th century, new areas of study in ecology emerged. The development of aerial photography and later the launching of satellites by the U.S. space program provided scientists with a new perspective of the surface of Earth through the use of remote sensing data. Ecologists began to explore spatial processes that linked adjacent communities and ecosystems through the new emerging field of **landscape ecology**. A new appreciation of the impact of changing land use on natural ecosystems led to the development of **conservation ecology**, which applies principles from different fields, from ecology to economics and sociology, to the maintenance of biological diversity. The application of principles of ecosystem development and function to the restoration and management of disturbed lands gave rise to **restoration ecology**, whereas understanding Earth as a system is the focus of the newest area of ecological study, **global ecology**.

Ecology has so many roots that it probably will always remain multifaceted—as the ecological historian Robert McIntosh calls it, "a polymorphic discipline." Insights from these many specialized areas of ecology will continue to enrich the science as it moves forward in the 21st century.

SUMMARY

Ecology 1.1

Ecology is the scientific study of the relationships between organisms and their environment. The environment includes the physical and chemical conditions and biological or living components of an organism's surroundings. Relationships include interactions with the physical world as well as with members of the same and other species.

Ecosystems 1.2

Organisms interact with their environment in the context of the ecosystem. Broadly, the ecosystem consists of two components, the living (biotic) and the physical (abiotic), interacting as a system.

Hierarchical Structure 1.3

Ecological systems may be viewed in a hierarchical framework, from individual organisms to the biosphere. Organisms of the same species that inhabit a given physical environment make up a population. Populations of different kinds of organisms interact with members of their own species as well as with individuals of other species. These interactions range from competition for shared resources to interactions that are mutually beneficial for the individuals of both species involved. Interacting populations make up a biotic community. The community plus the physical environment make up an ecosystem.

All communities and ecosystems exist in the broader spatial context of the landscape—an area of land (or water) composed of a patchwork of communities and ecosystems. Geographic regions having similar geological and climatic conditions support similar types of communities and ecosystems, referred to as biomes. The highest level of organization of ecological systems is the biosphere—the thin layer around Earth that supports all of life.

Ecological Studies 1.4

At each level in the hierarchy of ecological systems—from the individual organism to the biosphere—a different and unique set of patterns and processes emerges; subsequently, a different set of questions and approaches for studying these patterns and processes is required.

Scientific Method 1.5

All ecological studies are conducted by using the scientific method. All science begins with observation, from which questions emerge. The next step is the development of a hypothesis—a proposed answer to the question. The hypothesis must be testable through observation and experiments.

Models 1.6

From research data, ecologists develop models. Models allow us to predict some behavior or response using a set of explicit assumptions. They are abstractions and simplifications of natural phenomena. Such simplification is necessary to understand natural processes.

Uncertainty in Science 1.7

An inherent feature of scientific study is uncertainty; it arises from the limitation posed by focusing on only a small subset of nature, and it results in an incomplete perspective. Because we can develop any number of hypotheses that may be consistent with an observation, determining that experimental data are consistent with a hypothesis is not sufficient to prove that the hypothesis is true. The real goal of hypothesis testing is to eliminate incorrect ideas.

An Interdisciplinary Science 1.8

Ecology is an interdisciplinary science because the interactions of organisms with their environment and with one another involve physiological, behavioral, and physical responses. The study of these responses draws on such fields as physiology, biochemistry, genetics, geology, hydrology, and meteorology.

Individuals 1.9

The individual organism forms the basic unit in ecology. It is the individual that responds to the environment and passes genes to successive generations. It is the collective birth and death of individuals that determines the dynamics of populations, and the interactions among individuals of the same and different species that structures communities.

History Ecological Issues & Applications

Ecology has its origin in natural history and plant geography. Over the past century it has developed into a science that has its roots in disciplines as diverse as genetics and systems engineering.

STUDY QUESTIONS

1. How do ecology and environmentalism differ? In what way does environmentalism depend on the science of ecology?
2. Define the terms *population*, *community*, *ecosystem*, *landscape*, *biome*, and *biosphere*.
3. How might including the abiotic environment within the framework of the ecosystem help ecologists achieve the basic goal of understanding the interaction of organisms with their environment?
4. What is a hypothesis? What is the role of hypotheses in science?
5. An ecologist observes that the diet of a bird species consists primarily of large grass seeds (as opposed to smaller grass seeds or the seeds of other herbaceous plants found in the area). He hypothesizes that the birds are choosing the larger seeds because they have a higher concentration of nitrogen than do other types of seeds at the site. To test the hypothesis, the ecologist compares the large grass seeds with the other types of seeds, and the results clearly show that the large grass seeds do indeed have a much higher concentration of nitrogen. Did the ecologist prove the hypothesis to be true? Can he conclude that the birds select the larger grass seeds because of their higher concentration of nitrogen? Why or why not?
6. What is a model? What is the relationship between hypotheses and models?
7. Given the importance of ecological research in making political and economic decisions regarding current environmental issues such as global warming, how do you think scientists should communicate uncertainties in their results to policy makers and the public?

FURTHER READINGS

Classic Studies

Bates, M. 1956. *The nature of natural history*. New York: Random House.

A lone voice in 1956, Bates shows us that environmental concerns have a long history prior to the emergence of the modern environmental movement. A classic that should be read by anyone interested in current environmental issues.

McKibben, W. 1989. *The end of nature*. New York: Random House.

In this provocative book, McKibben explores the philosophies and technologies that have brought humans to their current relationship with the natural world.

McIntosh, R. P. 1985. *The background of ecology: Concept and theory*. Cambridge: Cambridge University Press.

McIntosh provides an excellent history of the science of ecology from a scientific perspective.

Current Research

Coleman, D. 2010. *Big ecology; the emergence of ecosystem science*. Berkeley, University of California Press.

History of the development of large-scale ecosystem research and its politics and personalities as told by one of the participants.

Edgerton, F. N. 2012. *Roots of ecology*. Berkeley: University of California Press.

This book explores the deep ancestry of the science of ecology from the early ideas of Herodotos, Plato, and Pliny, up through those of Linnaeus, Darwin, and Haeckel.

Golley, F. B. 1993. *A history of the ecosystem concept in ecology: More than the sum of its parts*. New Haven: Yale University Press.

Covers the evolution and growth of the ecosystem concept as told by someone who was a major contributor to ecosystem ecology.

Kingsland, S. E. 2005. *The evolution of American ecology, 1890–2000*. Baltimore: Johns Hopkins University Press.

A sweeping, readable review of the evolution of ecology as a discipline in the United States, from its botanical beginnings to ecosystem ecology as colored by social, economic, and scientific influences.

Savill, P. S., C. M. Perrins, K. J. Kirby, N. Fisher, eds. 2010. *Wytham Woods: Oxford's ecological laboratory*. Oxford: Oxford University Press.

A revealing insight into some of the most significant population ecology studies by notable pioneering population ecologists such as Elton, Lack, Ford, and Southwood.

Worster, D. 1994. *Nature's economy*. Cambridge: Cambridge University Press.

This history of ecology is written from the perspective of a leading figure in environmental history.

CHAPTER

2 Climate

As the sun rises, warming the morning air in this tropical rain forest on the island of Borneo, fog that has formed in the cooler night air begins to evaporate.

CHAPTER GUIDE

WHAT DETERMINES WHETHER a particular geographic region will be a tropical forest, a grassy plain, or a barren landscape of sand dunes? The aspect of the physical environment that most influences a particular ecosystem by placing the greatest constraint on organisms is climate. *Climate* is a term we tend to use loosely. In fact, people sometimes confuse climate with weather. **Weather** is the combination of temperature, humidity, precipitation, wind, cloudiness, and other atmospheric conditions occurring at a specific place and time. **Climate** is the long-term average pattern of weather and may be local, regional, or global.

The structure of terrestrial ecosystems is largely defined by the dominant plants, which in turn reflect the prevailing physical environmental conditions, namely climate (see Chapter 23). Geographic variations in climate, primarily temperature and precipitation, govern the large-scale distribution of plants and therefore the nature of terrestrial ecosystems. Here, we learn how climate determines the availability of thermal energy and water on Earth's surface and influences the amount of solar energy that plants may harness.

2.1 Surface Temperatures Reflect the Difference between Incoming and Outgoing Radiation

Solar radiation—the electromagnetic energy (**Figure 2.1**) emanating from the Sun—travels more or less unimpeded through the vacuum of space until it reaches Earth's atmosphere. Scientists conceptualize solar radiation as a stream of photons, or packets of energy, that—in one of the great paradoxes of science—behave either as waves or as particles, depending on how they are observed. Scientists characterize waves of energy in terms of their wavelength (λ), or the physical distances between successive crests, and their frequency (ν), or the number of crests that pass a given point per second. All objects emit radiant energy, typically across a wide range of wavelengths. The exact nature of the energy emitted, however, depends on the object's temperature (**Figure 2.2**). The hotter the object is, the more energetic the emitted photons and the shorter the wavelength. A hot surface such as that of the Sun (~5800°C) gives off primarily shortwave (solar) radiation. In contrast, cooler objects such as Earth's surface (average temperature of 15°C) emit radiation of longer wavelengths, or longwave (terrestrial) radiation.

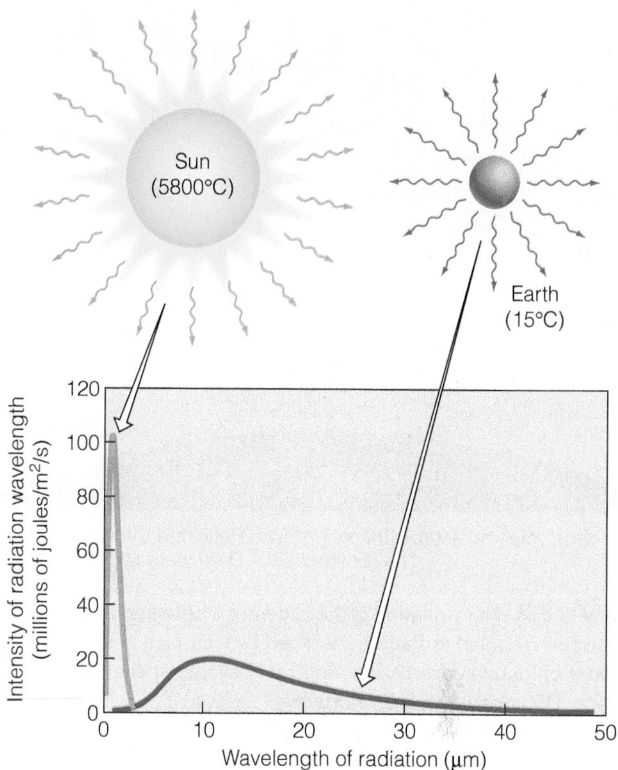

Figure 2.2 The wavelength of radiation emitted by an object is a function of its temperature. The Sun, with an average surface temperature of 5800°C, emits shortwave radiation as compared to Earth, with an average surface temperature of 15°C, which emits longwave radiation.

Some of the shortwave radiation that reaches the surface of our planet is reflected back into space. The quantity of shortwave radiation reflected by a surface is a function of its reflectivity, referred to as its *albedo*. Albedo is expressed as a proportion (0–1.0) of the shortwave radiation striking a surface that is reflected and differs for different surfaces. For example, surfaces covered by ice and snow have a high albedo (0.8–0.9), reflecting anywhere from 80 to 90 percent of incoming solar

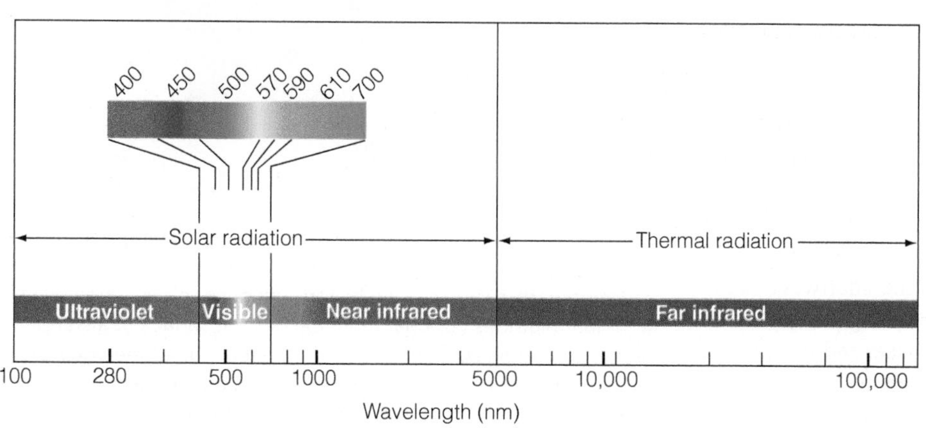

Figure 2.1 A portion of the electromagnetic spectrum, separated into solar (shortwave) and thermal (longwave) radiation. Ultraviolet, visible, and infrared light waves represent only a small part of the spectrum. To the left of ultraviolet radiation are X-rays and gamma rays (not shown).

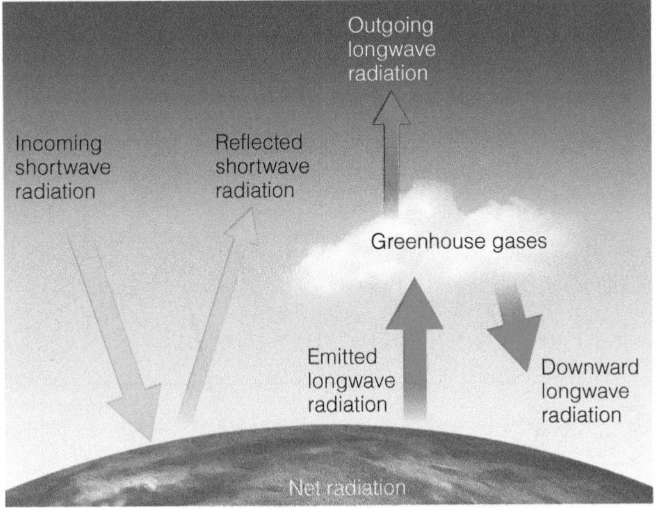

Figure 2.3 Net radiation is the difference between the amount of shortwave (solar) radiation absorbed by a surface and the amount of longwave radiation emitted back into space by that surface. LW, longwave; SW, shortwave.

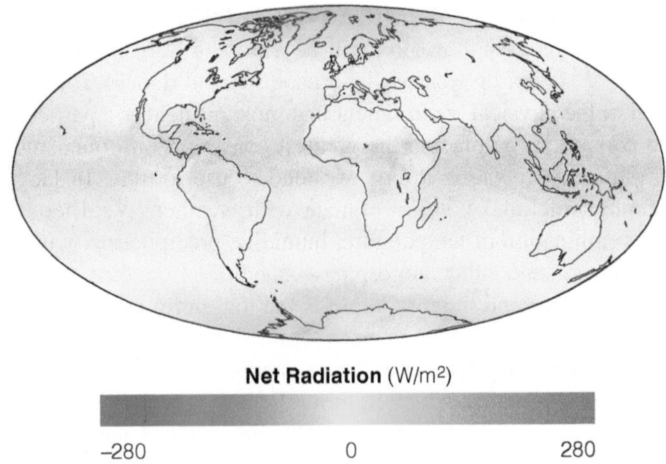

Figure 2.4 Global map of annual net radiation.

radiation, whereas a forest has a relatively low albedo (0.05), reflecting only 5 percent of sunlight. The global annual averaged albedo is approximately 0.30 (30 percent reflectance).

The difference between the incoming shortwave radiation and the reflected shortwave radiation is the net shortwave radiation absorbed by the surface. In turn, some of the energy absorbed by Earth's surface (both land and water) is emitted back out into space as terrestrial longwave radiation. The amount of energy emitted is dependent on the temperature of the surface. The hotter the surface, the more radiant energy it will emit. Most of the longwave radiation emitted by Earth's surface, however, is absorbed by water vapor and carbon dioxide in the atmosphere. This absorbed radiation is emitted downward toward the surface as longwave atmospheric radiation, which keeps near surface temperatures warmer than they would be without this blanket of gases. This is known as the "**greenhouse effect**," and gases such as water vapor and carbon dioxide that are good absorbers of longwave radiation are known as "**greenhouse gases**."

It is the difference between the incoming shortwave (solar) radiation and outgoing longwave (terrestrial) radiation that defines the **net radiation** (Figure 2.3) and determines surface temperatures. If the amount of incoming shortwave radiation exceeds the amount of outgoing longwave radiation, surface temperature increases. Conversely, surface temperature declines if the quantity of outgoing longwave radiation exceeds the incoming shortwave radiation (as is the case during the night). On average, the amount of incoming shortwave radiation intercepted by Earth and the quantity of longwave radiation emitted by the planet back into space balance, and the average surface temperature of our planet remains approximately 15°C. Note, however, from the global map of average annual surface net radiation presented in Figure 2.4 that there is a distinct latitudinal gradient of decreasing net surface radiation from the equator toward the poles. This decline is a direct function of the variation with latitude in the amount of shortwave radiation reaching the surface. Two factors influence this variation (Figure 2.5). First, at higher latitudes, solar radiation hits the surface at a steeper angle, spreading sunlight over a larger area. Second, solar radiation that penetrates the atmosphere at a steep angle must travel through a deeper layer of air. In the process, it encounters more particles in the atmosphere, which reflect more of the shortwave radiation back into space. The result of the decline in net radiation with latitude is a distinct gradient of decreasing mean annual temperature from the equator toward the poles (Figure 2.6).

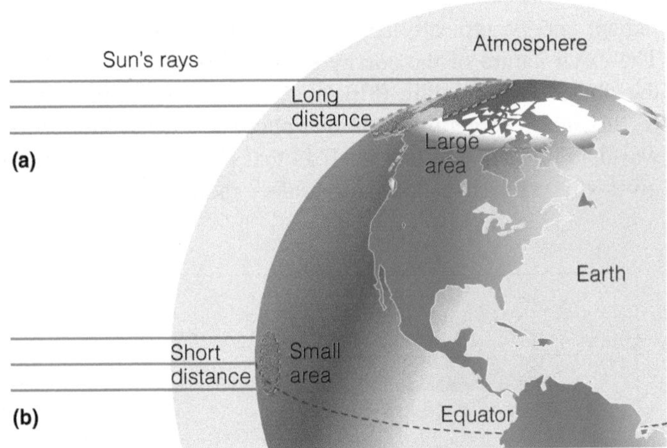

Figure 2.5 As one moves from the equator to the poles, there is a decrease in the average amount of solar (shortwave) radiation reaching Earth's surface. Two factors influence this variation. First, at higher latitudes (a), solar radiation hits the surface at a steeper angle, spreading sunlight over a larger area than at the equator (b). Second, solar radiation that penetrates the atmosphere at a steep angle must travel through a deeper layer of air.

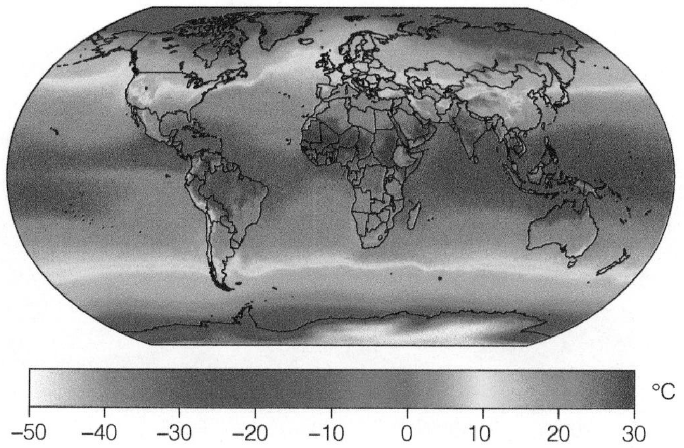
°C
−50 −40 −30 −20 −10 0 10 20 30

Figure 2.6 Global map of mean annual temperature (°C). Map based on annually averaged near-surface air temperature from 1961 to 1990.

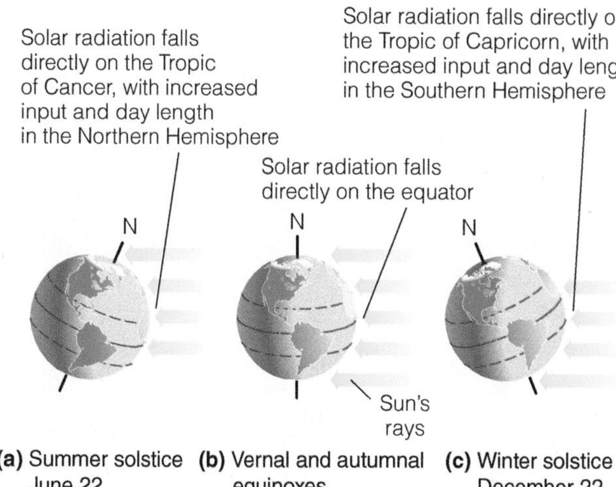
Solar radiation falls directly on the Tropic of Cancer, with increased input and day length in the Northern Hemisphere
Solar radiation falls directly on the Tropic of Capricorn, with increased input and day length in the Southern Hemisphere
Solar radiation falls directly on the equator
N
N
N
Sun's rays
(a) Summer solstice June 22
(b) Vernal and autumnal equinoxes
(c) Winter solstice December 22

Figure 2.7 Changes in the angle of the Sun and circle of illumination during Earth's yearly orbit (equinoxes and the winter and summer solstices are illustrated). Note that as a result of the 23.5° tilt of Earth on its north–south axis, the point of Earth's surface where the Sun is directly overhead migrates from the tropic of Cancer (23.5° N) to the tropic of Capricorn (23.5° S) over the course of the year.

2.2 Intercepted Solar Radiation and Surface Temperatures Vary Seasonally

Although the variation in shortwave (solar) radiation reaching Earth's surface with latitude can explain the gradient of decreasing mean annual temperature from the equator to the poles, it does not explain the systematic variation occurring over the course of a year. What gives rise to the seasons on Earth? Why do the hot days of summer give way to the changing colors of fall, or the freezing temperatures and snow-covered landscape of winter to the blanket of green signaling the onset of spring? The explanation is quite simple: it is because Earth does not stand up straight but rather tilts to its side.

Earth, like all planets, is subject to two distinct motions. While it orbits the Sun, Earth rotates about an axis that passes through the North and South Poles, giving rise to the brightness of day followed by the darkness of night (the diurnal cycle). Earth travels about the Sun in an ecliptic plane. By chance, Earth's axis of spin is not perpendicular to the ecliptic plane but tilted at an angle of 23.5°. As a result, as Earth follows its elliptical orbit about the Sun, the location on the surface where the Sun is directly overhead at midday migrates between 23.5° N and 23.5° S latitude over the course of the year (**Figure 2.7**).

At the vernal equinox (approximately March 21) and autumnal equinox (approximately September 22), the Sun is directly overhead at the equator (see Figure 2.7). At this time, the equatorial region receives the greatest input of shortwave (solar) radiation, and every place on Earth receives the same 12 hours each of daylight and night.

At the summer solstice (approximately June 22) in the Northern Hemisphere, solar rays fall directly on the Tropic of Cancer (23.5° N; see Figure 2.7). This is when days are longest in the Northern Hemisphere, and the input of solar radiation to the surface is the greatest. In contrast, the Southern Hemisphere experiences winter at this time. At winter solstice (about December 22) in the Northern Hemisphere, solar rays fall directly on the Tropic of Capricorn (23.5° S; see Figure 2.7). This period is summer in the Southern Hemisphere, whereas the Northern Hemisphere is enduring shorter days and colder temperatures. Thus, the summer solstice in the Northern Hemisphere is the winter solstice in the Southern Hemisphere.

In the equatorial region there is little seasonality (variation over the year) in net radiation, temperature, or day length. Seasonality systematically increases from the equator to the poles (**Figure 2.8**). At the Arctic and Antarctic circles (66.5° N and S, respectively), day length varies from 0 to 24 hours over the course of the year. The days shorten until the winter solstice, a day of continuous darkness. The days lengthen with spring, and on the day of the summer solstice, the Sun never sets.

2.3 Geographic Difference in Surface Net Radiation Result in Global Patterns of Atmospheric Circulation

As we discussed in the previous section, the average net radiation of the planet is zero; that is to say that the amount of incoming shortwave radiation absorbed by the surface is offset by the quantity of outgoing longwave radiation back into space. Otherwise, the average temperature of the planet would either increase or decrease. Geographically, however, this is not the case. Note from the global map of mean annual net radiation presented in Figure 2.4 that there are regions of positive (surplus) and negative (deficit) net radiation. In fact, there is a distinct latitudinal pattern of surface radiation illustrated in

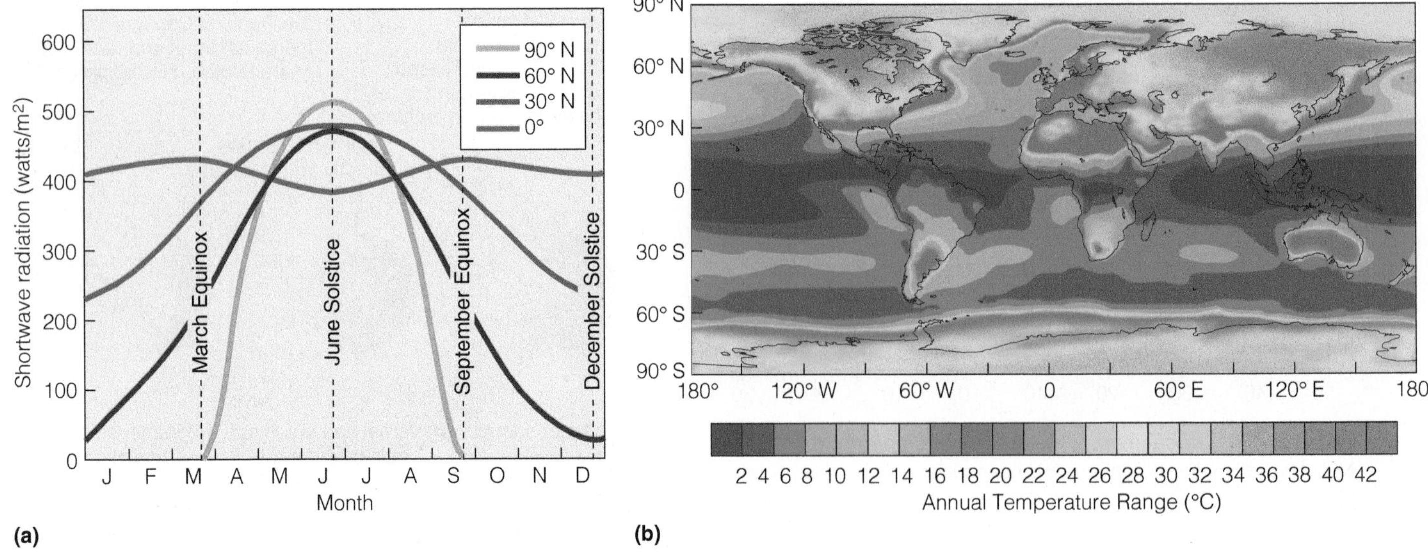

Figure 2.8 Two examples of changes in seasonality with latitude. (a) Annual variations in mean monthly solar (shortwave radiation) for different latitudes in the Northern Hemisphere. (b) Global map of annual temperature range, defined as the difference in temperature (°C) between the coldest and warmest month of the year (based on mean monthly temperatures for the period of 1979–2004).

Figure 2.9. Between 35.5° N and 35.5° S (from the equator to the midlatitudes), the amount of incoming shortwave radiation received over the year exceeds the amount of outgoing longwave radiation and there is a surplus. In contrast, from 35.5° N and S latitude to the poles (90° N and S), the amount of outgoing longwave radiation over the year exceeds the incoming shortwave radiation and there is a deficit. This imbalance in net radiation sets into motion a global scale pattern of the redistribution of thermal energy (heat) from the equator to the poles. Recall from basic physical sciences that energy flows from regions of higher concentration to regions of lower concentration, that is, from warmer regions to cooler regions. The primary mechanism of this planetary transfer of heat from the tropics (region of net radiation surplus) to the poles (region of net radiation deficit) is the process of convection, that is, the transfer of heat through the circulation of fluids (air and water).

As previously discussed, the equatorial region receives the largest annual input of solar radiation and greatest net radiation surplus. Air warmed at the surface rises because it is less dense than the cooler air above it. Air heated at the equatorial region rises to the top of the troposphere, establishing a zone of low pressure at the surface (**Figure 2.10**). This low atmospheric pressure at the surface causes air from the north and south to flow toward the equator (air moves from areas of higher pressure to areas of lower pressure). The resulting convergence of winds from the north and south in the region of the equator is called the *Intertropical Convergence Zone*, or *ITCZ*, for short.

The continuous column of rising air at the equator forces the air mass above to spread north and south toward the poles. As air masses move poleward, they cool, become heavier (more dense), and sink. The sinking air at the poles raises

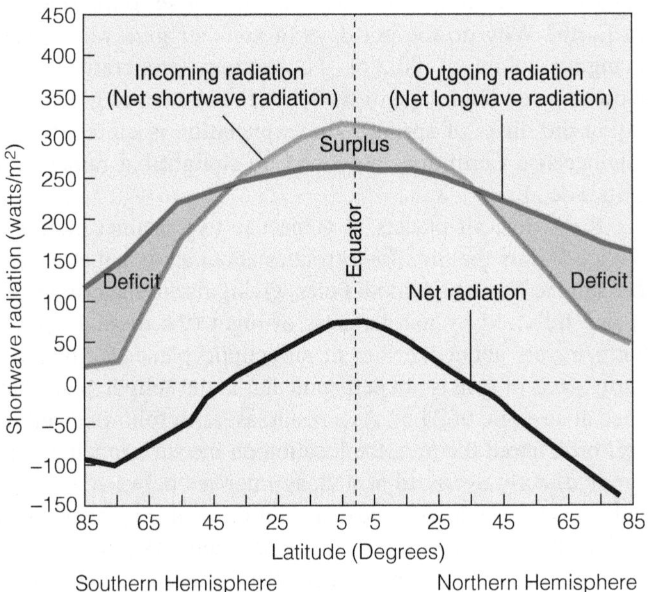

Figure 2.9 Variation in mean annual incoming shortwave radiation, outgoing longwave radiation, and net radiation as a function of latitude. Note that from the equator to approximately 35° N and S latitude, the amount of incoming shortwave radiation exceeds the amount of outgoing longwave radiation, and there is a net surplus of surface radiation (mean annual net radiation > 0). Conversely, there is a deficit (mean net radiation < 0) from 35° N and S to the poles (90° N and S). This gradient of net radiation drives the transport of heat from the tropics to the poles through the circulation of the atmosphere and oceans.

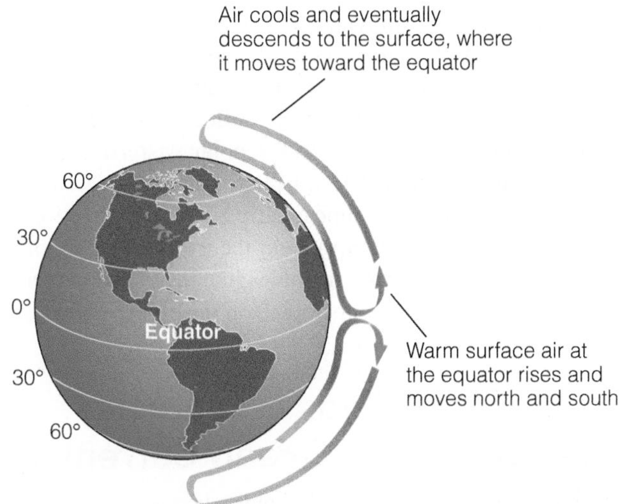

Figure 2.10 Circulation of air cells and prevailing winds on an imaginary, nonrotating Earth. Air heated at the equator rises and moves north and south creating a zone of low pressure at the surface. After cooling at the poles, it descends, creating a high pressure zone at the poles causing air to flow back toward the equator.

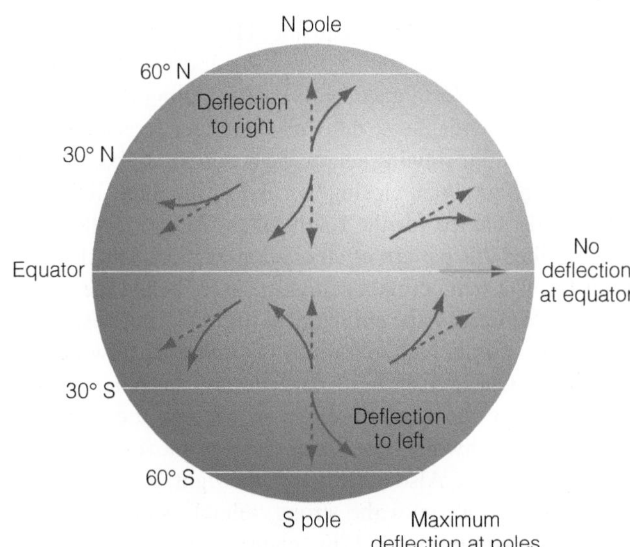

Figure 2.11 Effect of the Coriolis force on wind direction. The effect is absent at the equator, where the linear velocity is the greatest, 465 meters per second (m/s; 1040 mph). Any object on the equator is moving at the same rate. The Coriolis effect increases regularly toward the poles. If an object, including an air mass, moves northward from the equator at a constant speed, it speeds up because Earth moves more slowly (403 m/s at 30° latitude, 233 m/s at 60° latitude, and 0 m/s at the poles) than the object does. As a result, the object's path appears to deflect to the right or east in the Northern Hemisphere and to the left or west in the Southern Hemisphere.

surface air pressure, forming a high-pressure zone and creating a pressure gradient from the poles to the equator. The cooled, heavier air then flows toward the low-pressure zone at the equator, replacing the warm air rising over the tropics and closing the pattern of air circulation. If Earth were stationary and without irregular landmasses, the atmosphere would circulate as shown in Figure 2.10. Earth, however, spins on its axis from west to east. Although each point on Earth's surface makes a complete rotation every 24 hours, the speed of rotation varies with latitude (and circumference). At a point on the equator (its widest circumference at 40,176 km), the speed of rotation is 1674 km per hour. In contrast, at 60° N or S, Earth's circumference is approximately half that at the equator (20,130 km), and the speed of rotation is 839 km per hour. According to the law of angular motion, the momentum of an object moving from a greater circumference to a lesser circumference will deflect in the direction of the spin, and an object moving from a lesser circumference to a greater circumference will deflect in the direction opposite that of the spin. As a result, air masses and all moving objects in the Northern Hemisphere are deflected to the right (clockwise motion), and in the Southern Hemisphere to the left (counterclockwise motion). This deflection in the pattern of air flow is the **Coriolis effect**, named after the 19th-century French mathematician G. C. Coriolis, who first analyzed the phenomenon (**Figure 2.11**).

In addition to the deflection resulting from the Coriolis effect, air that moves poleward is subject to longitudinal compression, that is, poleward-moving air is forced into a smaller space, and the density of the air increases. These factors prevent a direct, simple flow of air from the equator to the poles. Instead, they create a series of belts of prevailing winds, named for the direction they come from. These belts break the simple flow of surface air toward the equator and they flow aloft to the poles into a series of six cells, three in each hemisphere. They produce areas of low and high pressure as air masses ascend from and descend toward the surface, respectively (**Figure 2.12**). To trace the flow of air as it circulates between the equator and poles, we begin at Earth's equatorial region, which receives the largest annual input of solar radiation.

Air heated in the equatorial zone rises upward, creating a low-pressure zone near the surface—the **equatorial low**. This upward flow of air is balanced by a flow of air from the

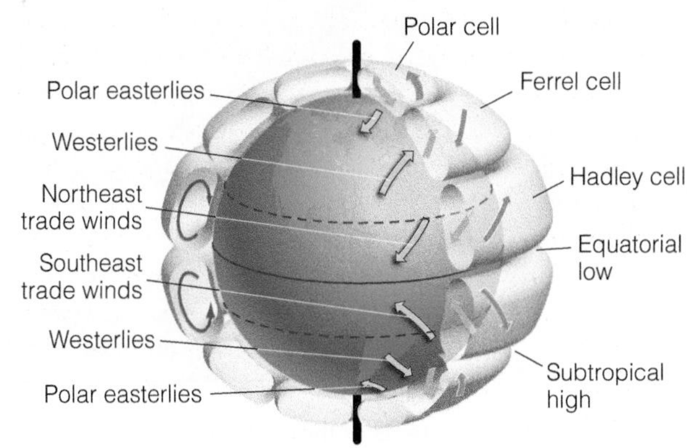

Figure 2.12 Belts and cells of air circulation about a rotating Earth. This circulation gives rise to the trade, westerly, and easterly winds.

north and south toward the equator (ITCZ). As the warm air mass rises, it begins to spread, diverging northward and southward toward the North and South Poles, cooling as it goes. In the Northern Hemisphere, the Coriolis effect forces air in an easterly direction, slowing its progress north. At about 30° N, the now-cool air sinks, closing the first of the three cells—the Hadley cells, named for the Englishman George Hadley, who first described this pattern of circulation in 1735. The descending air forms a semipermanent high-pressure belt at the surface that encircles Earth—the **subtropical high**. Having descended, the cool air warms and splits into two currents flowing over the surface. One moves northward toward the pole, diverted to the right by the Coriolis effect to become the prevailing **westerlies**. Meanwhile, the other current moves southward toward the equator. Also deflected to the right, this southward-flowing stream becomes the strong, reliable winds that were called **trade winds** by the 17th-century merchant sailors who used them to reach the Americas from Europe. In the Northern Hemisphere, these winds are known as the *northeast trades*. In the Southern Hemisphere, where similar flows take place, these winds are known as the *southeast trades*.

As the mild air of the westerlies moves poleward, it encounters cold air moving down from the pole (approximately 60° N). These two air masses of contrasting temperature do not readily mix. They are separated by a boundary called the *polar front*—a zone of low pressure (the **subpolar low**) where surface air converges and rises. Some of the rising air moves southward until it reaches approximately 30° latitude (the region of the subtropical high), where it sinks back to the surface and closes the second of the three cells—the Ferrel cell, named after U.S. meteorologist William Ferrel.

As the northward-moving air reaches the pole, it slowly sinks to the surface and flows back (southward) toward the polar front, completing the last of the three cells—the polar cell. This southward-moving air is deflected to the right by the Coriolis effect, giving rise to the **polar easterlies**. Similar flows occur in the Southern Hemisphere (see Figure 2.12).

This pattern of global atmospheric circulation functions to transport heat (thermal energy) from the tropics (the region of net radiation surplus) toward the poles (the regions of net radiation deficit), moderating temperatures at the higher latitudes.

2.4 Surface Winds and Earth's Rotation Create Ocean Currents

The global pattern of prevailing winds plays a crucial role in determining major patterns of surface water flow in Earth's oceans. These systematic patterns of water movement are called *currents*. In fact, until they encounter one of the continents, the major ocean currents generally mimic the movement of the surface winds presented in the previous section.

Each ocean is dominated by two great circular water motions, or **gyres**. Within each gyre, the ocean current moves clockwise in the Northern Hemisphere and counterclockwise in the Southern Hemisphere (**Figure 2.13**). Along the equator, trade winds push warm surface waters westward. When these waters encounter the eastern margins of continents, they split into north- and south-flowing currents along the coasts, forming north and south gyres. As the currents move farther from the equator, the water cools. Eventually, they encounter the westerly winds at higher latitudes (30–60° N

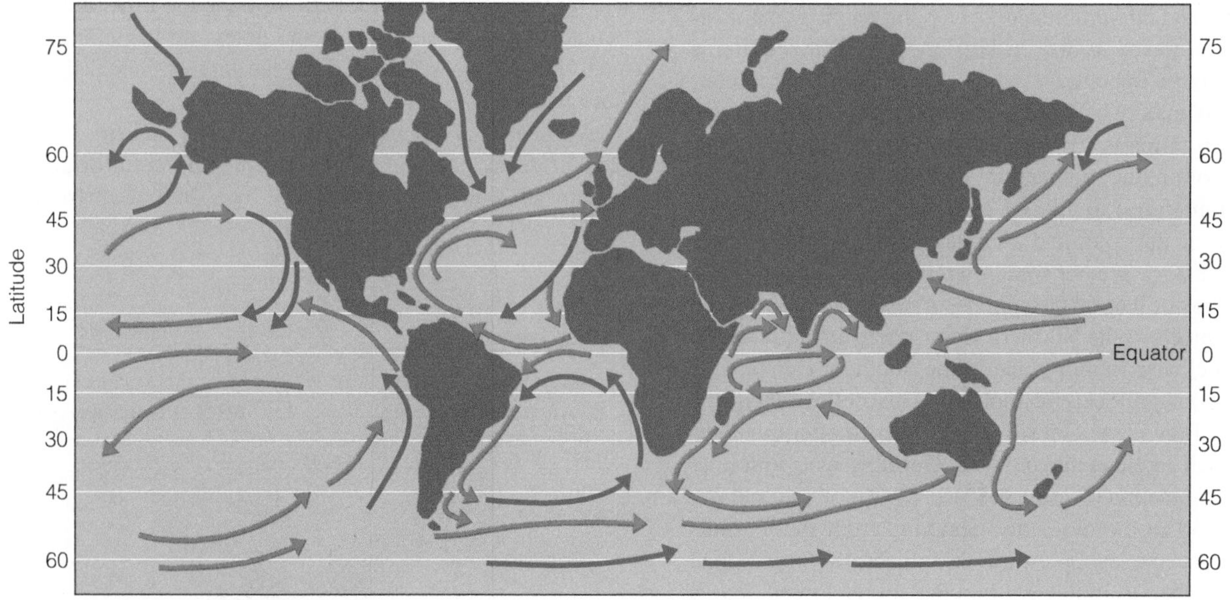

Figure 2.13 Ocean currents of the world. Notice how the circulation is influenced by the Coriolis force (clockwise movement in the Northern Hemisphere and counterclockwise movement in the Southern Hemisphere) and continental landmasses, and how oceans are connected by currents. Blue arrows represent cool water, and red arrows represent warm water.

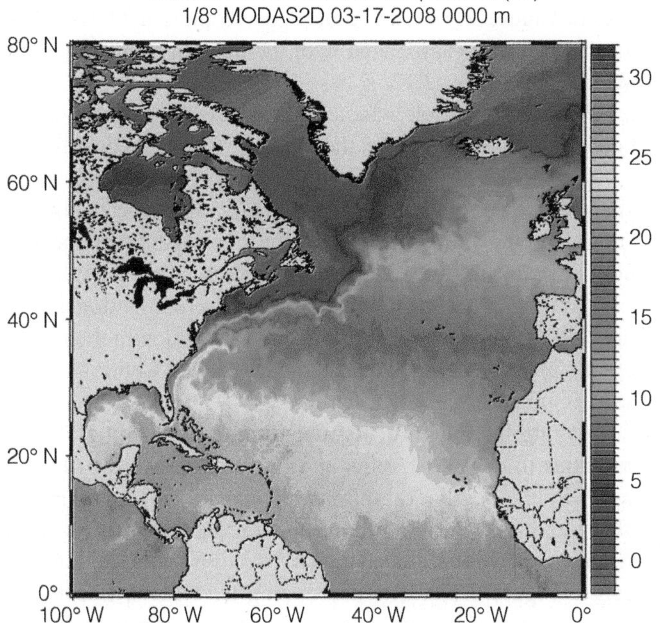

Figure 2.14 Color enhanced satellite image of the Gulf Steam current in the North Atlantic Ocean. Surface water temperatures increase from blue-green-yellow to red. The Gulf Stream carries warm tropical waters northward along the east coast of North America and into the cold waters of the North Atlantic moderating temperatures in Western Europe.

and 30–60° S), which produce eastward-moving currents. When these eastward-moving currents encounter the western margins of the continents, they form cool currents that flow along the coastline toward the equator. Just north of the Antarctic continent, ocean waters circulate unimpeded around the globe.

As with the patterns of global atmospheric circulation and winds, the gyres function to redistribute heat from the tropics northward and southward toward the poles (**Figure 2.14**).

2.5 Temperature Influences the Moisture Content of Air

Air temperature plays a crucial role in the exchange of water between the atmosphere and Earth's surface. Whenever matter, including water, changes from one state to another, energy is either absorbed or released. The amount of energy released or absorbed (per gram) during a change of state is known as *latent heat* (from the Latin *latens*, "hidden"). In going from a more ordered state (liquid) to a less ordered state (gas), energy is absorbed (the energy required to break bonds between molecules). While going from a less ordered to a more ordered state, energy is released. Evaporation, the transformation of water from a liquid to a gaseous state, requires 2260 joules (J) of energy per gram of liquid water to be converted to water vapor (1 joule is the equivalent of 1 watt of power radiated or dissipated for 1 second). Condensation, the transformation of water vapor to a liquid state, releases an equivalent amount of energy. When air comes into contact with liquid water, water molecules are freely exchanged between the air and the water's surface. When the evaporation rate equals the condensation rate, the air is said to be saturated. In the air, water vapor acts as an independent gas that has weight and exerts pressure. The amount of pressure that water vapor exerts independent of the pressure of dry air is called vapor pressure. Vapor pressure is typically defined in units of pascals (Pa). The water vapor content of air at saturation is called the *saturation vapor pressure*. The saturation vapor pressure, also known as the *water vapor capacity of air*, cannot be exceeded. If the vapor pressure exceeds the capacity, condensation occurs and reduces the vapor pressure. Saturation vapor pressure varies with temperature, increasing as air temperature increases (**Figure 2.15**). Having a greater quantity of thermal energy to support evaporation, warm air has a greater capacity for water vapor than does cold air.

The amount of water in a given volume of air is its absolute humidity. A more familiar measure of the water content of the air is **relative humidity**, or the amount of water vapor in the air expressed as a percentage of the saturation vapor pressure. At saturation vapor pressure, the relative humidity is 100 percent. If air cools while the actual moisture content (water vapor pressure) remains constant, then relative humidity increases as the value of saturation vapor pressure

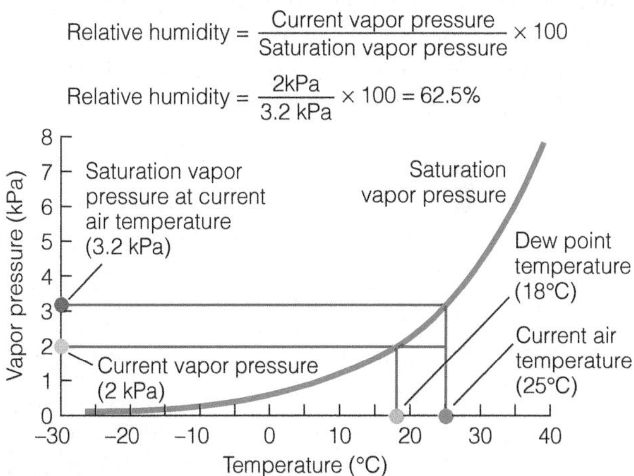

Figure 2.15 Saturation vapor pressure (VP) as a function of air temperature (saturation VP increases with air temperature). For a given air temperature, the relative humidity is the ratio of current VP to saturation VP (current VP/saturation VP) × 100. For a given VP, the temperature at which saturation VP occurs is called the dew point.

Interpreting Ecological Data

Q1. *Assume that the actual (current) water vapor pressure remains the same over the course of the day and that the current air temperature of 25°C in the above graph represents the air temperature at noon (12:00 p.m.). How would you expect the relative humidity to change from noon to 5:00 p.m.? Why?*

Q2. *What is the approximate relative humidity at 35°C? (Assume that actual water vapor pressure remains the same as in the above figure, 2 kilopascals [kPa].)*

declines. If the air cools to a point where the actual vapor pressure is equal to the saturation vapor pressure, moisture in the air will condense. This is what occurs when a warm parcel of air at the surface becomes buoyant and rises. As it rises, it cools, and as it cools, the relative humidity increases. When the relative humidity reaches 100 percent, water vapor condenses and forms clouds. As soon as particles of water or ice in the air become too heavy to remain suspended, precipitation falls. For a given water content of a parcel of air (vapor pressure), the temperature at which saturation vapor pressure is achieved (relative humidity is 100 percent) is called the **dew point temperature**. Think about finding dew or frost on a cool fall morning. As nightfall approaches, temperatures drop and relative humidity rises. If cool night air temperatures reach the dew point, water condenses and dew forms, lowering the amount of water in the air. As the sun rises, air temperature warms and the water vapor capacity (saturation vapor pressure) increases. As a result, the dew evaporates, increasing vapor pressure in the air.

2.6 Precipitation Has a Distinctive Global Pattern

By bringing together patterns of temperature, winds, and ocean currents, we are ready to understand the global pattern of precipitation. Precipitation is not evenly distributed across Earth (**Figure 2.16**). At first the global map of annual precipitation in Figure 2.16 may seem to have no discernible pattern or regularity. But if we examine the simpler pattern of variation in average rainfall with latitude (**Figure 2.17**), a general pattern emerges. Precipitation is highest in the region of the equator, declining as one moves north and south. The decline, however, is not continuous. Two troughs occur in the midlatitudes interrupting the general patterns of decline in precipitation from the equator toward the poles. The sequence of peaks and troughs seen in Figure 2.17 corresponds to the pattern of rising and falling air masses associated with the belts of prevailing winds presented in Figure 2.12.

As the warm trade winds move across the tropical oceans, they gather moisture. Near the equator, the northeasterly trade winds meet the southeasterly trade winds. This narrow region where the trade winds meet is the ITCZ, characterized by high amounts of precipitation. Where the two air masses meet, air piles up, and the warm humid air rises and cools. When the dew point is reached, clouds form, and precipitation falls as rain. This pattern accounts for high precipitation in the tropical regions of eastern Asia, Africa, and South and Central America (see Figure 2.16).

Having lost much of its moisture, the ascending air mass continues to cool as it splits and moves northward and southward. In the region of the subtropical high (approximately 30° N and S), where the cool air descends, two belts of dry climate encircle the globe (the two troughs at the midlatitudes seen in Figure 2.17). The descending air warms. Because the saturation vapor pressure rises, it draws water from the surface through evaporation, causing arid conditions. In these belts, the world's major deserts have formed (see Chapter 23).

As the air masses continue to move north and south, they once again draw moisture from the surface, but to a lesser degree because of the cooler surface conditions. Moving poleward, they encounter cold air masses originating at the poles (approximately 60° N and S). Where the surface air masses converge and rise, the ascending air mass cools and precipitation occurs (seen as the two smaller peaks in precipitation between 50° and 60° N and S in Figure 2.17). From this point on to the poles, the cold temperature and associated low-saturation vapor pressure function to restrict precipitation.

One other pattern is worth noting in Figure 2.17. In general, rainfall is greater in the Southern Hemisphere than in the

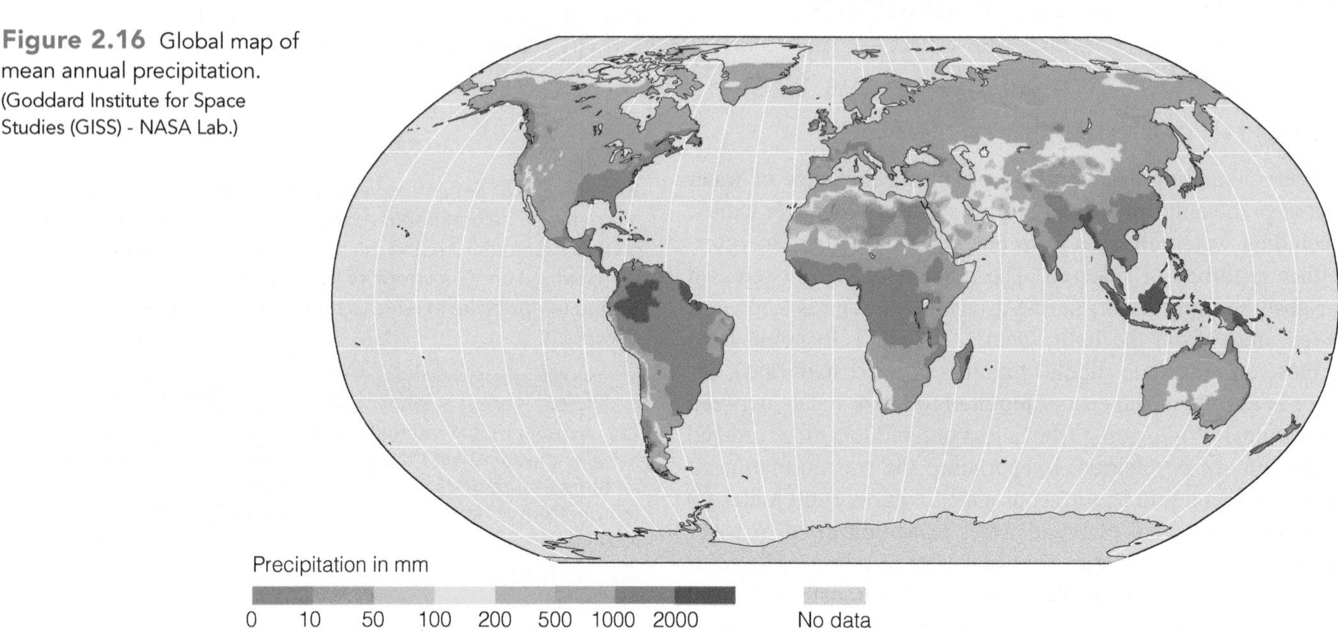

Figure 2.16 Global map of mean annual precipitation. (Goddard Institute for Space Studies (GISS) - NASA Lab.)

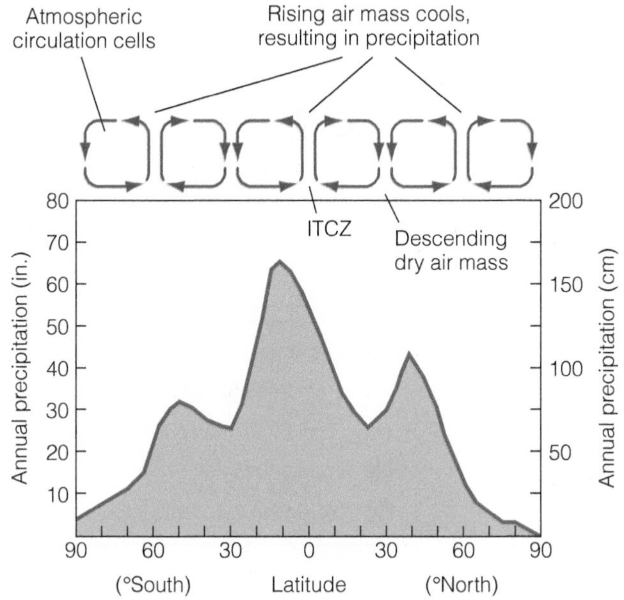

Figure 2.17 Variation in mean annual precipitation with latitude. The peaks in rainfall correspond to rising air masses, whereas the troughs are associated with descending dry air masses.

Northern Hemisphere (note the southern shift in the rainfall peak associated with the ITCZ). This is because the oceans cover a greater proportion of the Southern Hemisphere, and water evaporates more readily from the water's surface than from the soil and vegetation.

Missing from our discussion thus far is the temporal variation of precipitation over Earth. The temporal variation is directly linked to the seasonal changes in the surface radiation balance of Earth and its effect on the movement of global pressure systems and air masses. This is illustrated in seasonal movement north and south of the ITCZ, which follows the apparent migration of the direct rays of the Sun (**Figure 2.18**).

The ITCZ is not stationary but tends to migrate toward regions of the globe with the warmest surface temperature. Although tropical regions around the equator are always exposed to warm temperatures, the Sun is directly over the geographical equator only twice a year, at the spring and fall equinoxes. At the northern summer solstice, the Sun is directly over the Tropic of Cancer; at the winter solstice (which is summer in the Southern Hemisphere), the Sun is directly over the Tropic of Capricorn. As a result, the ITCZ moves poleward and invades the subtropical highs in northern summer; in the winter it moves southward, leaving clear, dry weather behind. As the ITCZ migrates southward, it brings rain to the southern summer. Thus, as the ITCZ shifts north and south, it brings on the wet and dry seasons in the tropics (**Figure 2.19**).

2.7 Proximity to the Coastline Influences Climate

At the continental scale, an important influence on climate is the relationship between land and water. Land surfaces heat and cool more rapidly than water as a result of differences in their specific heat. Specific heat is the amount of thermal energy necessary to raise the temperature of one gram of a substance by 1°C. The specific heat of water is much higher than that of land or air. It takes approximately four times the amount of thermal energy to raise the temperature of water by 1°C than land or air. As a result, land areas farther from the coast (or other large bodies of water) experience a greater seasonal variation in temperature than do coastal areas (**Figure 2.20**). This pattern is referred to as **continentality**. Annual differences of as much as 100°C (from 50°C to –50°C) have been recorded in some locations.

The converse effect occurs in coastal regions. These locations have smaller temperature ranges as a result of what is called a *maritime influence*. Summer and winter extremes are moderated by the movement onshore of prevailing westerly wind systems from the ocean. Ocean currents minimize seasonal variations in the surface temperature of the water. The moderated water temperature serves to moderate temperature changes in the air mass above the surface.

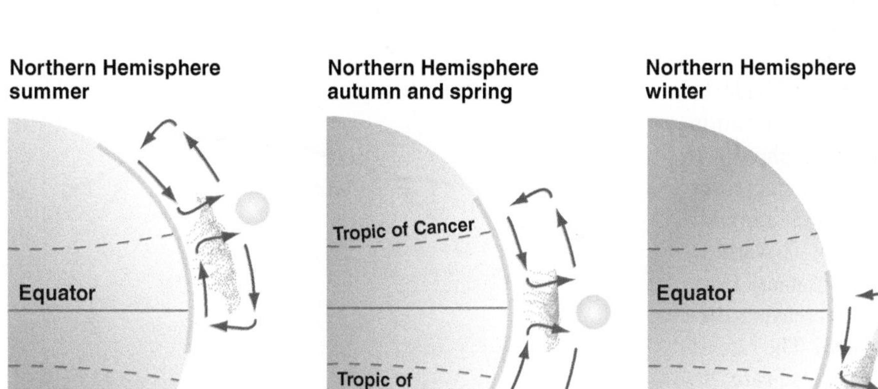

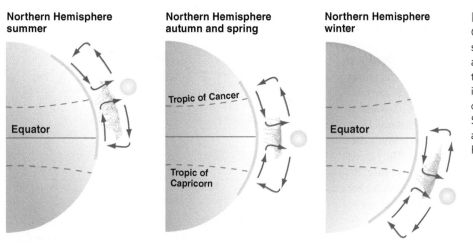

Figure 2.18 Shifts of the Intertropical Convergence Zone (ITCZ), producing seasonality in precipitation—rainy seasons and dry seasons. As the distance from the equator increases, the dry season is longer and the rainfall is less. These oscillations result from changes in the Sun's altitude between the equinoxes and the solstices, as diagrammed in Figure 2.7.

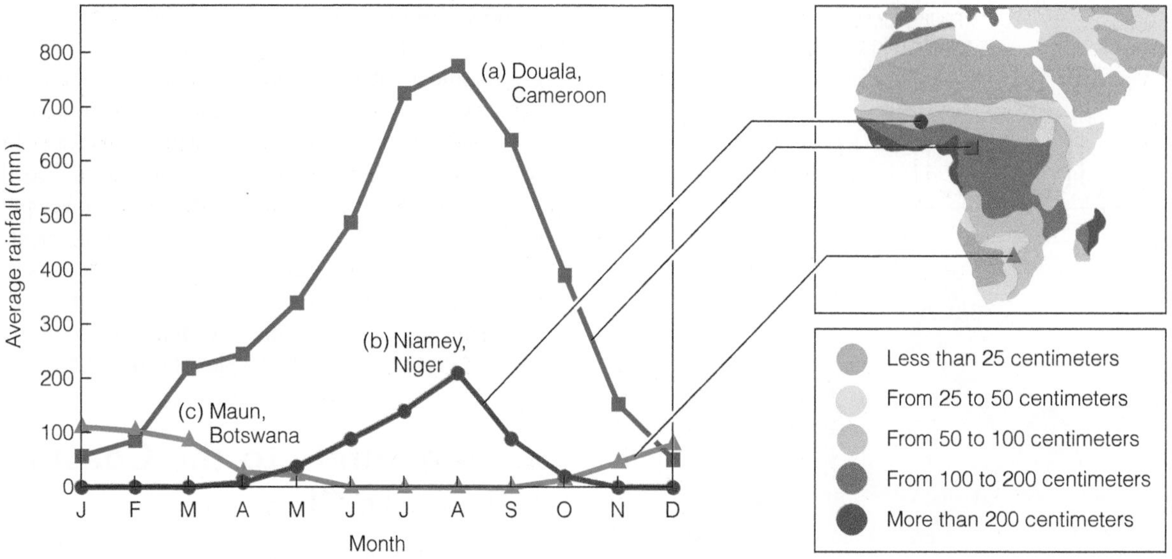

Figure 2.19 Seasonal variations in precipitation at three sites within the Intertropical Convergence Zone (ITCZ). Although site (a) shows a seasonal variation, precipitation exceeds 50 mm each month. Sites (b) and (c) are in the ITCZ regions that experience a distinct wet (summer) and dry (winter) season. The rainy season is six months out of phase for these two sites, reflecting the difference in the timing because the summer months occur at different times in the two hemispheres.

Proximity to large water bodies also tends to have a positive influence on precipitation levels. The interior of continents generally experience less precipitation than the coastal regions do. As air masses move inland from the coast, water vapor lost from the atmosphere through precipitation is not recharged (from surface evaporation) as readily as it is over the open waters of the ocean (note the gradients of precipitation from the coast to the interiors of North America and Europe/Asia in Figure 2.16). There are, however, notable exceptions to this rule, including the dry coast of southern California and the Arctic coastline of Alaska.

2.8 Topography Influences Regional and Local Patterns of Climate

Mountainous topography influences local and regional patterns of climate. Most obvious is the relationship between elevation and temperature. In the lower regions of the atmosphere (up to altitudes of approximately 12 km), temperature decreases with altitude at a fairly uniform rate because of declining air density and pressure. In addition, the atmosphere is warmed by conduction (transfer of heat through direct contact) from Earth's surface. So temperature declines with increasing distance from the conductive source (i.e., the surface). The rate of decline in temperature with altitude is called the **lapse rate**. So for the same latitude or proximity to the coast, locales at higher elevation will have consistently lower temperatures than those of lower elevation.

Mountains also influence patterns of precipitation. As an air mass reaches a mountain, it ascends, cools, relative

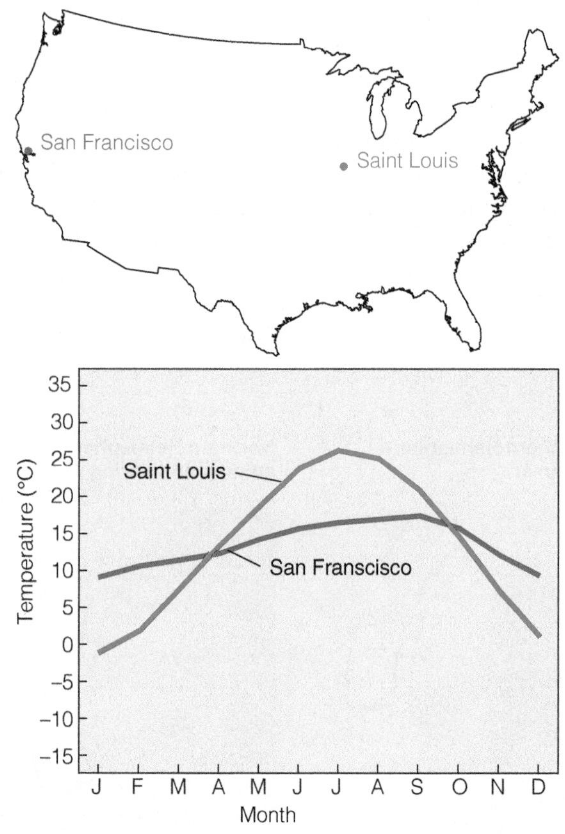

Figure 2.20 Pattern of mean monthly temperature (°C) for two locations in North America: San Francisco is located on the Pacific coast, whereas Saint Louis is in the middle of the continent.

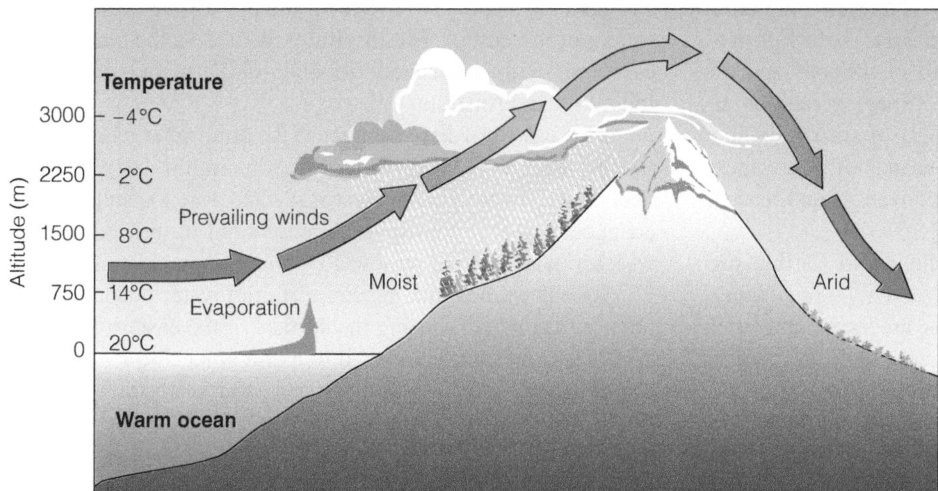

Figure 2.21 Formation of a rain shadow. Air is forced to go over a mountain. As it rises, the air mass cools and loses its moisture as precipitation on the windward side. The descending air, already dry, picks up moisture from the leeward side.

humidity rises (because of lower saturation vapor pressure). When the temperature cools to the dew point temperature, precipitation occurs at the upper altitudes of the windward side. As the now cool, dry air descends the leeward side, it warms again and relative humidity declines. As a result, the windward side of a mountain supports denser, more vigorous vegetation and different species of plants and associated animals than does the leeward side, where in some areas dry, desert-like conditions exist. This phenomenon is called a **rain shadow** (Figure 2.21). Thus, in North America, the westerly winds that blow over the Sierra Nevada and the Rocky Mountains, dropping their moisture on west-facing slopes, support vigorous forest growth. By contrast, the eastern slopes exhibit semidesert or desert conditions.

Some of the most pronounced effects of this same phenomenon occur in the Hawaiian Islands. There, plant cover ranges from scrubby vegetation on the leeward side of an island to moist, forested slopes on the windward side (Figure 2.22).

2.9 Irregular Variations in Climate Occur at the Regional Scale

The patterns of temporal variation in climate that we have discussed thus far occur at regular and predictable intervals: seasonal changes in temperature with the rotation of Earth around the Sun, and migration of the ITCZ with the resultant seasonality of rainfall in the tropics and monsoons in Southeast Asia. Not all features of the climate system, however, occur so regularly. Earth's climate system is characterized by variability at both the regional and global scales. The Little Ice Age, a period of cooling that lasted from approximately the mid-14th to the mid-19th century, brought bitterly cold winters to many parts of the Northern Hemisphere, affecting agriculture, health, politics, economics, emigration, and even art and literature. In the mid-17th century, glaciers in the Swiss Alps advanced, gradually engulfing farms and crushing entire villages. In 1780, New York Harbor froze, allowing people to walk from Manhattan to Staten Island. In fact, the image of a white Christmas evoked by Charles Dickens and the New England poets of the 18th and 19th centuries is largely a product of the cold and snowy winters of the Little Ice Age. But the climate has since warmed to the point that a white Christmas in these regions is becoming an anomaly.

The Great Plains region of central North America has undergone periods of drought dating back to the mid-Holocene period some 5000 to 8000 years ago, but the homesteaders of the early 20th century settled the Great Plains at a time of relatively wet summers. They assumed these moisture conditions were the norm, and they employed the agricultural methods they had used in the East. So they broke the prairie sod for crops, but the cycle of drought returned, and the prairie grasslands became a dust bowl (see Chapter 4, *Ecological Issues & Applications*).

These examples reflect the variability in Earth's climate systems, which operate on timescales ranging from decades

Figure 2.22 Rain shadow on the mountains of Maui, Hawaiian Islands. The windward, east-facing slopes intercept the trade winds and are cloaked with wet forest (left). Low-growing, shrubby vegetation is found on the dry side (right).

to tens of thousands of years, driven by changes in the input of energy to Earth's surface (see Section 2.1). Earth's orbit is not permanent. Changes occur in the tilt of the axis and the shape of the yearly path about the Sun. These variations affect climate by altering the seasonal inputs of solar radiation. Occurring on a timescale of tens of thousands of years, these variations are associated with the glacial advances and retreats throughout Earth's history (see Chapter 18).

Variations in the level of solar radiation to Earth's surface are also associated with sunspot activity—huge magnetic storms on the Sun. These storms are associated with strong solar emissions and occur in cycles, with the number and size reaching a maximum approximately every 11 years. Researchers have related sunspot activity, among other occurrences, to periods of drought and winter warming in the Northern Hemisphere.

Interaction between two components of the climate system, the ocean and the atmosphere, are connected to some major climatic variations that occur at a regional scale. As far back as 1525, historic documents reveal that fishermen off the coast of Peru recorded periods of unusually warm water. The Peruvians referred to these as El Niño because they commonly appear at Christmastime, the season of the Christ Child (Spanish: *El Niño*). Now referred to by scientists as the El Niño–Southern Oscillation (ENSO), this phenomenon is a global event arising from large-scale interaction between the ocean and the atmosphere. The Southern Oscillation, a more recent discovery, refers to an oscillation in the surface pressure (atmospheric mass) between the southeastern tropical Pacific and the Australian-Indonesian regions. When the waters of the eastern Pacific are abnormally warm (an El Niño event), sea level pressure drops in the eastern Pacific and rises in the west. The reduction in the pressure gradient is accompanied by a weakening of the low-latitude easterly trades.

Although scientists still do not completely understand the cause of the ENSO phenomenon, its mechanism has been well documented. Recall from Section 2.3 that the trade winds blow westward across the tropical Pacific (see Figure 2.12). As a consequence, the surface currents within the tropical oceans flow westward (see Figure 2.14), bringing cold, deeper waters to the surface off the coast of Peru in a process known as upwelling (see Section 3.8). This pattern of upwelling, together with the cold-water current flowing from south to north along the western coast of South America, results in this region of the ocean usually being colder than one would expect given its equatorial location (**Figure 2.23**).

As the surface currents move westward the water warms, giving the water's destination, the western Pacific, the warmest ocean surface on Earth. The warmer water of the western Pacific causes the moist maritime air to rise and cool, bringing abundant rainfall to the region (Figure 2.23; also see Figure 2.16). In contrast, the cooler waters of the eastern Pacific result in relatively dry conditions along the Peruvian coast.

During an El Niño event, the trade winds slacken, reducing the westward flow of the surface currents (see Figure 2.23). The result is a reduced upwelling and a warming of the surface waters in the eastern Pacific. Rainfall follows the warm water eastward, with associated flooding in Peru and drought in Indonesia and Australia.

This eastward displacement of the atmospheric heat source (latent heat associated with the evaporation of water; see Section 3.2) overlaying the warm surface waters results in large changes in global atmospheric circulation, in turn influencing weather in regions far removed from the tropical Pacific.

At other times, the injection of cold water becomes more intense than usual, causing the surface of the eastern Pacific to cool. This variation is referred to as La Niña (**Figure 2.24**). It results in droughts in South America and heavy rainfall, even floods, in eastern Australia.

2.10 Most Organisms Live in Microclimates

Most organisms live in local conditions that do not match the general climate profile of the larger region surrounding them. For example, today's weather report may state that the temperature is 28°C and the sky is clear. However, your weather forecaster is painting only a general picture. Actual conditions of specific environments will be quite different depending on whether they are underground versus on the surface, beneath vegetation or on exposed soil, or on mountain slopes or at the seashore. Light, heat, moisture, and air movement all vary greatly from one part of the landscape to another, influencing the transfer of heat energy and creating a wide range of localized climates. These microclimates define the conditions organisms live in.

On a sunny but chilly day in early spring, flies may be attracted to sap oozing from the stump of a maple tree. The flies are active on the stump despite the near-freezing air temperature because, during the day, the surface of the stump absorbs solar radiation, heating a thin layer of air above the surface. On a still day, the air heated by the tree stump remains close to the surface, and temperatures decrease sharply above and below this layer. A similar phenomenon occurs when the frozen surface of the ground absorbs solar radiation and thaws. On a sunny, late winter day, the ground is muddy even though the air is cold.

By altering soil temperatures, moisture, wind movement, and evaporation, vegetation moderates microclimates, especially areas near the ground. For example, areas shaded by plants have lower temperatures at ground level than do places exposed to the Sun. On fair summer days in locations 25 millimeters (mm; 1 inch) aboveground, dense forest cover can reduce the daily range of temperatures by 7°C to 12°C below the soil temperature in bare fields. Under the shelter of heavy grass and low plant cover, the air at ground level is completely calm. This calm is an outstanding feature of microclimates within dense vegetation at Earth's surface. It influences both temperature and humidity, creating a favorable environment for insects and other ground-dwelling animals.

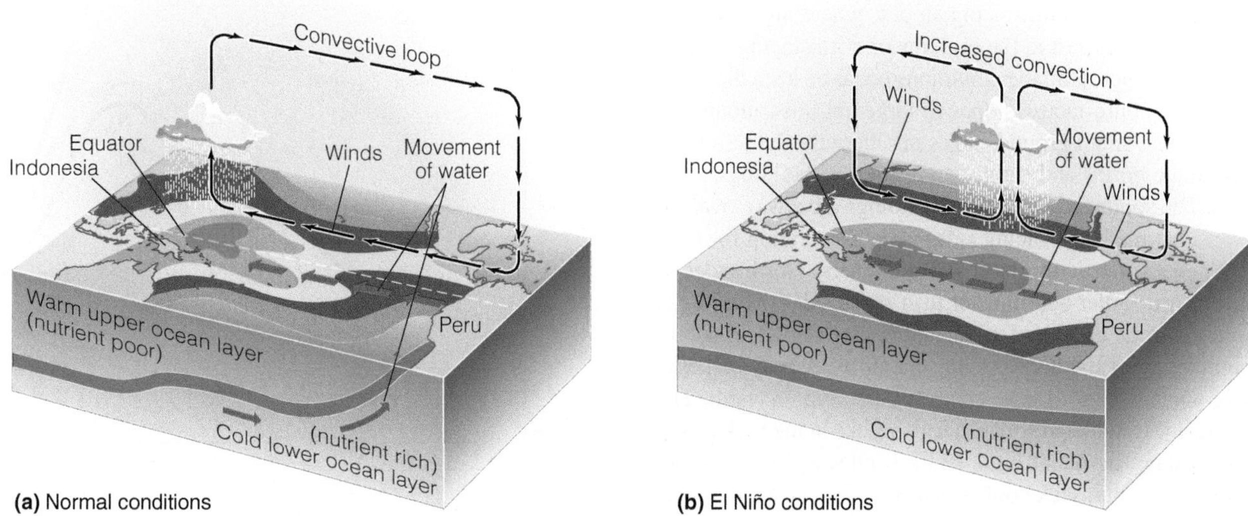

Figure 2.23 Schematic of the El Niño–Southern Oscillation (ENSO) that occurs off the western coast of South America. Under normal conditions, strong trade winds move surface waters westward (a). As the surface currents move westward, the water warms. The warmer water of the western Pacific causes the moist maritime air to rise and cool, bringing abundant rainfall to the region. Under ENSO conditions, the trade winds slacken, reducing the westward flow of the surface currents (b). Rainfall follows the warm water eastward, with associated flooding in Peru and drought in Indonesia and Australia.

Topography, particularly aspect (the direction that a slope faces), influences the local climatic conditions. In the Northern Hemisphere, south-facing slopes receive the most solar energy, whereas north-facing slopes receive the least (**Figure 2.25**). At other slope positions, energy received varies between these extremes, depending on their compass direction.

Different exposure to solar radiation at south- and north-facing sites has a marked effect on the amount of moisture and heat present. Microclimate conditions range from warm, dry, variable conditions on the south-facing slope to cool, moist, more uniform conditions on the north-facing slope. Because high temperatures and associated high rates of evaporation

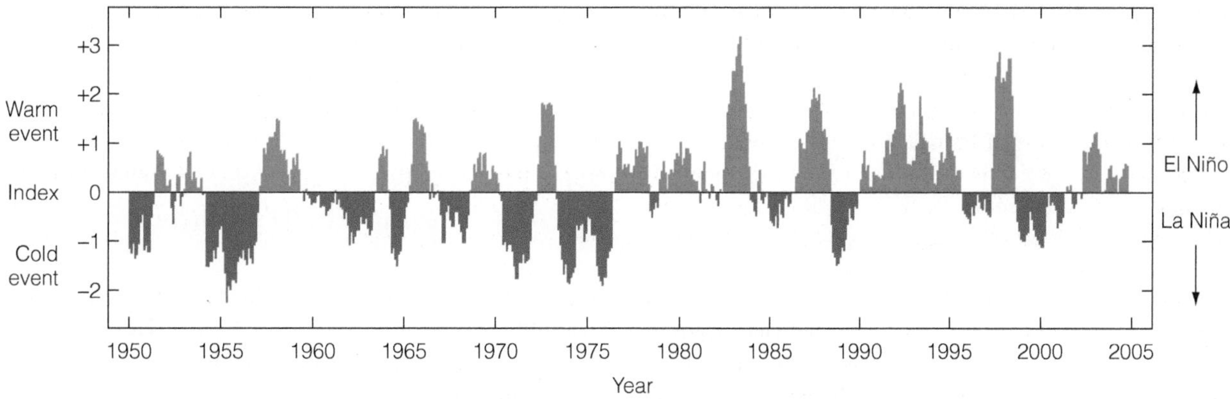

Figure 2.24 Record of El Niño–La Niña events during the second half of the 20th century. Numbers at the left of the diagram represent the ENSO index, which includes a combination of six factors related to environmental conditions over the tropical Pacific Ocean: air temperature, surface water temperature, sea-level pressure, cloudiness, and wind speed and direction. Warm episodes are in red; cold episodes are in blue. An index value greater than +1 represents an El Niño. A value less than −1 represents a La Niña.

draw moisture from soil and plants, the evaporation rate at south-facing slopes is often 50 percent higher, the average temperature is higher, and soil moisture is lower. Conditions are driest on the tops of south-facing slopes, where air movement is greatest, and dampest at the bottoms of north-facing slopes.

The same microclimatic conditions occur on a smaller scale on north- and south-facing slopes of large ant hills, mounds of soil, dunes, and small ground ridges in otherwise flat terrain, as well as on the north- and south-facing sides of buildings, trees, and logs. The south-facing sides of buildings are always warmer and drier than the north-facing sides—a consideration for landscape planners, horticulturists, and gardeners. North sides of tree trunks are cooler and moister than south sides, as reflected by more vigorous growth of moss on the north sides. In winter, the temperature of the north-facing side of a tree may be below freezing while the south side, heated by the Sun, is warm. This temperature difference may cause frost cracks in the bark as sap, thawed by day, freezes at night. Bark beetles and other wood-dwelling insects that seek cool, moist areas for laying their eggs prefer north-facing locations. Flowers on the south side of tree crowns often bloom sooner than those on the north side.

Microclimatic extremes also occur in depressions in the ground and on the concave surfaces of valleys, where the air is protected from the wind. Heated by sunlight during the day and cooled by terrestrial vegetation at night, this air often becomes stagnant. As a result, these sheltered sites experience lower nighttime temperatures (especially in winter), higher daytime temperatures (especially in summer), and higher relative humidity. If the temperature drops low enough, frost pockets form in these depressions. The microclimates of the frost pockets often display the same phenomenon, supporting different kinds of plant life than found on surrounding higher ground.

Although the global and regional patterns of climate discussed constrain the large-scale distribution and abundance of plants and animals, the localized patterns of microclimate define the actual environmental conditions sensed by the individual organism. This localized microclimate thus determines the distribution and activities of organisms in a particular region.

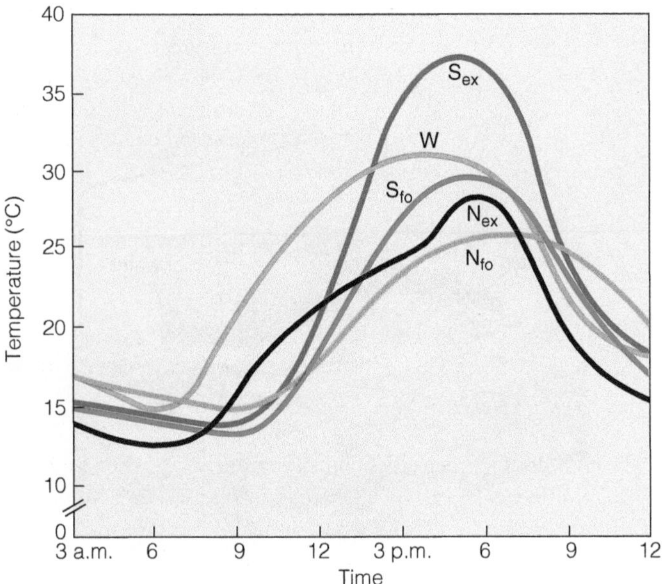

Key

W: Standard weather station on ridge
N_{fo}: Microclimate station on forested north-facing slope
N_{ex}: Microclimate station on exposed north-facing slope
S_{fo}: Microclimate station on forested south-facing slope
S_{ex}: Microclimate station on exposed south-facing slope

Figure 2.25 Diurnal changes in temperature (single clear day in August) recorded at five weather stations in Greer, West Virginia. Four of the stations provide data on the microclimate of a forested site on north-facing slope position (N_{fo}), an exposed (no vegetation cover) site on north-facing slope position (N_{ex}), a forested site on a south-facing slope (S_{fo}), and an exposed site on a south-facing slope (S_{ex}). The fifth station is located at the standard weather station position on the ridge top.

Interpreting Ecological Data

Q1. *Which of the two slope positions (north- or south-facing) has the higher maximum recorded temperatures (mid-afternoon)?*

Q2. *How does vegetation cover (forested vs. exposed slope) influence surface temperatures?*

ECOLOGICAL Issues & Applications

Rising Atmospheric Concentrations of Greenhouse Gases Are Altering Earth's Climate

Since the middle of the 19th century, direct measurements of surface temperature have been made at widespread locations around the world. These direct measures from instruments such as thermometers are referred to as the *instrumental record*. Besides these measurements made at the land surface, observations of sea surface temperatures have been made from ships since the mid-19th century. Since the late 1970s, both a network of instrumented buoys and Earth-observing satellites have been providing a continuous record of global observations for a wide variety of climate variables, supplementing the previous land- and ship-based instrumental records. What these various sources of data on the land and sea surface temperatures of our planet indicate is that Earth has been warming over the past 150 years (**Figure 2.26**).

Since the early 20th century, the global average surface temperature has increased by 0.74°C (±0.2°C). In addition, the 10 warmest years in the instrumental record since 1850 are, in descending order, 2010, 2005, 1998, 2003, 2013, 2002, 2006, 2009, 2007, and 2004. Analyses also indicate that global ocean heat content has increased significantly since the late 1950s. More than half of the increase in heat content has occurred in the upper 300 meters of the ocean; in this layer the temperature has increased at a rate of about 0.04°C per decade. Additional data examining trends on humidity, sea-ice extent, and snow cover likewise indicate a pattern of warming over the past century. What is the cause of this warming? The scientific consensus is that the warming is in large part a result of rising atmospheric concentrations of greenhouse gases. According to the most recent report of the Intergovernmental Panel on Climate Change (Report of Working Group I, 2013):

> Warming of the climate system is unequivocal, as is now evident from observations of increases in global average air and ocean temperatures.... Most of the observed increase in global average temperatures since the mid-20th century is very likely due to the observed increase in anthropogenic greenhouse gas concentrations.

Although human activities have increased the atmospheric concentration of a variety of greenhouse gases (e.g., methane [CH_4], nitrous oxide [N_2O]), the major concern is focused on carbon dioxide (CO_2). The atmospheric concentration of CO_2 has increased by more than 30 percent over the past 100 years. The evidence for this rise comes primarily from continuous observations of atmospheric CO_2 started in 1958 at Mauna Loa, Hawaii, by Charles Keeling (**Figure 2.27**) and from parallel records around the world. Evidence before the direct observations of 1958 comes from various sources, including the analysis of air bubbles trapped in the ice of glaciers in Greenland and Antarctica.

In reconstructing atmospheric CO_2 concentrations over the past 300 years, we see values that fluctuate between 280 and 290 parts per million (ppm) until the mid-1800s (see Figure 2.27). After the onset of the Industrial Revolution, the value increased steadily, rising exponentially by the mid-19th century onward. The change reflects the combustion of fossil fuels (coal, oil, and gas) as an energy source for industrialized nations (**Figure 2.28a**), as well as the increased clearing and burning of forests (primarily in the tropical regions; see **Figure 2.28b**).

Although there is an obvious correlation between rising atmospheric concentrations of CO_2 (and other greenhouse gases) and the observed increases in global temperature, what makes the scientific community so confident that the observed rise in global temperatures is a result of the greenhouse effect? One important factor is the actual pattern of warming itself. Recall from our discussion of the Earth's radiation balance, that surface temperature at any location or time reflects the net radiation balance, that is, the difference between incoming shortwave radiation and outgoing longwave radiation (Section 2.1). If incoming shortwave radiation exceeds outgoing longwave radiation, surface temperatures rise. Conversely, if outgoing longwave radiation exceeds incoming shortwave radiation, temperatures decline. It is this imbalance that accounts for the decline in mean annual temperatures with increasing latitude from the tropics (net radiation surplus) to the poles (net radiation deficit; see Figure 2.6). Likewise, it is the shift from surplus to deficit that results in the decline in surface temperatures from day to night (diurnal cycle) and from summer to winter (seasonal cycle). Since the influence of greenhouse gases on the radiation balance works through the absorption of outgoing longwave radiation, which is then emitted downward toward the surface instead, the net effect reduces cooling, that is, keeps the surface temperature warmer than it would otherwise be if the longwave radiation were lost to space. It therefore follows that the greater proportional warming from rising levels of greenhouse gases would occur in those places (i.e., polar) and times (i.e., winter and night)

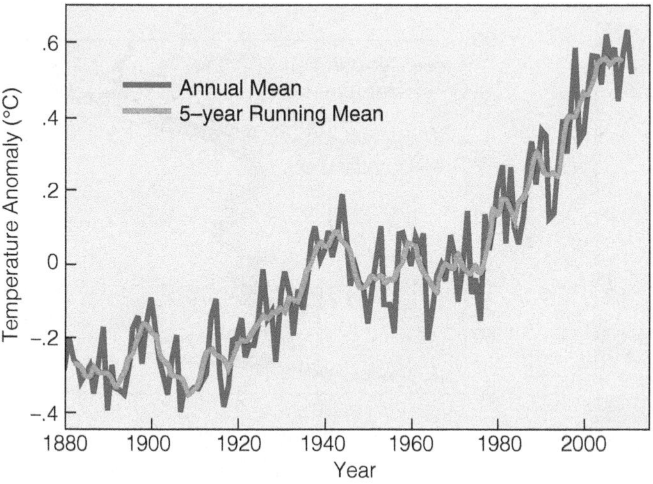

Figure 2.26 Global annual surface temperatures for the period of 1880 to 2012. Temperatures are expressed relative to the 1951 to 1980 mean (temperature anomaly plotted on *y*-axis is annual temperature – mean of annual temperatures over the period from 1951 to 1980). Data are from surface air measurements at meteorological stations and ship and satellite measurements for sea surface temperature. The five-year running mean (red line) is calculated for each year by averaging the sum of that year plus the preceding and following two years.

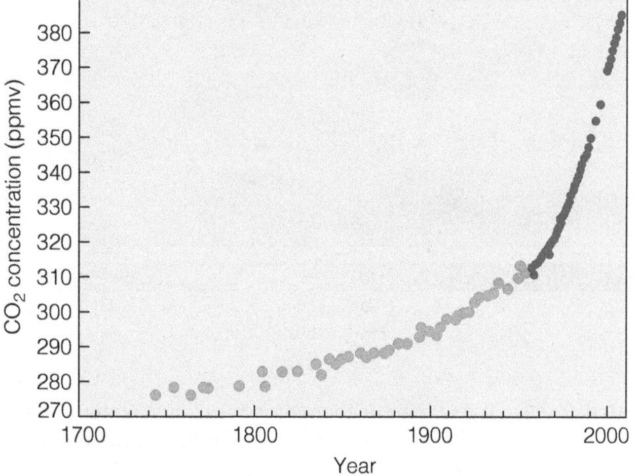

Figure 2.27 Historical record of average atmospheric carbon dioxide (CO_2) concentration over the past 300 years. Data collected prior to direct observation (1958 to present) are estimated from various techniques including analysis of air trapped in Antarctic ice sheets. ppmv, parts per million volume.

when and where temperatures are generally declining as a result of negative net radiation balance. An analysis of the patterns of warming over the past 50 years is in general agreement with this expectation.

The increase in global mean surface temperature illustrated in Figure 2.27 has not been the same at every location. The global map presented in **Figure 2.29a** shows the geographic patterns of surface temperature changes over the period from 1955 to 2005. Note that the greatest warming has occurred in the polar regions, particularly the Arctic (North America and Eurasia between 40 and 70° N). Although Earth's average temperature has risen 0.74°C during the 20th century, the Arctic is warming twice as fast as other parts of the world. In Alaska (U.S.) average temperatures have increased 3.0°C between 1970 and 2000. The warmer temperatures have caused other changes in the Arctic region such as melting of sea ice and continental ice sheets (Greenland). The reduction in ice cover potentially exacerbates the problem by reducing surface albedo and increasing the absorption of incoming shortwave radiation. In the Southern hemisphere, the Antarctic Peninsula has also undergone a great warming—five times the global average.

The changes in mean surface temperature presented in Figure 2.29a have been partitioned by season (December–February, March–May, June–August, and September–December.) in **Figure 2.29b**. The greatest observed warming over the last half century has occurred during the winter months. In effect, this represents a reduction of the normal pattern of cooling that occurs during the winter months as a result of the deficit in net radiation (deficit is reduced by increased absorption of outgoing longwave radiation). This pattern of

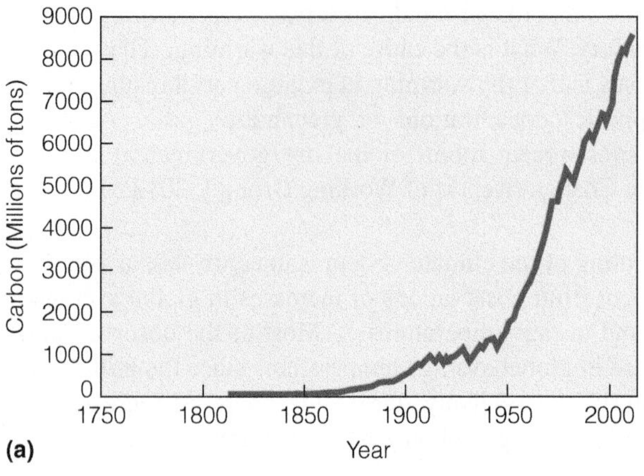

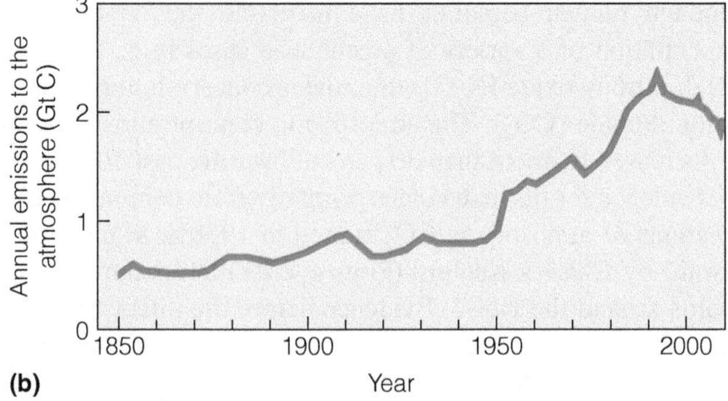

Figure 2.28 Historical record of global annual input of carbon dioxide (CO_2) to the atmosphere from the burning of fossil fuels since 1750 (a). Historical record of global annual input of CO_2 to the atmosphere from the clearing and burning of forest over the same period (b).

([b] Adapted from Houghton, J. *Global Warming: The Complete Briefing* [Cambridge University Press, 1997].)

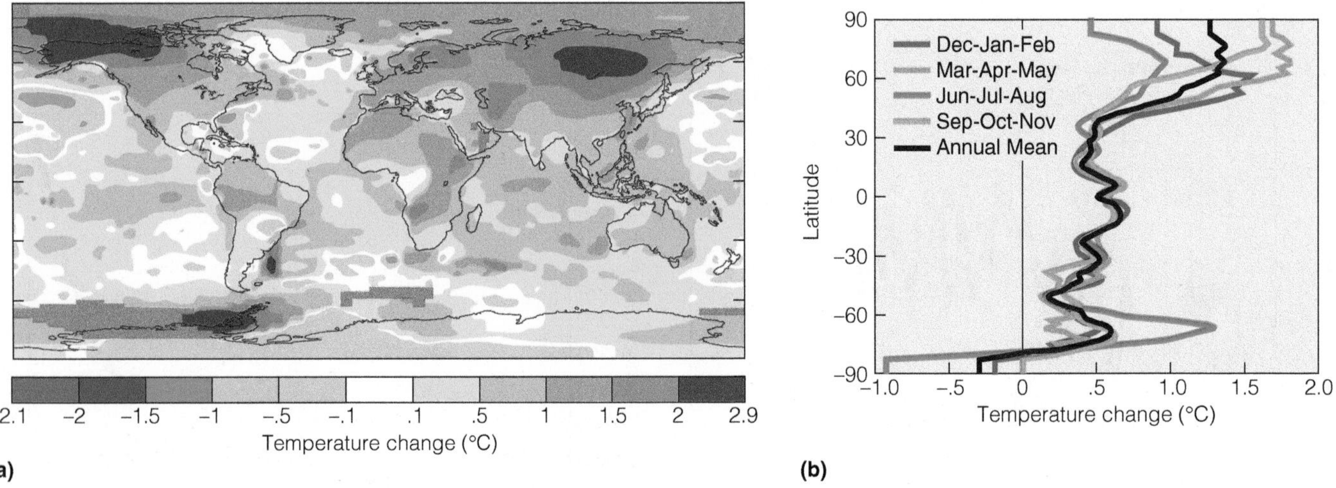

Figure 2.29 Global map of changes in mean surface temperature over the 50-year period from 1955 to 2005 (a). The changes in mean annual temperature shown in (a) are partitioned by season (December–February, March–May, June–August, and September–December) and averaged by latitude (b).

(NASA Goddard Institute for Space Studies 2005.)

Interpreting Ecological Data

Q1. *Based on the data provided in (b), which latitudes exhibit the greatest seasonal variations in surface temperature (Ts) change?*

Q2. *What accounts for the fact that the period of Jun–Aug in the arctic region (north of 60° N) shows the least warming, while the same period corresponds to the maximum temperature change in the Antarctic (south of 60° S)?*

winter warming becomes more apparent when the seasonal data are analyzed by latitude (see Figure 2.29b). The net result of winter warming is a reduction in the seasonal variations in temperature (differences between the warmest and coldest months).

Analyses of daily maximum and minimum land-surface temperatures from 1950 to 2000 show a decrease in the diurnal temperature range. On average, minimum temperatures are increasing at about twice the rate of maximum temperatures (0.2°C versus 0.1°C per decade). In other words, nighttime temperatures (minimum) have increased more than daytime temperatures (maximum) over this period.

These patterns of increasing surface temperatures over the past century have a major influence on the functioning of ecological systems, arranging the distribution of plant and animal species, the structure of communities, and the patterns of ecosystem productivity and decomposition. We will explore a variety of these issues in the *Ecological Issues & Applications* sections of the chapters that follow, and examine in more detail the current and future implication of global climate change in Chapter 27.

SUMMARY

Net Radiation 2.1

Earth intercepts solar energy in the form of shortwave radiation, some of which is reflected back into space. Earth emits energy back into space in the form of longwave radiation, a portion of which is absorbed by gases in the atmosphere and radiated back to the surface. The difference between incoming shortwave and outgoing longwave radiation is the net radiation. Surface temperatures are a function of net radiation.

Seasonal Variation 2.2

The amount of solar radiation intercepted by Earth varies markedly with latitude. Tropical regions near the equator receive the greatest amount of solar radiation, and high latitudes receive the least. Because Earth tilts on its axis, parts of Earth encounter seasonal differences in solar radiation. These differences give rise to seasonal variations in net radiation and temperature. There is a global gradient in mean annual temperature; it is warmest in the tropics and declines toward the poles.

Atmospheric Circulation 2.3

From the equator to the midlatitudes there is an annual surplus of net radiation, and there is a deficit from the midlatitudes to the poles. This latitudinal gradient of net radiation gives rise to global patterns of atmospheric circulation. The spin of Earth on its axis deflects air and water currents to the right in the Northern Hemisphere and to the left in the Southern Hemisphere. Three cells of global air flow occur in each hemisphere.

Ocean Currents 2.4

The global pattern of winds and the Coriolis effect cause major patterns of ocean currents. Each ocean is dominated by great circular water motions, or gyres. These gyres move clockwise in the Northern Hemisphere and counterclockwise in the Southern Hemisphere.

Atmospheric Moisture 2.5

Atmospheric moisture is measured in terms of relative humidity. The maximum amount of moisture the air can hold at any given temperature is called the saturation vapor pressure, which increases with temperature. Relative humidity is the amount of water in the air, expressed as a percentage of the maximum amount the air could hold at a given temperature.

Precipitation 2.6

Wind, temperature, and ocean currents produce global patterns of precipitation. They account for regions of high precipitation in the tropics and belts of dry climate at approximately 30° N and S latitude.

Continentality 2.7

Land surfaces heat and cool more rapidly than water; as a result, land areas farther from the coast experience a greater seasonal variation in temperature than do coastal areas. The interiors of continents generally receive less precipitation than the coastal regions do.

Topography 2.8

Temperature declines with altitude, so locations at higher elevations will have consistently lower temperatures that those of lower elevations. Mountainous topography influences local and regional patterns of precipitation. As an air mass reaches a mountain, it ascends, cools, becomes saturated with water vapor, and releases much of its moisture at upper altitudes of the windward side.

Irregular Variation 2.9

Not all temporal variation in regional climate occurs at a regular interval. Irregular variations in the trade winds give rise to periods of unusually warm waters off the coast of western South America. Referred to by scientists as El Niño; this phenomenon is a global event arising from large-scale interaction between the ocean and the atmosphere.

Microclimates 2.10

The actual climatic conditions that organisms live in vary considerably within one climate. These local variations, or microclimates, reflect topography, vegetative cover, exposure, and other factors on every scale. Angles of solar radiation cause marked differences between north- and south-facing slopes, whether on mountains, sand dunes, or ant mounds.

Climate Warming Ecological Issues & Applications

Over the past century the average surface temperature of the planet has been rising. The rise in surface temperature is related to increasing atmospheric concentrations of greenhouse gases caused by the burning of fossil fuels and clearing and burning of forests.

STUDY QUESTIONS

1. What is net radiation?
2. What is the greenhouse effect, and how does it influence the net radiation balance (and temperature) of Earth?
3. Why do equatorial regions receive more solar radiation than the polar regions? What is the consequence to latitudinal patterns of temperature?
4. The 23.5° tilt of Earth on its north–south axis gives rise to the seasons (review Figure 2.7). How would the pattern of seasons differ if the Earth's tilt were 90°? How would this influence the diurnal (night–day) cycle?
5. Why are the coastal waters of the southeastern United States warmer than the coastal waters off the southwestern coast? (Assume the same latitude. Hint: Look at the direction of the prevailing surface currents presented in Figure 2.13.)
6. The air temperature at noon on January 20 was 45°F, and the air temperature at noon on July 20 at the same location was 85°F. The relative humidity on both days was 75 percent. On which of these two days was there more water vapor in the air?
7. How might the relative humidity of a parcel of air change as it moves up the side of a mountain? Why?
8. What is the Intertropical Convergence Zone (ITCZ), and why does it give rise to a distinct pattern of seasonality in precipitation in the tropical zone?
9. What feature of global atmospheric circulation gives rise to the desert zones of the midlatitudes?
10. Which aspect of a slope, south- or north-facing, would receive the most solar radiation in the mountain ranges of the Southern Hemisphere?
11. Spruce Knob (latitude 38.625° N) in eastern West Virginia is named for the spruce trees dominating the forests at this site. Spruce trees are typically found in the colder forests of the more northern latitudes (northeastern United States and Canada). What does the presence of spruce trees at Spruce Knob tell you about this site?

FURTHER READINGS

Classic Studies

Geiger, R. 1965. *Climate near the ground.* Cambridge, MA: Harvard University Press.
A classic book on microclimate that continues to be a major reference on the subject.

Current Research

Ahrens, C. D. 2012. *Meteorology today: An introduction to weather, climate, and the environment.* 10th ed. Belmont, CA: Brooks/Cole.
An excellent introductory text on climate, clearly written and well illustrated.

Fagen, B. 2001. *The Little Ice Age: How climate made history, 1300–1850.* New York: Basic Books.
An enjoyable book that gives an overview of the effects of the Little Ice Age on human history.

Graedel, T. E., and P. J. Crutzen. 1997. *Atmosphere, climate and change.* New York: Scientific American Library.
A short introduction to climate written for the general public. Provides an excellent background for those interested in topics relating to air pollution and climate change.

Philander, G. 1989. "El Niño and La Niña." *American Scientist* 77:451–459.

Solomon, S., D. Qin, M. Manning, Z. Chen, M. Marquis, K. B. Averyt, M. Tignor and H. L. Miller (eds.). 2007. *Contribution of Working Group I to the Fourth Assessment Report of the Intergovernmental Panel on Climate Change.* UK: Cambridge University Press.
The most recent report of the IPCC outlining the state of the science regarding the issue of global climate change.

Suplee, C. 1999. "El Niño, La Niña." *National Geographic* 195:73–95.
These two articles provide a general introduction to the El Niño–La Niña climate cycle.

CHAPTER 3 The Aquatic Environment

A rainstorm over the ocean—a part of the water cycle.

CHAPTER GUIDE

WATER IS THE ESSENTIAL SUBSTANCE OF LIFE, the dominant component of all living organisms. About 75–95 percent of the weight of all living cells is water, and there is hardly a physiological process in which water is not fundamentally important.

Covering some 75 percent of the planet's surface, water is also the dominant environment on Earth. A major feature influencing the adaptations of organisms that inhabit aquatic environments is water salinity (see Section 3.5). For this reason, aquatic ecosystems are divided into two major categories: saltwater (or marine) and freshwater. These two major categories are further divided into a variety of aquatic ecosystems based on the depth and flow of water, substrate, and the type of organisms (typically plants) that dominate. We will explore the diversity of aquatic environments and the organisms that inhabit them later (Chapter 24). For now, we will examine the unique physical and chemical characteristics of water and how those characteristics interact to define the different aquatic environments and constrain the evolution of organisms that inhabit them.

3.1 Water Cycles between Earth and the Atmosphere

All marine and freshwater aquatic environments are linked, either directly or indirectly, as components of the **water cycle** (also referred to as the **hydrologic cycle**; Figure 3.1)—the process by which water travels in a sequence from the air to Earth and returns to the atmosphere.

Solar radiation, which heats Earth's atmosphere and provides energy for the evaporation of water, is the driving force behind the water cycle (see Chapter 2). **Precipitation** sets the water cycle in motion. Water vapor, circulating in the atmosphere, eventually falls in some form of precipitation. Some of the water falls directly on the soil and bodies of water. Some is intercepted by vegetation, dead organic matter on the ground, and urban structures and streets in a process known as **interception**.

Because of interception, which can be considerable, various amounts of water never infiltrate the ground but evaporate directly back to the atmosphere. Precipitation that reaches the soil moves into the ground by **infiltration**. The rate of infiltration depends on the type of soil, slope, vegetation, and intensity of the precipitation (see Section 4.8). During heavy rains when the soil is saturated, excess water flows across the surface of the ground as **surface runoff** or overland flow. At places, it concentrates into depressions and gullies, and the flow changes from sheet to channelized flow—a process that can be observed on city streets as water moves across the pavement into gutters. Because of low infiltration, runoff from urban areas might be as much as 85 percent of the precipitation.

Some water entering the soil seeps down to an impervious layer of clay or rock to collect as **groundwater** (see Figure 3.1). From there, water finds its way into springs and streams. Streams coalesce into rivers as they follow the topography of the landscape. In basins and floodplains, lakes and wetlands form. Rivers eventually flow to the coast, forming the transition from freshwater to marine environments.

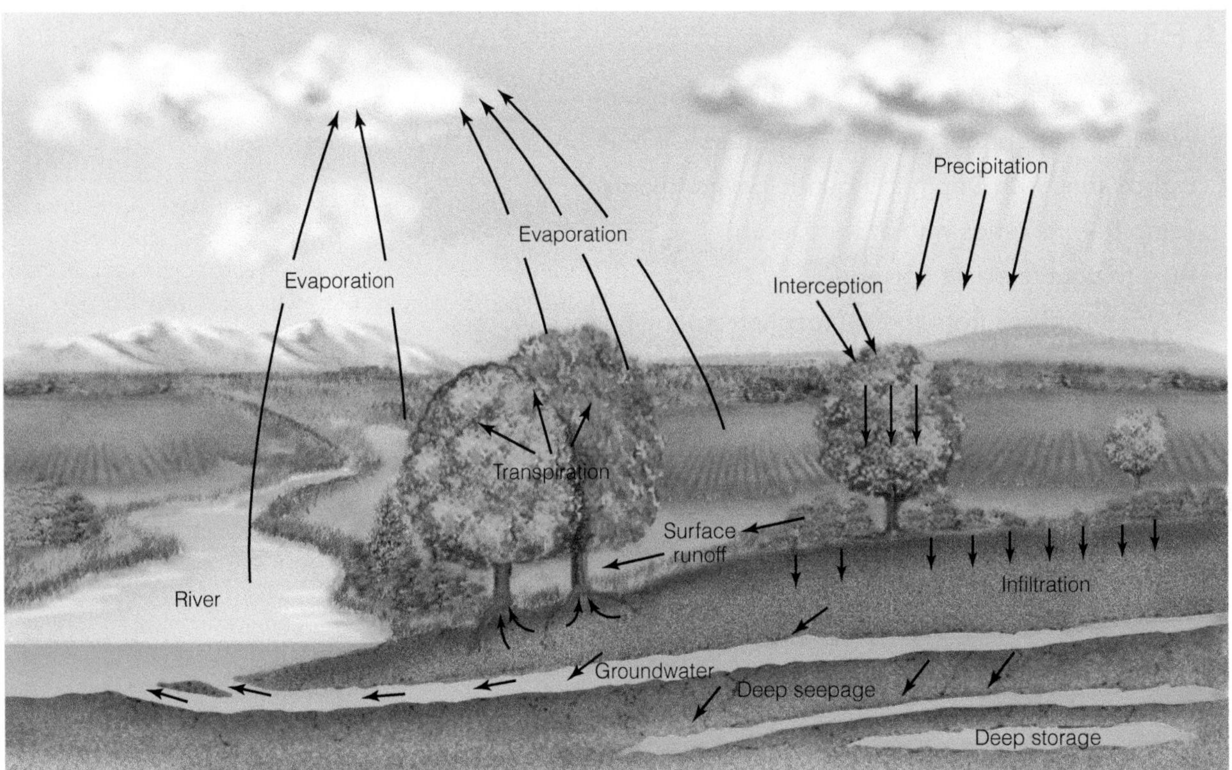

Figure 3.1 The water cycle on a local scale, showing major pathways of water movement.

Water remaining on the surface of the ground, in the upper layers of the soil, and collected on the surface of vegetation—as well as water in the surface layers of streams, lakes, and oceans—returns to the atmosphere by evaporation. The rate of evaporation is governed by how much water vapor is in the air relative to the saturation vapor pressure (relative humidity; see Section 2.5). Plants cause additional water loss from the soil. Through their roots, they take in water from the soil and lose it through their leaves and other organs in a process called transpiration. **Transpiration** is the evaporation of water from internal surfaces of leaves, stems, and other living parts (see Chapter 6). The total amount of evaporating water from the surfaces of the ground and vegetation (surface evaporation plus transpiration) is called **evapotranspiration**.

Figure 3.2 is a diagram of the global water cycle showing the various reservoirs (bodies of water) and fluxes (exchanges between reservoirs). The total volume of water on Earth is approximately 1.4 billion cubic kilometers (km^3) of which more than 97 percent resides in the oceans. Another 2 percent of the total is found in the polar ice caps and glaciers, and the third-largest active reservoir is groundwater (0.3 percent). Over the oceans, evaporation exceeds precipitation by some 40,000 km^3. A significant proportion of the water evaporated from the oceans is transported by winds over the land surface in the form of water vapor, where it is deposited as precipitation. Of the 111,000 km^3 of water that falls as precipitation on the land surface, only some 71,000 km^3 is returned to the atmosphere as evapotranspiration. The remaining 40,000 km^3 is carried as runoff by rivers and eventually returns to the oceans. This amount balances the net loss of water from the oceans to the atmosphere through evaporation that is eventually deposited on the continents (land surface) as precipitation (see Figure 3.2).

The relatively small size of the atmospheric reservoir (only 13 km^3) does not reflect its importance in the global water cycle. In Figure 3.2, note the large fluxes between the atmosphere, the oceans, and the land surface relative to the amount of water residing in the atmosphere at any given time (e.g., the size of atmospheric reservoir). The importance of the atmosphere in the global water cycle is better reflected by the turnover time of this reservoir. The turnover time is calculated by dividing the size of the reservoir by the rate of output (flux out). For example, the turnover time for the ocean is the size of the reservoir (1.37×10^6 km^3) divided by the rate of evaporation (425 km^3 per year) or more than 3000 years. In contrast, the turnover time of the atmospheric reservoir is approximately 0.024 year. That is to say, the entire water content of the atmosphere is replaced on average every nine days.

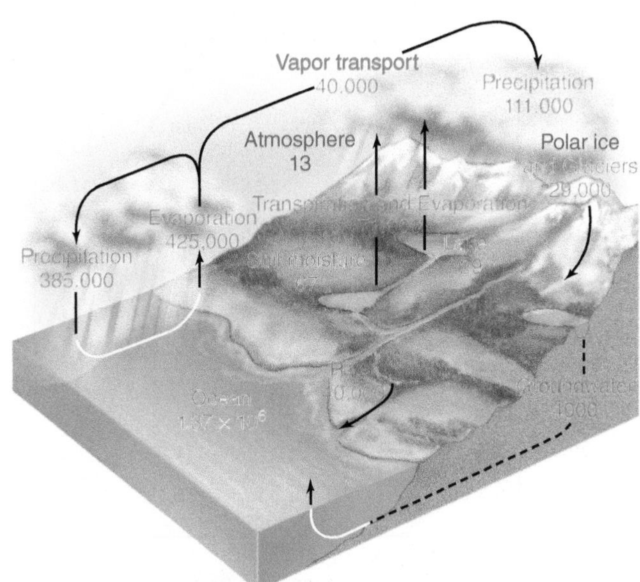

Figure 3.2 Global water cycle. Values for reservoirs (shown in blue) are in 10^8 km^3. Values for fluxes (shown in red) are in km^3 per year.

3.2 Water Has Important Physical Properties

The physical arrangement of its component molecules makes water a unique substance. A molecule of water consists of two atoms of hydrogen (H) joined to one atom of oxygen (O), represented by the chemical symbol H_2O. The H atoms are bonded to the O atom asymmetrically, such that the two H atoms are at one end of the molecule and the O atom is at the other (**Figure 3.3a**). The bonding between the two hydrogen atoms and the oxygen atom is via shared electrons (called a *covalent bond*), so that each H atom shares a single electron with the oxygen. The shared hydrogen atoms are closer to the oxygen atom than they are to each other. As a result, the side of the water molecule where the H atoms are located has a positive charge, and the opposite side where the oxygen atom is located has a negative charge, thus polarizing the water molecule (termed a *polar covalent bond*; **Figure 3.3b**).

Because of its polarity, each water molecule becomes weakly bonded with its neighboring molecules (**Figure 3.3c**). The positive (hydrogen) end of one molecule attracts the negative (oxygen) end of the other. The angle between the hydrogen atoms encourages an open, tetrahedral arrangement of water molecules. This situation, wherein hydrogen atoms act as connecting links between water molecules, is called **hydrogen bonding**. The simultaneous bonding of a hydrogen atom to the oxygen atoms of two different water molecules gives rise to a lattice arrangement of molecules (**Figure 3.3d**). These bonds, however, are weak in comparison to the bond between the hydrogen and oxygen atoms. As a result, they are easily broken and reformed.

Water has some unique properties related to its hydrogen bonds. One property is high **specific heat**—the number of calories necessary to raise the temperature of 1 gram of water 1 degree Celsius. The specific heat of water is defined as a value of 1, and other substances are given a value relative to that of water. Water can store tremendous quantities of heat energy with a small rise in temperature. As a result, great quantities of heat must be absorbed before the temperature of natural waters, such as ponds, lakes, and seas, rises just 1°C. These waters warm up slowly in spring and cool off just as slowly in the fall. This process prevents the wide seasonal fluctuations in the temperature of aquatic habitats so characteristic of

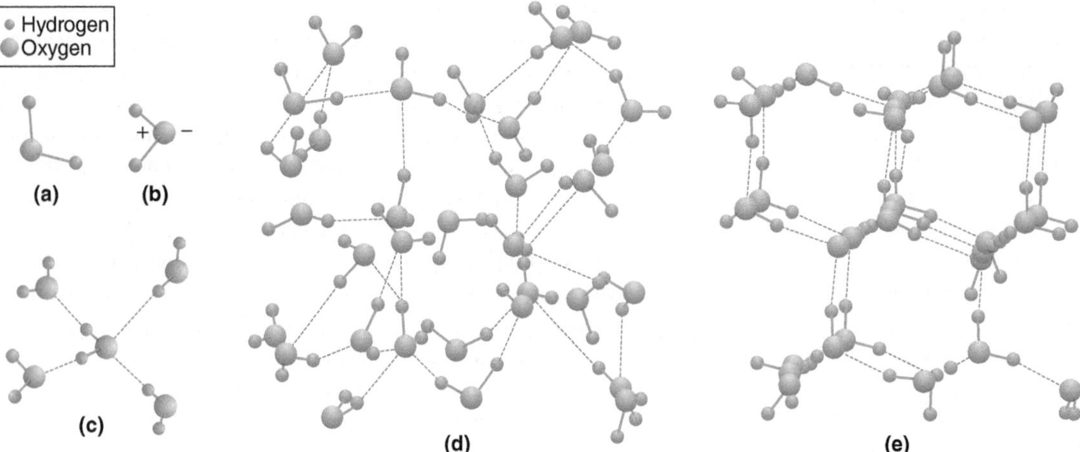

Figure 3.3 The structure of water. (a) An isolated water molecule, showing the angular arrangement of the hydrogen atoms. (b) Polarity of water. (c) Hydrogen bonds to one neighboring molecule of water. (d) The structure of liquid water. (e) The open lattice structure of ice.

air temperatures and moderates the temperatures of local and worldwide environments (see Section 2.7). The high specific heat of water is also important in the thermal regulation of organisms. Because 75–95 percent of the weight of all living cells is water, temperature variation is also moderated relative to changes in ambient temperature.

As a result of the high specific heat of water, large quantities of heat energy are required for it to change its state between solid (ice), liquid, and gaseous (water vapor) phases. Collectively, the energy released or absorbed in transforming water from one state to another is called latent heat (see Section 2.5). Removing only 1 calorie (4.184 joules [J]) of heat energy will lower the temperature of a gram of water from 2°C to 1°C, but approximately 80 times as much heat energy (80 calories per gram) must be removed to convert that same quantity of water at 1°C to ice (water's freezing point of 0°C). Likewise, it takes 536 calories to overcome the attraction between molecules and convert 1 gram (g) of water at 100°C into vapor, the same amount of heat needed to raise 536 g of water 1°C.

The lattice arrangement of molecules gives water a peculiar density–temperature relationship. Most liquids become denser as they are cooled. If cooled to their freezing temperature, they become solid, and the solid phase is denser than the liquid. This description is not true for water. Pure water becomes denser as it is cooled until it reaches 4°C (**Figure 3.4**). Cooling below this temperature results in a decrease in density. When 0°C is reached, freezing occurs and the lattice structure is complete—each oxygen atom is connected to four other oxygen atoms by means of hydrogen atoms. The result is a lattice with large, open spaces and therefore decreased density (see **Figure 3.3e**). When frozen, water molecules occupy more space than they do in liquid form. Because of its reduced density, ice is lighter than water and floats on it. This property is crucial to life in aquatic environments. The ice on the surface of water bodies insulates the waters below, helping to keep larger bodies of water from freezing solid during the winter months.

Because of hydrogen bonding, water molecules tend to stick firmly to one another, resisting external forces that would break their bonds. This property is called **cohesion**. In a body of water, these forces of attraction are the same on all sides. At the water's surface, however, conditions are different. Below the surface, molecules of water are strongly attracted to one another. Above the surface is the much weaker attraction between water molecules and air. Therefore, molecules on the surface are drawn downward, resulting in a surface that is taut like an inflated balloon. This condition, called **surface tension**, is important in the lives of aquatic organisms.

For example, the surface of water is able to support small objects and animals, such as the water striders (*Gerridae* spp.) and water spiders (*Dolomedes* spp.) that run across a pond's surface (**Figure 3.5**). To other small organisms, surface tension is a barrier, whether they wish to penetrate the water below or escape into

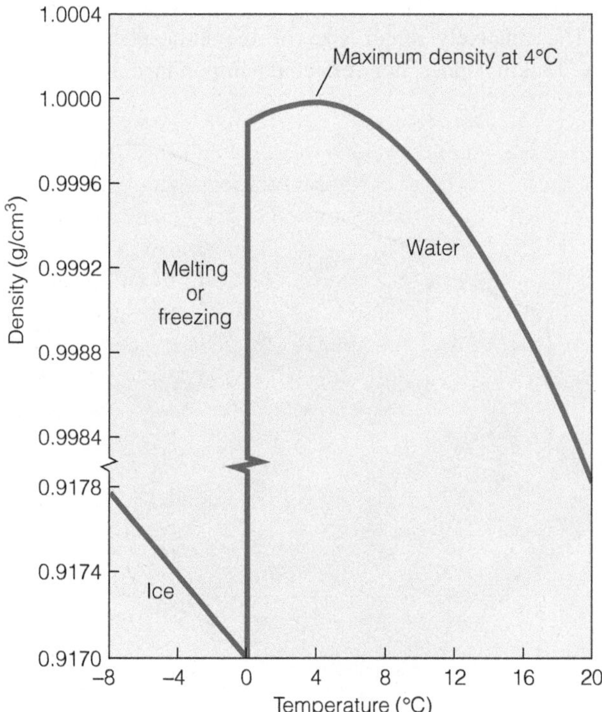

Figure 3.4 Density of pure water (and ice) as a function of temperature. The maximum density of water occurs at 4°C, and it declines dramatically as water changes from a liquid to a solid (ice).

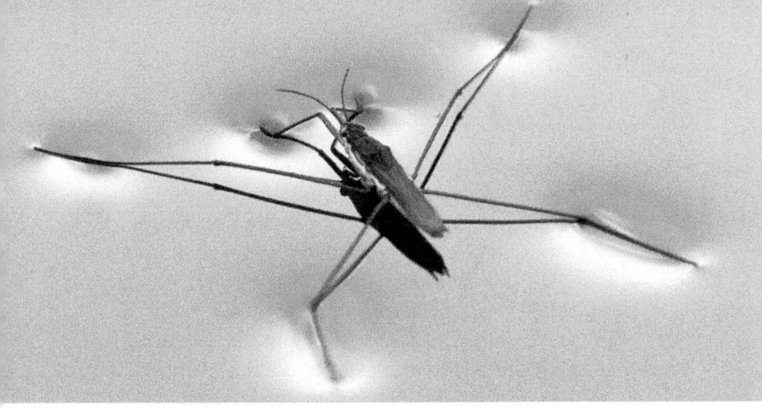

Figure 3.5 The property of surface tension allows the water strider (*Gerris remigis*) to glide across the water surface.

the air above. For some, the surface tension is too great to break; for others, it is a trap to avoid while skimming the surface to feed or to lay eggs. If caught in the surface tension, a small insect may flounder on the surface. The nymphs of mayflies (*Ephemeroptera* spp.) and caddis flies (*Trichoptera* spp.) that live in the water and transform into winged adults are hampered by surface tension when trying to emerge from the water. While slowed down at the surface, these insects become easy prey for fish.

Cohesion is also responsible for the viscosity of water. **Viscosity** is the property of a material that measures the force necessary to separate the molecules and allow an object to pass through the liquid. Viscosity is the source of frictional resistance to objects moving through water. This frictional resistance of water is 100 times greater than that of air. The streamlined body shape of many aquatic organisms, for example most fish and marine mammals, helps to reduce frictional resistance. Replacement of water in the space left behind by the moving animal increases drag on the body. An animal streamlined in reverse, with a short, rounded front and a rapidly tapering body, meets the least water resistance. The perfect example of such streamlining is the sperm whale (*Physeter catodon*; **Figure 3.6**).

Water's high viscosity relative to that of air is largely the result of its greater density. The density of water is about 860 times greater than that of air (pure water has a density of 1000 kilograms per cubic meter [kg/m^3]). Although the resulting viscosity of water limits the mobility of aquatic organisms, it also benefits them. If a body submerged in water weighs less than the water it displaces, it is subjected to an upward force called **buoyancy**. Because most aquatic organisms (plants and animals) are close to neutral buoyancy (their density is similar to that of water), they do not require structural material such as skeletons or cellulose to hold their bodies erect against the force of gravity. Similarly, when moving on land, terrestrial animals must raise their mass against the force of gravity with each step they take. Such movement requires significantly more energy than swimming movements do for aquatic organisms.

Figure 3.6 The body of the sperm whale (*Physeter catodon*) is streamlined in reverse, with a short, rounded front and a rapidly tapering body. This shape meets the least water resistance.

But water's greater density can profoundly affect the metabolism of marine organisms inhabiting the deeper waters of the ocean. Because of its greater density, water also undergoes greater changes in pressure with depth than does air. At sea level, the weight of the vertical column of air from the top of the atmosphere to the sea surface is 1 kilogram per square centimeter (kg/cm^2) or 1 atmosphere (atm). In contrast, pressure increases 1 atm for each 10 m in depth. Because the deep ocean varies in depth from a few hundred meters down to the deep trenches at more than 10,000 m, the range of pressure at the ocean bottom is from 20 atm to more than 1000 atm. Recent research has shown that both proteins and biological membranes are strongly affected by pressure, and animals living in the deep ocean have evolved adaptations that allow these biochemical systems to function under conditions of extreme pressure.

3.3 Light Varies with Depth in Aquatic Environments

When light strikes the surface of water, a certain amount is reflected back to the atmosphere. The amount of light reflected from the surface depends on the angle at which the light strikes the surface. The lower the angle, the larger the amount of light reflected. As a result, the amount of light reflected from the water surface will vary both diurnally and seasonally between the equator and the poles (see Section 2.1 and Figure 2.5 for a complete discussion).

The amount of light entering the water surface is further reduced by two additional processes. First, suspended particles, both alive and dead, intercept the light and either absorb or scatter it. The scattering of light increases the length of its path through the water and results in further attenuation. Second, water itself absorbs light (**Figure 3.7**). Moreover, water absorbs some wavelengths more than others. First to be absorbed are visible red light and infrared radiation in wavelengths greater than 750 nanometers (nm). This absorption reduces solar energy by half. Next, in clear water, yellow disappears, followed by green and violet, leaving only blue wavelengths to penetrate deeper water. A fraction of blue light is lost with increasing depth. In the clearest seawater, only about 10 percent of blue light reaches to more than 100 m in depth.

These changes in the quantity and quality of light have important implications for life in aquatic environments, both by directly influencing the quantity and distribution of productivity and by indirectly influencing the vertical profile of temperature with water depth (see Section 20.4 and Chapter 24). The lack of light in deeper waters of the oceans has resulted in various adaptations. Organisms of the deeper ocean (200–1000 m deep) are typically silvery gray or deep black, and organisms living in even deeper waters (below 1000 m) often lack pigment. Another adaptation is large eyes, which give these

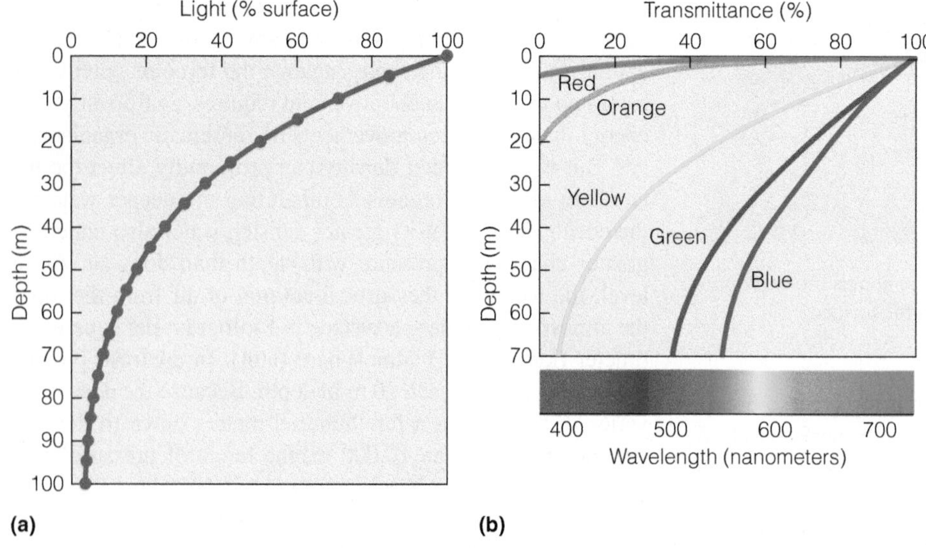

Figure 3.7 (a) Attenuation of incident light with water depth (pure water), expressed as a percentage of light at the water surface. Estimates assume a light extinction coefficient of $k_w = 0.035$ (see Quantifying Ecology 4.1, page 56). (b) The passage of light through water (transmittance) reduces the quantity of light and modifies its spectral distribution (see Figure 2.1). Red wavelengths are attenuated more rapidly than green and blue wavelengths.

Interpreting Ecological Data

Q1. *As you dive down in depth from the surface, which wavelength of light is the first to disappear? At approximately what depth would this occur?*

Q2. *Is it the shorter or longer wavelengths of visible light that penetrate the deepest into the water column? (Refer to Figure 2.1.)*

organisms maximum light-gathering ability. Many organisms have adapted organs that produce light through chemical reactions referred to as *bioluminescence* (see Section 24.10).

3.4 Temperature Varies with Water Depth

Surface temperatures reflect the balance of incoming and outgoing radiation (see Section 2.1). As solar radiation is absorbed in the vertical water column, the temperature profile with depth might be expected to resemble the vertical profile of light shown in Figure 3.7—that is, decreasing exponentially with depth. However, the physical characteristic of water density plays an important role in modifying this pattern (see Section 3.2, Figure 3.4).

As sunlight is absorbed in the surface waters, it heats up (**Figure 3.8**). Winds and surface waves mix the surface waters, distributing the heat vertically. Warm surface waters move downward, whereas the cooler waters below move up to the surface. As a result of this vertical mixing, heat is transported from the surface downward and the decline in water temperature with depth lags the decline in solar radiation. Below this mixed layer, however, temperatures drop rapidly. The region of the vertical depth profile where the temperature declines most rapidly is called the **thermocline**. The depth of the thermocline will depend on the input of solar radiation to the surface waters and on the degree of vertical mixing (wind speed and wave action). Below the thermocline, water temperatures continue to fall with depth but at a much slower rate. The result is a distinct pattern of temperature zonation with depth.

The difference in temperature between the warm, well-mixed surface layer and the cooler waters below the thermocline causes a distinctive difference in water density in these two vertical zones. The thermocline is located between an upper layer of warm, lighter (less dense) water called the **epilimnion** and a deeper layer of cold, denser water called the **hypolimnion** (see Figure 3.8; also see Section 21.10 and Figure 21.23). The density change at the thermocline acts as a physical barrier that prevents mixing of the upper (epilimnion) and lower (hypolimnion) layers.

Just as seasonal variation in the input of solar radiation to Earth's surface results in seasonal changes in surface temperatures (see Section 2.2), seasonal changes in the input of solar radiation to the water surface give rise to seasonal changes in the vertical profile of temperature in aquatic environments (**Figure 3.9**). Because of the relatively constant input of solar radiation to the water surface throughout the year, the thermocline is a permanent feature of tropical waters. In the waters of the temperate zone, a distinct thermocline exists during the summer months. By fall, conditions begin to change, and a turnabout takes place. Air temperatures and sunlight decrease, and the surface water of the epilimnion starts to cool. As it does, the water becomes denser and sinks, displacing the warmer water below to the surface, where it cools in turn. As the difference in water density between the epilimnion and hypolimnion continues to decrease, winds are able to mix the vertical profile to greater depths. This process continues until the temperature is uniform throughout the basin (see Figure 3.9). Now, pond and lake water circulate throughout the basin. This process of vertical circulation, called the *turnover*, is an important component of nutrient dynamics in open-water ecosystems (see Chapter 21). Stirred by wind, the process of vertical mixing may last until ice forms at the surface.

Then comes winter, and as the surface water cools to below 4°C, it becomes lighter again and remains on the surface. (Remember, water becomes lighter above and below 4°C; see Figure 3.4.) If the winter is cold enough, surface water freezes; otherwise, it remains close to 0°C. Now the warmest place in the pond or lake is on the bottom. In spring, the breakup of ice and heating of surface water with increasing inputs of solar radiation to the surface again causes the water to stratify.

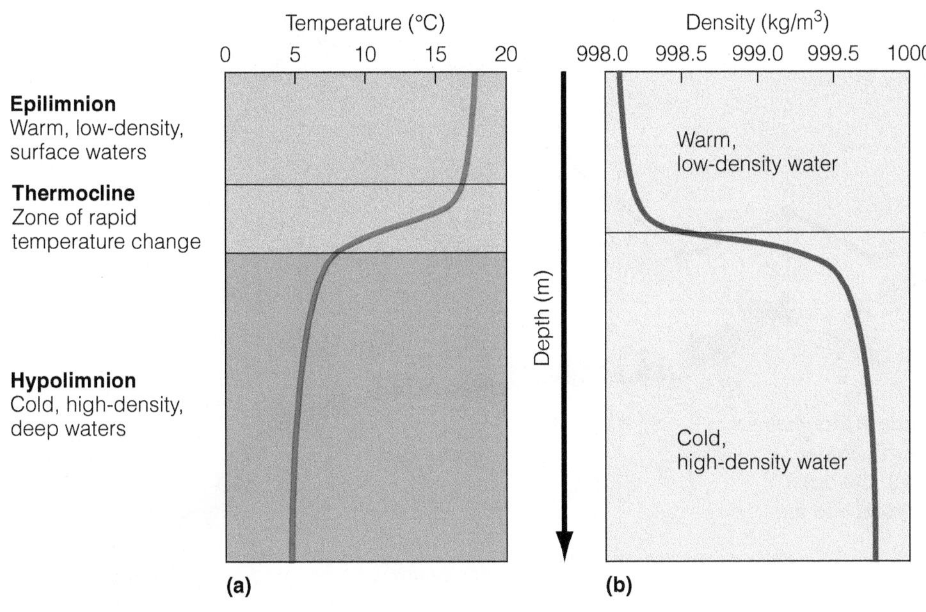

Figure 3.8 Temperature and density profiles with water depth for an open body of water such as a lake or pond. (a) The vertical profile of temperature might be expected to resemble the profile of light presented in Figure 3.7, but vertical mixing of the surface waters transports heat to the waters below. Below this mixed layer, temperatures decline rapidly in a region called the *thermocline*. Below the thermocline, temperatures continue declining at a slower rate. The vertical profile can therefore be divided into three distinct zones: epilimnion, thermocline, and hypolimnion. (b) The rapid decline in temperature in the thermocline results in a distinct difference in water density (see Figure 3.4) in the warmer epilimnion as compared to the cooler waters of the hypolimnion, leading to a two-layer density profile—warm, low-density surface water and cold, high-density deep water.

Because not all bodies of water experience such seasonal changes in stratification, this phenomenon is not necessarily characteristic of all deep bodies of water. In some deep lakes and the oceans, the thermocline simply descends during periods of turnover and does not disappear at all. In such bodies of water, the bottom water never becomes mixed with the top layer. In shallow lakes and ponds, temporary stratification of short duration may occur; in other bodies of water, stratification may exist, but the depth is not sufficient to develop a distinct thermocline. However, some form of thermal stratification occurs in all open bodies of water.

Water temperature (°C)
0 5 10 15 20
Water depth
Summer
Fall
Winter
Spring

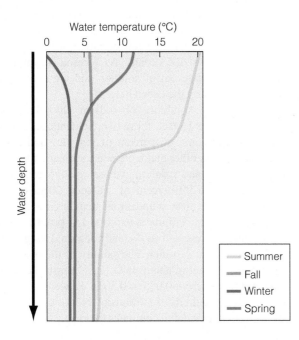

Figure 3.9 Seasonal changes in the vertical temperature profile (with water depth) for an open body of water such as a lake or pond. As air temperatures decline during the fall months, the surface water cools and sinks so that the temperature is uniform with depth. With the onset of winter, surface water further cools and ice may form on the surface. When spring arrives, the process reverses and the thermocline forms once again.

The temperature of a flowing body of water (stream or river), on the other hand, is variable (**Figure 3.10**). Small, shallow streams tend to follow, but lag behind, air temperatures. They warm and cool with the seasons but rarely fall below freezing in winter. Streams with large areas exposed to sunlight are warmer than those shaded by trees, shrubs, and high banks. That fact is ecologically important because temperature affects the stream community, influencing the presence or absence of cool- and warm-water organisms. For example, the dominant predatory fish shift from species such as trout and smallmouth bass, which require cooler water and more oxygen, to species such as suckers and catfish, which require warmer water and less oxygen (see Figure 24.13).

3.5 Water Functions as a Solvent

As you stir a spoonful of sugar into a glass of water, it dissolves, forming a homogeneous, or uniform, mixture. A liquid that is a homogeneous mixture of two or more substances is called a **solution**. The dissolving agent of a solution is the **solvent**, and the substance that is dissolved is referred to as the **solute**. A solution in which water is the solvent is called an **aqueous solution**.

Water is an excellent solvent that can dissolve more substances than can any other liquid. This extraordinary ability makes water a biologically crucial substance. Water provides a fluid that dissolves and transports molecules of nutrients and waste products, helps to regulate temperature, and preserves chemical equilibrium within living cells.

The solvent ability of water is largely a result of the bonding discussed in Section 3.2. Because the H atom is bonded to the O

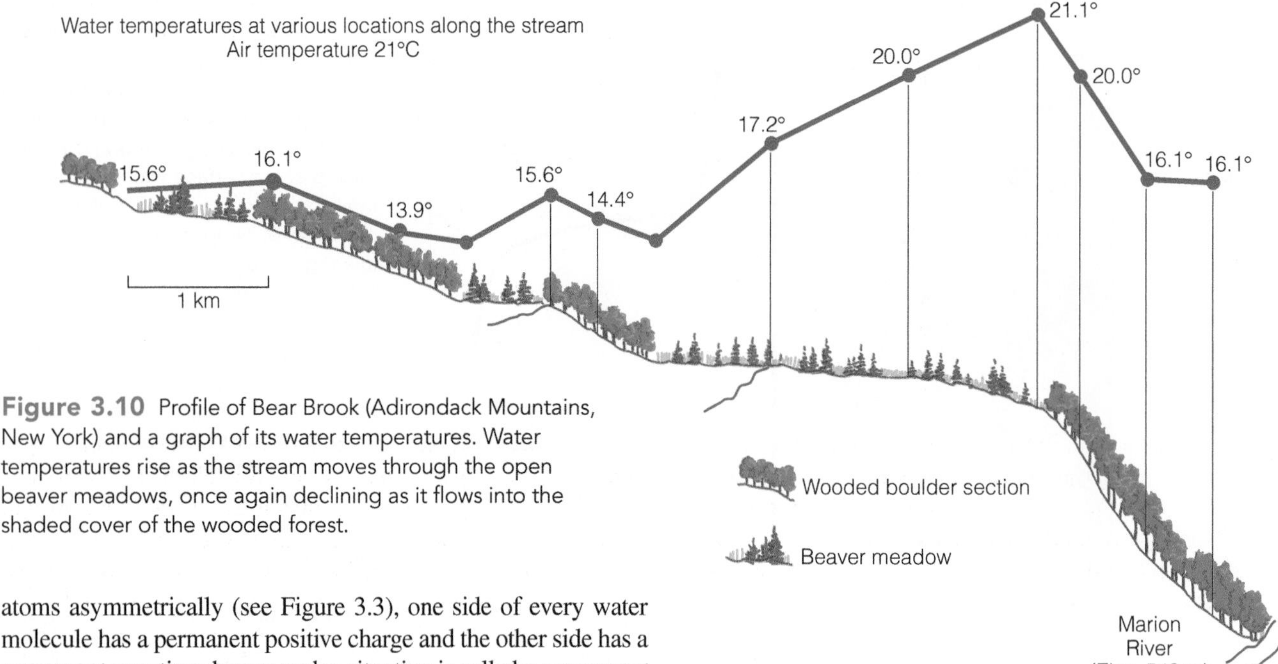

Figure 3.10 Profile of Bear Brook (Adirondack Mountains, New York) and a graph of its water temperatures. Water temperatures rise as the stream moves through the open beaver meadows, once again declining as it flows into the shaded cover of the wooded forest.

atoms asymmetrically (see Figure 3.3), one side of every water molecule has a permanent positive charge and the other side has a permanent negative charge; such a situation is called a *permanent dipole* (where *dipole* refers to oppositely charged poles). Because opposite charges attract, water molecules are strongly attracted to one another; they also attract other molecules carrying a charge.

Compounds that consist of electrically charged atoms or groups of atoms are called **ions**. Sodium chloride (table salt), for example, is composed of positively charged sodium ions (Na^+) and negatively charged chloride ions (Cl^-) arranged in a crystal lattice. When placed in water, the attractions between negative (oxygen atom) and positive (hydrogen atoms) charges on the water molecule (see Figure 3.3) and those of the sodium and chloride atoms are greater than the forces (ionic bonds) holding the salt crystals together. Consequently, the salt crystals readily dissociate into their component ions when placed in contact with water; that is, they dissolve.

The solvent properties of water are responsible for the presence of most of the minerals (elements and inorganic compounds) found in aquatic environments. When water condenses to form clouds, it is nearly pure except for some dissolved atmospheric gases. In falling to the surface as precipitation, water acquires additional substances from particulates and dust particles suspended in the atmosphere. Water that falls on land flows over the surface and percolates into the soil, obtaining more solutes. Surface waters, such as streams and rivers, pick up more solvents from the substances through and over which they flow. The waters of most rivers and lakes contain 0.01–0.02 percent dissolved minerals. The relative concentrations of minerals in these waters reflect the substrates over which the waters flow. For example, waters that flow through areas where the underlying rocks consist largely of limestone, composed primarily of calcium carbonate ($CaCO_3$), will have high concentrations of calcium (Ca^{2+}) and bicarbonate (HCO_3^-).

In contrast to freshwaters, the oceans have a much higher concentration of solutes. In effect, the oceans function as a large still. The flow of freshwaters into the oceans continuously adds to the solute content of the waters, as pure water evaporates from the surface to the atmosphere. The concentration of solutes, however, cannot continue to increase indefinitely. When the concentration of specific elements reaches the limit set by the maximum solubility of the compounds they form (grams per liter), the excess amounts will precipitate and be deposited as sediments. Calcium, for example, readily forms calcium carbonate ($CaCO_3$) in the waters of the oceans. The maximum solubility of calcium carbonate, however, is only 0.014 gram per liter of water, a concentration that was reached early in the history of the oceans. As a result, calcium ions continuously precipitate out of solution and are deposited as limestone sediments on the ocean bottom.

In contrast, the solubility of sodium chloride is high (360 grams per liter). In fact, these two elements, sodium and chlorine, make up some 86 percent of sea salt. Sodium and chlorine—along with other major elements such as sulfur, magnesium, potassium, and calcium, whose relative proportions vary little—compose 99 percent of sea salts (**Figure 3.11**). Determination of the most abundant element, chlorine, is used as an index of salinity. Salinity is expressed in **practical salinity units (psu)**, represented as ‰ and measured as grams of chlorine per kilogram of water. The salinity of the open sea is fairly constant, averaging about 35‰. In contrast, the salinity of freshwater ranges from 0.065 to 0.30‰. However, over geologic timescales (hundreds of millions of years), the salinity of the oceans has increased and continues to do so.

3.6 Oxygen Diffuses from the Atmosphere to the Surface Waters

Water's role as a solvent is not limited to dissolving solids. The surface of a body of water defines a boundary with the atmosphere. Across this boundary, gases are exchanged through

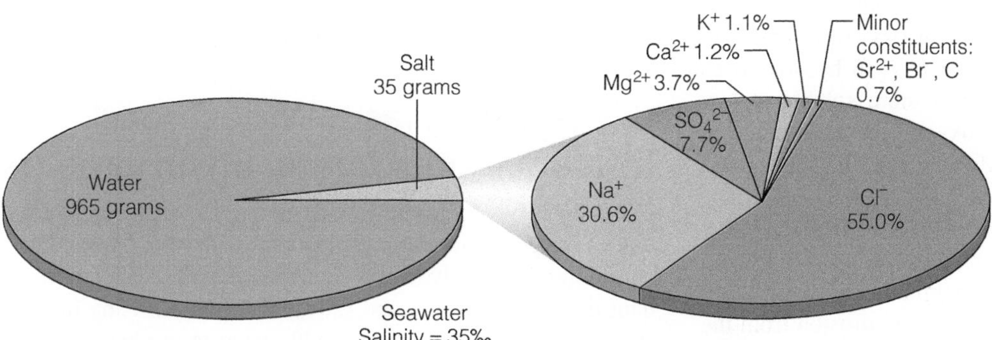

Figure 3.11 Composition of seawater of 35 practical salinity units (psu). Na^+, sodium; Cl^-, chlorine; SO_4^{-2}, sulfate; Mg^{2+}, magnesium; Ca^{2+}, calcium; K^+, potassium; Sr^{2+}, strontium; Br^-, bromine; C, carbon.

the process of diffusion. **Diffusion** is the general tendency of molecules to move from a region of high concentration to one of lower concentration. The process of diffusion results in a net transfer of two metabolically important gases, oxygen and carbon dioxide, from the atmosphere (higher concentration) into the surface waters (lower concentration) of aquatic environments.

Oxygen diffuses from the atmosphere into the surface water. The rate of diffusion is controlled by the solubility of oxygen in water and the steepness of the diffusion gradient (the difference in concentration between the air and the surface waters where diffusion occurs). The solubility of gases in water is a function of temperature, pressure, and salinity. The saturation value of oxygen is greater for cold water than warm water because the solubility (ability to stay in solution) of a gas in water decreases as the temperature rises. However, solubility increases as atmospheric pressure increases and decreases as salinity increases, which is not significant in freshwater.

Once oxygen enters the surface water, the process of diffusion continues, and oxygen diffuses from the surface to the waters below (because of their lower concentration). Water, with its greater density and viscosity relative to air, limits how quickly gases diffuse through it. Gases diffuse some 10000 times slower in water than in air. In addition to the process of diffusion, oxygen absorbed by surface water is mixed with deeper water by turbulence and internal currents. In shallow, rapidly flowing water and in wind-driven sprays, oxygen may reach and maintain saturation and even supersaturated levels because of the increase of absorptive surfaces at the air–water interface. Oxygen is lost from the water as temperatures rise, decreasing solubility, and through the uptake of oxygen by aquatic life.

During the summer, oxygen, like temperature (see Section 3.4), may become stratified in lakes and ponds. The amount of oxygen is usually greatest near the surface, where an interchange between water and atmosphere, further stimulated by the stirring action of the wind, takes place (**Figure 3.12**). Besides entering the water by diffusion from the atmosphere, oxygen is also a product of photosynthesis, which is largely restricted to the surface waters because of the limitations of available light (see Figure 3.7 and Chapter 6). The quantity of oxygen decreases with depth because of the oxygen demand of decomposer organisms living in the bottom sediments (Chapter 21). During spring and fall turnover, when water

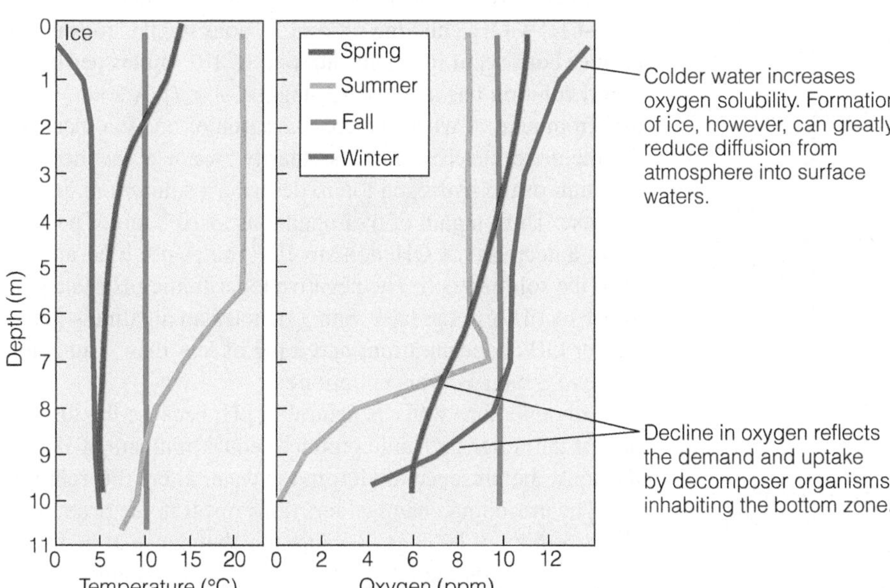

Figure 3.12 Oxygen stratification in Mirror Lake, New Hampshire, in winter, summer, and late fall. The late fall turnover results in uniform temperature as well as uniform distribution of oxygen throughout the lake basin. In summer, a pronounced stratification of both temperature and oxygen exists. Oxygen declines sharply in the thermocline and is nonexistent on the bottom because of its uptake by decomposer organisms in the sediments. In winter, oxygen levels are high in surface water reflecting higher solubility. Formation of ice during winter, however, can greatly reduce diffusion into surface waters.
(Adapted from Likens 1985.)

recirculates through the lake, oxygen becomes replenished in deep water. In winter, the reduction of oxygen in unfrozen water is slight because the demand for oxygen by organisms is reduced by the cold, and oxygen is more soluble at low temperatures. Under ice, however, oxygen depletion may be serious as a result of the lack of diffusion from the atmosphere to the surface waters.

As with ponds and lakes, oxygen is not distributed uniformly within the depths of the oceans (**Figure 3.13**). A typical oceanic oxygen profile shows a maximum amount in the upper 10–20 m, where photosynthetic activity and diffusion from the atmosphere often lead to saturation. With increasing depth, oxygen content declines. In the open waters of the ocean, concentrations reach a minimum value of 500–1000 m, a region referred to as the *oxygen minimum zone*. Unlike lakes and ponds, where the seasonal breakdown of the thermocline and resultant mixing of surface and deep waters result in a dynamic gradient of temperature and oxygen content, the limited depth of surface mixing in the deep oceans maintains the vertical gradient of oxygen availability year-round.

The availability of oxygen in aquatic environments characterized by flowing water is different. The constant churning of stream water over riffles and falls gives greater contact with the atmosphere; the oxygen content of the water is high, often near saturation for the prevailing temperature. Only in deep holes or in polluted waters does dissolved oxygen show any significant decline (see Chapter 24, *Ecological Issues & Applications*).

Even under ideal conditions, gases are not very soluble in water. Rarely is oxygen limited in terrestrial environments. In aquatic environments, the supply of oxygen, even at saturation levels, is meager and problematic. Compared with its concentration of 0.21 liter per liter in the atmosphere (21 percent by volume), oxygen in water reaches a maximum solubility of 0.01 liter per liter (1 percent) in freshwater at a temperature of 0°C. As a result, the concentration of oxygen in aquatic environments often limits respiration and metabolic activity.

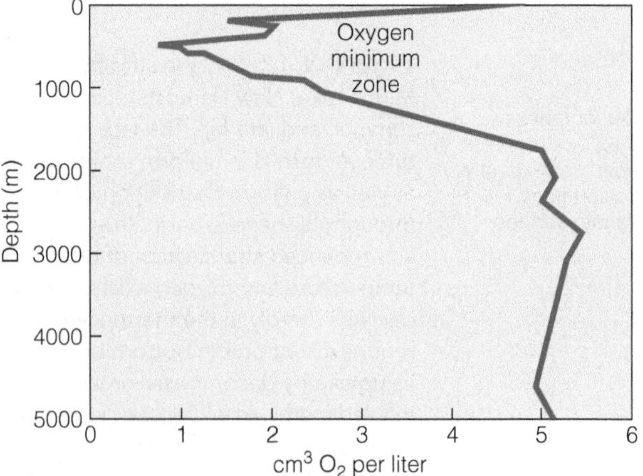

Figure 3.13 Vertical profile of oxygen with depth in the Atlantic Ocean. The oxygen content of the waters declines to a depth known as the oxygen minimum zone. Oxygen increase below this may be the result of the influx of cold, oxygen-rich waters that sank in the polar waters.

3.7 Acidity Has a Widespread Influence on Aquatic Environments

The solubility of carbon dioxide is somewhat different from that of oxygen in its chemical reaction with water. Water has a considerable capacity to absorb carbon dioxide, which is abundant in both freshwater and saltwater. Upon diffusing into the surface, carbon dioxide reacts with water to produce carbonic acid (H_2CO_3):

$$CO_2 + H_2O \rightleftharpoons H_2CO_3$$

Carbonic acid further dissociates into a hydrogen ion and a bicarbonate ion:

$$H_2CO_3 \rightleftharpoons HCO_3^- + H^+$$

Bicarbonate may further dissociate into another hydrogen ion and a carbonate ion:

$$HCO_3^- \rightleftharpoons H^+ + CO_3^{2-}$$

The carbon dioxide–carbonic acid–bicarbonate system is a complex chemical system that tends to stay in equilibrium. (Note that the arrows in the preceding equations go in both directions.) Therefore, if carbon dioxide (CO_2) is removed from the water, the equilibrium is disturbed and the equations will shift to the left, with carbonic acid and bicarbonate producing more CO_2 until a new equilibrium is reached.

The chemical reactions just described result in the production and absorption of free hydrogen ions (H^+). The abundance of hydrogen ions in solution is a measure of **acidity**. The greater the number of H^+ ions, the more acidic is the solution. **Alkaline** solutions are those that have a large number of OH^- (hydroxyl ions) and few H^+ ions. The measurement of acidity and alkalinity is called *pH*, calculated as the negative logarithm (base 10) of the concentration of hydrogen ions in solution. In pure water, a small fraction of molecules dissociates into ions: $H_2O \rightarrow H^+ + OH^-$, and the ratio of H^+ ions to OH^- ions is 1:1. Because both occur in a concentration of 10^{-7} moles per liter, a neutral solution has a pH of 7 [$-\log(10^{-7}) = 7$]. A solution departs from neutral when the concentration of one ion increases and the other decreases. Customarily, we use the negative logarithm of the hydrogen ion to describe a solution as an acid or a base. Thus, a gain of hydrogen ions to 10^{-6} moles per liter means a decrease of OH^- ions to 10^{-8} moles per liter, and the pH of the solution is 6. The negative logarithmic pH scale goes from 1 to 14. A pH greater than 7 denotes an alkaline solution (greater OH^- concentration) and a pH of less than 7 an acidic solution (greater H^+ concentration).

Although pure water is neutral in pH, because the dissociation of the water molecule produces equal numbers of H^+ and OH^- ions, the presence of CO_2 in the water alters this relationship. The preceding chemical reactions result in the production and absorption of H^+ ions. Because the abundance of hydrogen ions in solution is the measure of acidity, the dynamics of the carbon dioxide–carbonic acid–bicarbonate system directly affect the pH of aquatic ecosystems. In general, the carbon dioxide–carbonic acid–bicarbonate system functions as a buffer to keep the pH of water within a narrow range. It does this by absorbing hydrogen ions in the water when they are in excess (producing carbonic acid and bicarbonates) and producing them when they are in short supply (producing carbonate and bicarbonate ions). At neutrality (pH 7), most of the CO_2 is present as HCO_3^- (**Figure 3.14**). At a high pH, more CO_2 is present as CO_3^{2-} than at a low pH, where more CO_2 occurs in the free condition. Addition or removal of CO_2 affects pH, and a change in pH affects CO_2.

The pH of natural waters ranges between 2 and 12. Waters draining from watersheds dominated geologically by limestone ($CaCO_3$) have a much higher pH and are well buffered as compared to waters from watersheds dominated by acid sandstone and granite. The presence of the strongly alkaline ions sodium, potassium, and calcium in ocean waters results in seawater being slightly alkaline, usually ranging from 7.5 to 8.4.

The pH of aquatic environments can exert a powerful influence on the distribution and abundance of organisms. Increased acidity can affect organisms directly, by influencing physiological processes, and indirectly, by influencing the concentrations of toxic heavy metals. Tolerance limits for pH vary among plant and animal species, but most organisms cannot survive and reproduce at a pH below about 4.5. Aquatic organisms are unable to tolerate low pH conditions largely because acidic waters contain high concentrations of aluminum. Aluminum is highly toxic to many species of aquatic life and thus leads to a general decline in aquatic populations.

Aluminum is insoluble when the pH is neutral or basic. Insoluble aluminum is present in high concentrations in rocks, soils, and river and lake sediments. Under normal pH conditions, the aluminum concentrations of lake water are low; however, as the pH drops and becomes more acidic, aluminum begins to dissolve, raising the concentration in solution.

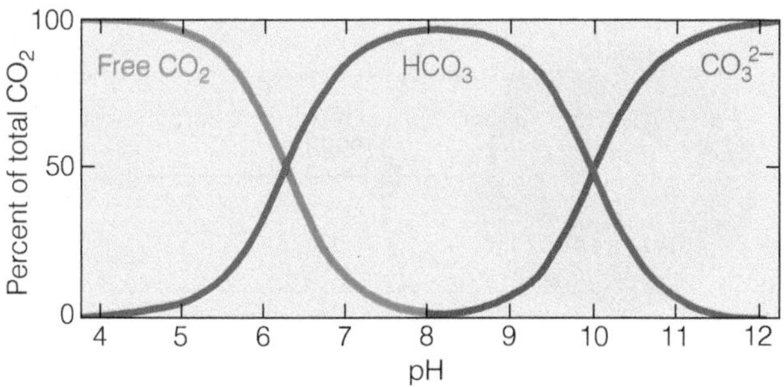

Figure 3.14 Theoretical percentages of carbon dioxide (CO_2) in each of the three forms present in water in relation to pH. At low values of pH (acidic conditions), most of the CO_2 is in its free form. At intermediate values (neutral conditions) bicarbonate dominates, whereas under alkaline conditions most of the CO_2 is in the form of carbonate ions.

Interpreting Ecological Data

Q1. *Under conditions of neutral pH, what is the relative abundance of the different forms of CO_2?*

Q2. *The current pH of the ocean is approximately 8.1. What is the dominant form of CO_2?*

3.8 Water Movements Shape Freshwater and Marine Environments

The movement of water—currents in streams and waves in an open body of water or breaking on a shore—determines the nature of many aquatic environments. The velocity of a current molds the character and structure of a stream. The shape and steepness of the stream channel, its width, depth, and roughness of the bottom, and the intensity of rainfall and rapidity of snowmelt all affect velocity. In fast streams, velocity of flow is 50 cm per second or higher (see Chapter 24, **Quantifying Ecology 24.1**). At this velocity, the current removes all particles less than 5 millimeters (mm) in diameter and leaves behind a stony bottom. High water volume increases the velocity; it moves bottom stones and rubble, scours the streambed, and cuts new banks and channels. As the gradient decreases and the width, depth, and volume of water increase, silt and decaying organic matter accumulate on the bottom. Thus, the stream's character changes from fast water to slow (**Figure 3.15**).

Wind generates waves on large lakes and on the open sea. The frictional drag of the wind on the surface of smooth water causes ripples. As the wind continues to blow, it applies more pressure to the steep side of the ripple, and wave size begins

Figure 3.15 (top) A fast mountain stream. The elevation gradient is steep, and fast-flowing water scours the stream bottom, leaving largely bedrock material. (bottom) In contrast, a slow-flowing stream meanders through a growth of willows. The relatively flat topography reduces the flow rate and allows finer sediments to build up on the stream bottom. These two streams represent different environmental conditions and subsequently support different forms of aquatic life (see Chapter 24).

to grow. As the wind becomes stronger, short, choppy waves of all sizes appear; as they absorb more energy, they continue to grow. When the waves reach a point where the energy supplied by the wind equals the energy lost by the breaking waves, they become whitecaps. Up to a certain point, the stronger the wind, the higher the waves.

The waves breaking on a beach do not contain water driven in from distant seas. Each particle of water remains largely in the same place and follows an elliptical orbit with the passage of the wave. As a wave moves forward, it loses energy to the waves behind and disappears, its place taken by another. The swells breaking on a beach are distant descendants of waves generated far out at sea.

As the waves approach land, they advance into increasingly shallow water. When the bottom of the wave intercepts the ocean floor, the wavelength shortens and the wave steepens until it finally collapses forward, or breaks. As the waves break onshore, they dissipate their energy, pounding rocky shores or tearing away sandy beaches in one location and building up new beaches elsewhere.

We have discussed the patterns of ocean currents, influenced by the direction of the prevailing winds and the Coriolis effect in Chapter 2 (see Section 2.4). As the warm surface currents of the tropical waters move northward and southward (see map of surface currents in Figure 2.13), they bring up deep, cold, oxygenated waters from below, a process known as **upwelling** (**Figure 3.16a**). A similar pattern occurs in coastal regions. Winds blowing parallel to the coast move the surface waters offshore. Water moving upward from the deep replaces this surface water, creating a pattern of coastal upwelling (**Figure 3.16b**).

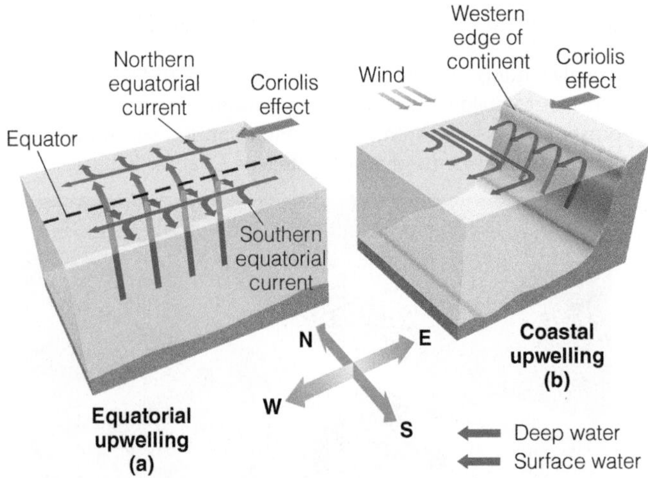

Figure 3.16 (a) Along the equator, the Coriolis effect acts to pull the westward-flowing currents to the north and south (purple solid arrows), resulting in an upwelling of deeper cold waters to the surface. (b) Along the western margins of the continents, the Coriolis effect causes the surface waters to move offshore (purple solid arrows). Movement of the surface waters offshore results in an upwelling of deeper, colder waters to the surface. Example shown is for the Northern Hemisphere.

3.9 Tides Dominate the Marine Coastal Environment

Tides profoundly influence the rhythm of life on ocean shores. Tides result from the gravitational pulls of the Sun and the Moon, each of which causes two bulges (tides) in the waters of the oceans. The two bulges caused by the Moon occur at the same time on opposite sides of Earth on an imaginary line extending from the Moon through the center of Earth (**Figure 3.17**). The tidal bulge on the Moon side is a result of gravitational attraction; the bulge on the opposite side occurs because the gravitational force there is less than at the Earth's center. As Earth rotates eastward on its axis, the tides advance westward. Thus, in the course of one daily rotation, Earth passes through two of the lunar tidal bulges, or high tides, and two of the lows, or low tides, at right angles (90° longitude difference) to the high tides.

The Sun also causes two tides on opposite sides of Earth, and these tides have a relation to the Sun like that of the lunar tides to the Moon. Because the Sun has a weaker gravitational pull than the Moon does, solar tides are partially masked by lunar tides—except for two times during the month: when the Moon is full and when it is new. At these times, Earth, Moon, and Sun are nearly in line, and the gravitational pulls of the Sun and the Moon are additive. This combination makes the high tides of those periods exceptionally large, with maximum rise and fall. These are the fortnightly spring tides, a name derived from the Saxon word *sprungen*, which refers to the brimming fullness and active movement of the water. When the Moon is at either quarter, its pull is at right angles to the pull of the Sun, and the two forces interfere with each other. At those times, the differences between high and low tides are exceptionally small. These are called the *neap tides*, from an old Scandinavian word meaning "barely enough."

Tides are not entirely regular, nor are they the same all over Earth. They vary from day to day in the same place, following

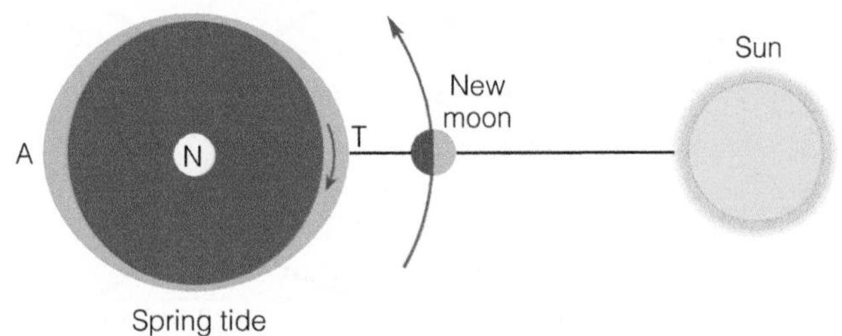

Figure 3.17 Tides result from the gravitational pull of the Moon. Centrifugal force applied to a kilogram of mass is 3.38 milligrams (mg). This force on a rotating Earth is balanced by gravitational force, except at moving points on Earth's surface that are directly aligned with the Moon. Thus, the centrifugal force at point N, the center of the rotating Earth, is 3.38 mg. Point T is directly aligned with the Moon. At this point, the Moon's gravitational force is 3.49 mg, a difference of 0.11 mg. Because the Moon's gravitational force is greater than the centrifugal force at T, the force is directed away from the Earth and causes a tidal bulge. At point A, the Moon's gravitational force is 3.27 mg, 0.11 mg less than the centrifugal force at N. This causes a tidal bulge on the opposite side of Earth.

the waxing and waning of the Moon. They may act differently in several localities within the same general area. In the Atlantic, semidaily tides are the rule. In the Gulf of Mexico and the Aleutian Islands of Alaska, the alternate highs and lows more or less cancel each other out, and flood and ebb follow one another at about 24-hour intervals to produce one daily tide. Mixed tides in which successive or low tides are of significantly different heights through the cycle are common in the Pacific and Indian oceans. These tides are combinations of daily and semidaily tides in which one partially cancels out the other.

Local tides around the world are inconsistent for many reasons. These reasons include variations in the gravitational pull of the Moon and the Sun as a result of the elliptical orbit of Earth, the angle of the Moon in relation to the axis of Earth, onshore and offshore winds, the depth of water, the contour of the shore, and wave action.

The area lying between the water lines of high and low tide, referred to as the **intertidal zone**, is an environment of extremes. The intertidal zone undergoes dramatic shifts in environmental conditions with the daily patterns of inundation and exposure. As the tide recedes, the uppermost layers of life are exposed to air, wide temperature fluctuations, intense solar radiation, and desiccation for a considerable period, whereas the lowest fringes of the tidal zone may be exposed only briefly before the high tide submerges them again. Temperatures on tidal flats may rise to 38°C when exposed to direct sunlight and drop to 10°C within a few hours when the flats are covered by water.

Organisms living in the sand and mud do not experience the same violent temperature fluctuations as those living on rocky shores do. Although the surface temperature of the sand at midday may be 10°C (or more) higher than that of the returning seawater, the temperature a few centimeters below the sand's surface remains almost constant throughout the year (see Section 25.3).

3.10 The Transition Zone between Freshwater and Saltwater Environments Presents Unique Constraints

Water from streams and rivers eventually drains into the sea. The place where freshwater mixes with saltwater is called an **estuary**. Temperatures in estuaries fluctuate considerably, both daily and seasonally. Sun and inflowing and tidal currents heat the water. High tide on the mudflats may heat or cool the water, depending on the season. The upper layer of estuarine water may be cooler in winter and warmer in summer than the bottom—a condition that, as in a lake, will cause spring and autumn turnovers (see Figures 3.9 and 3.12).

In the estuary, where freshwater meets the sea, the interaction of inflowing freshwater and tidal saltwater influences the salinity of the estuarine environment. Salinity varies vertically and horizontally, often within one tidal cycle (**Figure 3.18**).

Salinity may be the same from top to bottom or it may be completely stratified, with a layer of freshwater on top and a layer of dense, salty water on the bottom. Salinity is homogeneous when currents are strong enough to mix the water from top to bottom. The salinity in some estuaries is homogeneous at low tide, but at high tide a surface wedge of seawater moves upstream more rapidly than the bottom water. Salinity is then unstable, and density is inverted. The seawater on the surface tends to sink as lighter freshwater rises, and mixing takes place from the surface to the bottom. This phenomenon is known as **tidal overmixing**. Strong winds, too, tend to mix saltwater with freshwater in some estuaries, but when the winds are still, the river water flows seaward on a shallow surface over an upstream movement of seawater, more gradually mixing with the salt.

Horizontally, the least saline waters are at the river mouth and the most saline at the sea (see Figure 3.18). Incoming and outgoing currents deflect this configuration. In all estuaries of the Northern Hemisphere, outward-flowing freshwater and inward-flowing seawater are deflected to the right (relative to the axis of water flow from the river to ocean) because of Earth's rotation (Coriolis effect; see Section 2.3). As a result, salinity is higher on the left side; the concentration of metallic ions carried by rivers varies from drainage to drainage; and salinity and chemistry differ among estuaries. The portion of dissolved salts in the estuarine waters remains about the same as that of seawater, but the concentration varies in a gradient from freshwater to sea.

To survive in estuaries, aquatic organisms must have evolved physiological or behavioral adaptations to changes in salinity. Many oceanic species of fish are able to move inward during periods when the flow of freshwater from rivers is low and the salinity of estuaries increases. Conversely, freshwater fish move into the estuarine environment during periods of flood when salinity levels drop. Because of the stressful conditions that organisms face in the mixed zones of estuaries, there is often a relatively low diversity of organisms despite the high productivity found in these environments (see Chapter 24).

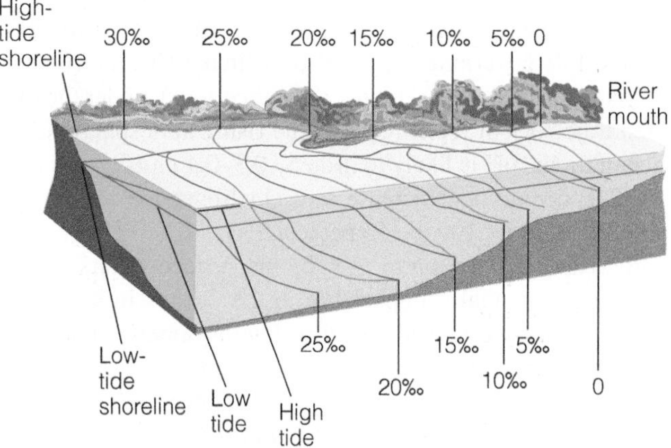

Figure 3.18 Vertical and horizontal stratification of salinity from the river mouth to the estuary at high tide (brown lines) and low tide (blue lines). At high tide, the incoming seawater increases the salinity toward the river mouth. At low tide, salinity is reduced. Salinity increases with depth because lighter freshwater flows over denser saltwater.

ECOLOGICAL Issues & Applications

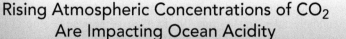

Rising Atmospheric Concentrations of CO_2 Are Impacting Ocean Acidity

The exchange of carbon dioxide (CO_2) between the atmosphere and the surface waters of the oceans is governed by the process of diffusion, with the net exchange moving CO_2 from higher concentrations (atmosphere) to lower concentrations (surface waters) (Section 3.7). Upon diffusing into the surface, the CO_2 reacts with the water to produce carbonic acid (H_2CO_3), which further dissociates into a hydrogen ion (H^+) and a bicarbonate ion (HCO_3^-). The bicarbonate may further dissociate into another hydrogen ion and a carbonate ion (CO_3^{2-}). In both of these chemical reactions, free hydrogen ions (H^+) are produced, the abundance of which is a measure of acidity. The greater the number of H^+ ions, the lower the value of pH and the more acidic the solution.

Under current ocean conditions, about 89 percent of the carbon dioxide dissolved in seawater takes the form of a bicarbonate ion, about 10 percent as a carbonate ion, and 1 percent as dissolved gas, and the pH of seawater on the surface of the oceans has remained relatively steady for millions of years at a value of about 8.2 (slightly basic—7.0 is neutral; see Figure 3.14). Since the height of the Industrial Revolution in the 19th century, however, atmospheric concentrations of CO_2 have been steadily rising as a result of the burning of fossil fuels (see Chapter 2, *Ecological Issues & Applications* and Chapter 27). As a consequence, the diffusion gradient of CO_2 between the atmosphere and oceans has increased, resulting in an increasing uptake of CO_2 into the surface waters. As a consequence, the pH of the surface waters of the oceans has fallen by about 0.1 pH unit from preindustrial times to today (**Figure 3.19**). Recall from Section 3.7 that the pH scale is logarithmic ($\log_{10}$; thus for every drop of 1 pH unit, hydrogen ion levels increase by a factor of 10), so this 0.1-unit drop in pH is equivalent to about a 25 percent increase in the ocean hydrogen ion concentration. According to estimates from the Intergovernmental Panel on Climate Change (IPCC; see Chapter 2, *Ecological Issues & Applications*), under the expected trajectory of fossil fuel use and rising atmospheric CO_2 concentrations, pH is likely to drop by 0.3–0.4 units by the end of the 21st century and increase ocean hydrogen ion concentration (or acidity) by 100 to 150 percent.

Increased absorption of CO_2 by the surface waters of the oceans can potentially impact life in the oceans in a variety of ways, both positive and negative. Photosynthetic algae and plants may benefit from higher CO_2 concentrations in the surface waters because elevated CO_2 may enhance rates of photosynthesis (see Chapter 6, *Ecological Issues & Applications*). On the other hand, one of the most important negative impacts of increasing ocean acidity relates to the process of calcification—the production of shells and plates out of calcium carbonate ($CaCO_3$)—which is important to the biology and survival of a wide range of marine organisms.

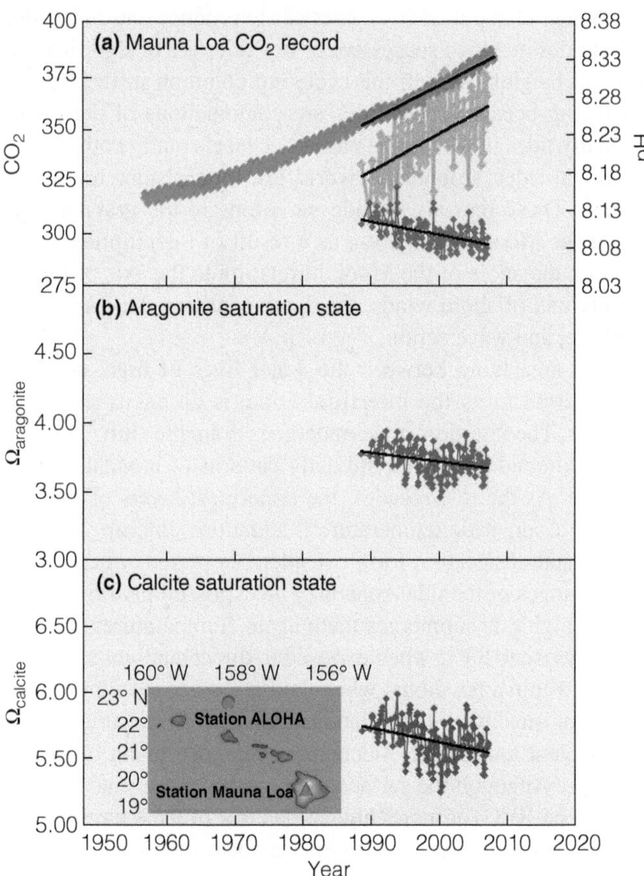

Figure 3.19 Time series of (a) atmospheric carbon dioxide (CO_2) at Mona Loa station in Hawaii (red), and CO_2 concentration (brown) and pH (blue) of surface ocean waters at Ocean station ALOHA (Pacific Ocean). The decline in pH has resulted in a decline in both the (b) aragonite and (c) calcite saturation state, that is, a decrease in mineral forms of calcium carbonate. (Adapted from Doney et al. 2009.)

$CaCO_3$ is formed in marine environments through the reaction of calcium and carbonate ions:

$$CO_3^{2-} + Ca^{2+} \rightleftharpoons CaCO_3$$

As with the chemical equations describing the formation of bicarbonate and carbonate ions from dissolved CO_2, the reaction involved in the formation of $CaCO_3$ proceeds in both directions. As sea water pH declines (acidity increases),

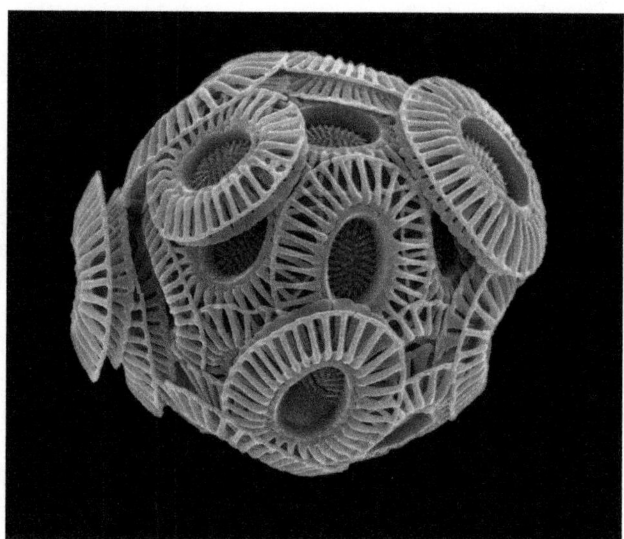

Figure 3.20 Calcium carbonate minerals are essential for the development of coccoliths (individual plates of calcium carbonate) in marine species such as this nanoplankton (the single-celled algae, *Emiliania huxleyi*).

carbonate ions (CO_3^{2-}) function like an antacid to neutralize the H^+, forming more bicarbonate ($CO_3^{2-} + H^+ \leftrightarrow HCO_3^-$). Therefore, declining pH results in an associated decline in carbonate ion concentrations (Figure 3.19a). This decline in carbonate ion concentration shifts the preceding equation in favor of the disassociation of $CaCO_3$ minerals into calcium and carbonate ions. The resulting decline in dissolved $CaCO_3$ minerals can have a significant impact on calcifying species, including oysters, clams, sea urchins, shallow water corals, deep sea corals, and calcareous plankton.

The process of calcification by marine organisms involves the precipitation of dissolved $CaCO_3$ into solid $CaCO_3$ structures, such as coccoliths (individual plates of $CaCO_3$ formed by single-celled algae; **Figure 3.20**). After they are formed, these structures are vulnerable to dissolution unless the surrounding seawater contains saturating concentrations of $CaCO_3$. As carbonate ions become depleted because of declining pH, seawater becomes undersaturated with respect to two $CaCO_3$ minerals vital for calcification, aragonite and calcite (Figure 3.19b and 3.19c). Current estimates suggest that the oceans are becoming undersaturated with respect to aragonite at the poles, where the cold and dense waters most readily absorb atmospheric CO_2, and that under projected rates of CO_2 emissions (IPCC), undersaturation would extend throughout the entire Southern Ocean ($<60°$ S) and into the subarctic Pacific by the end of the century (2100). These changes will threaten high-latitude aragonite secreting organisms including cold-water corals, which provide essential fish habitat, and shelled pteropods (free-swimming pelagic sea snails and sea slugs; **Figure 3.21**), an abundant food source for marine predators.

Table 3.1 Representative examples of impacts of ocean acidification on major groups of marine biota derived from experimental manipulation studies

		Response to increasing CO_2		
Major Group	Species Studied	Increasing	Decreasing	No response
Coccolithophores	4	2	1	1
Planktonic Foraminifera	2	2	-	-
Molluscs	4	4	-	-
Echinoderms	3	2	1	1
Tropical corals	11	11	-	-
Coralline red algae	1	1	-	-

(Data from Doney et al. 2009.)

In a review of experimental studies that have examined the response of marine calcifying species to elevated CO_2 conditions, Scott Doney of Woods Hole Oceanographic Institute and colleagues found that the degree of sensitivity varies among species, and some species may even show enhanced calcification at elevated CO_2 levels (**Table 3.1**). The researchers found, however, that in the vast majority of species, including every study published that has examined the calcification rates of coral species, elevated CO_2 concentrations and the associated decreasing aragonite saturation state had a negative effect on calcification rates. The researchers concluded that "ocean acidification impacts processes so fundamental to the overall structure and function of marine ecosystems that any significant changes could have far-reaching consequences for the oceans of the future."

Figure 3.21 Pterpod dissolving in water undersaturated with aragonite during a period of 45 days.
(Photograph by David Littschwager/National Geographic Society.)

SUMMARY

The Water Cycle 3.1

Water follows a cycle, traveling from the air to Earth and returning to the atmosphere. It moves through cloud formation in the atmosphere, precipitation, interception, and infiltration into the ground. It eventually reaches groundwater, springs, streams, and lakes from which evaporation takes place, bringing water back to the atmosphere in the form of clouds. The various aquatic environments are linked, either directly or indirectly, by the water cycle.

The largest reservoir in the global water cycle is the oceans, which contain more than 97 percent of the total volume of water on Earth. In contrast, the atmosphere is one of the smallest reservoirs but has a fast turnover time.

Properties of Water 3.2

Water has a unique molecular structure. The hydrogen atoms are located on the side of the water molecule that has a positive charge. The opposite side, where the oxygen atom is located, has a negative charge, thus polarizing the water molecule. Because of their polarity, water molecules become coupled with neighboring water molecules to produce a lattice-like structure with unique properties.

Depending on its temperature, water may occur in the form of a liquid, solid, or gas. It absorbs or releases considerable quantities of heat with a small rise or fall in temperature. Water has a high viscosity that affects its flow. It exhibits high surface tension, caused by a stronger attraction of water molecules for each other than for the air above the surface. If a body is submerged in water and weighs less than the water it displaces, it is subjected to the upward force of buoyancy. These properties are important ecologically and biologically.

Light 3.3

Both the quantity and quality of light change with water depth. In pure water, red and infrared light are absorbed first, followed by yellow, green, and violet; blue penetrates the deepest.

Temperature in Aquatic Environments 3.4

Lakes and ponds experience seasonal shifts in temperature. In summer there is a distinct vertical gradient of temperature, resulting in a physical separation of warm surface waters and the colder waters below the thermocline. When the surface waters cool in the fall, the temperature becomes uniform throughout the basin and water circulates throughout the lake. A similar mixing takes place in the spring when the water warms. In some deep lakes and the oceans, the thermocline simply descends during turnover periods and does not disappear at all.

Temperature of flowing water is variable, warming and cooling with the season. Within the stream or river, temperatures vary with depth, amount of shading, and exposure to sun.

Water as a Solvent 3.5

Water is an excellent solvent with the ability to dissolve more substances than any other liquid can. The solvent properties of water are responsible for most of the minerals found in aquatic environments. The waters of most rivers and lakes contain a relatively low concentration of dissolved minerals, determined largely by the underlying bedrock over which the water flows. In contrast, the oceans have a much higher concentration of solutes. As pure water evaporates from the surface to the atmosphere, the flow of freshwaters into the oceans continuously adds to the solute content of the waters.

The solubility of sodium chloride is high; together with chlorine, it makes up some 86 percent of sea salt. The concentration of chlorine is used as an index of salinity. Salinity is expressed in practical salinity units (psu; represented as ‰, measured as grams of chlorine per kilogram of water).

Oxygen 3.6

Oxygen enters the surface waters from the atmosphere through the process of diffusion. The amount of oxygen water can hold depends on its temperature, pressure, and salinity. In lakes, oxygen absorbed by surface water mixes with deeper water by turbulence. During the summer, oxygen may become stratified, decreasing with depth because of decomposition in bottom sediments. During spring and fall turnover, oxygen becomes replenished in deep water. Constant swirling of stream water gives it greater contact with the atmosphere and thus allows it to maintain a high oxygen content.

Acidity 3.7

The measurement of acidity is pH, the negative logarithm of the concentration of hydrogen ions in solution. In aquatic environments, a close relationship exists between the diffusion of carbon dioxide into the surface waters and the degree of acidity and alkalinity. Acidity influences the availability of nutrients and restricts the environment of organisms sensitive to acid situations.

Water Movement 3.8

Currents in streams and rivers as well as waves in open sea and breaking on ocean shores determine the nature of many aquatic and marine environments. The velocity of currents shapes the environment of flowing water. Waves pound rocky shores and tear away and build up sandy beaches. Movement of water in surface currents of the ocean affects the patterns of deep-water circulation. As the equatorial currents move northward and southward, deep waters move up to the surface, forming regions of upwelling. In coastal regions, winds blowing parallel to the coast create a pattern of coastal upwelling.

Tides 3.9

Rising and falling tides shape the environment and influence the rhythm of life in coastal intertidal zones.

Estuaries 3.10

Water from all streams and rivers eventually drains into the sea. The place where this freshwater joins and mixes with the salt is called an *estuary*. Temperatures in estuaries fluctuate considerably, both daily and seasonally. The interaction of inflowing freshwater and tidal saltwater influences the salinity of the estuarine environment. Salinity varies vertically and horizontally, often within one tidal cycle.

Ocean Acidification Ecological Issues & Applications

Rising atmospheric concentrations of carbon dioxide have resulted in increased concentrations in the surface waters of the oceans. The increased carbon dioxide concentrations of the surface waters have resulted in a decline in pH and reduced carbonate concentrations. The reduction in carbonate concentrations has reduced calcium carbonate mineral concentrations that are essential for calcifying marine species.

STUDY QUESTIONS

1. Draw a simple diagram of the water cycle and then describe the processes involved.
2. Why is there less seasonal variation in temperature in aquatic compared to terrestrial environments (at the same latitude)?
3. What property of water allows aquatic organisms to function with far fewer supportive structures (tissues) than terrestrial organisms have?
4. What is the fate of visible light in water?
5. What is the thermocline? What causes the development of a thermocline?
6. Explain why seasonal stratification of temperature and oxygen takes place in deep ponds and lakes.
7. What effect does increasing the carbon dioxide concentration of water have on its pH?
8. The concentration of which element is used to define the salinity of water?
9. What causes the upwelling of deeper, cold waters in the equatorial zone of the oceans?
10. What causes the tides?

FURTHER READINGS

Classic Studies

Hutchinson, G. E. 1957–1975. *A treatise on limnology. Vol. 1: Geography, physics, and chemistry*. New York: Wiley.
A classic reference.

Current Research

Doney, S. C., V. J. Farby, R. A. Feely, and J. A. Kleypas. 2009. "Ocean acidification: The other CO_2 problem." *Annual Review of Marine Science* 1:169–192.
This review paper provides an excellent overview of the issue of ocean acidification.

Garrison, T. 2012. *Oceanography: An invitation to marine science*. 8th ed. Belmont, CA: Brooks/Cole.
A clearly written and well-illustrated introductory text for those interested in more detail on the subject matter.

Hynes, H. B. N. 2001. *The ecology of running waters*. Caldwell, NJ: Blackburn Press.
A reprint of a classic and valuable work, this major reference continues to be influential.

McLusky, D. S. 2004. *The estuarine ecosystem*. 3rd ed. New York: Chapman & Hall.
Clearly describes the structure and function of estuarine ecosystems.

Nybakken, J. W., and M. D. Bertness. 2005. *Marine biology: An ecological approach*. 6th ed. San Francisco: Benjamin Cummings.
Chapters 1 and 6 provide an excellent introduction to the physical environment of the oceans.

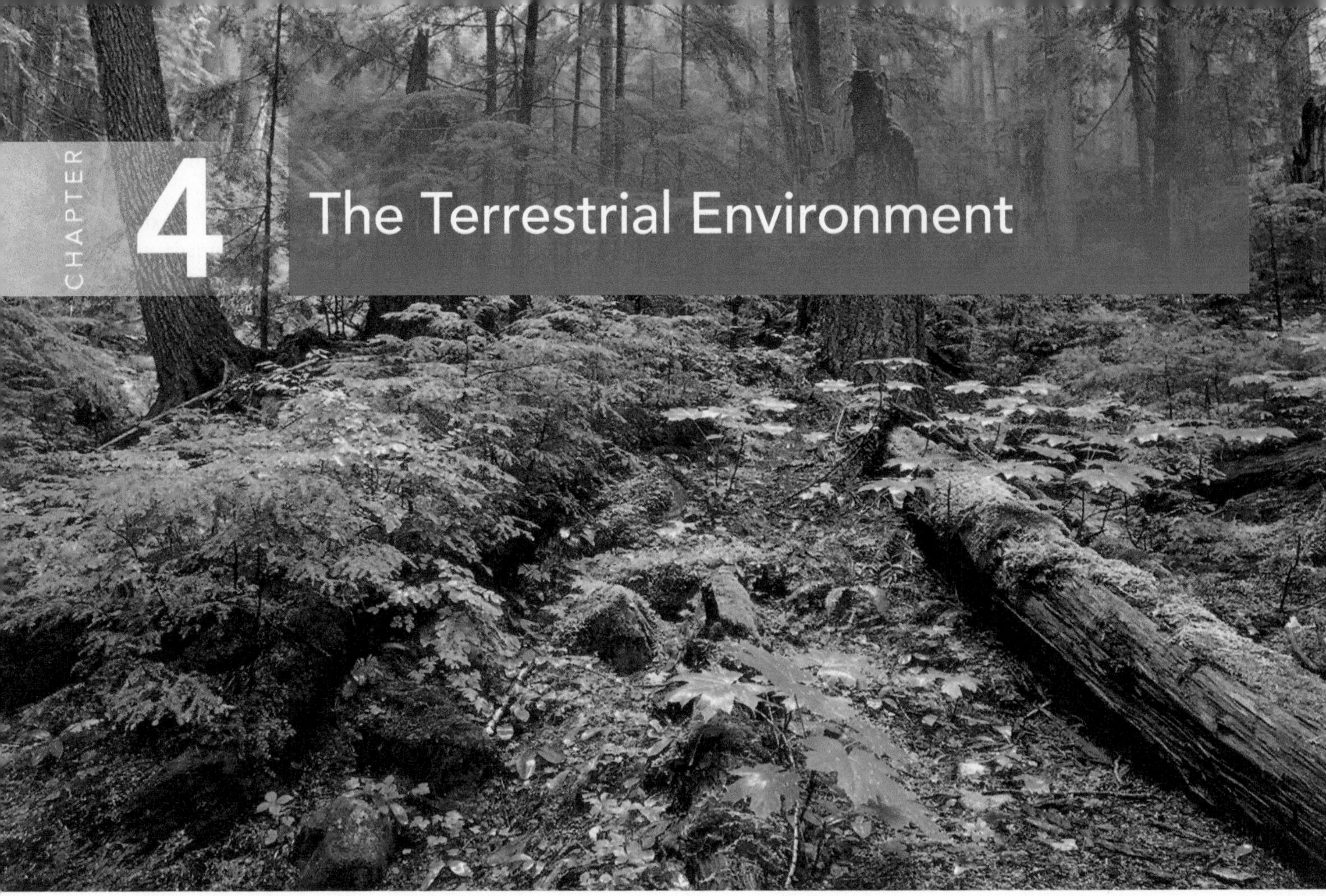

The input and decomposition of dead organic matter is a key factor in the development of forest soils.

CHAPTER GUIDE

OUR INTRODUCTION OF AQUATIC ENVIRONMENTS was dominated by discussion of the physical and chemical properties of water—characteristics such as depth, flow rate, and salinity (see Chapter 3). When considering the term *terrestrial environment*, however, people typically do not think of the physical and chemical characteristics of a place. What we most likely visualize is the vegetation: the tall, dense forests of the wet tropics; the changing colors of autumn in a temperate forest; or the broad expanses of grass that characterize the prairies. Animal life depends on the vegetation within a region to provide the essential resources of food and cover—and as such, the structure and composition of plant life constrain the distribution and abundance of animal life. But ultimately, as with aquatic environments, the physical and chemical features of terrestrial environments set the constraints for life. Plant life is a reflection of the climate and soils (as discussed in Chapter 6). Regardless of the suitability of plant life for providing essential resources, the physical conditions within a region impose the primary constraints on animal life as well (Chapter 7).

We will explore key features of the terrestrial environment that directly influence life on land. Life emerged from the water to colonize the land more than a billion years ago. The transition to terrestrial environments posed a unique set of problems for organisms already adapted to an aquatic environment. To understand the constraints imposed by the terrestrial environment, we must start by looking at the physical differences between the terrestrial and aquatic environments and at the problems these differences create for organisms making the transition from water to land.

4.1 Life on Land Imposes Unique Constraints

The transition from life in aquatic environments to life on land brought with it a variety of constraints. Perhaps the greatest constraint imposed by terrestrial environments is desiccation. Living cells, both plant and animal, contain about 75–95 percent water. Unless the air is saturated with moisture, water readily evaporates from the surfaces of cells via the process of diffusion (see Section 2.5). The water that is lost to the air must be replaced if the cell is to remain hydrated and continue to function. Maintaining this balance of water between organisms and their surrounding environment (referred to as an organism's **water balance**) has been a major factor in the evolution of life on land. For example, in adapting to the terrestrial environment, plants have evolved extensively specialized cells for different functions. Aerial parts of most plants, such as stems and leaves, are coated with a waxy cuticle that prevents water loss. While it reduces water loss, the waxy surface also prevents gas exchange (carbon dioxide and oxygen) from occurring. As a result, terrestrial plants have evolved pores on the leaf surface (stomata) that allow gases to diffuse from the air into the interior of the leaf (see Chapter 6).

To stay hydrated, an organism must replace water that it has lost to the air. Terrestrial animals can acquire water by drinking and eating. For plants, however, the process is passive. Early in their evolution, land plants evolved vascular tissues consisting of cells joined into tubes that transport water and nutrients throughout the plant body. The topic of water balance and the array of characteristics that plants and animals have evolved to overcome the problems of water loss are discussed in more detail later (see Chapters 6 and 7).

Desiccation is not the only constraint imposed by the transition from water to land. Because air is less dense than water, it results in a much lower drag (frictional resistance) on the movement of organisms; but it greatly increases the constraint imposed by gravitational forces. The upward force of buoyancy resulting from the displacement of water helps organisms in aquatic environments overcome the constraints imposed by gravity (see Section 3.2). In contrast, the need to remain erect against gravitational force in terrestrial environments results in a significant investment in structural materials such as skeletons (for animals) or cellulose (for plants). The giant kelp (*Macrocystis pyrifera*) inhabiting the waters off the coast of California is an excellent example (**Figure 4.1, left**). It grows in dense stands called *kelp forests*. Anchored to the bottom sediments, these kelp (macroalgae) can grow 100 feet or more toward the surface. The kelp are kept afloat by gas-filled bladders attached to each blade; yet when the kelp plants are removed from the water, they collapse into a mass. Lacking supportive tissues strengthened by cellulose and lignin, the kelp

Figure 4.1 (left) The giant kelp (*Macrocystis pyrifera*) inhabits the waters off the coast of California. Anchored to the bottom sediments, these kelp plants can grow 100 feet or more toward the surface despite their lack of supportive tissues. These kelp plants are kept afloat through the buoyancy of gas-filled bladders attached to each blade, yet when the kelp plants are removed from the water, they collapse into a mass. (right) In contrast, a redwood tree (*Sequoia sempervirens*) of comparable height allocates more than 80 percent of its biomass to supportive and conductive tissues that help the tree resist gravitational forces.

cannot support its own weight under the forces of gravity. In contrast, a tree of equivalent height inhabiting the coastal forest of California (**Figure 4.1, right**) must allocate more than 80 percent of its total mass to supportive and conductive tissues in the trunk (bole), branches, leaves, and roots.

Another characteristic of terrestrial environments is their high degree of variability, both in time and space. Temperature variations on land (air) are much greater than in water. The high specific heat of water prevents wide daily and seasonal fluctuations in the temperature of aquatic habitats (see Section 3.2). In contrast, such fluctuations are a characteristic of air temperatures (see Chapter 2). Likewise, the timing and quantity of precipitation received at a location constrains the availability of water for terrestrial plants and animals as well as their ability to maintain water balance. These fluctuations in temperature and moisture have both a short-term effect on metabolic processes and a long-term influence on the evolution and distribution of terrestrial plants and animals (see Chapters 6 and 7). Ultimately, the geographic variation in climate governs the large-scale distribution of plants and therefore the nature of terrestrial ecosystems (see Chapter 23).

4.2 Plant Cover Influences the Vertical Distribution of Light

In contrast to aquatic environments, where the absorption of solar radiation by the water itself results in a distinct vertical gradient of light, the dominant factor influencing the vertical gradient of light in terrestrial environments is the absorption and reflection of solar radiation by plants. When walking into a forest in summer, you will observe a decrease in light (**Figure 4.2a**). You can observe much the same effect if you examine the lowest layer in grassland or an old field (**Figure 4.2b**). The quantity and quality (spectral composition) of light that does penetrate the canopy of vegetation to reach the ground varies with both the quantity and orientation of the leaves.

The amount of light at any depth in the canopy is affected by the number of leaves above. As we move down through the canopy, the number of leaves above increases; so the amount of light decreases. However, because leaves vary in size and shape, the number of leaves is not the best measure of quantity. The quantity of leaves, or foliage density, is generally expressed as the leaf area. Because most leaves are flat, the leaf area is the surface area of one or both sides of the leaf. When the leaves are not flat, the entire surface area is sometimes measured. To quantify the changes in light environment with increasing area of leaves, we need to define the area of leaves per unit ground area (m^2 leaf area/m^2 ground area). This measure is the **leaf area index** ([LAI]; **Figure 4.3**). A LAI of 3 indicates a quantity of 3 m^2 of leaf area over each 1 m^2 of ground area.

The greater the LAI above any surface, the lower the quantity of light reaching that surface. As you move from the top of the canopy to the ground in a forest, the cumulative leaf area and LAI increase. Correspondingly, light decreases. The general relationship between available light and LAI is described by Beer's law (see **Quantifying Ecology 4.1**).

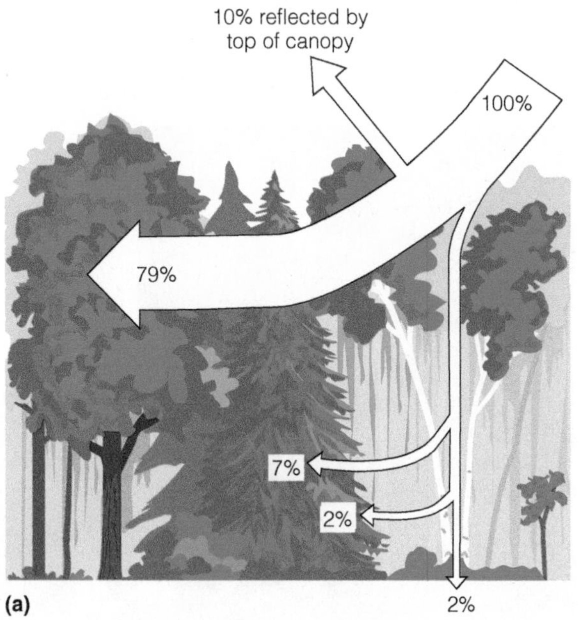

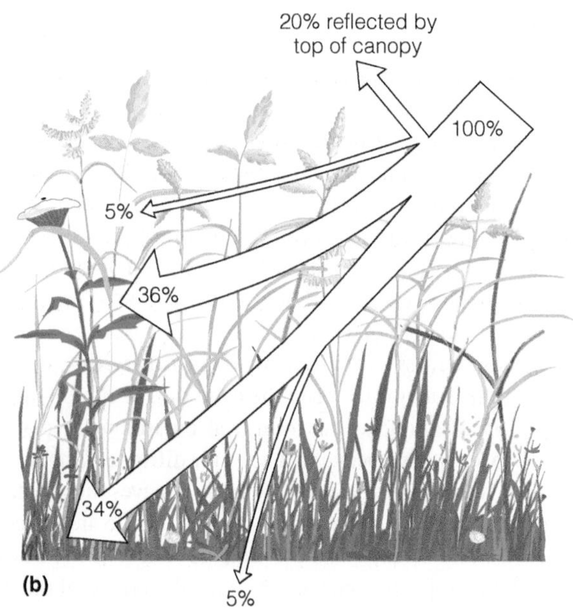

Figure 4.2 Absorption and reflection of light by the plant canopy. (a) A mixed conifer–deciduous forest reflects about 10 percent of the incident photosynthetically active radiation (PAR) from the upper canopy, and it absorbs most of the remaining PAR within the canopy. (b) A meadow reflects 20 percent of the PAR from the upper surface. The middle and lower regions, where the leaves are densest, absorb most of the rest. Only 2–5 percent of PAR reaches the ground.
(Adapted from Larcher 1980.)

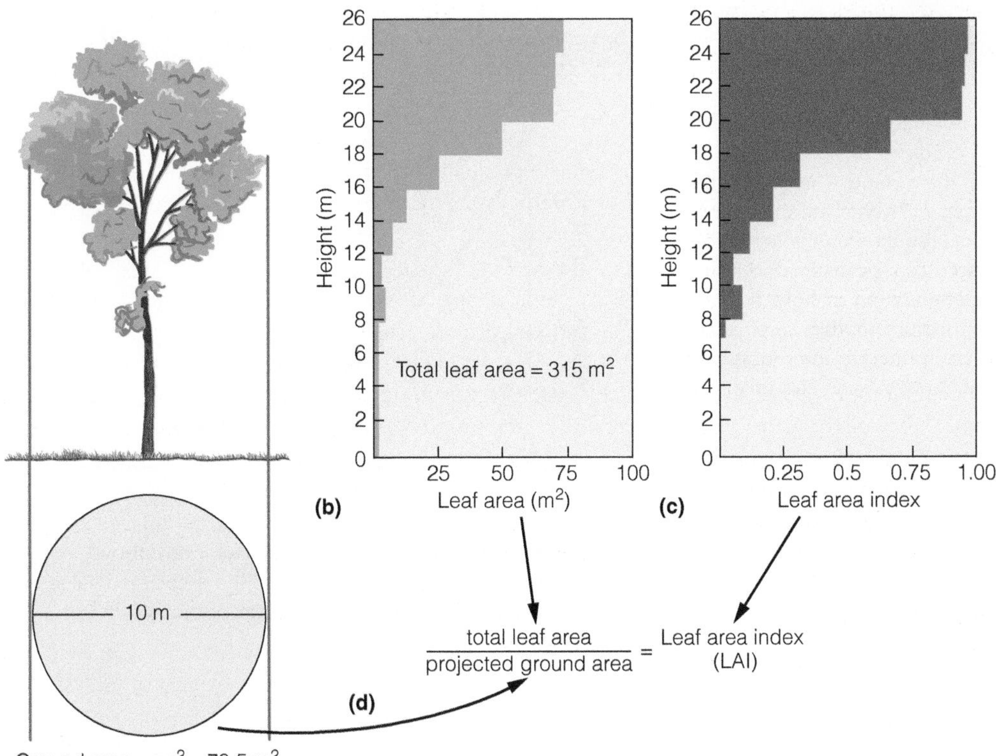

Figure 4.3 The concept of leaf area index (LAI). (a) A tree with a crown 10 m wide projects a circle of the same size on the ground. (b) Foliage density (area of leaves) at various heights above the ground. (c) Contributions of layers in the crown to the leaf area index. (d) Calculation of leaf area index (LAI). The total leaf area is 315 m^2. The projected ground area is 78.5 m^2. The LAI is 4.

In addition to the quantity of light, the spectral composition (quality) of light varies through the plant canopy. Recall that the wavelengths of approximately 400 to 700 nm make up visible light (Section 2.1 and Figure 2.1). These wavelengths are also known as photosynthetically active radiation (PAR) because they include the wavelengths used by plants as a source of energy in photosynthesis (see Chapter 6). The transmittance of PAR is typically less than 10 percent, whereas the transmittance of far-red radiation (730 nm) is much greater. As a result, the ratio of red (660 nm) to far-red radiation (R/FR ratio) decreases through the canopy. This shift in the spectral quality of light affects the production of phytochrome (a pigment that allows a plant to perceive shading by other plants), thus influencing patterns of growth and allocation (see Chapter 6, Section 6.8).

Besides the quantity of leaves, the orientation of leaves on the plant influences the attenuation of light through the canopy. The angle at which a leaf is oriented relative to the Sun changes the amount of light it absorbs. If a leaf that is perpendicular to the Sun absorbs 1.0 unit of light energy (per unit leaf area/time), the same leaf displayed at a 60-degree angle to the Sun will absorb only 0.5 units. The reason is that the same leaf area represents only half the projected surface area and therefore intercepts only half as much light energy (**Figure 4.4**). Thus, leaf angle influences the vertical distribution of light through the canopy as well as the total amount of light absorbed and reflected. The sun angle varies, however, both geographically (see Section 2.1) and through time at a given location (over the course of the day and seasonally). Consequently, different leaf angles are more effective at intercepting light in different locations and at different times. For example, in high-latitude environments, where sunlight angles are low, canopies having leaves that are displayed at an angle will absorb light more effectively (see Figure 2.5). Leaves that are displayed at an angle rather than perpendicular to the Sun are also typical of arid tropical environments. In these hot and dry environments,

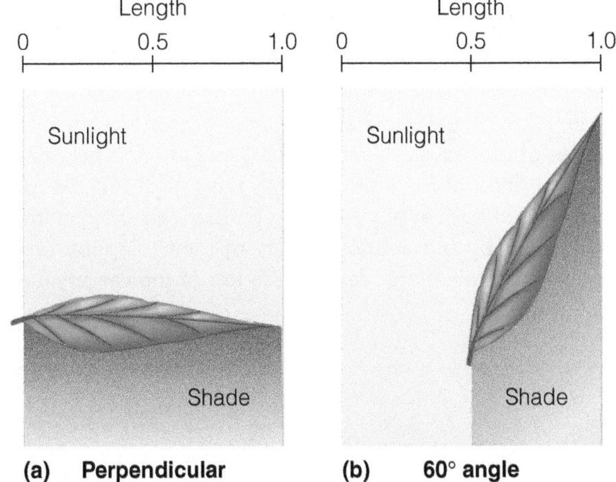

Figure 4.4 Influence of leaf orientation (angle) on the interception of light energy. If a leaf that is perpendicular to the source of light (a) intercepts 1.0 unit of light energy, the same leaf at an angle of 60 degrees relative to the light source will intercept only 0.5 unit (b). The reduction in intercepted light energy is a result of the angled leaf projecting a smaller surface area relative to the light source.

QUANTIFYING ECOLOGY 4.1 Beer's Law and the Attenuation of Light

Due to the absorption and reflection of light by leaves, there is a distinct vertical gradient of light availability from the top of a plant canopy to the ground. The greater the surface area of leaves, the less light will penetrate the canopy and reach the ground. The vertical reduction, or attenuation, of light through a stand of plants can be estimated using Beer's law, which describes the attenuation of light through a homogeneous medium. The medium in this case is the canopy of leaves. Beer's law can be applied to the problem of light attenuation through a plant canopy using the following relationship:

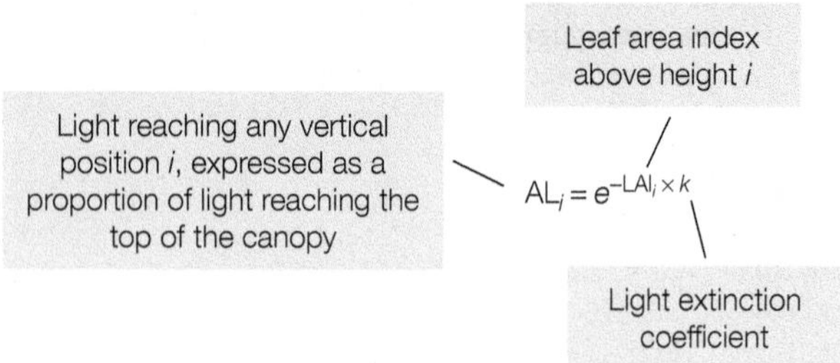

The subscript i refers to the vertical height of the canopy. For example, if i were in units of meters, a value of $i = 5$ refers to a height of 5 m above the ground. The value e is the natural logarithm (2.718). The light extinction coefficient, k, represents the quantity of light attenuated per unit of leaf area index (LAI) and is a measure of the degree to which leaves absorb and reflect light. The extinction coefficient will vary as a function of leaf angle (see Figure 4.4) and the optical properties of the leaves. Although the value of AL_i is expressed as a proportion of the light reaching the top of the canopy, the quantity of light at any level can be calculated by multiplying this value by the actual quantity of light (or photosynthetically active radiation) reaching the top of the canopy (units of μmol/m²/s).

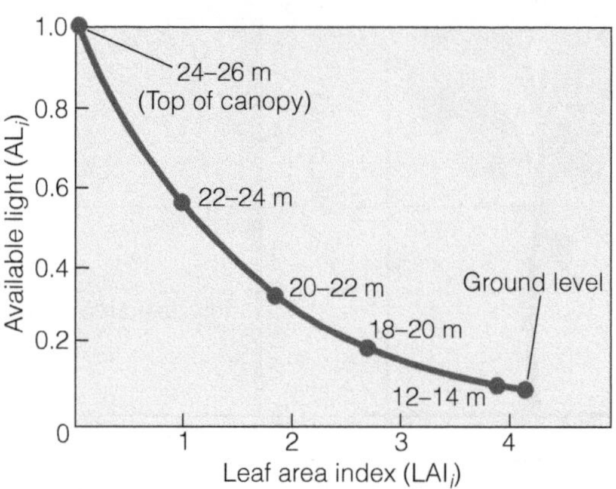

Figure 1 Relationship between leaf area index above various heights in the canopy (LAI_i) and the associated values of available light (AL_i), expressed as a proportion of PAR at the top of the canopy.

For the example presented in Figure 4.3, we can construct a curve describing the available light at any height in the canopy. In **Figure 1**, the light extinction coefficient has a value of $k = 0.6$ as an average value for a temperate deciduous forest. We label vertical positions from the top of the canopy to ground level on the curve. Knowing the amount of leaves (LAI) above any position in the canopy (i), we can use the equation to calculate the amount of light there.

The availability of light at any point in the canopy will directly influence the levels of photosynthesis (see Figure 6.2). The light levels and rates of light-limited photosynthesis for each of the vertical canopy positions are shown in the curve in **Figure 2**. Light levels are expressed as a proportion of values for fully exposed leaves at the top of the canopy (1500 μmol/m²/s). As one moves from the top of the canopy

angled leaves reduce light interception during midday, when temperatures and demand for water are at their highest.

Although light decreases downward through the plant canopy, some direct sunlight does penetrate openings in the crown and reaches the ground as sunflecks. Sunflecks can account for 70–80 percent of solar energy reaching the ground in forest environments (**Figure 4.5**).

In many environments, seasonal changes strongly influence leaf area. For example, in the temperate regions of the world, many forest tree species are deciduous, shedding their leaves

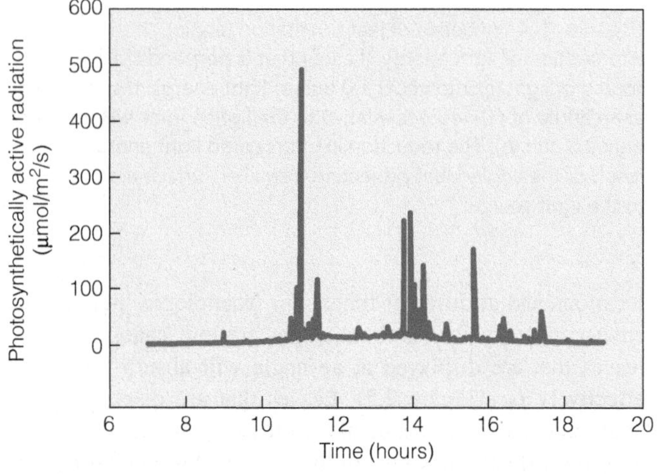

Figure 4.5 Changes in the availability of light (photosynthetically active radiation [PAR]) at ground level in a lowland rain forest in Mexico over the course of a day. The spikes result from sunflecks in an otherwise low-light.
(Adapted from Chazdon and Pearcy 1991.)

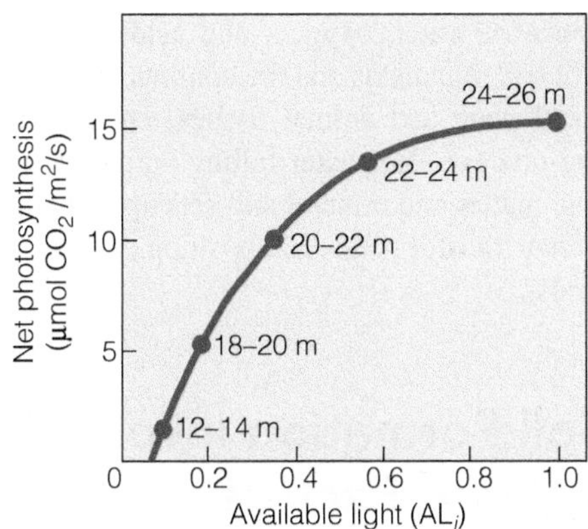

Figure 2 Relationship between available light (PAR) and rate of net photosynthesis at various heights in the canopy. Available light is expressed as the proportion of PAR at the top of the canopy (assumed to be 1500 μmol/m^2/s).

downward, the amount of light reaching the leaves and the corresponding rate of photosynthesis decline.

Beer's law can also be used to describe the vertical attenuation of light in aquatic environments, but applying the light extinction coefficient (k) is more complex. The reduction of light with water depth is a function of various factors: (1) attenuation by the water itself (see Section 3.3, Figure 3.7); (2) attenuation by phytoplankton (microscopic plants suspended in water), typically expressed as the concentration of chlorophyll (the light-harvesting pigment of plants) per volume of water (see Section 6.1); (3) attenuation by dissolved substances; and (4) attenuation by suspended particulates. Each of these factors has an associated light extinction coefficient, and the overall light extinction coefficient (k_T) is the sum of the individual coefficients:

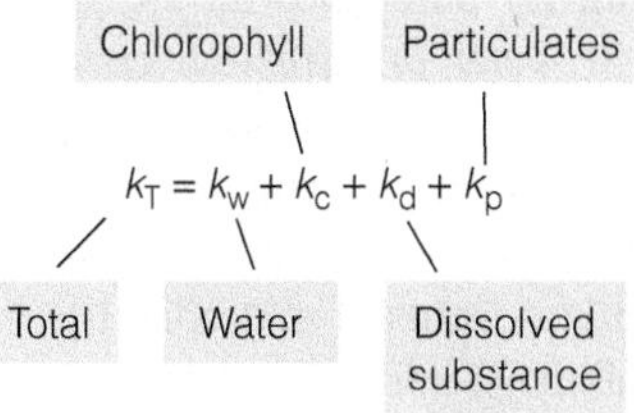

Whereas the light extinction coefficient for leaf area expresses the attenuation of light per unit of LAI, these values of k are expressed as the attenuation of light per unit of water depth (such as centimeter, meter, inches, or feet). Beer's law can then be used to estimate the quantity of light reaching any depth (z) by using the following equation:

$$AL_z = e^{-k_T z}$$

If the ecosystem supports submerged vegetation, such as kelp (see Figure 4.1), seagrass, or other plants that are rooted in the bottom sediments, the preceding equation can be used to calculate the available light at the top of the canopy. The equation describing the attenuation of light as a function of LAI can then be applied (combined) to calculate the further attenuation from the top of the plant canopy to the sediment surface.

1. If we assume that the value of k used to calculate the vertical profile of light in Figure 1 (k = 0.6) is for a plant canopy where the leaves are positioned horizontally (parallel to the forest floor), how would the value of k differ (higher or lower) for a forest where the leaves were oriented at a 60-degree angle? (See the example in Figure 4.4.)
2. In shallow-water ecosystems, storms and high wind can result in bottom sediments (particulates) being suspended in the water for some time before once again settling to the bottom. How would this situation affect the value of k_T and the attenuation of light in the water profile?

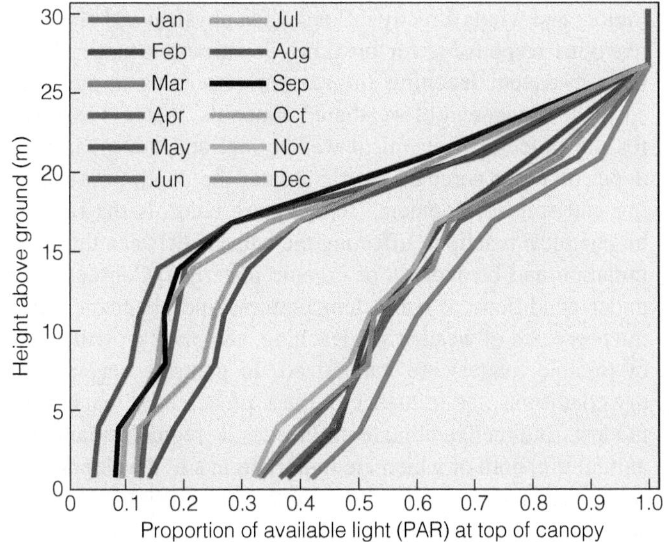

during the winter months. In these cases, the amount of light that penetrates a forest canopy varies with the season (**Figure 4.6**). In early spring in temperate regions, when leaves are just expanding, 20–50 percent of the incoming light may reach the forest floor. In other regions characterized by distinct wet and dry seasons, a similar pattern of increased light availability at the ground level occurs during the dry season (see Chapter 2).

Figure 4.6 Measured vertical profiles of photosynthetically active radiation (PAR) averaged monthly from 2003 to 2007 in Morgan-Monroe State Forest (Indiana, United States). Values are expressed as a proportion of PAR at the top of the canopy (27 m height). Note that the profiles form two distinct groups corresponding to the presence or absence of foliage in this temperate deciduous forest. (Adapted from Wenze Yanga, Wenge Ni-Meistera, Nancy Y. Kiangb, Paul R. Moorcroftc, Alan H. Strahler, Andrew Oliphante, *Agricultural and forest Meteorology* Volume 150, Issues 7–8, 15 July 2010, Pages 895–907.)

4.3 Soil Is the Foundation upon which All Terrestrial Life Depends

Soil is the medium for plant growth; the principal factor controlling the fate of water in terrestrial environments; nature's recycling system, which breaks down the waste products of plants and animals and transforms them into their basic elements; and a habitat to a diversity of animal life, from small mammals to countless forms of microbial life (see Chapter 21).

As familiar as it is, soil is hard to define. One definition says that soil is a natural product formed and synthesized by the weathering of rocks and the action of living organisms. Another states that soil is a collection of natural bodies of earth, composed of mineral and organic matter and capable of supporting plant growth. Indeed, one eminent soil scientist, Hans Jenny—a pioneer of modern soil studies—will not give an exact definition of soil. In his book *The Soil Resource*, he writes:

> Popularly, soil is the stratum below the vegetation and above hard rock, but questions come quickly to mind. Many soils are bare of plants, temporarily or permanently, or they may be at the bottom of a pond growing cattails. Soil may be shallow or deep, but how deep? Soil may be stony, but surveyors (soil) exclude the larger stones. Most analyses pertain to fine earth only. Some pretend that soil in a flowerpot is not a soil, but soil material. It is embarrassing not to be able to agree on what soil is. In this, soil scientists are not alone. Biologists cannot agree on a definition of life and philosophers on philosophy.

Of one fact we are sure. Soil is not just an abiotic environment for plants. It is teeming with life—billions of minute and not so minute animals, bacteria, and fungi. The interaction between the biotic and the abiotic makes the soil a living system.

Soil scientists recognize soil as a three-dimensional unit, or body, having length, width, and depth. In most places on Earth's surface, exposed rock has crumbled and broken down to produce a layer of unconsolidated debris overlaying the hard, unweathered rock. This unconsolidated layer, called the **regolith**, varies in depth from virtually nonexistent to tens of meters. This interface between rock and the air, water, and living organisms that characterizes the surface environment is where soil is formed.

4.4 The Formation of Soil Begins with Weathering

Soil formation begins with the weathering of rocks and their minerals. Weathering includes the mechanical destruction of rock materials into smaller particles as well as their chemical modification. **Mechanical weathering** results from the interaction of several forces. When exposed to the combined action of water, wind, and temperature, rock surfaces flake and peel away. Water seeps into crevices, freezes, expands, and cracks the rock into smaller pieces. Wind-borne particles, such as dust and sand, wear away at the rock surface. Growing roots of trees split rock apart.

Without appreciably influencing their composition, mechanical weathering breaks down rock and minerals into smaller particles. Simultaneously, these particles are chemically altered and broken down through **chemical weathering.** The presence of water, oxygen, and acids resulting from the activities of soil organisms and the continual addition of organic matter (dead plant and animal tissues) enhance the chemical weathering process. Rainwater falling on and filtering through this organic matter and mineral soil sets up a chain of chemical reactions that transform the composition of the original rocks and minerals.

4.5 Soil Formation Involves Five Interrelated Factors

Five interdependent factors are important in soil formation: parent material, climate, biotic factors, topography, and time. **Parent material** is the material from which soil develops. The original parent material could originate from the underlying bedrock; from glacial deposits (till); from sand and silt carried by the wind (eolian); from gravity moving material down a slope (colluvium); and from sediments carried by flowing water (fluvial), including water in floodplains. The physical character and chemical composition of the parent material are important in determining soil properties, especially during the early stages of development.

Biotic factors—plants, animals, bacteria, and fungi—all contribute to soil formation. Plant roots can function to break up parent material, enhancing the process of weathering, as well as stabilizing the soil surface and reducing erosion. Plant roots pump nutrients up from soil depths and add them to the surface. In doing so, plants recapture minerals carried deep into the soil by weathering processes. Through photosynthesis, plants capture the Sun's energy and transfer some of this energy to the soil in the form of organic carbon. On the soil surface, microorganisms break down the remains of dead plants and animals that eventually become organic matter incorporated into the soil (see Chapter 21). Climate influences soil development both directly and indirectly. Temperature, precipitation, and winds directly influence the physical and chemical reactions responsible for breaking down parent material and the subsequent **leaching** (movement of solutes through the soil) and movement of weathered materials. Water is essential for the process of chemical weathering, and the greater the depth of water percolation, the greater the depth of weathering and soil development. Temperature controls the rates of biochemical reactions, affecting the balance between the accumulation and breakdown of organic materials. Consequently, under conditions of warm temperatures and abundant water, the processes of weathering, leaching, and plant growth (input of organic matter) are maximized. In contrast, under cold, dry conditions, the influence of these processes is much more modest. Indirectly, climate influences a region's plant and animal life, both of which are important in soil development.

Topography, the contour of the land, can affect how climate influences the weathering process. More water runs off and less enters the soil on steep slopes than on level land; whereas water draining from slopes enters the soil on low and flat land. Steep slopes are also subject to soil erosion and soil creep—the downslope movement of soil material that accumulates on lower slopes and lowlands.

Time is a crucial element in soil formation: all of the factors just listed assert themselves over time. The weathering of rock material; the accumulation, decomposition, and mineralization of organic material; the loss of minerals from the upper surface; and the downward movement of materials through the soil all require considerable time. Forming well-developed soils may require 2000 to 20,000 years.

4.6 Soils Have Certain Distinguishing Physical Characteristics

Soils are distinguished by differences in their physical and chemical properties. Physical properties include color, texture, structure, moisture, and depth. All may be highly variable from one soil to another.

Color is one of the most easily defined and useful characteristics of soil. It has little direct influence on the function of a soil but can be used to relate chemical and physical properties. Organic matter (particularly humus) makes soil dark or black. Other colors can indicate the chemical composition of the rocks and minerals from which the soil was formed. Oxides of iron give a color to the soil ranging from yellowish-brown to red, whereas manganese oxides give the soil a purplish to black color. Quartz, kaolin, gypsum, and carbonates of calcium and magnesium give whitish and grayish colors to the soil. Blotches of various shades of yellowish-brown and gray indicate poorly drained soils or soils saturated by water. Soils are classified by color using standardized color charts (i.e., Munsell soil color charts).

Soil texture is the proportion of different-sized soil particles. Texture is partly inherited from parent material and partly a result of the soil-forming process. Particles are classified on the basis of size into gravel, sand, silt, and clay. Gravel consists of particles larger than 2.0 mm, but they are not part of the fine fraction of soil. Soils are classified based on texture by defining the proportion of sand, silt, and clay.

Sand ranges from 0.05 to 2.0 mm, is easy to see, and feels gritty. Silt consists of particles from 0.002 to 0.05 mm in diameter that can scarcely be seen by the naked eye; it feels and looks like flour. Clay particles are less than 0.002 mm and are too small to be seen under an ordinary microscope. Clay controls the most important properties of soils, including its water-holding capacity (see Section 4.8) and the exchange of ions between soil particles and soil solution (see Section 4.9). A soil's texture is the percentage (by weight) of sand, silt, and clay. Based on proportions of these components, soils are divided into texture classes (**Figure 4.7**).

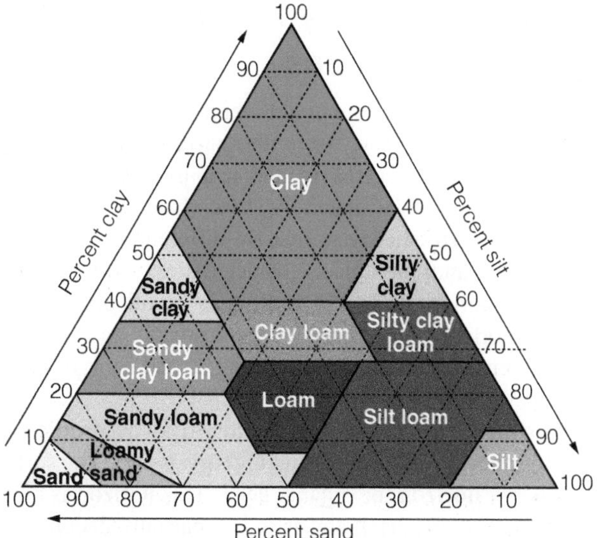

Figure 4.7 A soil texture chart showing the percentages of clay (less than 0.002 mm), silt (0.002–0.05 mm), and sand (0.05–2.0 mm) in the basic soil texture classes. For example, a soil with 60 percent sand, 30 percent silt, and 10 percent clay would be classified as a sandy loam.

Interpreting Ecological Data

Q1. *What is the texture classification for a soil with 60 percent silt, 35 percent clay, and 5 percent sand?*

Q2. *What is the texture classification for a soil with 60 percent clay and 40 percent silt?*

Soil texture affects pore space in the soil, which plays a major role in the movement of air and water in the soil and the penetration by roots. In an ideal soil, particles make up 50 percent of the soil's total volume; the other 50 percent is pore space. Pore space includes spaces within and between soil particles, as well as old root channels and animal burrows. Coarse-textured soils have large pore spaces that favor rapid water infiltration, percolation, and drainage. To a point, the finer the texture, the smaller the pores, and the greater the availability of active surface for water adhesion and chemical activity. Very fine-textured or heavy soils, such as clays, easily become compacted if plowed, stirred, or walked on. They are poorly aerated and difficult for roots to penetrate.

Soil depth varies across the landscape, depending on slope, weathering, parent materials, and vegetation. In grasslands, much of the organic matter added to the soil is from the deep, fibrous root systems of the grass plants. By contrast, tree leaves falling on the forest floor are the principal source of organic matter in forests. As a result, soils developed under native grassland tend to be several meters deep, and soils developed under forests are shallow. On level ground at the bottom of slopes and on alluvial plains, soils tend to be deep. Soils on ridgetops and steep slopes tend to be shallow, with bedrock close to the surface.

4.7 The Soil Body Has Horizontal Layers or Horizons

Initially, soil develops from undifferentiated parent material. Over time, changes occur from the surface down, through the accumulation of organic matter near the surface and the downward movement of material. These changes result in the formation of horizontal layers that are differentiated by physical, chemical, and biological characteristics. Collectively, a sequence of horizontal layers constitutes a **soil profile**. This pattern of horizontal layering, or **horizons**, is easily visible where a recent cut has been made along a road bank or during excavation for a building site (**Figure 4.8**).

The simplest general representation of a soil profile consists of four horizons: O, A, B, and C (**Figure 4.9**). The surface layer is the **O horizon**, or organic layer. This horizon is dominated by organic material, consisting of partially decomposed plant materials such as leaves, needles, twigs, mosses, and lichens. This horizon is often subdivided into a surface layer composed of undecomposed leaves and twigs (Oi), a middle layer composed of partially decomposed plant tissues (Oe), and a bottom layer consisting of dark brown to black, homogeneous organic material or the humus layer (Oa). This pattern of layering is easily seen by carefully scraping away the surface organic material on the forest floor. In temperate regions, the organic layer is thickest in the fall, when new leaf litter accumulates on the surface. It is thinnest in the summer after decomposition has taken place.

Figure 4.8 The pattern of horizontal layering or soil horizons is easily visible where a recent cut has been made along a road bank. This soil is relatively shallow, with the parent material close to the surface.

Below the organic layer is the **A horizon**, often referred to as the topsoil. This is the first of the layers that are largely composed of mineral soil derived from the parent materials. In this horizon, organic matter (humus) leached from above accumulates in the mineral soil. The accumulation of organic matter typically gives this horizon a darker color, distinguishing it from lower soil layers. Downward movement of water through this layer also results in the loss of minerals and finer soil particles, such as clay, to lower portions of the profile—sometimes giving rise to an **E horizon**, a zone or layer of maximum leaching, or eluviation (from Latin *ex* or *e*, "out," and *lavere*, "to wash") of minerals and finer soil particles to lower portions of the profile. Such E horizons are quite common in soils developed under forests, but because of lower precipitation they rarely occur in soils developed under grasslands.

Below the A (or E) horizon is the **B horizon**, also called the subsoil. Containing less organic matter than the A horizon, the B horizon shows accumulations of mineral particles such as clay and salts from the leaching from the topsoil. This process is called *illuviation* (from the Latin *il*, "in," and *lavere*, "to wash"). The B horizon usually has a denser structure than the A horizon, making it more difficult for plants to extend their roots downward. B horizons are distinguished on the basis of color, structure, and the kind of material that has accumulated as a result of leaching from the horizons above.

The **C horizon** is the unconsolidated material that lies under the subsoil and is generally made of original material from which the soil developed. Because it is below the zones of greatest biological activity and weathering and has not been sufficiently altered by the soil-forming processes, it typically

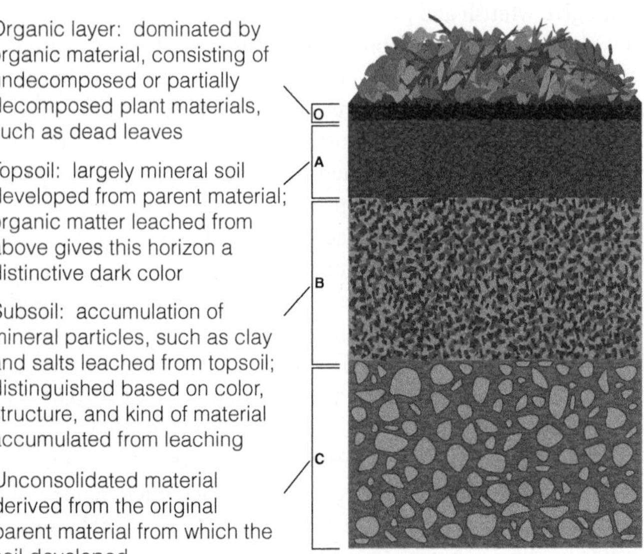
Organic layer: dominated by organic material, consisting of undecomposed or partially decomposed plant materials, such as dead leaves
Topsoil: largely mineral soil developed from parent material; organic matter leached from above gives this horizon a distinctive dark color
Subsoil: accumulation of mineral particles, such as clay and salts leached from topsoil; distinguished based on color, structure, and kind of material accumulated from leaching
Unconsolidated material derived from the original parent material from which the soil developed
O
A
B
C

Figure 4.9 A generalized soil profile. Over time, changes occur from the surface down, through the accumulation of organic matter near the surface and the downward movement of material. These changes result in the formation of horizontal layers, or horizons.

retains much of the characteristics of the parent materials from which it was formed. Below the C horizon lies the bedrock.

4.8 Moisture-Holding Capacity Is an Essential Feature of Soils

If you dig into the surface layer of a soil after a soaking rain, you should discover a sharp transition between wet surface soil and the dry soil below. As rain falls on the surface, it moves into the soil by infiltration. Water moves by gravity into the open pore spaces in the soil, and the size of the soil particles and their spacing determine how much water can flow in. Wide pore spacing at the soil surface increases the rate of water infiltration; so coarse soils have a higher infiltration rate than fine soils do.

If there is more water than the pore space can hold, we say that the soil is **saturated,** and excess water drains freely from the soil. If water fills all the pore spaces and is held there by internal capillary forces, the soil is at **field capacity** (physically defined as the water content at –0.33 bar suction pressure, or .0033 MPa). Field capacity is generally expressed as the percentage of the weight or volume of soil occupied by water when saturated compared to the oven-dried weight of the soil at a standard temperature. The amount of water a soil holds at field capacity varies with the soil's texture—the proportion of sand, silt, and clay. Coarse, sandy soil has large pores; water drains through it quickly. Clay soils have small pores and hold considerably more water. Water held between soil particles by capillary forces is **capillary water.**

As plants and evaporation from the soil surface extract capillary water, the amount of water in the soil declines. When the moisture level decreases to a point where plants can no longer extract water, the soil has reached the **wilting point** (physically defined as the water content at –15 bar suction pressure, or –1.5 MPa). The amount of water retained by the soil between field capacity and wilting point (or the difference between field capacity and wilting point) is the **available water capacity** (AWC), as shown in **Figure 4.10**. The AWC provides an estimate of the water available for uptake by plants. Although water still remains in the soil—filling up to 25 percent of the pore spaces—soil particles hold it tightly, making it difficult to extract.

Both the field capacity and wilting point of a soil are heavily influenced by soil texture. Particle size of the soil directly influences the pore space and surface area onto which water adheres. Sand has 30–40 percent of its volume in pore space, whereas clays and loams (see soil texture chart in Figure 4.7) range from 40 to 60 percent. As a result, fine-textured soils have a higher field capacity than sandy soils, but the increased surface area results in a higher value of the wilting point as well (see Figure 4.10). Conversely, coarse-textured soils (sands) have a low field capacity and a low wilting point. Thus, AWC is highest in intermediate clay loam soils.

The topographic position of a soil affects the movement of water both on and in the soil. Water tends to drain downslope, leaving soils on higher slopes and ridgetops relatively dry and creating a moisture gradient from ridgetops to streams.

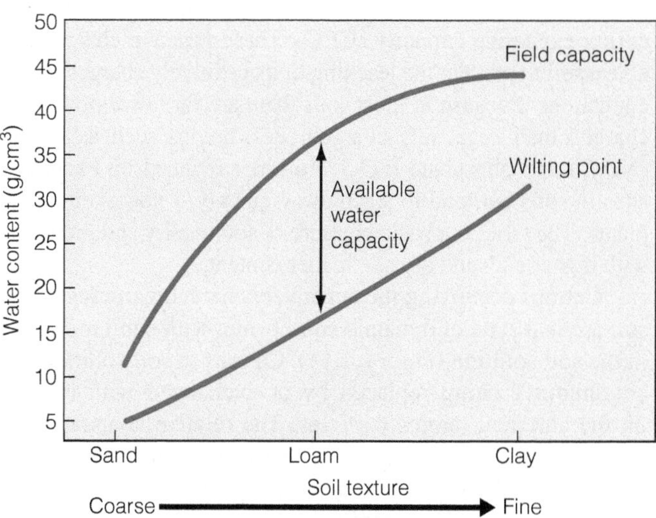

Figure 4.10 Water content of three different soils at wilting point (WP), field capacity (FC), and saturation. The three soils differ in texture from coarse-textured sand to fine-textured silty clay loam (see soil texture chart of Figure 4.7). Available water capacity (AWC) is defined as the difference between FC and WP. Both FC and WP increase from coarse- to fine-textured soils, and the highest AWC is in the intermediate-textured soils.

Interpreting Ecological Data

Q1. *Although fine-textured soils (silty clay loam) have a greater AWC, for this value to be achieved, the soil must be at or above FC. In arid regions, low and infrequent precipitation may keep soil water content below FC for most of the growing season. If the measured value of soil water content at a site is 10 g/cm^3, which soil texture (sand, silt, or clay) represented in Figure 4.10 would have the greatest soil water available for uptake by plants?*

Q2. *What if the value of soil water was 35 g/cm^3?*

4.9 Ion Exchange Capacity Is Important to Soil Fertility

Chemicals within the soil dissolve into the soil water to form a solution (see Section 3.5). Referred to as exchangeable nutrients, these chemical nutrients in solution are the most readily available for uptake and use by plants (see Chapter 6). They are held in soil by the simple attraction of oppositely charged particles and are constantly interchanging with the soil solution.

As described previously, an **ion** is a charged particle. Ions carrying a positive charge are **cations,** and ions carrying a negative charge are **anions.** Chemical elements and compounds exist in the soil solution both as cations, such as calcium (Ca^{2+}), magnesium (Mg^{2+}), and ammonium (NH_4^+), and as anions, such as nitrate (NO_3^-) and sulfate (SO_4^{2-}). The ability of these ions in soil solution to bind to the surface of soil particles depends on the number of negatively or positively charged sites within the soil. The total number of charged sites on soil particles within a volume of soil is called the **ion exchange capacity.** In most soils of the temperate zone, cation exchange predominates over anion exchange because of the prevalence of negatively charged particles in the soil, referred to as **colloids.** The total number of negatively charged sites, located on the leading edges of clay particles and soil organic matter (humus particles), is called the **cation exchange capacity** (CEC). These negative charges enable a soil to prevent the leaching of its positively charged nutrient cations. Because in most soils there are far fewer positively charged than negatively charged sites, anions such as nitrate (NO_3^-) and phosphate (PO_3^{4-}) are not retained on exchange sites in soils but tend to leach away quickly if not taken up by plants. The CEC is a basic measure of soil quality and increases with higher clay and organic matter content.

Cations occupying the negatively charged particles in the soil are in a state of dynamic equilibrium with similar cations in the soil solution (Figure 4.11). Cations in soil solution are continuously being replaced by or exchanged with cations on the clay and humus particles. The relative abundance of different ions on exchange sites is a function of their concentration in the soil solution and the relative affinity of each ion for the sites. In general, the physically smaller the ion and the greater its positive charge, the more tightly it is held. The lyotropic series places the major cations in order of their strength of bonding to the cation exchange sites in the soil:

$$Al^{3+} > H^+ > Ca^{2+} > Mg^{2+} > K^+ = NH_4^+ > Na^+$$

However, higher concentrations in the soil solution can overcome these differences in affinity.

Hydrogen ions added by rainwater, by acids from organic matter, and by metabolic acids from roots and microorganisms increase the concentration of hydrogen ions in the soil solution and displace other cations, such as Ca^{2+}, on the soil exchange sites. As more and more hydrogen ions replace other cations, the soil becomes increasingly acidic (see Section 3.7). Acidity is one of the most familiar of all chemical conditions in the soil. Typically, soils range from pH 3 (extremely acid) to pH 9 (strongly alkaline). Soils of more than pH 7 (neutral) are considered basic, and those of pH 5.6 or less are acid. As soil acidity increases, the proportion of exchangeable Al^{3+} increases, and Ca^{2+}, Na^+, and other cations decrease. High aluminum (Al^{3+}) concentrations in soil solution can be toxic to plants. Aluminum toxicity damages the root system first, making the roots short, thick, and stubby. The result is reduced nutrient uptake.

4.10 Basic Soil Formation Processes Produce Different Soils

Broad regional differences in geology, climate, and vegetation give rise to characteristically different soils. The broadest level of soil classification is the order. Each order has distinctive features, summarized in Figure 4.12, and its own distribution, mapped in Figure 4.13. Although a wide variety of processes are involved in soil formation (pedogenesis), soil scientists recognize five main soil-forming processes that give rise to these different classes of soils. These processes are laterization, calcification, salinization, podzolization, and gleization.

Laterization is a process common to soils found in humid environments in the tropical and subtropical regions. The hot, rainy conditions cause rapid weathering of rocks and minerals. Movements of large amounts of water through the soil cause heavy leaching, and most of the compounds and nutrients made available by the weathering process are transported out of the soil profile if not taken up by plants. The two exceptions to this

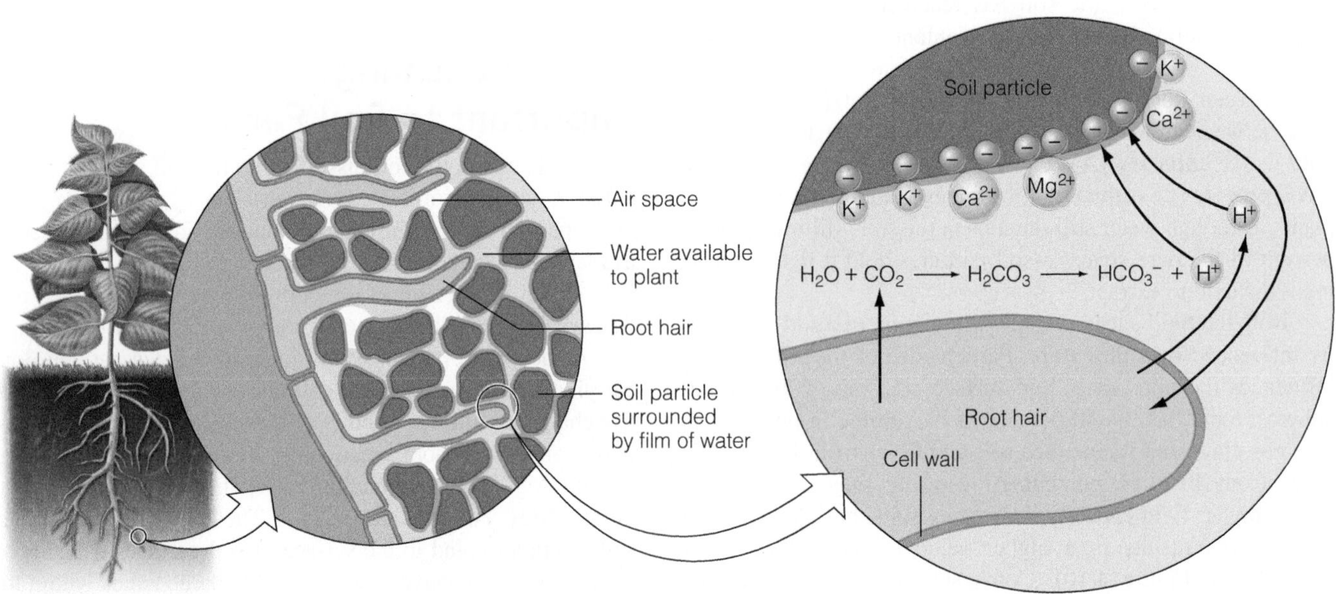

Figure 4.11 The process of cation exchange in soils. Cations occupying the negatively charged particles in the soil are in a state of dynamic equilibrium with similar cations in the soil solution. Cations in soil solution are continuously being replaced by or exchanged with cations on clay and humus particles. Cations in the soil solution are also taken up by plants and leached to ground and surface waters.

Entisol

Immature soils that lack vertical development of horizons; associated with recently deposited sediments

Aridisol

Develop in very dry environments; low in organic matter; high in base content; prone to the process of salinization

Vertisol

Dark clay soils that show significant expansion and contraction due to wetting and drying

Mollisol

Surface horizons dark brown to black with soft consistency; rich in bases; soils of semi-humid regions; prone to the process of calcification

Inceptisol

Young soils that are more developed than entisols; often shallow; moderate development of horizons

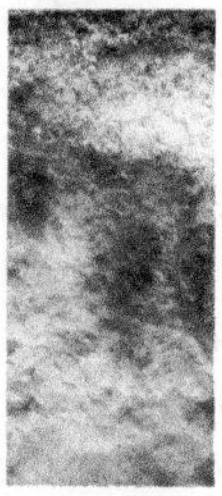

Spodosol

Light gray, whitish surface horizon on top of black or reddish B horizon; high in extractable iron and aluminum; formed through the process of podzolization

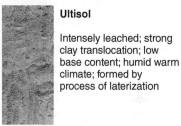

Alfisol

Shallow penetration of humus; translocation of clay; well-developed horizons

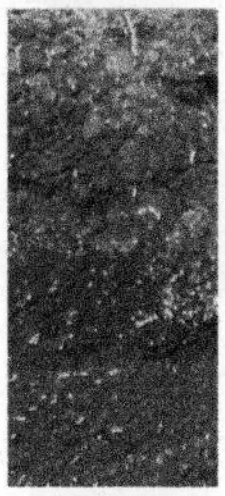

Histosol

High content of organic matter; formed in areas with poor drainage; bog and muck soils

Ultisol

Intensely leached; strong clay translocation; low base content; humid warm climate; formed by process of laterization

Andisol

Developed from volcanic parent material; not highly weathered; upper layers dark colored; low bulk density

Oxisol

Highly weathered soils with nearly featureless profile; red, yellow, or gray; rich in kalolinate, iron oxides, and often humus; in tropics and subtropics

Gelisol

Presence of permafrost or soil temperature of 0°C or less within 2 meters of the surface; formed through the process of gleization

Figure 4.12 Profiles and general description of the 12 major soil orders of the world.

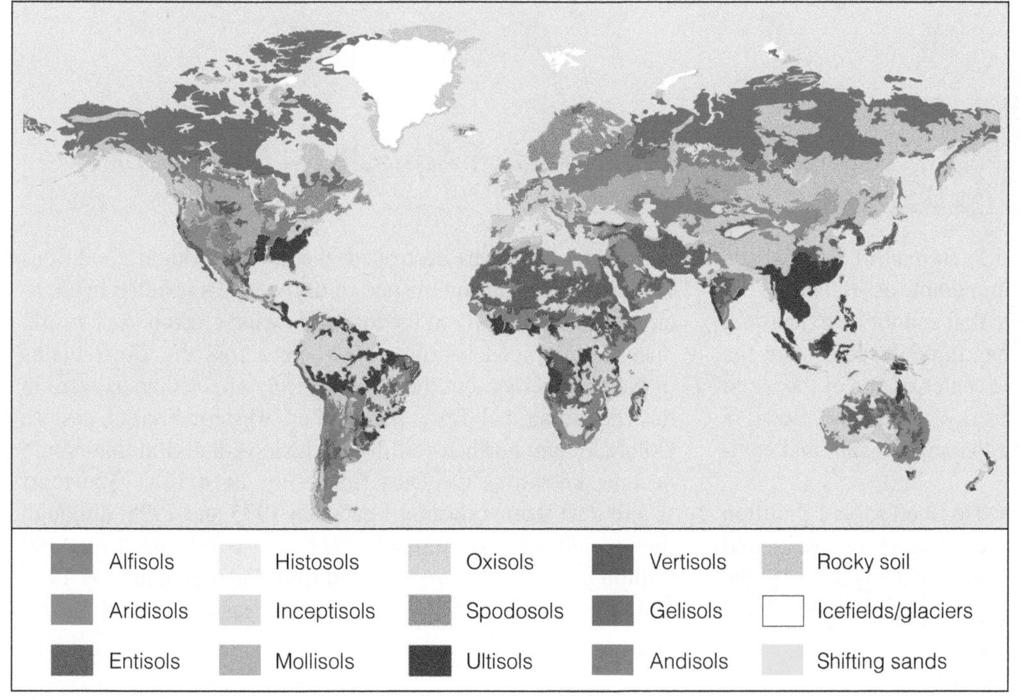

Figure 4.13 The world distribution of the 12 major soil orders shown in Figure 4.12. (Adapted from USGS, Soil Conservation Service.)

process are compounds of iron and aluminum. Iron oxides give tropical soils their unique reddish coloring (see Ultisol profile in Figure 4.12). Heavy leaching also causes these soils to be acidic because of the loss of other cations (other than H_1).

Calcification occurs when evaporation and water uptake by plants exceed precipitation. The net result is an upward movement of dissolved alkaline salts, typically calcium carbonate ($CaCO_3$), from the groundwater. At the same time, the infiltration of water from the surface causes a downward movement of the salts. The net result is the deposition and buildup of these deposits in the B horizon (subsoil). In some cases, these deposits can form a hard layer called *caliche* (**Figure 4.14 top**).

Salinization is a process that functions similar to calcification, only in much drier climates. It differs from calcification in that the salt deposits occur at or near the soil surface (**Figure 4.14 bottom**). Saline soils are common in deserts but may also occur in coastal regions as a result of sea spray. Salinization is also a growing problem in agricultural areas where irrigation is practiced.

Podzolization occurs in cool, moist climates of the mid-latitude regions where coniferous vegetation (e.g., pine forests) dominates. The organic matter of coniferous vegetation creates strongly acidic conditions. The acidic soil solution enhances the process of leaching, causing the removal of cations and compounds of iron and aluminum from the A horizon (topsoil). This process creates a sublayer in the A horizon that is composed of white- to gray-colored sand (see Spodosol profile in Figure 4.12).

Gleization occurs in regions with high rainfall or low-lying areas associated with poor drainage (waterlogged). The constantly wet conditions slow the breakdown of organic matter by decomposers (bacteria and fungi), allowing the matter to accumulate in upper layers of the soil. The accumulated organic matter releases organic acids that react with iron in the soil, giving the soil a black to bluish-gray color (see Gelisol profile in Figure 4.12 as an example of soil formed through the process of gleization).

Figure 4.14 (top) In arid regions, salinization occurs when salts (the white crust at the center of the photo) accumulate near the soil surface because of surface evaporation. (bottom) Calcification occurs when calcium carbonates precipitate out from water moving downward through the soil or from capillary water moving upward from below. The result is an accumulation of calcium in the B horizon (seen as the white soil layer in the photo).

These five processes represent the integration of climate and edaphic (relating to the soil) factors on the formation of soils, giving rise to the geographic diversity of soils that influence the distribution, abundance, and productivity of terrestrial ecosystems. (We will explore these topics further in Chapters 20, 21, and 23.)

ECOLOGICAL Issues & Applications — Soil Erosion Is a Threat to Agricultural Sustainability

In a report released in 1909, the U.S. Bureau of Soils stated "The soil is the one indestructible, immutable asset that the nation possesses. It is the one resource that cannot be exhausted; that cannot be used up." Yet less than three decades later, the loss of soil resources would be at the center of one of the worst environmental disasters in U.S. history—the Dust Bowl; a disaster that would have profound economic, social, and environmental costs.

Between 1909 and 1929 farmers had tilled some 13 million hectares of land in the Great Plains. In doing so they destroyed the sod—the grass-covered surface soil held together by the dense mat of fibrous roots. Once this protective cover of the native grassland was destroyed, the severe drought conditions and high winds during the period of the 1930s resulted in an increased susceptibility of the topsoil to wind erosion. As a result, dust storms raged nearly everywhere across the Great Plains of North America; but the most severely affected areas were in the Oklahoma and Texas panhandles, western Kansas, eastern Colorado, and northeastern New Mexico—a region that would become known as the Dust Bowl (**Figure 4.15a**). The most severe dust storms occurred between 1935 and 1938, although they would continue through 1941. It was estimated that 300 million tons of soil were removed from the region in May 1934 and spread over large portions of the eastern United States.

By 1935 an additional 850 million tons of topsoil were removed by wind erosion. It is estimated that by 1935 wind erosion had damaged 66 million hectares across 80 percent of the High Plains. By 1938 it was estimated that 12.5 inches of topsoil had been lost over an area of 4 million hectares and 6.5 cm had been lost over another 5.5 million hectares.

The storms generated by this environmental disaster darkened cities, buried homes and farm equipment, killed livestock, and represented a serious health risk (**Figure 4.15b** and **c**). Overall, the Dust Bowl rendered millions of acres of farmland virtually useless, left roughly half a million Americans homeless, and forced hundreds of thousands of people off the land. It also resulted in the most intense period of internal migration in U.S. history. Between 1932 and 1940, it is estimated that 2.5 million people abandoned the plains for other regions of the country.

In response to the environmental disaster of the Dust Bowl, U.S. president, Franklin Delano Roosevelt, established the Soil Erosion Service (later the Soil Conservation Service, and now the Natural Resources Conservation Service), which marked the first major federal commitment to the preservation of natural resources in private hands. Even more significantly, in 1935, the Prairie States Forestry Project was established. Under this federal project, nearly 220 million trees were planted, creating more than 18,000 miles of windbreaks on some 30,000 farms, which formed a "shelter belt" from the Texas Panhandle to the Canadian border.

Although the end of the drought, together with soil conservation efforts following the Dust Bowl, abated the dramatic dust storms that blackened the skies over North America, the problem of soil erosion on agricultural lands remains a serious environmental issue. Approximately 50 percent of Earth's land surface is devoted to agriculture, with about one-third planted in crops and two-thirds used for grazing. Of these two areas, cropland is more susceptible to erosion because the vegetation is most often removed and the soil tilled (plowed) before crops are planted. This functions to destabilize the soil surface, increasing rates of erosion resulting from both wind and water (**Figure 4.16a**). In addition, croplands are often left without vegetation cover between plantings (exposing the bare soil surface to erosion). According to David Pimentel of Cornell University, one of the leading experts in the study of agricultural ecology, currently about 80 percent of the world's agricultural land suffers moderate to severe soil erosion. Worldwide, erosion on cropland averages about 30 tons per hectare per year and ranges from 0.5 to 400 tons per hectare

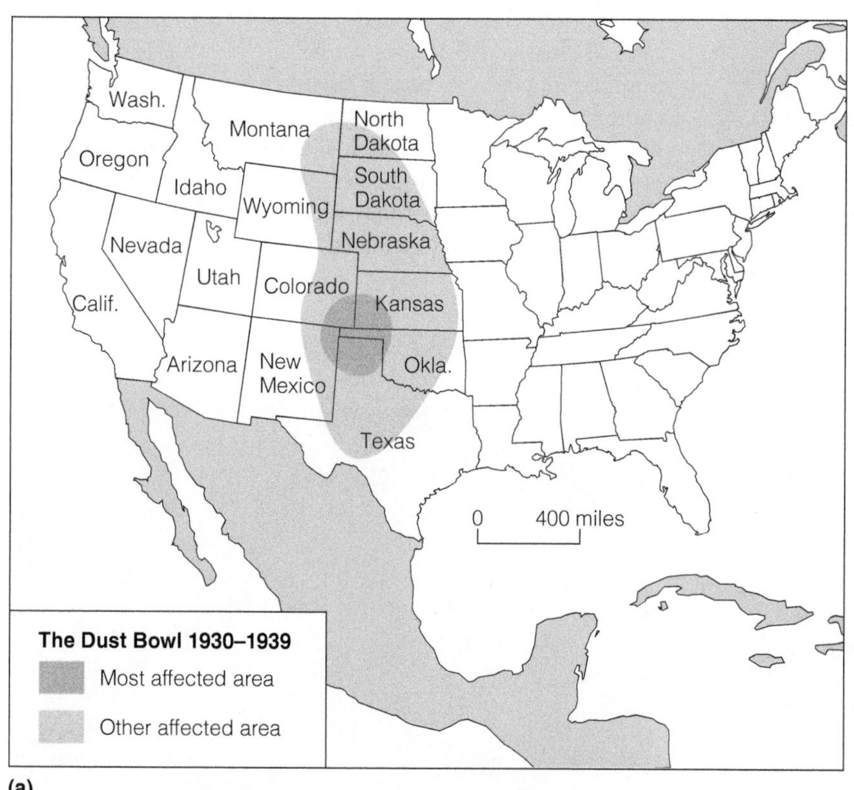

(a)

(b)

(c)

Figure 4.15 (a) Map of the region of the Great Plains known as the Dust Bowl. This represented the agricultural region most affected by soil erosion during the drought years from 1930 to 1939. Photographs of (b) dust storms (Boise City, Oklahoma, on April 15, 1935, and (c) their aftermath (Guymon, Oklahoma, March 29, 1937).

per year. As a result of soil erosion, during the past four decades about 30 percent of the world's arable land has become unproductive, much of which has been abandoned for agricultural use. Each year an estimated 10 million hectares of cropland worldwide are abandoned because of lack of productivity caused by soil erosion.

Rates of soil erosion on agricultural lands are influenced by a variety of factors. Topography of the landscape, patterns of rainfall and wind, and exposure all combine to influence the susceptibility of the soil surface to erosion. Soil structure influences the ease with which soils can be eroded. Soils with medium-to-fine texture (see Section 4.6) and low organic matter content are most easily eroded. Typically these soils have low water infiltration rates and are therefore susceptible to high rates of erosion by water and displacement by wind. Plant cover, both living and dead, greatly reduces rates of erosion by protecting the soil surface from exposure to agents of erosion.

Current estimates suggest that the degradation of agricultural lands alone will depress world food production by approximately 30 percent over the next 50 years, while during that same period the world population is predicted to exceed 9 billion (United Nations medium scenario; see Chapter 11, *Ecological Issues & Applications*). These forecasts point to the need to develop soil conservation techniques known to dramatically reduce soil erosion. For example, commercial corn production in the United States, which uses a practice of continuous crop production with annual plowing and removal of all plant materials at harvest, results in an average soil erosion rate of 44 tons per hectare per year. By using a practice of crop rotation in which a series of dissimilar/different types of crops are planted in the same area in sequential seasons (e.g., corn, wheat, and hay) erosion rates have been shown to decline to as little as 3 tons per hectare per year. No-till techniques, in which crops are planted directly in the soil without tilling or plowing the ground (**Figure 4.16b**), reduce average rates of erosion to 0.14 tons per hectare per year in corn fields. Similar reductions in rates of erosion have been measured with contour planting (plowing and planting row crops on a contour rather than up and down hill; **Figure 4.16c**) and the use of grass strips between crop rows. What all of these techniques share in common is that they serve to protect the soil surface from direct exposure to wind and rain.

Figure 4.16 (a) An agricultural field that has been recently tilled. As a result of the lack of cover, tilled fields have high rates of erosion from wind and water. Some methods being adopted in the effort to reduce soil erosion and promote sustainable agricultural practices are (b) contour and (c) no-till farming.

(a)

(b)

(c)

SUMMARY

Life on Land 4.1

Maintaining the balance of water between organisms and their surrounding environment has been a major influence on the evolution of life on land. The need to remain erect against the force of gravity in terrestrial environments results in a significant investment in structural materials. Variations in temperature and precipitation have both a short-term effect on metabolic processes and a long-term influence on the evolution and distribution of terrestrial plants and animals. The result is a distinct pattern of terrestrial ecosystems across geographic gradients of temperature and precipitation.

Light 4.2

Light passing through a canopy of vegetation becomes attenuated. The density and orientation of leaves in a plant canopy influence the amount of light reaching the ground. Foliage density is expressed as leaf area index (LAI), the area of leaves per unit of ground area. The amount of light reaching the ground in terrestrial vegetation varies with the season. In forests, only about 1–5 percent of light striking the canopy reaches the ground. Sunflecks on the forest floor enable plants to endure shaded conditions.

Soil Defined 4.3

Soil is a natural product of unconsolidated mineral and organic matter on Earth's surface. It is the medium for plant growth; the principal factor controlling the fate of water in terrestrial environments; nature's recycling system, which breaks down the waste products of plants and animals and transforms them into their basic elements; and a habitat to a diversity of animal life.

Weathering 4.4

Soil formation begins with the weathering of rock and minerals. In mechanical weathering, water, wind, temperature, and plants break down rock. In chemical weathering, the activity of soil organisms, the acids they produce, and rainwater break down primary minerals.

Soil Formation 4.5

Soil results from the interaction of five factors: parent material, climate, biotic factors, topography, and time. Parent material provides the substrate from which soil develops. Climate shapes soil development through temperature, precipitation, and its influence on vegetation and animal life. Biotic factors—vegetation, animals, bacteria, and fungi—add organic matter and mix it with mineral matter. Topography influences the amount of water entering the soil and the rates of erosion. Time is required to fully develop distinctive soils.

Distinguishing Characteristics 4.6

Soils differ in the physical properties of color, texture, and depth. Although color has little direct influence on soil function, it can be used to relate chemical and physical properties. Soil texture is the proportion of different-sized soil particles—sand, silt, and clay. A soil's texture is largely determined by the parent material but is also influenced by the soil-forming process. Soil depth varies across the landscape, depending on slope, weathering, parent materials, and vegetation.

Soil Horizons 4.7

Soils develop in layers called *horizons*. Four horizons are commonly recognized, although not all of them are necessarily present in any one soil: the O or organic layer; the A (sometimes E) horizon, or topsoil, characterized by accumulation of organic matter; the B horizon, or subsoil, in which mineral materials accumulate; and the C horizon, the unconsolidated material underlying the subsoil and extending downward to the bedrock.

Moisture-Holding Capacity 4.8

The amount of water a soil can hold is one of its important characteristics. When water fills all pore spaces, the soil is saturated. When a soil holds the maximum amount of water it can retain, it is at field capacity. Water held between soil particles by capillary forces is capillary water. When the moisture level is at a point where plants cannot extract water, the soil has reached wilting point. The amount of water retained between field capacity and wilting point is the available water capacity. The available water capacity of a soil is a function of its texture.

Ion Exchange 4.9

Soil particles, particularly clay particles and organic matter, are important to nutrient availability and the cation exchange capacity of the soil—the number of negatively charged sites on soil particles that can attract positively charged ions. Cations occupying the negatively charged particles in the soil are in a state of dynamic equilibrium with similar cations in the soil solution. Percent base saturation is the percentage of sites occupied by ions other than hydrogen.

Soil Formation Processes Form Different Soils 4.10

Broad regional differences in geology, climate, and vegetation give rise to characteristically different soils. The broadest level of soil classification is the order. Each order has distinctive features. Soil scientists recognize five main soil-forming processes that give rise to these different classes of soils. These processes are laterization, calcification, salinization, podzolization, and gleization.

Soil Erosion Ecological Issues & Applications

Soil erosion on agricultural lands is a serious environmental problem. The removal of natural vegetation and the plowing of the soil destabilizes the soil surface and greatly enhances erosion from wind and water. Sustainable practices such as contour and no-till farming can greatly reduce rates of soil loss.

STUDY QUESTIONS

1. Name two constraints imposed on organisms in the transition of life from aquatic to terrestrial environments.
2. Assume that two forests have the same quantity of leaves (leaf area index). In one forest, however, the leaves are oriented horizontally (parallel to the forest floor). In the other forest, the leaves are positioned at an angle of 60 degrees. How would the availability of light at the forest floor differ for these two forests at noon? In which forest would the leaves at the bottom of the canopy (lower in the tree) receive more light at mid-morning?
3. What is the general shape of the curve that describes the vertical attenuation of light through the plant canopy based on Beer's law? Why is it not a straight line (linear)?
4. What five major factors affect soil formation?
5. What role does weathering play in soil formation? What factors are involved in the process of weathering?
6. Use Figure 4.10 to answer this question: Which soil holds more moisture at field capacity: clay or sand? Which soil holds more moisture at wilting point: clay or sand? Which soil type has a greater availability of water for plant uptake when the water content of the soil is 3.0 in/ft (value on *y*-axis)?
7. What is the major factor distinguishing the O and A soil horizons?
8. Why do clay soils typically have a higher cation exchange capacity than do sandy soils?
9. How does pH influence the base saturation of a soil?
10. Why is the process of salinization more prevalent in arid areas? How does irrigation increase the process of salinization in agricultural areas?
11. What soil-forming process is dominant in the wet tropical regions? How does this process influence the availability of nutrients to plant roots in the A horizon?

FURTHER READINGS

Classic Studies

Wilde, S. A. 1946. *Forest Soils and Forest Growth.* Waltham, MA: Chronica Botanica.
A classic text on soils that was influential on the development of soil science.

Current Research

Brady, N. C., and R. W. Weil. 2008. *The nature and properties of soils.* 14th ed. Upper Saddle River, NJ: Prentice Hall.
The classic introductory textbook on soils. Used for courses in soil science.

Jenny, H. 1994. *Factors of soil formation.* Mineola, NY: Dover Publications.
A well-written and accessible book by one of the pioneers in soil science.

Kohnke, H., and D. P. Franzmeier. 1994. *Soil science simplified.* Prospect Heights, IL: Wavelength Press.
A well-written and illustrated overview of concepts and principles of soil science for the general reader. Provides many examples and applications of basic concepts.

Patton, T. R. 1996. *Soils: A new global view.* New Haven, CT: Yale University Press.
Presents a new view and approach to studying soil formation at a global scale.

CHAPTER 5

Adaptation and Natural Selection

A tenebrionid beetle (*Stenocara* spp.) perched on a sand dune in the Namib Desert of southwestern Africa. (Insert) Magnified view of the beetle's back shows fog droplets on the wing case (elytra).

CHAPTER GUIDE

THE NAMIB DESERT, stretching for 1200 miles along the southwest coast of Africa, is home to the highest sand dunes in the world. Rainfall is a rare event in the Namib. But each morning as the Sun rises, the cool, moist air of this coastal desert begins to warm, and the Namib becomes shrouded in fog. Here each morning, black thumbnail-sized beetles perform one of nature's more bizarre behaviors (see photos on chapter opener page). These tenebrionid beetles (*Stenocara* spp.) upend their bodies into a handstand. A beetle stays in this position as fog droplets collect on its back and then gradually roll down the wing case (called the *elytra*) into its mouth. By viewing the bumps on its back through an electron microscope, we can see a wax-coated carpet of tiny nodules covering the sides of the bumps as well as the valleys between them that aid in channeling water from the beetle's back to its mouth.

The tenebrionid beetles of the Namib Desert illustrate two important concepts: the relationship between structure and function, and how that relationship reflects adaptations of the organism to its environment. The structure of the beetle's back and the beetle's behavior of standing on its head in the morning fog serve the function of acquiring water, a scarce and essential resource in this arid environment. These characteristics represent adaptations to life in the unique environment of the Namib Desert. This same set of characteristics, however, are unlikely to be efficient for acquiring water in the desert regions of the continental interior, where morning fog may not form, or in wet environments, such as a tropical rain forest, where standing pools of water are readily available. Each environment presents a different set of constraints on processes relating to survival, growth, and reproduction. The set of characteristics that enable an organism to succeed in one environment typically preclude it from doing equally well under a different set of environmental conditions.

Prior to the mid-19th century this apparent match between species and their environment was seen as the work of the creator. As the late evolutionary ecologist Ernst Mayr of Harvard University so poetically wrote, examples such as the tenebrionid beetles of the Namib Desert served to illustrate the "wise laws that brought about the perfect adaptation of all organisms one to another and to their environment." Adaptation, after all, implied design—and design, a designer. Natural history was the task of cataloging the creations of the divine architect. By the mid-1800s, however, a revolutionary idea emerged that would forever change our view of nature.

> In considering the origin of species, it is quite conceivable that a naturalist... might come to the conclusion that species had not been independently created, but had descended, like varieties, from other species. Nevertheless, such a conclusion, even if well founded, would be unsatisfactory, until it could be shown how the innumerable species, inhabiting this world, have been modified, so as to acquire that perfection of structure and coadaptation which justly excites our admiration.

The pages that followed in Charles Darwin's *The Origin of Species,* first published on November 24, 1859, altered the history of science and brought into question a view of the world that had been held for millennia (**Figure 5.1**). Darwin put forward in those pages a mechanism to explain how the diversity of organisms inhabiting our world have acquired the features seemingly designed to enable them to survive and reproduce. He called it the *theory of natural selection.* Its beauty lay in its simplicity: the mechanism of natural selection is the simple elimination of "inferior" individuals.

Figure 5.1 Charles Darwin (1809–1882).

5.1 Adaptations Are a Product of Natural Selection

Stated more precisely, **natural selection** is the differential success (survival and reproduction) of individuals within the population that results from their interaction with their environment. As outlined by Darwin, natural selection is a product of two conditions: (1) that variation occurs among individuals within a population in some "heritable" characteristic, and (2) that this variation results in differences among individuals in their survival and reproduction as a result of their interaction with the environment. Natural selection is a numbers game. Darwin wrote:

> Among those individuals that do reproduce, some will leave more offspring than others. These individuals are considered more fit than the others because they contribute the most to the next generation. Organisms that leave few or no offspring contribute little or nothing to the succeeding generations and so are considered less fit.

The **fitness** of an individual is measured by the proportionate contribution it makes to future generations. Under a given set of environmental conditions, individuals having certain characteristics that enable them to survive and reproduce are selected for, eventually passing those characteristics on to the next generation. Individuals without those traits are selected against, failing to pass their characteristics on to future generations. In this way, the process of natural selection results in changes in the properties of populations of organisms over the course of generations, by a process known as **evolution**.

An **adaptation** is any heritable behavioral, morphological, or physiological trait of an organism that has evolved over a period of time by the process of natural selection such that it maintains or increases the fitness (long-term reproductive success) of an organism under a given set of environmental conditions. The concept of adaptation by natural selection is central to the science of ecology. The study of the relationship between organisms and their environment is the study of adaptations. Adaptations represent the characteristics (traits) that enable an organism to survive, grow, and reproduce under the prevailing environmental conditions. Adaptations likewise govern the interaction of the organism with other organisms, both of the same and different species. How adaptations enable an organism to function in the prevailing environment—and conversely, how those same adaptations limit its ability to successfully function in other environments—is the key to understanding the distribution and abundance of species, the ultimate objective of the science of ecology.

5.2 Genes Are the Units of Inheritance

By definition, adaptations are traits that are inherited—passed from parent to offspring. So to understand the evolution of adaptations, we must first understand the basis of inheritance: how characteristics are passed from parent to offspring and what forces bring about changes in those same characteristics through time (from generation to generation).

At the root of all similarities and differences among organisms is the information contained within the molecules of DNA (deoxyribonucleic acid). You will recall from basic biology that DNA is organized into discrete subunits—genes—that form the informational units of the DNA molecule. A **gene** is a stretch of DNA coding for a functional product (ribonucleic acid: RNA). The product is usually messenger RNA (mRNA) and mRNA ultimately results in the synthesis of a protein. The alternate forms of a gene are called **alleles** (derived from the term *allelomorphs*, which in Greek means "different form"). The process is called gene expression in which DNA is used in the synthesis of products such as proteins. All of the DNA in a cell is collectively called the **genome**.

Genes are arranged in linear order along microscopic, threadlike bodies called **chromosomes**. The position occupied by a gene on the chromosome is called the **locus** (Latin for *place*). In most multicellular organisms, each individual cell contains two copies of each type of chromosome (termed *homologous chromosomes*). In the process of asexual reproduction, both chromosomes are inherited from the single parent. In sexual reproduction, one is inherited from its mother through the ovum and one inherited from its father through the sperm. At any locus, therefore, every diploid individual contains two copies of the gene—one at each corresponding position in the homologous chromosomes. These two copies are the alleles of the gene in that individual. If the two copies of the gene are the same, then the individual is **homozygous** at that given locus. If the two alleles at the locus are different, then the individual is **heterozygous** at the locus. The pair of alleles present at a given locus defines the **genotype** of an individual; therefore, homozygous and heterozygous are the two main categories of genotypes.

5.3 The Phenotype Is the Physical Expression of the Genotype

The outward appearance of an organism for a given characteristic is its **phenotype**. The phenotype is the external, observable expression of the genotype. When an individual is heterozygous, the two different alleles may produce an individual with intermediate characteristics or one allele may mask the expression of the other (**Figure 5.2**). In the case in which one allele masks the expression of the other, the allele that is expressed is referred to as the **dominant allele**, whereas the allele that is masked is called the **recessive allele**. If the allele is recessive, it will only be expressed if the individual is homozygous for that allele (homozygous recessive). If the physical expression of the heterozygous individual is intermediate between those of the homozygotes, the alleles are said to be **incomplete dominance**, and each allele has a specific value (proportional effect) that it contributes to the phenotype.

Phenotypic characteristics that fall into a limited number of discrete categories, such as the example of flower color presented in Figure 5.2, are referred to as **qualitative traits**. Even though all genetic variation is discrete (in the form of alleles),

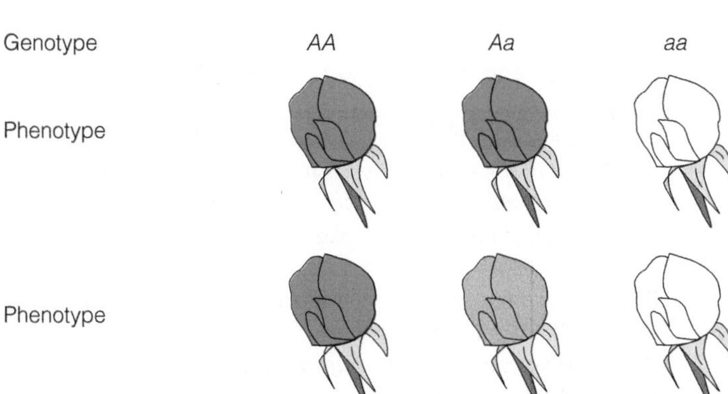

Figure 5.2 Example of different modes of gene expression. In this example, flower color is controlled by a single locus having two alternate alleles, *A* and *a*. The *A* allele codes for the production of red pigment, whereas the *a* allele does not (absence of pigment). In the first case, heterozygous individuals (*Aa*) exhibit the same phenotype as homozygous *AA* individuals, indicating that *A* is the dominant allele. The recessive allele (*a*) is expressed only in homozygous recessive (*aa*) individuals. In the second case, the heterozygous individuals are intermediate in form to the homozygotes. This is an example of incomplete dominance, and each allele has a proportional effect on the phenotype.

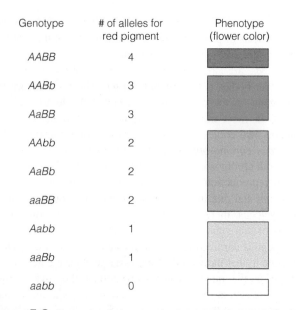

Genotype	# of alleles for red pigment	Phenotype (flower color)
AABB	4	
AABb	3	
AaBB	3	
AAbb	2	
AaBb	2	
aaBB	2	
Aabb	1	
aaBb	1	
aabb	0	

Figure 5.3 Example of phenotypic characteristics controlled by two loci. Assume that flower color is controlled by two genes, each having two alleles (*A:a* and *B:b*). Both the *A* and *B* alleles code for the production of red pigment, whereas the *a* and *b* alleles do not. There are nine possible genotypes, with the number of alleles coding for red pigment ranging from 4 (*AABB*) to 0 (*aabb*). The resulting phenotypes fall into five categories ranging from dark red through white depending on the number of alleles producing red pigment. The intermediate color (two alleles for red pigment) is the most abundant class. The number of possible phenotypes will increase as the number of loci (genes) controlling the phenotype increases.

most phenotypic traits have a continuous distribution. These traits, such as height or weight, are referred to as **quantitative traits**. The continuous distribution of most phenotypic traits occurs for two reasons. First, most traits have more than one gene locus affecting them. For example, if the phenotypic characteristic of flower color illustrated in Figure 5.2 is controlled by two loci rather than a single locus (each with two alleles—*A:a* and *B:b*), there are nine possible genotypes (**Figure 5.3**). In contrast to the three distinct flower colors (phenotypes) produced in the case of a single locus, there is now a range of flower colors varying in hue between dark red and white depending on the number of alleles coding for the production of red pigment (see Figure 5.2). The greater the number of loci, the greater is the range of possible phenotypes. The second factor influencing phenotypic variation is the environment.

5.4 The Expression of Most Phenotypic Traits Is Affected by the Environment

The expression of most phenotypic traits is influenced by the environment; that is to say, the phenotypic expression of the genotype is influenced by the environment. Because environmental factors themselves usually vary continuously—temperature, rainfall, sunlight, level of predation, and so on—the environment can cause the phenotype produced by a given genotype to vary continuously. To illustrate this point, we can use the example of flower color controlled by two loci presented previously (and in Figure 5.3). Pigment production during flower development can be affected by temperature. If temperatures below some optimal value or range function to reduce the expression of the *A* and *B* alleles in the production of red pigment, fluctuations in temperatures over the period of flower development in the population of plants will function to further increase the range of flower colors (shades between red and white) produced by the nine genotypes.

The ability of a genotype to give rise to different phenotypic expressions under different environmental conditions is termed **phenotypic plasticity**. The set of phenotypes expressed by a single genotype across a range of environmental conditions is referred to as the **norm of reaction** (**Figure 5.4**). Note that we are not talking about different genotypes adapted to different environmental conditions, but about a single genotype (set

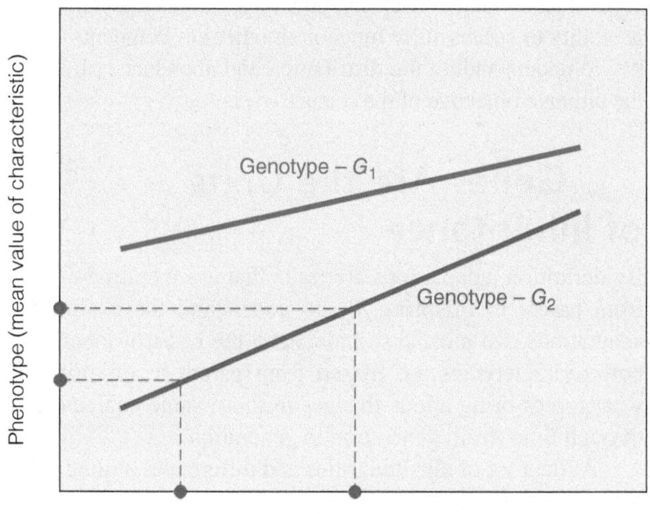

Figure 5.4 A hypothetical example of the norm of reaction: the range of phenotypes expressed by a given genotype in different environments. The norms of reaction for two genotypes, G_1 and G_2, along an environmental gradient are shown. The two lines represent the mean phenotypic characteristics (such as flower or body coloration) exhibited by two genotypes (G_1 and G_2) at any point along the environmental gradient (such as temperature). The dashed lines illustrate the change in phenotype for G_2 at two different points along the environmental gradient. The ability of a genotype to express different phenotypic characteristics under different environmental conditions is called *phenotypic plasticity*.

Interpreting Ecological Data

Q1. *Which of the two genotypes (G_1 or G_2) exhibits the greater norm of reaction?*

Q2. *What would the line look like for a genotype that did not exhibit phenotypic plasticity?*

Q3. *Is there any environment in which the two genotypes will express the same phenotype?*

Q4. *Is it possible for the two genotypes to exhibit the same phenotype?*

of alleles) capable of altering the development or expression of a phenotypic trait in response to the conditions encountered by the individual organism. The result is the improvement of the individual's ability to survive, grow, and reproduce under the prevailing environmental conditions (i.e., increase fitness). For example, the bodies of many species of insects change in color in response to the prevailing temperature during development (**Figure 5.5**). Development under colder temperatures typically results in darker coloration. Darker coloration most likely facilitates increased absorption of solar radiation, allowing them to compensate for the lower temperature (see Chapter 7 for discussion of thermoregulation in animals).

Some of the best examples of phenotypic plasticity occur among plants. The size of the plant, the ratio of reproductive tissue to vegetative tissue, and even the shape of the leaves may vary widely at different levels of nutrition, light, moisture, and temperature. An excellent illustration of phenotypic plasticity in plants is the work of Sonia Sultan of Wesleyan University. Sultan's research focuses on phenotypic plasticity in plant species in response to resource availability. In a series of greenhouse experiments, she examined the developmental response of the herbaceous annual *Polygonum lapathifolium* (common name *curlytop knotweed*) to different light environments. Sultan grew different individuals of the same genotype for eight weeks at two light levels: low light (20 percent available photosynthetically active radiation [PAR]) and high light (100 percent available PAR). Individuals of the same genotype grown under low-light conditions produced less biomass (slower growth rate), but produced far more photosynthetic leaf area per unit of biomass through changes in biomass allocation, morphology, and structure (**Figure 5.6**). Individuals grown under low-light conditions produced large, thin leaves and few branches. In contrast, the larger high-light plants grew narrow leaves on many more branches. This response is referred to as **developmental plasticity**. As such, these changes are irreversible. After the adult plant develops, these patterns of biomass allocation (proportions of leaf, stem, and root) will remain largely unchanged, regardless of any changes in the light environment.

In contrast to developmental plasticity, other forms of phenotypic plasticity in response to prevailing environmental

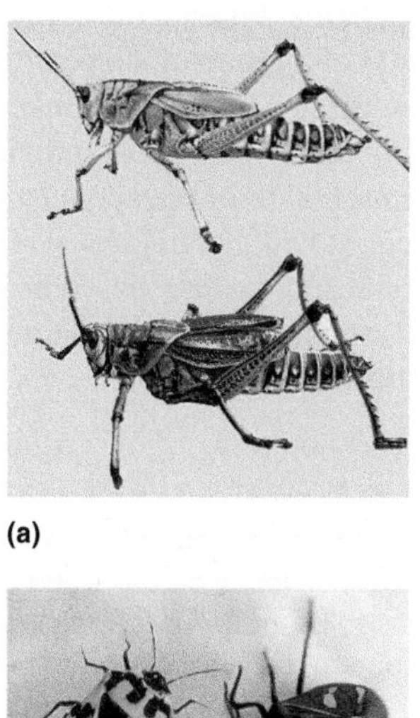

(a)

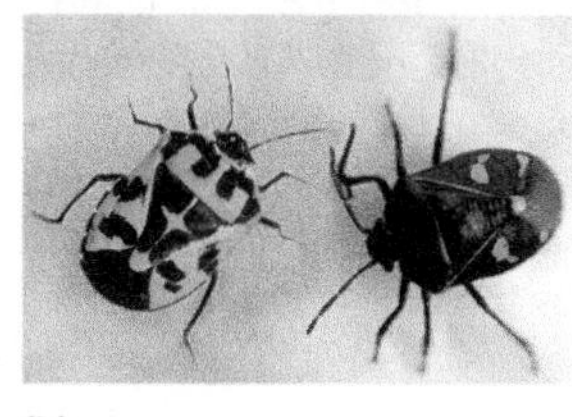

(b)

Figure 5.5 Examples of phenotypic plasticity in insects. The color of many insect species' bodies change in response to temperature during development. (a) Changes in body color in the lubber grasshopper (*Romalea microptera*) from south Florida reared at 35°C (*top*) and 25°C (*bottom*). (b) Changes in body color in harlequin bugs (*Murgantia histrionica*). Black individuals were reared at 22°C and yellow individuals at 30°C.
(From Whitman and Agrawal 2009.)

(a)

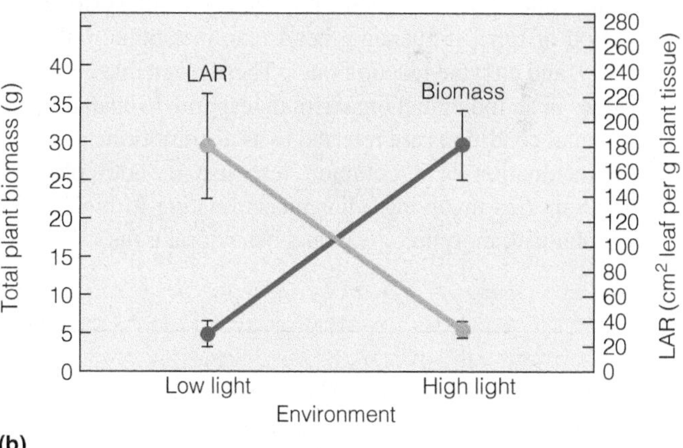

(b)

Figure 5.6 (a) Individuals of the same *Polygonum lapathifolium* genotype grown for eight weeks under low light (20 percent available PAR; *left*) and high light (100 percent available PAR; *right*). Low-light plants have large, thin leaves and few branches. The larger high-light grown plant has narrow leaves on many more branches and is more mature (developmentally). (b) Mean norms of reaction for 25 genotypes grown at low light (20 percent PAR) and high light (100 percent PAR). Plants grown under low light have less total biomass, but they produce far more photosynthetic leaf area per unit of total plant biomass (leaf area ratio: LAR) through changes in biomass allocation and leaf morphology.
([b] Based on Sultan, S.E. 2000. *Phenotypic plasticity for plant development, function and life history*. Trends in Plant Science 5: 537–542. Figure 2, pg. 538.)

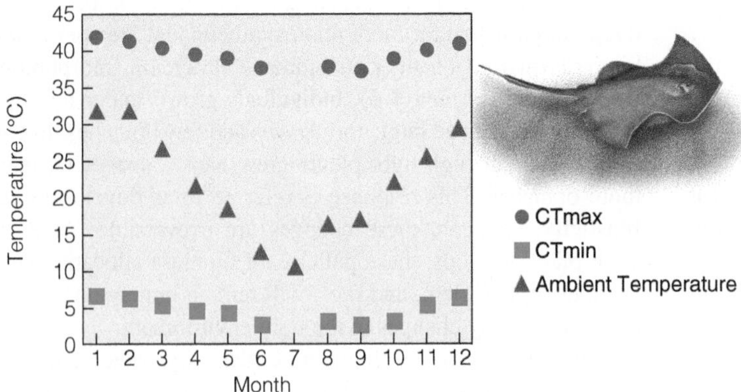

Figure 5.7 Example of seasonal acclimation to ambient temperatures. Plot of mean monthly ambient temperatures and corresponding monthly thermal maximum (*Tmax*) and minimum (*Tmin*), both in °C, for Atlantic stingray inhabiting St. John's Bay Florida. (Data from Fangue and Bennett 2003.)

conditions are reversible. For example, fish have an upper and lower limit of tolerance to temperature (see Chapter 7). They cannot survive at water temperatures above and below these limits. However, these upper and lower limits change seasonally as water temperatures warm and cool. This pattern of seasonal change in temperature tolerance is illustrated in the work of Nann Fangu and Wayne Bennett of the University of West Florida. Fangu and Bennett measured seasonal changes in the temperature tolerances of Atlantic stingrays (*Dasyatis sabina*) that inhabit shallow bays of the Florida coast. Their data for individuals inhabiting St. Josephs Bay on the Gulf Coast of Florida show a systematic shift in the critical minimum and maximum temperatures with seasonal changes in the ambient environmental (water) temperature (**Figure 5.7**). As water temperatures change seasonally, shifts in enzyme and membrane structure allow the individual's physiology to adjust slowly over a period of time, influencing heart rate, metabolic rate, neural activity, and enzyme reaction rates. These reversible phenotypic changes in an individual organism in response to changing environmental conditions are referred to as **acclimation**.

Acclimation is a common response in both plant and animal species involving adjustments relating to biochemical, physiological, morphological, and behavioral traits.

5.5 Genetic Variation Occurs at the Level of the Population

Adaptations are the characteristics of individual organisms—a reflection of the interaction of the genes and the environment. They are the product of natural selection. Although the process of natural selection is driven by the success or failure of individuals, the population—the collective of individuals and their alleles—changes through time, as individuals either succeed or fail to pass their genes to successive generations. For this reason, to understand the process of adaptation through natural selection, we must first understand how genetic variation is organized within the population.

A species is rarely represented by a single, continuous interbreeding population. Instead, the population of a species is typically composed of a group of subpopulations—local populations of interbreeding individuals, linked to each other in varying degrees by the movement of individuals (see Sections 8.2 and 19.7 for discussion of metapopulations). Thus, genetic variation can occur at two hierarchical levels, within subpopulations and among subpopulations. When genetic variation occurs among subpopulations of the same species, it is called **genetic differentiation**.

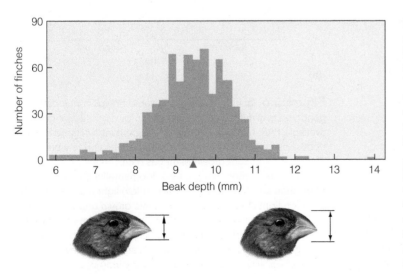

Figure 5.8 Variation in beak size (as measured by depth) in the population of Galápagos medium ground finch (*Geospiza fortis*) on the island of Daphne Major as estimated by individual birds sampled during 1976. The histogram represents the number of individuals that were sampled (*y*-axis) in each category (0.2 mm) of beak depth (*x*-axis). The estimate of the population mean is marked by the blue triangle. (Adapted from Grant 1999 after Boag and Grant 1984.)

Interpreting Ecological Data

Q1. What type of data do the original measures of beak depth represent? (See Chapter 1, Quantifying Ecology 1.1.)

Q2. How have the original measurements of beak depth been transformed for presentation purposes in Figure 5.8?

Q3. What is the range (maximum – minimum values) of beak depths observed for the sample of individuals presented in Figure 5.8? (Categories are in units of 0.2 mm.)

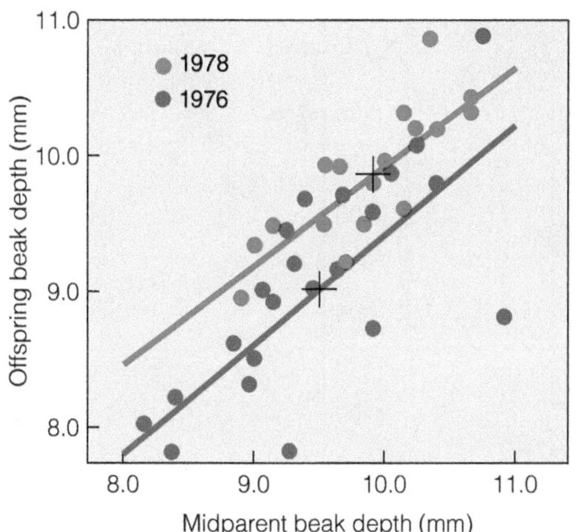

Figure 5.9 The relationship between the beak depth (size) of offspring and their parents in the medium ground finch (*Geospiza fortis*) population on Daphne Major. The *x*-axis represents midparent beak depth, which is the average beak depth for the two parent birds. The *y*-axis is the average beak depth of their offspring. The slope of the relationship (represented by the lines) is the estimate of heritability. The blue line and circles are data from 1976, and the red line and circles are data from 1978 (+ signs represent the average values). The results from the two years are consistent (nearly identical slopes); however, the average size of offspring was greater in 1978. Data from both years show a strong relationship between the beak depth of parents and their offspring.
(Adapted from Grant 1999 after Boag 1983.)

The sum of genetic information (alleles) across all individuals in the population is referred to as the **gene pool**. The gene pool represents the total genetic variation within a population. Genetic variation within a population can be quantified in several ways. The most fundamental measures are **allele frequency** and **genotype frequency**. The word *frequency* in this context refers to the proportion of a given allele or genotype among all the alleles or genotypes present at the locus in the population.

5.6 Adaptation Is a Product of Evolution by Natural Selection

We have defined evolution as changes in the properties of populations of organisms over the course of generations (Section 5.1). More specifically, phenotypic evolution can be defined as a change in the mean or variance of a phenotypic trait across generations as a result of changes in allele frequencies. In favoring one phenotype over another, the process of natural selection acts directly on the phenotype. But in doing so, natural selection changes allele frequencies within the population. Changes in allele frequencies from parental to offspring generations are a product of differences in relative fitness (survival and reproduction) of individuals in the parental generation.

The work of Peter Grant and Rosemary Grant provides an excellent documented example of natural selection. The Grants have spent more than three decades studying the birds of the Galápagos Islands, the same islands whose diverse array of animals so influenced the young Darwin when he was a naturalist aboard the expeditionary ship HMS *Beagle*. Among other events, the Grants' research documented a dramatic shift in a physical characteristic of finches inhabiting some of these islands during a period of extreme climate change.

Recall from our initial discussion in Section 5.1 that natural selection is a product of two conditions: (1) that variation occurs among individuals within a population in some heritable characteristic and (2) that this variation results in differences among individuals in their survival and reproduction. Figure 5.8 shows variation in beak size in Darwin's medium ground finch (*Geospiza fortis*) on the 40-hectare islet of Daphne Major, one of the Galápagos Islands off the coast of Ecuador. Heritability of beak size in this species was established by examining the relationship between the beak size of parents and their offspring (Figure 5.9).

Beak size is a trait that influences the feeding behavior of these seed-eating birds. Individuals with large beaks can feed on a wide range of seeds, from small to large, whereas individuals with smaller beaks are limited to feeding on smaller seeds (Figure 5.10).

During the early 1970s, the island received an average rainfall of between 127 and 137 millimeters (mm) per year, supporting an abundance of seeds and a large finch population

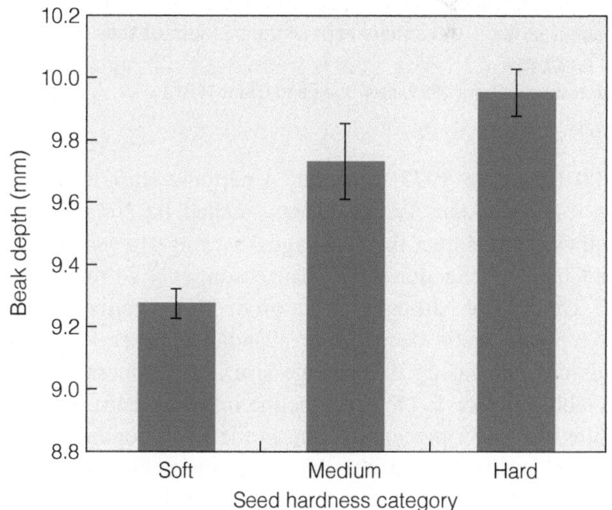

Figure 5.10 Beak depth of medium ground finches (*Geospiza fortis*) feeding on soft, medium, and hard seeds on Daphne Major in 1977. The bars represent the mean beak depth for birds feeding on the corresponding class of seeds, and the lines represent ±1 standard error. As can be seen, beak size has a direct influence on the hardness and size of seeds selected by individual birds.
(After Boag and Grant 1984.)

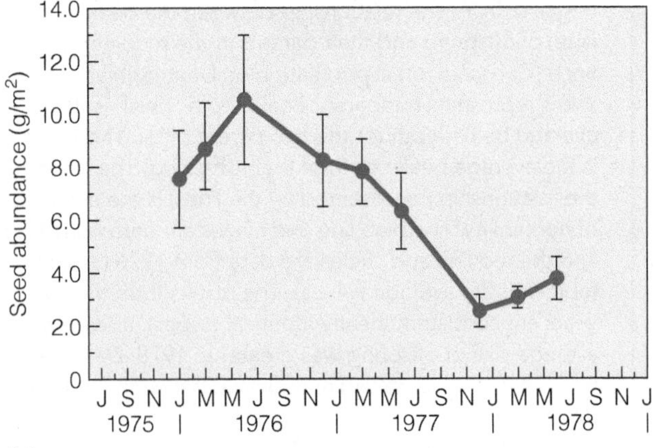

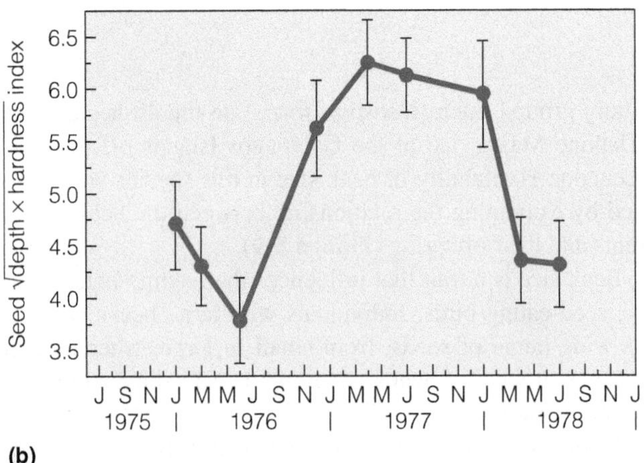

Figure 5.11 Changes in (a) seed abundance and (b) seed size and hardness on Daphne Major for the period of July 1975 to July 1978. Points represent mean values, and associated lines represent the 95 percent confidence intervals. Seed size and hardness index is the square root of the product of seed depth and hardness.
(Adapted from Grant 1999 after Boag and Grant 1981.)

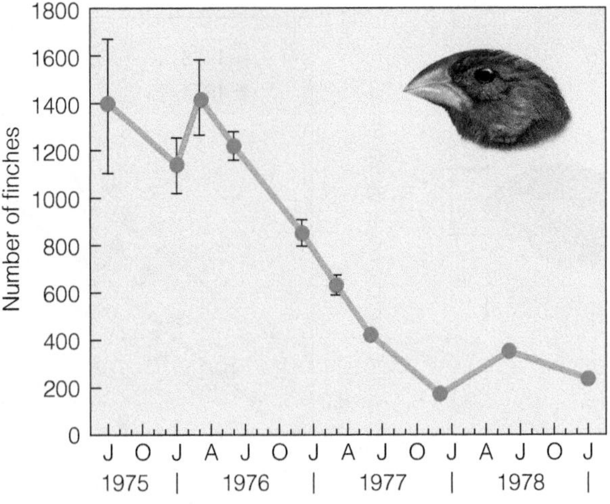

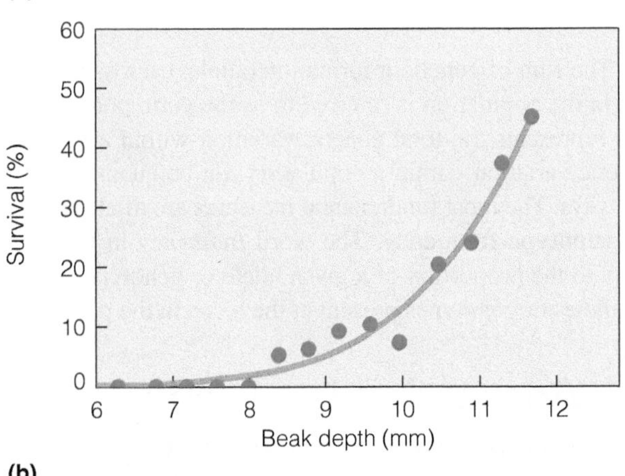

Figure 5.12 (a) Decline of the population of the medium ground finch on Daphne Major during the 1977 drought. Points represent mean estimates, and associated lines represent the 95 percent confidence interval. The population declined in the face of seed scarcity during a prolonged drought (Figure 5.11a). (b) Birds with larger beak size had a much greater rate of survival as a result of their ability to feed on the larger, harder seeds that comprised the majority of food resources during the drought period (see Figure 5.11b).
(Adapted from Grant 1999 after Boag and Grant 1981.)

(1500 birds). In 1977, however, a periodic shift in the climate of the eastern Pacific Ocean—called La Niña—altered weather patterns over the Galápagos, causing a severe drought (see Chapter 2, Section 2.9). That season, only 24 mm of rain fell. During the drought, seed production declined drastically. Small seeds declined in abundance faster than large seeds did, increasing the average size and hardness of seeds available (**Figure 5.11**). The decline in food (seed) resources resulted in an 85 percent decline in the finch population as a result of mortality and possible emigration (**Figure 5.12a**). Mortality, however, was not equally distributed across the population (**Figure 5.12b**). Small birds had difficulty finding food, whereas large birds, especially males with large beaks, had the highest rate of survival because they were able to crack large, hard seeds.

The graph in **Figure 5.12b** represents a direct measure of the differences in fitness (as measured by survival) among individuals in the population as a function of differences in phenotypic characteristics (beak size), the second condition for natural selection. The phenotypic trait that selection acts directly upon is referred to as the **target of selection**; in this example, it is beak size. The **selective agent** is the environmental cause of fitness differences among organisms with different phenotypes, or in this case, the change in food resources (abundance and size distribution of seeds).

The increased survival rate of individuals with larger beaks resulted in a shift in the distribution of beak size (phenotypes) in the population (**Figure 5.13**). This type of natural selection, in which the mean value of the trait is shifted toward one extreme over another (**Figure 5.14a**), is called **directional selection**. In other cases, natural selection may favor individuals near the population mean at the expense of the two extremes; this is referred to as **stabilizing selection** (**Figure 5.14b**). When natural selection favors both extremes simultaneously,

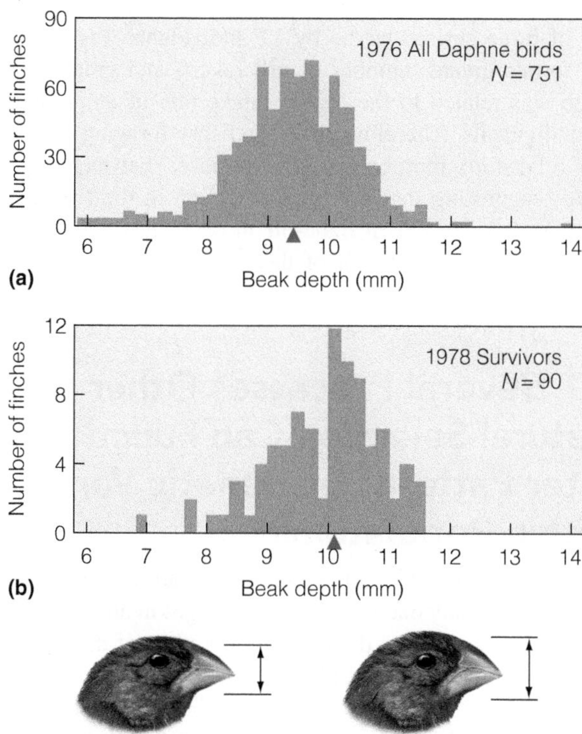

Figure 5.13 Distribution of beak depth for the population of medium ground finches inhabiting Daphne Major (a) before and (b) after natural selection. The estimate of mean beak depth for both census periods is shown by the blue triangles. Note the increase in the mean beak depth for the population resulting from the differential survival of individuals related to beak size as shown in Figure 5.12.

(Adapted from Grant 1999 after Boag and Grant 1984.)

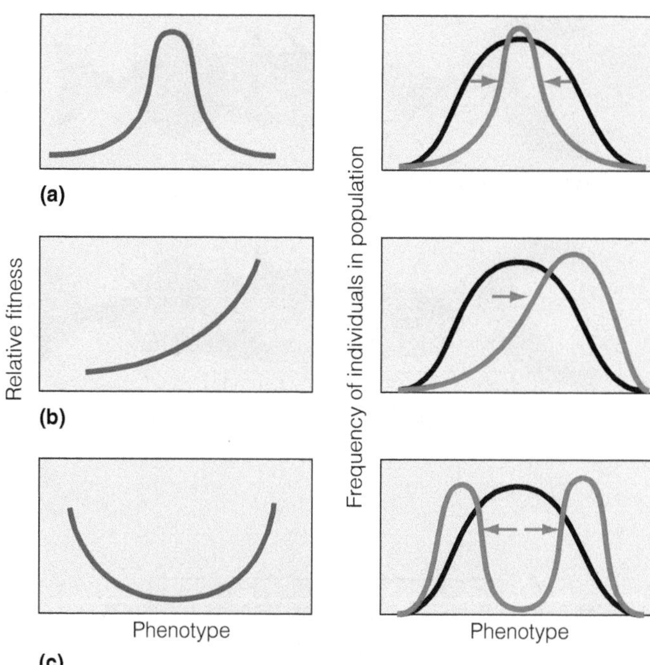

Figure 5.14 Three types of selection: curves in the left column represent the relative fitness of different phenotypes in the population under the three types of selection, whereas curves in the right column show the changes in the frequency of individuals in the population exhibiting different phenotypes under the corresponding three types of selection. Arrows represent the direction of change in the distribution of phenotypes in the population. (a) Under stabilizing selection, the mean phenotype in the population exhibits the highest relative fitness and the original distribution of phenotypes (black curve) is shifted to the center (mean value). (b) In directional selection, the distribution of phenotypes is shifted to one extreme. (c) In disruptive selection, the relative fitness is greatest for the extreme values of phenotype and the result is a bimodal distribution of phenotypes in the population.

although not necessarily to the same degree, it can result in a bimodal distribution of the characteristic(s) in the population (**Figure 5.14c**). Such selection, known as **disruptive selection**, occurs when members of a population are subject to different selection pressures.

The work of Beren Robinson of Guelph University in Canada provides an excellent example of disruptive selection. In studying the species of threespine stickleback (*Gasterosteus aculeatus*), which occupies Cranby Lake in the coastal region of British Columbia, Robinson found that individuals sampled from the open-water habitat (limnetic habitat) differed morphologically from individuals sampled from the shallower nearshore waters (benthic habitat). In a series of experiments, Robinson established that these individuals represented distinct phenotypes that are products of natural selection promoting divergence within the population. He initially established that morphological differences between the two forms were heritable, rather than an expression of phenotypic plasticity in response to the two different habitats or diets. He reared offspring of the two forms under identical laboratory conditions (environmental conditions and diet) and although there was some degree of phenotypic plasticity, differences in most characteristics remained between the two forms. On average, the benthic form (BF) had

Interpreting Ecological Data

Q1. *Figure 5.12b shows the survival of ground finches as a function of beak size during the period of drought. How does the graph in Figure 5.12b relate to this figure?*

Q2. *How do the patterns of relative fitness shown in the graphs on the left-hand column give rise to the corresponding patterns of selection illustrated by the arrows in the graphs shown in the right-hand column?*

(1) shorter overall body length, (2) deeper body, (3) wider mouth, (4) more dorsal spines, and (5) fewer gill rakers than did the limnetic form (LF) (**Figure 5.15a**).

The two habitats in the lake—benthic and limnetic—provide different food resources; so to determine the agent of selection that caused divergence within the population, Robinson conducted feeding trials in the laboratory to test for trade-offs in the foraging efficiency of the two forms on food resources found in the two habitats. The foraging success of individual fish was assessed in two artificial habitats, mimicking conditions in the limnetic and benthic environments.

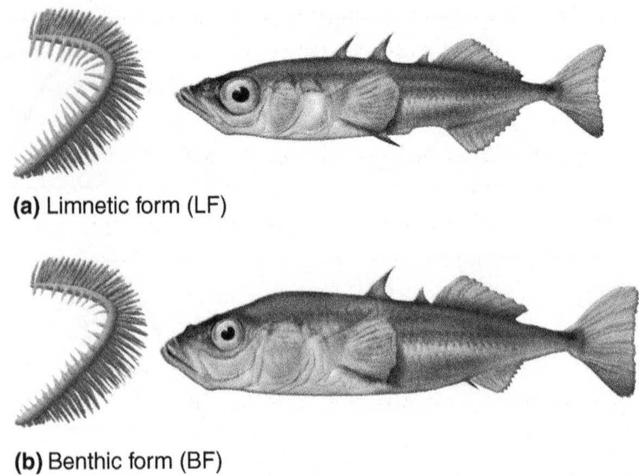

(a) Limnetic form (LF)

(b) Benthic form (BF)

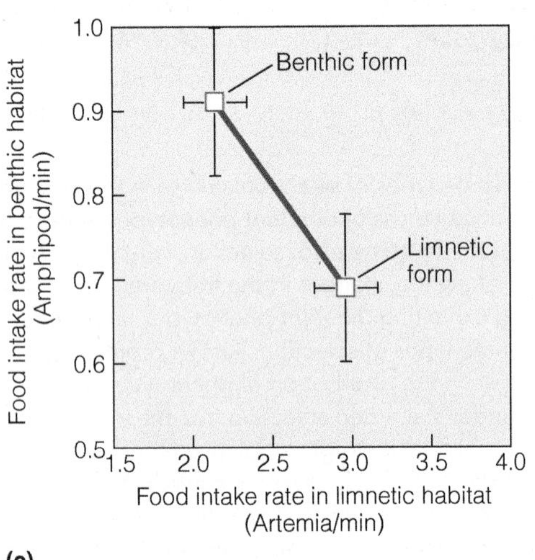

(c)

Figure 5.15 Morphological differences in the gill rakers of the stickleback species based on (a) limnetic form (phenotype) and (b) benthic form (phenotype) feeding habits. (c) Mean intake rate (and standard errors) of limnetic (open) and benthic (shallow) forms in open-water (food source: Artemia [brine shrimp larvae]) and shallow-water, benthic (food source: amphipod) feeding trials. ([a] Illustration by Laura Nagel, Queen's University. From Schulter 1993; [b] Adapted from Robinson 2000.)

Two food types were used in the trials. Brine shrimp larvae (*Artemia*), a common prey found in open water, were placed in the artificial limnetic habitats. Larger amphipods, fast-moving arthropods with hard exoskeletons that forage on dead organic matter on the sediment surface, were placed in the artificial benthic habitats.

Results of the foraging trials revealed distinct differences in the foraging success of the two morphological forms (phenotypes; **Figure 5.15b**). The LF individuals were most successful at foraging on the brine shrimp larvae. They had a higher consumption rate and required only half the number of bites to consume as compared to the BF individuals. In contrast, BF individuals had a higher intake rate for amphipods and on average consumed larger amphipods than did LF individuals.

Robinson was able to determine that the higher intake rate of brine shrimp larvae by LF individuals was related to this form's greater number of gill rakers, and greater mouth width was related to the higher intake rate of amphipods by BF individuals. Therefore, he found that foraging efficiency was related to morphological differences between the two forms, suggesting that divergent selection in the two distinct phenotypes represents a trade-off in characteristics related to the successful exploitation of these two distinct habitats and associated food resources.

5.7 Several Processes Other than Natural Selection Can Function to Alter Patterns of Genetic Variation within Populations

Natural selection is the only process that leads to adaptation because it is the only one in which the changes in allele frequency from one generation to the next are a product of differences in the relative fitness (survival and reproduction) of individuals in the population. Yet not all phenotypic characteristics represent adaptations, and processes other than natural selection can be important factors influencing changes in genetic variation (allele and genotype frequencies) within populations. For example, mutation is the ultimate source of the genetic variation that natural selection acts upon. **Mutations** are heritable changes in a gene or a chromosome. The word *mutation* refers to the process of altering a gene or chromosome as well as to the product, the altered state of the gene or chromosome. Mutation is a random force in evolution that produces genetic variation. Any altered phenotypic characteristic resulting from mutation may be beneficial, neutral, or harmful. Whether a mutation is beneficial depends on the environment. A mutation that enhances an organism's fitness in one environment could harm it in another. Most of the mutations that have significant effect, however, are harmful, but the harmful mutations do not survive long. Natural selection eliminates most deleterious genes from the gene pool, leaving behind only genes that enhance (or at least do not harm) an organism's ability to survive, grow, and reproduce in its environment.

Another factor that can directly influence patterns of genetic variation within a population is a change in allele frequencies as a result of random chance—a process known as **genetic drift**. Recall from basic biology that the recombination of alleles in sexual reproduction is a random process. The offspring produced in sexual reproduction, however, represent only a subset of the parents' alleles. If the parents have only a small number of offspring, then not all of the parents' alleles will be passed on to their progeny as a result of the random assortment of chromosomes at meiosis (the process of recombination). In effect, genetic drift is the evolutionary equivalent of sampling error, with each successive generation representing only a subset or sample of the gene pool from the previous generation.

In a large population, genetic drift will not affect each generation much because the effects of the random nature of the process will tend to average out. But in a small population, the effect could be rapid and significant. To illustrate this point, we can use the analogy of tossing a coin. With a single toss of the coin, the probability of each of the two possible outcomes, heads or tails, is equal, or 50 percent. Likewise, with a series of four coin tosses, the probability of the outcome being two heads and two tails is 50 percent. But each individual outcome in the coin tosses is independent; therefore, in a series of four coin tosses, there is also a probability of 0.0625, or 6.25 percent, that the outcome will be four heads. The probability of the outcome being all heads drops to 9.765×10^{-4} if the number of tosses is increased to 10, and this probability drops to 8.88×10^{-16} for 50 tosses. Likewise, the probability of heterozygous (*Aa*) individuals in the population producing only homozygous (either *aa* or *AA*) offspring under a system of random mating decreases with increasing population size.

Patterns of genetic variation within a population can also be influenced by the movement of individuals into, or out of, the population. Recall from the discussion of genetic variation in Section 5.5 that the population of a species is typically composed of a group of subpopulations—local populations of interbreeding individuals that are linked to one another in varying degrees by the movement of individuals (see Chapter 8). **Migration** is defined as the movement of individuals between local populations, whereas **gene flow** is the movement of genes between populations (see Chapter 8). Because individuals carry genes, the terms are often used synonymously; however, if an individual immigrates into a population but does not successfully reproduce, the new genes are not established in the population. Migration is a potent force in reducing the level of population differentiation (genetic differences among local populations; see Section 5.5).

One of the most important principles of genetics is that under conditions of random mating, and in the absence of the factors discussed thus far—natural selection, mutation, genetic drift, and migration—the frequency of alleles and genotypes in a population remains constant from generation to generation. In other words, no evolutionary change occurs through the process of sexual reproduction itself. This principle, referred to as the **Hardy–Weinberg principle**, is named for Godfrey Hardy and Wilhelm Weinberg, who each independently published the model in 1908 (see Quantifying Ecology 5.1). Mating is random when the chance that an individual mates with another individual of a given genotype is equal to the frequency of that genotype in the population. When individuals choose mates nonrandomly with respect to their genotype—or more specifically, select mates based on some phenotypic trait—the behavior is referred to as **assortative mating**. Perhaps the most recognized and studied form of assortative mating is female mate choice. Female mate choice is the behavior in which females exhibit a bias toward certain males as mates based on specific phenotypic traits (often secondary sex characteristics), such as body size or coloration (see Chapter 10, Section 10.11).

Positive assortative mating occurs when mates are phenotypically more similar to each other than expected by chance. Positive assortative mating is common, and one of the most widely reported examples relates to the timing of reproduction. Plants mate assortatively based on flowering time. In populations of plants with an extended flowering time, early flowering plants are often no longer flowering when late flowering plants are in bloom.

The genetic effect of positive assortative mating is an increase in the frequency of homozygotes with a decrease in the frequency of heterozygotes in the population. Think of a locus where *AA* individuals tend to be larger than *Aa,* which in turn are larger than *aa* individuals. With positive assortative mating, *AA* will mate with *AA*, and *aa* with other *aa*. All of these matings will produce only homozygous offspring. Even mating between *Aa* individuals will result in half of the offspring being homozygous. The genetic effects of positive assortative mating are only at the loci that affect the phenotypic characteristic by which the organisms are selecting mates.

Negative assortative mating occurs when mates are phenotypically less similar to each other than expected by chance. Though not as common as positive assortative mating, negative assortative mating results in an increase in the frequency of heterozygotes.

A special case of nonrandom mating is inbreeding. **Inbreeding** is the mating of individuals in the population that are more closely related than expected by random chance. Unlike positive assortative mating, inbreeding increases homozygosity at all loci. Inbreeding affects all loci equally because related individuals are genetically similar by common ancestry, and they are therefore more likely to share alleles throughout the genome than unrelated individuals.

Inbreeding can be detrimental. Offspring are more likely to inherit rare, recessive, deleterious genes. These genes can cause decreased fertility, loss of vigor, reduced fitness, reduced pollen and seed fertility in plants, and even death. These consequences are referred to as **inbreeding depression**.

As we have seen from the preceding discussion, nonrandom mating changes genotypic frequencies from one generation to the next, but assortative mating does not directly result in a change of allele frequencies within a population. The other three processes discussed—mutation, migration, and genetic drift, together with natural selection—alter the allele frequencies, and therefore result in a shift in the distribution of genotypes (and potentially phenotypes) within the population. As such, all four processes function as agents of evolution. However, natural selection is special among the four evolutionary processes because it is the only one that leads to adaptation. The other three can only speed up or slow the development of adaptations.

5.8 Natural Selection Can Result in Genetic Differentiation

The example of natural selection in the population of medium ground finches as described previously represents a shift in the distribution of phenotypes in the population inhabiting the island of Daphne Major in response to environmental changes

QUANTIFYING ECOLOGY 5.1 Hardy–Weinberg Principle

The Hardy–Weinberg principle states that both allele and genotype frequencies will remain the same in successive generations of a sexually reproducing population if certain criteria are met: (1) mating is random, (2) mutations do not occur, (3) the population is large, so that genetic drift is not a significant factor, (4) there is no migration, and (5) natural selection does not occur.

If we have only two alleles at a locus, designated as *A* and *a*, then the usual symbols for designating their frequencies are *p* and *q*, respectively. Because frequencies (proportions) must sum to 1, then:

$$p + q = 1 \text{ or } q = 1 - p$$

Genotypic frequencies are typically designated by uppercase letters. In the case of a locus with two alleles, *P* is the frequency of *AA*, *H* is the frequency of *Aa*, and *Q* is the frequency of *aa*. As with gene frequencies, genotype frequencies must sum to 1:

$$P + H + Q = 1$$

Given a population having the genotypic frequencies of

$$P = 0.64, H = 0.32, \text{ and } Q = 0.04$$

we can calculate the allele frequencies as follows:

$$p = P + H/2 = 0.64 + (0.32/2) = 0.8$$

$$q = Q + H/2 = 0.04 + (0.32/2) = 0.2$$

The frequency of heterozygous individuals (*H: Aa*) is divided by 2 because only one of the allele pair is *A* or *a*.

With a population consisting of the three genotypes just described (*AA*, *Aa*, and *aa*), there are six possible types of mating (**Table 1**). For example, the mating *AA* × *AA* occurs only when an *AA* female mates with an *AA* male, with the frequency of occurrence being $P \times P$ (or P^2) under the conditions of random mating. Similarly, an *AA* × *Aa* mating occurs when an *AA* female mates with an *Aa* male (proportion $P \times H$) or when an *Aa* female mates with an *AA* male (proportion $H \times P$). Therefore, the overall proportion of *AA* × *Aa* matings is $PH + HP = 2PH$. The frequencies of these and the other four types of matings are given in the second column of Table 1.

Table 1 Calculation of Offspring Genotype Frequencies for a Randomly Mating Population

(a) Parental Genotype Frequencies			(b) Parental Allele Frequencies	
P	H	Q	p	q
(*AA*)	(*Aa*)	(*aa*)	0.8	0.2
0.64	0.32	0.04		

(c) Mating Genotype		Frequency of Mating		Offspring Frequencies *AA*	*Aa*	*aa*
AA × *AA*	P^2	.64 × .64 =	.4096	1 .4096	0	0
AA × *Aa*	$2PH$	2 × .64 × .32 =	.4096	½ .2048	½ .2048	0
AA × *aa*	$2PQ$	2 × .64 × .04 =	.0512	0	1 .0512	0
Aa × *Aa*	H^2	.32 × .32 =	.1024	¼ .0256	½ .0512	¼ .0256
Aa × *aa*	$2HQ$	2 × .32 × .04 =	.0256	0	½ .0128	½ .0128
aa × *aa*	Q^2	.04 × .04 =	.0016	0	0	1 .0016
Total (frequency of mating):			1.0000			
Totals (next generation):				.64 P′	.32 H′	.04 Q′

that occurred over time (period of drought). This shift in the mean phenotype (beak size) reflects a change in genetic variation (allele and genotype frequencies) within the population. Natural selection can also function to alter genetic variation among local populations as a result of local differences in environmental conditions—the process of genetic differentiation (see Section 5.5).

Species having a wide geographic distribution often encounter a broader range of environmental conditions than do those species whose distribution is more restricted. The variation in environmental conditions can give rise to a corresponding variation in morphological, physiological, and behavioral characteristics (phenotypes). Significant differences often exist among local populations of a single species

To calculate the offspring genotypes produced by these matings, we must first examine the offspring produced by each of the six possible pairings of parental genotypes (Table 1). Because homozygous *AA* genotypes produce only *A*-bearing gametes (egg or sperm), and homozygous *aa* genotypes produce only *a*-bearing gametes, the mating of *AA* × *AA* individuals will produce only *AA* offspring, and likewise, the mating of *aa* × *aa* individuals will produce only offspring with genotype *aa*. In addition, the mating of *AA* × *aa* individuals will produce only heterozygous offspring (*Aa*).

In contrast to homozygous individuals, heterozygous individuals produce both *A*- and *a*-bearing gametes.

Therefore, the mating of a heterozygous individual with a homozygous (either *AA* or *aa*) or another heterozygous individual will produce offspring of all three possible genotypes (*AA*, *Aa*, and *aa*). The relative frequencies of offspring genotypes depend on the specific combination of parents (**Figure 1**). The offspring frequencies presented in Figure 1 are based on the assumption that an *Aa* heterozygote individual produces an equal number of *A*- and *a*-bearing gametes (referred to as Mendelian segregation).

Using the data presented in Figure 1 and the frequencies of the different types of matings in column 2 of Table 1, the genotype frequencies of the offspring, denoted as P' (*AA*), H' (*Aa*), and Q' (*aa*), are presented in column 3 of Table 1. The new genotype frequencies are calculated as the sum of the products shown at the bottom of Table 1. For each genotype, the frequency of each mating producing the genotype is multiplied by the fraction of the genotypes produced by that mating.

We can now calculate the allele frequencies ($p$and q) for the generation of offspring (designated as p' and q') using the formula presented previously:

$$p' = P' + H'/2 = 0.64 + (0.32/2) = 0.8$$

$$q' + Q' + H'/2 = 0.04 + (0.32/2) = 0.2$$

Note that both the genotype and allele frequencies of the offspring generation are the same as those of the parental generation.

In natural populations the assumptions of the Hardy–Weinberg principle are never fully met. Mating is not random, mutations do occur, individuals move between local populations, and natural selection does occur. All of these circumstances change the frequencies of genotypes and alleles from generation to generation, acting as evolutionary forces in a population. The beauty of the Hardy–Weinberg principle is that it functions as a null model, where deviations from the expected frequencies can provide insight into the evolutionary forces at work within a population.

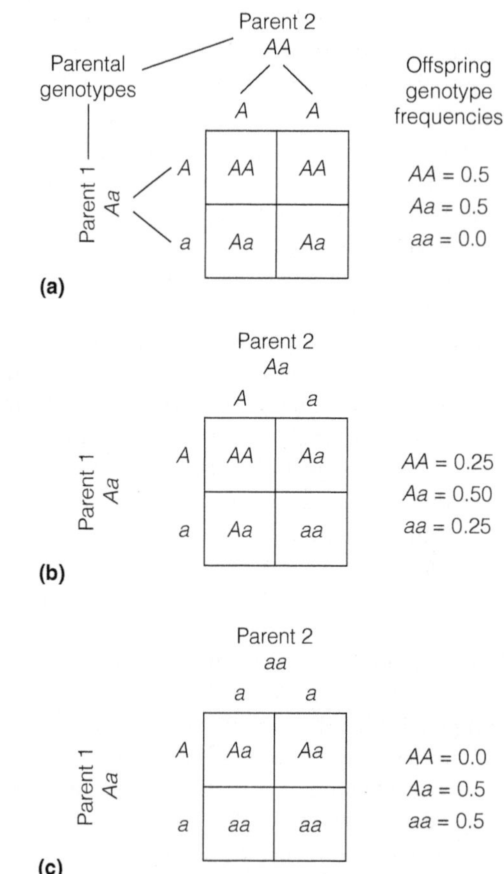

Figure 1 Proportion of offspring genotypes produced by heterozygous individuals mating with (a) homozygous *AA* individual, (b) another heterozygous individual, and (c) homozygous *aa* individual. The frequencies of genotypes produced from each of these matings are shown.

1. How would the frequency of heterozygotes in the population change if the frequency for the *A* allele in the example described was $p = 0.5$?

2. When the frequency of an allele is greater than 0.8, most of these alleles are contained in homozygous individuals (as illustrated in the preceding example). When the frequency of an allele is less than 0.1, in which genotype are most of these alleles?

inhabiting different regions. The greater the distance between populations, the more pronounced the differences often become as each population adapts to the locality it inhabits. The changes in phenotype across the landscape therefore reflect the changing nature of natural selection operating under the different local environmental conditions. Geographic variation within a species in response to changes in environmental conditions can result in the evolution of clines, ecotypes, and geographic isolates or subspecies.

A **cline** is a measurable, gradual change over a geographic region in the average of some phenotypic character, such as size and coloration. Clines are usually associated with an environmental gradient that varies in a continuous manner across the landscape, such as changes in temperature or moisture with elevation or latitude. Continuous variation in the phenotypic character across the species distribution results from gene flow from one population to another along the gradient. Because environmental constraints influencing natural selection vary along the gradient, any one population along the gradient will differ genetically to some degree from another—the difference increasing with the distance between the populations.

Clinal differences exist in size, body proportions, coloration, and physiological adaptations among animals. For example, the fence lizard (*Sceloporus undulatus*) is one of the most widely distributed species of lizards in North America, ranging throughout the eastern two-thirds of the United States and into northern Mexico. Across its range, the fence lizard exhibits a distinct gradient of increasing body size with latitude (**Figure 5.16**). Lizards from northern latitudes are larger than lizards from southern latitudes. Furthermore, lizards from higher elevations in geographically proximal areas exhibit larger body size than lizards from lower elevations. Thus, mean body size increases along an environmental gradient of decreasing mean annual temperature.

Similar clines are observed in plant species. Alicia Montesinos-Navarro of the University of Pittsburgh and colleagues examined phenotypic variation in *Arabidopsis thaliana*, a small annual flowering plant species that is native to Europe, Asia, and northwestern Africa. The researchers examined 17 natural populations that occupy an altitudinal gradient in the region of northeastern Spain. Along the gradient, precipitation increases, but maximum spring temperature and minimum winter temperature decrease with altitude. Examination of the local populations revealed a systematic variation in a variety of phenotypic characteristics. Aboveground mass, number of rosette leaves at bolting (a measure of size at reproduction), developmental time, and number of seeds and seed weight increased with altitude (**Figure 5.17**). Although these changes in phenotypic characteristics are clearly in response to the gradient of environmental conditions with altitude, how can the researchers be sure that the changes in phenotype represent changes in allele and genotype frequencies between populations rather than phenotypic plasticity?

Recall from Section 5.4 that phenotypic plasticity is the ability of a single genotype to produce different phenotypes under different environmental conditions (norms of reaction; see Figure 5.4). A common approach used to determine if observed phenotypic differences between local populations represent differences in allele frequencies (genetic differentiation) or phenotypic plasticity is the common garden experiment. In this experiment individuals (genotypes) from the different populations are grown under controlled environmental conditions—a common garden. If the phenotypic differences observed in the local populations are maintained in individuals grown in the common garden, the differences in phenotype represent genetic differences between the populations (genetic differentiation). If the individuals from the different populations no longer exhibit differences in phenotypic characteristics, then the differences observed in the local populations in their natural environments are a function of phenotypic plasticity. When Montesinos-Navarro and her colleagues grew genotypes from the 17 local populations under uniform controlled conditions (the common garden experiment), the phenotypic differences were maintained, revealing that the *A. thaliana* cline represents adaptations to local environmental conditions along the altitudinal gradient.

Clinal variation may show marked discontinuities. Such abrupt changes, or step clines, often reflect abrupt changes in local environments. Such variants are called ecotypes. An **ecotype** is a population adapted to its unique local environmental conditions (see this chapter, **Field Studies: Hopi Hoekstra**). For example, a population inhabiting a mountaintop may differ from a population of the same species in the valley below. This is the case with the weedy herbaceous annual *Diodia teres* (with the common name *poorjoe*) that occurs in a wide variety of habitats in eastern North America (**Figure 5.18**). In the southeastern United States, there are two distinct ecotypes: one occurs in inland agricultural fields and the other in coastal sand-dune habitats. The populations differ strikingly in morphology. Among many differences, the coastal

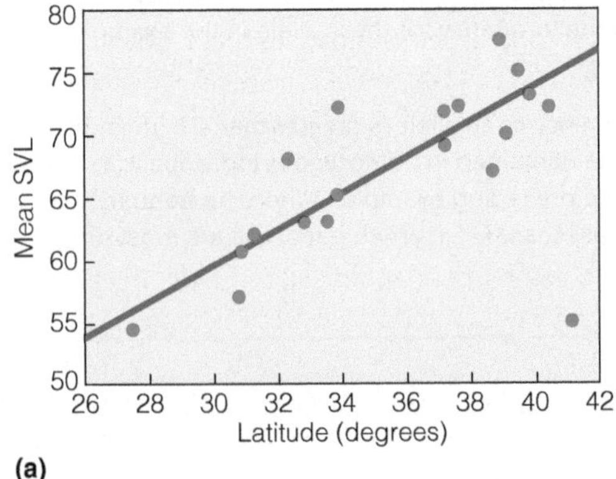

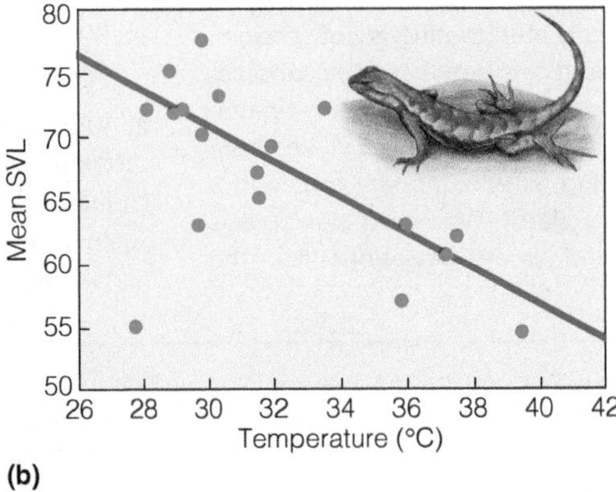

Figure 5.16 Example of a latitudinal cline in body size. Mean female body size, measured as snout-vent length (SVL), in 18 local populations of the fence lizard (*Sceloporus undulatus*) as related to latitude (a) and environmental temperature (b). Females at higher latitudes tended to be larger than those at lower latitudes, and females in colder environments tended to be larger than those in warmer environments.
(Adapted from Angilletta et al 2004.)

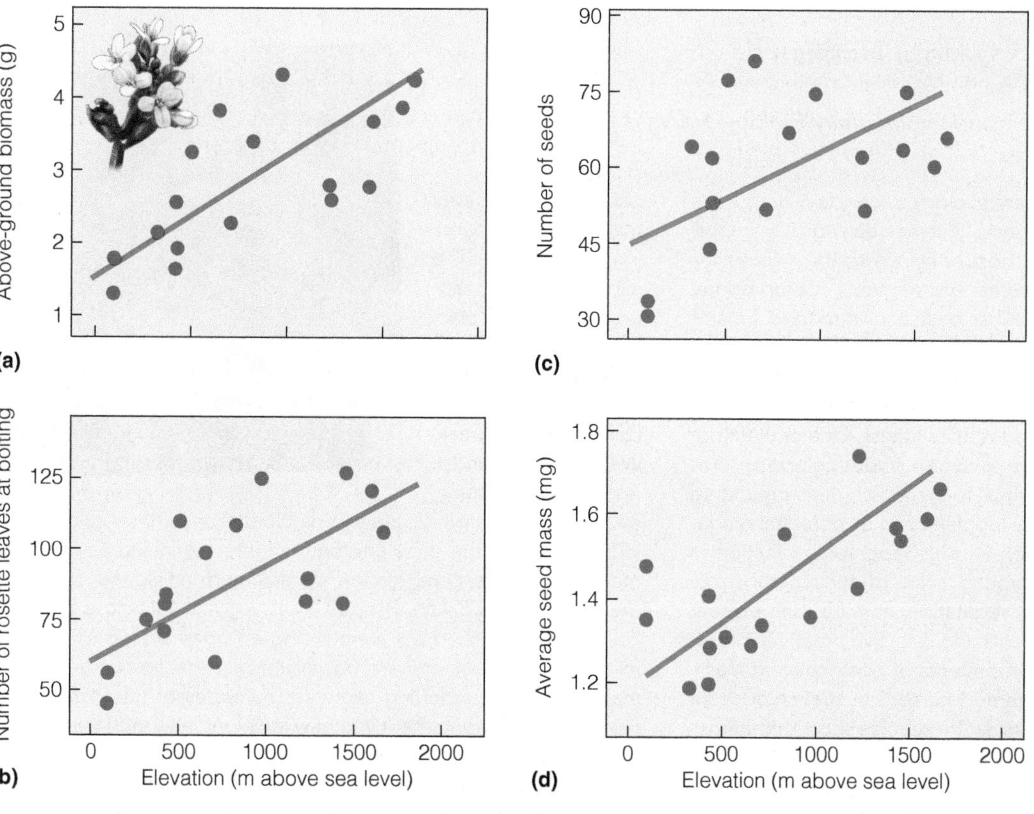

Figure 5.17 Example of altitudinal cline in *Arabidopsis thalania* on the Iberian Peninsula. Variations in mean (a) aboveground biomass, (b) rosette leaves at bolting (a measure of size at the onset of reproduction), (c) number of seeds produced per plant, and (d) average seed size as a function of altitude (meters above sea level) for 17 local populations.
(Adapted from Montesinos-Navarro et al. 2010.)

population has heavier stem pubescence (covered with short, soft, erect hairs), a more flattened growth habit, and larger seed size relative to the inland population. To understand the patterns of local adaptation acting on these two distinct ecotypes, Nicholas Jordan of the University of Minnesota undertook a series reciprocal transplant studies. In these studies, seeds from the two ecotypes were planted in each of the two distinct habitats (inland and coastal). By comparing patterns of survival, growth, and reproduction of the two ecotypes in the two different habitats, Jordan was able to analyze selection for and against native and introduced individuals. Results of the study reveal two important facts regarding the two ecotypes. First, phenotypic differences between the two ecotypes were maintained in both environments indicating that the ecotypes represent genetic differences between the two populations rather than phenotypic plasticity. Second, in each of the two habitats, the native ecotype performed better than the introduced ecotype in comparisons of survival, growth, and seed production. Each of the two ecotypes exhibited a greater relative fitness in its native habitat indicating that the phenotypic differences represent adaptations to the two different local environments (see Figure 5.18).

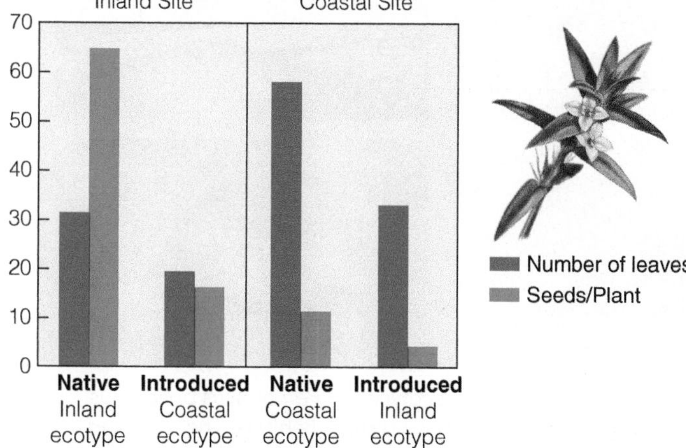

Figure 5.18 Estimates of fitness (plant size as measured by number of leaves and seeds produced per plant) for inland and coastal ecotypes of the annual plant *Diodia teres* planted at each of the two sites.
(Data from Jordan 1992.)

Although ecotypes typically represent distinct genetic populations (with respect to the phenotypic characteristics that relate to the local adaptations), gene flow occurs to varying degrees between adjacent populations, and often, zones of hybridization (mating between ecotypes) can be found. In some cases, however, geographic features such as rivers or mountain ranges that impede the movement of individuals (or gametes) can restrict gene flow between adjacent populations. For example, the southern Appalachian Mountains are noted for their diversity of salamanders. This diversity is fostered in part by a rugged terrain, an array of environmental conditions, and the limited ability of salamanders to disperse (**Figure 5.19**). Populations become isolated from one another, preventing a free flow of genes. One species of salamander, *Plethodon jordani*, formed a number of semi-isolated populations, each characteristic of a particular part of the mountains. These subpopulations make up **geographic isolates**, in which some extrinsic barrier—in the case of the salamanders, rivers and mountain ridges—prevents the free flow of genes

FIELD STUDIES *Hopi Hoekstra*

Department of Organismic and Evolutionary Biology, Harvard University, Cambridge, Massachusetts

A key focus of evolutionary ecology is on identifying traits (phenotypic characteristics) that are ecologically important and determining how those traits affect the relative fitness of individuals in different environments. Color is one phenotypic characteristic that has been shown to have a major influence on the way in which organisms interact with their environment. Color plays a central role in a wide variety of ecological processes relating to survival and reproduction and can therefore directly affect an individual's fitness. One of the more widely studied adaptive roles of color is cryptic coloration, that is, coloring that allows an animal to blend into the surrounding environment and therefore avoid detection by potential predators (see Chapter 14, Section 14.10). Geographic variation in habitat, such as the background color of surface substrate (soils, rocks, or snow cover), vegetation, or water, can present different adaptive environments and selective pressures resulting in localized differences in patterns of body color. In mammals, some of the most extreme variations in coat color occur in deer mice (genus *Peromyscus*). These mice occur throughout most of continental North America. One species of deer mouse that exhibits a large degree of variation in coat color over short geographic distances is *Peromyscus polionotus*. This species of *Peromyscus* occurs throughout the southeastern United States where it is commonly referred to as "oldfield mouse" because it inhabits abandoned agricultural fields. Understanding the evolution of variations in coat color within this species has been a focus of studies by the evolutionary ecologists Hopi Hoekstra and her students at Harvard University.

In the southeastern United States, oldfield mice (*P. polionotus*) typically occupy overgrown fields with dark soil and have a dark brown coat, which serves to camouflage the mice from predators (**Figure 1, left**). In the last few thousand years, however, these mice have colonized the sand dunes of Florida's coasts. Here the sands are lighter in color (white sands) than the inland soil, and there is much less vegetation cover. These beach-dwelling mice, known as "beach mice," have reduced pigmentation and have a much lighter color (compared to the inland populations) that blends well into the light-colored sand (**Figure 1, right**). Using a combination of field studies, classic genetics, and modern molecular biology, Hoekstra and her students are working to understand how, through changes in pigmentation genes, these mice have adapted to this new environment.

Although it may seem straightforward that being light colored would be a selective advantage over darker pigmentation on the white sand dunes of coastal Florida, evidence is needed to establish that differences in pigmentation among local populations represent adaptations. To accomplish this task, Hoekstra and colleague Sacha Vignieri undertook a series of field experiments to quantify the selective advantage of having pigmentation patterns that better match the surrounding environment. The experimental approach involved placing darker-colored mice from an inland, oldfield population (Lafayette Creek Wildlife Area, Florida) into the coastal dune environment (Topsail Hill State Park, Florida) and lighter-colored individuals from the dune environment into the inland environment. Rates of survival for the transplanted individuals could then be compared with those of local populations whose coat color better matched the surrounding environment. Using live mice for such an experiment, however, can present a number of major problems, including capturing hundreds of mice, transporting and releasing them into new locations, and then the difficulty of determining the fate of the mice over the experimental period. To avoid these problems, the researchers used a unique approach of creating model mice using nonhardening plasticine (a type of modeling clay). Although simple, this method has several advantages over using live mice. First, because plasticine preserves evidence of predation attempts (tooth, beak, or claw imprints), it is possible to quantify both predation rate and predator type. Additionally, using models, they were able to deploy a large number of individuals within a given environment. Finally, this experimental approach allowed them to focus on variation in a single trait of interest—coat color—controlling for other traits such as behavioral differences.

Figure 1 Habitats and associated ecotypes of *Peromyscus polionotus* in Florida. The ecotypes are found in two distinct habitats. The mainland ecotype (oldfield mice) inhabits old fields, which are vegetated and have dark loamy soil. These mice have a typical dark brown coat. In contrast, the beach ecotype (beach mice) occupies the coastal dunes, which have little vegetation and white sand. The beach mice largely lack pigmentation on their face, flanks, and tail.

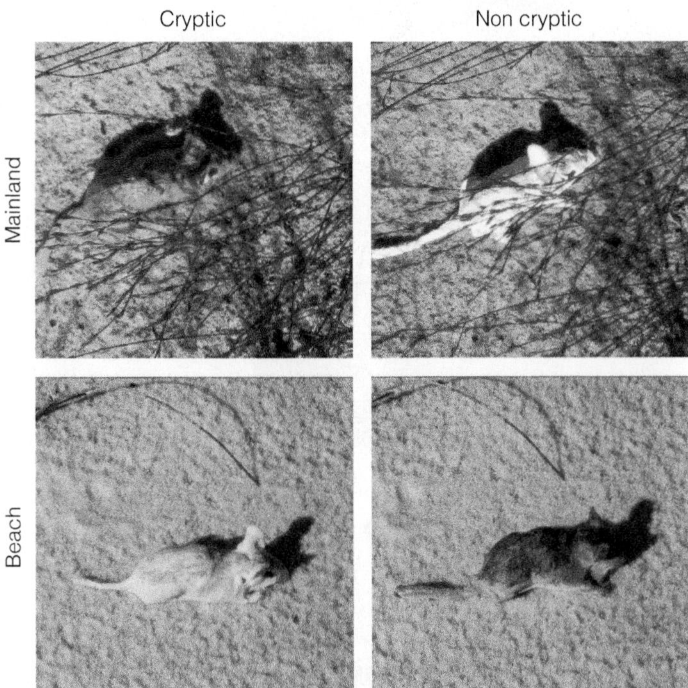

Figure 2 Examples of clay models of mice, painted to match the mainland (oldfield) and beach ecotypes, placed on the ground in both the mainland and beach habitats.

Some 250 models of *P. polionotus* were made, half of which were painted to mimic the coat color and pattern of the darker oldfield mouse and half the light-colored beach mouse (**Figure 2**). Each afternoon, the researchers set out the light and dark models in a straight line and in random order about 10 m apart in a habitat known to be occupied by either beach or mainland mice (and hence their natural predators). To determine the difference in color (brightness) between the model and the substrate on which it was placed, soil samples from around each deployed model were collected and measured for brightness (light reflection). The researchers would then return the following day and record which models showed evidence of predation. The shape of the imprint of the model left by the predator (beak or tooth marks) and the surrounding tracks gave clues as to the type of predator. By documenting predation events in both habitats, the researchers could determine how differences in color between the model and the background environment (soil) influenced rates of predation.

Results of the experiment revealed that models that were both lighter and darker than their local environment experienced a lower rate of survival (greater rate of predation) than models that were better matched in color to the soil on which they were placed (see Figure 2). Seventy five percent of all predation events occurred on mice that did not match their substrate, representing a large selective disadvantage. In the light-substrate beach environment, most attacked mice were dark, but some light models also were attacked. By analyzing the soil samples that were collected at the location where each model was placed, the researchers found that these light-colored models were all much lighter than their local substrate. In other words, selection acts against mice that are either too dark or too light relative to their background. This result demonstrates that in addition to predation acting as an agent of selection resulting in significant differences in pigmentation between inland and beach populations, there is also selection for subtle color phenotypes within a habitat.

In addition to establishing the role of natural selection in the evolution of phenotypic variation in color among local populations of *P. polionotus*, Hoekstra and her colleagues have also identified the genetic basis for these observed differences. Their work has revealed several interesting patterns. First, they have found that most of the differences in mouse fur color are caused by changes in just a handful of genes; this means that adaptation can sometimes occur via a few large mutational steps. For example, they identified a single DNA base-pair mutation in a pigment receptor, the presence or absence of which accounts for about 30 percent of the color differences between dark mainland mice and light-colored beach mice on the Florida coast. To date, this is one of the few examples of how a single DNA change can have a profound effect on the survival of individuals in nature.

Second, they have shown that the same adaptive characteristics can evolve by several different genetic pathways. Beach mice are not just restricted to Florida's Gulf Coast but are also found some 200 miles away on the Atlantic coast. They have shown that mice on the eastern coastal dunes have also evolved to possess light-colored fur but through different mechanisms. The pigment receptor mutation causing light color in the Gulf Coast mice is absent in the East Coast beach mice. Thus, similar evolutionary changes can sometimes follow different paths.

Bibliography

Hoekstra, H. E. 2010. From mice to molecules: "The genetic basis of color adaptation." In *In the light of evolution: Essays from the laboratory and field.* (Ed. J. B. Losos). Greenwood Village, CO: Roberts and Co. Publishers.

Hoekstra, H. E., R. J. Hirschmann, R. A. Bundey, P. A. Insel, and J. P. Crossland. 2006. A single amino acid mutation contributes to adaptive beach mouse color pattern. *Science* 313:101–104.

Mullen, L. M., S. N. Vignieri, J. A. Gore, and H. E. Hoekstra. 2009. Adaptive basis of geographic variation: Genetic, phenotypic and environmental differentiation among beach mouse populations. *Proceedings of the Royal Society B* 276:3809–3818.

Vignieri, S. N., J. Larson, and H. E. Hoekstra. 2010. The selective advantage of cryptic coloration in mice. *Evolution* 64:2153–2158.

1. Natural selection acts on phenotypic variation within a species or population. In the case of *Peromyscus polionotus*, what is the cause of this variation?
2. What is the "selective agent" in this example of natural selection?

among subpopulations. The degree of isolation depends on the efficiency of the extrinsic barrier, but rarely is the isolation complete. These geographic isolates are often classified as **subspecies**, a taxonomic term for populations of a species that are distinguishable by one or more characteristics. Unlike clines, for subspecies we can draw a geographic line separating the subpopulations into subspecies. Nevertheless, it is often difficult to draw the line between species and subspecies.

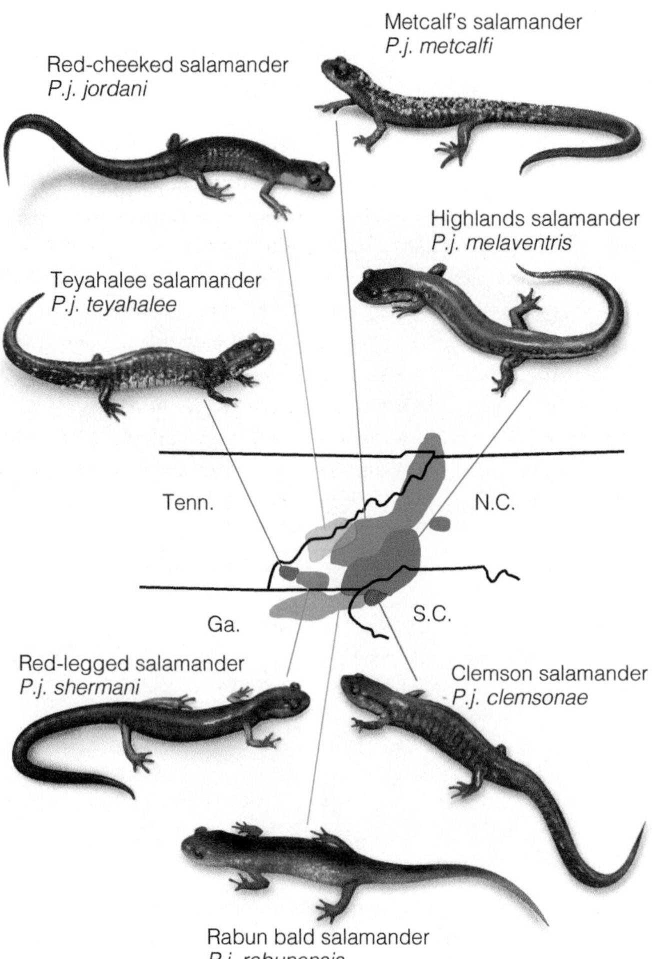

Figure 5.19 Geographical isolates in *Plethodon jordani* of the Appalachian highlands. These salamanders originated as a result of varying degrees of geographic isolation during the Pleistocene. Recent genetic studies have resulted in the members of the *Plethodon jordani* complex being classified as separate species, although reproductive isolation is not complete for any pair of these species.

5.9 Adaptations Reflect Trade-offs and Constraints

If Earth were one large homogeneous environment, perhaps a single phenotype, a single set of characteristics might bestow upon all living organisms the ability to survive, grow, and reproduce. But this is not the case. Environmental conditions that directly influence life vary in both space and time (Part One, The Physical Environment). Patterns of temperature, precipitation, and seasonality vary across Earth's surface, producing a diversity of unique terrestrial environments (Chapter 2). Likewise, variations in depth, salinity, pH, and dissolved oxygen define an array of freshwater and marine habitats (Chapter 3). Each combination of environmental conditions presents a unique set of constraints on the organisms that inhabit them—constraints on their ability to maintain basic metabolic processes that are essential to survival and reproduction. Therefore, as features of the environment change, so will the set of traits (phenotypic characteristics) that increase the ability of individuals to survive and reproduce. Natural selection will favor different phenotypes under different environmental conditions. This principle was clearly illustrated by the example of Darwin's medium ground finch in Section 5.6, in which a change in the resource base (abundance, size, and hardness of seeds) over time resulted in a shift in the distribution of phenotypes within the population, as well as the example of the two distinct phenotypes of the threespine stickleback adapted to the limnetic and benthic zones of lake ecosystems in the lakes of the Pacific Northwest of North America (also see this chapter, Field Studies: Hopi Hoekstra). Simply stated, the fitness of any phenotype is a function of the prevailing environmental conditions; the characteristics that maximize the fitness of an individual under one set of environmental conditions generally limit its fitness under a different set of conditions. The limitations on the fitness of a phenotype under different environmental conditions are a function of trade-offs imposed by constraints that can ultimately be traced to the laws of physics and chemistry.

This general but important concept of adaptive trade-offs is illustrated in the example of natural selection for beak size in the population of Darwin's medium ground finch (*G. fortis*) presented in Section 5.6. Recall from Figure 5.10 that the ability to use different seed resources (size and hardness) is related to beak size. Individuals with small beaks feed on the smallest and softest seeds, and individuals with larger beaks feed on the largest and hardest seeds. These differences in diet as a function of beak size reflect a trade-off in morphological characteristics (the depth and width of the beak) that allow for the effective exploitation of different seed resources. This pattern of trade-offs is even more apparent if we compare differences in beak morphology and the use of seed resources for the three most common species of Darwin's ground finch that inhabit Santa Cruz island in the Galápagos.

The distributions of beak size (phenotypes) for individuals of the three most common species of Darwin's ground finches are shown in **Figure 5.20a**. As their common names suggest, the mean value of beak size increases from the small (*Geospiza fuliginosa*) to the medium (*G. fortis*) and large (*Geospiza magnirostris*) ground finch. In turn, the proportions of various seed sizes in their diets (**Figure 5.20b**) reflect these differences in beak size, with the average size and hardness of seeds in the diets of these three populations increasing as a function of beak size. Small beak size restricts the ability of the smaller finch species (*G. fuliginosa*) to feed on larger, harder seed resources. In contrast, large beak size allows individuals of the largest species, *G. magnirostris*, to feed on a range of seed resources from small, softer seeds to larger, harder seeds. However, because they are less efficient at exploiting the smaller seed resources, these larger-beaked individuals restrict their diet to the larger, harder seeds. The profitability—defined as the quantity of food energy gained per unit of time spent handling these small seeds (see Section 14.7)—is extremely low for the larger birds and makes feeding on smaller seeds extremely inefficient for these individuals. This inefficiency is directly related to the greater metabolic (food energy) demands of the larger birds, which illustrates a second important concept regarding the role of constraints and trade-offs in the process of natural selection: individual phenotypic characteristics (such as beak size) often are components of a larger adaptive complex involving multiple traits and loci. The phenotypic trait of beak size in Darwin's ground finches is but one in a complex of interrelated morphological characteristics that determine the foraging behavior and diet of these birds. Larger beak size is accompanied by increased body size (length and weight) as well as specific changes in components of skull architecture and head musculature (**Figure 5.21**), all of which are directly related to feeding functions.

In summary, the beak size of a bird sets the potential range of seed types in the diet. This relationship between morphology and diet represents a basic trade-off that constrains the evolution of adaptations in Darwin's finches relating to their acquisition of essential food resources. In addition, the example of Darwin's ground finches illustrates how natural selection operates on genetic variation at the three levels we initially defined in Section 5.5: within a population, among subpopulations of the same species, and among different species. Natural selection operated in the local population of medium ground finches on the island of Daphne Major during the period of drought in the mid-1970s, increasing the mean beak size of birds in this population in response to the shift in the abundance and quality of seed resources. In addition, natural selection has resulted in differences in mean beak size between populations of the medium ground finch inhabiting the islands of Daphne Major and Santa Cruz (see Chapter 13, Figure 13.17). The larger mean beak size for the population on Santa Cruz is believed to be a result of competition from the population of small ground finch present on the island (*G. fuliginosa* does not occupy Daphne Major). The presence of the smaller species on the island has

Figure 5.20 (a) Frequency distribution of beak depths (upper mandible) of adult males in populations of the small (*Geospiza fuliginosa*), medium (*Geospiza fortis*), and large (*Geospiza magnirostris*) ground finch. Mean values are indicated by the solid triangles. (b) The differences in beak size among the three species result in differences in size of seeds in their diet, expressed as the proportion of various seed sizes in the diet.
(Adapted from Grant 1999 after Schluter 1982.)

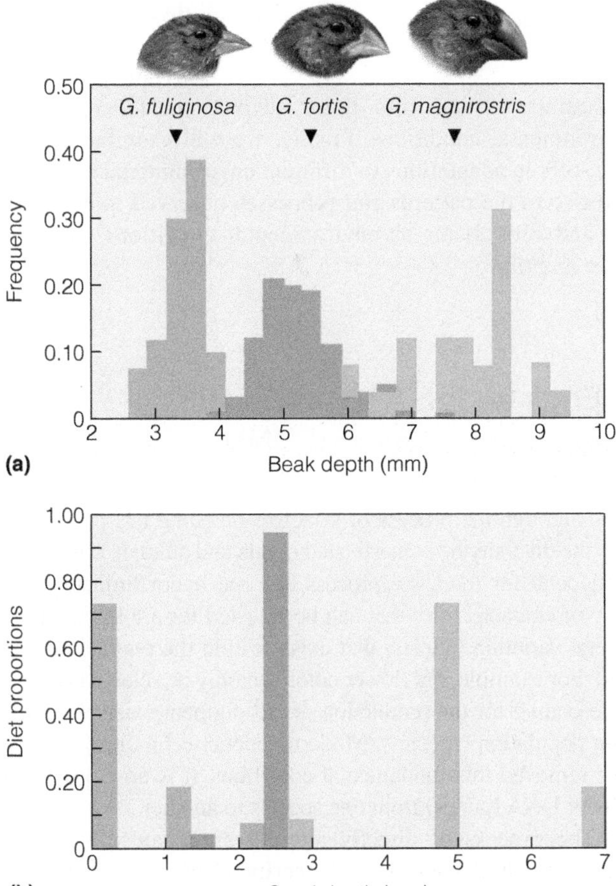

Figure 5.21 Beak structure and superficial jaw muscles of (a) large, (b) medium, and (c) small ground finches. Muscles identified by abbreviation are M. adductor mandibulae externus (superficial and medialis), M. depressor mandibulae, and M. pterygoideus ventralis lateralis. An individual's ability to exploit different seed resources (in size, hardness, etc.) is influenced by a complex of morphological and behavioral characteristics including those illustrated.
(Adapted from Grant 1999 after Bowman 1961.)

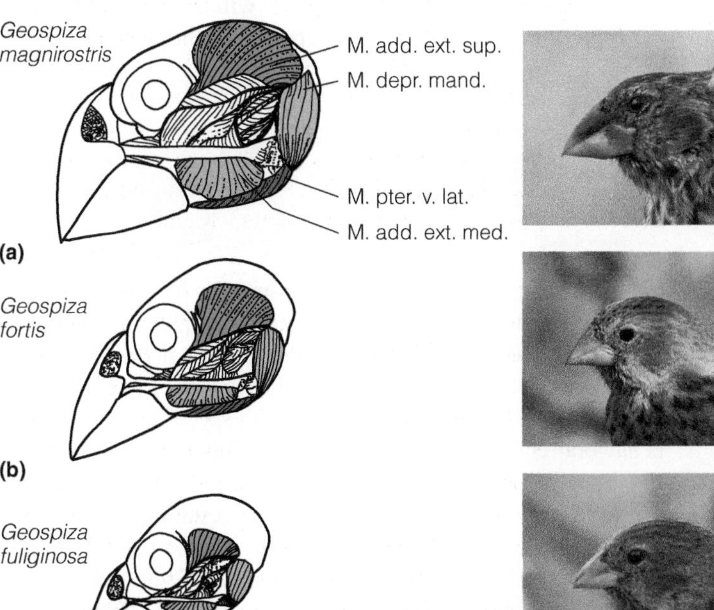

the effect of reducing the availability of smaller, softer seeds (see Figure 5.20) and increasing the relative fitness of *G. fortis* individuals with larger beaks that can feed on the larger, harder seeds (see Figure 5.10).

Population genetic studies have also shown that natural selection is the evolutionary force that has resulted in the genetic differentiation of various species of Darwin's finches inhabiting the Galápagos Islands (**Figure 5.22**). The process in which one species gives rise to multiple species that exploit different features of the environment, such as food resources or habitats, is called **adaptive radiation**. The different features of the environment exert the selection pressures that push the populations in various directions, and reproductive isolation, the necessary condition for speciation to occur, is often a by-product of the changes in morphology, behavior, or habitat preferences that are the actual targets of selection. In this way, the differences in beak size and diet among the three species of ground finch are magnified versions of the differences observed within a population, or among populations inhabiting different islands.

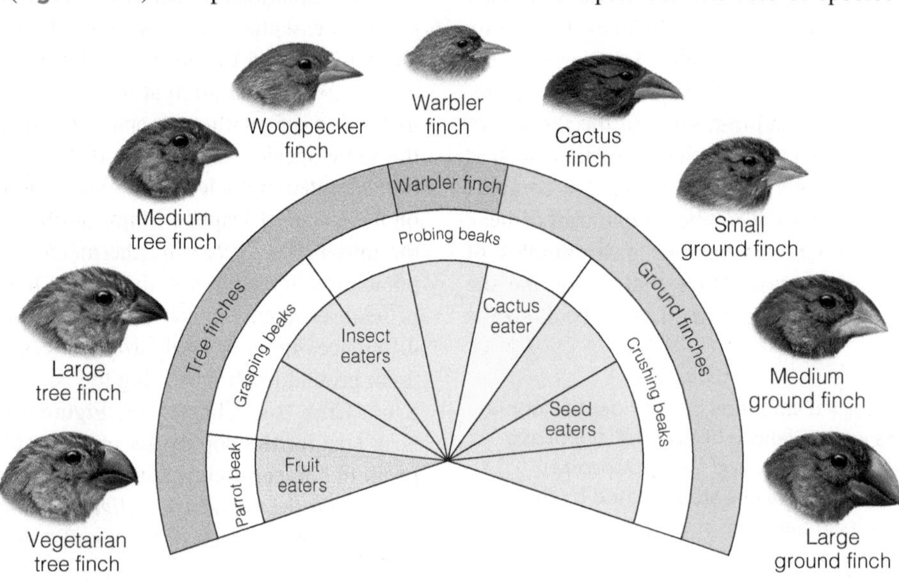

Figure 5.22 The concept of adaptive radiation is illustrated by the diversity of Darwin's ground finches inhabiting the Galápagos Islands. Genetic studies show that the various species of finches arise from a single ancestral species, but they have evolved a rich diversity of beak sizes and shapes. (From Patel 2006.)

In the chapters that follow, we will examine this basic principle of trade-offs as it applies to the adaptation of species and explore how the nature of adaptations changes with changing environmental conditions. We will explore various adaptations of plant and animal species, respectively, to key features of the physical environment that directly influence the basic processes of survival and assimilation in Chapters 6 and 7, and trade-offs involved in the evolution of life history characteristics (adaptations) relating to reproduction in Chapter 10. The role of species' interactions as a selective agent in the process of natural selection will be examined later in Part Four (Species Interactions).

Throughout our discussion, adaptation by natural selection is a unifying concept, a mechanism for understanding the distribution and abundance of species. We will explore the selective forces giving rise to the adaptations that define the diversity of species as well as the advantages and constraints arising from those adaptations under different environmental conditions. Finally, we will examine how the trade-offs in adaptations to different environmental conditions give rise to the patterns and processes observed in communities and ecosystems as environmental conditions change in space and time.

ECOLOGICAL Issues & Applications

Genetic Engineering Allows Humans to Manipulate a Species' DNA

For a millennia, humans have been using the process of selective breeding to modify the characteristics of plant and animal species. By selecting individuals that exhibit a desired trait, and mating them with individuals exhibiting the same trait (or traits), breeders produce populations with specific physical and behavioral characteristics (phenotypes). This process of **selective breeding** is analogous to natural selection—the differential fitness of individuals within the population resulting from differences in some heritable characteristic(s). Unlike natural selection, however, humans function as the agent of selection rather than the environment. Darwin referred to selective breeding as "artificial selection," and his understanding of this process was instrumental in his development of the idea of natural selection.

Although the process of selective breeding has provided us with the diversity of domesticated plants and animals upon which we depend for food, the process has one major limitation. The array of characteristics that can be selected for are limited to the genetic variation (alleles) that exists within the population (species). For example, red flower color can only be selected for if the allele coding for the production of red pigment exists within the plant population (species). Modern genetic techniques, however, have removed this fundamental constraint. It is now possible to transfer DNA (genes) from one species to another.

The process of directly altering an organism's genome is referred to as **genetic engineering**. The primary technology used in genetic engineering is genetic recombination; the development of recombinant DNA or rDNA by combining the genetic material from one organism into the genome of another organism (generally of a different species). The resulting modified gene is called a *transgene*, and the recipient of the recombinant DNA is called a *transgenic organism*.

The process of genetic engineering using rDNA requires the successful completion of a series of steps (**Figure 5.23**). DNA extraction is the first step in the process. To work with DNA, scientists must extract it from the donor organism (the organism that has the desired trait). During the process of DNA extraction, the complete sequence of DNA from the donor organism is extracted at once. The next task is to separate the single gene of interest from the rest of the DNA using specific enzymes that "cut" the desired segment of DNA from the larger strand. Copies of the gene can then produced using cloning techniques. Once the gene has been cloned, genetic engineers begin the third step, designing the gene to work once it is inside a recipient organism. This is done by using other enzymes that are capable of adding new segments called *promoters* (which start a sequence) and *terminators* (which stop a sequence). A promoter is a region of DNA that initiates the transcription of a particular gene (the first step of gene expression, in which a particular segment of DNA is copied into RNA). The terminator is a section of genetic sequence that marks the end of gene. The next step is to insert the new gene into the cell of the recipient organism. The process in which changes in a cell or organism are brought about through the introduction of new DNA is called *transformation*.

Transformation is accomplished through a variety of techniques, but two main approaches are used for plant species: the "gene gun" method and the Agrobacterium method. The gene gun method fires gold particles carrying the foreign DNA into plant cells. Some of these particles pass through the plant cell wall and enter the cell nucleus, where the transgene integrates itself into the plant chromosome. The Agrobacterium method involves the use of soil-dwelling bacteria known as *Agrobacterium tumefaciens* that cause crown gall disease in many plant species. This bacterium has a plasmid, or loop of nonchromosomal DNA, that contains tumor-inducing genes (T-DNA), along with additional genes that help the T-DNA integrate into the host genome. When the bacteria infect the plant, the plasmid is integrated into the plant's chromosomes, becoming part of the plant's genome. For genetic engineering purposes, the tumor-inducing part of the plasmid is removed so that it will not harm the plant. The desired gene from the donor organisms is then inserted into the bacteria's plasmid. The bacteria can now be used as a delivery system that will transfer the transgene into the plant.

An organism whose genetic material has been altered using genetic engineering techniques (including transgenic organisms) are commonly referred to as *genetically modified organisms* (GMOs). Genetic engineering has been used to produce a wide variety of GMOs. Organisms that have been genetically modified include microorganisms such as bacteria and yeast, insects, plants, fish, and mammals. GMOs are used in biological and medical research, production of pharmaceutical drugs, experimental medicine (e.g., gene therapy), but perhaps their most widespread application has been in agriculture (**Figure 5.24**). In agriculture, genetically engineered crops have been created that possess desirable traits such as resistance to pests or herbicides, increased nutritional value, or production of pharmaceuticals.

In addition to the ethical and health concerns that genetically modified crop species have raised, the practice of genetic engineering and the production of transgenic species (GMOs) has raised considerable concern among ecologists. There is little concern about gene transfer between major agricultural species such as corn, soybean, and rice and native plant populations as a result of the lack of close relatives capable of cross-pollination. However, other crop species, such as members of the genus *Brassica* (member of the mustard family) are represented by a variety of domesticated and wild (native) species and subspecies that are capable of cross-pollination. For example, *Brassica napus* (rapeseed) used in the production of rapeseed oil has been genetically modified to tolerate herbicides (used to kill weeds in agricultural fields), and the transfer of herbicide-tolerant traits by pollen to weedy relatives (other members of the genus *Bassica*) has been recorded. *Brassica* includes a number of important crop species such as turnips, cabbage, and broccoli.

A major application of genetic engineering in agriculture is the development of insect-resistant crop strains. Perhaps the most widely grown genetically modified crop plant is *Bt* corn. *Bacillus thuringiensis*, or *Bt*, is a common soil bacterium

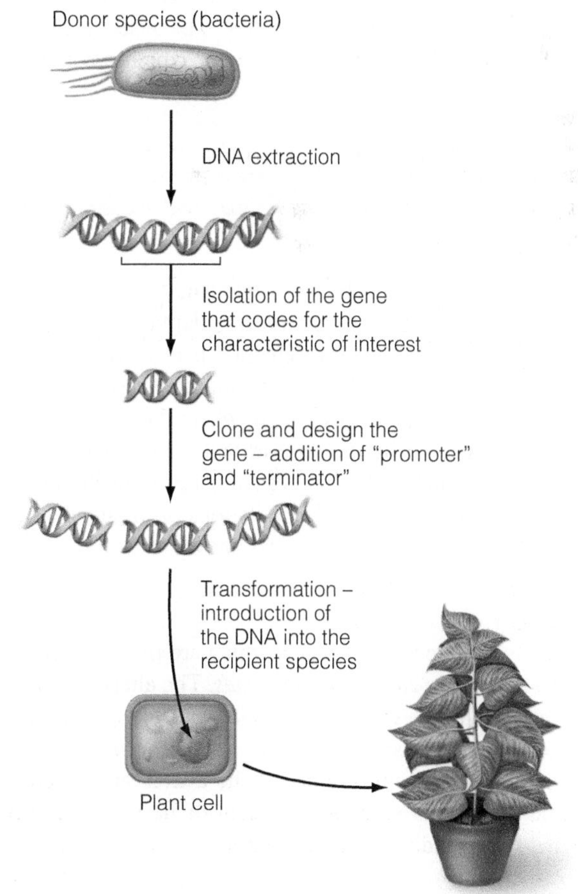

Figure 5.23 Steps involved in the process of genetic engineering and the production of a transgenic organism.

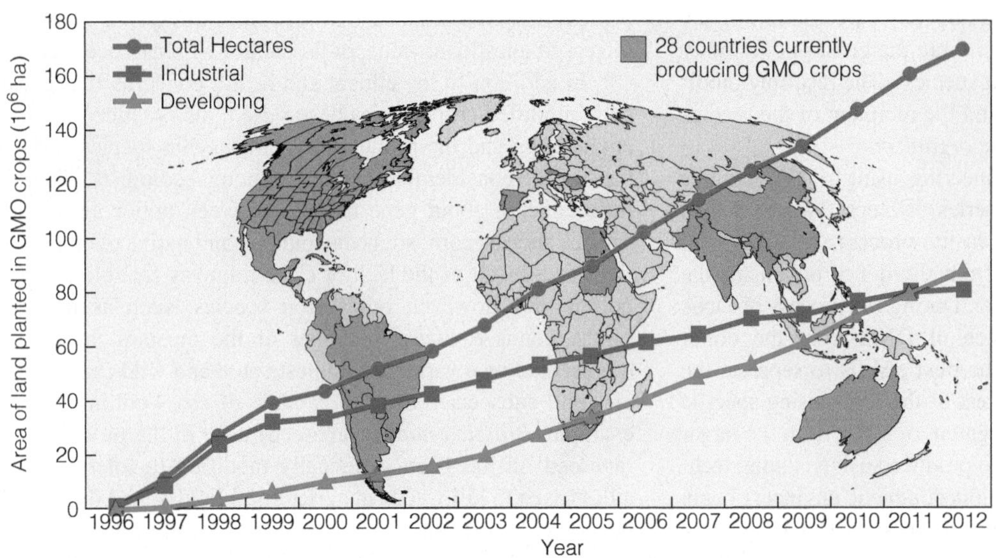

Figure 5.24 Changes in global area planted in genetically engineered crops for the period 1996–2012. Total area partitioning into industrial and developing countries. (Adapted from James 2012.)

whose genome contains genes for several proteins toxic to insects. For decades, *Bt* has been sprayed on fields as an organic pesticide. Starting in the mid-1990s, several varieties of corn were genetically engineered to incorporate *Bt* genes' encoding proteins, which are toxic to various insect pests. Some strains of *Bt* produce proteins that are selectively toxic to caterpillars, such as the southwestern corn borer, whereas others target mosquitoes, root worms, or beetles. The insecticide proteins are contained within the plant tissues, which are fatal to the pest species when ingested. Concerns have been raised over the potential impacts of *Bt* corn and other insect-resistant genetically modified crop species on nontarget insect species or to predators that feed on these insects.

The use of transgenes to confer disease resistance to crops represents another possible ecological risk. If genes that code for viral resistance are transferred to crops, there is a potential for transfer to wild plants, creating the potential for the natural development of new plant viruses of increased severity.

SUMMARY

Adaptation 5.1

Characteristics that enable an organism to thrive in a given environment are called *adaptations*. Adaptations are a product of natural selection. Natural selection is the differential fitness of individuals within the population that results from their interaction with their environment, where the fitness of an individual is measured by the proportionate contribution it makes to future generations. The process of natural selection results in changes in the properties of populations of organisms over the course of generations by a process known as *evolution*.

Genes 5.2

The units of heredity are genes, which are linearly arranged on threadlike bodies called *chromosomes*. The alternative forms of a gene are alleles. The pair of alleles present at a given locus defines the genotype. If both alleles at the locus are the same, the individual is homozygous. If the alleles are different, the individual is heterozygous. The sum of heritable information carried by the individual is the genome.

Phenotype 5.3

The phenotype is the physical expression of the genotype. The manner in which the genotype affects the phenotype is termed the *mode of gene action*. When heterozygous individuals exhibit the same phenotype as one of the homozygotes, the allele that is expressed is termed *dominant* and the masked allele is termed *recessive*. If the physical expression of the heterozygote is intermediate between the homozygotes, the alleles are said to be codominant.

Even though all genetic variation is discrete, most phenotypic traits have a continuous distribution because (1) most traits are affected by more than one locus, and (2) most traits are affected by the environment.

Phenotypic Plasticity 5.4

The ability of a genotype to give rise to a range of phenotypic expressions under different environmental conditions is termed *phenotypic plasticity*. The range of phenotypes expressed under different environmental conditions is termed the *norm of reaction*. If the phenotypic plasticity occurs during the growth and development of the individual and represents an irreversible characteristic, it is referred to as *developmental plasticity*. Reversible phenotypic changes in an individual organism in response to changing environmental conditions are referred to as *acclimation*.

Genetic Variation 5.5

Genetic variation occurs at three levels: within subpopulations, among subpopulations of the same species, and among different species. The sum of genetic information across all individuals in the population is the gene pool. The fundamental measures of genetic variation within a population are allele frequency and genotype frequency.

Natural Selection 5.6

Natural selection acts on the phenotype, but in doing so it alters both genotype and allele frequencies within the population. There are three general types of natural selection: directional selection, stabilizing selection, and disruptive selection. The target of selection is the phenotypic trait that natural selection acts upon, whereas the selective agent is the environmental cause of fitness differences among individuals in the population.

Processes Influencing Genetic Variation 5.7

Natural selection is the only evolutionary process that can result in adaptations; however, some processes can function to alter patterns of genetic variation from generation to generation. These include mutation, migration, genetic drift, and nonrandom mating. Mutations are heritable changes in a gene or chromosome. Migration is the movement of individuals between local populations. This movement results in the transfer of genes between local populations. Genetic drift is a change in allele frequency as a result of random chance.

Nonrandom mating on the basis of phenotypic traits is referred to as *assortative mating*. Assortative mating can be either positive (mates are more similar than expected by chance) or negative (dissimilar). A special case of nonrandom mating is inbreeding—the mating of individuals that are more closely related than expected by chance.

Genetic Differentiation 5.8

Natural selection can function to alter genetic variation between populations; this result is referred to as *genetic differentiation*. Species having a wide geographic distribution often encounter a broader range of environmental conditions than do species whose distribution is more restricted. The variation in environmental conditions often gives rise to a corresponding variation in many morphological, physiological, and behavioral characteristics as a result of different selective agents in the process of natural selection.

Trade-offs and Constraints 5.9

The environmental conditions that directly influence life vary in both space and time. Likewise, the objective of selection changes with environmental circumstances in both space and time. The characteristics enabling a species to survive, grow, and reproduce under one set of conditions limit its ability to do equally well under different environmental conditions.

Genetic Engineering Ecological Issues & Applications

Genetic engineering is the process of directly altering an organism's genome. The primary technology used in genetic engineering is genetic recombination—the combining of genetic material from one organism into the genome of another organism, generally of a different species. The result of this process is recombinant DNA (rDNA). The resulting modified gene is called a *transgene*, and the recipient of the rDNA is called a *transgenic organism*.

STUDY QUESTIONS

1. What is natural selection? What conditions are necessary for natural selection to occur?
2. Distinguish between the terms *gene* and *allele.*
3. What is the relationship between an individual's genotype and phenotype?
4. If the phenotype trait of an *Aa* heterozygous individual is the same as that of an *AA* homozygous individual, which allele is recessive?
5. Why are small populations more prone to variations in allele frequency from generation to generation as a result of genetic drift than are large populations?
6. How might genetic drift and inbreeding be important processes in the conservation of endangered species?
7. Why is natural selection the only process that can result in adaptation?
8. What is phenotypic plasticity?
9. Contrast developmental plasticity and acclimation.
10. Researchers studying populations of the seagrass *Thalassia testudinum* in the coastal waters of the Gulf of Mexico observed that the average leaf width in populations occupying the deeper waters offshore was significantly greater than for populations in the shallower nearshore waters. How might the researchers experimentally test whether these observed differences are a result of genetic differences between the populations (genetic differentiation) or a result of phenotypic plasticity? The nearshore waters are warmer and have higher values of PAR (light availability) than the deeper waters offshore.
11. David Reznick, an ecologist at the University of California at Riverside, studied the process of natural selection in populations of guppies (small freshwater fish) on the island of Trinidad. Reznick found that populations at lower elevations face the assault of predatory fish, whereas the populations at higher elevations live in peace because few predators can move upstream past the waterfalls. The average size of individuals in the higher-elevation waters is larger than the average size of guppies in the lower-elevation populations. Reznick hypothesized that the smaller size of individuals in the lower-elevation populations was a result of increased rates of predation on larger individuals; in effect, predation was selecting for smaller individuals in the population. To test this hypothesis, Reznick moved individuals from the lower elevations to unoccupied pools upstream, where predation was not a

factor. Eleven years in these conditions produced a population of individuals that were on average larger than the individuals of the downstream populations. Is the study by Reznick an example of natural selection (does it meet the necessary conditions)? If so, what type of selection does it represent (directional, stabilizing, or disruptive)? Can you think of any alternative hypotheses to explain why the average size of individuals may have shifted through time as a result of moving the population to the upstream (higher-elevation) environment?

12. Does the example of variation in body size of guppies from upstream and downstream populations presented in Question 11 represent a trade-off similar to that of variation in beak size of medium ground finches presented in Section 5.6? What is the selective agent in the example in Question 11?
13. What is the fundamental difference between selective breeding (artificial selection) and genetic engineering?
14. What might be some of the environmental concerns regarding genetically modified (transgenic) organisms?

FURTHER READINGS

Classic Studies

Darwin, C. 1859. *On the origin of species by means of natural selection, or the preservation of favoured races in the struggle for life*. London: John Murray.
Darwin's original text should be read by every student of ecology.

Ford, E. B. 1971. *Ecological genetics*. 3rd ed. London: Chapman and Hall.
A pioneering work that provides a good foundation in the study of ecological genetics as it relates to natural selection and adaptation.

Current Research

Conner, J. K., and D. L. Hartl. 2004. *A primer of ecological genetics*. Sunderland, MA: Sinauer Associates.
An excellent introduction to population and quantitative genetics for the ecologist.

Desmond, A., and J. Moore. 1991. *Darwin: The life of a tormented evolutionist*. New York: Norton.
Written by two historians, this book is an anthology providing an introduction to the man and his works.

Futuyma, D. J. 2013. *Evolution*. 3rd ed. Sunderland, MA: Sinauer Associates.
A comprehensive reference book for topics covered in the chapter as well as models of speciation.

Gotthard, K., and Nylin, S. 1995. "Adaptive plasticity and plasticity as an adaptation: A selective review of plasticity in animal morphology and life history." *Oikos* 74:3–17.
A review paper that examined phenotypic plasticity in animal species.

Gould, S. J. 1992. *Ever since Darwin: Reflections in natural history*. New York: Norton.
A collection of Gould's essays written for scientific journals. The first in a series of humorous and fun reading. See other collections, including *The Panda's Thumb*, *The Flamingo's Smile*, and *Dinosaur in a Haystack*.

Gould, S. J. 2002. *The structure of evolutionary theory*. Cambridge, MA: Belknap Press of Harvard University Press.
The last book written by Gould, the best, most respected popular science writer in the field of evolution. Although the book is more technical than others, Chapter 1 gives an excellent overview of evolutionary theory.

Grant, P. R., and B. R. Grant. 2000. "Non-random fitness variation in two populations of Darwin's finches." *Proceedings of the Royal Society of London Series B* 267:131–138.
An excellent source of additional information for those intrigued by the example of natural selection in Darwin's ground finch presented in the chapter.

Mayr, E. 2001. *What evolution is*. New York: Basic Books.
An excellent primer on the topics of natural selection and evolution by a leading figure in evolution. Wonderfully written and accessible to the general reader.

Reznick, D. N., F. H. Shaw, F. H. Rodd, and R. G. Shaw. 1997. "Evaluation of the rate of evolution in natural populations of guppies (*Poecilia reticulata*.") *Science* 275:1934–1937.
A beautifully designed experiment for evaluating the role of natural selection in the evolution of life history characteristics.

Warwick, S. I., H. J. Beckie, and L. M. Hall. 2009. "Gene flow, invasiveness and ecological impact of genetically modified crops." *Annal of the New York Academy of Sciences* 1168:72–79.
Provides an excellent discussion of the potential ecological impacts of genetically modified organisms.

Weiner, J. 1994. *The beak and the finch: A story of evolution in our time*. New York: Knopf.
Winner of the Pulitzer Prize. Gives the reader a firsthand view of scientific research in action.

CHAPTER 6

Plant Adaptations to the Environment

These Kokerboom trees in the desert region of Bloedkoppie, Namibia (in southwestern Africa), use the CAM photosynthetic pathway to conserve water in this harsh environment.

CHAPTER GUIDE

ALL LIFE ON EARTH is carbon based. This means all living creatures are made up of complex molecules built on a framework of carbon atoms. The carbon atom is able to bond readily with other carbon atoms, forming long, complex molecules. The carbon atoms needed to construct these molecules—the building blocks of life—are derived from various sources. The means by which organisms acquire and use carbon represent some of the most basic adaptations required for life. Humans, like all other heterotrophs, gain their carbon by consuming other organisms. However, the ultimate source of carbon from which life is constructed is carbon dioxide (CO_2) in the atmosphere.

Not all living organisms can use this abundant form of carbon directly. Only autotrophs can transform carbon in the form of CO_2 into organic molecules and living tissue. Autotrophs can be subdivided into chemoautotrophs and photoautotrophs, depending on how they derive the energy for their metabolism. Chemoautotrophs convert carbon dioxide into organic matter using the oxidation of inorganic molecules (such as hydrogen gas or hydrogen sulfide) or methane as a source of energy. Most chemotrophs are bacteria or archaea that live in hostile environments such as the hydrothermal vents of the deep sea floor (see Chapter 24). Photoautotrophs, the dominant form of autotrophs, use the Sun's energy to drive the process of converting CO_2 into simple organic compounds. That process, carried out by green plants, algae, and some types of bacteria, is photosynthesis. Photosynthesis is essential for the maintenance of life on Earth.

Although all green plants derive their carbon from photosynthesis, how organisms, from the most minute of flowering plants (members of the duckweed family—Lemnaceae) to the largest of trees, allocate the products of photosynthesis to the basic processes of growth and maintenance varies immensely. These differences represent the diverse outcomes of evolution that allow plants to acquire the essential resources of carbon, light, water, and mineral nutrients necessary to support the process of photosynthesis. In this chapter, we will examine the variety of adaptations that have evolved in plants that allow them to survive, grow, and reproduce across virtually the entire range of environmental conditions found on Earth.

First, let us review the process so essential to life on Earth—or as the author John Updike so poetically phrased it, "the lone reaction that counterbalances the vast expenditures of respiration, that reverses decomposition and death."

6.1 Photosynthesis Is the Conversion of Carbon Dioxide into Simple Sugars

Photosynthesis is the process by which energy from the Sun, in the form of shortwave radiation, is harnessed to drive a series of chemical reactions that result in the fixation of CO_2 into carbohydrates (simple sugars) and the release of oxygen (O_2) as a by-product. The portion of the electromagnetic spectrum that photosynthetic organisms use is between 400 and 700 nanometers (nm; roughly corresponding to the visible portion of the spectrum) and is referred to as photosynthetically active radiation (PAR).

The process of photosynthesis can be expressed in the simplified form shown here:

$$6CO_2 + 12H_2O \rightarrow C_6H_{12}O_6 + 6O_2 + 6H_2O$$

The net effect of this chemical reaction is the use of six molecules of water (H_2O) and the production of six molecules of oxygen (O_2) for every six molecules of CO_2 that are transformed into one molecule of sugar $C_6H_{12}O_6$. The synthesis of various other carbon-based compounds—such as complex carbohydrates, proteins, fatty acids, and enzymes—from these initial products occurs in the leaves as well as other parts of the plant.

Photosynthesis, a complex sequence of metabolic reactions, can be separated into two processes, often referred to as the light-dependent and light-independent reactions. The light-dependent reactions begin with the initial photochemical reaction in which chlorophyll (light-absorbing pigment) molecules within the chloroplasts absorb light energy. The absorption of a photon of light raises the energy level of the chlorophyll molecule. The excited molecule is not stable, and the electrons rapidly return to their ground state, thus releasing the absorbed photon energy. This energy is transferred to another acceptor molecule, resulting in a process called *photosynthetic electron transport*. This process results in the synthesis of adenosine triphosphate (ATP) from adenosine diphosphate (ADP) and of NADPH (the reduced form of nicotinamide adenine dinucleotide phosphate [NADP]) from $NADP^+$. The high-energy substance ATP and the strong reductant NADPH produced in the light-dependent reactions are essential for the second step in photosynthesis—the light-independent reactions.

In the light-independent reactions, CO_2 is biochemically incorporated into simple sugars. The light-independent reactions derive their name from the fact that they do not directly require the presence of sunlight. They are, however, dependent on the products of the light-dependent reactions and therefore ultimately depend on the essential resource of sunlight.

The process of incorporating CO_2 into simple sugars begins in most plants when the five-carbon molecule ribulose biphosphate (RuBP) combines with CO_2 to form two molecules of a three-carbon compound called *phosphoglycerate* (3-PGA).

CO_2 +	RuBP →	2 3-PGA
1-carbon molecule	5-carbon molecule	3-carbon molecule

This reaction, called *carboxylation*, is catalyzed by the enzyme **rubisco** (ribulose biphosphate carboxylase-oxygenase). The plant quickly converts the 3-PGA formed in this process into the energy-rich sugar molecule glyceraldehyde 3-phosphate (G3P). The synthesis of G3P from 3-PGA requires both ATP and NADPH—the high-energy molecule and reductant that are formed in the light-dependent reactions. Some of this G3P is used to produce simple sugars ($C_6H_{12}O_6$), starches, and other carbohydrates required for plant growth and maintenance; the remainder is used to synthesize new RuBP to continue the process. The synthesis of new RuBP from G3P requires additional ATP.

In this way, the availability of light energy (solar radiation) can limit the light-independent reactions of photosynthesis through its control on the production of ATP and NADPH required for the synthesis of G3P and the regeneration of RuBP. This photosynthetic pathway involving the initial fixation of CO_2 into the three-carbon PGAs is called the Calvin–Benson cycle, or C_3 cycle, and plants employing it are known as **C_3 plants** (Figure 6.1).

The C_3 pathway has one major drawback. The enzyme rubisco that drives the process of carboxylation also acts as an oxygenase; rubisco can catalyze the reaction between O_2 and RuBP. The oxygenation of RuBP results in the eventual release of CO_2 and is referred to as photorespiration (not to be confused with the process of cellular respiration discussed herein).

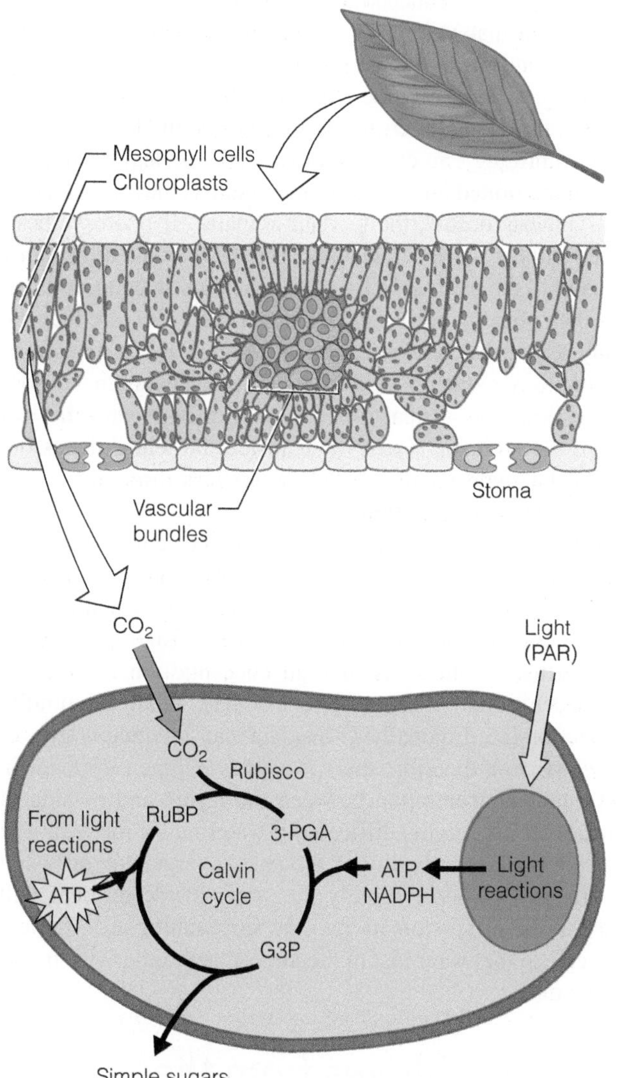

Figure 6.1 The process of photosynthesis occurs in the chloroplast within the mesophyll cells. A simple representation of the C_3 photosynthetic pathway, or Calvin cycle, is shown. Note the link between the light-dependent and light-independent reactions because the products of the light-dependent reactions (ATP and NADPH) are required to synthesize the energy-rich sugar molecule G3P and to regenerate RuBP (light-independent reactions). PAR, photosynthetically active radiation.

This competitive reaction to the carboxylation process reduces the efficiency of C_3 photosynthesis by as much as 25 percent.

Some of the carbohydrates produced in photosynthesis are used in the process of cellular respiration—the harvesting of energy from the chemical breakdown of simple sugars and other carbohydrates. The process of cellular respiration (also referred to as aerobic respiration) occurs in the mitochondria of all living cells and involves the oxidation of carbohydrates to generate energy in the form of ATP.

$$C_6H_{12}O_6 + 6O_2 \rightarrow 6CO_2 + 6H_2O + ATP$$

Because leaves both use CO_2 during photosynthesis and produce CO_2 during respiration, the difference in the rates of these two processes is the net gain of carbon, referred to as **net photosynthesis**.

$$\text{Net photosynthesis} = \text{Photosynthesis} - \text{Respiration}$$

The rates of photosynthesis and respiration, and therefore net photosynthesis, are typically measured in moles CO_2 per unit leaf area (or mass) per unit time ($\mu mol/m^2/s$).

6.2 The Light a Plant Receives Affects Its Photosynthetic Activity

Solar radiation provides the energy required to convert CO_2 into simple sugars. Thus, the availability of light (PAR) to the leaf directly influences the rate of photosynthesis (Figure 6.2). At night, in the absence of PAR, only respiration occurs and the net uptake of CO_2 is negative. The rate of CO_2 loss when the value of PAR is zero provides an estimate of the rate of

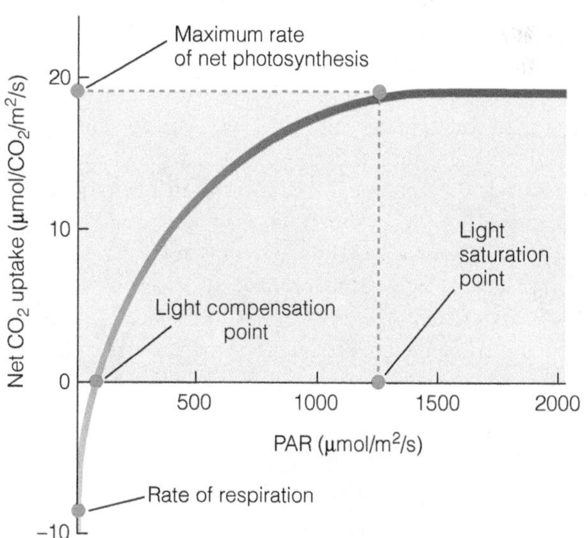

Figure 6.2 Response of net photosynthesis (y-axis) to available light (x-axis; PAR, photosynthetically active radiation). The plant's rate of photosynthesis increases as the light level increases up to a maximum rate known as the light saturation point. The light compensation point is the value of PAR at which the uptake of CO_2 for photosynthesis equals the loss of CO_2 in respiration. The value of net CO_2 exchange at PAR = 0 (dark) provides an estimate of the rate of respiration.

respiration. As the Sun rises and the value of PAR increases, the rate of photosynthesis likewise increases, eventually reaching a level at which the rate of CO_2 uptake in photosynthesis is equal to the rate of CO_2 loss in respiration. At that point, the rate of net photosynthesis is zero. The light level (value of PAR) at which this occurs is called the **light compensation point (LCP)**. As light levels exceed the LCP, the rate of net photosynthesis increases with PAR. Eventually, photosynthesis becomes light saturated. The value of PAR, above which no further increase in photosynthesis occurs, is referred to as the **light saturation point**. In some plants adapted to extremely shaded environments, photosynthetic rates decline as light levels exceed saturation. This negative effect of high light levels, called **photoinhibition**, can be the result of "overloading" the processes involved in the light-dependent reactions.

6.3 Photosynthesis Involves Exchanges between the Plant and Atmosphere

The process of photosynthesis occurs in specialized cells within the leaf called **mesophyll** cells (see Figure 6.1). For photosynthesis to take place within the mesophyll cells, CO_2 must move from the outside atmosphere into the leaf. In terrestrial (land) plants, CO_2 enters the leaf through openings on its surface called **stomata** (**Figure 6.3**) through the process of diffusion. Diffusion is the movement of a substance from areas of higher to lower concentration. CO_2 diffuses from areas of higher concentration (the air) to areas of lower concentration (the interior of the leaf). When the concentrations are equal, an equilibrium is achieved and there is no further net exchange.

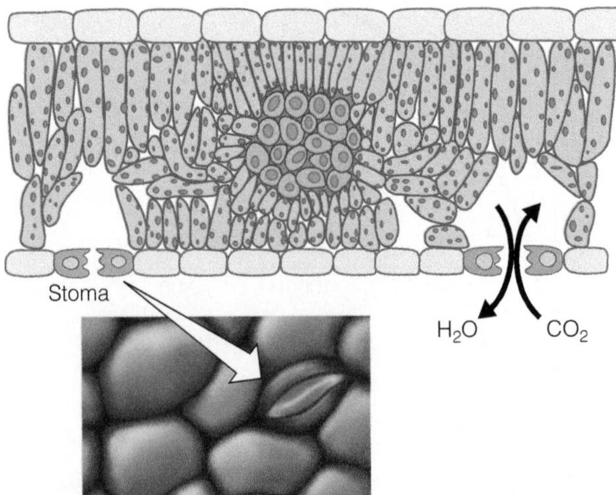

Figure 6.3 Cross section of a C_3 leaf, showing epidermal cells and stomata. In terrestrial (land) plants the stomata—openings on the surface of the leaf—allow for the exchange of CO_2 and water between the atmosphere and the leaf interior through the process of diffusion.

Two factors control the diffusion of CO_2 into the leaf: the diffusion gradient and stomatal conductance. The diffusion gradient is defined as the difference between the concentration of CO_2 in air adjacent to the leaf and the concentration of CO_2 in the leaf interior. Concentrations of CO_2 are often described in units of parts per million (ppm) of air. A CO_2 concentration of 400 ppm would be 400 molecules of CO_2 for every 1 million molecules of air. Stomatal conductance is the flow rate of CO_2 through the stomata (generally measured in units of $\mu mol/m^2/s$) and has two components: the number of stoma per unit leaf surface area (stomatal density) and aperture (the size of the stomatal openings). Stomatal aperture is under plant control, and stomata open and close in response to a variety of environmental and biochemical factors.

As long as the concentration of CO_2 in the air outside the leaf is greater than that inside the leaf and the stomata are open, CO_2 will continue to diffuse through the stomata into the leaf. So as CO_2 diffuses into the leaf through the stomata, why do the concentrations of CO_2 inside and outside the leaf not come into equilibrium? The concentration inside the leaf declines as CO_2 is transformed into sugar during photosynthesis. As long as photosynthesis occurs, the gradient remains. If photosynthesis stopped and the stomata remained open, CO_2 would diffuse into the leaf until the internal CO_2 equaled the outside concentration.

When photosynthesis and the demand for CO_2 are reduced for any reason (such as a reduction in light), the stomata tend to close, thus reducing flow into the leaf. The stomata close because they play a dual role. As CO_2 diffuses into the leaf through the stomata, water vapor inside the leaf diffuses out through the same openings. This water loss through the stomata is called **transpiration**.

As with the diffusion of CO_2 into the leaf, the rate of water diffusion out of the leaf will depend on the diffusion gradient of water vapor from inside to outside the leaf and the stomatal conductance (flow rate of water). Like CO_2, water vapor diffuses from areas of high concentration to areas of low concentration—from wet to dry. The relative humidity (see Section 2.5, Figure 2.15) inside a leaf is typically greater than 99 percent, therefore there is usually a large difference in water vapor concentration between the inside and outside of the leaf, resulting in the diffusion of water out of the leaf. The lower the relative humidity of the air, the larger the diffusion gradient and the more rapidly the water inside the leaf will diffuse through the stomata into the surrounding air. The leaf must replace the water lost to the atmosphere, otherwise it will wilt and die.

6.4 Water Moves from the Soil, through the Plant, to the Atmosphere

The force exerted outward on a cell wall by the water contained in the cell is called **turgor pressure**. The growth rate of plant cells and the efficiency of their physiological processes are highest when the cells are at maximum turgor—that is, when they are fully hydrated. When the water content of the cell declines, turgor pressure drops and water stress occurs, ranging from wilting to dehydration. For leaves to maintain maximum turgor, the water lost to the atmosphere in transpiration must be replaced by water taken up from the soil through the root system of the plant and transported to the leaves.

You may recall from basic physics that work—the displacement of matter, such as transporting water from the soil into the plant roots and to the leaves—requires the transfer of energy. The measure of energy available to do work is called Gibbs energy (*G*), named for the U.S. physicist Willard Gibbs, who first developed the concept in the 1870s. In the process of active transport, such as transporting water from the ground to an elevated storage tank using an electric pump, the input of energy to the system is in the form of electricity to the pump. The movement of water through the soil–plant–atmosphere continuum, however, is an example of passive transport, a spontaneous reaction that does not require an input of energy to the system. The movement of water is driven by internal differences in the Gibbs energy of water at any point along the continuum between the soil, plant, and atmosphere. The Second Law of Thermodynamics states that the transfer of energy (through either heat or work) always proceeds in the direction from higher to lower energy content (e.g., from hot to cold). Therefore, a gradient of decreasing energy content of the water between any two points along the continuum must exist to enable the passive movement of water between the soil, plant, and atmosphere.

The measure used to describe the Gibbs energy of water at any point along the soil–plant–atmosphere continuum is called *water potential* (ψ). Water potential is the difference in Gibbs energy per mole (the energy available to do work) between the water of interest and pure water (at a standard temperature and pressure). Plant physiologists have chosen to express water potential in terms of pressure, which has the dimensions of energy per volume, and is expressed in terms of Pascals (Pa = 1 Newton/m^2). By convention, pure water at atmospheric pressure has a water potential of zero and the addition of any solutes or the creation of suction (negative hydrostatic pressure) will function to lower the water potential (more negative values).

We can now examine the movement of water through the soil–plant–atmosphere continuum as a function of the gradient in water potential. As previously stated, the transfer of energy will always proceed from a region of higher energy content to a region of lower energy content, or in the case of water potential, from areas of higher water potential to areas of lower water potential. We can start with the exchange of water between the leaf and the atmosphere in the process of transpiration. When relative humidity of the atmosphere is 100 percent, the atmospheric water potential (ψ_{atm}) is zero. As values drop below 100 percent, the value of Gibbs energy declines, and ψ_{atm} becomes negative (Figure 6.4a). Under most physiological conditions, the air within the leaf is at or near saturation (relative humidity ~ 99 percent). As long as the relative humidity of the air is below 99 percent, a steep gradient of water potential between the leaf (ψ_{leaf}) and the atmosphere (ψ_{atm}) will drive the process of diffusion. Water vapor will move from the region of higher water potential (interior of the leaf) to the region of lower water potential (atmosphere)—that is, from a state of high to low Gibbs energy.

As water is lost to the atmosphere through the stomata, the water content of the cells decreases (turgor pressure drops) and in turn increases the concentration of solutes in the cell. This decrease in the cell's water content (and corresponding increase in solute concentration) decreases the water potential of the cells. Unlike the water potential of the atmosphere, which is determined only by relative humidity, several factors determine water potential within the plant. Turgor pressure (positive pressure) in the cell increases the plant's water potential. Therefore, a decrease in turgor pressure associated with water loss functions to decrease water potential. The component of plant water potential as a result of turgor pressure represents hydrostatic pressure and is represented as ψ_p.

Increasing concentrations of solutes in the cells are associated with water loss and will lower the water potential. This component of plant water potential is termed **osmotic potential** (ψ_π) because the difference in solute content inside and outside the cell results in the movement of water through the process of osmosis.

The surfaces of larger molecules, such as those in the cell walls, exert an attractive force on water. This tendency for water to adhere to surfaces reduces the Gibbs energy of the water molecules, reducing water potential. This component of water potential is called **matric potential** (ψ_m). The total water potential ψ at any point in the plant, from the leaf to the root, is the sum of these individual components:

$$\psi = \psi_p + \psi_\pi + \psi_m$$

The osmotic and matric potentials will always have a negative value, whereas the turgor pressure (hydrostatic pressure) component can be either positive or negative. Thus, the total potential can be either positive or negative, depending on the relative values of the individual components. Values of total water potential at any point along the continuum (soil, root, leaf, and atmosphere), however, are typically negative and the movement of water proceeds from areas of higher (zero or less negative) to lower (more negative) potential (from the region of higher energy to the region of lower energy). Therefore, the movement of water from the soil to the root, from the root to the leaf, and from the leaf to the atmosphere depends on maintaining a gradient of increasingly negative water potential at each point along the continuum (**Figure 6.4**).

$$\psi_{atm} < \psi_{leaf} < \psi_{root} < \psi_{soil}$$

Drawn by the low water potential of the atmosphere, water from the surface of and between the mesophyll cells within the leaf evaporates and escapes through the stomata. This gradient of water potential is transmitted into the mesophyll cells and on to the water-filled xylem (hollow conducting tubes throughout the plant) in the leaf veins. The gradient of increasingly negative water potential extends down to the fine rootlets in contact

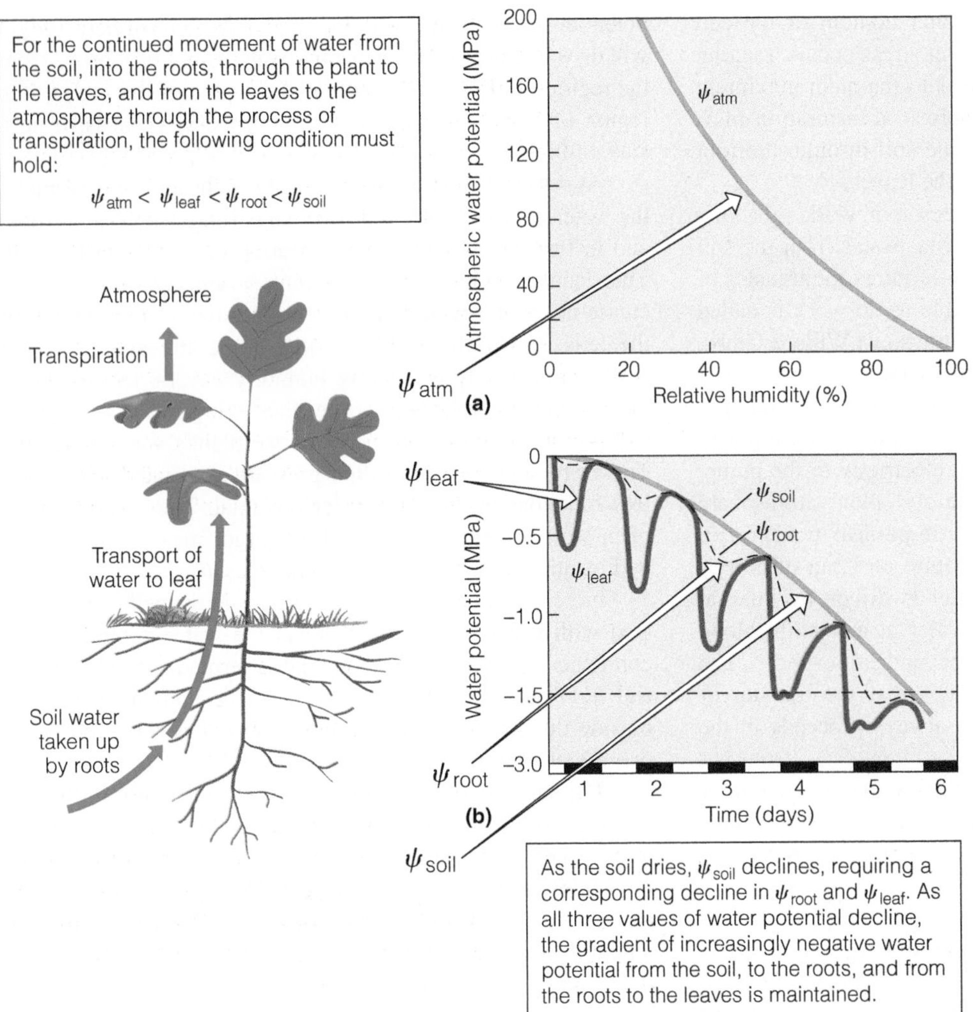

For the continued movement of water from the soil, into the roots, through the plant to the leaves, and from the leaves to the atmosphere through the process of transpiration, the following condition must hold:
$\psi_{atm} < \psi_{leaf} < \psi_{root} < \psi_{soil}$
Atmosphere
Transpiration
Transport of water to leaf
Soil water taken up by roots
ψ_{atm}
ψ_{leaf}
ψ_{root}
ψ_{soil}
(a)
Atmospheric water potential (MPa)
200
160
120
80
40
0
Relative humidity (%)
0
20
40
60
80
100
(b)
Water potential (MPa)
0
-0.5
-1.0
-1.5
-3.0
Time (days)
1
2
3
4
5
6
As the soil dries, ψ_{soil} declines, requiring a corresponding decline in ψ_{root} and ψ_{leaf}. As all three values of water potential decline, the gradient of increasingly negative water potential from the soil, to the roots, and from the roots to the leaves is maintained.

Figure 6.4 Transport of water along a gradient of water potential (ψ) from soil to leaves to air. Water moves from a region of high to low (more negative) water potential. (a) As the relative humidity of the air drops below 100 percent, a steep decline in atmospheric water potential maintains the gradient from the leaf to the air. (b) As the soil dries, the soil water potential becomes increasingly negative. This decline in soil water potential requires a corresponding decline in the water potential of the roots and leaves to maintain the gradient and flow of water. The graph depicts changes in the plant's leaf and root water potential in response to declining soil moisture (declining soil water potential) over a period of six days. Note the diurnal changes in leaf water potential. Leaf water potential is highest (least negative) at night when the stomata are closed.

with soil particles and pores. As water moves from the root and up through the stem to the leaf, the root water potential declines so that more water moves from the soil into the root.

Water loss through transpiration continues as long as (1) the amount of energy striking the leaf is enough to supply the necessary latent heat of evaporation (see Section 2.5), (2) moisture is available for roots in the soil, and (3) the roots are capable of maintaining a more negative water potential than that of the soil. At field capacity, water is freely available, and soil water potential (**ψ_{soil}**) is at or near zero (see Section 4.8). As water is drawn from the soil, the water content of the soil declines, and the soil water potential becomes more negative. As the water content of the soil declines, the remaining water adheres more tightly to the surfaces of the soil particles, and the matric potential becomes more negative. For a given water content, the matric potential of soil is influenced strongly by its texture (see Figure 4.10). Soils composed of fine particles, such as clays, have a higher surface area (per soil volume) for water to adhere to than sandy soils do. Clay soils, therefore, are characterized by more negative matric potentials for the same water content.

As soil water potential becomes more negative, the root and leaf water potentials must decline (become more negative) if the potential gradient is to be maintained. If precipitation does not recharge soil water, and soil water potentials continue to decline, eventually the gradient between the soil, root, and leaf cannot be maintained, and at that point, the stomata close to stop further water loss through transpiration. However, this closure also results in stopping further uptake of CO_2. The value of leaf water potential at which stomata close and net photosynthesis ceases varies among plant species (**Figure 6.5**) and reflects basic differences in their biochemistry, physiology, and morphology.

The rate of water loss varies with daily environmental conditions, such as humidity and temperature, and with the characteristics of plants. Opening and closing the stomata is probably the plant's most important means of regulating water loss. The trade-off between CO_2 uptake and water loss through the stomata results in a direct link between water availability in the soil and the plant's ability to carry out photosynthesis. To carry out photosynthesis, the plant must open its stomata; but when it does, it loses water, which it must replace to live. If water is scarce, the plant must balance the opening and closing of the stomata, taking up enough CO_2 while minimizing the loss of water. The ratio of carbon fixed

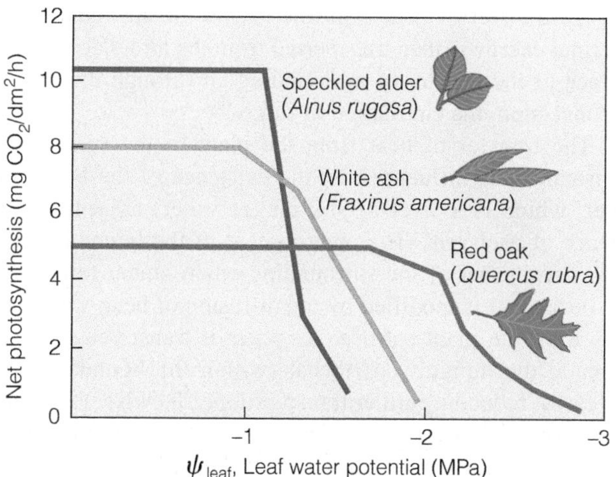

Figure 6.5 Changes in net photosynthesis (y-axis) as a function of leaf water potential for three tree species from the northeastern United States. The decline in net photosynthesis with declining leaf water potentials (more negative values) results primarily from stomatal closure. As the water content of the soil declines, the plant must reduce the leaf water potential to maintain the gradient so that water can move from the soil to the roots and from the roots to the leaves. The plant eventually reaches a point where it cannot further reduce leaf water potentials, and the stomata will close to reduce the loss of water.

(photosynthesis) per unit of water lost (transpiration) is called the **water-use efficiency**.

We can now appreciate the trade-off faced by terrestrial plants. To carry out photosynthesis, the plant must open the stomata to take up CO_2. But at the same time, the plant loses water through the stomata to the outside air—water that must be replaced through the plant's roots. If its access to water is limited, the plant must balance the opening and closing of stomata to allow for the uptake of CO_2 while minimizing water loss through transpiration. This balance between photosynthesis and transpiration is an extremely important constraint that has governed the evolution of terrestrial plants and directly influences the productivity of ecosystems under differing environmental conditions (see Chapter 20).

6.5 The Process of Carbon Uptake Differs for Aquatic and Terrestrial Autotrophs

A major difference in CO_2 uptake and assimilation by aquatic autotrophs (submerged plants, algae, and phytoplankton) versus terrestrial plants is the lack of stomata in aquatic autotrophs. CO_2 diffuses from the atmosphere into the surface waters and is then mixed into the water column. Once dissolved, CO_2 reacts with the water to form bicarbonate (HCO_3^-). This reaction is reversible, and the concentrations of CO_2 and bicarbonate tend toward a dynamic equilibrium (see Section 3.7). In aquatic autotrophs, CO_2 diffuses directly from the waters across the cell membrane. Once the CO_2 is inside the cell, photosynthesis proceeds in much the same way as outlined previously for terrestrial plants.

One difference between terrestrial and aquatic autotrophs is that some aquatic species can also use bicarbonate as a carbon source. However, the organism must first convert it to CO_2 using the enzyme carbonic anhydrase. This conversion can occur in two ways: (1) active transport of bicarbonate into the cell followed by conversion to CO_2 or (2) excretion of the enzyme into adjacent waters and subsequent uptake of converted CO_2 across the membrane. As CO_2 is taken up, its concentration in the waters adjacent to the organism decline. Because the diffusion of CO_2 in water is 10^4 times slower than in the air, it can easily become depleted (low concentrations) in the waters adjacent to the organism, reducing rates of uptake and photosynthesis. This constraint can be particularly important in still waters such as dense seagrass beds or rocky intertidal pools.

6.6 Plant Temperatures Reflect Their Energy Balance with the Surrounding Environment

Both photosynthesis and respiration respond directly to variations in temperature (**Figure 6.6**). As temperatures rise above freezing, both photosynthesis and respiration rates increase. Initially, photosynthesis increases faster than respiration. As temperatures continue to rise, the photosynthetic rate reaches a maximum related to the temperature response of the enzyme rubisco. As temperatures continue to rise, photosynthetic rate declines and respiration rate continues to increase. As temperatures rise further, even respiration declines as temperatures reach critical levels. The temperature response of net photosynthesis is the difference between the rate of carbon uptake in photosynthesis and the rate of carbon loss in respiration (see

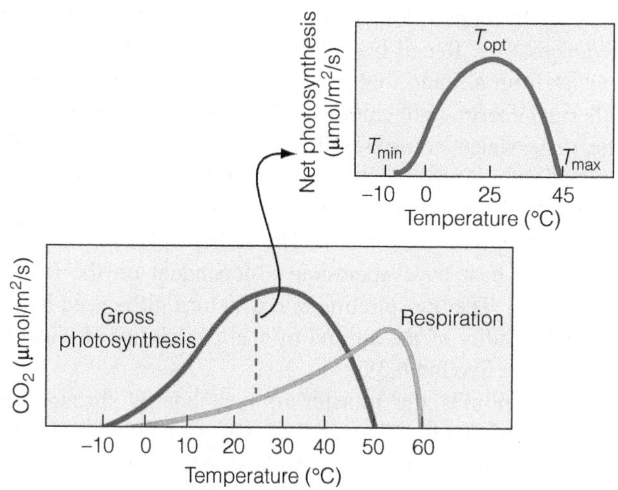

Figure 6.6 General response of the rates of photosynthesis and respiration to temperature. At any temperature, the difference between these two rates is the rate of net photosynthesis (net uptake rate of CO_2). Here, the optimal temperature for net photosynthesis is between 20°C and 30°C.

Figure 6.6). Three values describe the temperature response curve: T_{min}, T_{opt}, and T_{max}. The values T_{min} and T_{max} are, respectively, the minimum and maximum temperatures at which net photosynthesis approaches zero (meaning no net carbon uptake). T_{opt} is the temperature, or range of temperatures, over which net carbon uptake is at its maximum.

The temperature of the leaf, not the air, controls the rate of photosynthesis and respiration; and leaf temperature depends on the exchange of thermal energy between the leaf and its surrounding environment. Plants absorb both shortwave (solar) and longwave (thermal) radiation (see Section 2.1). Plants reflect some of this solar radiation and emit longwave radiation back to the atmosphere. The difference between the radiation a plant receives and the radiation it reflects and emits back to the surrounding environment is the net radiation balance of the plant (R_n). The net radiation balance of a plant is analogous to the concept of the radiation balance of Earth (see Chapter 2, Figure 2.3). Of the net radiation absorbed by the plant, some is used in metabolic processes and stored in chemical bonds—namely, in the processes of photosynthesis and respiration. This quantity is quite small, typically less than 5 percent of R_n. The remaining energy raises the temperature of the leaves and the surrounding air. On a clear, sunny day, the amount of energy plants absorb can raise internal leaf temperatures well above ambient (air or water temperature). Internal leaf temperatures may go beyond the optimum for photosynthesis and possibly reach critical levels (see Figure 6.6).

To maintain internal temperatures within the range of tolerance (positive net photosynthesis), plants must exchange thermal energy (heat) with the surrounding environment. The transfer of heat between the plant and the surrounding air (or water) is governed by the Second Law of Thermodynamics—thermal energy flows in only one direction, from areas of higher temperature to areas of lower temperature.

The primary means by which terrestrial plants dissipate heat are evaporation and convection; aquatic plants do so primarily by convection. Evaporation occurs in the process of transpiration. Recall from Chapter 3 that the phase change of water from a liquid to a gas (evaporation) requires an input of thermal energy (540 calories or 2260 joules per gram [g] of water). As waters transpires from the leaves of plants to the surrounding atmosphere through the stomata, the leaves lose thermal energy and their temperature declines through evaporative cooling (see Section 3.2). The ability of terrestrial plants to dissipate heat by evaporation is dependent on the rate of transpiration. The transpiration rate is in turn influenced by the relative humidity of the air and by the availability of water to the plant (see Section 6.3).

Convection is the transfer of heat energy through the circulation of fluids (Chapter 2), whereas conduction is the transfer of thermal energy through direct contact (between two objects). For convection to occur, the surface of the leaf must first transfer thermal energy between the adjacent molecules of air or water through conduction. The direction of this conductive exchange depends on the difference between the temperature of the leaf and the surrounding air. If the leaf temperature is higher than that of the surrounding air, there is a net transfer of heat from the leaf to the surrounding air. Thermal energy is then transported from the air adjacent to the surface of the leaf to the surrounding air through the process of convection, the circulation of fluids.

The transfer of heat from the plant to the surrounding environment is influenced by the existence of the **boundary layer**, which is a layer of still air (or water) adjacent to the surface of each leaf. The environment of the boundary layer differs from that of the surrounding environment (air or water) because it is modified by the diffusion of heat, water, and CO_2 from the plant surface. As water is transpired from the stomata, the humidity of the air within the boundary layer increases, reducing further transpiration. Likewise, as thermal energy (heat) is transferred from the leaf surface to the boundary layer, the air temperature of the boundary layer increases, reducing further heat transfer from the leaf surface. Under still conditions (no air or water flow), the boundary layer increases in thickness reducing the transfer of heat and materials (water and CO_2) between the leaf and the atmosphere (or water). Wind or water flow functions to reduce the size of the boundary layer, allowing for mixing between the boundary layer and the surrounding air (or water) and reestablishing the diffusion or temperature gradient between the leaf surface and the bulk air.

Leaf size and shape also influence the thickness and dynamics of the boundary layer, and therefore, the ability of plants to exchange heat through convection. Air tends to move more smoothly (laminar flow) over a larger surface than a smaller one, and as a result, the boundary layer tends to be thicker and more intact in larger leaves. Deeply lobed leaves, and small, compound leaves (**Figure 6.7**) tend to disrupt the flow of air, causing turbulence that functions to reduce the boundary layer and increase the exchange of heat and water.

The relative importance of evaporation and convection to the maintenance of leaf temperatures (dissipation of heat) is dependent on the physical environment. In locations where water is available, such as regions of high mean annual precipitation, most of the dissipation of heat can occur through transpiration (evaporation) as plants open stomata to support the uptake of CO_2. As conditions become drier, however, transpiration becomes limited and the average leaf size of species decreases, enhancing heat loss through convection (see Figure 6.18b).

6.7 Constraints Imposed by the Physical Environment Have Resulted in a Wide Array of Plant Adaptations

We have explored variation in the physical environment over Earth's surface: the salinity, depth, and flow of water; spatial and temporal patterns in climate (precipitation and temperature); variations in geology and soils (Part One). In all but the most extreme of these environments, autotrophs harness the energy of the Sun to fuel the conversion of CO_2 into glucose in

(a)

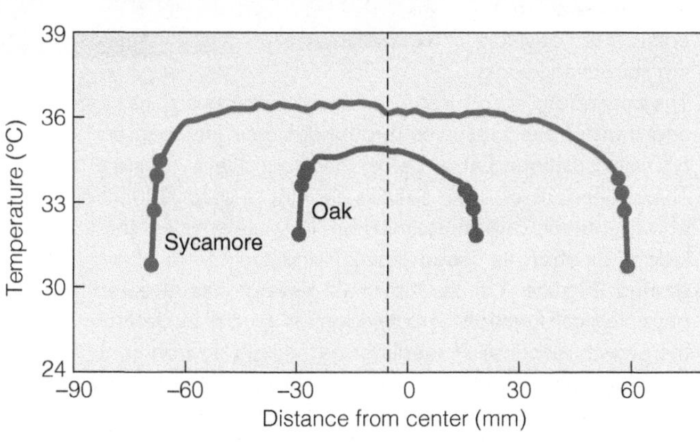

(b)

Figure 6.7 Example of the influence of leaf size and shape on leaf temperature and heat transfer. (a) Thermal image of nontranspiring leaves of sycamore and oak and (b) transects (labeled by white lines on leaf surfaces in part a) of temperature across leaf surface. Note that temperatures decrease from center to edge of leaf surface and the lower surface temperatures of smaller, lobed oak leaf as compared to sycamore. Conditions during measurements: mean wind speed = 0.6 m/s, air temperature = 30.2°C, PAR = 910 mmol/m^2/s.

([a] Based on Stokes, V.J., M.D. Morecrof, and J.I.L. Morison. 2006. *Boundary layer conductance for contrasting leaf shapes in a deciduous broadleaved forest canopy.* Agricultural and Forest Meteorology 139:40–54 Figure 3a, pg. 45, [b] Based on Stokes, V.J., M.D. Morecrof, and J.I.L. Morison. 2006. *Boundary layer conductance for contrasting leaf shapes in a deciduous broadleaved forest canopy.* Agricultural and Forest Meteorology 139:40–54 Figure 4a, pg. 46.)

the process of photosynthesis. To survive, grow, and reproduce, plants must maintain a positive carbon balance, converting enough CO_2 into glucose to offset the expenses of respiration (photosynthesis > respiration). To accomplish this, a plant must acquire the essential resources of light, CO_2, water, and mineral nutrients as well as tolerate other features of the environment that directly affect basic plant processes such as temperature, salinity, and pH. Although often discussed—and even studied as though they are independent of each other—the adaptations exhibited by plants to these features of the environment are not independent, for reasons relating to the physical environment and to the plants themselves.

Many features of the physical environment that directly influence plant processes are interdependent. For example, the light, temperature, and moisture environments are all linked through a variety of physical processes (Chapters 2–4). The amount of solar radiation not only influences the availability of light (PAR) required for photosynthesis but also directly influences the temperature of the leaf and its surroundings. In addition, air temperature directly affects the relative humidity, a key feature influencing the rate of transpiration and evaporation of water from the soil (see Section 2.5, Figure 2.15). For this reason, we see a correlation in the adaptations of plants to variations in these environmental factors. Plants adapted to dry, sunny environments must be able to deal with the higher demand for water associated with higher temperatures and lower relative humidity, and they tend to have characteristics such as smaller leaves and increased production of roots.

In other cases, there are trade-offs in the ability of plants to adapt to limitations imposed by multiple environmental factors, particularly resources. One of the most important of these trade-offs involves the acquisition of above- and belowground resources. Allocating carbon to the production of leaves and stems increases the plant's access to the resources of light and CO_2, but it is at the expense of allocating carbon to the production of roots. Likewise, allocating carbon to the production of roots increases access to water and soil nutrients but limits carbon allocation to the production of leaves. The set of characteristics (adaptations) that allow a plant to successfully survive, grow, and reproduce under one set of environmental conditions inevitably limits its ability to do equally well under different environmental conditions. We explore the consequences of this simple premise in the following sections.

6.8 Species of Plants Are Adapted to Different Light Environments

The amount of solar radiation reaching Earth's surface varies diurnally, seasonally, and geographically (Chapter 2, Section 2.3). However, a major factor influencing the amount of light (PAR) a plant receives is the presence of other plants through shading (see Section 4.2 and Chapter 4, **Quantifying Ecology 4.1**). Although the amount of light that reaches an individual plant varies continuously as a function of the area of leaves above it, plants live in one of two qualitatively different light environments—sun or shade—depending on whether they are overtopped by other plants. Plants have evolved to possess a range of physiological and morphological adaptations that allow individuals to survive, grow, and reproduce in these two different light environments (see this chapter, **Field Studies: Kaoru Kitajima**).

Plant species adapted to high-light environments are called **shade-intolerant** species, or sun-adapted species. Plant species adapted to low-light environments are called **shade-tolerant** species, or shade-adapted species. Shade-tolerant and shade-intolerant species differ across a wide variety of phenotypic characteristics that represent adaptations to sun and shade environments. One of the most fundamental differences between shade-intolerant and shade-tolerant plant species lies

FIELD STUDIES *Kaoru Kitajima*

Department of Botany, University of Florida, Gainesville, Florida

A major factor influencing the availability of light to a plant is its neighbors. By intercepting light, taller plants shade individuals below, influencing rates of photosynthesis, growth, and survival. Nowhere is this effect more pronounced than on the forest floor of a tropical rain forest, where light levels are often less than 1 percent of those recorded at the top of the canopy (see Section 4.2). With the death of a large tree, however, a gap is created in the canopy, giving rise to an "island" of light on the forest floor. With time, these gaps in the canopy eventually close because individuals grow up to the canopy from below or neighboring trees expand their canopies, which once again shades the forest floor.

How these extreme variations in availability of light at the forest floor have influenced the evolution of rain forest plant species has been the central research focus of plant ecologist Kaoru Kitajima of the University of Florida. Kitajima's work in the rain forests of Barro Colorado Island in Panama presents a story of plant adaptations to variations in the light environment that includes all life stages of the individual, from seed to adult.

Within the rain forests of Barro Colorado Island, the seedlings of some tree species survive and grow only in the high-light environments created by the formation of canopy gaps, whereas the seedlings of other species can survive for years in the shaded conditions of the forest floor.

In a series of experiments designed to determine shade tolerance based on patterns of seedling survival in sun and shade environments (see Figure 6.10), Kitajima noted that seed mass (weight) is negatively correlated with seedling mortality rates. Interestingly, large-seeded species not only had higher rates of survival in the shade but also exhibited slower rates of growth after germination. Intuitively, one might think that larger reserves of energy and nutrients within the seed (larger mass) would allow for a faster rate of initial development, but this was not the case. What role does seed size play in the survival and growth of species in different light environments? An understanding of these relationships requires close examination of how seed reserves are used.

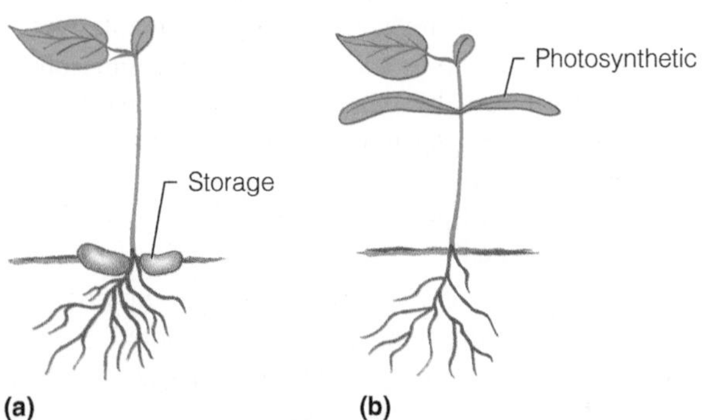

Figure 1 Two contrasting functional forms of cotyledons found in tropical rain forest trees: (a) storage and (b) photosynthetic.

The storage structure or structures within a seed are called the *cotyledon*. Upon germination, cotyledons transfer reserve materials (lipids, carbohydrates, and mineral nutrients) into developing shoots and roots. The cotyledons of some species serve strictly as organs to store and transfer seed reserves throughout their life span and are typically positioned at or below the ground level (**Figure 1a**). The cotyledons of other species develop a second function—photosynthetic carbon assimilation. In these species, the cotyledons function as "seed leaves" and are raised above the ground (**Figure 1b**). As Kitajima's research has revealed, the physiological function of cotyledons is crucial in determining the growth response of seedlings to the light environment.

Kitajima conducted an experiment involving tree species that differed in cotyledon function (photosynthetic and storage) and shade tolerance. Seedlings were raised from germination under two light levels: sun (23 percent full sun) and shade (2 percent full sun).

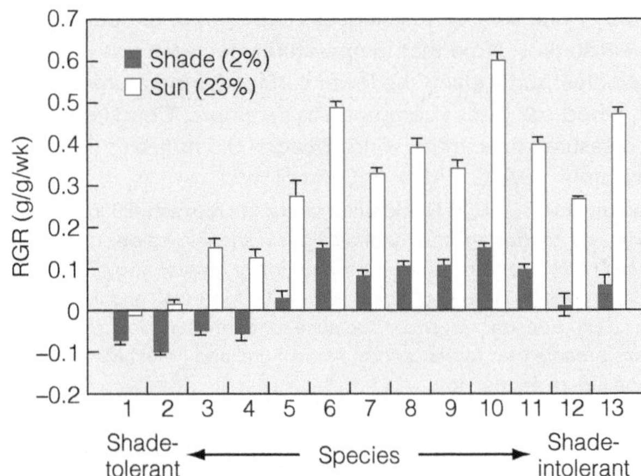

Figure 2 Relative growth rates (RGRs) of 13 tropical tree species grown as seedlings from germination to 10 weeks under controlled shade (2 percent) and sun (23 percent) conditions. Species are ranked based on their shade tolerance (survival in shade). Species codes: (1) *Aspidosperma cruenta*, (2) *Tachigalia versicolor*, (3) *Bombacopsis sessilis*, (4) *Platypodium elegans*, (5) *Lonchocarpus latifolius*, (6) *Lafoensia punicifolia*, (7) *Terminalia amazonica*, (8) *Cordia alliodora*, (9) *Pseudobombax septenatum*, (10) *Luehea seemannii*, (11) *Ceiba pentandra*, (12) *Cavanillesia platanifolia*, (13) *Ochroma pyramidale*
(Adapted from Kitajima 1994.)

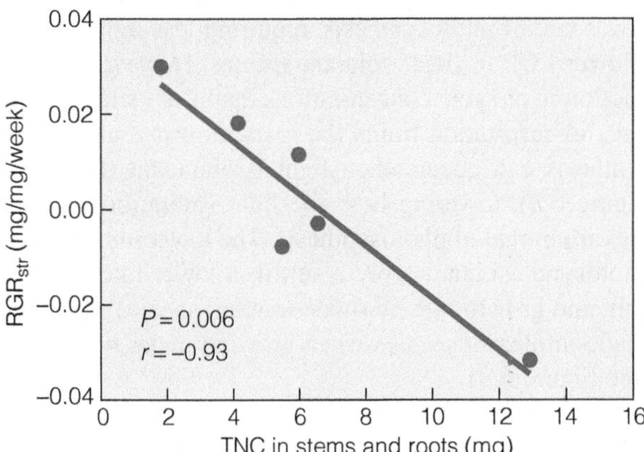

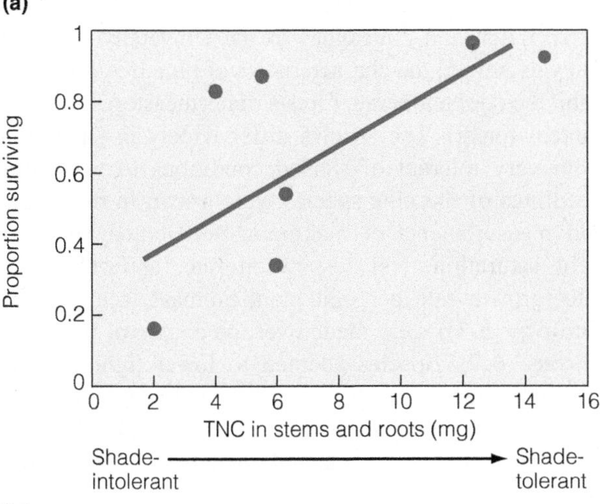

Figure 3 Relationship between initial carbohydrate pool size in stems and roots and (a) zero- to eight-week relative growth rate (RGR), and (b) first-year seedling survival under reduced light conditions. Each point is a species mean. (Data from Myers and Kitajima 2007.)

Changes in biomass and leaf area were recorded during a period of 40 days postgermination to determine when light began to affect seedling growth (determined for each species by the difference between individuals grown in shade and sun environments).

The smaller seeds of the shade-intolerant species had photosynthetic cotyledons and developed leaves earlier than did shade-tolerant species with their larger storage cotyledons. These differences reflect two different "strategies" in the use of initial seed reserves. Shade-intolerant species invested reserves in the production of leaves to bring about a rapid return (carbon uptake in photosynthesis), whereas shade-tolerant species kept seed reserves as storage for longer periods at the expense of growth rate.

Having used their limited seed reserves for the production of leaves, the shade-intolerant species responded to light availability earlier than did the shade-tolerant species. And without sufficient light, mortality was generally the outcome.

So the experiments revealed that larger seed storage in shade-tolerant species does not result in a faster initial growth under shaded conditions (**Figure 2**). Rather, these species (shade-tolerant) exhibit a conservative strategy of slow use of reserves over a prolonged period. In a later study, Kitajima established that the lower relative growth rate of shade tolerant seedlings is associated with an increased storage of nonstructural carbohydrates (sugars and starches) in the stem and roots (**Figure 3a**), enabling them to cope with periods of biotic (herbivory and disease) and abiotic (shade) stress. Results of the study show a significant positive relationship between seedling survival during the first year under shaded conditions and the storage of nonstructural carbohydrates (**Figure 3b**).

Whether shade tolerant or intolerant, once seedlings use up seed reserves, they must maintain a positive net carbon gain as a prerequisite for survival (see Section 6.7). What suites of seedling traits allow some species to survive better than others in the shade? To answer this question, Kitajima grew seedlings in the experimental sun and shade environments for an extended period beyond the reserve phase. Under both sun and shade conditions, shade-tolerant species had a greater proportional allocation to roots (relative to leaves), thicker leaves (lower SLA), and as a result, a lower photosynthetic surface area than did shade-intolerant species. As a result, the relative growth rates of shade-intolerant species were consistently greater than those of the shade-tolerant species, both in sun and shaded conditions (see Figure 2).

Whereas the characteristics exhibited by the shade-intolerant species reflect strong natural selection for fast growth within light gaps, shade-tolerant species appear adapted to survive for many years in the understory, where their ability to survive attacks by pathogens and herbivores is enhanced by their well-developed reserves within the root system.

Bibliography

Kitajima, K. 1994. "Relative importance of photosynthetic and allocation traits as correlates of seedling shade tolerance of 15 tropical tree species." *Oecologia* 98: 419–428.

Kitajima, K. 1996. Ecophysiology of tropical tree seedlings. In *Tropical forest plant ecophysiology* (S. Mulkey, R. L. Chazdon, and A. P. Smith, eds.), 559–596. New York: Chapman & Hall.

Kitajima, K. 2002. "Do shade-tolerant tropical tree seedlings depend longer on seed reserves?" *Functional Ecology* 16:433–444.

Myers, J. A., and Kitajima, K. 2007. "Carbohydrate storage enhances seedling shade and stress tolerance in a neotropical forest." *Journal of Ecology* 95:383–395.

1. What processes might create gaps in the forest canopy?

2. How might seed size influence the method of seed dispersal from the parent plant?

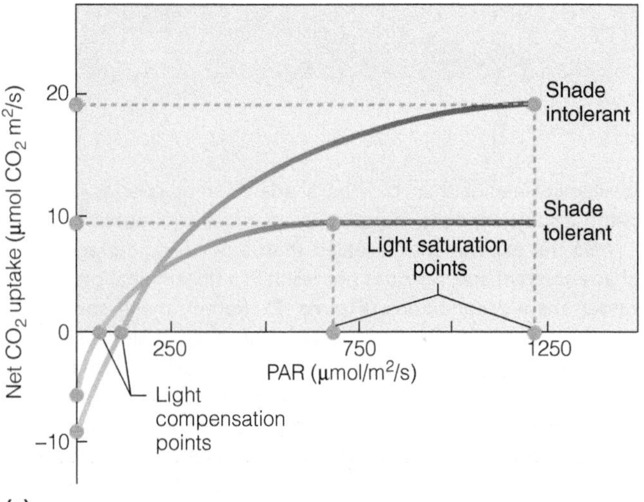

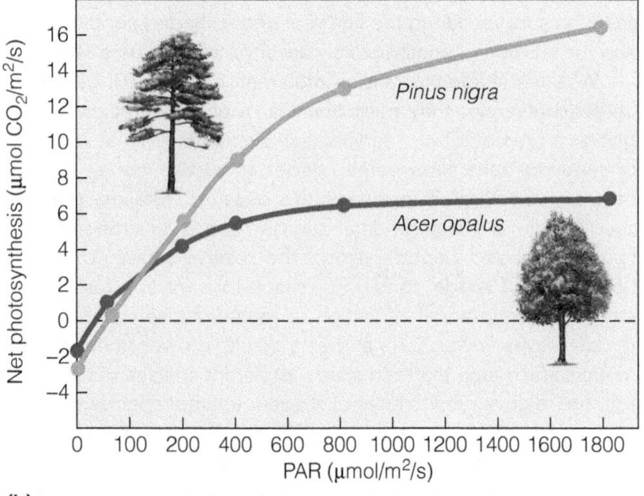

Figure 6.8 (a) General differences in the photosynthetic light response curves for shade-tolerant and shade-intolerant species. (b) Light response curves for two tree species that co-occur in the montane forests of the Mediterranean region. *Pinus nigra* is an example of a shade-intolerant tree species while *Acer opalus* is shade-tolerant.
(Data from Gomez-Aparicio et al. 2006.)

in their patterns of photosynthesis in response to varying levels of light availability. Shade-tolerant species tend to have a lower light saturation point and a lower maximum rate of photosynthesis than shade-intolerant species (**Figure 6.8**). These differences relate in part to lower concentrations of the photosynthetic enzyme rubisco (see Section 6.1) found in shade-tolerant plants. Plants must expend a large amount of energy and nutrients to produce rubisco and other components of the photosynthetic apparatus. In shaded environments, low light, not the availability of rubisco to catalyze the fixation of CO_2, limits the rate of photosynthesis. Shade-tolerant (shade-adapted) species produce less rubisco as a result. By contrast, production of chlorophyll, the light-harvesting pigment in the leaves, often increases. The reduced energy cost of producing rubisco and other compounds involved in photosynthesis lowers the rate of leaf respiration. Because the LCP is the value of PAR necessary to maintain photosynthesis at a rate that exactly offsets the loss of CO_2 in respiration (net photosynthesis = 0), the lower rate of respiration can be offset by a lower rate of photosynthesis, requiring less light. The result is a lower LCP in shade-tolerant species. However, the same reduction in enzyme concentrations that is associated with lower rates of respiration limits the maximum rate at which photosynthesis can occur when light is abundant (high PAR; see Figure 6.8), lowering both the light saturation point and the maximum rate of photosynthesis. The lower maximum rates of photosynthesis inevitably result in a lower rate of net carbon gain and growth rate by shade-tolerant species as compared to shade-intolerant species when growing under high light levels (see Figure 6.8).

The variations in photosynthesis, respiration, and growth rate that characterize plant species adapted to different light environments are illustrated in the work of plant ecologist Peter Reich and colleagues at the University of Minnesota. They examined the characteristics of nine tree species that inhabit the cool temperate forests of northeastern North America (boreal forest). The species differ widely in shade tolerance from very tolerant of shaded conditions to very intolerant. Seedlings of the nine species were grown in the greenhouse, and measurements of maximum net photosynthetic rate at light saturation, leaf respiration rate, and relative growth rate (growth rate per unit plant biomass; see **Quantifying Ecology 6.1**)) were made over the course of the experiment (**Figure 6.9**). Species adapted to lower light environments (shade-tolerant) are characterized by lower maximum rates of net photosynthesis, leaf respiration, and relative growth rate than are species adapted to higher light environments (shade-intolerant).

The difference in the photosynthetic characteristics between shade-tolerant and shade-intolerant species influences not only rates of net carbon gain and growth but also ultimately the ability of individuals to survive in low-light environments. This relationship is illustrated in the work of Caroline Augspurger of the University of Illinois. She conducted a series of experiments designed to examine the influence of light availability on seedling survival and growth for a variety of tree species, both shade-tolerant and shade-intolerant, that inhabit the tropical rain forests of Panama. Augspurger grew tree seedlings of each species under two levels of light availability. These two treatments mimic the conditions found either under the shaded environment of a continuous forest canopy (shade treatment) or in the higher light environment in openings or gaps in the canopy caused by the death of large trees (sun treatment). She continued the experiment for a year, monitoring the survival and growth of seedlings on a weekly basis. **Figure 6.10** presents the results for two contrasting species, shade-tolerant and shade-intolerant.

The shade-tolerant species (*Myroxylon balsamum*) showed little difference in survival and growth rates under sunlight and

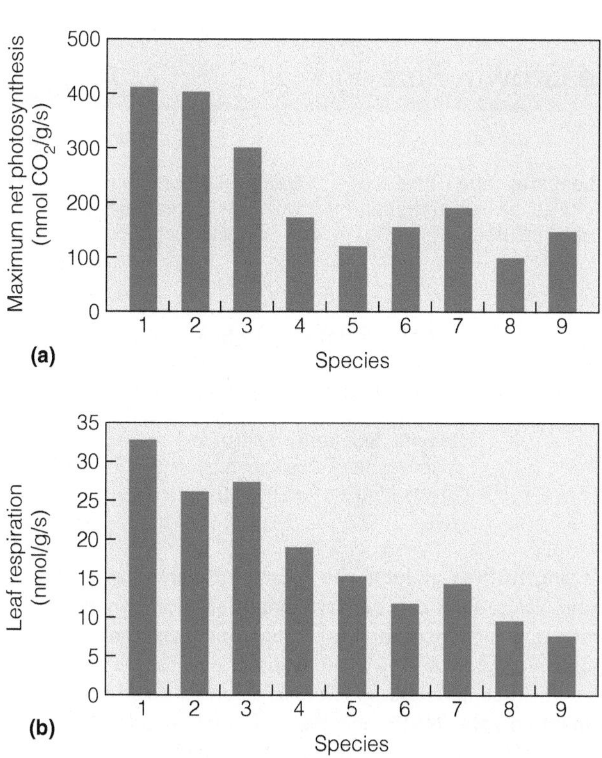

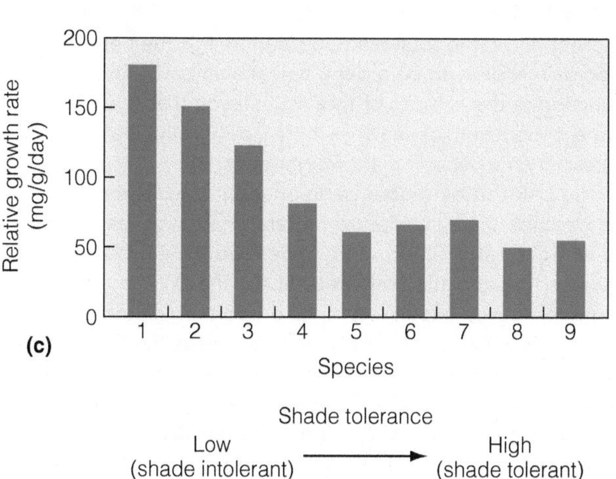

Figure 6.9 Differences in the rates of (a) light-saturated photosynthesis, (b) leaf respiration, and (c) relative growth rate for nine tree species that inhabit the forests of northeastern North America (boreal forest). Species are ranked from highest (shade-intolerant) to lowest (shade-tolerant) in tolerance to shade. Species codes: (1) *Populus tremuloides*, (2) *Betula papyrifera*, (3) *Betula allegheniensis*, (4) *Larix laricina*, (5) *Pinus banksiana*, (6) *Picea glauca*, (7) *Picea mariana*, (8) *Pinus strobus*, and (9) *Thuja occidentalis*.
(Adapted from Reich et al. 1998.)

Interpreting Ecological Data

Q1. *In general, how do net photosynthesis and leaf respiration vary with increasing shade tolerance for the nine boreal tree species? What does this imply about the corresponding pattern of gross photosynthesis with increasing shade tolerance for these species?*

Q2. *Based on the data presented in graphs (a) and (b), how would you expect the light compensation point to differ between* Populus tremuloides *and* Picea glauca*?*

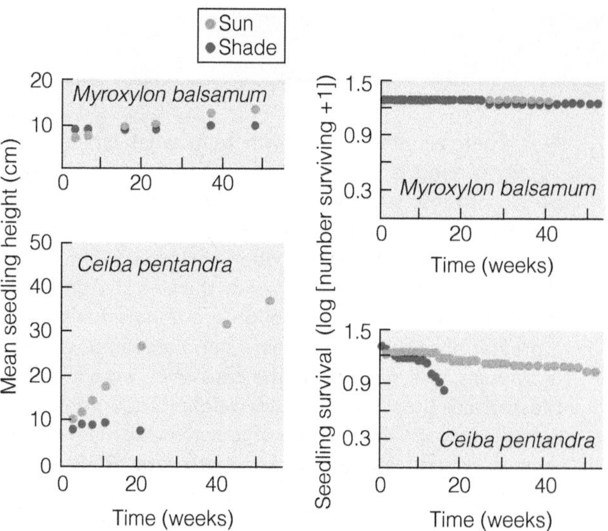

Figure 6.10 Seedling survival and growth over a period of one year for seedlings of two tree species on Barro Colorado Island, Panama, grown under sun and shade conditions. *Ceiba pentandra* is a shade-intolerant species; *Myroxylon balsamum* is shade-tolerant.
(Data from Augspurger 1982.)

shade conditions. In contrast, the survival and growth rates of the shade-intolerant species (*Ceiba pentandra*) were dramatically reduced under shade conditions. These observed differences are a direct result of the difference in the adaptations relating to photosynthesis and carbon allocation discussed previously. The higher rate of light-saturated photosynthesis resulted in a high growth rate for the shade-intolerant species in the high-light environment. The associated high rate of leaf respiration and LCP, however, reduced rates of survival in the shaded environment. By week 20 of the experiment, all individuals had died. In contrast, the shade-tolerant species was able to survive in the low-light environment. The low rates of leaf respiration and light-saturated photosynthesis that allow for the low LCP, however, limit rates of growth even in high-light environments.

In addition to the differences in photosynthesis and growth rate, shade-tolerant and shade-intolerant species also exhibit differences in patterns of leaf morphology. The ratio of surface area (measured in centimeters squared [cm^2]) to weight (g) for a leaf is called the specific leaf area (SLA; cm^2/g). The value of SLA represents the surface area of leaf produced per gram of biomass (or carbon) allocated to the production of leaves. Shade-tolerant species typically produce leaves with a greater specific leaf area. This difference in leaf structure effectively increases the surface area for the capture of light (the limiting resource) per unit of biomass allocated to the production of leaves. Marc Abrams and Mark Kubiske of Pennsylvania State University examined leaf morphology of 31 tree species inhabiting the forests of central Wisconsin. The researchers measured both SLA and leaf thickness for individuals of each species growing in full sunlight and in the shaded conditions of the

QUANTIFYING ECOLOGY 6.1 Relative Growth Rate

When we think of growth rate, what typically comes to mind is a measure of change in size during some period of time, such as change in weight during the period of a week (grams weight gain/week). However, this conventional measure of growth is often misleading when comparing individuals of different sizes or tracking the growth of an individual through time. Although larger individuals may have a greater absolute weight gain when compared with smaller individuals, this may not be the case when weight gain is expressed as a proportion of body weight (proportional growth). A more appropriate measure of growth is the mass-specific or relative growth rate. **Relative growth rate (RGR)** expresses growth during an observed period of time as a function of the size of the individual. This calculation is performed by dividing the increment of growth during some observed time period (grams [g] weight gain) by the size of the individual at the beginning of that time period (g weight gain/total g weight at the beginning of observation period) and then dividing by the period of time to express the change in weight as a rate (g/g/time).

Relative growth rate = weight gain during the period of observation per plant weight at the beginning of the observation period per time interval of observation period ($g/g^{-1}/time^{-1}$)

$$\text{RGR (g/g/time)} = \text{NAR (g/cm}^2\text{/time)} \times \text{LAR (cm}^2\text{/g)}$$

Net assimilation rate = weight gain during the period of observation per total area of leaves (leaf area) per time interval of observation period ($g/cm^2/time$)

Leaf area ratio = total area of leaves (leaf area) per total plant weight (cm^2/g)

Leaf area ratio = total area of leaves (leaf area) per total plant weight (cm^2/g)

Leaf weight ratio = total weight of leaves per total plant weight (g/g)

$$\text{LAR (cm}^2\text{/g)} = \text{LWR} \times \text{SLA}$$

Specific leaf area = total area of leaves (leaf area) per total weight of leaves (cm^2/g)

Using RGR to evaluate the growth of plants has an additional value; RGR can be partitioned into components reflecting the influences of assimilation (photosynthesis) and allocation on growth—the assimilation of new tissues per unit leaf area (g/centimeters squared [cm^2]/time) called the **net assimilation rate (NAR)**, and the leaf area per unit of plant weight (cm^2/g), called the **leaf area ratio (LAR)**.

The NAR is a function of the total gross photosynthesis of the plant minus the total plant respiration. It is the net assimilation gain expressed on a per-unit leaf area basis. The LAR is a function of the amount of that assimilation that is allocated to the production of leaves—more specifically, leaf area—expressed on a per-unit plant weight basis.

The LAR can be further partitioned into two components that describe the allocation of net assimilation to leaves, the **leaf weight ratio (LWR)**, and a measure of leaf density or thickness, the **specific leaf area (SLA)**. The LWR is the total weight of leaves expressed as a proportion of total plant weight (g leaves/g total plant weight), whereas the SLA is the area of leaf per gram of leaf weight. For the same tissue density, a thinner leaf would have a greater value of SLA.

The real value of partitioning the estimate of RGR is to allow for comparison, either among individuals of different

understory. Shade-tolerant species show a consistent pattern of higher SLA and lower leaf thickness than shade-intolerant species (**Figure 6.11**). Their data also illustrate a second important point regarding plant adaptations: plant species exhibit phenotypic plasticity in response to the light environment. Individuals of both shade-tolerant and shade-intolerant species exhibit an increase in SLA and a reduction in leaf thickness when growing under shaded conditions compared to open, sunny conditions (see Figure 6.11). The increased surface area of leaves in the shade functions to increase the photosynthetic surface area, partially offsetting the reduced rates of photosynthesis.

The dichotomy in adaptations between shade-tolerant and shade-intolerant species reflects a fundamental trade-off between characteristics that enable a species to maintain high rates of net photosynthesis and growth under high-light conditions and the ability to continue survival and growth under low-light conditions. The differences in biochemistry, physiology, and leaf morphology exhibited by shade-tolerant species reduce the amount of light required to survive and grow. However, these same characteristics limit their ability to maintain high rates of net photosynthesis and growth when light levels are high. In contrast, plants adapted to high-light environments (shade-intolerant species) can maintain high rates of net photosynthesis and growth under high-light conditions but at the expense of continuing photosynthesis, growth, and survival under shaded conditions.

species or among individuals of the same species grown under different environmental conditions. For example, the data presented in **Table 1** are the results of a greenhouse experiment in which seedlings of *Acacia tortilis* (a tree that grows on the savannas of southern Africa) were grown under two different light environments: full sun and shaded (50 percent full sun). Individuals were harvested at two times (at four and six weeks), and the total plant weight, total leaf weight, and total leaf area were measured. The mean values of these measures are shown in the table. From these values, estimates of RGR, LAR, LWR, and SLA were calculated. The values of RGR are calculated using the total plant weights at four and six weeks. NAR was then calculated by dividing the RGR by LAR. Because LAR varies through time (between weeks four and six), the average of LAR at four and six weeks was used to characterize LAR in estimating RGR. Note that the average size (weight and leaf area) of seedlings grown in the high-light environment is approximately twice that of seedlings grown in the shade. Despite this difference in size, and the lower light levels to support photosynthesis, the difference in RGR between sun- and shade-grown seedlings is only about 20 percent. By examining the components of RGR, we can see how the shaded individuals are able to accomplish this task. Low-light conditions reduced rates of photosynthesis, subsequently reducing NAR for the individuals grown in the shade. The plants compensated, however, by increasing the allocation of carbon (assimilates) to the production of leaves (higher LWR) and producing thinner leaves (higher SLA) than did the individuals grown in full sun. The result is that individuals grown in the shade have a greater LAR (photosynthetic surface area relative to plant weight), offsetting the lower NAR and maintaining comparable RGR.

These results illustrate the value of using the RGR approach for examining plant response to varying environmental conditions, either among individuals of the same species or among individuals of different species adapted to different environmental conditions. By partitioning the components of plant growth into measures directly related to morphology, carbon allocation, and photosynthesis, we can begin to understand how plants both acclimate and adapt to differing environmental conditions.

Table 1

	Week Four		Week Six	
	Sun	Shade	Sun	Shade
Leaf area (cm^2)	18.65	12.45	42	24
Leaf weight (g)	0.056	0.032	0.126	0.061
Stem weight (g)	0.090	0.058	0.283	0.138
Root weight (g)	0.099	0.043	0.239	0.089
Total weight (g)	0.245	0.133	0.648	0.288
LAR (cm^2/g)	75.998	93.750	64.854	83.304
SLA (cm^2/g)	334	392	332	394
LWR (g/g)	0.228	0.239	0.194	0.213
RGR (g/g/week)			0.471	0.382
NAR (g/cm^2/week)			0.007	0.004

1. When plants are grown under dry conditions (low water availability), there is an increase in the allocation of carbon to the production of roots at the expense of leaves. How would this shift in allocation influence the plant's LAR?
2. Nitrogen availability can directly influence the rate of net photosynthesis. Assuming no change in the allocation of carbon or leaf morphology, how would an increase in the rate of net photosynthesis resulting from an increase in nitrogen availability influence RGR? Which component of RGR would be influenced by the increase in net photosynthesis?

6.9 The Link between Water Demand and Temperature Influences Plant Adaptations

As with the light environment, a range of adaptations has evolved in terrestrial plants in response to variations in precipitation and soil moisture. As we saw in the previous discussion of transpiration (see Section 6.3), however, the demand for water is linked to temperature. As air temperature rises, the saturation vapor pressure will likewise rise, increasing the gradient of water vapor between the inside of the leaf and the outside air and subsequently the rate of transpiration (see Section 2.5). As a result, the amount of water required by the plant to offset losses from transpiration will likewise increase with temperature.

Plants exhibit both acclimation and developmental plasticity (both forms of phenotypic plasticity; see Section 5.4) in response to changes in water availability and demand on a variety of timescales. When the atmosphere or soil is dry, plants respond by partially closing the stomata and opening them for shorter periods of time. In the early period of water stress, a plant closes its stomata during the hottest part of the day when relative humidity is the lowest (**Figure 6.12**). It resumes normal activity in the afternoon. As water becomes scarcer, the plant opens its stomata only in the cooler, more humid conditions of morning. Closing the stomata reduces the loss of water

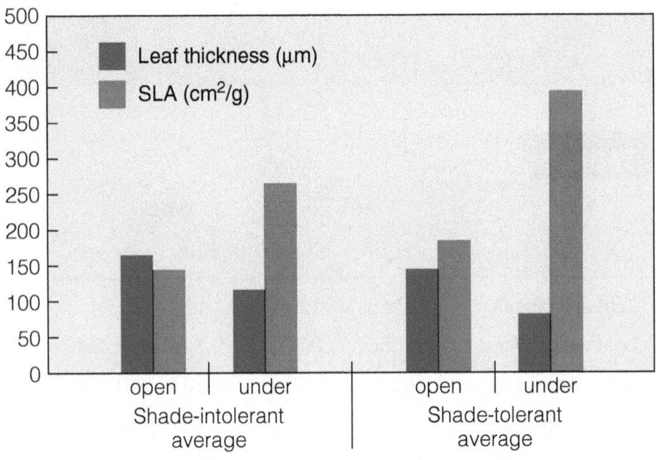

Figure 6.11 Differences in leaf morphology for shade-tolerant (average of seven species) and shade-intolerant (average of 11 species) tree species in central Wisconsin. Mean leaf thickness and specific leaf area (SLA is leaf area/leaf weight) are presented for leaves collected from both open (full sunlight) and under canopy (shaded) locations.

(Data from Abrams and Kubiske 1990.)

Interpreting Ecological Data

Q1. *How does specific leaf area (cm² of leaf area per gram of leaf weight) change for leaves grown in the shade as compared to the open (full sun)? Does the relationship differ for shade-tolerant and shade-intolerant species (same direction of change)?*

Q2. *How does the observed change in leaf thickness for leaves growing in the shade relate to the changes in specific leaf area (think about what a higher specific leaf area implies about leaf thickness)?*

Q3. *Are the observed changes in leaf morphology under shaded conditions an example of phenotypic plasticity?*

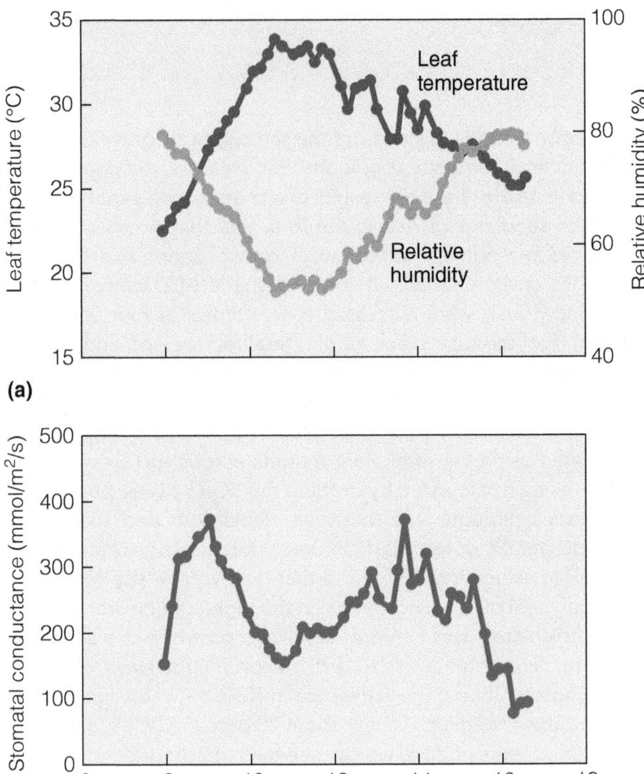

Figure 6.12 Diurnal changes in (a) temperature, relative humidity, and (b) stomatal conductance measured at the top of the canopy of *Fagus crenata* (Japanese beech).

(Data from Fukasawa et al. 2004.)

Interpreting Ecological Data

Q1. *How does relative humidity change with temperature? Why (see Section 2.5, Figure 2.15)?*

Q2. *How does stomatal conductance (the opening and closing of stomata) respond to changes in relative humidity?*

Q3. *What is the advantage to the plant of partially closing the stomata during the middle of the day? How would the decline in stomatal conductance influence net photosynthesis?*

through transpiration, but it also reduces CO_2 diffusion into the leaf and the dissipation of heat through evaporative cooling. As a result, the photosynthesis rate declines and leaf temperatures may rise. Some plant species, such as evergreen rhododendrons, respond to moisture stress by an inward curling of the leaves. Others show it in a wilted appearance caused by a lack of turgor in the leaves. Leaf curling and wilting allow leaves to reduce water loss and heat gain by reducing the surface area exposed to solar radiation.

Prolonged moisture stress inhibits the production of chlorophyll, causing the leaves to turn yellow or, later in the summer, to exhibit premature autumn coloration. As conditions worsen, deciduous trees may prematurely shed their leaves—the oldest ones dying first. Such premature shedding can result in dieback of twigs and branches.

Plants also exhibit developmental plasticity in response variations in the availability of water to meet the demands of transpiration. During development, plants respond to low soil water availability by increasing the allocation of carbon to the production of roots while decreasing the production of leaves (**Figure 6.13a**). By increasing its production of roots, the plant can explore a larger volume and depth of soil for extracting water. The reduction in leaf area decreases the amount of solar radiation the plant intercepts as well as the surface area that is losing water through transpiration. The combined effect is to increase the uptake of water per unit leaf area while reducing the total amount of water that is lost to the atmosphere through transpiration.

The decline in leaf area with decreasing water availability is actually a combined effect of reduced allocation of carbon to the production of leaves (**Figure 6.13b**) and changes in leaf morphology (size and shape). The leaves of plants grown under reduced water conditions tend to be smaller and thicker (lower specific leaf area; see Section 6.8) than those of individuals growing in more mesic (wet) environments (**Figure 6.13c**).

On an evolutionary timescale, a wide array of adaptations has evolved in plant species in response to variations in the availability of water relative to demand. In some species of

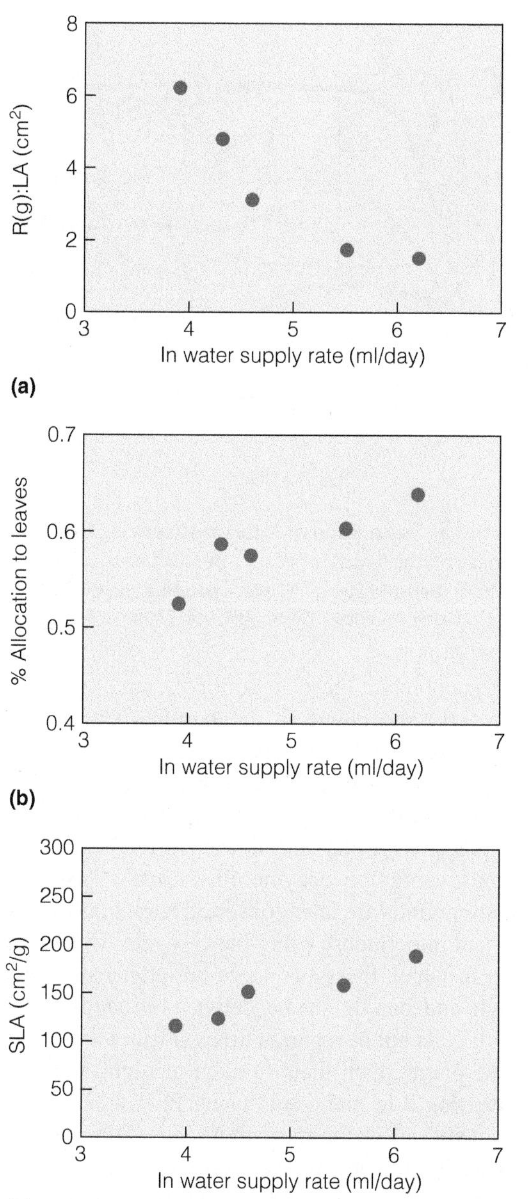

Figure 6.13 Relationship between plant water availability and the ratio of (a) root mass (mg) to leaf area (cm^2), (b) allocation to leaves, and (c) specific leaf area (SLA) for broadleaf peppermint (*Eucalyptus dives*) seedlings grown in the greenhouse. Each point on the graph represents the average value for plants grown under the corresponding water treatment. As water availability decreases, plants allocate more carbon to producing roots than to producing leaves. This increased allocation to roots increases the surface area of roots for the uptake of water, whereas the decline in leaf area decreases water loss through transpiration.
(Adapted from Austin et al. 2010.)

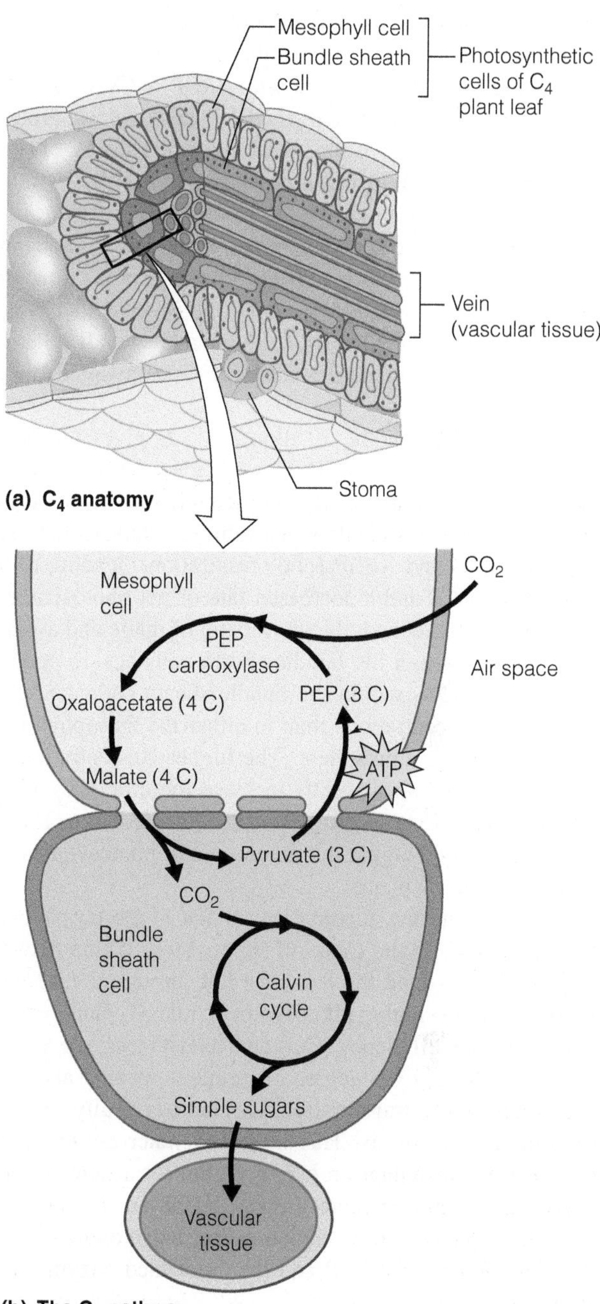

Figure 6.14 The pathway of photosynthesis. Different reactions take place in the mesophyll and bundle sheath cells. Compare to the C_3 pathway (see Figure 6.1). OAA, oxaloacetate; PEP, phosphoenolpyruvate.

plants, referred to as C_4 plants and CAM plants, a modified form of photosynthesis has evolved that increases water-use efficiency in warmer and drier environments. The modification involves an additional step in the conversion (fixation) of CO_2 into sugars.

In C_3 plants, the capture of light energy and the transformation of CO_2 into sugars occur in the mesophyll cells (see Section 6.1). The products of photosynthesis move into the vascular bundles, part of the plant's transport system, where they can be transported to other parts of the plant. In contrast, plants possessing the **C_4 photosynthetic pathway** have a leaf anatomy different from that of C_3 plants (see Figure 6.3). C_4 plants have two distinct types of photosynthetic cells: the mesophyll cells and the **bundle sheath cells**. The bundle sheath cells surround the veins or vascular bundles (**Figure 6.14**). C_4 plants divide photosynthesis between the mesophyll and the bundle sheath cells.

In C_4 plants, CO_2 reacts with phosphoenolpyruvate (PEP), a three-carbon compound, within the mesophyll cells. This is in contrast to the initial reaction with RuBP in C_3 plants. This reaction is catalyzed by the enzyme **PEP carboxylase**, producing oxaloacetate (OAA) as the initial product. The OAA is then rapidly transformed into the four-carbon molecules of malic and aspartic acids, from which the name C_4 photosynthesis is derived. These organic acids are then transported to the bundle sheath cells (see Figure 6.14). There, enzymes break down the organic acids to form CO_2, reversing the process that is carried out in the mesophyll cells. In the bundle sheath cells, the CO_2 is transformed into sugars using the C_3 pathway involving RuBP and rubisco.

The extra step in the fixation of CO_2 gives C_4 plants certain advantages. First, PEP does not react with oxygen, as does RuBP. This eliminates the inefficiency that occurs in the mesophyll cells of C_3 plants when rubisco catalyzes the reaction between O_2 and RuBP (photorespiration), leading to the production of CO_2 and a decreased rate of net photosynthesis (see Section 6.1). Second, the conversion of malic and aspartic acids into CO_2 within the bundle sheath cells acts to concentrate CO_2. The CO_2 within the bundle sheath cells can reach much higher concentrations than in either the mesophyll cells or the surrounding atmosphere. The higher concentrations of CO_2 in the bundle sheath cells increase the efficiency of the reaction between CO_2 and RuBP catalyzed by rubisco. The net result is generally a higher maximum rate of photosynthesis in C_4 plants than in C_3 plants.

To understand the adaptive advantage of the C_4 pathway, we must go back to the trade-off in terrestrial plants between the uptake of CO_2 and the loss of water through the stomata. Resulting from the higher photosynthetic rate, C_4 plants exhibit greater water-use efficiency (CO_2 uptake/H_2O loss; see Section 6.4). That is, for a given degree of stomatal opening and associated water loss in transpiration, C_4 plants typically fix more carbon in photosynthesis. This increased water-use efficiency can be a great advantage in hot, dry climates where water is a major factor limiting plant growth. However, it comes at a price. The C_4 pathway has a higher energy expenditure because of the need to produce PEP and the associated enzyme, PEP carboxylase.

The C_4 photosynthetic pathway is not found in algae, bryophytes, ferns, gymnosperms (includes conifers, cycads, and ginkgos), or the more primitive flowering plants (angiosperms). C_4 plants are mostly grasses native to tropical and subtropical regions and some shrubs characteristic of arid and saline environments, such as *Larrea* (creosote bush) and *Atriplex* (saltbush) that dominate regions of the desert southwest in North America. The distribution of C_4 grass species in North America reflects the advantage of the C_4 photosynthetic pathway under warmer and drier conditions (**Figure 6.15**). The proportion of grass species that are C_4 increases from north to south, reaching a maximum in the southwest.

In the hot deserts of the world, environmental conditions are even more severe. Solar radiation is high, and water is scarce. To counteract these conditions, a small group of desert plants, mostly succulents in the families Cactaceae (cacti), Euphorbiaceae, and Crassulaceae, use a third type of photosynthetic pathway—crassulacean acid metabolism (CAM). The **CAM pathway** is similar to the C_4 pathway in that CO_2 initially reacts with PEP and is transformed into four-carbon compounds using the enzyme PEP carboxylase. The four-carbon compounds are later converted back into CO_2, which is transformed into glucose using the C_3 cycle. Unlike C_4 plants, however, in which these two steps are physically separate (in mesophyll and bundle sheath cells), both steps occur in the mesophyll cells but at separate times (**Figure 6.16**).

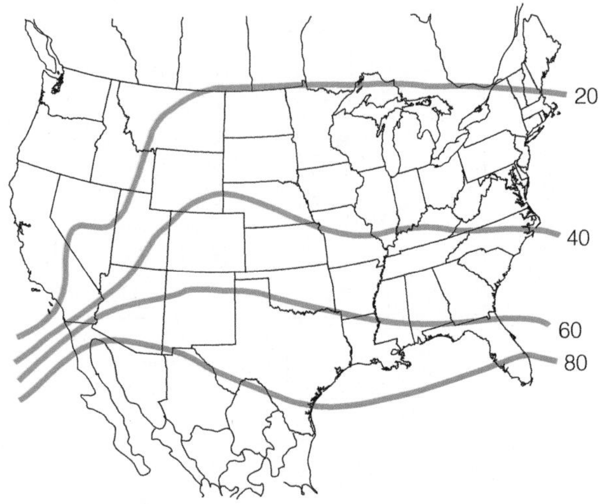

Figure 6.15 Percentage of total grass species that are C_4. Values on isopleths (lines) represent percentages.
(Adapted from Teeri and Stone, "Climatic patterns and the distribution of C4 grasses in North America," Oecologia, vol. 23, no. 1 1976, Fig. 1.)

CAM plants open their stomata at night, taking up CO_2 and converting it to malic acid using PEP, which accumulates in large quantities in the mesophyll cells. During the day, the plant closes its stomata and reconverts the malic acid into CO_2, which it then fixes using the C_3 cycle. Relative to both C_3 and C_4 plants, the CAM pathway is slow and inefficient in the fixation of CO_2. But by opening their stomata at night when temperatures are lowest and relative humidity is highest, CAM plants dramatically reduce water loss through transpiration and increase water-use efficiency.

In addition to adaptations relating to modifications of the photosynthetic pathway, plants adapted to different soil moisture environments exhibit a variety of physiological and morphological characteristics that function to allow them to either tolerate or avoid drought conditions. Plant species adapted to xeric conditions typically have a lower stomatal conductance (lower number and size of stomata) than species adapted to more mesic conditions. This results in a lower rate of transpiration but also functions to decrease rates of photosynthesis. Because of the higher diffusion gradient of water relative to CO_2, the reduction in stomatal conductance functions to increase water-use efficiency.

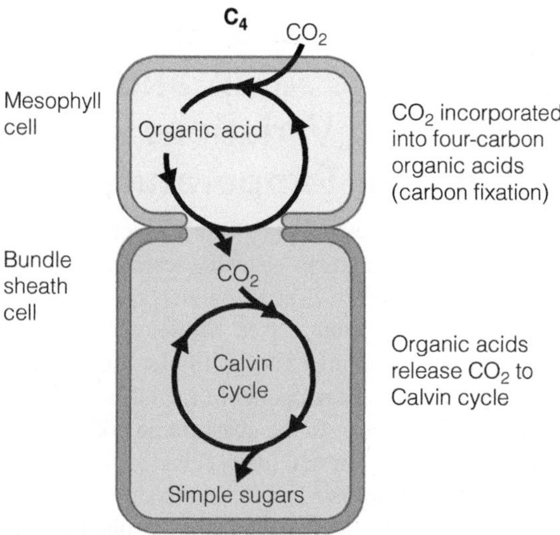

(a) Spatial separation of steps. In C_4 plants, carbon fixation and the Calvin cycle occur in different types of cells.

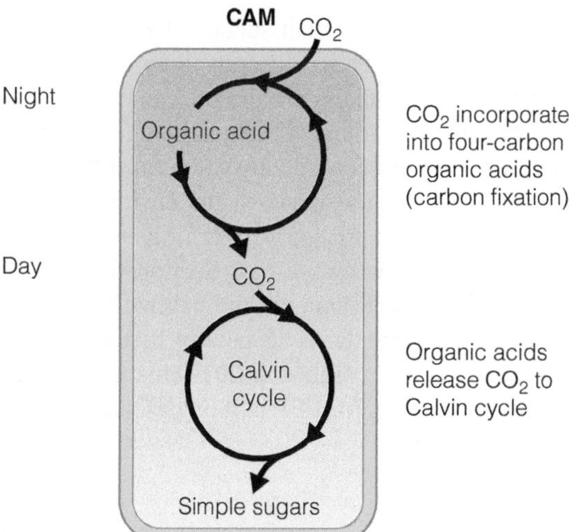

(b) Temporal separation of steps. In CAM plants, carbon fixation and the Calvin cycle occur in the same cells at different times.

Figure 6.16 Photosynthesis in CAM plants. (a) At night, the stomata open, the plant loses water through transpiration, and CO_2 diffuses into the leaf. CO_2 is stored as malate in the mesophyll to be used in photosynthesis by day. (b) During the day, when stomata are closed, the stored CO_2 is refixed in the mesophyll cells using the C_3 cycle.

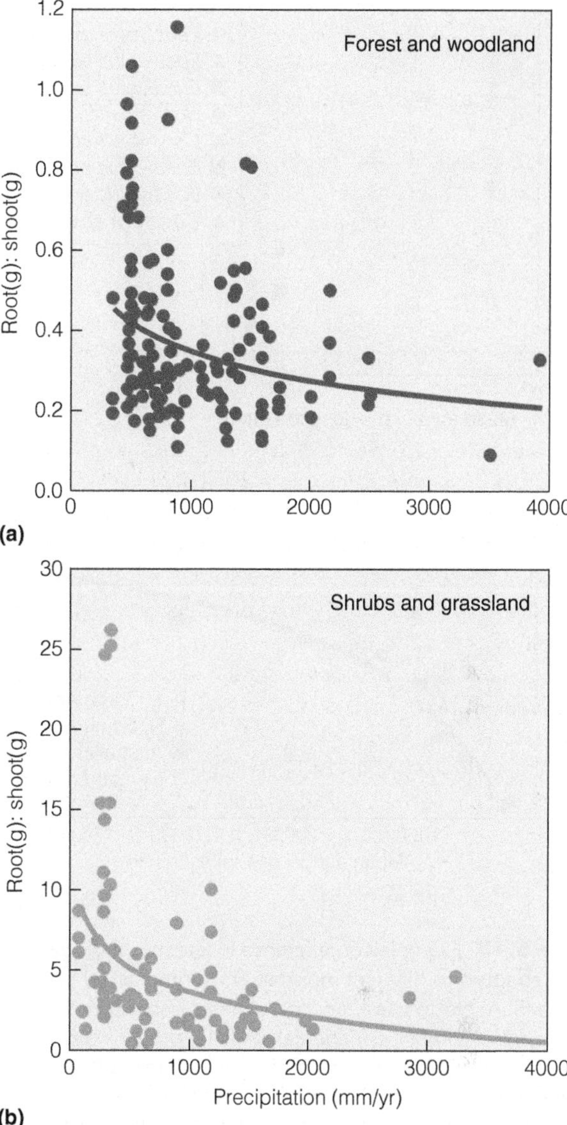

Figure 6.17 Root-to-shoot ratio as a function of mean annual precipitation for (a) forest and woodland ecosystems, and (b) shrub and grassland ecosystems.
(Data from Mokany et al. 2006.)

Plant species adapted to drier conditions tend to have a greater allocation of carbon to the production of roots relative to aboveground tissues (greater ratio of roots to shoots), particularly leaves (**Figure 6.17**). This pattern of carbon allocation allows the plant to explore a larger volume and depth of soil for extracting water. The decline in leaf area in more xeric environments is actually a combined effect of reduced allocation of carbon to the production of leaves and changes in leaf morphology (size and shape). The leaves of plant species adapted to xeric conditions tend to be smaller and thicker (lower specific leaf area; see Section 6.9) than those of species adapted to more mesic environments (**Figure 6.18**). In some plants, the leaves are small, the cell walls are thickened, the stomata are tiny, and the vascular system for transporting water is dense. Some species have leaves covered with hairs that scatter incoming solar radiation, whereas others have leaves coated with waxes and resins that reflect light and reduce its absorption. All these structural features function to reduce the amount of energy striking the leaf, enhance the dissipation of heat through convection (see Section 6.6, Figure 6.7), and thus,

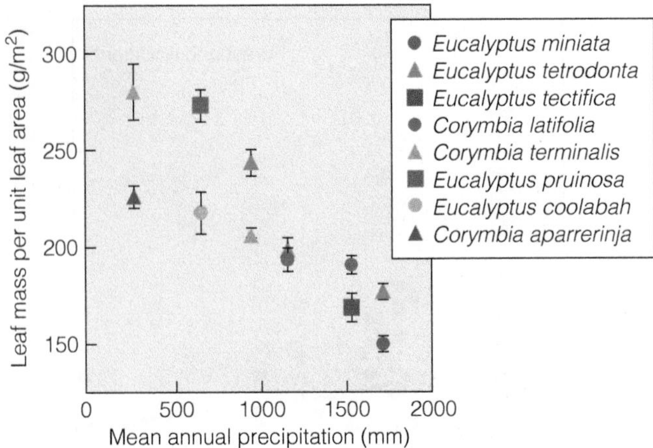

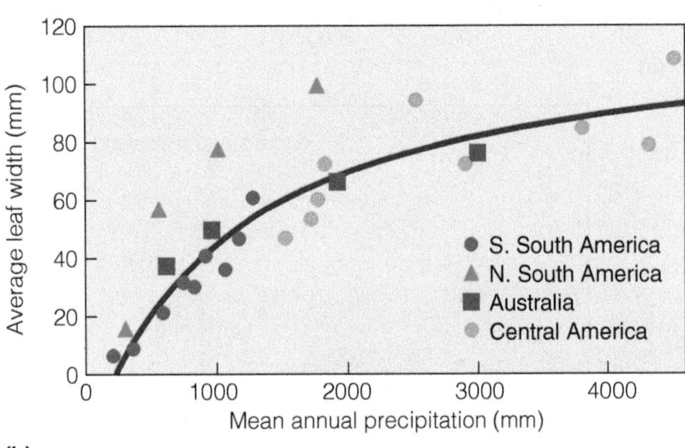

Figure 6.18 Examples of difference in leaf morphology for species adapted to different moisture environments. (a) Specific leaf area (SLA) plotted as a function of mean annual precipitation for tree species along a continental rainfall gradient in northern Australia. Each symbol represents the mean of the species. Error bars are one standard error. (Data from Cernusaka et al. 2011.) (b) Changes in average leaf width as a function of mean annual rainfall for species occupying different tropical forest ecosystems in South America, Central America, and Australia.

([a] Based on Cernusaka, L.A., L.B. Hutleya, J. Beringerc, J.A.M. Holtumd, and B.L. Turnere. 2011. *Agricultural and Forest Meteorology* 151: 1462–1470 Figure 2a, pg. 1465.)

reduce the loss of water through transpiration. In tropical regions with distinct wet and dry seasons, some species of trees and shrubs have evolved the characteristic of dropping their leaves at the onset of the dry season (see Section 2.6). These plants are termed *drought deciduous*. In these species, leaf senescence occurs as the dry season begins, and new leaves are grown just before the rainy season begins.

Although the decrease in leaf area and corresponding increase in biomass allocated to roots observed for plant species adapted to reduced water availability functions to reduce transpiration and increase the plant's ability to acquire water from the soil, this shift in patterns of allocation has consequences for plant growth. The reduced leaf area decreases carbon gain from photosynthesis resulting in a reduction in plant growth rate.

6.10 Plants Exhibit Both Acclimation and Adaptation in Response to Variations in Environmental Temperatures

As sessile organisms, terrestrial plants are subject to wide variations in temperature on a number of spatial scales and timescales. As we discussed in Chapter 2, at a continental to global scale, temperatures vary with latitude (see Section 2.2). At a local to regional scale, temperatures vary with elevation, slope, and aspect. Seasonal changes in temperature are influenced by both latitude and position relative to the coast (large bodies of water; see Section 2.7), whereas diurnal (daily) changes in temperature occur everywhere. These patterns of temperature variation are consistent and predictable, and evolution has resulted in a variety of adaptations that enable plants to cope with these variations.

When examined across a range of plant species inhabiting different thermal environments, T_{min}, T_{opt}, and T_{max} (see Figure 6.6) tend to match the prevailing environmental temperatures. Species adapted to cooler environments typically have a lower T_{min}, T_{opt}, and T_{max} than species that inhabit warmer climates (**Figure 6.19**). These differences in the temperature response of net photosynthesis are directly related to a variety of biochemical and physiological adaptations that act to shift the temperature responses of photosynthesis and respiration toward the prevailing temperatures in the environment. These differences are most pronounced between plants using the C_3 and C_4 photosynthetic pathways (see Section 6.9). C_4 plants inhabit warmer, drier environments and exhibit higher optimal temperatures for photosynthesis (generally between 30°C and 40°C) than do C_3 plants (**Figure 6.20**). This is in large part because of the higher T_{opt} for PEP carboxylase as compared to rubisco (see Section 6.9).

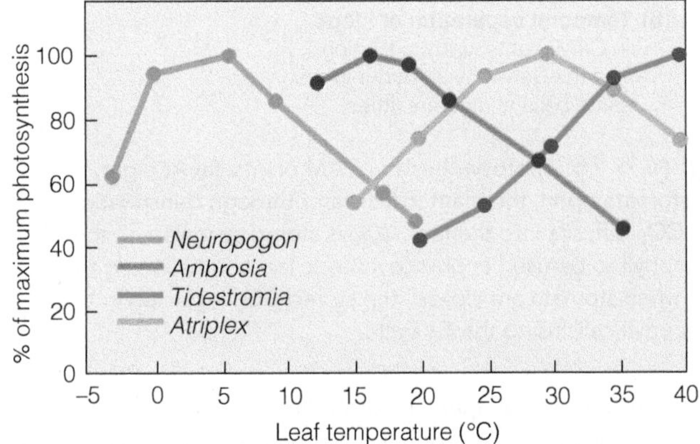

Figure 6.19 Relationship between net photosynthesis and temperature for various terrestrial plant species from dissimilar thermal habitats: *Neuropogon acromelanus* (Arctic lichen), *Ambrosia chamissonis* (cool, coastal dune plant), *Atriplex hymenelytra* (evergreen desert shrub), and *Tidestromia oblongifolia* (summer-active desert perennial).

(Adapted from Mooney et al. 1976.)

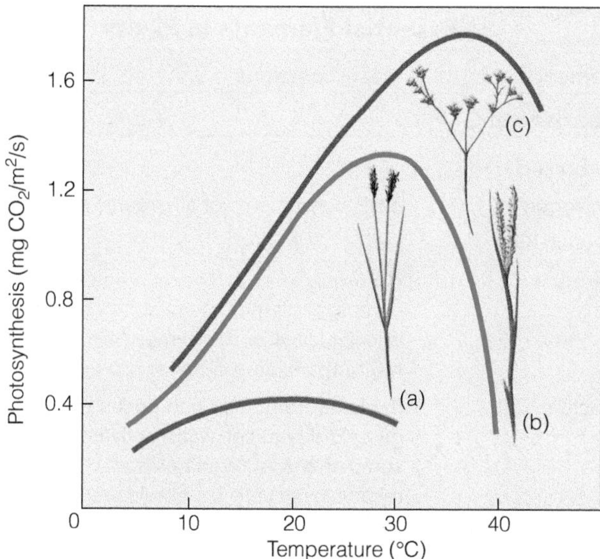

Figure 6.20 Effect of change in leaf temperature on the photosynthetic rates of C_3 and C_4 plants. (a) A C_3 plant, the north temperate grass *Sesleria caerulea*, exhibits a decline in the rate of photosynthesis as the temperature of the leaf increases. (b) A C_4 north temperate grass, *Spartina anglica*. (c) A C_4 shrub of the North American hot desert, *Tidestromia oblongifolia* (Arizona honeysweet). The maximum rate of photosynthesis for the C_4 species occurs at higher temperatures than for the C_3 species.

(Adapted from Bjorkman, O. [1973] "Comparative studies on photosynthesis in higher plants" page 53 in A.C. Geise, ed., Photophysiology [Academic Press].)

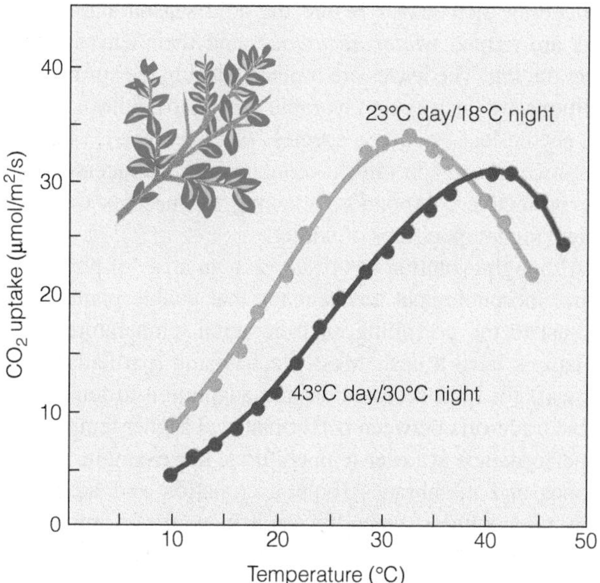

Figure 6.21 Relationship between temperature and net photosynthesis for cloned plants of big saltbush (*Atriplex lentiformis*) grown under two different day/night temperature regimes. The shift in T_{opt} corresponds to the temperature conditions under which the plants were grown.

(Adapted from Pearcy 1977.)

Although species from different thermal habitats exhibit different temperature responses for photosynthesis and respiration, these responses are not fixed. When individuals of the same species are grown under different thermal conditions in the laboratory or greenhouse, divergence in the temperature response of net photosynthesis is often observed (**Figure 6.21**). In general, the range of temperatures over which net photosynthesis is at its maximum shifts in the direction of the thermal conditions under which the plant is grown. That is to say, individuals grown under cooler temperatures exhibit a lowering of T_{opt}, whereas those individuals grown under warmer conditions exhibit an increase in T_{opt}. This same shift in the temperature response can be observed in individual plants in response to seasonal shifts in temperature (**Figure 6.22**). These modifications in the temperature response of net photosynthesis are a result of the process of acclimation—reversible phenotypic changes in response to changing environmental conditions (see Section 5.4).

In addition to the influence of temperature on plant carbon balance, periods of extreme heat or cold can directly damage plant cells and tissues. Plants that inhabit seasonally cold environments, where temperatures drop below freezing for periods of time, have evolved several adaptations for survival. The ability to tolerate extreme cold, referred to as frost hardening, is a genetically controlled characteristic that varies among species as well as among local populations of the same species. In seasonally changing environments, plants develop frost hardening through the fall and achieve maximum hardening in winter. Plants acquire frost hardiness—the turning of cold-sensitive cells into hardy ones—through the formation or addition of protective compounds in the cells. Plants synthesize and distribute substances such as sugars, amino acids, and other compounds that function as antifreeze, lowering the temperature at which freezing occurs. Once growth starts in spring, plants lose this tolerance quickly and are susceptible to frost damage in late spring.

Producing the protective compounds that allow leaves to survive freezing temperatures requires a significant expenditure of energy and nutrients. Some species avoid these costs

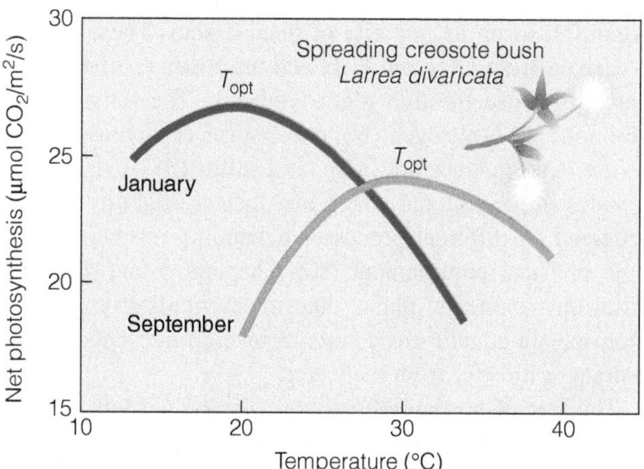

Figure 6.22 Seasonal shift in the relationship between net photosynthesis and temperature for creosote bush (*Larrea divaricata*) shrubs growing in the field. Note that T_{opt} shifts to match the prevailing temperatures.

(Adapted from Mooney et al [1976] "Photosynthetic capacity of in situ Death Valley plants" Carnegie Institute Yearbook 75:310–413.)

by shedding their leaves before the cold season starts. These plants are termed *winter deciduous*, and their leaves senesce during the fall. The leaves are replaced during the spring, when conditions are once again favorable for photosynthesis. In contrast, needle-leaf evergreen species—such as pine (*Pinus* spp.) and spruce (*Picea* spp.) trees—contain a high concentration of these protective compounds, allowing the needles to survive the freezing temperatures of winter.

Although evolution has resulted in an array of physiological and morphological mechanisms that enable plant species to adjust to the prevailing environmental temperatures, these adaptations have a cost. Most mechanisms (particularly biochemical) for both acclimation and adaptation to temperature involve trade-offs between performance at higher temperatures and performance at lower temperatures. For example, shifts of enzymes and membranes (both acclimation and adaptation) to low temperatures generally result in poor performance (or maladaptation) to high temperatures, that is, shifts in T_{min} are associated with a corresponding shift in T_{max}. In addition, reductions in T_{opt} are typically associated with a decline in maximum rates of net photosynthesis and growth.

6.11 Plants Exhibit Adaptations to Variations in Nutrient Availability

Plants require a variety of chemical elements to carry out their metabolic processes and to synthesize new tissues (**Table 6.1**). Thus, the availability of nutrients has many direct effects on plant survival, growth, and reproduction. Some of these elements, known as **macronutrients**, are needed in large amounts. Other elements are needed in lesser, often minute quantities. These elements are called **micronutrients**, or trace elements. The prefixes *micro–* and *macro–* refer only to the quantity of nutrients needed, not to their importance to the organism. If micronutrients are lacking, plants fail as completely as if they lacked nitrogen, calcium, or any other macronutrient.

Of the macronutrients, carbon (C), hydrogen (H), and oxygen (O) form the majority of plant tissues. These elements are derived from CO_2 and H_2O and are made available to the plant as glucose through photosynthesis. The remaining six macronutrients—nitrogen (N), phosphorus (P), potassium (K), calcium (Ca), magnesium (Mg), and sulfur (S)—exist in varying states in the soil and water, and their availability to plants is affected by different processes depending on their location in the physical environment (see Chapters 3 and 4). In terrestrial environments, plants take up nutrients from the soil. Autotrophs in aquatic environments take up nutrients from the substrate or directly from the water.

The rate of nutrient absorption (uptake per unit root) depends on concentrations in the external solution (soil or water; **Figure 6.23a**). As the availability (concentration) of nutrients at the root surface declines, the rate of absorption declines, which eventually results in a decline in tissue nutrient concentrations (**Figure 6.23b**). In the case of nitrogen, the decrease in leaf concentrations has a direct effect on maximum rates of photosynthesis (**Figure 6.23c**) through a reduction in the production of rubisco and chlorophyll (see Section 6.1). In

Table 6.1 Essential Elements in Plants

Element	Major Functions
Macronutrients	
Carbon (C) Hydrogen (H) Oxygen (O)	Basic constituents of all organic matter
Nitrogen (N)	Used only in a fixed form: nitrates, nitrites, and ammonium; component of chlorophyll and enzymes (such as rubisco); building block of protein
Calcium (Ca)	In plants, combines with pectin to give rigidity to cell walls; activates some enzymes; regulates many responses of cells to stimuli; essential to root growth
Phosphorus (P)	Component of nucleic acids, phospholipids, adenosine triphosphate (ATP), and several enzymes
Magnesium (Mg)	Essential for maximum rates of enzymatic reactions in cells; integral part of chlorophyll; involved in protein synthesis
Sulfur (S)	Basic constituent of protein
Potassium (K)	Involved in osmosis and ionic balance; activates many enzymes
Micronutrients	
Chlorine (Cl)	Enhances electron transfer from water to chlorophyll in plants
Iron (Fe)	Involved in the production of chlorophyll; is part of the complex protein compounds that activate and carry oxygen and transport electrons in mitochondria and chloroplasts
Manganese (Mn)	Enhances electron transfer from water to chlorophyll and activates enzymes in fatty-acid synthesis
Boron (B)	Fifteen functions are ascribed to boron in plants, including cell division, pollen germination, carbohydrate metabolism, water metabolism, maintenance of conductive tissue, and translation of sugar
Copper (Cu)	Concentrates in chloroplasts, influences photosynthetic rates, and activates enzymes
Molybdenum (Mo)	Essential for symbiotic relationship with nitrogen-fixing bacteria
Zinc (Zn)	Helps form growth substances (auxins); associated with water relationships; active in formation of chlorophyll; component of several enzyme systems
Nickel (Ni)	Necessary for enzyme functioning in nitrogen metabolism

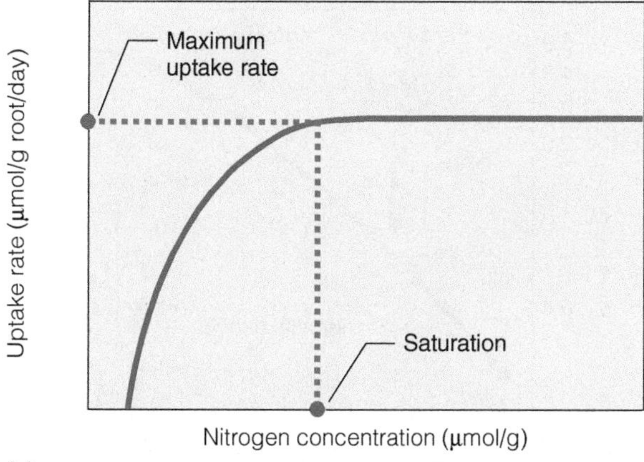

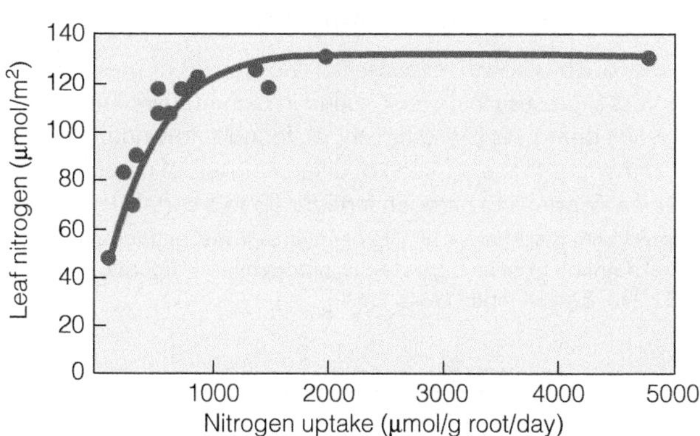

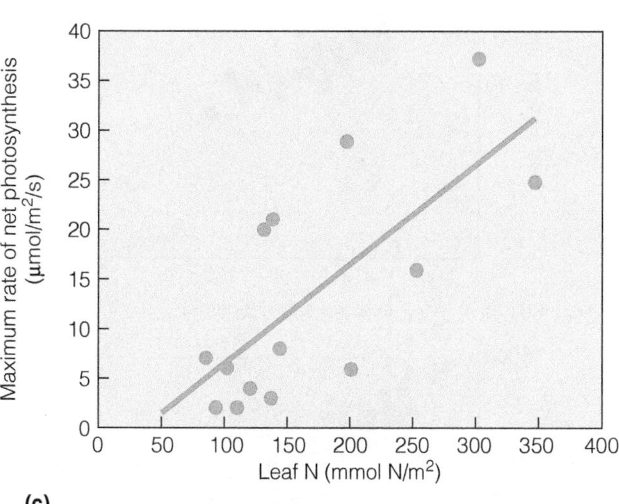

Figure 6.23 (a) Uptake of nitrogen by plant roots increases with concentration in soil until the plant arrives at maximum uptake. (b) Influence of root nitrogen uptake on leaf nitrogen concentrations. (c) Influence of leaf nitrogen concentrations on maximum observed rates of net photosynthesis for a variety of species from differing habitats.

(Adapted from Woodward and Smith 1994.)

fact, more than 50 percent of the nitrogen content of a leaf is in some way involved directly with the process of photosynthesis, with much of it tied up in these two compounds.

In response to reduced nutrient availability, carbon is allocated to root growth at the expense of shoot growth, resulting in an increase in the ratio of roots to shoots (**Figure 6.24**).

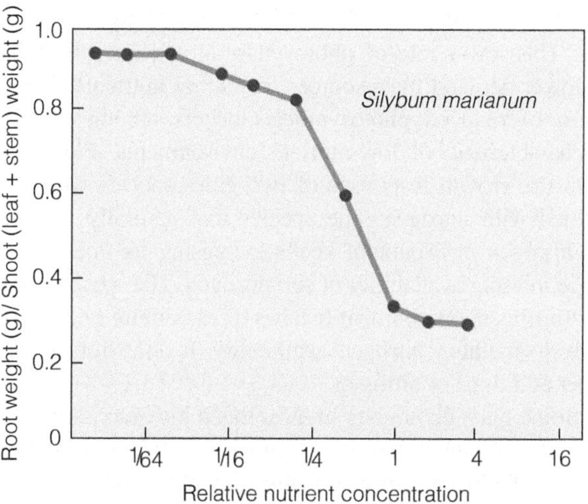

Figure 6.24 Changes in allocation to roots and shoots (root-to-shoot ratio) for *Silybum marianum* (milk thistle) grown along a gradient of nutrient availability. Nutrient levels (x-axis) represent 11 different nutrient treatments, ranging from 1/64 to 16 times the recommended concentration of standard greenhouse nutrient solution.

(Data from Austin et al. 1986.)

The increased production of roots is an example of phenotypic plasticity and allows the plant to compensate for the decrease in nutrient absorption per-unit root by increasing the root area and soil volume from which nutrients are absorbed. Despite the shift in patterns of carbon allocation to compensate for reduced nutrient availability, decreased rates of photosynthesis and reduced allocation to leaves resulting from increased allocation to roots inevitably lead to a reduction in growth rate.

We have seen that geology, climate, and biological activity alter the availability of nutrients in the soil (see Chapter 4). Consequently, some environments are relatively rich in nutrients, and others are poor. How do plants from low-nutrient environments succeed?

Species that inhabit low nutrient environments exhibit a wide array of phenotypic characteristics that enable them to survive, grow, and reproduce under reduced nutrient levels in the soil or water. Compared with species from more fertile soils, species characteristic of infertile soils usually exhibit a low absorption rate. Consequently, in comparison with species from high-nutrient environments, species from infertile soils absorb considerably less nutrient under high-nutrient conditions but similar quantities, and in some cases, even more nutrients at extremely low availability. In addition, plants adapted to low-nutrient environments generally have a greater allocation of carbon to the production of roots and subsequently have a higher ratio of roots to shoots.

Because plants require nutrients for synthesizing new tissue, in physiological terms, it is growth that creates demand for nutrients. Conversely, as we have seen, the plant's uptake rate of the nutrients directly influences its growth rate. This relationship may seem circular, but the important point is that not all plants have the same inherent (maximum potential) rate of growth. In Section 6.8 (see Figure 6.8), we saw how shade-tolerant plants have an inherently lower rate of photosynthesis and growth than shade-intolerant plants do, even under high-light conditions. This lower rate of photosynthesis and growth translates to a lower demand for resources, including nutrients. The same pattern of reduced photosynthesis occurs among plants that are characteristic of low-nutrient environments. **Figure 6.25** shows the growth responses of two grass species when soil is enriched with nitrogen. The species that naturally grows in a high-nitrogen environment keeps increasing its rate of growth with increasing availability of soil nitrogen. The species native to a low-nitrogen environment reaches its maximum rate of growth at low to medium nitrogen availability. It does not respond to further additions of nitrogen.

Some plant ecologists suggest that a low maximum growth rate is an adaptation to a low-nutrient environment. One advantage of slower growth is that the plant can avoid stress under low-nutrient conditions. A slow-growing plant can still maintain optimal rates of photosynthesis and other metabolic processes crucial for growth under low-nutrient availability. In contrast, a plant with an inherently high rate of growth will show signs of stress.

Another hypothesized adaptation to low-nutrient environments is leaf longevity (**Figure 6.26**). Leaf production has a cost to the plant. This cost can be defined in terms of the carbon and other nutrients required to grow the leaf. At a low rate of photosynthesis, a leaf needs a longer time to "pay back" the cost of its production. As a result, plants inhabiting low-nutrient environments tend to have longer-lived leaves. A good example is the dominance of pine species on nutrient poor, sandy soils in the coastal region of the southeastern United States. In contrast to deciduous tree species, which shed their leaves every year, these pines have needles that live for up to three years.

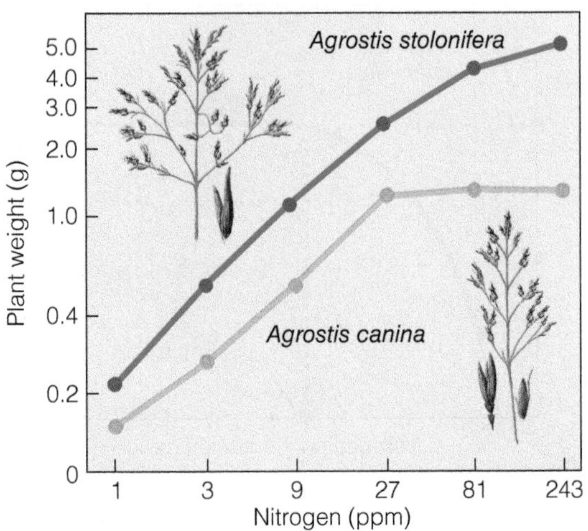

Figure 6.25 Growth responses of two species of grass—creeping bentgrass (*Agrostis stolonifera*), found in high-nutrient environments, and velvet bentgrass (*Agrostis canina*), found in low-nutrient environments—to the addition of different levels of nitrogen fertilizer. *A. canina* responds to nitrogen fertilizer up to a certain level only. (Adapted from Bradshaw et al., "Experimental investigations into the mineral nutrition of several grasses: IV. nirtogen level" Journal of Ecology, Vol. 52, No. 3 [November 1964], Fig.1.)

6.12 Plant Adaptations to the Environment Reflect a Trade-off between Growth Rate and Tolerance

As we have seen in the preceding discussion, plant adaptations to the abiotic (physical and chemical) environment represent a fundamental trade-off between phenotypic characteristics that enable high rates of photosynthesis and plant growth in high resource/energy environments and the ability to tolerate (survive, grow, and reproduce) low resource/energy conditions (**Figure 6.27**). The basic physiological processes, particularly photosynthesis, function optimally under warm temperatures and adequate supplies of light, water, and mineral nutrients. As environmental temperatures get colder and supplies of essential resources decline, plants respond through a variety of mechanisms that function to both increase access to the limiting resource or enhance the ability of the plant to function under the reduced resource/energy conditions. For example, species adapted to low-light environments exhibit lower rates of respiration that enable the maintenance of positive rates of photosynthesis under low-light levels (reduced LCP), but at the same time these characteristics reduce maximum rates of photosynthesis and plant growth under high-light levels. Likewise,

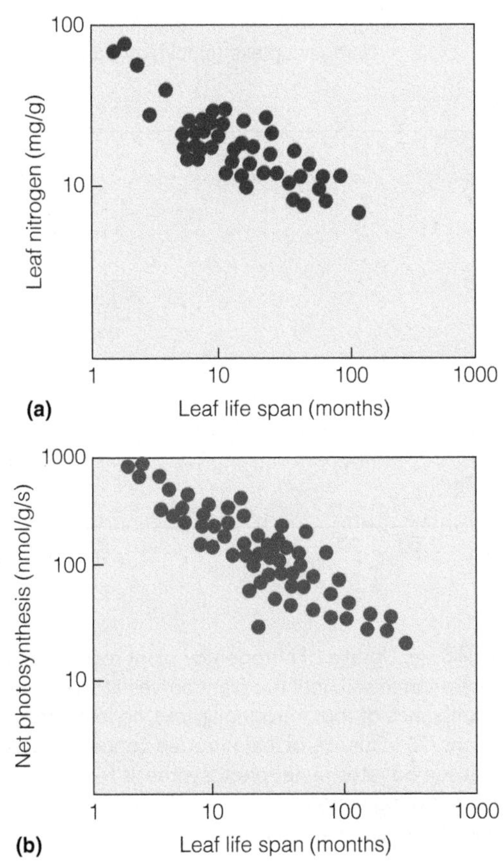

Figure 6.26 Relationship between (a) leaf longevity (life span) and leaf nitrogen concentration and (b) leaf longevity and net photosynthetic rate (maximum) for a wide variety of plants from different environments. Each data point represents a single species. Species having longer-lived leaves tend to have lower leaf nitrogen concentrations and subsequently lower rates of photosynthesis. (Adapted from Reich et al. 1982.)

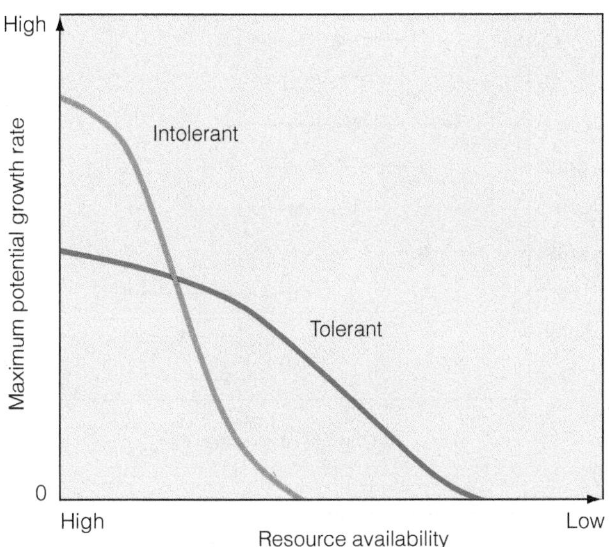

Figure 6.27 Trade-off in the strategies of high maximum potential growth rate under high resource availability (intolerant strategy) and the ability to continue survival and growth under low resource availability (tolerant strategy). Compare general pattern with Figures 6.5, 6.8, and 6.25.
(Adapted from Smith and Huston 1989.)

characteristics that enable plant species to successfully grow in arid and drought-prone environments, such as increased production of root, reduced leaf area, and smaller leaf size, enhance its ability to access water and reduce rates of water loss through transpiration; however, these same characteristics limit growth rates under mesic conditions.

What has emerged in our discussion is a general pattern of evolutionary constraints and trade-offs (costs and benefits) such that the set of phenotypic characteristics that enhance an organism's relative fitness under one set of environmental conditions inevitably limit its relative fitness under different environmental conditions. The set of phenotypic characteristics that enhance a species' carbon gain (photosynthesis and plant growth) under high resource/energy environments limit its tolerance (survival and growth) of low resource/energy conditions. Conversely, the phenotypic characteristics that enable a species to survive, grow, and reproduce under low resource/energy conditions limit its ability to maximize growth rate in high resource/energy environments. This basic concept will provide a foundation for later discussions of the interactions of plant species under different environmental conditions (e.g., competition) and how patterns of plant species distribution and abundance change across the landscape.

ECOLOGICAL Issues & Applications — Plants Respond to Increasing Atmospheric CO_2

In Chapter 2 we discussed that atmospheric concentrations of CO_2 have been rising exponentially since the mid-19th century from preindustrial levels of approximately 280 ppm to current levels of 400 ppm (as of June 2013; Chapter 2, *Ecological Issues & Applications*). In addition to influencing the planet's energy balance (Section 2.1) and the pH of the oceans (Chapter 3, *Ecological Issues & Applications*), rising atmospheric concentrations of CO_2 have a direct influence on terrestrial plants.

Recall that CO_2 diffuses from the air into the leaf through the stomatal openings (see Section 6.3). The rate of diffusion is a function of two factors: the diffusion gradient (the difference in CO_2 concentration between the air and the leaf interior) and stomatal conductance. Therefore, for a given stomatal conductance, an increase in the CO_2 concentration of the air will increase the diffusion gradient, subsequently increasing the movement of CO_2 into the leaf interior. In turn, the increased concentration of CO_2 within the leaf (mesophyll cells) will result in a greater rate of photosynthesis. The higher rates of diffusion and photosynthesis under elevated atmospheric concentrations of CO_2 have been termed the **CO_2 fertilization effect**.

The increased rate of photosynthesis under elevated CO_2 is in large part a result of reduced photorespiration (Section 6.1). The higher internal concentrations of CO_2 increase the affinity of rubisco to catalyze the reaction of RuBP with CO_2 (photosynthesis) rather than with O_2 (photorespiration). Because photorespiration can reduce photosynthetic rates by as much as 25 percent, the reduction, or even elimination of photorespiration under elevated CO_2 greatly enhances potential rates of net photosynthesis. Because C_4 plants avoid photorespiration (see discussion of C_4 pathway in Section 6.9) they do not exhibit the same increase in photosynthesis under elevated CO_2 (**Figure 6.28**).

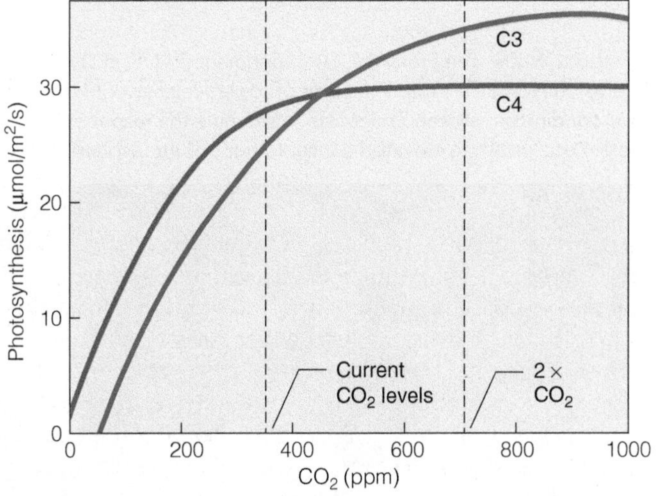

Figure 6.28 Net photosynthesis in relation to atmospheric CO_2 concentration. Although the specific values will vary for different species, in general, species with the C_3 photosynthetic pathway will show a greater relative benefit from elevated CO_2 than will C_4 species.

A second observed response of plants to elevated CO_2 is a reduction in stomatal conductance. Recall that stomatal conductance has two components: the number of stoma per unit area (stomatal density) and aperture (the size of the stomatal openings). In the short term, the observed decrease in stomatal conductance under elevated CO_2 is caused by a reduction in the aperture (partial closure of the stomata); in the long term, developmental plasticity has been shown to result in a decline in the stomatal density. As with the partial closure of stomata in response to decreased relative humidity (see Section 6.3 and Figure 6.12), this decrease in stomatal conductance functions to reduce rates of transpiration to a greater degree than CO_2 uptake and photosynthesis, and therefore results in an increase in water-use efficiency (ratio of photosynthesis to transpiration; see Section 6.3).

Most of our fundamental understanding about the response of plants to elevated CO_2 has come from experiments in controlled environments, greenhouses, and open-top chambers. However, because these techniques can alter the environment surrounding the plant, the use of free-air CO_2 enrichment (FACE) experiments—in which plants are grown at elevated CO_2 in the field under fully open-air conditions—provide scientists with the best estimates of how plants will respond to increasing atmospheric concentrations of CO_2 in natural ecosystems (**Figure 6.29**). Elizabeth Ainsworth and Alistair Rogers of the University of Illinois conducted a meta-analysis of the results of FACE experiments and summarized the current understanding of the response of plant species to elevated CO_2 concentrations. Averaged across all plant species grown at elevated CO_2 (567 ppm) in FACE experiments, stomatal conductance was reduced by 22 percent (**Figure 6.30a**). There was significant variation among different plant groups. On average, trees, shrubs, and forbs showed a lower percentage decrease in stomatal conductance as compared to C_3 and C_4 grasses and herbaceous crop species. Although studies have shown a reduction in stomatal density in a wide variety of species when grown under elevated CO_2, the observed decrease in stomatal conductance in the FACE experiments was not significantly influenced by a change in stomatal density.

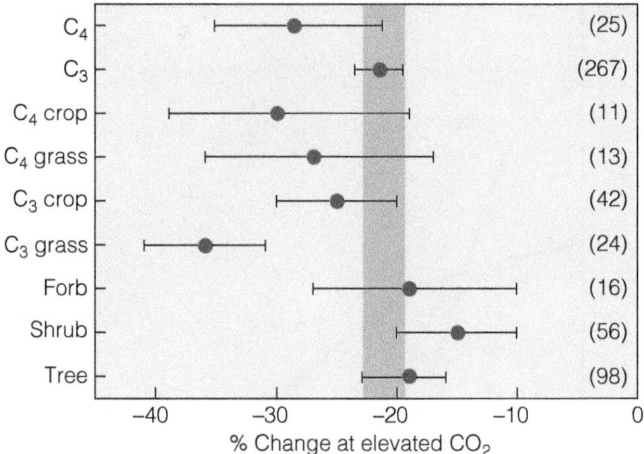

(a)

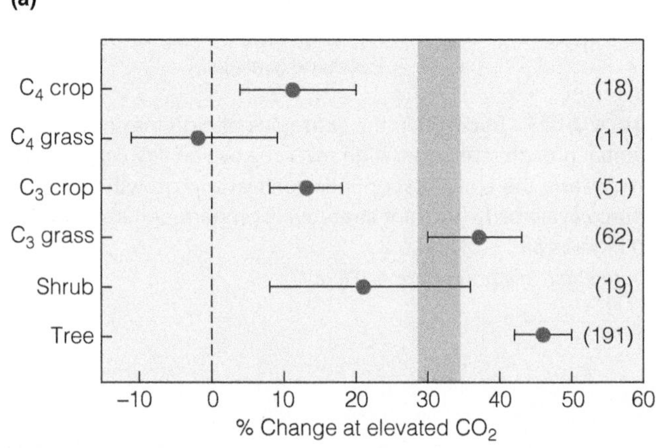

(b)

Figure 6.30 Meta-analysis of the response of (a) stomatal conductance (gs) and (b) light-saturated rates of net photosynthesis to elevated CO_2 in free-air CO_2 enrichment (FACE) experiments. The ambient and elevated CO_2 for all studies averaged 366 and 567 μmol/mol, respectively. The grey bar represents the overall mean and 95 percent confidence interval of all measurements. The symbol represents the mean response (±95 percent CI) of C_3 and C_4 species and different functional groups. (Data from Ainsworth and Rogers 2007.)

Figure 6.29 The Free Air CO_2 Experiment (FACE) at Duke Forest in North Carolina. The circle of towers releases CO_2 into the surrounding air, allowing scientists to examine the response of the forest ecosystem to elevated concentrations of atmospheric CO_2.

Elevated CO_2 stimulated light-saturated photosynthetic rates (see Figure 6.2) in C_3 plants grown in FACE experiments by an average of 31 percent (**Figure 6.30b**). The magnitude of increase in photosynthetic rates, however, varied with plant type and environment. Trees showed the largest response to elevated CO_2, whereas shrub species showed the smallest response. There was a surprising increase in photosynthetic rates of C_4 crop species, however; this stimulation of photosynthesis at elevated CO_2 was an indirect effect of reduced stomatal conductance. The reduction in stomatal conductance is associated with improved soil water status (**Figure 6.31**) as a result of reduced transpiration. Increased

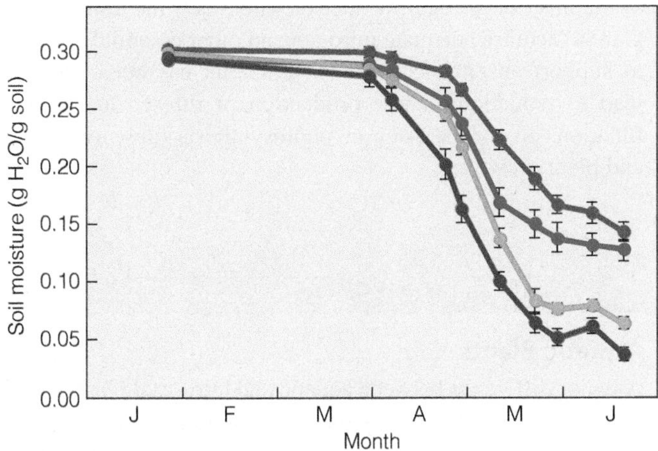

Figure 6.31 Response of soil moisture to increased CO_2 with open-topped chambers in California grassland in 1993. Soil moisture (g H_2O/g soil) at depths of 0.0–0.15 m (circles) and 0.15–0.30 m (squares) is shown. Dashed lines and open symbols represent ambient CO_2 concentrations, and solid lines and closed symbols represent elevated CO_2 concentration. The higher soil water content under elevated CO_2 is a result of reduced stomatal conductance and transpiration rates, extending the period where water is available for plant growth.
(Data from Field et al. 1995.)

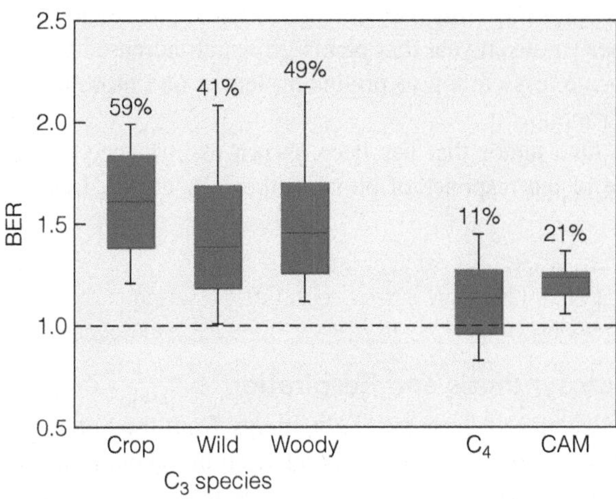

Figure 6.32 Distribution of biomass enhancement ratio (BER) for several functional types of species. BER is the ratio of biomass growth at elevated and ambient levels of CO_2. Distributions are based on 280 C_3, 30 C_4, and 6 CAM species. C_3 species were separated into three groups: crop, wild herbaceous, and woody species. Boxes indicate the distribution of the range of observation. Line represents median value; lower box, 25th percentile; and upper box, 75th percentile. Error bars give 10th and 90th percentile.
(Data from Poorter and Pérez-Soba 2002.)

Interpreting Ecological Data

Q1. How does soil moisture at both soil depths (0.0–0.15 m [circles] and 0.15–0.30 m [squares]) differ between ambient and elevated CO_2 chambers during the month of June?

Q2. How is the increase in soil moisture under elevated CO_2 a result of reduction in stomatal conductance and lower rates of transpiration?

Q3. How might the increased soil moisture under elevated CO_2 during the summer months affect net photosynthesis of plants (in addition to the direct enhancement of net photosynthesis by elevated CO_2)?

rates of photosynthesis in the C_4 crops sorghum and maize (corn) were associated with improved water status or were limited to periods of low rainfall.

The effects of long-term exposure to elevated CO_2 on plant growth and development, however, may be more complicated. Plant ecologists Hendrik Poorter and Marta Pérez-Soba of Utrecht University in the Netherlands reviewed the results from more than 600 experimental studies examining the growth of plants at elevated CO_2 levels. These studies examined a wide variety of plant species representing all three photosynthetic pathways: C_3, C_4, and CAM (Section 6.9). Their results revealed that C_3 species respond most strongly to elevated CO_2, with an average increase in biomass of 47 percent (**Figure 6.32**). Data on the response of CAM species were limited, but the mean response for the six species reported was 21 percent. The C_4 species examined also responded positively to elevated CO_2, with an average increase of 11 percent.

On average within C_3 species, crop species show the highest biomass enhancement (59 percent) and wild herbaceous plants the lowest (41 percent). Most of the experiments with woody species were conducted with seedlings, therefore covering only a small part of their life cycle. The growth stimulation of woody plants was on average 49 percent.

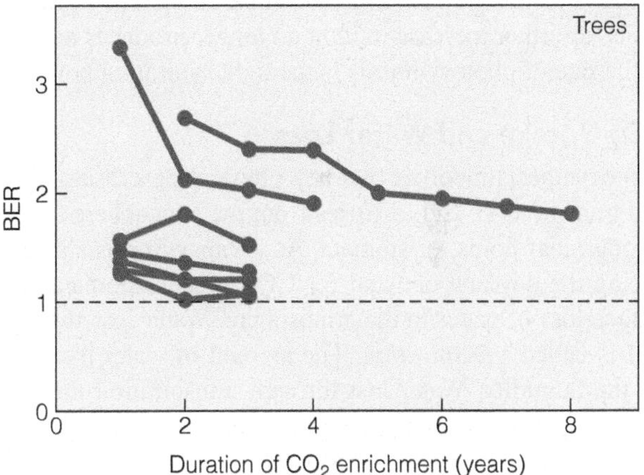

Figure 6.33 Time course of biomass enhancement ratio (BER) resulting from elevated CO_2. BER is the ratio of biomass growth at elevated and ambient levels of CO_2. Each line represents the results of an experiment with a different tree species.
(Data from Poorter and Pérez-Soba 2002.)

In some longer-term studies, the enhanced effects of elevated CO_2 levels on plant growth have been short-lived (**Figure 6.33**). Some plants produce less of the photosynthetic enzyme rubisco at elevated CO_2, reducing photosynthesis to rates comparable to those measured at lower CO_2

concentrations; this phenomenon is known as *downregulation.* Other studies reveal that plants grown at increased CO_2 levels allocate less carbon to producing leaves and more to producing roots.

One factor that has been shown to influence the magnitude of the response of photosynthesis to elevated CO_2 is the availability of nitrogen. Under elevated CO_2 the ability of the plant to acquire adequate nitrogen and other essential resources to support an enhanced growth potential has been shown to lead to reductions in the production of rubisco and thereby functions to reduce rates of photosynthesis (downregulation) and plant growth.

SUMMARY

Photosynthesis and Respiration 6.1

Photosynthesis harnesses light energy from the Sun to convert CO_2 and H_2O into glucose. A nitrogen-based enzyme, rubisco, catalyzes the transformation of CO_2 into sugar. Because the first product of the reaction is a three-carbon compound, this photosynthetic pathway is called C_3 photosynthesis. Cellular respiration releases energy from carbohydrates to yield energy, H_2O, and CO_2. The energy released in this process is stored as the high-energy compound ATP. Respiration occurs in the living cells of all organisms.

Photosynthesis and Light 6.2

The amount of light reaching a plant influences its photosynthetic rate. The light level at which the rate of CO_2 uptake in photosynthesis equals the rate of CO_2 loss as a result of respiration is called the *light compensation point.* The light level at which a further increase in light no longer produces an increase in the rate of photosynthesis is the light saturation point.

CO_2 Uptake and Water Loss 6.3

Photosynthesis involves two key physical processes: diffusion and transpiration. CO_2 diffuses from the atmosphere to the leaf through leaf pores, or stomata. As photosynthesis slows down during the day and demand for CO_2 lessens, stomata close to reduce loss of water to the atmosphere. Water loss through the leaf is called *transpiration.* The amount of water lost depends on the humidity. Water lost through transpiration must be replaced by water taken up from the soil.

Water Movement 6.4

Water moves from the soil into the roots, up through the stem and leaves, and out to the atmosphere. Differences in water potential along a water gradient move water along this route. Plants draw water from the soil, where the water potential is the highest, and release it to the atmosphere, where it is the lowest. Water moves out of the leaves through the stomata in transpiration, and this reduces water potential in the roots so that more water moves from the soil through the plant. This process continues as long as water is available in the soil. This loss of water by transpiration creates moisture conservation problems for plants. Plants need to open their stomata to take in CO_2, but they can conserve water only by closing the stomata.

Aquatic Plants 6.5

A major difference between aquatic and terrestrial plants in CO_2 uptake and assimilation is the lack of stomata in submerged aquatic plants. In aquatic plants, there is a direct diffusion of CO_2 from the waters adjacent to the leaf across the cell membrane.

Plant Energy Balance 6.6

Leaf temperatures affect both photosynthesis and respiration. Plants have optimal temperatures for photosynthesis beyond which photosynthesis declines. Respiration increases with temperature. The internal temperature of all plant parts is influenced by heat gained from and lost to the environment. Plants absorb longwave and shortwave radiation. They reflect some of it back to the environment. The difference is the plant's net radiation balance. The plant uses some of the absorbed radiation in photosynthesis. The remainder must be either stored as heat in the plant and surrounding air or dissipated through the processes of evaporation (transpiration) and convection.

Interdependence of Plant Adaptations 6.7

A wide range of adaptations has evolved in plants in response to variations in environmental conditions. The adaptations exhibited by plants to these features of the environment are not independent for reasons relating to the physical environment and to the plants themselves.

Plant Adaptations to High and Low Light 6.8

Plants exhibit a variety of adaptations and phenotypic responses (phenotypic plasticity) in response to different light environments. Shade-adapted (shade-tolerant) plants have low photosynthetic, respiratory, metabolic, and growth rates. Sun plants (shade-intolerant) generally have higher photosynthetic, respiratory, and growth rates but lower survival rates under shaded conditions. Leaves in sun plants tend to be small, lobed, and thick. Shade-plant leaves tend to be large and thin.

Alternative Pathways of Photosynthesis 6.9

The C_4 pathway of photosynthesis involves two steps and is made possible by leaf anatomy that differs from C_3 plants. C_4 plants have vascular bundles surrounded by chlorophyll-rich bundle sheath cells. C_4 plants fix CO_2 into malate and aspartate in the mesophyll cells. They transfer these acids to the bundle sheath cells, where they are converted into CO_2. Photosynthesis then follows the C_3 pathway. C_4 plants are characterized by high water-use efficiency (the amount of carbon fixed per unit of water transpired). Succulent desert plants, such as cacti, have a third type of photosynthetic pathway, called CAM. CAM plants open their stomata to take in CO_2 at night, when the humidity is high. They convert CO_2 to malate, a four-carbon compound. During the day, CAM plants close their stomata, convert malate back to CO_2, and follow the C_3 photosynthetic pathway.

Adaptations to Temperature 6.10

Plants exhibit a variety of adaptations to extremely cold as well as hot environments. Cold tolerance is mostly genetic and varies among species. Plants acquire frost hardiness through the formation or addition of protective compounds in the cell, where these compounds function as antifreeze. The ability to tolerate high air temperatures is related to plant moisture balance.

Plant Adaptations to Nutrient Availability 6.11

Terrestrial plants take up nutrients from soil through the roots. As roots deplete nearby nutrients, diffusion of water and nutrients through the soil replaces them. Availability of nutrients directly affects a plant's survival, growth, and reproduction. Nitrogen is important because rubisco and chlorophyll are nitrogen-based compounds essential to photosynthesis. Uptake of nitrogen and other nutrients depends on availability and demand. Plants with high nutrient demands grow poorly in low-nutrient environments. Plants with lower demands survive and grow, slowly, in low-nutrient environments. Plants adapted to low-nutrient environments exhibit lower rates of growth and increased longevity of leaves.

Trade-off between Growth and Tolerance 6.12

Plant adaptations to the abiotic environment represent a fundamental trade-off between phenotypic characteristics that enable high rates of photosynthesis and plant growth in high resource/energy environments and the ability to tolerate (survive, grow, and reproduce) under low resource/energy conditions.

Plant Response to Elevated CO_2 Ecological Issues & Applications

Plants exhibit two primary responses to CO_2: an increase in photosynthesis and a reduction in stomatal conductance. The increase in photosynthesis occurs primarily in C_3 plant species and is a response to reduced photorespiration. The decrease in stomatal conductance functions to increase water-use efficiency. Increased rates of photosynthesis result in an increase in growth rates.

STUDY QUESTIONS

1. How does the fact that the enzyme rubisco catalyzes the reaction between RuBP and O_2 as well as the reactions between RuBP and CO_2 affect the rate of net photosynthesis?
2. How would an increase in the rate of cellular respiration influence the rate of net photosynthesis?
3. What is the role of light (PAR) in the process of photosynthesis?
4. In the relationship between net photosynthesis and available light (PAR) shown in Figure 6.2, there is a net loss of CO_2 by the leaf at levels of light below the light compensation point (LCP). Why does this occur? Based on this relationship, how do you think net photosynthesis varies over the course of the day?
5. How does diffusion control the uptake of CO_2 and the loss of water from the leaf?
6. How does the availability of water to a plant constrain the rate of photosynthesis?
7. What is the advantage of the C_4 photosynthetic pathway as compared to the conventional C_3 pathway? How might these advantages influence where these plant species are found?
8. What is the advantage of a lower LCP for plant species adapted to low-light environments? What is the cost of maintaining a low LCP?
9. How do plants growing in shaded environments respond developmentally (through phenotypic plasticity) to increase their photosynthetic surface area?
10. How does a decrease in soil water availability influence the allocation of carbon (photosynthates) to the production of roots?
11. What is the basis for the relationship between leaf nitrogen concentration and rate of net photosynthesis shown in Figure 6.23c?
12. How could increased leaf longevity (longer-lived leaves) function as an adaptation to low-nutrient environments?
13. For a given stomatal conductance, how will an increase in atmospheric concentrations of CO_2 possibly influence the rate of photosynthesis? How might this affect the water-use efficiency of the plant?

FURTHER READINGS

Classic Studies

Bjorkman, O., and P. Holmgren. 1963. "Adaptability of the photosynthetic apparatus to light intensity in ecotypes from exposed and shaded habitats." *Physiologia Plantarum* 16:889–914.

One of the early, classic studies of plants' adaptation to variations in the light environment.

Chapin, F. S. 1980. "The mineral nutrition of wild plants." *Annual Review of Ecology and Systematics* 11:233–260.

This is an excellent review of the role of nutrients in controlling plant growth in natural ecosystems.

Grime, J. P. 1971. *Plant strategies and vegetative processes*. New York: Wiley.

An excellent integrated overview of plant adaptations to the environment. This book describes how the various features of a plant's life history, from seed to adult, reflect adaptations to different habitats and constraints imposed on plant survival, growth, and reproduction.

Schulze, E. D., R. H. Robichaux, J. Grace, P. W. Randel, and J. R. Ehleringer. 1987. "Plant water balance." *Bioscience* 37:30–37.

A good introduction to plant water balance that is well written, well illustrated, and not too technical. For students who wish to expand their understanding of the topic, it complements the materials presented in this chapter.

Woodward, F. I. 1987. *Climate and plant distribution*. Cambridge, UK: Cambridge University Press.

An excellent overview of plant energy and water balance, as well as plant adaptations to climate. It is easy to read, well referenced, and concise.

Current Research

Ainsworth, E. A., and S. P. Long. 2005. "What have we learned from 15 years of free-air CO_2 enrichment (FACE)? A meta-analytic review of the responses of photosynthesis, canopy properties and plant production to rising CO_2." *New Phytologist* 165:351–372.

Lambers, H., F. S. Chapin III, and T. L. Pons, 2008. *Plant physiological ecology*. 2nd ed. New York: Springer.

For more information, read this technical (but well written, illustrated, and organized) book that delves further into the processes presented in this chapter.

Larcher, W. 2003. *Physiological plant ecology*. 4th ed. New York: Springer-Verlag.

An excellent reference on plant ecophysiology. Like the previous text, this is a fine reference book for more information on the materials presented in the chapter. Less technical, but also less comprehensive.

CHAPTER 7 Animal Adaptations to the Environment

A male brown lemur (*Eulemur fulvus*) feeds on fruits in the canopy of a rain forest in Madagascar (Africa). This omnivore feeds on fruits, leaves, and insects in the forest canopy, where it lives in small, cohesive groups of 3–12 individuals.

CHAPTER GUIDE

ALL AUTOTROPHS, WHETHER MICROSCOPIC phytoplankton or the giant sequoia trees of the western United States, derive their energy from the same process—photosynthesis. The story is quite different for animals. Because heterotrophic organisms derive their energy, and most of their nutrients, by consuming organic compounds contained in plants and animals, they encounter literally hundreds of thousands of different types of potential food items—packaged as the diversity of plant and animal species inhabiting Earth. For this reason alone, animal adaptation is a much more complex and diverse topic than that of plants (presented in Chapter 6). However, several key processes are common to all animals: acquiring and digesting food, absorbing oxygen, maintaining body temperature and water balance, and adapting to systematic variation in light and temperature (the diurnal and seasonal cycles). The aquatic and terrestrial environments also impose several fundamentally different constraints on animal adaptation. In our discussion, we examine the variety of adaptations that have evolved in animals that enable them to maintain the basic metabolic processes that allow them to survive, grow, and reproduce in the diversity of environments existing on Earth. In doing so, we will focus on the benefits and constraints imposed by specific adaptations and how the trade-offs involved influence the organisms' success under different environmental conditions.

7.1 Size Imposes a Fundamental Constraint on the Evolution of Organisms

Living organisms occur in a wide range of sizes (**Figure 7.1**). The smallest animals are around 2–10 micrograms [μg], and the largest living animals are mammals (the blue whale weighing more than 100,000 kilograms (kg) in marine environments and the African elephant at 5000 kg on land). Each taxonomic group of animals has its own particular size range, largely as a result of morphological and physiological constraints. Some groups such as Bryozoa (aquatic colonial animals) contain species all within one or two orders of magnitude, whereas mammals are hugely variable in size. The smallest mammal is a species of shrew weighing only about 2 grams (g) fully grown—or about 100 million (10^8) times less than the blue whale.

Size has consequences for structural and functional relationships in animals, and as such, presents a fundamental constraint on adaptation. Most morphological and physiological features change as a function of body size in a predictable way—by a process known as **scaling**.

Geometrically similar objects, such as cubes or spheres, are referred to as being *isometric* (Greek for "having equal measurement"). The surface area (SA) and volume (V) of isometric objects are related to their linear dimensions (length = l) to the second and third power, respectively. For example, the surface area of a square is l^2, where l is the length of each side. Therefore the surface area of a cube of length l is $6l^2$ (six sides). In contrast, the volume of the cube is l^3 (**Figure 7.2a**).

Bryozoa are aquatic colonial animals, with individuals as small as 0.5 mm and 5 μg.

The largest mammal in marine environments is the blue whale weighs more than 100,000 kg.

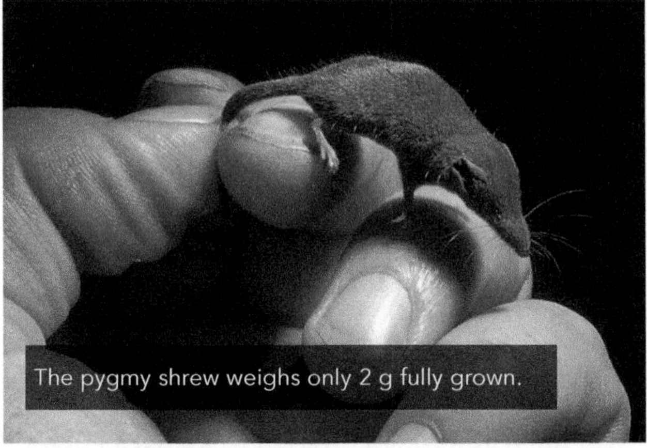
The pygmy shrew weighs only 2 g fully grown.

The African elephant weighs 5000 kg.

Figure 7.1 Living organisms occur in a wide range of body size.

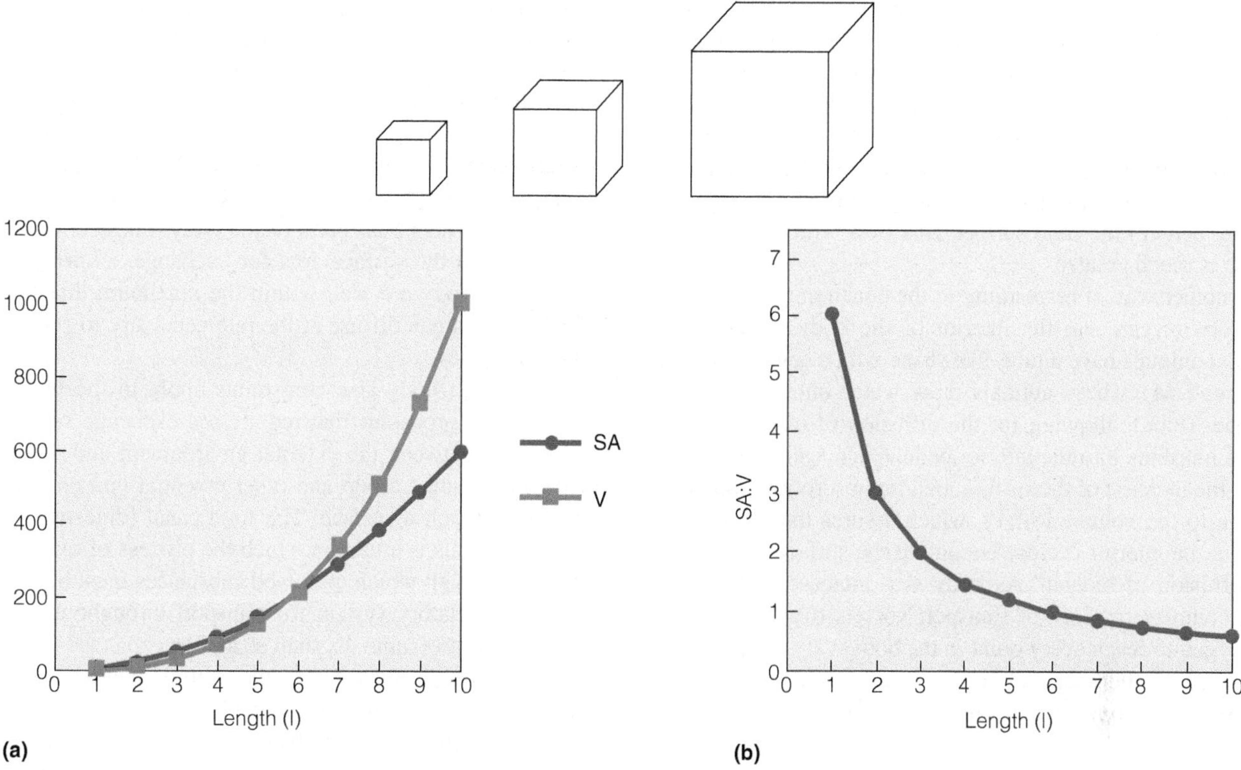

Figure 7.2 (a) The surface area (*SA*) of a cube increases as a square of length ($6l^2$), while the volume (*V*) increases as a function of l^3. (b) As a result, the ratio of surface area to volume (*SA*:*V*) decreases with increasing length (*l*). This relationship holds true for any isometric object and constrains the ability of organisms to exchange energy and matter with the external environment with increasing body size.

Interpreting Ecological Data

Q1. *The volume of a cube of length 4 is 64 (or 4^3), and the surface area is 96 (or 6×4^2). Now consider a three-dimensional rectangle having the following dimensions of length (l), height (h), and width (w): l = 16, h = 2, w = 2. The volume is $l \times h \times w = 64$. The surface area is $4(l \times h) + 2(w \times h)$. Calculate the surface area. How does the SA:V ratio differ for these two objects with the same volume?*

Q2. *Which of the two objects (cube or three-dimensional rectangle) would be a more efficient design for exchanging substances between the surface and the body interior?*

An important consequence of the characteristics of isometric scaling is the relationship between the surface area and volume. If the ratio of surface area to volume (*SA*:*V*) is plotted against length (*l*) for a square, there is an inverse relationship between *SA*:*V* and *l* (**Figure 7.2b**); smaller bodies have a larger surface area relative to their volume than do larger objects of the same shape.

This relationship between surface area and volume imposes a critical constraint on the evolution of animals. The range of biochemical and physiological processes associated with basic metabolism (assimilation and respiration) requires the transfer of materials and energy between the organism's interior and its exterior environment. For example, most organisms depend on oxygen (O_2) to maintain the process of cellular respiration (see Section 6.1). Every living cell in the body, therefore, requires that oxygen diffuse into it to function and survive. Oxygen is a relatively small molecule that readily diffuses across the cell surface; in a matter of seconds, it can penetrate into a millimeter (mm) of living tissue. So the center of a spherical organism that is 1 mm in radius is close enough to the surface that as the organism uses oxygen in the process of respiration, its oxygen is replenished by a steady diffusion from the surface in contact with the external environment (air or water).

Now imagine a spherical organism with the radius of a golf ball: approximately 21 mm. It would now take more than an hour for oxygen to diffuse into the center. Although the layers of cells just below the surface would receive adequate oxygen, the continuous depletion of oxygen as it diffused toward the center and the greater distance over which oxygen would have to diffuse, would result in the death of the interior cells (and eventually the organism) because of oxygen depletion.

The problem is that, as the size (length or radius) of the organism increases, the surface area of the body across which oxygen diffuses into the organism decreases relative to the interior volume of the body that requires the oxygen (the *SA*:*V* ratio decreases as shown in Figure 7.2b). So how can animals respond to this constraint so that an adequate flow of oxygen may reach the entire interior of the body in larger organisms?

A more complex, convoluted, or wrinkled surface, as shown in **Figure 7.3**, functions to increase the surface area of an object having the same volume as the golf-ball-shaped organism. The difference is that now (1) no point on the interior of the organism is more than a few millimeters from the surface, and (2) the total surface area over which oxygen can diffuse is much greater.

Another way of responding to the constraint is to actively transport oxygen into the interior of the body. Many of the smallest animals have a tube-like shape with a central chamber (**Figure 7.4a**). These animals draw water into their interior chamber (tube), allowing for the diffusion of oxygen and essential nutrients into the interior cells. Once again, the end result is the increase of the surface area for absorption (diffusion) relative to the volume (*SA*:*V*), which assures that every point (cell) in the interior is close enough to the surface to allow for the diffusion of oxygen. As body size increases, however, a more complex network of transport vessels (tubes) is needed for oxygen to reach every point in the body.

Much of the shape of larger organisms is governed by the transport of oxygen and other essential substances to cells in the interior of the body. To allow for this, a complex set of anatomical structures has evolved in animals. Lungs function as interior chambers that bring oxygen close to blood vessels, where it can be transferred to molecules of hemoglobin for transport throughout the body. A circulatory system with a heart functioning as a pump assures that oxygen-containing blood is actively transported into the minute vessels or capillaries that permeate all parts of the body. These complex systems increase the surface area for exchange, assuring that all cells in the body are well within the maximum distance over which oxygen can diffuse at the rate necessary to support cellular respiration.

The same body size constraints apply to the wide range of metabolic processes that require the exchange of materials and energy between the external environment and the interior of the organism. Carbon and other essential nutrients must be taken in through a surface. The food canal (digestive system) in most animals is a tube in which the process of digestion occurs and through which dissolved substances must be absorbed into the circulatory system for transport throughout the body. In the smallest of animals, such as the Bryozoa (see Figure 7.1) or tube worms (**Figure 7.4b**), the central chamber into which water is drawn also functions as the food canal, where digestion occurs and substances are absorbed. Waste products then exit through the opening as water is expelled. In larger animals, the food canal is a tube extending from the mouth to the anus. As food travels through the tube it is broken down, and essential nutrients and amino acids are absorbed and transported into the circulatory system. The greater the surface area of the food

Figure 7.3 One way to increase the ratio of surface area to volume (*SA*:*V*) for an object of a given size (volume) is to alter the shape. Objects (a) and (b) have the same volume; but by creating a more complex, convoluted surface, object (b) has a much greater surface area (and *SA*:*V*) than does object (a). Note the similarity between the body form of the sea anemone (c) and object (b).

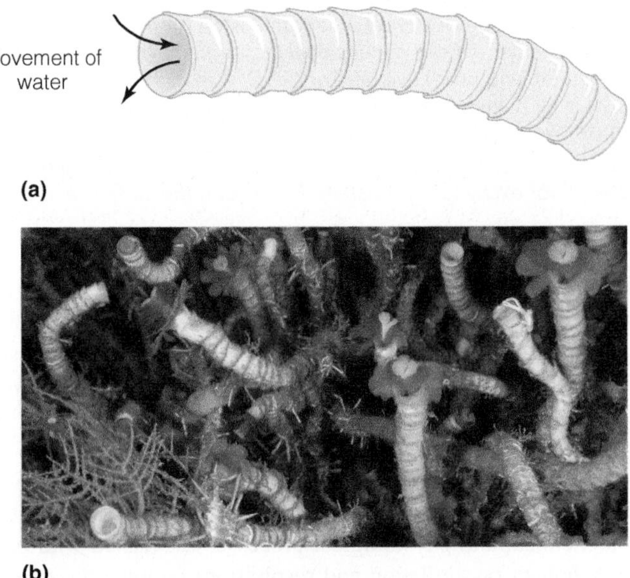

Figure 7.4 (a) A tube-shaped body with a central chamber greatly increases the surface area of an organism, providing a greater exchange surface with the external environment and reducing the distance from the surface to any point within the body interior. Note the similarity in design of (a) with that of tube-shaped organisms such as the tube worms shown in (b).

canal, the greater its ability to absorb food. Because surface area increases as the square of length, the larger the animal (which increases as a cube), the greater the surface area of its food canal must be to maintain a constant ratio of surface area to volume.

From these simple examples, it should be clear that greater body size requires complex changes in the organism's structure. These changes represent adaptations that maintain the relationship between the volume (or mass) of living cells that must be constantly supplied with essential resources from the outside environment and the surface area through which these exchanges occur.

We will examine various adaptations relating to the ability of animals to maintain the exchange of essential nutrients (food), oxygen, water, and thermal energy (heat) with the external environment. We will also consider how those adaptations are constrained by both body size and the physical environments in which the animals live (Section 7.11).

7.2 Animals Have Various Ways of Acquiring Energy and Nutrients

The diversity of potential energy sources in the form of plant and animal tissues requires an equally diverse array of physiological, morphological, and behavioral characteristics that enable animals to acquire (Figure 7.5) and assimilate these resources. There are many ways to classify animals based on the resources they use and how they exploit them. The most general of these classifications is the division based on how animals use plant and animal tissues as sources of food. Animals that feed exclusively on plant tissues are classified as **herbivores**. Those that feed exclusively on the tissues of other animals are classified as **carnivores**, and those that feed on both plant and animal tissues are called **omnivores**. In addition, animals that feed on dead plant and animal matter, called *detritus*, are detrital feeders, or **detritivores** (see Chapter 21). Each of these four feeding groups has characteristic adaptations that allow it to exploit its particular diet.

Herbivory

Because plants and animals have different chemical compositions, the problem facing herbivores is how to convert plant tissue to animal tissue. Animals are high in fat and proteins, which they use as structural building blocks. Plants are low in proteins and high in carbohydrates—many of them in the form of cellulose and lignin in cell walls, which have a complex structure and are difficult to break down (see Chapter 21). Nitrogen is a major constituent of protein. In plants, the ratio of carbon to nitrogen is about 50:1. In animals, the ratio is about 10:1.

Herbivores are categorized by the type of plant material they eat. Grazers feed on leafy material, especially grasses. Browsers feed mostly on woody material. Granivores feed on seeds, and frugivores eat fruit. Other types of herbivorous animals, such as avian sapsuckers (*Sphyrapicus* spp.) and sucking insects such as aphids, feed on plant sap; hummingbirds, butterflies, and a variety of moth and ant species feed on plant nectar (nectivores).

Grazing and browsing herbivores, with some exceptions, live on diets high in cellulose (complex carbohydrates made up of hundreds or thousands of simple sugar molecules). In doing so, they face several dietary problems. Their diets are rich in carbon but low in protein. Most of the carbohydrates are locked in indigestible cellulose, and the proteins exist in chemical compounds. Lacking the enzymes needed to digest cellulose, herbivores depend on specialized bacteria and protists living in their digestive tracts. These bacteria and protozoans digest cellulose and proteins, and they synthesize fatty acids, amino acids, proteins, and vitamins.

The highest-quality plant food for herbivores, vertebrate and invertebrate, is high in nitrogen in the form of protein. As the nitrogen content of their food increases, the animals' assimilation of plant material improves, increasing growth, reproductive success, and survival. Nitrogen is concentrated in the growing tips, new leaves, and buds of plants. Its content declines as leaves and twigs mature and become senescent. Herbivores have adapted to this period of new growth. Herbivorous insect larvae are most abundant early in the growing season, and they complete their growth before the leaves

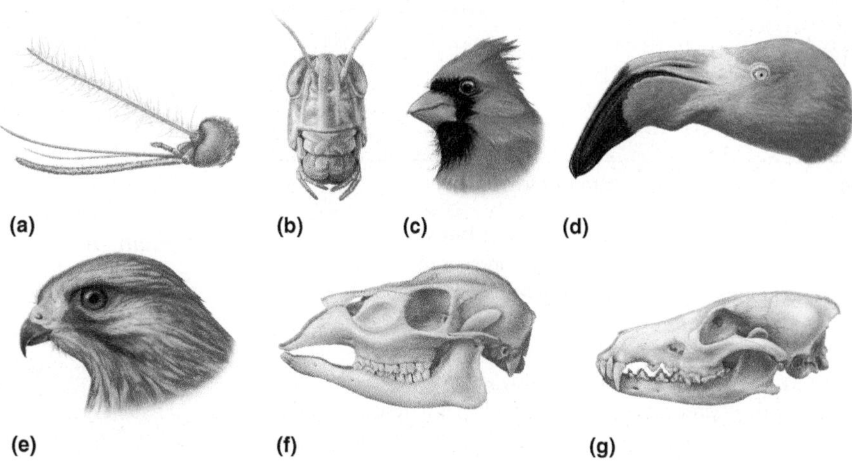

Figure 7.5 Mouthparts reflect how organisms obtain their food. (a) Piercing mouthparts of a mosquito. (b) Chewing mouthparts of a grasshopper. (c) Strong, conical bill of a seed-eating bird. (d) Straining bill of a flamingo. (e) Tearing beak of a hawk. (f) Grinding molars of an herbivore, a deer. (g) Canine and shearing teeth of a carnivorous mammal, the coyote.

mature. Many vertebrate herbivores, such as deer, give birth to their young at the start of the growing season, when the most protein-rich plant foods are available for their growing young.

Although availability and season strongly influence food selection, both vertebrate and invertebrate herbivores do show some preference for the most nitrogen-rich plants, which they probably detect by taste and odor. For example, beavers show a strong preference for willows (*Salix* spp.) and aspen (*Populus* spp.), two species that are high in nitrogen content. Chemical receptors in the nose and mouth of deer encourage or discourage consumption of certain foods. During drought, nitrogen-based compounds are concentrated in certain plants, making them more attractive and vulnerable to herbivorous insects. However, preference for certain plants means little if they are unavailable. Food selection by herbivores reflects trade-offs between quality, preference, and availability (see this chapter, **Field Studies: Martin Wikelski**).

Carnivory

Herbivores are the energy source for carnivores—the flesh eaters. Unlike herbivores, carnivores are not faced with problems relating to digesting cellulose or to the quality of food. Because the chemical composition of the flesh of prey and the flesh of predators is quite similar, carnivores encounter no problem in digesting and assimilating nutrients from their prey. Their major problem is obtaining enough food.

Among the carnivores, quantity is more important than quality. Carnivores rarely have a dietary problem because they consume animals that have resynthesized and stored protein and other nutrients from plants in their tissues.

Omnivory

Omnivores feed on both plants and animals. The food habits of many omnivores vary with the seasons, stages in the life cycle, and their size and growth rate. The red fox (*Vulpes vulpes*), for example, feeds on berries, apples, cherries, acorns, grasses, grasshoppers, crickets, beetles, and small rodents. The black bear (*Ursus americanus*) feeds heavily on vegetation—buds, leaves, nuts, berries, tree bark—supplemented with bees, beetles, crickets, ants, fish, and small- to medium-sized mammals.

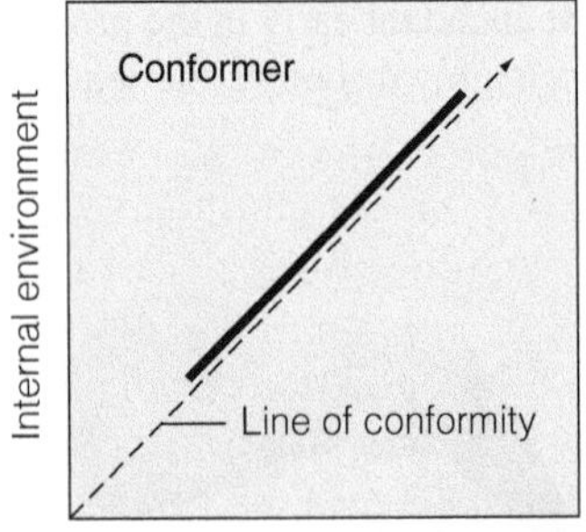

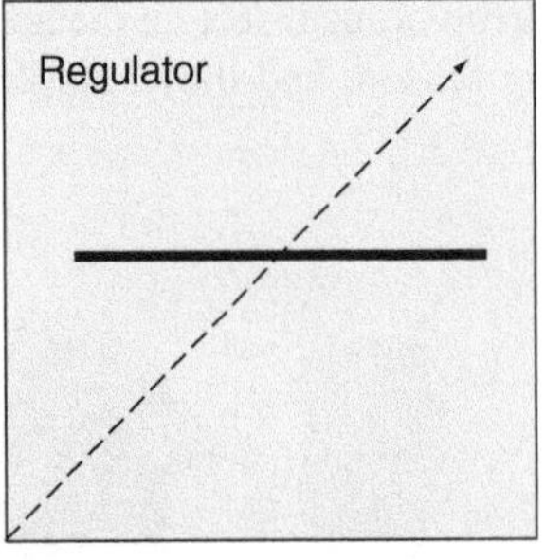

Figure 7.6 Contrast in the relationship between ambient (environmental) and body (internal) environment for an organism that is a (a) conformer and a (b) regulator. The internal body environment (temperature, oxygen, pH, solute concentration, etc.) of conformer organisms parallels the ambient conditions (line of conformity is a straight line where body environment = external environment), whereas a regulator organism maintains relatively constant internal body conditions regardless of external environmental conditions.

The means of food resource acquisition functions as a major selective agent in the process of natural selection, directly influencing the physiology, morphology, and behavior of animal species. From the specific behaviors and morphologies necessary to locate, capture, and consume different food resources (see Figures 5.10, 5.15, 5.20, 5.21, and 7.5 for specific examples), to the different enzymes and digestive systems necessary to break down and extract essential nutrients from the plant and animal tissues upon which they feed, the means of acquiring food resources has been a major force in the evolution of animal diversity.

7.3 In Responding to Variations in the External Environment, Animals Can Be either Conformers or Regulators

Some environments change little on timescales relevant to living organisms, such as the deep waters of the oceans. However, the majority of environments on our planet vary on a wide range of timescales. Regular annual, lunar, and daily cycles (see Chapters 2 and 3) present organisms with predictable changes in environmental conditions, whereas changes on a much shorter timescale of hours, minutes, or seconds as a result of weather are much less predictable. When an animal is confronted with changes in its environment, it can respond in one of two ways: conformity or regulation.

In some species, changes in external environmental conditions induce internal changes in the body that parallel the external conditions (**Figure 7.6a**). Such animals, called **conformers**, are unable to maintain consistent internal conditions such as body fluid salinity or levels of tissue oxygen. Echinoderms such as the starfish, for example, are osmoconformers whose internal body fluids quickly come to equilibrium with the seawater that surrounds them. The degree to which conformers can survive in changing environments depends largely on the tolerance of their body tissues to internal changes brought about by the changes in the external environment.

Conforming largely involves changes at the physiological and biochemical levels. If the internal conditions are allowed to vary widely, be it in terms of temperature, salinity, or oxygen supply, then tissues and cells will need to have biochemical systems in place that can continue to function under the new conditions. In extreme conditions, changes in these systems must be sufficient enough to keep the animal functional, even if at a low level, to avoid potentially irreversible damages, such as freezing, hypoxia (lack of oxygen), or osmotic water loss. Typically the biochemical and physiological changes that occur are simple and energetically inexpensive but carry the cost of reduced activity and growth.

Regulators, as their name implies, use a variety of biochemical, physiological, morphological, and behavioral mechanisms to regulate their internal environments over a broad range of external environmental conditions (**Figure 7.6b**). For example, in contrast to an osmoconformer, an osmoregulator maintains the ion concentrations of its body fluids within a limited range of values when faced with changes in the ion concentration of the surrounding water.

In contrast to conformity, regulation may require substantial and energetically expensive changes in biochemistry, physiology, morphology, and behavior. Behavior is often the first line of defense; however, behavior is augmented by substantial physiological and biochemical adjustments.

The strategies of conformity and regulation, therefore, have different costs and benefits. The benefit of conformity is a low energetic expenditure associated with mechanisms that maintain internal environmental conditions, but it results in reduced activity, growth, and reproduction as environmental conditions deviate from those that optimize the function of cells, tissues, and organs. In contrast, regulation is generally expensive. For example, regulation of body temperature in terrestrial animals may account for as much as 90 percent of their total energy budget. The benefit, however, is in the level of performance and the greatly extended range of environmental conditions over which activity can be maintained (see Section 7.11).

Although conformity and regulation represent two distinct strategies for coping with variations in the external environment, a single species may exhibit a different strategy under different environmental conditions or during different activities (**Figure 7.7**). Extreme environmental conditions may exceed the ability of a species to regulate internal conditions, resulting in conformity with external environmental conditions (see Section 7.12). In addition, a species may be a regulator with respect to one feature of the environment, such as oxygen, but a conformer with respect to another, such as temperature.

7.4 Regulation of Internal Conditions Involves Homeostasis and Feedback

Organisms that maintain their internal environment within narrow limits need some means of regulating internal conditions relative to the external environment, including body temperature, water balance, pH, and the amounts of salts in fluids and tissues. For example, the human body must maintain internal temperatures within a narrow range around 37°C. An increase or decrease of only a few degrees from this range could prove fatal. The maintenance of a relatively constant internal environment in a varying external environment is called **homeostasis**.

Whatever the processes involved in regulating an organism's internal environment, homeostasis depends on negative feedback—meaning that when a system deviates from the normal or desired state, referred to as the **set point**, mechanisms function to restore the system to that state. All feedback systems consist of a parameter or variable that is the focus of regulation (e.g., temperature or oxygen) and three components: receptor, integrator, and effector (**Figure 7.8**). The receptor measures the internal environment for the variable and transfers the information to the integrator. The integrator evaluates the information from the receptor (compares to set point) and determines whether action must be taken by the effector. The effector functions to modify the internal environment (the variable being regulated).

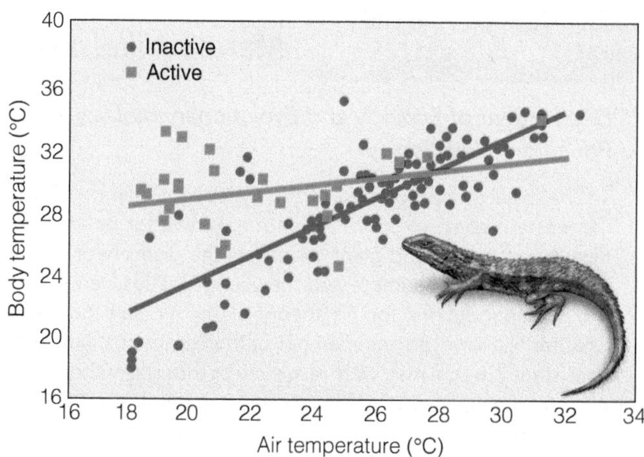

Figure 7.7 The relationship between air temperature and body temperature for the girdled lizard (*Cordylus macropholis*). The temperature of active lizards falls within a narrow range, regardless of air temperature, suggesting some form of regulation; however, the temperatures of inactive lizards depends on (conforms to) air temperature.
(Adapted from Bauwens et al. 1999.)

The thermostat that controls the temperature in your home is an example of a negative feedback system (see Figure 7.8). If we wish the temperature of the room to be 20°C (68°F), we set that point on the thermostat. When the temperature of the room air falls below that point, a temperature-sensitive device within the thermostat trips the switch that turns on the furnace. When the room temperature reaches the set point, the thermostat responds by shutting off the furnace. Should the thermostat fail to function properly and not shut off the furnace, then the furnace would continue to heat, the temperature would continue to rise, and the furnace would ultimately overheat, causing either a fire or a mechanical breakdown.

Among animals, the control of homeostasis is both physiological and behavioral. An example is temperature regulation in humans (see Figure 7.8). The normal temperature, or set point, for humans is 37°C. When the temperature of the environment rises, sensory mechanisms in the skin detect the change. They send a message to the brain, which automatically relays the message to receptors that increase blood flow to the skin, induces sweating, and stimulates behavioral responses. Water excreted through the skin evaporates, cooling the body. When the environmental temperature falls below a certain point, another reaction takes place. This time it reduces blood flow and causes shivering, an involuntary muscular exercise that produces more heat. If the environmental temperature

FIELD STUDIES Martin Wikelski

Department of Ecology and Evolutionary Biology, Princeton University, Princeton, New Jersey

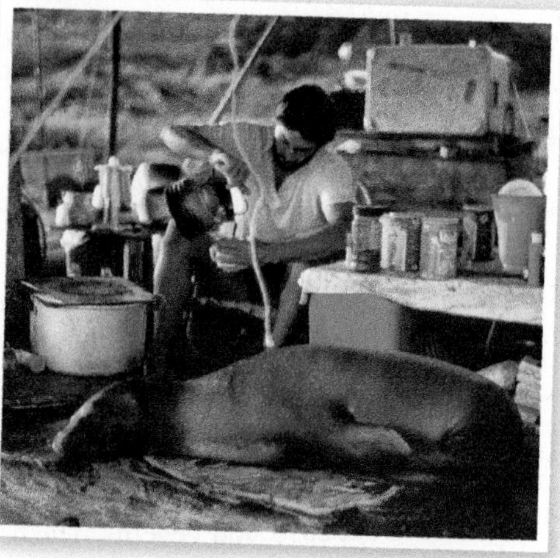

The isolated archipelago of the Galápagos Islands off the western coast of South America is known for its amazing diversity of animal and plant life. It was the diversity of life on these islands that so impressed the young Charles Darwin and laid the foundations for his theory of natural selection (see Chapter 5). However, one member of the Galápagos fauna has consistently been met with revulsion by historic visitors: the marine iguana, *Amblyrhynchus cristatus*. Indeed, even Darwin himself commented on this "hideous-looking creature."

Marine iguanas are widely distributed throughout the Galápagos Islands, and individuals of different populations vary dramatically in size (both length and weight). Because of these variations, many of the iguana populations were long considered separate species, yet modern genetic studies have confirmed that all of the populations are part of a single species. What could possibly account for the marked variation in body size among populations? This question has been central to the research of University of Konstanz ecologist Martin Wikelski. Studies by Wikelski and his colleagues during the past decade have revealed an intriguing story of the constraints imposed by variations in the environment of the Galápagos on the evolution of these amazing creatures.

In a series of studies, Wikelski and his colleagues have examined differences in body size between two populations of marine iguanas that inhabit the islands of Santa Fe and Genovesa. The populations of these two islands differ markedly in body size (as measured by the snout-vent length), with an average body length of 25 cm (maximum body weight of 900 g) for adult males on Genovesa as compared to 40 cm (maximum body weight of 3500 g) for adult males on the island of Santa Fe. Wikelski hypothesized that these differences reflected energetic constraints on the two populations in the form of food supply.

Marine iguanas are herbivorous reptiles that feed on submerged intertidal and subtidal algae (seaweed) along the rocky island shores, referred to as *algae pastures*. To determine the availability of food for iguana populations, Wikelski and colleagues measured the standing biomass and productivity of pastures in the tidal zones of these two islands. Their results show that the growth of algae pastures correlates with sea surface temperatures. Waters in the tidal zone off Santa Fe (the more southern island) are cooler than those off Genovesa, and as a result, both the length of algae plants and the productivity of pastures are five times greater off Santa Fe than Genovesa.

By examining patterns of food intake and growth of marked individuals on the two islands, Wikelski was able to demonstrate that food intake limits growth rate and subsequent body size in marine iguanas, which in turn depends on the availability of algae (**Figure 1**). Body size differences between members of the two island populations can be explained by differences in food availability.

Temporal variations in climate and sea surface temperatures also influence food availability for the marine iguanas across the Galápagos Islands. Marine iguanas can live for up to 30 years, and environmental conditions can change dramatically within an individual's lifetime. El Niño events usually recur at intervals of three to seven years but were more prevalent in the decade of the 1990s (see Section 2.9). During El Niño years in the Galápagos, sea surface temperatures increase from an average of 18°C to a maximum of 32°C as cold ocean currents and cold-rich upwellings are disrupted. As a result, green and red algal species—the preferred food of marine iguanas—disappear and are replaced by the brown algae, which the iguanas find hard to digest. Up to 90 percent of marine iguana populations on islands can die of starvation as a result of these environmental changes.

In studying patterns of mortality during the El Niño events of the 1990s, Wikelski observed the highest mortality rate among larger individuals. This higher mortality rate was directly related to observed differences in foraging efficiency with body size. Wikelski and colleagues determined that although larger individuals have a higher daily intake of food, smaller individuals have a higher food intake per unit body mass, a result of higher foraging efficiency (food intake per bite per gram body mass). Large iguanas on both islands showed a marked decline in body mass during the El Niño events. The result is a strong selective pressure against large body size during these periods of food shortage (**Figure 2**).

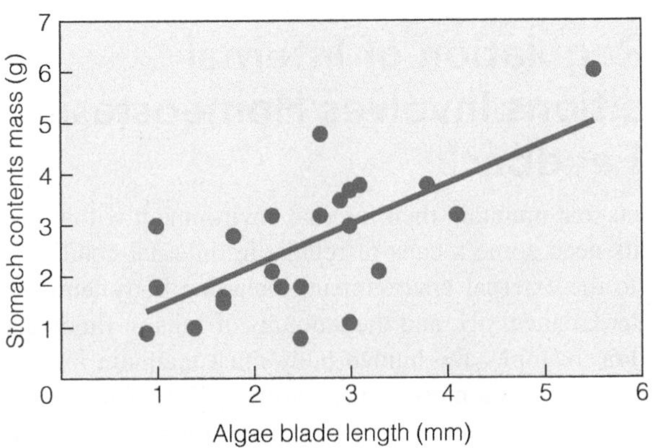

Figure 1 The food intake (dry mass in stomach) for iguanas of a given length (200–250 snout-vent length) from both study islands increased with increasing length of the algae pasture (estimated by average algae blade length) in the intertidal zone. (Adapted from Wikelski et al. 1997.)

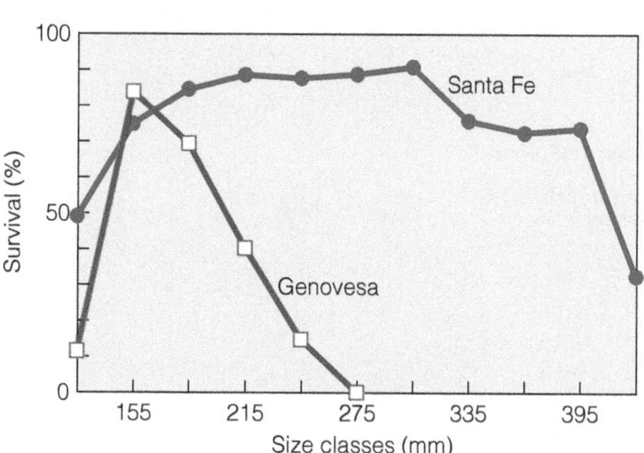

Figure 2 Survival of individually marked animals on Genovesa (squares) and Santa Fe (dots).
(Adapted from Wikelski and Trillmich 1997.)

Perhaps the most astonishing result of Wikelski's research is that the marine iguanas exhibit an unusual adaptation to the environmental variations caused by El Niño. Change in body length is considered to be unidirectional in vertebrates, but Wikelski repeatedly observed shrinkage of up to 20 percent in the length of individual adult iguanas. This shrinking coincided with low food availability resulting from El Niño events.

Shrinking did not occur equally across all size classes. Wikelski found an inverse relationship between the initial body size of individuals and the observed change in body length during the period of food shortage—larger individuals shrank less than smaller individuals.

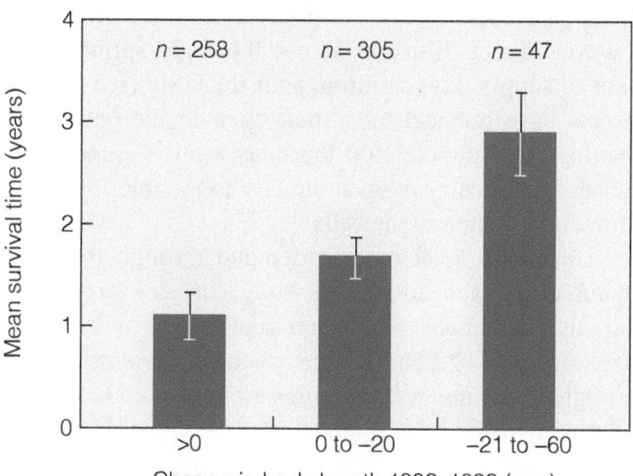

Figure 3 Relation between change in body length and survival time for adult iguanas on the island of Santa Fe during the 1997–1998 El Niño cycle. Values of *n* refer to sample size.
(Data from Wikelski and Thom 2000.)

The Galápagos marine iguana (*Amblyrhynchus cristatus*).

Shrinkage was found to influence survival. Large adult individuals that shrank more survived longer because their foraging efficiency increased and their energy expenditure decreased (**Figure 3**).

Given the disadvantage of larger body size during periods of low resource availability, what factors were selecting for larger body size in the marine iguana? What is the advantage of being big? Marine iguanas do not compete for food, either with other iguanas or other species of marine herbivores, and their potential predators are not size specific, so these factors were discounted as selective agents influencing body size. Instead Wikelski found that larger body size benefits males in attracting mates. Male iguanas establish display territories, and females select males for mating. Wikelski found that females favor larger males, and therefore larger males have greater reproductive success and relative fitness. It appears that the evolution of body size in the marine iguana is a continuous battle (trade-off) between the advantage of large body size in reproductive success and the disadvantage of large body size during regular periods of resource shortage.

Bibliography

Wikelski, M. 2005. "Evolution of body size in Galápagos marine iguanas." *Proceedings of the Royal Society B* 272:1985–1993.

Wikelski, M., and C. Thom. 2000. "Marine iguanas shrink to survive El Niño." *Nature* 403:37–38.

Wikelski, M., V. Carrillo, and F. Trillmich. 1997. "Energy limits on body size in a grazing reptile, the Galápagos marine iguana." *Ecology* 78:2204–2217.

Wikelski, M. and F. Trillmich. 1997. "Body size and sexual size dimorphism in marine iguanas fluctuate as a result of opposing natural and sexual selection: An island comparison." *Evolution* 51:922–936.

1. Does the mortality of iguanas during El Niño events represent a case of natural selection? Which of the three models of selection best describes the pattern of natural selection?
2. If the iguanas could not shrink during the period of resource shortage, how do you think the El Niño events would influence natural selection?

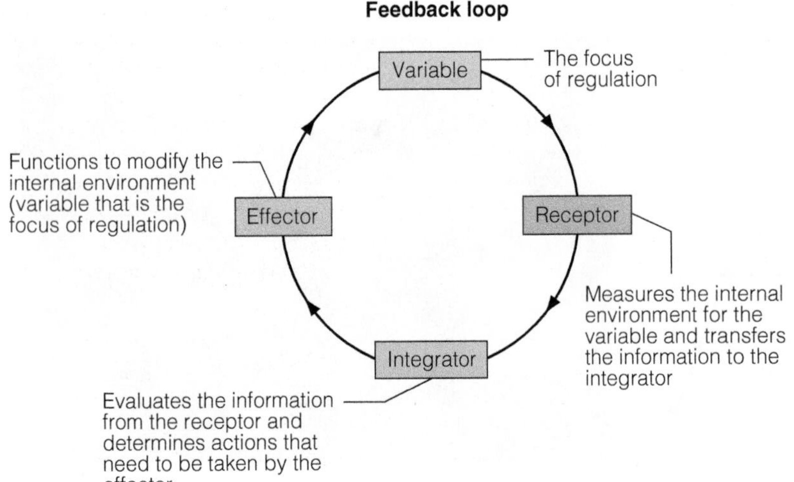

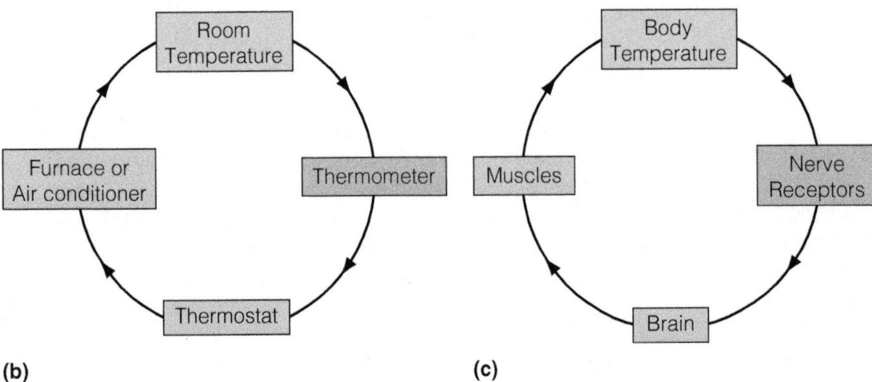

Figure 7.8 The role of negative feedback in the process of homeostasis. (a) Generalized structure of a negative feedback and examples of the concept applied to (b) temperature regulation in a room and (c) the human body.

becomes extreme, the homeostatic system breaks down. When it gets too warm, the body cannot lose heat fast enough to maintain normal temperature. Metabolism speeds up, further raising body temperature, until death results from heatstroke. If the environmental temperature drops too low, metabolic processes slow down, further decreasing body temperature until death by freezing ensues.

7.5 Animals Require Oxygen to Release Energy Contained in Food

Animals obtain their energy from organic compounds in the food they eat; and they do so primarily through aerobic respiration, which requires oxygen (see Section 6.1). Most organisms are oxygen regulators, maintaining their own oxygen consumption even when external (ambient) oxygen levels drop below normal. Oxygen conformity in which oxygen consumption decreases in proportion to decreasing ambient oxygen concentrations is found, however, in some smaller aquatic organisms.

Oxygen is easily available in the atmosphere for terrestrial animals. However, for aquatic animals, oxygen may be limiting and its acquisition problematic (see Section 3.6). Differences between terrestrial and aquatic animals in the means of acquiring oxygen reflect the availability of oxygen in the two environments. Minute terrestrial organisms take in oxygen by diffusion across the body surface. With increasing body size, however, direct diffusion across the body surface is insufficient to supply oxygen throughout the body (see Section 7.1). Insects have tracheal tubes that open to the outside through openings (or spiracles) on the body wall (**Figure 7.9a**). The tracheal tubes carry oxygen directly to the interior of the body allowing diffusion to the cells.

Unable to meet oxygen demand through the direct diffusion of oxygen across the body surface, larger terrestrial animals (mammals, birds, and reptiles) have some form of lungs (**Figure 7.9b**). Unlike tracheal systems that branch throughout the insect body, lungs are restricted to one location. Structurally, lungs have innumerable small sacs that increase surface area across which oxygen readily diffuses into the bloodstream. Amphibians take in oxygen through a combination of lungs and vascularized skin (containing blood vessels). Lungless salamanders are an exception; they live in a moist environment and take in oxygen directly through the skin.

In aquatic environments, organisms must take in oxygen from the water or gain oxygen from the air in some way. Marine

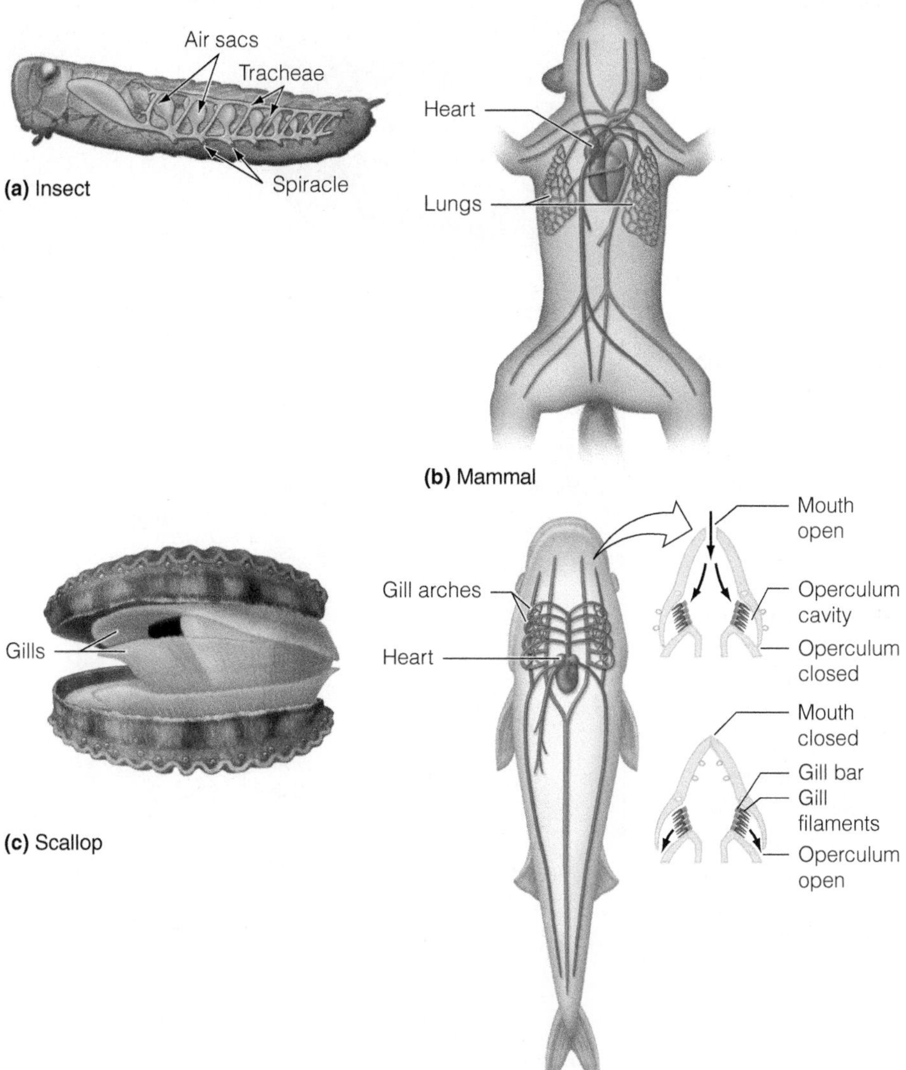

Figure 7.9 Respiratory systems. (a) The tracheal system and spiracles of an insect (grasshopper). Air enters the tracheal tubes through spiracles—openings on the body wall. (b) The lungs of a mammal. Location of gills on (c) scallop (marine invertebrate), and (d) fish. Water flows into the mouth as it opens (gill openings in closed position) and flows out over the gills as it closes (gill openings in open position).

mammals such as whales and dolphins come to the surface to expel carbon dioxide and take in air containing oxygen to the lungs. Some aquatic insects rise to the surface to fill the tracheal system with air. Others, like diving beetles, carry a bubble of air with them when submerged. Held beneath the wings, the air bubble contacts the spiracles of the beetle's abdomen.

A number of smaller aquatic animals are oxygen conformers, particularly sedentary marine invertebrates, most cnidarians (corals, jellyfish, and sea anemones), and echinoderms (starfish and sea urchins). Most, however, are oxygen regulators, and as with terrestrial animals, the mechanisms controlling oxygen uptake are related to size. Minute aquatic animals, zooplankton, take up oxygen from the water by diffusion across the body surface. Larger aquatic animals have gills, that is, outfoldings of the body surface that are suspended in the water and across which oxygen can diffuse. The gills of many aquatic invertebrates, such as starfish, are simple in shape and distributed over much of their body. In others, such as the crayfish or sea scallop, gills are restricted to specific regions of the body (**Figure 7.9c**). Fish, the major aquatic vertebrates, pump water through their mouth. The water flows over gills and exits through the back of the gill covers (**Figure 7.9d**). The close contact with and the rapid flow of water over the gills allows for exchanges of oxygen and carbon dioxide between water and the gills.

7.6 Animals Maintain a Balance between the Uptake and Loss of Water

Living cells, both plant and animal, contain about 75–95 percent water. Water is essential for virtually all biochemical reactions within the body, and it functions as a medium for excreting metabolic wastes and for dissipating excess heat through evaporative cooling. For an organism to stay properly hydrated, these water losses must be offset by the uptake of water from the external environment. This balance between the uptake and loss of water with the surrounding environment is referred to as an organism's *water balance* (see Section 4.1).

Terrestrial animals have three major ways of gaining water and solutes: directly by drinking and eating and indirectly by producing metabolic water in the process of respiration (see Section 6.1). They lose water and solutes through urine, feces, evaporation from the skin, and from the moist air they exhale. Some birds and reptiles have a salt gland, and all birds and reptiles have a cloaca—a common receptacle for the digestive, urinary, and reproductive tracts. They reabsorb water from the cloaca back into the body proper. Mammals have kidneys capable of producing urine with high ion concentrations.

In arid environments, animals, like plants, face a severe problem of water balance. Survival depends on either evading the drought or by avoiding its effects. Animals of semiarid and desert regions may evade drought by leaving the area during the dry season and moving to areas where permanent water is available. Many of the large African ungulates (**Figure 7.10**) and many birds use this strategy.

Many animals that inhabit arid regions avoid the effects of drought by entering a period of physiological inactivity (dormancy) termed **estivation**. During hot, dry periods the spadefoot toad (*Scaphiopus couchi*) of the southern deserts of the United States remains below ground in a state of estivation and emerges when the rains return (**Figure 7.11**). Some invertebrates inhabiting ponds that dry up in summer, such as the flatworm *Phagocytes vernalis*, develop hardened casings and remain in them for the dry period. Other aquatic or semiaquatic animals retreat deep into the soil until they reach the level of groundwater. Many insects undergo **diapause**, a stage of arrested development in their life cycle from which they emerge when conditions improve.

Other animals remain active during the dry season but reduce respiratory water loss. Some small desert rodents lower the temperature of the air they breathe out. Moist air from the lungs passes over cooled nasal membranes, leaving condensed water on the walls. As the rodent inhales, this water humidifies and cools the warm, dry air.

There are other approaches to the problem. Some small desert mammals reduce water loss by remaining in burrows by day and emerging by night. Many desert mammals, from kangaroos to camels, extract water from the food they eat—either directly from the moisture content of the plants or from metabolic water produced during respiration—and produce highly concentrated urine and dry feces. Some desert mammals can tolerate a certain degree of dehydration. Desert rabbits may withstand water losses of up to 50 percent and camels of up to 27 percent of their body weight.

Unlike terrestrial animals, aquatic animals face the constant exchange of water with the external environment through the process of osmosis. As in the discussion of passive transport of water in plants, osmotic pressure moves water through cell membranes from the side of greater water concentration to the side of lesser water concentration (see Section 6.4). Aquatic organisms living in freshwater are **hyperosmotic**; they have a higher salt concentration in their bodies than does the surrounding water. Consequently, water moves inward into the body, whereas salts move outward. Their problem is the prevention of uptake, or the removal of excess water, and replacement of salts lost to the external environment. Because of the large disparity between the osmotic concentration of the freshwater and body fluids (e.g., blood), osmoconformity is not an option in freshwater environments. Freshwater fish maintain

(a)

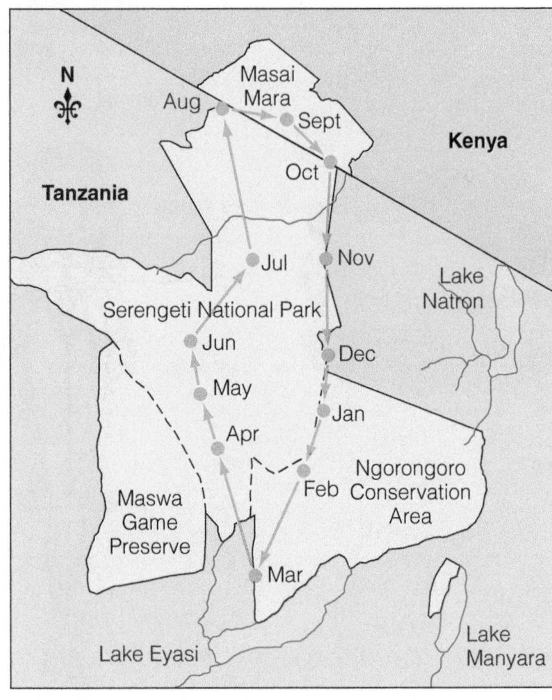

(b)

Figure 7.10 (a) Many of the large ungulate species in the semiarid regions of Africa, such as the wildebeest shown here, migrate over the course of the year, following the seasonal shift in rainfall. (b) The changing distribution of wildebeest populations in the contiguous Serengeti, Masai Mara, and Ngorongoro Conservation areas in East Africa. This seasonal pattern of migration gives these species consistent access to food (grass production) and water.

Figure 7.11 During periods of extreme heat and drought, the spadefoot toad (*Scaphiopus couchi*) of the southern deserts of the United States remains belowground in a state of dormancy called *estivation* and emerges when the rains return.

osmotic balance by absorbing and retaining salts in special cells in the gills and by producing copious amounts of watery urine (**Figure 7.12a**). Amphibians balance the loss of salts through the skin by absorbing ions directly from the water and transporting them across the skin and gill membranes. In the terrestrial phase, amphibians store water from the kidneys in the bladder. If circumstances demand it, they can reabsorb the water through the bladder wall.

The constraint imposed upon marine organisms is opposite of that faced by freshwater organisms. These organisms are **hypoosmotic**; they have a lower salt concentration in their bodies than does the surrounding water. When the concentration of salts is greater outside the body than within, organisms tend to dehydrate. Osmosis draws water out of the body into the surrounding environment. In marine and brackish environments, organisms have to inhibit water loss by osmosis through the body wall and prevent an accumulation of salts in the body (see Chapter 3).

Marine animals have evolved a variety of mechanisms that function to regulate water balance. Some animals are **isosmotic**; their body fluids have the same osmotic pressure as the surrounding seawater. For example, the bodies of invertebrates such as tunicates, jellyfish, many mollusks, and sea anemones are unable to actively adjust the amount of water in their tissues. These animals are osmoconformers, and their bodies gain water and lose ions until they are isosmotic to the surrounding water. In contrast, others function as osmoregulators, employing a variety of mechanisms to maintain constant salt concentration in their body. Marine bony (teleost) fish absorb saltwater into the gut. They secrete magnesium and calcium through the kidneys and pass these ions off as a partially crystalline paste. In general, fish excrete sodium and chloride, major ions in seawater, by pumping the ions across special membranes in the gills (**Figure 7.12b**). This pumping process is one type of active transport, moving salts against the concentration gradient, but it has a high energy cost. Sharks and rays retain enough urea to maintain a slightly higher concentration of solute in the body than exists in surrounding seawater. Birds of the open sea and sea turtles can consume seawater because they possess special salt-secreting nasal glands. Seabirds of the order Procellariiformes (e.g., albatrosses, shearwaters, and petrels) excrete fluids in excess of 5 percent salt from these glands. Petrels forcibly eject the fluids through the nostrils; other species drip the fluids out of the internal or external nares. In marine mammals, the kidney is the main route for elimination of salt; porpoises have highly developed kidneys to eliminate salt loads rapidly.

7.7 Animals Exchange Energy with Their Surrounding Environment

In principle, an animal's energy balance is the same as that described for a plant (see Section 6.6). Animals, however, differ significantly from plants in their thermal relations with the environment. Animals can produce significant quantities of

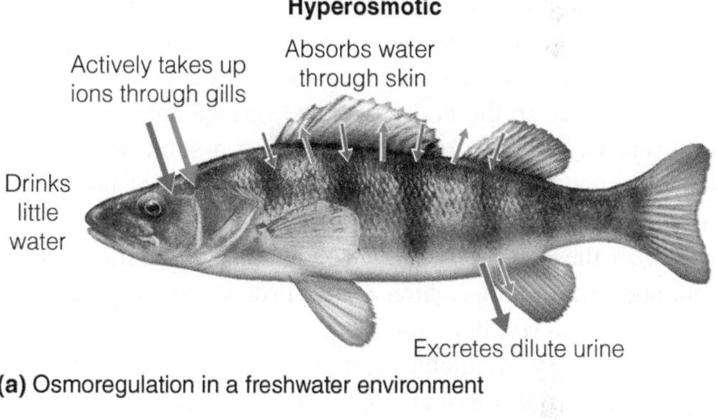

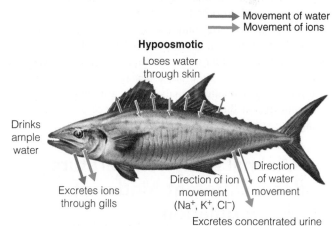

Figure 7.12 Exchanges in water and solutes involved in the process of osmoregulation in (a) freshwater and (b) saltwater (marine) fish species. Freshwater fish are hyperosmotic; their tissues have a greater solute concentration than the surrounding water. Marine fish are hypoosmotic; their tissues have a lower solute concentration than the surrounding water. Note directions of net movement of water and ions into and out of the bodies.

heat by metabolism, and their mobility allows them to seek out or escape heat and cold.

Body structure influences the exchange of heat between animals and the external environment. Consider a simple thermal model of an animal body (**Figure 7.13**). The interior or core of the body must be regulated within a defined range of temperature. In contrast, the temperature of the environment surrounding the animal's body varies. The temperature at the body's surface, however, is not the same as the air or water temperature in which the animal lives. Rather, it is the temperature at a thin layer of air (or water) called the *boundary layer*, which lies at the surface just above and within hair, feathers, and scales (see Section 6.6).

Therefore, body surface temperature differs from both the air (or water) and the core body temperature. Separating the body core from the body surface are layers of muscle tissue and fat, across which the temperature gradually changes from the core temperature to the body surface temperature. This layer of insulation influences the organism's **thermal conductivity**; that is, the ability to conduct or transmit heat.

To maintain its core body temperature, the animal must balance gains and losses of heat to the external environment. It does so through changes in metabolic rate and by heat exchange. The core area exchanges heat (produced by metabolism and stored in the body) with the surface area by conduction, that is, the transfer of heat through a solid. Influencing this exchange are the thickness and conductivity of fat and the movement of blood to the surface. The surface layer exchanges heat with the environment by conduction, convection, radiation, and evaporation, which are all influenced by the characteristics of skin and body covering.

External environmental conditions heavily influence how animals confront thermal stress. Because air has a lower specific heat and absorbs less solar radiation than water does (Section 3.2), terrestrial animals face more radical and dangerous changes in their thermal environment than do aquatic animals. Incoming solar radiation can produce lethal heat. The loss of radiant heat to the air, especially at night, can result in deadly cold. Aquatic animals live in a more stable energy environment, but they have a lower tolerance for temperature changes (Section 7.9).

7.8 Animal Body Temperature Reflects Different Modes of Thermoregulation

Different animal species exhibit different ranges of body temperature in their natural environments. In some, body temperature varies; these species are referred to as **poikilotherms** (from the Greek *poikilos* meaning "changeable"). In others species, termed **homeotherms** (from the Greek *homoeo* meaning "same"), body temperature is constant or nearly constant. These terms, *poikilotherm* and *homeotherm*, are not, however, synonymous with conformers and regulators discussed in Section 7.3. In fact, probably the only true thermoconformers are those animals that live in environments, such as the deep regions of the oceans, that have little to no variations in ambient temperature, and body temperature is virtually identical to the unchanging water temperature. Whether poiklotherms or homeotherms, all animals exhibit some degree of regulation.

Although environmental temperatures vary widely, both temporally and spatially, in most organisms body temperature is regulated through behavior, physiology, and morphology. The term *thermoregulation* does not merely refer to an organism's internal temperature differing from that of the surrounding environment; rather regulation implies maintaining the average body temperature or variations in body temperature within certain bounds. This requires mechanisms for the organism to sense and respond to its thermal environment. There are two categories of thermal regulation that emphasize the source of thermal energy used to influence body temperature: ectothermy and endothermy.

Ectothermy is the process of maintaining body temperature through the exchange of thermal energy with the surrounding environment. Species that use this mechanism of thermoregulation are called *ectotherms*. In contrast, **endothermy** is the process of maintaining body temperature through internally generated metabolic heat. Species that use this mechanism of thermoregulation are called *endotherms*. Although in practice all animals generate some internal heat as a function of metabolic processes, and all animals use external sources of thermal energy to modify body temperatures (such as basking in the sun or seeking shade), these two categories

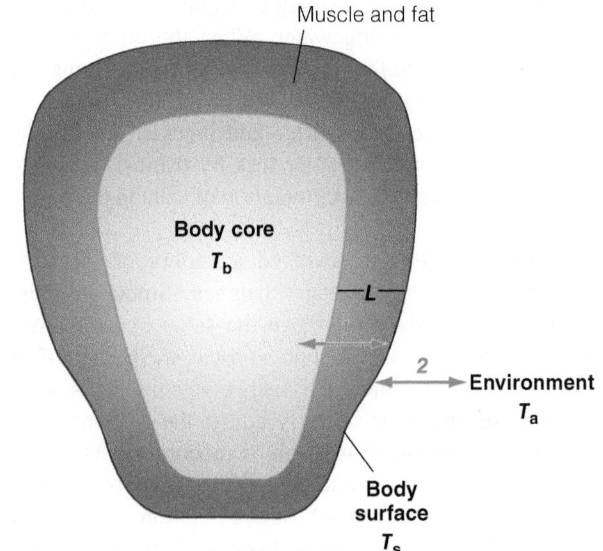

Figure 7.13 Temperatures in an animal body. Object represents an idealized cross-section of an animal body. Body core temperature is T_b, environmental temperature is T_a, surface temperature is T_s, and L is the thickness of the outer layer of the body. The exchange of heat between the body core and body surface (1) is a function of the difference in temperature (T_b–T_s) and the thermal conductivity, which is influenced by the insulating properties of muscle and fat (L). The transfer of heat from the body surface to the surrounding air or water (2) is influenced by the difference in temperature (T_s–T_e) and the insulating properties of any surface structures (e.g., hair or feathers).

are largely distinct. Endothermic species have the special ability to raise their metabolic activity markedly in excess of their immediate needs, using the resulting metabolic heat to maintain body temperature. In contrast, ectothermic species lack this ability and depend on external sources.

So how is the classification of animals based on variations in body temperature (poikilotherm and homeotherm) related to the classification of species based on primary means of temperature regulation (ectotherm and endotherm)? Although some species that inhabit environments where the thermal environment is fairly constant, such as the cold deep waters of the ocean or the litter layer of a tropical rain forest, may exhibit little if any variations in body temperature, the term *homeotherm* is generally applied to endothermic animals that maintain a constant body temperature through metabolic processes (endothermy). The only animals that fall within the category of endothermic homeotherms are birds and mammals. All other animals are typically classified as poikilotherms.

To simplify our proceeding discussion, we will discuss mechanisms in thermoregulation in terms of the two categories of animals based on variations in body temperature: poikilotherms, who use primarily ectothermy, and homeotherms, who primarily regulate body temperature using endothermy.

7.9 Poikilotherms Regulate Body Temperature Primarily through Behavioral Mechanisms

The performance (common measures include locomotion, growth, development, fecundity, and survivorship) of poikilotherms varies as a function of body temperature. As with plants (see Section 6.6, Figure 6.6), each species has minimum and maximum temperatures at which performance approaches zero (T_{min} and T_{max}) and a temperature or range of temperatures over which performance is optimal (T_{opt}; **Figure 7.14**). Likewise, the relationship between body temperature and performance varies among species and is correlated to the temperature characteristics of the environments they inhabit. Jonathon Stillman and George Somero of Stanford University examined the upper thermal tolerance limits (T_{max}) of 20 species of porcelain crabs, genus *Petrolisthes*, from intertidal and subtidal habitats (see Chapter 25, Figure 25.1) throughout the eastern Pacific. The researchers found that the upper thermal tolerance limit (T_{max}) was positively correlated with surface water temperature and maximum temperature in the microhabitats in which the species were found (**Figure 7.15**).

Poikilotherms have a low metabolic rate and a high ability to exchange heat between body and environment (high thermal conductivity; see Figure 7.13). During normal activities, poikilotherms carry out aerobic respiration. Under stress and while pursuing prey, the poikilotherms' inability to supply sufficient oxygen to the body requires that much of their energy production come from anaerobic respiration, in which oxygen is not used. This process depletes stored energy and accumulates lactic acid in the muscles. (Anaerobic respiration can occur in the muscles of marathon runners and other athletes, causing leg cramps.) Anaerobic respiration metabolism limits poikilotherms to short bursts of activity and results in rapid physical exhaustion.

To maintain body temperature in the "preferred" or optimal range, terrestrial and amphibious poikilotherms rely largely on behavioral thermoregulation. They seek out appropriate microclimates where environmental temperatures allow for body temperatures to approach optimal values. Insects such as butterflies, moths, bees, dragonflies, and damselflies bask in the sun to raise their body temperature to the level necessary to become highly active. When they become too warm, these animals seek the shade. Semiterrestrial frogs, such as bullfrogs (*Rana catesbeiana*) and green frogs (*Rana clamitans*), exert considerable control over their body temperature. By basking in the sun, frogs can raise their body temperature as much as 10°C above ambient temperature. Because of associated

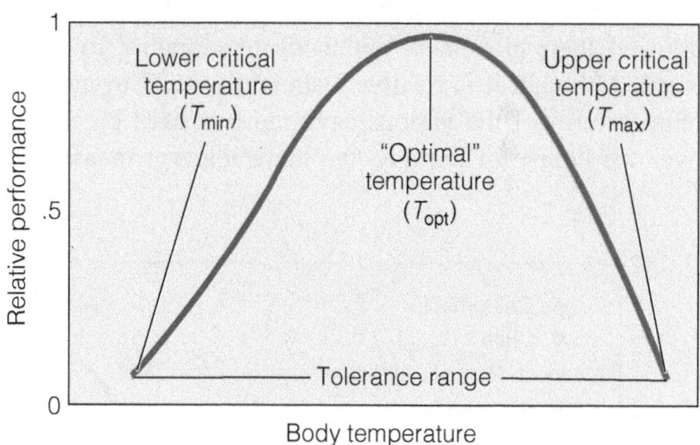

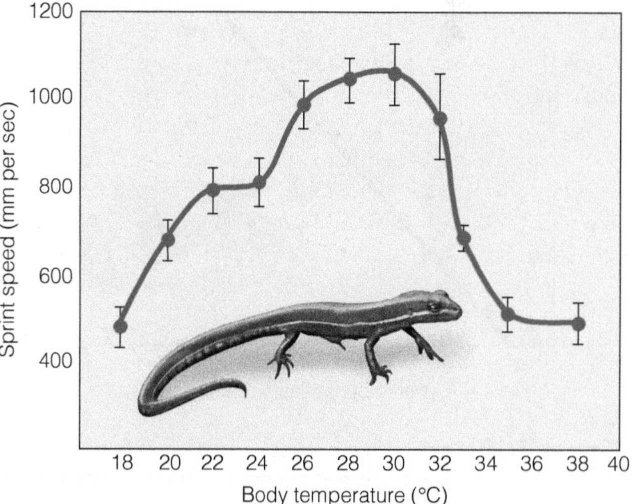

Figure 7.14 (a) Hypothetical performance curve of an ectotherm as a function of body temperature (note similarity to temperature response curve for plants). (b) A performance curve relating sprint speed of the white-striped grass lizard (*Takydromus wolteri*) at different body temperatures.

([b] Data from Chen, X., and X. Xub X. Jia. 2003. "Influence of body temperature on food assimilation and loco motor performance in white-striped grass lizards,"Takydromus (Lacertidae). *Journal of Thermal Biology* 28:385–391Fig. 3, pg. 389.)

evaporative water losses, such amphibians must be either near or partially submerged in water. By changing position or location or by seeking a warmer or cooler substrate, amphibians can maintain body temperatures within a narrow range.

Lizards raise and lower their bodies and change body shape to increase or decrease heat conduction between them and the rocks or soil they rest on. They also seek sunlight or shade or burrow into the soil to adjust their temperatures. Desert beetles, locusts, and scorpions exhibit similar behavior. They raise their legs to reduce contact between their body and the ground, minimizing conduction and increasing convection by exposing body surfaces to the wind.

The work of Gabriel Blouin-Demers and Patrick Weatherhead of Carleton University (Ottawa, Canada) illustrates the role that behavior plays in the thermoregulation of snakes. The researchers conducted a series of studies to examine how the body temperatures of individual black rat snakes (*Elaphe obsoleta*) varied on a daily basis under field conditions. Individual snakes were implanted with sensors that allowed their movement and body temperature to be monitored. Although it is relatively straightforward to monitor the temperatures of the various environments used by the snakes over the course of the day, the more relevant measure is the body temperature that occurs when the snake occupies each of these environments, referred to as the **operative environmental temperature**. For example, the body temperature of a snake lying on bare soil would not be the same as either the air temperature at the soil surface or the surface temperature of the soil. As presented in Section 7.7, the temperature of the snake is influenced by the physical characteristics of the snake (body shape, color, and thermal conductivity) and the exchange of heat between the snake and the surrounding environment. To better estimate the range of body temperatures that each environment represented, the researchers used physical models of a black rat snake constructed from painted copper tubing that matched the reflectance and conductance properties of the snake's body. The preferred (selected) body temperature(s) of the black rat snakes was established using thermal gradients in the laboratory.

Daily variations in average body temperature for the month of July are presented in **Figure 7.16**. By selecting a variety of microhabitats (rocks, bare ground, forest, fields, in the open or in the shade; **Figure 7.17**), individuals were able to maintain their preferred temperature during most of the active period of the day regardless of variations in the operative environmental temperatures. Both the thermal environment and the behavior of the snakes determined the daily pattern of body temperature.

When faced with longer-term changes in environmental temperatures, such as seasonal changes, poikilotherms are

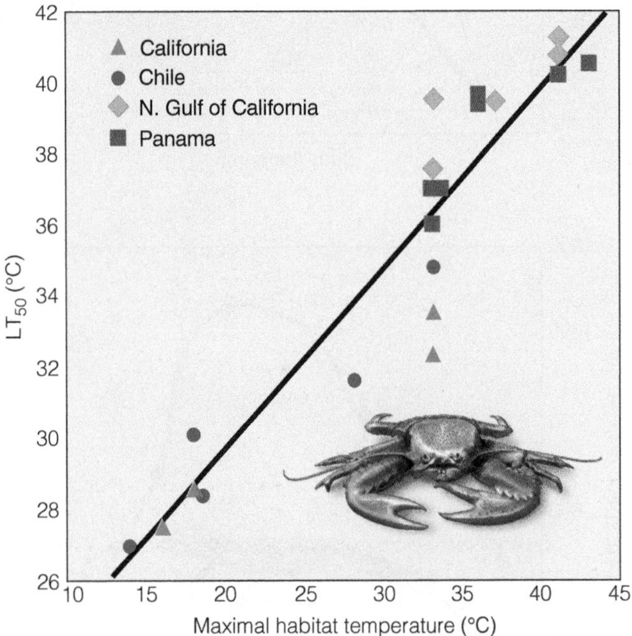

Figure 7.15 Relationship between maximum habitat temperature and upper thermal tolerance limits (T_{max}) for 20 species of porcelain crabs of the genus *Petrolisthes*, from intertidal and subtidal habitats throughout the eastern Pacific (geographic locations noted by symbols). Each symbol represents a different crab species. Maximal habitat temperature is the maximum recorded water temperature at the geographic location and microhabitat inhabited by the species. T_{max} represents the LT_{50} (lethal temperature); the water temperature at which 50 percent of test crabs died.
(Adapted from Stillman and Somero 2000.)

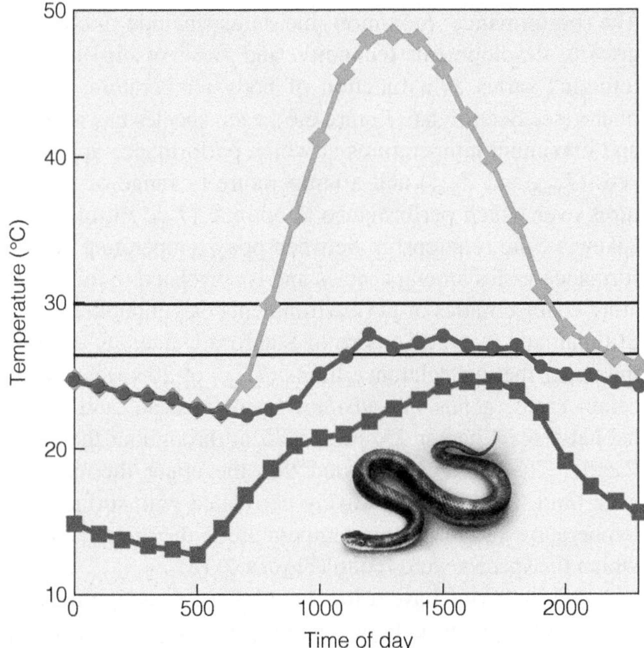

Figure 7.16 Hourly mean body temperature (T_b; circles) of black rat snakes in Ontario and hourly mean maximum (To_{max}; diamonds) and minimum (To_{min}; squares) operative environmental temperatures for the month of July. The preferred body temperature range (T_{opt}) for black snakes in Ontario is represented by the horizontal solid lines.
(Adapted from Blouin-Demers and Weatherhead 2001.)

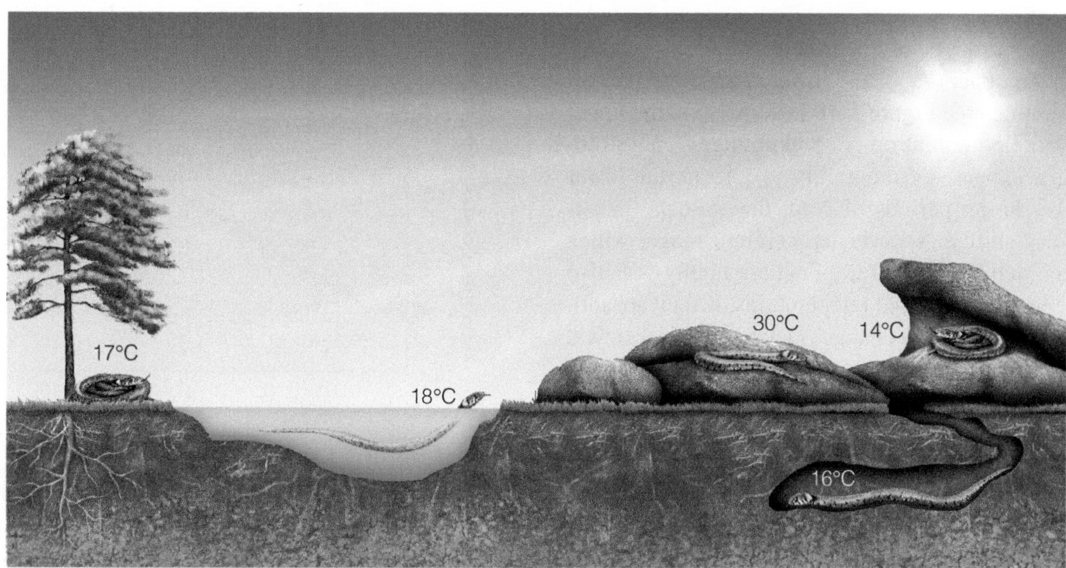

Figure 7.17 Microclimates selected by black snakes to regulate body temperature during the summer months.
(Adapted from Pearson et al. 1993.)

able to undergo the process of temperature acclimation (see Chapter 5, Section 5.4). Acclimation allows an animal's relationship between body temperature and performance to shift. For example, under acclimation, an animal's metabolic reactions in cold temperatures are increased to a level that is closer to that of warm-acclimated individuals, even though their body temperatures are that of the environment (**Figure 7.18**). This type of thermal acclimation involves specific biochemical changes (such as shifts in enzyme systems).

The thermal conductivity of water is approximately 25 times greater than air, meaning that heat is transferred 25 times faster than in air. For this reason, animals in water reach an equilibrium with their surrounding environment much faster than terrestrial animals. As a result, it is much more difficult for the body temperature of aquatic animals to be independent of the surrounding water temperature. Aquatic poikilotherms, when completely immersed, maintain no appreciable difference between their body temperature and the surrounding water. Aquatic poikilotherms are poorly insulated. Any heat produced in the muscles moves to the blood and on to the gills and skin, where heat transfers to the surrounding water by convection. Exceptions are sharks and tunas, which use a form of countercurrent exchange—a blood circulation system that allows them to keep internal temperatures higher than external ones. (See Section 7.13 and Figure 7.23 for discussion of countercurrent heat exchange.) Because seasonal water temperatures are relatively stable, fish and aquatic invertebrates maintain a constant temperature within any given season. They adjust seasonally to changing temperatures by acclimation or physiological adjustment to a change in environmental conditions (for an example of seasonal acclimation see Chapter 5, Section 5.4 and Figure 5.7). They undergo these physiological changes over a period of time. Because water temperature

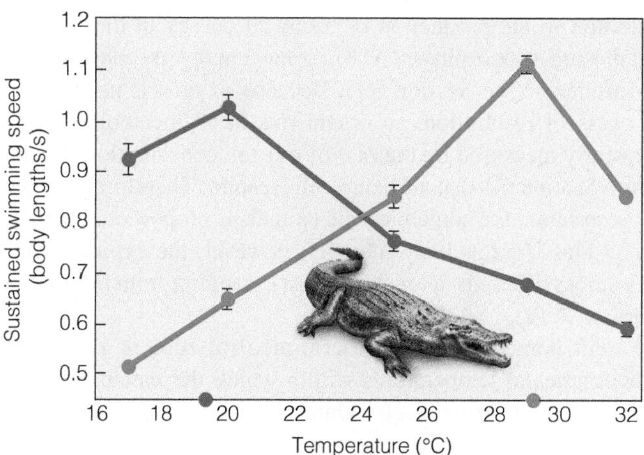

Figure 7.18 Example of thermal acclimation. The maximal sustained swimming performance of estuarine crocodiles (*Crocodylus porosus*) shifts significantly with acclimation and coincides with mean body temperatures within each treatment (blue line, cold acclimation; red line, warm acclimation). Water temperatures for cold (blue) and warm (red) water treatment are 19.5°C and 29.2°C, respectively (shown on x-axis).
(Data from Glanville and Seebacher 2006.)

Interpreting Ecological Data

Q1. *How does the temperature at which the maximum sustained swimming speed occurs differ for each of the two water acclimation treatments? How do these differences relate to the water temperatures to which the crocodiles were acclimated (acclimation treatment temperatures)?*

Q2. *How would you describe the trade-off that occurs between temperature acclimation and swimming speed for the crocodiles (costs and benefits of acclimation to the two different temperatures)?*

changes slowly through the year, aquatic poikilotherms may adjust slowly. The process of thermal acclimation involves changes in both the upper (T_{max}) and lower (T_{min}) limits of tolerance to temperature (see Figure 5.4). If they live at the upper end of their tolerable thermal range, poikilotherms' physiologies adjust at the expense of the ability to tolerate the lower range. Similarly, during periods of cold, the animals' physiological functions shift to a lower temperature range, which would have been debilitating before. Fish are highly sensitive to rapid change in environmental temperatures. If they are subjected to a sudden temperature change (faster than biochemical and physiological adjustments can occur), they may die of thermal shock.

7.10 Homeotherms Regulate Body Temperature through Metabolic Processes

Homeothermic birds and mammals meet the thermal constraints of the environment by being endothermic. Their body temperature is maintained by the oxidization of glucose and other energy-rich molecules in the process of respiration. The process of oxidation is not 100 percent efficient, and in addition to the production of chemical energy in the form of adenosine triphosphate (ATP), some energy is converted to heat energy (see Section 6.1). Because oxygen is used in the process of respiration, an organism's basal metabolic rate is typically measured by the rate of oxygen consumption. Recall from Section 6.1 that all living cells respire. Therefore, the rate of respiration for homeothermic animals is proportional to their body mass (grams body mass$^{0.75}$; however, the exponent varies across different taxonomic groups, ranging from 0.6 to 0.9; **Figure 7.19**).

For homeotherms, the **thermoneutral zone** is a range of environmental temperatures within which the metabolic rates are minimal (**Figure 7.20**). Outside this zone, marked by upper and lower critical temperatures, metabolic rate increases.

Maintenance of a high body temperature is associated with specific enzyme systems that operate optimally within a high temperature range, with a set point of about 40°C. Because efficient cardiovascular and respiratory systems bring oxygen to their tissues, homeotherms can maintain a high level of energy production through aerobic respiration (high metabolic rates). Thus, they can sustain high levels of physical activity for long periods. Independent of external temperatures, homeotherms can exploit a wider range of thermal environments. They can generate energy rapidly when the situation demands, escaping from predators or pursuing prey.

To regulate the exchange of heat between the body and the environment, homeotherms use some form of insulation—a covering of fur, feathers, or body fat (see Figure 7.13). For mammals, fur is a major barrier to heat flow, but its insulation value varies with thickness, which is greater in large mammals than in small ones. Small mammals are limited in the amount of fur they carry because a thick coat would reduce their ability to move. The thickness of mammals' fur changes with the season, a form of acclimation (see Section 5.4). Aquatic mammals—especially of Arctic regions—and Arctic and Antarctic birds such as auklets (Alcidae) and penguins (Spheniscidae) have a heavy layer of fat beneath the skin. Birds reduce heat loss by fluffing the feathers and drawing the feet into them, making the body a round, feathered ball. Some Arctic birds, such as ptarmigan (*Lagopus* spp.), have feathered feet—unlike most birds, which have scaled feet that function to lose heat.

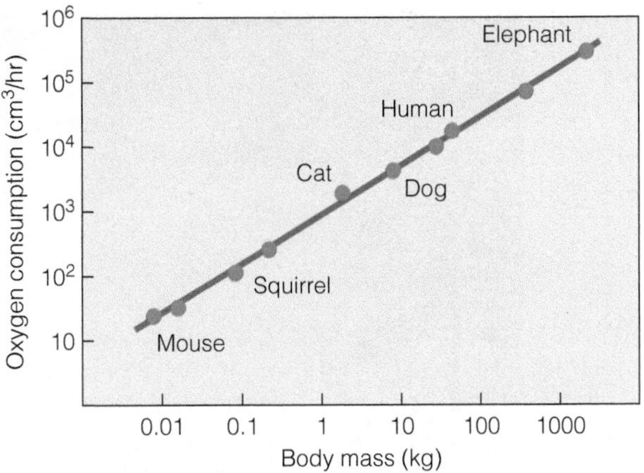

Figure 7.19 Relationship between metabolic rate, as measured by oxygen consumption per hour, and body mass (kg) for a range of mammal species. Both variables (metabolic rate and body mass) are plotted on a logarithmic scale ($\log_{10}$).
(After Schmidt-Nielsen, K. *Animal Physiology: Adaptation and Environment*, 5th Edition 1997.)

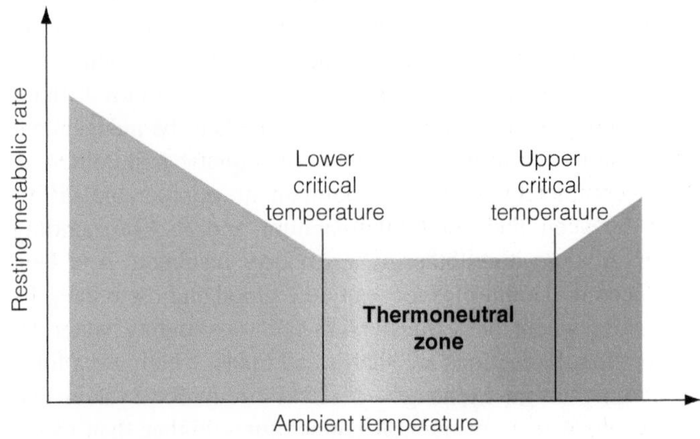

Figure 7.20 General resting metabolic response of homeotherms to changes in ambient temperature. For temperatures within the thermal neutral zone, resting metabolic rate changes little with a change in ambient temperature. Beyond these limits, however, metabolic rate increases markedly with either an increase or decrease in ambient temperature as a result of feedback mechanisms (see Section 7.4 and Figure 7.8).
(After Schmidt-Nielsen 1997.)

Although the major function of insulation is to keep body heat in, it also keeps heat out. In a hot environment, an animal must either rid itself of excess body heat or prevent heat from being absorbed in the first place. One way is to reflect solar radiation from light-colored fur or feathers. Another way is to grow a heavy coat of fur that heat does not penetrate. Large mammals of the desert, notably the camel, use this method. The outer layers of hair absorb heat and return it to the environment.

Some insects—notably moths, bees, and bumblebees—have a dense, furlike coat over the thoracic region that serves to retain the high temperature of flight muscles during flight. The long, soft hairs of caterpillars, together with changes in body posture, act as insulation to reduce convective heat exchange.

When insulation fails, many animals resort to shivering, which is a form of involuntary muscular activity that increases heat production. Many species of small mammals increase heat production without shivering by burning (oxidizing) highly vascular brown fat. Found about the head, neck, thorax, and major blood vessels, brown adipose tissue (fat) is particularly prominent in hibernators, such as bats and groundhogs (*Marmota monax*).

Many species employ evaporative cooling to reduce the body heat load. Birds and mammals lose some heat through the evaporation of moisture from the skin. When their body heat is above the upper critical temperature, evaporative cooling is accelerated by sweating and panting. Only certain mammals have sweat glands—in particular, horses and humans. Panting in mammals and gular fluttering in birds function to increase the movement of air over moist surfaces in the mouth and pharynx. Many mammals, such as pigs, wallow in water and wet mud to cool down.

7.11 Endothermy and Ectothermy Involve Trade-offs

Prime examples of the trade-offs involved in the adaptations of organisms to their environment are endothermy and ectothermy, which are the two alternative approaches to regulation of body temperature in animals. Each strategy has advantages and disadvantages that enable the organisms to excel under different environmental conditions. For example, endothermy allows homeotherms to remain active regardless of variations in environmental temperatures, whereas environmental temperatures largely dictate the activity of poikilotherms (ectothermy). However, the freedom of activity enjoyed by homeotherms comes at a great energy cost. The maintenance of internal body temperature in homeotherms requires a high metabolic rate, and heat lost to the surrounding environment must be continuously replaced by additional heat generated through respiration. As a result, metabolic costs weigh heavily against homeotherms. In contrast, ectotherms, not needing to burn calories to provide metabolic heat, allocate more of their energy intake to biomass production than to metabolic needs. Ectotherms, therefore, require fewer calories (food) per gram of body weight. A homeotherm must take in some 20 times more food energy than a poikilotherm of equal body mass. Because they do not depend on internally generated body heat, ectotherms can curtail metabolic activity in times of food and water shortage and temperature extremes. Low energy demands enable some terrestrial poikilotherms to colonize areas with limited food and water.

One of the most important features influencing its ability to regulate body temperature is an animal's size. Poikilotherms (ectothermy) absorb heat across their body's surface but must absorb enough energy to heat the entire body mass (volume). Therefore, the ratio of surface area to volume ($SA{:}V$) is a key factor controlling the uptake of heat and the maintenance of body temperature. As an organism's size increases, the $SA{:}V$ ratio decreases (see Figure 7.2b). Because the organism must absorb sufficient energy across its surface to warm the entire body mass, the amount of energy or the period of time required to raise body temperature increases. For this reason, ectothermy imposes a constraint on maximum body size for poikilotherms and restricts the distribution of the larger poikilotherms to the warmer, aseasonal regions of the subtropics and tropics. For example, large reptiles such as alligators, crocodiles, iguanas, komodo dragons, anacondas, and pythons are all restricted to warm tropical environments.

The constraint that size imposes on homeotherms (endothermy) is opposite that presented earlier for poikilotherms. For homeotherms, it is the body mass (or volume) that produces heat through respiration, while heat is lost to the surrounding environment across the body surface. The smaller the organism, the larger the $SA{:}V$ ratio, therefore, the greater the relative heat loss to the surrounding environment. To maintain a constant body temperature, the heat loss must be offset by increased metabolic activity (respiration). Thus, small homeotherms have a higher mass-specific metabolic rate (metabolic rate per unit body mass; **Figure 7.21**) and consume more food energy per unit of body weight than do large ones. Small shrews (*Sorex* spp.), for example, ranging in weight from 2 to 29 g (see Figure 7.1), require a daily amount of food (wet weight) equivalent to their own body weight. Therefore, small animals must spend most of their time seeking and eating food. The mass-specific metabolic rate (respiration rate per gram of body weight) of small endotherms rises so rapidly that below a certain size, they do not meet their energy demands. On average, 2 g is about as small as an endotherm may be and still maintain a metabolic heat balance; however, this minimum constraint depends on the thermal environment. Some shrews and hummingbirds undergo daily torpor (see Section 7.13) to reduce their metabolic needs. As a result of the conflicting metabolic demands of body temperature and growth, most young birds and mammals are born in an altricial state, meaning they are blind, naked, helpless, and begin life as ectotherms. They depend on the body heat of their parents to maintain their body temperature, which allows most of these young animals' energy to be allocated to growth.

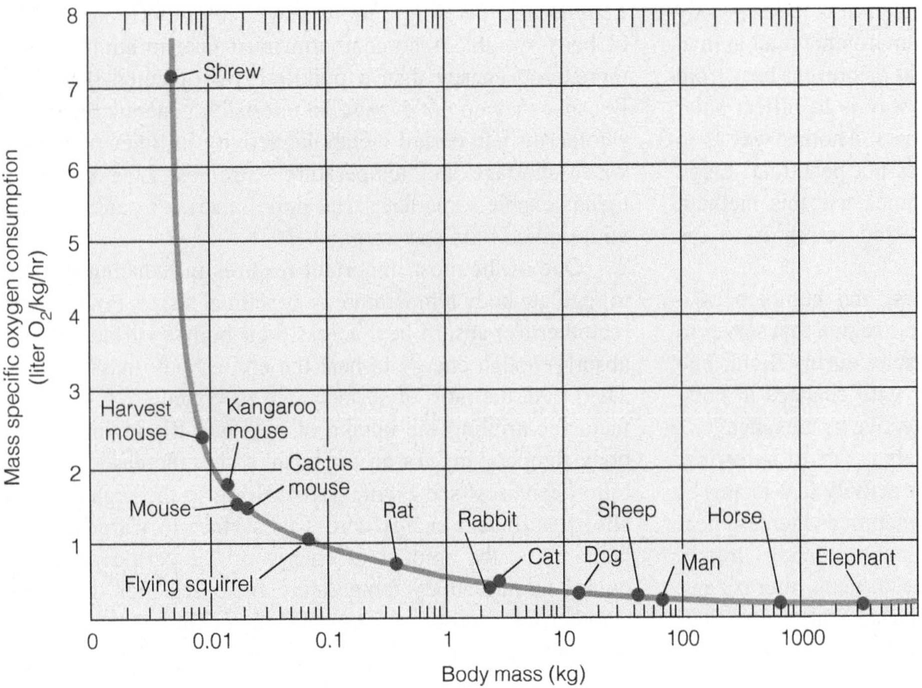

Figure 7.21 Observed relationship between metabolic rate (oxygen consumption) per unit body mass (mass-specific metabolic rate) and body mass for various mammal species. Mass-specific metabolic rate increases with decreasing body mass. Body mass is plotted on a logarithmic scale (log_{10}).
(Adapted from Schmidt-Nielson 1979.)

Interpreting Ecological Data

Q1. How does the variable plotted on the y-axis of this graph (mass-specific metabolic weight) differ from the variable plotted on the y-axis of Figure 7.19?

Q2. What does the graph imply about the rates of cellular respiration for a mouse compared to a horse?

Q3. How would the graph differ if the y-axis was plotted on a logarithmic scale (log_{10})?

7.12 Heterotherms Take on Characteristics of Ectotherms and Endotherms

Species that sometimes function as homeotherms while at other times as poikilotherms are called temporal **heterotherms**. At different stages of their daily and seasonal cycle or in certain situations, these animals take on characteristics of endotherms or ectotherms. They can undergo rapid, drastic, repeated changes in body temperature.

Insects are ectothermic and poikilothermic; yet in the adult stage, most species of flying insects are heterothermic. When flying, they have high rates of metabolism, with heat production as great as or greater than that of homeotherms. They reach this high metabolic state in a simpler way than do homeotherms because they are not constrained by the uptake and transport of oxygen through the lungs and vascular system. Insects take in oxygen by demand through openings in the body wall and transport it throughout the body in a tracheal system (see Section 7.5).

Temperature is crucial to the flight of insects. Most cannot fly if the temperature of the body muscles is less than 30°C, nor can they fly if muscle temperature is higher than 44°C. This constraint means that an insect must warm up to take off, and it must get rid of excess heat in flight. With wings beating up to 200 times per second, flying insects can produce a prodigious amount of heat.

Some insects, such as butterflies and dragonflies, warm up by orienting their bodies and spreading their wings to the sun. Most warm up by shivering their flight muscles in the thorax. Moths and butterflies vibrate their wings to raise thoracic temperatures above ambient temperatures. Bumblebees pump their abdomens without any external wing movements. They do not maintain any physiological set point, and they cool down to ambient temperatures when not in flight.

To reduce metabolic costs during periods of inactivity, some small homeothermic animals become heterothermic and enter into torpor daily. Daily **torpor** is the dropping of body temperature to approximately ambient temperature for a part of each day, regardless of season.

Some birds, such as the common poorwill (*Phalaenoptilus nuttallii*) and hummingbirds (Trochilidae), and small mammals, such as bats, pocket mice, kangaroo mice, and white-footed mice, undergo daily torpor. Such daily torpor seems to have evolved as a way to reduce energy demands over that part of the day when the animals are inactive, allowing them to save the energy that would otherwise be used to maintain a high (normal) body temperature. Nocturnal mammals, such as bats, go into torpor by day; and diurnal animals, such as hummingbirds, go into torpor by night. As the animal goes into torpor, its body temperature falls steeply and oxygen consumption drops (**Figure 7.22**). With the relaxation of homeothermic responses, the body temperature declines to within a few degrees of ambient temperature. Arousal returns the body temperature to normal rapidly as the animal renews its metabolic heat production.

To escape the rigors of long, cold winters, some heterothermic mammals go into a long, seasonal torpor called **hibernation**. Hibernation is characterized by the cessation of activity and controlled hypothermia (reduction of body temperature). Homeothermic regulation is relaxed, and the body

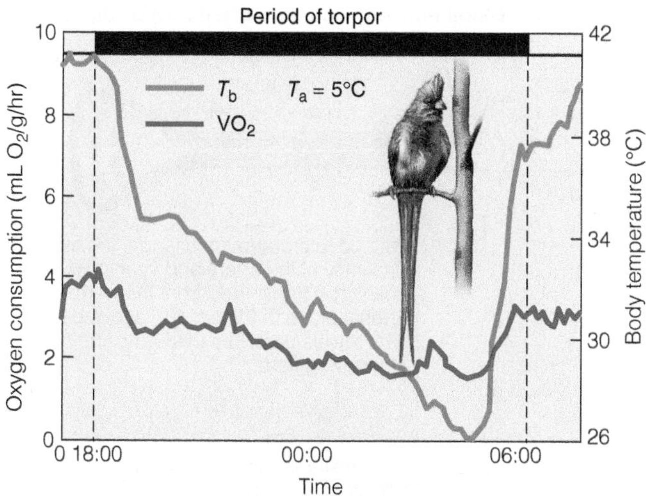

Figure 7.22 Period of torpor in a White-backed Mousebird (*Colius colius*) under laboratory conditions (air temperature [T_a] = 5°C). The onset of entry (18:00 hr) and arousal (06:00 hr) phases of torpor are noted by vertical dashed lines. Body temperature (T_b) and oxygen consumption (VO_2) drop during the period of torpor.
(Adapted from McKechnie and Lovegrove 2001.)

temperature is allowed to approach ambient temperature. Heart rate, respiration, and total metabolism fall, and body temperature sinks below 10°C. Associated with hibernation are high blood levels of carbon dioxide and an associated decrease in blood pH (increased acidity). This state, called *acidosis*, lowers the threshold for shivering and reduces the metabolic rate. Hibernating homeotherms, however, are able to rewarm spontaneously using only metabolically generated heat.

Among homeotherms, entrance into hibernation is a controlled process difficult to generalize from one species to another. Some hibernators, such as the groundhog (*Marmota monax*), feed heavily in late summer to store large fat reserves from which they will draw energy during hibernation. Others, like the chipmunk (*Tamias striatus*), lay up a store of food instead. All hibernators, however, convert to a means of metabolic regulation different from that of the active state. Most hibernators rouse periodically and then drop back into torpor. The chipmunk, with its large store of seeds, spends much less time in torpor than do species that store large amounts of fat.

Although popularly said to hibernate, black bears (*Ursus americanus*), grizzly bears (*Ursus arctos*), and female polar bears (*Ursus maritimus*) do not. Instead, they enter a unique winter sleep from which they easily rouse. They do not enter extreme hypothermia but allow body temperatures to decline only a few degrees below normal. The bears do not eat, drink, urinate, or defecate, and females give birth to and nurse young during their sleep; yet they maintain a metabolism that is near normal. To do so, the bears recycle urea, normally excreted in urine, through the bloodstream. The urea is degraded into amino acids that are reincorporated in plasma proteins.

Hibernation provides selective advantages to small homeotherms. For them, the maintenance of high body temperature during periods of cold and limited food supply is too costly. It is far less expensive to reduce metabolism and allow the body temperature to drop. Doing so eliminates the need to seek scarce food resources to maintain higher body temperatures.

7.13 Some Animals Use Unique Physiological Means for Thermal Balance

Because of an animal's limited tolerance for heat, storing body heat does not seem like a sound option to maintain thermal balance in the body. But certain mammals, especially the camel, oryx, and some gazelles, do just that. The camel, for example, stores body heat by day and dissipates it by night, especially when water is limited. Its temperature can fluctuate from 34°C in the morning to 41°C by late afternoon. By storing body heat, these animals of dry habitats reduce the need for evaporative cooling and thus reduce water loss and food requirements.

Many ectothermic animals of temperate and Arctic regions withstand long periods of below-freezing temperatures in winter through supercooling and developing a resistance to freezing. **Supercooling** of body fluids takes place when the body temperature falls below the freezing point without actually freezing. The presence of certain solutes in the body that function to lower the freezing point of water influences the amount of supercooling that can take place (see Chapter 3). Some Arctic marine fish, certain insects of temperate and cold climates, and reptiles exposed to occasional cold nights employ supercooling by increasing solutes, notably glycerol, in body fluids. Glycerol protects against freezing damage, increasing the degree of supercooling. Wood frogs (*Rana sylvatica*), spring peepers (*Hyla crucifer*), and gray tree frogs (*Hyla versicolor*) can successfully overwinter just beneath the leaf litter because they accumulate glycerol in their body fluids.

Some intertidal invertebrates of high latitudes and certain aquatic insects survive the cold by freezing and then thawing out when the temperature moderates. In some species, more than 90 percent of the body fluids may freeze, and the remaining fluids contain highly concentrated solutes. Ice forms outside the shrunken cells, and muscles and organs are distorted. After thawing, they quickly regain normal shape.

To conserve heat in a cold environment and to cool vital parts of the body under heat stress, **countercurrent heat exchange** has evolved in some animals (**Figure 7.23**). For example, the porpoise (*Phocaena* spp.), swimming in cold Arctic waters, is well insulated with blubber. It could experience an excessive loss of body heat, however, through its uninsulated flukes and flippers. The porpoise maintains its body core temperature by exchanging heat between arterial (coming from the lungs) and venous (returning to the lungs) blood in these

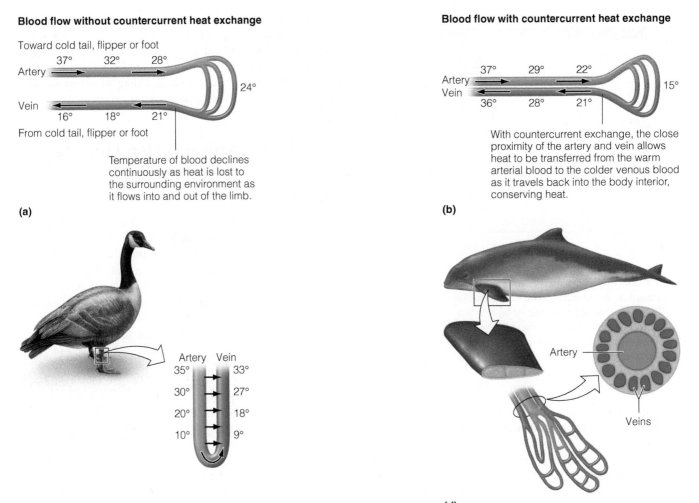

Blood flow without countercurrent heat exchange

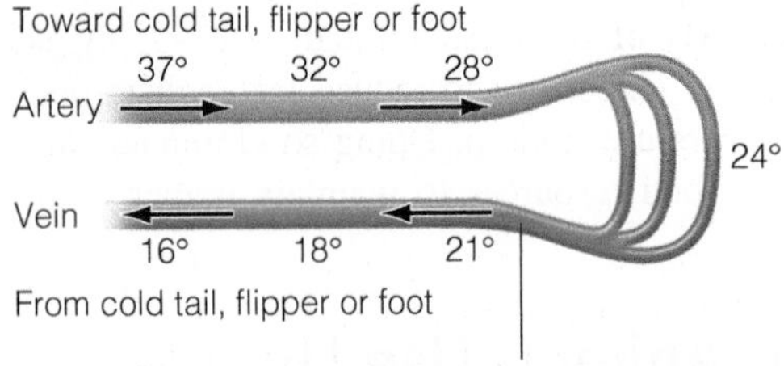

Temperature of blood declines continuously as heat is lost to the surrounding environment as it flows into and out of the limb.

(a)

Blood flow with countercurrent heat exchange

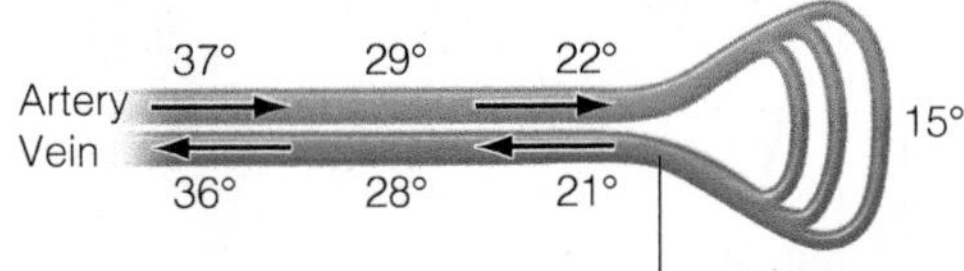

With countercurrent exchange, the close proximity of the artery and vein allows heat to be transferred from the warm arterial blood to the colder venous blood as it travels back into the body interior, conserving heat.

(b)

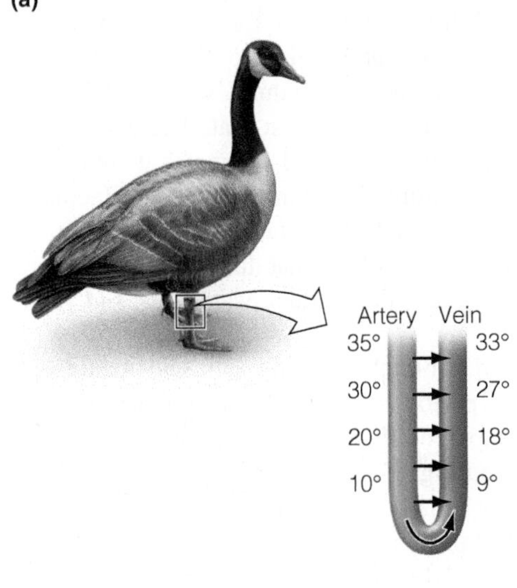

(c)

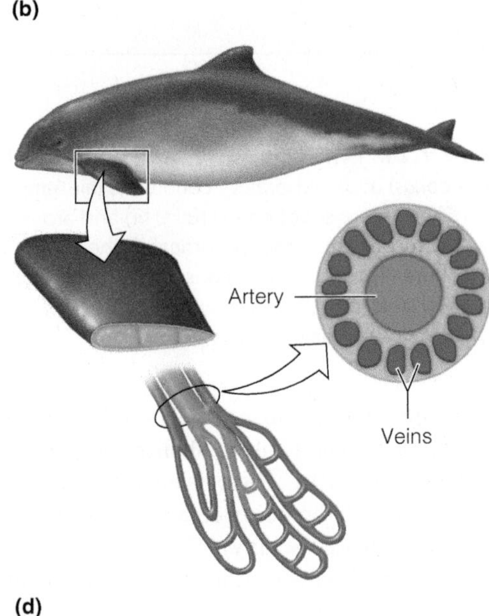

(d)

Figure 7.23 A model of countercurrent flow in the limb of an animal, showing hypothetical temperature changes in the blood (a) in the absence and (b) in the presence of countercurrent heat exchange. Examples of countercurrent exchange in the (c) legs of birds, and (d) flippers of porpoises.

structures (see Figure 7.23). Veins completely surround arteries, which carry warm blood from the heart to the extremities. Warm arterial blood loses its heat to the cool venous blood returning to the body core. As a result, little body heat passes to the environment. Blood entering the flippers cools, whereas blood returning to the deep body warms. In warm waters, where the animals need to get rid of excessive body heat, blood bypasses the heat exchangers. Venous blood returns unwarmed through veins close to the skin's surface to cool the body core. Such vascular arrangements are common in the legs of mammals and birds as well as in the tails of rodents, especially the beaver (*Castor canadensis*).

Many animals have arteries and veins divided into small, parallel, intermingling vessels that form a discrete vascular bundle or net known as a **rete**. In a rete, countercurrent heat exchange occurs as blood flows in opposite directions. Countercurrent heat exchange can also keep heat out. The oryx (*Oryx beisa*), an African desert antelope exposed to high daytime temperatures, experiences elevated body temperatures yet keeps the highly heat-sensitive brain cool by a rete in its head. The external carotid artery passes through a cavernous sinus filled with venous blood that is cooled by evaporation from the moist mucous membranes of the nasal passages (**Figure 7.24**). Arterial blood passing through the cavernous sinus cools on the way to the brain, lowering the temperature of the brain by 2°C to 3°C compared to the body core.

Countercurrent heat exchangers are not restricted to homeotherms. Certain poikilotherms that assume some degree of endothermism employ the same mechanism. The swift, highly predaceous tuna (*Thunnus* spp.) and the mackerel shark (*Isurus tigris*) possess a rete in a band of dark muscle tissue used for sustained swimming effort. Metabolic heat produced in the muscle warms the venous blood, which gives up heat to the adjoining newly oxygenated blood returning from the

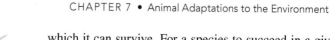

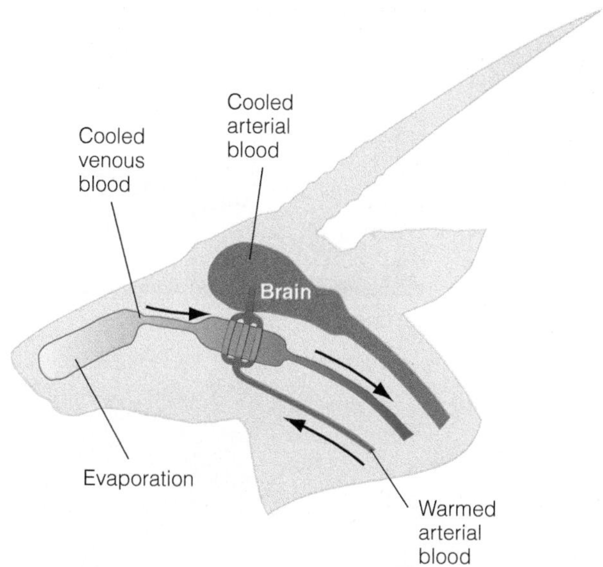

Figure 7.24 Arterial blood passes in small arteries through a pool of venous blood that is cooled by evaporation as it drains from the nasal region and into the pool.

gills. Such a countercurrent heat exchange increases the power of the muscles because warm muscles contract and relax more rapidly. Sharks and tuna maintain fairly constant body temperatures, regardless of water temperatures.

7.14 An Animal's Habitat Reflects a Wide Variety of Adaptations to the Environment

One of the most fundamental factors defining the relationship between an organism and the environment is the place where it is found, its **habitat**. The millions of species that inhabit our planet are not found everywhere, nor are they distributed at random across Earth's environments. There is a correspondence between the different species of organisms and the environments in which they are found. Each species on our planet occupies a unique geographic area where its members live and reproduce. So why are some species found in one location—one habitat—but not in another?

The most fundamental constraint on the distribution of species is the ability of the environment to provide essential resources and environmental conditions capable of sustaining basic life processes. For a species to persist in a given location, it must have the physiological potential to survive and reproduce in that habitat. As we have seen, physical and chemical (abiotic) features of the environment, such as pH, temperature, or salinity have a direct influence on basic physiological processes necessary to sustain life. Because physiological processes proceed at different rates under different conditions, any one organism has only a limited range of conditions under which it can survive. For a species to succeed in a given location, however, it must do more than survive. A species' habitat must provide the wide array of resources necessary to sustain growth and reproduction. The environment must provide essential food resources, cover from potential predators, areas for successful reproduction (courtship, mating, nesting, etc.), and a substrate for a wide array of life activities. As such, a species' habitat is a reflection of the wide variety of adaptations relating to these processes (e.g., feeding and mating behaviors). For plants and sessile animal species, there is only the hope that gametes or individuals dispersed by a variety of means (wind, water, animals, etc.) arrive at a site that is suitable for successful establishment. In mobile animal species, however, individuals actively choose a specific location to inhabit. The process of selecting a specific location to inhabit is called **habitat selection**.

Given the importance of habitat selection on an organism's fitness, how are organisms able to assess the suitability of an area in which to settle? What do they seek in a living place? Such questions have been intriguing ecologists for many years. Habitat selection has been most widely studied in birds, particularly in species that defend breeding territories—areas of habitat that the individual defends against other individuals and in which it carries out its life activities (feeding, mating, and rearing of offspring). (See Section 11.10 for discussion of territoriality.) The advantage of studying habitat selection in territorial species is that territories can be delineated, and the features of the habitat within the territory can be described and contrasted with the surrounding environment.

Of particular importance is the contrast between those areas that have been chosen as habitats and adjacent areas that have not. Using this approach, a wide variety of studies examining the process of habitat selection in birds has demonstrated a strong correlation between the selection of an area as habitat and structural features of the vegetation. These studies suggest that habitat selection most likely involves a hierarchical approach. Birds appear initially to assess the general features of the landscape: the type of terrain; presence of lakes, ponds, streams, and wetland; gross features of vegetation such as open grassland, shrubby areas, and type and extent of forest. Once in a broad general area, the birds respond to more specific features, such as the structural configuration of the vegetation, particularly the presence or absence of various vertical layers such as shrubs, small trees, tall canopy, and degree of patchiness (**Figure 7.25**). Frances James, an avian ecologist at Florida State University, coined the term *niche gestalt* to describe the vegetation profile associated with the breeding territory of a particular species.

In addition to the physical structure of the vegetation, the actual plant species present can be important. Certain species of plants might produce preferred food items, such as seeds or fruits, or influence the type and quantity of insects available as food for insectivorous birds.

The structural features of the vegetation that define its suitability for a given species may be related to a variety of

(a) Yellowthroat

(b) Hooded warbler

(c) Ovenbird

Figure 7.25 Vegetation structure characterizing the habitat of three warbler species. (a) Common yellowthroat (*Geothlypis trichas*), a bird of shrubby margins of woodland and wetlands and brushy fields. (b) Hooded warbler (*Wilsonia citrina*), a bird of small forest openings. (c) Ovenbird (*Seiurus aurocapillus*), an inhabitant of deciduous or mixed conifer–deciduous forests with open forest floor. The labels 9–10 m, 18–30 m, and 15–30 m refer to height of vegetation.
(Adapted from James 1971.)

specific needs, such as food, cover, and nesting sites. The lack of an adequate nesting site may prevent an individual from occupying an otherwise suitable habitat. Animals require sufficient shelter to protect themselves and young from enemies and adverse weather conditions. Cavity-nesting animals require suitable dead trees or other structures in which they can access cavities (**Figure 7.26**).

Habitat selection is a common behavioral characteristic of a wide array of vertebrates other than birds; fish, amphibians, reptiles, and mammals furnish numerous examples. Garter snakes (*Thamnophis elegans*) living along the shores of Eagle Lake in the sagebrush-ponderosa pine ecosystems of northeastern California select rocks of intermediate thickness (20–30 cm) over thinner and thicker rocks as their retreat sites. Shelter under thin rocks becomes lethally hot; shelter under thicker rocks does not allow the snakes to warm their bodies to preferred temperatures (see Figure 7.17). Insects, too, cue in on habitat features. Thomas Whitham of the University of Utah studied habitat selection by the gall-forming aphid *Pemphigus betae*, which parasitizes the narrow-leaf cottonwood (*Populus angustifolia*). He found that aphids select the largest leaves to colonize and discriminate against small leaves. Beyond that, they select the best positions on the leaf. Occupying this particular habitat, which provides the best food source, produces individuals with the highest fitness.

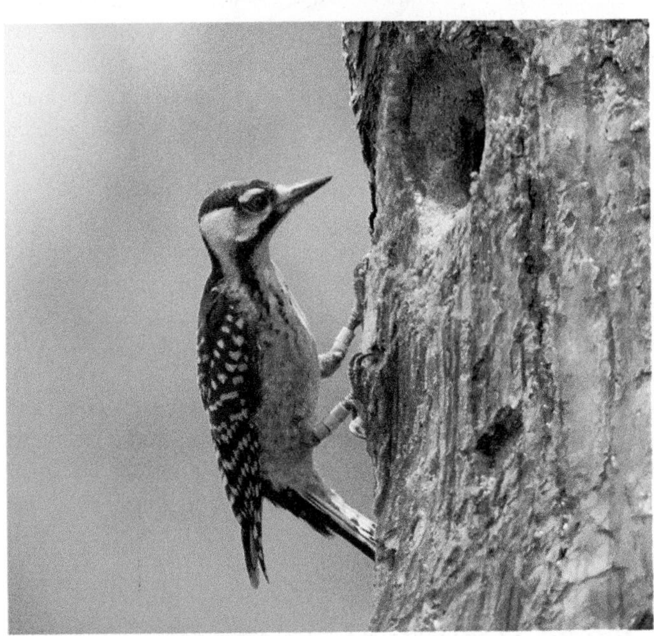

Figure 7.26 The red-cockaded woodpecker (*Picoides borealis*) is a cavity-nesting species that depends on habitat that contains large, standing, dead trees for nesting sites.

ECOLOGICAL Issues & Applications
Increasing Global Temperature Is Affecting the Body Size of Animals

Body size is one of the most important phenotypic traits of animals, influencing virtually all physiological and ecological processes (Section 7.1). Variation in body size, both geographically and through time, is a common phenomenon and assumed to be a product of adaptation through natural selection (see Section 5.6). For example, temperatures have a direct effect on an animal's energy balance, and the relationship between body size and heat exchange (see Section 7.11) is the basis for Bergmann's rule. Bergmann's rule states that for endotherms, body size for a species tends to increase with decreasing mean annual temperature. The result is a cline in body size with latitude (for discussion of clines, see Section 5.8): increasing body size with increasing latitude. Similar changes in body size in response to temperature have been observed over time. For example, Ross Secord of the University of Nebraska and colleagues examined shifts in body size in the earliest known horses (family Equidae) during the Paleocene-Eocene Thermal Maximum (PETM) approximately 56 million years ago. A high-resolution record of continental climate and body size of fossil horses shows a directional body size decrease of approximately 30 percent over the first 130,000 years of the PETM (a period of warming), followed by a 76 percent increase in the recovery phase of the PETM as temperatures cooled.

Given the strong selective (evolutionary) influence that temperature has on body size, could patterns of recent climate warming over the past century (0.6°C increase in mean global temperature; see discussion in Chapter 2, *Ecological Issues & Applications*) as a result of human activities have influenced the body size of animals? Numerous studies have documented recent changes in animal body size for local populations over the timescale of decades to a century that are correlated to changes in local temperature. For example, Celine Teplitsky of the University of Helsinki (Finland) and colleagues examined data on mean body mass of red-billed gulls (*Larus novaehollandiae scopulinus*) from New Zealand over a 47-year period (1958–2004). Results of their analyses show that mean body mass had decreased over time as ambient temperatures increased (**Figure 7.27**). Similar patterns have been observed from a wide array of bird and mammal species. Some of the most pronounced changes have occurred in animal species that inhabit the northern latitudes, where the largest changes in temperature have occurred over recent decades (see Chapter 2, *Ecological Issues & Applications*). Yoram Yom-Tov of Tel Aviv University (Israel) and colleagues examined variations in body size of the stone marten (*Martes foina*) collected in Denmark between 1858 and 1999. Analyses show that skull size (and by implication body size) had two periods of decrease and that these two periods coincided with the periods of increase in ambient temperature (of 0.7°C and 0.55°C, respectively). The changes in temperature for the region of Denmark during these periods are equivalent to the observed global rise in mean global temperature during the 20th century (approximately 0.6°C).

Although numerous studies have found evidence of decreasing body size with increasing temperatures across various species of endotherms—a pattern consistent with the hypothesis that smaller body size is more energetically efficient under a warmer climate—other studies have observed increases in body size with rising temperatures. For example, in contrast to the pattern of decreasing body size in the stone marten that he observed in Denmark, Yoram Yom-Tov observed an increase in average body size with increasing ambient temperature for Eurasian otter (*Lutra lutra*) collected in Sweden between 1962 and 2008. To understand the apparent discrepancy in these studies, it is important to understand that temperature has direct and indirect effects on animals through a variety of processes other than thermoregulation that may complicate the story. For example, temperature can have a significant influence on both food availability and nutrition. In cold regions, the direct effect of an increase in temperature may be a reduction in the cost of body maintenance, thus enabling animals to divert energy toward growth, which results in an increase in body size. Increases in temperature can also have an effect on the availability of food resources. In the case of the observed increase in body size in the Eurasian otter in Sweden, Yom-Tov found that increasing temperature reduced the length of time of ice cover on freshwater lakes in Sweden, thus increasing the access of otter to food resources (fish and invertebrates). Similarly, elevated plant growth and thus food availability under warmer climate conditions is suggested as an explanation for increase in the body size of the masked shrew (*Sorex cinereus*) and the American marten (*Martes americana*) in Alaska, and the weasel (*Mustela nivalis*) and stoat (*Mustela ereminea*) in Sweden.

The story that emerges from these and many other studies is that rising global temperatures are having an impact on the performance and fitness of animal species. As we shall see later in Chapter 27, just as with body size, the variety of responses to a warming climate are as diverse and complex as the array of biological processes that respond both directly and indirectly to temperature and other features of the climate system.

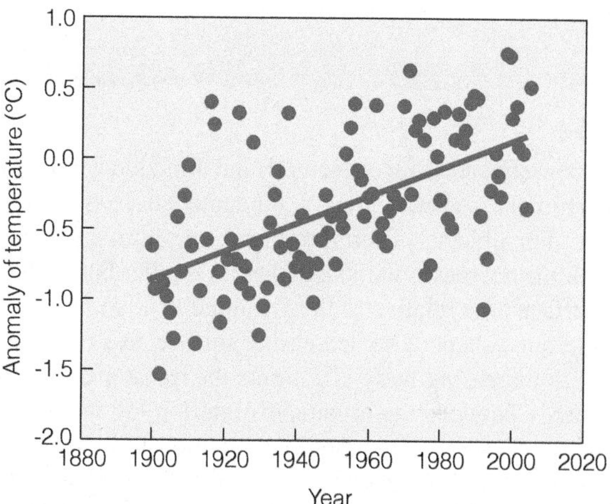

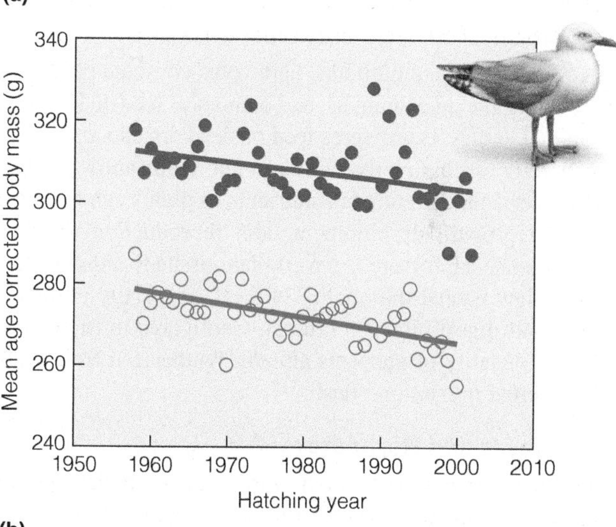

Figure 7.27 Trends in (a) mean annual temperature and (b) mean body mass for male (filled symbol) and female (open symbols) red-billed gulls in New Zealand for the period of 1958–2004. Anomaly of temperature is the difference between mean annual temperature and the average annual temperature for the period of 1958–2001. Body mass has been corrected for age. (Adapted from Teplitsky et al. 2008.)

SUMMARY

Body Size 7.1

Size has consequences for structural and functional relationships in animals; as such, it is a fundamental constraint on adaptation. For objects of similar shape, the ratio of surface area to volume decreases with size; that is, smaller bodies have a larger surface area relative to their volume than do larger objects of the same shape. The decreasing surface area relative to volume with increasing body size limits the transfer of materials and energy between the organism's interior and its exterior environment. An array of adaptations function to increase the surface area and enable adequate exchange of energy and materials between the interior cells and the external environment.

Acquisition of Energy and Nutrients 7.2

To acquire energy and nutrients, herbivores consume plants, carnivores consume other animals, and omnivores feed on both plant and animal tissues. Detritivores feed on dead organic matter.

Directly or indirectly, animals get their nutrients from plants. Low concentrations of nutrients in plants can adversely affect the growth, development, and reproduction of plant-eating animals. Herbivores convert plant tissue to animal tissue. Among plant eaters, the quality of food, especially its protein content and digestibility, is crucial. Carnivores must secure a sufficient quantity of nutrients already synthesized from plants and converted into animal flesh.

Conformers and Regulators 7.3

When an animal is confronted with changes in the environment, it can respond in one of two ways: conformity or regulation. In conformers, changes in the external environmental conditions induce internal changes in the body that parallel external conditions. Conformity largely involves changes at the physiological and biochemical level that are simple and energetically inexpensive but are accompanied by reduced activity and growth. Regulators use a variety of biochemical, physiological, morphological, and behavioral mechanisms that regulate their internal environments. Regulation is energetically expensive, but the benefit is a high level of performance and a greatly extended range of environmental conditions over which an activity can be maintained.

Regulation of Internal Conditions 7.4

To confront daily and seasonal environmental changes, organisms must maintain some equilibrium between their internal and external environment. Homeostasis is the maintenance of a relatively constant internal environment in a variable external environment through negative feedback responses. Through various sensory mechanisms, an organism responds physiologically or behaviorally to maintain an optimal internal environment relative to its external environment. Doing so requires an exchange between the internal and external environments.

Oxygen Uptake 7.5

Animals generate energy by breaking down organic compounds primarily through aerobic respiration, which requires oxygen. Differences between terrestrial and aquatic animals in their means of acquiring oxygen reflect the availability of oxygen in the two environments. Most terrestrial animals have some form of lungs, whereas most aquatic animals use gills to transfer gases between the body and the surrounding water.

Water Balance 7.6

Terrestrial animals must offset water loss from evaporation, respiration, and waste excretion by consuming or conserving water. Terrestrial animals gain water by drinking, eating, and producing metabolic water. Animals of arid regions may reduce water loss by becoming nocturnal, producing highly concentrated urine and feces, becoming hyperthermic during the day, using only metabolic water, and tolerating dehydration.

Aquatic animals must prevent the uptake of, or rid themselves of, excess water. Freshwater fish maintain osmotic balance by absorbing and retaining salts in special cells in the body and by producing copious amounts of watery urine. Many marine invertebrates' body cells maintain the same osmotic pressure as that in seawater. Marine fish secrete excess salt and other ions through kidneys or across gill membranes.

Energy Exchange 7.7

Animals maintain a fairly constant internal body temperature, known as the body core temperature. They use behavioral and physiological means to maintain a heat balance in a variable environment. Layers of muscle fat and surface insulation of scales, feathers, and fur insulate the animal body core against environmental temperature changes. Terrestrial animals face a more dynamic and often more threatening thermal environment than do aquatic animals.

Thermal Regulation 7.8

Animals fall into three major groups relating to temperature regulation: poikilotherms, homeotherms, and heterotherms. Poikilotherms, so named because they have variable body temperatures influenced by ambient temperatures, are ectothermic. Animals that depend on internally produced heat to maintain body temperatures are endothermic. They are called homeotherms because they maintain a rather constant body temperature independent of the environment. Many animals are heterotherms that function either as endotherms or ectotherms, depending on external circumstances.

Poikilotherms 7.9

Poikilotherms gain heat from and lose heat to the environment. Poikilotherms have low metabolic rates and high thermal conductance. Environmental temperatures control their rates of metabolism. Poikilotherms are active only when environmental temperatures are moderate; they are sluggish when temperatures are cool. They have, however, upper and lower limits of tolerable temperatures. Most aquatic poikilotherms maintain no appreciable difference between body temperature and water temperature.

Poikilotherms use behavioral means of regulating body temperature. They exploit variable microclimates by moving into warm, sunny places to heat up and by seeking shaded places to cool off. Many amphibians move in and out of water. Insects and desert reptiles raise and lower their bodies to reduce or increase conductance from the ground or for convective cooling. Desert animals enter shade or spend the heat of day in underground burrows.

Homeotherms 7.10

Homeotherms maintain high internal body temperature through the oxidiziation of glucose and other energy-rich molecules. They have high metabolic rates and low thermal conductance. Body insulation of fat, fur, feathers, scales, and furlike covering on many insects reduces heat loss from the body. A few desert mammals employ heavy fur to keep out desert heat and cold. When insulation fails during the cold, many homeotherms resort to shivering and burning fat reserves. For homeotherms, evaporative cooling by sweating, panting, and wallowing in mud and water is an important way of dissipating body heat.

Trade-offs in Thermal Regulation 7.11

The two approaches to maintaining body temperature, ectothermy and endothermy, involve trade-offs. Unlike poikilotherms, homeotherms are able to remain active regardless of environmental temperatures. For homeotherms, a high rate of aerobic metabolism comes at a high energy cost. This cost places a lower limit on body size. Because of the low metabolic cost of ectothermy, poikilotherms can curtail metabolic activity in times of food and water shortage and temperature extremes. Their low energy demands enable some terrestrial poikilotherms to colonize areas of limited food and water.

Heterotherms 7.12

Based on environmental and physiological conditions, heterotherms take on the characteristics of endotherms or ectotherms. Some normally homeothermic animals become ectothermic and drop their body temperature under certain environmental conditions. Many poikilotherms, notably insects, must increase their metabolic rate to generate heat before they can take flight. Most accomplish this feat by vibrating their wings or wing muscles or by basking in the sun. After flight, their body temperatures drop to ambient temperatures.

During environmental extremes, some animals enter a state of torpor to reduce the high energy costs of staying warm or cool. Their metabolism, heartbeat, and respiration slows, and their body temperature decreases. Hibernation (seasonal torpor during winter) involves a complete rearrangement of metabolic activity so that it runs at a very low level. Heartbeat, breathing, and body temperature are all greatly reduced.

Unique Physiological Means to Maintain Thermal Balance 7.13

Many homeotherms and heterotherms employ countercurrent circulation, the exchange of body heat between arterial and venous blood reaching the extremities. This exchange reduces heat loss through body parts or cools blood flowing to such vital organs as the brain.

Some desert mammals use hyperthermia to reduce the difference between body and environmental temperatures. They store up body heat by day, then release it to the cool desert air by night. Hyperthermia reduces the need for evaporative cooling and thus conserves water. Some cold-tolerant poikilotherms use supercooling, the synthesis of glycerol in body fluids, to resist freezing in winter. Supercooling takes place when the body temperature falls below freezing without freezing body fluids. Some intertidal invertebrates survive the cold by freezing, then thawing with warmer temperatures.

Habitat 7.14

The place where an animal is found is called its habitat. A species' habitat must provide the wide array of resources necessary to sustain growth and reproduction. The environment must provide essential food resources, cover from potential predators, areas for successful reproduction (courtship, mating, nesting, etc.), and a substrate for a wide array of life activities. In mobile animals, the behavioral process of selecting a location to occupy is called habitat selection.

Climate Change and Body Size Ecological Issues & Applications

Bergmann's rule states that for endotherms, body size for a species tends to increase with decreasing mean annual temperature. Studies have documented recent changes in animal body size for local populations over the timescale of decades to a century that are correlated to changes in local temperature. Recent global warming has resulted in both increases and decreases in the average size of different animal species. Decreases in body size have been related to the benefit of smaller body size in thermal balance, whereas increases in body size have been associated with increases in food availability under warmer climates.

STUDY QUESTIONS

1. What constraints are imposed by a diet of plants as compared to one of animal tissues?
2. What is the difference between conformers and regulators?
3. What is homeostasis?
4. What are the three components of a negative feedback system?
5. How does body size influence the ability to supply oxygen to all of the living cells in an animal's body?
6. Contrast the constraints imposed by freshwater and marine environments on the maintenance of water balance for fish.
7. How does the size and shape of an animal's body influence its ability to exchange heat with the surrounding environment?

8. In Figure 7.21, why does mass-specific metabolic rate (metabolic rate per unit weight) of mammals increase with decreasing body mass?
9. Why are the largest species of poikilotherms found in the tropical and subtropical regions?
10. How might you expect the average size of mammal species to vary from the tropics to the polar regions? Why?
11. Why might it be easier to capture a snake in the early morning rather than the afternoon?
12. Why do homeotherms typically have a greater amount of body insulation than do poikilotherms?
13. What behaviors help poikilotherms maintain a fairly constant body temperature during their season of activity?
14. How does supercooling enable some insects, amphibians, and fish to survive freezing conditions?
15. How does countercurrent circulation work, and what function does it serve?
16. Distinguish between hibernation and torpor.
17. Consider a population of fish living below a power plant that is discharging heated water. The plant shuts down for three days in the winter. How would that affect the fish?

FURTHER READINGS

Classic Studies

Bogert, C. M. 1949. "Thermoregulation in reptiles, a factor in evolution." *Evolution* 3:195–211.
Classic study of thermoregulation in ectotherms that discusses the difference between ectothermy and conformity.

Cowles, R. B., and C. M. Bogert.1944. "A preliminary study of the thermal requirements of desert reptiles." *Bulletin of the American Museum of Natural History* 83:265–296.
One of the classic studies on thermoregulation reptiles by a pioneer in the biology of temperature.

Stevenson, R. D. 1985. "The relative importance of behavioral and physiological adjustments controlling body temperature in terrestrial ectotherms." *The American Naturalist* 126:362–386.
Early study in the importance of different mechanisms involved in thermoregulation in ectotherms.

Recent Research

Angilletta, M. J. 2009. *Thermal adaptation*. New York: Oxford University Press.
This book provides an overview of the "state of the science" regarding adaptations to the thermal environment. Clear and concise in its presentation, it provides a weather of research examples on the different responses of animal species to variations in the thermal environment.

Bonner, J. T. 2006. *Why size matters*. Princeton, NJ: Princeton University Press.
This short book provides an excellent overview of the constraints imposed by body size in the evolution and ecology of animals from one of the leading figures in the field.

French, A. R. 1988. "The patterns of mammalian hibernation." *American Scientist* 76:569–575.
This article provides an excellent, easy-to-read, and well-illustrated overview of the concept of hibernation.

Heinrich, B. 1996. *The thermal warriors: Strategies of insect survival*. Cambridge: Harvard University Press.
This enjoyable book describes the variety of strategies insects use to heat their bodies. It is full of strange and wonderful examples of evolution in the world of insects.

Heinrich, B. 2003. *Winter world*. New York. Harper Collins.
Fascinating account of evolutionary adaptations of animals to survive the cold temperatures of winter.

Lee, R. E., Jr. 1989. "Insect cold-hardiness: To freeze or not to freeze." *Bioscience* 39:308–313.
This article offers an excellent overview of the diversity of adaptations that allow insects to respond to seasonal variations in temperature. It includes a good discussion of supercooling and cold hardening in insects.

Randall, D., W. Burggren, and K. French. 2001. *Eckert Animal Physiology*. 5th ed. New York: W. H. Freeman.
This classic animal physiology text focuses on comparative examples that illustrate the general principles of physiology at all levels of organization.

Schmidt-Neilsen, K. 1997. *Animal physiology: Adaptation and environment*. 5th ed. New York: Cambridge University Press.
This comprehensive text gives more information about animal physiology and the range of adaptations that allow animals to cope with various terrestrial and aquatic environments.

Storey, K. B., and J. M. Storey. 1996. "Natural freezing survival in animals." *Annual Review of Ecology and Systematics* 27:365–386.
An excellent, if somewhat technical, review of the adaptations that allow animals to cope with freezing temperatures.

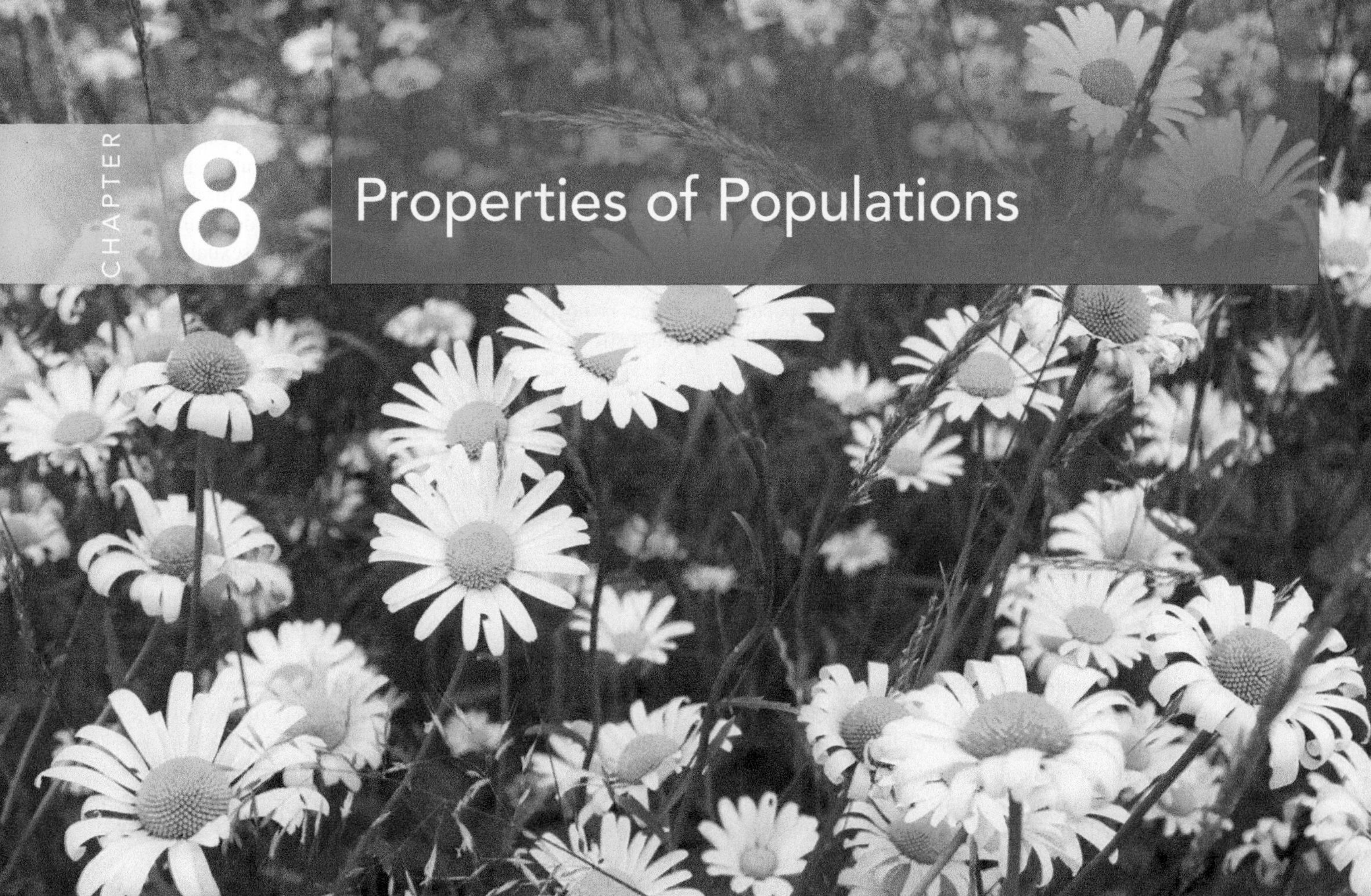

This field of daisies represents a population—a group of individuals of the same species inhabiting a given area.

CHAPTER GUIDE

8.1 Organisms May Be Unitary or Modular

8.2 The Distribution of a Population Defines Its Spatial Location

8.3 Abundance Reflects Population Density and Distribution

8.4 Determining Density Requires Sampling

8.5 Measures of Population Structure Include Age, Developmental Stage, and Size

8.6 Sex Ratios in Populations May Shift with Age

8.7 Individuals Move within the Population

8.8 Population Distribution and Density Change in Both Time and Space

ECOLOGICAL Issues & Applications Invasive Species

AS AN INDIVIDUAL, how do you perceive the world? Most of us regard a friend, a neighborhood maple tree, a daisy in a field, a squirrel in the park, or a bluebird nesting in the backyard as individuals. Rarely do we consider such individuals as part of a larger unit—a population. Although the term *population* has many different meanings and uses, for biologists and ecologists it has a specific definition. A population is a group of individuals of the same species that inhabit a given area. This definition has two important features. First, by requiring that individuals be of the same species, the definition suggests the potential (in sexually reproducing organisms) for interbreeding among members of the population. As such, the population is a genetic unit. It defines the gene pool, the focus of evolution (see Chapter 5). Second, the population is a spatial concept, requiring a defined spatial boundary—for example, the population of Darwin's ground finch inhabiting the Island of Daphne Major in the Galápagos Islands (Chapter 5, Section 5.6).

Populations have unique features because they are an aggregate of individuals. Populations have structure, which relates to characteristics of the collective, such as density (the number of individuals per unit area), proportion of individuals in various age classes, and spacing of individuals relative to each other. Populations also exhibit dynamics—a pattern of continuous change through time that results from the birth, death, and movement of individuals. In this chapter we explore the basic features used to describe the structure of populations, and in doing so we set the foundation for examining the dynamics of population structure in subsequent chapters (Chapters 9 and 11).

8.1 Organisms May Be Unitary or Modular

A population is considered to be a group of individuals, but what constitutes an individual? For most of us, defining an individual would seem to be no problem. We are individuals, and so are dogs, cats, spiders, insects, fish, and so on throughout much of the animal kingdom. What defines us as individuals is our unitary nature. Form, development, growth, and longevity of unitary organisms are predictable and determinate from conception on. The zygote, formed through sexual reproduction, grows into a genetically unique organism (see Chapter 5). There is no question about recognizing an individual. This simplistic view of an individual breaks down, however, when the organism is modular rather than unitary.

In modular organisms, the zygote (the genetic individual) develops into a unit of construction, a module, which then produces further, similar modules. Most plants are modular in that they develop by branching, repeated units of structure. The fundamental unit of aboveground construction is the leaf with its axillary bud and associated internode of the stem. As the bud develops and grows, it produces further leaves, each bearing buds in their axils. The plant grows by accumulating these modules (**Figure 8.1**). The growth of the root system is also modular, growing through the process of branching and providing a continuous connection between above- and belowground modules. There are, however, a variety of growth forms produced by modular growth in plants; some plants spread their modules laterally as well as vertically. For example, some species produce specialized stems that either grow above the surface of the substrate, referred to as *stolons* (**Figure 8.2a**), or below the surface of the substrate, referred to as *rhizomes* (**Figure 8.2b**). These plants can produce new vertical stems and associated root systems from these laterally growing stolons or rhizomes. Similarly, some plants sprout new stems from the surface roots, which are called *suckers* (**Figure 8.2c**). In these laterally growing forms, the new modules may cover a considerable area and appear to be individuals. The plant produced by sexual reproduction, thus arising from a zygote, is a genetic individual, or **genet**. Modules produced asexually by the genet are **ramets**. These ramets are clones—genetically identical modules—and are collectively referred to as a *clonal colony*.

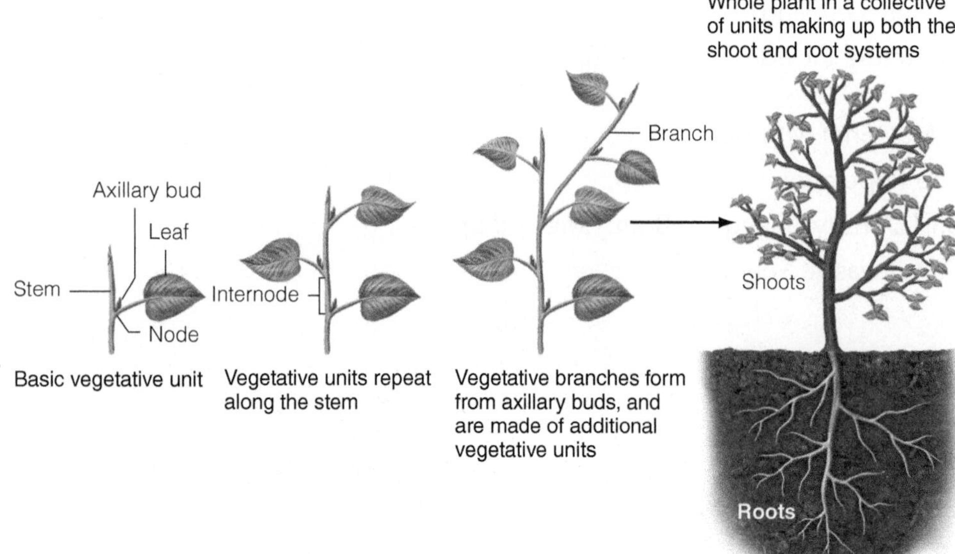

Whole plant in a collective of units making up both the shoot and root systems
Branch
Axillary bud
Leaf
Stem
Internode
Node
Shoots
Roots
Basic vegetative unit
Vegetative units repeat along the stem
Vegetative branches form from axillary buds, and are made of additional vegetative units

Figure 8.1 Illustration of modular growth in a plant. The fundamental unit of aboveground construction is the leaf with its axillary bud and associated internode of the stem. As the bud develops and grows, it produces further leaves, each bearing buds in their axils. The plant grows by accumulating these modules. The growth of the root system is also modular, growing through the process of branching and providing a continuous connection between above- and belowground modules.

Figure 8.2 Three types of lateral modular growth in plants that produce ramets. (a) Stolons are modified stems that develop and extend laterally above the soil surface from which new ramets (above ground stems and root systems) are produced. (b) Rhizomes are similar to stolons but grow below the ground or sediment. (c) In woody plants (shrubs and trees), such as an aspen tree (*Populus tremuloides*), ramets develop from root suckers.

(a)

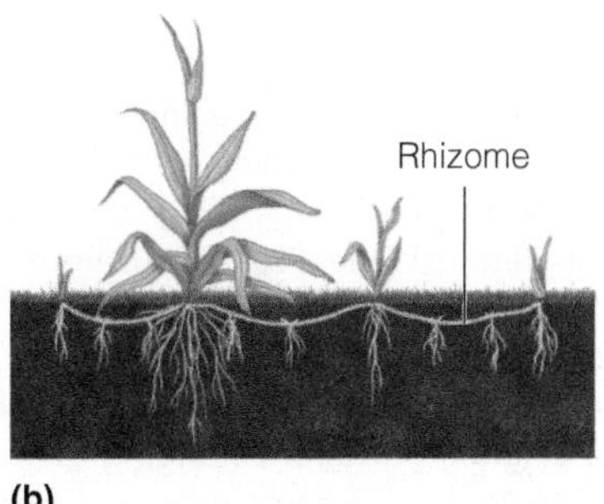

(b)

(c)

The ramets may remain physically linked or they may separate. The connections between the modules may die or rot away, so that the products of the original zygote become physiologically independent units. These ramets can produce seeds (through sexual reproduction) and their own lateral extensions or ramets (asexual reproduction).

Whether living independently or physically linked to the original individual, all ramets are part of the same genetic individual. Thus, by producing ramets, the genet can cover a relatively large area and considerably extend its life. Some modules die, others live, and new ones appear.

Plants are the most obvious group of modular organisms; however, modular organisms include animals, such as corals, sponges, and bryozoans (**Figure 8.3**), as well as many protists and fungi.

Technically, to study populations of modular organisms, we must recognize two levels of population structure: the module (ramet) and the individual (genet). As such, characterizing the population structure of a modular species presents special problems. For practical purposes, ramets are often counted as—and function as—individual members of the population. Modern genetic techniques, however, have allowed ecologists to determine the structure of these populations in terms of genets and ramets, quantifying the patterns of genetic diversity (see this chapter, **Field Studies: Filipe Alberto**).

8.2 The Distribution of a Population Defines Its Spatial Location

The **distribution** of a population describes its spatial location, the area over which it occurs. Distribution is based on the presence and absence of individuals. If we assume that

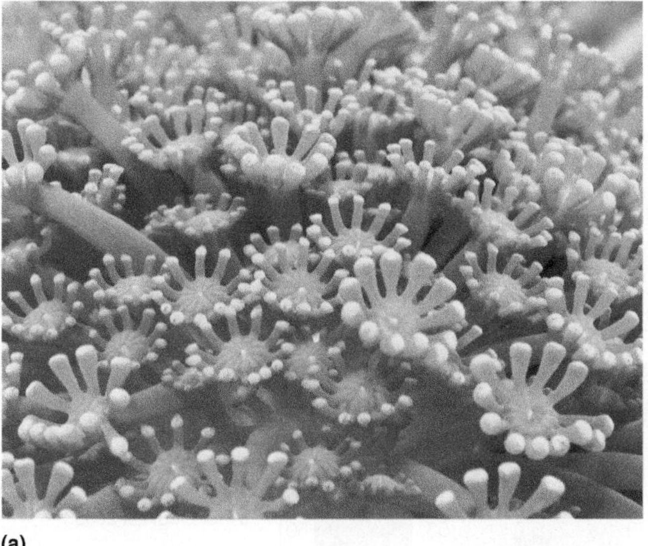
(a)

(b)

Figure 8.3 Examples of clonal animal species: (a) corals and (b) sponges.

FIELD STUDIES *Filipe Alberto*

Department of Biological Sciences, University of Wisconsin–Milwaukee

Numerous plant species reproduce both sexually and asexually. For example, many grass species form dense mats of ramets through the growth of rhizomes or stolons, yet also produce new offspring by flowering and through seed production (new genets). For plant ecologists, this dual strategy of reproduction presents a critical problem in understanding the structure of plant populations. In theory, a field of grass occupied by a species exhibiting both reproductive strategies (sexual and asexual) could consist of only a single genetic individual (genet) and all of the apparent "individual" grass plants could be genetically identical ramets produced by a single parent plant. In contrast, the field could consist of a diversity of genets, each with an associated clonal colony of ramets that intermingle with ramets from adjacent clones.

Although impossible only a few decades ago, the development of new genetic technologies now enables ecologists to analyze the genetic structure of a population, and one of the "new generation" of ecologists that is engaged in this new field of molecular ecology is Filipe Alberto of the University of Wisconsin. The focus of Alberto's research is to understand the population structure of marine plants and algae that inhabit the shallow waters of coastal environments. One of the species that has been the focus of Alberto's work is the seagrass *Cymodocea nodosa*. *C. nodosa* is common throughout the Mediterranean and Atlantic coast of Africa, where it forms meadows in the shallow nearshore waters (see Figure 18.8). It colonizes disturbed areas where it plays an important role in the stabilization of sediments. The species is dioecious (having separate male and female individuals) reproducing both sexually as well as vegetatively (asexually) through the extension of rhizomes. It exhibits fast clonal growth, with extension rates up to 2 m per year.

In one study, Alberto and colleagues examined the genetic structure of a population of *C. nodosa* inhabiting Cádiz Bay along the southeastern coast of Spain. Within the bay, the researchers established a grid of 20 × 38 m with a grid spacing of 2 m, yielding a total of 220 sampling units (see **Figure 1**). For each sampling unit the researchers collected three to five shoots belonging to the same rhizome (genet) for genetic analysis. This sampling scheme allowed them to analyze the spatial pattern of relatedness (genetic similarity) among shoots at any point within the grid.

Relatedness (similarity in genotypes) of sampled shoots was determined through the use of microsatellites, tandem repeats of one to six nucleotides found along a strand of DNA (see Section 5.2). Tandem repeats occur in DNA when a pattern of two or more nucleotides is repeated and the repetitions are directly adjacent to each other (tandem); for example: ACACACACAC. The repeated sequence is often simple, consisting of two (as with the preceding example of AC), three, or four nucleotides (di-, tri-, and tetranucleotide repeats, respectively). Microsatellites are ideal for population studies because, first, they are typically abundant in all species, and secondly, they are highly polymorphic—that is, for a given microsatellite locus on the DNA, there are typically many different forms of the microsatellite—different alleles. Each sequence with a specific number of repeated nucleotides is designated as an allele. So, a locus with six repeats is one allele (ACACACACACAC or AC_6), whereas the same locus in another individual that contains nine repeats is another (different) allele. For the analysis of relatedness in the *C. nodosa*

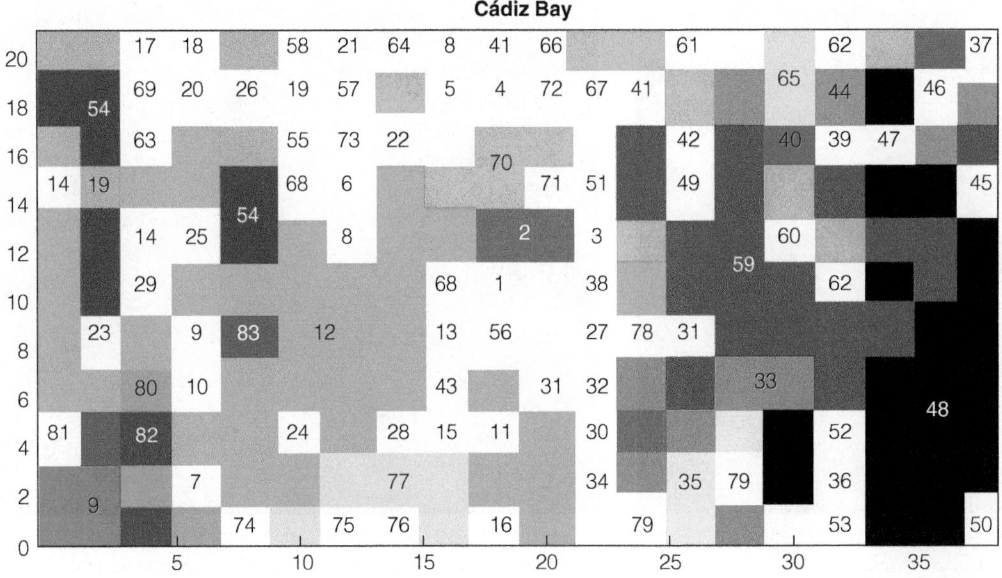

Figure 1 Sampling grid of *Cymodocea nodosa* in Cádiz Bay. The minimum distance between consecutive sampling points was 2 m. The numbers shown code the different genets; different patterns are used to represent each of the clones with more than one copy. Some sampling sites had no cover; these are represented by blank grid units. (After Gouveia 2005.)

population, Alberto used nine different microsatellites that he had identified in a previous genetic study of the species.

The researchers identified 41 different alleles (across the nine microsatellites) and a total of 83 different genotypes in the sample grid. The 83 different genotypes represent 83 genetically unique individuals that were produced through sexual reproduction. The number of different genotypes also corresponds to the number of clones (where clone is defined as a colony of ramets originating from the same genet) in the population. A map of the genotypes on the sampling grid is presented in Figure 1. The result shows an extremely skewed distribution of clone sizes (**Figure 2**), with a median clone dimension of 3.6 m. Despite the genetic richness of the meadow (83 different genotypes) and the resulting presence of many unique smaller clones, the meadow (sample grid) is spatially dominated by a few large clones.

To determine how the two strategies of reproduction (sexual and asexual) influence the genetic structure of the population across the landscape, Alberto undertook a spatial analysis of relatedness using the data presented in Figure 1. With a plant species that can reproduce asexually through the lateral extension of rhizomes, one would assume that there is a high probability that adjacent shoots are ramets from the same genet (they belong to the same clone). But how would the probability of two shoots belonging to the same clone change for greater and greater distance between them? To answer this question, Alberto calculated the probability of clonal identity for the different distance classes (recall that the sample points are on a grid of 2 m). The probability of clonal identity F_r is defined as the probability that a randomly chosen pair of shoots separated by a distance r belong to the same clone (genetically identical).

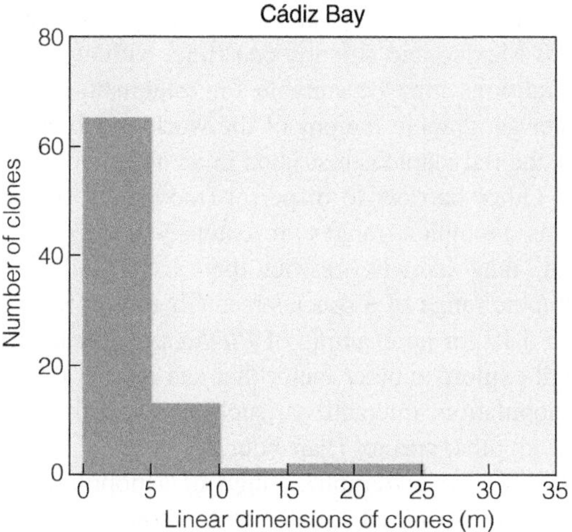

Figure 2 Distribution of the linear dimensions of *Cymodocea nodosa* clones for Cádiz Bay. The linear dimension is the minimum clone size, which is estimated as the distance between the farthest clonemates (sampled shoots belonging to the same clone).
(After Gouveia 2005.)

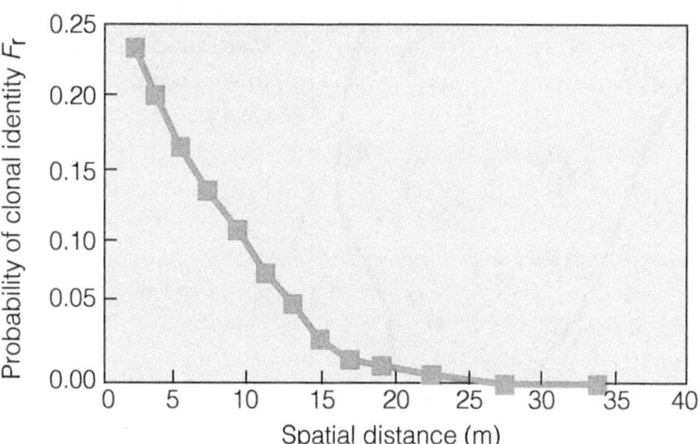

Figure 3 Analysis of *Cymodocea nodosa* clonal structure using the probability of clonal identity, F_r. The spatial distance where probability of clonal identity approaches zero provides an estimate the radius of the clonal influence on the genetic structure of the population.
(After Gouveia 2005.)

The calculated values of F_r for each distance class (0–2 m; 2–4; 4–6; 6–8; ...) is presented in **Figure 3**. The probability of clonal identity (F_r) declines with increasing distance, from approximately 25 percent in the first distance class (0–2 m) to reach zero (meaning no pairs shared the same genotype) at a distance of 25–30 m. This distance (25–30 m) corresponds to the dimensions of the largest clones found in the populations (see Figure 3). The results indicate that in this population of seagrass, sampling shoots at an interval of 30 m or more will assure that the samples represent unique genotypes—genetically unique individuals—rather than ramets.

References

Alberto, F., S. Massa, P. Manent, E. Diaz-Almela, S. Arnaud-Haond, C. M. Duarte, and E.A. Serrão. 2008. "Genetic differentiation and secondary contact zone in the seagrass *Cymodocea nodosa* across the Mediterranean-Atlantic transition region." *Journal of Biogeography* 35:1279–1294.

Alberto, F., L. Gouveia, S. Arnaud-Haond, J. L. Pérez-Lloréns, C. M. Duarte, and E. A. Serrão. 2005. "Within-population spatial genetic structure, neighborhood size and clonal subrange in the seagrass *Cymodocea nodosa*." *Molecular Ecology* 14:2669–2681.

Alberto, F., L. Correia, C. Billot, C. Duarte, and E. Serrão. 2003. "Isolation and characterization of microsatellite markers for the seagrass *Cymodocea nodosa*." *Molecular Ecology Notes* 3:397–399.

1. How might a change in the relative importance of asexual versus sexual reproduction change the distribution of clone sizes presented in Figure 1?
2. The results presented in Figure 3 show the change in probability of clonal identity with distance. How might an increase in the importance of sexual reproduction over asexual reproduction influence the result presented in Figure 3?

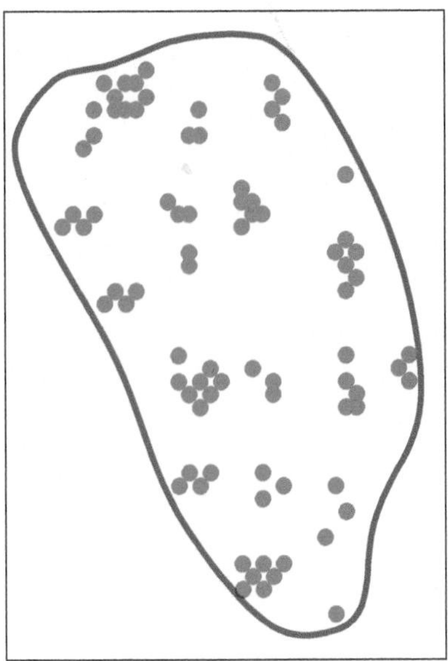

Figure 8.4 A hypothetical population. Each red dot represents an individual organism. The blue line defines the population distribution, or the area in which the population occurs.

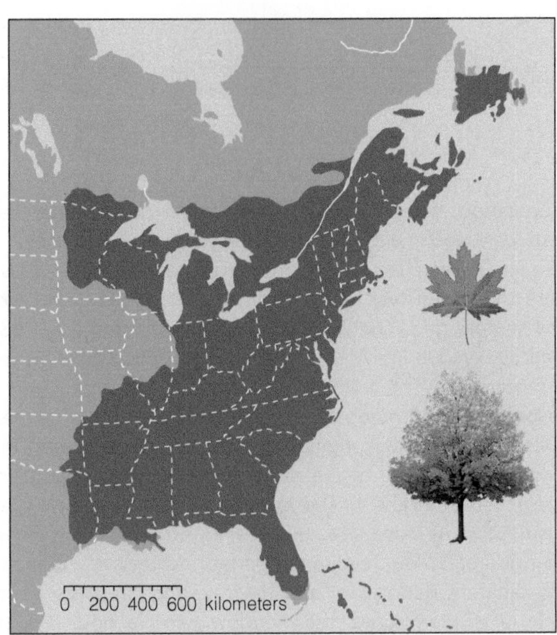

Figure 8.5 Red maple (*Acer rubrum*), one of the most abundant and widespread trees in eastern North America, thrives on a wider range of soil types, texture, moisture, acidity, and elevation than does any other forest species in North America. The northern extent of its range coincides with the minimum winter temperature in southeastern Canada.

each red dot in **Figure 8.4** represents an individual's position within a population on the landscape, we can draw a line (shown in blue) defining the population distribution—a spatial boundary within which all individuals in the population reside. When the defined area encompasses all the individuals of a species, the distribution describes the population's **geographic range**.

Population distribution is influenced by a number of factors. We introduced the concept of habitat—the place or environment where an organism lives—in Chapter 7 (Section 7.14). Each species has a range of abiotic environmental and resource conditions under which it can survive, grow, and reproduce. The primary factor influencing the distribution of a population is the occurrence of suitable environmental and resource conditions—habitat suitability. Red maple (*Acer rubrum*), for example, is the most widespread of all deciduous trees of eastern North America (**Figure 8.5**). The northern limit of its geographic range coincides with the area in southeastern Canada where minimum winter temperatures drop to –40°C. Its southern limit is the Gulf Coast and southern Florida. Dry conditions halt its westward range. Within this geographic range, the tree grows under a wide variety of soil types, soil moisture, acidity, and elevations—from wooded swamps to dry ridges. Thus, the red maple exhibits high tolerance to temperature and other environmental conditions. In turn, this high tolerance allows a widespread geographic range.

A species with a geographically widespread distribution, such as red maple, is referred to as **ubiquitous**. In contrast, a species with a distribution that is restricted to a particular locality or localized habitat is referred to as **endemic**. Many endemic species have specialized habitat requirements. For example, the shale-barren evening primrose (*Oenothera argillicola*; **Figure 8.6**) is a member of the evening primrose family (Onagraceae). This species is adapted to hot, shale-barren environments that form when certain types of shale form outcrops on south- to southwest-facing slopes of the Allegheny Mountains. Most members of this group of plants are listed as endangered or threatened because they are found in these specific habitats only from southern Pennsylvania through West Virginia to southern Virginia, where shale barrens are formed.

The geographic distribution of red maple in Figure 8.5 illustrates another important factor limiting the distribution of a population: geographic barriers. Although this tree species occupies several islands south of mainland Florida, the southern and eastern limits of its geographic range correspond to the Gulf of Mexico and Atlantic coastline. Although environmental conditions may be suitable for establishment and growth in other geographic regions of the world (such as Europe and Asia), the red maple is restricted in its ability to colonize those areas. Other barriers to dispersal (movement of individuals), such as mountain ranges or extensive areas of unsuitable habitat, may likewise restrict the spread and therefore the geographic range of a species (see Chapter 5, Section 5.8 and Figure 5.19 for an example of *Plethodon* salamanders). Later, we will explore another factor that can restrict the distribution of a population: interactions, such as competition and predation, with other species (Part Four).

Within the geographic range of a population, individuals are not distributed equally. Individuals occupy only those areas that can meet their requirements (suitable habitat). Because organisms respond to a variety of environmental factors, they can inhabit only those locations where all factors fall within their range of tolerance (see Chapter 7, Section 7.14). As a result, we can describe the distribution

Figure 8.6 (a) The distribution map of the species shows that it is found in only three states in the eastern United States (Pennsylvania, West Virginia, and Virginia). Within these states it is only found only in the Allegheny Mountain range. (b) The shale-barren evening primrose (*Oenothera argillicola*), an endemic species found only on shale outcrops (c) on south-facing mountain slopes.

of a population at various spatial scales. For example, in **Figure 8.7**, the distribution of the moss *Tetraphis pellucida* is described at several different spatial scales, ranging from its geographic distribution at a global scale to the location of individuals within a single clump occupying the stump of a dead conifer tree. This species of moss can grow only in areas in which the temperature, humidity, and pH are suitable; different factors may be limiting at different spatial scales. At the continental scale, the suitability of climate (temperature and humidity) is the dominant factor. Within a particular area, distribution of the moss is limited to microclimates along stream banks, where conifer trees are abundant. Within a particular locality, it occupies the stumps of conifer trees where the pH is sufficiently acidic.

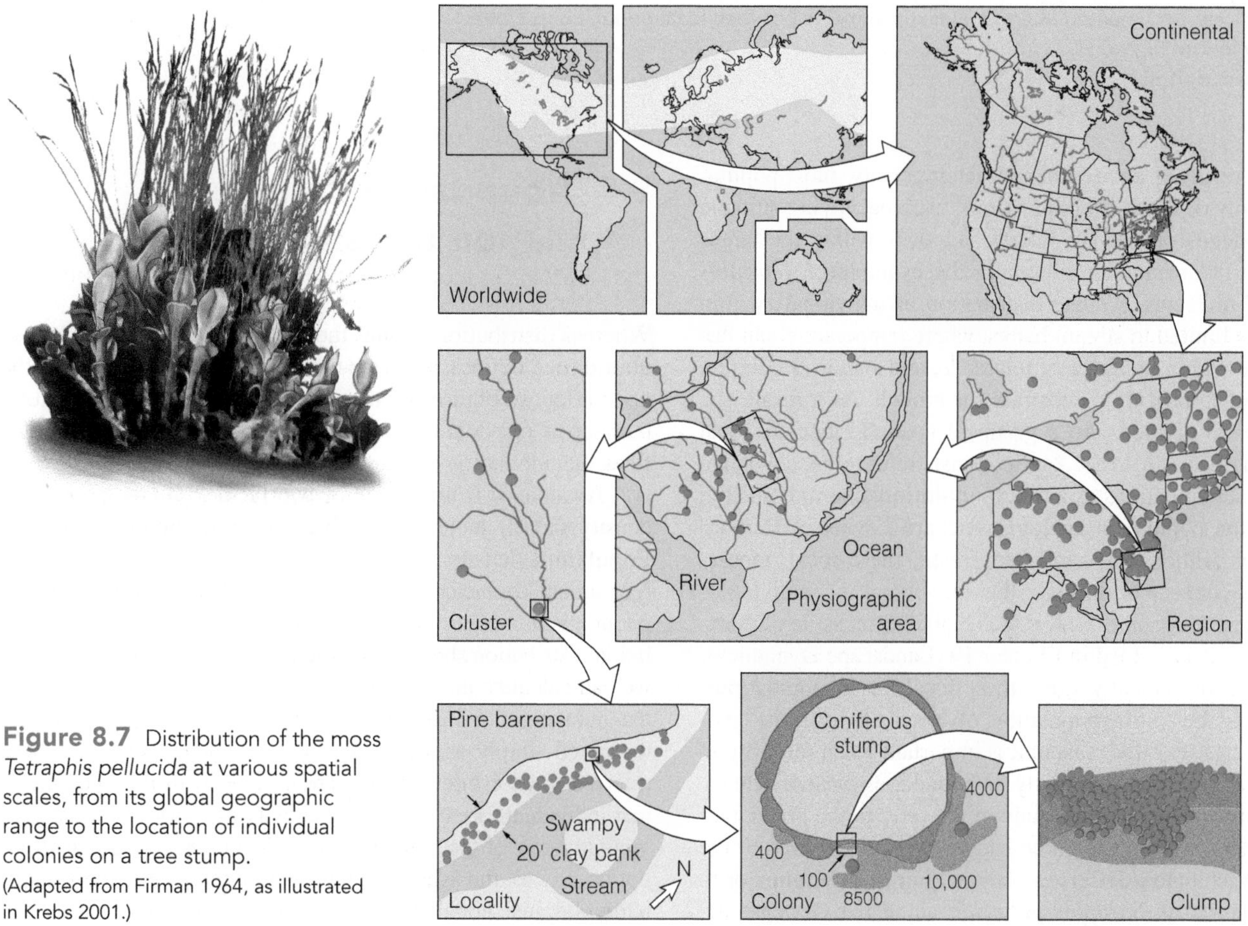

Figure 8.7 Distribution of the moss *Tetraphis pellucida* at various spatial scales, from its global geographic range to the location of individual colonies on a tree stump.
(Adapted from Firman 1964, as illustrated in Krebs 2001.)

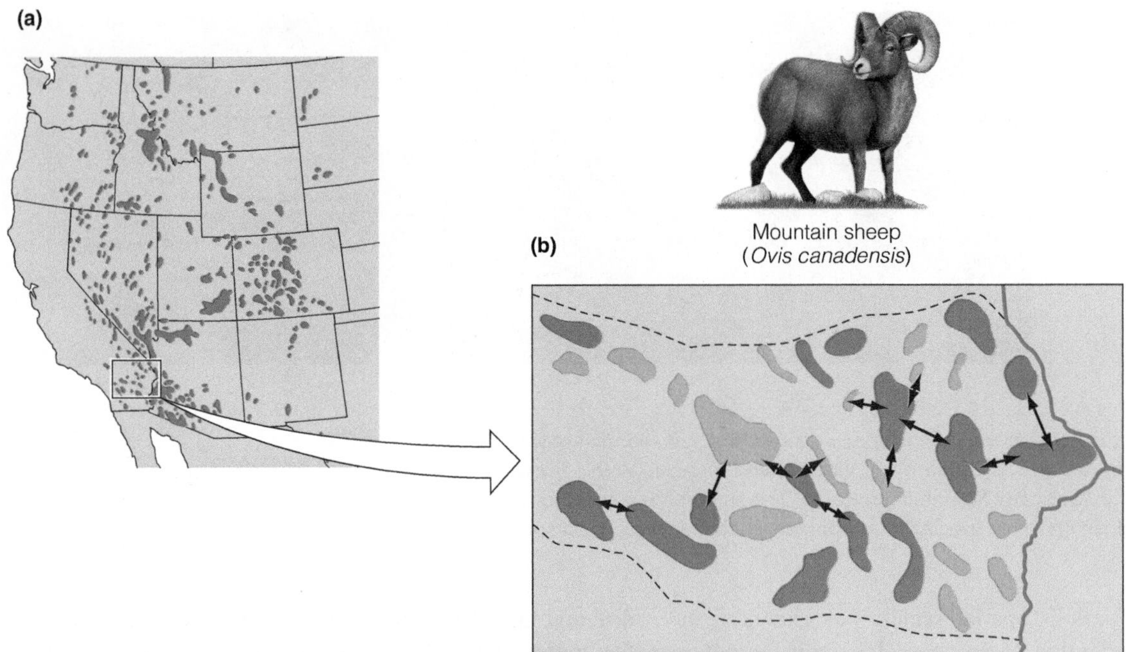

Figure 8.8 Example of a metapopulation (a collective of local populations linked by the dispersal of individuals) of mountain sheep (*Ovis canadensis*) in southern California. (a) The southern California metapopulation exists within the larger geographic distribution of the mountain sheep (dark areas on map of western United States). (b) For the southern California metapopulation, shaded areas indicate mountain ranges with resident populations, whereas unshaded areas are suitable areas of habitat not currently occupied by resident populations. Arrows indicate documented intermountain movements between local populations. Dotted lines show fenced highways.
(After Bleich et al. 1990.)

As a result of environmental heterogeneity, most populations are divided into subpopulations, each occupying suitable habitat patches of various shapes and sizes within the larger landscape of unsuitable habitat. In the example of *Tetraphis* presented in Figure 8.7, the distribution of individuals within a region is limited to stream banks, where temperature and humidity are within its range of tolerance, and stands of conifers are present to provide a substrate for growth. As a result, the population is divided into a group of spatially discrete local subpopulations, (**Figure 8.8**). Ecologists refer to the collective of local subpopulations as a **metapopulation**, a term coined in 1970 by the population ecologist Richard Levins of Harvard University. Although spatially separated, these local populations are connected through the movement of individuals among them (Section 8.7). A more detailed discussion of metapopulations is presented in Chapter 19 (Landscape Dynamics).

Ecologists typically study these local, or subpopulations, rather than the entire population of a species over its geographic range. For this reason, it is important when referring to a population to define explicitly its boundaries (spatial extent). For example, an ecological study might refer to the population of red maple trees in the Three Ridges Wilderness Area of the George Washington–Jefferson National Forest in Virginia or to the population of *Tetraphis pellucida* along the Oswagatchie River in in the Adirondack Mountains of New York.

8.3 Abundance Reflects Population Density and Distribution

Whereas distribution defines the spatial extent of a population, **abundance** defines its size—the number of individuals in the population. In Figure 8.4, the population abundance is the total number of red dots (individuals) within the blue line that defines the population distribution.

Abundance is a function of two factors: (1) the population density and (2) the area over which the population is distributed. **Population density** is the number of individuals per unit area (per square kilometer [km], hectare [ha], or square meter [m]), or per unit volume (per liter or m^3). By placing a grid over the population distribution shown in Figure 8.4, as is done in **Figure 8.9**, we can calculate the density for any given grid cell by counting the number of red dots that fall within its boundary. Density measured simply as the number of individuals per unit area is referred to as **crude density**. The trouble with this measure is that individuals are typically not equally numerous over the geographic range of the population (see Section 8.2). Individuals do not occupy all the available space within the population's distribution because not all areas are suitable. As a result, density can vary widely from location to location (as in Figure 8.9).

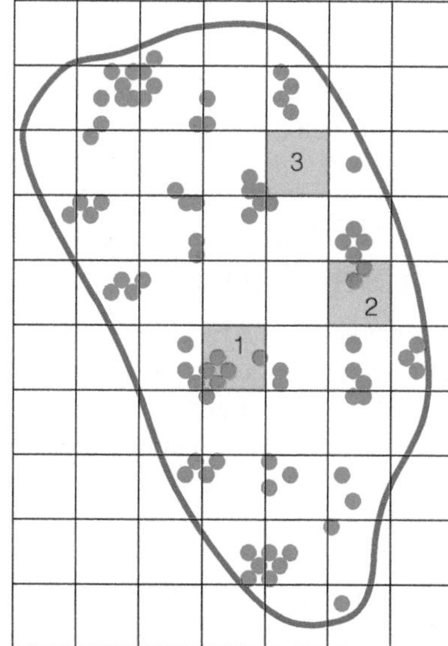

Figure 8.9 We can use the hypothetical population shown in Figure 8.4 to distinguish abundance from density. Abundance is defined as the total number of individuals in the population (red dots). Population density is defined as the number of individuals per unit area. The grid divides the distribution into quadrats of equal size. If we assume that each grid cell is 1 m^2, the density of grid cell 1 is 5 individuals/m^2, the density of grid cell 2 is 2 individuals/m^2, and the density of the third grid cell is zero (unoccupied).

How individuals are distributed within a population—in other words, their spatial position relative to each other—has an important bearing on density. Individuals of a population may be distributed randomly, uniformly, or in clumps (aggregated; **Figure 8.10**). Individuals may be distributed randomly if each individual's position is independent of those of the others. In contrast, individuals distributed uniformly are more or less evenly spaced. A uniform distribution usually results from some form of negative interaction among individuals, such as competition, which functions to maintain some minimum distance among members of the population (see Chapter 11). Uniform distributions are common in animal populations where individuals defend an area for their own exclusive use (territoriality) or in plant populations where severe competition exists for belowground resources such as water or nutrients (Figure 8.11; see also Figures 11.17 and 11.19).

The most common spatial distribution is clumped, in which individuals occur in groups. Clumping results from a variety of factors. For example, suitable habitat or other resources may be distributed as patches on the larger landscape. Some species form social groups, such as fish that move in schools or birds in flocks (see Figure 8.10). Plants that reproduce asexually form clumps, as ramets extend outward from the parent plant (see Figure 8.1). The distribution of humans is clumped because of social behavior, economics, and geography, reinforced by the growing development of urban areas during the past century. In the example presented in Figure 8.9, the individuals within the population are clumped; as a result, the density varies widely between grid cells.

Figure 8.10 Patterns of the spatial distribution for individuals within a population: random, uniform, and clumped. Examples include (a) random, flowering plants in meadow; (b) uniform, nesting shorebirds; and (c) clumped, school of marine fish.

(a)

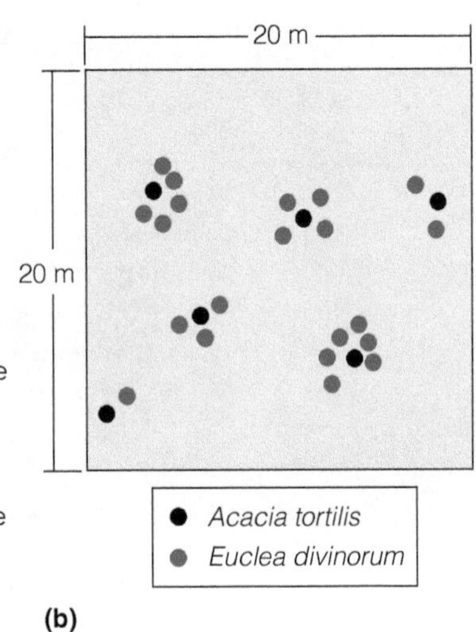

(b)

Figure 8.11 Spatial distribution of the shrub *Euclea divinorum*, inhabiting the savannas of southern Africa. Individuals are clumped under the canopies of *Acacia tortilis* trees, as seen in both (a) the photograph, and in (b) the mapping of individuals within a sample plot. The clumps, however, are uniformly spaced because of the uniform spacing of *A. tortilis* trees on the landscape.

As with geographic distribution (see Section 8.2), the spatial distribution of individuals within the population can also be described at multiple spatial scales. For example, the distribution of the shrub *Euclea divinorum*, found in the savanna ecosystems of Southern Africa, is clumped (**Figure 8.11a**). The clumps of *Euclea* are associated with the canopy cover of another plant that occupies the savanna: trees of the genus *Acacia* (**Figure 8.11b**). The clumps, however, are uniformly spaced, reflecting the uniform distribution of *Acacia* trees on the landscape. The regular distribution of *Acacia* trees is a function of competition among neighboring individuals for water (see Section 11.10). In the example of *Tetraphis* presented in Figure 8.7, the spatial distribution of individuals is clumped at two different spatial scales. Populations are concentrated in long bands or strips along the stream banks, leaving the rest of the area unoccupied. Within these patches, individuals are further clumped in patches corresponding to the distribution of conifer stumps.

To account for patchiness, ecologists often refer to **ecological density**, which is the number of individuals per unit of available living space. For example, in a study of bobwhite quail (*Colinus virginianus*) in Wisconsin, biologists expressed density as the number of birds per mile of hedgerow (the birds' preferred habitat), rather than as birds per hectare (**Figure 8.12**). Ecological densities are rarely estimated because determining what portion of a habitat represents living space is typically a difficult undertaking.

8.4 Determining Density Requires Sampling

Population size (abundance) is a function of population density and the area that is occupied (geographic distribution). In other words, population size = density × area. But how is density determined? When both the distribution (spatial extent) and abundance are small—as in the case of many rare or endangered species—a complete count may be possible. Likewise, in some habitats that are unusually open, such as antelope living on an open plain or waterfowl concentrated in a marsh, density may be determined by a direct count of all individuals. In most cases, however, density must be estimated by sampling the population.

A method of sampling used widely in the study of populations of plants and sessile (attached) animals involves quadrats, or sampling units (**Figure 8.13**). Researchers divide the area of study into subunits, in which they count animals or plants of concern in a prescribed manner, usually counting individuals in only a subset or sample of the subunits (as in Figure 8.9). From these data, they determine the mean density of the units sampled. Multiplying the mean value by the total area provides an estimate of population size (abundance). The accuracy of estimates of density derived from population sampling can be influenced by the manner in which individuals are spatially distributed

(a)

(b)

Figure 8.12 (a) The bobwhite quail (*Colinus virginianus*) inhabits the (b) hedgerow habitats (forested strips of land bordering the agricultural fields) on this agricultural landscape in Wisconsin (United States). The estimate of population density for this species is different if expressed as number per acre as compared to number per area of hedgerow, the latter being an estimate of ecological density.

Figure 8.13 This researcher is estimating population density of marsh plants by positioning quadrats (the 1 m^2 frame seen in the photo) at random locations across the landscape (study area). Density can then be estimated by averaging the results of the replicate samples.

within the population (Section 8.3). The estimate of density can also be influenced by the choice of boundaries or sample units. If a population is clumped—concentrated into small areas—and the population density is described in terms of individuals per square kilometer, the average number of individuals per unit area alone does not adequately represent the spatial variation in density that occurs within the population (**Figure 8.14**). In this case, it is important to report an estimate of variation or provide a confidence interval for the estimate of density. In cases where clumping is a result of habitat heterogeneity (habitat is clumped), ecologists may choose to use the index of ecological density for the specific areas (habitats) in which the species is found (for example, stream banks in Figure 8.7).

For mobile populations, animal ecologists must use other sampling methods. Capturing, marking, and recapturing individuals within a population—known generally as mark-recapture—is the most widely used technique to estimate animal populations (**Figure 8.15**). There are many variations of this technique, and entire books are devoted to various methods of application and statistical analysis. Nevertheless, the basic concept is simple.

Capture-recapture or mark-recapture methods are based on trapping, marking, and releasing a known number of marked animals (M) into the population (N). After giving the marked individuals an appropriate period of time to once again mix with the rest of the population, some individuals are again captured from the population (n). Some of the individuals caught in this second period will be carrying marks (recaptured, R), and others will not. If we assume that the ratio of marked to sampled individuals in the second sample (n/R) represents the

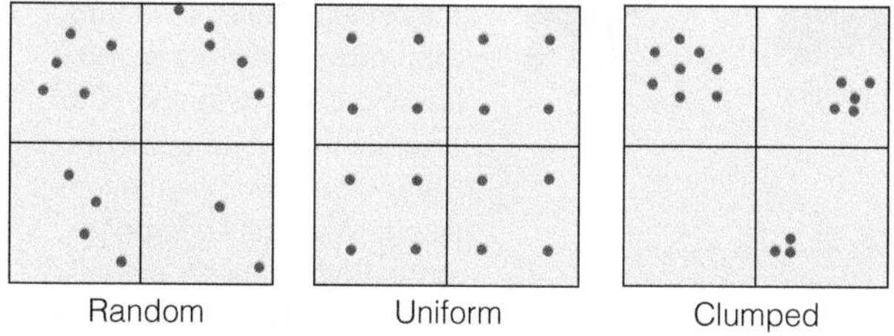

Figure 8.14 The difficulty of sampling. Each area contains a population of 16 individuals. We divide each area into four sampling units and choose one at random. Our estimates of population size will be quite different depending on which unit we select. For the random population, the estimates are 20, 20, 16, and 8. For the uniform population, any sampling unit gives a correct estimate (16). For the clumped population, the estimates are 32, 20, 0, and 12.

ratio for the entire population (N/M), we can compute an estimate of the population using the following relationship:

$$\frac{N}{M} = \frac{n}{R}$$

The only variable that we do not know in this relationship is N. We can solve for N by rearranging the equation as follows:

$$N = \frac{nM}{R}$$

For example, suppose that in sampling a population of rabbits, a biologist captures and tags 39 rabbits from the population. After their release, the ratio of the number of rabbits in the entire population (N) to the number of tagged or marked rabbits (M) is N/M. During the second sample period, the biologist captures 15 tagged rabbits (R) and 19 unmarked ones—a total of 34 (n). The estimate of population size, N, is calculated as

$$N = \frac{nM}{R} = \frac{(34 \times 39)}{15} = 88$$

This simplest method, the single mark–single recapture, is known as the Lincoln–Petersen index of relative population size.

As with any method of population estimation, the accuracy of the Lincoln–Peterson index depends on a number of assumptions. First, the method assumes that the sampling is random, that is, each individual in the population has an equal probability of being captured. Secondly, the marked individuals must distribute themselves randomly throughout the population so that the second sample will accurately represent the population. Last, the ratio of marked and unmarked individuals must not change between the sampling periods. This is especially important if the method of marking individuals influences their survival, as in the case of highly visible marks or tags that increase their visibility to predators.

For work with most animals, ecologists find that a measure of relative density or abundance is sufficient. Methods involve

Figure 8.15 Researchers are using mark-recapture sampling to estimate the population of Bristle-thighed Curlews (*Numenius tahitiensis*) on their breeding grounds in the Andreafsky Wilderness of the Yukon Delta National Wildlife Refuge. Captured birds are marked with leg flags as part of the effort to estimate population size and track survival rates of adults.

observations relating to the presence of organisms rather than to direct counts of individuals. Techniques include counts of vocalizations, such as recording the number of drumming ruffed grouse heard along a trail, counts of animal scat seen along a length of road traveled, or counts of animal tracks, such as may be left by a number of opossums crossing a certain dusty road. If these observations have some relatively constant relationship to total population size, the data can be converted to the number of individuals seen per kilometer or heard per hour. Such counts, called *indices of abundance*, cannot function alone as estimates of actual density. However, a series of such index figures collected from the same area over a period of years depicts trends in abundance. Counts obtained from different areas during the same year provide a comparison of abundance between different habitats. Most population data on birds and mammals are based on indices of relative abundance rather than on direct counts.

8.5 Measures of Population Structure Include Age, Developmental Stage, and Size

Abundance describes the number of individuals in the population but provides no information on their characteristics—that is, how individuals within the population may differ from one another. Unless each generation reproduces and dies in a single season, not overlapping the next generation (such as annual plants and many insects), the population will have an age structure: the number or proportion of individuals in different age classes. Because reproduction is restricted to certain age classes and mortality is most prominent in others, the relative proportions of each age group bear on how quickly or slowly populations grow (see Chapter 9).

Populations can be divided into three ecologically important age classes or stages: prereproductive, reproductive, and postreproductive. We might divide humans into young people, working adults, and senior citizens. How long individuals remain in each stage depends largely on the organism's life history (see Chapter 10). Among annual species, the length of the prereproductive stage has little influence on the rate of population growth (see Chapter 9). In organisms with variable generation times, the length of the prereproductive period has a pronounced effect on the population's rate of growth. The populations of short-lived organisms often increase rapidly, with a short span between generations. Populations of long-lived organisms, such as elephants and whales, increase slowly and have a long span between generations.

Determining a population's age structure requires some means of obtaining the ages of its members. For humans, this task is not a problem, but it is for wild populations. Age data for wild animals can be obtained in several ways, and the method varies with the species (**Figure 8.16**). The most accurate, but most difficult, method is to mark young individuals in a population and follow their survival through time (see discussion of life table, Chapter 9). This method requires a large number of marked individuals and a lot of time. For this reason, biologists may use other, less-accurate methods. These methods include examining a representative sample of individual carcasses to determine their ages at death. A biologist might look for the wear and replacement of teeth in deer and other ungulates, growth rings in the cementum of the teeth of carnivores and ungulates, or annual growth rings in the horns of mountain sheep. Among birds, observations of plumage changes and wear in both living and dead individuals can separate juveniles from subadults (in some species) and adults. Aging of fish is most commonly accomplished by counting rings deposited annually (annuli) on hard parts including scales, otoliths (ear bones), and spines.

Studying the age structure of plant populations can prove even more difficult. The major difficulty lies in determining

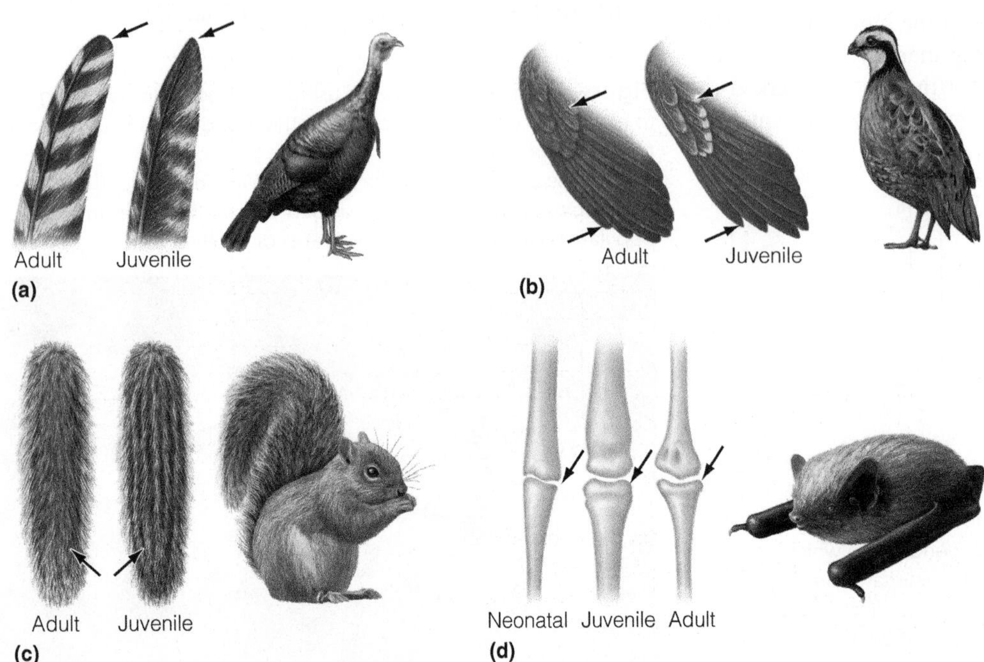

Figure 8.16 Examples of methods used by ecologists for determining the age of birds and mammals. (a) The leading primary feather on an adult wild turkey is rounded, whereas that of a juvenile is pointed. (b) Besides having a pointed leading primary feather, the juvenile bobwhite quail has buff-colored primary coverts (see arrows). (c) Differences in the color bars on the tail of gray squirrels distinguish adults from juveniles. Juveniles have much more distinctive bands of white and black along the tail edge. (d) Researchers can detect differences in the bone structure of bats' wings and determine age by feeling the wing bones of living individuals.

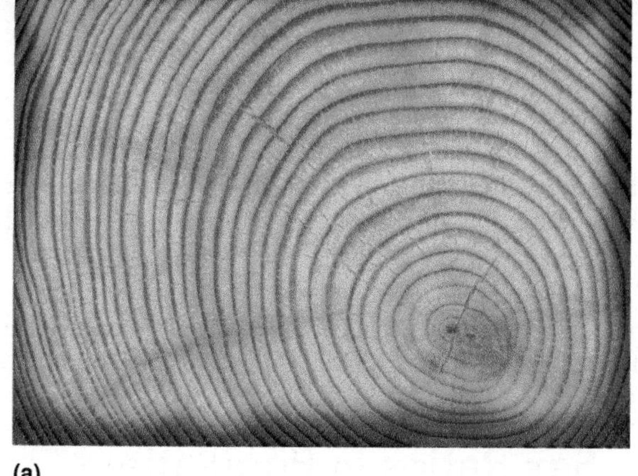

(a)

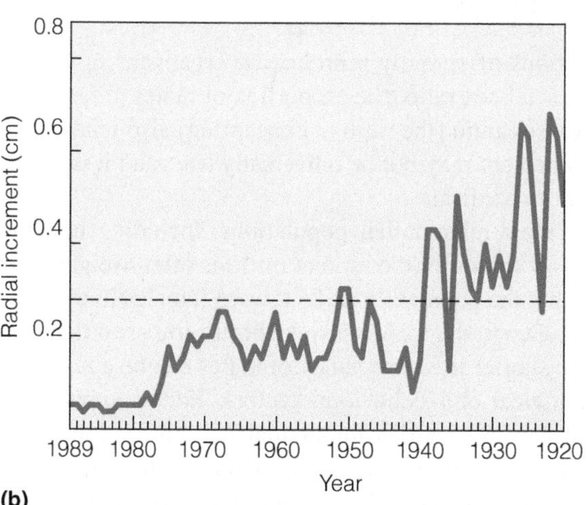

(b)

Figure 8.17 (a) Cross section of tree trunk showing annual growth rings. By measuring the width of each ring, a pattern of radial growth through time can be established. (b) Example of time series of radial increments for an American beech tree (*Fagus grandifolia*) in central Virginia. The dashed line is the overall average for the tree over time.

the age of plants and whether the plants are genetic individuals (genets) or ramets (Section 8.1).

The approximate ages of trees in which growth is seasonal can be determined by counting annual growth rings (**Figure 8.17**), a procedure called *dendrochronology*. But given the time and expense necessary to collect and analyze samples of tree rings, forest ecologists have tried to employ size (diameter of the trunk at breast height, or dbh) as an indicator of age on the assumption that diameter increases with age—the greater the diameter, the older the tree. Such assumptions, it was discovered, were valid for dominant canopy trees; but with their growth suppressed by lack of light, moisture, or nutrients, smaller understory trees, seedlings, and saplings add little to their diameters. Although their diameters suggest youth, small trees are often the same age as large individuals in the canopy.

Attempts to age nonwoody plants have met with less success. The most accurate method of determining the age structure of short-lived herbaceous plants is to mark individual seedlings and follow them through their lifetimes. The results of recent studies, however, suggest that annual growth rings form in the root tissues (secondary root xylem) of many perennial herbaceous plants and can be used successfully in the analysis of age structure in this group of plants. However, the use of size or developmental stage classes is often more appropriate than age in describing the structure of plant (and some animal) populations. As we shall see in Chapter 9, size or developmental stage often provides a better indicator of patterns of mortality and reproduction necessary for predicting patterns of population dynamics.

Once the age structure of a population has been determined, it can be represented graphically in the form of an age pyramid. Age pyramids (**Figure 8.18**) are snapshots of the

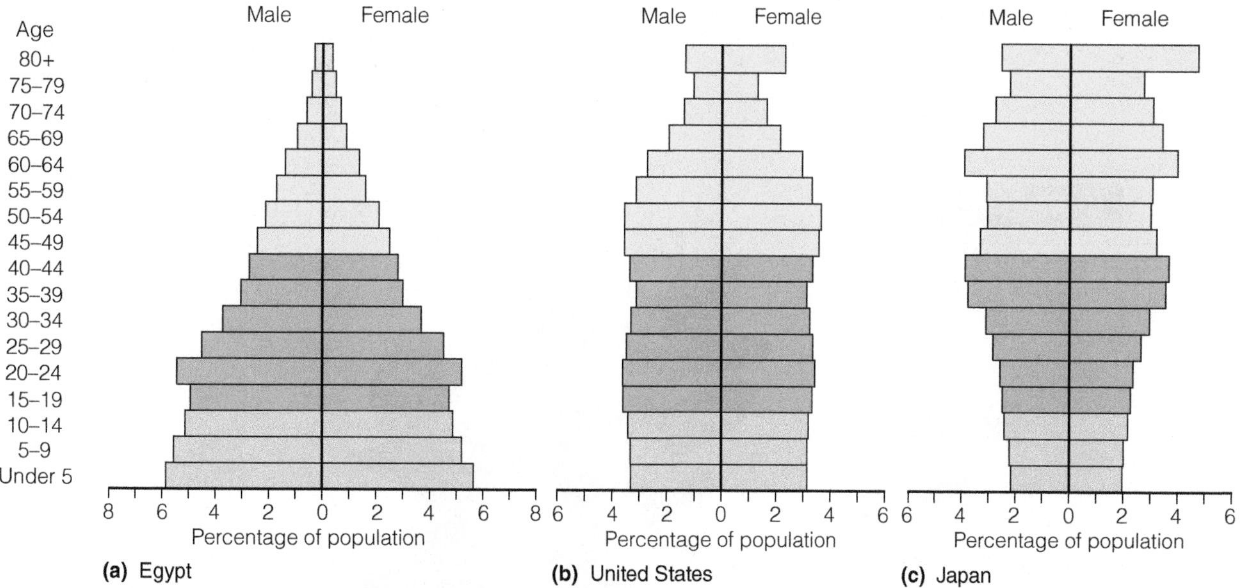

Figure 8.18 Age pyramids for three human populations. (a) Egypt shows an expanding population. A broad base of young will enter the reproductive age classes. (b) The United States has a less-tapered age pyramid approaching zero growth. The youngest age classes are no longer the largest. (c) The age pyramid for Japan is characteristic of a population that is aging with a negative growth rate.

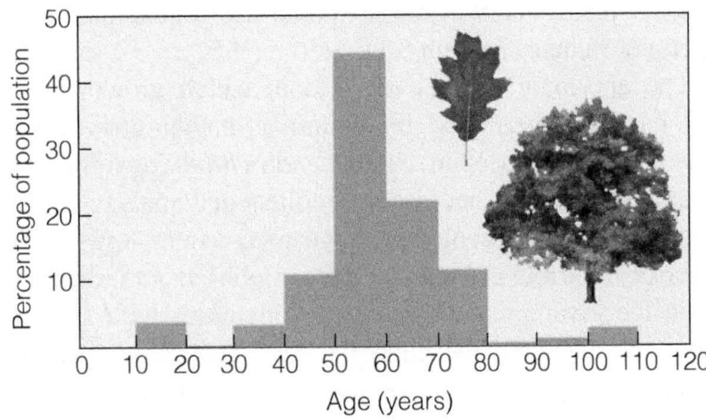

Figure 8.19 Age structure of a red oak (*Quercus rubra*) forest in the Ozark Highlands of Missouri. The forest is dominated by trees in the 30- to 80-year age classes. There has been little recruitment of new trees in the past 30 years.
(Based on data from Loewenstein et al. 2000.)

age structure of a population at some period in time, providing a picture of the relative sizes of different age groups in the population. As we shall see, the age structure of a population is a product of the age-specific patterns of mortality and reproduction (Chapter 9). In many plant populations, the distribution of age classes is often highly skewed (**Figure 8.19**). In forests, for example, dominant overstory trees can inhibit the establishment of seedlings and growth and survival of juvenile trees. One or two age classes dominate the site until they die or are removed, allowing trees in young age classes access to resources such as light, water, and nutrients so they can grow and develop.

When size or developmental stage classes are used as descriptors of population structure, size and stage classes can be used to construct pyramids similar to those for age (**Figure 8.20**).

8.6 Sex Ratios in Populations May Shift with Age

Populations of sexually reproducing organisms in theory tend toward a 1:1 sex ratio (the proportion of males to females). The primary sex ratio (the ratio at conception) also tends to be 1:1. This statement may not be universally true, and it is, of course, difficult to confirm.

In most mammalian populations, including humans, the secondary sex ratio (the ratio at birth) is often weighted toward males, but the population shifts toward females in the older age groups. Generally, males have a shorter life span than females do. The shorter life expectancy of males can be a result of both physiological and behavioral factors. For example, in many animal species, rivalries among males occur for dominant positions in social hierarchies or for the acquisition of mates (see Section 10.11). Among birds, males tend to outnumber females because of increased mortality of nesting females, which are more susceptible to predation and attack (**Figure 8.21**).

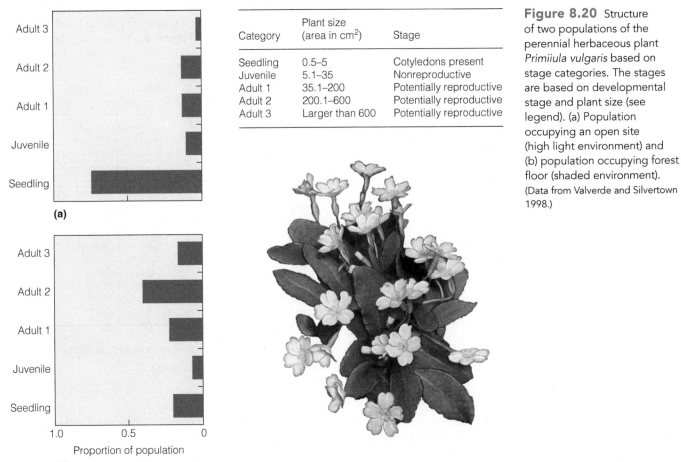

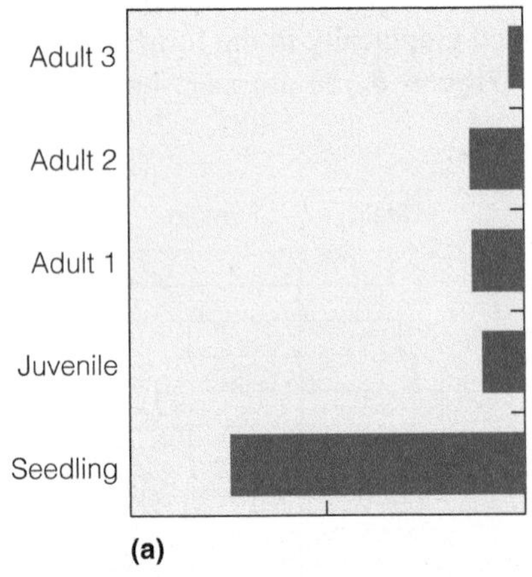

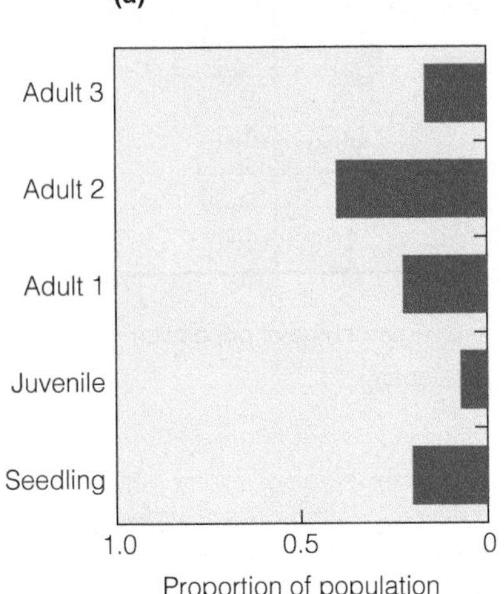

Category	Plant size (area in cm^2)	Stage
Seedling	0.5–5	Cotyledons present
Juvenile	5.1–35	Nonreproductive
Adult 1	35.1–200	Potentially reproductive
Adult 2	200.1–600	Potentially reproductive
Adult 3	Larger than 600	Potentially reproductive

Figure 8.20 Structure of two populations of the perennial herbaceous plant *Primiiula vulgaris* based on stage categories. The stages are based on developmental stage and plant size (see legend). (a) Population occupying an open site (high light environment) and (b) population occupying forest floor (shaded environment).
(Data from Valverde and Silvertown 1998.)

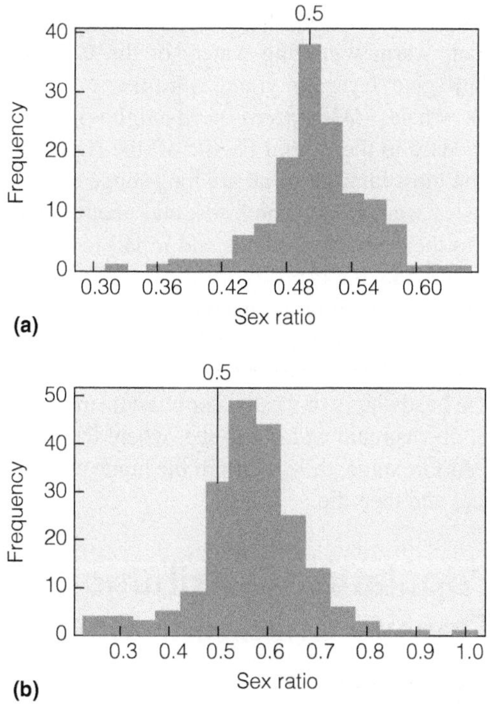

Figure 8.21 Frequency distribution of published estimates of sex ratio (male-to-female) for (a) offspring (eggs and nestlings: 114 studies) and (b) adults (201 studies) from a wide range of bird species.
(Adapted from Donald 2007.)

8.7 Individuals Move within the Population

At some stage in their lives, most organisms are mobile to some degree. The movement of individuals directly influences their local density. The movement of individuals in space is referred to as **dispersal**, although the term *dispersal* most often refers to the more specific movement of individuals away from one another. When individuals move out of a subpopulation, it is referred to as **emigration**. When an individual moves from another location into a subpopulation, it is called **immigration**. The movement of individuals among subpopulations within the larger geographic distribution is a key process in the dynamics of metapopulations and in maintaining the flow of genes between these subpopulations (see Chapters 5 and 19).

Many organisms, especially plants, depend on passive means of dispersal involving gravity, wind, water, and animals. The distance these organisms travel depends on the agents of dispersal. Seeds of most plants fall near the parent, and their density falls off quickly with distance (**Figure 8.22**). Heavier seeds, such as the acorns of oaks (*Quercus* spp.), have a much shorter dispersal range than do the lighter wind-carried seeds of maples (*Acer* spp.), birch (*Betula* spp.), milkweed (Asclepiadaceae), and dandelions (*Taxaxacum officinale*). Some plants, such as cherries and viburnums (*Viburnum* spp.), depend on active carriers such as particular birds and mammals to disperse their seeds by eating the fruits and carrying the seeds to some distant point. These seeds pass through the animals' digestive tract and are deposited in their feces. Other plants possess seeds armed with spines and hooks that catch on the fur of mammals, the feathers of birds, and the clothing of humans. In the example of the clumped distribution of *E. divinorum* shrubs (see Figure 8.11), birds disperse seeds of this species. The birds feed on the fruits and deposit the seeds in their feces as they perch atop the *Acacia* trees. In this way, the seeds are dispersed across the landscape, and the clumped distribution of the *E. divinorum* is associated with the use of *Acacia* trees as bird perches.

For mobile animals, dispersal is active, but many others depend on a passive means of transport, such as wind and moving water. Wind carries the young of some species of spiders, larval gypsy moths, and cysts of brine shrimp (*Artemia salina*). In streams, the larval forms of some invertebrates disperse downstream in the current to suitable habitats. In the oceans, the dispersal of many organisms is tied to the movement of currents and tides.

Dispersal among mobile animals may involve young and adults, males and females; there is no hard-and-fast rule about who disperses. The major dispersers among birds are usually the young. Among rodents, such as deer mice (*Peromyscus maniculatus*) and meadow voles (*Microtus pennsylvanicus*), subadult males and females make up most of the dispersing

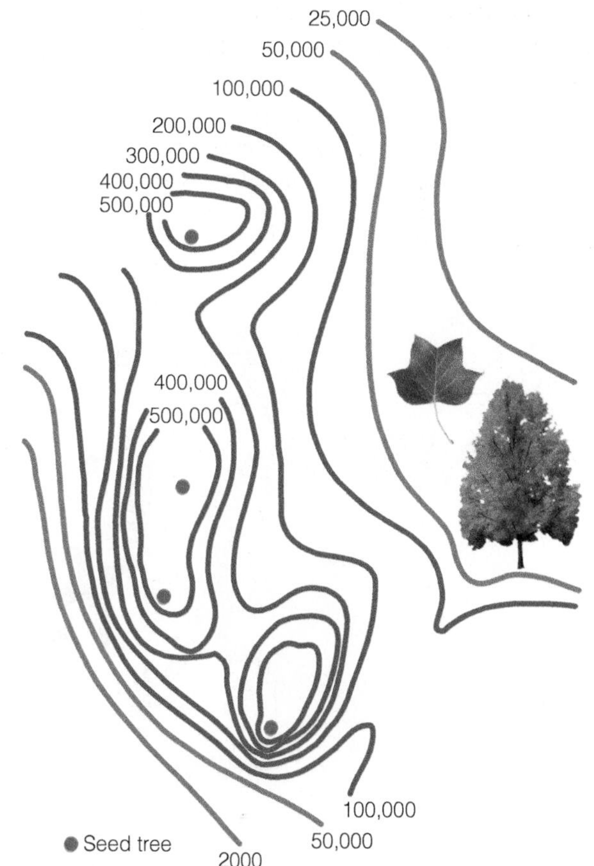

Figure 8.22 Pattern of annual seedfall of yellow poplar (*Liriodendron tulipifera*). Lines define areas of equal density of seeds. With this wind-dispersed species, seedfall drops off rapidly away from the parent trees.
(Adapted from Engle 1960.)

individuals. Crowding, temperature change, quality and abundance of food, and photoperiod all have been implicated in stimulating dispersal in various animal species (Chapter 11, Section 11.8).

Often, the dispersing individuals are seeking vacant habitat to occupy. As a result, the distance they travel depends partly on the density of surrounding subpopulations and the availability of suitable unoccupied areas. The dispersal of individuals is a key feature in the dynamics of metapopulations (see Figure 8.8), where colonization involves the movement of individuals from occupied habitat patches (existing local populations) to unoccupied habitat patches to form new local populations. The role of dispersal and colonization in metapopulation dynamics is discussed further in Chapter 19 (Section 19.7).

Unlike the one-way movement of animals in the processes of emigration and immigration, **migration** is a round trip. The repeated return trips may be daily or seasonal. Zooplankton in the oceans, for example, move down to lower depths by day and move up to the surface by night. Their movement appears to be related to a number of factors including predator avoidance. Bats leave their daytime roosting places in caves and trees, travel to their feeding grounds, and return by daybreak. Other migrations are seasonal, either short range or long range. Earthworms annually make a vertical migration deeper into the soil to spend the winter below the freezing depths and move back to the upper soil when it warms in spring. Elk (*Cervus canadensis*) move down from their high mountain summer ranges to lowland winter ranges. On a larger scale, caribou (*Rangifer tarandus*) move from the summer calving range in the arctic tundra to the boreal forests for the winter, where lichens are their major food source. Gray whales (*Eschrichtius robustus*) move down from the food-rich arctic waters in summer to their warm wintering waters of the California coast, where they give birth to young (**Figure 8.23**). Similarly, humpback whales (*Megaptera novaeangliae*) migrate from northern oceans to the central Pacific off the Hawaiian Islands. Perhaps the most familiar of all are long-range and short-range migrations of waterfowl, shorebirds, and neotropical migrants in spring to their nesting grounds and in fall to their wintering grounds (see Chapter 11, **Field Studies: T. Scott Sillett**).

Another type of migration involves only one return trip. Such migrations occur among Pacific salmon (*Oncorhynchus* spp.) that spawn in freshwater streams. The young hatch and grow in the headwaters of freshwater coastal streams and rivers and travel downstream and out to sea, where they reach sexual maturity. At this stage, they return to the home stream to spawn (reproduce) and then die.

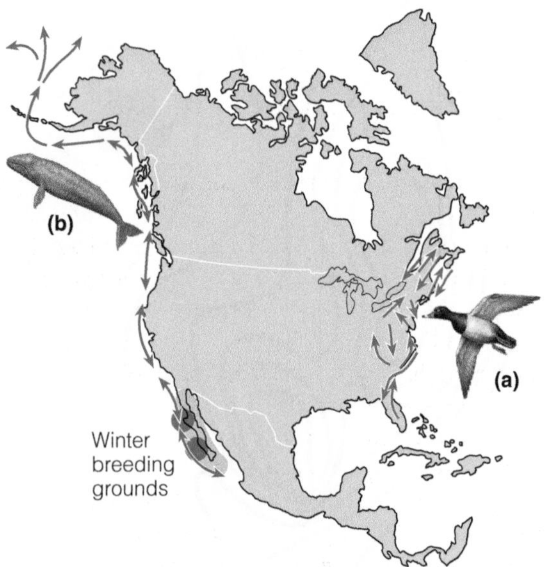

Figure 8.23 Migratory pathways of two vertebrates. (a) Ring-necked ducks (*Aythya collaris*) breeding in the northeast migrate in a corridor along the coast to wintering grounds in South Carolina and Florida. (b) The gray whale (*Eschrichtius robustus*) summers in the Arctic and Bering seas; it winters in the Gulf of California and the waters off Baja California.

8.8 Population Distribution and Density Change in Both Time and Space

Dispersal has the effect of shifting the spatial distribution of individuals and consequently the localized patterns of population density. Emigration may cause density in some areas to decline, whereas immigration into other areas increases the density of subpopulations or even establishes new subpopulations in habitats that were previously unoccupied.

In some instances, dispersal can result in the shift or expansion of a species' geographic range. The role of dispersal in range expansion is particularly evident in populations that have been introduced to a region where they did not previously exist. A wide variety of species have been introduced, either intentionally or unintentionally, into regions outside their geographic distribution. As the initial population becomes established, individuals disperse into areas of suitable habitat, expanding their geographic distribution as the population grows. A map showing the spread of the gypsy moth (*Lymantria dispar*) in the eastern United States after its introduction in 1869 is shown in **Figure 8.24**. The story of the introduction of this species is presented in the following section (see this chapter, *Ecological Issues & Applications*).

In other cases, the range expansion of a population has been associated with temporal changes in environmental conditions, shifting the spatial distribution of suitable habitats. Such is the case of the shift in the distribution of tree populations in eastern North America as climate has changed during the past 20,000 years (see Section 18.9, Figure 18.25). Examples of predicted changes in the distribution of plant and animal populations resulting from future human-induced changes in Earth's climate are discussed later in Chapter 27. Although the movement of individuals within the population results in a changing pattern of distribution and density through time, the primary factors driving the dynamics of population abundance are the demographic processes of birth and death. The processes of birth and death, and the resulting changes in population structure, are the focus of our attention in the following chapter.

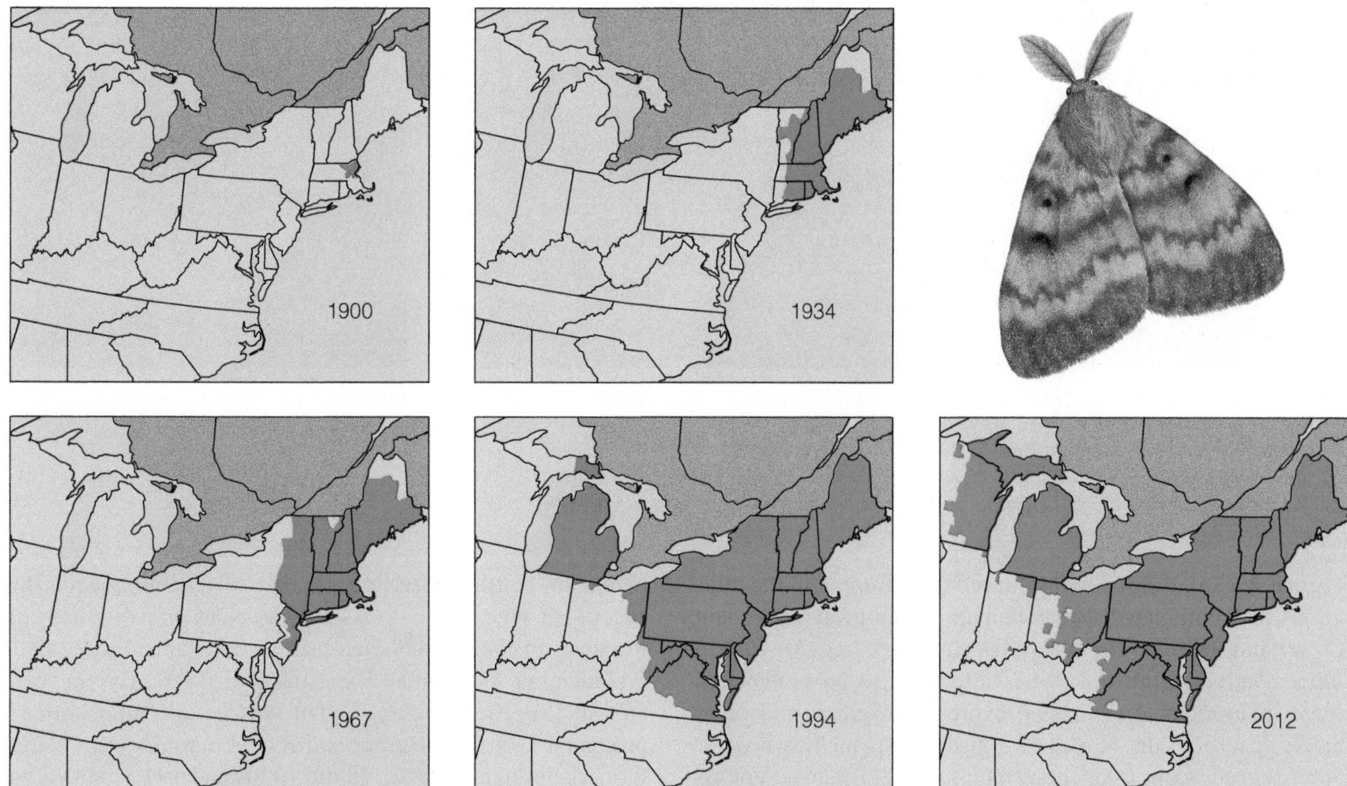

Figure 8.24 Spread of the gypsy moth (*Lymantria dispar*) in the United States after its introduction to Massachusetts in 1869. Note how dispersal since its original introduction has expanded the geographic distribution of this population. See this chapter, *Ecological Issues & Applications*.
(Adapted from Liebhold 1992, as illustrated in Krebs 2001.)

ECOLOGICAL Issues & Applications: Humans Aid in the Dispersal of Many Species, Expanding Their Geographic Range

Dispersal is a key feature of the life histories of all species, and a diversity of mechanisms have evolved to allow plant and animal species to move across the landscape and seascape. In plants, seeds and spores can be dispersed by wind, water, or through active dispersal by animals (see Section 15.14). In animals, the dispersal of fertilized eggs, particularly in aquatic environments, can result in the dispersal of offspring across significant distances. But dispersal typically involves the movement—either active or passive—of individuals, both juvenile and adult. In recent centuries, however, a new source of long-distance dispersal has led to the redistribution of species at a global scale: dispersal by humans.

Humans are increasingly moving about the world. As they do so, they may either accidentally or intentionally introduce plants and animals to places where they have never occurred (outside their geographic range). Although many species fail to survive in their new environments, others flourish. Freed from the constraints of their native competitors, predators and parasites, they successfully establish themselves and spread. These nonnative (nonindigenous) plants and animals are referred to as **invasive species**.

Sometimes these introductions are harmless, but often the introduced organisms negatively affect native species and ecosystems. In the past few centuries, many plants and animals, especially insects, have been introduced accidentally by accompanying imported agricultural and forest products. The seeds of weed species are unintentionally included in shipments of imported crop seeds or on the bodies of domestic animals. Or seed-carrying soil from other countries is often loaded onto ships as ballast and then dumped in another country in exchange for cargo. Major forest insect pests such as the Asian longhorned beetle (*Anoplophors galbripennis*) are hitchhikers on wooden shipping containers and pallets.

Humans have also introduced nonnative plants intentionally for ornamental and agricultural purposes. Most of these introduced plants do not become established and reproduce, but some do. On the North American continent, the ornamental perennial herb purple loosestrife (*Lythrum salicaria*;

(a)

(b)

(c)

Figure 8.25 Three invasive plant species that have significant negative impacts on native communities in North America: (a) the perennial herb purple loosestrife (*Lythrum salicaria*), (b) Australian paperbark tree (*Melaleuca quinquenervia*), and (c) cheatgrass (*Bromus tectorum*).

Figure 8.25a), originally introduced from Europe in the mid-1800s, has eliminated native wetland plants to the detriment of wetland wildlife. The Australian paperbark tree (*Melaleuca quinquenervia*; Figure 8.25b), introduced as an ornamental plant in Florida, is displacing cypress, sawgrass, and other native species in the Florida Everglades, drawing down water and fostering more frequent or intense fires. The most notorious plant invader in the United States is cheatgrass (*Bromus tectorum*; Figure 8.25c), a winter annual accidentally introduced from Europe into Colorado in the 1800s. It arrived in the form of packing material and possibly crop seeds. It spread explosively across overgrazed rangeland and winter wheat fields in the Pacific Northwest and the Intermountain Region. By 1930 it became the dominant grass, replacing native vegetation. Cheatgrass is highly flammable, and densely growing populations provide ample fuel that increases fire intensity and often decreases the time intervals between fires (fire frequency). One of the most widely spread invasive plants in North America is kudzu (*Pueraria montana*), a species of vine native to Asia. This plant was originally introduced to the United States as an ornamental vine at the Philadelphia Centennial Exposition of 1876. By the early part of the 20th century, kudzu was being enthusiastically promoted as a fodder crop, and rooted cuttings were sold to farmers through the mail. In the 1930s and 1940s, kudzu was propagated and promoted by the Soil Conservation Service as a means of holding soil on the swiftly eroding gullies of the deforested southern landscape, especially in the Piedmont regions of Alabama, Georgia, and Mississippi. By the 1950s, however, kudzu was recognized as a pest and removed from the list of species acceptable for use under the Agricultural Conservation Program, and in 1998 it was listed by the United States Congress as a Federal Noxious Weed. Although it spreads slowly, kudzu completely covers all other vegetation, blanketing trees with a dense canopy through which little light can penetrate (Figure 8.26). Estimates of kudzu infestation

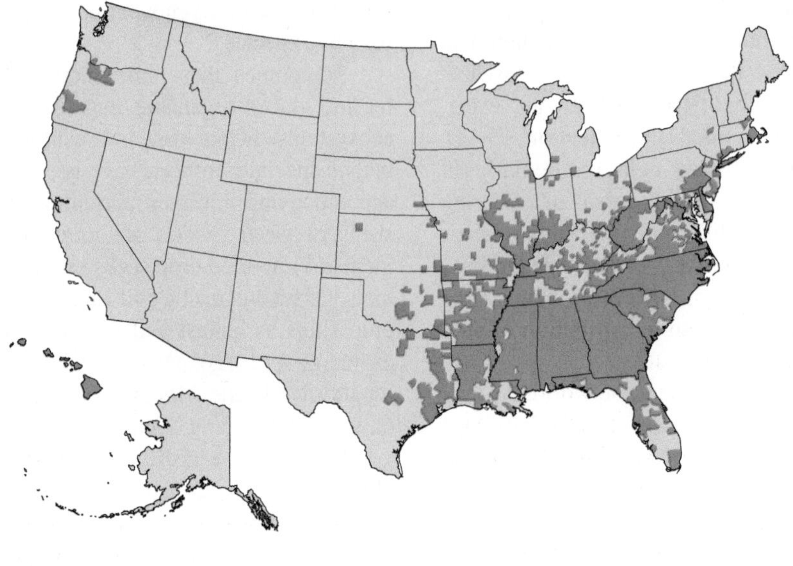

(a)

(b)

Figure 8.26 (a) Map of United States counties where kudzu is currently found. (b) Kudzu vines have blanketed the ground, shrubs, and trees.

(a)

(b)

Figure 8.27 An oak forest (a) that has been completely defoliated by gypsy moth caterpillars (b) in summer.

in the southeastern United States vary greatly, from as low as 2 million to as high as 7 million acres.

The most damaging introduced insect pest of the eastern United States hardwood forests is the gypsy moth (*Lymantria dispar*). The gypsy moth is found mainly in the temperate regions of the world, including central and southern Europe, northern Africa, central and southern Asia, and Japan. Leopold Trouvelot, a French astronomer with an interest in insects, originally introduced the species into Medford, Massachusetts in 1869. As part of an effort to begin a commercial silk industry, Trouvelot wanted to develop a strain of silk moth that was resistant to disease. However, several gypsy moth caterpillars escaped from Trouvelot's home and established themselves in the surrounding areas. Some 20 years later, the first outbreak of gypsy moths occurred, and despite all control efforts since that time, the gypsy moth has persisted and extended its range (see Figure 8.24). In the United States, the gypsy moth has rapidly moved north to Canada, west to Wisconsin, and south to North Carolina. Gypsy moth caterpillars defoliate millions of acres of trees annually in the United States (**Figure 8.27**). In the forests of eastern North America, annual losses to European gypsy moths are estimated at $868 million, and the Asian strain that has invaded the Pacific Northwest has already necessitated a $20 million eradication campaign.

The problem that invasive species present is not restricted to terrestrial environments. More than 139 nonindigenous aquatic species that affect native plant and animal species have invaded the Great Lakes by way of global shipping. Most notorious is the zebra mussel (*Dreissena polymorpha*; **Figure 8.28**) native to the lakes of southern Russia. The species was introduced from the ballast of ships traversing the St. Lawrence Seaway. Since it first appeared in 1988, the zebra mussel has spread to most eastern river systems (**Figure 8.29**). In addition to their impact on wildlife, zebra mussels colonize water intake pipes, severely restricting the water flow to power plants or other municipal or private facilities.

The San Francisco Bay Area is occupied by 96 nonnative invertebrates, from sponges to crustaceans. Exotic fish, introduced purposefully or accidentally, have been responsible for 68 percent of fish extinctions in North America during the past 100 years and for the population decline of 70 percent of the fish species listed as endangered.

Figure 8.28 Zebra mussels, so called because of the striping on the shell, cover solid substrate, water intake pipes, and the shells of native mussels.

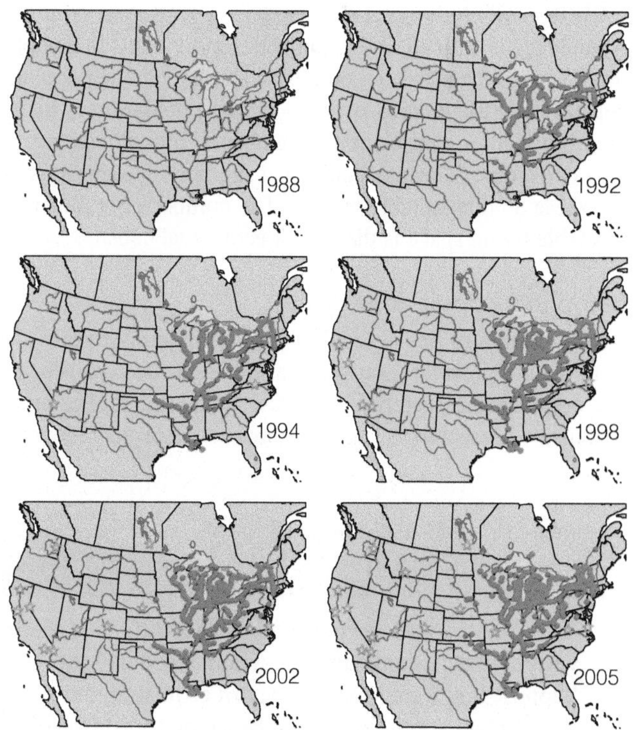

Figure 8.29 Spread of the zebra mussel (*Dreissena polymorpha*) from its original introduction site in 1988.
(U.S. Geological Survey, Nonindigenous Aquatic Species Program, 2007. U.S. Geological Survey, Department for the Interior http://nas.er.usgs.gov/.)

SUMMARY

Unitary and Modular Organisms 8.1

A population is a group of individuals of the same species living in a defined area. Populations are characterized by distribution, abundance, density, and age structure. Most animal populations are made up of unitary individuals with a definitive growth form and longevity. In most plant populations, however, organisms are modular. These plant populations may consist of sexually produced parent plants and asexually produced stems arising from roots. A similar population structure occurs in animal species that exhibit modular growth.

Distribution 8.2

The distribution of a population describes its spatial location, or the area over which it occurs. The distribution of a population is influenced by the occurrence of suitable environmental conditions. Within the geographic range of a population, individuals are not distributed equally throughout the area. Therefore, the distribution of individuals within the population can be described as a range of different spatial scales.

Individuals within a population are distributed in space. If the spacing of each individual is independent of the others, then the individuals are distributed randomly; if they are evenly distributed, with a similar distance among individuals, it is a uniform distribution. In most cases, individuals are grouped together in a clumped or aggregated distribution.

Abundance 8.3

Abundance is defined as the number of individuals in a population. Abundance is a function of two factors: (1) the population density and (2) the area over which the population is distributed. Population density is the number of individuals per unit area or volume. Because landscapes are not homogeneous, not all of the area is suitable habitat. The number of organisms in available living space is the true or ecological density.

Sampling Populations 8.4

Determination of density and dispersion requires careful sampling and appropriate statistical analysis of the data. For sessile organisms, researchers often use sample plots. For mobile organisms, researchers use capture-recapture techniques or determine relative abundance using indicators of animal presence, such as tracks or feces.

Age, Stage, and Size Structure 8.5

The number or proportion of individuals within each age class defines the age structure of a population. Individuals making up the population are often divided into three ecological periods: prereproductive, reproductive, and postreproductive. Populations can also be characterized by the number of individuals in defined classes of size or stage of development.

Sex Ratios 8.6

Sexually reproducing populations have a sex ratio that tends to be 1:1 at conception and birth but often shifts as a function of sex-related differences in mortality.

Dispersal 8.7

At some stage of their life cycles, most individuals are mobile. For some organisms, such as plants, dispersal is passive and dependent on various dispersal mechanisms. For mobile organisms, dispersal can occur for a variety of reasons, including the search for mates and unoccupied habitat. For some species, dispersal is a systematic process of movement between areas in a process called *migration*.

Population Dynamics 8.8

Dispersal has the effect of shifting the spatial distribution of individuals and as a result the localized patterns of population density. Although the movement of individuals within the population results in a changing pattern of distribution and density through time, the primary factors driving the dynamics of population abundance are the demographic processes of birth and death.

Invasive Species Ecological Issues & Applications

Humans have either accidentally or intentionally introduced plant and animal species to places outside their geographic range. Sometimes these introductions are harmless, but often the introduced organisms negatively affect the populations of native species and ecosystems.

STUDY QUESTIONS

1. How does asexual reproduction make it difficult to define what constitutes an individual within a population?
2. Suppose you were given the task of estimating the density of two plant species in a field. Based on the life histories of the two species, you expect that the spatial distribution of one of the species is approximately uniform, whereas the other is likely to be clumped. How might your approach to estimating the density of these two species differ?
3. The age structure of a population can provide insight into whether the population is growing or declining. The presence of a large number of individuals in the young age classes relative to the older age classes often indicates a growing population. In contrast, a large proportion

of individuals in the older age classes relative to the young age classes suggests a population in decline (see Figure 8.18). What factors might invalidate this interpretation? When might a large number of individuals in the young age classes relative to the older age classes not indicate a growing population?

4. Modern humans are a highly mobile species. Think of three locations in your local community that might be used as areas for estimating the population density. How might the daily movement pattern of people in your community change the estimate of density at these locations during the course of the day?

FURTHER READINGS

Classic Studies

Cook, R. E. 1983. "Clonal plant populations." *American Scientist* 71:244–253.
An introduction to the nature of modular growth in plants and its implications for the study of plant populations.

Elton, C. S. 1958. *The ecology of invasions by animals and plants*. London: Methuen and Co. Ltd.
In this classic text, Elton sounds an early warning regarding the ecological consequences of introducing nonnative species.

Shelford, V. E. 1931. "Some concepts of bioecology." *Ecology* 12:455–467.
Classic article in which the author develops the concept of environmental tolerances and their role in defining the geographic distribution of populations.

Recent Research

Brown, J. H., D. W. Mehlman, and G. C. Stevens. 1995. "Spatial variation in abundance." *Ecology* 76:2028–2043.
An excellent overview of the factors influencing geographic patterns of abundance in populations.

Gaston, K. J. 1991. "How large is a species' geographic range?" *Oikos* 61:434–438.
This article explores the methods used to define a species' geographic range and how range size is influenced by various aspects of a species' life history.

Gompper, M. E. 2002. "Top carnivores in the suburbs? Ecological and conservation issues raised by colonization of Northeastern North America by coyotes." *Bioscience* 52:185–190.
Fascinating story of the coyote's dispersal and enormous range expansion.

Krebs, C. J. 1999. *Ecological methodology.* 2nd ed. San Francisco: Benjamin Cummings.
Essential reading for those interested in the sampling of natural populations. The text provides an excellent introduction with many illustrative examples.

Laliberte, A. S., and W. J. Ripple. 2004. "Range contractions of North American carnivores and ungulates." *Bioscience* 54:123–138.
In contrast to the Gompper article that presents a case of range expansion, this article explores examples of range contraction of many of the large mammals that were once widely distributed across North America.

Mack, R., and W. M. Lonsdale. 2001. "Humans as global plant dispersers: Getting more than we bargained for." *Bioscience* 51:95–102.
Tells the story of how humans are acting as agents of species dispersal and illustrates some of the unexpected consequences. A good follow-up to the *Ecological Issues* essay in this chapter.

Mooney, H. A. and R. J. Hobbs. 2000. *Invasive species in a changing world.* Washington, D.C.: Island Press.
This edited volume provides an overview of invasive species in freshwater, marine, and terrestrial ecosystems and examines the potential influence of global climate change to exacerbate the invasive species problem.

CHAPTER 9 Population Growth

Elephant (*Loxodonta africana*) herd in Amboseli National Park, Kenya, Africa. This small group is composed of adult females as well as juveniles of various ages. Because African elephants are largely restricted to national parks and conservation areas, these local populations function as closed populations, with no immigration or emigration.

CHAPTER GUIDE

- 9.1 Population Growth Reflects the Difference between Rates of Birth and Death
- 9.2 Life Tables Provide a Schedule of Age-Specific Mortality and Survival
- 9.3 Different Types of Life Tables Reflect Different Approaches to Defining Cohorts and Age Structure
- 9.4 Life Tables Provide Data for Mortality and Survivorship Curves
- 9.5 Birthrate Is Age-Specific
- 9.6 Birthrate and Survivorship Determine Net Reproductive Rate
- 9.7 Age-Specific Mortality and Birthrates Can Be Used to Project Population Growth
- 9.8 Stochastic Processes Can Influence Population Dynamics
- 9.9 A Variety of Factors Can Lead to Population Extinction

ECOLOGICAL Issues & Applications Habitat Loss and Extinction

THE TERM *POPULATION GROWTH* refers to how the number of individuals in a population increases or decreases over time. This growth is controlled by the rate at which new individuals are added to the population through the processes of birth and immigration and the rate at which individuals leave the population through the processes of death and emigration (**Figure 9.1**). We refer to populations in which immigration or emigration occurs as *open populations.* Those in which movement into and out of the population does not occur (or is not a significant influence on population growth) are referred to as *closed populations.*

In this chapter we will explore the process of population growth under the conditions in which population dynamics are a function only of demographic processes relating to birth and death (that is, the conditions in which either the population is closed [no immigration or emigration] or the rates of immigration and emigration are equal). We will relax this assumption to examine how through the process of dispersal, the interactions among subpopulations influence the dynamics of local populations as well as the overall dynamics of the larger metapopulation in Chapter 19 (Landscape Ecology).

We begin our discussion by examining how the changes in the size of a population over time reflect the difference between the rates of birth and death.

9.1 Population Growth Reflects the Difference between Rates of Birth and Death

Suppose we were to undertake an experiment in which we monitor a population of an organism that has a very simple life cycle, such as a population of freshwater hydra (**Figure 9.2**) growing in an aquarium in the laboratory. Most reproduction is asexual, involving a process termed *budding,* in which a new hydra develops as a bud from the parent (see Figure 9.2). If we assume only asexual reproduction, then all individuals are capable of reproduction and produce a single offspring at a time.

We define the population size at a given time (t) as $N(t)$, where N represents the number of individuals. Let us assume that the initial population is small, $N(0) = 100$ (where 0 refers to time zero at the start of the experiment), so that the food supply within the aquarium is much more than is needed to support the current population. How will the population change over time?

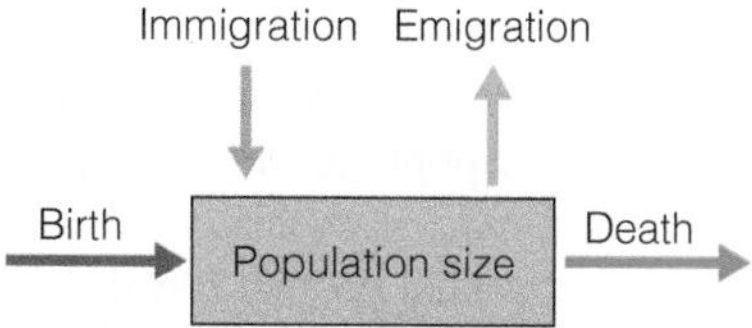

Figure 9.1 Simple representation of population growth. The change in the population size over time is the difference in the arrival of new individuals through the processes of birth and immigration and the loss of individuals through the processes of death and emigration.

Figure 9.2 The freshwater hydra reproduces asexually by budding.

Because no emigration or immigration is allowed by the lab setting, the population is closed. The number of hydra will increase as a result of new "births." Additionally, the population will decrease as a result of some hydra dying. Because the processes of birth and death in this population are continuous—no defined period of synchronized birth or death—we can observe the number of births that occur over some appropriate time interval. Given the rates of budding (reproduction) for freshwater hydra, an appropriate time unit (t) would be one day. We can define the number of new hydra produced by budding over the period of one day as B and the number of hydra dying over the same day as D. In our hypothetical experiment, let us assume that the initial population produced 40 new individuals (births) over the first day ($B = 40$) and that 10 of the original 100 hydra died ($D = 10$). The population size at the end of day 1, $N(1)$, can then be calculated from the initial population size, $N(0)$, and the observed numbers of births (B) and deaths (D):

$$N(0) + B - D = N(1) \quad \text{or} \quad 100 + 40 - 10 = 130$$

But what if we now want to predict what the population will be the following day, $N(2)$? How could we use the measures of B and D to determine the number of births and deaths that will occur in our population that is now composed of 130 hydra? Although B and D represent the measure of birth and death in the population, the actual values are dependent on the initial population size, $N(0) = 100$. For example, if the initial population size was 200 rather than 100, we could assume that the values of B and D would be twice as large. If we wish to calculate an estimate of birthrate that is independent of the initial population size we need to divide the number of hydra born during the day by the initial population size: $B/N(0)$ or 40/100. We can now define the resulting value 0.4 as b, which is the per capita birthrate (per capita meaning per individual). The per capita birthrate is the average number of births per individual during the time period t (one day). Likewise, we can calculate the per capita death rate as $D/N(0)$ of $10/100 = 0.1$. The advantage of expressing the observed values of birth (B) and death (D) for the population as per capita rates (b and d) is that if we assume they are constant (do not change over time), we can use b and d to predict the growth of the population over time regardless of the population size $N(t)$.

If we start with $N(t)$ hydra at time t, then to calculate the total number of hydra reproducing over the following day $(t + 1)$, we must simply multiply the per capita birthrate (b) by the total number of hydra at time t $[N(t)]$, which is $bN(t)$. The number of hydra dying over the time interval is calculated in a similar manner: $dN(t)$.

The population size at the next time period $(t + 1)$ would then be

$$N(t + 1) = N(t) + bN(t) - dN(t)$$

Applied to the hydra population:

$$N(0) = 100$$

$$N(1) = 100 + 0.4(100) - 0.1(100) = 130$$

$$N(2) = 130 + 0.4(130) - 0.1(130) = 169$$

The resulting pattern of population size as a function of time is shown in **Figure 9.3** and is referred to as **geometric population growth**.

We can calculate the rate of change in the population (the population growth rate) by subtracting $N(t)$ from both sides of the preceding equation:

$$N(t + 1) - N(t) = bN(t) - dN(t)$$

or

$$N(t + 1) - N(t) = (b - d)N(t)$$

Applied to the hydra population over the first two days:

$$N(1) - N(0) = 0.4(100) - 0.1(100) = 30$$

$$N(2) - N(1) = 0.4(130) - 0.1(130) = 39$$

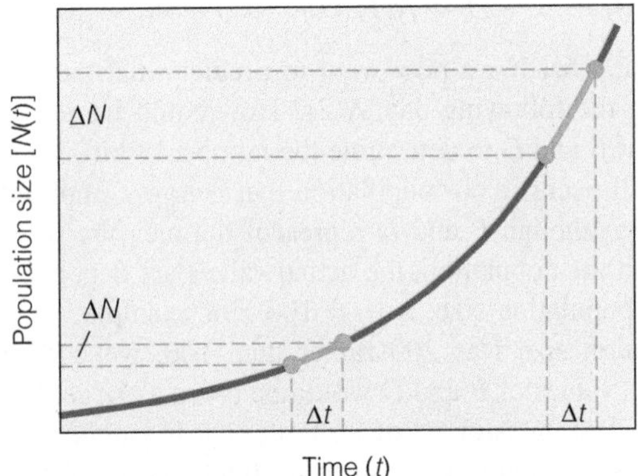

Figure 9.3 Change in the size of a hypothetical population of hydra over time (green line). The change in population size, ΔN, for a given time interval, Δt, differs as a function of time (t), as indicated by the slope of the line segments shown in orange.

The term on the left side of the equation $[N(t + 1) - N(t)]$ is the change in the population size N over the time interval $[(t + 1) - t]$. If we represent the change in population size as ΔN and the change in time (the time interval) as Δt—the mathematical symbol Δ refers to a "change" in the associated variable—we can rewrite the equation for the rate of population change in a simplified form:

$$\frac{\Delta N}{\Delta t} = (b - d)N(t)$$

Because per capita birthrates and death rates, b and d, are constants (fixed values), we can simplify the equation even further by defining a new parameter $r = (b - d)$. The value r is the per capita growth rate.

$$\frac{\Delta N}{\Delta t} = rN(t)$$

Thus, the population growth rate ($\Delta N/\Delta t$) defines the unit change in population size per unit change in time, or the slope of the relationship between $N(t)$ and t (the "rise" over the "run") presented in Figure 9.3. Note that because the pattern of population growth is an upward sloping curve, the rate of population change depends on the time interval being viewed (see Figure 9.3 and preceding calculations). With a population that is growing geometrically, the rate of population growth is continuously increasing as the population size increases.

It is important to remember that the model of geometric growth that we have developed for the hydra population can *only* predict changes in population size on discrete time intervals of one day. This is because the estimates of b and d (and therefore, r) were estimated using observations of birth and death over a one-day period. For populations, such as the hydra, where birth and death are occurring continuously (not daily intervals), population ecologists often represent the processes of birth and death as instantaneous rates and population growth as a continuous process rather than on defined time steps (such as one day). The model is then presented as a differential equation:

$$\frac{dN}{dt} = rN$$

The term $\Delta N/\Delta t$ is replaced by dN/dt to express that Δt (the time interval) approaches a value of zero, and the rate of change becomes instantaneous. The value $\boldsymbol{r}$ is now the instantaneous per capita rate of growth (sometimes called the *intrinsic rate of population growth*), and the resulting equation is referred to as the model of **exponential population growth** (in contrast to geometric population growth based on discrete time steps).

The model of exponential growth ($dN/dt = rN$) predicts the rate of population change over time. If we wish to define the equation to predict population size, $N(t)$, under conditions of exponential growth $[N(t)$ at any given value of $t]$, it is necessary to integrate the differential equation presented previously. The result is:

$$N(t) = N(0)e^{rt}$$

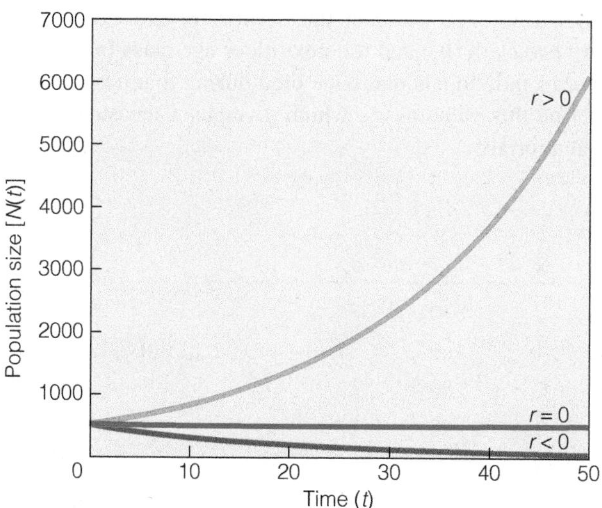

Figure 9.4 Examples of exponential growth under differing values of *r*, which is the instantaneous per capita growth rate. When $r > 0$ ($b > d$), the population size increases exponentially; for values of $r < 0$ ($b < d$), there is an exponential decline. When $r = 0$ ($b = d$), there is no change in population size through time.

where $N(0)$ is the initial population size at $t = 0$, and e is the base of the natural logarithms; its value is approximately 2.72.

Examples of exponential growth for differing values of *r* are shown in **Figure 9.4**. Note that when $r = 0$ (when $b = d$), there is no change in population size. For values of $r > 0$ (when $b > d$) the population increases exponentially, whereas values of $r < 0$ (when $b < d$) result in an exponential decline in the population. As with the pattern of geometric population growth, exponential growth results in a continuously accelerating (or decelerating) rate of population growth as a function of population size.

Exponential (or geometric) growth is characteristic of populations inhabiting favorable environments at low population densities, such as during the process of colonization and establishment in new environments. An example of a population undergoing exponential growth is the rise of the reindeer herd introduced on St. Paul, one of the Pribilof Islands, Alaska (**Figure 9.5**). In the fall of 1911, the United States government introduced 25 reindeer on the island of St. Paul to provide the native residents with a sustained source of fresh meat. Over the next 30 years, the original herd of 4 males and 21 females grew to a herd of more than 2000 individuals.

The whooping crane (*Grus americana*) provides another example of a population exhibiting exponential growth (**Figure 9.6**). At the time of European settlement of North America, the population of the whooping crane was estimated at more than 10,000. That number dropped to between 1300 and 1400 individuals by 1870, and by 1938, only 15 birds existed. The species was declared endangered in 1967, and thanks to conservation efforts, the 2011 population was estimated at more than 300 individuals. The whooping crane breeds in the Northwestern Territories of Canada and migrates to overwinter on the Texas coast at the Aransas National Wildlife Refuge. Counts of the entire population from the period of 1938 to 2013 have provided the data presented in Figure 9.6. (For an example of exponential population growth for humans, see Chapter 10, *Ecological Issues & Applications*).

Figure 9.5 Exponential growth of the St. Paul reindeer (*Rangifer tarandus*) herd following introduction in 1910. (Adapted from Scheffer 1951.)

9.2 Life Tables Provide a Schedule of Age-Specific Mortality and Survival

As we established in the previous section, change in population abundance over time is a function of the rates of birth and death, as represented by the per capita growth rate *r*. But how do ecologists estimate the per capita growth rate of a population? For the hydra population, where all individuals can be treated as identical, the rates of birth and death for the population were estimated by counting the number of individuals in the population either giving birth or dying per unit of time. This simple approach was possible because each individual within the

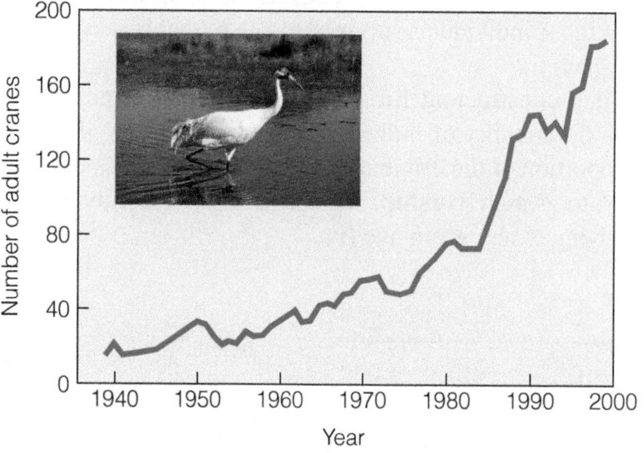

Figure 9.6 Pattern of exponential population growth of the whooping crane, an endangered species that has recovered from near-extinction in 1941. Population estimates are based on an annual count of adult birds on the wintering grounds at Aransas National Wildlife Refuge.
(Data from Binkley and Miller 1983, and Cannon 1996.)

population could be treated as identical with respect to its probability of giving birth or dying. But when birthrates and death rates vary with age, a different approach must be used.

To obtain a clear and systematic picture of mortality and survival within a population, ecologists use an approach involving the construction of life tables. The **life table** is simply an age-specific account of mortality. Life insurance companies use this technique, first developed by students of human populations, as the basis for evaluating age-specific mortality rates. Now, however, population ecologists are using life tables to examine systematic patterns of mortality and survivorship within animal and plant populations.

The construction of a life table begins with a **cohort**, which is a group of individuals born in the same period of time. For example, data presented in the following table represent a cohort of 530 gray squirrels (*Sciurus carolinensis*) from a population in northern West Virginia that was the focus of a decade-long study. The fate of these 530 individuals was tracked until all had died some six years later. The first column of numbers, labeled x, represents the age classes; in this example, the age classes are in units of years. The second column, n_x, represents the number of individuals from the original cohorts that are alive at the specified age (x).

x	n_x
0	530
1	159
2	80
3	48
4	21
5	5

Of the original 530 individuals born (age 0), only 159 survived to an age of 1 year, whereas of those 159 individuals, only 80 survived to age 2. Only 5 individuals survived to age 5, and none of those individuals survived to age 6 (that is why there is no age class 6).

When constructing life tables, it is common practice to express the number of individuals surviving to any given age as a proportion of the original cohort size (n_x/n_0). This value, l_x, referred to as **survivorship**, represents the probability at birth of surviving to any given age (x).

x	n_x	l_x	
0	530	1.00	$n_0 / n_0 = 530 / 530$
1	159	0.30	$n_1 / n_0 = 159 / 530$
2	80	0.15	$n_2 / n_0 = 80 / 530$
3	48	0.09	
4	21	0.04	
5	5	0.01	

The difference between the number of individuals alive for any age class (n_x) and the next older age class (n_{x+1}) is the number of individuals that have died during that time interval. We define this value as d_x, which gives us a measure of age-specific mortality.

x	n_x	d_x	
0	530	371	$n_0 - n_1 = 530 - 159$
1	159	79	$n_1 - n_2 = 159 - 80$
2	80	32	
3	48	27	
4	21	16	
5	5	5	

The number of individuals that died during any given time interval (d_x) divided by the number alive at the beginning of that interval (n_x) provides an **age-specific mortality rate,** q_x.

x	n_x	d_x	q_x	
0	530	371	0.70	$d_0 / n_0 = 371 / 530$
1	159	79	0.50	$d_1 / n_1 = 79 / 159$
2	80	32	0.40	
3	48	27	0.55	
4	21	16	0.75	
5	5	5	1.00	

A complete life table for the cohort of gray squirrels, including all of the preceding calculations, is presented in **Table 9.1**. In addition, the calculation of age-specific life expectancy, e_x, which is the average number of years into the future that an individual of a given age is expected to live, is presented in **Quantifying Ecology 9.1**.

Table 9.1 Gray Squirrel Life Table

x	n_x	l_x	d_x	q_x
0	530	1.0	371	0.7
1	159	0.3	79	0.5
2	80	0.15	32	0.4
3	48	0.09	27	0.55
4	21	0.04	16	0.75
5	5	0.01	5	1.0

QUANTIFYING ECOLOGY 9.1 Life Expectancy

Most of us are not familiar with the concept of life tables; however, almost everyone has heard or read statements like, "The average life expectancy for a male in the United States is 72 years." What does this mean? What is life expectancy? Life expectancy (*e*) typically refers to the average number of years an individual is expected to live from the time of its birth. Life tables, however, are used to calculate age-specific life expectancies (e_x), or the average number of years that an individual of a given age is expected to live into the future. We can use the life table for the cohort of female gray squirrels presented in Table 9.1 to examine the process of calculating age-specific life expectancies for a population.

The first step in estimating e_x is to calculate L_x using the n_x column of the life table. L_x is the average number of individuals alive during the age interval *x* to *x* + 1. It is calculated as the average of n_x and n_{x+1}. This estimate assumes that mortality within any age class is distributed evenly over the year.

x	n_x	L_x	
0	530	344.5	$= (n_0 + n_1)/2 = (530 + 159)/2 = 344.5$
1	159	119.5	
2	80	64.0	$= (n_2 + n_3)/2 = (80 + 48)/2 = 64$
3	48	34.5	
4	21	13.0	
5	5	2.5	$= (n_5 + n_6)/2 = (5 + 0)/2 = 2.5$

Next, the L_x values are used to calculate T_x, which is the total years lived into the future by individuals of age class *x* in the population. This value is calculated by summing the values of L_x cumulatively from the bottom of the column to age *x*.

x	L_x	T_x	
0	344.5	578.0	$= L_0 + L_1 + L_2 + L_3 + L_4 + L_5 = 344.5 + 119.5 + 64 + 34.5 + 13 + 2.5 = 578$
1	119.5	233.5	
2	64.0	114.0	
3	34.5	50.0	
4	13.0	15.5	$= L_4 + L_5 = 13 + 2.5 = 15.5$
5	2.5	2.5	$= L_5 = 2.5$

In the example of the gray squirrel, the value of T_0 is 578. This means that the 530 individuals in the cohort lived a total of 578 years (some only 1 year, whereas others lived to age 5).

The life expectancy for each age class (e_x) is then calculated by dividing the value of T_x by the corresponding value of n_x. In other words, it is calculated by dividing the total number of years lived into the future by individuals of age *x* by the total number of individuals in that age group.

x	n_x	T_x	e_x	
0	530	578.0	1.09	$= T_0 / n_0 = 578 / 530 = 1.09$
1	159	233.5	1.47	
2	80	114.0	1.43	$= T_2 / n_2 = 114 / 80 = 1.43$
3	48	50.0	1.06	
4	21	15.5	0.75	
5	5	2.5	0.50	

Note that life expectancy changes with age. On average, at birth gray squirrel individuals can expect to live for only 1.09 years. However, for those individuals that survive past their first birthday, life expectancy increases to 1.47. Life expectancy remains high for age class 2 and then declines for the remainder of the age classes.

1. Why does life expectancy increase for those individuals that survive to age 1 (1.47 as compared to 1.09 for newborn individuals)?
2. Which would have a greater influence on the life expectancy of a newborn (age 0): a 20 percent decrease in mortality rate for individuals of age class 0 (x = 0) or a 20 percent decrease in the mortality rate of age class 4 (x = 4) individuals? Why?

9.3 Different Types of Life Tables Reflect Different Approaches to Defining Cohorts and Age Structure

There are two basic kinds of life tables. The first type is the **cohort** or **dynamic life table**. This is the approach used in constructing the gray squirrel life table presented in Table 9.1. The fate of a group of individuals, born at a given time, is followed from birth to death—for example, a group of individuals born in the year 1955. A modification of the dynamic life table is the **dynamic composite life table**. This approach constructs a cohort from individuals born over several time periods instead of just one. For example, you might follow the fate of individuals born in 1955, 1956, and 1957.

The second type of life table is the **time-specific life table**. This approach does not involve following a single or group of

Table 9.2 Life Table of a Sparse Gypsy Moth Population in Northeastern Connecticut

x	n_x	l_x	d_x	q_x
Eggs	450	1.000	135	0.300
Instars I–III	315	0.700	258	0.819
Instars IV–VII	57	0.127	33	0.582
Prepupae	24	0.053	1	0.038
Pupae	23	0.051	7	0.029
Adults	16	0.036	0	1.000

(Source: Data from R. W. Campbell 1969.)

cohorts, but rather it is constructed by sampling the population in some manner to obtain a distribution of age classes during a single time period. Although it is much easier to construct, this type of life table requires some crucial assumptions. First, it must be assumed that each age class was sampled in proportion to its numbers in the population. Second, it must be assumed that the age-specific mortality rates (and birthrates) have remained constant over time.

Most life tables have been constructed for long-lived vertebrate species having overlapping generations (such as humans). Many animal species, especially insects, live through only one breeding season. Because their generations do not overlap, all individuals belong to the same age class. For these species, we obtain the values of n_x by observing a natural population several times over its annual season, estimating the size of the population at each time. For many insects, the n_x values can be obtained by estimating the number surviving from egg to adult. If records are kept of weather, abundance of predators and parasites, and the occurrence of disease, death from various causes can also be estimated.

Table 9.2 represents the fate of a cohort from a single gypsy moth egg mass. The age interval, or x column, indicates the different life stages, which are of unequal duration. The n_x column indicates the number of survivors at each stage. The d_x column gives an accounting of deaths in each stage.

Table 9.3 Life Table for a Natural Population of *Sedum smallii*

x	l_x	d_x	q_x
Seed produced	1.000	0.16	0.160
Available	0.840	0.630	0.750
Germinated	0.210	0.177	0.843
Established	0.033	0.009	0.273
Rosettes	0.024	0.010	0.417
Mature plants	0.014	0.014	1.000

(Source: Data from Sharitz and McCormick 1973.)

In plant demography, the life table is most useful in studying three areas: (1) seedling mortality and survival, (2) population dynamics of perennial plants marked as seedlings, and (3) life cycles of annual plants. An example of the third type is **Table 9.3**, showing a life table for the annual elf orpine (*Sedum smallii*). The time of seed formation is the initial point in the life cycle. The l_x column indicates the proportion of plants alive at the beginning of each stage; the d_x column indicates the proportion dying, rather than the actual number of individuals (as in the other examples).

9.4 Life Tables Provide Data for Mortality and Survivorship Curves

Although we can graphically display data from any of the columns in a life table, the two most common approaches are the construction of (1) a mortality curve based on the q_x column and (2) a survivorship curve based on the l_x column. A mortality curve plots mortality rates in terms of q_x as a function of age. Mortality curves for the life tables presented in Table 9.1 (gray squirrel) and Table 9.3 (*S. smallii*) are shown in **Figure 9.7**. For the gray squirrel cohort (Figure 9.7a),

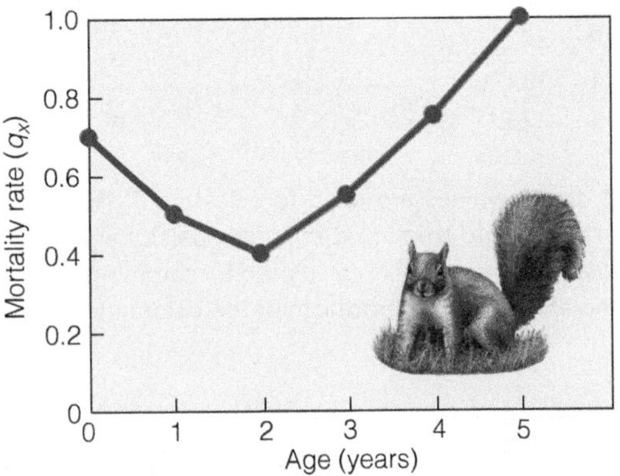

(a)

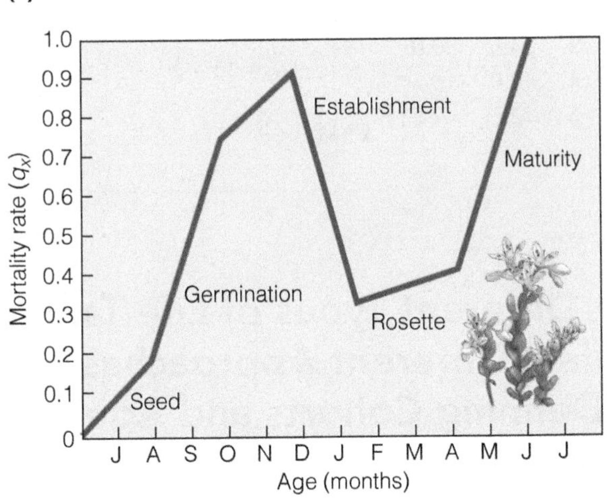

(b)

Figure 9.7 Examples of mortality curves for (a) gray squirrel (*Sciurus carolinensis*) population based on Table 9.1 and (b) *Sedum smallii* based on Table 9.3.

the curve consists of two parts: a juvenile phase, in which the rate of mortality is high, and a post-juvenile phase, in which the rate decreases with age until mortality reaches some low point, after which it increases again. For plants, the mortality curve may assume various patterns, depending on whether the plant is annual or perennial and how we express the age structure. Mortality rates for the *Sedum* population (Figure 9.7b) are initially high, declining once seedlings are established.

Survivorship curves plot the l_x from the life table against time or age class (x). The time interval is on the horizontal axis, and survivorship is on the vertical axis. Survivorship (l_x) is typically plotted on a $\log_{10}$ scale. Survivorship curves for the life tables presented in Table 9.1 (gray squirrel) and Table 9.3 (*S. smallii*) are shown in **Figure 9.8**.

Life tables and survivorship curves are based on data obtained from one population of the species at a particular time and under certain environmental conditions. They are like snapshots. For this reason, survivorship curves are useful for comparing one time, area, or sex with another (**Figure 9.9**).

Survivorship curves fall into three general idealized types (**Figure 9.10**). When individuals tend to live out their physiological life span, survival rate is high throughout the life span, followed by heavy mortality at the end. With this type of survivorship pattern, the curve is strongly convex, or Type I.

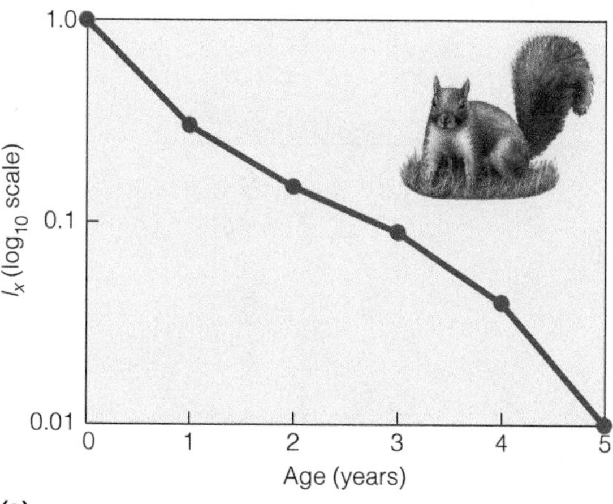

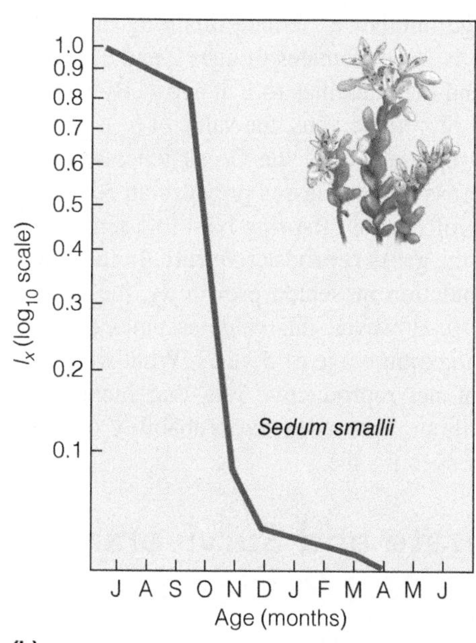

Figure 9.8 Survivorship curve for (a) gray squirrel based on Table 9.1 and (b) *Sedum smallii* based on Table 9.3.

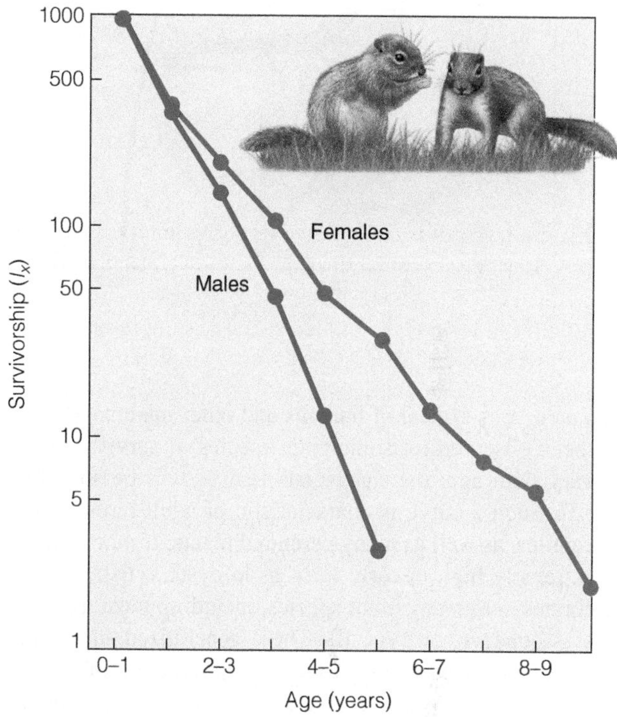

Figure 9.9 Comparison of survivorship curves for male and female of Belding's ground squirrels (*Spermophilus beldingi*) at Tioga Pass, in the central Sierra Nevada of California. (Adapted from Sherman 1984.)

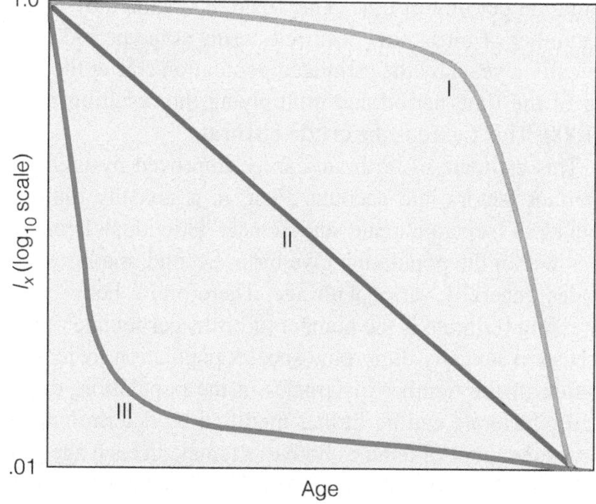

Figure 9.10 The three basic types of survivorship curves.

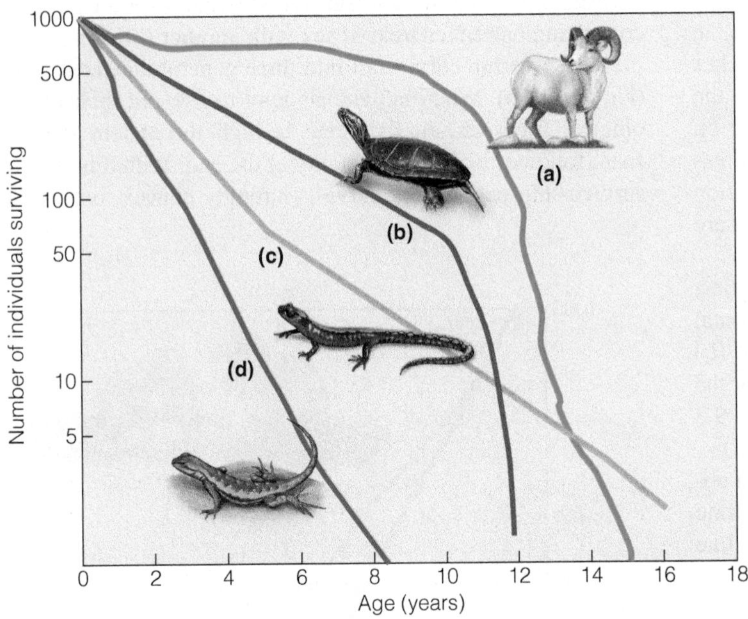

Figure 9.11 Examples of survivorship curves for four animal species: (a) Dall sheep (*Ovis dalli*), (b) painted turtle (*Chrysemys picta*), (c) Allegheny Mountain dusky salamander (*Desmognathus ochrophaeus*), and (d) striped plateau lizard (*Sceloporus virgatus*). The Dall sheep population exhibits a typical Type I survivorship curve, whereas the striped plateau lizard exhibits a Type II survivorship curve. The survivorship curves of the other two species are hybrids, combining features of the different standard survivorship curves (Types I, II, and III) at different times during their life cycles. (Data from: (a) Deevey 1947, (b) Tinkle et al. 1981, (c) Tilley 1980, (d) Vinegar 1975.)

Such a curve is typical of humans and other mammals and has also been observed for some plant species. If survival rates do not vary with age, the survivorship curve will be straight, or Type II. Such a curve is characteristic of adult birds, rodents, and reptiles, as well as many perennial plants. If mortality rates are extremely high in early life—as in oysters, fish, many invertebrates, and many plant species, including most trees—the curve is concave, or Type III. These generalized survivorship curves are idealized models to which survivorship of a species can be compared. Many survivorship curves show components of these three generalized types at different times in the life cycle of a species (**Figure 9.11**).

9.5 Birthrate Is Age-Specific

A standard convention in demography (the study of populations) is to express birthrates as births per 1000 individuals of a population per unit of time. This figure is obtained by dividing the number of births that occurred during some period of time (typically a year) by the estimated population size at the beginning of the time period and multiplying the resulting number by 1000. This figure is the **crude birthrate**.

This estimate of birthrate can be improved by taking two important factors into account. First, in a sexually dimorphic population (separate male and female individuals), only females within the population give birth. Second, the birthrate of females generally varies with age. Therefore, a better way of expressing birthrate is the number of births per female of age x. Because in sexually dimorphic species population increase is a function of the number of females in the population, the age-specific birthrate can be further modified by determining only the mean number of females born to a female in each age group, b_x. Following is the table of **age-specific birthrates** for the gray squirrel population used to construct the life table (Table 9.1):

x	b_x
0	0
1	2
2	3
3	3
4	2
5	0
Σ	10

At age 0, females produce no young; thus, the value of b_x is 0. The average number of female offspring produced by a female of age 1 is 2. For females of ages 2 and 3, the b_x value increases to 3 and then declines to 2 at age 4. By age 5 the females no longer reproduce; thus, the value of b_x is 0.

The sum—represented by the Greek letter sigma, Σ—of the b_x values across all age classes provides an estimate of the average number of female offspring born to a female over her lifetime; this is the **gross reproductive rate**. In the example of the squirrel population presented previously, the gross reproductive rate is 10. However, this value assumes that a female survives to the maximum age of 5 years. What we really need is a measure of net reproductive rate that incorporates the age-specific birthrate as well as the probability of a female's surviving to any specific age.

9.6 Birthrate and Survivorship Determine Net Reproductive Rate

We can use the gray squirrel population as the basis for constructing a fecundity, or fertility, table (**Table 9.4**). The **fecundity table** uses the survivorship column, l_x, from the life

Table 9.4 Gray Squirrel Fecundity Table

x	l_x	b_x	l_xb_x
0	1.0	0.0	0.00
1	0.3	2.0	0.60
2	0.15	3.0	0.45
3	0.09	3.0	0.27
4	0.04	2.0	0.08
5	0.01	0.0	0.00
Σ		10.0	1.40

table together with the age-specific birthrates (b_x) described previously. Although b_x may initially increase with age, survivorship (l_x) in each age class declines. To adjust for mortality, we multiply the b_x values by the corresponding l_x (the survivorship values). The resulting value, l_xb_x, is the mean number of females born in each age group, adjusted for survivorship.

Thus, for 1-year-old females, the b_x value is 2; but when adjusted for survival (l_x), the value drops to 0.6. For age 2, the b_x is 3; but l_xb_x drops to 0.45, reflecting poor survival of adult females. The values of l_xb_x are summed over all ages at which reproduction occurs. The result represents the **net reproductive rate**, R_0, defined as the average number of females that are produced during a lifetime by a newborn female. If the R_0 value is 1, on average females will replace themselves in the population (produce one daughter). If the R_0 value is less than 1, the females do not replace themselves. If the value is greater than 1, females are more than replacing themselves. For the gray squirrel, an R_0 value of 1.4 suggests a growing population of females. Note the significant difference between the gross and net reproductive rates (10 and 1.4, respectively). The difference reflects the fact that only a small proportion of the females born will survive to the maximum age and produce 10 female offspring.

Because the value of R_0 is a function of the age-specific patterns of birth and survivorship, it is a product of the life history characteristics: the allocation of resources to reproduction, the timing of reproduction, the trade-off between the size and number of offspring produced, and the degree of parental care. The net reproductive rate (R_0), therefore, provides a means of evaluating both the individual (fitness) and the population consequences of specific life history characteristics. We will discuss this topic in detail in the following chapter (Chapter 10).

9.7 Age-Specific Mortality and Birthrates Can Be Used to Project Population Growth

Age-specific mortality rates (q_x) from the life table together with the age-specific birthrates (b_x) from the fecundity table can be combined to project changes in the population into the future. To simplify the process, the values for age-specific mortality are converted to age-specific survival. If q_x is the proportion of individuals alive at the beginning of an age class that die before reaching the next age class, then $1 - q_x$ is the proportion that survive to the next age class (**Table 9.5**) and is designated as s_x. With age-specific values of s_x and b_x, we can project the growth of a population by constructing a population projection table.

We can illustrate the construction of a **population projection table** by using data from Table 9.5 and a hypothetical population of squirrels introduced into an unoccupied oak forest. Because females form the reproductive units of the population, we follow only the females in constructing the table; we are assuming that values of q_x presented in Table 9.1 are the same for both males and females. The year of establishment is designated as year 0. The introduced population of female squirrels consists of 20 juveniles (age 0) and 10 adults (age 1), giving a total population of $N(0) = 30$. The following table gives age-specific birthrates (b_x), survival rates (s_x), and number of females in each age class (x) at year 0. These values can now be used to project the population at year 1. We begin by projecting the fate of the initial population of 30 squirrels.

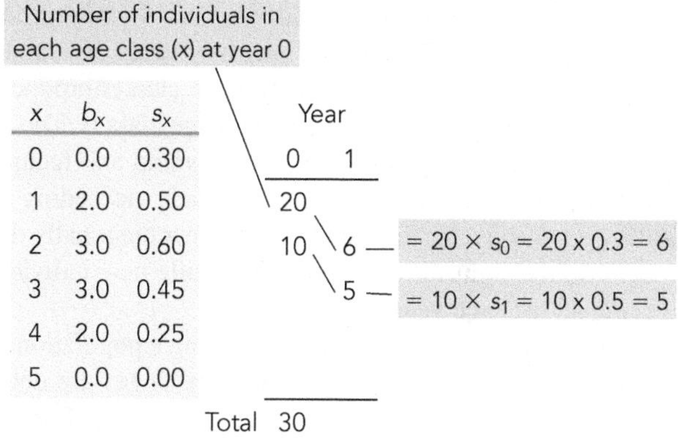

Not all of the squirrels in the initial population (year 0) will survive into the next year (year 1). The survival of these two age groups is obtained by multiplying the number of each by the s_x value. Because the s_x of the 1-year-old females is 0.5, we know

Table 9.5 Age-Specific Survival and Birthrates for Squirrel Population

x	l_x	q_x	s_x	b_x
0	1.0	0.7	0.3	0.0
1	0.3	0.5	0.5	2.0
2	0.15	0.4	0.6	3.0
3	0.09	0.55	0.45	3.0
4	0.04	0.75	0.25	2.0
5	0.01	1.0	0.0	0.0

that 5 of the original 10 individuals (10 × 0.5 = 5) survive to year 1 and reach age 2. The s_x value of age 0 is 0.3, so that only 6 of the original 20 females less than 1 year of age in year 0 survive (20 × 0.3 = 6) to year 1 and become 1-year-olds.

In year 1, we now have six 1-year-olds and five 2-year-olds, and we are now ready to calculate reproduction (recruitment into age class 0).

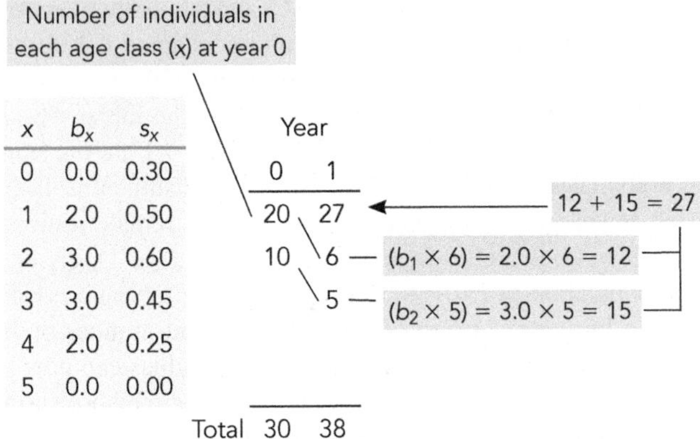

The b_x value of the six 1-year-olds is 2, so they produce 12 offspring. The five 2-year-olds have a b_x value of 3, so they produce 15 offspring. Together, the two age classes produce 27 young for year 1, and they now make up age class 0. The total population for year 1 [$N(1)$] is 38. Survivorship and fecundity are determined in a similar manner for each succeeding year (**Table 9.6**). Survival is tabulated year by year diagonally down the table to the right through the years, while new individuals are being added each year to age class 0.

As we can see, the process of calculating a population projection table from a life table in which life stages are defined by age (or age classes) is relatively straightforward. For many populations, such as perennial plants or fish, however, survival and birth are better described in terms of size rather than age, and rates are expressed as the probability of survival or number of offspring produced per individual of a given size. For these species, the process of developing a population projection table is similar to that presented previously if the size of an individual increases continuously through time. If, however, the size of an individual can either increase or decrease from one time to the next, as is the case with most perennial herbaceous plants, a more complex approach must be taken (see **Quantifying Ecology 9.2**).

Given the population projection table presented in Table 9.6, we can calculate the **age distribution** for each successive year—the proportion of individuals in the various age classes for any one year—by dividing the number in each age class (x) by the total population size for that year [$N(t)$] (see Section 8.5). In comparing the age distribution of the squirrel population over time (**Table 9.7**), we observe that the population attains an unchanging or **stable age distribution** by year 7. From that year on, the proportions of each age group in the population remain the same year after year, even though the population [$N(t)$] is increasing. Another piece of information that can be derived from the population projection shown in Table 9.6 is an estimate of population growth. By dividing the total number of individuals in year $t + 1$, $N(t + 1)$, by the total number of individuals in the previous year, $N(t)$, we can arrive at the **finite multiplication rate**—λ (Greek letter lambda)—for each time period.

$$N(t + 1)/N(t) = \lambda$$

The rate λ has been calculated for each time interval and is shown at the bottom of each column (year) in Table 9.6. Note that initially λ varies between years, but once the population has achieved a stable age distribution, the value of λ remains constant. Values of λ greater than 1.0 indicate a population that is growing, whereas values less than 1.0 indicate a population in decline. A value of $\lambda = 1.0$ indicates a stable population size—neither increasing nor decreasing through time.

The population projection table demonstrates two important concepts of population growth: (1) the rate of population growth, as estimated by λ, is a function of the age-specific rates of survival (s_x) and birth (b_x), and (2) the constant rate of

Table 9.6 Population Projection Table, Squirrel Population

	Year (t)										
Age	0	1	2	3	4	5	6	7	8	9	10
0	20	27	34.1	40.71	48.21	58.37	70.31	84.8	101.86	122.88	148.06
1	10	6	8.1	10.23	12.05	14.46	17.51	21.0	25.44	30.56	36.86
2	0	5	3.0	4.05	5.1	6.03	7.23	8.7	10.50	12.72	15.28
3	0	0	3.0	1.8	2.43	3.06	3.62	4.4	5.22	6.30	7.63
4	0	0	0	1.35	0.81	1.09	1.38	1.6	1.94	2.35	2.83
5	0	0	0	0	0.33	0.20	0.27	0.35	0.40	0.49	0.59
Total $N(t)$	30	38	48.2	58.14	68.93	83.21	100.32	120.85	145.36	175.30	211.25
Lambda	λ	1.27	1.27	1.21	1.19	1.21	1.20	1.20	1.20	1.20	1.20

QUANTIFYING ECOLOGY 9.2 Life History Diagrams and Population Projection Matrices

The construction of life tables and their use in the development of population projection tables are important approaches in studying the dynamics of age-structured populations. We can represent the steps involved in the construction of the population projection table presented in Table 9.6 graphically using a life history diagram.

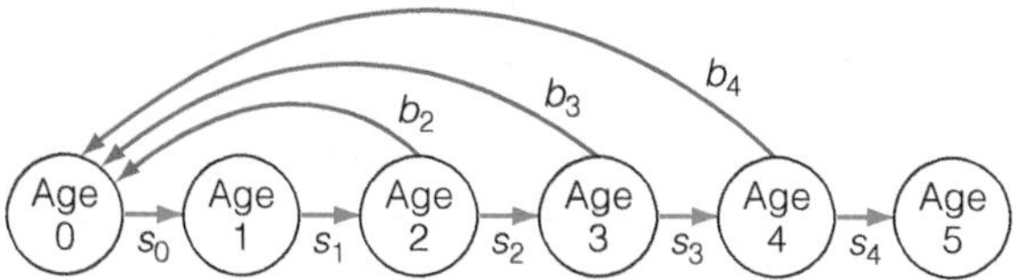

Each of the circles represents an age class. The red arrows pointing from each age class to the next represent the values of s_x, the age-specific values of survival in Table 9.5. The blue arrows from age classes 1 to 4 that point to age class 0 are the age specific birth rates, b_x.

Although the life history diagram provides a convenient way of graphically representing the age-specific rates of survival and birth from a life table, the real value of this approach is that it summarizes the changes that occur in populations (species) that exhibit a more complex pattern of transition between different developmental stages or size classes. For example, in perennial herbaceous plant species, birthrates (seed or seedling production) are a function of size rather than age. In addition, the die-back and regrowth each year of aboveground tissues (stem, leaves, and flowers) can result in an individual remaining in the same size class (or developmental stage) for multiple years or even reverting to a smaller size class. The population structure of the perennial herb *Prinmila vulgaris* presented in Figure 8.20 provides an example.

In their analysis of the population growth of the forest herb *Prinmila vulgaris,* Teresa Valverde and Jonathan Silverston represent the population structure using five developmental stage classes based on reproductive stage and size (see legend to figure). The resulting life history diagram is presented here.

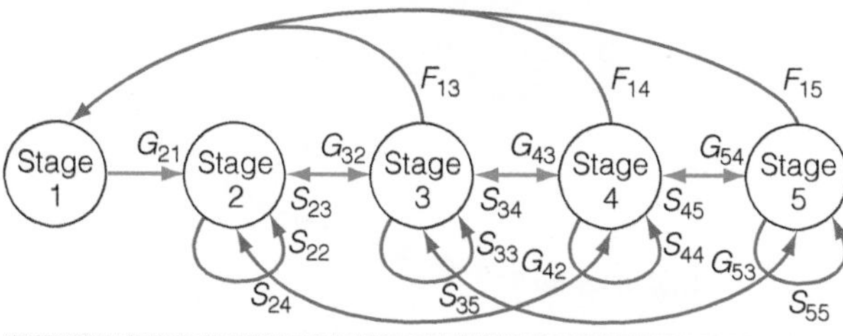

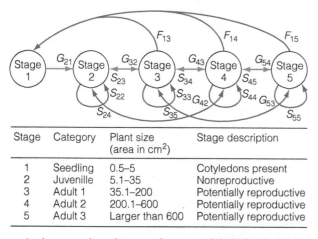

Stage	Category	Plant size (area in cm^2)	Stage description
1	Seedling	0.5–5	Cotyledons present
2	Juvenille	5.1–35	Nonreproductive
3	Adult 1	35.1–200	Potentially reproductive
4	Adult 2	200.1–600	Potentially reproductive
5	Adult 3	Larger than 600	Potentially reproductive

In the preceding diagram, the arrows labeled with the letter *F* represent fecundity or birthrates (the number of seedlings produced by the various adult stage classes: Adults 1, 2, and 3). These values are equivalent to the b_x values used in the analysis of the grey squirrel population (see Table 9.5). The arrows labeled with the letter *G* represent the probability of an individual in that stage class growing into the next larger stage class the following year. These values are equivalent to the s_x values in Table 9.5. The arrows labeled *S* represent the probability that an individual of a given stage class will either stay in the same age class or revert into a previous stage class the following year (smaller in size than the previous year). These values do not exist for a population that is described in terms of age classes because an individual cannot revert to a previous age class.

Because of the greater complexity of possible transitions, a system of subscripts is required to identify the transitions. Each transition has a two-number subscript, *i* and *j*, that represents the probability that an individual of stage class *j* at year *t* will move into stage class *i* the following year (calculated as *t* + 1). For example, there are three possible transitions for individuals currently in the Stage Class 3. The transition G_{43} is the probability that an individual in Stage Class 3 will move into Stage Class 4 the following year, whereas the transition S_{23} is the probability that the individual will revert to Stage Class 2 the following year. The transition labeled S_{33} is the probability that the individual will remain in Stage Class 3 the following year.

The values of fecundity (*F*) and the transition probabilities (*G* and *S*) are organized in the form of a population projection matrix.

The top row of the matrix contains the values of fecundity for each of the stage classes (F_{ij}). The other elements of the matrix contain the transition probabilities between stage classes (G_{ij} and S_{ij}). The elements of the matrix that have a value of zero represent rates or transitions that are either not possible or do not exist for the population under study.

Combined with estimates of the current population structure (number of individuals in each of the stage classes), the population projection matrix can be used to project patterns of population growth and structure in the future, just as was done in Table 9.6 for the grey squirrel population. The procedure involves matrix multiplication—multiplying the preceding matrix by a vector representing the current population structure. To further investigate this technique and apply it to predict patterns of population growth for the population of *P. vulgaris* studied by Teresa Valverde and Jonathan Silverston and discussed previously, go to Analyzing Ecological Data at www.masteringbiology.com.

Stage at time *t*+1 \ Stage at time *t*	1	2	3	4	5
1	0	0	F_{13}	F_{14}	F_{15}
2	G_{21}	S_{22}	S_{23}	S_{24}	0
3	0	G_{32}	S_{33}	S_{34}	S_{35}
4	0	G_{42}	G_{43}	S_{44}	S_{45}
5	0		G_{53}	G_{54}	S_{55}

Table 9.7 Approximation of Stable Age Distribution, Squirrel Population

	Proportion in Each Age Class for Year										
Age	0	1	2	3	4	5	6	7	8	9	10
0	.67	.71	.71	.71	.69	.70	.70	.70	.70	.70	.70
1	.33	.16	.17	.17	.20	.17	.17	.18	.18	.18	.18
2		.13	.06	.07	.06	.07	.07	.07	.07	.07	.07
3			.06	.03	.03	.04	.04	.03	.03	.03	.03
4				.02	.01	.01	.01	.01	.01	.01	.01
5					.01	.01	.01	.01	.01	.01	.01

increase of the population from year to year and the stable age distribution are results of survival and birthrates for each age class that are constant through time.

Given a stable age distribution in which λ does not vary, λ can be used as a multiplier to project population size into the future $(t + 1)$. This can be shown simply by multiplying both sides of the equation for λ shown previously by the current population size, $N(t)$, giving:

$$N(t + 1) = N(t)\lambda$$

We can predict the population size at year 1 by multiplying the initial population size $N(0)$ by λ, and for year 2 by multiplying $N(1)$ by λ:

$$N(1) = N(0)\lambda$$

$$N(2) = N(1)\lambda$$

Note that by substituting $N(0)\lambda$ for $N(1)$, we can rewrite the equation predicting $N(2)$ as:

$$N(2) = [N(0)\lambda]\lambda = N(0)\lambda^2$$

In fact, we can use λ to project the population at any year into the future using the following general form of the relationship developed previously:

$$N(t) = N(0)\lambda^t$$

For our squirrel population, we can multiply the population size at year 0 — $N(0) = 30$ — by $\lambda = 1.2$, which is the value derived from the population projection table, to obtain a population size of 36 for year 1. If we again multiply 36 by 1.20, or the initial population size 30 by λ^2 (1.20^2), we get a population size of 43 for year 2; and if we multiply the initial population size of 30 by λ^{10}, we arrive at a projected population size of 186 for year 10 (**Figure 9.12**). These population sizes do not correspond exactly to the population sizes calculated in the population projection table because λ fluctuates above and below the eventual value attained at stable age distribution. Only after the population achieves a stable age distribution does the λ value of 1.20 project future population size.

The equation $N(t) = N(0)\lambda^t$ describes a pattern of population growth (see Figure 9.12) similar to that presented for the exponential growth model developed in Section 9.1. Recall that when described over discrete time intervals, however, the pattern of growth is termed *geometric population growth* (see discussion in Section 9.1). In this example, the time interval (Δt) is 1 year, the interval (x) used in constructing the life and fecundity tables from which λ is derived.

Note that the equation predicting population size through time using the finite growth multiplier λ is similar to the corresponding equation describing conditions of exponential growth developed in Section 9.1:

$$N(t) = N(0)\lambda^t \text{ (geometric population growth)}$$

$$N(t) = N(0)e^{rt} \text{ (exponential population growth)}$$

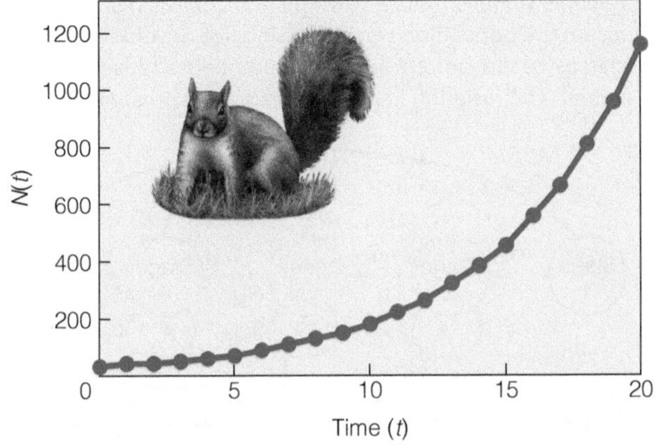

Figure 9.12 Change in population size [$N(t)$] with time for the gray squirrel population represented in Table 9.1. Population estimates are based on the discrete model of geometric population growth [$N(t) = N(0)\lambda^t$] with an initial population density [$N(0)$] of 30 and a finite growth multiplier of $\lambda = 1.2$ from the population projection table (Table 9.6).

In fact, the two equations (finite and continuous) illustrate the relationship between λ and *r:*

$$\lambda = e^r \text{ or } r = \ln \lambda$$

For the gray squirrel population, we can calculate the value of $r = \ln(1.20)$, or 0.18.

Unlike the original calculation of r for the hydra population in Section 9.1, this estimate of the per capita population growth rate does not assume that all individuals in the population are identical. It is derived from λ, which as we have seen is an estimate of population growth based on the age-specific patterns of birth and death for the population. This estimate does, however, assume that the age-specific rates of birth and death for the population are constant; that is, they do not change through time. It is this assumption that results in the population converging on a stable age distribution and constant value of λ.

The geometric and exponential models developed thus far provide an important theoretical framework for understanding the demographic processes governing the dynamics of populations. But nature is not constant; systematic and stochastic (random) processes, both internal (demographic) and external (environmental), can influence population dynamics.

9.8 Stochastic Processes Can Influence Population Dynamics

Thus far we have considered population growth as a deterministic process. Because the rates of birth and death are assumed to be constant for a given set of initial conditions — values of r or λ and $N(0)$ — both the exponential and geometric models of population growth will predict only one exact outcome. Recall, however, that the age-specific values of survival and birth in the life and fecundity tables (Tables 9.1 and 9.4) represent probabilities and averages derived from the cohort or population under study. For example, the values of b_x are the average number of females produced by a female of that age group. For the 1-year-old females, the average value is 2.0; however, some female squirrels in this age class may have given birth to four female offspring, whereas others may not have given birth at all. The same holds true for the age-specific survival rates (s_x), which represent the probability of a female of that age surviving to the next age class. For example, in Table 9.5 the probability of survival for a 1-year-old female gray squirrel is 0.5—the same probability of getting a heads or tails in a coin toss. Although survival (and mortality) is expressed as a probability, it is a discrete event for any individual—it either survives to the next year or not, just as the outcome of a single coin toss will be either heads or tails. If we toss a coin 10 times, however, we expect to get on average an outcome of 5 heads and 5 tails. This is in fact what we assume when we multiply the probability of survival (0.5) by the number of females in an age class (10) to project the number surviving to the next year (5) in Table 9.6. But each individual outcome in the 10 coin tosses is independent, and there is a possibility of getting 4 heads and 6 tails (probability $p = 0.2051$), or even 0 heads and 10 tails ($p = 9.765 \times 10^{-4}$). The same is true for the probability of survival when applied to individuals in a specific age class. The realization that population dynamics represent the combined outcome of many individual probabilities has led to the development of probabilistic, or stochastic, models of population growth. These models allow the rates of birth and death to vary about the mean estimate represented by the values of b_x and s_x.

The stochastic (or random) variations in birthrates and death rates occurring in populations from year to year are called **demographic stochasticity**, and they cause populations to deviate from the predictions of population growth based on the deterministic models discussed in this chapter.

Besides demographic stochasticity, random variations in the environment, such as annual variations in climate (temperature and precipitation) or the occurrence of natural disasters, such as fire, flood, and drought, can directly influence birthrates and death rates within the population. Such variation is referred to as **environmental stochasticity**. We will discuss the role of environmental stochasticity in controlling the growth of populations later in Chapter 11 (Section 11.13).

9.9 A Variety of Factors Can Lead to Population Extinction

When deaths exceed births, populations decline. R_0 becomes less than 1.0, and r becomes negative. Unless the population reverses the trend, it may become so low that it declines toward extinction (see Figure 9.4).

Small populations—because of their greater vulnerability to demographic and environmental stochasticity (see Section 9.8) and loss of genetic variability (see Section 5.7)—are more susceptible to extinction than larger populations (we shall explore this issue at greater length in Chapter 11). However, a wide variety of factors can lead to population extinction regardless of population size.

Extreme environmental events, such as droughts, floods, or extreme temperatures (heat waves or frosts), can increase mortality rates and reduce population size. Should the environmental conditions exceed the bounds of tolerance for the species, the event could well lead to extinction (for examples of environmental tolerances see Figures 5.7, 6.6, and 7.14a). Changes in regional and global climate over the past century (see Chapter 2, *Ecological Issues & Applications*) have led to a decline in many plant and animal species, and projected future climate change could result in the extinction of many species (see Chapter 27).

A severe shortage of resources, caused by either environmental extremes (as discussed previously) or overexploitation, could result in a sharp population decline and possible extinction should the resource base not recover in time to allow for adequate reproduction by survivors. In the example of

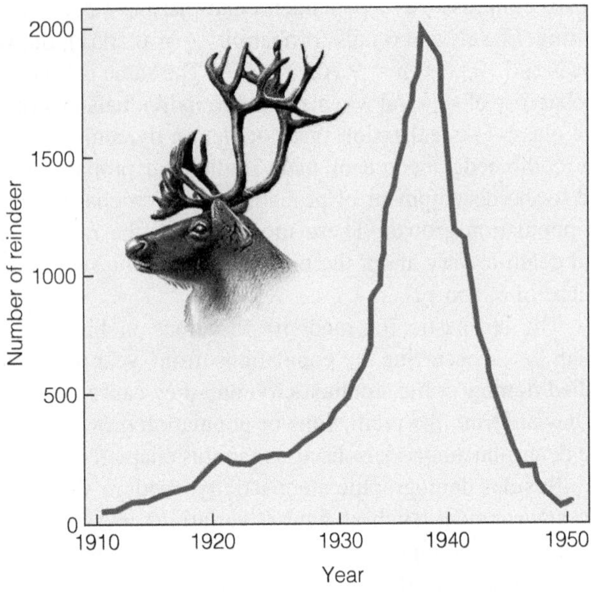

Figure 9.13 After the initial period of exponential growth of the St. Paul reindeer herd presented in Figure 9.5, overexploitation of resources degraded the habitat, resulting in a sharp decline in population numbers.
(Adapted from Scheffer 1951.)

exponential growth in the population of reindeer introduced on St. Paul in 1911 (see Figure 9.5), the reindeer overgrazed their range so severely that the herd plummeted from its high of more than 2000 in 1938 to 8 animals in 1950 (**Figure 9.13**). The decline produced a curve typical of a population that exceeds the resources of its environment. Growth stops abruptly and declines sharply in the face of environmental deterioration. From a low point, the population may recover to undergo another phase of exponential growth or it may decline to extinction.

As we discussed in Chapter 8 (*Ecological Issues & Applications*), when a nonnative species (invasive species) is introduced to an ecosystem through human activity, the resulting interactions with species in the community can often be detrimental. The introduction of a novel predator, competitor, or parasite (disease) can increase mortality rates, having a devastating effect on the target population and causing population decline or even extinction.

ECOLOGICAL Issues & Application — The Leading Cause of Current Population Declines and Extinctions Is Habitat Loss

A multitude of ecological studies over the past several decades have documented a pattern of population decline and extinction for an ever-growing number of plant and animal species across the planet (**Figure 9.14**). The primary cause of current population extinctions is the loss of habitat as a result of human activities. The cutting of forests, draining of wetlands, clearing of lands for agriculture, and damming of rivers have resulted in a significant decline in the available habitat for many species and are currently the leading causes of species extinctions on a global scale.

Freshwater ecosystems have been particularly vulnerable to habitat destruction over the past century. In the United States alone, there are approximately 75,000 dams impounding some 970,000 km of river, or about 17 percent of the rivers in the nation (**Figure 9.15**). Dams remove sections of turbulent river and create standing bodies of water (lakes and reservoirs),

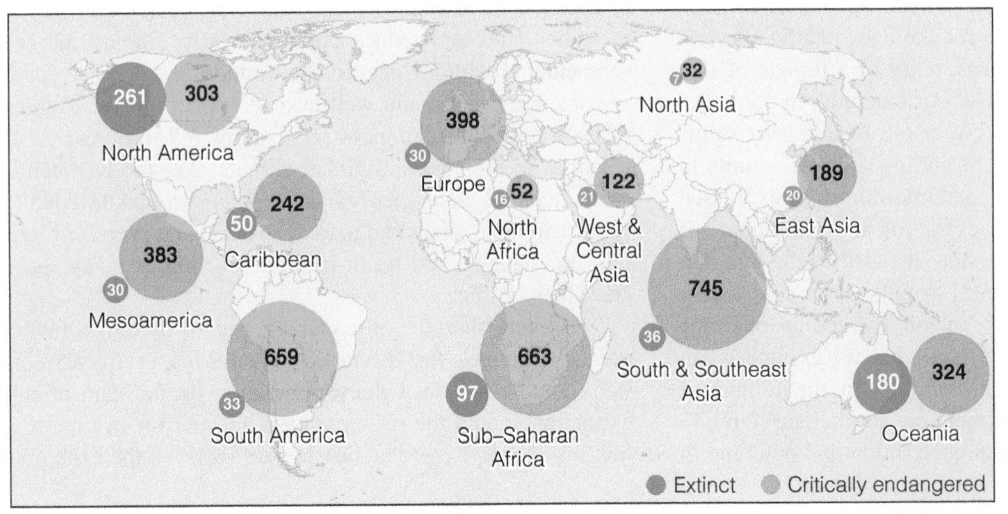

Figure 9.14 Global map showing the numbers of recent extinctions and critically endangered species for various geographic regions.
(Data from IUCN.)

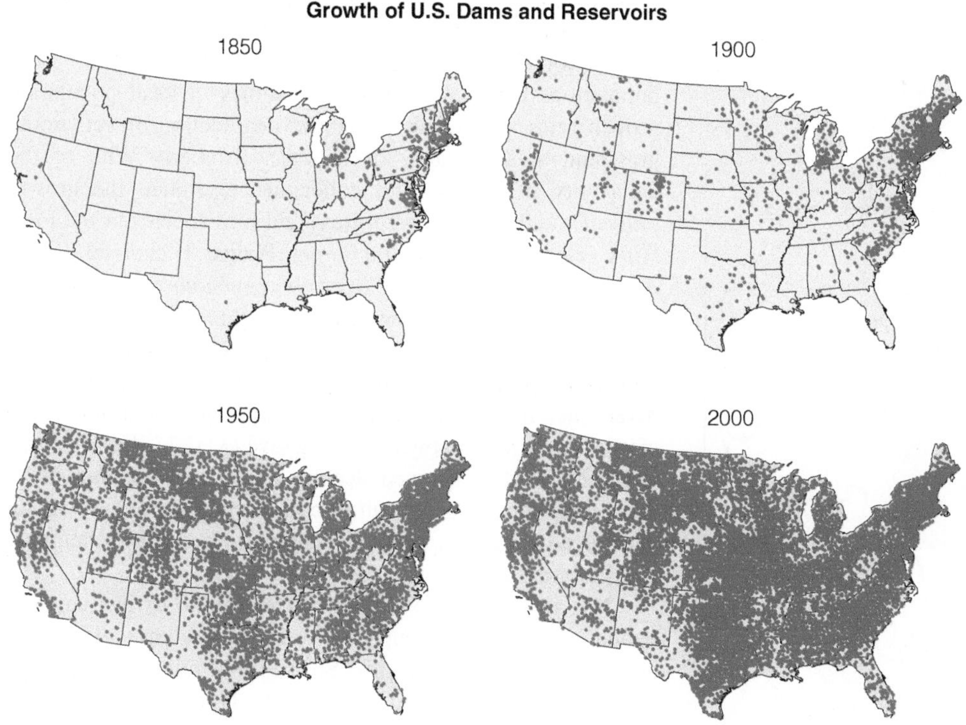

Figure 9.15 The growth of dams and reservoirs in the United States during the period of 1850–2000. (Data from Syvitski and Kettner. 2011.)

affecting flow rates, temperature and oxygen levels, and sediment transport. Dams have a particularly negative impact on migratory species, restricting movement upriver to breeding areas. As a result of the degradation of freshwater habitats, between the years 1900 and 2010, 57 species and subspecies of North American freshwater fish have become extinct (Figure 9.16).

Birds are perhaps the most extensively monitored group of terrestrial species in North America over the past 50 years and therefore provide some of the best examples of population declines resulting from human activity and land-use change. The North American Breeding Bird Survey (a joint effort between the United States Geological Service and the Canadian Wildlife Service) has conducted annual surveys in the United States and Canada since its initial launch in 1966. These data provide a basis for evaluating population trends that can be related to changes in land use and habitat decline over the same period.

Data from the Breeding Bird Survey show that one of the most negatively impacted groups of birds over the past 50 years has been species that inhabit the grassland habitats of the Great Plains of central North America. Beginning in the latter half of the 19th century, the expansion of agriculture west of the Mississippi River has led to the decline of native grasslands (prairies) as land has been converted to cropland (Figure 9.17). More than 80 percent of North American grasslands have been converted to agriculture or other land uses. The loss of habitat has led to a steady decline in grassland bird

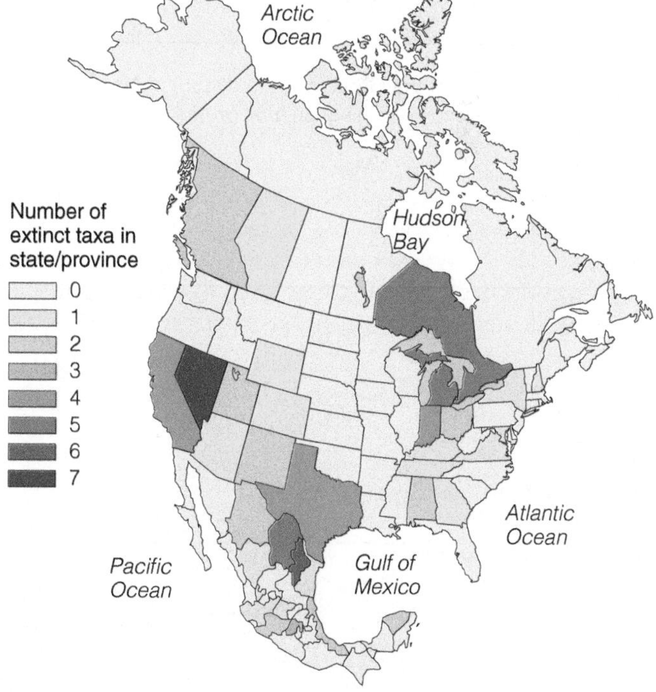

Figure 9.16 Map of North America showing the number of extinct freshwater fish species and subspecies by state (United States) and province (Canada). Data cover extinctions recorded during the period from 1900 to 2010.
(Data from Burkhead 2012, Map from USGS.)

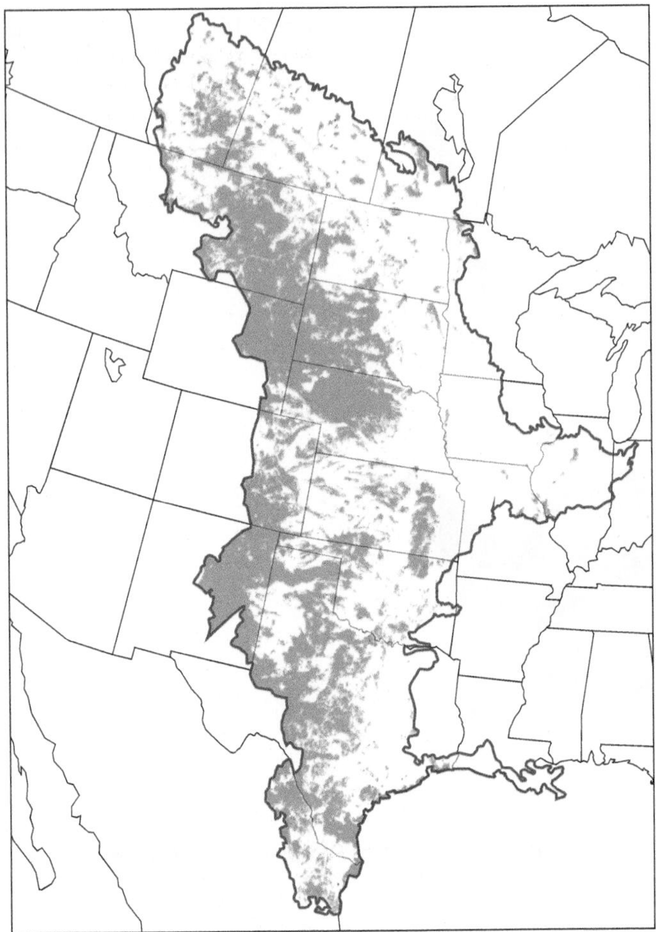

Figure 9.17 Map of the Great Plains (outlined) of North America. Areas currently not in agricultural production (not tilled) are shaded in green.
(Data from Ostlie and Haferman. 1999.)

species in both the United States and Canada (**Figure 9.18**). The observed decline in populations is not only a result of a reduction in the area of native grasslands to support local populations but also because of declines in local population growth rates as a result of habitat degradation on remaining grassland areas (quality of habitat). Kimberly With of the University of Kansas and colleagues examined the growth rates of the three dominant grassland bird species in the Flint Hills region of the central Great Plains: Dickcissel (*Spiza Americana*), Grasshopper Sparrow (*Ammodramus savannarum*), and Eastern Meadowlark (*Sturnella magna*). Using estimates of annual population growth rate (λ; see Section 9.7) for numerous local populations across the region, the researchers determined that the mean annual growth rates for all three species on remaining habitats are negative ($\lambda < 1$). These results indicate that the observed decline in regional populations of these three species is a result of both decreasing area of habitat and negative growth rates for populations on the remaining areas of grassland.

Not all species are equally susceptible to extinction from habitat decline. One group of species that are particularly vulnerable is migratory species. Species that migrate seasonally depend on two or more distinct habitat types in different geographic regions (see Section 8.7). If either of these habitats is altered or destroyed, the species will not persist. The more than 120 species of neotropical birds that migrate each year between the temperate zone of eastern North America and the tropics of Central and South America (and the islands of the Caribbean) depend on suitable habitat in both locations (**Figure 9.19**) as well as stopover habitat in between. An analysis of the North American Breeding Bird Survey data for Canada during the period of 1970 to 2010 shows a decline of more than 50 percent in the populations of bird species that breed in Canada (spring and summer months) and migrate to South America for the

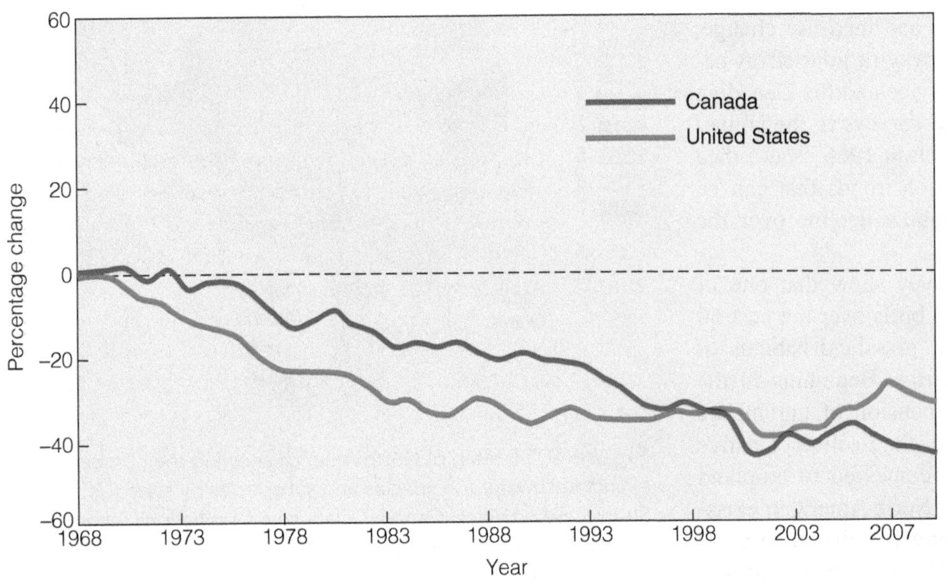

Figure 9.18 Percentage change in the populations of grassland bird species in Canada and the United States based on the North American Breeding Bird Survey.
(Data from North American Bird Conservation Initiative, U.S. Committee, 2009; The state of the Birds, United States of America, 2009, U.S. Department of Interior, pg. 8; North American Bird Conservation Initiative, Canada, Ottawa, Canada, pg 14.)

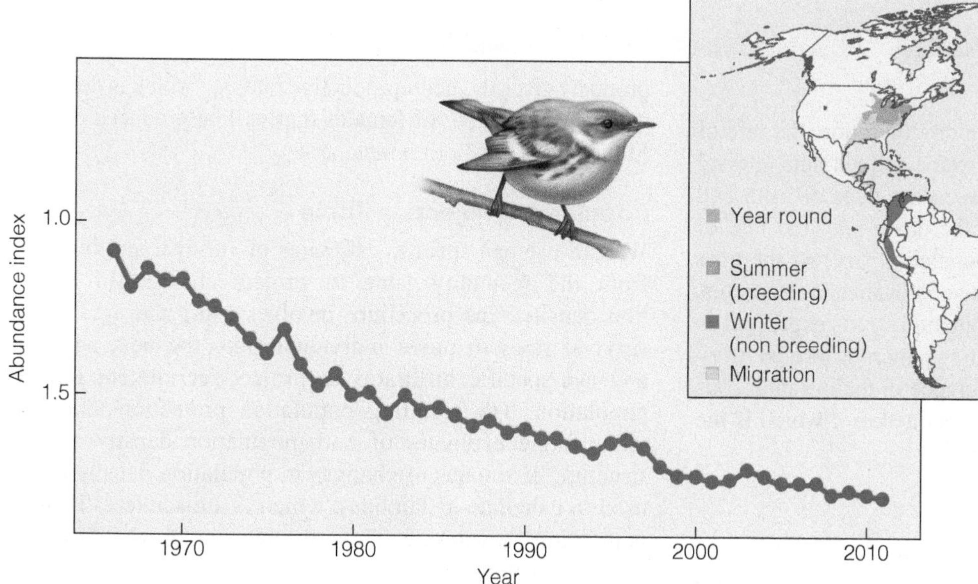

Figure 9.19 Decline in the breeding population of the cerulean warbler (*Setophaga cerulea*). This species is a neotropical migrant that breeds in the northeastern region of North America during the spring and summer months and migrates to northern South America during the winter (see range map). The decline in the breeding population is associated with a decline in forest habitat in South America. (Data from U.S. Department of Interior 2012.)

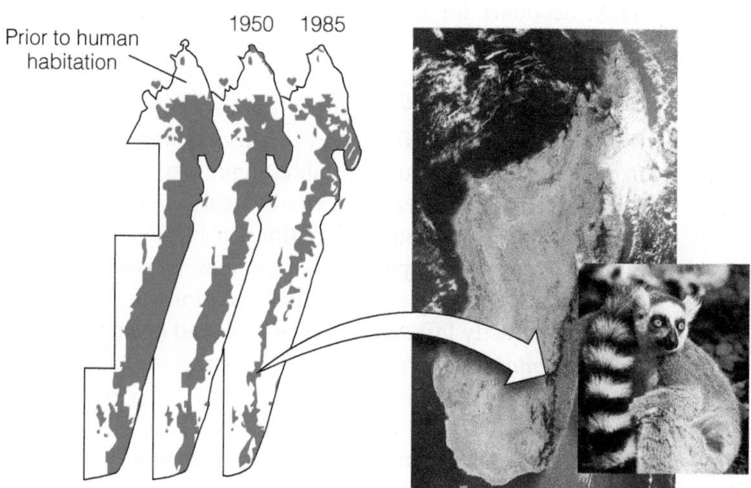

Figure 9.20 The series of maps represent the decline of rain forest in eastern Madagascar from the time of humans' arrival to 1985. The photograph of Madagascar was taken from a space shuttle. The dark areas shown on the inland regions of the east coast represent existing rain forest, which once extended to the coast. The ring-tailed lemur (*Lemur catta*) is one of more than 100 known species of lemur on the island. Currently, 24 lemur species are listed as critically endangered. In the 2000 years since humans first arrived in Madagascar, at least 17 species and 8 genera are believed to have become extinct.

winter. The primary reason for this decline in migratory bird populations is the destruction of rain forest habitats in South America.

The species most vulnerable to extinction are endemics, which are species found only in a particular locality or localized habitat. Endemic species are particularly susceptible to extinction because of their limited geographic distribution (see Chapter 8, Section 8.2 and Figure 8.6). Environmental changes, disturbances, or human activities within their limited range could result in a complete loss of habitat for the species. For example, the island of Madagascar off the east coast of Africa is home to a diverse flora and fauna, of which approximately 90 percent are endemics and found only on Madagascar. The majority of these species inhabit the island's tropical rain forest habitats, which have declined in extent steadily over the past 50 years (**Figure 9.20**). More than 90 percent of the original rain forest has been cleared, and as a result, Madagascar has the largest percentage of plant and animal species listed as threatened or endangered compared to any other geographic region in the world. For example, lemurs are a group of primates endemic to the island of Madagascar that depend on the rain forest habitat. As a result of forest clearing and habitat loss, 91 percent of the known lemur species are threatened. Twenty-three of the species are now considered "critically endangered," 52 are "endangered," and 19 are listed as "vulnerable" on the International Union for Conservation of Nature's (IUCN) Red List of Threatened Species. At least 17 species and 8 genera are believed to have become extinct in the 2000 years since humans first arrived in Madagascar.

SUMMARY

Population Growth 9.1

In a population with no immigration or emigration, the rate of change in population size over time over a defined time interval is a function of the difference between the rates of birth and death. When the birthrate exceeds the death rate, the rate of population change increases with population size. As the time interval over which population change is evaluated decreases, approaching zero, the change in population size is expressed as a continuous function, and the resulting pattern is termed *exponential population growth.* The difference in the instantaneous per capita rates of birth and death is defined as r, which is the instantaneous per capita growth rate.

Life Table 9.2

Mortality and its complement, survivorship, are best analyzed by means of a life table—an age-specific summary of mortality. By following the fate of a cohort of individuals until all have died, we can calculate age-specific estimates of mortality and survival.

Types of Life Tables 9.3

We can construct a cohort or dynamic life table by following one or more cohorts of individuals over time. A time-specific life table is constructed by sampling the population in some manner to obtain a distribution of age classes during a single time period.

Mortality and Survivorship Curves 9.4

From the life table, we derive both mortality curves and survivorship curves. They are useful for comparing demographic trends within a population and among populations under different environmental conditions and for comparing survivorship among various species. Survivorship curves fall into three major types: Type I, in which individuals tend to live out their physiological life span; Type II, in which mortality and thus survivorship are constant through all ages; and Type III, in which the survival rate of the young is low.

Birthrate 9.5

Birth has the greatest influence on population increase. Like mortality rate, birthrate is age-specific. Certain age classes contribute more to the population than others do.

Net Reproductive Rate 9.6

The fecundity table provides data on the gross reproduction, b_x, and survivorship, l_x, of each age class. The sum of these products gives the net reproductive rate, R_0, which is defined as the average number of females that will be produced during a lifetime by a newborn female.

Population Projection Table 9.7

We can use age-specific estimates of survival and birthrates from the fecundity table to project changes in population density. The procedure involves using the age-specific survival rates to move individuals into the next age class and age-specific birthrates to project recruitment into the population. The resulting population projection table provides future estimates of both population density and age structure. Estimates of changes in population density can be used to calculate λ (lambda), which is a discrete estimate of population growth rate. This estimate can be used to predict changes in population size through time (geometric growth model). In addition, λ can be used to estimate r, which is the instantaneous per capita growth rate. The estimate of r, based on λ, accounts for differences in the age-specific rates of birth and death.

Stochastic Processes 9.8

Because the age-specific values of survival and birth derived from the life and fecundity tables represent average values (probabilities), actual values for individuals within the population can vary. The random variations in birthrates and death rates that occur in populations from year to year are called *demographic stochasticity.* Random variations in the environment that directly influence rates of birth and death are termed *environmental stochasticity.*

Extinction 9.9

A variety of factors can result in a population declining to extinction, including environmental stochasticity, the introduction of new species, and habitat destruction.

Habitat Loss and Extinction Ecological Issues & Applications

The primary cause of species extinctions is habitat destruction resulting from the expansion of human populations and activities. Declining populations are a result of both a reduction in the area of habitat available to support populations and a decline in the growth rate of populations that inhabit remaining areas of habitat. The latter is the result of the degradation of remaining habitats because of human activities.

STUDY QUESTIONS

1. What is the difference between a discrete ($\Delta N/\Delta t$) and continuous (dN/dt) model of population growth? What is the difference between geometric and exponential growth?
2. What is a life table, and what information (data) is required to construct one?
3. How do the gross reproductive rate and the net reproductive rate differ?
4. How can the value of net reproductive rate for a population (R_0) be used to assess whether a population is increasing or declining?
5. To use a life table to project population growth, what assumption must be made regarding the age-specific rates of survival (s_x)?
6. How is the finite growth multiplier (λ) calculated from the population projection table?
7. How can the finite growth multiplier be used to predict future values of population density, $N(t)$?
8. What environmental factors might result in random yearly variations in the rates of survival and birth within a population?
9. Identify two factors that could possibly cause a population to decline to extinction. How might population size influence the impact of these factors?
10. Why are small populations more susceptible to extinction from demographic stochasticity than larger populations?

FURTHER READINGS

Classic Studies

Deevey, E. S., Jr. 1947. "Life tables for natural populations of animals." *Quarterly Review of Biology* 22:283–314.
A classic, pioneering article about life tables.

Kaufman, L., and K. Mallory, eds. 1986. *The last extinction.* Cambridge, MA: MIT Press.
An excellent overview of the causes of global extinction.

Recent Research

Begon, M., and M. Mortimer. 1996. *Population ecology: A unified study of animals and plants*. 3rd ed. Oxford: Blackwell Scientific Publications.
This well-written and organized introductory text on the ecology of populations is an excellent resource for those wishing to read further about the structure and dynamics of natural populations.

Burkhead, N. M. 2012. "Extinction rates in North American freshwater fishes, 1900–2010." *BioScience* 62: 798–808.
This article presents an overview of the recent extinctions of freshwater fish in North America and an analysis of factors responsible for their decline.

Carey, J. R. 2001. "Insect biodemography." *Annual Review of Entomology* 46:79–110.
A review of life tables of insect populations.

Gotelli, N. J. 2008. *A primer of ecology*. 4th ed. Sunderland, MA: Sinauer Associates.
This text explains in detail the most common mathematical models in population and community ecology.

North American Bird Conservation Initiative, U.S. Committee. 2009. *The state of the birds, United States of America, 2009*. Washington, DC: U.S. Department of Interior. 36 pages.

North American Bird Conservation Initiative Canada. 2012. *The state of Canada's birds, 2012*. Ottawa, Canada: Environment Canada. 36 pages.
The preceding two reports present an excellent overview of the decline and current status of bird populations in North America.

CHAPTER 10 Life History

A female Alaskan brown bear (*Ursus arctos*) with her three cubs. Litter sizes range from one to four cubs, born between the months of January and March. The cubs remain with their mothers for at least 2 1/2 years, so the most frequently females can breed is every 3 years.

CHAPTER GUIDE

AN ORGANISM'S LIFE HISTORY is its lifetime pattern of growth, development, and reproduction. At what age or size do individuals mature and begin to reproduce? How many offspring are produced during their lifetimes? How long do individuals live? Life history characteristics are traits that affect and are reflected in the life table of an organism. Previously, we discussed how the growth of a population (population dynamics) is a function of age-specific patterns of mortality (survivorship) and fecundity (birthrate; Chapter 9). Here, we will see that these age-specific patterns are a product of evolution by natural selection. We have explored how natural selection functions to shape the physiology, morphology, and behavior of organisms, resulting in specific adaptations that allow an organism to survive, grow, and reproduce in a given environment (Part Two). Likewise, we saw that the same adaptations also function to limit the organism to do equally well under different environmental conditions. Adaptation is a story of constraints and trade-offs, when a beneficial change in one trait is associated with a detrimental change in another. We will extend the discussion of this framework of constraints and trade-offs that seemingly guide the process of evolution to understand how the age-specific patterns of reproduction and mortality discussed in Chapter 9 evolve.

10.1 The Evolution of Life Histories Involves Trade-offs

If reproductive success (the number of offspring that survive to reproduce) is the measure of fitness, imagine designing an organism with the objective of maximizing its fitness. It would reproduce as soon as possible after birth, and it would reproduce continuously, producing large numbers of large offspring that it would nurture and protect. Yet such an organism is not possible. Each individual has a limited amount of resources that it can allocate to specific tasks. Its allocation to one task reduces its resources available for other tasks. Thus, allocation to reproduction reduces the amount of resources available for growth. Should an individual reproduce early in life or delay reproduction? For a given allocation of resources to reproduction, should an individual produce many small offspring or fewer and larger offspring? Each possible action has both benefits and costs. Thus, organisms face trade-offs in life history characteristics related to reproduction, just as they do in the adaptations related to carbon, water, and energy balance (discussed in Chapters 6 and 7). These trade-offs involve modes of reproduction; age at reproduction; allocation to reproduction; number and size of eggs, young, or seeds produced; and timing of reproduction. These trade-offs are imposed by constraints of physiology, energetics, and the prevailing physical and biotic environment—the organism's habitat. As such, the evolution of an organism's life history reflects the interaction between intrinsic and extrinsic factors. Extrinsic ecological factors such as the physical environment and the presence of predators or competitors directly influence age-specific rates of mortality and survivorship. Intrinsic factors relating to phylogeny (the evolutionary history of the species), patterns of development, genetics, and physiology impose constraints resulting in trade-offs among traits. In our discussion, we explore these trade-offs and the diversity of solutions that have evolved to assure success at the one essential task for continuation of life on our planet, reproduction.

10.2 Reproduction May Be Sexual or Asexual

In Chapter 5, we explored how genetic variation among individuals within a population arises from the shuffling of genes and chromosomes in sexual reproduction. In sexual reproduction between two diploid individuals, the individuals produce haploid (one-half the normal number of chromosomes) gametes—egg and sperm—that combine to form a diploid cell, or zygote, that has a full complement of chromosomes. Because the possible number of gene recombinations is enormous, recombination is an immediate and major source of genetic variability among offspring. However, not all reproduction is sexual. Many organisms reproduce asexually (see Section 8.1). Asexual reproduction produces offspring without the involvement of egg and sperm. It takes many forms, but in all cases, the new individuals are genetically the same as the parent. Strawberry plants spread by stolons, modified lateral stems from which new roots and vertical stems sprout (see Figure 8.2). The one-celled paramecium reproduces by dividing in two. Hydras, coelenterates that live in freshwater (see Figure 9.2), reproduce by budding—a process by which a bud pinches off as a new individual. In spring, wingless female aphids emerge from eggs that have survived the winter and give birth to wingless females without fertilization, a process called **parthenogenesis** (Greek *parthenos,* "virgin"; Latin *genesis,* "to be born").

Organisms that rely heavily on asexual reproduction revert occasionally to sexual reproduction. Many of these reversions to sexual reproduction are induced by an environmental change at some time in their life cycle. During warmer parts of the year, hydras turn to sexual reproduction to produce eggs that lay dormant over the winter and from which young hydras emerge in the spring to mature and reproduce asexually. After giving birth to several generations of wingless females, aphids produce a generation of winged females that migrate to different food plants, become established, and reproduce parthenogenetically. Later in the summer, these same females move back to the original food plants and give birth to true sexual forms—winged males and females that lay eggs rather than give birth to young.

Each form of reproduction, asexual and sexual, has its trade-offs. The ability to survive, grow, and reproduce indicates that an organism is adapted to the prevailing environmental conditions. Asexual reproduction produces offspring that are genetically identical to the parent and are, therefore, adapted to the local environment. Because all individuals are capable of reproducing, asexual reproduction results in a potential for high population growth. However, the cost of asexual reproduction is the loss of genetic recombination that increases variation among offspring. Low genetic variability among individuals in the population means that the population

Figure 10.1 Floral structure in (a) dioecious plant (separate male and female individuals), (b) hermaphroditic plant possessing bisexual flowers, and (c) monoecious plant possessing separate male and female flowers.

responds more uniformly to a change in environmental conditions than does a sexually reproducing population. If a change in environmental conditions is detrimental, the effect on the population can be catastrophic.

In contrast, the mixing of genes and chromosomes that occurs in sexual reproduction produces genetic variability to the degree that each individual in the population is genetically unique. This genetic variability produces a broader range of potential responses to the environment, increasing the probability that some individuals will survive environmental changes. But this variability comes at a cost. Each individual can contribute only one-half of its genes to the next generation. It requires specialized reproductive organs that, aside from reproduction, have no direct relationship to an individual's survival. Production of gametes (egg and sperm), courtship activities, and mating are energetically expensive. The expense of reproduction is not shared equally by both sexes. The eggs (ovum) produced by females are much larger and energetically much more expensive than sperm produced by males. As we shall examine in the following sections, this difference in energy investment in reproduction between males and females has important implications in the evolution of life history characteristics.

10.3 Sexual Reproduction Takes a Variety of Forms

Sexual reproduction takes a variety of forms. The most familiar involves separate male and female individuals. It is common to most animals. Plants with that characteristic are called **dioecious** (Greek *di*, "two," and *oikos*, "home"; **Figure 10.1a**).

In some species, individual organisms possess both male and female organs. They are **hermaphrodites** (Greek *hermaphroditos*). In plants, individuals can be hermaphroditic by possessing bisexual flowers with both male organs, stamens, and female organs, ovaries (**Figure 10.1b**). Such flowers are termed *perfect*. Asynchronous timing of the maturation of pollen and ovules reduces the chances of self-fertilization. Other plants are **monoecious** (Greek *mono*, "one," and *oikos*, "home"). They possess separate male and female flowers on the same plant (**Figure 10.1c**). Such flowers are called *imperfect*. This strategy of sexual reproduction can be an advantage in the process of colonization. A single self-fertilized hermaphroditic plant can colonize a new habitat and reproduce, establishing a new population; this is what self-fertilizing annual weeds do that colonize disturbed sites.

Among animals, hermaphroditic individuals possess the sexual organs of both males and females (both testes and ovaries), a condition common in invertebrates such as earthworms (**Figure 10.2**). In these species, referred to as **simultaneous hermaphrodites**, the male organ of one individual is mated with the female organ of the other and vice versa. The result is that a population of hermaphroditic individuals is in theory able to produce twice as many offspring as a population of unisexual individuals.

Other species are **sequential hermaphrodites**. Animals—such as some mollusks and echinoderms—and some plants

Figure 10.2 Hermaphroditic earthworms mating.

Figure 10.3 Parrotfishes (Scaridae) that inhabit coral reefs exhibit sex change. When a large dominant male mating with a harem of females is removed (by a predator or experimenter), within days, the largest female in the harem becomes a dominant male and takes over the missing male's function.

may be males during one part of their life cycle and females in another part. Some fish may be females first, then males. Sex change usually takes place as individuals mature or grow larger. A change in the sex ratio of the population stimulates sex change among some animals. Removing individuals of the other sex initiates sex reversal among some species of marine fish (**Figure 10.3**). Removal of females from a social group among some coral reef fish stimulates males to change sex and become females. In other species, removal of males stimulates a one-to-one replacement of males by sex-reversing females. Among the mollusks, the Gastropoda (snails and slugs) and Bivalvia (clams and mussels) have sex-changing species. Almost all of these species change from male to female.

Plants also can undergo sex change. One such plant is jack-in-the-pulpit (*Arisaema triphyllum*), a clonal herbaceous plant found in the woodlands of eastern North America (**Figure 10.4**). Jack-in-the-pulpit may produce male flowers one year, an asexual vegetative shoot the next, and female flowers the next. Over its life span, a jack-in-the-pulpit may produce both sexes as well as an asexual vegetative shoot but in no particular sequence. Usually an asexual stage follows a sex change. Sex change in jack-in-the-pulpit appears to be triggered by the large energy cost of producing female flowers. Jack-in-the-pulpit plants generally lack sufficient resources to produce female flowers in successive years; male flowers and pollen are much cheaper to produce than female flowers and subsequent fruits.

Figure 10.4 The jack-in-the-pulpit becomes asexual, male, or female depending on energy reserves. The plant gets its name from the flower stem, or spadix, enclosed in a hoodlike sheath. This fruiting plant is in the female stage.

10.4 Reproduction Involves Both Benefits and Costs to Individual Fitness

To understand how trade-offs function to influence natural selection requires an understanding of the balance between benefits and costs associated with a phenotypic trait. If the objective of reproduction is to maximize the relative fitness of the individual, then the benefit of increasing the number of offspring produced would seem obvious. Yet a central tenet of life history theory is that the behavioral, physiological, and energetic activities involved in reproduction extract some sort of cost to future reproductive success in the form of reduced survival, fecundity, or growth.

There are many examples of various activities involved in reproduction that increase an individual's probability of mortality in addition to the direct physiological costs of reproduction. Activities associated with the acquisition of a mate (see Sections 10.11 and 10.12), defense of a breeding territory (see Chapter 11, Section 11.10), and the feeding and protection of young can reduce the probability of future survival.

The work of Tim Cutton-Block of Cambridge University provides an example of the costs of reproduction in terms of increased probability of future survival. In the development of life tables for a population of red deer in central Scotland, he examined differences in the age-specific patterns of mortality for females—referred to as milk hinds—who have reared a calf to weaning age and those who have not—referred to as yeld hinds (**Figure 10.5**). The higher reproductive costs to milk hinds associated with the care and feeding (lactation) of calves result in higher mortality rates than those observed for yeld hinds (Figure 10.5a).

Reproduction can also directly reduce an individual's ability to produce future offspring. The current reproductive expenditure might leave the individual with insufficient energy resources to produce the same number of offspring during future periods of reproduction (Figure 10.5b). For example, studies by Sveinn Hanssen of the University of Tromso in Norway have shown that current reproduction results in reduced future fecundity in eider ducks. The common eider (*Somateria mollissima*) is a long-lived sea duck whose females do not eat during the incubation period. As a result, the reproductive effort of the female results in an increased loss of body mass and reduced immune function.

In a four-year study, Richard Primack and Pamela Hall of Boston University examined the costs of reproduction in

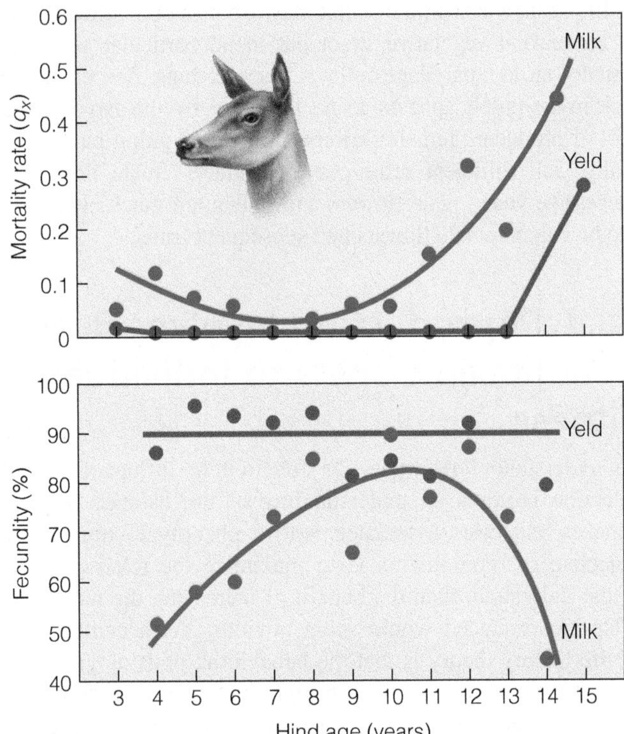

Figure 10.5 Example of the costs of reproduction. (a) Age-specific mortality rates for red deer hinds (milk hinds) that have reared a calf to weaning and those that failed to produce or rear a calf (yeld hinds) show the effects of breeding on survival (measured as overwinter mortality). Increase in mortality rate is associated with declining condition (measured in terms of kidney fat) resulting from the energetic costs of lactation. (b) Milk hinds are less likely to conceive the following year than yeld hinds. (Clutton-Brock 1984.)

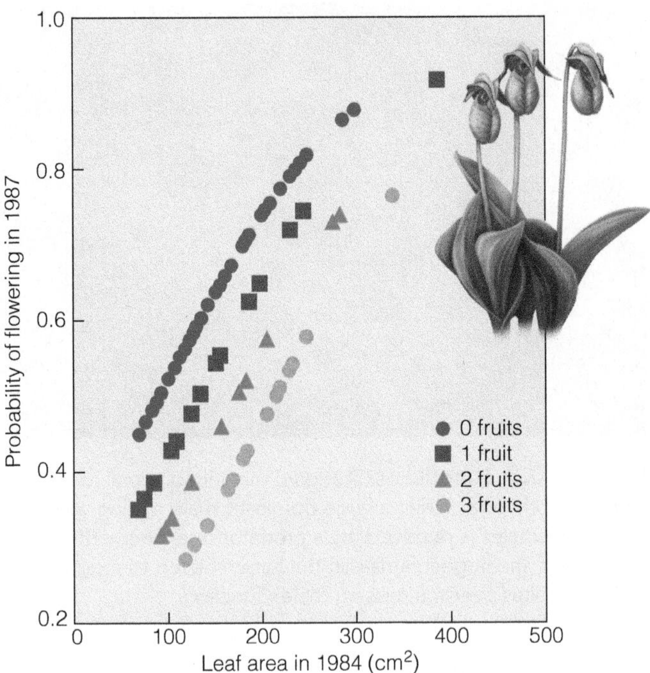

Figure 10.6 Example of cost of reproduction in the pink lady's slipper orchid (*Cypripedium acaule*): the probability of a plant flowering in 1987 based on its initial size (leaf area in cm^2 in 1984) and allocation to reproduction (the number of fruits produced for the period of 1984–1986). The probability of flowering increases with plant size, but for a given size plant, the probability of flowering declines as a function of previous allocation to reproduction (number of fruits produced during the period of 1984–1986). (Adapted from Primack and Hall 1990.)

Interpreting Ecological Data

Q1. What is the approximate difference in the probability of flowering in 1987 for individuals with a leaf area of 200 cm^2 that produced zero fruits and three fruits during the period from 1984 to 1986? What does this tell you about the impact of the costs of past reproduction on future prospects of reproduction?

Q2. According to the preceding figure, the probability of an individual with leaf area of 100 cm^2 that produced zero fruits over the past three years (1984–1986) flowering in the following year (1987) is approximately 0.5 (or 50 percent). How large would an individual that bore three fruits over the past three years have to be to have the same probability of flowering?

the pink lady's slipper orchid (*Cypripedium acaule*). In two eastern Massachusetts populations, the researchers randomly assigned plants to be hand pollinated (increased reproduction) or left as controls, and the treatments were repeated in four successive years. By the third and fourth years of the study, the high cost of reproduction resulted in a lower growth and flowering rate of hand-pollinated plants in comparison with the control plants. For an average-sized plant, the production of fruit in the current year results in an estimated 10–13 percent decrease in leaf area and a 5–16 percent decrease in the probability of flowering in the following year. Increased allocation of resources to reproduction relative to growth diminished future fecundity (**Figure 10.6**).

Allocation to reproduction has been shown to reduce allocation to growth in a wide variety of plant and animal species (**Figure 10.7**). In many species, there is a direct relationship between body size and fecundity (**Figure 10.8**). As a result, an individual reproducing earlier in age will produce fewer offspring per reproductive period than an individual that postpones reproduction in favor of additional growth.

The act of reproduction at a given age, therefore, has potential implications to both age-specific patterns of mortality (survivorship) and fecundity (birthrate) moving forward. For this reason, the age at which reproduction begins—the age at maturity—is a key aspect of the organism's life history.

10.5 Age at Maturity Is Influenced by Patterns of Age-Specific Mortality

When should an organism begin the process of reproduction? Some species begin reproduction early in their life cycle, whereas others have a period of growth and development before

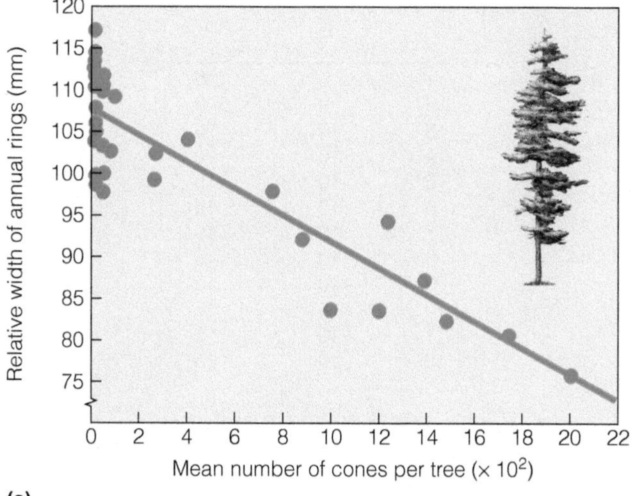

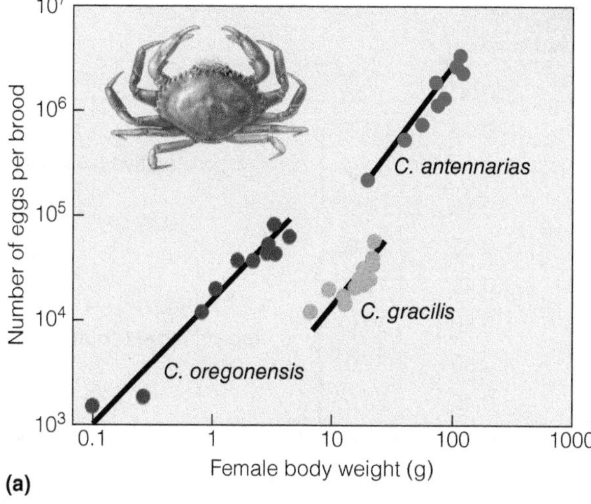

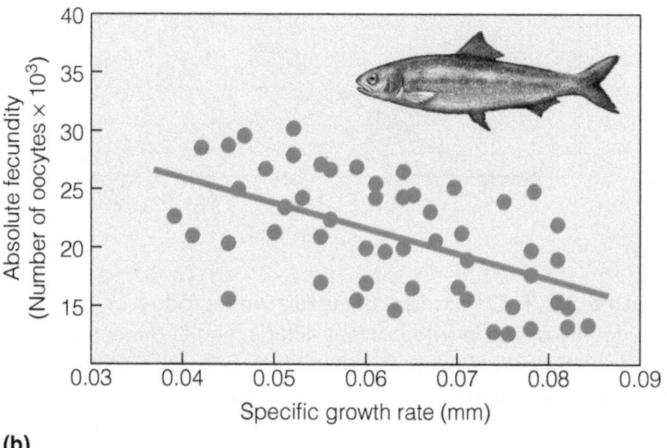

Figure 10.7 (a) Douglas fir trees (*Pseudotsuga menziesii*) exhibit an inverse relationship between the allocation to reproduction (number of cones produced) and annual growth (as measured by radial growth). (b) The relationship between specific growth rate (natural log of the increase in total body length between February and June) and absolute fecundity (measured as the number of oocytes) of three-year-old round sardinella.
([a] Adapted from Eis 1965; [b] Adapted from Tsikliras et al. 2007.)

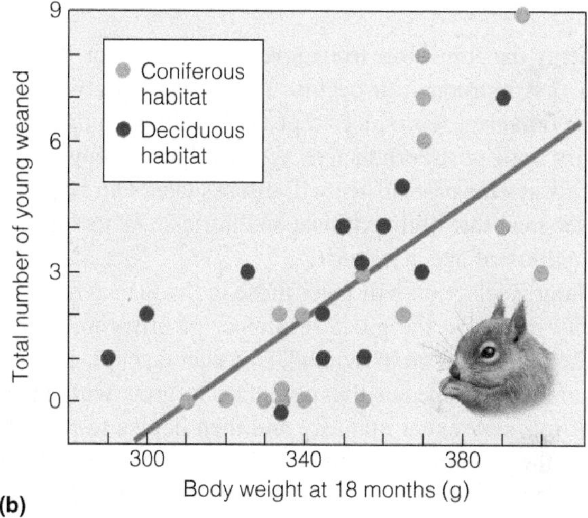

Figure 10.8 (a) Relationship between number of eggs per brood and body weight for three crab species of the genus *Cancer*. (b) Lifetime reproductive success of the European red squirrel is correlated with body weight in the first winter as an adult (age of 18 months).
([a] Data from Hines 1991; [b] Adapted from Wauters and Dhondf 1989.)

the onset of reproduction. If natural selection functions to maximize the relative fitness of the individual, then the age and size at maturity are optimized when the difference between the costs and benefits of maturation at different ages and sizes is maximized. That is when the "payoff" in the fitness of the individual is greatest. An important component of this evolution is the age-specific pattern of mortality because it both shapes and is shaped by the age-specific expenditures of reproductive effort.

To explore how natural selection can function to influence the age at maturity, let's return to think about the age-specific patterns of survival and fecundity that we developed in the previous chapter, that is, the patterns of survival and fecundity that determine the trajectory of population growth (**Table 10.1**; see also Section 9.6). Recall that the first column labeled x shows the age or age class of individuals in the population. The column labeled s_x shows the age-specific survival rates (the probability of an individual of age x surviving to age $x + 1$), and column b_x represents the average number of female offspring produced by an individual female of age x. The three columns have been divided into three distinct age categories relating to reproduction: prereproductive, reproductive, and postreproductive. Prereproductive age categories represent juveniles, whereas the reproductive and postreproductive categories are referred to as adults. The age of maturity then

Table 10.1

x	s_x	b_x	
0	0.05	00.0	Prereproductive (Juvenile)
1	0.05	00.0	
2	0.10	00.0	
3	0.25	25.0	Reproductive (Adult)
4	0.45	30.0	
5	0.50	30.0	
6	0.50	25.0	
7	0.45	20.0	
8	0.40	15.0	
9	0.20	00.0	Postreproductive (Adult)
10	0.00	00.0	

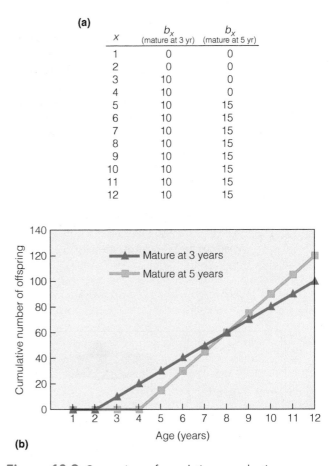

(a)

x	b_x (mature at 3 yr)	b_x (mature at 5 yr)
1	0	0
2	0	0
3	10	0
4	10	0
5	10	15
6	10	15
7	10	15
8	10	15
9	10	15
10	10	15
11	10	15
12	10	15

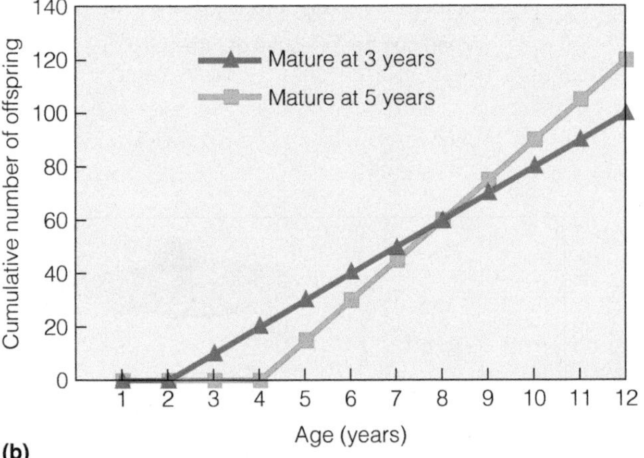

(b)

Figure 10.9 Comparison of cumulative reproduction over the lifetime of two individuals that differ in age at maturity. Both individuals continue to grow only until the onset of maturity. (a) One individual matures at age three and has a fixed fecundity rate of 10 offspring per year. The other individual matures at age five, and the additional allocation to growth and increased body size results in a 50 percent increase in fecundity (average of 15 offspring per year). (b) Until age eight, the individual that matures at age three has a greater accumulated fecundity. From age eight forward, however, the individual that matures at age five has the greater accumulated reproduction.

represents the transition from juvenile to adult, or the age at which first reproduction occurs. In our example, we assume that the organism reproduces repeatedly following the onset of maturity until postreproductive age is achieved; however, this is not always the case, as we will discuss later. Our objective is to understand that both extrinsic and intrinsic factors influence the evolution of age at maturity.

Natural selection will favor those individuals whose age at maturity results in the greatest number of offspring produced over the lifetime of an individual. Consider a simple hypothetical example of a species that continues to grow with age only until it reaches sexual maturity and then begins to reproduce. As with the examples presented in Figure 10.8, assume that fecundity increases with body weight—the larger the individual female, the greater the number of offspring produced per time period (reproductive event). Now assume that individuals within the population vary in the age at which they achieve maturity. As a result of differences in body weight, a female that begins to reproduce at age 3 will produce 10 offspring per year over the duration of her lifetime, whereas a female that delays reproduction until age 5 will have a 50 percent greater fecundity, or 15 offspring per year (**Figure 10.9**). Therefore, we can calculate the cumulative number of offspring produced at any point in each female's life by summing the number of offspring from the onset of maturity to that age (see Figure 10.9). Note in Figure 10.9 that the female that delayed maturity until year 5 has produced a greater number of offspring over her lifetime. Thus, natural selection should favor delayed maturity. However, this conclusion assumes that the females live to their maximum age (12 years). In fact, before age eight the female that matured early has a greater cumulative number of offspring, and it is only if females survive past year eight that the strategy of delayed maturity increases fitness. Recall the difference between gross and net reproductive rate presented previously (Section 9.6). The value obtained by summing the values in the b_x column as was done in this example is a measure of gross reproductive rate, and the strategy of delayed maturity is clearly the winner in terms of fitness. However, the true measure of reproductive rate is net reproductive rate (R_0), as it considers both the age-specific values of fecundity (b_x) and the age-specific values of survivorship (l_x). If survival beyond age eight is an improbable event for this species, then the strategy of early maturity results in the greater fitness.

As the preceding hypothetical example demonstrates, the primary fitness advantage of delaying maturity is the larger initial body size obtained by individuals when they first reproduce. The primary cost of delaying reproduction (late maturity) is the increased risk of death before reproduction, or death before the advantage of increased fecundity as a result of delayed maturity are fully realized—in this example, death before age eight. If one assumes that natural selection acts on age-specific potential of producing future offspring, then age at maturity can be predicted from the mean juvenile and adult survival rates (s_x column in Table 10.1). Decreases in the ratio of adult-to-juvenile survival (low survival for adults relative to juveniles) appear to favor reductions in age at maturity. As external factors (those unassociated with reproduction) increase adult mortality, selection would be expected to favor genotypes that mature earlier (before those ages), thus increasing their probability of contributing genes to future generations.

Laboratory studies and comparative data from natural populations, as well as a number of long-term experiments in which patterns of mortality have been manipulated, support the prediction that natural selection favors earlier maturation when adult survival is reduced, and conversely, that it favors delayed maturation when, relative to adult survival, juvenile survival is reduced. David Reznick and colleagues at the University of California–Riverside conducted a long-term experiment on guppies in Trinidad in which the predictions relating to age-specific patterns of mortality and age at maturity are supported. Local populations of guppies on the island differ in their life history characteristics, and differences among populations are closely associated with the identity of the predator species living in their habitat. Predator species alter age-specific survival because they are size specific in their choice of prey. *Crenicichla alta* (a cichlid), the main predator at one set of the localities, preys predominantly on larger guppies from sexually mature size classes. At other localities, *Rivulus hartii* (a killifish) is the main predator. *Rivulus* feeds primarily on small guppies from immature size classes. Guppies from localities with *Crenicichla* mature at an earlier age than do guppies from localities with *Rivulus*. To prove that differences in age-specific patterns of survival result from different patterns of predation causing differences in age at maturity, Reznick and colleagues transplanted guppies from a site with *Crenicichla* to a site that contained *Rivulus*, but no guppies. This manipulation released the guppies from selective predation on adults and exposed them to selective predation on juveniles.

Eleven years (30–60 generations) after the shift in predation-induced mortality from adults to juveniles, guppies responded to the increase in the ratio of adult-to-juvenile survival with significantly increased age at maturity (from 48 to 58 days). The increased age at maturity was accompanied by a larger average size at age of maturity for females and the production of fewer, but larger offspring (**Figure 10.10**).

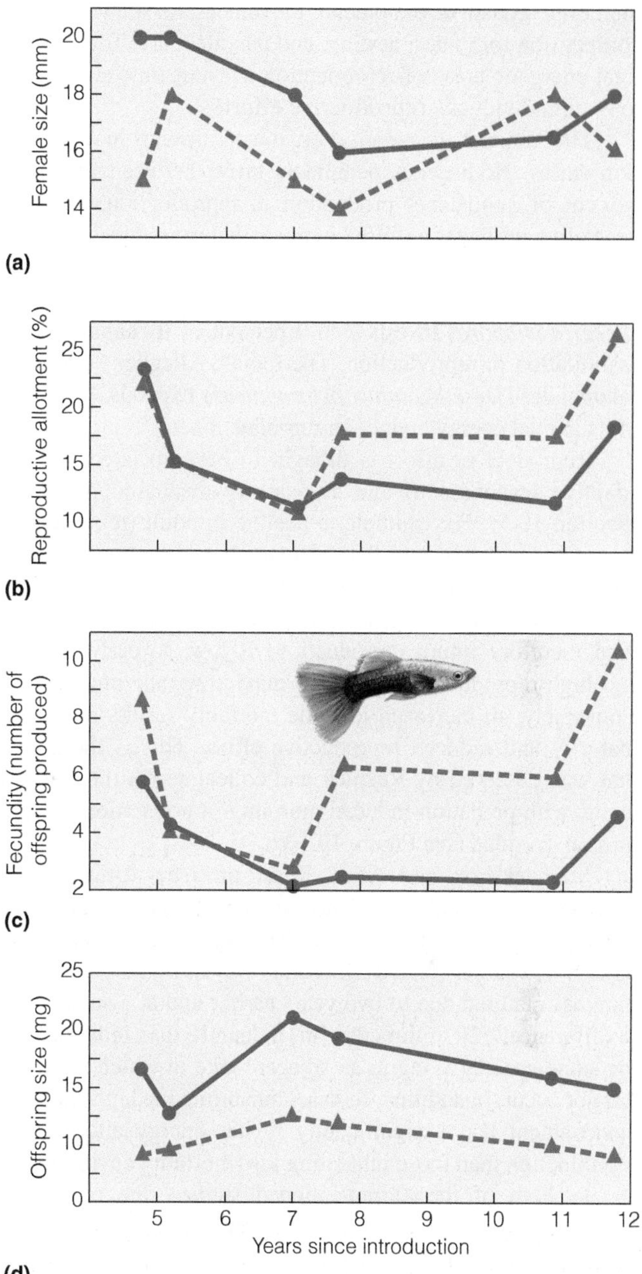

Figure 10.10 Experiment in which researchers shifted predation-induced mortality from adults to juveniles in a population of guppies on the island of Trinidad. The dashed line represents the control group (high adult mortality relative to juveniles), and the solid line represents the introduction of predators that shifted mortality from adults to juveniles. Over the 11-year experiment (*x*-axis: years since introduction), guppies responded to the increase in the ratio of adult to juvenile survival with significantly increased age at maturity (from 48 to 58 days). The increased age at maturity was accompanied by: (a) larger average size at age of maturity for females, (b) decrease in the allocation to reproduction, (c) reduction in fecundity (number of offspring produced), and (d) an increase in the average size of offspring produced.
(Adapted from Reznick et al. 1990.)

10.6 Reproductive Effort Is Governed by Trade-offs between Fecundity and Survival

Fecundity is the number of offspring produced per unit of time (b_x), but the energetic costs of reproduction include a wide variety of physiological and behavioral activities in addition to the energy and nutrient demands of the reproductive event, including gonad development, movement to spawning area, competition for mates, nesting, and parental care. Together, the total energetic costs of reproduction per unit time are referred to as an individual's **reproductive effort**.

The amount of energy organisms invest in reproduction varies. Herbaceous perennials invest between 15 and 20 percent of annual net production in reproduction, including vegetative propagation. Wild annuals that reproduce only once expend 15 to 30 percent, most grain crops—25 to 30 percent, and corn and barley—35 to 40 percent. The common lizard (*Lacerta vivipara*) invests 7 to 9 percent of its annual energy assimilation in reproduction. The female Allegheny Mountain salamander (*Desmognathus ochrophaeus*) expends 48 percent of its annual energy budget on reproduction.

Reproductive effort is thought to be associated with the adaptive responses to age at maturity discussed previously (Section 10.5). For example, a decline in adult (reproductive) survival rate relative to that of juveniles (prereproductive) is predicted to favor genotypes that mature earlier in life and increase reproductive effort. The probability of future survival (and therefore future reproduction) is low, so early maturity and high reproductive effort will maximize individual fitness. Conversely, an increased juvenile mortality results in delayed maturity and reduced reproductive effort. This is the pattern that was observed by Reznick and colleagues in their experiments with predation-induced mortality in populations of guppies on Trinidad (see Figure 10.10a).

Michael Fox and Allen Keast of Trent University in Canada found that pumpkinseed fish (*Lepomis gibbosus*) inhabiting two shallow ponds that experienced major winterkills (weather-related mortality events during winter months) matured one to two years earlier and at a smaller size (a difference >20 millimeters [m] in length) than individuals of the same species living in an adjacent lake in which winterkill did not occur. In addition, females inhabiting the high-mortality environment had a significantly higher energy allocation to reproduction than those inhabiting low-mortality environments.

In both of the aforementioned studies, the researchers found that variations in allocation to reproduction were related to patterns of mortality caused by extrinsic factors (predation or extreme temperatures). Patterns of mortality, however, are also influenced directly by reproductive effort. For example, in the study of red deer presented in Figure 10.5, allocation to reproduction resulted in an increased female mortality rate. Therefore, allocation to reproduction at any time during the life of an individual involves trade-offs between current benefits from the production of offspring and costs in terms of potential reduction in future reproduction. Natural selection functions to optimize the trade-offs between present and future reproduction.

An optimized life history is one wherein conflicts between the competing demands for survival and reproduction are resolved to the advantage of the individual in terms of fitness. To explore this relationship, we can examine how fecundity and survival vary as a function of allocation to reproduction at any given time period in an individual's life.

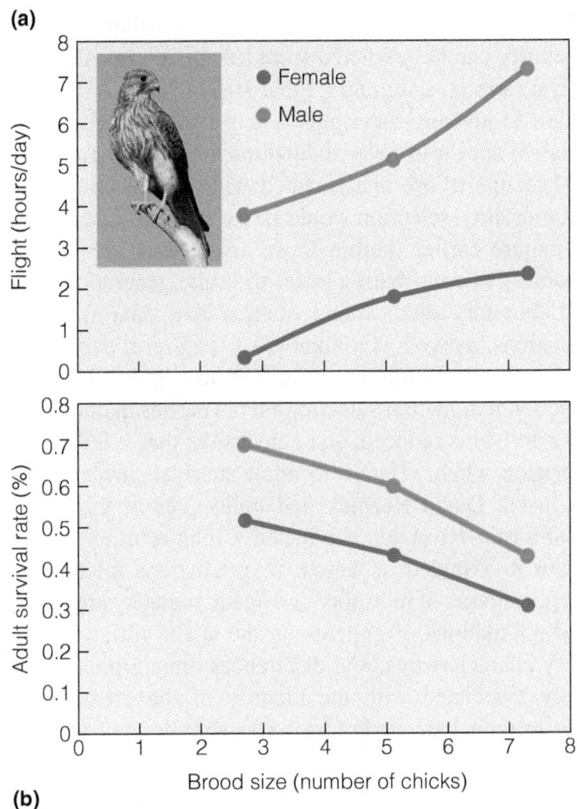

Figure 10.11 Results from an experiment involving the manipulation of brood size in breeding pairs of European kestrels. (a) The amount of time spent foraging (flight time) by both the male and female parent increased as a function of the brood size (number of chicks in the nest). (b) As a result of the increase in energy expenditure, the survival rate of parents decreased with increasing brood size.
(Data from Dijkstra et al., "Brood size manipulations in the kestrel (Falco tinnunculus): Effects on offspring and parent survival" Journal of Animal Ecology Vol. 59, no. 1, February, 1990.)

Cor Dijkstra and colleagues at the University of Groningen in the Netherlands examined the trade-off between fecundity and survival in the European kestrel (*Falco tinnunculus*), which is a predatory bird that feeds on small mammals. Both parents provide food for the brood (offspring); therefore, reproductive allocation associated with the feeding of young can be approximated by the time (hours/day) spent in flight (hunting activity). The brood size for nesting pairs of kestrels in the study area averaged five chicks.

The researchers divided the nesting population in the study area into two groups: a nonmanipulated control group and a group in which brood size in the nests was manipulated. Starting when the nestlings were 5 to 10 days old, the researchers removed two nestlings from selected nests, thus reducing brood size by two. These chicks were then transferred to other nests, increasing the size of the brood by two. Brood enlargement forced an increase in daily hunting activities of both parents (**Figure 10.11a**). Despite the increase, food

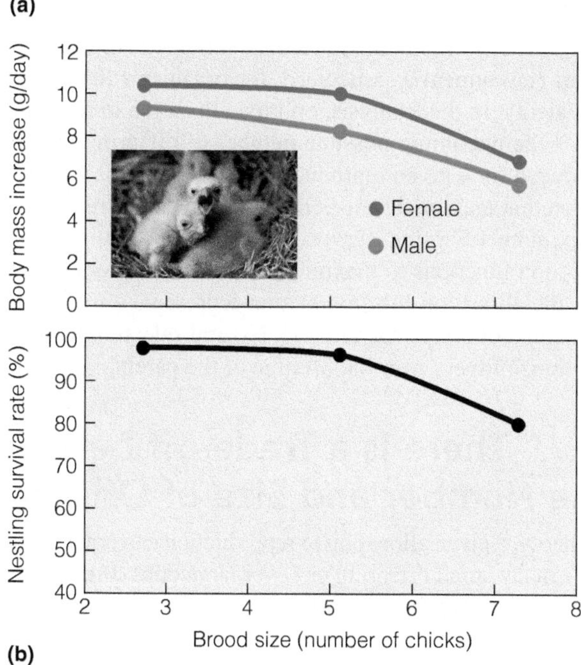

Figure 10.12 Results from an experiment involving the manipulation of brood size in breeding pairs of European kestrels. (a) The body mass of both male and female chicks decreased as the number of chicks in the nest (brood size) increased. (b) As a result, nestling survival rate declined with increasing brood size. (Adapted from Dijkstra et al. 1990.)

intake per chick declined with increased brood size. This decline resulted in reduced growth rate of the nestlings and increased nestling mortality (**Figure 10.12**). In addition, brood enhancement resulted in enhanced weight loss in the female parent, and survival of the parents was negatively correlated with the experimental change in brood size (**Figure 10.11b**). Increased allocation to reproduction (energy expenditure to the feeding and caring of offspring) resulted in a reduction in the probability of future survival of parents and, therefore, future reproduction.

The responses of offspring and parental survival to brood enhancement in the study by Dijkstra reveal two patterns that are essential to understanding how natural selection functions to optimize reproductive effort. First, as reproductive effort increased, the number of offspring increased, but the probability of offspring survival decreased. Therefore, for any given value of reproductive effort, current reproductive success is the product of the two: number of offspring produced multiplied by the probability of their survival. As a result of the inverse relationship between the number of offspring and their probability of survival (**Figure 10.13**), the resulting pattern of current reproductive success is one of diminishing returns, with each additional unit of reproductive allocation returning a decreasing benefit in terms of current fecundity (**Figure 10.14**). Second, as reproductive effort increased, parental survival decreased, once again with each additional

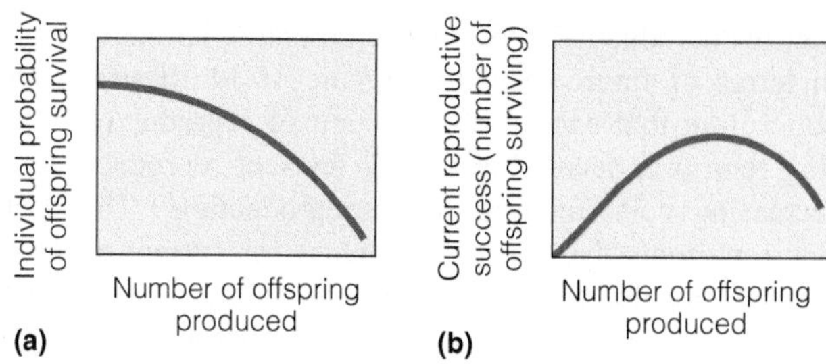

Figure 10.13 (a) As the number of offspring produced per reproductive effort increases, there is a corresponding decline in their probability of survival. As a result, (b) the current reproductive success (calculated as the number of offspring produced times their probability of survival: $x \times y$) shows a pattern of diminishing return, reaching a maximum value at intermediate vales of reproductive effort.

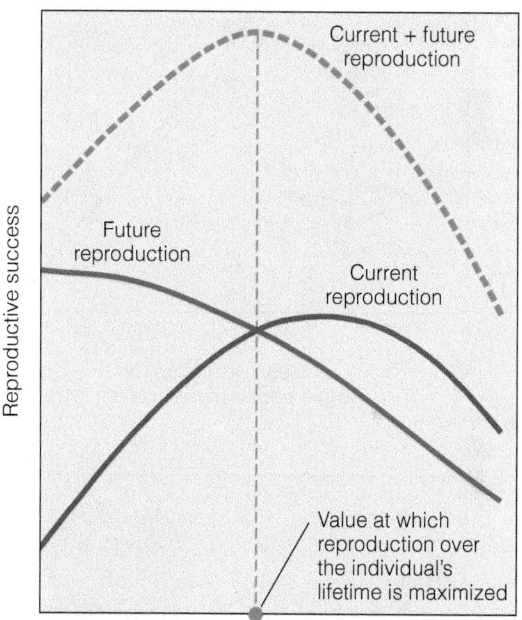

Figure 10.14 If we assume that future reproductive success follows the general pattern of parental survival with current reproductive success seen in Figure 10.11b (decreasing with increasing number of offspring), then we can plot predicted current (from Figure 10.13) and future reproductive success of parents as a function of current allocation to reproduction. By summing the values of current and future reproduction for any value of current reproductive allocation (dashed red line), we can see that reproduction over an individual's lifetime (current + future reproduction) is maximized at intermediate values of reproductive allocation.

Interpreting Ecological Data

Q1. *How would a decrease in the probability of offspring survival in Figure 10.13a change the optimal value for current reproductive allocation?*

Q2. *How would a decrease in adult survival in Figure 10.11b change the optimal value for current reproductive allocation?*

unit of reproductive allocation representing an increased cost in terms of future fecundity. Figure 10.14 illustrates these patterns in that each additional unit of reproductive allocation returns a decreasing benefit (current reproduction) and increasing cost (reduced future reproduction). The dashed line represents the sum of the values for current and future reproduction for any given allocation to reproduction (value along *x*-axis). The dashed line reaches its maximum value at intermediate values of reproductive allocation. Fitness of the parent is often maximized at intermediate reproductive effort (investment), particularly for organisms that reproduce repeatedly. In this analysis, optimal allocation to reproduction is not the maximum possible number of offspring that can be produced for a given reproductive event, and it is not the allocation that maximizes the benefit in terms of current fecundity (maximum offspring survival; see Figure 10.14). If natural selection functions to maximize the relative fitness of the parent, the allocation to reproduction represents a trade-off with parental, not offspring, survival. Natural selection functions to maximize fitness over the lifetime of the parent.

10.7 There Is a Trade-off between the Number and Size of Offspring

In theory, a given allocation to reproduction can potentially produce many small offspring or fewer large ones (**Figure 10.15**). The number of offspring affects the parental investment each receives. If the parent produces a large number of young, it can afford only minimal investment in each one. In such cases,

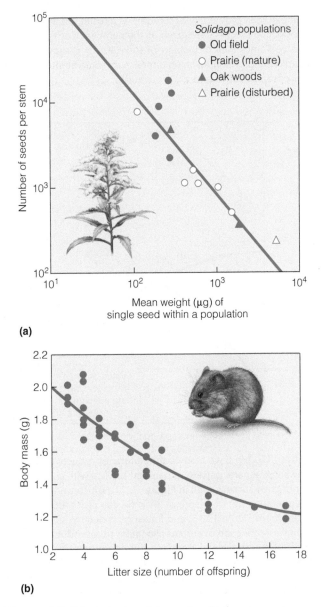

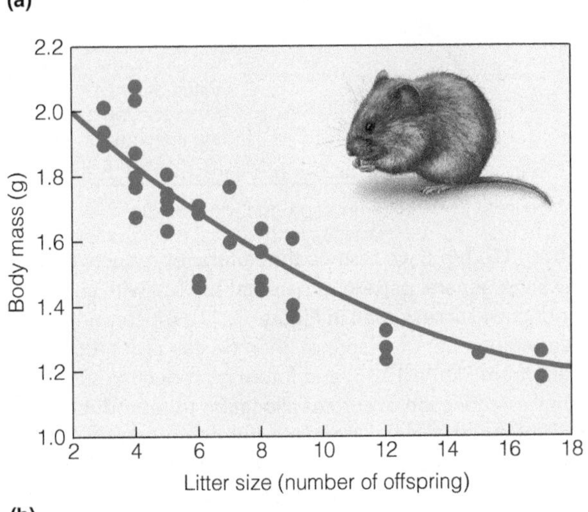

Figure 10.15 Examples of the trade-off between number of offspring produced and the average size of each offspring. (a) Inverse relationship between mean weight of individual seeds and the number of seeds produced per stem for populations of goldenrod (*Solidago* spp.) in a variety of habitats. (b) Inverse relationship between litter size (number of offspring per litter) and average body mass (g) of offspring at birth for litters of bank vole (*Clethrionomys glareolus*).
([a] Adapted from Werner and Platt 1976; [b] Adapted from Oksanen et al. 2003.)

Figure 10.16 The spotted (*Ambystoma maculatum*) and redback (*Plethodon cinereus*) salamanders found in eastern North America provide an example of contrasting life history strategies. (top) The spotted salamander lays a large number of eggs that form an egg mass, which it then abandons. (bottom) In contrast, the redback lays only few eggs, which it guards until they hatch.

animals provide no parental care, and plants store little food energy in seeds. Such organisms usually inhabit disturbed sites, unpredictable environments, or places such as the open ocean where opportunities for parental care are difficult at best. By dividing energy for reproduction among as many young as possible, these parents increase the chances that some young will successfully settle and reproduce in the future.

Parents that produce few young are able to expend more energy on each. The amount of energy varies with the number, size, and maturity of individuals at birth. Some organisms expend less energy during incubation. The young are born or hatched in a helpless condition and require considerable parental care. These animals, such as young mice or nestling American robins (*Turdus migratorius*), are **altricial**. Other animals have longer incubation or gestation, so the young are born in an advanced stage of development. They are able to move about and forage for themselves shortly after birth. Such young are called **precocial**. Examples are gallinaceous birds, such as chickens and turkeys, and ungulate mammals, such as cows and deer.

The degree of parental care varies widely. Some species of fish, such as cod (*Gadus morhua*), lay millions of floating eggs that drift freely in the ocean with no parental care. Other species, such as bass, lay eggs in the hundreds and provide some degree of parental care. Among amphibians, parental care is most prevalent among tropical toads and frogs and some species of salamanders. The spotted (*Ambystoma maculatum*) and redback (*Plethodon cinereus*) salamanders found in eastern North America provide such an example of contrasting life history strategies relating to the number of young produced and parental care (Figure 10.16). The spotted salamander is found under logs and piles of damp leaves in deciduous forest habitats. During the month of February, individuals migrate to ponds and other small bodies of water to reproduce. After mating, females lay up to 250 eggs in large, compact, gelatinous masses that are attached to twigs just below the water surface. After mating, adults leave the water and provide no parental care of eggs or young. In contrast, the redback salamander occupies similar habitats in mixed coniferous–deciduous forests. After mating, females lay 4 to 10 eggs, which are deposited in a cluster within the crevice of a rotting log or stump. The female then curls about the egg cluster, guarding it until the larvae hatch.

Among reptiles, which rarely exhibit parental care, crocodiles are an exception. They actively defend the nest and young for a considerable time. Invertebrates exhibit parental care to varying degrees. Octopi, crustaceans (such as lobsters, crayfish, and shrimp), and certain amphipods brood and defend eggs. Parental care is best developed among the social insects: bees, wasps, ants, and termites. Social insects perform all functions of parental care, including feeding, defending, heating and cooling, and sanitizing.

How a given investment in reproduction is allocated, the number and size of offspring produced, and the care and defense provided all interact in the context of the environment to determine the return to the individual in terms of increased fitness (see Quantifying Ecology 10.1).

10.8 Species Differ in the Timing of Reproduction

How should an organism invest its allocation to reproduction through time? Thus far we have focused on age-structured populations in which reproduction begins with the onset of maturity and continues over some period of time until either reproduction ceases (postreproductive period) or senescence occurs. Organisms that produce offspring more than once over their lifetime are called **iteroparous**. Iteroparous organisms include most vertebrates, perennial herbaceous plants, shrubs, and trees. As we have explored, the timing of reproduction for iteroparous species involves trade-offs. Early reproduction means earlier maturity, less growth, reduced fecundity per reproductive period, reduced survivorship, and reduced potential for future reproduction. Later reproduction means increased growth, later maturity, and increased survivorship but less time for reproduction. In effect, to maximize contributions to future generations, an organism balances the benefits of immediate reproduction and future reproductive prospects, including the cost to fecundity (total offspring produced) and its own survival (Section 10.6).

Another approach to reproduction is to initially invest all energy in growth, development, and energy storage, followed by one massive reproductive effort, and then death. In this strategy, an organism sacrifices future prospects by expending all its energy in one suicidal act of reproduction. Organisms exhibiting this mode of reproduction are called **semelparous**.

Semelparity is employed by most insects and other invertebrates, by some species of fish (notably, salmon), and by many plants. It is common among annuals, biennials, and some species of bamboos. Many semelparous plants, such as ragweed (*Ambrosia* spp.), are small, short-lived, and found in ephemeral or disturbed habitats. For them, it would not be efficient, fitness-wise, to hold out for future reproduction because their chances of success are slim. They gain maximum fitness by expending all their energy in one bout of reproduction.

Other semelparous organisms, however, are long-lived and delay reproduction. Mayflies (*Ephemeroptera*) may spend several years as larvae before emerging from the surface of the water for an adult life of several days devoted to reproduction. Periodical cicadas spend 13 to 17 years belowground before they emerge as adults to stage a single outstanding exhibition of reproduction. Some species of bamboo delay flowering for 100 to 120 years, produce one massive crop of seeds, and die. Hawaiian silverswords (*Argyroxiphium* spp.) live 7 to 30 years before flowering and dying. In general, the fitness of species that evolved semelparity must increase enough to compensate for the loss of repeated reproduction.

As we have seen, optimal reproductive effort per unit time (per reproductive event) is the balance between current and future reproduction that functions to maximize the individual's (parental) fitness. Within this framework, semelparity implies that one single maximum reproductive effort followed by death represents the optimal strategy for the individual in the context of its environment (external constraints). It follows that iteroparity evolved through a change in conditions

QUANTIFYING ECOLOGY 10.1 Interpreting Trade-offs

Many of the life history characteristics discussed in this chapter involve trade-offs, and understanding the nature of trade-offs involves the analysis of costs and benefits for a particular trait.

One trade-off in reproductive effort discussed in Section 10.7 involves the number and size of offspring produced. The graph in **Figure 1** is similar to the one presented in Figure 10.15, showing the trade-off relationship between seed size and the number of seeds produced per plant. The example assumes a fixed allocation (100 units); therefore, the number of seeds produced per plant declines with increasing seed size.

Based on this information alone, it would appear that the best strategy for maximizing reproductive success would be to produce small seeds, thereby increasing the number of offspring produced. However, we must also consider any benefits to reproductive success that might vary as a function of seed size. The reserves of energy and nutrients associated with large seed size have been shown to increase the probability of successful establishment, particularly for seedlings in low-resource environments. For example, J. A. Ramírez-Valiente of the Center for Forestry Research (CIFOR: Madrid, Spain) found that average seed size in local populations of cork oak (*Quercus suber*) in Spain increases with decreasing precipitation (**Figure 2**), and that increased seed size was positively related to seedling survival under dry conditions (decreasing precipitation) (also see an example of the relationship between seed size and seedling survival in shade in Chapter 6, Field Studies: Kaoru Kitajima). A generalized relationship between seed size and seedling survival for two different environments (wet and dry) is plotted in **Figure 3**. In both environments, survival increases with seed size; however, in dry environments, the probability of survival declines dramatically with decreasing seed size.

By multiplying the number of seeds produced by the probability of survival, we can now calculate the expected reproductive success (the number of surviving offspring produced per plant) for plants producing seeds of a given size in both the wet and dry environments (**Figure 4**).

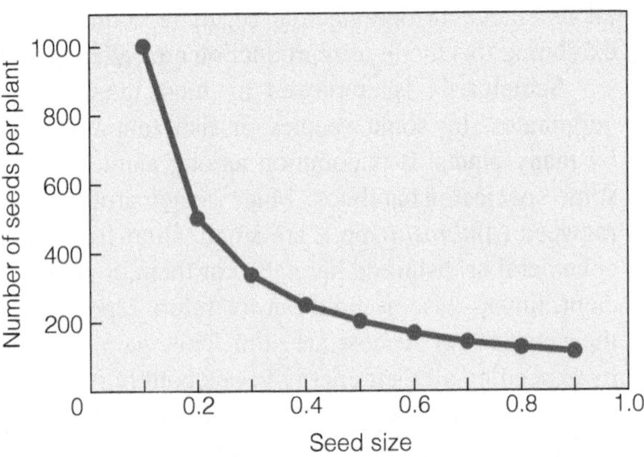

Figure 1

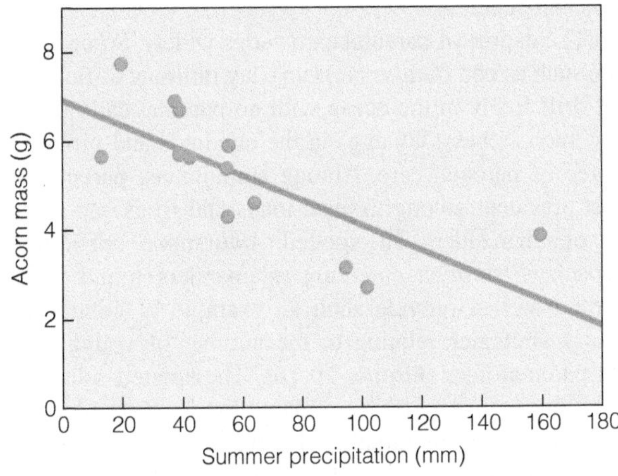

Figure 2 Relationships between average seed size (acorn mass) and summer precipitation for 13 populations of cork oak (*Quercus suber*) in Spain.

(Based on J.A. Rami ́rez-Valiente, F. Valladare, L. Gil and I. Aranda. 2009. Population differences in juvenile survival under increasing drought are mediated by seed size in cork oak (Quercus suber L.). Forest Ecology and Management 257:1676–1683 Fig. 2b, pg. 1679.)

such that less than the maximum possible reproductive effort is optimal on the first reproductive attempt and the organism survives to reproduce during future time intervals. What type of change in conditions might bring about the shift from semelparity to iteroparity? If the external environment imposes a high adult mortality relative to juvenile mortality, and if individuals reach maturity, chances are that they will not survive much longer; therefore, future reproductive expectations are bleak. Under these conditions, semelparity would be favored. If the opposite holds true and juvenile mortality is high compared to adult mortality, an individual has a good chance of surviving into the future once it survives to maturity; hence, prospects of future reproduction are good. Under these conditions iteroparity is favored.

10.9 An Individual's Life History Represents the Interaction between Genotype and the Environment

Natural selection acts on phenotypic variation among individuals within the population and variation in life history characteristics, such as age at maturity, allocation to reproduction, and the average number and size of offspring produced, is common

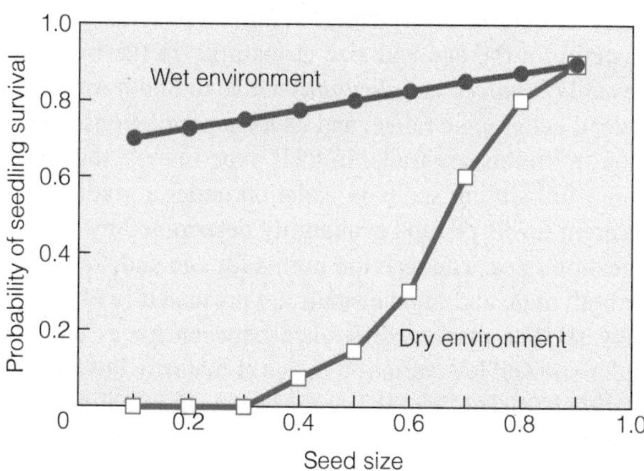

Figure 3

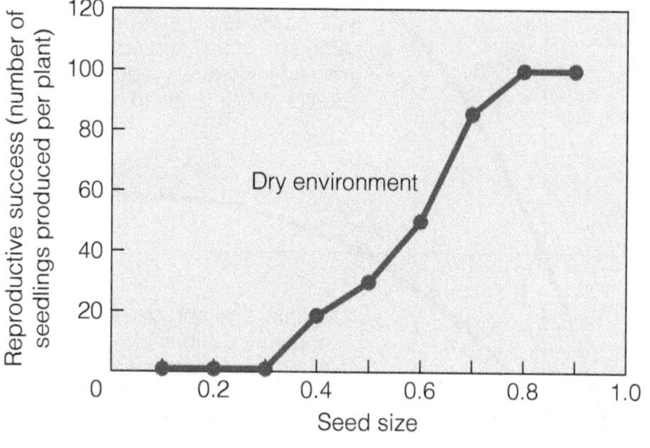

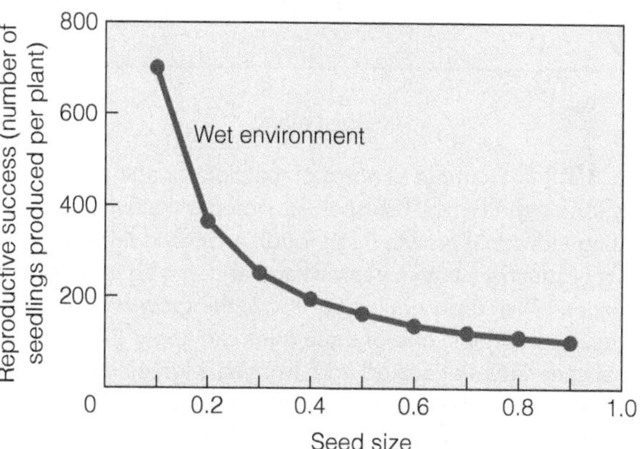

Figure 4

In wet environments, where all seed sizes have comparable probabilities of survival, the strategy of producing many small seeds results in the highest reproductive success and fitness. In contrast, the greater probability of survival makes the strategy of producing large seeds the most successful in dry environments, even though far fewer seeds are produced.

Interpreting the trade-offs observed in life history characteristics, such as the one illustrated between seed size and the number of seeds produced, requires understanding how those trade-offs function in the context of the environment (both biotic and abiotic) in which the species lives. Costs and benefits of a trait can change as the environmental conditions change. The diversity of life history traits exhibited by species is testimony that there is no single "best" solution for all environmental conditions.

1. In the example just presented, natural selection should favor plants producing small seeds in wet environments and plants that produce larger seeds in dry environments, resulting in a difference in average seed size in these two environments. What might you expect in an environment where annual rainfall is relatively high during most years (wet) but in which periods of drought (dry) commonly persist for several years?
2. The seeds of shade-tolerant plant species are typically larger than those of shade-intolerant species. How might this difference reflect a trade-off in life history characteristics relating to successful reproduction in sun and shade environments? See the discussion of shade tolerance in Chapter 6 and the Field Studies feature in that chapter.

among individuals within a population (see Chapter 5). The observed phenotypic variation within populations can arise from two sources: genotypic variation among individuals and interactions between the genotype and environment. Recall that most phenotypic traits are influenced by the environment; that is to say, the phenotypic expression of the genotype is influenced by the environment (see Chapter 5, Section 5.4). The ability of a genotype to give rise to different phenotypic expressions under different environmental conditions is termed *phenotypic plasticity*, and the set of phenotypes expressed by a single genotype across a range of environmental conditions is referred to as the *norm of reaction* (see Figure 5.4). Just as with the examples of phenotypic plasticity related to physiological, morphological, and behavioral characteristics involved in the thermal, energy, and water balance of plants and animals, the characteristics related to life history also exhibit reaction norms as a result of interactions between genes and environment (see Chapters 6 and 7). One life history trait that has received a great deal of focus regarding response to environmental variation is the relationship between age and size at maturity.

Let's begin by examining the expected patterns of size and age at maturity for a given genotype. **Figure 10.17** shows the graph of a growth curve for a hypothetical fish species that under the best of conditions can mature at two years of age and

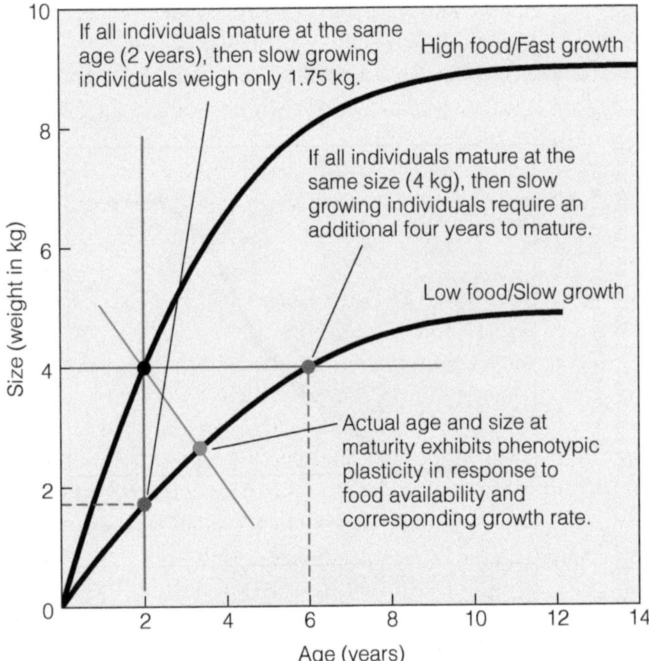

Figure 10.17 Example of norm of reaction for age and size at maturity for a hypothetical fish species. Assume that when food is not limiting individual growth (high food), an individual matures (begins reproduction) at two years of age and at 4 kg in weight (black circle). When food availability is low, the growth rate slows. Under these conditions, one of three rules can apply. First, the individual can mature at a fixed age: two years (green line and circle). As a result of the reduced growth rate because of a lack of food, the individual will only weigh 1.5 kg. If reproductive effort is related to body weight, this will result in a reduction in fecundity. Second, the individual can mature at a fixed size: 4 kg (blue line and circle). Under this rule, the individual will not mature until age six. Third, the individual may exhibit phenotypic plasticity in the relationship between size and age at maturity, exhibiting a compromise between age and size at maturity (red line and circle).

at a weight of 4 kilograms (kg). Now let us change the environmental conditions by reducing the availability of food. The result is a slower rate of growth (Figure 10.17). The question now becomes when would the initiation of reproduction (onset of maturity) maximize fitness for the slow-growing fish? There are three possible ways to go. First, individuals could always mature at the same size (blue line in Figure 10.17). The problem with this approach is that it now requires an additional four years to reach maturity, increasing the probability of mortality before the individual has the opportunity to breed (blue dashed line). The second approach is to always mature at the same age (green line in Figure 10.17). This approach also presents a downside; now the individual would weigh only 1.75 kg and smaller individuals produce fewer offspring (green dashed line). Somewhere between these two approaches is a compromise between the increased costs represented by the increased risk of mortality and that of reduced fecundity. The species could evolve to possess a norm of reaction for age and size at maturity (red line in Figure 10.17). The optimal solution for any growth rate would depend on the relationship between size and fecundity and the age-specific patterns of juvenile mortality.

Nicolas Tamburi and Pablo Martin of the Universidad Nacional del Sur in Argentina examined patterns of phenotypic plasticity in the age and size at maturity in the freshwater applesnail (*Pomacea canaliculata*) native to South America. It has a broad geographic range, and its local populations exhibit variation in life history traits. In their experiments, the researchers reared full sibling snails in isolation under a gradient of seven different levels of food availability determined by size-specific ingestion rates. The reaction norms for age and size at maturity for both male and female snails are presented in **Figure 10.18**. They show a marked difference between males and females. Males showed less variation in age at maturity but a wide variation in shell size. Size is largely irrelevant in gaining access to females, and male fitness can be maximized through fast maturation at the expense of size at maturity. In contrast, a minimum size is required for females to reach maturity, so there is a much greater variation in age at maturity rather than size. In both cases, the reaction norms reflect a trade-off between age and size at maturity. The differences between the reaction norms of males and females reflect basic differences in the trade-offs between the sexes as they relate to fitness.

10.10 Mating Systems Describe the Pairing of Males and Females

In all sexually reproducing species there is a social framework involving the selection of mates. The pattern of mating between males and females in a population is called the **mating system** (also see Chapter 5). The structure of mating systems in animal species ranges from **monogamy**, which involves the formation of a lasting pair bond between one male and one female, to **promiscuity**, in which males and females mate with one or many of the opposite sex and form no pair bond. The primary mating systems in plants are **outcrossing** (cross-fertilization in which pollen from one individual fertilizes the ovum of another) and **autogamy** (self-fertilization). However, a mixed mating system, in which plants use both outcrossing and autogamy, is common.

The mating system of a species has direct relevance to its life history because it influences allocation to reproduction, particularly in males. Competition among males for mates, courtship behavior, territorial defense, and parental care (feeding and protection of offspring) can represent a significant component of reproductive allocation. In addition, we shall see that the degree of parental care differs among mating systems and parental care has a direct effect on offspring survival. As such, a mating system is both influenced by and influences age-specific patterns of fecundity and mortality.

Monogamy is most prevalent among birds and rare among mammals, except for several carnivores, such as foxes (*Vulpes* spp.) and weasels (*Mustela* spp.), and a few herbivores, such as the beaver (*Castor* spp.), muskrat (*Ondatra zibethica*), and

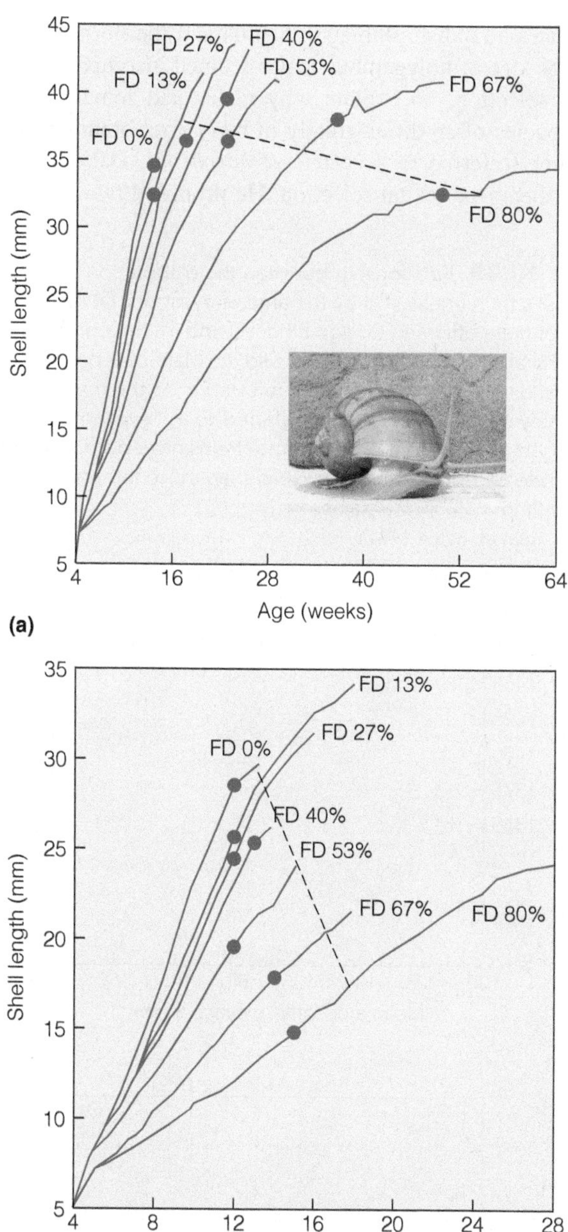

Figure 10.18 Mean growth curves for (a) females and (b) males of the freshwater applesnail *Pomacea canaliculata* grown under a gradient of food deprivation (FD; from FD 0 percent to FD 80 percent). Dots indicate the age and size at sexual maturity. The dashed line represents the general trend in the norm of reaction—phenotypic plasticity in the age and size at maturity in response to food availability (interaction between genotype and environment). (Adapted from Tamburi and Martin 2009.)

prairie vole (*Microtus ochrogaster*). Monogamy exists mostly among species in which cooperation by both parents is needed to raise the young successfully. Most species of birds are seasonally monogamous (during the breeding season) because most young birds are helpless at hatching and need food, warmth, and protection. The avian mother is no better suited than the father to provide these needs. Instead of seeking other mates, the male can increase his fitness by continuing his investment in the young. Without him, the young carrying his genes may not survive. Among mammals, the situation is different. The females lactate (produce milk), which provides food for the young. Males often contribute little or nothing to the survival of the young, so it is to their advantage in terms of fitness to mate with as many females as possible. Among the mammalian exceptions, the male provides for the female and young and defends the territory (area defended for exclusive use and access to resources; see Section 11.10 for discussion). Both males and females regurgitate food for the weaning young.

Monogamy, however, has another side. Among many species of monogamous birds, such as bluebirds (*Sialia sialis*), the female or male may "cheat" by engaging in extra-pair copulations while maintaining the reproductive relationship with the primary mate and caring for the young. By engaging in extra-pair relationships, the female may increase her fitness by rearing young sired by two or more males. The male increases his fitness by producing offspring with several females.

Polygamy is the acquisition of two or more mates by one individual. It can involve one male and several females or one female and several males. A pair bond exists between the individual and each mate. The individual having multiple mates—male or female—is generally not involved in caring for the young. Freed from parental duty, the individual can devote more time and energy to competition for more mates and resources. The more unevenly such crucial resources as food or quality habitat are distributed, the greater the opportunity for a successful individual to control the resource and several mates.

The number of individuals of the opposite sex an individual can monopolize depends on the degree of synchrony in sexual receptivity. For example, if females in the population are sexually active for only a brief period, as with the white-tailed deer, the number a male can monopolize is limited. However, if females are receptive over a long period of time, as with elk (*Cervus elaphus*), the size of a harem a male can control depends on the availability of females and the number of mates the male has the ability to defend.

Environmental and behavioral conditions result in various types of polygamy. In **polygyny**, an individual male pairs with two or more females. In **polyandry**, an individual female pairs with two or more males. Polyandry is interesting because it is the exception rather than the rule. This system is best developed in three groups of birds: the jacanas (Jacanidae; **Figure 10.19**), phalaropes (*Phalaropus* spp.), and some sandpipers (Scolopacidae). The female competes for and defends resources essential for the male and the males themselves. As in polygyny, this mating system depends on the distribution and defensibility of resources, especially quality habitat. The female produces multiple clutches of eggs, each with a different male. After the female lays a clutch, the male begins incubation and becomes sexually inactive.

The nature and evolution of male–female relationships are influenced by environmental conditions, especially the availability and distribution of resources and the ability of

Figure 10.19 An example of polyandry. The male African jacana (*Actophilornis africanus*) is shown defending the young. After the female lays a clutch, the male incubates the eggs and cares for the young while the female seeks additional mates.

individuals to control access to resources. If the male has no role in feeding and protecting the young and defends no resources available to them, the female gains no advantage by remaining with him. Likewise, the male gains no increase in fitness by remaining with the female. If the habitat were sufficiently uniform, so that little difference existed in the quality of territories held by individuals, selection would favor monogamy because female fitness in all habitats would be nearly the same. However, if the habitat is diverse, with some parts more productive than others, competition may be intense, and some males will settle on poorer territories. Under such conditions, a female may find it more advantageous to join another female in the territory of the male defending a rich resource than to settle alone with a male in a poorer territory. Selection under those conditions will favor a polygamous relationship, even though the male may provide little aid in feeding the young.

10.11 Acquisition of a Mate Involves Sexual Selection

For females, the production and care of offspring represents the largest component of reproduction expenditure. For males, however, the acquisition of a mate is often the major energetic expenditure that influences fitness.

The flamboyant plumage of the peacock (**Figure 10.20c**) presented a troubling problem for Charles Darwin. Its tail feathers are big and clumsy and require a considerable allocation of energy to grow. They are also conspicuous and present a hindrance when a peacock is trying to escape predators. In the theory of natural selection, what could account for the peacock's tail? Of what possible benefit could it be (see Chapter 5)?

In his book *The Descent of Man and Selection in Relation to Sex,* published in 1871, Darwin observed that the elaborate and often outlandish plumage of birds and the horns, antlers, and large size of polygamous males seemed incompatible with natural selection. To explain why males and females of the same species often differ greatly in body size, ornamentation, and color (referred to as *sexual dimorphism*), Darwin developed a theory of sexual selection. He proposed two processes

Figure 10.20 Relationship between the elaboration of the father's train (measured by the area of eyespots) (a) mean weight of male offspring (at age 84 days) and (b) their probability of survival after introduction to the wild. (c) Male peacock in courtship display. The tail feathers, referred to as the train, are elaborately colored with marking referred to as "eyespots." The offspring of males with more elaborate trains have both a higher growth rate and probability of survival than those fathered by males with less elaborate trains.
(Adapted from M. Petrie 1994.)

(a)

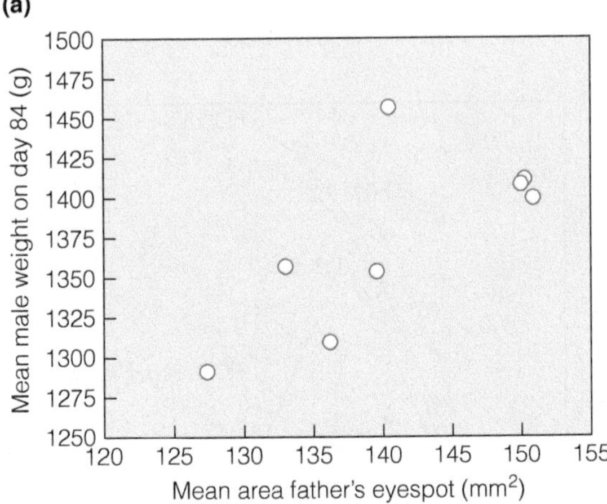

(b)

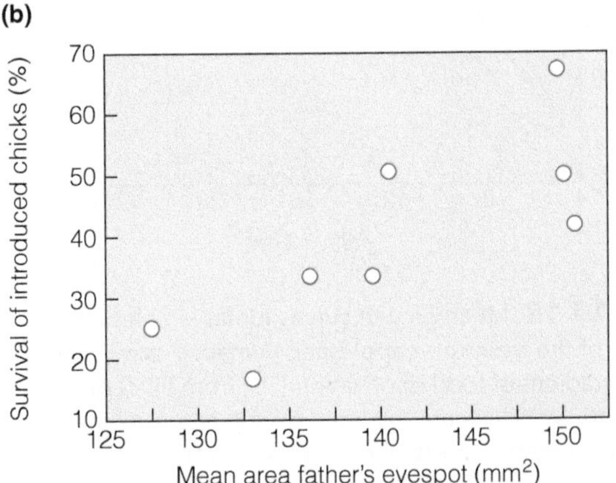

(c)

to account for these differences between the sexes: intrasexual selection and intersexual selection.

Intrasexual selection involves male-to-male (or in some cases, female-to-female) competition for the opportunity to mate. It leads to exaggerated secondary sexual characteristics such as large size, aggressiveness, and organs of threat, such as antlers and horns (**Figure 10.21**), that aid in competition for access to mates.

Intersexual selection involves the differential attractiveness of individuals of one sex to another (see this chapter, **Field Studies: Alexandra L. Basolo**). In the process of intersexual selection, the targets of selection are characteristics in males such as bright or elaborate plumage, vocalizations used in sexual displays, and the elaboration of some of the same characteristics related to intrasexual selection (such as horns and antlers). It is a form of assortative mating in which the female selects a mate based on specific phenotypic characteristics (see Section 5.7). There is intense rivalry among males for female attention. In the end, the female determines the winner, selecting an individual as a mate. The result is an increase in relative fitness for those males that are chosen, shifting the distribution of male phenotypes in favor of the characteristics on which female choice is based (see Chapter 5). But do characteristics such as bright coloration, elaborate plumage, vocalizations, or size really influence the selection of males by females of the species?

Marion Petrie of the University of Newcastle, England, conducted some experiments to examine intersexual selection in peacocks (*Pavo cristatus*). She measured characteristics of the tail feathers (referred to as the *train*) of male peacocks chosen by females as mates over the breeding season. Her results show that females selected males with more elaborate trains. In particular, she found a positive correlation between the number of eyespots a male had on his train (see Figure 10.20c) and the number of females he mated with. She then altered the tail feathers from a group of males with elaborate trains and found that reduction in the number of eyespots led to a reduction in mating success.

However, the train itself is not what is important; it is what the elaborate tail feathers imply about the individual. The large, colorful, and conspicuous tail makes the male more vulnerable to predation, or in many other ways, reduces the male's probability of survival. A male that can carry these handicaps and survive shows his health, strength, and genetic superiority. Females showing preference for males with elaborately colored tail feathers produce offspring that carry genes for high viability. Thus, the selective force behind the evolution of exaggerated secondary sexual characteristics in males is preferred by females. In fact, later experiments by Petrie found that the offspring of female peacocks that mated with males having elaborate tail feathers had higher rates of survival and growth than did the offspring of those paired with males having less elaborate trains (**Figures 10.20a** and **10.20b**). A similar mechanism may be at work in the selection of male birds with bright plumage. One hypothesis proposes that only healthy males can develop bright plumage. There is evidence from some species that males with low parasitic infection have the brightest plumage. Females selecting males based on differences in the brightness of plumage are in fact selecting males that are the most disease resistant (for example, see Section 15.7).

In some animal species, male vocalizations play an important role in courtship behavior, and numerous studies have found evidence of female mate choice based on the complexity of a male's song. In aviary studies, Myron Baker and colleagues at the University of Trondheim (Norway) found that female great tits (*Parus major*) were more receptive of males with more varied or elaborate songs (**Figure 10.22**).

In a 20-year study of song sparrows (*Melospiza melodia*) inhabiting Mandarte Island, British Columbia (Canada), Jane M. Reid of Cambridge University (England) and colleagues found that males with larger song repertoires were more likely to mate and that repertoire size predicted overall measures of male and offspring fitness. Males with larger song repertoires contributed more independent offspring—those hatching and reaching independence from parental care—and recruited offspring into the breeding population on the island; furthermore,

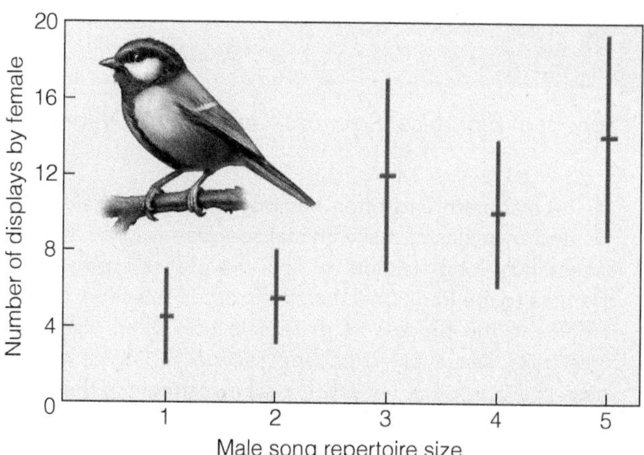

Figure 10.22 Mean (± standard error) number of copulation–solicitation displays given by 11 female great tits (*Parus major*) as a function of the complexity of the male song. Complexity of song was measured in terms of repertoire size, ranging from one to five song types. Females responded more frequently to male courtship songs that included a greater number of song types.
(Adapted from Baker et al. 1986.)

Figure 10.21 This bull elk is bugling a challenge to other males in a contest to control a harem.

FIELD STUDIES Alexandra L. Basolo

School of Biological Sciences, University of Nebraska–Lincoln

The elaborate and often flamboyant physical traits exhibited by males of many animal species—bright coloration, exceedingly long feathers or fins—have always presented a dilemma to the traditional theory of natural selection. Because females in the process of mate selection often favor these male traits, sexual selection (see Section 10.11) will reinforce these characteristics. However, male investment in these traits may also reduce the amount of energy available for other activities that are directly related to individual fitness, such as reproduction, foraging, defense, predator avoidance, and growth. The effect of such trade-offs in energy allocation on the evolution of animal traits is the central focus of ecologist Alexandra L. Basolo's research, which is changing how behavioral ecologists think about the evolution of mate selection.

Basolo's research focuses on the small freshwater fishes of the genus *Xiphophorus* that inhabit Central America. One group of species within this genus, the swordtail fish, exhibits a striking sexual dimorphism in the structure of the caudal fin. Males have a colorful, elongated caudal appendage, which is termed the *sword* (**Figure 1**), which is absent in females. This appendage appears to play no role other than as a visual signal to females in the process of mate selection. To test the hypothesis that this trait results partly from female choice (intersexual selection), Basolo undertook a series of laboratory experiments to determine if females exhibited a preference for male sword length. Her test subject was the green swordtail, *Xiphophorus helleri,* shown in Figure 1. These experiments allowed females to choose between a pair of males differing in sword length. Five tests with different pairs of males were conducted in which the sword differences between paired males varied. Female preference was measured by scoring the amount of time a female spent in association with each male.

Results of the experiments revealed that females preferred males with longer swords. The greater the difference in sword length between two males, the greater was the difference in time that the female spent with them (**Figure 2**). The results suggest that sexual selection through female choice will influence the relative fitness of males. The benefit of having a long sword is the increased probability of mating. But what is the cost? Locomotion accounts for a large part of the energy budget of fish, and the elongated caudal fin (sword) of the male swordtails may well influence the energetic cost of swimming. The presence of the sword increases mating success (via female choice) but may well negatively affect swimming activities.

To evaluate the costs associated with sword length, Basolo undertook a series of experiments using another species of swordtail, the Montezuma swordtail (*Xiphophorus montezumae*) found in Mexico. Like the green swordtail, the

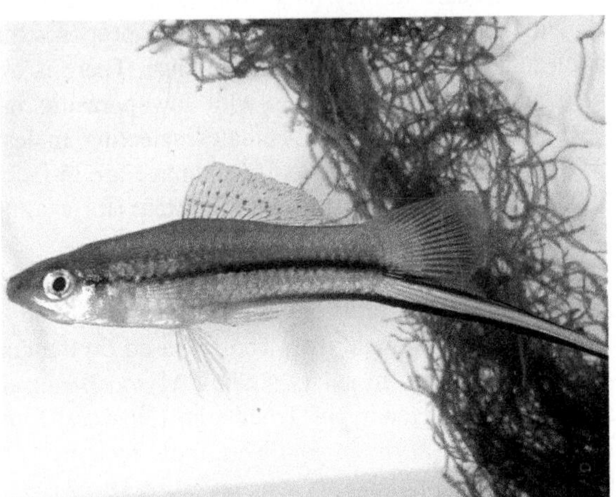

Figure 1 The green swordtail, *Xiphophorus helleri.* Males (shown in the photo) have a colorful, elongated appendage on the lower caudal fin (tail), which is termed the *sword* and functions as a visual cue to attract females during courtship.

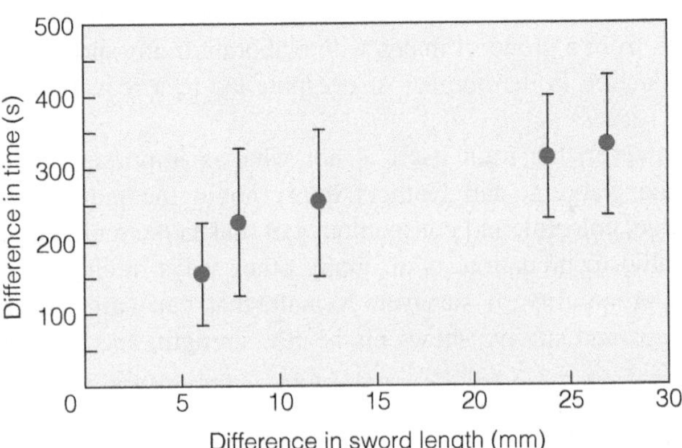

Figure 2 Relationship between the difference in sword length between the two test males and the difference in the time spent with the male having the longer sword. (Adapted from Basolo 1990b.)

Montezuma swordtail males have an asymmetric caudal fin as a result of an extended sword, and the presence of this sword increases mating success. The experiments were designed to quantify the metabolic costs of the sword fins during two types of swimming—routine and courtship—for males with and without sword fins. Males having average-length sword fins were chosen from the population. For some of these males, the sword was surgically removed (excised). Comparisons were then made between males with and without swords for both routine (no female present) and courtship (female present) swimming. Male courtship behavior involves a number of active maneuvers. Routine swimming by males occurs in the absence of females, whereas the presence of females elicits courtship-swimming behavior.

Basolo placed test males into a respirometric chamber—a glass chamber instrumented to measure the oxygen content of the water continuously. For a trial where a female was present, the female was suspended in the chamber in a cylindrical glass tube having a separate water system. During each trial, water was sampled from the chamber for oxygen content to determine the rate of respiration. Higher oxygen consumption indicates a higher metabolic cost (respiration rate).

Results of the experiments show a significant energy cost associated with courtship behavior (**Figure 3**). A 30 percent increase in net cost was observed when females were present for both groups (males with and without swords) as a result of increased courtship behavior. However, the energy cost for males with swords was significantly higher than that for males without swords for both routine and courtship swimming behavior (Figure 3). Thus, although sexual selection via female choice favors long swords, males with longer swords experience higher metabolic costs during swimming, suggesting that sexual and natural selection have opposing effects on sword evolution.

The cost of a long sword to male swordtails extends beyond the energetics of swimming. Other studies have shown that more conspicuous males are more likely to be attacked by predators than are less conspicuous individuals. In fact, Basolo and colleague William Wagner have found that in green swordtail populations that occur sympatrically (together) with predatory fish, the average sword length of males in the population is significantly shorter than in populations where predators are not present. These results suggest that although sexual selection favors longer swords, natural selection may have an opposing effect on sword length in populations with predators.

Despite the cost, both in energy and probability of survival, the sword fin of the male swordtails confers an advantage in the acquisition of mates that must offset the energy and survival costs in terms of natural selection.

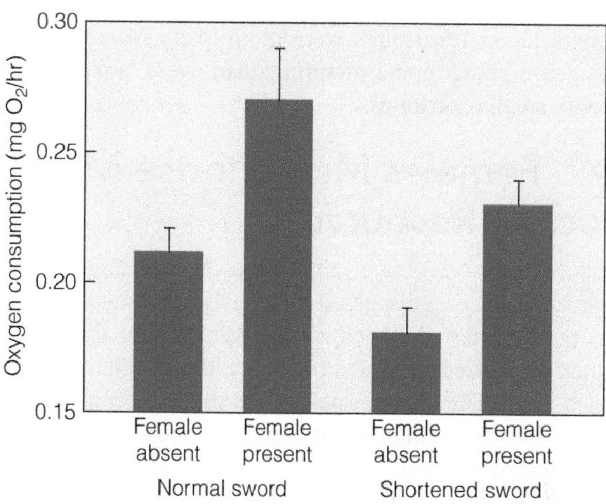

Figure 3 Mean oxygen consumption for males with intact (normal) swords and shortened (excised) swords in the absence and presence of females. Female-absent results represent routine swimming. Female-present results represent male-courtship swimming.
(Adapted from Basolo 2003.)

Bibliography

Basolo, A. L. 1990a. "Female preference predates the evolution of the sword in swordtail fish." *Science* 250:808–810.

Basolo, A. L. 1990b. "Female preference for male sword length in the green swordtail, *Xiphophorus helleri*." *Animal Behaviour* 40:332–338.

Basolo, A. L., and G. Alcaraz. 2003. "The turn of the sword: Length increases male swimming costs in swordtails." *Proceedings of the Royal Society of London* 270: 1631–1636.

Basolo, A. L., and W. E. Wagner Jr. 2004. "Covariation between predation risk, body size and fin elaboration in the green swordtail, *Xiphophorus helleri*." *Biological Journal of the Linnean Society* 83:87–100.

1. In her experiments examining the costs and benefits of swords on the caudal fins of male fish, did Basolo actually quantify differences in fitness associated with this characteristic?

2. Often, sexual selection favors characteristics that appear to reduce the probability of survival for the individual. Does this not run counter to the idea of natural selection presented in Chapter 5?

those males also contributed more independent and recruited grand-offspring to the island population (**Figure 10.23**). This was because these males lived longer and reared a greater proportion of hatched chicks to independence from parental care, not because females mated to males with larger repertoires laid or hatched more eggs. Furthermore, independent offspring of males with larger repertoires were more likely to recruit and then to leave more grand-offspring than were offspring of males with small repertoires.

10.12 Females May Choose Mates Based on Resources

A female exhibits two major approaches in choosing a mate. In the sexual selection discussed previously, the female selects a mate for characteristics such as exaggerated plumage or displays that are indirectly related to the health and quality of the male as a mate. The second approach is that the female bases her choice on the availability of resources such as habitat or food that improve fitness.

Numerous studies have shown that in some species, mate choice by females appears to be associated with the acquisition of resources, usually a defended high quality habitat or territory (see Section 11.10). Ethan Temeles of Amherst College (Amherst, Massachusetts) and J. John Kress of the National Museum of Natural History (Smithsonian Institution, Washington, D.C.) found that female purple-throated carib hummingbirds (*Eulampis jugularis*) on the island of Dominica in the West Indies preferred to mate with males that had high standing crops of nectar on their flower territories (**Figure 10.24**). A male's ability to maintain high nectar standing crops on his territory not only depended on the number of flowers in his territory but also on his ability to enhance his territory through the prevention of nectar losses to intruders.

Andrew Balmford and colleagues at Cambridge University (England) examined the distribution of females across male territories to assess mate choice in puku (*Kobus vardoni*) and topi (*Damaliscus lunatus*), which are two species of grazing antelope in southern Africa. In both species, males defend areas (territories) in which they have exclusive use of resources. Both species are polygamous, and visitation to territories by females was found to be a good predictor of where females tended to mate. In both species, female choice (visitation rate) was correlated with the quality of forage in different territories, indicating that female choice was influenced by the quality of defended resources.

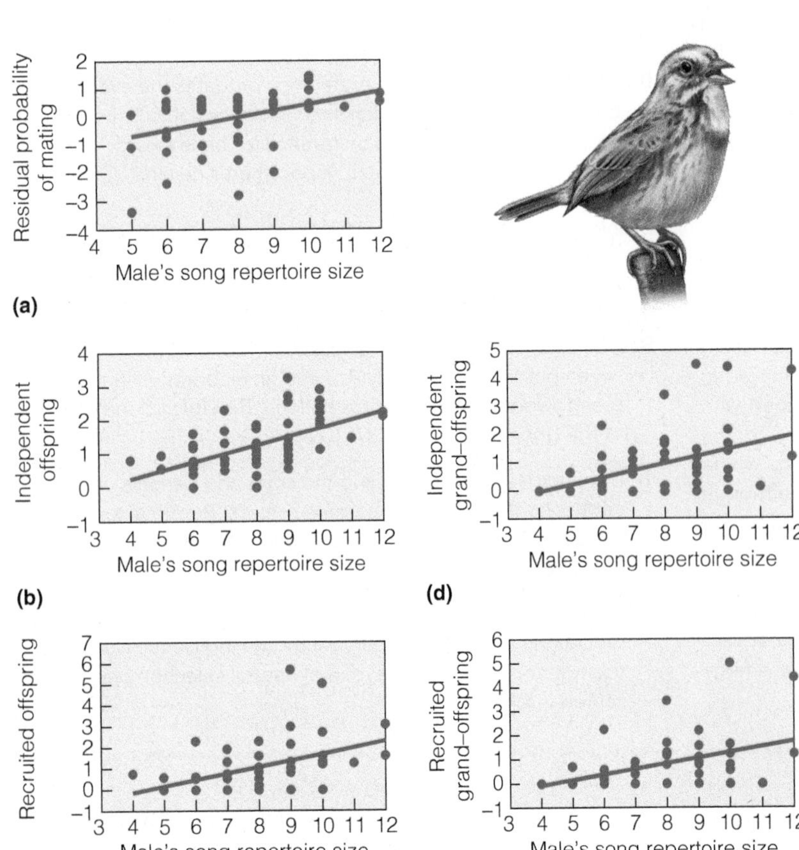

Figure 10.23 (a) Relationship between song repertoire size and probability of mating for male song sparrows (*Melospiza melodia*) inhabiting Mandarte Island, British Columbia, Canada. Probability of mating is measured as the residual probability of mating during the month of April after controlling for territory size (residuals from linear regression between probability of mating and territory size). Relationship between a male song sparrow's song repertoire size and relative fitness measures (b) total offspring raised to independence, (c) total offspring recruited into island population, (d) total independent grand-offspring, and (e) total grand-offspring recruited into island population.

([a] Based on Reid, Jane M., et al., Song repertoire size predicts initial mating success in male song sparrows, Melospiza Melodia, Animal Behaviour, 2004, 68, 1055–1063. Figure 2a pg. 1059.)

Interpreting Ecological Data

Q1. *In Figures 10.23b–e, the lines show the general trend of the measure of male fitness plotted on the y-axis with an increase in male song repertoire size (x-axis). Notice that the spread of points (values of y) for a given male song repertoire size increases as the repertoire size increases. (Males with small song repertoire size have low fitness, and males with large song repertoire size can either have low or high fitness.) However, if we focus on the maximum value fitness (y) for any repertoire size it is clear that maximum fitness increases with male song repertoire size. What might account for the spread of points (values of y) for a given male song repertoire size to increase as the repertoire size increases?*

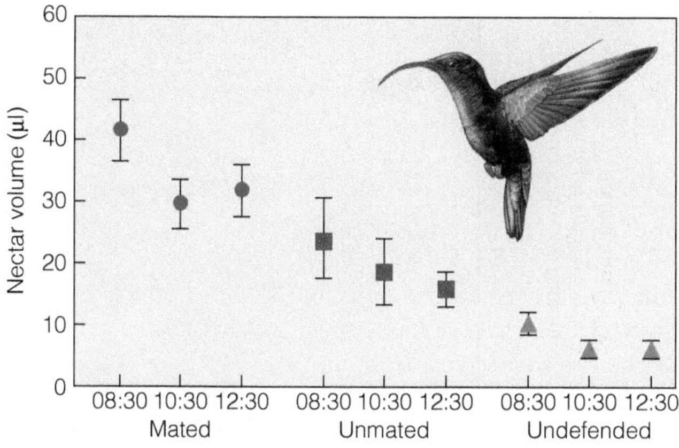

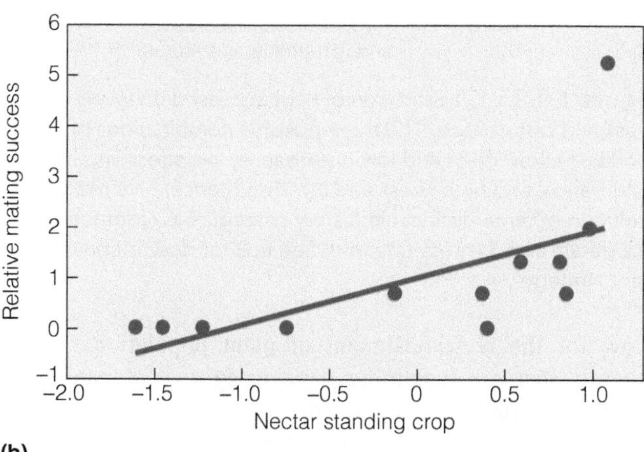

Figure 10.24 Relationship between resource availability of territories and mating success of territorial males of purple throated carib hummingbirds (*Eulampis jugularis*). (a) Mean standing crops of nectar (± standard error; μl/flower) on territories of mated, unmated, and undefended areas. (b) Relationship between nectar standing crop on territory and relative mating success (copulation rate per male/mean copulation rate). Both variables have been log(e) transformed.
(Adapted from Temeles and Kress 2009.)

10.13 Patterns of Life History Characteristics Reflect External Selective Forces

Nature presents us with a richness of form and function in the diversity of life that inhabits our planet. Some species are large, and others are small. Some mature early, and others mature later in their lives. Some organisms produce only a few offspring over their lifetime, whereas other species produce thousands in a single reproductive event. Some organisms fit an entire lifetime into a single season, and others live for centuries. Are the characteristics exhibited by any given species a random assemblage of these traits, or is there a discernable pattern? What we have seen so far is that these characteristics, which define the life history of an individual, are not independent of one another. These characteristics are products of evolution by natural selection, the possible outcomes molded by the external environment, and constrained by trade-offs relating to fundamental physiological and developmental processes. Ecologists have long recognized that the set of characteristics that define a species' life history covaries, forming what appear to be distinctive "suites" of characteristics that seem to be a product of broad categories of selective forces. A number of empirical models have developed to account for the observed covariation among life history traits. One such model is the fast–slow continuum hypothesis.

The fast–slow continuum hypothesis emphasizes the selective forces imposed by mortality at different stages of the life cycle. Under this scheme, species can be arranged along a continuum from those experiencing high adult mortality levels to those experiencing low adult mortality. This differential mortality is responsible for the evolution of contrasting life histories on either end of the continuum. Species undergoing high adult mortality are expected to have a shorter life cycle (longevity) with faster development rates, early maturity, and higher fecundity than those experiencing low adult mortality. This approach has proven accurate in predicting patterns of life history characteristics in many groups of species and is generally consistent with the patterns presented in the preceding sections.

Other approaches to understanding the observed correlations among life history traits have focused on constraints imposed by the abiotic environment. If the life history characteristics and mating system exhibited by a species are the products of evolution, would they not reflect adaptations to the prevailing environmental conditions under which natural selection occurred? If this is the case, do species inhabiting similar environments exhibit similar patterns of life history characteristics? Do life history characteristics exhibit patterns related to the habitats that species occupy?

One way of classifying environments (or species habitats) relates to their variability in time. We can envision two contrasting types of habitats: (1) those that are variable in time or short-lived and (2) those that are relatively stable (long-lived and constant) with few random environmental fluctuations. The ecologists Robert MacArthur of Princeton University, E. O. Wilson of Harvard University, and later E. Pianka of the University of Texas used this dichotomy to develop the concept of *r*- and *K*-selection.

The theory of *r*- and *K*-selection predicts that species adapted to these two different environments will differ in life history traits such as size, fecundity, age at first reproduction, number of reproductive events during a lifetime, and total life span. Species popularly known as ***r*-strategists** are typically short-lived. They have high reproductive rates at low population densities, rapid development, small body size, large number of offspring (with low survival), and minimal parental care. They make use of temporary habitats. Many inhabit unstable or unpredictable environments that can cause catastrophic mortality independent of population density. Environmental resources are rarely limiting. They exploit noncompetitive situations. Some *r*-strategists, such as weedy species, have means of wide dispersal, are good colonizers, and respond rapidly to disturbance.

***K*-strategists** are competitive species with stable populations of long-lived individuals. They have a slower growth rate at low populations, but they maintain that growth rate at high densities. *K*-strategists can cope with physical and biotic pressures. They possess both delayed and repeated reproduction and have a larger body size and slower development. They produce few seeds, eggs, or young. Among animals, parents care for the young; among plants, seeds possess stored food that gives the seedlings a strong start. Mortality relates more to density than to unpredictable environmental conditions. They are specialists—efficient users of a particular environment—but their populations are at or near carrying capacity (maximum sustainable population size) and are resource-limited. These qualities, combined with their lack of means for wide dispersal, make *K*-strategists poor colonizers of new and empty habitats.

The terms *r* and *K* used to characterize these two contrasting strategies related to the parameters of the logistic model of population growth (presented in Chapter 11); *r* is the per capita rate of growth, and *K* is the carrying capacity (maximum sustainable population size). Using the classification of *r* and *K*, strategies for comparing species across a wide range of sizes is of limited value. For example, the correlation among body size, metabolic rate, and longevity in warm-blooded organisms results in species with small body size generally being classified as *r* species and those with large body size as *K* species (see Chapter 7). The concept of *r* species and *K* species is most useful in comparing organisms that are either taxonomic or functionally similar.

The plant ecologist J. Phillip Grime of the University of Sheffield, England, used a framework similar to that used by MacArthur and Wilson to develop a life history classification for plants. Grime's life history classification of plants is based on the assumption that any habitat can be classified into one of two categories: stress and disturbance. Stress is defined as conditions that restrict plant growth and productivity, such as shortages of light, water, mineral nutrients, or suboptimal temperatures (see Chapter 6). Disturbance is associated with the partial or total destruction of plant biomass that arises from the activity of herbivores, pathogens, or natural disasters such as wind, fire, or flooding. When the four permutations of high and low stress couple with high and low disturbance, it is apparent that only three are suitable as habitat for plants, because in highly disturbed environments, stress does not

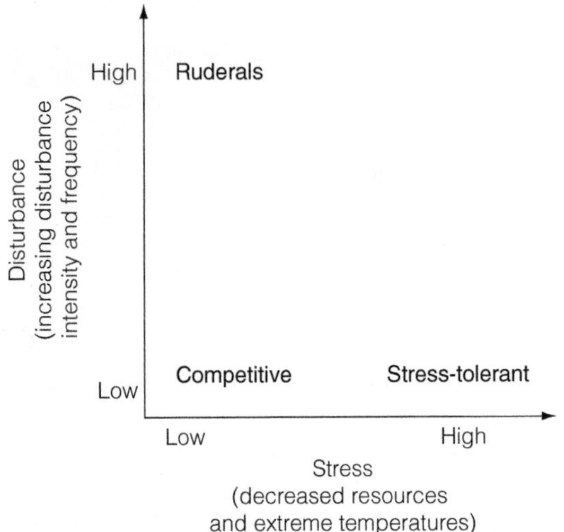

Figure 10.25 Classification of habitats based on levels of stress and disturbance. The three possible combinations of habitat conditions (low stress and low disturbance, low stress and high disturbance, and high stress and low disturbance) give rise to the evolution of three distinct life history strategies: *C* (competitive), *R* (ruderal), and *S* (stress-tolerant). See text for description of each plant strategy.

allow for the reestablishment of plant populations. Grime suggests that the remaining three categories of habitat are associated with the evolution of distinct types of plant life history strategies—*R*, *C*, and *S* (**Figure 10.25**). Species exhibiting the *R*, or ruderal, strategy rapidly colonize disturbed sites. These species are typically small in stature and short-lived. Allocation of resources is primarily to reproduction, with characteristics allowing for a wide dispersal of seeds to newly disturbed sites. Predictable habitats with abundant resources favor species that allocate resources to growth, favoring resource acquisition and competitive ability (*C* species). Habitats in which resources are limited favor stress-tolerant species (*S* species) that allocate resources to maintenance. These three strategies form the end points of a triangular classification system that allows for intermediate strategies, depending on resource availability (levels of stress) and frequency of disturbance (**Figure 10.26**).

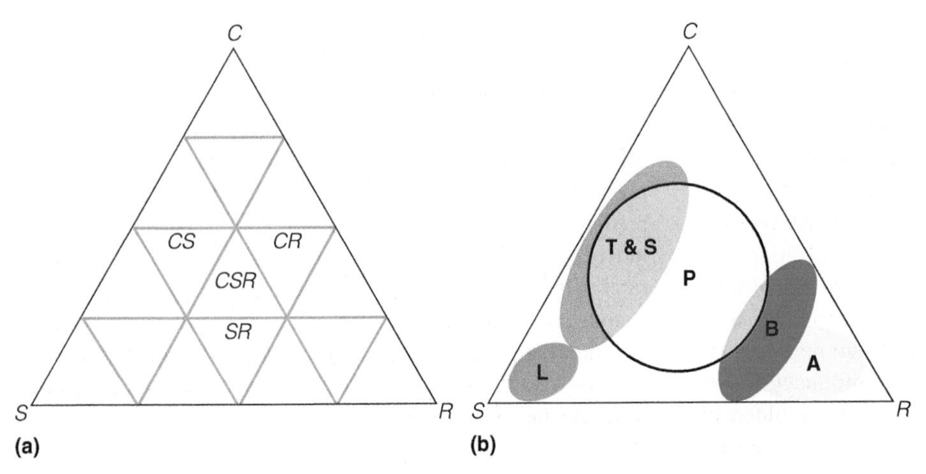

Figure 10.26 Grime's model of life history variation in plants based on three primary strategies: ruderals (*R*), competitive (*C*), and stress-tolerant (*S*). (a) These primary strategies define the three points of the triangle. Intermediate strategies are defined by combinations of these three (e.g., *CS*, *CR*, *CSR*, and *SR*). (b) Grime's assessment of life history strategies of most trees and shrubs (T & S), lichens (L), biennial herbs (B), perennial herbs (P), and annual herbs (A). (Adapted from Grime, J.P. "Evidence for the existence of three primary strategies in plants and its relevance to ecological and evolutionary theory," American Naturalist, Vol. 111, 1977.)

ECOLOGICAL Issues & Applications

The Life History of the Human Population Reflects Technological and Cultural Changes

The history of the human population presents what appears to be a classic example of exponential population growth, yet on closer inspection, what emerges is a story of a species that has redefined its life history through a series of technological, cultural, and economic changes over the past two centuries.

With the end of the last glacial period (~18,000 BP) and the development of agriculture some 10,000 years ago, human demographers estimate that the human population was approaching 5 million. By 1 AD the population had risen to approximately 250 million, and it would take until the beginning of the 19th century before that number would reach a billion. By the 19th century, the human population entered a period of expansive growth, rising to 2 billion by 1930. Adding the next billion would take only 30 years (1960), and on October 31, 2011, the United Nations officially declared that the human population had reached 7 billion.

So what are the prospects for the future? The United Nations' projection of future population growth shows the global population continuing to expand over the next several decades before peaking near 10 billion later in the 21st century. Although this may appear as an astonishingly large number, it represents a significant decline in population growth rates moving forward and the possibility of population numbers stabilizing in the foreseeable future.

When combined with projections of future growth over the next century, it becomes apparent the human population is not following a continuous pattern of exponential growth. Rather, the graph of the human population presented in **Figure 10.27** suggests three distinct periods, or phases, of population growth in modern time (19th century forward). In phase 1, the period before the early 20th century, population growth is slow and steady. By the early 20th century, however, Phase 2 began, which was a period of dramatic exponential growth. This period of growth continued until the latter part of the 20th century when the population growth rate began to slow. We have now entered Phase 3 as the population growth rate declines and the population potentially peaks at 10 billion.

What has caused these three phases? Why did the population growth rate explode in the early 20th century, and what caused it to decline as the 20th century came to a close? These three phases of population growth are the central components of what human demographers—ecologists who study the human population—refer to as the **demographic transition**. The demographic transition describes the transition from high birthrates and death rates to low birthrates and death rates as countries move from a preindustrial to an industrialized social and economic system (**Figure 10.28**).

Phase 1 is associated with premodern times and is characterized by a balance between high rates of birth and death. This was the situation of the human population before the late 18th century. This balance between birthrate and death rate resulted in a slow growth rate (<0.05 percent). Death rates were high because of the lack of sanitation, knowledge of disease prevention and cures, and occasional food shortages (usually climate related). The infant mortality rate in the United Kingdom and the United States during the 18th century was as high as 500 per 1000, or one in every two infants born. With high child mortality rates, there was little incentive in rural societies to control fertility.

By the early 19th century death rates began to decline, first in Europe and then in other countries, over the next 100 years. The decline in death rates would lead to Phase 2 of the transition characterized by exponential growth as the population growth rate rose (difference between birthrate and death rate increased; see Figure 10.28).

The decline was a result of improved food supply and sanitation (particularly water supplies). This decline gained momentum in the early 20th century with significant improvements in public health. Improved sanitation and the identification of causes of and cures for infectious diseases led to a dramatic decline in death rates as the 20th century progressed. The greatest reduction in death rates was realized by children; infant mortality rates declined steadily in the 20th century (**Figure 10.29**). The reduced infant mortality rates had a swift and dramatic effect on population growth rates. Increases in life expectancy for older individuals (post-reproductive) has little effect on population growth rates; in contrast, increased survival rate of infants results in those individuals entering the reproductive ages and adding to overall population growth (see Section 9.7).

Phase 3 of the transition moved the population toward stability through a decline in birthrates. This phase began as

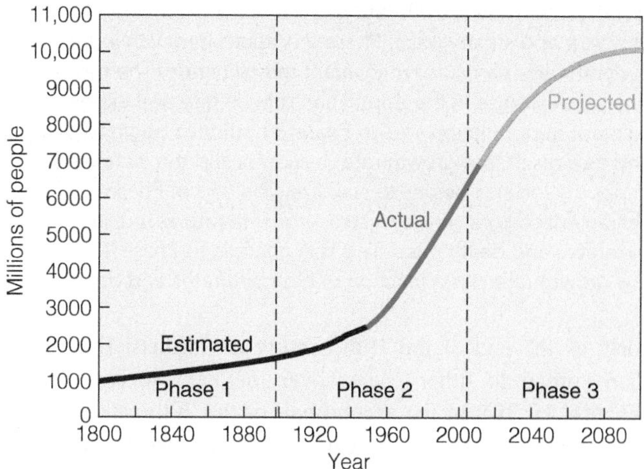

Figure 10.27 Pattern of human population growth for period of 1800–2100. Projected population estimates are based on United Nations' medium population scenario. The pattern of population growth can be divided into three distinct phases. Phase 1 is a period of slow and steady growth before the 20th century. Phase 2 is the onset of exponential growth over the 20th century, and Phase 3 begins a period of decline in growth rates and population stabilization.
(Data from United Nations.)

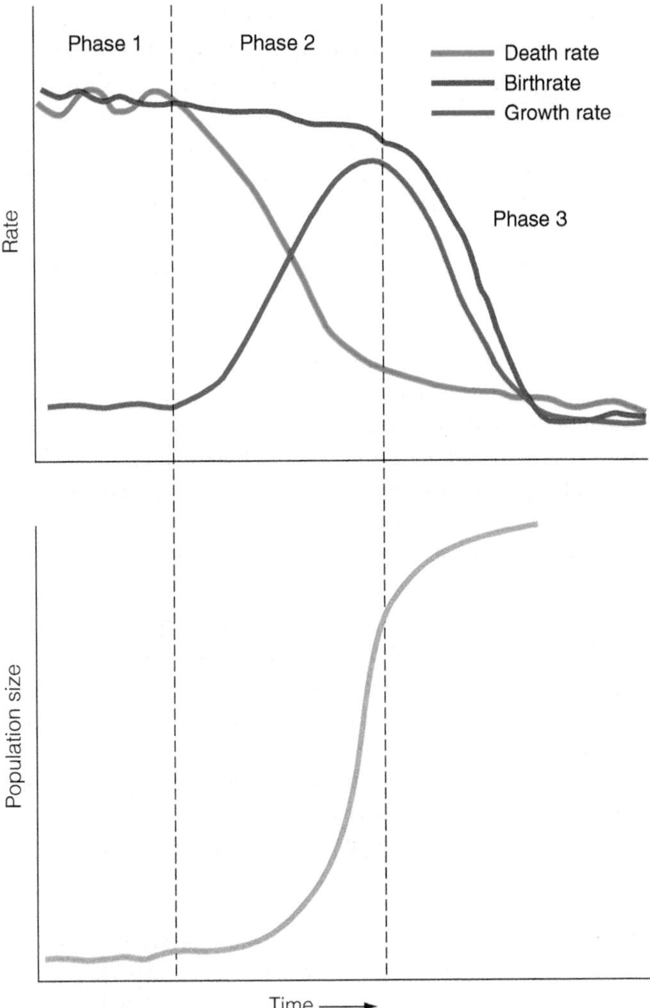

Figure 10.28 Representation of the demographic transition. During Phase1 both birthrates and death rates are high, and as a result, population growth rate is low and the population increases at a slow and steady pace. Phase 2 is associated with a decline in death rate, particularly the infant mortality rate. The result is a dramatic increase in the population growth rate and exponential increase in population size. In Phase 3 birthrates begin to decline, and as a result, the growth rate decreases and the increase in population size begins to stabilize. The end of Phase 3 is characterized by a slow and steady growth rate as a result of low birthrates and death rates. This is in contrast to Phase 1 where the low growth rate was a function of high birthrates and death rates.

early as the end of the 19th century in northern Europe and then spread to other places over the next several decades (**Figure 10.30**). In the second half of the 20th century, birthrates declined, and by the early 1960s, the world population growth rate peaked at more than 2 percent and has been declining ever since. The number of individuals added to the global population each year peaked in the 1990s.

There are a number of factors that contributed to this decline, although many of them remain speculative and are the focus of continued research by social scientists. In rural areas where children played an important role in farm life, the continued decline in infant mortality meant that at some point parents realized the need to control family size. Likewise, increased urbanization changed the traditional value of large family size that was essential to farm life. The increased role of women in the workforce and the improvements to contraception in the second half of the 20th century led to even further declines in birthrates, and by the early 1960s the world population growth rate had peaked at slightly more than 2 percent.

So begins Phase 3 of the transition. The global population growth rate has been declining since its peak in the 1960s and the number of new individuals added to the global population each year peaked in the 1990s. Demographic transition describes the patterns of human population growth for all regions of the planet, but the timing of the transition has differed for different regions. The more industrialized economies began the transition earlier, with many countries in Western Europe and

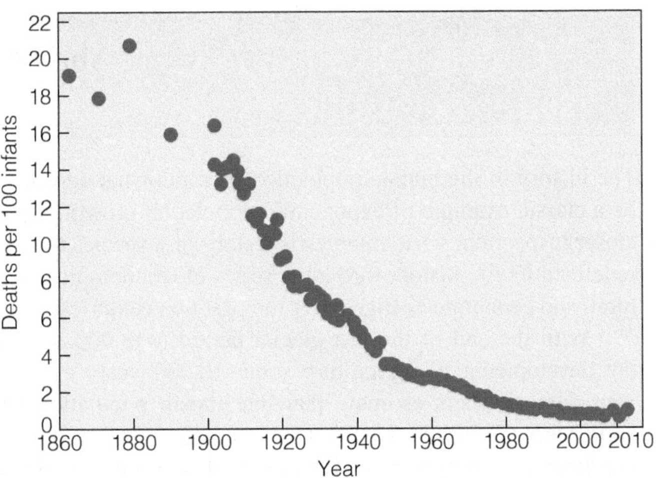

Figure 10.29 Infant mortality rates for the United States for the period of 1860–2012 expressed in deaths per 100 infants. (Data from U.S. Center for Disease Control.)

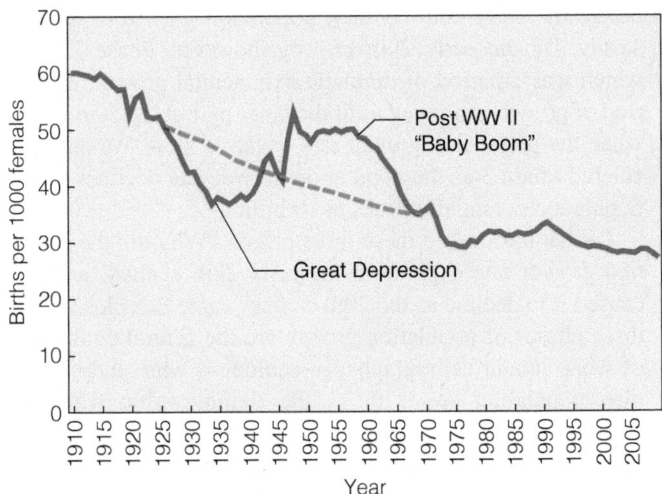

Figure 10.30 Birthrate in the United States over the period of 1909–2010. Birthrate expressed as births per 1000 females. The dashed red line follows the general downward trend over the 20th century for purposes of noting two important periods in United States history where birthrates deviated significantly from the overall trend.
(Data from U.S. Center for Disease Control.)

Asia, such as Poland, Germany, and Japan, now exhibiting a negative growth rate. In contrast, many of the developing countries of the world are still in the mid to latter phases of Phase 2, exhibiting growth rates that still exceed 2 percent.

From an ecological perspective, the amazing point of the demographic transition is that it represents a modification of the life history of our species, not as a result of natural selection but by means of "social evolution." Humans have dramatically altered age-specific patterns of birth and death through changes in technology and cultural changes that have occurred as the social structure has changed from rural agrarian to industrial urbanized society.

SUMMARY

Trade-offs 10.1

Organisms face trade-offs in life history characteristics related to reproduction. Trade-offs are necessitated by the constraints of physiology, energetics, and the prevailing physical and biotic environment. Trade-offs involve conflicting demands on resources or negative correlation among traits.

Asexual and Sexual Reproduction 10.2

Fitness is an organism's ability to leave behind reproducing offspring. Organisms that contribute the most offspring to the next generation are the fittest. Reproduction can be asexual or sexual. Asexual reproduction, or cloning, results in new individuals that are genetically the same as the parent. Sexual reproduction combines egg and sperm in a diploid cell, or zygote. Sexual reproduction produces genetic variability among offspring.

Forms of Sexual Reproduction 10.3

Sexual reproduction takes a variety of forms. Plants with separate males and females are called *dioecious*. An organism with both male and female sex organs is hermaphroditic. Plant hermaphrodites have bisexual flowers, or if they are monoecious, separate male and female flowers on the same individual. Some plants and animals change sex.

Benefits and Costs 10.4

The behavioral, physiological, and energetic activities involved in reproduction represent a cost to future reproductive success of the parent in the form of reduced survival, fecundity, or growth.

Age at Maturity 10.5

Natural selection favors the age at maturity that results in the greatest number of offspring produced over the lifetime of an individual. Environmental factors that result in reduced adult survival select for earlier maturation, and conversely, environmental factors that result in reduced juvenile survival relative to that of adults select for delayed maturation.

Reproductive Effort 10.6

Optimal reproductive effort represents a trade-off between current and future reproduction. Allocation to current reproduction functions to increase current fecundity but reduces parental survival, resulting in decreased future reproduction. Fitness is often maximized by an intermediate reproductive effort, particularly for organisms that reproduce repeatedly over their life spans.

Number and Size of Offspring 10.7

Organisms that produce many offspring have a minimal investment in each offspring. They can afford to send a large number into the world with a chance that a few will survive. By so doing, they increase parental fitness but decrease the fitness of the young. Organisms that produce few young invest considerably more in each one. The fitness of the young of such organisms is increased at the expense of the fitness of the parents.

Timing of Reproduction 10.8

To maximize fitness, an organism balances immediate reproductive efforts against future prospects. One alternative, semelparity, is the investment of maximum energy in a single reproductive effort. The other alternative, iteroparity, is the allocation of less energy to repeated reproductive efforts.

Reaction Norms 10.9

The characteristics related to life history, such as age at maturity, exhibit reaction norms (phenotypic plasticity) as a result of the interaction between genes and the environment.

Mating Systems 10.10

The pattern of mating between males and females in a population is the mating system. In animal species, mating systems range from monogamy to promiscuity.

Sexual Selection 10.11

In general, males compete with males for the opportunity to mate with females, but females finally choose mates. Sexual selection favors traits that enhance mating success, even if it handicaps the male by making him more vulnerable to predation. Male competition represents intrasexual selection, whereas intersexual selection involves the differential attractiveness of individuals of one sex to the other. By choosing the best males, females ensure their own fitness.

Resources and Mate Selection 10.12

Females may also choose mates based on the acquisition of resources, usually a defended territory or habitat. By choosing a male with a high-quality territory, the female may increase her fitness.

Life History Strategies 10.13

The set of characteristics that define a species' life history covary, forming what appear to be distinctive suites of characteristics that seem to be a product of broad categories of selective forces. A number of hypotheses have been developed to explain these patterns. The fast–slow continuum hypothesis says that species can be arranged along a continuum from high to low adult mortality. High adult mortality results in selection for a shorter life cycle, faster development rates, and higher fecundity than populations experiencing low adult mortality. Another hypothesis is based on two contrasting types of habitats: those that are variable in time or short-lived and those that are relatively stable. The former habitat type creates selection pressure for short life cycle, fast development, and high reproductive rates, and high fecundity, and the latter for longevity, delayed maturity, and lower reproductive effort distributed over a longer period of time.

Human Life History Ecological Issues & Applications

The history of the human population is described by the transition from high birthrates and death rates to low birthrates and death rates as countries move from a preindustrial to an industrialized social and economic system This dynamic is known as the *demographic transition* and represents a modification of the life history of our species, not as a result of natural selection, but by means of "social evolution."

STUDY QUESTIONS

1. What are some of the costs associated with reproduction, and how might they function to limit future reproduction by the individual?
2. How might an increase in adult (reproductive ages) mortality influence selection for age at maturity? Given your answer(s) to Question 1, what might be a consequence (cost) associated with this change in age at maturity?
3. What is the trade-off between the number and size of offspring produced for a given reproductive effort? How is this influenced by the level of parental care?
4. What type(s) of environment would favor plant species with a strategy of producing many small seeds rather than a few large ones?
5. What conditions favor semelparity over iteroparity?
6. Distinguish between monogamy and polygamy as well as between polygyny and polyandry.
7. How might female preference for a male trait (sexual selection), such as coloration or body size, drive selection in a direction counter to that of natural selection?
8. Contrast intrasexual selection and intersexual selection.
9. What is the difference between *r*-selected and *K*-selected organisms? Which strategy would you expect to be more prevalent in unpredictable environments (high stochastic variation in conditions)?
10. How have patterns of birthrates and death rates for the human population changed over the past two centuries, and how have these changes influenced patterns of population growth?

FURTHER READINGS

Classic Studies

Cole, L. C. 1954. "The population consequences of life history phenomena." *Quarterly Review of Biology* 29:103–137.
This classic work sets out the mathematical framework for examining the population consequences of variations in life history traits.

Lack, D. 1947. "The significance of clutch size." *Ibis* 89:302–352.
A classic study on the evolution of life history characteristics, examining factors relating to variations in clutch size in birds.

Recent Research

Alcock, J. 2009. *Animal behavior: An evolutionary approach.* 9th ed. Sunderland, TMS: Sinauer.
This text is an excellent treatment of topics covered in this chapter. It is a good reference for students who want to pursue specific topics relating to behavioral ecology.

Andersson, M., and Y. Iwasa. 1996. "Sexual selection." *Trends in Ecology and Evolution* 11:53–58.
An excellent but technical review of sexual selection.

Krebs, J. R., and N. D. Davies. 1993. *An introduction to behavioral ecology.* 3rd ed. Oxford: Blackwell Scientific.
This text provides a comprehensive discussion of behavioral topics that are covered in this chapter.

Stearns, S. C. 1992. *The evolution of life histories.* Oxford: Oxford University Press.
This book explores the link between natural selection and life history. It does an excellent job of illustrating how both biotic and abiotic factors interact to influence the evolution of specific life history traits.

CHAPTER 11 Intraspecific Population Regulation

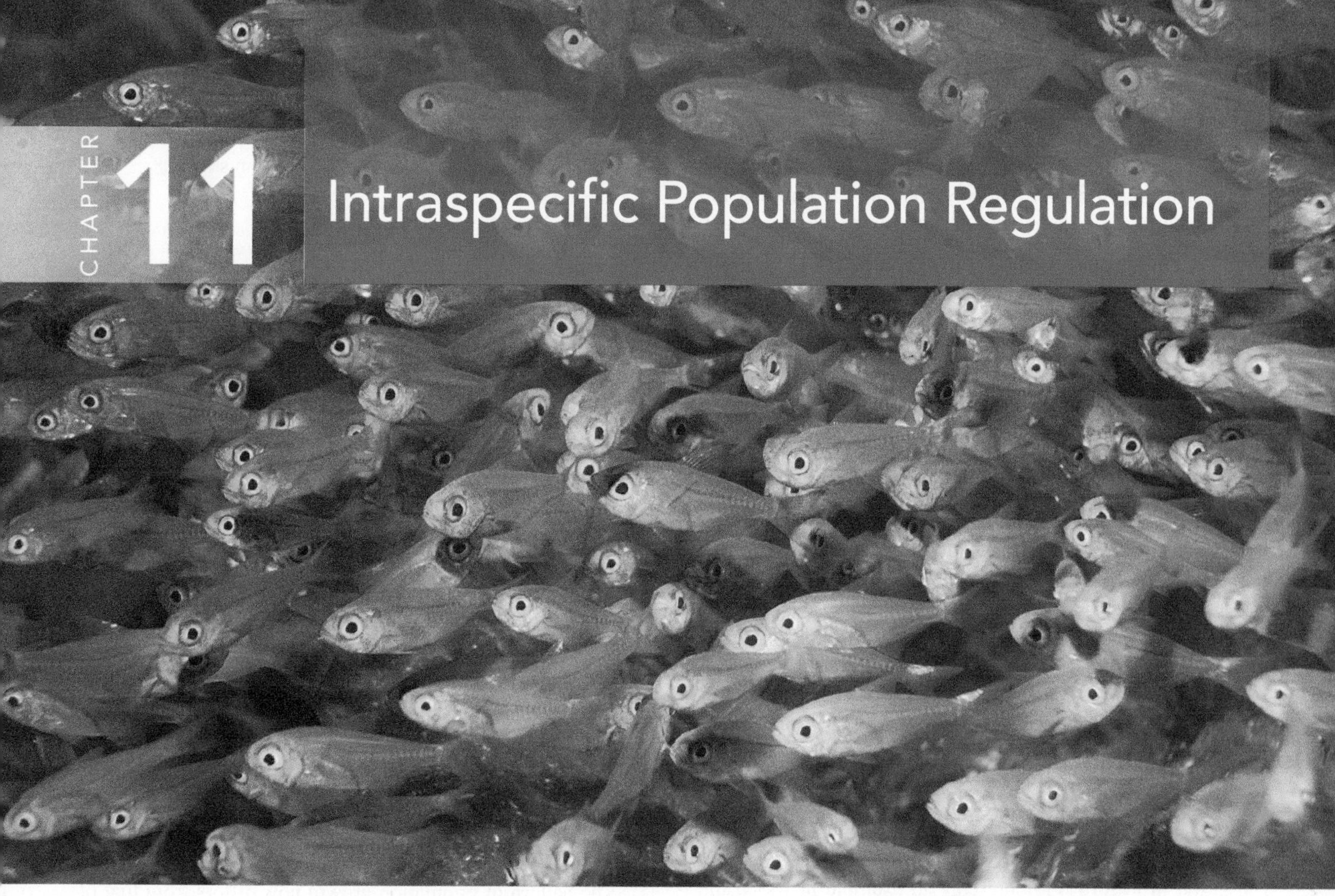

The pygmy sweep (*Parapriacanthus ransonetti*) form densely packed schools in small openings in the coral knolls of the South Pacific. This species feeds at night on small plankton by using bioluminescent organs located at the back of their pectoral fins.

CHAPTER GUIDE

11.1 The Environment Functions to Limit Population Growth

11.2 Population Regulation Involves Density Dependence

11.3 Competition Results When Resources Are Limited

11.4 Intraspecific Competition Affects Growth and Development

11.5 Intraspecific Competition Can Influence Mortality Rates

11.6 Intraspecific Competition Can Reduce Reproduction

11.7 High Density Is Stressful to Individuals

11.8 Dispersal Can Be Density Dependent

11.9 Social Behavior May Function to Limit Populations

11.10 Territoriality Can Function to Regulate Population Growth

11.11 Plants Preempt Space and Resources

11.12 A Form of Inverse Density Dependence Can Occur in Small Populations

11.13 Density-Independent Factors Can Influence Population Growth

ECOLOGICAL Issues & Applications Conservation of Populations

NO POPULATION CONTINUES TO GROW indefinitely. In particular, populations that exhibit exponential growth eventually confront the limits of the environment. As a population's density changes, interactions mediated by the environment occur among members of the population and tend to regulate the population's size. These interactions include a wide variety of mechanisms relating to physiological, morphological, and behavioral adaptations.

11.1 The Environment Functions to Limit Population Growth

The exponential model of population growth that we developed in Chapter 9 [$dN/dt = (b - d)N$] is based on several assumptions about the environment in which the population is growing. The model assumes that essential resources (space, food, etc.) are unlimited and that the environment is constant, but this is not the case because the environment is not constant, and resources are limited. As the density of a population increases, demand for resources increases. If the rate of consumption exceeds the rate at which resources are replenished, then the resource base will shrink. Shrinking resources and the potential for an unequal distribution of those resources results in increased mortality, decreased fecundity, or both. The simplest form of representing changes in birthrates and death rates with increasing population is a straight line (linear function). The graph in **Figure 11.1** presents an example in which the per capita birthrate (b) decreases with increasing population size (N). Conversely, the per capita death rate (d) increases

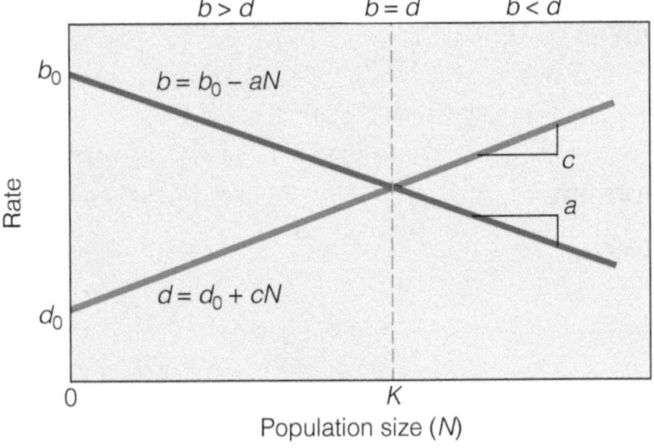

Figure 11.1 Rates of birth (b) and death (d), represented as a linear function of population size N. The values b_0 and d_0 represent the ideal birth and death rates (respectively) under conditions where the population size is near zero and resources are not limiting. The values a and c represent the slopes of the lines describing changes in birthrates and death rates as a function of N (respectively). The population density where $b = d$ and population growth is zero is defined as K, the carrying capacity. For values of N above K, b is less than d and the population growth rate is negative. For values of N below K, b is greater than d, and the population growth rate is positive.

with population size. We can describe the line representing the change in birthrate as a function of population size as:

$$b = b_0 - aN$$

In this equation, b_0 is the intercept (value of b when N is near zero), and a is the slope of the line ($\Delta b/\Delta N$; see Figure 11.1). The intercept, b_0, represents the birthrate achieved under ideal conditions (uncrowded with no resource limitation), whereas b is the actual birthrate, which is reduced as a function of crowding. The maximum birthrate, b_0, is the value used in the exponential model of population growth (see Section 9.1). Likewise, we can represent the change in death rate as a function of population size as:

$$d = d_0 + cN$$

Again, the constant d_0 is the death rate when the population size is close to zero (no crowding or resource limitation), and the constant c represents the increase in death rate with increasing population size (slope of the line shown in Figure 11.1).

We can now rewrite the exponential model of population growth developed in Section 9.1 [$dN/dt = (b - d)N$] to include the variations in the rates of birth and death as a function of population size presented previously:

$$\frac{dN}{dt} = [(b_0 - aN) - (d_0 + cN)]N$$

The pattern of population growth now differs from that of the original exponential model. As N increases, the birthrate ($b_0 - aN$) declines, the death rate ($d_0 + cN$) increases, and the result is a slowing of the rate of population growth. If the value of d exceeds that of b, population growth is negative, and population size declines (see Figure 11.1). When the birthrate (b) is equal to the death rate (d), the rate of population change is zero ($dN/dt = 0$). The value of population size at which the birthrate is equal to the death rate ($b = d$) represents the maximum sustainable population size under the prevailing environmental conditions. We can solve for this value by setting the equation for population growth equal to zero and solving for N (see **Quantifying Ecology 11.1**). The result is:

$$N = (b_0 - d_0)/(a + c)$$

Because b_0, d_0, a, and c are constants, this value of N represents a constant—a single value at which $b = d$ and the population growth rate is zero ($dN/dt = 0$). We define this unique value of N as the **carrying capacity**, represented by the letter K. The carrying capacity is the maximum sustainable population size for the prevailing environment. It is a function of the supply of resources (e.g., food, water, space, etc.).

We can now rewrite the equation for population growth that includes the rates of birth (b) and death (d) that vary with population size using the value of carrying capacity, K, defined previously:

$$\frac{dN}{dt} = rN\left(1 - \frac{N}{K}\right)$$

QUANTIFYING ECOLOGY 11.1 Defining the Carrying Capacity (K)

We can solve for the population size at which $b = d$ by setting the equation for population growth developed in Section 11.1 equal to zero and solving for N:

$$\frac{dN}{dt} = [(b_0 - aN) - (d_0 + cN)]N = 0$$

$$(b_0 - aN)N - (d_0 + cN)N = 0$$

(move the term for death rate to the right side of the equation)

$$(b_0 - aN)N = (d_0 + cN)N$$

(then divide both sides by N)

$$b_0 - aN = d_0 + cN$$

(move d_0 to the left side of the equation and aN to the right side)

$$b_0 - d_0 = aN + cN$$

(rearrange the right-hand side of the equation)

$$(b_0 - d_0) = (a + c)N$$

[divide both sides by $(a + c)$]

$$\frac{(b_0 - d_0)}{(a + c)} = N$$

Because b_0, d_0, a, and c are constants, we can define a new constant:

$$K = \frac{(b_0 - d_0)}{(a + c)}$$

This new value, K, is the carrying capacity; the value of N at which $(b_0 - aN) = (d_0 + cN)$ and therefore $dN/dt = 0$.

In this form, referred to as the **logistic model of population growth**, the per capita growth rate, r, is defined as $b_0 - d_0$. The derivation of the logistic equation is presented in **Quantifying Ecology 11.2**.

The logistic model effectively has two components: the original exponential term (rN) and a second term $(1 - N/K)$ that functions to reduce population growth as the population size approaches the carrying capacity. When the population density (N) is low relative to the carrying capacity (K), the term $(1 - N/K)$ is close to 1.0, and population growth follows the exponential model (rN). However, as the population grows and N approaches K, the term $(1 - N/K)$ approaches zero, slowing population growth. Should the population density exceed K, population growth becomes negative and population density declines toward carrying capacity.

As with the exponential growth model, we can use the rules of calculus to integrate the logistic growth equation and express population size as a function of time:

$$N(t) = \frac{K}{1 + \left\{\frac{[K - N(0)]}{N(0)}\right\}e^{-rt}}$$

The graph of population size (N) through time for the logistic model is shown in **Figure 11.2a**. When the population is small, it increases rapidly but at a rate slightly lower than that predicted by the exponential model. The rate of population growth (dN/dt) is at its highest when $N = K/2$ (called the *inflection point*) and then decreases as it approaches the carrying capacity (K; **Figure 11.2b**). This is in contrast to the exponential model, in which the population growth rate increases linearly with population size.

As an illustration of logistic growth, we can return to the example of the gray squirrel population (Chapter 9). We assume a carrying capacity of 200 individuals ($K = 200$). Given the value of $r = 0.18$ calculated from the life table and an initial population size of 30 individuals, the predicted patterns

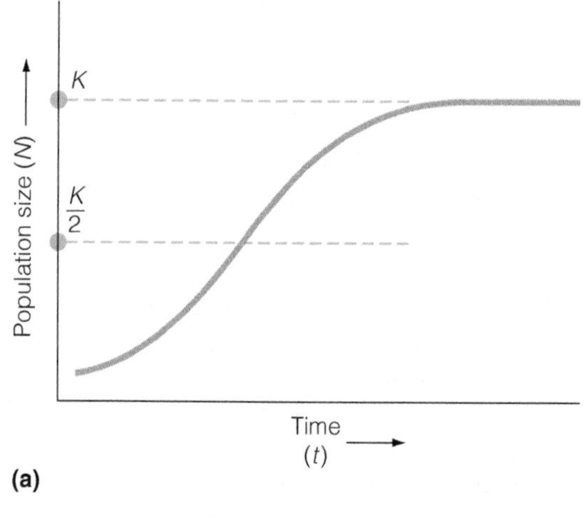

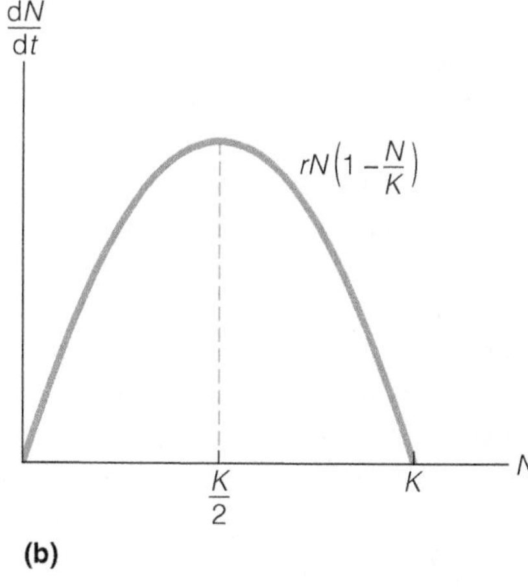

Figure 11.2 (a) Change in population size (N) through time as predicted by the logistic model of population growth [$dN/dt = rN(1 - N/K)$]. Initially (low values of N), the population grows exponentially; as N increases, the rate of population growth decreases, eventually reaching zero as the population size approaches the carrying capacity (K). (b) The relationship between the rate of population growth, dN/dt, and population size, N, takes the form of a parabola, reaching a maximum value at a population size of $N = K/2$.

QUANTIFYING ECOLOGY 11.2 The Logistic Model of Population Growth

We can derive the logistic population growth by beginning with the equation that allows the rates of birth (b) and death (d) to vary as a function of population size, as outlined in Section 11.1:

$$\frac{dN}{dt} = (b - d)N$$

Because $b = b_0 - aN$ and $d = d_0 + cN$, we rewrite the equation as follows:

$$\frac{dN}{dt} = [(b_0 - aN) - (d_0 + cN)]N$$

After rearranging the terms (see p. 221), we have

$$\frac{dN}{dt} = [(b_0 - d_0) - (a + c)N]N$$

Next, we multiply by $(b_0 - d_0)/(b_0 - d_0)$. This term is equal to 1.0, so it only simplifies the equation further:

$$\frac{dN}{dt} = \frac{(b_0 - d_0)}{(b_0 - d_0)}[(b_0 - d_0) - (a + c)N]N$$

$$\frac{dN}{dt} = [(b_0 - d_0)]\left[\frac{(b_0 - d_0)}{(b_0 - d_0)} - \frac{(a + c)}{(b_0 - d_0)}N\right]N$$

Because we have defined $r = (b_0 - d_0)$ in Section 11.1, we have:

$$\frac{dN}{dt} = rN\left[1 - \frac{(a + c)}{(b_0 - d_0)}N\right]$$

Note that $(a + b)/(b_0 - d_0) = 1/K$, as shown on page 221.

Making the appropriate substitution, we have:

$$\frac{dN}{dt} = rN\left[1 - N\left(\frac{1}{K}\right)\right]$$

$$\frac{dN}{dt} = rN\left(1 - \frac{N}{K}\right)$$

This is the equation for the logistic model of population growth.

1. The equation is sometimes presented in an alternative but equivalent form: $dN/dt = rN[(K - N)/K]$. Show algebraically how this equation is equivalent to the one presented previously.

of population growth under both the exponential and logistic models of population growth are shown in **Figure 11.3**.

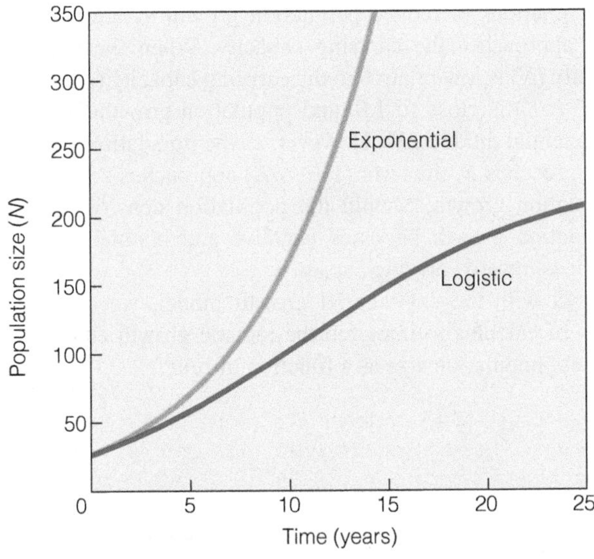

Figure 11.3 Predictions of the exponential and logistic population growth models for the gray squirrel population from Tables 9.1 and 9.6; $r = 0.18$, $K = 200$, and $N(0) = 30$.

11.2 Population Regulation Involves Density Dependence

The concept of carrying capacity suggests a negative feedback between population increase and resources available in the environment. As population density increases, the per capita availability of resources declines. The decline in per capita resources eventually reaches some crucial level at which it functions to regulate population growth. Implicit in this model of population regulation is **density dependence**.

Density-dependent effects influence a population in proportion to its size. They function to slow the rate of population growth with increasing population density by increasing the rate of mortality (termed **density-dependent mortality**), decreasing the rate of fecundity (**density-dependent fecundity**), or both. In the case of the logistic growth model, density-dependent mortality and fecundity are incorporated by varying the rates of birth (b) and death (d), expressed through the value of the carrying capacity, K (**Figure 11.4**).

Mechanisms of density-dependent population regulation may include factors other than the direct effects of resource availability. For example, population density can influence patterns of predation or the spread of disease and parasites (see Chapters 14 and 15).

Other factors that can directly influence rates of birth and death function independently of population density. If some environmental factor such as adverse weather conditions affects the population regardless of the number of individuals, or if the proportion of individuals affected is the same at any density, then the influence is referred to as **density independent**.

In the following sections of this chapter, we will explore the variety of interactions among individuals within a population that can influence the rates of birth and death, and therefore, the rates of density-dependent population growth.

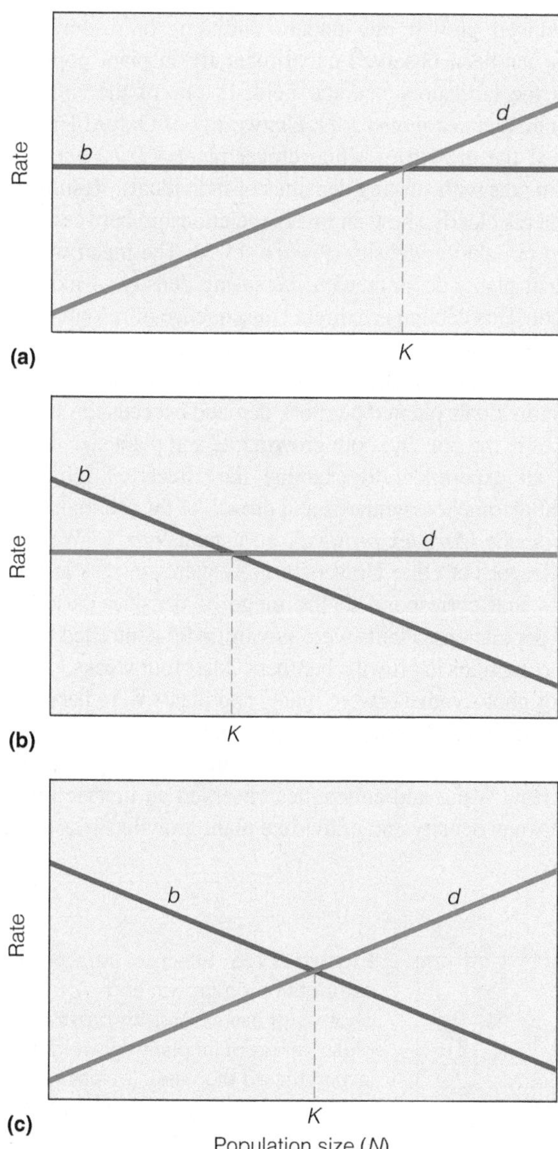

Figure 11.4 Regulation of population size in three situations. (a) Birthrate *(b)* is independent of population density, as indicated by the horizontal line. Only the death rate *(d)* increases with population size. At *K*, equilibrium is maintained by increasing mortality. (b) The situation is reversed. Mortality is independent, but birthrate declines with population size. At *K*, a decreasing birthrate maintains equilibrium. (c) Full density-dependent regulation. Both birthrate and mortality are density dependent. Fluctuations in either one hold the population at or near *K*.

11.3 Competition Results When Resources Are Limited

Implicit in the concept of carrying capacity is competition among individuals for essential resources. **Competition** occurs when individuals use a common resource that is in short supply relative to the number seeking it. Competition among individuals of the same species is referred to as **intraspecific competition**. As long as the availability of resources does not impede the ability of individuals to survive, grow, and reproduce, no competition exists. When resources are insufficient to satisfy all individuals, the means by which they are allocated has a marked influence on the welfare of the population.

When resources are limited, a population may exhibit one of two responses: scramble competition or contest competition. **Scramble competition** occurs when growth and reproduction are depressed equally across individuals in a population as the intensity of competition increases. **Contest competition** takes place when some individuals claim enough resources while denying others a share. Generally, under the stress of limited resources, a species exhibits only one type of competition. Some are scramble species and others are contest species. One species may practice both types of competition at different stages in the life cycle. For example, some insects endure scramble competition during their larval stages until the population declines. The adult stage is then characterized by contest competition.

The outcomes of scramble and contest competition vary. At its extreme, scramble competition can lead to all individuals receiving insufficient resources for survival and reproduction, resulting in local extinction. In contest competition, only a fraction of the population suffers—the unsuccessful individuals. The survival, growth, and reproduction of individuals that successfully compete for the limited resources all function to sustain the population.

In many cases, competing individuals do not directly interact with one another. Instead, individuals respond to the level of resource availability that is depressed by the presence and consumption of other individuals in the population. For example, large herbivores such as zebras grazing on the savannas of Africa may influence one another not through direct interactions but by reducing the amount of grass available as food. Similarly, as a tree in the forest takes up water through its roots, it decreases the remaining amount of water in the soil for other trees. In these cases, competition is termed **exploitation**.

In other situations, however, individuals interact directly with one another, preventing others from occupying a habitat or accessing resources within it. For example, most bird species actively defend the area around their nest during the breeding season, denying other individuals access to the site and its resources (see Section 11.10). In this case, competition is termed **interference**.

11.4 Intraspecific Competition Affects Growth and Development

Because the intensity of intraspecific competition is usually density dependent, it increases gradually, and at first affects growth and development. Later, it affects individual survival and reproduction.

As population density increases toward a point at which resources are insufficient to provide for all individuals in the population, some (contest competition) or all individuals (scramble competition) reduce their intake of resources. That reduction slows the rate of growth and development. The result is an inverse relationship between population density and individual growth, referred to as **density-dependent growth**. Examples of this inverse relationship between density and individual growth rate have been observed in a wide variety of plant and animal populations.

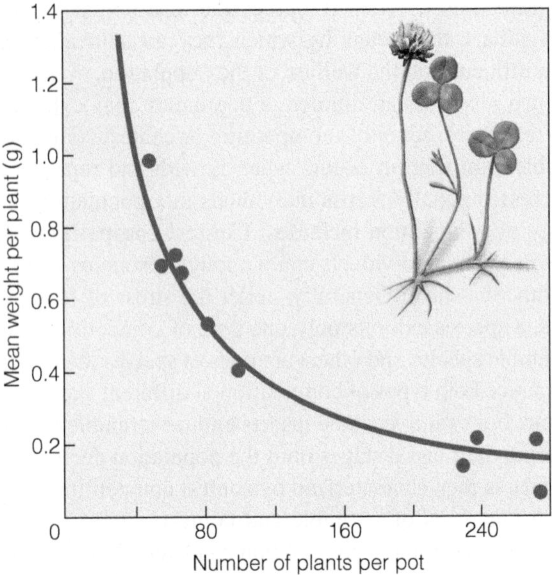

Figure 11.5 Effect of population density on the growth of individuals. The growth rate and subsequent weight of white clover (*Trifolium repens*) plants declines markedly with increasing density of individuals planted in the pot.
(Adapted from Chatworthy 1960.)

Reduced growth rate under conditions of resource competition has been observed experimentally in plant populations in both the laboratory and the field. In one of the earliest experiments, plant ecologist J. N. Clatworthy of Oxford University examined the growth of white clover plants (*Trifolium repens*) grown in pots with varying densities of individuals. Results of the experiments clearly show an inverse relationship between growth rate and population density (**Figure 11.5**). The mean weight of individual plants declines with increasing density of individuals in the pot. This decline is a direct consequence of resource limitation. At low densities, all individuals are able to acquire sufficient resources to meet demands for growth. As the density is increased (more individuals planted per pot), demand exceeds the supply of resources in the pot, and both growth rate and plant size decline.

In an experiment to examine the effects of intraspecific competition on photosynthesis and growth of the salt marsh spear-leaved orache (*Atriplex prostrat*), an annual forb, Li-Wen Wang and colleagues at Ohio University grew plants in pots at varying densities that correspond to the range of densities observed in natural populations. Plants were grown under controlled environmental conditions in growth chambers. After four weeks, measurements of photosynthesis were made, and plants were harvested to provide measures of tissue dry weight, leaf area, and plant height.

As with the results of the previous experiment by Clatworthy, Wang and colleagues observed an inverse relationship between density and individual plant growth (**Figure 11.6**).

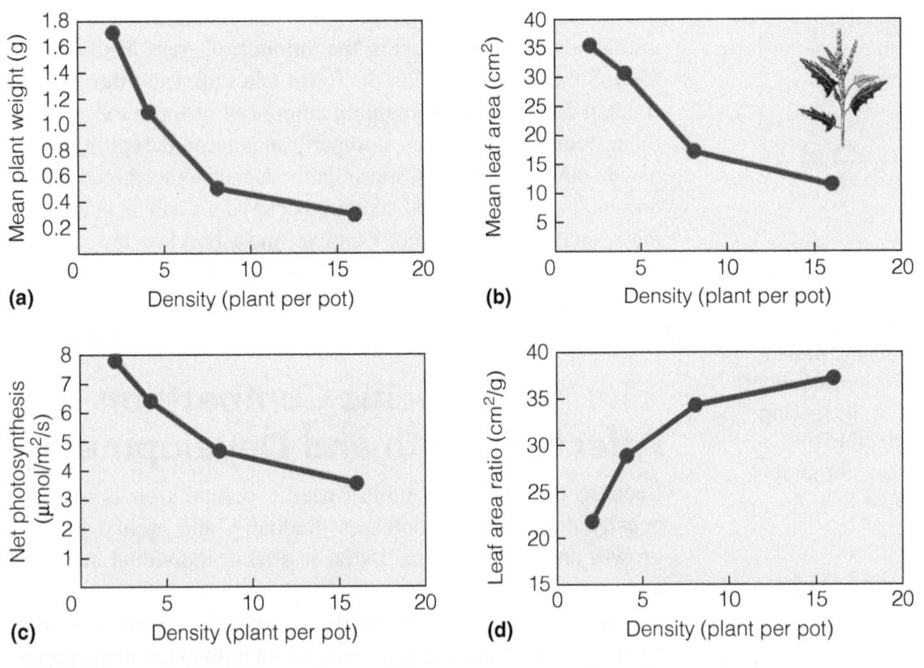

Figure 11.6 Effects of intraspecific competition on growth and photosynthesis of *Atriplex prostrata*. Mean values of (a) plant dry weight (accumulated biomass), (b) leaf area, and (c) net photosynthetic rate plotted as a function of varying densities of individuals per pot in the experiments. Besides the general pattern of reduced photosynthesis and growth experienced under increasing competition, individual plants exhibit a shift in patterns of carbon allocation, as illustrated by the increase in leaf area ratio (leaf area cm^2-to-plant weight g) with increasing density of plants per pot (d). The increase in leaf area ratio is characteristic of reduced light availability during plant development.
(Data from Wang et al. 2005.)

Interpreting Ecological Data

Q1. *What is the approximate difference in mean plant weight between individuals grown at the density of 2 plants per pot and those grown at 8 plants per pot? How does this difference in mean plant weight compare to that between plants grown at 8 and 16 plants per pot?*

Q2. *Why is the decline in mean plant weight with increasing density not linear (straight line)? (Hint: Calculate the total biomass in each pot by multiplying the density [2, 4, 8, and 16 plants per pot] by mean plant weight [1.7, 1.1, 0.5, and 0.3 g] for each treatment.)*

High plant density caused an 80 percent reduction in mean plant weight (Figure 11.6a) and a 72 percent reduction in leaf area (Figure 11.6b) over values observed at the lowest plant densities. A 50 percent reduction in net photosynthetic rate paralleled the observed growth inhibition with increasing plant density (Figure 11.6c), indicating that the growth inhibition caused by intraspecific competition is mainly the result of a decline in the net rate of carbon uptake. In addition to the decline in plant weight and leaf area with increasing density, the data illustrate a corresponding shift in carbon allocation and plant morphology. The average leaf area ratio of individual plants (leaf area-to-total plant weight) increased with increasing plant density (Figure 11.6d), reflecting a relative increase in the allocation of photosynthates (total plant mass) to the production of leaf area (photosynthetic surface). As you may recall from the discussion of plant response to variations in the light environment, an increase in leaf area ratio is indicative of reduced light availability, strongly suggesting that the observed reductions in photosynthesis and growth with increasing plant density are a result of shading and competition for light resources (see Section 6.8 and Chapter 6, **Quantifying Ecology 6.1**).

Similar patterns of density-dependent growth to those reported for plants have been observed among populations of ectothermic (poikilothermic) vertebrates. Rick Relyea of the University of Pittsburgh examined the effects of intraspecific competition on experimental populations of wood frog tadpoles (*Rana sylvatica*). In addition to the influence of density on individual growth rate, Relyea examined the impacts of intraspecific competition on phenotypic variations in behavior and morphology (phenotypic plasticity; see Section 5.4 for discussion). Relyea found a decrease in the average individual growth rate with increasing population density (**Figure 11.7a**). The reductions in growth rate, however, were accompanied by distinct changes in behavior and morphology in response to the competitive environment. Individuals reared in higher-density populations exhibited an increase in activity (time spent in movement; **Figure 11.7b**) and generally developed longer bodies (**Figure 11.7c**), shorter tails, and wider mouths, which is a clear example of competition-induced phenotypic plasticity (see Section 5.4). Increased activity was shown to increase resource acquisition, and previous studies have found that the observed shifts in morphology are associated with enhanced growth under competitive environments, suggesting that both behavioral and developmental responses are adaptive.

11.5 Intraspecific Competition Can Influence Mortality Rates

In addition to suppressing the growth of individuals, competition for resources at high population densities can function to reduce survival (**Figure 11.8**). In turn, mortality functions to increase per capita resource availability, allowing for increased growth of the surviving individuals. This link between density-dependent mortality, resource availability, and growth rate is particularly apparent in organisms that exhibit indeterminate growth rates that respond strongly to resource availability, such as plants. In an experiment aimed at exploring this relationship, the late plant ecologist Kyoji Yoda planted seeds of horseweed (*Erigeron canadensis*) at a density of 100,000 seeds per square meter. As the seedlings grew, competition for the limited

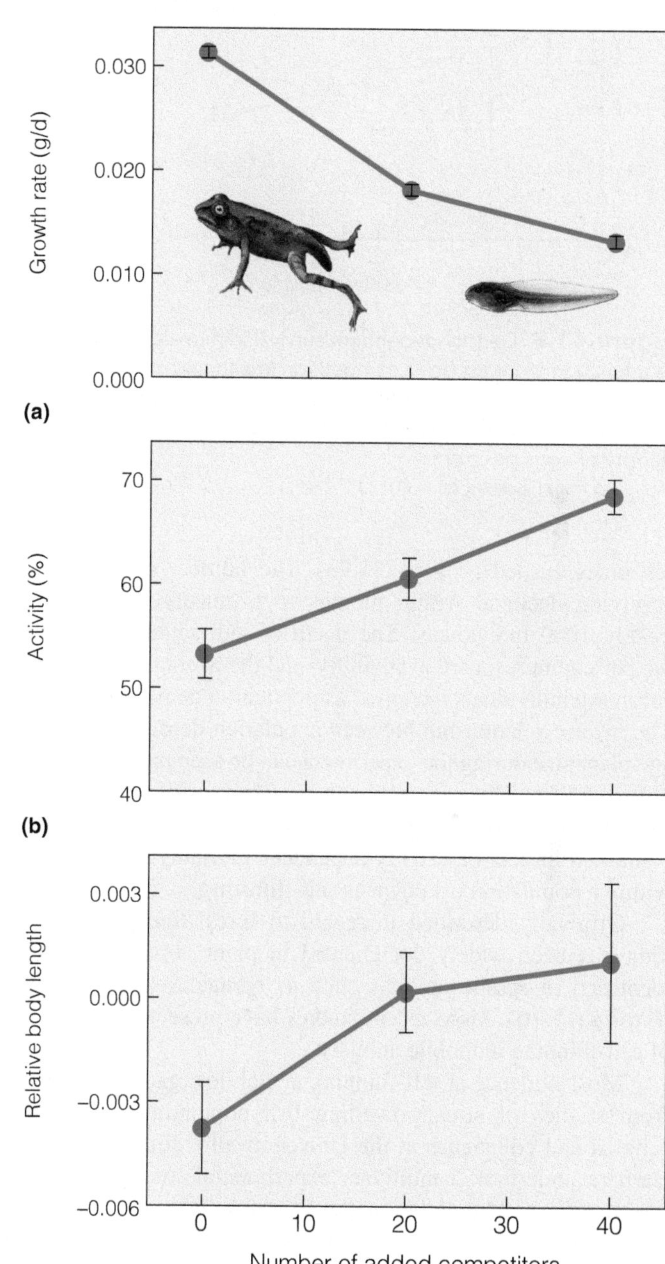

Figure 11.7 Mean (a) growth rate, (b) activity, and (c) relative body length (±SE) of larval wood frogs as a function of density of intraspecific competitors. Activity was measured by determining the proportion (percentage) of observed individuals in motion during the observation period. The measure of relative body length is the residual of the regression of body length as a function of body mass. This approach removes differences in body length among individuals purely as a result of differences in overall size (body mass). Mean values (and estimates of variance) for the residuals were then calculated for each treatment, as presented in graph (c). (Adapted from Relyea 2002.)

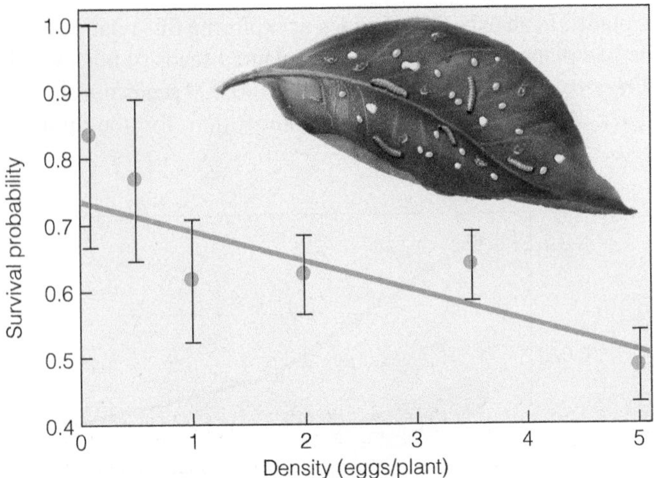

Figure 11.8 Density-dependent survival in monarch butterflies as a function of intraspecific competition (mean and ±SE). The probability of survival from egg to eclosion (from egg, to larva, and eventual emergence from pupal case) as a function of the density of eggs per plant.
(Adapted from Flockhart et al. 2012.)

resources ensued (**Figure 11.9a**). The number of seedlings surviving declined within months to a density of approximately 1000 individuals. The death of individuals increased the per capita resource availability, and the average size of the surviving individuals increased as population density declined. The inverse relationship between population density and average plant size during the experiment can be seen more easily in **Figure 11.9b**. This progressive decline in density and increase in biomass (growth) of remaining individuals caused by the combined effects of density-dependent mortality and growth within a population is known as **self-thinning**.

Originally described in regard to forest trees, self-thinning has been widely documented in plant populations and identified in sessile animals such as barnacles and mussels (**Figure 11.10**). More recent studies have presented evidence of self-thinning in mobile animals.

Most evidence of self-thinning in mobile organisms comes from studies of stream-dwelling fish populations. Thomas Jenkins and colleagues at the University of California–Santa Barbara undertook a multiyear experimental study to examine the effects of population density on individual growth of brown trout (*Salmo trutta*) in two stream ecosystems in the Sierra Nevada mountains (Sierra Nevada Aquatic Research Laboratory). Their data reveal an inverse relationship between average mass of individuals and density of surviving brown trout over the study period (**Figure 11.11**) that is similar to the general patterns shown in Figures 11.9 and 11.10 for populations of plants and sessile animals. The reduction in density of brown trout through time was the result of intraspecific competition for limited food resources as mean body size (and associated demand for food resources) increased.

Ernest Keeley of the University of British Columbia conducted experiments in artificial stream channels, manipulating density of competitors and food abundance to examine

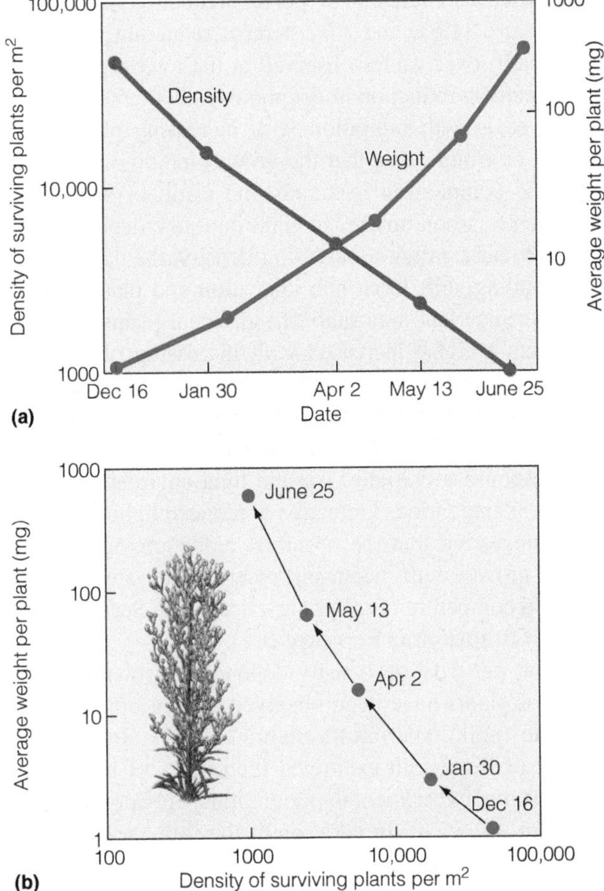

Figure 11.9 (a) Changes in the number of surviving individuals and average plant size (weight in milligrams) through time for an experimental population of horseweed (*Erigeron canadensis*). (b) Data from (a) replotted to show the relationship between population density and average plant weight. Competition results in mortality, which in turn increases the per capita availability of resources, resulting in increased growth for the survivors.
(Adapted from Yoda, K. et al, "self-thinning in overcrowded pure stands under cultivated and natural conditions" Journal of Biology Vol. 14, Osaka City University, 1963.)

density-dependent growth and mortality in populations of steelhead trout (*Oncorhynchus mykiss*). His results illustrate a pattern of decreasing growth and increasing mortality with increasing levels of per capita food competition, resulting either from an increase in population density for a given level of food abundance or from reduced food abundance for a given population density (**Figure 11.12**).

11.6 Intraspecific Competition Can Reduce Reproduction

Besides directly influencing the survival and growth of individuals, competition within a population can reduce fecundity. The timing of the response depends on the nature of the population, and the mechanisms by which competition influences reproductive rate can vary with species. Harp seals (*Phoca groenlandica*) become sexually mature when they reach approximately

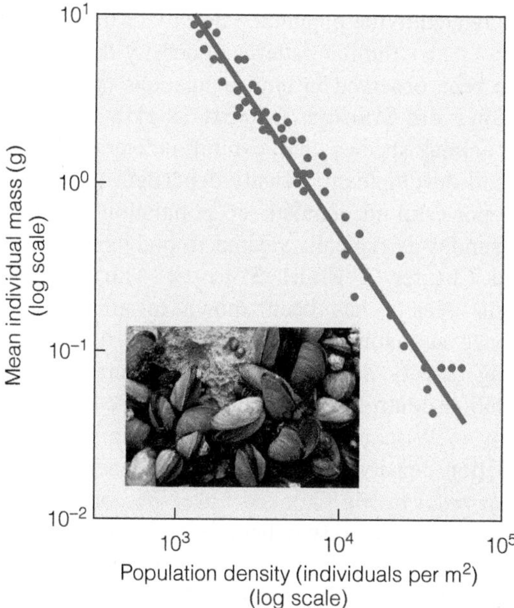

Figure 11.10 Relationship between mean individual mass (g) and population density (individuals per m^2) for the black mussel, *Choromytilus meridionalis*.
(Adapted from Hughes and Griffiths 1988.)

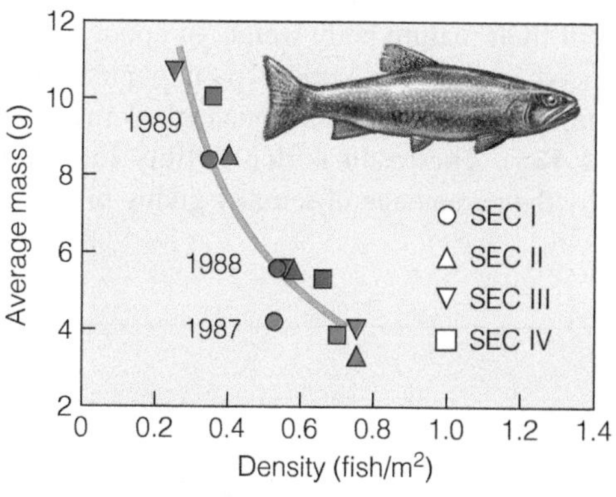

Figure 11.11 Self-thinning in a population of brown trout inhabiting Convict Creek in the Sierra Nevada Mountain range. Average body mass of surviving brown trout is expressed as a function of the total density of individuals in different stream sections as measured during the fall censuses 1987–1989. Data points represent means of from 118 to 699 individuals. Each symbol represents the subpopulation occupying a specific stream section. The color red denotes data from 1987; the color blue, from 1988; the color green, from 1989.
(Adapted from Jenkins et al. 1999.)

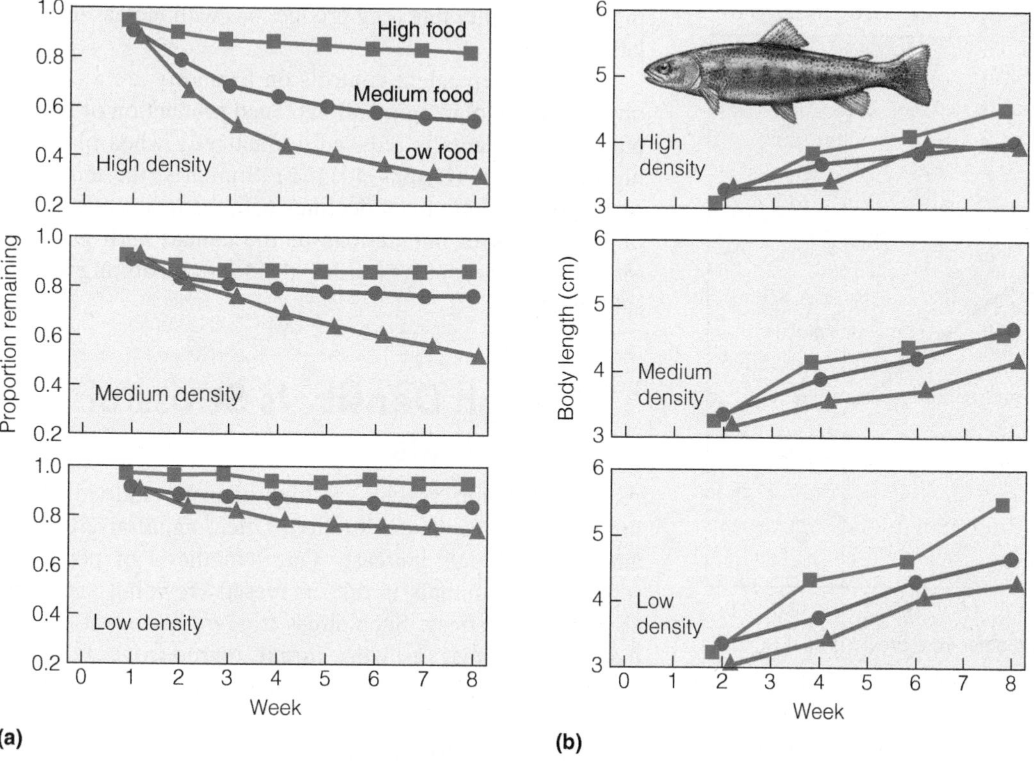

Figure 11.12 Patterns of (a) mortality (as measured by proportion of initial individuals remaining) and (b) growth (as measured by body length) of juvenile steelhead trout in stream channels over an eight-week experimental period. Experiments examined the influence of intraspecific competition by manipulating density of competitors and abundance of food. Initial densities of individuals in the experimental stream channels were set at 32 (low density), 64 (medium density), or 128 (high density) individuals per m^2. Food abundance was varied by providing either 0.3 (low food), 0.6 (medium food), or 1.2 (high food) grams/m^2/day.
(Adapted from Keeley 2001.)

Interpreting Ecological Data

Q1. *How does mortality differ for the high-density experimental populations under low, medium, and high food abundance? How do these differences among the three food abundance treatments change as population density is reduced (medium and low population densities)?*

Q2. *How does growth (body length) differ for the high-density experimental populations under low, medium, and high food abundance? How do these differences among the three food abundance treatments change as population density is reduced (medium and low population densities)?*

Q3. *Why do the greatest differences in mortality among the three food abundance treatments occur at high population density, whereas the greatest differences in growth rate among the three food abundance treatments occur at low population density?*

87 percent of their mature body weight of about 120 kilograms (kg). Reduced weight gain under high population densities increase the mean age at which females become reproductive (**Figure 11.13a**). The result is that fertility in harp seals, as measured by the percentage of females giving birth to young is inversely related to the population density of the previous year (**Figure 11.13b**). Similar patterns of density-dependent fecundity have been observed in bird populations (see this chapter, Figure 4, in **Field Studies: T. Scott Sillett**).

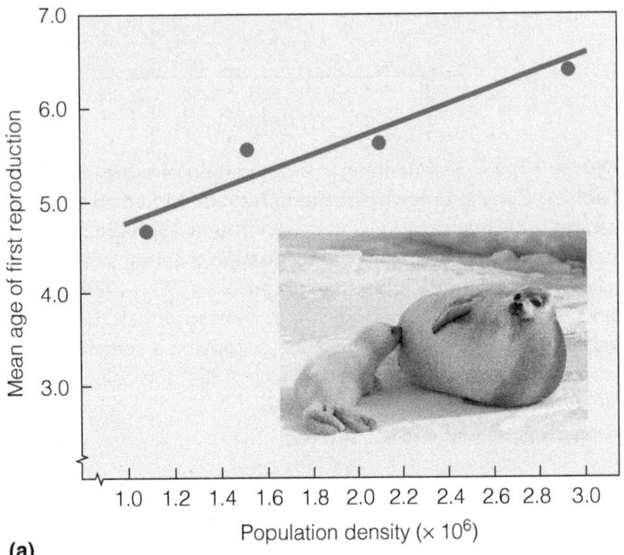

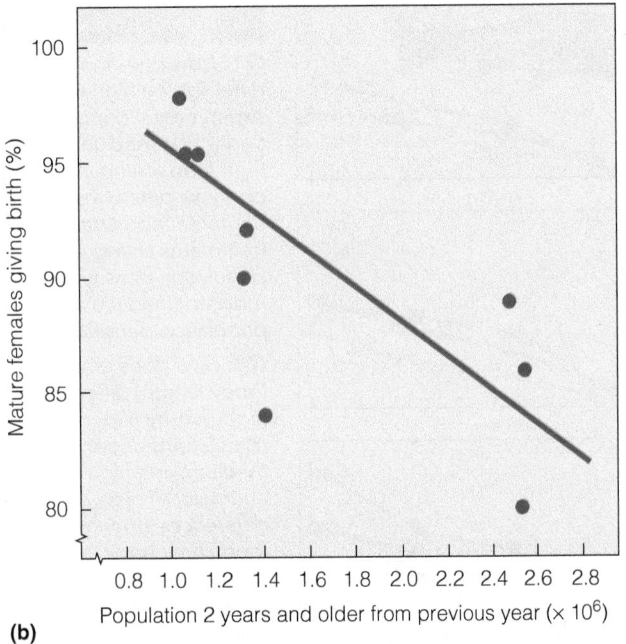

Figure 11.13 (a) The mean age of sexual maturity (whelping) for female harp seals (*Phoca groenlandica*) increases with increasing population density. Sexual maturity is related to weight more than to age. Seals arrive at sexual maturity when they reach 87 percent of average adult body weight. Seals attain this weight at an earlier age when population density is low. (b) As a result, fertility is density dependent. As the seal population (measured by the number of individuals two years old and older during the previous year) increases, the percentage of females giving birth to young decreases markedly.
(Adapted from Lett et al. 1981.)

For animal species that exhibit indeterminate rates of growth and development, density-dependent growth is a potentially powerful mechanism of population regulation because fecundity is typically related to body size (see Section 10.4 and Chapter 7, **Field Studies: Martin Wikelski**). Population density has been shown to affect patterns of growth, age at maturity, and fecundity in fish populations, suggesting density dependence as an important mechanism of population control. Amy Schueller and colleagues at the University of Wisconsin–Stevens Point examined the effects of population density on maturity and fecundity of walleyes (*Sander vitreus*) in Big Crooked Lake, Wisconsin, over a period of six years. The researchers found a significant positive relationship between age of maturity (reproduction) and adult walleye population density (**Figure 11.14a**) similar to that observed for the harp seal population shown in Figure 11.13. The onset of sexual maturity is size dependent (**Figure 11.14b**), and the delayed maturity observed with increasing population density is a direct result of the inverse relationship between individual growth rates and population density (**Figure 11.14c**). In addition to delayed maturity, Schueller observed a decrease in mean fecundity rate (egg production) with increasing population density.

Density-dependent controls on fecundity are a common observation in plant populations. Seed production of individual soybean plants is reduced dramatically when planted at higher densities (**Figure 11.15a**). Similarly, the number of seeds produced per plant declines with the increasing density of individuals in populations of the annual herb glasswort (*Salicornia europaea;* **Figure 11.15b**) inhabiting coastal salt marshes.

11.7 High Density Is Stressful to Individuals

As a population reaches a high density, individual living space can become restricted. Often, aggressive contacts among individuals increase. One hypothesis of population regulation in animals is that increased crowding and social contact cause stress. Such stress triggers hormonal changes that can suppress growth, curtail reproductive functions, and delay sexual activity. They may also suppress the immune system and break down white blood cells, increasing vulnerability to disease. In mammals, social stress among pregnant females may increase mortality of the young in the fetal stage (unborn) and cause inadequate lactation, stunting the growth and development of nursing young. Thus, stress results in decreased births and increased infant mortality. Such population-regulating effects have been confirmed in confined laboratory populations of several species of mice and to a lesser degree in enclosed wild populations of woodchucks (*Marmota monax*) and rabbits (*Oryctolagus cuniculus*). Evidence of the effects of stress in free-ranging wild animals, however, is difficult to obtain.

Pheromones are perfume-like chemical substances that are produced and released into the environment by an animal that affect the behavior or physiology of other individuals of its species. In social insects, pheromones released by the queen have been identified as an important mechanism in controlling the development and reproduction of colony members. Pheromones present in the urine of adult rodents can encourage or inhibit reproduction. One study by Adrianne Massey and John Vendenbergh of North Carolina State University involved wild female house mice (*Mus musculus*) confined to grassy areas surrounded by roadways that prohibited dispersal. One group lived in a high-density population and the other in a low-density population. Urine from females of each population was absorbed onto filter paper and placed in laboratory cages with juvenile test females. Exposing juvenile females to urine from high-density populations delayed puberty, whereas exposing females to urine from low-density populations did not. The results suggest that pheromones contained in the urine of adult females in high-density populations function to delay puberty and help slow further population growth.

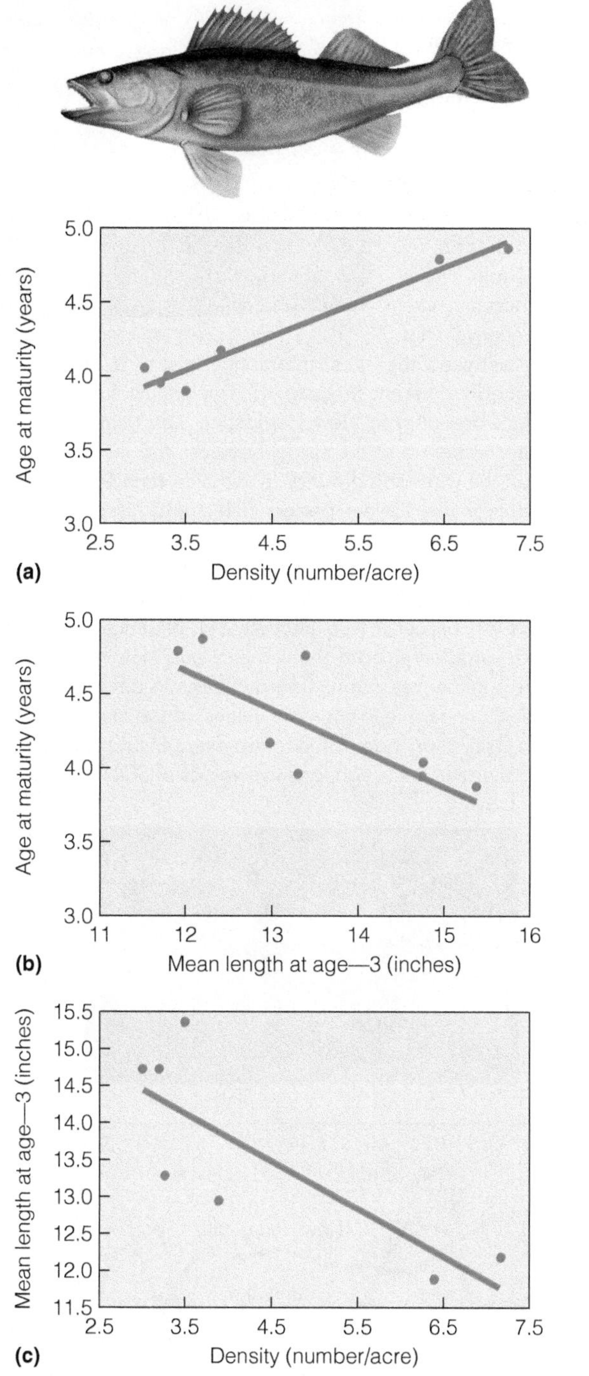

Figure 11.14 Influence of population density on growth rate and age at sexual maturity for female walleye in Big Crooked Lake, Wisconsin. (a) The result is a positive relationship between age at maturity and population density for female walleye in the lake ecosystem. (b) The age at which walleye achieve sexual maturity is size dependent, therefore increasing with decreasing growth rate. (c) Individual walleye growth, as measured by the mean length at three years of age, is inversely related to population density. (Adapted from Schueller et al. 2005.)

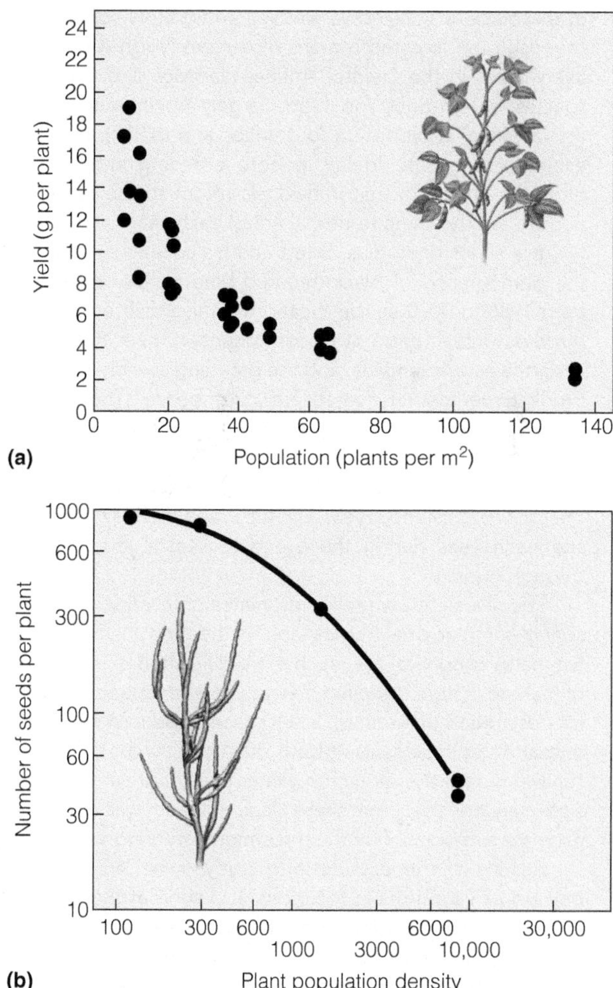

Figure 11.15 Two examples of density-dependent effects on fecundity in plants. (a) Grain (seed) production of individual soy bean plants declines with increasing density of plants. (b) Number of seeds produced per plant declines with increasing population density in the annual herb glasswort (*Salicornia europaea*). ([a] Adapted from Fery and Janick 1971; [b] adapted from Ball et al. 2000.)

FIELD STUDIES T. Scott Sillett

Smithsonian Migratory Bird Center, National Zoological Park, Washington, D.C.

One of the most studied and publicized conservation issues today is the decline of migratory songbirds, particularly neotropical migrants, which are species that migrate between breeding grounds in the temperate zone of North America and that winter in the tropics of Central and South America (see Chapter 9, *Ecological Issues & Applications*). Determining the factors responsible for the decline of these species, however, is difficult because birds spend parts of their annual cycle in different geographic locations. Furthermore, events during one period of the annual cycle are likely to influence populations in subsequent stages.

Understanding the factors influencing the population dynamics of migratory bird species is the research focus of avian ecologist T. Scott Sillett of the Smithsonian Migratory Bird Center. Sillett's research focuses on the population dynamics of the black-throated blue warbler, a migratory songbird that breeds in the forested regions of eastern North America and overwinters in the Greater Antilles (Jamaica and Cuba). The species is territorial and feeds largely on insects, primarily the larvae of Lepidoptera (butterflies and moths). Individuals exhibit strong site fidelity in both breeding and wintering grounds—meaning that individuals return to the same locations (often the same territorial areas) each year.

In a series of studies, Sillett and his colleagues quantified the demography of black-throated blue warbler populations from 1986 to 2000 at two locations during their annual cycle: the overwinter period at Copse Mountain, near Bethel Town in northwestern Jamaica, and the breeding season at Hubbard Brook Experimental Forest, New Hampshire. The Jamaican site was visited twice annually, first at the beginning of the overwinter period in autumn (October) and again at the end of the winter (March), before the start of the spring migration, in the following calendar year. Sillett studied the New Hampshire site each year during the breeding season from mid-May through August.

The research team made estimates of fecundity during the spring–summer breeding season at the New Hampshire site. Estimates of survival are much more difficult, requiring the use of mark-recapture sampling over multiple years (see Section 9.4 for description of method). Sillett's team used mist nets to capture and tag individuals at both sites to allow for estimates of survival during the six-month winter and three-month summer stationary periods. From these data, the team was also able to estimate survivorship for the three-month migratory period.

Results of the decade-long study reveal an interesting pattern of variations in fecundity and survival that are influenced by both density-dependent and density-independent factors. When averaged over the decade-long study, the survival rates of birds at both locations (Jamaica and New Hampshire) did not differ significantly. Average monthly values during the summer (0.99, New Hampshire) and winter (0.93, Jamaica) stationary periods (when the birds were in residence) were high. In contrast, monthly survival probability during the periods of migration ranged from 0.77 to 0.81. Thus, apparently mortality rates are at least 15 times higher during the migration (0.19 to 0.23) compared to the stationary periods (0.01 to 0.07), and more than 85 percent of annual mortality occurred during migration.

Although the mean monthly values of survival did not vary between the sites, interannual patterns of survival were markedly different (**Figure 1**). The annual survival rate of warblers breeding in New Hampshire was relatively constant over the period of the study; however, the probability of survival on the wintering grounds in Jamaica varied more than threefold over the 10-year period. Sillett and his colleagues found that the difference in the interannual patterns of survival is related to different impacts of climate variation on the birds at the two locations. The El Niño–Southern Oscillation (ENSO) discussed in Chapter 2 (see Section 2.9) influences patterns of climate variation at both the Jamaica and New Hampshire sites, but it influences population dynamics in different ways. Sillett used annual mean monthly values of the standardized Southern Oscillation Index (SOI) to represent ENSO conditions for each calendar year. High, positive values of SOI indicate La

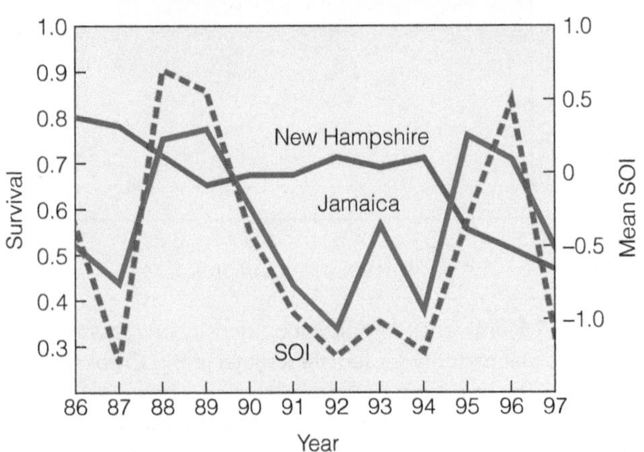

Figure 1 Comparison of annual survival estimates for Jamaica and New Hampshire to mean monthly values of the Southern Oscillation Index (SOI).

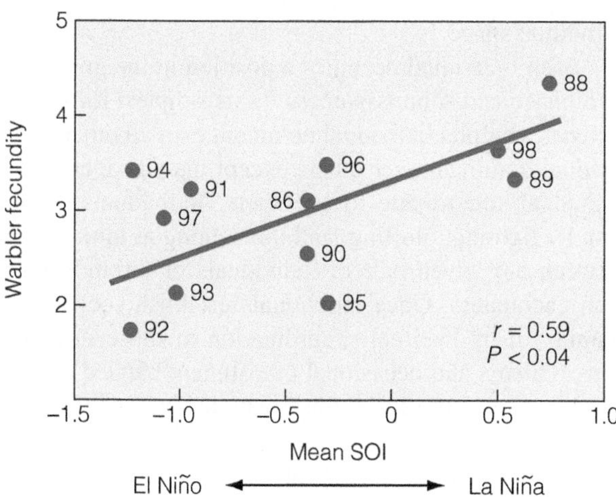

Figure 2 Relationship between the mean monthly Southern Oscillation Index (SOI) and warbler fecundity, measured as the mean number of offspring fledged per warbler pair per year at the New Hampshire site. Numbers at each point represent the calendar year.

Niña conditions, and low, negative values indicate El Niño conditions (Figure 2). Survival at the Jamaica site was low in El Niño years and high in La Niña years (see Figure 2). During El Niño years in Jamaica, reduced rainfall led to decreased food availability for warblers in the winter dry season and, hence, to lower survival. La Niña years, in contrast, tend to be wetter with increased food availability and higher survival rates.

In contrast to Jamaica, the ENSO cycle had little effect on survival rates at the New Hampshire site during the breeding season. The ENSO cycle did, however, significantly influence fecundity rates. Both the mean number of young fledged per warbler pair (**Figure 2**) and the mass at fledging were lower in El Niño years relative to La Niña years. The influence of ENSO on fecundity was the result of the influence of climate variation on food abundance. Prey biomass (lepidopteran larvae) was low during El Niño (dry) years and high during La Niña (wet) years.

In addition to the influence of the ENSO cycle, Sillett and his colleagues observed density-dependent regulation of fecundity rates at the New Hampshire site over the period of study. Using a series of experiments in which population density of areas of the study site were manipulated through the removal of breeding pairs, the researchers observed an increase in territory size, time spent foraging by male birds, and the number of young fledged per territory under reduced population density. The result is an inverse relationship between fecundity and population density (**Figure 3**). Sillett and colleagues found that the influence of population density on fecundity and population growth rate was greatest during El Niño years when food resources were depressed because of climate conditions.

Sillett and colleagues have shown that the population dynamics of the black-throated blue warbler is driven by a combination of density-independent (ENSO) and density-dependent factors. In El Niño years, high mortality (low survival) in the wintering grounds (Jamaica) and low fecundity in the breeding grounds (New Hampshire) function to reduce the density of the breeding population. In subsequent La Niña years, increasing rates of survival during the winter, combined with high rates of fecundity, cause the population to rise. As the density of the breeding population increases, density-dependent factors (crowding) control fecundity rates and population growth.

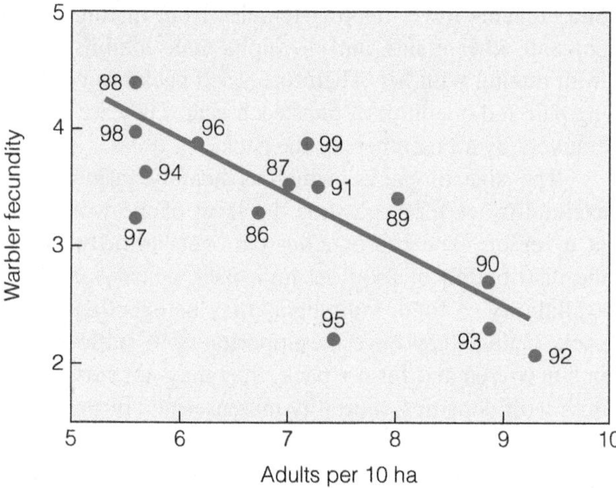

Figure 3 Relationship between population density (adults per 10 ha) and warbler fecundity, measured as the mean number of offspring fledged per warbler pair per year at the New Hampshire site. Numbers at each point represent the calendar year.

Bibliography

Rodenhouse, N. L., T. S. Sillett, P. J. Doran, and R. T. Holmes. 2003. "Multiple density-dependence mechanisms regulate a migratory bird population during the breeding season." *Proceedings of the Royal Society of London* 270:2105–2110.

Sillett, T. S., R. T. Holmes, and T. W. Sherry. 2000. "Impacts of a global climate cycle on population dynamics of a migratory songbird." *Science* 288:2040–2042.

Sillett, T. S., and R. T. Holmes. 2002. "Variation in survivorship of a migratory songbird throughout its annual cycle." *Journal of Animal Ecology* 71:296–308.

Sillett, T.S., N.L. Rodenhouse, and R. T. Holmes. 2004. "Experimentally reducing neighbor density affects reproduction and behavior in a migratory songbird." *Ecology* 85:2467–2477.

1. Suppose that environmental conditions associated with El Niño resulted in a decreased survival in Jamaica during the winter but increased fecundity rates in New Hampshire. How would the population dynamics of the species differ from that discussed?
2. How does population density influence the birthrate at the New Hampshire site?

11.8 Dispersal Can Be Density Dependent

Instead of coping with stress, some animals disperse, leaving the population to seek vacant habitats (Section 8.7). Although dispersal is most apparent when population density is high, it occurs all the time. Some individuals leave the parent population whether it is crowded or not. There is no hard-and-fast rule about who disperses.

When a lack of resources resulting from high population density forces some individuals to disperse, the ones to go are usually sub-adults driven out by adult aggression. The odds are that such individuals will perish, although a few may arrive at some suitable area and successfully become established. Because dispersal under conditions of high population density is a response to overpopulation, this type of dispersal does not function as an effective means of population regulation. More important to population regulation is dispersal when density is low or increasing, well before the local population reaches the point of overexploiting resources.

For example, ecologists Dominique Berteaux and Stan Boutin of McGill University (Quebec, Canada) studied dispersal in a red squirrel (*Tamiasciurus hudsonicus*) population in the Yukon in Canada. They found that every year, a fraction of older reproductive females left their home areas during the summer months when food availability was high. The dispersal of female adults increased the survival of their juvenile offspring that remained on the home area during the winter months when food resources are scarce.

Some dispersing individuals, especially juveniles, can maximize their probability of survival and reproduction only if they leave their birthplace. When intraspecific competition at home is intense, dispersers can relocate to habitats where resources are more accessible, breeding sites more available, and competition less intense. Further, the disperser reduces the risk of inbreeding (see Section 5.7). At the same time, dispersers incur risks that come with living in unfamiliar terrain.

Does dispersal actually regulate a population? Although dispersal is often positively correlated with population density, no relationship that can be generalized exists between the proportion of the population leaving and its increase or decrease. Dispersal may not function as a regulatory mechanism, but it contributes strongly to population expansion (see Section 8.7). It also aids in the persistence of local populations (see Chapter 19).

11.9 Social Behavior May Function to Limit Populations

Intraspecific competition can express itself in social behavior, or the degree to which individuals of the same species tolerate one another. Social behavior appears to be a mechanism that limits the number of animals living in a particular habitat, having access to a common food supply, and engaging in reproductive activities.

Many species of animals live in groups with some kind of social organization. In some populations, the group structure is crucial to acquiring resources (as with predators that hunt in packs) and maintaining defense. The organization, however, is often based on aggressiveness, intolerance, and the dominance of one individual by another. Two opposing forces are at work. One is mutual attraction of individuals; the other is a negative reaction against crowding, that is, the need for individual space.

Each individual occupies a position in the group based on dominance and submissiveness. In its simplest form, the group includes an alpha individual dominant over all others, a beta individual dominant over others except the alpha, and an omega individual subordinate to all others. Individuals settle social rank by fighting, bluffing, and threatening at initial encounters between any given pair of individuals or through a series of such encounters. Once individuals establish social rank, they maintain it by habitual subordination of those in lower positions. Threats and occasional punishment handed out by those of higher rank reinforce this relationship. Such organization stabilizes and formalizes intraspecific (contest) competitive relationships and resolves disputes with a minimum of fighting and wasted energy.

Social dominance plays a role in population regulation when it affects reproduction and survival in a density-dependent manner. An example is the wolf. Wolves live in small groups, called packs, of 6 to 12 or more individuals. The pack is an extended kin group consisting of a mated pair, one or more juveniles from the previous year that do not become sexually mature until the second year, and several related nonbreeding adults.

The pack has two social hierarchies, one headed by an alpha female and the other headed by an alpha male, who is the leader of the pack and to which all other members defer. Below the alpha male is the beta male, who is closely related, often a full brother, and who has to defend his position against pressure from males below him in the social hierarchy.

Mating within the pack is rigidly controlled. The alpha male (occasionally the beta male) mates with the alpha female. She prevents lower-ranking females from mating with the alpha and other males, and the alpha male inhibits other males from mating with her. Therefore, each pack has one reproducing pair and one litter of pups each year. They are reared cooperatively by all members of the pack.

The size of packs, which is heavily influenced by the availability of food, governs the level of the wolf population in a region. The reproducing pair gets priority in terms of the distribution of food. At high pack density and decreased availability of food, individuals may be expelled or leave the pack. Unless they have the opportunity to settle successfully in a new area and form a pack, they may not survive. Thus, at high wolf densities, mortality increases and birthrates decline. When the population of wolves is low, sexually mature males and females leave the pack, settle in unoccupied habitat, and establish their own packs that have an alpha (reproducing) female. In this case, nearly every sexually mature female reproduces, and the wolf population increases. At very low densities, however, females may have difficulty in locating males to establish a pack with, and so they fail to reproduce or even survive.

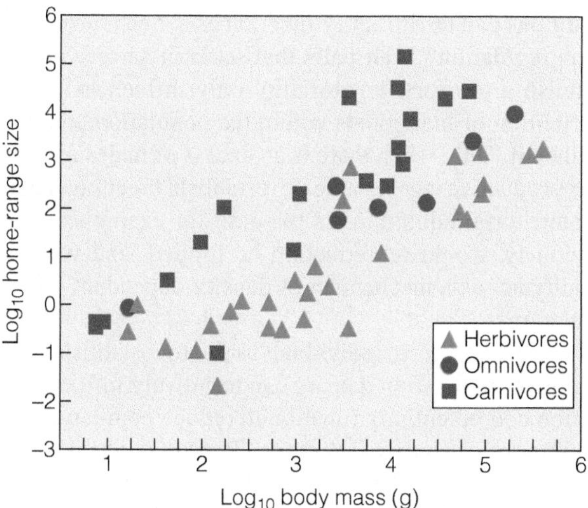

Figure 11.16 Relationship between the size of home range and body weight of North American mammals. For a given body mass, the home range of carnivores is larger than that of herbivores because the home range of a carnivore must be large enough to support a population of the prey (other animals) that it feeds on.
(Adapted from Harestad A.S. and Bunnell, F.L. "Home range and body weight: A reevaluation" Ecology, vol. 60, no. 2 April 1979. Figure 1.)

11.10 Territoriality Can Function to Regulate Population Growth

The area that an animal normally uses during a year is its **home range**. Overall size of the home range varies with the available food resources, mode of food gathering, body size, and metabolic needs. Among mammal species, the home-range size is related to body size (**Figure 11.16**), which reflects the link between body size and energy requirements (food resources; see Chapter 7, Field Studies: Martin Wikelski). In general, carnivores require a larger home range than herbivores and omnivores of the same size. Males and adults have larger home ranges than do females and juveniles.

Although the home range is not defended, aggressive interactions may influence the movements of individuals within another's home range. Some species, however, defend a core area of the home range against others. If the animal defends any part of its home range, we define that part as a **territory**, which is a defended area. If the animal defends its entire home range, its home range and territory are the same.

By defending a territory, the individual secures sole access to an area of habitat and the resources it contains. The defense of a territory involves well-defined behavioral patterns: song and call, intimidation displays such as spreading wings and tail in birds and baring fangs in mammals, attack and chase, and marking with scents that evoke escape and avoidance in rivals. As a result, territorial individuals tend to occur in more or less regular patterns of distribution (see Section 8.3 and **Figure 11.17**).

The total area available divided by the average size of the territory determines the number of territorial owners a habitat can support. When the available area is filled, owners evict excess individuals, denying them access to resources and potential mates. These individuals make up a floating reserve of potential breeders. The existence of such a reserve of potentially breeding adults has been described for several bird species, including the red grouse (*Lagopus lagopus*) in Scotland, the Australian magpie (*Gymnorhina tibicen*), Cassin's auklet (*Ptychoramphus aleuticus*) of Alaska, and the white-crowned sparrow (*Zonotrichia leucophrys*) of California. Studies of a banded (marked for identification) population

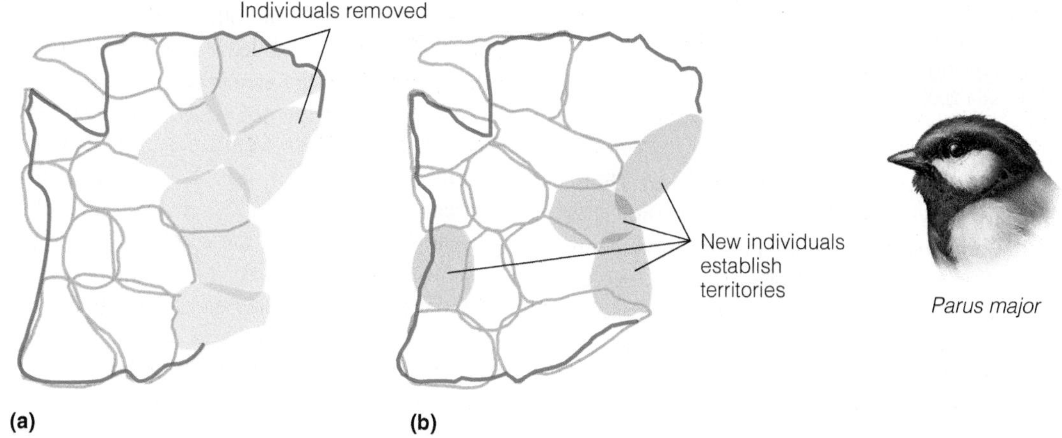

Figure 11.17 Replacement of removed individuals and settlement of vacant territories in an oak woodland. Lines represent the territory boundaries of breeding pairs of great tits (*Parus major*). Six pairs were removed between March 19 and March 24 (1969), shown as the shaded areas on map (a). Within three days, some of the resident pairs had shifted (and in some cases expanded) the boundaries of their territories, and four new pairs had taken up the new territories shown on map (b). After this short period of adjustment, the territories again formed a complete mosaic over the woodland.
(Adapted from Krebs 1971.)

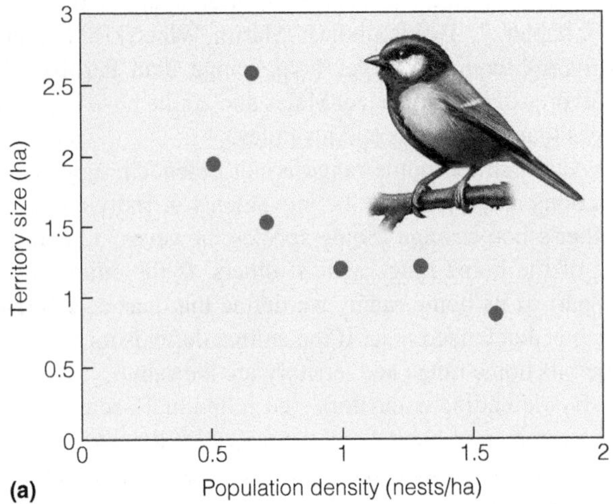

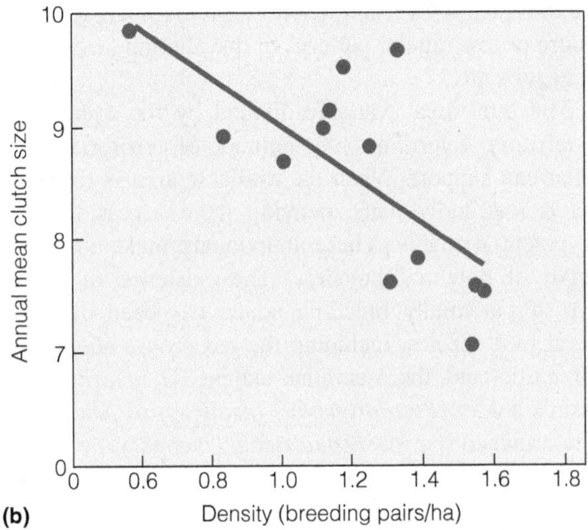

Figure 11.18 (a) Territory size as a function of breeding density, and (b) annual mean clutch size for a population of great tits (*Parus major*) occupying a woodland in central Netherlands. (Data from Both and Visser 2000.)

of white-crowned sparrows by Lewis Petrinvich and Thomas Patterson of the University of California–Riverside indicated a surplus of potential breeding individuals. In fact, 24 percent of the individuals holding territories had been floaters (no territory) for a period ranging from two to five years before acquiring a territory. Floaters quickly replaced territory holders that disappeared during the breeding season.

Ecologist John Krebs of the University of Oxford conducted a field experiment in which breeding pairs of great tits (*Parus major*) were removed from their territories in an English oak woodland (see Figure 11.17). The pairs were quickly replaced by new birds, largely first-year individuals that moved into the vacated territories from adjacent areas of suboptimal habitat, such as hedgerows.

Territoriality functions to limit access to the defended area by other individuals in the population, but under what conditions can territoriality function as a mechanism of population regulation? If all pairs that settle in an area are able to establish a territory, territoriality only influences the spatial distribution of individuals within the population and does not regulate it. Only when there is an excess of males and females of reproductive age unable to establish breeding territories (nonterritorial adults), as is the case for examples presented previously, would reproduction be limited, and would territoriality act as a mechanism of density-dependent population regulation.

Even for those individuals able to establish breeding territories, population density can negatively influence reproduction and potentially function to reduce population growth. In a three-year study of great tits (*P. major*) inhabiting woodlands in central Netherlands, ecologists Christian Booth and Marcel Visser (Netherlands Institute of Ecology) found an inverse relationship between territory size and population density (**Figure 11.18a**). Increased population density (from one location to another, or from one year to the next) resulted in increased competition and smaller territory size for males that were successful in acquiring a territory. The resulting reduction in territory size negatively impacted the probability that a territorial pair nested, the growth rate of their chicks, the number of fledglings that were recruited into the breeding population, and the survival of the territorial adults. In addition, population density had a direct negative effect on clutch size (number of eggs produced by females; **Figure 11.18b**).

11.11 Plants Preempt Space and Resources

Plants are not territorial in the same sense that animals are, but plants can capture and hold onto space. Plants from dandelions to trees occupy a certain amount of space and exclude individuals of their own and other species. When a dandelion plant spreads its rosettes of leaves on the ground, it eliminates all other plants from the area it covers. Plants also establish zones of resource depletion associated with their canopy (leaves) and root systems. Taller individuals intercept light, shading the ground below and limiting successful establishment to species that can tolerate the reduced availability of light (see Sections 4.2 and 6.8). Likewise, the uptake of water and nutrients from the soil limits availability to individuals with overlapping rooting zones. This phenomenon in the plant world is analogous to territoriality in animals, especially if we accept as an alternative definition of territoriality: individual organisms spaced out more than we would expect from a random occupancy of suitable habitat. In fact, the presence of a uniform distribution is often used as an indication that competition is occurring within plant populations (see Figure 8.11).

Ecologists Thomas Smith and Peter Goodman of the University of Witwatersrand (South Africa) examined the role of competition for belowground resources (water and nutrients) in the spatial distribution of trees in the *Acacia* savannas of Mkuzi Game Reserve in Kwazulu Natal, South Africa.

Through a statistical analysis of nearest neighbor distances (distance between an individual tree and its nearest neighboring tree), the researchers established that adult *Acacia* trees were regularly distributed within local populations (see Figure 8.11), and through a series of stand-thinning experiments (the removal of nearest neighbors) the researchers were able to establish that the observed regular distribution of trees is a result of overlap in the rooting zones and competition for belowground resources. When the researchers compared the distribution of *Acacia* trees for a series of locations in Mkuzi Game Reserve where soil moisture varied significantly as a function of soil texture (see Section 4.8, Figure 4.10), they found that trees maintained a regular spatial distribution. But the average nearest neighbor distance (for a given combined size or neighboring trees) decreased with increasing soil moisture (**Figure 11.19**), which indicated a decrease in the intensity of competition between neighboring plants.

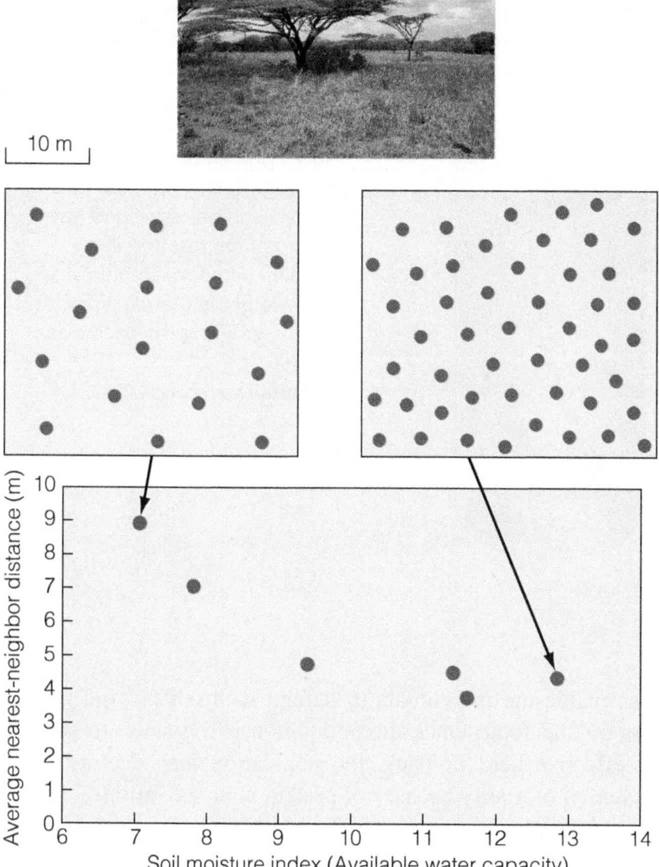

Figure 11.19 Relationship between soil moisture availability (measure as available water capacity; see Chapter 4, Figure 4.10) and average nearest neighbor distance for *Acacia tortilis* trees at six sites in Mkuzi Game Reserve, Kwazulu Natal, South Africa. Trees at all sites exhibit a regular spatial distribution. Inserts are representations of the spatial distribution of trees at two of the sites illustrating the differences in nearest neighbor distance with increasing soil water availability.
(Data from Smith and Goodman 1986.)

Interpreting Ecological Data

Q1. *The preceding figure is based on trees of the same average size. That is to say, for a given tree size, the average distance between neighboring trees decreases as the availability of soil water increases (trees are closer together at a wetter site than a drier site). If we assume that the size of the root system of trees does not change with soil moisture, what does the relationship tell us about the average overlap in the roots of neighboring trees as soil water availability increases?*

Q2. *Based on the discussion of plant response to water availability in Chapter 6 (Section 6.9), how might the size of the root systems of* Acacia *trees change with increasing soil water availability (as indicated on the x-axis)?*

Plant ecologist James Cahill of the University of Alberta has conducted a variety of experiments that have quantified the relationship between overlap in the rooting zones of neighboring plants and belowground competition by using root exclusion tubes made of PVC pipe. The exclusion tubes are placed vertically into the soil to separate roots of the target plant, which is planted inside the tube, from the roots of neighboring individuals in the population that naturally occur surrounding it. Differing quantities of neighboring roots are then allowed to access the soil within the tubes by placing differing numbers of holes through the sides of the tube before installing it. Results of these experiments have demonstrated a clear relationship between the growth of target plants and the degree of overlap with the biomass of neighboring individuals (**Figure 11.20**).

Because of their longevity, some plants, especially trees, occupy space for a long time, preventing invasion by individuals of the same or other species. Plants successful in capturing space increase their fitness at the expense of others.

11.12 A Form of Inverse Density Dependence Can Occur in Small Populations

In contrast to the model of declining rates of birth and survival with increasing population size presented in Section 11.1 (see Figure 11.1), density-dependent mechanisms have also been identified that function to reduce rates of birth and survival at low population densities. This is referred to as the **Allee effect**, which is named for the ecologist W. C. Allee, who first proposed this mechanism of population regulation. Small populations can be susceptible to a variety of factors that directly influence the rates of survival and birth that result from life history characteristics related to mating, reproduction, and defense. The result is a form of inverse density dependence, where at low population size, birthrate declines or mortality increases; and below some minimum population density, the rate of population growth is negative (**Figure 11.21**).

Among species that are widely dispersed, such as large cats, finding a mate may be impossible once the population density falls below a certain point. Many insect species use chemical odors or pheromones to communicate with and attract mates. As population density falls, there is less probability

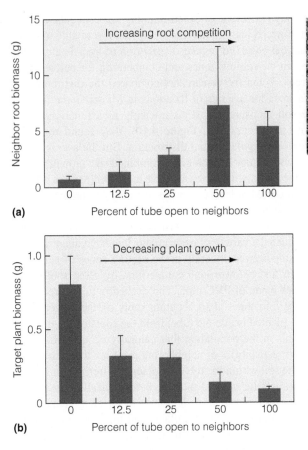

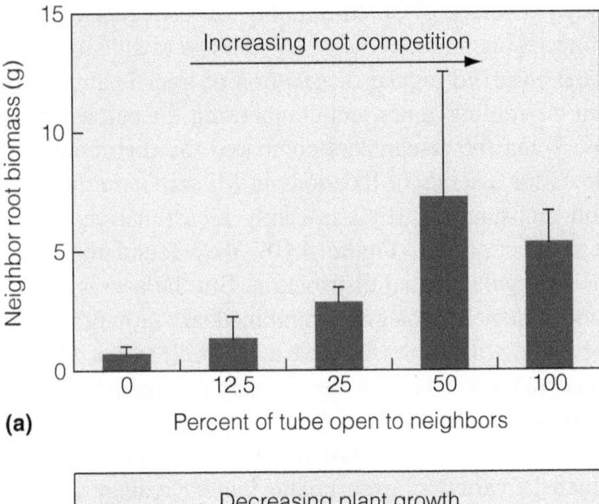

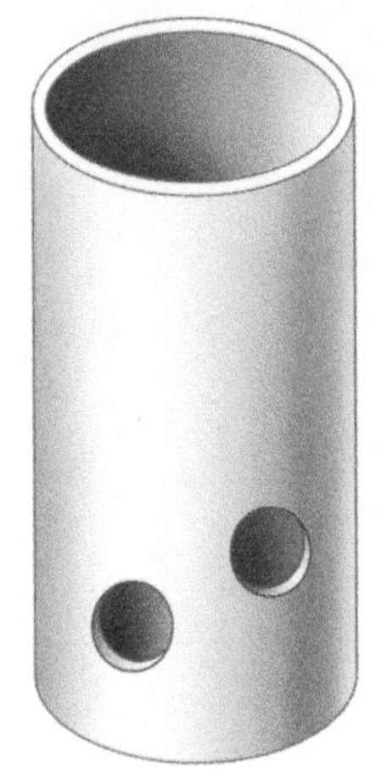

Figure 11.20 Results of an experiment controlling root competition among neighboring plants by using root exclusion tubes made out of PVC pipe. By drilling holes, the investigator controlled access to soil in the tube by neighboring plants. (a) The amount of root biomass from neighboring plants that grew into the tubes increased as a function of the number of openings in the tube (percentage of tube open to neighbors, x-axis). (b) The increase in belowground (root) competition resulted in a decline in the mean biomass (dry weight in grams) of target plants grown in the exclusion tubes.
(Adapted from Cahill 2000.)

that an individual's chemical message will reach a potential mate, and reproductive rates may decrease. Similarly, as a plant population declines and individuals are more widely scattered, the distance between plants increases and pollination may become less likely.

Ecologists Erine Hackney and James McGraw of West Virginia University examined the reproductive limitations imposed by small population size on American ginseng (*Panax quinquefolius*), a perennial herbaceous plant species inhabiting the deciduous forest ecosystems of eastern North America. Wild populations of American ginseng have historically been harvested for their medicinal value, and at some locations have been reduced to populations of only a few dozen individuals. Hackney and McGraw established experimental populations of varying density using cultivated plants. Fruit production per plant declined with decreasing population size (**Figure 11.22**). The reduced pattern of per-plant fertility was the result of reduced visitation by pollinators. Similar studies have confirmed that smaller populations of flowering plants are less apparent to potential pollinator species and therefore have a lower frequency of visitation (see Chapter 15).

In other cases, small population size may result in the breakdown of social structures in species that practice facilitation or cooperative behaviors relating to mating, foraging, or defense. Species that aggregate into groups on communal courtship grounds (called *leks*) are particularly susceptible to disruption of mating behavior and reproduction with declining population size. Many gregarious species live in herds or packs that enable the individuals to defend themselves from predators or find food. Once the population is too small to sustain an effective herd or pack, the population may decline from increased mortality because of predation or starvation.

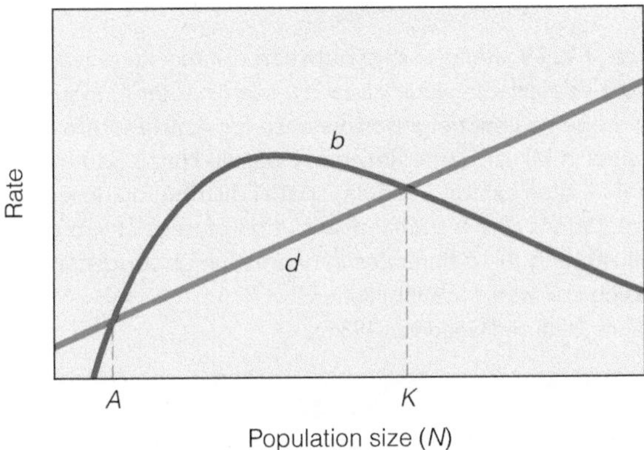

Figure 11.21 The Allee effect provides an example of how density dependence can operate in small populations. In this example, the per capita birthrate (*b*) declines at low population size (or density) because of the increased difficulty of finding a mate (see example in Figure 11.22). In theory, a low-density equilibrium would be sustained at population size *A*, where the birthrate equals the death rate. However, given the susceptibility of small populations to demographic and environmental stochasticity, the population could be driven below the low-density equilibrium, and thus continue to decline to extinction.

(a)

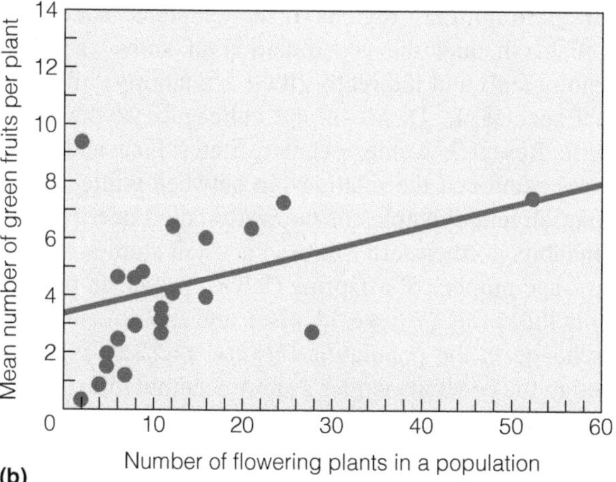

(b)

Figure 11.22 Example of the Allee effect of reduced fecundity rate at low population density in the perennial herbaceous plant (a) American ginseng. (b) There is a reduction in fruit production per plant at small population sizes in the 26 planted experimental populations. The line represents the general trend as defined by the linear regression $y = 3.36 + 0.077x$, where x is the number of flowering plants in population and y is the mean number of green fruits per plant.
(Adapted from Hackney and McGraw 2001.)

The work of Justin Brashares of the University of California–Berkeley and colleagues provides an example of the Allee effect as it relates to cooperative defense. The researchers examined the effects of declining population size on the behavior and population ecology of the critically endangered Vancouver Island marmot (*Marmota vancouverensis*). Like other highly social squirrels, the Vancouver Island marmot historically occurred in dense colonies where members benefited from group vigilance, anti-predator alarm calls, communal burrow maintenance, and access to mates in neighboring colonies. Between 1973 and 2006, however, the estimated wild population of marmots declined from at least 300 individuals to fewer than 35. As a result, the average colony size declined from 8.3 animals per colony in 1973 to 3.6 by 2006. The 90 percent decline in population size also drove increased isolation of colonies from a mean intercolony distance of roughly 3.5 km in the early 1980s to more than 20 km in 2006. The researchers found that observed reductions in colony size and proximity resulted

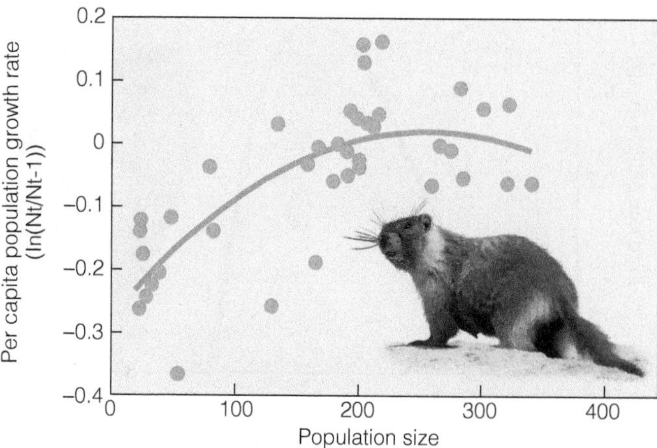

Figure 11.23 Example of the Allee effect in the Vancouver Island marmot population, an endangered species. The population exhibits strong inverse density dependence with per capita population growth rate declining with population size over the period of 1970 to 2007.
(Adapted from Brashares et al 2010.)

in a reduction in social interactions and a 10-fold increase in the time spent in anti-predator vigilance. As a result of the increased time allocated to predator defense, marmots showed an 86 percent decline in feeding rate and entered hibernation on average 20 days later than historical records indicated for previous populations on the island. These changes in time allocation, movement, and social behavior resulted in declines in survival and reproduction. The net effect reflected a strong inverse density dependence in per capita population growth for the period from 1970 to 2007 (**Figure 11.23**).

11.13 Density-Independent Factors Can Influence Population Growth

We have seen that population growth and fecundity are heavily influenced by density-dependent responses. But there are other, often overriding influences on population growth that do not relate to density. These influences are termed **density independent**. Factors such as temperature, precipitation, and natural disasters (fire, flood, and drought) may influence the rates of birth and death within a population (see this chapter, Field Studies: T. Scott Sillett) but do not regulate population growth because regulation implies feedback.

If environmental conditions exceed an organism's limits of tolerance, the result can be disastrous, affecting growth, maturation, reproduction, survival, and movement. The resulting increase in mortality rates can even lead to the extinction of local populations. For terrestrial poikilotherms (see Chapter 7, Section 7.8), variations in temperature can have a direct effect on population growth rates (**Figure 11.24**).

Pronounced changes in population growth often correlate directly with variations in moisture and temperature. For example, outbreaks of spruce budworm (*Choristoneura fumiferana*) are usually preceded by five or six years of low rainfall

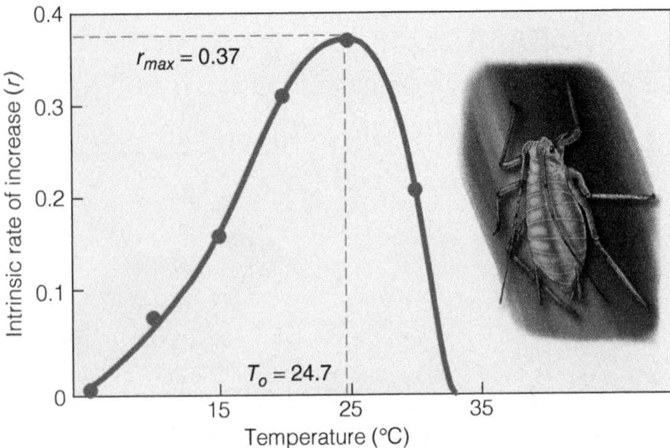

Figure 11.24 Relationship between population growth rate (r) and body temperature for laboratory populations of the turnip aphid (*Hyadaphis pseudobrassicae*). The thermal optimum (T_o) and maximum rate of population growth (r_{max}) are indicated. (Data from DeLoach 1974 as presented in Frazier et al. 2006.)

and drought. Outbreaks end when wet weather returns. Such density-independent effects can occur on a local scale where topography and microclimate influence the fortunes of local populations.

In desert regions, a direct relationship exists between precipitation and rate of increase in certain rodents and birds. Merriam's kangaroo rat (*Dipodomys merriami*) occupies lower elevations in the Mojave Desert. The kangaroo rat has the physiological capacity to conserve water and survive long periods of aridity. However, it does require the prevailing patterns of seasonal moisture availability to be sufficient to stimulate the growth of herbaceous desert plants in fall and winter. The kangaroo rat becomes reproductively active in January and February when plant growth, stimulated by fall rains, is green and succulent. Herbaceous plants provide a source of water, vitamins, and food for pregnant and lactating females. If rainfall is scant, annual plants fail to develop, and reproduction by kangaroo rats is low. This close relationship between population dynamics, seasonal rainfall, and success of winter annuals is also apparent in other rodents and birds occupying similar desert habitats.

In more northern regions of the temperate zone, winters can be harsh, and the accumulation of snow can directly (physiological) and indirectly (food availability) affect many animal species. L. D. Mech and colleagues at the Patuxent Wildlife Research Center (United States Fish and Wildlife Service) examined the relationship between winter snow accumulation and dynamics of the white-tailed deer population that inhabits northeastern Minnesota. Their studies reveal that the average number of offspring (fawns) produced per female (doe) in the spring (**Figure 11.25a**), and subsequently the annual change in the population (**Figure 11.25b**), is inversely related to the previous winter's snow accumulation.

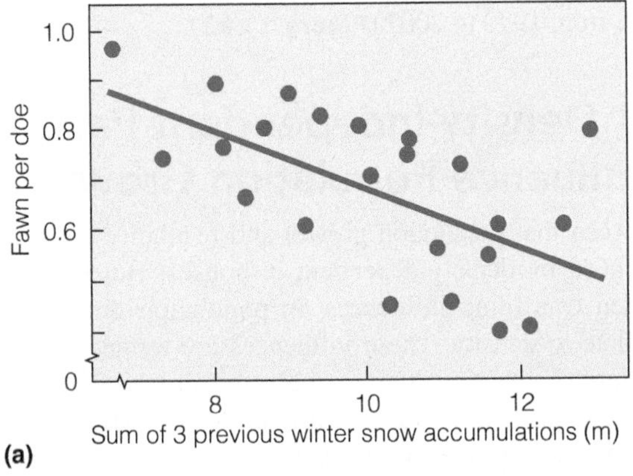

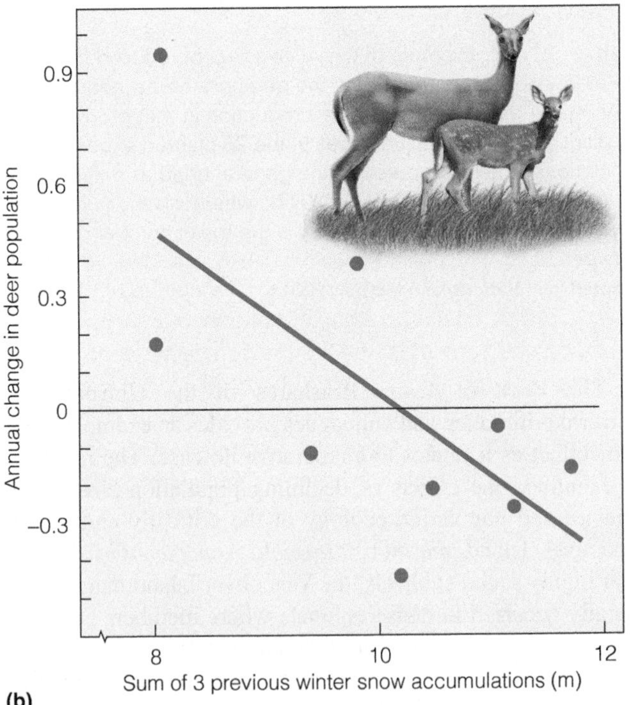

Figure 11.25 Relationship between the sum of the previous three winter month snow accumulations in northeastern Minnesota and the population of white-tailed deer. (a) Fecundity (fawn-to-doe ratio). (b) Percentage of annual change in next winter's population. (Adapted from Mech et al. 1987.)

ECOLOGICAL Issues & Applications

The Conservation of Populations Requires an Understanding of Minimum Viable Population Size and Carrying Capacity

Human activities and associated habitat loss have resulted in population decline for an ever-increasing number of plant and animal species (see Chapter 9, *Ecological Issues & Applications*). Often, these populations are restricted to protected areas (nature reserves, etc.), and an adequate conservation plan requires the preservation of as many individuals as possible within the greatest possible area. But with limited land and resources, conservation ecologists must address a number of fundamental questions relating to the minimum number of individuals and the physical area of suitable habitat necessary to sustain a population.

The number of individuals needed to ensure that a species persists in a viable state must be large enough to cope with chance variations in demographic processes (births and deaths), environmental changes, genetic drift, and catastrophes (see Chapters 5 and 9). In addition, for some species, the Allee effect (see Section 11.12 and Figure 11.21) can impose a direct constraint on the minimum size required for population survival. Conservation ecologist M. L. Shaffer defined the number of individuals necessary to ensure the long-term survival of a species as the **minimum viable population (MVP)**. Shaffer defined the MVP for any given population in any given habitat as the "smallest isolated population having a 99 percent chance of remaining extant for 1000 years despite the foreseeable effects of demographic and environmental stochasticity, and natural catastrophes." Although Shaffer realized the tentative nature of this definition, the key point is that MVP size allows a quantitative assessment of how large a population must be to assure long-term survival.

Genetic models suggest that for vertebrate species, populations with an actual size of less than 1000 are highly vulnerable to extinction. For species exhibiting extreme variation in population size, such as invertebrates and annual plants, it has been suggested that MVPs of 10,000 individuals are required.

In fact, the actual MVP for a species depends on the life history of the species (longevity, mating system, etc.) and the ability of individuals to disperse among habitat patches. Despite the difficulty in quantifying MVP for a given species, the concept is of paramount importance in conserving species and maintaining biological diversity.

David Reed of Macquire University (New South Wales, Australia) and colleagues used population viability analysis to estimate MVPs for 102 vertebrate species for which suitable data were available (2 amphibians, 28 birds, 1 fish, 53 mammals, and 18 reptiles). Population viability analysis (PVA) is a statistical method of risk assessment that brings together species' characteristics and environmental variability to forecast population dynamics and extinction risk. Each PVA is individually developed for a target population or species, and consequently, each PVA is unique. The researchers defined a MVP size as one with a 99 percent probability of persistence for 40 generations. The mean and median estimates of MVP were 7316 and 5816 adults, respectively. MVPs did not differ significantly among major taxa, or with geographic region (latitude), but were negatively correlated with population growth rate (**Figure 11.26**).

Once a MVP size has been established for a species, the area required to support that population must be considered. This requires an understanding of the carrying capacity of the habitat, that is, the maximum sustainable population for a given area (and the resources that the area supplies; see Section 11.1). The area of suitable habitat necessary for maintaining the MVP is called the **minimum dynamic area (MDA)**. An estimation of MDA for a species begins with an understanding of the home-range size of the individuals, family groups, or colonies. Recall that the area (home range) requirement of an individual of a species increases with body size (see Figure 11.16). In addition, for a given body size, the home-range requirement of a carnivore is larger than that of an herbivore. With knowledge of the area requirement per individual of a species, and an estimate of MVP, the area necessary to sustain a viable population can be established (**Figure 11.27**). For large carnivores, the area required to sustain an MVP can be enormous. Wildlife biologist Reed Noss estimated that to preserve a population of 1000 grizzly bears would require an area of 2 million km^2. This is why most large carnivore populations (such as the African lion, Asian tiger, and gray wolf of North America) are endangered, and why they are restricted to only the largest public lands and nature reserves.

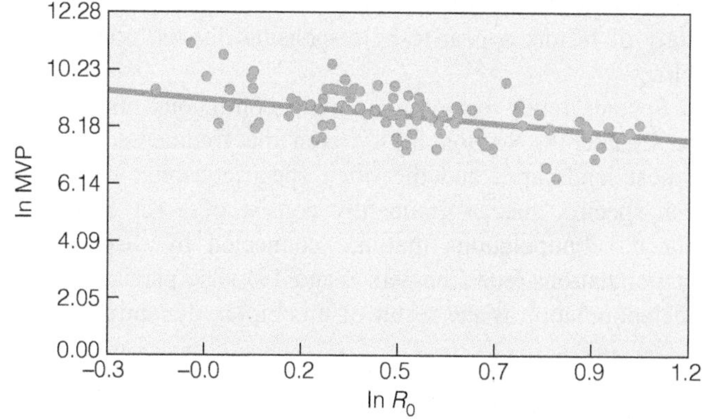

Figure 11.26 The minimum viable population (MVP; defined as the minimum number of adults required for a 99 percent chance of a population persisting for 40 generations) as a function of the population growth rate (R_0). Values of both MVP and R_0 are $\log_e$ transformed (natural log). Each point represents one of the 102 species examined in the study.

(Adapted from Reed, D. H., J. J. O'Grady, B. W. Brook, J. D. Balloc, and R. Frankham. 2003. Estimates of minimum viable population sizes for vertebrates and factors influencing those estimates. Biological Conservation 113:23–34. Fig. 3, pg. 28.)

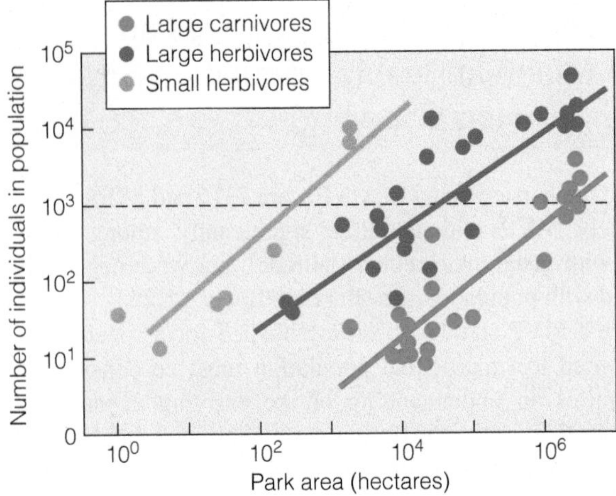

Figure 11.27 Large parks contain larger populations of each species than small parks; only the largest parks may contain long-term viable populations of many large vertebrate species. Each symbol represents an animal population in a park. If the minimum viable population (MVP) of a species is 1000 (dashed line), parks of at least 100 ha will be needed to protect small herbivores. Parks of more than 10,000 hectares will be needed to support populations of large herbivores, and parks of at least 1 million hectares will be needed to protect populations of large carnivores.
(Adapted from Schonewald-Cox 1983.)

One of the best-documented cases of MVP size comes from Joel Berger's (University of Montana) study of the persistence of bighorn sheep (*Ovis canadensis*) populations in the deserts of the southwestern United States. The study, which examined 120 populations, found that all populations with 50 individuals or fewer became extinct within 50 years. In contrast, virtually all of the populations with 100 or more individuals persisted over that same period (**Figure 11.28**). No single cause was identified for the local extinctions; rather, a wide variety of factors appear to be responsible for the population declines.

Species rarely occur in a single contiguous population (see Chapter 8, Section 8.2). Given the fragmented nature of most landscapes and the often specific habitat needs of given species, species frequently consist of a set of semi-isolated subpopulations that are connected by dispersal—metapopulations (see Chapters 8 and 19). The persistence of a metapopulation is the result of a complex dynamic among subpopulations. The rates of birth, death, immigration, and emigration of each subpopulation interact with the size and spatial arrangement of habitat patches to determine dynamics of the metapopulation as a whole, which is a topic we will explore in detail in Chapter 19.

(a)

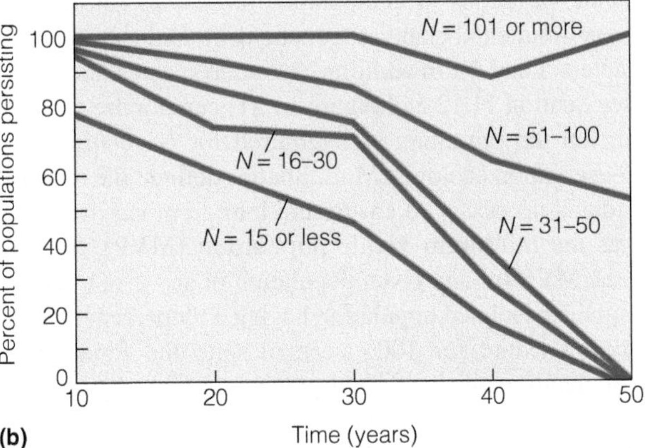

(b)

Figure 11.28 Relationship between the size of a population of bighorn sheep, shown in (a), and the percentage of populations that persist over time. In (b), (N) represents population size. Almost all populations of 100 or more sheep persisted beyond 50 years, whereas populations with fewer than 50 individuals did not.
(Adapted from Berger 1990.)

SUMMARY

Logistic Population Growth 11.1

Because resources are limited, exponential growth cannot be sustained indefinitely. The maximum population size that can be sustained for a particular environment is termed the *carrying capacity* (K). The logistic model of population growth incorporates the concept of carrying capacity into the previously developed model of exponential growth. The result is a decrease in the rate of population growth as the population size approaches the carrying capacity.

Density-Dependent Regulation 11.2

Populations do not increase indefinitely. As resources become less available to an increasing number of individuals, birthrates decrease, mortality increases, and population growth slows.

If the population declines, mortality decreases, births increase, and population growth speeds up. Positive and negative feedbacks function to regulate the population.

Competition 11.3

Intraspecific competition occurs when resources are in short supply. Competition can take two forms: scramble and contest. In scramble competition, growth and reproduction are depressed equally across individuals as competition increases. In contest competition, dominant individuals claim sufficient resources for growth and reproduction. Others produce no offspring or perish. Competition can involve interference among individuals or indirect interactions via exploitation of resources.

Competition, Growth, and Development 11.4

Competition for scarce resources can decrease or retard growth and development. Up to a point, plants respond to competition by modifying form and size.

Competition and Mortality 11.5

A common response to high population density is reduced survival. Mortality functions to increase resource availability for the remaining individuals, allowing for increased growth.

Competition and Reproduction 11.6

High population density and competition can also function to delay reproduction in animals and reduce fecundity in both plants and animals.

Density and Stress 11.7

In animals, the stress of crowding may cause delayed reproduction, abnormal behavior, and reduced ability to resist disease and parasitic infections; in plants, crowding results in reduced growth and production of seeds.

Dispersal 11.8

Dispersal is a constant phenomenon in populations. When dispersal occurs in response to the overexploitation of resources, or crowding, it is not a mechanism of population regulation. It can, however, regulate populations if the rate of dispersal increases in response to population growth.

Social Behavior 11.9

Intraspecific competition may be expressed through social behavior. The degree of tolerance can limit the number of animals in an area and access of some animals to essential resources. A social hierarchy is based on dominance. Dominant individuals secure most of the resources. Shortages are borne by subdominant individuals. Such social dominance may function as a mechanism of population regulation.

Territoriality 11.10

Social interactions influence the distribution and movement of animals. The area an animal normally covers in its life cycle is its home range. The size of a home range is influenced by body size.

If the animal or a group of animals defends a part or all of its home range as its exclusive area, it exhibits territoriality. The defended area is its territory. Territoriality is a form of contest competition in which part of the population is excluded from reproduction. The nonreproducing individuals act as a floating reserve of potential breeders, available to replace losses of territory holders. In such a manner, territoriality can act as a population-regulating mechanism.

Plants Preempt Space 11.11

Plants are not territorial in the same sense as animals are, but they do hold on to space, excluding other individuals of the same or smaller size. Plants capture and hold space by intercepting light, moisture, and nutrients.

Inverse Density Dependence 11.12

Small populations can be susceptible to a variety of factors that directly influence the rates of survival and birth that result from life history characteristics related to mating, reproduction, and defense. The result is a form of inverse density dependence, where at low population size, birthrate declines or mortality increases; and below some minimum population density, the rate of population growth is negative.

Density-Independent Factors 11.13

Density-independent influences, such as weather, affect but do not regulate populations. They can reduce populations to the point of local extinction. Their effects, however, do not vary with population density. Regulation implies feedback.

Conservation of Populations Ecological Issues & Applications

The loss of habitat as a result of human activities has caused the populations of many species to decline to levels where their future existence is threatened. The conservation of these species depends on an understanding of the minimum viable population size necessary for the survival of future generations and the area of habitat necessary to sustain these populations.

STUDY QUESTIONS

1. What is the difference between the exponential and logistic models of population growth?
2. What is the definition of carrying capacity (K)? How is it used in the framework of the logistic model to introduce a mechanism of density-dependent population regulation?
3. We have seen from many examples in Chapter 11 that competition among individuals within a population can result in an inverse relationship between population density and the growth (see Figures 11.8 and 11.9) and reproduction (see Figures 11.14 and 11.15) of individuals

within the population. Besides the inverse correlation between population density and growth and reproduction, what condition must hold for the researcher to establish that competition is responsible for these relationships?

4. Competition can function as a mechanism of density-dependent population regulation. How might scramble and contest competition differ in their effect on population growth (regulation)?
5. Distinguish between home range and territory.
6. How does the relationship between body size and home-range size influence the estimates of population density and abundance discussed in Chapter 8?
7. What condition must hold true for territoriality to function as a density-dependent mechanism regulating population growth (and density)?
8. How might social dominance within a population function to regulate population growth?
9. How might the Allee effect be a factor in the conservation of endangered species?
10. Years of below-average rainfall in Kruger National Park (South Africa) cause a decline in the growth and productivity of grasses. Mortality rates in populations of herbivores such as the African buffalo then increase, resulting in a decline in population density. Based on the information provided, would you conclude the annual variations in rainfall within Kruger function as a density-dependent mechanism regulating growth of the buffalo population? Is there any additional information that would cause you to change your answer?

FURTHER READINGS

Classic Studies

Lack, D. 1954. *The natural regulation of animal numbers.* London: Oxford University Press.
A classic text on population regulation that should be read by those interested in the historical development of ideas in population ecology.

Shaffer, M. L. 1981. "Minimum population sizes for species conservation." *BioScience* 31:131–134.
This article introduces the concept of minimum viable population size in conservation ecology.

Sinclair, A. R. E. 1977. *The African buffalo: A study of resource limitations of populations.* Chicago: University of Chicago Press.
A classic study of intraspecific population regulation. Provides an example of the interaction between various factors that together function to regulate the population of this large herbivore.

Recent Research

Barrett, G. W., and E. P. Odum. 2000. "The twenty-first century: The world at carrying capacity." *Bioscience* 50:363–368.
A discussion on the relationships among global human population growth, economic growth, and world carrying capacity.

Courchamp, F., L. Berec, and J. Gascoigne. 2008. *Allee effects in ecology and conservation.* New York: Oxford University Press.
A concise and clearly written text that provides an overview of the topic, providing examples for plant and animal populations in both marine and terrestrial environments.

Murdoch, W. W. 1994. "Population regulation in theory and practice." *Ecology* 75:271–287.
This excellent article contrasts the theory of population regulation as understood by ecologists against the difficulties of experimentally detecting regulation in natural populations.

Newton, I. 1998. *Population limitation in birds.* London: Academic Press Limited.
Chapters 2 through 5 present an excellent, readable overview of population regulation.

Stephens, P. A., and W. J. Sutherland. 1999. "Consequences of the Allee effect for behavior, ecology and conservation." *Trends in Ecology and Evolution* 14:401–405.
An extensive, but technical, review of the concept of the Allee effect on population regulation.

Turchin, P. 1999. "Population regulation: A synthetic view." *Oikos* 84:160–163.

Wolff, J. O. 1997. "Population regulation in mammals: An evolutionary perspective." *Journal of Animal Ecology* 66:1–13.
The above two references provide an excellent discussion of the broader issues relating to the regulation of natural populations.

CHAPTER 12 Species Interactions, Population Dynamics, and Natural Selection

In the species interaction known as *brood parasitism*, female birds of one species lay their eggs in an active nest of another bird species. When the eggs hatch, the adoptive parents feed and care for the nestlings. In this example, a Reed Warbler (*Acrocephalus scripaceus*) is feeding a Common Cuckoo (*Cuculos canous*) chick, which hatched from an egg deposited in the Warbler nest by a female Cuckoo.

CHAPTER GUIDE

THUS FAR, WE HAVE EXAMINED INTERACTIONS among organisms only as they pertain to individuals within the same population or species (intraspecific interactions). We examined interactions that relate to reproduction (selecting mates and care and defense of offspring, Chapter 10). We also examined how competition for limited resources among individuals can function to limit population growth (Chapter 11). But species cannot be viewed in isolation. Species occupying the same physical area—be it a lake, stream, forest, or field—interact in many ways. Central to these interactions is the need to acquire the basic resources necessary for growth and reproduction. Although different plant species that co-occur within a habitat may differ in their specific needs for certain essential nutrients, all plants require the same resources of water, light, carbon dioxide, and other essential nutrients. As a result, competition for these resources within a habitat can become intense, with acquisition by individuals of one species reducing availability to individuals of other species.

Among heterotrophic organisms, the range of potential interactions expands. Because heterotrophic organisms derive energy and nutrients from consuming organic matter, the very act of feeding involves interaction between species: interaction between predator (the consumer) and prey (the consumed). When different predator species feed on the same species as prey, there is also the potential for competition.

For some species, other organisms provide habitat as well as nourishment. Many microorganisms, such as bacteria and fungi, take up residence in or on other organisms. They draw their energy and nutrients from the host organism. This is the interaction between parasite and host.

Not all interactions among species are negative (i.e., involving winners and losers). Interactions that are mutually beneficial to the parties involved are ubiquitous and relate to nutrition, shelter, defense, and reproduction.

In this chapter, we develop a simple classification of species interactions based on a qualitative description (neutral, positive, or negative) of how individuals of one population influence individuals of the population with which they are interacting. We then show how species interactions influence the probabilities of survival and reproduction of individuals within each population. Finally, we explore how species interactions influence their respective populations on two timescales: (1) by influencing the demographic processes of birth and death—with interspecific interactions playing a central role in population dynamics—and (2) by differentially influencing the survival and reproduction of individuals within the population these same interactions can also function as agents of natural selection. In functioning as agents of natural selection, species interactions play a central role in the process of evolution. In the chapters that follow, we will examine specific interactions in more detail, developing quantitative models of how interactions between two species influence their linked population dynamics.

Table 12.1 Population Interactions between Individuals of Two Species (A and B)

Type of Interaction	Response: Species A	Response: Species B
Neutral	0	0
Mutualism	+	+
Commensalism	+	0
Competition	−	−
Amensalism	−	0
Predation	+	−
Parasitism	+	−
Parasatoidism	+	−

12.1 Species Interactions Can Be Classified Based on Their Reciprocal Effects

If we designate the positive effect of one species on another as +, a detrimental effect as −, and no effect as 0, we can use this qualitative description of the different ways in which populations of two species interact to develop a classification of possible interactions between two co-occurring species (**Table 12.1**). When neither of the two populations affects the other, the relationship is (00), or neutral. If the two populations mutually benefit, the interaction is (++), or positive, and the relationship is called **mutualism** (Chapter 15). When one species maintains or provides a condition that is necessary for the welfare of another but does not affect its own well-being, the relationship (+0) is called **commensalism**. For example, the trunk or limb of a tree provides the substrate on which an epiphytic orchid grows (**Figure 12.1**). The arrangement benefits

Figure 12.1 The trunk or limb of a tree provides the substrate on which an epiphytic orchid grows. This is an example of commensalism (+0) in which the orchid benefits from the interaction while the effect on the tree is neutral.

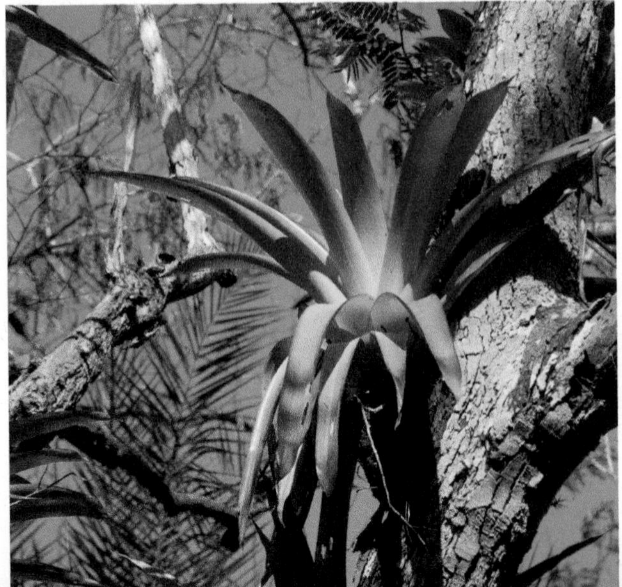

the orchid, which gets nutrients from the air and moisture from aerial roots, whereas the tree is unaffected.

When the relationship is detrimental to the populations of both species (−−), the interaction is termed **competition** (Chapter 13). In some situations, the interaction is (−0). One species reduces or adversely affects the population of another, but the affected species has no influence in return. This relationship is **amensalism**. It is considered by many ecologists as a form of asymmetric competition, such as when taller plant species shade species of smaller stature.

Relationships in which one species benefits at the expense of the other (+−) include predation, parasitism, and parasitoidism (see Chapter 14 for more information on predation and Chapter 15 for more information on parasitism and parasitoidism). **Predation** is the process of one organism feeding on another, typically killing the prey. Predation always has a negative effect on the individual prey. In **parasitism**, one organism feeds on the other but rarely kills it outright. The parasite and host live together for some time. The host typically survives, although its fitness is reduced. **Parasitoidism**, like predation, kills the host eventually. Parasitoids, which include certain wasps and flies, lay eggs in or on the body of the host. When the eggs hatch, the larvae feed on it. By the time the larvae reach the pupal stage, the host has succumbed.

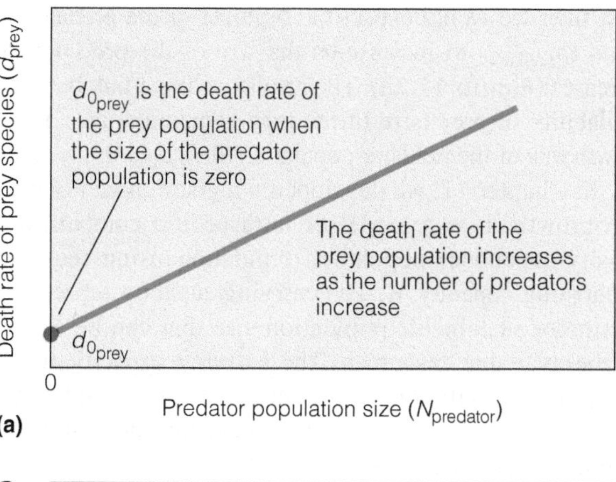

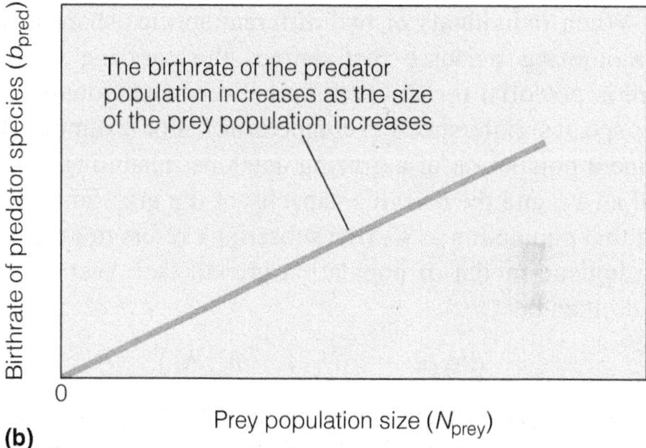

Figure 12.2 Examples of interactions between predator and prey populations that directly or indirectly influence their respective growth rates (difference between rates of birth and death). (a) Relationship between predator population size (N_{prey}) and the rate of mortality (d_{prey}) for the prey population. The prey population has a rate of mortality in the absence of predators (d_0). As the predator population increases, the probability of an individual prey being captured and consumed increases. The result is an increase in the mortality rate of the prey population as the predator population increases. (b) Relationship between the birthrate of predator population (b_{pred}) and the size of the prey population (N_{prey}). As the size of the prey population increases, the rate at which individual predators are able to capture and consume prey increases. This increase in food supply results in an increase in the reproductive output of individual predators and subsequently an increase in the birthrate of the predator population.

12.2 Species Interactions Influence Population Dynamics

The varieties of species interactions outlined in the previous section typically involve the interaction of individual organisms. A predator captures a prey or a bacterium infects a host organism. Yet through their beneficial or detrimental effects on the individuals involved, these interactions influence the collective properties of birth and death at the population level, and in doing so, influence the dynamics of the respective populations. For example, by capturing and killing individual prey, predators function as agents of mortality. We might therefore expect that as the number of predators ($N_{predator}$) in an area increases, the number of prey captured and killed will likewise increase. If we assume the simplest case of a linear relationship, we can represent the influence of changes in the predator population ($N_{predator}$) on the death rate of the prey population (d_{prey}) as shown in **Figure 12.2a**. As the number of predators in the population ($N_{predator}$) increases, the probability of an individual in the prey population (N_{prey}) being captured and killed increases. Subsequently, the death rate of the prey population increases. The net effect is a decline in the growth rate of the prey population. Note the similarity in the functional relationship presented in Figure 12.2a with the example of density-dependent population control presented earlier (Chapter 11, Figure 11.1). Previously, we examined how an increase in population size can function as a negative feedback on population growth by increasing the mortality rate or decreasing the birthrate (density-dependent population regulation; Section 11.2 and Figure 11.4). The relationship shown in Figure 12.2a expands the concept of density-dependent population regulation to include the interaction *between* species. As the population of predators increases, there is a subsequent decline in the population of prey as a direct result of the prey's increased rate of mortality.

A similar approach can be taken to evaluate the positive effects of species interactions. In the example of predation, whereas the net effect of predation on the prey is negative, the predator benefits from the capture and consumption of prey. Prey provides basic food resources to the predator and directly influences its ability to survive and reproduce. If we assume that the ability of a predator to capture and kill prey increases as the number of potential prey increase (N_{prey}), and that the reproductive fitness of a predator is directly related to its consumption of prey, then we would expect the birthrate of the predator population ($b_{predator}$) to increase as the size of the prey population increases (**Figure 12.2b**). The result is a direct link between the availability of prey (size of the prey population, N_{prey}) and the growth rate of the predator population ($dN_{predator}/dt$).

In Chapter 11, we developed a logistic model of population growth. It is a model of intraspecific competition and density-dependent population regulation using the concept of carrying capacity, *K*. The carrying capacity represents the maximum sustainable population size that can be supported by the available resources. The carrying capacity functions to regulate population growth in that as the population size approaches *K*, the population growth rate approaches zero ($dN/dt = 0$).

When individuals of two different species share a common limiting resource that defines the carrying capacity, there is potential for competition between individuals of the two species (interspecific competition). For example, let's define a population of a grazing antelope inhabiting a grassland as N_1, and the carrying capacity of the grassland to support that population as K_1 (the subscript 1 refers to *species 1*). The logistic model of population growth (see Section 11.1) would then be:

$$dN_1/dt = r_1N_1(1 - N_1/K_1)$$

Now let's assume that a second species of antelope inhabits the same grassland, and to simplify the example, we assume that individuals of the second species—whose population we define as N_2—have the same body size and exactly the same rate of food consumption (grazing of grass) as do individuals of the first species. As a result, when we evaluate the role of density-dependent regulation on the population of *species 1* (N_1), we must now also consider the number of individuals of *species 2* (N_2) because individuals of both species feed on the grass that defines the carrying capacity of *species 1* (K_1). The new logistic model for *species 1,* will be:

$$dN_1/dt = r_1N_1(1 - (N_1 + N_2)/K_1)$$

For example, if the carrying capacity of the grassland for *species 1* is 1000 individuals ($K_1 = 1000$)—because *species 2* draws on the exact same resource in exactly the same manner—the combined carrying capacity of the grassland is also 1000. If there are 250 individuals of *species 2* ($N_2 = 250$) living on the grassland, it effectively reduces the carrying capacity for *species 1* from 1000 to 750 (**Figure 12.3a**). The population growth rate of *species 1* now depends on the population sizes of both *species 1* and 2 relative to the carrying capacity (**Figure 12.3b**). Although we have defined the two antelope species as being identical in their use of the limiting resource that defines the carrying capacity, this is not always the case. In reality, it is necessary to evaluate the overlap in resource use and quantify the equivalency of one species to another (see **Quantifying Ecology 12.1**).

In all cases in which individuals of two species interact, the nature of the interaction can be classified qualitatively as neutral, positive, or negative, and the influence of the specific interaction can be evaluated in terms of its impact on the survival or reproduction of individuals within the populations. In the discussion that follows, we develop quantitative models to examine how the diversity of species interactions outlined in Table 12.1 influence the combined population dynamics of the species involved (Chapters 13, 14, and 15). In all cases, these models involve quantifying the per capita effect of interacting individuals on the birthrates and death rates of the respective populations.

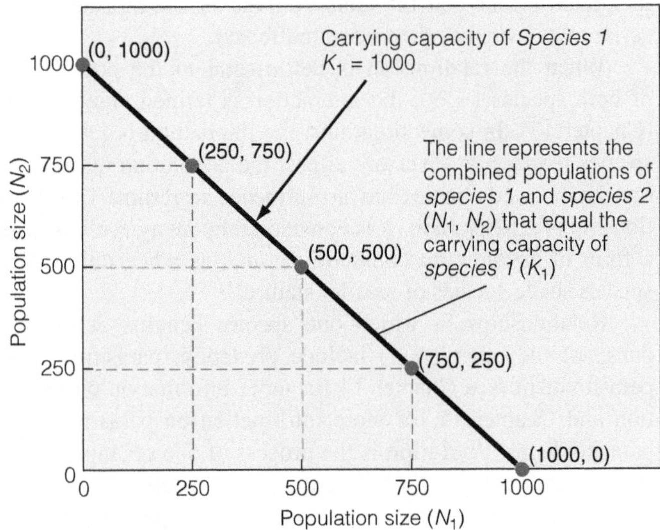

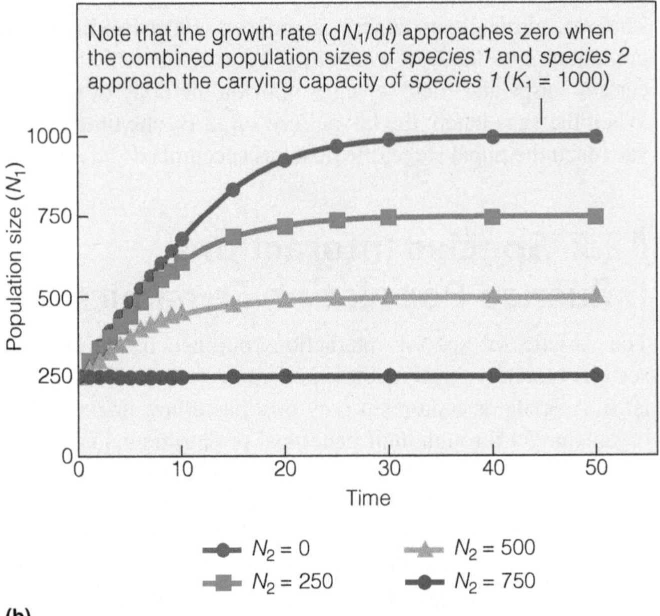

Figure 12.3 Influence of interspecific competition on carrying capacity. (a) If we assume that the carrying capacity of *species 1* (K_1) is 1000 individuals and that individuals of *species 2* (N_2) are identical to *species 1* (N_1) in their use of the shared, limited food resource, the presence of each individual of *species 2* will effectively reduce the carrying capacity for *species 1* by one individual. (b) The effect will be a reduction in the growth rate of *species 1* for given population sizes of *species 2*. (Growth rates based on logistic model with $r = 0.15$ and initial population size of N_1 at 250.)

QUANTIFYING ECOLOGY 12.1 Incorporating Competitive Interactions in Models of Population Growth

When individuals of two different species (represented as populations N_1 and N_2) share a common limiting resource that defines the carrying capacity for each population (K_1 and K_2), there is potential for competition between individuals of the two species (interspecific competition). Thus, the population density of both species must be considered when evaluating the role of density-dependent regulation on each population. In Section 12.2, we gave the example of two species of antelope that share the common limiting food resource of grass. We assumed that individuals of the two species were identical in their food selection and the rate at which they feed, therefore, with respect to the carrying capacity of the grassland, individuals of the two species are equivalent to each other; that is, in resource consumption one individual of *species 1* is equivalent to one individual of *species 2*. As a result, when evaluating the growth rate of *species 1* using the logistic model of population growth, it is necessary to include the population sizes of both species relative to the carrying capacity (see Figure 12.4):

$$dN_1/dt = r_1N_1(1 - (N_1 + N_2)/K_1)$$

However, two species, even closely related species, are unlikely to be identical in their use of resources. So it is necessary to define a conversion factor that can equate individuals of *species 2* to individuals of *species 1* as related to the consumption of the shared limited resource. This is accomplished by using a competition coefficient, defined as α, that quantifies individuals of *species 2* in terms of individuals of *species 1* as related to the consumption of the shared resource. Using the example of two antelope species, let us now assume that both species still feed on the same resource (grass), however, individuals of *species 2* have on average only half the body mass of individuals of *species 1* and therefore consume grass at only half the rate of *species 1*. Now an individual of *species 2* is only equivalent to one-half an individual of *species 1* with respect to the use of resources. In this case, $\alpha = 0.5$, and we can rewrite the logistic model for *species 1* shown previously as:

$$dN_1/dt = r_1N_1(1 - (N_1 + \alpha N_2)/K_1)$$

Because in Section 12.2 we defined the carrying capacity of the grassland for *species 1* as $K_1 = 1000$, we can substitute the values of α and K_1 in the preceding equation:

$$dN_1/dt = r_1N_1(1 - (N_1 + 0.5N_2)/1000)$$

Now the growth rate of *species 1* (dN_1/dt) approaches zero as the combined populations of *species 1* and *2*, represented as $N_1 + 0.5N_2$, approach a value of 1000 (the value of K_1).

We have considered how to incorporate the effects of competition from *species 2* into the population dynamics of *species 1* using the competition coefficient α, but what about the effects of *species 1* on *species 2*? The competition for food resources (grass) will also function to reduce the availability of resources to *species 2*. We can take the same approach and define a conversion factor that can equate individuals of *species 1* to individuals of *species 2*, defined as β. Because individuals of *species 1* consume twice as much resource (grass) as individuals of *species 2*, it follows that an individual of *species 1* is equivalent to 2 individuals of *species 2*; that is, $\beta = 2.0$. It also follows that if individuals of *species 2* require only half the food resources as individuals of *species 1*, then the carrying capacity of the grassland for *species 2* should be twice that for *species 1*; that is, $K_2 = 2000$. The logistic growth equation for *species 2* is now:

$$dN_2/dt = r_2N_2(1 - (N_2 + \beta N_1)/K_2)$$

or, substituting the values for β and K_2

$$dN_2/dt = r_2N_2(1 - (N_2 + 2.0N_1)/2000)$$

We now have a set of equations that can be used to calculate the growth of the two competing species that considers their interaction for the limiting food resource. We explore this approach in more detail in the following chapter (Chapter 13).

In the example of the two hypothetical antelope species presented previously, the estimation of the competition coefficients (α and β) appear simple and straightforward. Both species are identical in their diet and differ only in the rate at which they consume the resource (which is defined as a simple function of their relative body masses). In reality, even closely related species drawing on a common resource (such as grazing herbivores) differ in their selection (preferring one group of grasses of herbaceous plants over another), foraging behavior, timing of foraging, and other factors that influence the nature of their relative competitive effects on each other. As such, quantifying species interactions, such as resource competition, can be a difficult task, as we shall see in the following chapter (Chapter 13, Interspecific Competition).

12.3 Species Interactions Can Function as Agents of Natural Selection

For a number of reasons, the interaction between two species will not influence all individuals within the respective populations equally. First, interactions among species involve a diverse array of physiological processes and behavioral activities that are influenced by phenotypic characteristics (physiological, morphological, and behavioral characteristics of the individuals). Secondly, these phenotypic characteristics vary among individuals within the populations (see Chapter 5). Therefore, the variations among individuals within the populations will result in differences in the nature and degree of interactions that occur. For example, imagine a species of seed-eating bird that feeds on the seeds of a single plant species. Individuals of the plant species exhibit a wide degree of variation in the size of seeds that they produce. Some individuals produce smaller seeds, whereas others produce larger seeds (**Figure 12.4a**), and seed size is a heritable characteristic (genetically determined). Seed size is important to the birds because the larger the seed, the thicker the seed coat, and the more difficult it is for a bird to crush the seed with its bill. If the seed coat is not broken, the seed passes through the digestive system undigested and provides no food value to the bird. As a result, birds actively select smaller seeds in their diet (**Figure 12.4c**). In doing so, the birds are decreasing the reproductive success of individual plants that produce small seeds while increasing the relative fitness of those individuals that produce larger seeds. The net effect is a shift in the distribution of phenotypes in the plant population to individuals that produce larger, harder seeds (**Figure 12.4d**). In this situation, the bird population (and pattern of seed predation) is functioning as an agent of natural selection, increasing the relative fitness of one phenotype over another (see Section 5.6). Over time, the result represents a directional change in the genetic structure of the population (gene frequencies), that is, the process of evolution (Chapter 5).

In this example, the predator functions as an agent of natural selection, decreasing the reproduction for certain phenotypes (small seed-producing individuals) within the plant population and increasing the relative fitness of other phenotypes (large seed-producing individuals). But the shifting distribution of phenotypes within the plant population and the resulting change in the distribution of food resources will in turn have a potential influence on the predator population (**Figure 12.4b**). The directional selection for increased seed size within the plant population decreases the relative abundance of smaller seeds, effectively decreasing the availability of food resources for birds with smaller bill sizes. If the birds with smaller bills are unable to crack the larger seeds, these individuals will experience a decreased probability of survival and reproduction, which increases the relative fitness of individuals with larger bill size. The shift in the distribution of phenotypes in the plant population, itself a function of selective pressures imposed by the bird population, now functions as an agent of natural selection in the predator (bird) population. The result is a shift in the distribution of phenotypes and associated gene frequencies within the bird population toward larger bill size (**Figure 12.4e**). This process in which two species undergo reciprocal evolutionary change through natural selection is called **coevolution**.

Unlike adaptation to the physical environment, adaptation in response to the interaction with another species can produce reciprocal evolutionary responses that either thwart (counter) these adaptive changes, as in the previous example, or in mutually beneficial interactions, magnify (reinforce) their effect. An example of the latter can be found in the relationship between flowering plants and their animal pollinators. Many species of flowering plants require the transfer of pollen from one individual to another for successful fertilization and reproduction (outcrossing; **Figure 12.5**). In some plant species, this is accomplished through passive transport by the wind, but many plants depend on animals to transport pollen between flowers. By attracting animals, such as insects or birds, to the flower, pollen is spread. When the animal comes into contact with the

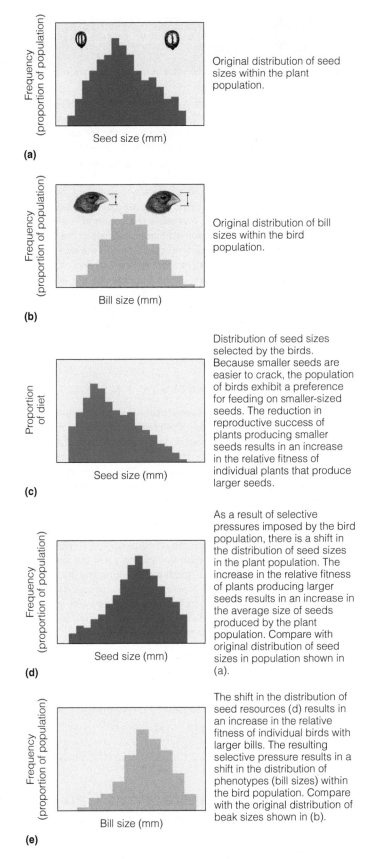

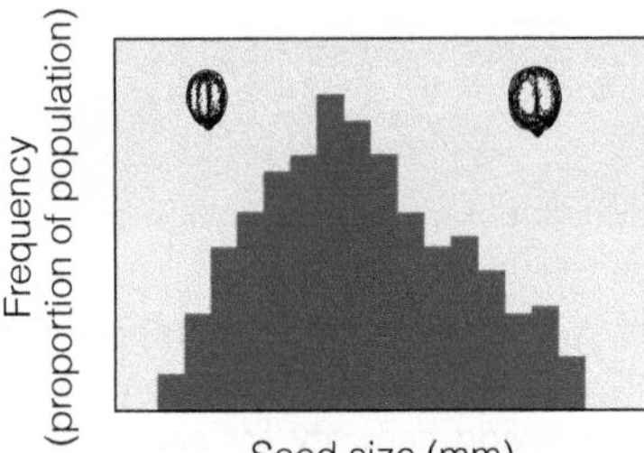

Original distribution of seed sizes within the plant population.

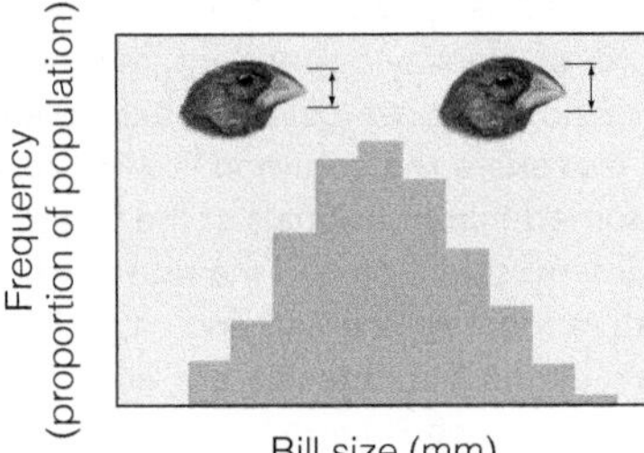

Original distribution of bill sizes within the bird population.

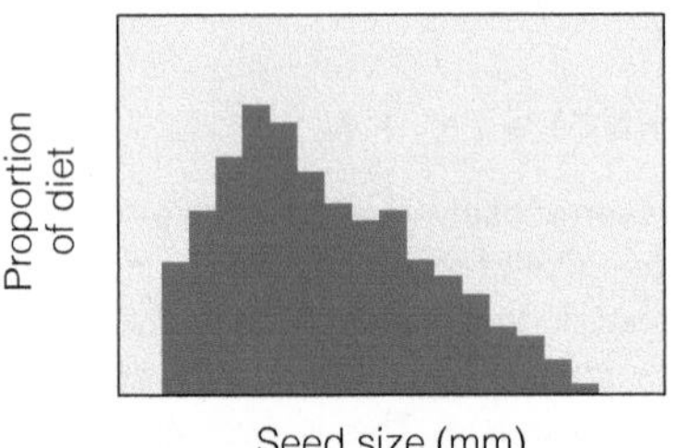

Distribution of seed sizes selected by the birds. Because smaller seeds are easier to crack, the population of birds exhibit a preference for feeding on smaller-sized seeds. The reduction in reproductive success of plants producing smaller seeds results in an increase in the relative fitness of individual plants that produce larger seeds.

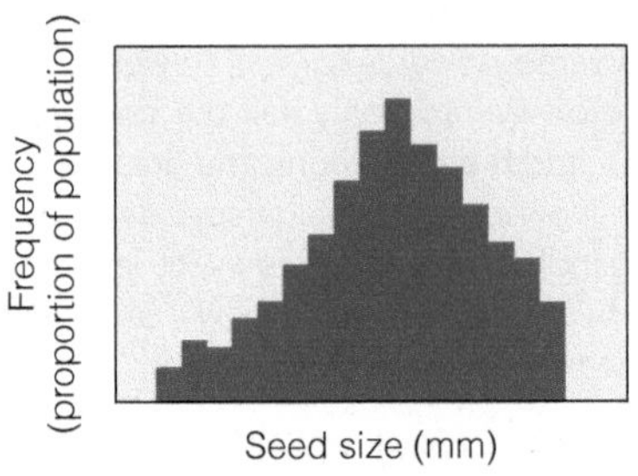

As a result of selective pressures imposed by the bird population, there is a shift in the distribution of seed sizes in the plant population. The increase in the relative fitness of plants producing larger seeds results in an increase in the average size of seeds produced by the plant population. Compare with original distribution of seed sizes in population shown in (a).

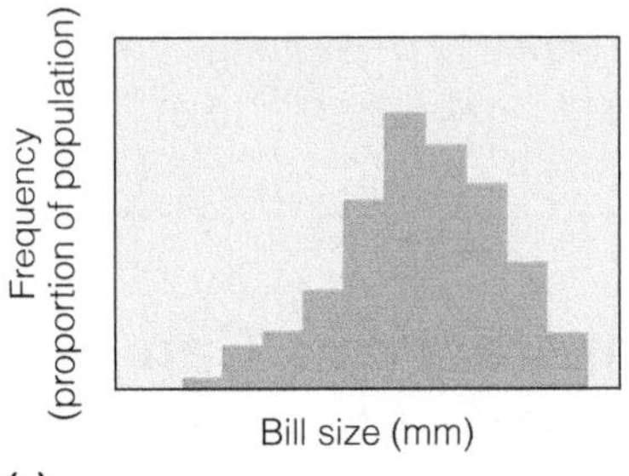

The shift in the distribution of seed resources (d) results in an increase in the relative fitness of individual birds with larger bills. The resulting selective pressure results in a shift in the distribution of phenotypes (bill sizes) within the bird population. Compare with the original distribution of beak sizes shown in (b).

Figure 12.4 Example of the coevolution of a predator (seed-eating bird) and its prey (seeds).

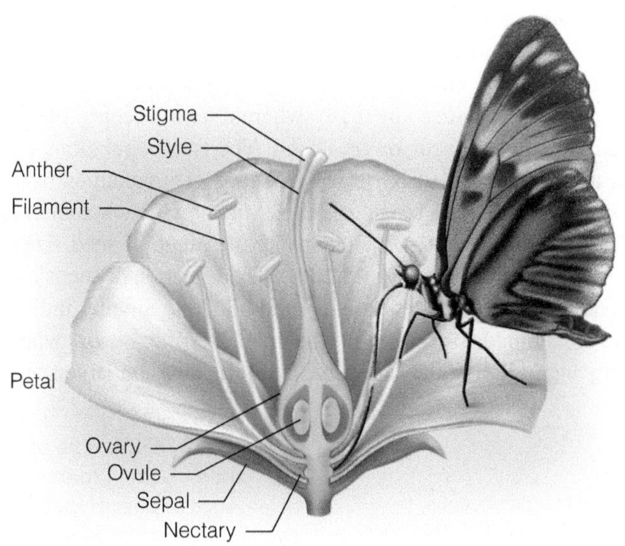

Figure 12.5 Cross-sectional illustration of flower structure. For pollination and reproduction to occur, pollen from the anther must be transported and deposited on the stigma of another individual. As an animal pollinator gains access to nectar located in the nectary, pollen is deposited on its body. The pollen is then transported to other flowers as the animal continues to forage.

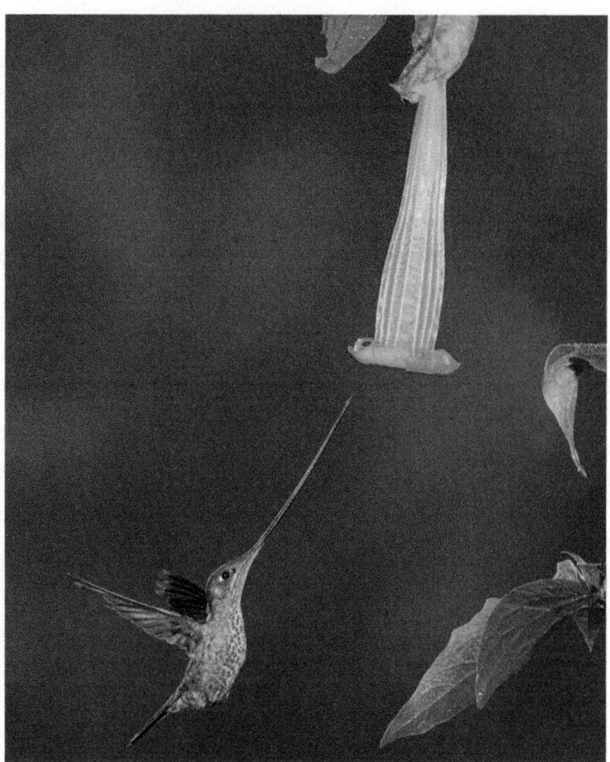

Figure 12.6 The bill of the hummingbird is adapted for extracting nectar from flowers. In this example, the extremely long bill of the sword-billed hummingbird (*Ensifera ensifera*) is uniquely adapted for extracting nectar from the long floral tubes of the passion flower (*Passiflora mixa*) growing in the highlands of the Andes Mountains of South America.

flower, pollen is deposited on its body, which is then transferred to another individual as the animal travels from flower to flower. This process requires the plant species to possess some mechanism to attract the animal to the flower. A wide variety of characteristics has evolved in flowering plants that function to entice animals through either signal or reward. Signals can involve brightly colored flowers or scents. The most common reward to potential pollinators is nectar, a sugar-rich liquid produced by plants, which serves no purpose for the individual plant other than to attract potential pollinators. Nectar is produced in glands called nectaries, which are most often located at the base of the floral tube (see Figure 12.5).

The relationship between nectar-producing flowers and nectar-feeding birds provides an excellent example of the magnification of reciprocal evolutionary responses—coevolution—resulting from a mutually beneficial interaction. The elongated bill of hummingbirds distinguishes them from other birds and is uniquely adapted to the extraction of nectar (**Figure 12.6**). Their extremely long tongues are indispensable in gaining nectar from long tubular flowers. Let us assume a species of hummingbird feeds on a variety of flowering plants within a tropical forest but prefers the flowers of one plant species in particular because it produces larger quantities of nectar. Thus, the reward to the hummingbird for visiting this species is greater than that of other plant species in the forest. Now assume that flower size (an inherited characteristic) varies among individuals within the plant population and that an increase in nectar production is associated with elongation of the floral tube (larger flower size). Individual plants with larger flowers and greater nectar production would have an increased visitation rate by hummingbirds. If this increase in visitation rate results in an increase in pollination and reproduction, the net effect is an increase in the relative fitness of individuals that produce larger flowers, shifting the distribution of phenotypes within the plant population. The larger flower size and longer floral tube, however, make it more difficult to gain access to the nectar. Individual hummingbirds with longer bills are more efficient at gaining access, and bill size varies among individuals within the population. With increased access to nectar resources, the relative fitness of longer-billed individuals increases at the expense of individuals with shorter bills. In addition, any gene mutation that results in increasing bill length with be selected for because it will increase the fitness of the individual and its offspring (assuming that they exhibit the phenotype). The genetic changes that are occurring in each population are reinforced and magnified by the mutually beneficial interaction between the two species. The plant characteristic of nectar production is reinforced and magnified by natural selection in the form of improved pollination success by the plant and reproductive success by the hummingbird. In turn, the increased flower size and associated nectar production functions as a further agent of natural selection in the bird population, resulting in an increase in average bill size (length). One consequence of this type of coevolutionary process is specialization, wherein changes in phenotypic characteristics of the species involved function to limit the ability of the species to carry out the same or similar interactions with other species. For example, the increase in bill size in the hummingbird population will function to limit its ability to efficiently forage on plant species that produce smaller flowers, restricting its feeding to the subset of flowering plants within the tropical forest that produces large flowers with long floral tubes (see Figure 12.6). In the extreme case, the interaction can become obligate, where the degree of specialization in phenotypic characteristics results in the two species being dependent on each other for survival and successful reproduction. We will examine the evolution of obligate species interactions in detail later (Chapter 15).

Unlike the case of mutually beneficial interactions in which natural selection functions to magnify the intensity of the interaction, interactions that are mutually negative to the species involved can lead to the divergence in phenotypic characteristics that function to reduce the intensity of interaction. Such is the case when the interaction involves competition for essential resources. Consider the case wherein two species of seed-eating birds co-occur on an island. The two populations differ in average body and bill size, yet the two populations overlap extensively in the range of these phenotypic characteristics (**Figure 12.7a**) and subsequently in the range of seed sizes on which they forage (**Figure 12.7b**). The selection of seeds by individual birds is related to body and bill size. Smaller individuals are limited to feeding on the smaller, softer seeds, whereas only larger individuals are capable of cracking the larger, harder

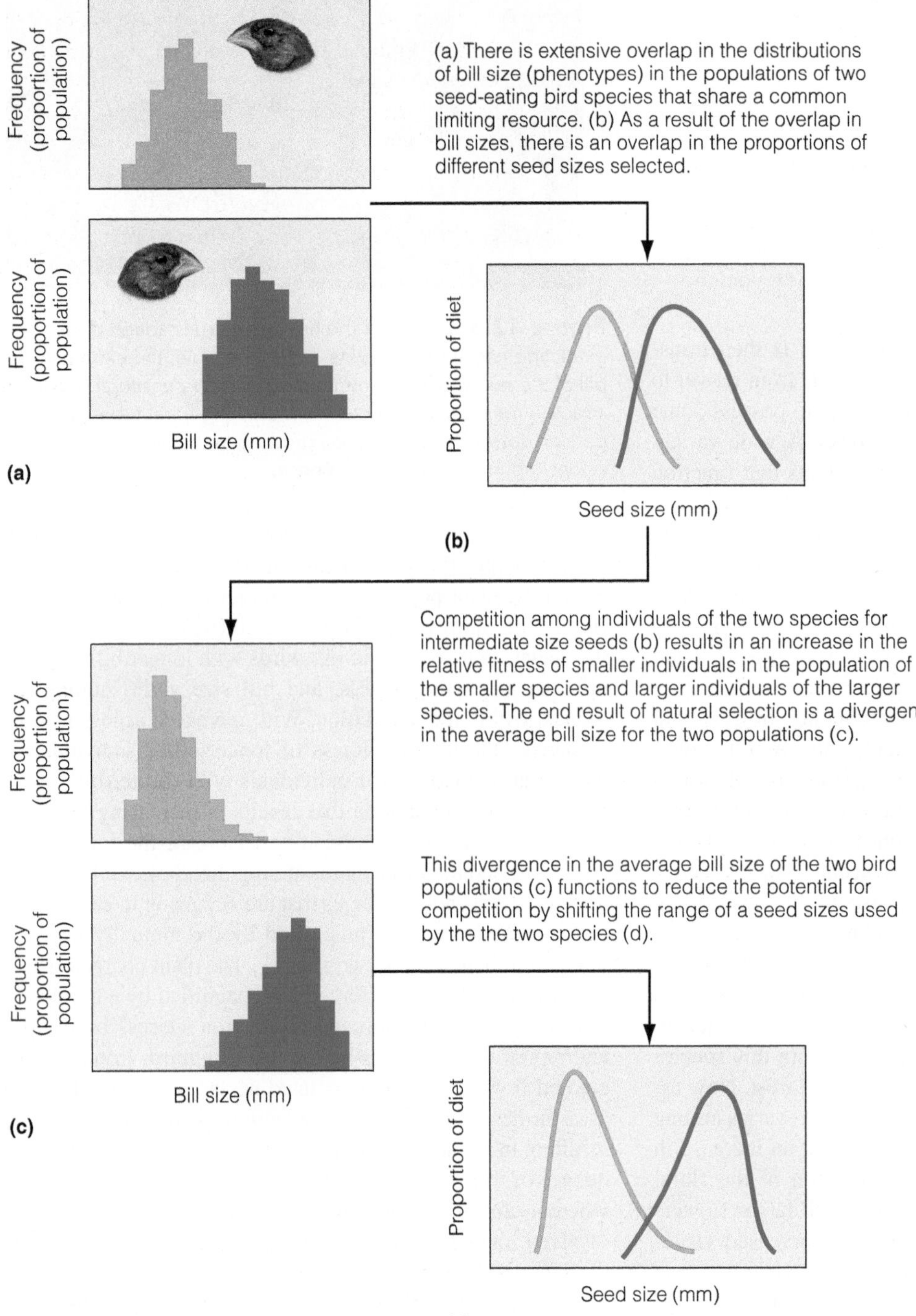
Frequency (proportion of population)
(a) There is extensive overlap in the distributions of bill size (phenotypes) in the populations of two seed-eating bird species that share a common limiting resource. (b) As a result of the overlap in bill sizes, there is an overlap in the proportions of different seed sizes selected.
Frequency (proportion of population)
Bill size (mm)
(a)
Proportion of diet
Seed size (mm)
(b)
Competition among individuals of the two species for intermediate size seeds (b) results in an increase in the relative fitness of smaller individuals in the population of the smaller species and larger individuals of the larger species. The end result of natural selection is a divergence in the average bill size for the two populations (c).
Frequency (proportion of population)
This divergence in the average bill size of the two bird populations (c) functions to reduce the potential for competition by shifting the range of a seed sizes used by the the two species (d).
Frequency (proportion of population)
Bill size (mm)
(c)
Proportion of diet
Seed size (mm)
(d)

Figure 12.7 Example of coevolution leading to the divergence of phenotypes in two competing species.

seeds. Although larger birds are able to feed on smaller seeds, it is energetically inefficient; therefore, their foraging is restricted to relatively larger seeds (see Section 5.8 for an example).

Seed resources on the island are limited relative to the populations of the two species, hence, competition is often intense for the intermediate-sized seeds for which both species forage. If competition for intermediate-sized seeds functions to reduce the fitness of individual birds that depend on these resources, the result would be reduced survival and reproductive rates for larger individuals of the smaller species and smaller individuals of the larger species (**Figure 12.7c**). This result represents a divergence in the average body and bill size for the two populations that functions to reduce the potential for competition between the two species (**Figure 12.7d**).

12.4 The Nature of Species Interactions Can Vary across Geographic Landscapes

We have examined how natural selection can result in genetic differentiation, that is, genetic differences among local populations. Species with wide geographic distributions generally encounter a broader range of physical environmental conditions than species whose distribution is more restricted. The variation in physical environmental conditions often gives rise to a corresponding variation in phenotypic characteristics. As a result, significant genetic differences can occur among local populations inhabiting different regions (see Section 5.8 for examples). In a similar manner, species with wide geographic distributions are more likely to encounter a broader range of biotic interactions. For example, a bird species such as the warbling vireo (*Vireo gilvus*) that has an extensive geographic range in North America, extending from northern Canada to Texas and from coast to coast, is more likely to encounter a greater diversity of potential competitors, predators, and pathogens than will the cerulean warbler (*Dendroica cerula*), whose distribution is restricted to a smaller geographic region of the eastern United States (see Figure 17.2 for distribution maps). Changes in the nature of biotic interactions across a species geographic range can result in different selective pressures and adaptations to the local biotic environment. Ultimately, differences in the types of species interactions encountered by different local populations can result in genetic differentiation and the evolution of local ecotypes similar to those that arise from geographic variations in the physical environment (see Section 5.8 for examples of the latter). The work of Edmund Brodie Jr. of Utah State University presents an excellent example.

Brodie and colleagues examined geographic variation among western North American populations of the garter snake (*Thamnophis sirtalis*) in their resistance to the neurotoxin tetrodotoxin (TTX). The neurotoxin TTX is contained in the skin of newts of the genus *Taricha* on which the garter snakes feed (**Figure 12.8a**). These newts are lethal to a wide range of potential predators, yet to garter snakes having the TTX-resistant phenotype, the neurotoxin is not fatal. Both the toxicity of newts (TTX concentration in their skin) and the TTX resistance of garter snakes vary geographically (Figure 12.8b). Previous studies have established that TTX resistance in the garter snake is highly heritable (passed from parents to offspring), so if TTX resistance in snakes has coevolved in response to toxicity of the newt populations on

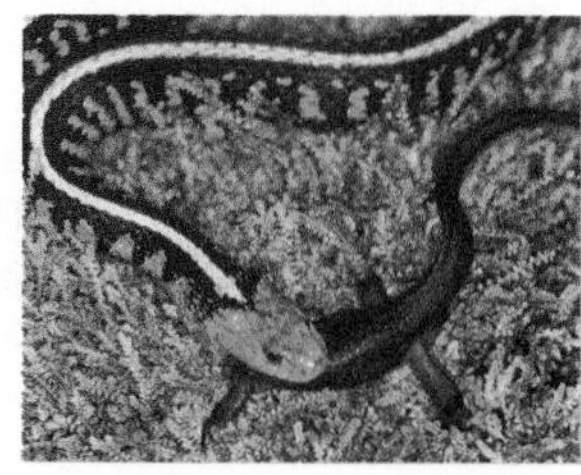
(a)

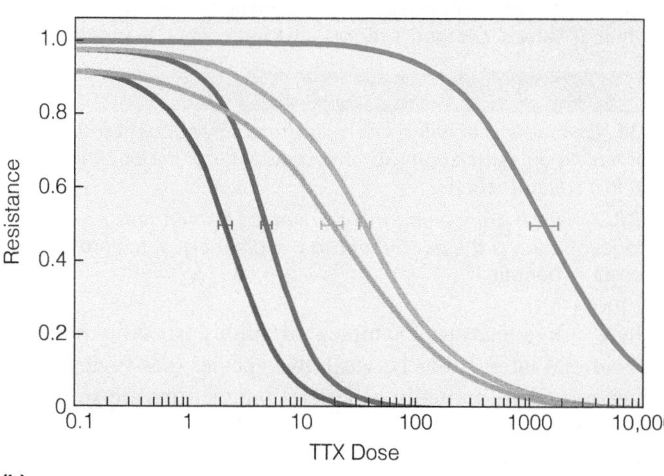

(b)

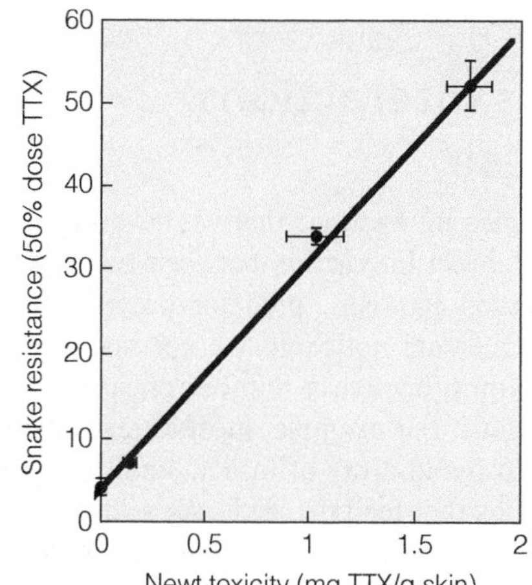

(c)

Figure 12.8 (a) Garter snake feeding on newt. (b) Dose-response curves for five representative populations of garter snake. The administered tetrodotoxin (TTX) dose is shown on the *x*-axis, and the corresponding average response is given on the *y*-axis. Dose response is measured as proportion of baseline activity (measured as sprint speed in laboratory in absence of TTX). The 50 percent doses are marked on each of the five dose-response curves. These values correspond to the dose at which sprint speed is reduced by 50 percent from baseline. (c) Relationship between 50 percent dose response (snake resistance) and newt toxicity (TTX concentration of skin) for the five populations shown in (b).
(Adapted from Brodie et al. 2002.)

which they feed, it is possible that levels of TTX resistance exhibited by local populations of garter snakes will vary as a function of the toxicity of newts on which they feed. The strength of selection for resistance would vary as a function of differences in selective pressure (the toxicity of the newts).

To test this hypothesis, the researchers examined TTX resistance in more than 2900 garter snakes from 40 local populations throughout western North America, as well as the toxicity of newts at each of the locations. The researchers found that the level of TTX resistance in local populations varies with the presence of toxic newts. Where newts are absent or nontoxic (as is the case on Vancouver Island, British Columbia), snakes are minimally resistant to TTX. In contrast, levels of TTX resistance increased more than a thousand-fold with increasing toxicity of newts (see **Figure 12.8b**). Brodie and his colleagues found that for local populations, the level of resistance to TTX varies as a direct function of the levels of TTX in the newt population on which they prey (**Figure 12.8c**). The resistance and toxicity levels match almost perfectly over a wide geographic range, reflecting the changing nature of natural selection across the landscape.

In some cases, even the qualitative nature of some species interactions can be altered when the background environment is changed. For example, mycorrhizal fungi are associated with a wide variety of plant species (see Chapter 15, Section 15.11). The fungi infect the plant root system and act as an extension of the root system. The fungi aid the plant in the uptake of nutrients and water, and in return, the plant provides the fungi with a source of carbon. In environments in which soil nutrients are low, this relationship is extremely beneficial to the plant because the plant's nutrient uptake and growth increase. (**Figure 12.9a**). Under these conditions, the interaction between plant and fungi is mutually beneficial. In environments in which soil nutrients are abundant, however, plants are able to meet nutrient demand through direct uptake of nutrients through their root system. Under these conditions, the fungi are of little if any benefit to the plant; however, the fungi continue to represent an energetic cost to the plant, reducing its overall net carbon balance and growth (**Figure 12.9b**). Across the landscape, the interaction between plant and fungi changes from mutually beneficial (++) to parasitic (+−) with increasing soil nutrient availability.

12.5 Species Interactions Can Be Diffuse

The examples of species interactions that we have discussed thus far focus on the direct interaction between two species. However, most interactions (e.g., predator–prey, competitors, mutually beneficial) are not exclusive nor involve only two species. Rather, they involve a number of species that form diffuse associations. For example, most terrestrial communities are inhabited by an array of insect, small mammal, reptile, and bird species that feed on seeds. As a result, there is a potential for competition to occur among any number of species that draw on this limited food resource. Similarly,

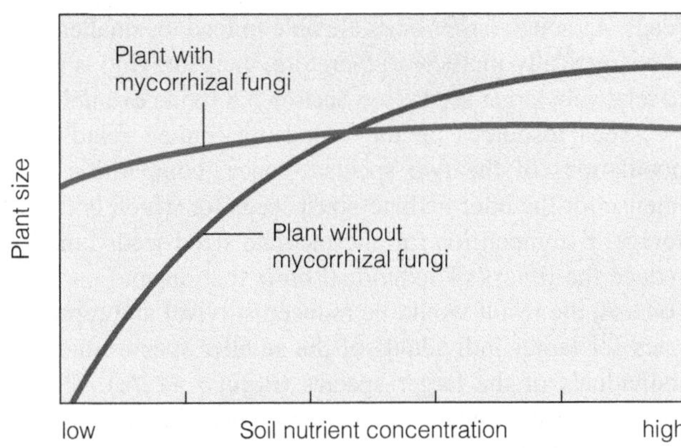

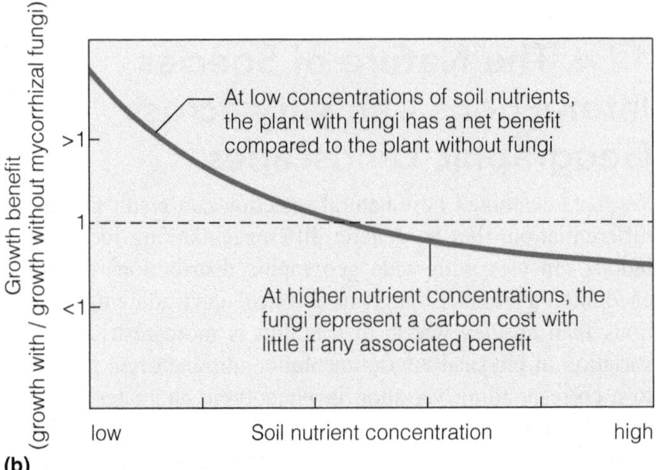

Figure 12.9 Cost-benefit curves for individual plants with and without mycorrhizal fungi associated with the root system across a gradient of soil nutrient concentration. Mycorrhizal fungi provide increased access to soil nutrients but cost the plant carbon resources. (a) At low soil nutrient concentrations, the presence of the fungi increases growth rate and plant size compared to plants without fungi. (b) As nutrient availability in the soil increases, the impact of fungi switches from a net benefit to a net cost. The carbon cost to the plant in supporting the fungi yields little benefit but costs the plant carbon that could otherwise be used for growth. Under high soil nutrient concentration, the plants without fungi have a higher growth rate and size. Dashed line (value of 1) represents no net benefit. Values greater than 1 represent a net growth benefit of mycorrhizal association (+ + interaction), whereas values less than 1 represent a net cost (+ − interaction).

Interpreting Ecological Data

Q1. *Given the preceding figure, is there a net benefit to the plant of having an association with mycorrhizal fungi under conditions of low soil nutrients?*

Q2. *At which point along the gradient of soil nutrient concentration is the net benefit to the plant equal to zero (costs = benefits)?*

there are numerous examples of highly specific mutually beneficial interactions between two species (see Figure 12.6); however, most mutually beneficial interactions are somewhat diffuse. In plant-pollinator interactions, most plants are pollinated by multiple animal species, and each animal species

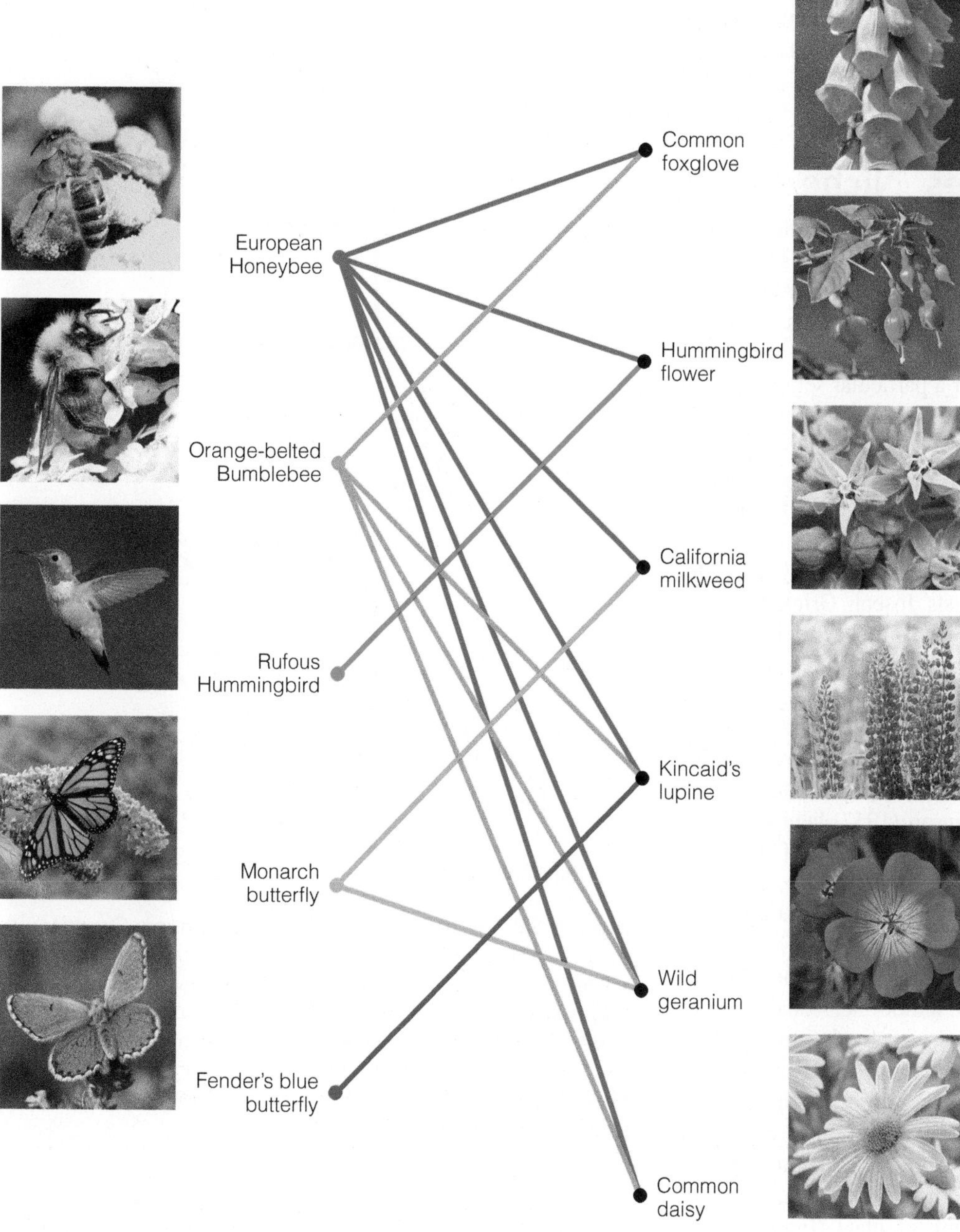

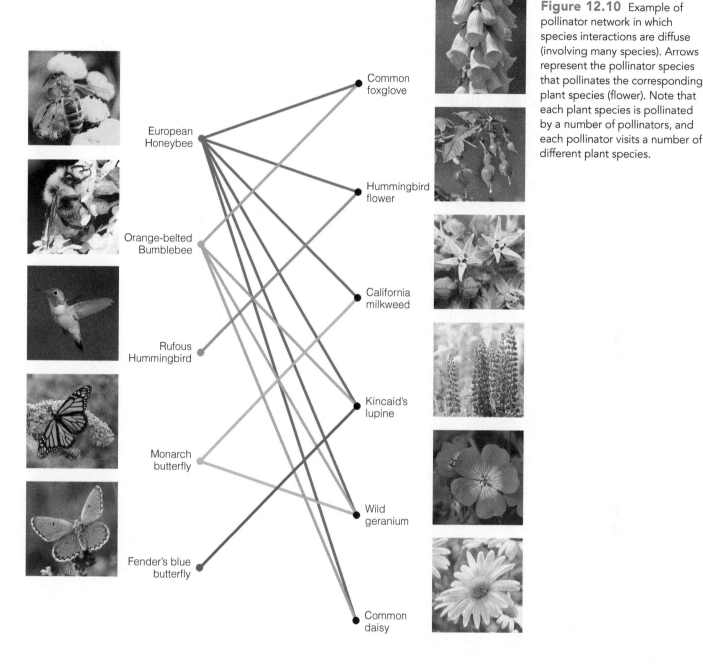

Figure 12.10 Example of pollinator network in which species interactions are diffuse (involving many species). Arrows represent the pollinator species that pollinates the corresponding plant species (flower). Note that each plant species is pollinated by a number of pollinators, and each pollinator visits a number of different plant species.

pollinates multiple plant species. For example, honey bees (*Apis melifera*) are known to visit the flowers of hundreds of plant species, and white mangrove (*Laguncularia racemosa*) is visited by more than 50 different insect species. Species of plants and pollinators form pollination networks, and the resulting selective forces that reinforce the mutually beneficial interactions are likewise diffuse (**Figure 12.10**). This process in which a network of species undergoes reciprocal evolutionary change through natural selection is referred to as *diffuse coevolution.*

In diffuse coevolution, groups of species interact with other groups of species, leading to natural selection and evolutionary changes that cannot be identified as examples of specific, pairwise coevolution between two species. For example, the evolution of resistance to the neurotoxin TTX by garter snakes presented in the previous section (see Figure 12.8) is in response to TTX concentrations in the skin of newts of the genus *Taricha* on which they prey. This genus consists of three species and four subspecies of western newts, so the evolution of resistance by snake populations is not in response to its interaction with a single species but rather a group of closely related species that all produce the neurotoxin and on which they feed. Likewise, the evolution of toxicity by members of the genus *Taricha* provides a defense mechanism to avoid predation by an array of vertebrate predators, not just a single species of predator.

In the chapters that follow, we will explore an array of examples of co-evolution. Some represent highly specialized co-adaptations between two species in which the interaction has become obligate (essential to the survival of the two species involved), whereas others represent the result of generalized relationships between groups of species—diffuse relationships between competitors, predator and prey, or mutualists.

12.6 Species Interactions Influence the Species' Niche

The diversity of species inhabiting our planet reflect different evolutionary solutions to the same basic processes of assimilation and reproduction, and that the characteristics that distinguish each species often reflect adaptations (products of natural selection) that allow individuals of that species to survive, grow, and reproduce under a particular set of environmental conditions (see Part Two). As such, each species may be described in terms of the range of physical and chemical conditions under which it persists (survives and reproduces) and the array of essential resources it uses. This characterization of a species is referred to as its **ecological niche**.

The concept of the ecological niche was originally developed independently by ecologists Joseph Grinnell (1917, 1924) and Charles Elton (1927), who proposed slightly different meanings for the term. Grinnell's definition centered on the concept of habitat (see Section 7.14, Figure 7.25) and the limitations imposed by the physical environment (as discussed in Chapters 6 and 7), whereas Elton emphasized the role of the species in the context of the community (species interactions). The limnologist G. Evelyn Hutchinson (1957) later expanded the concept of the niche by proposing the idea of the niche as a multidimensional space called a *hypervolume,* in which each axis (dimension) is defined by a variable relating to the specific resource need or environmental factor that is essential for a species' survival and successful reproduction. We can begin to visualize this concept of a multidimensional niche by modeling a three-dimensional one—a niche defined by three resources or environmental variables: temperature, salinity, and pH (Figure 12.11). For each axis there is a range of values (conditions) that permit a species to survive and reproduce (or in Hutchinson's own words, "for the population to persist indefinitely"). For example, in Chapters 6 and 7 we presented numerous examples of the response of plant (Figures 5.19–5.22) and animal (Figures 7.14 and 7.18) species to variation in environmental temperature. Each of these figures represents a description of the species' niche for the single dimension (variable) of environmental temperature. Likewise, the distribution of seed sizes used by the three species of Darwin's ground finch inhabiting the Galapagos Islands presented in Figure 5.20 represents a description of the species' niches for the single dimension of food resource size.

Hutchinson referred to this hypervolume that defines the environmental conditions under which a species can survive and reproduce as the **fundamental niche**. The fundamental niche, sometimes referred to as the *physiological niche,* provides a description of the set of environmental conditions under which a species can persist. As we have discussed in the previous sections, however, a population's response to the environment may be modified by interactions with other species.

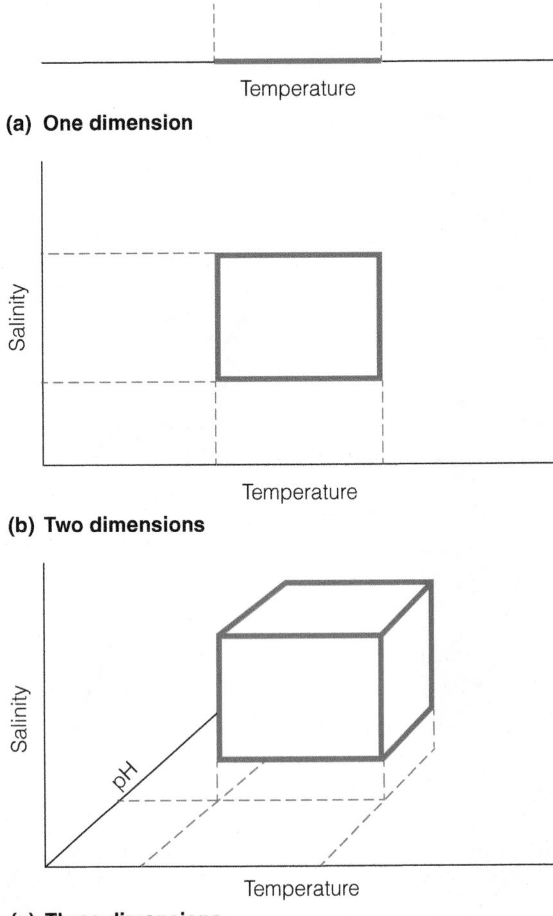

Figure 12.11 An illustration of niche dimension. Consider three variables composing the species' niche: temperature, salinity, and pH. (a) A one-dimensional niche involving only temperature. (b) A second dimension, salinity, has been added. Enclosing that space we have a two-dimensional niche. (c) Adding a third dimension, pH, and enclosing all three gives a three-dimensional niche space, or volume. A fourth variable would create a hypervolume.

Hutchinson recognized that interactions such as competition may restrict the environment in which a species can persist and referred to the portion of the fundamental niche that a species actually exploits as a result of interactions with other species as the **realized niche** (Figure 12.12).

An illustration of the difference between a species' fundamental and realized niche is provided in the work of J. B. Grace and R. G. Wetzel of the University of Michigan. Two species of cattail (*Typha*) occur along the shorelines of ponds in Michigan. One species, *Typha latifolia* (wide-leaved cattail), dominates in the shallower water, whereas *Typha angustifolia* (narrow-leaved cattail) occupies the deeper water farther from shore. When these two species grew along the water depth gradient in the absence of the other species, a comparison of the results with their natural distributions revealed how competition influences their realized niche (Figure 12.13). Both species can survive in shallow waters, but only the narrow-leaved cattail,

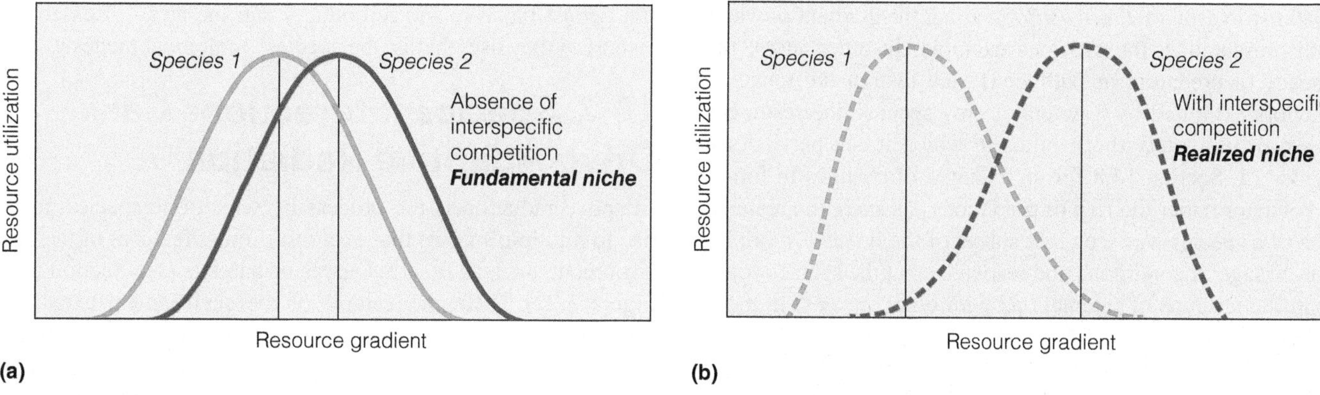

Figure 12.12 Comparison of (a) fundamental and (b) realized niche for two species along a resource gradient. The fundamental niches (solid curves) represent resource use in the absence of the other species (no competition), whereas the realized niches (dashed curves) represent resource use in the presence of competing species. Competition is a function of overlap in resource use. Note that competition results in a shift in resource use (vertical lines). Shift in use reflects reduced availability of the shared resource.

T. angustifolia, can grow in water deeper than 80 centimeters (cm). When the two species grow together along the same gradient of water depth, their distributions, or realized niches, change. Even though *T. angustifolia* can grow in shallow waters (0–20 cm depth) and above the shoreline (−20 to 0 cm depth), in the presence of *T. latifolia* it is limited to depths of 20 cm or deeper. Individuals of *T. latifolia* outcompete individuals of *T. angustifolia* for the resources of nutrients, light, and space, limiting the distribution of *T. angustifolia* to the deeper waters. Note that the maximum abundance of *T. angustifolia* occurs in the deeper waters, where *T. latifolia* is not able to survive.

As originally proposed, the concept of realized niche focused on how the fundamental niche of a species is restricted as a result of negative interactions with other species. Competition can function to restrict the range of resources or environmental conditions used by a species, as in the example

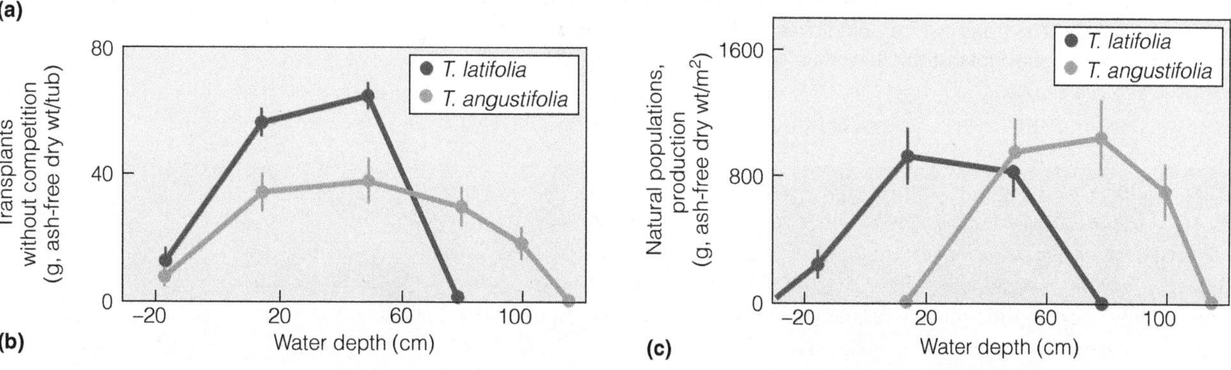

Figure 12.13 Distribution of two species of cattail (*Typha latifolia* and *Typha angustifolia*): (a) along a gradient of water depth; (b) grown separately in an experiment; (c) growing together in natural populations. The response of the two species in the absence of competition (b) reflects their fundamental niche (physiological tolerances). The response of each species is altered by the presence of the other (c). They are forced to occupy only their realized niches.
(Adapted from Grace and Wetzel 1981.)

of the distribution of *T. angustifolia* along the gradient of water depth presented in the previous example. In other cases, the presence of predators or pathogens may restrict the range of behaviors exhibited by a potential prey species, the resources it uses, or ultimately the habitats in which it can persist (see Chapter 14, Section 14.8 for an example of changes in foraging behavior under the risk of predation). As such, the realized niche of a species was seen as a subset of the broader, more inclusive range of conditions and resources that the species could use in the absence of interactions with other species. In more modern times, however, ecologists have come to appreciate the importance of positive interactions, particularly mutually beneficial interactions, and how this class of interactions can modify the species' fundamental niche. By either directly or indirectly enhancing the probabilities of survival and reproduction of individuals in the participating populations, interactions that are either beneficial to one species and neutral to the other (commensalism), or mutually beneficial to both (mutualism), can function to expand the range of environmental conditions or resources under which a species can persist. In this case, the realized niche of the species is greater (more expansive) than that of its fundamental niche. For example, nitrogen-fixing *Rhizobium* bacteria associated with the root systems of certain plant species provide a direct source of mineral nitrogen to the plant, enabling it to persist in soils that have low mineral nitrogen content (see Section 15.11 for a detailed discussion of this mutualistic interaction). In the absence of interaction with the bacteria, the plants are restricted to a narrower range of soils that have higher availability of mineral nitrogen.

Although the realized niche is by definition a product of species interactions, over evolutionary time, biotic interactions can play a critical role in defining a species' fundamental niche. The previous discussion of species' adaptation to the environment focused almost exclusively on the role of the physical and chemical environments as agents of natural selection (see Part Two). We now have seen, however, that species interactions also function as agents of natural selection, and phenotypic characteristics often reflect adaptations to these selective pressures. As such, over evolutionary timescales, species interactions can have a major role in determining the range of physical and chemical conditions under which species can persist (survive and reproduce) and the array of essential resources they use, that is, the species' ecological niches.

12.7 Species Interactions Can Drive Adaptive Radiation

Adaptive radiation is the process by which one species gives rise to multiple species that exploit different features of the environment, such as food resources or habitats (see Section 5.9, Figure 5.22). Different features of the environment exert the selective pressures that push populations in various directions (phenotypic divergence); reproductive isolation, the necessary condition for speciation to occur, is often a by-product of the changes in morphology, behavior, or habitat preferences that are the actual objects of selection. Likewise, variations among local populations in biotic interactions can result in phenotypic divergence and therefore have the potential to function as mechanisms of adaptive radiation. Resource competition is often inferred as a primary factor driving phenotypic divergence. For example, species of the globeflower fly *Chiastocheta* present a unique case of adaptive radiation as a result of resource competition. At least six sister species of the genus *Chiastocheta* lay their eggs (oviposition) on the fruits of the globeflower, *Trollius europaeus* (**Figure 12.14**); however, the different species of globeflower flies differ in the timing of their egg laying. One species lays its eggs in 1-day-old flowers, whereas all the other species sequentially deposit their eggs throughout the flower life span. In a series of field experiments, Laurence Despres and Mehdi Cherif of Université Joseph Fourier (Grenoble, France) found evidence that supports the hypothesis that the evolutionary divergence of species of *Chiastocheta* was a result of disruptive selection on the timing of egg laying (reproduction). The researchers established that intense intraspecific competition occurs within each of the species, but differences in the timing of egg laying and larval development functions to minimize competition among species (the concept of resource partitioning will be examined in Chapter 13).

Although numerous studies have illustrated the role of competitive interactions in adaptive radiation, the importance of other interactions, such as mutualism or predation, remain largely unexplored. The research of Patrik Nosil and Bernard

Figure 12.14 (a) The globeflower fly (genus *Chiastocheta*) presents a unique case of adaptive radiation, with at least six coexisting species whose larvae develop in fruits of the globeflower (*Trollius europaeus*). (b) These species all feed on seeds and differ in their timing of egg laying. One species lays its eggs in 1-day-old flowers, whereas all the other species sequentially deposit their eggs throughout the flower life span.

(a)

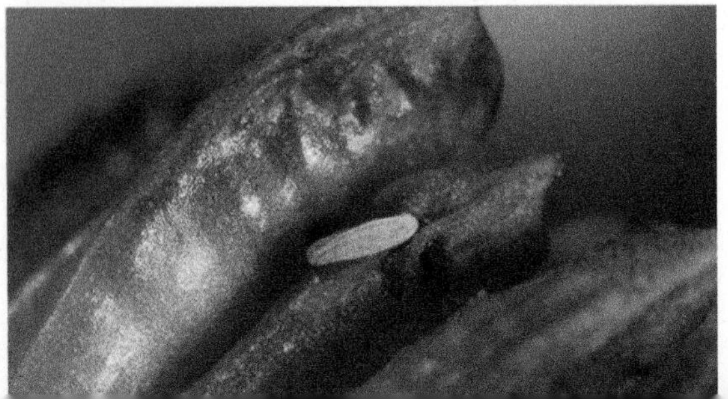

(b)

Crespi of Simon Fraser University (British Columbia, Canada), however, has shown that adaptive radiation can result from divergent adaptations to avoid predators. Nosil and Crespi's research focused on two ecotypes (populations of the same species adapted to their local environments) of the stick insect *Timema cristinae* (see Section 5.8 and Chapter 5, **Field Studies: Hopi Hoekstra** for discussion of ecotypes). *Timema* walking sticks are wingless insects inhabiting southwestern North America. Individuals feed and mate on the host plants on which they reside. The two distinct ecotypes of *Timema* are adapted to feeding on different host plants, *Ceanothus* and *Adenostoma.* The two host plants differ strikingly in foliage form, with *Ceanothus* plants being relatively large and tree-like with broad leaves and *Adenostoma* plants being small and shrub-like with thin, needle-like leaves (**Figure 12.15**).

The two *Timema* ecotypes differ in 11 quantitative traits (see Figure 12.15), comprising aspects of color, color pattern, body size, and body shape. These differences between the two ecotypes appears to be a result of divergent selection. The different traits exhibited by each of the ecotypes appear to provide crypsis (avoidance of observation) from avian predators on the respective host-plant species. Field experiments were conducted to determine how differences in phenotypic traits influenced the survival rates of the two ecotypes on the two plant species. Each of the two *Timema* ecotypes was placed on each of the two host-plant species. The results of the experiment clearly indicated that the direction and magnitude of divergence in traits represent adaptations that function to reduce rates of predation on *Timema* on their respective host-plant species. The ecotypes of *T. cristinae*, like the example of the limnetic and benthic ecotypes of sticklebacks examined in Chapter 5, can be considered to represent an early stage of adaptive radiation because studies indicate that reproductive isolation is not complete (see Section 5.6, Figure 5.15).

Figure 12.15 Contrasting phenotypic traits of two ecotypes of the walking stick *Timema cristinae* that are associated with two host plants, *Ceanothus* and *Adenostoma*. The insects feed and reproduce on the host plants. The phenotypic differences in the two ecotypes reflect adaptations that aid in avoiding detection by avian predators on the two morphologically different plant species. The two ecotypes represent an early stage of adaptive radiation. (Adapted from Nosil and Crespi 2006.)

ECOLOGICAL Issues & Application: Urbanization Has Negatively Impacted Most Species while Favoring a Few

As we will see in the chapters that follow, species interactions are ubiquitous in nature and play a fundamental role in the structuring of ecological communities. Perhaps no other interaction, however, has as great an impact on the diverse array of plants and animals that inhabit our planet as their interaction with the human species.

As we first presented in Chapter 9 (*Ecological Issues & Applications*), the primary cause of population declines and recent species extinctions is habitat loss as a result of human activities—namely, changing land-use patterns. There are two major land-use changes that are responsible for habitat loss in terrestrial environments: expanding agriculture and urbanization.

According to the Food and Agricultural Organization (FAO) United Nations' statistics, at present some 11 percent (1.5 billion hectares) of the globe's land surface (13.4 billion ha) is used in crop production (arable land and land under permanent crops), and even more land (3.2 to 3.6 billion ha) is used to raise livestock. Together, agricultural lands account for almost 40 percent of Earth's land surface. The negative impacts of the expansion of agriculture to meet the needs of the growing human population have been central to the discussion of the decline of biological diversity on our planet, a topic we will examine in more detail in Chapter 26. The increasing urbanization of the human population over the past century (**Figure 12.16**), however, has led to the emergence of a new field of ecology—**urban ecology**—to study the ecology of organisms in the context of the urban environment.

Ecology has historically focused on "pristine" natural environments; however, by as early as the 1970s, many ecologists began turning their attention toward ecological interactions taking place in urban environments. What has emerged is a picture of species interactions dominated by humans, which negatively impacts most species and benefits only a few.

Estimates of urban land area vary widely from 0.5 to slightly more than 2.0 percent of the world's land, depending on the criteria used to define urban development. Historically, cities have been compact areas with high population densities that grew slowly in their physical extent. Today, however, urban areas are expanding twice as fast as their populations. According to the United States Census Bureau, about 30 percent of the U.S. population currently lives in cities, whereas another 50 percent lives in the suburbs. More than 5 percent of the total surface area of the United States is covered by urban and other developed areas; this is more than the land covered by the combined totals of national and state parks.

The expansion of urbanization produces some of the greatest local extinction rates and frequently eliminates the large majority of native species. Eyal Shochat of Arizona State University's Global Institute of Sustainability and colleagues used data from Phoenix, Arizona, and Baltimore, Maryland, to contrast the distribution of species in these two urban areas as compared to the surrounding natural ecosystems. Their findings show a general pattern of decline in the number of species in urban environments as compared to both surrounding agricultural and natural ecosystems (**Figure 12.17**).

Species vary in their ability to adapt to the often drastic physical changes along the gradient from rural to urban habitat. Moving from the rural landscape of natural ecosystems and cultivated lands into the suburban landscape, one moves through a heterogeneous mixture of residential areas, commercial centers, and the managed vegetation of parks and cemeteries. The main cause for the loss of species in these suburban environments is habitat alteration. Yet in contrast to the decline in the number of species, both suburban areas and urban centers are usually characterized by higher population densities of resident species as compared to adjacent natural lands. For example, in a study of population of northern cardinals (*Cardinalis cardinalis*) in the metropolitan area of Columbus, Ohio, and surrounding forested landscape of central Ohio, Lionel Leston and Amanda Rodewald of Ohio State University found that birds were four times more abundant in urban than rural forests. Their research showed that food abundance was as much as four times greater in the urban habitat as compared to the forests of the surrounding region because exotic vegetation, refuse, and bird feeders may all provide food sources for birds in these urban environments.

Some mammals, such as raccoons (*Procyon lotor*), skunks (*Mephitis mephitis*), and rabbits (*Sylvilagus* spp.) have also benefited from the spread of the suburban landscape, finding

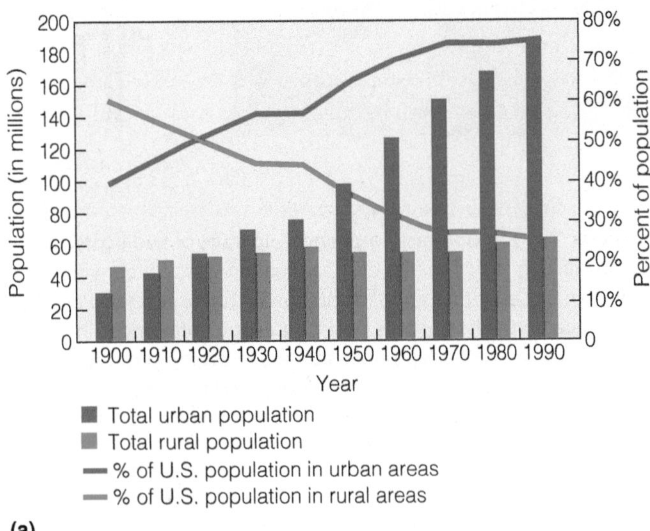

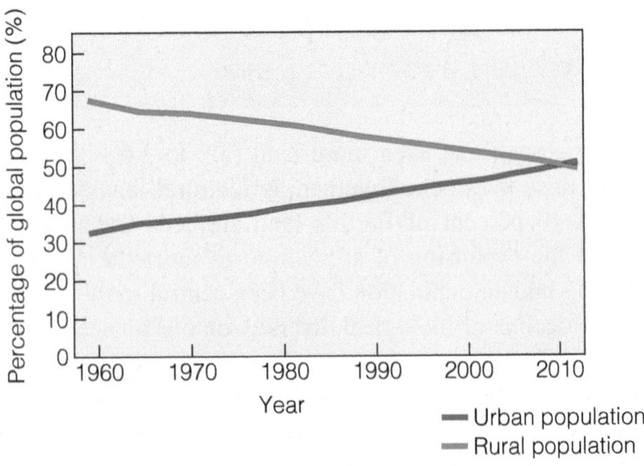

Figure 12.16 (a) Growth of rural and urban population in the United States over the 20th century. Essentially all population growth over the 20th century has been in cities, increasing the urban population fraction from 40 percent in 1900 to more than 75 percent in the 1990s. (b) A similar trend has been seen globally.

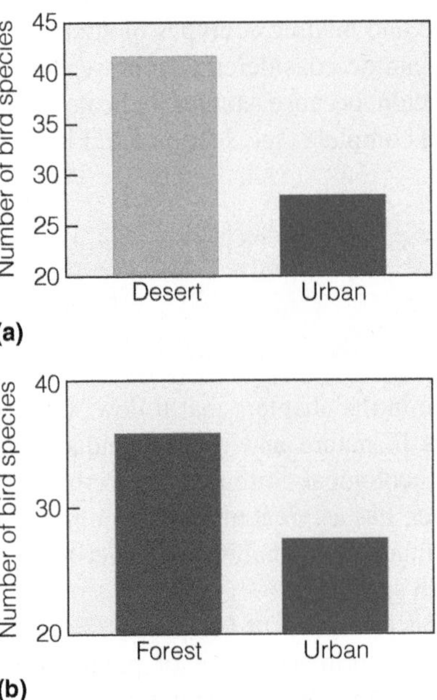

Figure 12.17 Comparison of the number of bird species occurring in (a) Phoenix, Arizona, as compared to the surrounding Sonoran desert ecosystem, and (b) Baltimore, Maryland, as compared to the surrounding forest ecosystems. In both studies, the comparison of the number of species in the urban and rural areas was corrected to account for area that was in the census. (Adapted from Shochat et al. 2010.)

shelter beneath sheds and porches, and an abundance of food—for raccoons, garbage; for skunks, insects and larvae on lawns and in gardens; and for rabbits, an abundance of high quality food plants in gardens and flowerbeds. Larger species, rapidly adapting to human presence, are moving into the suburban landscape and dramatically increasing in number. White-tailed deer (*Odocoileus virginiaus*), carriers of Lyme disease, find an abundance of forage on grass, shrubs, and gardens. Resident Canada geese (*Branta canadensis*), attracted to large open areas of grass—including golf courses and parks—create both a nuisance and health problems. In recent years, coyotes (*Canis latrans*), attracted by garbage and small prey including rodents and pets (cats and small dogs), are becoming more common in suburban areas. Even black bears (*Ursus americanus*) are attracted to backyard bird feeders and dumpsters in suburban areas adjacent to forested, rural landscapes.

In addition to increased abundance and predictability of food resources, recent research indicates that a reduction in predator populations in urban environments favors resident species. Evidence has been gathered that supports the idea that urban environments are safer for some species than are rural habitats. Both birds and squirrels in urban environments benefit from reduced nest predation and are able to spend a greater proportion of their time foraging compared with individuals in the surrounding natural ecosystems, indicating that the urban habitat is less risky than the surrounding rural habitats.

Species adapted to habitats along the suburban gradient drop out as they come to urban centers where habitat changes sharply. Vegetation is limited to scattered parks, some tree-lined streets, and vacant lots. Species that benefit from the habitat provided by these core urban centers are often referred to as "urban exploiters." Among plants, urban exploiters tend to be ruderal species (see discussion of plant life history classification in Section 10.13) that can tolerate high levels of disturbance. Examples include wind-dispersed weeds (grasses and annuals) that colonize abandoned lots and properties, and plants that can grow in and around pavement.

Bird species that thrive in urban habitats are often adapted to nesting in environments that are similar to the cityscape. For example, species that use cliff-like rocky areas, such as the rock dove (pigeons, *Columba livia*) and peregrine falcon (*Falco peregrinus*), are "pre-adapted" to using the barren concrete edifices of urban buildings, whereas cavity-nesting species, such as the house sparrow (*Passer domesticus*), house finch (*Haemorhous mexicanus*), and European starling (*Sturnus vulgaris*) are able to inhabit human dwellings.

Mammalian urban exploiters consist of species that are able to find shelter in human dwellings and exploit the rich food source provided by refuse, such as the house mouse (*Mus musculus*), the black rat (*Rattus rattus*), and brown rat (Norway rat: *Rattus norvegicus*).

Urban environments typically have more in common with other cities than with adjacent natural ecosystems, so species that flourish in urban habitats are often not native to the region. Rather, these species tend to disperse from city to city, typically with assistance—either intentionally or unintentionally—from humans (see Chapter 8, *Ecological Issues & Applications*). Species such as rock doves, starlings, house sparrows, Norway rats, and the house mouse are found in all cities in Europe and North America. As a result, many studies have found that the number (and proportion) of non-native species tends to increase as you move from rural habitats toward urban centers. In general, the proportion of species that is non-native goes from less than a few percent in rural areas to more than 50 percent at the urban core.

This combination of negative interactions with the majority of native species—while enhancing a small subset of often non-native species, which we have manipulated to serve our needs, facilitated through dispersal, or created urban environments in which their populations flourish—is resulting in what urban and conservation ecologists refer to as *biotic homogenization,* which is the gradual replacement of regionally distinct ecological communities with cosmopolitan communities that reflect the increasing global activity of humans.

SUMMARY

Classification 12.1

By designating the positive effect of one species on another as +, a detrimental effect as −, and no effect as 0, we can develop a classification of possible interactions between two co-occurring species: (00) neutral; (0+) commensalism; (++) mutualism; (0−) amensalism; (−−) competition; (+−) predation, parasitism, or parasitoidism.

Population Dynamics 12.2

Species interactions typically involve the interaction of individual organisms within the respective populations. By influencing individuals' probabilities of survival or reproduction, interactions influence the collective properties of birth and death at the population level, and in doing so, influence the dynamics of the respective populations.

Natural Selection 12.3

Phenotypic variations among individuals within the populations can result in differences in the nature and degree of interactions that occur. These phenotypic differences may influence the relative fitness of individuals within the populations in the degree of interaction, resulting in the process of natural selection. The process in which two species undergo reciprocal evolutionary change through natural selection is called *coevolution.* Mutually beneficial interactions typically serve to reinforce the phenotypic changes that result from the species interaction, and mutually detrimental interactions typically result in phenotypic changes that function to reduce the intensity of interaction.

Geographic Variation 12.4

Species with wide geographic distributions are more likely to encounter a broader range of biotic interactions. Changes in the nature of biotic interactions across a species' geographic range can result in different selective pressures and adaptations to the local biotic environment. Ultimately, differences in the types of species interactions encountered by different local populations can result in genetic differentiation and the evolution of local ecotypes.

Diffuse Interactions 12.5

Most interactions are not exclusive involving only two species but rather involve a number of species that form diffuse associations.

Niche 12.6

The range of physical and chemical conditions under which a species can persist and the array of essential resources it uses define its ecological niche. The ecological niche of a species in the absence of interactions with other species is referred to as the *fundamental niche.* The species' realized niche is its ecological niche as modified by its interactions with other species within the community. Species interactions can function to either restrict or expand the fundamental niche of a species dependent on whether the interaction is detrimental or beneficial.

Adaptive Radiation 12.7

Variations among local populations in biotic interactions can result in phenotypic divergence and therefore have the potential to function as mechanisms of adaptive radiation, if the divergence in phenotypic characteristics results in reproductive isolation.

Urban Ecology Ecological Issues & Applications

Urban ecology is the study of the ecology of organisms in the context of the urban environment. Increased urbanization has led to a decline in habitat and loss of many native species, while providing habitat for other, often non-native species.

STUDY QUESTIONS

1. Contrast amensalism and competition. What do ecologists mean when they refer to amensalism as a form of asymmetrical competition?
2. What does a researcher need to establish to distinguish whether an interaction between two species is an example of commensalism or mutualism?
3. Do you think all species interactions influence the population dynamics of the species involved? Can you construct an example wherein a population of predators may not have a negative impact on the population growth rate of a prey species?
4. Do you think that all positive interactions necessarily lead to an increase in the relative fitness of the individuals involved?
5. If a predator affects all phenotypes within the prey population equally, does the interaction function as an agent of natural selection?
6. If a phenotypic characteristic exhibited by a prey species functions to help individuals avoid being detected by predators, can we assume that the characteristic is a product of coevolution?
7. Take note of the animal and plant species that are most common in the area of your campus or around your home. Are these species different from the animals and plants that you would associate with the natural ecosystems that define the landscape in your region (forest, desert, grassland, etc.)? What features of your suburban or urban surroundings make the environment suitable for these species?

FURTHER READINGS

Classic Studies

Elton, C. S. 1927. *Animal Ecology.* New York: Macmillan Co.
One of the two classic works that introduced the concept of ecological niche to the field of ecology. A new edition was published in 2001 by University of Chicago Press.

Grinnell, J. 1917. "The niche-relationships of the California Thrasher." *Auk* 34:427–433.
One of the two classic articles that introduced the concept of ecological niche to the field of ecology.

Hutchinson, G. E. 1957. "Concluding remarks." *Cold Spring Harbor Symposia on Quantitative Biology* 22 (2):415–427.
In this article, Hutchinson defines the modern concept of the ecological niche.

Current Research

Bascompte, J. 2010. "Structure and dynamics of ecological networks." *Science* 329:765–766.
An excellent overview of the concept of diffuse interactions in ecological communities.

Chase J. M., and M. A. Leibold. 2003. *Ecological niches: Linking classical and contemporary approaches.* Chicago: University of Chicago Press.

Futuyma, D. J., and M. Slatkin. 1983. *Coevolution.* Sunderland, MA: Sinauer Associates Inc.
This now-classic volume is a compilation of contributed chapters that presents a wide array of studies examining the process of coevolution.

Gaston, K. J. (ed.). 2010. *Urban ecology.* Cambridge: Cambridge University Press.
An edited volume that provides an excellent overview of the diversity of research on the ecology of urban environments.

Holt, R. D. 2009. "Bringing the Hutchinsonian niche into the 21st century: Ecological and evolutionary perspectives." *Proceedings of the National Academy of Sciences* 106:19659–19665.
This manuscript provides a modern perspective on the concept of the ecological niche.

Lindberg, A. B., and J. M. Olesen. 2001. "The fragility of extreme specialization: *Passiflora mixta* and its pollinating hummingbird *Ensifera ensifera.*" *Journal of Tropical Ecology* 17:323–329.
An excellent example of the evolution of specialization in the coevolution of plants and pollinators.

McKinney, M. L. 2002. "Urbanization, biodiversity and conservation." *BioScience* 52:883–890.
This article provides an excellent overview of the impacts of urbanization on habitat and species composition along the rural–urban gradient of landscapes.

Schluter, D. 2000. "Introduction to the Symposium: Species Interactions and Adaptive Radiation." *The American Naturalist* 156:S1–S3.
This article provides an excellent introduction to the role of species interaction in the process of speciation.

Thompson, J. N. 2005. *The geographic mosaic of coevolution.* Chicago: University of Chicago Press.
This volume provides a discussion of how the biology of species provides the raw material for long-term coevolution and evaluates how local coadaptation forms the basic module of coevolutionary change.

Travis, J. 1996. "The significance of geographic variation in species interactions." *American Naturalist* 148 (Supplement):S1–S8.
This work explores how interspecific interactions vary across the landscape and explores the implications to both evolutionary processes and the structure of ecological communities.

CHAPTER 13 Interspecific Competition

A yellow-necked mouse (*Apodemus flavicollis*) feeds on an acorn on a forest floor in South Lower Saxony, Germany. This mouse is only one of an array of species within the forest that depend on acorns as a food resource.

CHAPTER GUIDE

IN CHAPTER 3 OF *THE ORIGIN OF SPECIES*, Charles Darwin wrote: "as more individuals are produced than can possibly survive, there must in every case be a struggle for existence, either one individual with another of the same species, or with the individuals of distinct species, or with the physical conditions of life." The concept of interspecific competition is a cornerstone of evolutionary ecology. Darwin based his idea of natural selection on competition, the "struggle for existence." Because it is advantageous for individuals to avoid this struggle, competition has been regarded as the major force behind species divergence and specialization (Chapter 5).

13.1 Interspecific Competition Involves Two or More Species

A relationship that affects the populations of two or more species adversely (– –) is interspecific competition. In interspecific competition, as in intraspecific competition, individuals seek a common resource in short supply (see Chapter 11). But in interspecific competition, the individuals are of two or more species. Both kinds of competition may take place simultaneously. In the deciduous forest of eastern North America, for example, gray squirrels compete among themselves for acorns during a year when oak trees produce fewer acorns. At the same time, white-footed mice, white-tailed deer, wild turkey, and blue jays vie for the same resource. Because of competition, one or more of these species may broaden the base of their foraging efforts. Populations of these species may be forced to turn away from acorns to food that is less in demand.

Like intraspecific competition, interspecific competition takes two forms: exploitation and interference (see Section 11.3). As an alternative to this simple dichotomous classification of competitive interactions, Thomas Schoener of the University of California–Davis proposed that six types of interactions are sufficient to account for most instances of interspecific competition: (1) consumption, (2) preemption, (3) overgrowth, (4) chemical interaction, (5) territorial, and (6) encounter.

Consumption competition occurs when individuals of one species inhibit individuals of another by consuming a shared resource, such as the competition among various animal species for acorns. Preemptive competition occurs primarily among sessile organisms, such as barnacles, in which the occupation by one individual precludes establishment (occupation) by others. Overgrowth competition occurs when one organism literally grows over another (with or without physical contact), inhibiting access to some essential resource. An example of this interaction is when a taller plant shades those individuals below, reducing available light (as discussed in Chapter 4, Section 4.2). In chemical interactions, chemical growth inhibitors or toxins released by an individual inhibit or kill other species. **Allelopathy** in plants, in which chemicals produced by some plants inhibit germination and establishment of other species, is an example of this type of species interaction. Territorial competition results from the behavioral exclusion of others from a specific space that is defended as a territory (see Section 11.10). Encounter competition results when nonterritorial meetings between individuals negatively affect one or both of the participant species. Various species of scavengers fighting over the carcass of a dead animal provide an example of this type of interaction.

13.2 The Combined Dynamics of Two Competing Populations Can Be Examined Using the Lotka–Volterra Model

In the early 20th century, two mathematicians—the American Alfred Lotka and the Italian Vittora Volterra—independently arrived at mathematical expressions to describe the relationship between two species using the same resource (consumption competition). Both men began with the logistic equation for population growth that we developed previously in Chapter 11:

$$\textit{Species 1}: \quad dN_1/dt = r_1N_1(1 - N_1/K_1)$$

$$\textit{Species 2}: \quad dN_2/dt = r_2N_2(1 - N_2/K_2)$$

Next, they both modified the logistic equation for each species by adding to it a term to account for the competitive effect of one species on the population growth of the other. For *species 1*, this term is αN_2, where N_2 is the population size of *species 2*, and α is the **competition coefficient** that quantifies the per capita effect of *species 2* on *species 1*. Similarly, for *species 2*, the term is βN_1, where β is the per capita competition coefficient that quantifies the per capita effect of *species 1* on *species 2*. The competition coefficients can be thought of as factors for converting an individual of one species into the equivalent number of individuals of the competing species, based on their shared use of the resources that define the carrying capacities (see Chapter 12, Section 12.2 and Figure 12.3, and Quantifying Ecology 12.1). In resource use, an individual of *species 1* is equal to β individuals of *species 2*. Likewise, an individual of *species 2* is equivalent to α individuals of *species 1*. These terms (α and β), in effect, convert the population size of the one species into the equivalent number of individuals of the other. For example, assume *species 1* and *species 2* are both grazing herbivores that feed on the exact same food resources (grasses and other herbaceous plants). If individuals of *species 2* have, on average, twice the body mass as individuals of *species 1* and consume food resources at twice the rate, with respect to the food resources, an individual of *species 2* is equivalent to two individuals of *species 1* (that is, $\alpha = 2.0$). Likewise, consuming food resources at only half the rate as *species 2*, an individual of *species 1* is equivalent to one-half an individual of *species 2* (that is, $\beta = 0.5$).

Now we have a pair of equations that consider both intraspecific and interspecific competition.

$$\textit{Species 1}: \quad dN_1/dt = r_1N_1(1 - (N_1 + \alpha N_2)/K_1) \qquad (1)$$

$$\textit{Species 2}: \quad dN_2/dt = r_2N_2(1 - (N_2 + \beta N_1)/K_2) \qquad (2)$$

As you can see, in the absence of interspecific competition—either α or $N_2 = 0$ in Equation (1) and β or $N_1 = 0$ in Equation (2)—

the population of each species grows logistically to equilibrium at K, the respective carrying capacity. In the presence of competition, however, the picture changes.

For example, the carrying capacity for *species 1* is K_1, and as N_1 approaches K_1, the population growth (dN_1/dt) approaches zero. However, *species 2* is also vying for the limited resource that determines K_1, so we must consider the impact of *species 2*. Because α is the per capita effect of *species 2* on *species 1*, the total effect of *species 2* on *species 1* is αN_2, and as the combined population $N_1 + \alpha N_2$ approaches K_1, the growth rate of *species 1* approaches zero as well. The greater the population size of the competing species (N_2), the greater the reduction in the growth rate of *species 1* is (see discussion in Section 12.2 and Figure 12.3).

The simplest way to examine the possible outcomes of competition using the Lotka–Volterra equations presented is a graphical approach in which we first define the zero-growth isocline for each of the two competing species. The **zero-growth isocline** represents the combined values of population size for *species 1* (N_1) and *species 2* (N_2) at which the population growth rate of the respective species is zero ($dN/dt = 0$). This occurs when the combined population sizes are equal to the carrying capacity of that species (see Figure 12.3). We can begin by defining the zero-growth isocline for *species 1* (**Figure 13.1a**). The two axes in the graph shown in Figure 13.1a define the population size of *species 1* (x-axis, N_1) and *species 2* (y-axis, N_2). We must now solve for the combined values of N_1 and N_2 at which the growth rate of *species 1* is equal to zero ($dN_1/dt = 0$). This occurs when: $(1 - (N_1 + \alpha N_2)/K_1) = 0$ or $K_1 = N_1 + \alpha N_2$ (see Equation 1). In effect, we are determining the combined values of N_1 and N_2 that equal the carrying capacity of *species 1* (K_1). This task is made simple because $K_1 = N_1 + \alpha N_2$ represents a line and all that is necessary to draw the line is to solve for two points. The two simplest solutions are to solve for the two intercepts (where the line intersects the two axes). The x-intercept occurs when $N_2 = 0$, giving us a value of $N_1 = K_1$. The y-intercept occurs when $N_1 = 0$, giving us a value of $\alpha N_2 = K_1$, or $N_2 = K_1/\alpha$. Given these two points (values for N_1, N_2), we can draw the line defining the zero isocline for *species 1* (Figure 13.1a). For any combined value of N_1, N_2 along this line, $N_1 + \alpha N_2 = K_1$ and $dN_1/dt = 0$. For combinations of (N_1, N_2) that fall below the line (toward the origin: 0, 0), $N_1 + \alpha N_2 < K_1$ and the population of *species 1* can continue to grow. An increase in the population of *species 1* is represented by a green horizontal arrow pointing to the right. The arrow is horizontal because the x-axis represents the population of *species 1*. For combinations of N_1 and N_2 that fall above the line, $N_1 + \alpha N_2 > K_1$, the population growth rate is negative (as represented by the green horizontal line pointing to the left), and the population size declines until it reaches the line.

We can take this same approach and define the zero isocline for *species 2* (**Figure 13.1b**). The x-intercept is $N_2 = 0$ and $N_1 = K_2/\beta$, and the y-intercept is $N_2 = K_2$ and $N_1 = 0$. As with the zero-growth isocline for *species 1*, for combinations of N_1 and N_2 that fall below the line, $N_2 + \beta N_1 < K_2$ and the population of *species 2* can continue to grow. The yellow vertical arrow pointing up represents an increase in the population of *species 2*. The arrow is vertical because the x-axis represents the population of *species 2*. For combinations of (N_1, N_2) that fall above the line, $N_2 + \beta N_1 > K_2$, the population growth rate is negative (yellow vertical arrow pointing down), and the population size declines until it reaches the line (see Figure 13.1b). We can now combine the two zero-growth isoclines onto a single graph and examine the combined population dynamics of the two species for different values of N_1 and N_2.

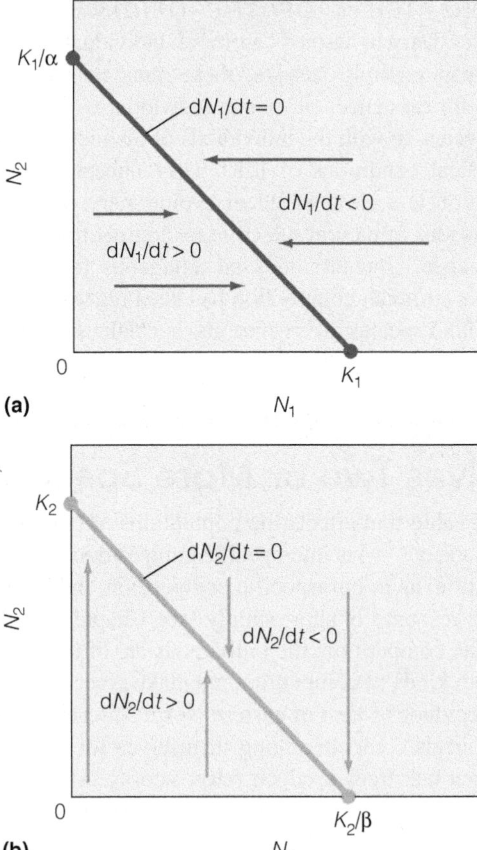

Figure 13.1 The Lotka–Volterra model of competition between two species: zero-growth isoclines for *species 1* (a) and *species 2* (b). The zero-growth isocline for each species is defined as the combinations of (N_1, N_2) for which $dN/dt = 0$ (zero population growth). In the area (combined values of [N_1, N_2]) below the line, population growth is positive and the population increases (as indicated by the arrows); for combined values of (N_1, N_2) above the line, the population decreases.

13.3 There Are Four Possible Outcomes of Interspecific Competition

To interpret the combined dynamics of the two competing species, their isoclines must be drawn on the same x–y graph. Although there are an infinite number of isoclines that can be constructed by using different values of K_1, K_2, α, and β, there are only four qualitatively different ways in which to plot the isoclines. These four possible outcomes are shown in **Figure 13.2**. In the first case (Figure 13.2a), the isocline of *species 1* is parallel to, and lies completely above, the isocline of *species 2*. In this case, the isoclines define three areas of the graph. In the lower left-hand area of the graph (point A), the combined values of N_1 and N_2 are below the zero-growth isoclines for both species, and the populations of both species can increase. The green horizontal arrow representing *species 1* points right, indicating an increase in the population of *species 1*, whereas the orange vertical arrow representing *species 2* points up, indicating an increase in the population of *species 2*. The next point representing the combined values of N_1 and N_2 must therefore lie somewhere between the two arrows and is represented by the black arrow pointing away from the origin. In the upper right-hand corner of the graph, the combined values of N_1 and N_2 are above the zero-growth isoclines for both species. In this case, the populations of both species decline (black arrow points toward the origin).

In the interior region between the two isoclines, the dynamics of the two populations diverge. Here (at point C) the combined values of N_1 and N_2 are below the isocline for *species 1*, so its population increases in size, and the green horizontal arrow points to the right. However, this region is above the isocline for *species 2*, so its population is declining, and the yellow vertical arrow is pointing down. The black arrow now points down and toward the right, which takes the populations toward the carrying capacity of *species 1* (K_1). Note that this occurs regardless of where the initial point (N_1, N_2) lies within this region. If the isocline of *species 1* lies above the isocline for *species 2*, *species 1* is the more competitive species and *species 2* is driven to extinction ($N_2 = 0$).

In the second case (Figure 13.2b), the situation is reversed. The zero-growth isocline for *species 2* lies above the isocline for *species 1*, and therefore *species 2* "wins" leading to the extinction of *species 1* ($N_1 = 0$). Note that in the interior region (between the isoclines), the combined values of N_1 and N_2 are now below the isocline for *species 2* allowing its population to grow (yellow vertical arrow pointing up), whereas it is above the isocline for *species 1*, causing its population to decline (green horizontal arrow pointing to the left). The result is a movement of the populations toward the upper left (see black arrow), the carrying capacity of *species 2* (K_2).

In the remaining two cases (Figures 13.2c and 13.2d), the isoclines of the two species cross, dividing the graph into four regions, but the outcomes of competition for the two cases are quite different. As with the previous two cases, we determine the outcomes by plotting the arrows, indicating changes in the two populations within each of the regions. However, the point where the two isoclines cross represents an equilibrium point, a combined value of N_1 and N_2 for which the growth of both *species 1* and *species 2* is zero. At this point, the combined population sizes of the two species are equal to the carrying capacities of both species ($N_1 + \alpha N_2 = K_1$ and $N_2 + \beta N_1 = K_2$).

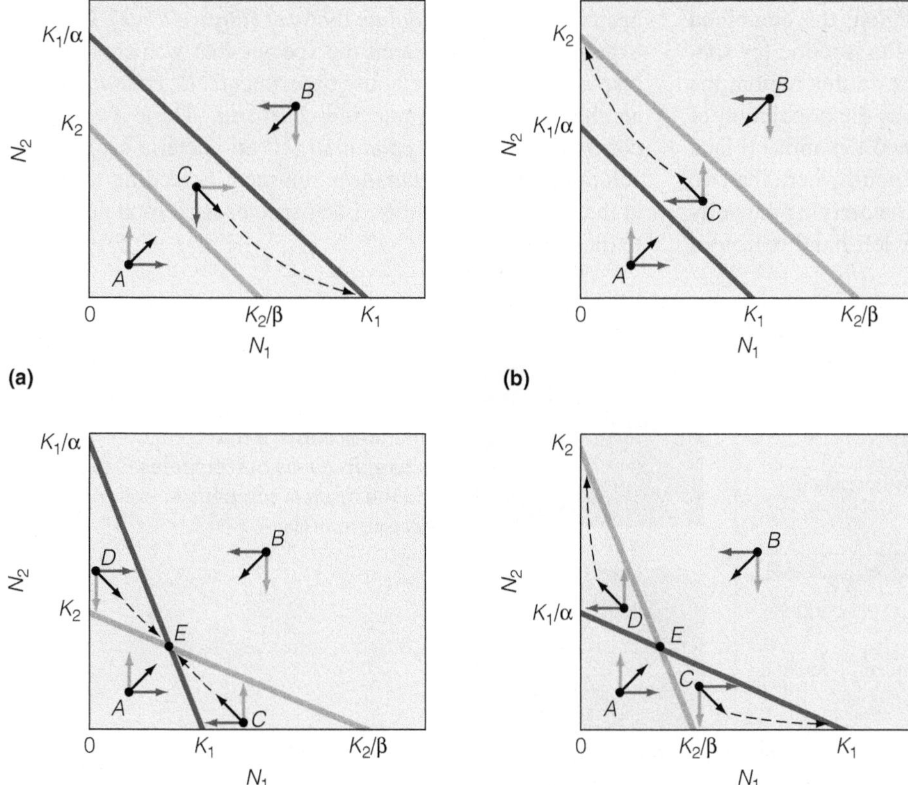

Figure 13.2 Four possible arrangements of the zero-growth isoclines for *species 1* and *species 2* using the Lotka-Volterra model of competition. In case (a), the isocline of *species 1* falls outside the isocline of *species 2*. *Species 1* always wins, leading to the extinction of *species 2*. In case (b), the situation is the reverse of (a). (c) Each species inhibits the growth of its own population more than that of the other by intraspecific competition. The species coexist. (d) The isoclines cross. Each species inhibits the growth of the other more than its own growth. The more abundant species often wins.

The third case is presented in Figure 13.2c. The region closest to the origin (point A) is below the isocline of both species, and therefore the growth of both populations is positive and the arrows point outward. The upper right-hand region (point B) is above the isoclines for both species, so both populations decline and the arrows point inward toward the axes and origin. In the bottom right-hand region of the graph (point C), we are above the isocline for *species 1*, but below the isocline for *species 2*. In this region, the population of *species 1* declines (green horizontal arrow points to left), whereas the population of *species 2* increases (yellow vertical arrow points up). As a result, the combined dynamics (black arrow) point toward the center of the graph where the two isoclines intersect. The upper left-hand region of the graph (point D) is above the isocline for *species 2* but below the isocline for *species 1*. In this region, the population of *species 2* declines, and the population of *species 1* increases. Again, the combined dynamics (black arrow) point toward the center of the graph where the two isoclines intersect. The fact that the arrows in all four regions of the graph point to where the two isoclines intersect indicates that this point (combined values of N_1 and N_2) represents a "stable equilibrium." The equilibrium is stable when no matter what the combined values of N_1 and N_2 are, both populations move toward the equilibrium value.

In the fourth case (Figure 13.2d), the isoclines cross, but in a different manner than in the previous case (Figure 13.2c). Again, both populations increase in the region of the graph closest to the origin (point A). Likewise, both populations decline in the upper right-hand region (point B). However, the dynamics differ in the remaining two regions of the graph. In the lower right-hand region, the combined values of N_1 and N_2 (point C) are below the isocline for *species 1* but above the isocline for *species 2*. In this region, the population of *species 1* decreases, whereas the population of *species 2* continues to grow. The combined dynamics (black arrow) move away from the equilibrium point where the two isoclines intersect (point E) and toward the carrying capacity of *species 1* (K_1 on x-axis). In the upper left-hand region of the graph, the combined values of N_1 and N_2 (point D) are below the isocline for *species 2* but above the isocline for *species 1*. In this region of the graph, the combined dynamics (black arrow) move away from the equilibrium point where the two isoclines intersect (point E) and toward the carrying capacity of *species 2* (K_2 on y-axis). This case represents an "unstable equilibrium." If the combined values of N_1 and N_2 are displaced from the equilibrium (point E), the populations move into one of the two regions of the graph that will eventually lead to one species excluding the other (driving it to extinction: $N = 0$). Which of the two species will "win" is difficult to predict and depends on the initial population values (N_1 and N_2) and the growth rates of the populations (r_1 and r_2).

13.4 Laboratory Experiments Support the Lotka–Volterra Model

The theoretical Lotka–Volterra equations stimulated studies of competition in the laboratory, where under controlled conditions an outcome is more easily determined than in the field. One of the first to study the Lotka–Volterra competition model experimentally was the Russian biologist G. F. Gause. In a series of experiments published in the mid-1930s, he examined competition between two species of *Paramecium*, *Paramecium aurelia* and *Paramecium caudatum*. *P. aurelia* has a higher rate of population growth than *P. caudatum* and can tolerate a higher population density. When Gause introduced both species to one tube containing a fixed amount of bacterial food, *P. caudatum* died out (**Figure 13.3**). In another experiment, Gause reared the species that was competitively displaced in the previous experiment, *P. caudatum*, with another species, *Paramecium bursaria*. These two species coexisted because *P. caudatum* fed on bacteria suspended in solution, whereas *P. bursaria* confined its feeding to bacteria at the bottom of the tubes. Each species used food unavailable to the other.

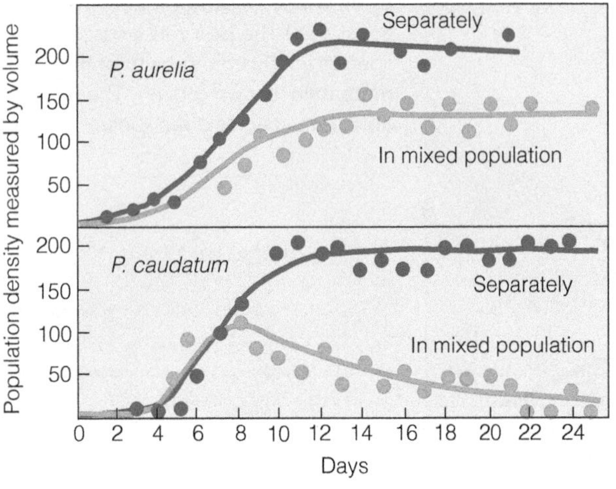

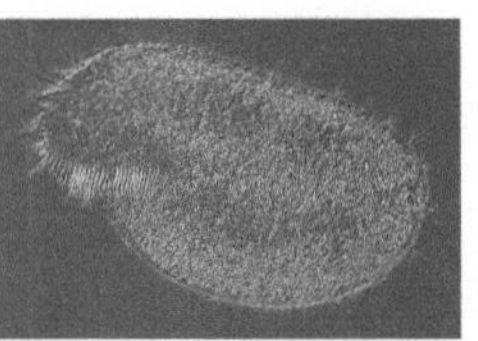

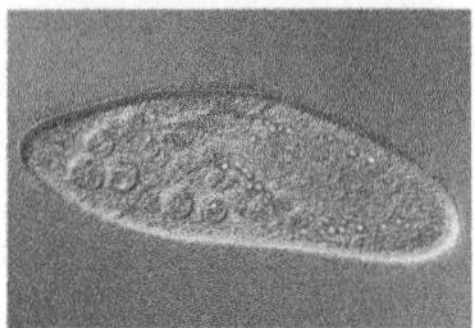

Figure 13.3 Competition experiments with two ciliated protozoans, *Paramecium aurelia* and *Paramecium caudatum*, grown separately and in a mixed culture. In a mixed culture, *P. aurelia* outcompetes *P. caudatum*, and the result is competitive exclusion. (Adapted from Gause 1934.)

In the 1940s and 1950s, Thomas Park at the University of Chicago conducted several classic competition experiments with laboratory populations of flour beetles. He found that the outcome of competition between *Tribolium castaneum* and *Tribolium confusum* depended on environmental temperature, humidity, and fluctuations in the total number of eggs, larvae, pupae, and adults. Often, the outcome of competition was not determined until many generations had passed.

In a much later study, ecologist David Tilman of the University of Minnesota grew laboratory populations of two species of diatoms, *Asterionella formosa* and *Synedra ulna.* Both species require silica for the formation of cell walls. The researchers monitored population growth and decline as well as the level of silica in the water. When grown alone in a liquid medium to which silica was continually added, both species kept silica at a low level because they used it to form cell walls. However, when grown together, the use of silica by *S. ulna* reduced the concentration to a level below that necessary for *A. formosa* to survive and reproduce (**Figure 13.4**). By reducing resource availability, *S. ulna* drove *A. formosa* to extinction.

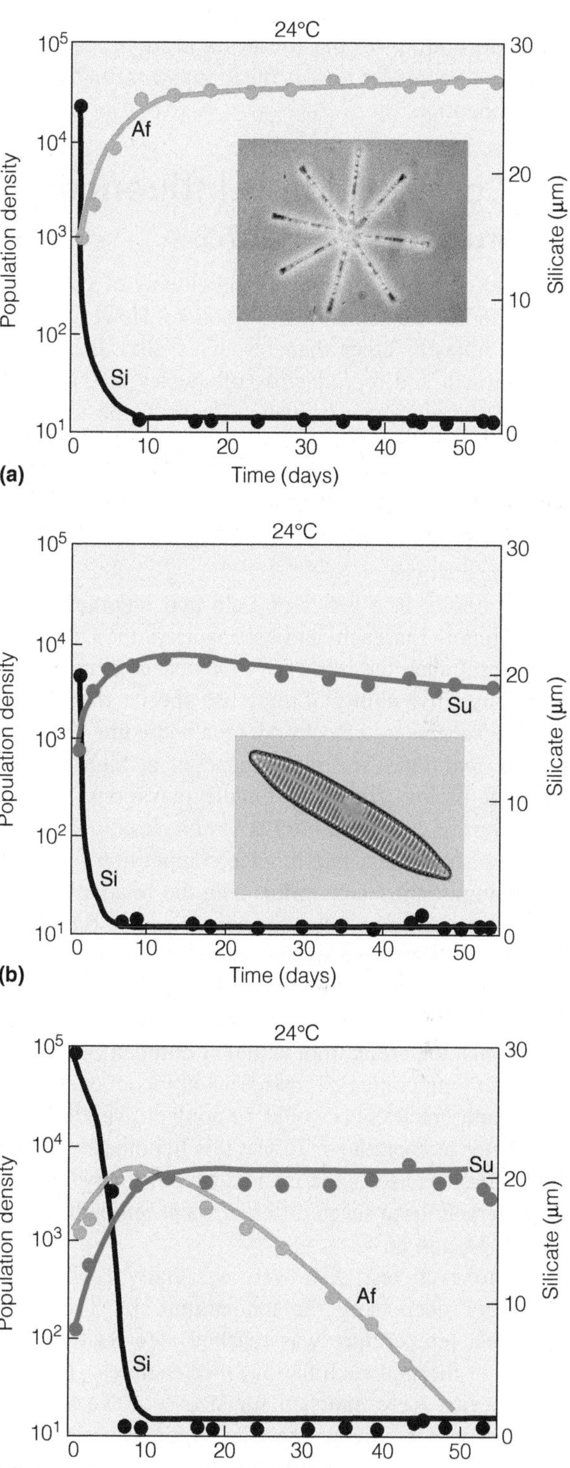

Figure 13.4 Competition between two species of diatom, *Asterionella formosa* (Af) and *Synedra ulna* (Su), for silicate (Si). (a, b) Grown alone in a culture flask, both species reach a stable population that keeps silicate at a constant low level. *Synedra* draws silicate lower. (c) When the two are grown together, *Synedra* reduces silicate to a point at which *Asterionella* dies out. (Adapted from Tilman et al. 1981.)

13.5 Studies Support the Competitive Exclusion Principle

In three of the four situations predicted by the Lotka–Volterra equations, one species drives the other to extinction. The results of the laboratory studies just presented tend to support the mathematical models. These and other observations have led to the concept called the **competitive exclusion principle**, which states that "complete competitors" cannot coexist. Complete competitors are two species (non-interbreeding populations) that live in the same place and have exactly the same ecological requirements (see concept of fundamental niche in Chapter 12, Section 12.6). Under this set of conditions, if population A increases the least bit faster than population B, then A will eventually outcompete B, leading to its local extinction.

Competitive exclusion, then, invokes more than competition for a limited resource. The competitive exclusion principle involves assumptions about the species involved as well as the environment in which they exist. First, this principle assumes that the competitors have exactly the same resource requirements. Second, it assumes that environmental conditions remain constant. Such conditions rarely exist. The idea of competitive exclusion, however, has stimulated a more critical look at competitive relationships in natural situations. How similar can two species be and still coexist? What ecological conditions are necessary for coexistence of species that share a common resource base? The resulting research has identified a wide variety of factors affecting the outcome of interspecific competition, including environmental factors that directly influence a species' survival, growth, and reproduction but are not consumable resources (such as temperature or pH), spatial and temporal variations in resource availability, competition for multiple limiting resources, and resource partitioning. In the following sections, we examine each topic and consider how it functions to influence the nature of competition.

13.6 Competition Is Influenced by Nonresource Factors

Interspecific competition involves individuals of two or more species vying for the same limited resource. However, features of the environment other than resources also directly influence the growth and reproduction of species (see Chapters 6 and 7) and therefore can influence the outcome of competitive interactions. For example, environmental factors such as temperature, soil or water pH, relative humidity, and salinity directly influence physiological processes related to growth and reproduction, but they are not consumable resources that species compete over.

For example, in a series of field and laboratory experiments, Yoshinori Taniguchi and colleagues at the University of Wyoming examined the influence of water temperature on the relative competitive ability of three fish species that show longitudinal replacement in Rocky Mountain streams. Brook trout (*Salvelinus fontinalis*) are most abundant at high elevations, brown trout (*Salmo trutta*) at middle elevations, and creek chub (*Semotilus atromaculatus*) at lower elevations. Previous studies have shown that interference competition for foraging sites is an important factor influencing the relative success of individuals at sites where the species co-occur. Based on the distribution of these three species along elevation gradients in the Rocky Mountain streams and differences in physiological performance with respect to temperature, the researchers hypothesized that the brook trout would be competitively superior at cold water temperatures, brown trout at moderate water temperatures, and creek chub would be competitively superior at warmer water temperatures. To test this hypothesis, Taniguchi and his colleagues used experimental streams to examine competitive interactions at seven different water temperatures: 3, 6, 10, 22, 22, 24, and 26°C.

Prior to each test, fish were thermally acclimated by increasing or decreasing the temperature by 1°C per day until the test temperature was reached (see Section 7.9 for discussion of thermal acclimation). For each test, individuals of each species were matched for size (<10%) and placed in the experimental stream together. Aggressive interactions and food intake were monitored. Competitive superiority was based on which species consumed the most food items because food intake is considered a limiting factor for these drift-feeding, stream fishes.

Patterns of food consumption clearly show changes in the relative competitive abilities of the three fish species across the gradient of water temperatures (**Figure 13.5**). At 3°C, brook trout exhibited the highest rate of food consumption, although differences between the two trout species were minimal below 20°C, and both trout species consumed significantly more food than creek chub. However, as temperature increased, food consumption by creek chub increased. At 24°C, food intake

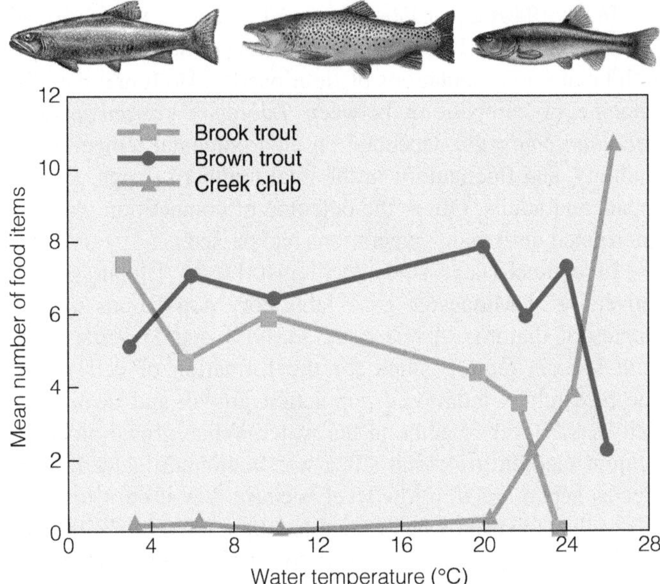

Figure 13.5 Food consumption in relation to water temperature for brook trout, brown trout, and creek chub during competition trials in experimental streams.
(Adapted from Taniguchi et al. 1998.)

by brook trout dropped to zero, whereas intake rate of brown trout still exceeded that of creek chub. At 26°C, the rate of food intake reversed for the two species and food intake by creek chub exceeded that of brown trout. In an additional series of experiments, the researchers were able to establish that the observed patterns of food intake during the competition trials were a result of differences in competitive ability and no changes in appetite because of water temperature.

The transition in competitive ability from 24 to 26°C in the laboratory experiments are in agreement with the transition in dominance from trout species to creek chub at a similar temperature range in the field. The results of Taniguchi and his colleagues provide a clear example of temperature mediation of competitive interactions. The relative competitive abilities of the three fish species for limiting food resources are directly influenced by abiotic conditions, that is, water temperature.

A similar case of competitive ability being influenced by nonresource factors is illustrated in the work of Susan Warner of Pennsylvania State University. Warner and her colleagues examined the effect of water pH (acidity) on interspecific competition between two species of tadpoles (*Hyla gratiosa* and *Hyla femoralis*). The two species overlap broadly in their geographic distribution, yet differ in their responses to water acidity. The researchers conducted experiments using two levels of water pH (4.5 and 6.0) and varying levels of population densities to examine the interactions of pH and population density on both intra- and interspecific competition. The results of the experiments indicated that interspecific interactions were minimal at low water pH (4.5); however, at higher water pH (6.0), interspecific competition from *H. fermoralis* caused decreased survival and an increased larval period for *H. gratiosa*. The latter resulted in decreased size at metamorphosis for *H. gratiosa* individuals.

13.7 Temporal Variation in the Environment Influences Competitive Interactions

When one species is more efficient at exploiting a shared, limiting resource, it may be able to exclude the other species (see Section 13.2). However, when environmental conditions vary through time, the competitive advantages may also change. As a result, no one species reaches sufficient density to displace its competitors. In this manner, environmental variation allows competitors to coexist whereas under constant conditions, one would exclude the other.

The work of Peter Dye of the South African Forestry Research Institute provides an example of shifting competitive ability resulting from temporal variation in resource availability in the grasslands of southern Africa. He examined annual variations in the relative abundance of grass species occupying a savanna community in southwest Zimbabwe. From 1971 to 1981, the dominant grass species shifted from *Urochloa mosambicensis* to *Heteropogon contortus* (**Figure 13.6a**). This observed shift in dominance was a result of yearly variations in rainfall (**Figure 13.6b**). Rainfall during the 1971–1972 and 1972–1973 rainy seasons was much lower than average. *U. mosambicensis* can maintain higher rates of survival and growth under dry conditions than can *H. contortus,* making it a better competitor under conditions of low rainfall. With the return to higher rainfall during the remainder of the decade, *H. contortus* became the dominant grass species. Annual rainfall in this semiarid region of southern Africa is highly variable, and fluctuations in species composition such as those shown in Figure 13.6 are a common feature of the community.

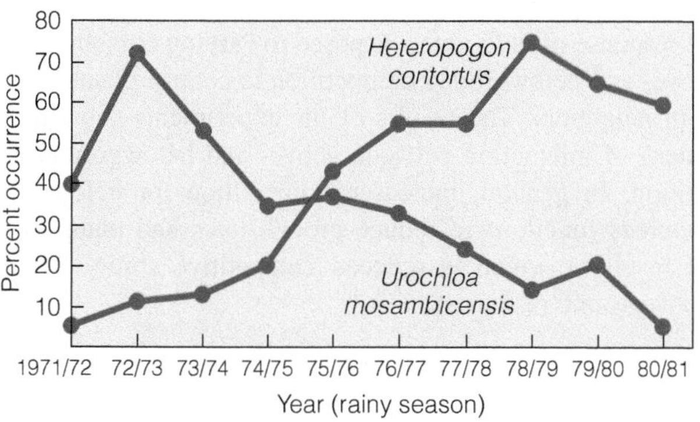

(a)

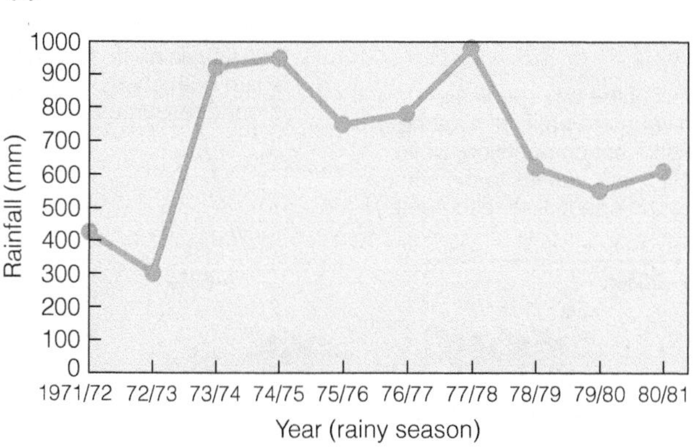

(b)

Figure 13.6 (a) Shift in the dominant grass species in a savanna community in southwest Zimbabwe during the period 1971–1981. The shift is in response to changing patterns of precipitation during the same period. (b) *Urochloa mosambicensis* was able to compete successfully under the drier conditions during the 1971–1972 and 1972–1973 rainy seasons. With the increase in rainfall beginning in the 1973–1974 season, *Heteropogon contortus* came to dominate the site.
(Adapted from Dye and Spear 1982.)

Peter Adler (Utah State University) and colleagues observed a similar pattern for a prairie grassland site at Hays, Kansas, in the Great Plains region of North America. Adler and colleagues examined the role of interannual climate variability on the relative abundance of prairie grasses over a period of 30 years (1937–1968). The researchers found that year-to-year variations in climate correlated with interannual variations in species performance. The year-to-year variations in the relative competitive abilities of the species functioned to buffer species from competitive exclusion.

Besides shifting the relative competitive abilities of species, variation in climate can function as a density-independent limitation on population growth (see Section 11.13). Periods of drought or extreme temperatures may depress populations below carrying capacity. If these events are frequent enough relative to the time required for the population to recover (approach carrying capacity), resources may be sufficiently abundant during the intervening periods to reduce or even eliminate competition.

13.8 Competition Occurs for Multiple Resources

In many cases, competition between species involves multiple resources and competition for one resource often influences an organism's ability to access other resources. One such example is the practice of interspecific territoriality, where competition for space influences access to food and nesting sites (see Section 11.10).

A wide variety of bird species in the temperate and tropical regions exhibit interspecific territoriality. Most often, this practice involves the defense of territories against closely related species, such as the gray (*Empidonax wrightii*) and dusky (*Empidonax oberholseri*) flycatchers of the western United States. Some bird species, however, defend their territories against a much broader range of potential competitors. For example, the acorn woodpecker (*Melanerpes formicivorus*) defends territories against jays and squirrels as well as other species of woodpeckers. Strong interspecific territorial disputes likewise occur among brightly colored coral reef fish.

Competition among plants provides many examples of how competition for one resource can influence an individual's ability to exploit other essential resources, leading to a combined effect on growth and survival. R. H. Groves and J. D. Williams examined competition between populations of subterranean clover (*Trifolium subterraneum*) and skeletonweed (*Chondrilla juncea*) in a series of greenhouse experiments. Plants were grown both in monocultures (single populations) and in mixtures (two populations combined). The investigators used a unique experimental design to determine the independent effects of competition for aboveground (light) and belowground (water and nutrients) resources (see Section 11.11). In the monocultures, plants were grown in pots, allowing for the canopies (leaves) and roots to intermingle. In the two-species mixtures (**Figure 13.7**), three different approaches were used: (1) plants of both species were grown in the same pot, allowing their canopies and roots to intermingle, (2) plants of both species were grown in the same pot allowing their roots to overlap, but with their canopies separated, (3) the plant species were grown in separate pots with their canopies intermingled, but not allowing the roots to overlap.

Clover was not significantly affected by the presence of skeletonweed; however, the skeletonweed was affected in all three treatments where the two populations were grown together. When the roots were allowed to intermingle, the biomass (dry weight of the plant population) of skeletonweed was reduced by 35 percent compared to the biomass of the species when grown as a monoculture. The biomass was reduced by 53 percent when the canopies were intermingled. When both the canopies and roots were intermingled, the biomass was reduced by 69 percent, indicating an interaction in the competition for aboveground and belowground resources. Clover plants were the superior competitors for both aboveground and belowground resources, resulting in a combined effect of competition for these two resources (see Sections 11.11 and 18.4). This type of interaction has been seen in a variety of laboratory and field experiments. The species with the faster growth rate grows taller than the slower-growing species, reducing its available light, growth, and demand for belowground resources. The result is increased access to resources and further growth by the superior competitor.

In a series of field studies, James Cahill of the University of Alberta (Canada) examined the interactions between competition for above- and belowground resources in an old field grassland community in Pennsylvania. With an experimental design in the field similar to that used by Groves and Williams in the greenhouse, Cahill grew individual plants with varying degrees of interaction with the roots of neighboring plants through the use of root exclusion tubes made of PVC pipe. He planted the target plant inside an exclusion tube that was placed vertically into the soil to separate roots of the target plant from the roots of other individuals in the population that naturally surround it. He controlled the degree of belowground competition by drilling varying numbers of holes in the PVC pipe that allowed access to the soil volume from neighboring plants (see Section 11.11 and Figure 11.20 for further description of method). Cahill varied the level of aboveground competition by tying back the aboveground neighboring vegetation. In total, he created 16 combinations of varying intensities of above- and belowground interaction with neighboring plants. This experimental design allowed Cahill to compare the response of individuals exposed to varying combinations of above- and belowground competition to control plants isolated from neighbors. The results of his experiments show a clear pattern of interaction between above- and belowground competition. In general, increased competition for belowground resources functions to reduce growth rates and plant stature, the result of which is reduced competitive ability for light (aboveground resource).

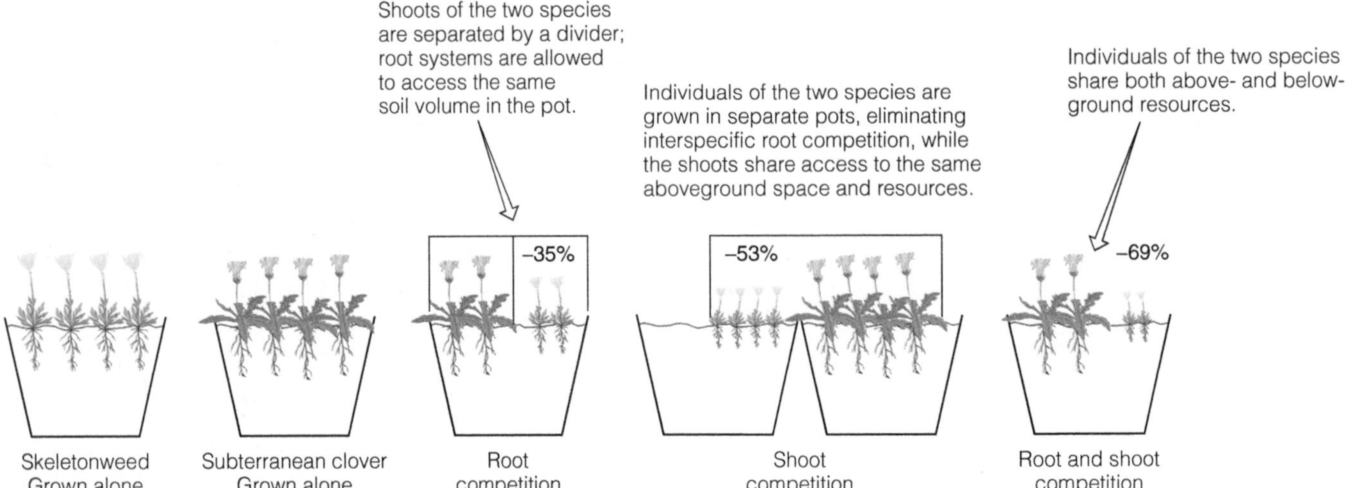

Figure 13.7 Experimental design used to examine above- and belowground competition between subterranean clover and skeletonweed. The growth of subterranean clover was not impacted by the presence of skeletonweed; however, skeletonweed growth was reduced as a result of competition from clover. Values represent the percent reduction in growth of skeletonweed in the various treatments.
(Adapted from Groves and Williams 1975.)

13.9 Relative Competitive Abilities Change along Environmental Gradients

As environmental conditions change, so do the relative competitive abilities of species. Shifts in competitive ability can result either from changes in the carrying capacities of species (values of *K;* see **Quantifying Ecology 13.1**) related to a changing resource base or from changes in the physical environment that interact with resource availability.

Many laboratory and field studies have examined the outcomes of competition among plant species across experimental gradients of resource availability. Mike Austin and colleagues at the Commonwealth Scientific and Industrial Research Organization (CSIRO) research laboratory in Canberra, Australia, have conducted several greenhouse studies to explore the changing nature of interspecific competition among plant species across experimental gradients of nutrient availability. In one such experiment, the researchers examined the response of six species of thistle along a gradient of nutrient availability (application of nutrient solution). Plants were grown both in monoculture (single species) and mixture (all six species) under 11 different nutrient treatments, ranging from 1/64 to 16 times the recommended concentration of standard greenhouse nutrient solution. After 14 weeks, the plants were harvested, and their dry weights were determined. Responses of the six species along the nutrient gradient for monoculture and mixture experiments are shown in **Figure 13.8**.

Two important results emerged from the experiment. First, when grown in mixture, the response of each species along the resource gradient differed from the pattern observed when grown in isolation—interspecific competition directly influenced the patterns of growth for each species. Second, the relative competitive abilities of the species changed along the nutrient gradient. This result was easily seen when the response of each species in the mixed-species experiments was expressed on a relative basis. The relative response of each species across the gradient was calculated by dividing the biomass (dry weight) value for each species at each nutrient level by the value of the species that achieved the highest biomass at that level. The relative performance of each species at each nutrient

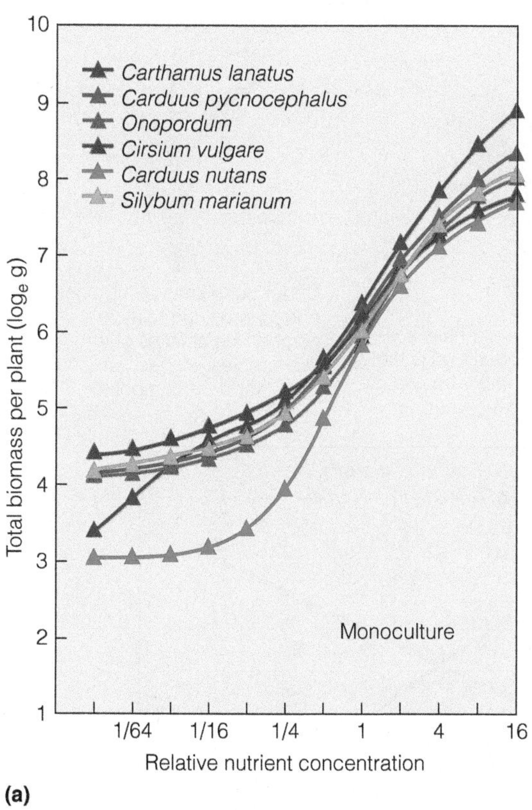

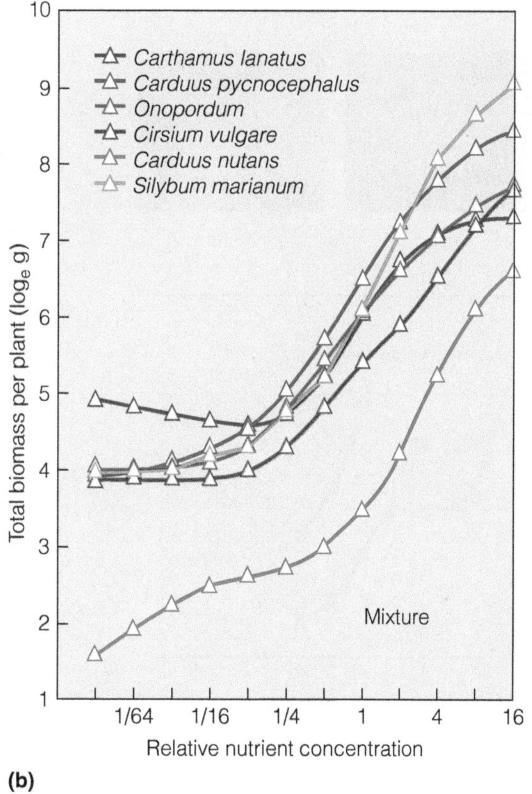

Figure 13.8 Response of six thistle species to an experimental gradient of nutrient availability: (a) single-species populations (monocultures) and (b) mixed populations. (Adapted from Austin et al. 1986.)

Interpreting Ecological Data

Q1. Which of the three species of thistle included in the graph had the highest biomass production under the 1/64 nutrient treatment? What does this imply about this species' competitive ability under low nutrient availability relative to other thistle species?

Q2. Using relative biomass production at each treatment level as an indicator of competitive ability, which thistle species is the superior competitor under the standard concentration of nutrient solution (1.0)?

Q3. At which nutrient level is the relative biomass of the three species most similar (smallest difference in the biomass of the three species)?

level then ranged from 0 to 1.0. Relative responses of the three dominant thistle species along the nutrient gradient are shown in **Figure 13.9**. Note that *Carthamus lanatus* was the superior competitor under low nutrient concentrations, *Carduus pycnocephalus* at intermediate values, and *Silybum marianum* at the highest nutrient concentrations.

In a series of field experiments, Richard Flynn and colleagues at the University of KwaZulu-Natal (South Africa) examined trade-offs in competitive ability among five perennial C_4 grass species at different levels of soil fertility and disturbance. Soil fertility treatments were established through the application of different levels of fertilizer, whereas varying levels of clipping were used to simulate disturbance resulting from grazing by herbivores. Individuals of the five grass species were grown in both monoculture and mixtures at each treatment level. The results of their experiments show a pattern of changing relative competitive abilities of the species along the gradients of soil fertility and disturbance (**Figure 13.10**). Moreover, in some of the results there were clear interactions between soil fertility and disturbance on competitive outcomes.

Field studies designed to examine the influence of interspecific competition across an environmental gradient often reveal that multiple environmental factors interact to influence the response of organisms across the landscape. In New England salt marshes, the boundary between frequently flooded low marsh habitats and less frequently flooded high marsh habitats is characterized by striking plant zonation in which monocultures of the cordgrass *Spartina alterniflora* (smooth cordgrass) dominate low marsh habitats, whereas the high marsh habitat is generally dominated by *Spartina patens* (**Figure 13.11a**). The gradient from high to low marsh is characterized by changes in nutrient availability as well as increasing physical stress relating to waterlogging, salinity, and oxygen availability in the soil and sediments. In a series of field experiments, ecologist Mark Bertness of Brown University found that *S. patens* individuals transplanted into the low marsh zone (dominated by *S. alterniflora*) were severely stunted with or without *S. alterniflora* neighbors, that is, with or without competition (**Figure 13.11b**). In contrast, *S. alterniflora* transplants grew vigorously in the high marsh (zone dominated by *S. patens*) when neighbors were removed (without competition), but were excluded from the high marsh when *S. patens* was present, that is, with competition (**Figure 13.11c**). Bertness also observed that *S. alterniflora* rapidly invaded the high marsh habitats in the absence of *S. patens*.

Figure 13.9 Differences in the response of three thistle species grown at different relative nutrient levels. Nutrient levels (x-axis) represent 11 different nutrient treatments, ranging from 1/64 to 16 times the recommended concentration of standard greenhouse nutrient solution. Response (y-axis) is measured as the normalized ecological response of the three dominant thistle species when grown as a mixture (interspecific competition). This expression is achieved by dividing the biomass (dry weight) value for each species at each nutrient level by the value of the species that achieved the highest biomass at that level (see results of experiments presented in Figure 14.7b). The relative performance of each species at each nutrient level then ranges from 0 to 1.0. (Adapted from Austin et al. 1986.)

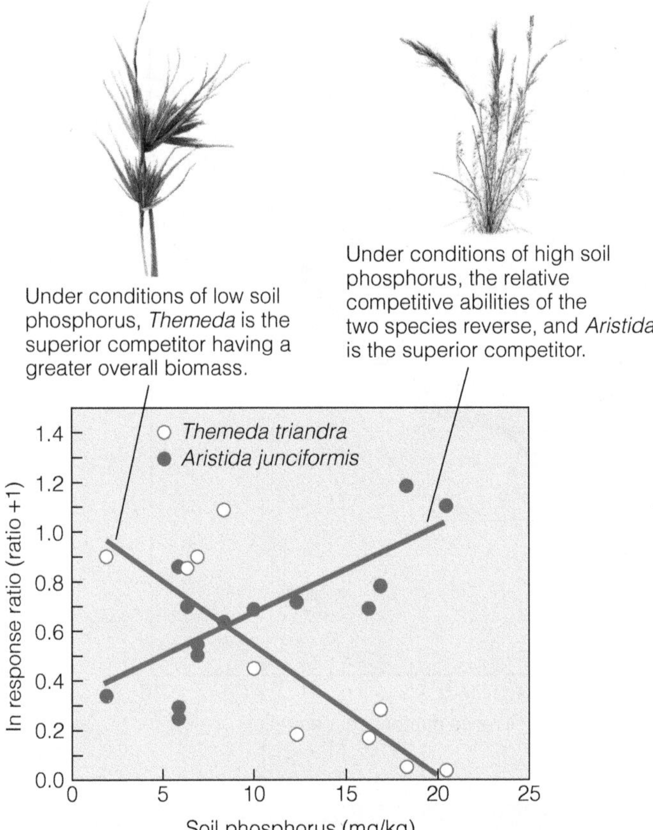

Figure 13.10 Patterns of response ratio (mixture biomass/monoculture biomass) for two grass species (*Themeda triandra* and *Aristida junciformis*) grown along a gradient of soil phosphorus. Plants were grown both in monoculture and mixtures (both species present) along the gradient, and the response ratio reflects the relative competitive abilities of the two species at the varying levels of soil phosphorus. (Fynn et al. 2005.)

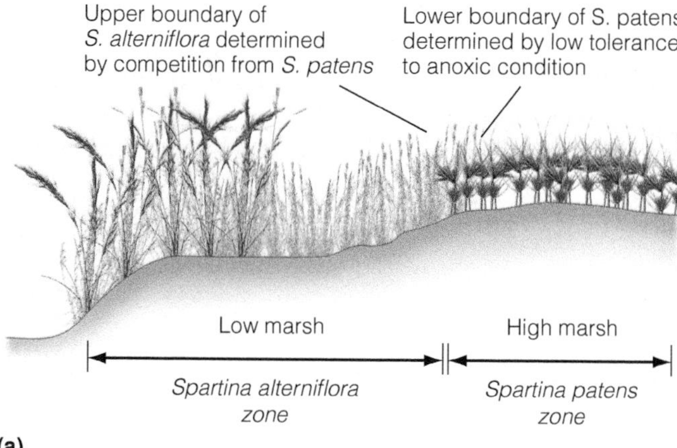

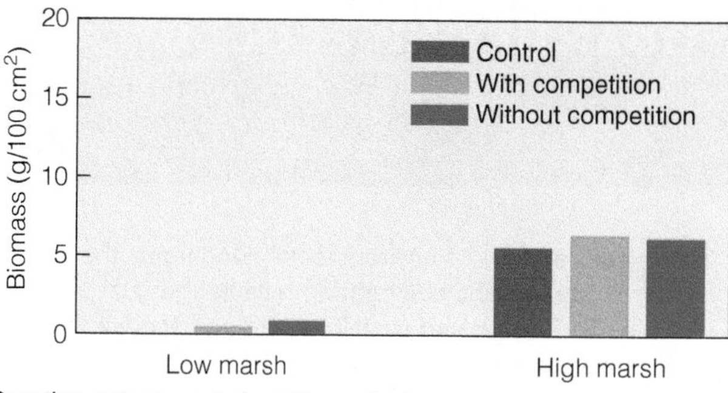

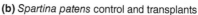

(b) *Spartina patens* control and transplants

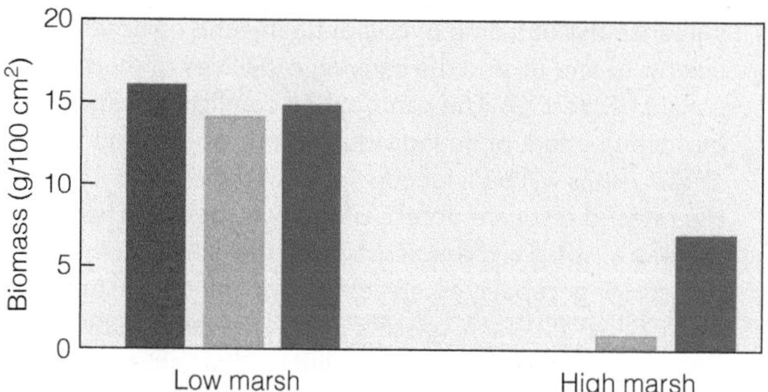

(c) *Spartina alterniflora* control and transplants

Figure 13.11 (a) Zonation of the dominant perennial plant species (*Spartina alterniflora* and *Spartina patens*) in a New England salt-marsh community. Results of transplant experiments of (b) *S. patens* and (c) *S. alterniflora* into each of the two zones (low marsh and high marsh) show that the upper border of *S. alterniflora* is a function of competition, whereas the lower border of *S. patens* is a function of the species' ability to tolerate the physical stress associated with salinity, waterlogging, and low oxygen concentrations in the sediments.
(Adapted from Bertness 1991.)

Interpreting Ecological Data

Q1. Based on the results of the transplant experiment presented in (c), are individuals of S. alterniflora able to grow in the upper zone of the marsh where S. patens dominates (S. patens zone) in the absence of competition?

Q2. Based on the results of the transplant experiment presented in (b), how does competition from S. alterniflora effect the growth of S. patens in the upper zone of the marsh (S. patens zone)?

He concluded that *S. alterniflora* dominates the physically stressful low marsh habitats because of its ability to persist in anoxic (low oxygen) soils, but it is competitively excluded from the high marsh by *S. patens*. *S. patens* is limited to high marsh habitats as a result of its inability to tolerate the harsh physical conditions in frequently flooded low marsh habitats.

Chipmunks furnish a striking example of the interaction of competition and tolerance to physical stress in determining species distribution along an environmental gradient. In this case, physiological tolerance, aggressive behavior, and restriction to habitats in which one organism has competitive advantage all play a part. On the eastern slope of the Sierra Nevada live four species of chipmunks: the alpine chipmunk (*Tamias alpinus*), the lodgepole chipmunk (*Tamias speciosus*), the yellow-pine chipmunk (*Tamias amoenus*), and the least chipmunk (*Tamias minimus*). Each of these species has strongly overlapping food requirements, and each species occupies a different altitudinal zone (**Figure 13.12**).

The line of contact between chipmunks is determined partly by interspecific aggression. Aggressive behavior by the

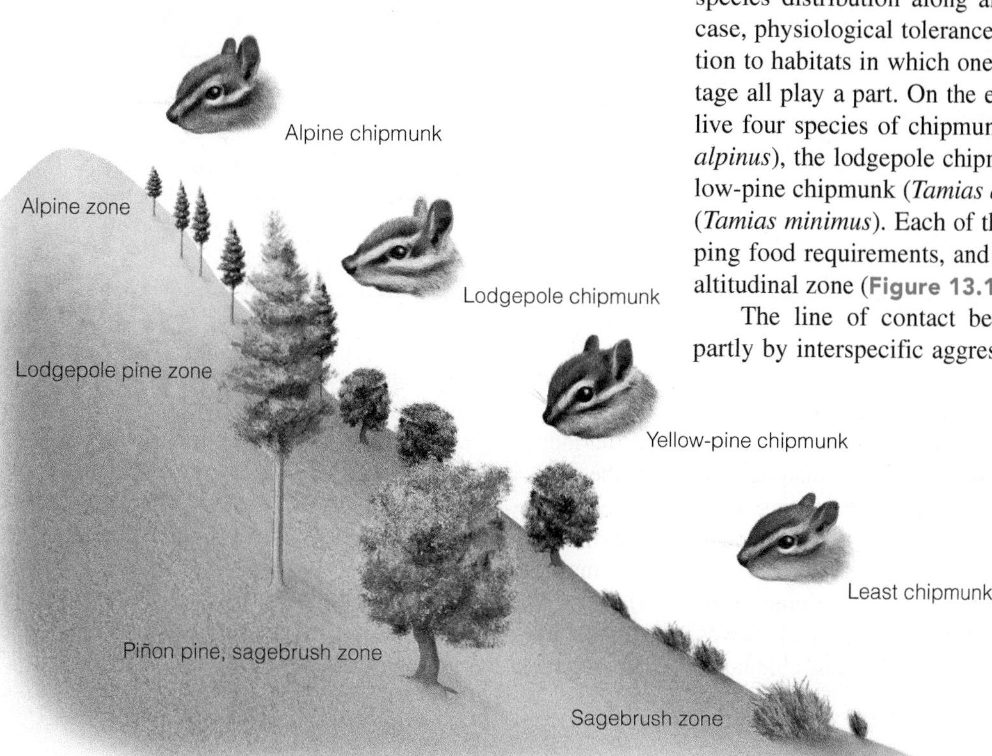

Figure 13.12 A transect of the Sierra Nevada in California, 38° N latitude, showing vegetation zonation and the altitudinal ranges of four species of chipmunks (*Eutamias*) on the east slope.
(Adapted from Heller-Gates 1971.)

QUANTIFYING ECOLOGY 13.1 Competition under Changing Environmental Conditions: Application of the Lotka–Volterra Model

Under any set of environmental conditions, the outcome of interspecific competition reflects the relative abilities of the species involved to gain access and acquire the essential resources required for survival, growth, and reproduction. As we have seen in the analysis of interspecific competition using the Lotka–Volterra equations, two factors interact to influence the outcome of competition—the competition coefficients (α and β), and the carrying capacities of the species involved (K_1 and K_2). The competition coefficients represent the per capita effect of an individual of one species on the other. These values will be a function of both the overlap in diets and the rates of resource uptake of the two species. These values, therefore, reflect characteristics of the species. In contrast, the carrying capacities are a function of the resource base (availability) for each species in the prevailing environment. Changes in environmental conditions that influence resource availability, therefore, influence the relative carrying capacities of the species and can directly influence the nature of competition.

Consider, for example, two species (*species 1* and *2*) that draw on the same limiting food resource: seeds. The diets of the two species are shown in **Figure 1a**. Note that the overlap in diet of the two species is symmetric. If the rate of food intake (seeds eaten per unit time) is the same, we can assume that the competition coefficients are the same. For this example, let us assume a value of 0.5 for both α and β.

Now let's assume that the size distribution of seeds and their abundance vary as a function of environmental conditions. For example, in **Figure 1b** the average seed size increases from environment A to B and C. As the size distribution of seeds changes, so will the carrying capacity (K) for each species. Now assume that the carrying capacities of the two species vary as shown in the following table.

	Environment		
	A	B	C
Species 1 (K)	225	150	75
Species 2 (K)	75	150	225

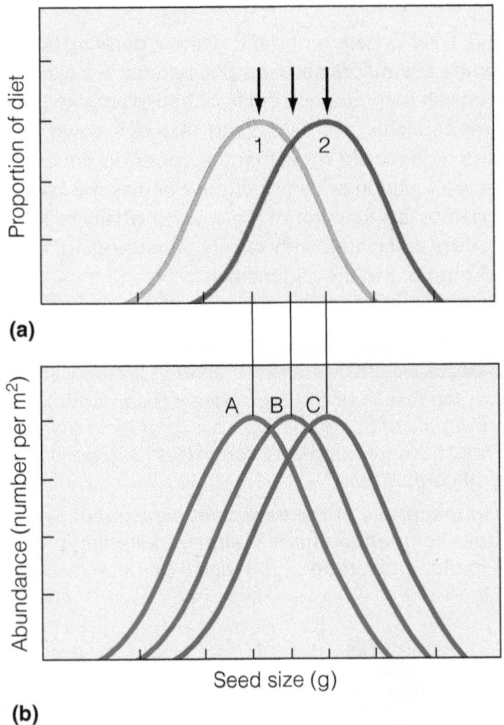

Figure 1 (a) Proportion of different seed sizes making up the diets of two species (*species 1* and *2*). (b) Relative abundances of the different seed sizes in three different environments (A, B, and C).

dominant yellow-pine chipmunk determines the upper range of the least chipmunk. Although the least chipmunk can occupy a full range of habitats from sagebrush desert to alpine fields, it is restricted in the Sierra Nevada to sagebrush habitat. Physiologically, it is more capable of handling heat stress than the others, enabling it to inhabit extremely hot, dry sagebrush. In a series of field experiments, ecologist Mark Chappell of Stanford University found that when the yellow-pine chipmunk is removed from its habitat, the least chipmunk moves into vacated open pine woods. However, if the least chipmunk is removed from the sagebrush habitat, the yellow-pine chipmunk does not invade the habitat. The aggressive behavior of the lodgepole chipmunk in turn determines the upper limit of the yellow-pine chipmunk. The lodgepole chipmunk is restricted to shaded forest habitat because it is vulnerable to heat stress. Most aggressive of the four, the lodgepole chipmunk also may limit the downslope range of the alpine chipmunk. Thus, the range of each chipmunk is determined both by aggressive exclusion and by its ability to survive and reproduce in a habitat hostile to the other species.

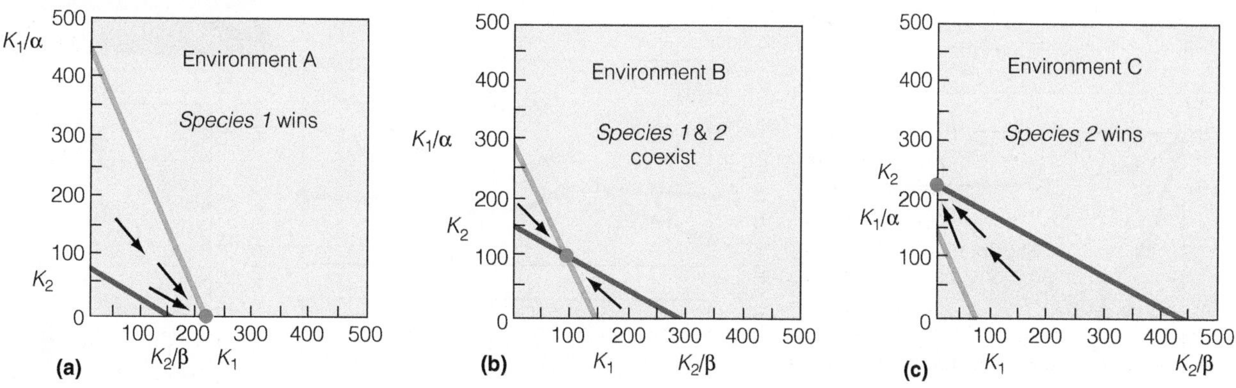

Figure 2 Outcomes for the competitive interactions of the two species in the three different environments (parameters of Lotka–Volterra equations given in text).

We can now explore the changing nature of competition between the two species at these three sites by using the Lotka–Volterra equations and the graphical analysis of the zero-growth isoclines presented in Section 13.2. Construction of the zero-growth isoclines in each of the three environments requires the values presented in the following table.

Environment	K_1	K_1/α	K_2	K_2/β
A	225	450	75	150
B	150	300	150	300
C	75	250	225	450

The outcomes of competition in these three environments using the Lotka–Volterra equations are graphically depicted in **Figure 2**. In environment A, *species 1* wins; in environment B, *species 1* and 2 coexist; and in environment C, *species 2* wins. As the relative abundance of different seed sizes changes from one environment to another, the resulting changes in carrying capacities of the two species shift the nature of interspecific competition.

1. Suppose that the competition coefficients were not the same for both species, but rather $\alpha = 0.5$ and $\beta = 0.25$. How would this influence the outcome of competition in environment A?
2. What factor or factors might cause the relative values of the competition coefficients (α and β) to change in the three different environments?

13.10 Interspecific Competition Influences the Niche of a Species

Previously, we defined the ecological niche of a species as the range of physical and chemical conditions under which it can persist (survive and reproduce) and the array of essential resources it uses and drew the distinction between the concepts of fundamental and realized niche (Chapter 12, Section 12.6). The fundamental niche is the ecological niche in the absence of interactions with other species, whereas the realized niche is the portion of the fundamental niche that a species actually exploits as a result of interactions with other species. As preceding examples have illustrated, competition may force species to restrict their use of space, range of foods, or other resource-oriented activities. As a result, species do not always occupy that part of their fundamental niche where conditions yield the highest growth rate, reproductive rate, or fitness. The work of Jessica Gurevitch of the University of New York–Stony Brook illustrates this point well. Gurevitch examined the role of interspecific competition on the local distribution of *Stipa*

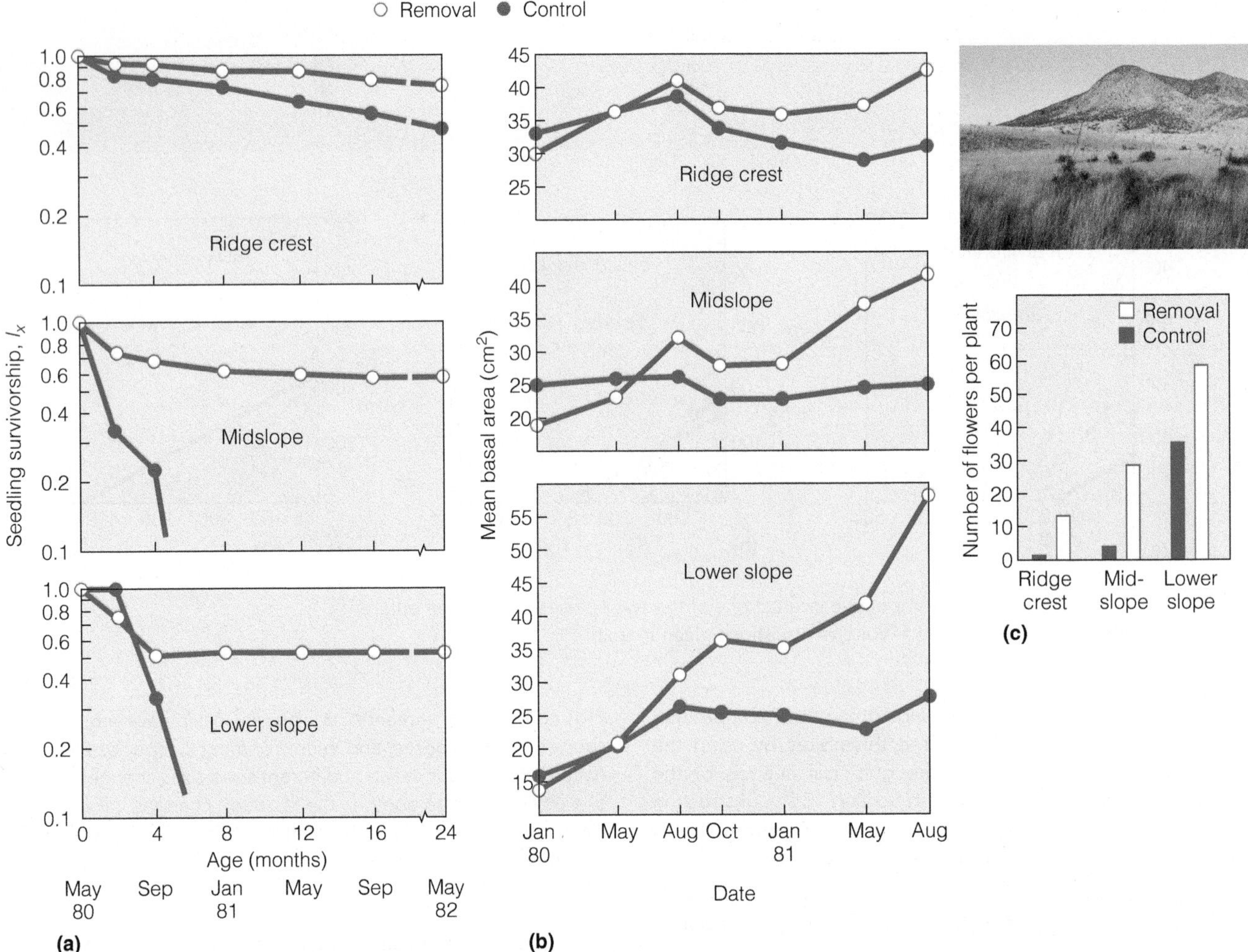

Figure 13.13 Response of *Stipa neomexicana* plants in three different habitats (ridge crest, midslope, and lower slope). Results of both treatment (neighboring plants removed) and control (neighbors *not* removed) plants are shown for (a) seedling survival, (b) mean growth rate, and (c) flowers produced per plant. Under natural conditions, distribution of *Stipa* is restricted to the ridge-crest habitats as a result of competition from other grass species. (Adapted from Gurevitch 1986.)

Interpreting Ecological Data

Q1. How does the influence of interspecific competition on seedling survival of Stipa differ between the ridge-crest and lower-slope habitats?

Q2. Experiment results show that Stipa individuals can effectively grow at the lower slope even under conditions of interspecific competition, as indicated by values of mean basal area in part (b). Based on the results in Figure 13.13, what part(s) of the Stipa life cycle are most heavily influenced by interspecies competition, and how would these limitations affect distribution of the species on the landscape?

neomexicana, a C_3 perennial grass found in the semiarid grassland communities of southeastern Arizona. *Stipa* is found only on the dry ridge crests where grass cover is low, rather than in moister, low-lying areas below the ridge crests where grass cover is greater. In a series of experiments, Gurevitch removed neighboring plants from individual *Stipa* plants in ridge-crest, midslope, and lower-slope habitats. She compared the survival, growth, and reproduction of these plants with control individuals (whose neighboring plants were not removed). Her results clearly show that *Stipa* has a higher growth rate, produces more flowers per plant, and has higher rates of seedling survival in midslope and lower-slope habitats (**Figure 13.13**). But its population density in these habitats is limited by competition with more successful grass species. Thus, *Stipa* distribution (or realized niche) is limited to suboptimal habitats because of interspecific competition.

Much of the evidence for competition comes from studies, such as the one just presented, demonstrating the contraction of a fundamental niche in the presence of a competitor. Conversely, when a species' niche expands in response to the removal of a competitor, the result is termed **competitive release**. Competitive release may occur when a species invades an island that is free of potential competitors, moves into habitats it never occupied on a mainland, and becomes more abundant. Such expansion may also follow when a competing species is removed from a community, allowing remaining species to move into microhabitats they previously could not occupy. Such was the case with the distribution of cattails along the gradient of water depth discussed previously, where in the absence of competition from *Typha latifoli*, the distribution of *Typha angustifolia* expanded to areas above the shoreline (expressed as negative values of water depth; see Figure 12.13).

An example of competitive release in a lake ecosystem is presented by Daniel Bolnick and colleagues at the University of Texas. Bolnick and his colleagues tested for short-term changes in the feeding niche of the three-spine stickleback (*Gasterosteus aculeatus*) after experimentally manipulating the presence or absence of two interspecific competitors: juvenile cut-throat trout (*Oncorhynchus clarki*) and prickly sculpin (*Cottus asper*). Direct examination of stomach contents of sculpin and trout reveals overlap with stickleback diets. Sculpin are exclusively benthic feeders, whereas juvenile trout feed at the surface and in the water column. In contrast, stickleback feed in both microhabitats. The experiment consisted of 20 experimental enclosures (made of netting) in Blackwater Lake on northern Vancouver Island, British Columbia. Five replicate blocks of four enclosures each were distributed along the shoreline of the lake. Sticklebacks collected from similar habitats nearby were placed in the enclosures. The enclosures in each of the blocks were assigned to one of four treatments: (1) competition with sculpin and trout present, (2) release from sculpin with trout present, (3) release from trout with sculpin present, and (4) total release with no competitors. The experimental treatments were left undisturbed for 15 days, after which all sticklebacks were removed, and the researchers identified (to the lowest feasible taxonomic level) and counted prey in the stomach of each stickleback. The diversity of prey species in the diet of the sticklebacks in each treatment was used as a measure of niche breadth. Results of the experiment reveal no significant change in the niche breadth (diversity of prey consumed) for the stickleback population when released from competition from sculpin. When released from competition from juvenile cut-throat trout, however, the researchers observed a significant expansion of niche breadth for the stickleback population (**Figure 13.14**).

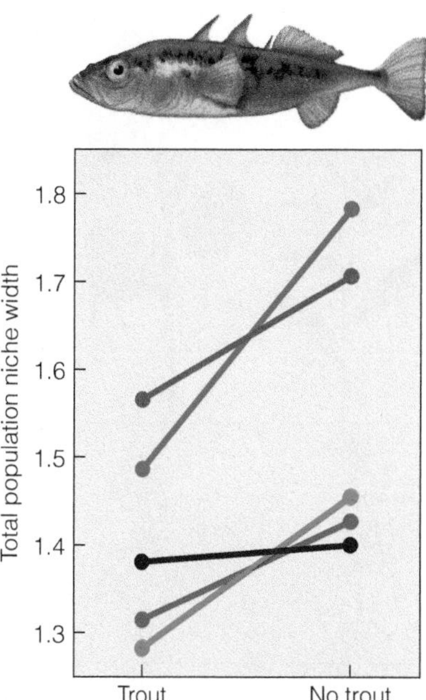

Figure 13.14 Example of niche expansion in a population of three-spine stickleback (*Gasterosteus aculeatus*) with the removal of its competitor, juvenile cut-throat trout (*Oncorhynchus clarki*). Total population niche width represents the diversity of prey species in the diet of the stickleback. When food competition from trout was removed, the diversity of the prey consumed by stickleback increased significantly. Each of the five lines represents the average response of stickleback in one of the five replicate experimental blocks (see text for discussion).
(Bolnick et al. 2010.)

13.11 Coexistence of Species Often Involves Partitioning Available Resources

All terrestrial plants require light, water, and essential nutrients such as nitrogen and phosphorus. Consequently, competition between various co-occurring species is common. The same is true for the variety of insect-feeding bird species inhabiting the canopy of a forest, large mammalian herbivores feeding on grasslands, and predatory fish species that make the coral reef their home. How is it that these diverse arrays of potential competitors can coexist in the same community? The competitive exclusion principle introduced in Section 13.5 suggests that if two species have identical resource requirements, then one species will eventually displace the other. But how different do two species have to be in their use of resources before competitive exclusion does not occur (or conversely, how similar can two species be in their resource requirements and still coexist)?

We have seen that the coexistence of competitors is associated with some degree of "niche differentiation"—differences in the range of resources used or environmental tolerances—in the species' fundamental niches. Observations of similar species sharing the same habitat suggest that they coexist by partitioning available resources. Animals use different kinds and sizes of food, feed at different times, or forage in different areas. Plants require different proportions of nutrients or have different tolerances for light and shade. Each species exploits a portion of the resources unavailable to others, resulting in differences among co-occurring species that would not be expected purely as a result of chance.

Field studies provide many reports of apparent resource partitioning. One example involves three species of annual plants growing together on prairie soil abandoned one year after plowing. Each plant exploits a different part of the soil

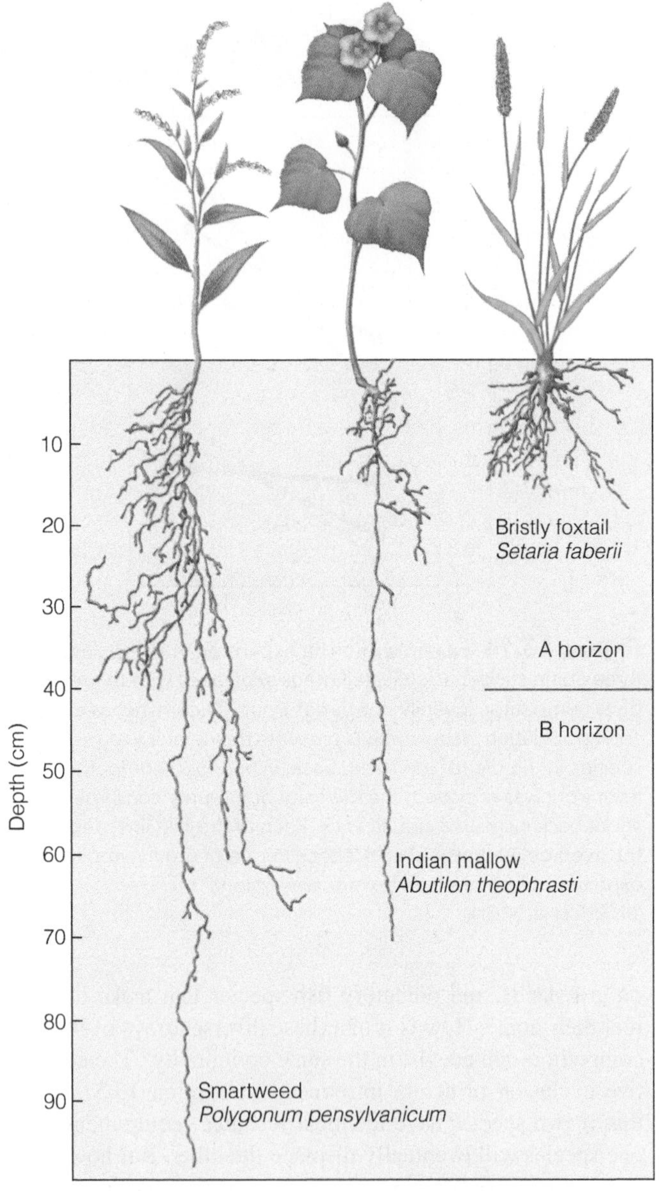

Figure 13.15 Vertical partitioning of the prairie soil resource at different levels by three species of annual plants, one year after disturbance.

resource (**Figure 13.15**). Bristly foxtail (*Setaria faberii*) has a fibrous, shallow root system that draws on a variable supply of moisture. It recovers rapidly from drought, takes up water rapidly after a rain, and carries on a high rate of photosynthesis even when partially wilted. Indian mallow (*Abutilon theophrasti*) has a sparse, branched taproot extending to intermediate depths, where moisture is adequate during the early part of the growing season but is less available later on. The plant is able to carry on photosynthesis at low water availability (Section 6.10). The third species, smartweed (*Polygonum pensylvanicum*), has a taproot that is moderately branched in the upper soil layer and develops mostly below the rooting zone of other species, where it has a continuous supply of moisture.

Apparent resource partitioning is also common among related animal species that share the same habitat and draw on a similar resource base. Tamar Dayan, at Tel Aviv University, examined possible resource partitioning in a group of coexisting species of wild cats inhabiting the Middle East. Dayan and colleagues examined differences among species in the size of canine teeth, which are crucial to wild cats in capturing and killing their prey. For these cats, there is a general relationship between the size of canine and the prey species selected. Dayan found clear evidence of systematic differences in the size of the canine teeth, not only between male and female individuals within each of the species (sexual dimorphism) but also among the three coexisting cat species (**Figure 13.16**; see also Chapter 10). The pattern observed suggests an exceptional regularity in the spacing of species along the axis defined by the average size of canine teeth (*x*-axis in Figure 13.16). Dayan and colleagues hypothesize that intraspecific and interspecific competition for food has resulted in natural selection favoring the observed differences, thereby reducing the overlap in the types and sizes of prey that are taken.

The patterns of resource partitioning discussed previously are a direct result of differences among co-occurring species in specific physiological, morphological, or behavioral adaptations that allow individuals access to essential resources while at the same time function to reduce competition (see Chapter 5). Because the adaptations function to reduce competition, they are often regarded as a product of coevolutionary forces (see Chapter 12, Sections 12.3 and Section 12.6 for discussion and

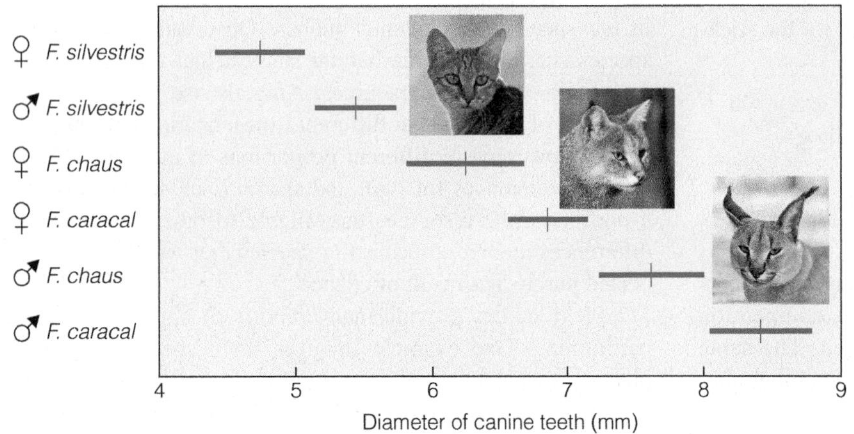

Figure 13.16 Size (diameter) of canine teeth for small cat species that co-occur in Israel. Note the regular pattern of differences in size between species. Size is correlated with the size of prey selected by the different species.
(Adapted from Dayan et al. 1990.)

example of coevolution driven by competition). Although patterns of resource partitioning observed in nature are consistent with the hypothesis of phenotypic divergence arising from coevolution between competing species, it is difficult to prove that competition functioned as the agent of natural selection that resulted in the observed differences in resource use (observed differences in fundamental niches of the species). Differences among species may relate to adaptation for the ability to exploit a certain environment or range of resources independent of competition. Differences among species have evolved over a long period of time, and we have limited or no information about resources and potential competitors that may have influenced natural selection. This issue led Joseph Connell, an ecologist at the University of California–Santa Barbara, to refer to the hypothesis of resource portioning as a product of coevolution between competing species as the "ghosts of competition past." Unable to directly observe the role of past competition on the evolution of characteristics, some of the strongest evidence supporting the role of "competition past" comes from studies examining differences in the characteristics of subpopulations of a species that face different competitive environments. A good example is the work of Peter Grant and Rosemary Grant, of Princeton University, involving two Darwin's finches of the Galápagos Islands. The Grants studied the medium ground finch (*Geospiza fortis*) and the small ground finch (*Geospiza fuliginosa*), both of which feed on an overlapping array of seed sizes—for further discussion and illustrations, see Section 5.9. On the large island of Santa Cruz, where the two species of finch coexist, the distribution of beak sizes (phenotypes) of the two species does not overlap. Average beak size is significantly larger for *G. fortis* than for the smaller *G. fuliginosa* (**Figure 13.17a**). On the adjacent—and much smaller—islands of Los Hermanos and Daphne Major, the two species do not coexist, and the distributions of beak sizes for the two species are distinctively different from the patterns observed on Santa Cruz. The medium ground finch is allopatric (lives separately) on the island of Daphne Major, and the small ground finch is allopatric on Los Hermanos. Populations of each species on these two islands possess intermediate and overlapping distributions of beak sizes (**Figures 13.17b** and **13.17c**). These patterns suggest that on islands where the two species coexist, competition for food results in natural selection favoring medium ground finch individuals with a large beak size that can effectively exploit larger seeds while also favoring small ground finch individuals that feed on smaller seeds. The outcome of this competition was a shift in feeding niches. When the shift involves features of the species' morphology, behavior, or physiology, it is referred to as **character displacement**.

The preceding example suggests that the competing species on the island of Santa Cruz exhibit character displacement as a result of coevolutionary forces—that is, divergence in phenotypic traits relating to the exploitation of a shared and limited resource. However, until recently, the process of character displacement had never been documented by direct observational data. The first direct evidence of character displacement is provided by the work of Peter and Rosemary Grant on the population of *G. fortis* inhabiting the small island of Daphne Major.

Before 1982, *G. fortis* (medium ground finch) was the only species of ground finch inhabiting the island of Daphne Major. The situation changed in 1982 when a new competitor species emigrated from the larger adjacent islands—the large ground finch, *Geospiza magnirostris* (see Section 5.9 and Figure 5.20). *G. magnirostris* is a potential competitor on the island as a result of diet overlap with *G. fortis*. *G. magnirostris* feeds primarily on seeds of the herbaceous forb, Jamaican feverplant (*Tribulus cistoides*). The seeds are contained within a hard seed coat and exposed when a finch cracks or tears away the woody outer coating. Large-beaked members of the *G. fortis* population also feed on these seeds; in fact, during the 1976–1977 drought, the survival of the population depended on this seed resource (see Section 5.6 for a discussion of natural selection in this population).

Initially, the population of *G. magnirostris* on Daphne Major was too small in relation to the food supply to have anything but a small competitive effect on *G. fortis*. From 1982 to 2003, however, the population increased. Then little rain fell on the island during 2003 and 2004, and populations of both finch species declined dramatically as a result of declining food resources. During this period, *G. magnirostris* depleted the supply of large seeds from the Jamaican feverplant, causing the *G. fortis* population to depend on the smaller seed resources on

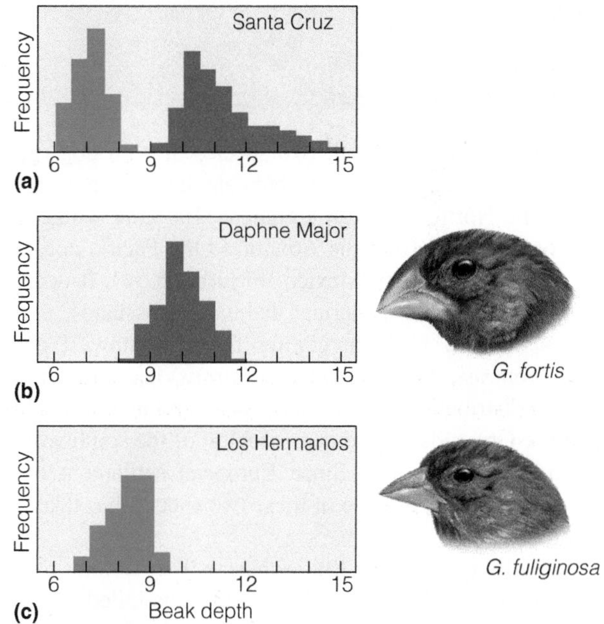

Figure 13.17 Apparent character displacement in beak size in two populations of Galápagos finches—the medium ground finch (*Geospiza fortis*) and the small ground finch (*Geospiza fuliginosa*). (a) On the large island of Santa Cruz, where the two finch species coexist, distribution of beak sizes (phenotypes) for these species does not overlap. Average beak size is significantly larger for *G. fortis* than for the smaller *G. fuliginosa*. In contrast, on the smaller islands of (b) Daphne Major and (c) Los Hermanos, where the two species do not co-occur, distribution of beak sizes for these species is intermediate and overlapping.
(Adapted from Grant and Grant 1986.)

the island. The result of this shift in resource availability because of competition from *G. magnirostris* was that during 2004 and 2005, *G. fortis* experienced strong directional selection against individuals with large beaks. The resulting decrease in the average beak size of the *G. fortis* population provides a clear example of the coevolutionary process of character displacement.

13.12 Competition Is a Complex Interaction Involving Biotic and Abiotic Factors

Demonstrating interspecific competition in laboratory "bottles" or the greenhouse is one thing; demonstrating competition under natural conditions in the field is another. In the field, researchers (1) have little control over the environment, (2) have difficulty knowing whether the populations are at or below carrying capacity, and (3) lack full knowledge of the life history requirements or the subtle differences between the species.

In the previous sections, we reviewed an array of studies examining the role of competition in the field. Perhaps the most common are removal experiments, in which one of the potential competitors is removed and the response of the remaining species is monitored. These experiments might appear straightforward, yielding clear evidence of competitive influences. But removing individuals may have direct and indirect effects on the environment that are not intended or understood by the investigators and that can influence the response of the remaining species. For example, removing (neighboring) plants from a location may increase light reaching the soil surface, soil temperatures, and evaporation. The result may be reduced soil moisture and increased rates of decomposition, influencing the abundance of belowground resources. These sometimes "hidden treatment effects" can hinder the interpretation of experimental results.

As we have seen in previous sections, competition is a complex interaction that seldom involves the interaction between two species for a single limiting resource. Interaction between species involves a variety of environmental factors that directly influence survival, growth, and reproduction; these factors vary in both time and space. The outcome of competition between two species for a specific resource under one set of environmental conditions (temperature, salinity, pH, etc.) may differ markedly from the outcome under a different set of environmental conditions. As we shall see in the following chapters, competition is only one of many interactions occurring between species—interactions that ultimately influence population dynamics and community structure.

ECOLOGICAL Issues & Applications

Is Range Expansion of Coyote a Result of Competitive Release from Wolves?

Before European settlement, two species of wild dog (genus *Canis*) were among the most abundant large carnivores occupying the North American continent. The gray wolf, *Canis lupus,* once ranged from the Atlantic to the Pacific coast and from Alaska to northern Mexico (**Figure 13.18**). It occurred in virtually all North American habitats (grasslands, eastern deciduous forest, northern conifer forest, southwest desert, etc.). In contrast, the coyote (*Canis latrans*) had a much more restricted distribution to the prairie grassland and desert habitats of the Great Plains and desert region of the southwest and Mexico (**Figure 13.19**). Since European settlement of the continent, however, the fate of these two species has taken different paths.

As early as 1630, the Massachusetts Bay Colony paid an average month's salary for any wolf that was killed. Bounties like this continued until the last wolf in the Northeast was killed around 1897. The fate of the wolf population in other areas of its range was similar. Settlers moving westward depleted the populations of bison, deer, elk, and moose on which the wolves preyed. Wolves then turned to attacking sheep and cattle, and to protect livestock, ranchers and government agencies began an eradication campaign. Bounty programs initiated in the 19th century continued as late as 1965. Wolves were trapped, shot, dug from their dens, and hunted with dogs. Poisoned animal carcasses were left out for wolves, a practice that also killed eagles, ravens, foxes, bears, and other animals that fed on the tainted carrion. By the time wolves were protected by the Endangered Species Act of 1973, only a few hundred remained in extreme northeastern Minnesota and a small number on Isle Royale, Michigan.

In contrast to the gray wolf, the coyote did not originally occur in eastern North America, and with the westward expansion of settlement into the Great Plains, the coyote was perceived as less of a threat to farmers and ranchers. By the turn of the 20th century, it began to take advantage of newly open habitat that agriculture and logging had created, and its distribution expanded eastward. There were two main waves of colonization, northern and southern (Figure 13.19). The northern wave occurred first—coyote were reported in Michigan in about 1900, in southern Ontario by 1919, and in northern New York in the late 1930s. Most of the southeast was not colonized until the 1960s. Whereas the gray wolf population has been virtually eliminated in the continental United States, the range of the coyote has expanded to cover most of the areas once occupied by wolves, and coyote now occupy virtually every habitat in eastern North America (compare Figures 13.18 and 13.19) from forests, wooded areas, grassland, and agricultural land to suburban areas.

The concurrent expansion of the coyote with the decline of the wolf population in North America has caused ecologists to question whether the two occurrences are linked in some way. In North American ecosystems where gray wolves occur, interactions with other large carnivores are common, with competition being most intense with species having a similar ecology. Interference competition (see Section 13.1) occurs between the wolves and coyotes, with wolves limiting coyote access to resources by direct aggression. Field studies in regions where wolves and coyotes overlap indicate that coyotes are excluded from wolf territories and that wolves will go out of their way to kill coyotes. One of the leading hypotheses put forward to explain the dramatic range expansion of the coyote is that the eradication of the gray wolf from its former range may have reduced the competitive pressures limiting coyotes to their former range: range expansion is a result of "competitive release" (see Section 13.10). Now as a result of recent conservation efforts, ecologists are able to test this hypothesis directly.

Thanks to conservation efforts, the gray wolf is beginning to make a comeback. The wolf's comeback within the United States is as a result of its listing under the Endangered Species Act, which provided protection from unregulated killing and resulted in increased scientific research, along with reintroduction and management programs. As of 2013 about 2200 wolves live in Minnesota, 8 on Lake Superior's Isle Royale, about 650 in Michigan's Upper Peninsula, and at least 800 in Wisconsin. In the northern Rocky Mountains, the U.S. Fish and Wildlife Service reintroduced gray wolves into Yellowstone National Park and U.S. Forest Service lands in central Idaho in 1995 and 1996. The reintroduction was successful, and as of 2013 there were at least 1650 wolves in the northern Rocky Mountains of Montana, Idaho, and Wyoming. These reintroductions of wolves into areas now occupied by coyotes have enabled ecologists to directly examine the role of competition on the populations of the two carnivores and test the hypothesis that the range expansion of the coyote in the United States is in part the result of competitive release from wolves.

Kim Berger and Eric Gese of Utah State University used data collected on wolf and coyote distribution and abundance to test the hypothesis that interference competition with wolves limits the distribution and abundance of coyotes in two regions of the Northern Rocky Mountains in which wolves have been recently reintroduced. From August 2001 to August 2004, the two researchers gathered data on cause-specific mortality and survival rates of coyotes captured at wolf-free and wolf-abundant sites in Grand Teton National Park (GTNP), and data on population densities of both species at three study areas across the Greater Yellowstone Ecosystem (GYE), to determine whether competition with wolves is sufficient to reduce coyote densities in these areas.

Berger and Gese found that although coyotes were the numerically dominant predator, across the GYE, densities varied spatially and temporally as a function of wolf abundance. Mean coyote densities were 33 percent lower at wolf-abundant sites in GTNP, and densities declined 39 percent in

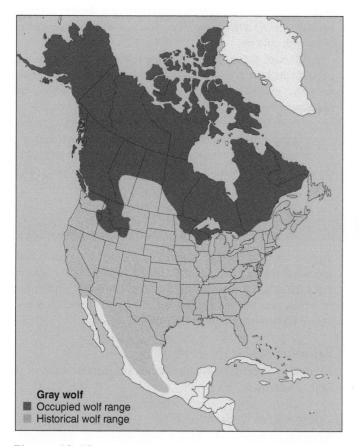

Figure 13.18 Map of North America showing historical and current geographic distribution of gray wolf.

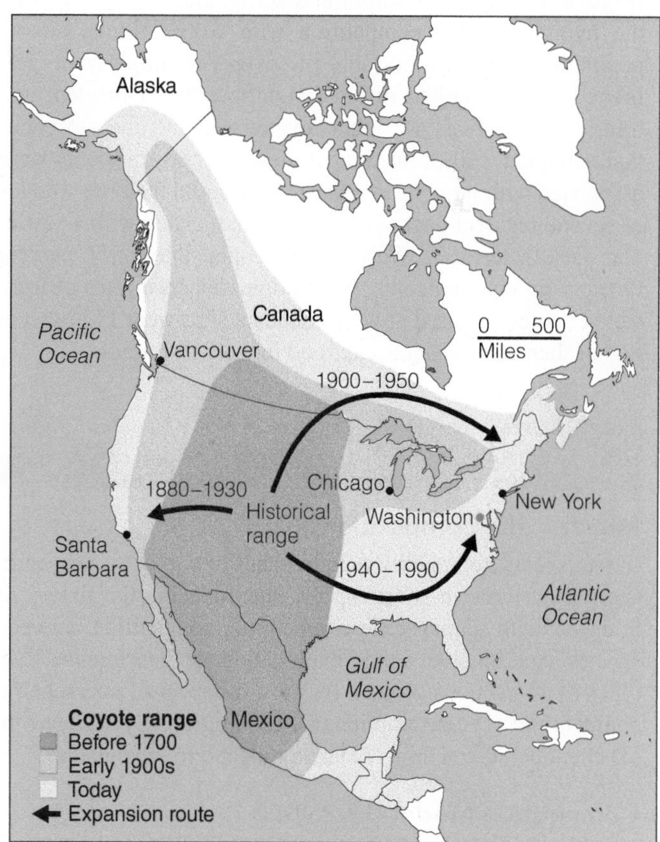

Figure 13.19 Map of North America showing historical geographic range of coyote and dramatic east and westward expansion of range beginning in the early 20th century.

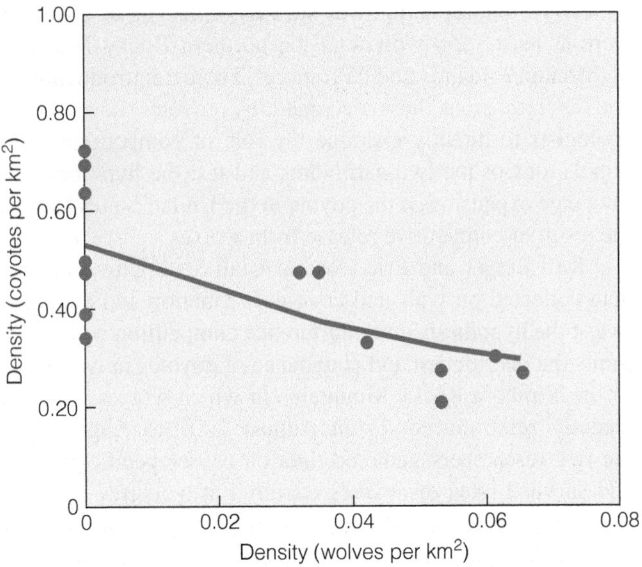

Figure 13.20 Negative relationship between coyote and wolf densities for three study areas: Grand Teton National Park (GTNP), Lamar River Valley (LRV), and Northern Madison Study Area (NMSA) in the Greater Yellowstone Ecosystem 1991–2005. (Adapted from Berger and Gese 2007.)

Yellowstone National Park following wolf reintroduction. A strong negative relationship between coyote and wolf densities (**Figure 13.20**), both within and across study sites, supports the hypothesis that competition with wolves limits coyote populations. Overall mortality of coyotes resulting from wolf predation was low but differed significantly for resident and transient individuals. Resident coyotes were members of packs that defended well-defined territories, whereas transients were associated with larger areas that encompassed the home ranges of several resident packs but were not associated with a particular pack or territory. Wolves were responsible for 56 percent of transient coyote deaths. In addition, dispersal rates of transient coyotes captured at wolf-abundant sites were 117 percent higher than for transients captured in wolf-free areas.

Figure 13.21 Interaction between wolf and coyote (interference competition) at the site of an elk carcass (wolf kill) in Yellowstone National Park.

The work by Jerod Merkle and colleagues at the Yellowstone Wolf Project (Yellowstone Center for Resources, Yellowstone National Park) provides a detailed picture of the nature of competitive interactions between wolves and coyotes in areas where wolves have been reintroduced. In a series of field studies, the researchers examined interference competition between gray wolves and coyotes in Yellowstone National Park using radio-collared wolves (**Figure 13.21**). Merkle and colleagues documented 337 wolf–coyote interactions from 1995 to 2007. The majority (75 percent) of interactions occurred at the sites of wolf-killed ungulate carcasses (elk, buffalo, moose, mule deer, etc.) with coyotes attempting to scavenge. Wolves initiated the majority of encounters (85 percent), generally outnumbered coyotes (39 percent), and dominated (91 percent) most interactions. Wolves typically (79 percent) chased coyotes without physical contact; however, 7 percent of encounters resulted in a coyote death. Interactions decreased over time, suggesting coyote adaptation or a decline in coyote density. The results clearly show that wolves dominate interactions with coyotes.

Although data are limited to the few regions in which wolf populations have been successfully introduced, when combined with the results of studies of wolf–coyote interactions and population studies for regions of North America where these two species naturally co-occur (regions of Minnesota and Canada), a consistent picture emerges that the dramatic range expansion of coyote over the past century is as a result, at least in part, of the decline of wolf populations throughout most of its former range.

SUMMARY

Interspecific Competition 13.1

In interspecific competition, individuals of two or more species share a resource in short supply, thus reducing the fitness of both. As with intraspecific competition, competition between species can involve either exploitation or interference. Six types of interactions account for most instances of interspecific competition: (1) consumption, (2) preemption, (3) overgrowth, (4) chemical interaction, (5) territorial, and (6) encounter.

Competition Model 13.2–13.3

The Lotka–Volterra equations describe four possible outcomes of interspecific competition. *Species 1* may outcompete *species 2*; *species 2* may outcompete *species 1*. Both of these outcomes represent competitive exclusion. The other two outcomes involve coexistence. One is unstable equilibrium, in which the species that was most abundant at the outset usually outcompetes the other. A final possible outcome is stable equilibrium, in which two species coexist but at a lower population level than if each existed without the other.

Experimental Tests 13.4

Laboratory experiments with species interactions support the Lotka–Volterra model.

Competitive Exclusion 13.5

Experiment results led to the formulation of the competitive exclusion principle—two species with exactly the same ecological requirements cannot coexist. This principle has stimulated critical examinations of competitive relationships outside the laboratory, especially of how species coexist and how resources are partitioned.

Nonresource Factors 13.6

Environmental factors such as temperature, soil or water pH, relative humidity, and salinity directly influence physiological processes related to growth and reproduction but are not consumable resources that species compete over. By differentially influencing species within a community, these nonresource factors can influence the outcome of competition.

Environmental Variability 13.7

Environmental variability may give each species a temporary advantage. It allows competitors to coexist, whereas under constant conditions one would exclude the other.

Multiple Factors 13.8

In many cases, competition between species involves multiple resources. Competition for one resource often influences an organism's ability to access other resources.

Environmental Gradients 13.9

As environmental conditions change, so may the relative competitive ability of species. Shifts in competitive ability can result either from changes in the carrying capacities related to a changing resource base or from changes in the physical environment that interact with resource availability. Natural environmental gradients often involve the covariation of multiple factors—both resource and nonresource factors—such as salinity, temperature, and water depth.

Niche 13.10

A species' fundamental niche compresses or shifts when competition restricts the species' type of food or habitat. In some cases, the realized niche may not provide optimal conditions for the species. In the absence of competition, the species may experience competitive release, and its niche may expand.

Resource Partitioning 13.11

Many species that share the same habitat coexist by partitioning available resources. When each species exploits a portion of the resources unavailable to others, competition is reduced. Resource partitioning is often viewed as a product of the coevolution of characteristics that function to reduce competition. Interspecific competition can reduce the fitness of individuals. If certain phenotypes within the population function to reduce competition with individuals of other species, those individuals will encounter less competition and increased fitness. The result is a shift in the distribution of phenotypes (characteristics) within the competing population(s). When the shift involves features of the species' morphology, behavior, or physiology, it is referred to as *character displacement.*

Complexity of Competition 13.12

Competition is a complex interaction that seldom involves the interaction between two species for a single limiting resource. Competition involves a variety of environmental factors that directly influence survival, growth, and reproduction—factors that vary in both time and space.

Wolves and Coyotes Ecological Issues & Applications

The decline of gray wolf populations throughout much of North America have been paralleled by a dramatic expansion in the range of coyotes. Evidence from areas in which wolves have been reintroduced suggests that the expansion of coyotes was in part a result of competitive release from wolf populations over the past century.

STUDY QUESTIONS

1. What condition(s) must be established before a researcher can definitively state that two species are competing for a resource? Is establishing that two species overlap in their use of a resource enough to determine that interspecific competition is occurring?
2. In analyzing the Lotka–Volterra model of interspecific competition in Section 13.3, four outcomes of competition between two species were identified. In three of the cases, one species outcompetes another, driving its population to zero. In the fourth case, the two species coexist. What condition is necessary for this outcome?
3. Environmental factors such as temperature and salinity are not consumable resources, but species that occupy the same habitat often differ in their response to these factors. How might environmental factors such as temperature and salinity influence the outcome of competition between two species that occupy the same habitat?
4. In Chapter 12, Figure 12.6 presents a hypothetical case where two bird species overlap in their use of food resources (seeds). Let us assume that because of interspecific competition, the two species cannot co-occur in the same habitat (say *species 1* succeeds). Now assume that the distribution of seed sizes in their habitat changes from year to year depending on rainfall, and that the average seed size varies from *a* to *b* (Figure 12.19). How might this temporal variation in resources influence the outcome of competition between the two species?
5. In the experiment conducted by Austin and colleagues (Figures 13.8 and 13.9), relative competitive abilities of the three dominant thistle species change under different nutrient availability. Based on the discussion of plant adaptations to nutrient availability in Chapter 6 (see Section 6.12), how do you think these three species might differ (morphology, maximum rates of photosynthesis, etc.)?

6. Resource partitioning as a result of natural selection is often interpreted as the cause of observed differences in the characteristics of closely related species that occupy the same area, such as the example of the cats in Figure 13.16, or the Galápagos finches in Figure 13.17. What other factors unrelated to competition might account for the observed differences in the bill size of the finches? Think about possible differences in the use of habitats within the area.
7. Distinguish between niche shift and character displacement. Which of these two concepts best describes the example of the Galápagos finches in Figure 13.17? Why?

FURTHER READINGS

Classic Studies

Connell, J. H. 1983. "On the prevalence and relative importance of interspecific competition: Evidence from field experiments." *American Naturalist* 122:661–696.
This article provides a review of field experiments that have examined the role of competition.

Gause, G. F. 1932. "Experimental studies on the struggle for existence: Mixed population of two species of yeast." *Journal of Experimental Biology* 9:389–402.

Gause, G. F. 1934. *The struggle for existence.* Baltimore, MD: Williams & Wilkins.
Gause's works present the early experiments on which the competitive exclusion hypothesis is based.

Schoener, T. W. 1983. "Field experiments on interspecific competition." *American Naturalist* 122:240–285.
A companion article to the one by Connell.

Werner, E. E., and J. D. Hall. 1976. "Niche shifts in sunfishes: Experimental evidence and significance." *Science* 191:404–406.
A classic work in the study of how competition influences the species' niche.

Wiens, J. A. 1977. "On competition and variable environments." *American Scientist* 65:590–597.
This article discusses the influence of temporal variation in environmental conditions on the outcome of competition. It provides a contrast to the papers by Connell and Schoener.

Current Research

Bazzaz, F. A. 1996. *Plants in changing environments: Linking physiological, population, and community ecology.* New York: Cambridge University Press.
An excellent overview of competition in plant populations, linking species characteristics and competitive interactions. Also illustrates the shifting nature of competitive interactions as environmental conditions change in time and space.

Grace, J. B., and D. Tilman. 1990. *Perspectives on plant competition.* San Diego: Academic Press.
An excellent overview of the concept of competition in plant communities. Provides a wealth of illustrated examples from laboratory, greenhouse, and field experiments.

Gurevitch, J. L., L. Murrow, A. Wallace, and J. J. Walch. 1992. "Meta-analysis of competition in field experiments." *American Naturalist* 140:539–572.
An update of the article by Connell.

Keddy, P. 1989. *Competition.* London: Chapman and Hall.
An outstanding review of theoretical developments in the area of competition. Also provides a wide array and review of experimental research aimed at examining the importance of competition in structuring ecological communities.

CHAPTER 14 Predation

Lioness attacking a female Kudu in Etosha National Park, Namibia, Africa.

CHAPTER GUIDE

WHEN THE POET ALFRED, LORD TENNYSON wrote "Tho' Nature, red in tooth and claw," he was no doubt seeking to evoke savage images of predation. The term *predator* brings to mind lions on the African savannas or the great white shark cruising coastal waters; however, *predation* is defined more generally as "the consumption of one living organism by another." Although all heterotrophic organisms acquire their energy by consuming organic matter, predators are distinguished from scavengers and decomposers because they feed on living organisms (see Chapter 21). As such, they function as agents of mortality with the potential to regulate prey populations. Likewise, being the food resource, the prey population has the potential to influence the growth rate of the predator population. These interactions between predator and prey species can have consequences for community structure and serve as agents of natural selection, influencing the evolution of both predator and prey.

14.1 Predation Takes a Variety of Forms

The broad definition of predation as the consumption of one living organism (the prey) by another (the predator) excludes scavengers and decomposers. Nevertheless, this definition results in the potential classification of a wide variety of organisms as predators. The simplest classification of predators is represented by the categories of heterotrophic organisms presented previously, which are based on their use of plant and animal tissues as sources of food: carnivores (carnivory—consumption of animal tissue), herbivores (herbivory—consumption of plant or algal tissue), and omnivores (omnivory—consumption of both plant and animal tissues); see Chapter 7. Predation, however, is more than a transfer of energy. It is a direct and often complex interaction of two or more species: the eater and the eaten. As a source of mortality, the predator population has the potential to reduce, or even regulate, the growth of prey populations. In turn, as an essential resource, the availability of prey may function to regulate the predator population. For these reasons, ecologists recognize a functional classification, which provides a more appropriate framework for understanding the interconnected dynamics of predator and prey populations and which is based on the specific interactions between predator and prey.

In this functional classification of predators, we reserve the term *predator*, or *true predator*, for species that kill their prey more or less immediately upon capture. These predators typically consume multiple prey organisms and continue to function as agents of mortality on prey populations throughout their lifetimes. In contrast, most herbivores (grazers and browsers) consume only part of an individual plant. Although this activity may harm the plant, it usually does not result in mortality. Seed predators and planktivores (aquatic herbivores that feed on phytoplankton) are exceptions; these herbivores function as true predators. Like herbivores, parasites feed on the prey organism (the host) while it is still alive and although harmful, their feeding activity is generally not lethal in the short term. However, the association between parasites and their host organisms has an intimacy that is not seen in true predators and herbivores because many parasites live on or in their host organisms for some portion of their life cycle. The last category in this functional classification, the parasitoids, consists of a group of insects classified based on the egg-laying behavior of adult females and the development pattern of their larvae. The parasitoid attacks the prey (host) indirectly by laying its eggs on the host's body. When the eggs hatch, the larvae feed on the host, slowly killing it. As with parasites, parasitoids are intimately associated with a single host organism, and they do not cause the immediate death of the host.

In this chapter we will use the preceding functional classifications, focusing our attention on the two categories of true predators and herbivores. (From this point forward, the term *predator* is used in reference to the category of true predator). We will discuss the interactions of parasites and parasitoids and their hosts later, focusing on the intimate relationship between parasite and host that extends beyond the feeding relationship between predator and prey (Chapter 15).

We will begin by exploring the connection between the hunter and the hunted, developing a mathematical model to define the link between the populations of predator and prey. The model is based on the same approach of quantifying the per capita effects of species interactions on rates of birth and death within the respective populations that we introduced previously (Chapter 13, Section 13.2). We will then examine the wide variety of subjects and questions that emerge from this simple mathematic abstraction of predator–prey interactions.

14.2 Mathematical Model Describes the Interaction of Predator and Prey Populations

In the 1920s, Alfred Lotka and Vittora Volterra turned their attention from competition (see Section 13.2) to the effects of predation on population growth. Independently, they proposed mathematical statements to express the relationship between predator and prey populations. They provided one equation for the prey population and another for the predator population.

The population growth equation for the prey population consists of two components: the exponential model of population growth ($dN/dt = rN$; see Chapter 9) and a term that represents mortality of prey from predation. Mortality resulting from predation is expressed as the per capita rate at which predators consume prey (number of prey consumed per predator per unit time). The per capita consumption rate by predators is assumed to increase linearly with the size of the prey population (**Figure 14.1a**) and can therefore be represented as cN_{prey}, where c represents the capture efficiency of the predator, defined by the slope of the relationship shown in Figure 14.1a. (Note that the greater the value of c, the greater the number of prey captured and consumed for a given prey population size, which means that the predator is more efficient at capturing prey.) The total rate of predation (total number of prey captured per unit time) is the product of the per capita rate of consumption (cN_{prey}) and the number of predators (N_{pred}), or $(cN_{\text{prey}})N_{\text{pred}}$. This value represents a source of mortality for the prey population and must be subtracted from the rate of

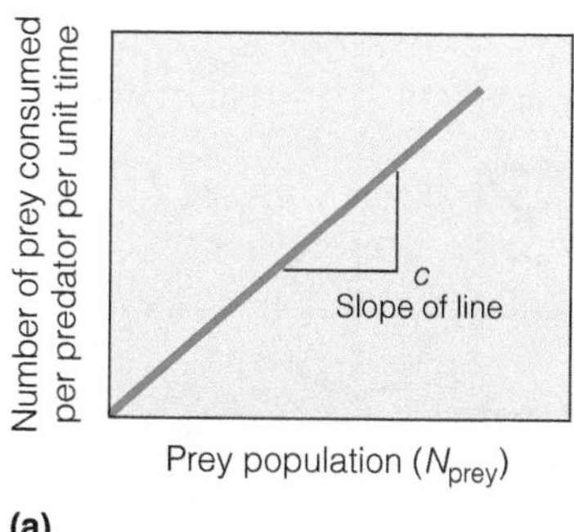

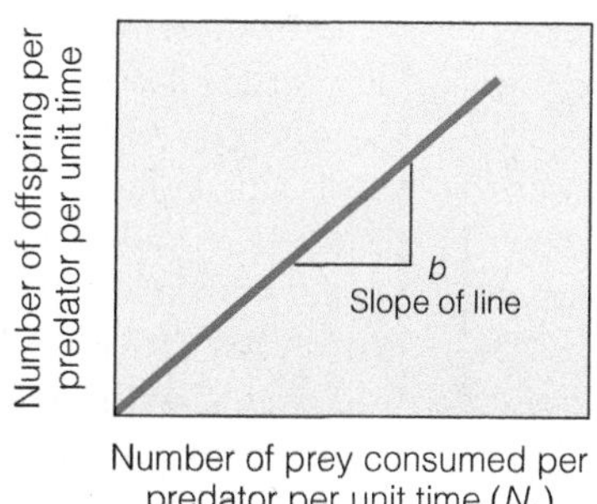

Figure 14.1 (a) Relationship between prey population (x-axis) and the per capita rate of predation (y-axis). The slope of the relationship ($\Delta y/\Delta x$) defined as c, represents the "efficiency of predation." (b) Relationship between the per capita rate of predation (x-axis) and the per capita rate or predator reproduction (y-axis). The slope of the relationship, b, represents the efficiency with which food is converted into predator population growth (reproduction).

population increase represented by the exponential model of growth. The resulting equation representing the rate of change in the prey population (dN_{prey}/dt) is:

$$dN_{prey}/dt = rN_{prey} - (cN_{prey})N_{pred}$$

The equation for the predator population likewise consists of two components: one representing birth and the other death of predators. The predator mortality rate is assumed to be a constant proportion of the predator population and is therefore represented as dN_{pred}, where d is the per capita death rate (this value is equivalent to the per capita death rate in the exponential model of population growth developed in Chapter 9). The per capita birthrate is assumed to be a function of the amount of food consumed by the predator, the per capita rate of consumption (cN_{prey}), and increases linearly with the per capita rate at which prey are consumed (**Figure 14.1b**). The per capita birthrate is therefore the product of b, the efficiency with which food is converted into population growth (reproduction), which is defined by the slope of the relationship shown in Figure 14.1b, and the rate of predation (cN_{prey}), or $b(cN_{prey})$. The total birthrate for the predator population is then the product of the per capita birthrate, $b(cN_{prey})$, and the number of predators, N_{pred}: $b(cN_{prey})N_{pred}$. The resulting equation representing the rate of change in the predator population is:

$$dN_{pred}/dt = b(cN_{prey})N_{pred} - dN_{pred}$$

The Lotka–Volterra equations for predator and prey population growth therefore explicitly link the two populations, each functioning as a density-dependent regulator on the other. Predators regulate the growth of the prey population by functioning as a source of density-dependent mortality. The prey population functions as a source of density-dependent regulation on the birthrate of the predator population. To understand how these two populations interact, we can use the same graphical approach used to examine the outcomes of interspecific competition (Chapter 13, Section 13.2).

In the absence of predators (or at low predator density), the prey population grows exponentially ($dN_{prey}/dt = rN_{prey}$). As the predator population increases, prey mortality increases until eventually the mortality rate resulting from predation, $(cN_{prey})N_{pred}$, is equal to the inherent growth rate of the prey population, rN_{prey}, and the net population growth for the prey species is zero ($dN_{prey}/dt = 0$). We can solve for the size of the predator population (N_{pred}) at which this occurs:

$$cN_{prey}N_{pred} = rN_{prey}$$

$$cN_{pred} = r$$

$$N_{pred} = \frac{r}{c}$$

Simply put, the growth rate of the prey population is zero when the number of predators is equal to the per capita growth rate of the prey population (r) divided by the efficiency of predation (c).

This value therefore defines the zero-growth isocline for the prey population (**Figure 14.2a**). As with the construction of the zero-growth isoclines in the analysis of the Lotka–Volterra competition equations (see Section 13.2, Figure 13.1), the two axes of the graph represent the two interacting populations. The x-axis represents the size of the prey population (N_{prey}), and the y-axis represents the predator population (N_{pred}). The prey zero-growth isocline is independent of the prey population size (N_{prey}) and is represented by a line parallel to the x-axis at a point along the y-axis represented by the value $N_{pred} = r/c$. For values of N_{pred} below the zero-growth isocline, mortality resulting from predation, $(cN_{prey})N_{pred}$, is less than the inherent growth rate of the prey population (rN_{prey}), so population growth is positive and the prey population increases, as represented by the green horizontal arrow pointing to the right. If the predator population exceeds this value, mortality resulting from predation, $(cN_{prey})N_{pred}$, is greater than the inherent growth rate of the prey population (rN_{prey}) and the growth rate of the prey becomes negative. The corresponding decline in the size of the prey population is represented by the green arrow pointing to the left.

Likewise, we can define the zero-growth isocline for the predator population by examining the influence of prey population size on the growth rate of the predator population. The growth rate of the predator population is zero ($dN_{pred}/dt = 0$) when the rate of predator increase (resulting from the consumption of prey) is equal to the rate of mortality:

$$b(cN_{prey})N_{pred} = dN_{pred}$$

$$bcN_{prey} = d$$

$$N_{prey} = \frac{d}{bc}$$

The growth rate of the predator population is zero when the size of the prey population (N_{prey}) equals the per capita mortality rate of the predator (d) divided by the product of the efficiency of predation (c) and the ability of predators to convert the prey consumed into offspring (b). Note that these are the two factors that determine the per capita predator birthrate for a given prey population (N_{prey}). As with the prey population, we can now use this value

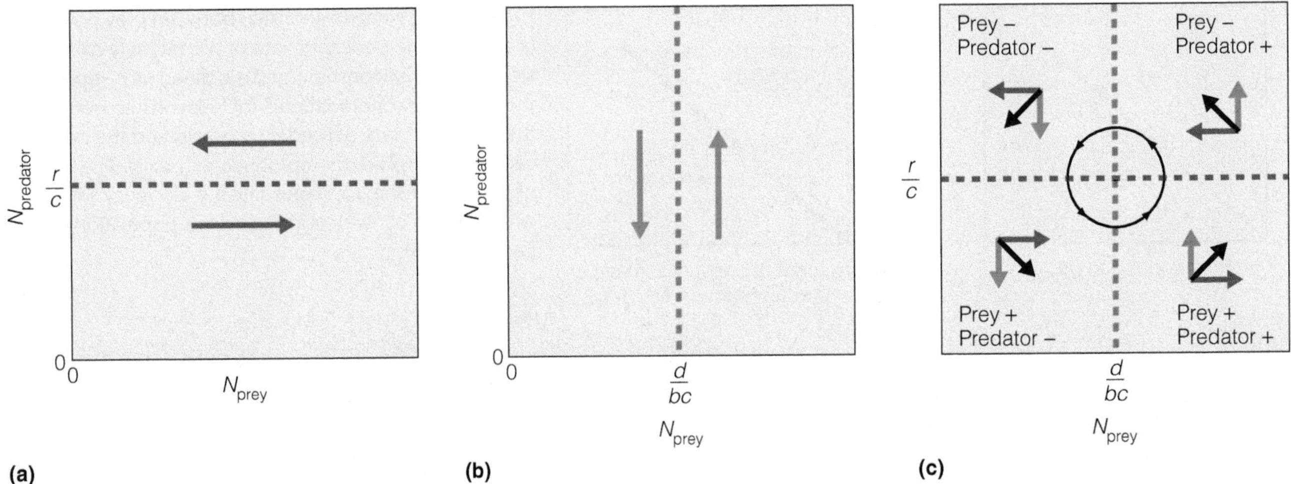

Figure 14.2 The zero-growth isoclines ($dN/dt = 0$) for predator and prey populations under the Lotka–Volterra model. (a) The zero isocline for the prey population is defined by the value $N_{pred} = r/c$. Note the zero-growth isocline is independent of the prey population size. For value of $N_{pred} < r/c$, the prey population increases (horizontal green arrow pointing to right). For values of $N_{pred} > r/c$, the prey population decreases (horizontal green arrow pointing to left). (b) The zero isocline for the predator population is defined by the value $N_{prey} = d/bc$. Note the zero-growth isocline is independent of the predator population size. For value of $N_{prey} > d/bc$, the predator population increases (vertical red arrow pointing up). For values of $N_{prey} < d/bc$, the predator population decreases (vertical red arrow pointing down). (c) The combined zero isoclines provide a means of examining the combined population trajectories of the predator and prey populations. The black arrows represent the combined population trajectory. A minus sign indicates population decline, and a plus sign indicates population increase. This trajectory shows the cyclic nature of the predator–prey interaction.

to define the zero-growth isocline for the predator population (**Figure 14.2b**). The predator zero-growth isocline is independent of the predator population size (N_{pred}) and is represented by a line parallel to the y-axis at a point along the x-axis (represented by the value $N_{prey} = d/bc$). For values of N_{prey} to the left of the zero-growth isocline (toward the origin) the rate of birth in the predator population, $b(cN_{prey})N_{pred}$, is less than the rate of mortality, dN_{pred}, and the growth rate of the predator population is negative. The corresponding decline in population size is represented by the red arrow pointing downward. For values of N_{prey} to the right of the predator zero-growth isocline, the population birthrate is greater than the mortality rate and the population growth rate is positive. The increase in population size is represented by the vertical red arrow pointing up.

As we did in the graphical analysis of competitive interactions (see Section 13.3, Figure 13.2), the two zero-growth isoclines representing the predator and prey populations can be combined to examine changes in the growth rates of two interacting populations for any combination of population sizes (**Figure 14.2c**). When plotted on the same set of axes, the zero-growth isoclines for the predator and prey populations divide the graph into four regions. In the lower right-hand region, the combined values of N_{prey} and N_{pred} are below the prey zero-isocline (green dashed line), so the prey population increases, as represented by the green arrow pointing to the right. Likewise, the combined values lie above the zero-growth isocline for the predator population so the predator population increases, as represented by the red arrow pointing upward. The next value of (N_{prey}, N_{pred}) will therefore be within the region defined by the green and red arrows represented by the black arrow. The combined dynamics indicated by the black arrow point toward the upper right region of the graph. For the upper right-hand region, combined values of N_{prey} and N_{pred} are above the prey isocline, so the prey population declines as indicated by the green horizontal arrow pointing left. The combined values are to the right of the predator isocline, so the predator population increases as indicated by the vertical red arrow pointing up. The black arrow indicating the combined dynamics points toward the upper left-hand region of the graph. In the upper left-hand region of the graph, the combined values of N_{prey} and N_{pred} are above the prey isocline and to the left of the predator isocline so both populations decline. In this case, the combined dynamics (black arrow) point toward the origin. In the last region of the graph, the lower left, the combined values of N_{prey} and N_{pred} are below the prey isocline and to the left of the predator isocline. In this case, the prey population increases and the predator population declines. The combined dynamics point in the direction of the lower left-hand region of the graph, completing a circular, or cyclical, pattern, where the combined dynamics of the predator and prey populations move in a counterclockwise pattern through the four regions defined by the population isoclines.

14.3 Predator–Prey Interaction Results in Population Cycles

The graphical analysis of the combined dynamics of the predator (N_{pred}) and prey (N_{prey}) populations using the zero-growth isoclines presented in Figure 14.2c reveal a cyclical pattern that represents the changes in the two populations through time (**Figure 14.3a**). If we plot the changes in the predator and prey populations as a function of time, we see that the two populations rise and fall in oscillations (**Figure 14.3b**) with the predator population lagging behind the prey population. The oscillation occurs because as the predator population increases, it consumes more and more prey until the prey population begins to decline. The declining prey population no longer supports the large predator population. The predators now face a food shortage, and many of them starve or fail to reproduce. The predator population declines sharply to a point where the reproduction of prey more than balances its losses through predation. The prey population increases, eventually followed by an increase in the population of predators. The cycle may continue indefinitely. The prey is never quite destroyed; the predator never completely dies out.

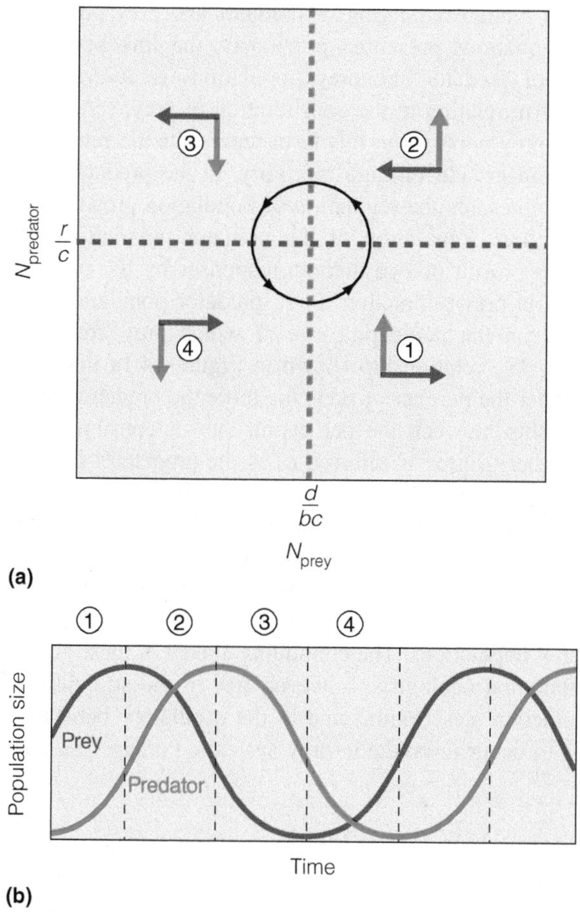

Figure 14.3 By plotting the implied changes in size for both the predator and prey populations through time as indicated in (a) for each of the four regions of the graph, we can see that (b) the two populations continuously cycle out of phase with each other, and the density of predators lags behind that of prey. The sections defined by the dashed lines and labeled by numbers 1–4 correspond to the four regions in (a).

Interpreting Ecological Data

Q1. In Figure 14.3a, which of the four quadrants (regions of the graph) correspond to the following conditions?

$$N_{prey} < d/bc \text{ and } N_{pred} > r/c$$

Q2. In which of the four labeled regions of Figure 14.3b does the mortality rate of predators exceed their birthrate as a result of low prey density?

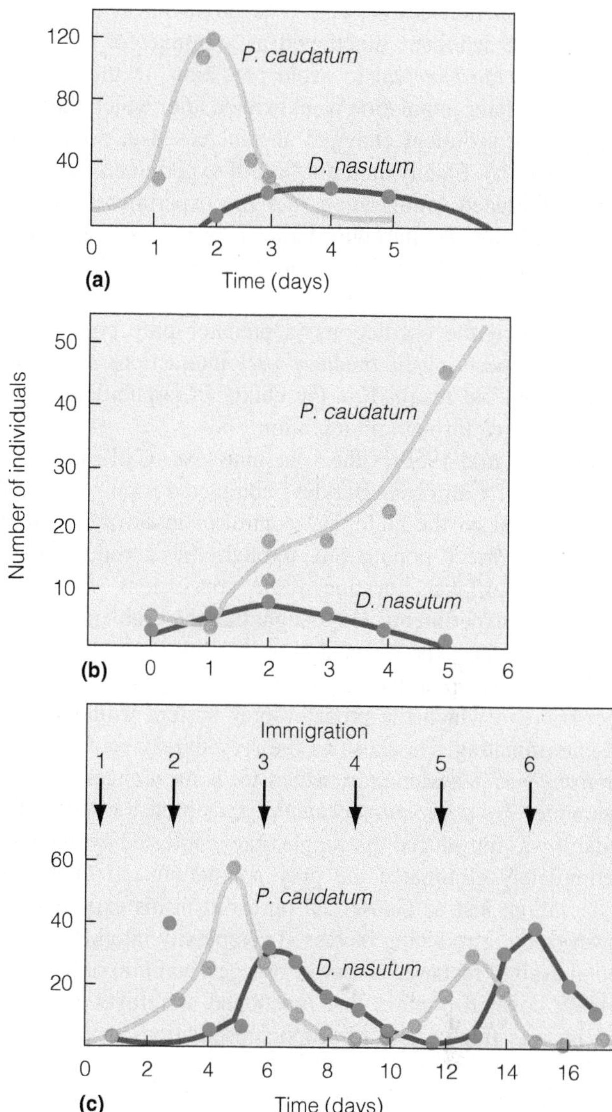

Figure 14.4 Outcome of Gause's experiments of predator–prey interactions between the protozoans *Paramecium caudatum* and *Didinium nasutum* in three microcosms: (a) oat medium without sediment, (b) oat medium with sediment, and (c) with immigration.
(Data from Gause 1934.)

How realistic are the predictions of the Lotka–Volterra model of predator–prey interactions? Do predator–prey cycles actually occur, or are they just a mathematical artifact of this simple model? The Russian biologist G. F. Gause was the first to empirically test the predictions of the predator–prey models in a set of laboratory experiments conducted in the mid-1930s. Gause raised protozoans *Paramecium caudatum* (prey) and *Didinium nasutum* (predator) together in a growth medium of oats. In these initial experiments, *Didinium* always exterminated the *Paramecium* population and then went extinct as a result of starvation (**Figure 14.4a**). To add more complexity to the experimental design, Gause added sediment to the oat medium. The sediment functioned as a refuge for the prey, allowing the *Paramecium* to avoid predation. In this experiment the predator population went extinct, after which the prey hiding in the sediment emerged and increased in population (**Figure 14.4b**). Finally, in a third set of experiments in which Gause introduced immigration into the experimental design (every third day he introduced one new predator and prey individual to the populations), the populations produced the oscillations predicted by the model (**Figure 14.4c**). Gause concluded that the oscillations in predator–prey populations are not a property of the predator–prey interactions suggested by the model but result from the ability of populations to be "supplemented" through immigration.

In the mid-1950s, the entomologist Carl Huffaker (University of California–Berkley) completed a set of experiments focused on the biological control of insect populations (controlling insect populations through the introduction of predators). Huffaker questioned the conclusions drawn by Gause in his experiments. He thought that the problem was the simplicity of the experiment design used by Gause. Huffaker sought to develop a large and complex enough laboratory experiment in which the predator–prey system would not be self-exterminating. He chose as the prey the six-spotted mite, *Eotetranychus sexmaculatus*, which feeds on oranges and another mite, *Typhlodromus occidentalis*, as predator. When the predator was introduced to a single orange infested by the prey, it completely eliminated the prey population and then died of starvation, just as Gause had observed in his experiments. However, by introducing increased complexity into his experimental design (rectangular tray of oranges, addition of barriers, partially covered oranges that functioned as refuges for prey, etc.) he was finally able to produce oscillations in predator–prey populations (**Figure 14.5**).

These early experiments show that predator–prey cycles can result from the direct link between predator and prey populations as suggested by the Lotka–Volterra equations (Section 14.2), but only by introducing environmental heterogeneity—which is a factor not explicitly considered in the model. As we shall see as our discussion progresses, environmental heterogeneity is a key feature of the natural environment that influences species interactions and community structure. However, these laboratory experiments do confirm that predators can have a significant effect on prey populations, and likewise, prey populations can function to control the dynamics of predators.

14.4 Model Suggests Mutual Population Regulation

The Lotka–Volterra model of predator–prey interactions assumes a mutual regulation of predator and prey populations. In the equations presented previously, the link between the growth of predator and prey populations is described by a single term relating to the consumption of prey: $(cN_{prey})N_{pred}$. For the prey population, this term represents the regulation of population growth through mortality. In the predator population, it represents the regulation of population growth through reproduction. Regulation of the predator population growth is a direct result of two distinct responses by the predator to changes in prey population. First, predator population growth depends on the per capita rate at which prey are captured (cN_{prey}). The relationship shown in Figure 14.1a implies that the greater the number of prey, the more the predator eats. The relationship between the per capita rate of consumption and the number of prey is referred to as the predator's **functional response**. Second, this increased consumption of prey results in an increase in predator reproduction $[b(cN_{prey})]$, referred to as the predator's **numerical response**.

This model of predator–prey interaction has been widely criticized for overemphasizing the mutual regulation of predator and prey populations. The continuing appeal of these equations to population ecologists, however, lies in the straightforward mathematical descriptions and in the oscillatory behavior that seems to occur in predator–prey systems. Perhaps the greatest

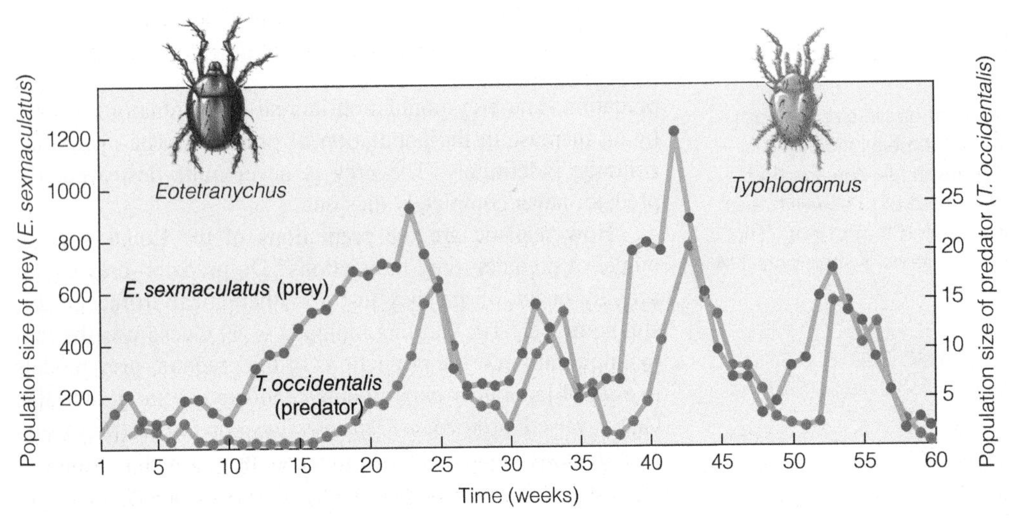

Figure 14.5 Three oscillations in predator and prey populations when the predatory mite *Typhlodromus occidentalis* preyed on the orange-feeding six-spotted mite, *Eotetranychus sexmaculatus*, in a complex environment.
(Data from Huffaker 1958.)

value of this model is in stimulating a more critical look at predator–prey interactions in natural communities, including the conditions influencing the control of prey populations by predators. A variety of factors have emerged from laboratory and field studies, including the availability of cover (refuges) for the prey (as in the experiments discussed in Section 14.3), the increasing difficulty of locating prey as it becomes scarcer, choice among multiple prey species, and evolutionary changes in predator and prey characteristics (coevolution). In the following sections, we examine each of these topics and consider how they influence predator–prey interactions.

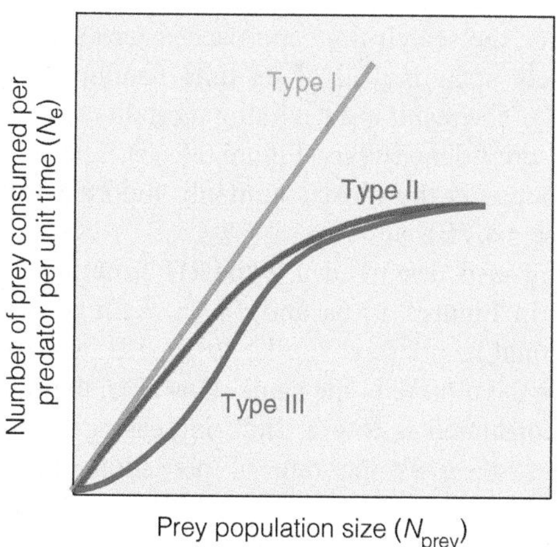

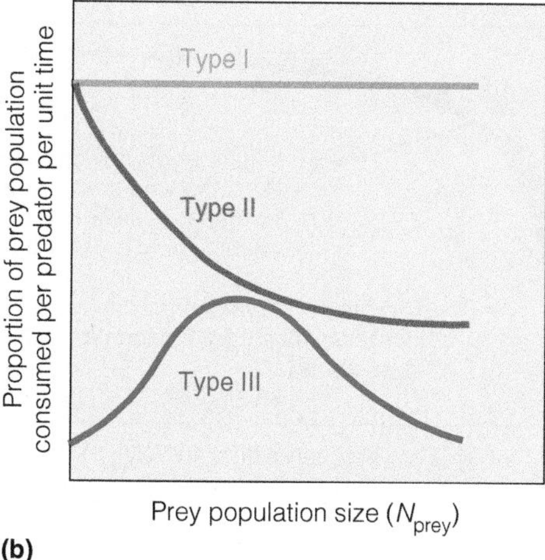

Figure 14.6 Three types of functional response curves that relate the per capita rate of predation (N_e) to prey population size. Curves are expressed in terms of (a) number of prey consumed per predator, and (b) the proportion of prey population consumed per predator. Type I: The number of prey taken per predator increases linearly as prey population size increases. Expressed as a proportion of the prey population, the rate of predation is constant, independent of prey population size. Type II: The predation rate rises at a decreasing rate to a maximum level. Expressed as a proportion of prey population, the rate of predation declines as the prey population grows. Type III: The rate of predation is low at first and then increases in a sigmoid fashion, approaching an asymptote. Plotted as a proportion of prey population, the rate of predation is low at low prey population size, rising to a maximum before declining as the rate of predation reaches its maximum.

14.5 Functional Responses Relate Prey Consumed to Prey Density

The English entomologist M. E. Solomon introduced the idea of functional response in 1949. A decade later, the ecologist C. S. Holling explored the concept in more detail, developing a simple classification based on three general types of functional response (**Figure 14.6**). The functional response is the relationship between the per capita predation rate (number of prey consumed per predator per unit time, N_e) and prey population size (N_{prey}) shown in Figure 14.1a. How a predator's rate of consumption responds to changes in the prey population is a key factor influencing the predator's ability to regulate the prey population.

In developing the predatory prey equations in Section 14.2, we defined the per capita rate of predation as cN_{prey}, where c is the "efficiency" of predation, and N_{prey} is the size of the prey population. This is what Holling refers to as a **Type I functional response.** In the Type I functional response, the number of prey captured per unit time by a predator (per capita rate of predation, N_e) increases linearly with increasing number of prey (N_{prey}; Figure 14.6a). The rate of prey mortality as a result of predation (proportion of prey population captured per predator per unit time) for the Type I response is constant, equal to the efficiency of predation (c), as in Figure 14.6b.

The Type I functional response is characteristic of passive predators, such as filter feeders that extract prey from a constant volume of water that washes over their filtering apparatus. A range of aquatic organisms, from zooplankton (**Figure 14.7a**) to blue whales, exhibit this feeding bahavior. Filter feeders capture prey that flow through and over their filtering system, so for a given rate of water flow over their feeding apparatus, the rate of prey capture will be a direct function of the density of prey per volume of water.

The Type I functional response is limited in its description of the response of predators to prey abundance for two reasons. First, it assumes that predators never become satiated, that is, the per capita rate of consumption increases continuously with increasing prey abundance. In reality, predators will become satiated ("full") and stop feeding. Even for filter feeders, there will be a maximum amount of prey that can be captured (filtered) per unit time above that it can no longer increase regardless of the increase in prey density (see Figure 14.7a). Secondly, even in the absence of satiation, predators will be limited by the handling time, that is, the time needed to chase, capture, and consume each prey item. By incorporating the constraint of handling time, the response of the per capita rate of predation (N_e) to increasing prey abundance (N_{prey}) now exhibits what Holling refers to as a **Type II functional response**. In the Type II functional response, the per capita rate of predation (N_e) increases in a decelerating

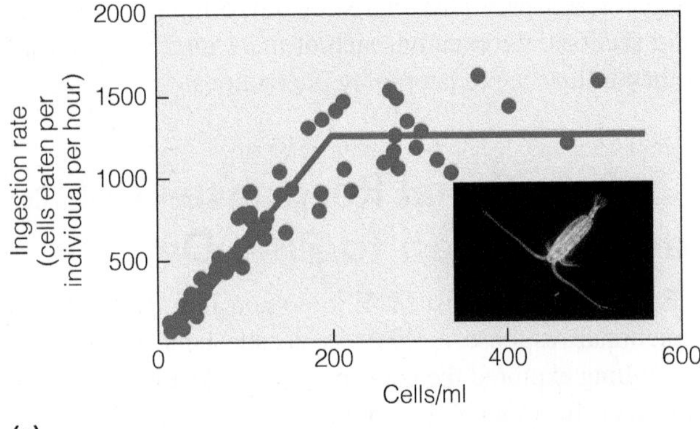

(a)

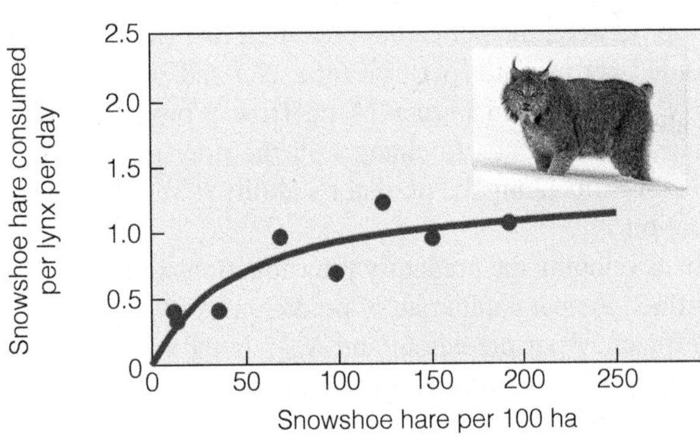

(b)

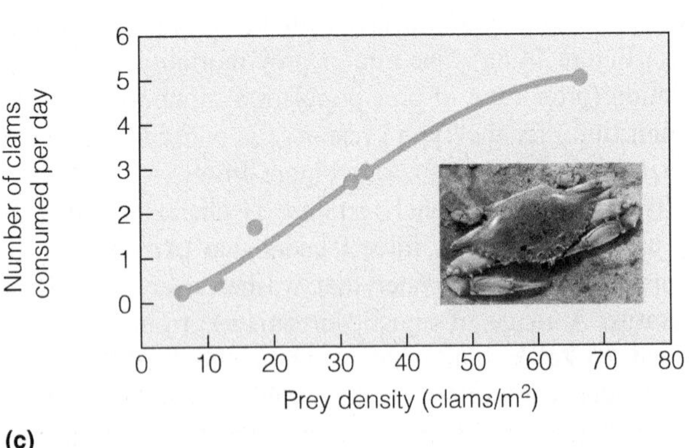

(c)

Figure 14.7 Three examples of functional response curves relating the per capita rate of predation (y-axis) to prey density (x-axis). (a) Type I functional response of the marine copepod *Calanus* (zooplankton filter feeder) feeding on *Coscinodiscus angstii*. The x-axis represents prey population size as cells per ml of water, and the y-axis is ingestion rate (cells eaten per individual per hour). (b) Type II functional response of Canadian lynx (*Lynx canadensis*) feeding on snowshoe hare (*Lepus americanus*) at a site in the southwest Yukon Territory, Canada. (c) Type III functional response of blue crabs (*Callinectes sapidus*) feeding on the clam (*Mya arenaria*). ([a] Frost 1972; [b] O'Donoghue et al. 1997; [c] Seitz et al. 2001.)

fashion, reaching a maximum rate at some high prey population size (see Figure 14.6a). The reason that the value of N_e approaches an asymptote is related to the predator's time budget (**Figure 14.8**; for a mathematical derivation of the Type II functional response, see **Quantifying Ecology 14.1**).

We can think of the total amount of time that a predator spends feeding as T. This time consists of two components: time spent searching for prey, T_s, and time spent handling the prey once it has been encountered, T_h. The total time spent feeding is then: $T = T_s + T_h$. Now as prey abundance (N_{prey}) increases, the number of prey captured (N_e) during the time period T increases (because it is easier to find a prey item as the prey become more abundant); however, the handling time (T_h) also increases (because it has captured more prey to handle), decreasing the time available for further searching (T_s). Handling time (T_h) will place an upper limit on the number of prey a predator can capture and consume in a given time (T). At high prey density, the search time approaches zero and the predator is effectively spending all of its time handling prey (T_h approaches T). The result is a declining mortality rate of prey with increasing prey density (see Figure 14.6b). The Type II functional response is the most commonly reported for predators (see **Figure 14.7b**).

Holling also described a **Type III functional response**, illustrated in Figures 14.6a and **14.7c**. At high prey density, this functional response is similar to Type II, and the explanation for the asymptote is the same. However, the rate at which prey are consumed is low at first, increasing in an *S*-shaped (sigmoid) fashion as the rate of predation approaches the

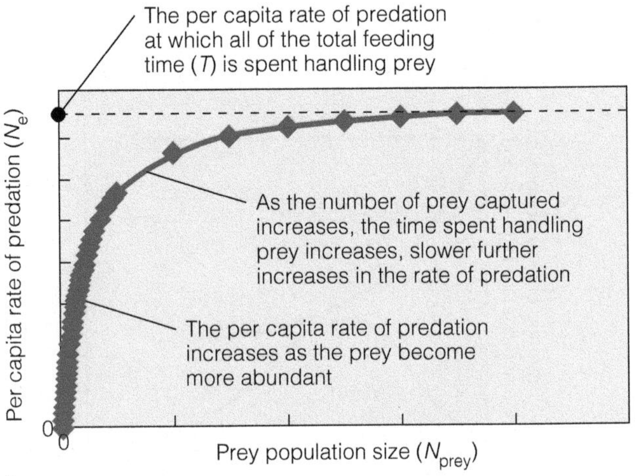

(a)

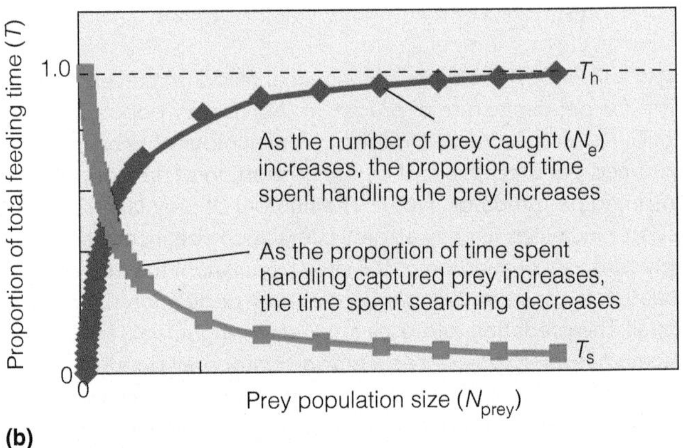

(b)

Figure 14.8 (a) The Type II functional response is a result of (b) the predator allocation of total feeding time (T) between searching (T_s) and handling prey that are being consumed (T_h).

QUANTIFYING ECOLOGY 14.1 Type II Functional Response

The Type I functional response suggests a form of predation in which all of the time allocated to feeding is spent searching (T_s). In general, however, the time available for searching is shorter than the total time associated with consuming the N_e prey because time is required to "handle" the prey item. Handling includes chasing, killing, eating, and digesting. (Type I functional response assumes no handling time below the maximum rate of ingestion.) If we define t_h as the time required by a predator to handle an individual prey item, then the time spent handling N_e prey will be the product $N_e t_h$. The total time (T) spent searching and handling the prey is now:

$$T = T_s + (N_e t_h)$$

By rearranging the preceding equation, we can define the search time as:

$$T_s = T - N_e t_h$$

For a given total foraging time (T), search time now varies, decreasing with increasing allocation of time to handling.

We can now expand the original equation describing the type I functional response $[N_e - (cN_{\text{prey}})T_s]$ by substituting the equation for T_s just presented. This includes the additional time constraint of handling the N_e prey items:

$$N_e = c(T - N_e t_h)N_{\text{prey}}$$

Note that N_e, the number of prey consumed during the time period T, appears on both sides of the equation, so to solve for N_e, we must rearrange the equation.

$$N_e = c(N_{\text{prey}}T - N_{\text{prey}}N_e t_h)$$

Move c inside the brackets, giving:

$$N_e = cN_{\text{prey}}T - N_e cN_{\text{prey}}t_h$$

Add $N_e cN_{\text{prey}}t_h$ to both sides of the equation, giving:

$$N_e + N_e cN_{\text{prey}}t_h = cN_{\text{prey}}T$$

Rearrange the left-hand side of the equation, giving:

$$N_e(1 + cN_{\text{prey}}t_h) = cN_{\text{prey}}T$$

Divide both sides of the equation by $(1 + cN_{\text{prey}}t_h)$, giving:

$$N_e = \frac{cN_{\text{prey}}T}{(1 + cN_{\text{prey}}t_h)}$$

We can now plot the relationship between N_e and N_{prey} for a given set of values for c, T, and t_h. (Recall that the values of c, T, and t_h are constants.)

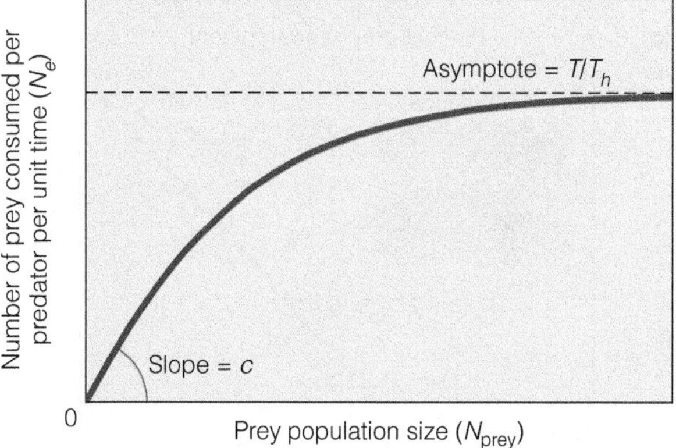

Figure 1 Relationship between the density of prey population (x-axis) and the per capita rate of prey consumed (y-axis) for the model of predator functional response presented above that includes both search (T_s) and handling ($T_h = N_e t_h$) time ($T = T_s + T_h$). At low prey density, the number of prey consumed is low, as is handling time. As prey density increases, the number of prey consumed increases; a greater proportion of the total foraging time (T) is spent handling prey, reducing time available for searching. As the handling time approaches the total time spent foraging, the per capita rate of prey consumed approaches an asymptote. The resulting curve is referred to as a Type II functional response.

maximum value. In the Type III functional response, mortality rate of the prey population is negligible at low prey abundance, but as the prey population increases (as indicated by the upward sweep of the curve), the mortality rate of the population increases in a density-dependent fashion (Figure 14.6b). However, the regulating effect of predators is limited to the interval of prey density where mortality increases. If prey density exceeds the upper limit of this interval, then mortality resulting from predation starts to decline.

Several factors may result in a Type III response. Availability of cover (refuge) that allows prey to escape predators may be an important factor. If the habitat provides only a limited number of hiding places, it will protect most of the prey population at low density, but the susceptibility of individuals will increase as the population grows.

Another reason for the sigmoidal shape of the Type III functional response curve may be the predator's **search image**, an idea first proposed by the animal behaviorist L. Tinbergen. When a new prey species appears in the area, its risk of becoming selected as food by a predator is low. The predator has not yet acquired a search image—a way to recognize that species as a potential food item. Once the predator has captured an individual, it may identify the species as a desirable prey. The predator then has an easier time locating others of the same kind. The more adept the predator becomes at securing a particular prey item, the more intensely it concentrates on it. In

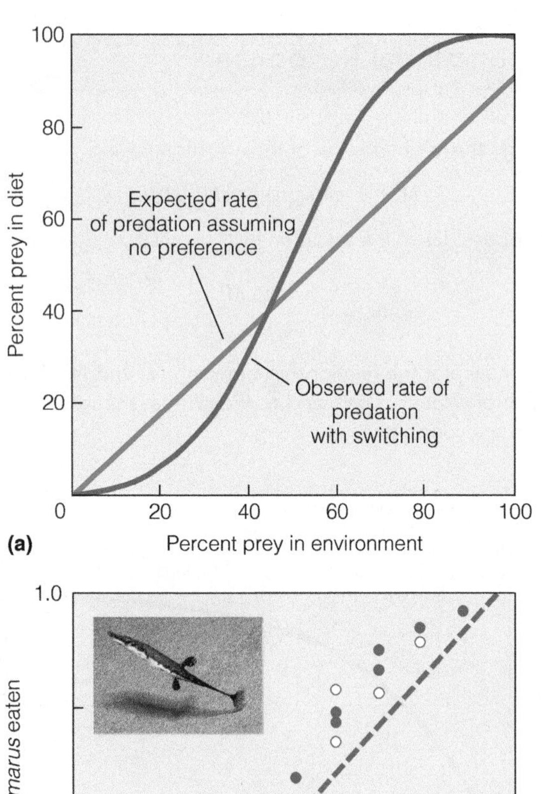

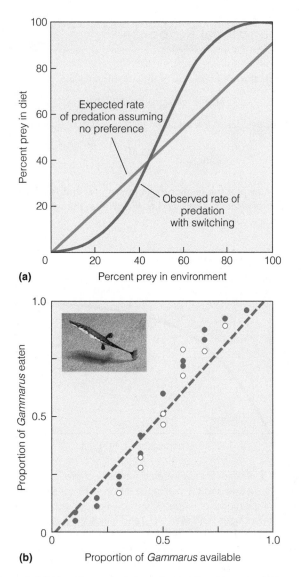

Figure 14.9 (a) A model of prey switching. The straight line represents the expected rate of predation assuming no preference by the predator. The prey are eaten in a fixed proportion to their relative availability (percentage of total prey available to predator in environment). The curved line represents the change in predation rate observed in the case of prey switching. At low densities, the proportion of the prey species in the predator's diet is less than expected based on chance (and on its proportional availability to other prey species). Over this range of abundance, the predator is selecting alternative prey species. When prey density is high, the predator takes more of the prey than expected. Switching occurs at the point where the lines cross. The habit of prey switching results in a Type III functional response between a predator and its prey species. (b) Example of frequency-dependent predation (prey switching) by sticklebacks (*Spinachia spinachia*) fed on mixtures of *Gammarus* and *Artemia*. Proportion of *Gammarus* in the diet is plotted as a function of the proportion available. Dotted line represents frequency-independent predation. Closed symbols denote trials with increasing availability of *Gammarus*, open symbols decreasing availability of *Gammarus* prey.
(Hughes and Croy 1993.)

time, the number of this particular prey species becomes so reduced or its population becomes so dispersed that encounters between it and the predator lessen. The search image for that prey item begins to wane, and the predator may turn its attention to another prey species.

A third factor that can result in a Type III functional response is the relative abundance of different, alternative prey species. Although a predator may have a strong preference for a certain prey, in most cases it can turn to another, more abundant prey species that provides more profitable hunting. If rodents, for example, are more abundant than rabbits and quail, foxes and hawks will concentrate on rodents.

Ecologists call the act of turning to more abundant, alternate prey **switching** (Figure 14.9a). In switching, the predator feeds heavily on the more abundant species and pays little attention to the less abundant species. As the relative abundance of the second prey species increases, the predator turns its attention to that species.

The point in prey abundance when a predator switches depends considerably on the predator's food preference. A predator may hunt longer and harder for a palatable species before turning to a more abundant, less palatable alternate prey. Conversely, the predator may turn from the less desirable species at a much higher level of abundance than it would from a more palatable species.

In a series of laboratory experiments, Roger Hughes and M. I. Croy of the University of Wales (Great Britain) examined prey switching in 15-spined stickleback (*Spinachia spinachia*) feeding on two prey species: amphipod (*Gammarus locusta*) and brine shrimp (*Aremia* spp.). In all experiments, fish showed the sigmoid response to changing relative abundances of prey, typical of switching (Figure 14.9b). The researchers found that a combination of changing attack efficiency and search image formation contributed to the observed pattern of prey switching.

Although simplistic, the model of functional response developed by Holling has been a valuable tool. It allows ecologists to explore how various behaviors—exhibited by both the predator and prey species—influence predation rate and subsequently predator and prey population dynamics. Because the model explicitly addresses the principle of time budget in the process of predation, this framework has been expanded to examine questions relating to the efficiency of foraging, a topic we will return to in Section 14.7.

14.6 Predators Respond Numerically to Changing Prey Density

As the density of prey increases, the predator population growth rate is expected to respond positively. A numerical response of predators can occur through reproduction by predators (as suggested by the conversion factor *b* in the Lotka–Volterra equation for predators) or through the movement of predators into areas of high prey density (immigration). The latter is referred to as an

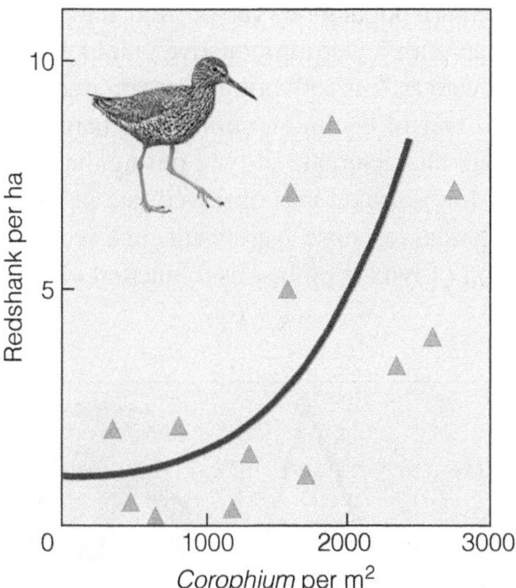

Figure 14.10 Aggregative response in the redshank (*Tringa totanus*). The curve plots the density of the redshank in relation to the average density of its arthropod prey (*Corophium* spp.).
(Data from Hassel and May 1974.)

aggregative response (**Figure 14.10**). The tendency of predators to aggregate in areas of high prey density can be a crucial feature in determining a predator population's ability to regulate prey density. Aggregative response is important because most predator populations grow slowly in comparison to those of their prey.

Marc Salamolard of the Center for Biological Studies (French National Center for Scientific Research) and colleagues provide an example of how these two components of numerical response (immigration and increased reproduction) can combine to influence the response of a predator population to changes in prey abundance. Salamolard quantified the functional and numerical responses of Montagu's harrier (*Circus pygargus*), a migratory raptor, to variations in abundance of its main prey, the common vole (*Microtus arvalis*). The researchers monitored variations in the vole population over a 15-year period and the response of the harrier population to this variable food supply. This predatory bird species exhibits a Type II functional response; the per capita rate of predation increases with increasing prey density up to some maximum (see **Figure 14.11a**). The researchers were able to provide a number of measures relating to the bird's numerical response. The breeding density of birds increases with increasing prey. This increase in predator density is a result of an increase in the number of nesting pairs occupying the area and represents an aggregative response density (**Figure 14.11b**). In addition, the mean brood size of nesting pairs (mean number of chicks at fledging) also increased (**Figure 14.11c**). The net result is an increase in the growth rate of the predator population in response to an increase in the abundance of prey (vole population).

The work of Włodzimierz Jędrzejewski and colleagues at the Mammal Research Institute of the Polish Academy of Sciences provides an example where the numerical response of the predator population is dominated by reproductive effort. Jędrzejewski

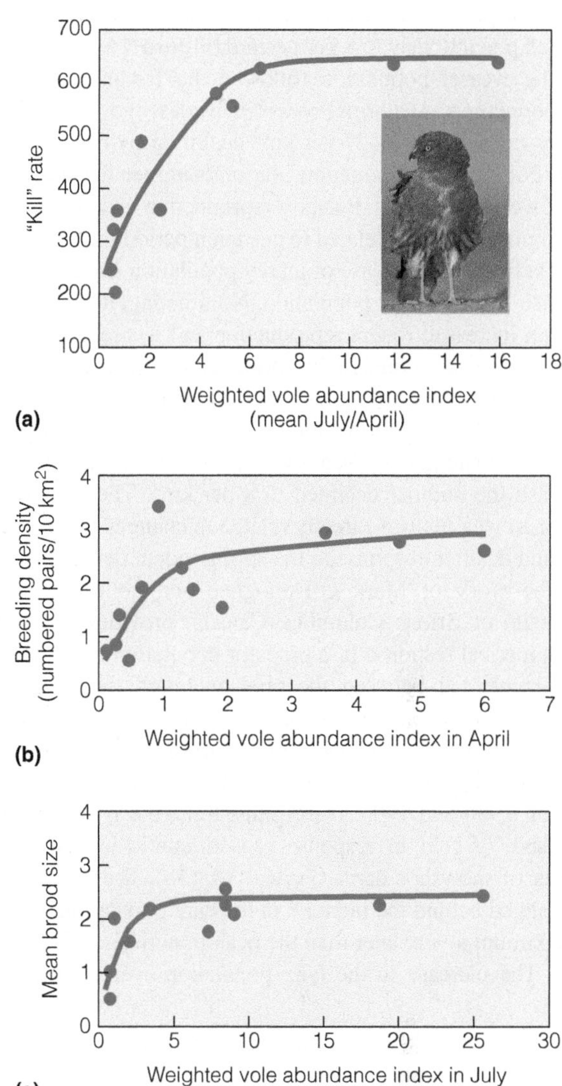

Figure 14.11 Functional and numerical responses of Montagu's harriers (*Circus pygargus*). (a) Functional response is represented as the "kill rate," defined as the number of common voles preyed on by a breeding pair of Montagu's harriers over the breeding season. Numerical response expressed as (b) the breeding density of harriers (aggregative response) and (c) mean brood size (number of chicks fledged per nest) expressed as a function of vole abundance. The latter is a measure of increased reproductive success per breeding pair.
(Salamolard et al. 2000.)

examined the response of a weasel (*Mustela nivalis*) population to the density of two rodents, the bank vole (*Clethrionomys glareolus*) and the yellow-necked mouse (*Apodemus flavicollis*), in Białowieża National Park in eastern Poland in the early 1990s. During that time, the rodents experienced a two-year irruption in population size brought about by a heavy crop of oak, hornbeam, and maple seeds. The abundance of food stimulated the rodents to breed throughout the winter. The long-term average population density was 28–74 animals per hectare. During the irruption, the rodent population reached nearly 300 per hectare and then declined precipitously to 8 per hectare (**Figure 14.12**).

The weasel population followed the fortunes of the rodent population. At normal rodent densities, the winter weasel density ranged from 5–27 per km^2 declining by early spring to 0–19. Following reproduction, the midsummer density rose to 42–47 weasels per km^2. Because reproduction usually requires a certain minimal time (related to gestation period), a lag typically exists between an increase of a prey population and a numerical response by a predator population. No time lag, however, exists between increased rodent reproduction and weasel reproductive response. Weasels breed in the spring, and with an abundance of food they may have two litters or one larger litter. Young males and females breed during their first year of life. During the irruption, the number of weasels grew to 102 per km^2 and during the crash the number declined to 8 per km^2. The increase and decline in weasels was directly related to changes in the rates of birth and death in response to the spring rodent density.

The work of Mark O'Donoghue and colleagues at the University of British Columbia (Canada) provides an example of a numerical response of a predator population in which there is a distinct lag between the prey and predator populations. The researchers monitored populations of Canadian lynx (*Lynx canadensis*) and their primary prey, the snowshoe hare (*Lepus americanus*) at a site in the southwest Yukon Territory, Canada, between 1986 and 1995. During this time, the lynx population increased 7.5-fold in response to a dramatic increase in the number of snowshoe hares (**Figure 14.13a**). The abundance of lynx lagged behind the increase in the hare population, reaching its maximum a year later than the peak in numbers of snowshoe hares. The increase in the lynx population eventually led to a decline in the hare population. The decline in the number of lynx was associated with lower reproductive output and high emigration rates. Few to no kits (offspring) were produced by lynx after the second winter of declining numbers of hares. High emigration rates were characteristic of lynx during the cyclic peak and decline, and low survival was observed late in the decline. The delayed numerical response (lag) results in a cyclic pattern when the population of lynx is plotted as a function of size of the prey

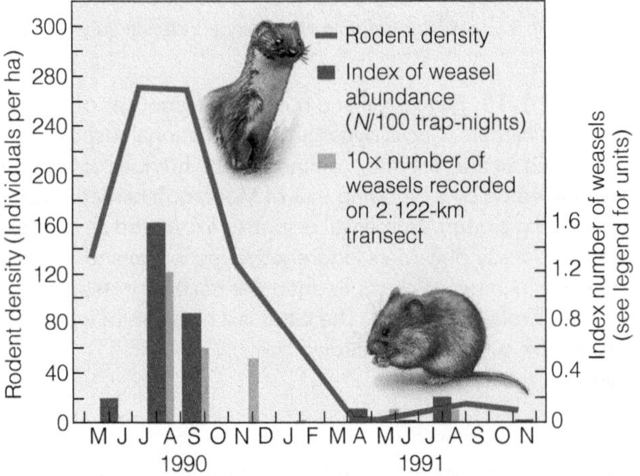

Figure 14.12 Numerical response of weasels (predators) to an irruption and crash in the populations of forest rodents (prey). The x-axis represents time of year (month); the y-axes represent the density of the rodent population (left axis) and an index of the predator population density (right axis). Brown bars represent weasel data from live trapping; orange bars represent data from captures, visual observations, and radio tracking. The reproductive rate of the weasel population is high enough to allow the population to track increases in the prey population during 1990.
(Adapted from Jędrzejewski et al. 1995.)

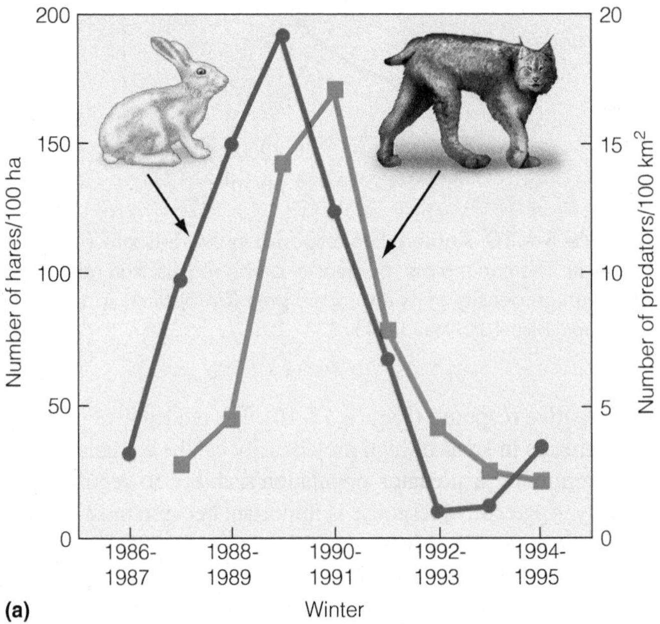

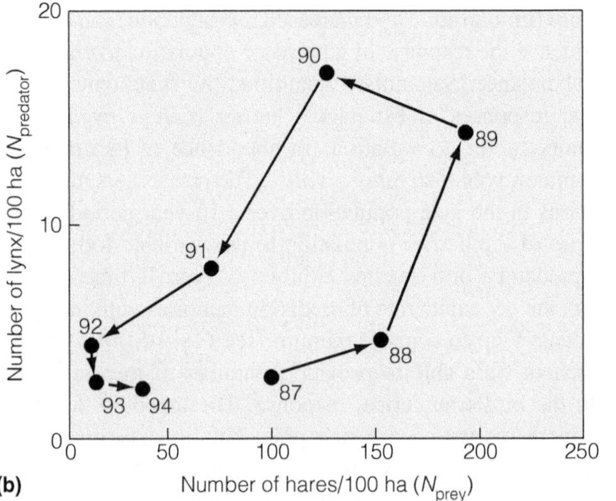

Figure 14.13 Numerical response of Canadian lynx (predator) to changes in snowshoe hare population (prey). (a) Population estimates (expressed as density per 100 km^2 and 100 ha respectively) from 1986 through 1995 in the southwest Yukon, Canada, based on autumn and later winter counts. (b) When data from (a) are plotted on axes representing the size of predator and prey populations, the lynx population (predator) responds to changes in the hare population (prey) following the counterclockwise joint population trajectory (cycle) predicted by the Lokta–Volterra equations (see Figure 14.2c).
(Adapted from O'Donoghue et al 1997.)

population (**Figure 14.13b**), as was observed in the analysis of the Lotka–Volterra model in Section 14.2 (see Figure 14.2c).

14.7 Foraging Involves Decisions about the Allocation of Time and Energy

Thus far, we have discussed the activities of predators almost exclusively in terms of foraging. But all organisms are required to undertake a wide variety of activities associated with survival, growth, and reproduction. Time spent foraging must be balanced against other time constraints such as defense, avoiding predators, searching for mates, or caring for young. This trade-off between conflicting demands has led ecologists to develop an area of research known as **optimal foraging theory**. At the center of optimal foraging theory is the hypothesis that natural selection favors "efficient" foragers, that is, individuals that maximize energy or nutrient intake per unit of effort. Efficient foraging involves an array of decisions: what food to eat, where and how long to search, and how to search. Optimal foraging theory approaches these decisions in terms of costs and benefits. Costs can be measured in terms of the time and energy expended in the act of foraging, and benefits should be measured in terms of fitness. However, it is extremely difficult to quantify the consequences of a specific behavioral choice on the probability of survival and reproduction. As a result, benefits are typically measured in terms of energy or nutrient gain, which is assumed to correlate with individual fitness.

One of the most active areas of research in optimal foraging theory has focused on the composition of animal diets—the process of choosing what to eat from among a variety of choices. We can approach this question using the framework of time allocation developed in the simple model of function response in Section 14.5, where the total time spent foraging (T) can be partitioned into two categories of activity: searching (T_s) and handling (T_h). Here we will define the search time for a single prey (per capita search time) as t_s, and the handling time for a single captured prey as t_h (capital letters refer to total search and handling time during a given period of hunting or feeding, T).

For simplicity, consider a predator hunting in a habitat that contains just two kinds of prey: P_1 and P_2. Assume that the two prey types yield E_1 and E_2 units of net energy gain (benefits), and they require t_{h_1} and t_{h_2} seconds to handle (costs). Profitability of the two prey types is defined as the net energy gained per unit handling time: E_1/t_{h_1} and E_2/t_{h_2}. Now suppose that P_1 is more profitable than P_2: $E_1/t_{h_1} > E_2/t_{h_2}$. Optimal foraging theory predicts that P_1 would be the preferred prey type because it has a greater profitability.

This same approach can be applied to a variety of prey items within a habitat. Behavioral ecologist Nicholas B. Davies of the University of Cambridge examined the feeding behavior of the pied wagtail (*Motacilla alba*) in a pasture near Oxford, England. The birds fed on various dung flies and beetles attracted to cattle droppings. Potential prey types were of various sizes: small, medium, and large flies and beetles. The wagtails showed a decided preference for medium-sized prey (**Figure 14.14a**). The size of the prey selected corresponded to the prey the birds could handle most profitably (E/t_h; **Figure 14.14b**). The birds virtually ignored smaller prey. Although easy to handle (low value of t_h), small prey did not return sufficient energy (E), and large prey required too much time and effort to handle relative to the energy gained.

The simple model of optimal foraging presented here provides a means for evaluating which of two or more potential prey types is most profitable based on the net energy gain per

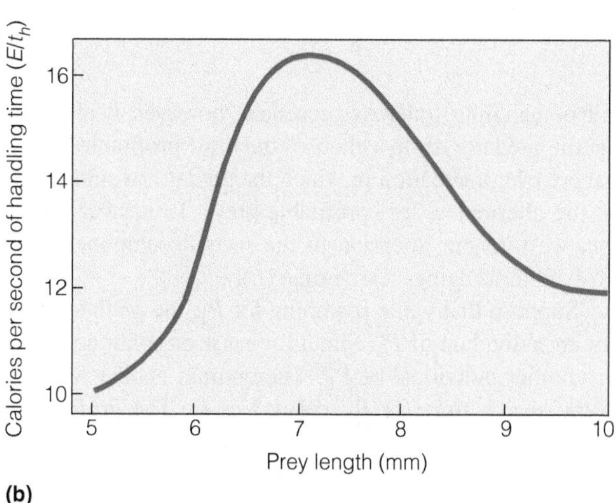

Figure 14.14 (a) Relationship between prey length (x-axis) and frequency of prey in the diet of pied wagtails (y-axis) foraging in a pasture. Individuals show a definite preference for medium-sized prey, which are taken in disproportionate amounts compared to the frequencies of prey available. (b) The prey size (x-axis) chosen by pied wagtails represented the optimal size, providing maximum energy per handling time (E/t_h: y-axis).
(Adapted from Davies 1977.)

QUANTIFYING ECOLOGY 14.2 A Simple Model of Optimal Foraging

Faced with a variety of potential food choices, predators make decisions regarding which types of food to eat and where and how long to search for food. But how are these decisions made? Do predators function opportunistically, pursuing prey as they are encountered, or do they make choices and pass by potential prey of lesser quality (energy content) while continuing the search for more preferred food types? If the objective is to maximize energy intake (energy gain per unit time), a predator should forage in a way that maximizes benefits (energy gained from consuming prey) relative to costs (energy expended). This concept of maximizing energy intake is the basis of models of optimal foraging.

Any food item has a benefit (energy content) and a cost (in terms of time and energy involved in search and acquisition). The benefit–cost relationship determines how much profit a particular food item represents. The profitability of a prey item is the ratio of its energy content (E) to the time required for handling the item (t_h), or E/t_h.

Let us assume that a predator has two possible choices of prey, P_1 and P_2. The two prey types have energy contents of E_1 and E_2 (units of kilojoules [kJ]) and take t_{h_1} and t_{h_2} seconds to handle. The searching time for the two prey types are t_{s_1} and t_{s_2} in seconds. We will define P_1 as the most profitable prey type (greater value of E/t_h).

As the predator searches for P_1, it encounters an individual of P_2. Should the predator capture and eat P_2 or continue to search for another individual of P_1? Which decision—capture P_2 or continue to search—would be the more profitable and maximize the predator's energy intake? This is the basic question posed by optimal foraging theory, and the solution depends on the search time for P_1.

The profitability of capturing and eating P_2 is E_2/t_{h_2} and the profitability of continuing the search, capturing, and eating another individual of P_1 is $E_1/(t_{h_1} + t_{s_1})$. Notice that the decision to ignore P_2 and continue the search carries the additional cost of the average search time for P_1, t_{s_1}. Therefore, the optimal solution, the decision that will yield the greater profit, is based on the following conditions:

If:

$$E_2/t_{h_2} > E_1/(t_{h_1} + t_{s_1})$$

then capture and eat P_2.

If:

$$E_2/t_{h_2} < E_1/(t_{h_1} + t_{s_1})$$

then ignore P_2 and continue to search for P_1.

Therefore, if the search time for P_1 is short, the predator will be better off continuing the search; if the search time is long, the most profitable decision is to capture and consume P_2.

The benefit–cost trade-off for the optimal choice in prey selection is best understood through an actual example. David Irons and colleagues at Oregon State University examined the foraging behavior of glaucous-winged gulls (*Larus glaucescens*) that forage in the rock intertidal habitats of the Aleutian Islands, Alaska. Data on the abundance of three prey types (urchins, chitons, and mussels) in three intertidal zones (A, B, and C) are presented in the table. Mean densities of the three prey types in numbers per m^2 are given for the three zones. Average energy content (E), handling time (t_h), and search time (t_s) for each of the three prey types are also listed in the table.

unit of handling time. As presented, however, it also implies that the predator always chooses the most profitable prey item. Is there ever a situation in which the predator would choose to eat the alternative, less profitable prey? To answer this question, we turn our attention to the second component of time involved in foraging, search time (t_s).

Suppose that while searching for P_1, the predator encounters an individual of P_2. Should it eat it or continue searching for another individual of P_1? The optimal choice will depend on the search time for P_1, defined as t_{s_1}. The profitability of consuming the individual of P_2 is E_2/t_{h_2}; the alternative choice of continuing to search, capture, and consume an individual of P_1 is $E_1/(t_{h_1} + t_{s_1})$, which now includes the additional time cost of searching for another individual of P_1 (t_{s_1}). If $E_2/t_{h_2} > E_1/(t_{h_1} + t_{s_1})$, then according to optimal foraging theory, the predator would eat the individual of P_2. If this condition does not hold true, then the predator would continue searching for P_1. Testing this hypothesis requires the researcher to quantify the energy value and search and handling times of the various potential prey items. An example of this simple model of optimal prey choice is presented in **Quantifying Ecology 14.2**.

A wealth of studies examines the hypothesis of optimal prey choice in a wide variety of species and habitats, and patterns of prey selection generally follow the rules of efficient foraging. But the theory as presented here fails to consider the variety of other competing activities influencing a predator's time budget and the factors other than energy content that may influence prey selection. One reason that a predator consumes a varied diet is that its nutritional requirements may not be met by eating a single prey species (see Chapter 7).

14.8 Risk of Predation Can Influence Foraging Behavior

Most predators are also prey to other predatory species and therefore face the risk of predation while involved in their routine activities, such as foraging. Habitats and foraging areas vary in their foraging profitability and their risk of predation. In deciding whether to feed, the forager must balance its potential energy gains against the risk of being eaten. If predators are about, then it may be to the forager's advantage not to visit a most profitable, but predator-prone, area

In feeding preference experiments, where search and handling time were not a consideration, chitons were the preferred prey type and the obvious choice for maximizing energy intake. However, the average abundance of urchins across the three zones is greater than that of chitons. As a gull happens upon an urchin while hunting for chitons, should it capture and eat the urchin or continue to search for its preferred food? Under conditions of optimal foraging, the decision depends on the conditions outlined previously. The profit gained by capturing and consuming the urchin is E/t_h = (7.45 kJ/8.3 s), or 0.898. In contrast, the profit gained by ignoring the urchin and searching, capturing, and consuming another chiton is $E/(t_h + t_s)$ = [24.52 kJ/(3.1 s + 37.9 s)] or 0.598. Because the profit gained by consuming the urchin is greater than the profit gained by ignoring it and continuing the search for chitons, it would make sense for the gull to capture and eat the urchin.

What about a gull foraging in intertidal zone A that happens upon a mussel? The profit gained by capturing and eating the mussel is (1.42/2.9), or 0.490, and the profit gained by continuing the search for a chiton remains [24.52 kJ/(3.1 s + 37.9 s)] or 0.598. In this case, the gull would be better off ignoring the mussel and continuing the search for chitons.

We now know what the gulls "should do" under the hypothesis of optimal foraging. But do they in fact forage optimally as defined by this simple model of benefits and costs? If gulls are purely opportunistic, their selection of prey in each of the three zones would be in proportion to their relative abundances. Irons and colleagues, however, found that the relative preferences for urchins and chitons were in fact related to their profitability (E/t_h); mussels, however, were selected less frequently than predicted by their relative value of E.

1. How would reducing the energy content of chitons by half (to 12.26 kJ) influence the decision whether the gull should capture and eat the mussel or continue searching for a chiton in the example presented?
2. Because the gulls do not have the benefit of the optimal foraging model in deciding whether to select a prey item, how might natural selection result in the evolution of optimal foraging behavior?

Prey Type	Density Zone A	Density Zone B	Density Zone C	Energy (kJ/individual)	Handling Time (s)	Search Time (s)
Urchins	0.0	3.9	23.0	7.45	8.3	35.8
Chitons	0.1	10.3	5.6	24.52	3.1	37.9
Mussels	852.3	1.7	0.6	1.42	2.9	18.9

and to remain in a less profitable but more secure part of the habitat. Many studies report how the presence of predators affects foraging behavior. In one such study, Jukka Suhonen of the University of Jyväskylä (Finland) examined the influence of predation risk on the use of foraging sites by willow tits (*Parus montanus*) and crested tits (*Parus cristatus*) in the coniferous forests of central Finland. During the winter months, flocks of these two bird species forage in spruce, pine, and birch trees. The major threat to their survival is the Eurasian pygmy owl (*Glaucidium passerinum*). The owl is a diurnal ambush, or sit-and-wait hunter, that pounces downward on its prey. Its major food is voles, and when vole populations are high, usually every three to five years, the predatory threat to these small passerine birds declines. When vole populations are low, however, the small birds become the owl's primary food. During these periods, the willow and crested tits forsake their preferred foraging sites on the outer branches and open parts of the trees, restricting their foraging activity to the denser inner parts of spruce trees that provide cover and to the tops of the more open pine and leafless birch trees.

14.9 Coevolution Can Occur between Predator and Prey

By acting as agents of mortality, predators exert a selective pressure on prey species (see Chapter 12, Section 12.3). That is, any characteristic that enables individual prey to avoid being detected and captured by a predator increases its fitness. Natural selection functions to produce "smarter," more evasive prey (fans of the *Road Runner* cartoons should already understand this concept). However, failure to capture prey results in reduced reproduction and increased mortality of predators. Therefore, natural selection also produces "smarter," more skilled predators. As characteristics that enable them to avoid being caught evolve in prey species, more effective means of capturing prey evolve in predators. To survive as a species, the prey must present a moving target that the predator can never catch. This view of the coevolution between predator and prey led the evolutionary biologist Leigh Van Valen to propose the Red Queen hypothesis. In Lewis Carroll's *Through the Looking Glass, and What Alice Found There*, there is a scene in the Garden of Living Flowers in which everything is continuously moving. Alice is surprised to see that no matter how fast she moves, the world around her remains motionless—to which the Red Queen responds, "Now, here, you see, it takes all the running you can do, to keep in the same place." So it is with prey species. To avoid extinction at the hands of predators, prey must evolve means of avoiding capture; they must keep moving just to stay where they are.

14.10 Animal Prey Have Evolved Defenses against Predators

Animal species have evolved a wide range of characteristics to avoid being detected, selected, and captured by predators. These characteristics are collectively referred to as **predator defenses**.

Chemical defense is widespread among many groups of animals. Some species of fish release alarm pheromones (chemical signals) that, when detected, induce flight reactions in members of the same and related species. Arthropods, amphibians, and snakes employ odorous secretions to repel predators. For example, when disturbed, the stinkbug (*Cosmopepla bimaculata*) discharges a volatile secretion from a pair of glands located on its back (**Figure 14.15a**). The stinkbug can control the amount of fluid released and can reabsorb the fluid into the gland. In a series of controlled experiments, Bryan Krall and colleagues at Illinois State University have found that the secretion deters feeding by both avian and reptile predators.

Many arthropods possess toxic substances, which they acquire by consuming plants and then store in their own bodies. Other arthropods and venomous snakes, frogs, and toads synthesize their own poisons.

Prey species have evolved numerous other defense mechanisms. Some animals possess **cryptic coloration**, which includes colors and patterns that allow prey to blend into the background of their normal environment (**Figure 14.15b**). Such protective coloration is common among fish, reptiles, and many ground-nesting birds. **Object resemblance** is common among insects. For example, walking sticks (Phasmatidae) resemble twigs (**Figure 14.15c**), and katydids (Pseudophyllinae) resemble leaves. Some animals possess eyespot markings,

Figure 14.15 Animals have evolved a variety of defenses against predators.

(a)
The stinkbug discharges a volatile secretion that discourages predators.
(b)
The flounder uses cryptic coloration to avoid detection by both predators and potential prey.
(c)
The walking stick, which feeds on the leaves of deciduous trees, strongly resembles a twig.
(d)
The white-tailed deer has a characteristic white tail patch that serves as an alarm and a distraction while the deer flees predators.
(e)
(f)
The bright and distinctive coloration patterns of the monarch butterfly and the strawberry poison dart frog warn potential predators of these species' toxicity.

Figure 14.16 The warning coloration of (a) the poisonous coral snake (*Micrurus fulvius*) is mimicked by (b) the nonvenomous scarlet king snake (*Lampropeltis triangulum*).

which intimidate potential predators, attract the predators' attention away from the animal, or delude them into attacking a less vulnerable part of the body. Associated with cryptic coloration is **flashing coloration**. Certain butterflies, grasshoppers, birds, and ungulates, such as the white-tailed deer, display extremely visible color patches when disturbed and put to flight. The flashing coloration may distract and disorient predators; in the case of the white-tailed deer, it may serve as a signal to promote group cohesion when confronted by a predator (**Figure 14.15d**). When the animal comes to rest, the bright or white colors vanish, and the animal disappears into its surroundings.

Animals that are toxic to predators or use other chemical defenses often possess **warning coloration**, or **aposematism**, that is, bold colors with patterns that may serve as warning to would-be predators. The black-and-white stripes of the skunk, the bright orange of the monarch butterfly, and the yellow-and-black coloration of many bees and wasps and some snakes may serve notice of danger to their predators (**Figures 14.15e** and **14.15f**). All their predators, however, must have an unpleasant experience with the prey before they learn to associate the color pattern with unpalatability or pain.

Some animals living in the same habitats with inedible species sometimes evolve a coloration that resembles or mimics the warning coloration of the toxic species. This type of mimicry is called **Batesian mimicry** after the English naturalist H. E. Bates, who described it when observing tropical butterflies. The mimic, an edible species, resembles the inedible species, called the model. Once the predator has learned to avoid the model, it avoids the mimic also. In this way, natural selection reinforces the characteristic of the mimic species that resembles that of the model species.

Most discussions of Batesian mimicry concern butterflies, but mimicry is not restricted to Lepidoptera and other invertebrates. Mimicry has also evolved in snakes with venomous models and nonvenomous mimics (**Figure 14.16**). For example, in eastern North America, the scarlet king snake (*Lampropeltis triangulum*) mimics the eastern coral snake (*Micrurus fulvius*) and in southwestern North America, the mountain kingsnake (*Lampropeltis pyromelana*) mimics the western coral snake (*Micruroides euryxanthus*). Mimicry is not limited to color patterns. Some species of nonvenomous snakes are acoustic mimics of rattlesnakes. The fox snake (*Elaphe vulpina*) and the pine snake of eastern North America, the bull snake of the Great Plains, and the gopher snake of the Pacific States, all subspecies of *Pituophis melanoleucus*, rapidly vibrate their tails in leafy litter to produce a rattle-like sound.

Another type of mimicry is called *Müllerian*, after the 19th-century German zoologist Fritz Müller. With **Müllerian mimicry**, many unpalatable or venomous species share a similar color pattern. Müllerian mimicry is effective because the predator must only be exposed to one of the species before learning to stay away from all other species with the same warning color patterns. The black-and-yellow striped bodies of social wasps, solitary digger wasps, and caterpillars of the cinnabar moths warn predators that the organism is inedible (**Figure 14.17**). All are unrelated species with a shared color pattern that functions to keep predators away.

Figure 14.17 Example of Müllerian mimicry. The black-and-yellow striped bodies of (a) social wasps (Vespidae), (b) solitary digger wasps (Sphecidae), and (c) caterpillars of the cinnabar moths (*Callimorpha jacobaeae*) warn predators that these organisms are inedible. All are unrelated species with a shared color pattern that functions to keep predators away.

(a)

(b)

(c)

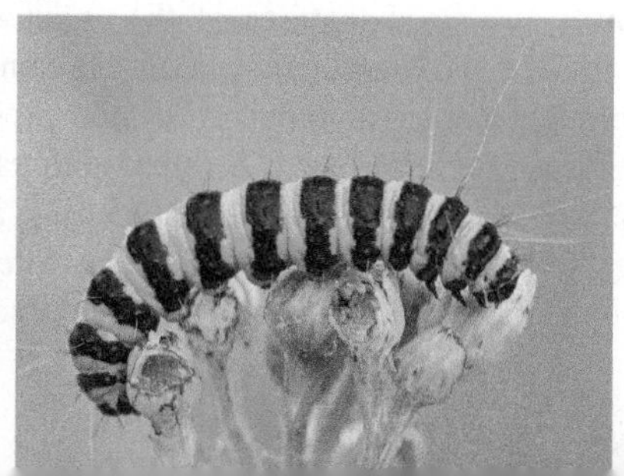

Some animals employ **protective armor** for defense. Clams, armadillos, turtles, and many beetles all withdraw into their armor coats or shells when danger approaches. Porcupines, echidnas, and hedgehogs possess quills (modified hairs) that discourage predators.

Still other animals use **behavioral defenses**, which include a wide range of behaviors by prey species aimed at avoiding detection, fleeing, and warning others of the presence of predators. Animals may change their foraging behavior in response to the presence of predators, as in the example of the willow and crested tits (see Section 14.8). Some species give an alarm call when a predator is sighted. Because high-pitched alarm calls are not species specific, they are recognized by a wide range of nearby animals. Alarm calls often bring in numbers of potential prey that mob the predator. Other behavioral defenses include distraction displays, which are most common among birds. These defenses direct the predator's attention away from the nest or young.

For some prey, living in groups is the simplest form of defense. Predators are less likely to attack a concentrated group of individuals. By maintaining tight, cohesive groups, prey make it difficult for any predator to obtain a victim (**Figure 14.18**). Sudden, explosive group flight can confuse a predator, leaving it unable to decide which individual to follow.

A subtler form of defense is the timing of reproduction so that most of the offspring are produced in a short period. Prey are thus so abundant that the predator can take only a fraction of them, allowing a percentage of the young to escape and grow to a less-vulnerable size. This phenomenon is known as **predator satiation**. Periodic cicadas (*Magicicada* spp.) emerge as adults once every 13 years in the southern portion of their range in North America and once every 17 years in the northern portion of their range, living the remainder of the period as nymphs underground. Though these cicadas emerge only once every 13 or 17 years, a local population emerges somewhere within their range virtually every year. When emergence occurs, the local density of cicadas can number in the millions of individuals per hectare. Ecologist Kathy Williams of San Diego State University and her colleagues tested the effectiveness of predator satiation during the emergence of periodic cicadas in northwest Arkansas. Williams found that the first cicadas emerging in early May were eaten by birds, but avian predators quickly became satiated. Birds consumed 15–40 percent of the cicada population at low cicada densities but only a small proportion as cicada densities increased (**Figure 14.19**). Williams's results demonstrated that, indeed, the synchronized, explosive emergences of periodic cicadas are an example of predator satiation.

The predator defenses just discussed fall into two broad classes: permanent and induced. Permanent, or **constitutive defenses**, are fixed features of the organism, such as object resemblance and warning coloration. In contrast, defenses that are brought about, or induced, by the presence or action of predators are referred to as **induced defenses**. Behavioral defenses are an example of induced defenses, as are chemical defenses such as alarm pheromones that, when detected, induce flight reactions. Induced defenses can also include shifts in physiology or morphology, representing a form of phenotypic plasticity (see this chapter, **Field Studies: Rick A. Relyea**).

Figure 14.18 Musk ox (*Ovibos moschatus*) form a circle, each facing outward, to present a combined defense when threatened by predators.

14.11 Predators Have Evolved Efficient Hunting Tactics

As prey have evolved ways of avoiding predators, predators have evolved better ways of hunting. Predators use three general methods of hunting: ambush, stalking, and pursuit. Ambush hunting means lying in wait for prey to come along. This method is typical of some frogs, alligators, crocodiles, lizards, and certain insects. Although ambush hunting has a low frequency of success, it requires minimal energy. Stalking, typical of herons and some cats, is a deliberate form of hunting with a quick attack. The predator's search time may be great, but pursuit time is minimal. Pursuit hunting, typical of many hawks, lions, wolves, and insectivorous bats, involves minimal search time because the predator usually knows the location of

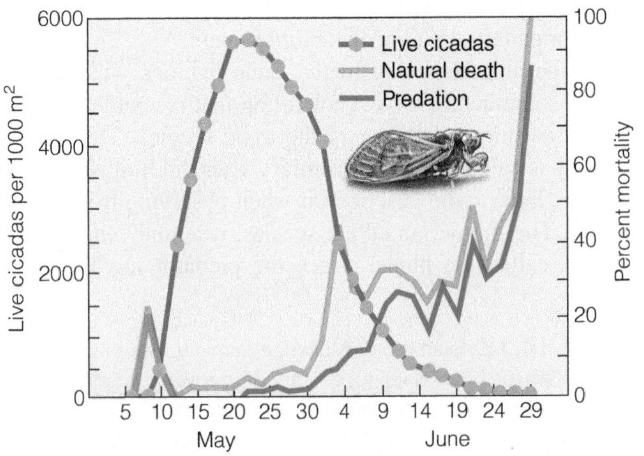

Figure 14.19 Estimated daily population density of periodic cicadas (*Magicicada*) on a study site in Arkansas (left *y*-axis) and estimated daily mortality resulting from bird predation and natural causes (right *y*-axis). Maximum cicada density occurred around May 24, and maximum predation occurred around June 10. At the height of predation, most of the cicadas had already emerged and escaped bird predation.

(Adapted from Williams et al. 1993.)

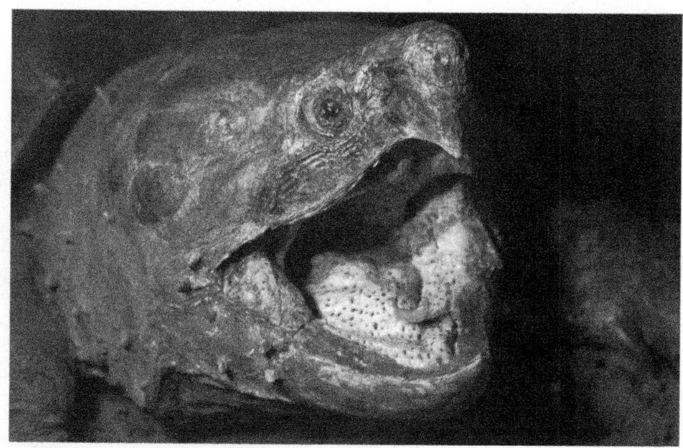

Figure 14.20 The alligator snapping turtle uses a combination of cryptic coloration and mimicry to avoid detection and attract prey. By lying motionless on the bottom with its mouth wide open, it wiggles its worm-shaped tongue (see bottom of mouth) to attract and ambush potential prey.

the prey, but pursuit time is usually great. Stalkers spend more time and energy encountering prey. Pursuers spend more time capturing and handling prey.

Predators, like their prey, may use cryptic coloration to blend into the background or break up their outlines (**Figure 14.20**). Predators use deception by resembling the prey. Robber flies (*Laphria* spp.) mimic bumblebees, their prey (**Figure 14.21**). The female of certain species of fireflies imitates the mating flashes of other species to attract males of those species, which she promptly kills and eats. Predators may also employ chemical poisons, as do venomous snakes, scorpions, and spiders. They may form a group to attack large prey, as lions and wolves do.

14.12 Herbivores Prey on Autotrophs

Although the term *predator* is typically associated with animals that feed on other animals, herbivory is a form of predation in which animals prey on autotrophs (plants and algae). Herbivory is a special type of predation because herbivores typically do not kill the individuals they feed on. Because the ultimate source of food energy for all heterotrophs is carbon fixed by plants in the process of photosynthesis (see Chapter 6), autotroph–herbivore interactions represent a key feature of all communities.

Figure 14.21 The robber fly (*Laphria* spp.) illustrates aggressive mimicry. It mimics the bumblebee (*Megabombus pennsylvanicus*), on which it preys.

If you measure the amount of biomass actually eaten by herbivores, it may be small—perhaps 6–10 percent of total plant biomass present in a forest community or as much as 30–50 percent in grassland communities (see Chapter 20, Section 20.12). In years of major insect outbreaks, however, or in the presence of an overabundance of large herbivores, consumption is considerably higher (**Figure 14.22**). Consumption, however, is not necessarily the best measure of the impact of herbivory within a community. Grazing on plants can have a subtler impact on both plants and herbivores.

The removal of plant tissue—leaf, bark, stems, roots, and sap—affects a plant's ability to survive, even though the plant may not be killed outright. Loss of foliage and subsequent loss of roots will decrease plant biomass, reduce the vigor of the plant, place it at a competitive disadvantage with surrounding vegetation, and lower its reproductive effort. The effect is especially strong in the juvenile stage, when the plant is most vulnerable and least competitive with surrounding vegetation.

A plant may be able to compensate for the loss of leaves with the increase of photosynthesis in the remaining leaves. However, it may be adversely affected by the loss of nutrients, depending on the age of the tissues removed. Young leaves are dependent structures—importers and consumers of nutrients drawn from reserves in roots and other plant tissues. Grazing herbivores, both vertebrate and invertebrate, often concentrate on younger leaves and shoots because they are lower in structural carbon compounds such as lignins, which are difficult to digest and provide little if any energy (see Section 21.4). By selectively feeding on younger tissues, grazers remove considerable quantities of nutrients from the plant.

Plants respond to defoliation with a flush of new growth that drains nutrients from reserves that otherwise would go

(a)

(b)

Figure 14.22 Examples of the impact of high rates of herbivory. (a) Contrast between heavily grazed grassland in southeast Africa and an adjacent area where large herbivores have been excluded. (b) Intense predation on oaks by gypsy moths in the forests of eastern North America.

FIELD STUDIES *Rick A. Relyea*

Department of Biological Sciences, University of Pittsburgh

Ecologists have long appreciated the influence of predation on natural selection. Predators select prey based on their sizes and shapes, thereby acting as a form of natural selection that alters the range of phenotypes within the population. In doing so, predators alter the genetic composition of the population (gene pool), which determines the range of phenotypes in future generations. Through this process, many of the mechanisms of predator avoidance discussed in Section 14.10 are selected for in prey populations. In recent years, however, ecologists have discovered that predators can have a much broader influence on the characteristics of prey species through nonlethal effects. For example, presence of a predator can change the behavior of prey, causing them to reduce activity (or hide) to avoid being detected. This change in behavior can reduce foraging activity. In turn, changes in the rate of food intake can influence prey growth and development, resulting in shifts in their morphology (size and shape of body). This shift in the phenotype of individual prey, induced by the presence and activity of predators, is termed *induction* and represents a form of phenotypic plasticity (see Section 5.4).

The discovery that predators can influence the characteristics (phenotype) of prey species through natural selection and induction presents a much more complex picture of the role of predation in evolution. Although ecologists are beginning to understand how natural selection and induction function separately, little is known about how these two processes interact to determine the observed range of phenotypes within a prey population. Thanks to the work of ecologist Rick Relyea, however, this picture is becoming much clearer.

Relyea's research is conducted in wading pools that are constructed to serve as experimental ponds. In one series of experiments, Relyea explored the nature of induced changes in behavior and morphology in prey (gray tree frog tadpoles, *Hyla versicolor*) by introducing caged predators (dragonfly larvae, *Anax longipes*) into the experimental ponds (**Figure 1**). The tadpoles can detect waterborne chemicals produced by the predators, allowing Relyea to simulate the threat of predation to induce changes in the tadpoles while preventing actual predation. By comparing the characteristics of tadpoles in control ponds (no predator present) and in ponds with caged predators, he was able to examine the responses induced by the presence of predators.

Results of the experiments reveal that induction by predatory chemical cues altered the tadpoles' behavior. They became less active in the presence of predators (**Figure 2**). Reduced activity makes prey less likely to encounter predators and improves their probability of survival. The predators' presence also induced a shift in the morphology of tadpoles—a form of phenotypic plasticity. Tadpoles raised in the experimental ponds in which predators were present have a greater tail depth and shorter overall body length than do individuals raised in the absence of predators (control ponds; **Figure 3**). Interestingly, previous studies showed that tadpoles with deeper tails and shorter bodies escape dragonfly predators better than tadpoles with the opposite morphology. Therefore, the induced morphological responses that were observed in Relyea's experiments are adaptive; they are a form of phenotypic plasticity that functions to increase the survival of individual tadpoles. To assess the heritability of traits and trait plasticities, Relyea conducted artificial crosses of adults, reared their progeny in predator and no-predator environments, and then quantified tadpole behavior (activity), morphology (body and tail shape), and life history (mass and development). Results of the study found that predator-induced traits were heritable, however, the magnitude of heritability varied across traits and environments. Interestingly, several traits had significant heritability for plasticity, suggesting a potential for selection to act on phenotypic plasticity per se. Relyea's experiments clearly show that predators can induce changes in prey phenotype and that the induced changes are heritable and result from natural selection.

The experiments discussed here focus on only one life stage in the development of the tree frog: the larval (tadpole) stage. But how might these changes in morphology early in development affect traits later in life? As the tadpoles metamorphose into adult frogs, they have drastically different morphologies and occupy different habitats. To answer this question, Relyea conducted an experiment to examine how differences in the morphology of wood frog tadpoles

Figure 1 Dragonfly larvae feeding on tadpole.

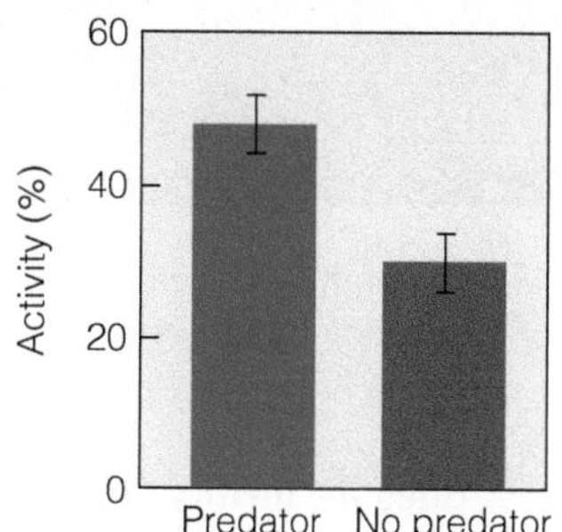

Figure 2 Activity of gray tree frog tadpoles when reared either in the absence (green bar) or presence (blue bar) of caged predators. Go to QUANTIFY it! at www.masteringbiology.com to use data to calculate standard deviation.
(Data from Relyea 2002.)

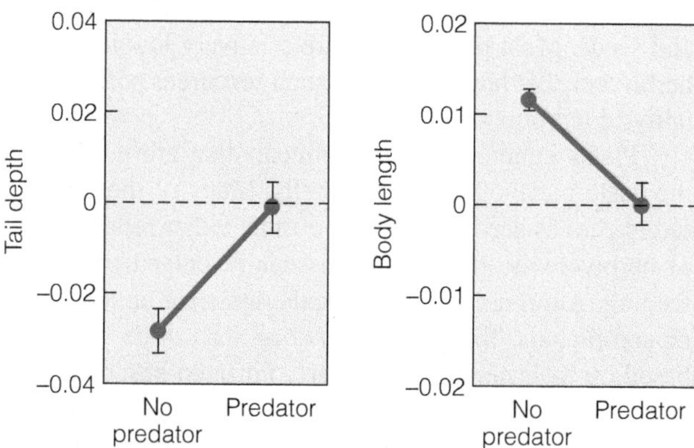

Figure 3 Relative morphology of tree frog tadpoles when reared in either the presence or absence of caged predators. All values are relative to the mean, with negative values below and positive values above the mean value of the respective characteristic (see Figure 11.7 for complete description of statistical analyses).
(Adapted from Relyea 2002.)

(*Rana sylvatica*), induced by the presence of predators, subsequently affected the morphology of the adult frog later in development.

As in previous experiments, tadpoles reared with caged predators developed relatively deeper tail fins and had shorter bodies, lower mass, and longer developmental times than did tadpoles reared without predators. Adult frogs that emerged from the tadpoles exposed to predators (and exhibiting these induced changes during the larval stage) exhibited no differences in mass but developed relatively large hindlimbs and forelimbs and narrower bodies as compared to individuals

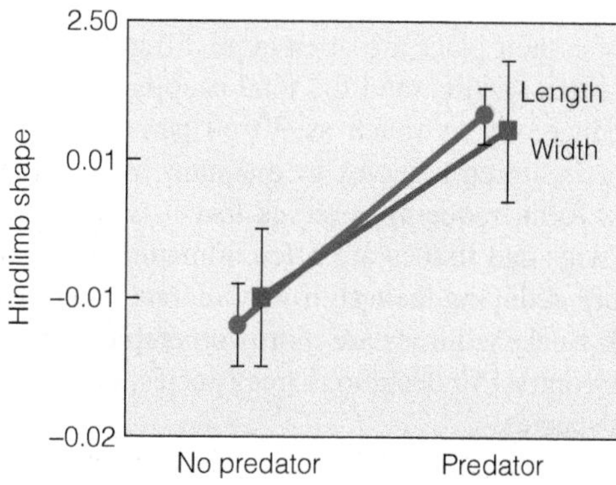

Figure 4 Relative morphology of adult wood frogs that developed from tadpoles reared either in the presence or absence of caged predators. All values are relative to the mean (see Figure 3).
(Adapted from Reylea 2001.)

emerging from environments where predators were absent (**Figure 4**). These results clearly show that predator-induced shifts in traits early in development can subsequently alter traits later in development.

Bibliography

Relyea, R. 2001. "The lasting effects of adaptive plasticity: Predator-induced tadpoles become long-legged frogs." *Ecology* 82:1947–1955.

Relyea, R. 2002a. "The many faces of predation: How induction, selection, and thinning combine to alter prey phenotypes." *Ecology* 83:1953–1964.

Relyea, R. 2002b. "Local population differences in phenotypic plasticity: Predator induced changes in wood frog tadpoles." *Ecological Monographs* 72:77–93.

Relyea, R. 2005. "The heritability of inducible defenses in tadpoles." *Journal of Evolutionary Biology* 18:856–866.

1. Theory predicts that phenotypic plasticity evolves when alternative phenotypes are favored in different environments. How does this prediction relate to the patterns of phenotypic plasticity observed by Relyea? What are the different environments that may function as a selective force to reinforce the benefit of phenotypic plasticity?

2. In the last experiment described, induced shifts in morphology observed during the larval stage (tadpole) influenced the morphology of these individuals at the adult stage. What type of experiment might determine if these induced changes in phenotype influence the fitness of adult frogs?

to growth and reproduction. For example, Anurag Agrawal of the University of Toronto found that herbivory by longhorn beetles (*Tetraopes* spp.) reduced fruit production and mass of milkweed plants (*Asclepias* spp.) by as much as 20–30 percent.

If defoliation of trees is complete (Figure 14.22a), as often happens during an outbreak of gypsy moths (*Lymantria dispar*) or fall cankerworms (*Alsophila pometaria*), leaves that regrow in their place are often quite different in form. The leaves are often smaller, and the total canopy (area of leaves) may be reduced by as much as 30–60 percent. In addition, the plant uses stored reserves to maintain living tissue until new leaves form, reducing reserves that it will require later. Regrown twigs and tissues are often immature at the onset of cold weather, reducing their ability to tolerate winter temperatures. Such weakened trees are more vulnerable to insects and disease. In contrast to deciduous tree species, defoliation kills coniferous species.

Browsing animals such as deer, rabbits, and mice selectively feed on the soft, nutrient-rich growing tips (apical meristems) of woody plants, often killing the plants or changing their growth form. Burrowing insects, like the bark beetles, bore through the bark and construct egg galleries in the phloem–cambium tissues. In addition to phloem damage caused by larval and adult feeding, some bark beetle species carry and introduce a blue stain fungus to a tree that colonizes sapwood and disrupts water flow to the tree crown, hastening tree death.

Some herbivores, such as aphids, do not consume tissue directly but tap plant juices instead, especially in new growth and young leaves. Sap-sucking insects can decrease growth rates and biomass of woody plants by as much as 25 percent.

Grasses have their meristems, the source of new growth, close to the ground. As a result, grazers first eat the older tissue and leave intact the younger tissue with its higher nutrient concentration. Therefore, grasses are generally tolerant of grazing, and up to a point, most benefit from it. The photosynthetic rate of leaves declines with leaf age. Grazing stimulates production by removing older tissue functioning at a lower rate of photosynthesis, increasing the light availability to underlying young leaves. Some grasses can maintain their vigor only under the pressure of grazing, even though defoliation reduces sexual reproduction. Not all grasses, however, tolerate grazing. Species with vulnerable meristems or storage organs can be quickly eradicated under heavy grazing.

14.13 Plants Have Evolved Characteristics that Deter Herbivores

Most plants are sessile; they cannot move. Thus, avoiding predation requires adaptations that discourage being selected by herbivores. The array of characteristics used by plants to deter herbivores includes both structural and other defenses. Structural defenses, such as hairy leaves, thorns, and spines, can discourage feeding (**Figure 14.23**), thereby reducing the amount of tissues removed by herbivores.

Figure 14.23 Thorns of this *Acacia* tree can deter herbivores or reduce levels of defoliation by browsers.

For herbivores, often the quality rather than the quantity of food is the constraint on food supply. Because of the complex digestive process needed to break down plant cellulose and convert plant tissue into animal flesh, high-quality forage rich in nitrogen is necessary (see Chapter 7, Section 7.2). If the nutrient content of the plants is not sufficient, herbivores can starve to death on a full stomach. Low-quality foods are tough, woody, fibrous, and indigestible. High-quality foods are young, soft, and green or they are storage organs such as roots, tubers, and seeds. Most plant tissues are relatively low in quality, and herbivores that have to live on such resources suffer high mortality or reproductive failure.

Plants contain various chemicals that are not involved in the basic metabolism of plant cells. Many of these chemicals, referred to as **secondary compounds**, either reduce the ability of herbivores to digest plant tissues or deter herbivores from feeding. Although these chemicals represent an amazing array of compounds, they can be divided into three major classes based on their chemical structure: nitrogen-based compounds, terpenoids, and phenolics. Nitrogen-based compounds include alkaloids such as morphine, atropine, nicotine, and cyanide. Terpenoids (also called isoprenoids) include a variety of essential oils, latex, and plant resins (many spices and fragrances contain terpenoids). Phenolics are a general class of aromatic compounds (i.e., contain the benzene ring) including the tannins and lignins.

Some secondary compounds are produced by the plant in large quantities and are referred to as **quantitative inhibitors**. For example, tannins and resins may constitute up to 60 percent of the dry weight of a leaf. In the vacuoles of their leaves, oaks and other species contain tannins that bind with proteins and inhibit their digestion by herbivores. Between 5–35 percent of the carbon contained in the leaves of terrestrial plants occurs in the form of lignins—complex, carbon-based molecules that are impossible for herbivores to digest, making the nitrogen and other essential nutrients bound in these compounds unavailable to the herbivore. These types of compounds reduce digestibility and thus potential energy gain from food (see Section 7.2).

Other secondary compounds that function as defenses against herbivory are present in small to minute quantities and are referred to as **qualitative inhibitors**. These compounds are toxic, often causing herbivores to avoid their consumption. This category of compounds includes cyanogenic compounds (cyanide) and alkaloids such as nicotine, caffeine, cocaine, morphine, and mescaline that interfere with specific metabolic pathways of physiological processes. Many of these compounds, such as pyrethrin, have become important sources of pesticides.

Although the qualitative inhibitors function to protect against most herbivores, some specialized herbivores have developed ways of breaching these chemical defenses. Some insects can absorb or metabolically detoxify the chemical substances. They even store the plant poisons to use them in their own defense, as the larvae of monarch butterflies do, or in the production of pheromones (chemical signals). Some beetles and certain caterpillars sever veins in leaves before feeding, stopping the flow of chemical defenses.

Some plant defenses are constitutive, such as structural defenses or quantitative inhibitors (tannins, resins, or lignins) that provide built-in physical or biological barriers against the attacker. Others are active, induced by the attacking herbivore. These induced responses can be local (occur at the site of the attack) or can extend systematically throughout the plant. Often, these two types of defenses are used in combination. For example, when attacked by bark beetles carrying an infectious fungus in their mouthparts, conifer trees release large amounts of resin (constitutive, quantitative defense) from the attack sites that flows out onto the attackers, entombing the beetles. Meanwhile, the tree mobilizes induced defenses against the pathogenic fungus that the intruder has deposited at the wound site.

In another kind of plant–insect interaction, some plants appear to "call for help," attracting the predators of their predators. Parasitic and predatory arthropods often prevent plants from being severely damaged by killing herbivores as they feed on the plants. Recent studies show that a variety of plant species, when injured by herbivores, emit chemical signals to guide natural enemies to the herbivores. It is unlikely that the herbivore-damaged plants initiate the production of chemicals solely to attract predators. The signaling role probably evolved secondarily from plant responses that produce toxins and deterrents against herbivores. For example, in a series of controlled laboratory studies, Ted Turlings and James Tumlinson, researchers at the Agricultural Research Service of the U.S. Department of Agriculture, found that corn seedlings under attack by caterpillars release several volatile terpenoid compounds that function to attract parasitoid wasps (*Cotesia marginiventris*) that then attack the caterpillars. Experiment results showed that the induced emission of volatiles is not limited to the site of damage but occurs throughout the plant. The systematic release of volatiles by injured corn seedlings results in a significant increase in visitation by the parasitoid wasp.

Various hypotheses have been put forward to explain why different types of defenses that help in the avoidance of herbivores have evolved in plants. A feature common to all of these hypotheses is the trade-off between the costs and benefits of defense. The cost of defense in diverting energy and nutrients from other needs must be offset by the benefits of avoiding predation.

14.14 Plants, Herbivores, and Carnivores Interact

In our discussion thus far, we have considered herbivory on plants and carnivory on animals as two separate topics, linked only by the common theme of predation. However, they are linked in another important way. Plants are consumed by herbivores, which in turn are consumed by carnivores. Therefore, we cannot really understand an herbivore–carnivore system without understanding plants and their herbivores, nor can we understand plant–herbivore relations without understanding predator–herbivore relationships. All three—plants, herbivores, and carnivores—are interrelated. Ecologists are beginning to understand these three-way relationships.

A classic case (**Figure 14.24**) is the three-level interaction of plants, the snowshoe hare (*Lepus americanus*), and its predators—lynx (*Felis lynx*), coyote (*Canis latrans*), and horned owl (*Bubo virginianus*). The snowshoe hare inhabits the high-latitude forests of North America. In winter, it feeds on the buds of conifers and the twigs of aspen, alder, and willow, which are termed *browse*. Browse consists mainly of smaller stems and young growth rich in nutrients. The hare–vegetation interaction becomes critical when the amount of essential browse falls below that needed to support the

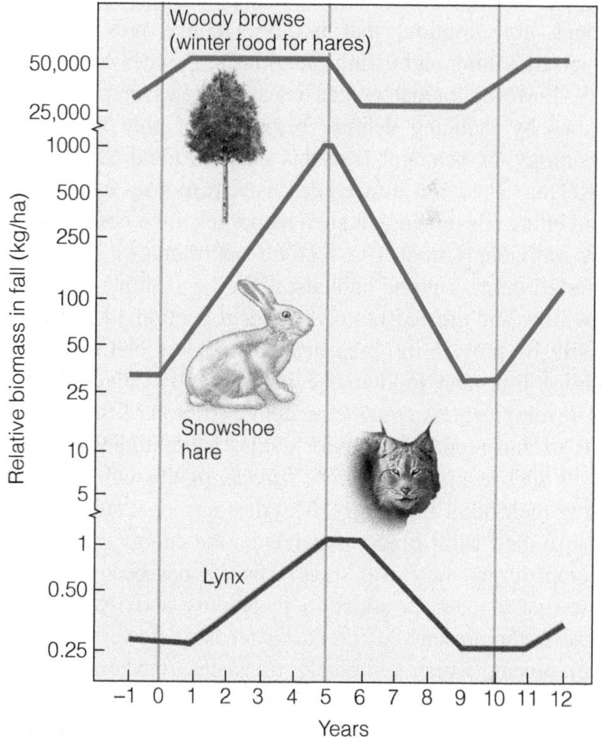

Figure 14.24 The three-way interaction of woody vegetation, snowshoe hare, and lynx. Note the time lag between the cycles of the three populations. Abundance of organisms is based on biomass (kg/ha). Note change in scale of *y*-axis for the different organisms.
(Data from Keith 1983.)

population over winter (approximately 300 grams [g] per individual per day). Excessive browsing when the hare population is high reduces future woody growth, bringing on a food shortage.

The shortage and poor quality of food lead to malnutrition, parasite infections, and chronic stress. Those conditions and low winter temperatures weaken the hares, reducing reproduction and making them extremely vulnerable to predation. Intense predation causes a rapid decline in the number of hares. Now facing their own food shortage, the predators fail to reproduce, and populations decline. Meanwhile, upon being released from the pressures of browsing by hares, plant growth rebounds. As time passes, with the growing abundance of winter food as well as the decline in predatory pressure, the hare population starts to recover and begins another cycle. Thus, an interaction between predators and food supply (plants) produces the hare cycle and, in turn, the hare cycle affects the population dynamics of its predators (see Figure 14.13).

14.15 Predators Influence Prey Dynamics through Lethal and Nonlethal Effects

The ability of predators to suppress prey populations has been well documented. Predators can suppress prey populations through consumption; that is, they reduce prey population growth by killing and eating individuals. Besides causing mortality, however, predators can cause changes in prey characteristics by inducing defense responses in prey morphology, physiology, or behavior (see this chapter, Field Studies: Rick A. Relyea). Predator-induced defensive responses can help prey avoid being consumed, but such responses often come at a cost. Prey individuals may lose feeding opportunities by avoiding preferred but risk-prone habitats, as in the example of foraging by willow and crested tits presented in Section 14.8. Reduced activity by prey in the presence of predators can reduce prey foraging time and food intake, subsequently delaying growth and development. A convincing demonstration of the long-term costs of anti-predator behavior comes from studies of aquatic insects such as mayflies (*Baetis tricaudatus*), which do not feed during their adult life stages. Mayflies are ideal study subjects because their adult fitness depends on the energy reserves they develop during the larval stage. Thus, it has been possible to show that a marked reduction in feeding activity by mayfly larvae in the presence of predators leads to slower growth and development, which ultimately translates into smaller adults that produce fewer eggs (**Figure 14.25**).

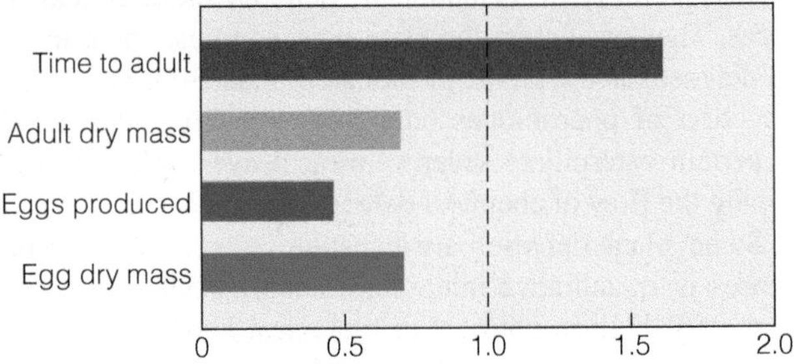

Figure 14.25 Consequences of reduced activity in the mayfly *Baetis tricaudatus*. For each measure (time to adult, dry mass of adults, number of eggs produced, and dry mass of eggs), the ratio (x-axis) of its average value in the presence of predators (fish) relative to that in the absence of predators is given. A ratio smaller than 1.0 represents a reduced average value in the presence of predators, whereas values greater than 1.0 represent a greater average value in the presence of predators.
(Adapted from Scrimgeour and Culp 1994, as illustrated in Lima 1998.)

Interpreting Ecological Data

Q1. *Based on the results of the experimental study presented in Figure 14.25, how does the reduced activity of larval mayflies in the presence of predators influence the time required for larvae to develop into adult mayflies?*

Q2. *How does the presence of predators and associated reduction in activity during the larval stage influence the fitness of adult mayflies? Explain the variables you used to draw your conclusions about adult fitness.*

Predator-induced defensive responses can potentially influence many aspects of prey population regulation and dynamics, given the negative reproductive consequences of anti-predator behavior. Translating behavior decisions to population-level consequences, however, can be difficult. But research by Eric Nelson and colleagues at the University of California–Davis has clearly demonstrated an example of reduction in prey population growth resulting from predator-induced changes in prey behavior. Nelson and colleagues studied the interactions between herbivorous and predatory insects in fields of alfalfa (*Medicago sativa*). Pea aphids (*Acyrthosiphon pisum*) feed by inserting their mouthparts into alfalfa phloem tissue, and they reproduce parthenogenetically (asexual reproduction through the development of an unfertilized ovum) at rates of 4 to 10 offspring per day. A suite of natural enemies attacks the aphids, including damsel bugs (*Nabis* spp.). The aphids respond to the presence of foraging predators by interrupting feeding and walking away from the predator or dropping off the plant. The costs suffered by the aphids because of their defensive behavior may include increased mortality or reduced reproduction.

Damsel bugs feed by piercing aphids with a long proboscis and ingesting the body contents. Damsel bugs, therefore, influence prey in two ways: first by consuming aphids and second by disturbing their feeding behavior. In a series of controlled experiments, Nelson was able to distinguish between the effects of these two influences by surgically removing the mouthparts (proboscises) of some damsel bugs, therefore making them unable to kill and feed on aphids. By exposing aphids to these damsel bugs, the researchers were able to test the predators' ability to suppress aphid population growth through behavioral mechanisms only. Normal predators that were able to consume and disturb the aphids caused the greatest reduction in aphid population growth; however, nonconsumptive predators also strongly reduced aphid population growth (**Figure 14.26**). These field experiments clearly demonstrated that predators reduce population growth partly through predator-induced changes in prey behavior and partly through direct mortality (consuming prey individuals).

An array of specific behavioral, morphological, and physiological adaptations influence the relationship between a predator and its prey, making it difficult to generalize about the influence of predation on prey populations. Nonetheless, many laboratory and field studies offer convincing evidence that predators can significantly alter prey abundance. Whereas the influence of competition on community structure is somewhat obscure, the influence of predation is more demonstrable. Because all heterotrophs derive their energy and nutrients from consuming other organisms, the influence of predation can be more readily noticed throughout a community. As we shall see later in our discussion, the direct influence of predation on the population density of prey species can have the additional impact of influencing the interactions among prey species, particularly competitive relationships (Chapter 17).

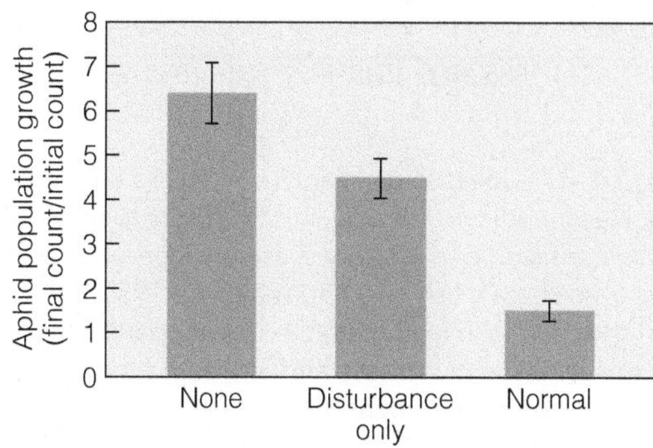

Figure 14.26 Per capita population growth rate (*y*-axis) of pea aphids in field cages containing either no predatory damsel bugs (none), damsel bugs with proboscis removed (disturbance only), and normal damsel bugs (normal). Bars represent the mean population growth for each treatment, where population growth was measured as the ratio of final to initial population size. Vertical lines associated with each bar represent ±1 standard error of the mean.

Go to QUANTIFY it! at www.masteringbiology.com to perform confidence intervals and *t*-tests.

(Adapted from Nelson et al. 2004.)

ECOLOGICAL Issues & Applications

Sustainable Harvest of Natural Populations Requires Being a "Smart Predator"

Although the advent of agriculture some 10000 years ago reduced human dependence on natural populations of plants and animals as a food source, more than 80 percent of the world's commercial catches of fish and shellfish is from the harvest of naturally occurring populations in the oceans (71 percent) and inland freshwaters (10 percent). When humans exploit natural fish populations as a food resource, they are effectively functioning as predators. So what effect is predation by humans having on natural fish populations? Unfortunately, in most cases it is a story of overexploitation and population decline. The cod fishery of the North Atlantic provides a case in point.

For 500 hundred years the waters of the Atlantic Coast from Newfoundland to Massachusetts supported one of the greatest fisheries in the world. The English explorer John Cabot in 1497 discovered and marveled at the abundance of cod off the Newfoundland Coast. Upon returning to Britain, he told of seas "swarming with fish that could be taken not only with nets but with baskets weighted down with stone." Some cod were five to six feet long and weighed up to 200 pounds. Cabot's news created a frenzy of exploitative fishery. Portuguese, Spanish, English, and French fishermen sailed to Newfoundland, and by 1542 the French sailed no fewer than 60 ships, each making two trips a year. In the 1600s, England took control of Newfoundland and its waters and established numerous coastal posts where English merchants salted and dried cod before shipping it to England. So abundant were the fish that the English thought nothing could seriously affect this seemingly inexhaustible resource.

Catches remained rather stable until after World War II, when the demand for fish increased dramatically and led to intensified fishing efforts. Large factory trawlers that could harvest and process the catch at sea replaced smaller fishing vessels. Equipped with sonar and satellite navigation, fishing fleets could locate spawning schools. They could engulf

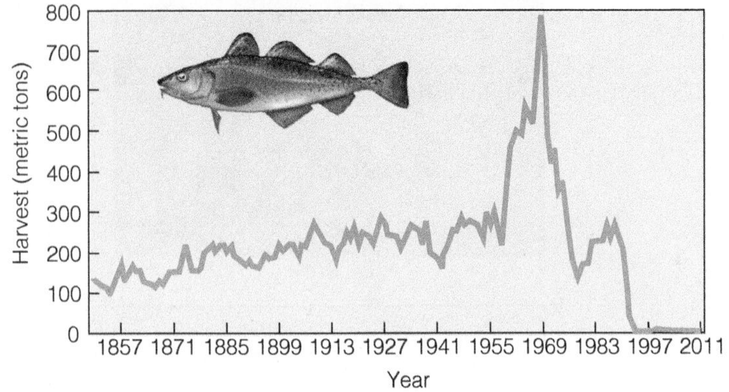

Figure 14.27 Collapse of Atlantic cod stocks off the East Coast of Newfoundland, Canada, in 1992. Prior to the 1950s, the fishery consisted of migratory seasonal fleets and resident near-shore small-scale fisheries. From the late 1950s, offshore bottom trawlers began exploiting the deeper part of the ocean, leading to a large catch increase and a strong decline in the cod populations. The establishment of quotas in the early 1970s, and the establishment by the Canadian government of an exclusive fishing zone in 1977, ultimately failed to stop the decline.

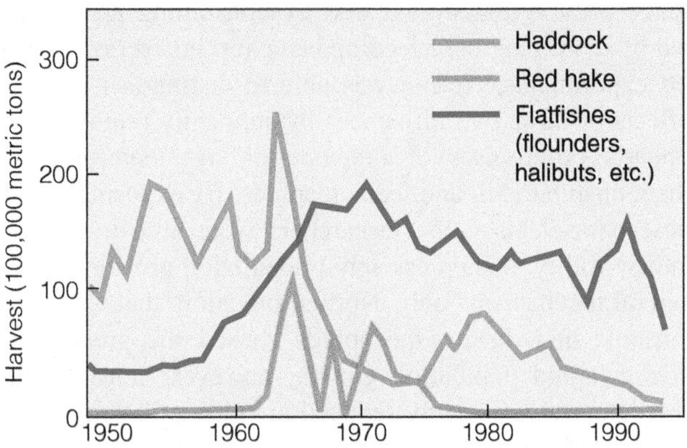

Figure 14.28 Commercial harvests in the Northwest Atlantic of some important fish species for the period 1950–1995. (Source: United Nations Food and Agriculture Organization (FAO), Fishstat-PC (FAO, Rome, 1997).)

schools with huge purse nets and sweep the ocean floor clean of fish and all associated marine life. In the 1950s, annual average catch off the coast of Newfoundland was 300,000 metric tons (MT) of cod, but by the 1960s the catch had almost tripled (**Figure 14.27**). In 15 years from the mid-1950s through the 1960s, 200 factory ships off Newfoundland took as many northern cod as were caught over the prior 250-year span since Cabot's arrival.

The cod fishery could not endure such intense exploitation. By 1978 the catch had declined to less than a quarter of the harvest just a decade before. To protect their commercial interests in the fishery, the Canadian and U.S. governments excluded all foreign fisheries in a zone extending 200 miles. But instead of capitalizing on this opportunity to allow the fish populations to recover, the Canadian government provided the industry with subsidies to build huge factory trawlers. After a brief surge in catches during the 1980s, in 1992 the North Atlantic Canadian cod fishery collapsed (see Figure 14.27).

The story of the North Atlantic cod fishery is an example of the rate of predation exceeding the ability of the prey population to recover; and unlike natural predator–prey systems, there is no negative feedback on the predator population. (Despite the economic consequences of the collapse of the fishery, humans do not exhibit a numerical response to declining fish populations). Unfortunately, the story of the North Atlantic cod fishery is not unique (**Figure 14.28**). Often following the collapse of one fishery, the industry shifts to another species, and the pattern of overexploitation repeats itself. Over the past decades, however, there has been a growing effort toward the active scientific management of fisheries resources to ensure their continuance. The goal of fisheries science is to provide for the long-term sustainable harvesting of fish populations based on the concept of sustainable yield. The amount of resources (fish) harvested per unit of time is called the **yield**. **Sustainable yield** is the yield that allows for populations to recover to their pre-harvest levels. The population of fish will be reduced by a given harvest, but under sustainable management, the yield should not exceed the ability of natural population growth (reproduction) to replace the individuals harvested, allowing the level of harvest (yield) to be sustained through time.

A central concept of sustainable harvest in fisheries management is the logistic model of population growth (Chapter 11, see Section 11.1). Under conditions of the logistic model, growth rate (overall numbers of new organisms produced per year) is low when the population is small (Figure 14.28). It is also low when a population nears its carrying capacity (K) because of density-dependent processes such as competition for limited resources. Intermediate-sized populations have the greatest growth capacity and ability to produce the most harvestable fish per year. The key insight of this model is that fisheries can optimize harvest of a particular species by keeping the population at an intermediate level and harvesting the species at a rate equal to its annual growth rate (**Figure 14.29**). This strategy is called the **maximum sustainable yield**.

In effect, the concept of sustainable yield is an attempt at being a "smart predator." The objective is to maintain the prey population at a density where the production of new individuals just offsets the mortality represented by harvest. The higher the rate of population increase, the higher will be the rate of harvest that produces the maximum sustainable yield. Species characterized by a high rate of population growth often lose much of their production to a high density-independent mortality, influenced by variation in the physical environment such as temperature (see Section 11.13). The management objective for these species is to reduce "waste" by taking all individuals that otherwise would be lost to natural mortality. Such species are difficult to manage, however, because populations can be depleted if annual patterns of reproduction are interrupted as a result of environmental conditions. An example is the Pacific sardine (*Sardinops sagax*). Exploitation of the Pacific sardine population in the 1940s and 1950s shifted the age structure of the population to younger age classes. Before exploitation, reproduction was distributed among the first five age classes (years). In the exploited population, this pattern of

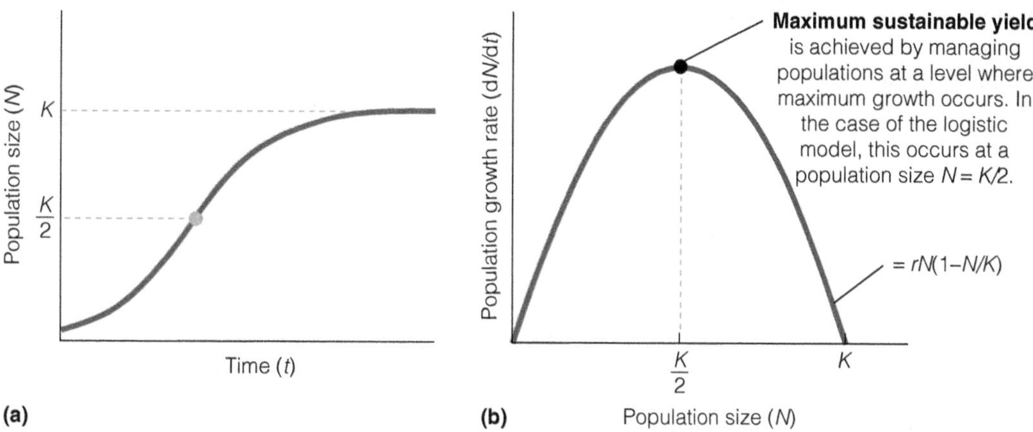

Figure 14.29 Assuming the growth rate of the fish population follows the logistic model presented in Chapter 11 [$dN/dt = rN(1 - N/K)$], (a) in the absence of fishing, the population will grow to carrying capacity, K. (b) The relationship between the rate of population growth, dN/dt, and population size, N, takes the form of a parabola, reaching a maximum value at a population size of $N = K/2$. Maximum sustainable yield occurs when the population growth rate is at its maximum.

reproduction shifted, and close to 80 percent of reproduction was associated with the first two age classes. Two consecutive years of environmentally induced reproductive failure (a result of natural climate variations associated with El Niño–Southern Oscillation [ENSO]; see Chapter 2) caused a population collapse the species never recovered from.

Sustainable yield requires a detailed understanding of the population dynamics of the fish species. Recall that the intrinsic rate of population growth, r, is a function of the age-specific birthrate and mortality rate (Chapter 9). Unfortunately, the usual approach to maximum sustained yield more often than not fails to consider adequately the sex ratio, size and age class structure, size and age-dependent rates of mortality and reproduction, and environmental uncertainties—all data that is difficult to obtain. Adding to the problem is the common-property nature of the resource; because it belongs to no one, it belongs to everyone to use as each of us sees fit.

Perhaps the greatest problem with sustainable harvest models is that they fail to incorporate the most important component of population exploitation: economics. Once commercial exploitation begins, the pressure is on to increase it to maintain the underlying economic investment. Attempts to reduce the rate of exploitation meet strong opposition. People argue that reduction will mean unemployment and industrial bankruptcy—that, in fact, the harvest effort should increase. This argument is shortsighted. An overused resource will fail, and the livelihoods it supports will collapse, because in the long run the resource will be depleted. The presence of abandoned fish processing plants and rusting fishing fleets support this view. With conservative, sustainable exploitation, the resource can be maintained.

SUMMARY

Forms of Predation 14.1

Predation is defined generally as the consumption of all or part of one living organism by another. Forms of predation include carnivory, parasitoidism, cannibalism, and herbivory.

Model of Predation 14.2

A mathematical model that links the two populations through the processes of birth and death can describe interactions between predator and prey. Predation represents a source of mortality for the prey population, whereas the reproduction of the predator population is linked to the consumption of prey.

Population Cycles 14.3

The models of predator–prey interactions predict oscillations of predator and prey populations, with the predator population lagging behind that of the prey population.

Mutual Population Regulation 14.4

The results of the models assume mutual regulation of predator and prey populations. The growth rate of the prey population is influenced by the per capita consumption of prey by the predator population. The relationship between the per capita rate of consumption and the number of prey is referred to as the predator's functional response. This increased consumption of prey results in an increase in predator reproduction referred to as the predator's numerical response.

Functional Response 14.5

There are three types of functional responses. In Type I, the number of prey affected increases linearly. In Type II, the number of prey affected increases at a decreasing rate toward a maximum value. The Type II response is a function of allocation of feeding time by predators between the activities of searching for prey and handling prey (chasing, capturing, killing, consuming, etc.). In Type III, the number of prey consumed increases sigmoidally as the density of prey increases.

Numerical Response 14.6

A numerical response is the increase of predators with an increased food supply. Numerical response may involve an aggregative response: the influx of predators to a food-rich area. More important, a numerical response involves a change in the growth rate of a predator population through changes in fecundity.

Optimal Foraging 14.7

Central to the study of predation is the concept of optimal foraging. This approach to understanding the foraging behavior of animals assumes that natural selection favors "efficient" foragers, that is, individuals that maximize their energy or nutrient intake per unit of effort. Decisions are based on the relative profitability of alternative prey types, defined as the energy gained per unit of handling time. An optimal diet includes the most efficient size of prey for handling and net energy return.

Foraging Behavior and Risk of Predation 14.8

Most predators are also prey to other predatory species and thus face the risk of predation while involved in their routine activities, such as foraging. If predators are about, it may be to the forager's advantage not to visit a most profitable but predator-prone area and to remain in a less profitable but more secure part of the habitat.

Coevolution of Predator and Prey 14.9

Prey species evolve characteristics to avoid being caught by predators. Predators have evolved their own strategies for overcoming these prey defenses. This process represents a coevolution of predator and prey in which each functions as an agent of natural selection on the other.

Predator Defenses 14.10

Chemical defense in animals usually takes the form of distasteful or toxic secretions that repel, warn, or inhibit would-be attackers. Cryptic coloration and behavioral patterns enable prey to escape detection. Warning coloration declares that the prey is distasteful or disagreeable. Some palatable species mimic unpalatable species for protection. Armor and aggressive use of toxins defend some prey. Alarms and distraction displays help others. Another form of defense is predator satiation wherein prey species produce many young at once so that predators can take only a fraction of them. Predator defenses can be classified as permanent or induced.

Predator Evolution 14.11

Predators have evolved different methods of hunting that include ambush, stalking, and pursuit. Predators also employ cryptic coloration for hiding and aggressive mimicry for imitating the appearance of prey.

Herbivory 14.12

Herbivory is a form of predation. The amount of plant or algal biomass actually eaten by herbivores varies between communities. Plants respond to defoliation with a flush of new growth, which draws down nutrient reserves. Such drawdown can weaken plants, especially woody ones, making them more vulnerable to insects and disease. Moderate grazing may stimulate leaf growth in grasses up to a point. By removing older leaves less active in photosynthesis, grazing stimulates the growth of new leaves.

Herbivore Defenses 14.13

Plants affect herbivores by denying them palatable or digestible food or by producing toxic substances that interfere with growth and reproduction. Certain specialized herbivores are able to breach the chemical defenses. They detoxify the secretions, block their flow, or sequester them in their own tissues as a defense against predators. Defenses can be either permanent (constitutive) or induced by damage inflicted by herbivores.

Vegetation–Herbivore–Carnivore Systems 14.14

Plant–herbivore and herbivore–carnivore systems are closely related. An example of a three-level feeding interaction is the cycle of vegetation, hares, and their predators. Malnourished hares fall quickly to predators. Recovery of hares follows recovery of plants and decline in predators.

Lethal and Nonlethal Influences 14.15

Besides influencing prey population directly through mortality, predators can cause changes in prey characteristics by inducing defense responses in prey morphology, physiology, or behavior. Reduced activity by prey in the presence of predators can reduce foraging time and food intake, subsequently delaying growth and development. The net result can be a reduction in the growth rate of the prey population.

Fisheries Management Ecological Issues & Applications

The harvesting of natural fish populations often leads to overexploitation and population decline. Management practices based on sustainable yield attempt to limit harvests to levels at which natural recruitment (reproduction) offsets mortality resulting from fishing activities.

STUDY QUESTIONS

1. The Lotka–Volterra model of predator–prey dynamics suggests mutual control between predator and prey populations that results in the two populations oscillating through time (see Figure 14.2). Why does the predator population lag behind the prey population?
2. What is a functional response in predation? What component of the Lotka–Volterra model of predator–prey dynamics presented in Section 14.2 represents the functional response?
3. Distinguish among Type I, Type II, and Type III functional responses. Which type of functional response is

included in the Lotka–Volterra model of predator–prey dynamics presented in Section 14.2?

4. In Hindu mythology, Brahma created a large and monstrous creature that grew rapidly and devoured everything in its path. In reality, predators are much more selective about what they eat. What factors appear to be important in determining what a predator selects to eat among the array of potential prey?
5. Optimal foraging theory suggests that a predator selects among possible prey based on their relative profitability (energy gained per unit of energy expended). Do you think that predators directly evaluate the profitability of potential prey items before selecting or rejecting them? If not, how might a foraging strategy evolve?
6. What is a numerical response? Which terms in the Lotka–Volterra model of predator–prey dynamics presented in Section 14.2 relate to the numerical response of predators to prey?
7. How would changes in the value of b (the efficiency with which consumption of prey is converted into predator population growth) influence the lag between predator and prey populations (see Figure 14.2c)?
8. How can predators function as agents of natural selection in prey populations?
9. Distinguish between permanent and induced predator defenses. Provide an example of each in both plants and animals.
10. How can prey function as agents of natural selection in predator populations?

FURTHER READINGS

Classic Studies

Gause, G. F., N. P. Smaragdova, and A. A. Witt. 1936. "Further studies of interaction between predators and prey." *Journal of Animal Ecology* 5:1–18.
The original article describing the laboratory predator–prey experiments presented in the chapter.

Holling, C. S. 1959. "Some characteristics of simple types of predation and parasitism." *Canadian Entomologist* 91:385–398.
The classic article in which the concept of functional response is developed.

Current Research

Clark, T. W., A. P. Curlee, S. C. Minta, and P. M. Kareiva (Eds.) 1999. *Carnivores in ecosystems: The Yellowstone experience.* New Haven: Yale University Press.
Covers interactions among canids and smaller predators and the three species systems of vegetation, herbivores, and predators.

Danell, K., R. Bergstrom, P. Duncan, and J. Pastor. (Eds.) 2006. *Large herbivore ecology, ecosystem dynamics, and conservation.* New York: Cambridge University Press.
Excellent synthesis of the impact of large herbivores on plant community structure and long-term plant diversity.

Hay, M. E. 1991. "Marine-terrestrial contrasts in the ecology of plant chemical defenses against herbivores." *Trends in Ecology and Evolution* 6:362–365.
This review article, an excellent overview of plant chemical defenses, contrasts the differences in plant chemical defenses that have evolved in terrestrial and aquatic plant species. It also considers how adaptations to the abiotic environment constrain the adaptation of plant defenses.

Krebs, C. J., R. Boonstra, S. Boutin, and A. R. E. Sinclair. 2001. "What drives the 10-year cycle of snowshoe hares?" *Bioscience* 51:25–35.
An excellent, easy-to-read overview of factors involved in the predator–prey cycle discussed in Section 14.3.

Pauly, D., V. Christensen, S. Guenette, T. J. Pitcher, U. R. Sumaila, C. J. Walters, R. Watson, and D. Zeller. 2002. "Towards sustainability in world fisheries." *Nature* 418: 689–695.

Tollrian, R., and C. D. Harvell. 1999. *The ecology and evolution of inducible defenses.* Princeton, NJ: Princeton University Press.
An extensive, excellent review of the defenses exhibited by prey species (both plant and animal) that are induced by the presence and activity of predators. The text provides a multitude of illustrated examples from recent research in the growing area of coevolution between predator and prey.

Williams, K., K. G. Smith, and F. M. Steven. 1993. "Emergence of the 13-year periodic cicada: Phenology, mortality, and predator satiation." *Ecology* 74:1143–1152.
A classic example of the consequences of predator satiation on the functional response of predators.

CHAPTER 15 Parasitism and Mutualism

Corn (*Zea mays*) infected with corn smut (*Ustilago maydis*).

CHAPTER GUIDE

IN CHAPTER 12 WE INTRODUCED THE CONCEPT of coevolution, the process in which two species undergo reciprocal evolutionary change through natural selection. Competition between two species can lead to character displacement and resource partitioning (Chapter 13, Section 13.11). Prey have evolved means of defense against predators, and predators have evolved ways to breach those defenses (Chapter 14, Sections 14.10 and 14.11) in an evolutionary "game" of adaptation and counteradaptation.

The process of coevolution is even more evident in the interactions between parasites and their hosts. The parasite lives on or in the host organism for some period of its life, in a relationship referred to as symbiosis (Greek *sym*, "together," and *bios*, "life"). **Symbiosis**, as defined by the eminent evolutionary biologist Lynn Margulis, is the "intimate and protracted association between two or more organisms of different species." This definition does not specify whether the result of the association between the species is positive, negative, or benign. It therefore includes a wide variety of interactions in which the fate of individuals of one species depends on their association with individuals of another.

Some symbiotic relationships benefit both species involved, as in the case of mutualism. In other symbiotic relationships, however, one species benefits at the expense of the other. Such is the case of the parasitic relationship, in which the host organism is the parasite's habitat as well as its source of nourishment. For the parasite, this is an obligatory relationship; the parasite requires the host organism for its survival and reproduction. To minimize the negative impact of parasites, host species have evolved a variety of defense mechanisms.

15.1 Parasites Draw Resources from Host Organisms

Parasitism is a type of symbiotic relationship between organisms of different species. One species—the parasite—benefits from a prolonged, close association with the other species—the host—which is harmed. Parasites increase their fitness by exploiting host organisms for food, habitat, and dispersal. Although they draw nourishment from the tissues of the host organism, parasites typically do not kill their hosts as predators do. However, the host may die from secondary infection or suffer reduced fitness as a result of stunted growth, emaciation, modification of behavior, or sterility. In general, parasites are much smaller than their hosts, are highly specialized for their mode of life, and reproduce more quickly and in greater numbers than their hosts.

The definition of parasitism just presented may appear unambiguous. But as with predation the term *parasitism* is often used in a more general sense to describe a much broader range of interactions (see Section 14.1). Interactions between species frequently satisfy some, but not all, parts of this definition because in many cases it is hard to demonstrate that the host is harmed. In other cases, there may be no apparent specialization by the parasite or the interaction between the organisms may be short-lived. For example, because of the episodic nature of their feeding habits, mosquitoes and hematophagic (blood-feeding) bats are typically not considered parasitic. *Parasitism* can also be used to describe a form of feeding in which one animal appropriates food gathered by another (the host), which is a behavior termed *cleptoparasitism* (literally meaning "parasitism by theft"). An example is the brood parasitism practiced by many species of cuckoo (Cuculidae). Many cuckoos use other bird species as "babysitters"; they deposit their eggs in the nest of the host species, which raise the cuckoo young as one of their own (see Chapter 12 opening photograph). In the following discussion, we use the narrower definition of *parasite* as given in the previous paragraph, which includes a wide range of organisms—viruses, bacteria, protists, fungi, plants, and an array of invertebrates, among them arthropods. A heavy load of parasites is termed an **infection**, and the outcome of an infection is a **disease**.

Parasites are distinguished by size. Ecologically, parasites may be classified as microparasites and macroparasites. **Microparasites** include viruses, bacteria, and protists. They are characterized by small size and a short generation time. They develop and multiply rapidly within the host and are the class of parasites that we typically associate with the term *disease*. The infection generally lasts a short time relative to the host's expected life span. Transmission from host to host is most often direct, although other species may serve as carriers.

Macroparasites are relatively large. Examples include flatworms, acanthocephalans, roundworms, flukes, lice, fleas, ticks, fungi, rusts, and smuts. Macroparasites have a comparatively long generation time and typically do not complete an entire life cycle in a single host organism. They may spread by direct transmission from host to host or by indirect transmission, involving intermediate hosts and carriers.

Although the term *parasite* is most often associated with heterotrophic organisms such as animals, bacteria, and fungi, more than 4000 species of parasitic plants derive some or all of their sustenance from another plant. Parasitic plants have a modified root—the haustorium—that penetrates the host plant and connects to the vascular tissues (xylem or phloem). Parasitic plants may be classified as holoparasites or hemiparasites based on whether they carry out the process of photosynthesis. **Hemiparasites**, such as most species of mistletoe (**Figure 15.1**), are photosynthetic plants that contain chlorophyll when mature and obtain water, with its dissolved nutrients, by connecting to the host xylem. **Holoparasites**, such as broomrape and dodder (**Figure 15.2**), lack chlorophyll and are thus nonphotosynthetic. These plants function as heterotrophs that rely totally on the host's xylem and phloem for carbon, water, and other essential nutrients.

Parasites are extremely important in interspecific relations. In contrast with the species interactions of competition and predation, however, it was not until the late 1960s that ecologists began to appreciate the role of parasitism in population dynamics and community structure. Parasites have dramatic effects when they are introduced to host populations that have not evolved to possess defenses against them. In such cases, diseases sweep through and decimate the population.

(a) (b)

Figure 15.1 (a) Mistletoe (*Viscum album*) in poplar tree. Mistletoes are hemiparasites. (b) Although capable of photosynthesis, they penetrate the host tree, extracting water and nutrients.

15.2 Hosts Provide Diverse Habitats for Parasites

Hosts are the habitats of parasites, and the diverse arrays of parasites that have evolved exploit every conceivable habitat on and within their hosts. Parasites that live on the host's skin, within the protective cover of feathers and hair, are **ectoparasites**. Others, known as **endoparasites**, live within the host. Some burrow beneath the skin. They live in the bloodstream, heart, brain, digestive tract, liver, spleen, mucosal lining of the stomach, spinal cord, nasal tract, lungs, gonads, bladder, pancreas, eyes, gills of fish, muscle tissue, or other sites. Parasites of insects live on the legs, on the upper and lower body surfaces, and even on the mouthparts.

Parasites of plants also divide up the habitat. Some live on the roots and stems; others penetrate the roots and bark to live in the woody tissue beneath. Some live at the root collar, commonly called a crown, where the plants emerge from the soil. Others live within the leaves, on young leaves, on mature leaves, or on flowers, pollen, or fruits. A major problem for parasites, especially parasites of animals, is gaining access to and escaping from the host. Parasites can enter and exit host animals through various pathways including the mouth, nasal passages, skin, rectum, and urogenital system; they travel to their point of infection through the pulmonary, circulatory, or digestive systems.

Figure 15.2 Squawroot (*Conopholis americana*), a member of the broomrape family, is a holoparasite on the roots of oak.

For parasites, host organisms are like islands that eventually disappear (die). Because the host serves as a habitat enabling their survival and reproduction, parasites must escape from one host and locate another, which is something that they cannot do at will. Endo-macroparasites can escape only during a larval stage of their development, known as the infective stage, when they must make contact with the next host. The process of transmission from one host to another can occur either directly or indirectly and can involve adaptations by parasites to virtually all aspects of feeding, social, and mating behaviors in host species.

15.3 Direct Transmission Can Occur between Host Organisms

Direct transmission occurs when a parasite is transferred from one host to another without the involvement of an intermediate organism. The transmission can occur by direct contact with a carrier, or the parasite can be dispersed from one host to another through the air, water, or other substrate. Microparasites are more often transmitted directly, as in the case of influenza (airborne) and smallpox (direct contact) viruses and the variety of bacterial and viral parasites associated with sexually transmitted diseases.

Many important macroparasites of animals and plants also move from infected to uninfected hosts by direct transmission. Among internal parasites, the roundworms (*Ascaris*) live in the digestive tracts of mammals. Female roundworms lay thousands of eggs in the host's gut that are expelled with the feces, where they are dispersed to the surrounding environment (water, soil, ground vegetation). If they are swallowed by a host of the correct species, the eggs hatch in the host's intestines, and the larvae bore their way into the blood vessels and come to rest in the lungs. From there they ascend to the mouth, usually by causing the host to cough, and are swallowed again to reach the stomach, where they mature and enter the intestines.

The most important debilitating external parasites of birds and mammals are spread by direct contact. They include lice, ticks, fleas, botfly larvae, and mites that cause mange. Many of these parasites lay their eggs directly on the host; but fleas just lay their eggs and their larvae hatch in the host's nests and bedding, and from there they leap onto nearby hosts.

Some parasitic plants also spread by direct transmission; notably those classified as holoparasites, such as members of the broomrape family (Orobanchaceae). Two examples are squawroot (*Conopholis americana*), which parasitizes the roots of oaks (see Figure 15.2), and beechdrops (*Epifagus virginiana*), which parasitizes mostly the roots of beech trees. Seeds of these plants are dispersed locally; upon germination, their roots extend through the soil and attach to the roots of the host plant.

Some fungal parasites of plants spread through root grafts. For example, *Fomes annosus*, an important fungal infection of white pine (*Pinus strobus*), spreads rapidly through pure stands of the tree when roots of one tree grow onto (and become attached to) the roots of a neighbor.

15.4 Transmission between Hosts Can Involve an Intermediate Vector

Some parasites are transmitted between hosts by an intermediate organism, or vector. For example, the black-legged tick (*Ixodes scapularis*) functions as an arthropod vector in the transmission of Lyme disease, which is the major arthropod-borne disease in the United States. Named for its first noted occurrence at Lyme, Connecticut, in 1975, the disease is caused by a bacterial spirochete, *Borrelia burgdorferi*. It lives in the bloodstream of vertebrates, from birds and mice to deer and humans. The spirochete depends on the tick for transmission from one host to another (see this chapter's *Ecological Issues & Applications*).

Malaria parasites infect a wide variety of vertebrate species, including humans. The four species of protists parasites (*Plasmodium*) that cause malaria in humans are transmitted to the bloodstream by the bite of an infected female mosquito of the genus *Anopheles* (**Figure 15.3**; see this chapter, *Ecological Issues & Applications*). Mosquitoes are known to transmit more than 50 percent of the approximately 102 arboviruses (a contraction of "arthropod-borne viruses") that can produce disease in humans, including dengue and yellow fever.

Insect vectors are also involved in the transmission of parasites among plants. European and native elm bark beetles (*Scolytus multistriatus* and *Hylurgopinus rufipes*) carry spores of the fungi *Ophiostoma ulmi* that spreads the devastating Dutch elm disease from tree to tree. Mistletoes (*Phoradendron* spp.) belong to a group of plant parasites known as hemiparasites (see Figure 15.1) that, although photosynthetic, draw water and nutrients from their host plant.

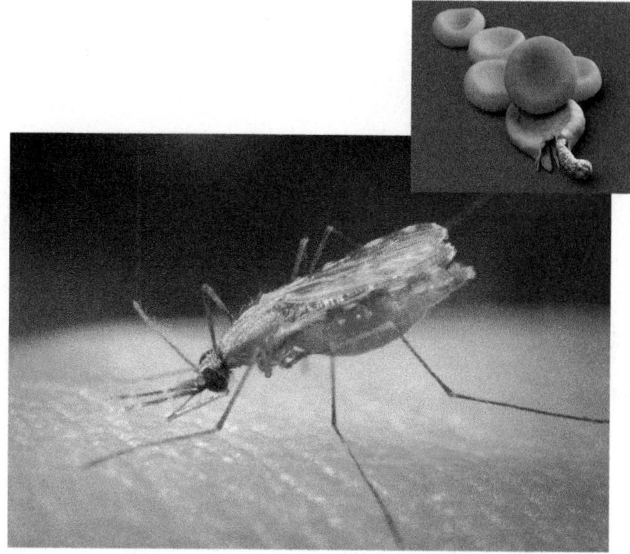

(a)

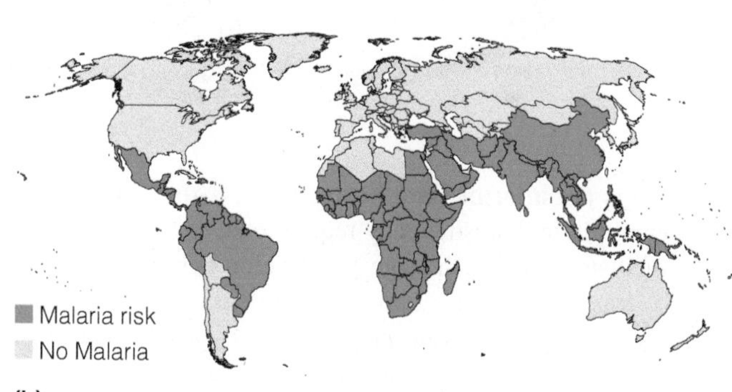

(b)

Figure 15.3 (a) Malaria is a recurring infection produced in humans by protozoan parasites transmitted by the bite of an infected female mosquito of the genus *Anopheles*. Insert shows *Plasmodium falciparum* parasite infecting two blood cells. (b) Today, more than 40 percent of the world's population is at risk, and more than 1 million are killed each year by malaria.

Transmission of mistletoes between host plants is linked to seed dispersal. Birds feed on the mistletoe fruits. The seeds pass through the digestive system unharmed and are deposited on trees where the birds perch and defecate. The sticky seeds attach to limbs and send out rootlets that embrace the limb and enter the sapwood.

15.5 Transmission Can Involve Multiple Hosts and Stages

Previously, we introduced the concept of life cycle—the phases associated with the development of an organism, typically divided into juvenile (or prereproductive), reproductive, and postreproductive phases (Chapter 10). Some species of parasites cannot complete their entire life cycle in a single host species. The host species in which the parasite becomes an adult and reaches maturity is referred to as the **definitive host**.

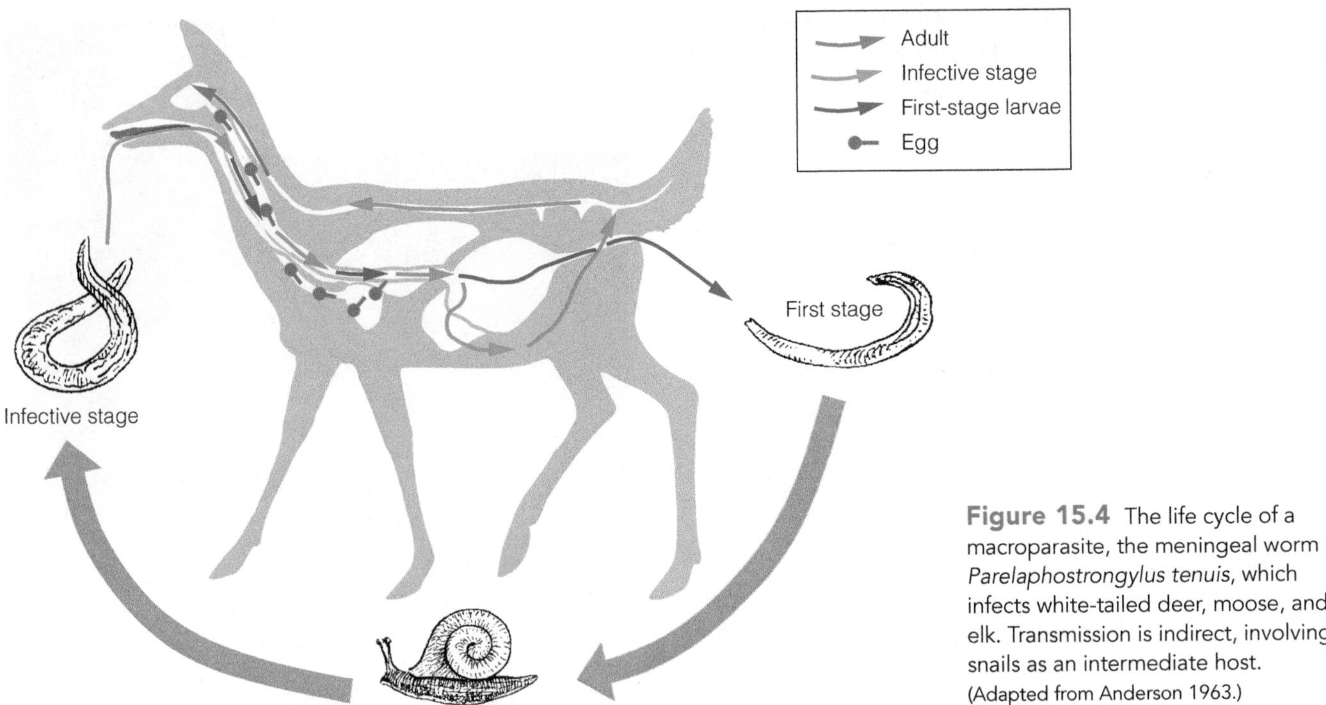

Figure 15.4 The life cycle of a macroparasite, the meningeal worm *Parelaphostrongylus tenuis*, which infects white-tailed deer, moose, and elk. Transmission is indirect, involving snails as an intermediate host. (Adapted from Anderson 1963.)

All others are **intermediate hosts**, which harbor some developmental phase. Parasites may require one, two, or even three intermediate hosts. Each stage can develop only if the parasite can be transmitted to the appropriate intermediate host. Thus, the dynamics of a parasite population are closely tied to the population dynamics, movement patterns, and interactions of the various host species.

Many parasites, both plant and animal, use this form of indirect transmission and spend different stages of the life cycle with different host species. Figure 15.4 shows the life cycle of the meningeal worm (*Parelaphostrongylus tenuis*), which is a parasite of the white-tailed deer in eastern North America. Snails or slugs that live in the grass serve as the intermediate host species for the larval stage of the worm. The deer picks up the infected snail while grazing. In the deer's stomach, the larvae leave the snail, puncture the deer's stomach wall, enter the abdominal membranes, and travel via the spinal cord to reach spaces surrounding the brain. Here, the worms mate and produce eggs. Eggs and larvae pass through the bloodstream to the lungs, where the larvae break into air sacs and are coughed up, swallowed, and passed out with the feces. The snails acquire the larvae as they come into contact with the deer feces on the ground. Once within the snail, the larvae continue to develop to the infective stage.

15.6 Hosts Respond to Parasitic Invasions

Just as the coevolution of predators and prey has resulted in the adaptation of defense mechanisms by prey species, host species likewise exhibit a range of adaptations that minimize the impact of parasites. Some responses are mechanisms that reduce parasitic invasion. Other defense mechanisms aim to combat parasitic infection once it has occurred.

Some defensive mechanisms are behavioral, aimed at avoiding infection. Birds and mammals rid themselves of ectoparasites by grooming. Among birds, the major form of grooming is preening, which involves manipulating plumage with the bill and scratching with the foot. Both activities remove adults and nymphs of lice from the plumage. Deer seek dense, shaded places where they can avoid deerflies, which are common to open areas.

If infection should occur, the first line of defense involves the inflammatory response. The death or destruction (injury) of host cells stimulates the secretion of histamines (chemical alarm signals), which induce increased blood flow to the site and cause inflammation. This reaction brings in white blood cells and associated cells that directly attack the infection. Scabs can form on the skin, reducing points of further entry. Internal reactions can produce hardened cysts in muscle or skin that enclose and isolate the parasite. An example is the cysts that encase the roundworm *Trichinella spiralis* (Nematoda) in the muscles of pigs and bears and that cause trichinosis when ingested by humans in undercooked pork.

Plants respond to bacterial and fungal invasion by forming cysts in the roots and scabs in the fruits and roots, cutting off fungal contact with healthy tissue. Plants react to attacks on leaf, stem, fruit, and seed by gall wasps, bees, and flies by forming abnormal growth structures unique to the particular gall insect (Figure 15.5). Gall formation exposes the larvae of some gall parasites to predation. For example, John Confer and Peter Paicos of Ithaca College (New York) reported that the conspicuous, swollen knobs of the goldenrod ball gall (Figure 15.5d) attract the downy woodpecker (*Picoides pubescens*), which excavates and eats the larva within the gall.

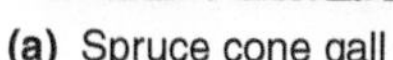

(a) Spruce cone gall

(b) Oak succulent gall

(c) Oak bullet gall

(d) Goldenrod ball gall

Figure 15.5 Galls are a growth response to an alien substance in plant tissues. In this case, the presence of a parasitic egg stimulates a genetic transformation of the host's cells. (a) The spruce cone gall caused by the aphid *Adelges abietis.* (b) The oak succulent gall induced by the gall wasp *Andricus paloak.* (c) The oak bullet gall caused by the gall wasp *Disholcaspis globulus.* (d) The goldenrod ball gall, a stem gall induced by a gallfly, *Eurosta solidaginis.*

The second line of defense is the immune response (or immune system). When a foreign object such as a virus or bacteria—termed an *antigen* (a contraction of "antibody-generating")—enters the bloodstream, it elicits an immune response. White cells called lymphocytes (produced by lymph glands) produce antibodies. The antibodies target the antigens present on the parasite's surface or released into the host and help to counter their effects. These antibodies are energetically expensive to produce. They also are potentially damaging to the host's own tissues. Fortunately, the immune response does not have to kill the parasite to be effective. It only has to reduce the feeding, movements, and reproduction of the parasite to a tolerable level. The immune system is extremely specific, and it has a remarkable "memory." It can "remember" antigens it has encountered in the past and react more quickly and vigorously to them in subsequent exposures.

The immune response, however, can be breached. Some parasites vary their antigens more or less continuously. By doing so, they are able to keep one jump ahead of the host's response. The result is a chronic infection of the parasite in the host. Antibodies specific to an infection normally are composed of proteins. If the animal suffers from poor nutrition and its protein deficiency is severe, normal production of antibodies is inhibited. Depletion of energy reserves breaks down the immune system and allows viruses or other parasites to become pathogenic. The ultimate breakdown in the immune system occurs in humans infected with the human immunodeficiency virus (HIV)—the causal agent of AIDS—which is transmitted sexually, through the use of shared needles, or by infected donor blood. The virus attacks the immune system itself, exposing the host to a range of infections that prove fatal.

15.7 Parasites Can Affect Host Survival and Reproduction

Although host organisms exhibit a wide variety of defense mechanisms to prevent, reduce, or combat parasitic infection, all share the common feature of requiring resources that the host might otherwise have used for some other function. Given that organisms have a limited amount of energy, it is not surprising that parasitic infections function to reduce both growth and reproduction. Joseph Schall of the University of Vermont examined the impact of malaria on the western fence lizard (*Sceloporus occidentalis*) inhabiting California. Clutch size (number of eggs produced) is approximately 15 percent smaller in females infected with malaria compared with non-infected individuals (**Figure 15.6**). Reproduction is reduced because infected females are less able to store fat during the summer, so they have less energy for egg production the following spring. Infected males likewise exhibit numerous reproductive pathologies. Infected males display fewer courtship and territorial behaviors, have altered sexually dimorphic coloration, and have smaller testes.

Parasitic infection can reduce the reproductive success of males by impacting their ability to attract mates. Females of

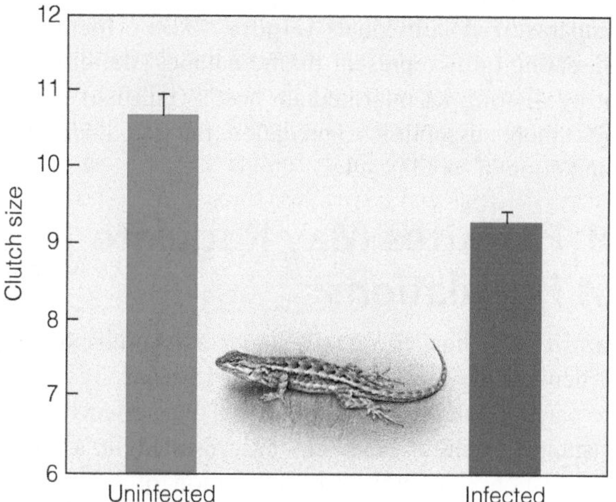

Figure 15.6 Effects of malaria (*Plasmodium mexicanum*) on clutch size of western fence lizards (*Scleroporus occidentalis*). Data average over period of 1978–1982.
(Data from Schall 1983.)

many species choose mates based on the secondary sex characteristics, such as bright and ornate plumage of male birds (see discussion of intrasexual selection in Chapter 10). Full expression of these characteristics can be limited by parasite infection, thus reducing the male's ability to successfully attract a mate. For example, the bright red color of the male zebra finch's beak depends on its level of carotenoid pigments, which are the naturally occurring chemicals that are responsible for the red, yellow, and orange coloration patterns in animals as well as in foods such as carrots. Birds cannot synthesize carotenoids and must obtain them through the diet. Besides being colorful pigments, carotenoids stimulate the production of antibodies and absorb some of the damaging free radicals that arise during the immune response. In a series of laboratory experiments, Jonathan Blount and colleagues from the University of Glasgow (Scotland) found that only those males with the fewest parasites and diseases can devote sufficient carotenoids to producing bright red beaks and therefore succeed in attracting mates and reproducing.

Although most parasites do not kill their host organisms, increased mortality can result from a variety of indirect consequences of infection. One interesting example is when the infection alters the behavior of the host, increasing its susceptibility to predation. Rabbits infected with the bacterial disease tularemia (*Francisella tularensis*), transmitted by the rabbit tick (*Haemaphysalis leporis-palustris*), are sluggish and thus more vulnerable to predation. In another example, ecologists Kevin Lafferty and Kimo Morris of the University of California–Santa Barbara observed that killifish (*Fundulus parvipinnis*; **Figure 15.7a**) parasitized by trematodes (flukes) display abnormal behavior such as surfacing and jerking. In a comparison of parasitized and unparasitized populations, the scientists found that the frequency of conspicuous behaviors displayed by individual fish is related to the intensity of parasitism (**Figure 15.7b**). The abnormal behavior of the infected killifish attracts fish-eating birds. Lafferty and Morris found that heavily parasitized fish were preyed on more frequently than unparasitized individuals (**Figure 15.7c**). Interestingly, the fish-eating birds represent the trematodes' definitive host, so that by altering its intermediate host's (killifish) behavior, making it more susceptible to predation, the trematode ensures the completion of its life cycle.

(a)

Uninfected
Infected
Conspicuous behaviors
Intensity of infection
0 500 1000 1500 2000 2500
0 5 10 15 20 25 30 35

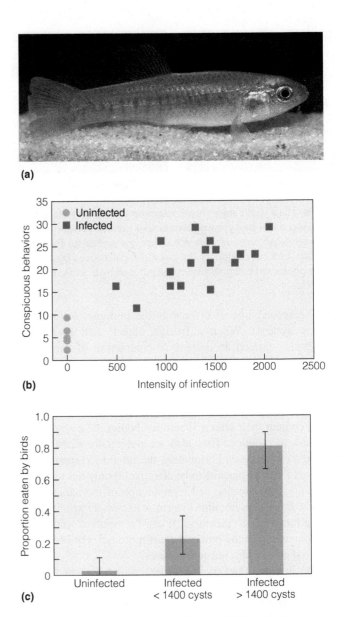

(b)

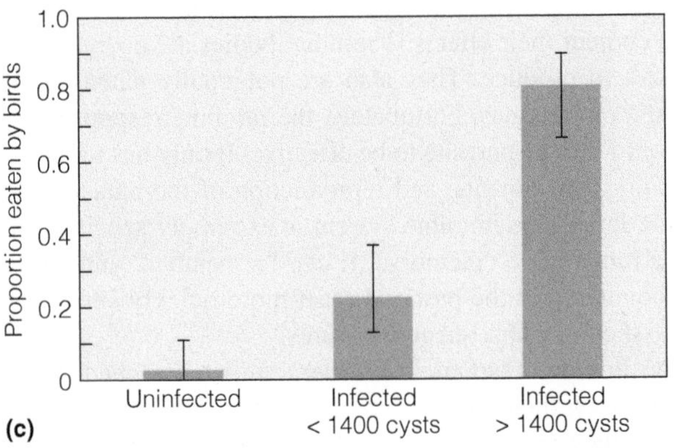

(c)

Figure 15.7 Infection of the California killifish (a) by the trematode parasite causes abnormal behavior that increases the susceptibility of individuals to predation. (b) The frequency of conspicuous behaviors each fish displayed within a 30-minute period (*y*-axis) in relation to the intensity of parasitism (number of cysts per fish brain). In the parasitized population (squares), the number of conspicuous behaviors increased with parasite intensity. All unparasitized fish (circles) had smaller numbers of conspicuous behaviors than parasitized fish. (c) A comparison of the proportion of fish eaten by birds after 20 days (*y*-axis), showing that heavily parasitized fish were preyed on more frequently than unparasitized fish. Vertical lines in bars represent 95 percent confidence intervals. (Adapted from Lafferty and Morris 1996.)

15.8 Parasites May Regulate Host Populations

For parasite and host to coexist under a relationship that is hardly benign, the host needs to resist invasion by eliminating the parasites or at least minimizing their effects. In most circumstances, natural selection has resulted in a level of immune response in which the allocation of metabolic resources by the host species minimizes the cost of parasitism yet does not unduly impair its own growth and reproduction. Conversely, the parasite gains no advantage if it kills its host. A dead host means dead parasites. The conventional wisdom about host–parasite evolution is that virulence is selected against, so that parasites become less harmful to their hosts and thus persist. Does natural selection work this way in parasite–host systems?

Natural selection does not necessarily favor peaceful coexistence of hosts and parasites. To maximize fitness, a parasite should balance the trade-off between virulence and other components of fitness such as transmissibility. Natural selection may yield deadly (high virulence) or benign (low virulence) parasites depending on the requirements for parasite reproduction and transmission. For example, the term *vertical transmission* is used to describe parasites transmitted directly from the mother to the offspring during the perinatal period (the period immediately before or after birth). Typically, parasites that depend on this mode of transmission cannot be as virulent as those transmitted through other forms of direct contact between adult individuals because the recipient (host) must survive until reproductive maturity to pass on the parasite. The host's condition is important to a parasite only as it relates to the parasite's reproduction and transmission. If the host species did not evolve, the parasite might well be able to achieve some optimal balance of host exploitation. But just as with the coevolution of predator and prey, host species do evolve (see discussion of the Red Queen hypothesis in Section 14.9). The result is an "arms race" between parasite and host.

Parasites can have the effect of decreasing reproduction and increasing the probability of host mortality, but few studies have quantified the effect of a parasite on the dynamics of a particular plant or animal population under natural conditions. Parasitism can have a debilitating effect on host populations, a fact that is most evident when parasites invade a population that has not evolved to possess defenses. In such cases, the spread of disease may be virtually density independent, reducing populations, exterminating them locally, or restricting distribution of the host species. The chestnut blight (*Cryphonectria parasitica*), introduced to North America from Europe, nearly exterminated the American chestnut (*Castanea dentata*) and removed it as a major component of the forests of eastern North America. Dutch elm disease, caused by a fungus (*Ophiostoma ulmi*) spread by beetles, has nearly removed the American elm (*Ulmus americana*) from North America and the English elm (*Ulmus glabra*) from Great Britain. Anthracnose (*Discula destructiva*), a fungal disease, is decimating flowering dogwood (*Cornus florida*), an important understory tree in the forests of eastern North America. Rinderpest, a viral disease of domestic cattle, was introduced to East Africa in the late 19th century and subsequently decimated herds of African buffalo (*Syncerus caffer*) and wildebeest (*Connochaetes taurinus*). Avian malaria carried by introduced mosquitoes has eliminated most native Hawaiian birds below 1000 m (the mosquito cannot persist above this altitude).

On the other hand, parasites may function as density-dependent regulators on host populations. Density-dependent regulation of host populations typically occurs with directly transmitted endemic (native) parasites that are maintained in the population by a small reservoir of infected carrier individuals. Outbreaks of these diseases appear to occur when the host population density is high; they tend to reduce host populations sharply, resulting in population cycles of host and parasite similar to those observed for predator and prey (see Section 14.2). Examples are distemper in raccoons and rabies in foxes, both of which are diseases that significantly control their host populations.

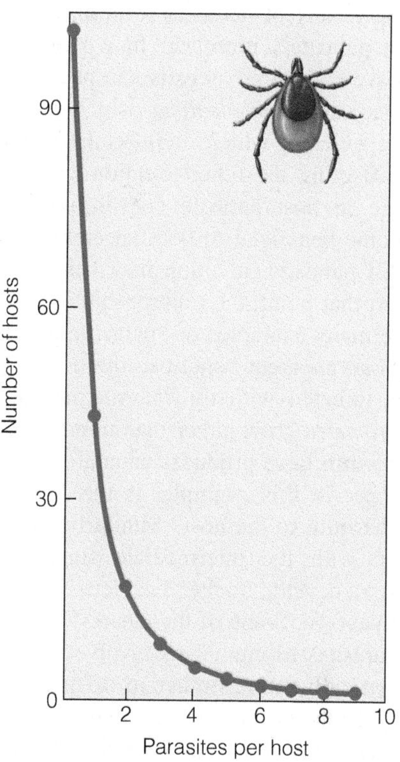

Figure 15.8 Clumped distribution of the shrew tick (*Ixodes trianguliceps*) on a population of the European field mouse (*Apodemus sylvaticus*). Most individuals in the host population carry no ticks. A few individuals carry most of the parasite load. (Adapted from Randolph 1975.)

In other cases, the parasite may function as a selective agent of mortality, infecting only a subset of the population. Distribution of macroparasites, especially those with indirect transmission, is highly clumped. Some individuals in the host population carry a higher load of parasites than others do (**Figure 15.8**). These individuals are most likely to succumb to parasite-induced mortality, suffer reduced reproductive rates, or both. Such deaths often are caused not directly by the macroparasites, but indirectly by secondary infection. In a study of reproduction, survival, and mortality of bighorn sheep (*Ovis canadensis*) in south-central Colorado, Thomas Woodard and colleagues at Colorado State University found that individuals may be infected with up to seven different species of lungworms (Nematoda). The highest rates of infection occur in the spring when lambs are born. Heavy lungworm infections in the lambs bring about a secondary infection—pneumonia—that kills them. The researchers found that such infections can sharply reduce mountain sheep populations by reducing reproductive success.

15.9 Parasitism Can Evolve into a Mutually Beneficial Relationship

Parasites and their hosts live together in a symbiotic relationships in which the parasite derives its benefit (habitat and food resources) at the expense of the host organism. Host species have evolved a variety of defenses to minimize the negative impact of the parasite's presence. In a situation in which adaptations have countered negative impacts, the relationship may be termed *commensalism*, which is a relationship between two species in which one species benefits without significantly affecting the other (Section 12.1, Table 12.1). At some stage in host–parasite coevolution, the relationship may become beneficial to both species. For example, a host tolerant of parasitic infection may begin to exploit the relationship. At that point, the relationship is termed *mutualism*. There are many examples of "parasitic relationships" in which there is an apparent benefit to the host organism. For example, rats infected with the intermediate stages of the tapeworm *Spirometra* grow larger than uninfected rats do because the tapeworm larva produces an analogue of vertebrate growth hormone. In this example, is the increased growth beneficial or harmful to the host? Similarly, many mollusks, when infected with the intermediate stages of digenetic flukes (Digenea), develop thicker, heavier shells that could be deemed an advantage. Some of the clearest examples of evolution from parasites to mutualists involve parasites that are transmitted vertically from mother to offspring (see discussion in Section 15.8). Theory predicts that vertically transmitted parasites are selected to increase host survival and reproduction because maximization of host reproductive success benefits both the parasite and host. This prediction has been supported by studies examining the effects of *Wolbachia*, a common group of bacteria that infect the reproductive tissues of arthropods. Investigations of the effects of *Wolbachia* on host fitness in the wasp *Nasonia vitripennis* have shown that infection increases host fitness and that infected females produce more offspring than do uninfected females. Similar increases in fitness have been reported for natural populations of fruit flies (*Drosophila*).

Mutualism is a relationship between members of two species in which the survival, growth, or reproduction is enhanced for individuals of both species. Evidence, however, suggests that often this interaction is more of a reciprocal exploitation than a cooperative effort between individuals. Many classic examples of mutualistic associations appear to have evolved from species interactions that previously reflected host–parasite or predator–prey interactions. In many cases of apparent mutualism, the benefits of the interaction for one or both of the participating species may be dependent on the environment (see Section 12.4). For example, many tree species have the fungal mycorrhizae associated with their roots (see Section 15.11). The fungi obtain organic nutrients from the plant via the phloem, and in nutrient-poor soil the trees seem to benefit by increased nutrient uptake, particularly phosphate by the fungus. In nutrient-rich soils, however, the fungi appear to be a net cost rather than benefit; this seemingly mutualistic association appears much more like a parasitic invasion by the fungus. Depending on external conditions, the association switches between mutualism and parasitism (see further discussion of example in Section 12.4, Figure 12.9).

15.10 Mutualisms Involve Diverse Species Interactions

Mutualistic relationships involve many diverse interactions that extend beyond simply acquiring essential resources. Thus, it is important to consider the different attributes of mutualistic relationships and how they affect the dynamics of the populations involved. Mutualisms can be characterized by a number of variables: the benefits received, the degree of dependency, the degree of specificity, and the duration of the intimacy.

Mutualism is defined as an interaction between members of two species that serves to benefit both parties involved, and the benefits received can include a wide variety of processes. Benefits may include provision of essential resources such as nutrients or shelter (habitat) and may involve protection from predators, parasites, and herbivores, or they may reduce competition with a third species. Finally, the benefits may involve reproduction, such as dispersal of gametes or zygotes.

Mutualisms also vary in how much the species involved in the mutualistic interaction depend on each other. Obligate mutualists cannot survive or reproduce without the mutualistic interaction, whereas facultative mutualists can. In addition, the degree of specificity of mutualism varies from one interaction to another, ranging from one-to-one, species-specific associations (termed *specialists*) to association with a wide diversity of mutualistic partners (*generalists*). The duration of intimacy in the association also varies among mutualistic interactions. Some mutualists are symbiotic, whereas others are free living (nonsymbiotic). In symbiotic mutualism, individuals coexist and their relationship is more often obligatory; that is, at least one member of the pair becomes totally dependent on the other. Some forms of mutualism are so permanent and obligatory that the distinction between the two interacting organisms becomes blurred. Reef-forming corals of the tropical waters provide an example. These corals secrete an external skeleton composed of calcium carbonate. The individual coral animals, called *polyps*, occupy little cups, or corallites, in the larger skeleton that forms the reef (**Figure 15.9**). These corals have single-celled, symbiotic algae in their tissues called *zooxanthellae*. Although the coral polyps are carnivores, feeding on zooplankton suspended in the surrounding water, they acquire only about 10 percent of their daily energy requirement from zooplankton. They obtain the remaining 90 percent of their energy from carbon produced by the symbiotic algae through photosynthesis. Without the algae, these corals would not be able to survive and flourish in their nutrient-poor environment (see this chapter, **Field Studies: John J. Stachowicz**). In turn, the coral provides the algae with shelter and mineral nutrients, particularly nitrogen in the form of nitrogenous wastes.

Lichens are involved in a symbiotic association in which the fusion of mutualists has made it even more difficult to distinguish the nature of the individual. Lichens (**Figure 15.10**) consist of a fungus and an alga (or in some cases cyanobacterium) combined within a spongy body called a *thallus*. The

(a)

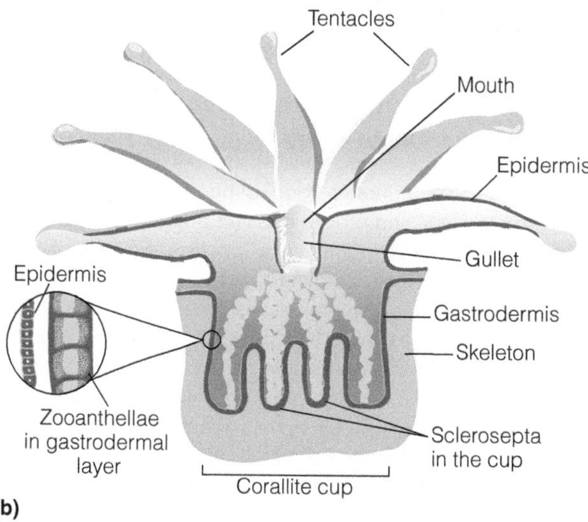

(b)

Figure 15.9 (a) Photograph showing individual polyps of the great star (*Montastrea cavernosa*) coral. (b) Anatomy of a coral polyp, showing the location of the symbiotic zooanthellae.

(a)

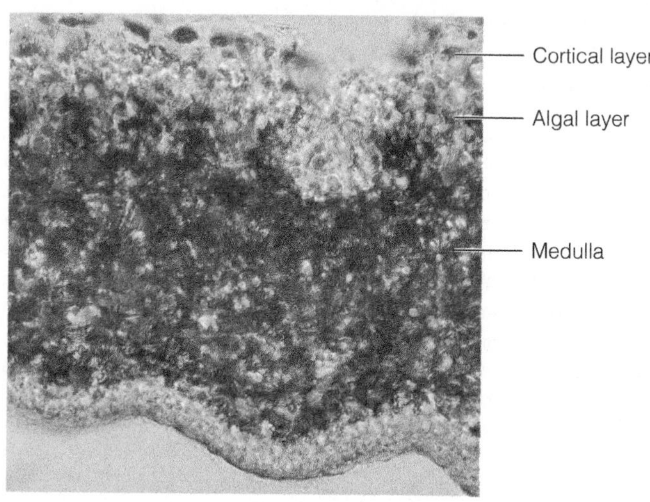

(b)

Figure 15.10 (a) Lichens are composite organisms consisting of a fungus and a photosynthetic organism (algae or cyanobacteria) growing together in a symbiotic relationship. (b) The lichen body, or thallus, typically consists of a cortex or outer layer, an algal layer, and the medulla consisting of the fungal filaments.

alga supplies food to both organisms, and the fungus protects the alga from harmful light intensities, produces a substance that accelerates photosynthesis in the alga, and absorbs and retains water and nutrients for both organisms. There are about 25,000 known species of lichens, each composed of a unique combination of fungus and alga.

In nonsymbiotic mutualism, the two organisms do not physically coexist, yet they depend on each other for some essential function. Although nonsymbiotic mutualisms may be obligatory, most are not. Rather, they are facultative, representing a form of mutual facilitation. Pollination in flowering plants and seed dispersal are examples. These interactions are generally not confined to two species, but rather involve a variety of plants, pollinators, and seed dispersers.

In the following sections, we explore the diversity of mutualistic interactions. The discussion centers on the benefits derived by mutualists: acquisition of energy and nutrients, protection and defense, and reproduction and dispersal.

15.11 Mutualisms Are Involved in the Transfer of Nutrients

The digestive system of herbivores is inhabited by a diverse community of mutualistic organisms that play a crucial role in the digestion of plant materials. The chambers of a ruminant's stomach contain large populations of bacteria and protists that carry out the process of fermentation (see Section 7.2). Inhabitants of the rumen are primarily anaerobic, adapted to this peculiar environment. Ruminants are perhaps the best studied but are not the only example of the role of mutualism in animal nutrition. The stomachs of virtually all herbivorous mammals and some species of birds and lizards rely on the presence of a complex microbial community to digest cellulose in plant tissues.

Mutualistic interactions are also involved in the uptake of nutrients by plants. Nitrogen is an essential constituent of

FIELD STUDIES *John J. Stachowicz*

Section of Evolution and Ecology, Center for Population Biology, University of California–Davis

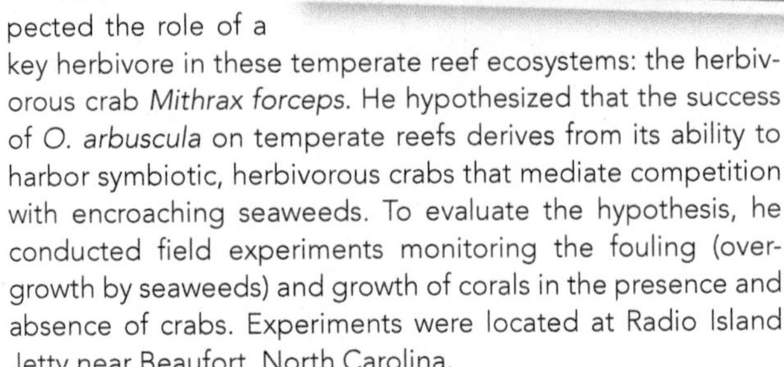

Facilitative, or positive, interactions are encounters between organisms that benefit at least one of the participants and cause harm to neither. Such interactions are considered *mutualisms*, in which both species derive benefit from the interaction. Ecologists have long recognized the existence of mutualistic interactions, but there is still far less research on positive interactions than on competition and predation. Now, however, ecologists are beginning to appreciate the ubiquitous nature of positive interactions and their importance in affecting populations and in the structuring of communities. The research of marine ecologist John Stachowicz has been at the center of this growing appreciation of the importance of facilitation.

Stachowicz works in the shallow-water coastal ecosystems of the southeastern United States. The large colonial corals and calcified algae that occupy the warm subtropical waters of this region provide a habitat for a diverse array of invertebrate and vertebrate species. In well-lit habitats, corals and calcified algae (referred to as coralline algae) grow slowly relative to the fleshy species of seaweed. The persistence of corals appears to be linked to the high abundance of herbivores that suppress the growth of the seaweeds, which grow on and over the coral and coralline algae and eventually cause their death. In contrast, the relative cover of corals is generally low in habitats such as reef flats and seagrass beds, where herbivory is less intense.

Stachowicz hypothesized that mutualism plays an influential role in the distribution of coral species. Although corals are typically associated with the colorful and diverse coral reef ecosystems of the tropical and subtropical coastal waters, many temperate and subarctic habitats support corals, and some tropical species occur where temperatures drop to 10°C or below for certain months of the year. One such species is the coral *Oculina arbuscula*.

O. arbuscula occurs as far north as the coastal waters of North Carolina, forming dense aggregations in poorly lit habitats where seaweeds are rare or absent. In certain areas of the coastal waters, however, *O. arbuscula* does co-occur with seaweeds on natural and artificial reefs. It is the only coral in this region with a structurally complex branching morphology that provides shelter for a species-rich epifauna. More than 300 species of invertebrates are known to live among the branches of *Oculina* colonies.

How can *O. arbuscula* persist in the well-lit, shallow-water systems? In well-lit habitats, corals grow slowly relative to seaweeds, and the persistence of coral reefs appears to be tightly linked to high abundance of herbivores that prevent seaweed from growing on and over the corals. When herbivorous fish or sea urchins are naturally or experimentally removed from tropical reefs, seaweed biomass increases dramatically and corals are smothered. In contrast, on the temperate reefs of North Carolina, herbivorous fish are less abundant than in the tropics, and the standing biomass of seaweed is typically much higher. On these reefs, herbivorous fish and urchins also alter the species composition of the seaweed community by selectively removing their preferred species, but they do not diminish the total seaweed biomass. The dependence of corals on positive interactions with herbivores may thus explain why corals are generally uncommon in temperate latitudes.

Stachowicz suspected the role of a key herbivore in these temperate reef ecosystems: the herbivorous crab *Mithrax forceps*. He hypothesized that the success of *O. arbuscula* on temperate reefs derives from its ability to harbor symbiotic, herbivorous crabs that mediate competition with encroaching seaweeds. To evaluate the hypothesis, he conducted field experiments monitoring the fouling (overgrowth by seaweeds) and growth of corals in the presence and absence of crabs. Experiments were located at Radio Island Jetty near Beaufort, North Carolina.

In these experiments, metal stakes were driven into substrate, and one coral (which had previously been weighed) was fastened to each stake. A single crab was then placed on a subset of the corals, and the remainder was left vacant. At the end of the experiment, all seaweed (and other epiphytic growth) was removed from the corals, dried, and weighed. After removal of the seaweeds, the corals were reweighed to measure growth.

To determine if association with *O. arbuscula* reduced predation on *M. forceps*, Stachowicz tethered crabs both with and without access to coral. He checked each tether after 1 and 24 hours to see if crabs were still present.

The experiment results clearly demonstrated a mutually beneficial association between *O. arbuscula* and *M. forceps*. The coral shelters the crab from predators (**Figure 1**); in turn, the

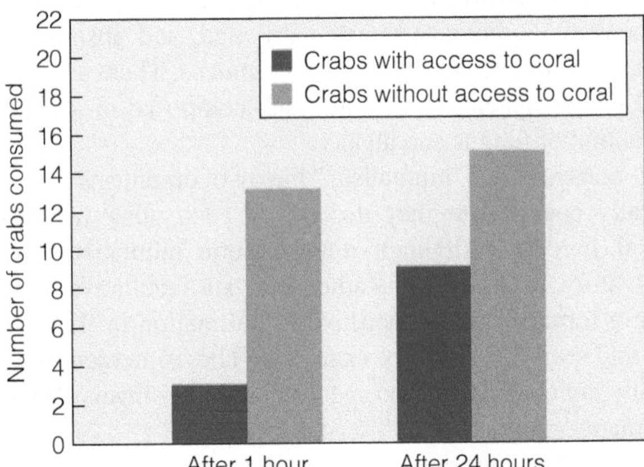

Figure 1 Predation on *Mithrax forceps* with and without access to corals.
(Adapted from Stachowicz 1999.)

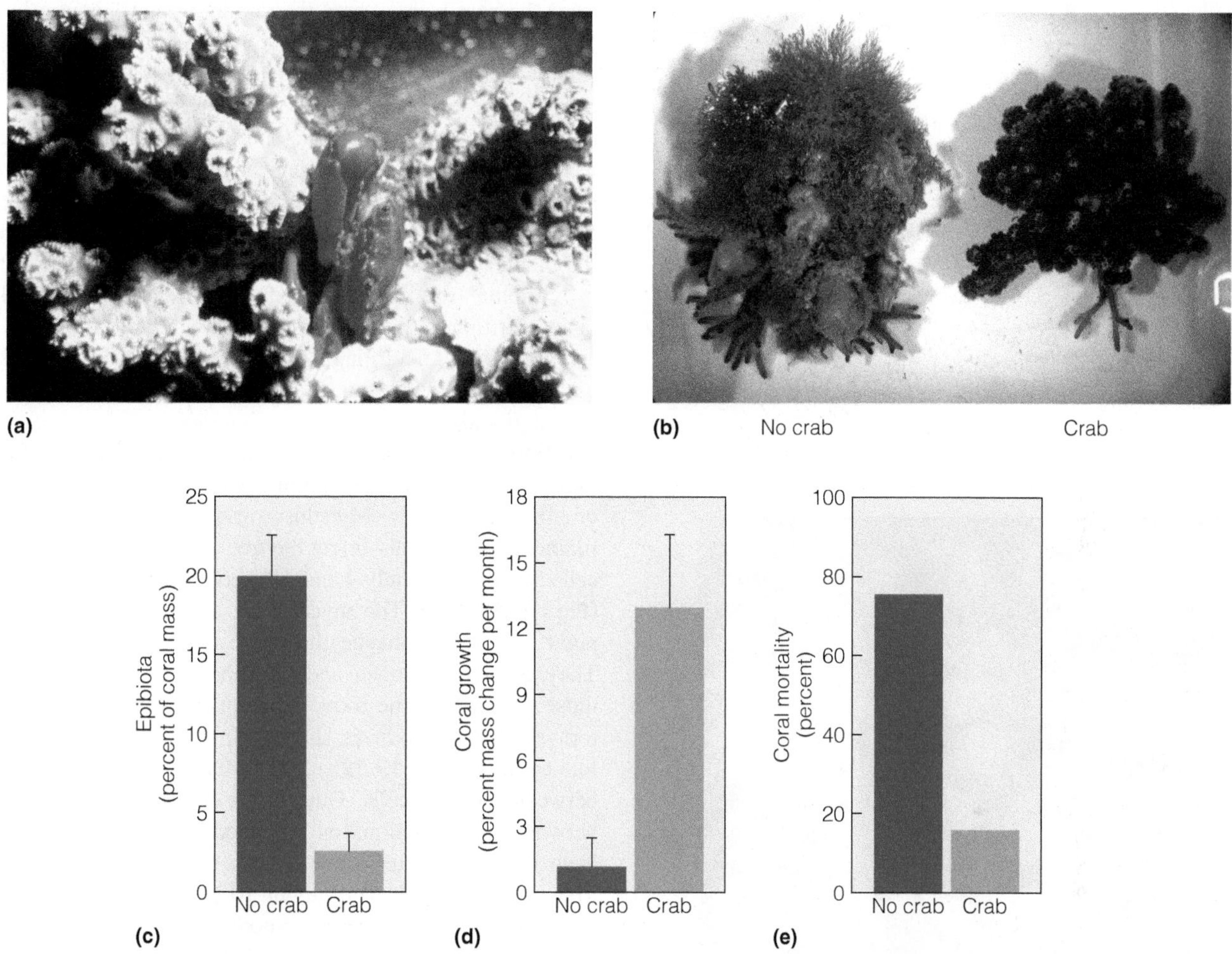

Figure 2 (a) The crab (*Mithrax forceps*) hiding within the branches of the coral (*Oculina arbuscula*). (b) When corals are grown in the field without crabs, they are smothered by a dense cover of seaweed. By reducing overgrowth (c), the crab increases coral growth (d) and survival (e). (Adapted from Stachowicz 2001.)

crab defends the coral from overgrowth by encroaching competitors, thus enhancing coral growth and survivorship (**Figure 2**).

Interactions between *O. arbuscula* and *M. forceps* have population and community implications extending beyond these two species. The crab directly alters the benthic community, enhancing the growth and survival of its host and ensuring the persistence of the diverse community associated with the coral structure.

Bibliography

Bruno, J. F., J. J. Stachowicz, and M. D. Bertness. 2003. "Inclusion of facilitation into ecological theory." *Trends in Ecology & Evolution* 18:119–125.

Stachowicz, J. 2001. "Mutualism, facilitation, and the structure of ecological communities." *BioScience* 51:235–246.

Stachowicz, J., and M. Hay. 1996. "Facultative mutualism between an herbivorous crab and a coralline alga: Advantages of eating noxious seaweeds." *Oecologia* 105:337–387.

Stachowicz, J., and M. Hay. 1999. "Mutualism and coral persistence: The role of herbivore resistance to algal chemical defense." *Ecology* 80:2085–2101.

1. Would you classify the relationship between corals and crabs described here as an example of facilitation or obligate mutualism? Can you determine this based on the information provided? If not, what additional information would you need?

2. In the results presented in Figures 1 and 2, has Stachowicz demonstrated that the mutualistic interactions directly influence the fitness of the two species involved?

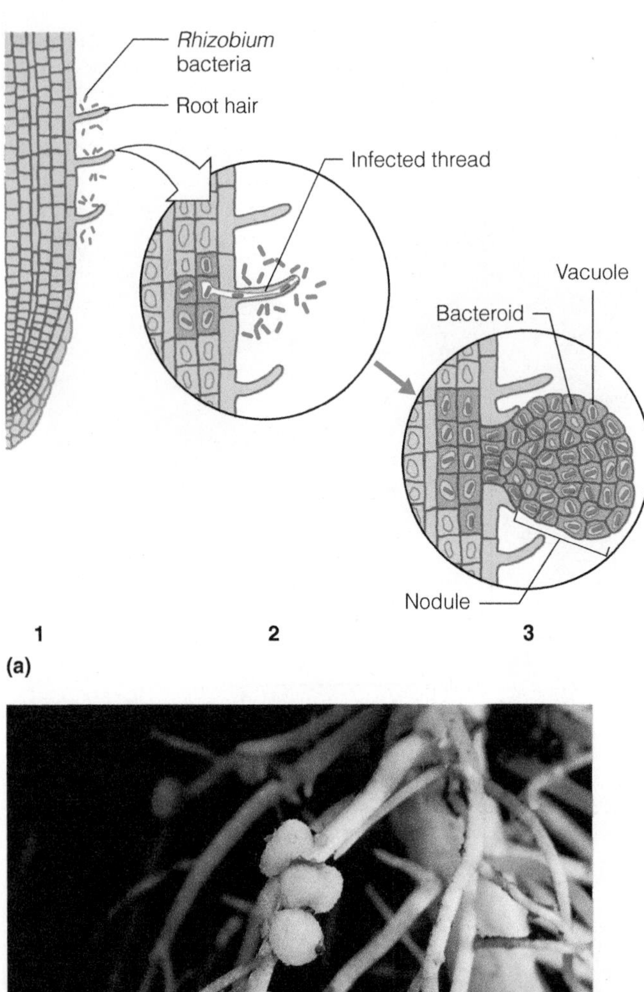

Figure 15.11 Mutualistic association between Rhizobium bacteria and plant roots (a). (1) The process of infection begins when *Rhizobium* bacteria come into contact with the root hairs of the host plant. The presence of the roots in the soil stimulates the bacteria to reproduce, thus larger and larger numbers of bacteria are produced. (2) The bacteria gradually form an infection thread that allows the bacteria to enter root cells of the plant via the root hairs. (3) Bacteria in the root cells gradually grow and develop into structures known as bacteroids. The bacteroids are able to absorb atmospheric N_2 and convert it to nitrogen in the form of ammonia (NH_3). During the infection process, the bacteria stimulate cell division in the root cells. This results in the eventual formation of the structures known as root nodules (b).

protein, a building block of all living material. Although nitrogen is the most abundant constituent of the atmosphere—approximately 79 percent in its gaseous state—it is unavailable to most life. It must first be converted into a chemically usable form. One group of organisms that can use gaseous nitrogen (N_2) is the nitrogen-fixing bacteria of the genus *Rhizobium*. These bacteria (called rhizobia) are widely distributed in the soil, where they can grow and multiply. But in this free-living state, they do not fix nitrogen. Legumes—a group of plant species that include clover, beans, and peas—attract the bacteria through the release of exudates and enzymes from the roots. Rhizobia enter the root hairs, where they multiply and increase in size. This invasion and growth results in swollen, infected root hair cells, which form root nodules (**Figure 15.11**). Once infected, rhizobia within the root cells reduce gaseous nitrogen to ammonia (a process referred to as nitrogen fixation). The bacteria receive carbon and other resources from the host plant; in return, the bacteria contribute fixed nitrogen to the plant, allowing it to function and grow independently of the availability of mineral (inorganic) nitrogen in the soil (see Chapter 6, Section 6.11).

Another example of a symbiotic relationship involving plant nutrition is the relationship between plant roots and mycorrhizal fungi. The fungi assist the plant with the uptake of nutrients and water from the soil. In return, the plant provides the fungi with carbon, a source of energy.

Endomycorrhizae have an extremely broad range of hosts; they have formed associations with more than 70 percent of all plant species. Mycelia—masses of interwoven fungal filaments in the soil—infect the tree roots. They penetrate host cells to form a finely bunched network called an arbuscule (**Figure 15.12a**). The mycelia act as extended roots for the plant but do not change the shape or structure of the roots. They draw in nitrogen and phosphorus at distances beyond those reached by the roots and root hairs. Another form, ectomycorrhizae, produces shortened, thickened roots that look like coral (**Figure 15.12b**). The threads of the fungi penetrate between the root cells. Outside the root, they develop into a network that functions as extended roots. Ectomycorrhizae have a more restricted range of hosts than do endomycorrhizae. They are associated with about 10 percent of plant families, and most of these species are woody.

Together, either ecto- or endomycorrhizae are found associated with the root systems of the vast majority of terrestrial plant species and are especially important in nutrient-poor soils. They aid in the decomposition of dead organic matter and the uptake of water and nutrients, particularly nitrogen and phosphorus, from the soil into the root tissue (see Sections 21.7 and 6.11).

15.12 Some Mutualisms Are Defensive

Other mutualistic associations involve defense of the host organism. A major problem for many livestock producers is the toxic effects of certain grasses, particularly perennial ryegrass and tall fescue. These grasses are infected by symbiotic endophytic fungi that live inside plant tissues. The fungi (Clavicipitaceae and Ascomycetes) produce alkaloid compounds in the tissue of the host grasses. The alkaloids, which impart a bitter taste to the grass, are toxic to grazing mammals, particularly domestic animals, and to a number of insect herbivores. In mammals, the alkaloids constrict small blood vessels in the brain, causing convulsions, tremors, stupor, gangrene of

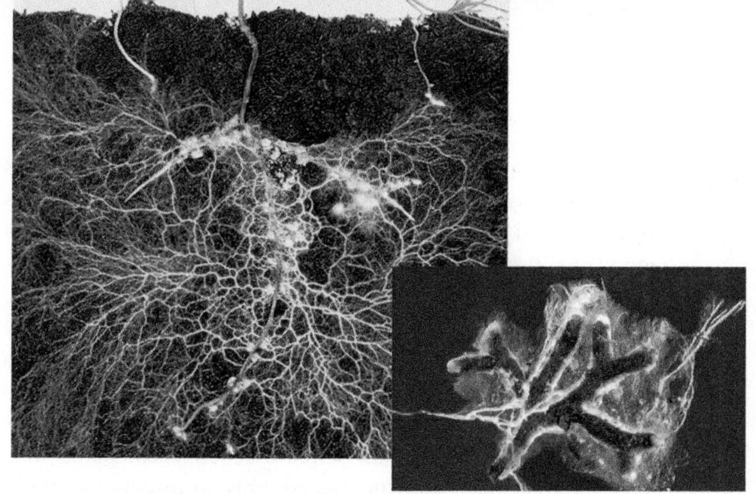

(a)

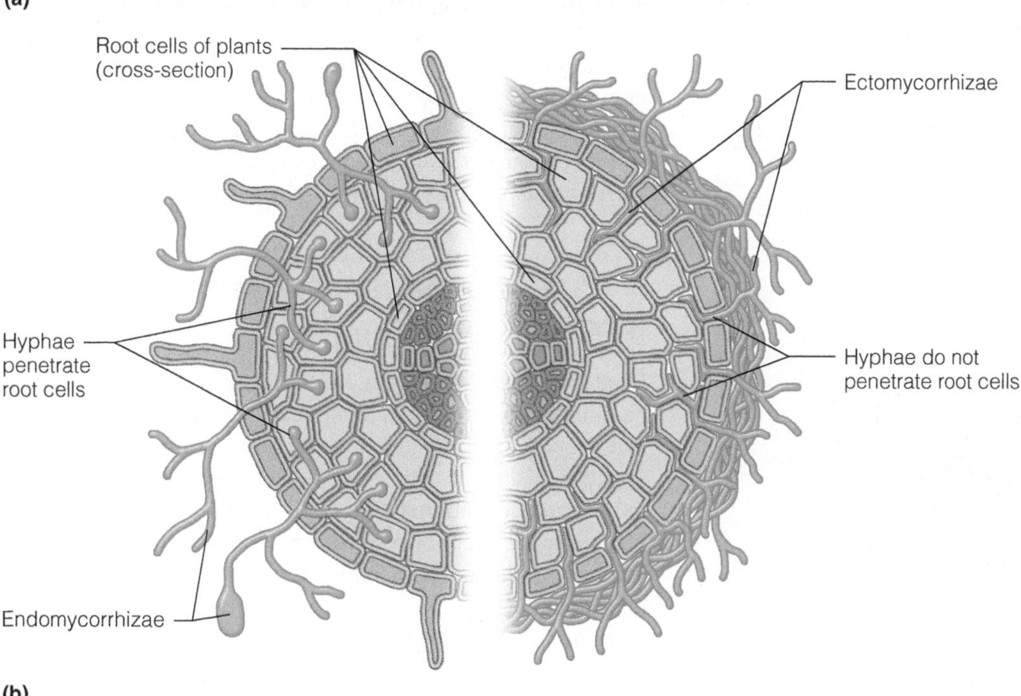

(b)

Figure 15.12 (a) Mycorrhizae on root system of tree seedling. Much of what appears as plant roots are actually fungi (see insert). (b) There are two broad types of mycorrhizal fungi. Endomycorrhizae grow within tree roots, and fungal hyphae enter the cells. Ectomycorrhizae form a mantle of fungi about the tips of rootlets. Hyphae invade the tissues of rootlets between the cells.

the extremities, and death. At the same time, these fungi seem to stimulate plant growth and seed production. This symbiotic relationship suggests a defensive mutualism between plant and fungi. The fungi defend the host plant against grazing. In return, the plant provides food to the fungi in the form of photosynthates (products of photosynthesis).

A group of Central American ant species (*Pseudomyrmex* spp.) that live in the swollen thorns of acacia (*Vachellia* spp.) trees provides another example of defensive mutualism. Besides providing shelter, the plants supply a balanced and almost complete diet for all stages of ant development. In return, the ants protect the plants from herbivores. At the least disturbance, the ants swarm out of their shelters, emitting repulsive odors and attacking the intruder until it is driven away.

Perhaps one of the best-documented examples of a defensive or protective mutualistic association is the cleaning mutualism found in coral reef communities between cleaner shrimp or cleaner fishes and a large number of fish species. Cleaner fishes and shrimp obtain food by cleaning ectoparasites and diseased and dead tissue from the host fish (**Figure 15.13a**). In so doing, they benefit the host fish by removing harmful and unwanted materials.

Cleaning mutualism also occurs in terrestrial environments. The red-billed oxpecker (**Figure 15.13b**) of Africa is a bird that feeds almost exclusively by gleaning ticks and other parasites from the skin of large mammals such as antelope, buffalo, rhinoceros, or giraffe (also domestic cattle). It has always been assumed that these birds significantly reduce the number of ticks on the host animal, yet a recent study by ecologist Paul Weeks of Cambridge University brings into question whether this relationship is indeed mutualistic. In a series of field experiments, Weeks found that changes in adult tick load of cattle were unaffected by excluding the birds. In addition, oxpeckers will peck a vulnerable area (often an ear) and drink blood when parasites are not available.

15.13 Mutualisms Are Often Necessary for Pollination

The goal of cross-pollination is to transfer pollen from the anthers of one plant to the stigma of another plant of the same species (see Figure 12.3). Some plants simply release their pollen in the wind. This method works well and costs little when plants grow in large homogeneous stands, such as grasses and

(a)

(b)

Figure 15.13 Examples of cleaning mutualism. (a) Bluehead wrasse (small fish at eel's mouth) participating in cleaning symbiosis with a moray eel (*Muraenidae*). The cleaner fish obtains food by cleaning ectoparasites from the host fish. (b) The red-billed oxpecker of Africa feeds almost exclusively by gleaning ticks and other parasites from the skin of large mammals such as the impala shown here.

pine trees often do. Wind dispersal can be unreliable, however, when individuals of the same species are scattered individually or in patches across a field or forest. In these circumstances, pollen transfer typically depends on insects, birds, and bats.

Plants entice certain animals by color, fragrances, and odors, dusting them with pollen and then rewarding them with a rich source of food: sugar-rich nectar, protein-rich pollen, and fat-rich oils (Section 12.3, Figure 12.5). Providing such rewards is expensive for plants. Nectar and oils are of no value to the plant except as an attractant for potential pollinators. They represent energy that the plant might otherwise expend in growth.

Nectivores (animals that feed on nectar) visit plants to exploit a source of food. While feeding, the nectivores inadvertently pick up pollen and carry it to the next plant they visit. With few exceptions, the nectivores are typically generalists that feed on many different plant species. Because each species flowers briefly, nectivores depend on a progression of flowering plants through the season.

Many species of plants, such as blackberries, elderberries, cherries, and goldenrods, are generalists themselves. They flower profusely and provide a glut of nectar that attracts a diversity of pollen-carrying insects, from bees and flies to beetles. Other plants are more selective, screening their visitors to ensure some efficiency in pollen transfer. These plants may have long corollas, allowing access only to insects and hummingbirds with long tongues and bills and keeping out small insects that eat nectar but do not carry pollen. Some plants have closed petals that only large bees can pry open. Orchids, whose individuals are scattered widely through their habitats, have evolved a variety of precise mechanisms for pollen transfer and reception. These mechanisms assure that pollen is not lost when the insect visits flowers of other species.

15.14 Mutualisms Are Involved in Seed Dispersal

Plants with seeds too heavy to be dispersed by wind depend on animals to carry them some distance from the parent plant and deposit them in sites favorable for germination and seedling establishment. Some seed-dispersing animals on which the plants depend may be seed predators as well, eating the seeds for their own nutrition. Plants depending on such animals produce a tremendous number of seeds during their reproductive lives. Most of the seeds are consumed, but the sheer number ensures that a few are dispersed, come to rest on a suitable site, and germinate (see concept of predator satiation, Section 14.10).

For example, a mutualistic relationship exists between wingless-seeded pines of western North America (whitebark pine [*Pinus albicaulis*], limber pine [*Pinus flexilis*], southwestern white pine [*Pinus strobiformis*], and piñon pine [*Pinus edulis*]) and several species of jays (Clark's nutcracker [*Nucifraga columbiana*], piñon jay [*Gymnorhinus cyanocephalus*], western scrub jay [*Aphelocoma californica*], and Steller's jay [*Cyanocitta stelleri*]). In fact, there is a close correspondence between the ranges of these pines and jays. The relationship is especially close between Clark's nutcracker and the whitebark pine. Research by ecologist Diana Tomback of the University of Colorado–Denver has revealed that only Clark's nutcracker has the morphology and behavior appropriate to disperse the seeds significant distances away from the parent tree. A bird can carry in excess of 50 seeds in cheek pouches and caches them deep enough in the soil of forest and open fields to reduce their detection and predation by rodents.

Seed dispersal by ants is prevalent among a variety of herbaceous plants that inhabit the deserts of the southwestern United States, the shrublands of Australia, and the deciduous

Figure 15.14 Bleeding hearts, trilliums, and several dozen other plants have appendages on their seeds that contain oils attractive to ants. The ants carry the seeds to their nest, where the elaiosomes are removed and consumed as food, leaving the seeds unharmed.

forests of eastern North America. Such plants, called **myrmecochores**, have an ant-attracting food body on the seed coat called an **elaiosome** (**Figure 15.14**). Appearing as shiny tissue on the seed coat, the elaiosome contains certain chemical compounds essential for the ants. The ants carry seeds to their nests, where they sever the elaiosome and eat it or feed it to their larvae. The ants discard the intact seed within abandoned galleries of the nest. The area around ant nests is richer in nitrogen and phosphorus than the surrounding soil, providing a good substrate for seedlings. Further, by removing seeds far from the parent plant, the ants significantly reduce losses to seed-eating rodents. Plants may enclose their seeds in a nutritious fruit attractive to fruit-eating animals—the frugivores (**Figure 15.15**). Frugivores are not seed predators. They eat only the tissue surrounding the seed and, with some exceptions, do not damage the seed. Most frugivores do not depend exclusively on fruits, which are only seasonally available and deficient in proteins.

To use frugivorous animals as agents of dispersal, plants must attract them at the right time. Cryptic coloration, such as green unripened fruit among green leaves, and unpalatable texture, repellent substances, and hard outer coats discourage consumption of unripe fruit. When seeds mature, fruit-eating animals are attracted by attractive odors, softened texture, increasing sugar and oil content, and "flagging" of fruits with colors.

Most plants have fruits that can be exploited by an array of animal dispersers. Such plants undergo quantity dispersal; they scatter a large number of seeds to increase the chance that various consumers will drop some seeds in a favorable site. Such a strategy is typical of, but not exclusive to, plants of the temperate regions, where fruit-eating birds and mammals rarely specialize in one kind of fruit and do not depend exclusively on fruit for sustenance. The fruits are usually succulent and rich in sugars and organic acids. They contain small seeds with hard seed coats resistant to digestive enzymes, allowing the seeds to pass through the digestive tract unharmed. Such seeds may not germinate unless they have been conditioned or scarified by passage through the digestive tract. Large numbers of small seeds may be dispersed, but few are deposited on suitable sites.

Figure 15.15 The frugivorous cedar waxwing (*Bombycilla cedrorum*) feeds on the red berries of mountain ash (*Sorbus*).

In tropical forests, 50–75 percent of the tree species produce fleshy fruits whose seeds are dispersed by animals. Rarely are these frugivores obligates of the fruits they feed on, although exceptions include many tropical fruit-eating bats.

15.15 Mutualism Can Influence Population Dynamics

Mutualism is easy to appreciate at the individual level. We grasp the interaction between an ectomycorrhizal fungus and its oak or pine host, we count the acorns dispersed by squirrels and jays, and we measure the cost of dispersal to oaks in terms of seeds consumed. Mutualism improves the growth and reproduction of the fungus, the oak, and the seed predators. But what are the consequences at the population and community levels?

Mutualism exists at the population level only if the growth rate of *species 1* increases with the increasing density of *species 2*, and vice versa (see **Quantifying Ecology 15.1**). For symbiotic mutualists where the relationship is obligate, the influence is straightforward. Remove species 1 and the population of species 2 no longer exists. If ectomycorrhizal spores fail to infect the rootlets of young pines, the fungi do not develop. If the young pine invading a nutrient-poor field fails to acquire a mycorrhizal symbiont, it does not grow well, if at all.

Discerning the role of facultative (nonsymbiotic) mutualisms in population dynamics can be more difficult. As discussed in Sections 15.13 and 15.14, mutualistic relationships are common in plant reproduction, where plant species often depend on animal species for pollination, seed dispersal, or germination. Although some relationships between pollinators and certain flowers are so close that loss of one could result in the extinction of the other, in most cases the effects are subtler and require detailed demographic studies to determine the consequences on species fitness.

When the mutualistic interaction is diffuse, involving a number of species—as is often the case with pollination systems (see discussion of pollination networks in Section 12.5) and seed dispersal by frugivores—the influence of specific species–species interactions is difficult to determine. In other situations, the mutualistic relationship between two

QUANTIFYING ECOLOGY 15.1 A Model of Mutualistic Interactions

The simplest model of a mutualistic interaction between two species is similar to the basic Lotka–Volterra model as described in Chapter 13 for two competing species. The crucial difference is that rather than negatively influencing each other's growth rate, the two species have positive interactions. The competition coefficients α and β are replaced by positive interaction coefficients, reflecting the per capita effect of an individual of species 1 on species 2 (α_{12}) and the effect of an individual of species 2 on species 1 (α_{21}).

$$\text{Species 1: } \frac{dN_1}{dt} = r_1N_1\left(\frac{K_1 - N_1 + \alpha_{21}N_2}{K_1}\right)$$

$$\text{Species 2: } \frac{dN_2}{dt} = r_2N_2\left(\frac{K_2 - N_2 + \alpha_{12}N_1}{K_2}\right)$$

All of the terms are analogous to those used in the Lotka–Volterra equations for interspecific competition, except that $\alpha_{21}N_2$ and $\alpha_{12}N_1$ are added to the respective population densities (N_1 and N_2) rather than subtracted.

This model describes a facultative, rather than obligate, interaction because the carrying capacities of the two species are positive, and each species (population) can grow in the absence of the other. In this model, the presence of the mutualist offsets the negative effect of the species' population on the carrying capacity. In effect, the presence of the one species increases the carrying capacity of the other.

To illustrate this simple model, we can define values for the parameters r_1, r_2, K_1, K_2, α_{21}, and α_{12}.

$$r_1 = 3.22,\ K_1 = 1000,\ \alpha_{12} = 0.5$$

$$r_2 = 3.22,\ K_2 = 1000,\ \alpha_{21} = 0.6$$

As with the Lotka–Volterra model for interspecific competition, we can calculate the zero isocline for the two mutualistic species that are represented by the equations presented two paragraphs above. The zero isocline for species 1 is solved by defining the values of N_1 and N_2, where ($K_1 - N_1 + \alpha_{21}N_2$) is equal to zero. As with the competition model, because the equation is a linear function, we can define the line (zero isocline) by solving for only two points. Likewise, we can solve for the species 2 isocline. The resulting isoclines are shown in **Figure 1**.

Note that, unlike the possible outcomes with the competition equations, the zero isoclines extend beyond the carrying capacities of the two species (K_1 and K_2), reflecting that the carrying capacity of each species is effectively increased by the presence of the mutualist (other species, see Figure 14.2). If we use the equations to project the density of the two populations through time (**Figure 2**), each species attains a higher density in the presence of the other species than when they occur alone (in the absence of the mutualist).

1. On the graph displaying the zero isoclines shown in Figure 1, plot the four points listed and indicate the direction of change for the two populations.

   $(N_1, N_2) = 500, 500$
   $(N_1, N_2) = 3500, 3000$
   $(N_1, N_2) = 3000, 1000$
   $(N_1, N_2) = 1000, 3000$

2. What outcome do the isoclines indicate for the interaction between these two species?

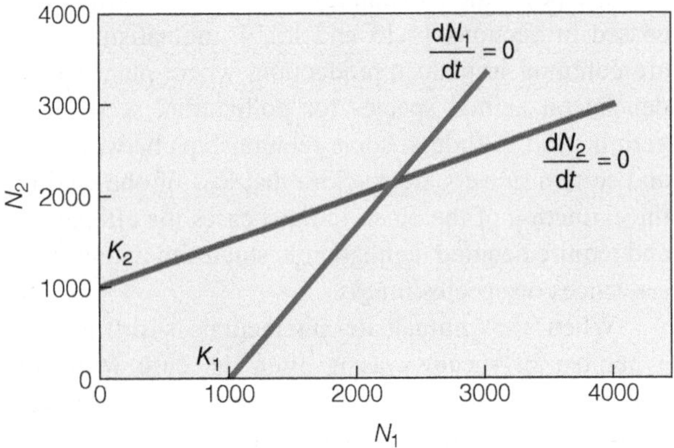

Figure 1 Zero isoclines for species 1 (N_1) and species 2 (N_2) based on the modified Lotka–Volterra model for two mutualistic species presented in text.

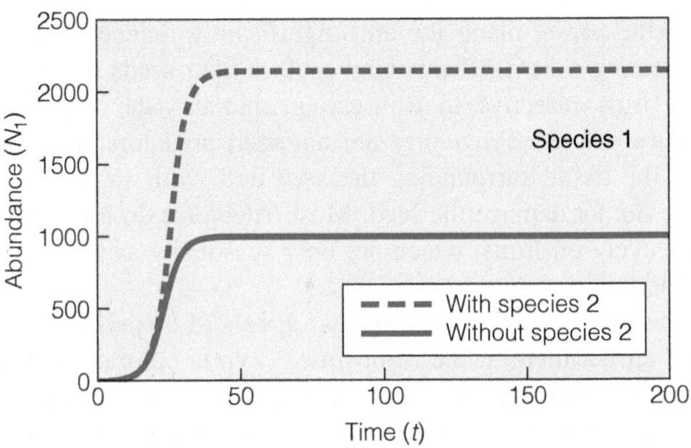

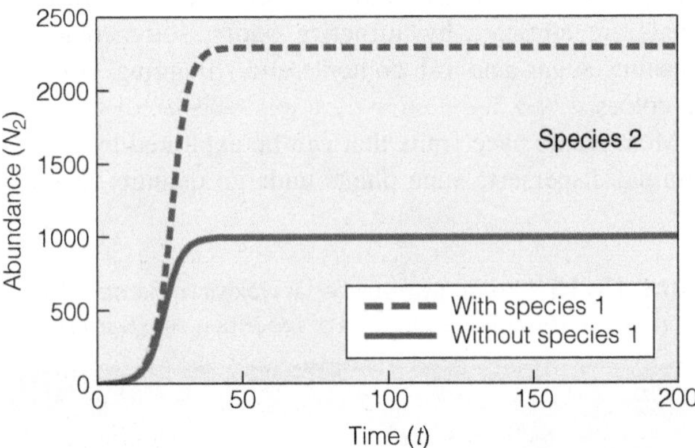

Figure 2 Population trajectories for a pair of facultative mutualists using the equations presented in text. Parameters for the Lotka–Volterra equations are presented in table form in the text. Population projections are shown for simulations of the equations both with and without the presence of the mutualist. (Adapted from Morin 1999.)

species may be mediated or facilitated by a third species, much the same as for vector organisms and intermediate hosts in parasite–host interactions. Mutualistic relationships among conifers, mycorrhizae, and voles in the forests of the Pacific Northwest as described by ecologist Chris Maser of the University of Puget Sound (Washington) and his colleagues are one such example (**Figure 15.16**). To acquire nutrients from the soil, the conifers depend on mycorrhizal fungi associated with the root system. In return, the mycorrhizae depend on the conifers for energy in the form of carbon (see Section 15.10). The mycorrhizae also have a mutualistic relationship with voles that feed on the fungi and disperse the spores, which then infect the root systems of other conifer trees.

Perhaps the greatest limitation in evaluating the role of mutualism in population dynamics is that many—if not most—mutualistic relationships arise from indirect interaction in which the affected species never come into contact. Mutualistic species influence each other's fitness or population growth rate indirectly through a third species or by altering the local environment (habitat modification)—topics we will revisit later (Chapter 17). Mutualism may well be as significant as either competition or predation in its effect on population dynamics and community structure.

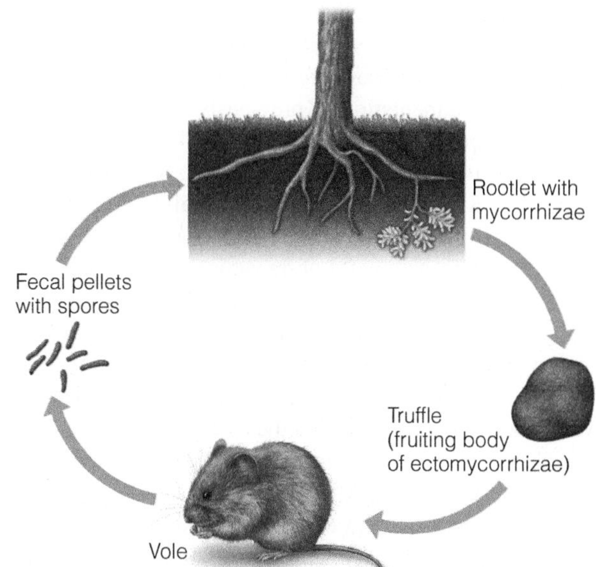

Figure 15.16 A mutualism involving three species and both symbiotic and nonsymbiotic interactions. Voles eat truffles, the belowground fruiting bodies of some mycorrhizae. The spores become concentrated in fecal pellets. The voles disperse the spores to locations where they can infect new host plants.

ECOLOGICAL Issues & Applications

Land-use Changes Are Resulting in an Expansion of Infectious Diseases Impacting Human Health

The cutting and clearing of forests to allow for the expansion of agriculture and urbanization has long been associated with declining plant and animal populations and the reduction of biological diversity resulting from habitat loss (see Chapters 9 and 12, *Ecological Issues & Applications*); however, recent research is showing that these land-use changes are directly impacting human health because they facilitate the expansion of infectious diseases. In many regions of the world, forest clearing has altered the abundance or dispersal of pathogens—parasites causing disease in the host organisms—by influencing the abundance and distribution of animal species that function as their hosts and vectors. One of the best-documented cases of forest clearing impacting the transmission of an infectious disease involves Lyme disease, which is an infectious disease that has been dramatically increasing in the number of reported cases in North America (see Section 15.4). New estimates indicate that Lyme disease is 10 times more common than previous national counts indicated, with approximately 300,000 people, primarily in the Northeast, contracting the disease each year.

Lyme disease is caused by the bacterial parasite *Borrelia burgdorferi*, which, in eastern and central North America, is transmitted by the bite of an infected blacklegged tick (*Ixodes scapularis*). The ticks have a four-stage life cycle: egg, larvae, nymph, and adult (**Figure 15.17**). Larval ticks hatch uninfected; however, they feed on blood, and if they feed on an organism infected by the *Borrelia burgdorferi* bacteria, they too can become infected and later transmit the bacteria to people. Whether a larval tick will acquire an infection and subsequently molt into an infected nymph depends largely on the species of host on which it feeds. The larval ticks may feed on a wide variety of host species that carry the bacterial parasite, including birds, reptiles, and mammals. However, not all host species are equally likely to transmit the infection to the feeding tick. One species with high rates of transmission to larval ticks that feed on its blood is the white-footed mouse (*Peromyscus leucopus*), which infects between 40 and 90 percent of feeding tick larvae. It is at this point in the story that human activity comes into play.

Human activities in the northeastern United States have resulted in the fragmentation of what was once a predominantly

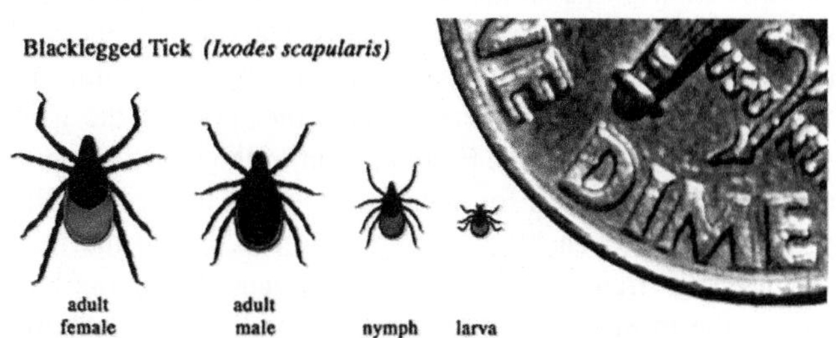

Figure 15.17 Life stages of blacklegged tick: egg, larva, nymph, and adult (female and male). United States dime shown for comparison of size.
(United States Center for Disease Control.)

forested landscape. Fragmentation involves both a reduction in the total forested area as well as a reduction in the average size of remaining forest patches (see Chapter 19). One key consequence of the fragmentation of previously continuous forest is a reduction in species diversity (Section 19.4). However, certain species thrive in highly fragmented landscapes. One such organism is the white-footed mouse, the small mammal species with high transmission rates of the bacterial parasite *B. burgdorferi* to their primary vector of transmission to humans, larval blacklegged ticks. White-footed mice reach unusually high densities in small forest fragments, which is most likely a result of decreased abundance of both predators and competitors. Could forest fragmentation and associated increases in the populations of white-footed mice in the Northeast be responsible for the increased transmission of Lyme disease in this region? To address this question, Brian Allen of Rutgers University and colleagues Felicia Keesing and Richard Ostfeld undertook a study to examine the impact of forest clearing and fragmentation in southeastern New York State on the potential for transmission of Lyme disease. The researchers hypothesized that small forest patches (<2 hectares [ha]) have a higher density of infected nymphal blacklegged ticks than larger patches (2–8 ha). To test this hypothesis Allen and his colleagues sampled tick density and *B. burgdorferi* infection prevalence in forest patches, ranging in size from 0.7 to 7.6 ha. The researchers found both an exponential decline in the density of nymphal ticks, as well as a significant decline in the nymphal infection prevalence with increasing size of forest patches (**Figure 15.18**). The consequence was a dramatic increase in the density of infected nymphs, and therefore in Lyme disease risk, with decreasing size of forest patches. Forest clearing and fragmentation clearly lead to a potential increase in the transmission of Lyme disease.

An additional factor resulting from forest clearing and fragmentation in the region is an increase in the population of white-tailed deer, the primary host species for the adult ticks. Adult ticks feed on white-tailed deer, after which the female tick drops her eggs to the ground for the cycle to begin once again. Together, the increases in white-footed mice and white-tailed deer population in the Northeast that have resulted from alterations of the landscape have dramatically increased the populations of ticks, and the transmission rate of the bacterial pathogen that causes Lyme disease.

Forest clearing has had a similar impact on the rise of vector-borne infectious disease in the tropical regions. Deforestation in the Amazon rainforest has been linked to an increase in the prevalence of malaria. Malaria is a recurring infection produced in humans by protists parasites transmitted by the bite of an infected female mosquito of the genus *Anopheles* (Section 15.4). Forty percent of the world's population is currently at risk for malaria, and more than two million people are killed each year by this disease. Of all the forest species that transmit diseases to humans, mosquitoes are among the most sensitive to environmental changes resulting from deforestation. Their survival, population density, and geographic distribution are dramatically influenced by small changes in environmental conditions, such as temperature, humidity, and the availability of suitable breeding sites. The main

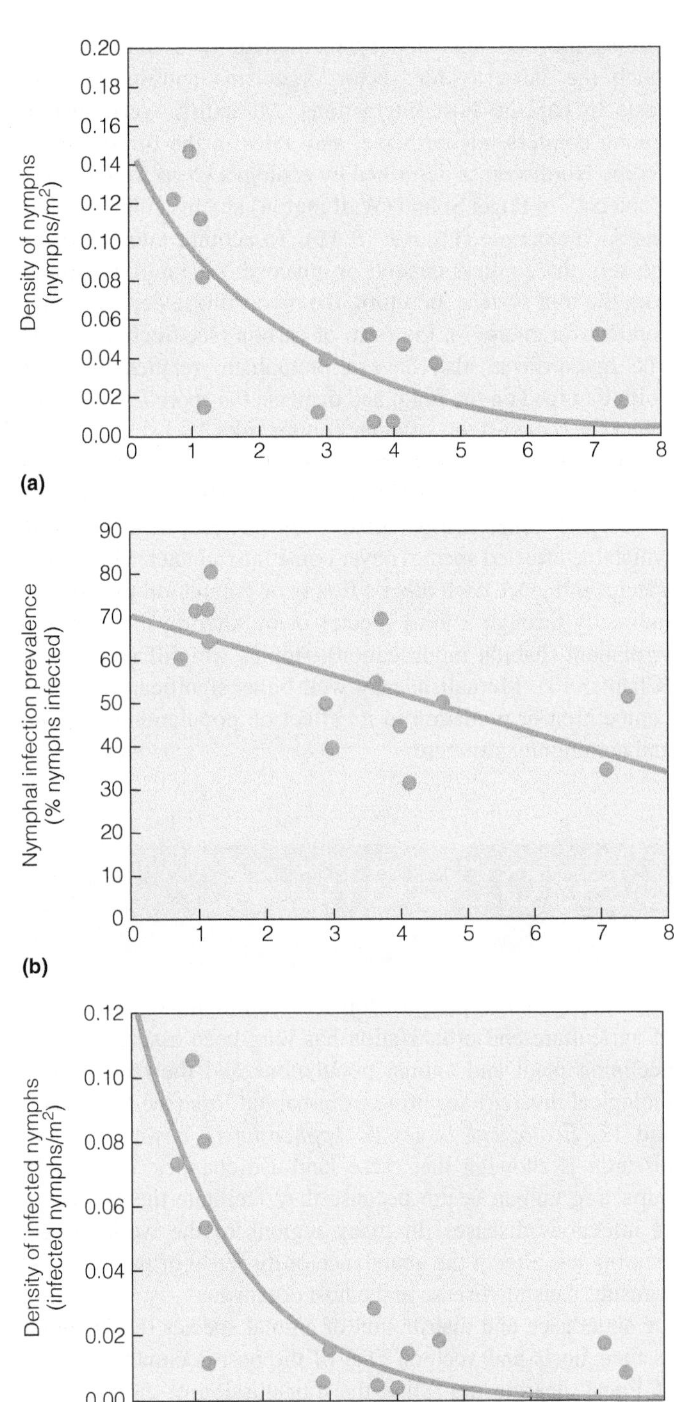

Figure 15.18 Relationship between measures of Lyme disease risk and forest patch area in a fragmented landscape in Dutchess County, New York. (a) Density of nymphal ticks (*y*-axis) decreases exponentially with increasing size of forest patch. (b) The percent of nymphal ticks in the population infected with the bacteria that causes Lyme disease decreases with increasing size of forest patch. (c) Density of infected nymphal ticks decreases with increasing forest patch size.
(Adapted from Allan et al. 2003.)

vectors of malaria in the Amazon, *Anopheles darlingi* mosquitoes, seek out larval habitat in partially sunlit areas, with clear water of neutral pH and aquatic plant growth. *A. darlingi* prefers to lay its eggs in water surrounded by short vegetation, so the abundance of this mosquito species has been enhanced by forest clearing in the Amazon region.

To examine the impact of tropical rainforest clearing on malaria, Amy Vittor of Stanford University and colleagues conducted a year-long study focused on a region of the Peruvian Amazon to examine the influence of forest clearing on the abundance of *A. darlingi*, and the rates at which they fed on humans in areas with varying degrees of forest clearing. The researchers found that the likelihood of finding *A. darlingi* larvae doubled in breeding sites with <20 percent forest compared with sites with 20–60 percent forest, and the likelihood increased sevenfold when compared with sites with >60 percent forest (**Figure 15.19**). As a result, deforested sites had a biting rate that was approximately 300 times higher than the rate of areas that were predominantly forested. Their results indicate that *A. darlingi* is both more abundant and displays significantly increased human-biting activity in areas that have undergone deforestation.

A similar pattern was observed by Sarah Olson of the University of Wisconsin and colleagues who examined the role of forest clearing on the transmission of malaria in the Amazon Basin of Brazil. The researchers found that after adjusting for population, access to health care and district size, a 4.3 percent increase in deforestation between 1997 and 2000 was associated with a 48 percent increase in malaria risk.

The impacts of forest clearing and changing land-use patterns are not limited to the enhancement of pathogen populations and their vectors. Land-use change and expansion of human populations into forest areas is resulting in the exposure of humans and domestic animal populations to pathogens not previously encountered but that naturally occur in wildlife. The result has been the emergence of new and often deadly parasites and associated diseases. There is also potential for changes in the distribution of pathogens and their vectors as a result of changing climate conditions (see Chapter 2, *Ecological Issues & Applications*), a subject we will address later in Chapter 27.

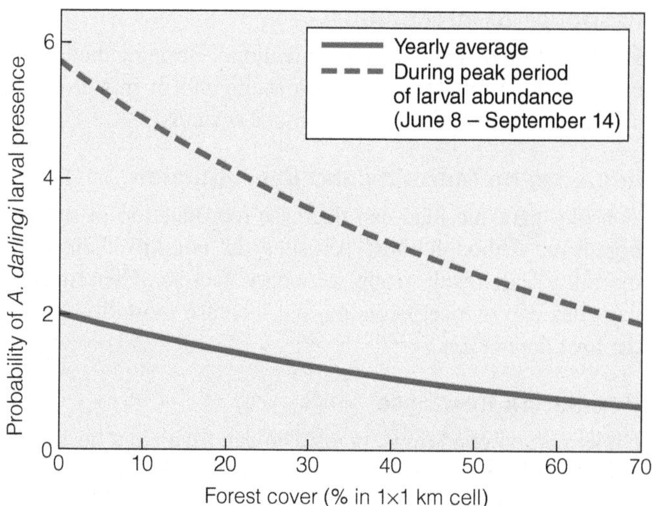

Figure 15.19 The predicted probability of *Anopheles darlingi* larval (the vector for transmission of malaria to humans) presence by the percent of forest cover. The likelihood of *A. darlingi* presence decreases from ~2 percent in areas with very little forest cover to less than 1 percent in areas with more than 50 percent forest cover. During the peak *A. darlingi* period, the likelihood decreases from ~6 to 2 percent.
(Adapted from Vittor et al 2009.)

SUMMARY

Characteristics of Parasites 15.1

Parasitism is a symbiotic relationship between individuals of two species in which one benefits from the association, wherease the other is harmed. Parasitic infection can result in disease. Microparasites include viruses, bacteria, and protozoa. They are small, have a short generation time, multiply rapidly in the host, tend to produce immunity, and spread by direct transmission. They are usually associated with dense populations of hosts. Macroparasites are relatively large and include parasitic worms, lice, ticks, fleas, rusts, smuts, fungi, and other forms. They have a comparatively long generation time, rarely multiply directly in the host, persist with continual reinfection, and spread through both direct and indirect transmission.

Parasite–Host Relationships 15.2

Parasites exploit every conceivable habitat in host organisms. Many are specialized to live at certain sites, such as in plant roots or an animal's liver. Parasites must (1) gain entrance to and (2) escape from the host. Their life cycle revolves about these two problems.

Direct Transmission 15.3

Transmission for many species of parasites occurs directly from one host to another. It occurs either through direct physical contact or through the air, water, or another substrate.

Indirect Transmission 15.4

Some parasites are transmitted between hosts by means of other organisms, called *vectors*. These carriers become intermediate hosts of some developmental or infective stage of the parasite.

Intermediate Hosts 15.5

Other species of parasites require more than one type of host. Indirect transmission takes them from definitive to intermediate to definitive host. Indirect transmission often depends on the feeding habits of the host organisms.

Response to Infection 15.6

Hosts respond to parasitic infections through behavioral changes, inflammatory responses at the site of infection, and subsequent activation of their immune systems.

Influence on Mortality and Reproduction 15.7

A heavy parasitic load can decrease reproduction of the host organism. Although most parasites do not kill their hosts, mortality can result from secondary factors. Consequently, parasites can reduce fecundity and increase mortality rates of the host population.

Population Response 15.8

Under certain conditions, parasitism can regulate a host population. When introduced to a population that has not developed defense mechanisms, parasites can spread quickly, leading to high rates of mortality and in some cases to virtual extinction of the host species.

Predation to Mutualism 15.9

Mutualism is a positive reciprocal relationship between two species that may have evolved from predator–prey or host–parasite relationships. Where adaptations have countered the negative impacts of predators or parasites, the relationship is termed *commensalism*. Where the interaction is beneficial to both species, the interaction is termed *mutualism*.

Mutualistic Relationships 15.10

Mutualistic relationships involve diverse interactions. Mutualisms can be characterized by a wide number of variables relating to the benefits received, degree of dependency of the interaction, degree of specificity, and duration and intimacy of the association.

Nutrient Uptake 15.11

Symbiotic mutualisms are involved in the uptake of nutrients in both plants and animals. The chambers of a ruminant's stomach contain large populations of bacteria and protozoa that carry out the process of fermentation. Some plant species have a mutualistic association with nitrogen-fixing bacteria that infect and form nodules on their roots. The plants provide the bacteria with carbon, and the bacteria provide nitrogen to the plant. Fungi form mycorrhizal associations with the root systems of plants, assisting in the uptake of nutrients. In return, they derive energy in the form of carbon from the host plant.

Mutualisms Involving Defense 15.12

Other mutualistic associations are associated with defense of the host organism.

Pollination 15.13

Nonsymbiotic mutualisms are involved in the pollination of many species of flowering plants. While extracting nectar from the flowers, the pollinator collects and exchanges pollen with other plants of the same species. To conserve pollen, some plants have morphological structures that permit only certain animals to reach the nectar.

Seed Dispersal 15.14

Mutualism is also involved in seed dispersal. Some seed-dispersing animals that the plant depends on may be seed predators as well, eating the seeds for their own nutrition. Plants depending on such animals must produce a tremendous number of seeds to ensure that a few are dispersed, come to rest on a suitable site, and germinate. Alternatively, plants may enclose their seeds in a nutritious fruit attractive to frugivores (fruit-eating animals). Frugivores are not seed predators. They eat only the tissue surrounding the seed and, with some exceptions, do not damage the seed.

Population Dynamics 15.15

Mutualistic relationships, both direct and indirect, may influence population dynamics in ways that we are just beginning to appreciate and understand.

Deforestation and Disease Ecological Issues & Applications

Land-use changes associated with human activities have led to an increase in the transmission of infectious diseases. In many regions of the world, forest clearing has altered the abundance or dispersal of pathogens by influencing the abundance and distribution of animal species that function as their hosts and vectors.

STUDY QUESTIONS

1. In both predation and parasitism, one organism (species) derives its energy and nutrients from consuming another organism. How do these two processes differ?
2. For the parasite trematode discussed in Section 15.7, infection begins as snails grazing on algae incidentally ingest worm eggs. The eggs hatch into worms that prevent a snail's own reproduction. Instead, the infected snail nourishes the growing larval worms, which eventually develop into a free-swimming stage and leave the snails to seek their second, or intermediate, host—the California killifish. In traveling to the fish's brain, the worm causes the fish to behave differently from other killifish; it moves about jerkily near the water's surface. This behavior attracts predators like herons. The heron, in turn, becomes the host to the adult worm. The adult trematode takes up final residence in the bird's gut, releasing thousands of eggs that are deposited by way of bird droppings back into the salt marsh, completing the parasite's life cycle. How might such a complex life cycle have evolved?
3. How might a patchy or clumped distribution of hosts affect the spread of parasites? What spatial distribution of

hosts (random, uniform, or clumped) would present the greatest difficulty in transmitting parasites from host to host?

4. What is mutualism? Look up some examples of relationships that have been identified as mutualisms, and examine them critically. Are they in fact mutualistic?
5. Distinguish among symbiosis, obligate, and facultative mutualism.
6. Is fruit predation a chancy way of distributing seeds? Why?
7. Is mutualism reciprocal exploitation, or are the two species acting together for mutual benefit?

FURTHER READINGS

Classic Studies

Boucher, D. H., ed. 1988. *The biology of mutualism*. New York, Oxford University Press.
This volume provides a multitude of examples describing the range of mutualistic relationships covered in this chapter.

Futuyma, D. J., and M. Slatkin, eds. 1983. *Coevolution.* Sunderland, MA: Sinauer Associates.
This edited volume provides an excellent overview of research in the area of coevolution.

Muscatine, L., and J. W. Porter. 1977. "Reef corals: Mutualistic symbioses adapted to nutrient-poor environments." *BioScience* 27:454–460.
An excellent introduction to the ecology of corals and the symbiotic relationships that are the foundation of the most diverse aquatic ecosystems on our planet.

Price, P. W. 1980. *Evolutionary biology of parasites*. Princeton NJ: Princeton University Press.
A classic that provided the foundation for the study of interaction between parasites and their hosts.

Recent Research

Barth, F. G. 1991. *Insects and flowers: The biology of a partnership.* Princeton, NJ: Princeton University Press.
Although technical, this monograph is an excellent overview of the ecology of plant–pollinator interactions.

Bronstein, J. L. 2009. "The evolution of facilitation and mutualism." *Journal of Ecology* 97:1160–1170.
This article explores the parallels and differences between facilitation and mutualism.

Bruno, J. F., J. J. Stachowicz, and M. D. Bertness. 2003. "Inclusion of facilitation into ecological theory." *Trends in Ecology & Evolution* 18:119–125.
Provides an excellent discussion of species interactions that fall within the broader classification of facilitation.

Dobson, A. P., and E. R. Carper. 1996. "Infectious diseases and human population history." *BioScience* 46:115–125.
An interesting and well-written review of the role of infectious disease in the history of the human population.

Handel, S. N., and A. J. Beattie. 1990. "Seed dispersal by ants." *Scientific American* 263:76–83.
This review provides an excellent discussion of the mutualistic relationship between ants and plants.

Hatcher, M. J., and A. M. Dunn. 2011. *Parasites in ecological communities: From interactions to ecosystems.* New York: Cambridge University Press.
Investigates the role parasites play in influencing interactions both detrimental and beneficial between competitors, predators, and their prey.

Moore, J. 1984. "Parasites that change the behavior of their host." *Scientific American* 250:108–115.
An excellent introduction to this fascinating area of research.

Stachowicz, J. 2001. "Mutualism, facilitation, and the structure of ecological communities." *BioScience* 51:235–246.
An excellent review of the role of positive interactions on population dynamics and community structure. Contains many well-illustrated examples of current research in this growing area of research.

Coral reefs inhabiting the shallow waters of subtropical and tropical oceans are among the most biologically diverse communities on our planet.

CHAPTER GUIDE

IN WALKING THROUGH A FOREST or swimming along a coral reef, we see a collection of individuals of different species; plants and animals that form local populations. Sharing environments and habitats, these species of plants and animals interact in various ways. The group of species that occupy a given area, interacting either directly or indirectly, is called a **community**. This definition embraces the concept of community in its broadest sense. It is a spatial concept—the collective of species occupying a place within a defined boundary. Because ecologists generally do not study the entire community, the term *community* is often used in a more restrictive sense. It refers to a subset of the species, such as a plant, bird (avian), small mammal, or fish community. This use of *community* suggests relatedness or similarity among the members in their taxonomy, response to the environment, or use of resources.

The definition of community also recognizes that species living in close association may interact. They may compete for a shared resource, such as food, light, space, or moisture. One may depend on the other as a source of food. They may provide mutual aid, or they may not directly affect one another at all.

Like a population, a community has attributes that differ from those of its components and that have meaning only within the collective. These attributes include the number of species, their relative abundances, the nature of their interactions, and their physical structure (defined primarily by the growth forms of the plant components of the community). In our discussion, we examine the properties defining the structure of the community. In the chapters that follow, we will turn our attention to the processes influencing community structure and dynamics.

16.1 Biological Structure of Community Defined by Species Composition

The biological structure of a community is defined by its species composition, that is, the set of species present and their relative abundances. For example, **Table 16.1** contains samples representing tree species composition of two forest communities in northern West Virginia. For each forest, the first column provides the number of individuals of each species, and the second column provides a measure of their **relative abundance** expressed as the proportion each species contributes to the total number of individuals of all species within the community. Relative abundance in Table 16.1 is calculated as:

$$p_i = n_i/N$$

where p_i is the proportion of individuals in the community belonging to species i, n_i is the number of individuals belonging to species i, and N is the total number of individuals of all species in the community. These values have then been converted to percentages.

The sample from the first forest community consists of 256 individuals representing 24 species. Two species—yellow poplar and white oak—make up nearly 44 percent of the total number of individuals. The four next most abundant trees—black oak, sugar maple, red maple, and American beech—each make up a little more than 5 percent of the total. Nine species range from 1.2 percent to 4.7 percent, and the 9 remaining species as a group represent about 0.4 percent each. The second forest presents a somewhat different picture. This community consists of 274 individuals representing 10 species, of which two species—yellow poplar and sassafras—make up almost 84 percent of the total number of individuals in the tree community.

A common method for examining the patterns of relative abundance within communities involves plotting the relative abundance of each species against rank, where rank is defined by the order of species from the most to the least abundant (the tree species in Table 16.1 are presented in order from the most to the least abundant). Thus, the most abundant species is plotted first along the x-axis, with the corresponding value on the y-axis being its value of relative abundance. This process is continued until all species are plotted. The resulting graph is called a **rank-abundance diagram**. **Figure 16.1** depicts the rank-abundance curves for the two forest communities presented in Tables 16.1.

The rank-abundance diagram illustrates two features of community structure: species richness and species evenness. **Species richness** is a count of the number of species occurring within the community, and is typically denoted by the symbol S. **Species evenness** refers to the equitability in the distribution of individuals among the species. The maximum species evenness would occur if each species in the community was equally abundant.

The rank-abundance curves presented in Figure 16.1 shows that the two forest communities from Table 16.1 differ in both species richness and how individuals are apportioned among the species (evenness). The first forest community has

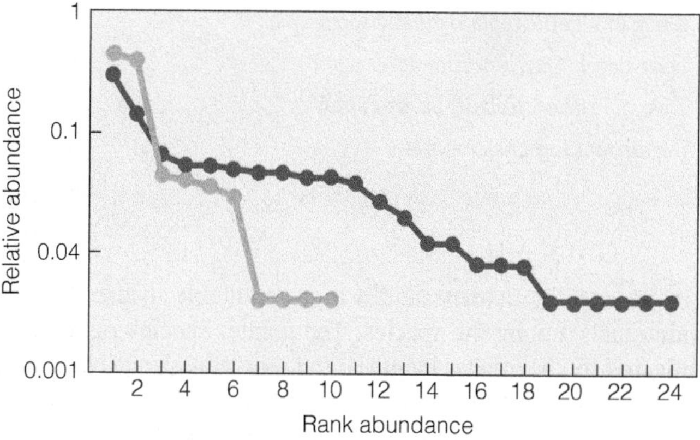

Figure 16.1 Rank-abundance curves for the two forest communities described in Table 16.1. Rank abundance is the species ranking based on relative abundance, ranked from the most to least abundant (x-axis). Relative abundance (y-axis) is expressed on a $\log_{10}$ axis. The first forest community in Table 16.1 (brown line) has a higher species richness (length of curve) and evenness (slope of curve) than the second forest community (orange line).

Interpreting Ecological Data

Q1. *How does the slope of the rank-abundance curve vary with increasing species evenness? Why?*

Q2. *What would the rank-abundance curve look like for a forest community consisting of 10 species, in which all of the 10 tree species are equally abundant?*

Table 16.1 Structure of Two Deciduous Forest Stands in Northern West Virginia

	Stand 1		Stand 2	
Species	Number of Individuals	Relative Abundance (Total Individuals, %)	Number of Individuals	Relative Abundance (Total Individuals, %)
Yellow poplar (*Liriodendron tulipifera*)	76	29.7	122	44.5
White oak (*Quercus alba*)	36	14.1		
Black oak (*Quercus velutina*)	17	6.6		
Sugar maple (*Acer saccharum*)	14	5.4	1	0.4
Red maple (*Acer rubrum*)	14	5.4	10	3.6
American beech (*Fagus grandifolia*)	13	5.1	1	0.4
Sassafras (*Sassafras albidum*)	12	4.7	107	39.0
Red oak (*Quercus rubra*)	12	4.7	8	2.9
Mockernut hickory (*Carya tomentosa*)	11	4.3		
Black cherry (*Prunus serotina*)	11	4.3	12	4.4
Slippery elm (*Ulmus rubra*)	10	3.9		
Shagbark hickory (*Carya ovata*)	7	2.7	1	0.4
Bitternut hickory (*Carya cordiformis*)	5	2.0		
Pignut hickory (*Carya glabra*)	3	1.2		
Flowering dogwood (*Cornus florida*)	3	1.2		
White ash (*Fraxinus americana*)	2	0.8		
Hornbeam (*Carpinus carolinia*)	2	0.8		
Cucumber magnolia (*Magnolia acuminate*)	2	0.8	11	4.0
American elm (*Ulmus americana*)	1	0.4		
Black walnut (*Juglans nigra*)	1	0.4		
Black maple (*Acer nigra*)	1	0.4		
Black locust (*Robinia pseudoacacia*)	1	0.4		
Sourwood (*Oxydendrum arboreum*)	1	0.4		
Tree of heaven (*Ailanthus altissima*)	1	0.4		
Butternut (*Juglans cinerea*)			1	0.4
	256	**100.0**	**274**	**100.0**

greater species richness and a more equitable distribution of individuals among the species. The greater species richness is reflected by the greater length of the rank-abundance curve (24 species compared to 10 in the second community). The more equitable distribution of individuals among the species (species evenness) is indicated by the more gradual slope of the rank-abundance curve. If each species was equally abundant, the rank-abundance curve would be a straight line parallel to the x-axis at the value $1.0/S$ on the y-axis (where S is species richness, the number of species in the community).

16.2 Species Diversity Is Defined by Species Richness and Evenness

Although the graphical procedure of rank-abundance diagrams can be used to visually assess (interpret) differences in the biological structure of communities, these diagrams offer no means of quantifying the observed differences. The simplest quantitative measure of community structure is the index of species richness (S). However, species richness does not account for differences in the relative abundance of species within the community. For example, two communities may both be inhabited by the same number of species and therefore have the same value of species richness; yet in one community the vast majority of individuals may be of a single species, whereas in the other community the individuals may be more equally distributed among the various species (greater evenness; **Figure 16.2**). Ecologists have addressed this shortcoming by developing mathematical indices of **species diversity**, which consider both the number and relative abundance of species within the community.

One of the simplest and most widely used indices of species diversity is the Simpson's index. The term *Simpson's diversity index* can actually refer to any one of three closely related indexes.

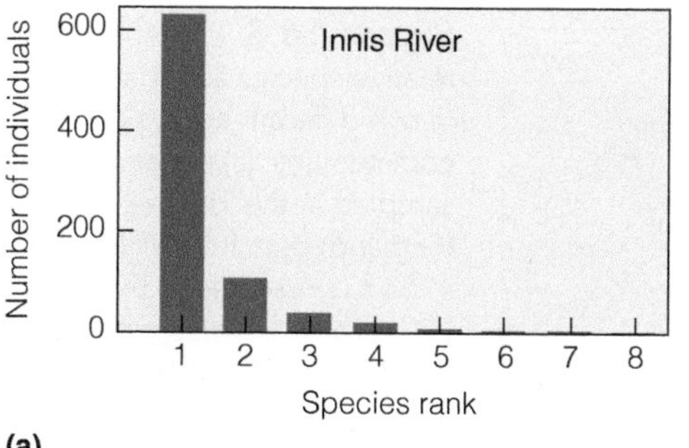

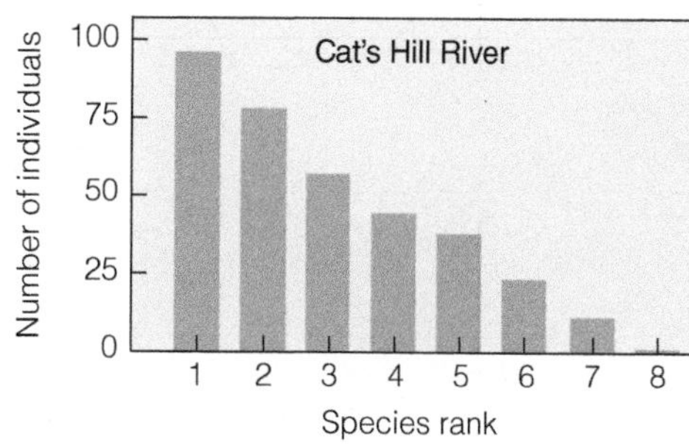

(a)

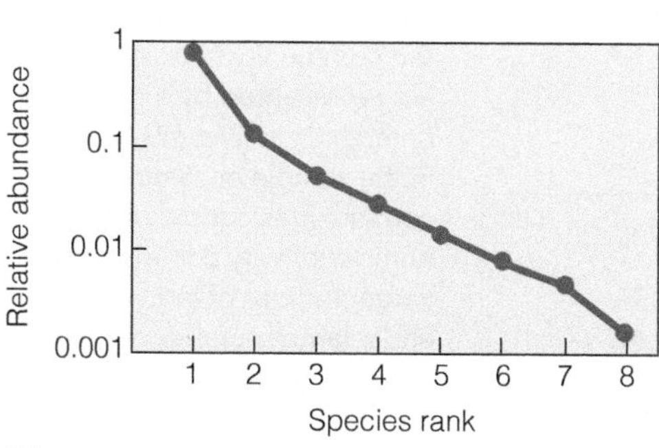

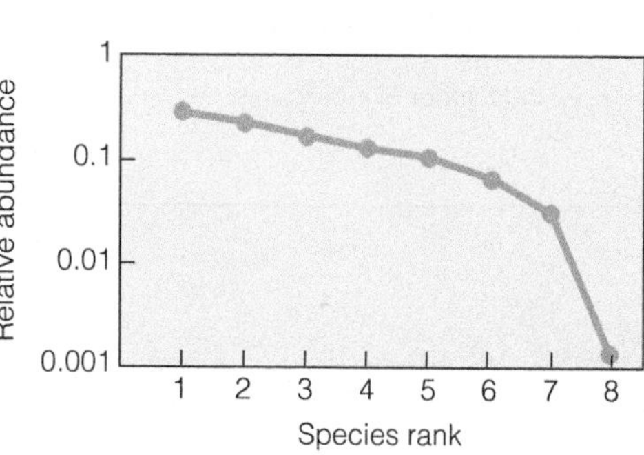

(b)

Figure 16.2 Patterns of fish species diversity for two rivers on the Island of Trinidad provide an example of two communities with equal species richness ($S = 8$) but different evenness. (a) The abundance of the eight species of fish in the Innis River and Cat's Hill River. (b) The same data are presented in the form of rank-abundance curves. The greater evenness of the Cat's Hill River fish community is evident from the shallower slope of the rank-abundance curve. The most abundant species in the Cat's River community accounts for 28 percent of total individuals as compared to 76 percent for the Innis River.

(© Pearson Education Inc.)

Simpson's index (*D*) measures the probability that two individuals randomly selected from a sample will belong to the same species (category):

$$D = \sum p_i^2$$

Where p_i is the proportion of the total individuals in the community represented by species i (relative abundance). The value of D ranges between 0 and 1. In the absence of diversity, where only one species is present, the value of D is 1. As both species richness and evenness increase, the value approaches 0.

Because the greater the value of D, the lower the diversity, D is often subtracted from 1 to give:

$$\text{Simpson's index of diversity} = 1 - D$$

The value of this index also ranges between 0 and 1, but now the value increases with species diversity. In this case, the index represents the probability that two individuals randomly selected from a sample will belong to different species.

The most common way to use the Simpson's index is to take the reciprocal of D:

$$\text{Simpson's reciprocal index} = 1/D$$

The lowest possible value of this index is 1, representing a community containing only one species. The higher the value, the greater is the species diversity. The maximum value of the reciprocal index is the number of species in the community, the value of species richness (S). For example, there are 10 tree species in the second forest community presented in Table 16.1, so the maximum possible value of the index is 10. Because S is the maximum value of the index, a measure of evenness (E_D) can be calculated as the ratio of the reciprocal index ($1/D$) divided by S:

$$E_D = \frac{(1/D)}{S}$$

Values of evenness (E_D) range from 0 to 1, with a value of 1 representing complete evenness (all species equally abundant).

Because the Simpson's index actually refers to three related but different indexes, it is important to identify which is being used and reported.

Another widely used index of diversity that also considers both species richness and evenness is the **Shannon index** (also called the Shannon–Weiner index).

The Shannon index (H) is then computed as:

$$H = -\Sigma(p_i)(\ln p_i)$$

Where p_i is the proportion of the total individuals in the community represented by species i, and ln is the natural logarithm.

In the absence of diversity, where only one species is present, the value of H is 0. The maximum value of the index, which occurs when all species are present in equal numbers, is $H_{max} = \ln S$, where S is the total number of species (species richness). As with the Simpson's index of diversity, the maximum value of the Shannon index (H_{max}) can be used to calculate an index of species evenness (E_H):

$$E_H = H/H_{max}$$

As with the Simpson's index, values of evenness (E_H) range from 0 to 1, with a value of 1 representing complete evenness (all species equally abundant).

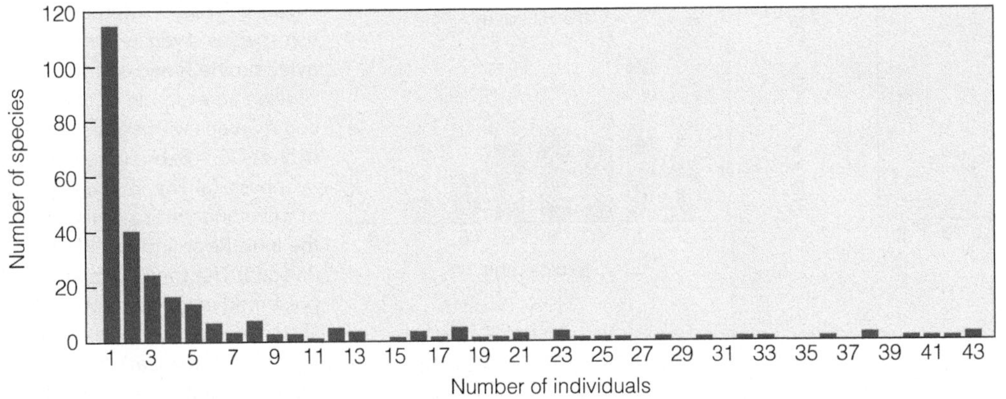

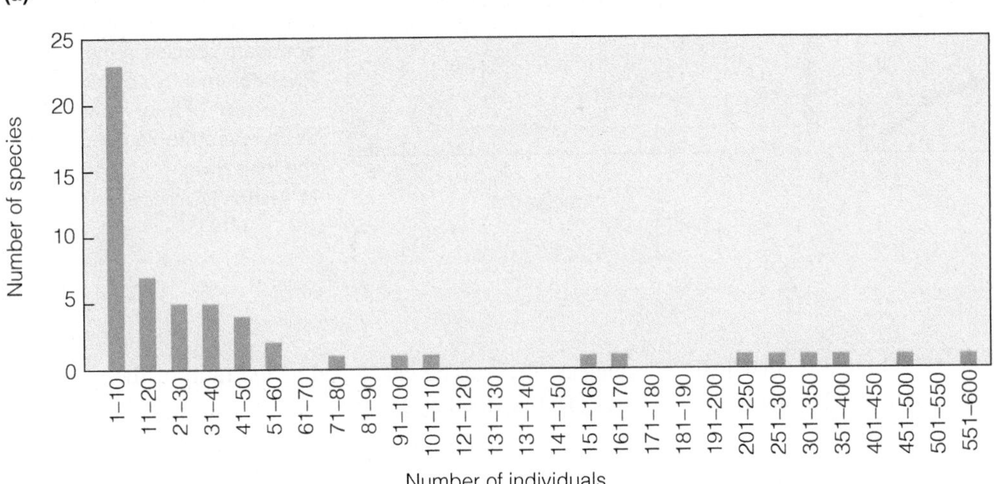

Figure 16.3 Patterns of relative species abundance in two different ecological communities: (a) aquatic beetles sampled in the Thames River (England; data from Williams 1963 as presented in Magurran 2004) , and (b) fish species in the Wabash River (Indiana, United States; data from Edgell and Long 2009). Graphs plot the number of species in the community (*y-axis*) that are represented by a given number of individuals (*x-axis*) in the sample or samples. Both communities consist of a small number of very common species (large number of individuals) and a larger number of relatively rare species (small number of individuals).

16.3 Dominance Can Be Defined by a Number of Criteria

Although the numbers of tree species occurring in the two forest communities (species richness) presented in Tables 16.1 differ more than twofold, the two communities share a common feature. Both communities are composed of a few common tree species with high population density, whereas the remaining tree species are relatively rare and at low population density. This is a characteristic of most communities (**Figure 16.3**). When a single or few species predominate within a community, those species are referred to as **dominants**.

Dominance is the converse of diversity. In fact, the basic Simpson index, D, is often used as a measure of dominance. Recall that values of D range from 0 to 1, where 1 represents complete dominance; that is, only one species is present in the community.

Dominant species are usually defined separately for different taxonomic or functional groups of organisms within the community. For example, yellow poplar is a dominant tree species in both of the forest communities just discussed, but we could likewise identify the dominant herbaceous plant species within the forest or the dominant species of bird or small mammal.

Dominance typically is assumed to mean the greatest in number. But in populations or among species in which individuals vary widely in size, abundance alone is not always a sufficient indicator of dominance. In a forest, for example, the small or understory trees can be numerically superior, yet a few large trees that overshadow the smaller ones will account for most of the biomass (living tissue). For example, the species composition of trees in a forest community in central Virginia is presented in **Table 16.2**. When the structure of the forest is quantified in terms of relative abundance (percentage of total individuals in community) two species—red maple and dogwood—account for approximately 60 percent of individuals in the forest. When the structure of the community is quantified in terms of relative biomass (percentage of total biomass in community), however, the picture of dominance that emerges is quite different. Now white oak, which accounts for less than 9 percent of the individuals, accounts for approximately 60 percent of the total biomass, and the two numerically dominant tree species (red maple and dogwood) account for slightly more than 10 percent. This discrepancy between relative abundance and relative biomass occurs because a few large white oak trees that make up the forest canopy account for the majority of the biomass, and the much larger number of smaller red maple and dogwood occupy the understory (see Figure 16.12 for description and graphic of vertical structure of forest). In such a situation, we may wish to define dominance

Table 16.2 Structure of Deciduous Forest Stand in Central Virginia

Species	Number of Individuals	Relative Abundance	Relative Biomass
Red maple *Acer rubrum*	30	33.0	6.2
Dogwood *Cornus florida*	24	26.4	4.6
White Oak *Quercus alba*	8	8.8	58.5
Tulip poplar *Liriodendron tulipifera*	6	6.6	12.3
Red Oak *Quercus rubra*	6	6.6	7.6
Mockernut hickory *Carya tomentosa*	5	5.5	2.2
Virginia pine *Pinus virginiana*	4	4.4	6.3
Cedar *Juniperus virginiana*	2	2.2	0.5
Beech *Fagus grandifolia*	2	2.2	0.9
Blackgum *Nyssa sylvatica*	1	1.1	0.2
Black cherry *Prunus serotina*	1	1.1	0.2
Sweetgum *Liquidambar styraciflua*	1	1.1	0.4
American hornbeam *Carpinus carolinia*	1	1.1	0.2
	91	**100.0**	**100.0**

Species composition is expressed in terms of both relative abundance (percentage of total individuals) and relative biomass (percentage of total stand biomass). Biomass estimates based on species-specific allometric equations relating tree diameter (at 1.5 m height) to biomass.

based on some combination of characteristics that include both the number and size of individuals.

Because dominant species typically achieve their status at the expense of other species in the community, they are often the dominant competitors under the prevailing environmental conditions. For example, the American chestnut tree (*Castanea dentata*) was a dominant component of oak–chestnut forests in eastern North America until the early 20th century. At that time, the chestnut blight introduced from Asia decimated chestnut tree populations. Since then a variety of species—including oaks, hickories, and yellow poplar—have taken over the chestnut's position in the forest. As we shall see, however, processes other than competition can also be important in determining dominance within communities (Chapter 17).

16.4 Keystone Species Influence Community Structure Disproportionately to Their Numbers

Relative abundance is just one measure, based only on numerical supremacy, of a species' contribution to the community. Other, less-abundant species, however, may play a crucial role in the function of the community. A species that has a disproportionate impact on the community relative to its abundance is referred to as a **keystone species**.

Keystone species function in a unique and significant manner, and their effect on the community is disproportionate to their numerical abundance. Their removal initiates changes in community structure and often results in a significant loss of diversity. Their role in the community may be to create or modify habitats or to influence the interactions among other species. One organism that functions as a keystone species by creating habitat is the coral *Oculina arbuscula*, which occurs along the eastern coast of the United States as far north as the coastal waters of North Carolina. It is the only coral in this region with a structurally complex, branching morphology that provides shelter for a species-rich epifauna (organisms that live on and among the coral). More than 300 species of invertebrates are known to live among the branches of *Oculina* colonies, and many more are reported to complete much of their life cycle within the coral (see Chapter 15, **Field Studies John J. Stachowicz**).

In other cases, keystone herbivores may modify the local community through their feeding activities. An excellent example is the role of the African elephant in the savanna communities of southern Africa. This herbivore feeds primarily on a diet of woody plants (browse). Elephants are destructive feeders that often uproot, break, and destroy the shrubs and trees they feed on (**Figure 16.4a**). Reduced density of trees and shrubs favors the growth and production of grasses. This change in the composition of the plant community is to the elephant's disadvantage, but other herbivores that feed on the grasses benefit from it. In a study of the influence of tree cover on grass productivity and local densities of large herbivore populations in the savanna communities of East Africa (Kenya), Corrina Riginos of the University of California–Davis found that both grass productivity and large herbivore density increase with decreasing tree cover (**Figures 16.4b** and **16.4c**). In addition to benefiting grazing herbivores, the destruction of trees creates a variety of habitats for smaller vertebrate species. Ecologist Robert Pringle of Stanford University found that by damaging trees and increasing their structural complexity, browsing elephants create refuges used by arboreal lizards. In a study conducted at the Mpala Research Center in central Kenya, Pringle found that lizard density increased with the density of trees damaged by elephants. Daniel Parker of Rhodes University in South Africa found a similar influence of elephant feeding on community diversity. In a comparison of paired sites with and without elephants in savanna grassland communities of the

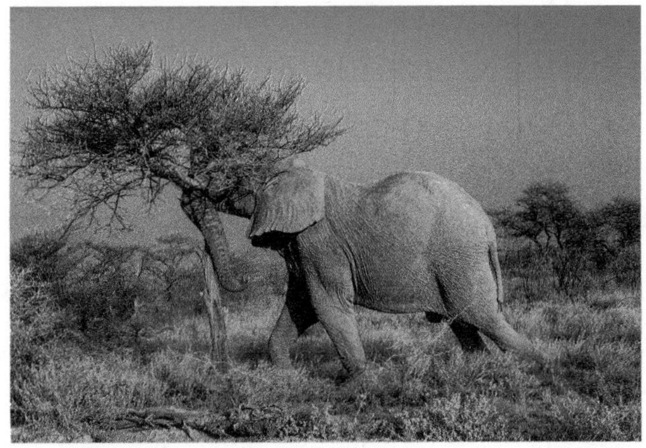

(a)

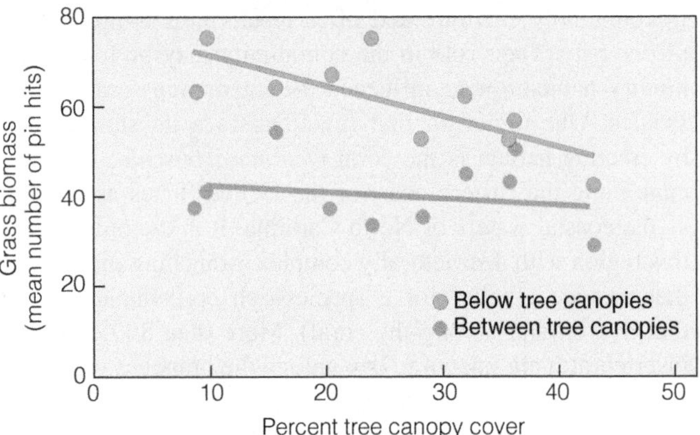

(b)

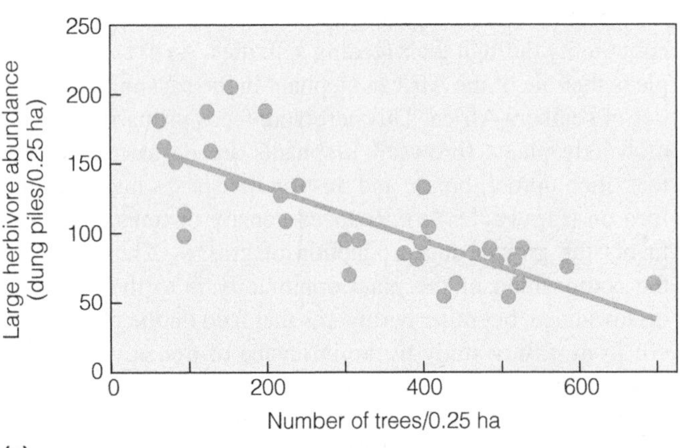

(c)

Figure 16.4 Elephants function as a keystone species in the savanna communities of Africa. (a) Elephant browsing damages trees and reduces tree density. (b) Reduced tree density functions to both increase the productivity of grasses (both in the open and under the remaining trees), and increase the abundance of large herbivores (c). Grass biomass in (b) is measured using 10-point pin frame at 10-m intervals along four 50-m transects (for a total of 240 pins per plot), which is strongly correlated with biomass. Herbivore abundance in (c) based on dung counts, which have been shown to be highly correlated to both areal and ground animal counts.

(Adapted from Riginos and Grace 2008 and Riginos et al. 2009.)

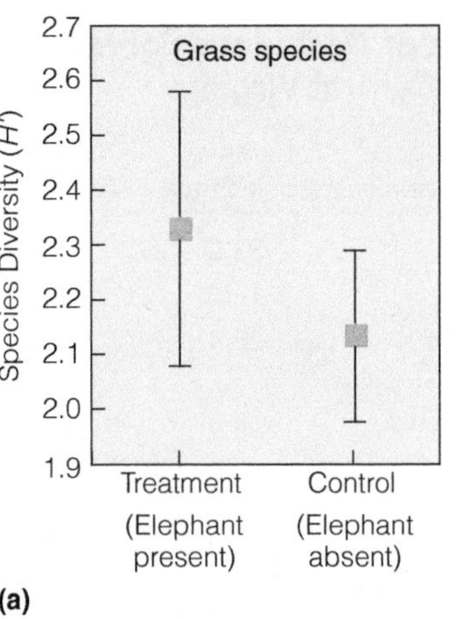

(a)

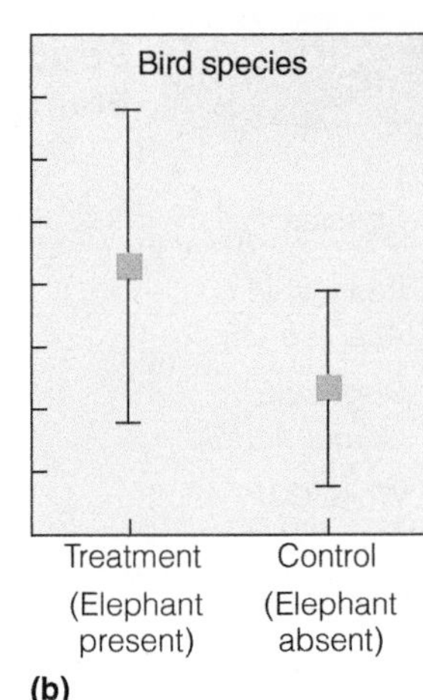

(b)

Figure 16.5 Comparison of (a) grass species diversity and (b) bird species diversity for savanna grassland sites in the Eastern Cape region of South Africa where elephants are absent (control) and where elephants are present (treatment). Species diversity is measured by the Shannon index (see Section 16.2). Boxes represent mean values for control and treatment plots. Bars represent ±1 standard error.

(Data from Parker 2008.)

Eastern Cape region of South Africa, Parker found an increase in both grass and bird species diversity in those sites inhabited by elephants (**Figure 16.5**). In addition, Parker found that insect and small mammal communities also appeared to benefit from elephant foraging through the modification of habitats.

Predators often function as keystone species within communities (see Section 17.4 for further discussion of keystone predators). For example, sea otters (*Enhydra lutris*) are a keystone predator in the kelp bed communities found in the coastal waters of the Pacific Northwest. Sea otters eat urchins, which feed on kelp. The kelp beds provide habitat to a wide diversity of other species. Since the 1970s, however, there has been a dramatic decline in sea otter populations. In a study of sea otter populations in the Aleutian Island of Alaska, James Estes and colleagues at the United States Geological Survey and the University of California–Santa Cruz found that sea otter populations are declining as a result of increased predation by killer whales (*Orcinus orca*). With the decline of sea otters, the sea urchin population has increased dramatically (**Figure 16.6**). The result is overgrazing of the kelp beds and a loss of habitat for the many species inhabiting these communities.

16.5 Food Webs Describe Species Interactions

Perhaps the most fundamental process in nature is that of acquiring the energy and nutrients required for assimilation. The species interactions discussed earlier—predation, parasitism, competition, and mutualism—are all involved in acquiring these essential resources (Part Four). For this reason, ecologists studying the structure of communities often focus on the feeding relationships among the component species, or how species

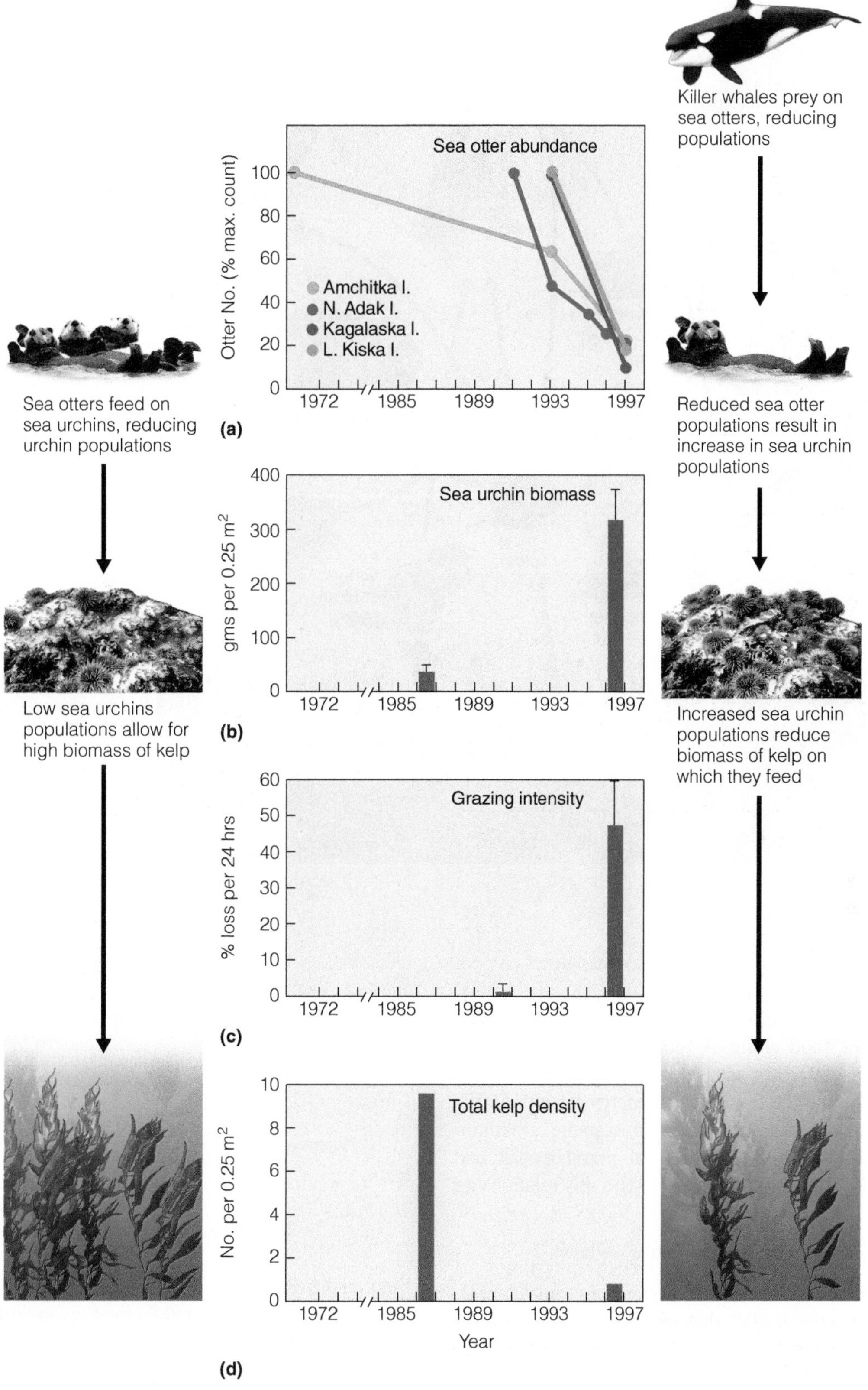

Figure 16.6 Sea otter function as a keystone predator species in the coastal kelp communities of the North Pacific; however, their role as top predator has changed over the past several decades. Increased predation of otter by killer whales in the 1990s resulted in a (a) decline in sea otter abundance at several islands in the Aleutian archipelago and concurrent changes in (b) sea urchin biomass, (c) grazing intensity, and (d) kelp density measured from kelp forests at Adak Island. The proposed mechanisms of change are portrayed in the marginal cartoons.
(From Estes et al. 1998.)

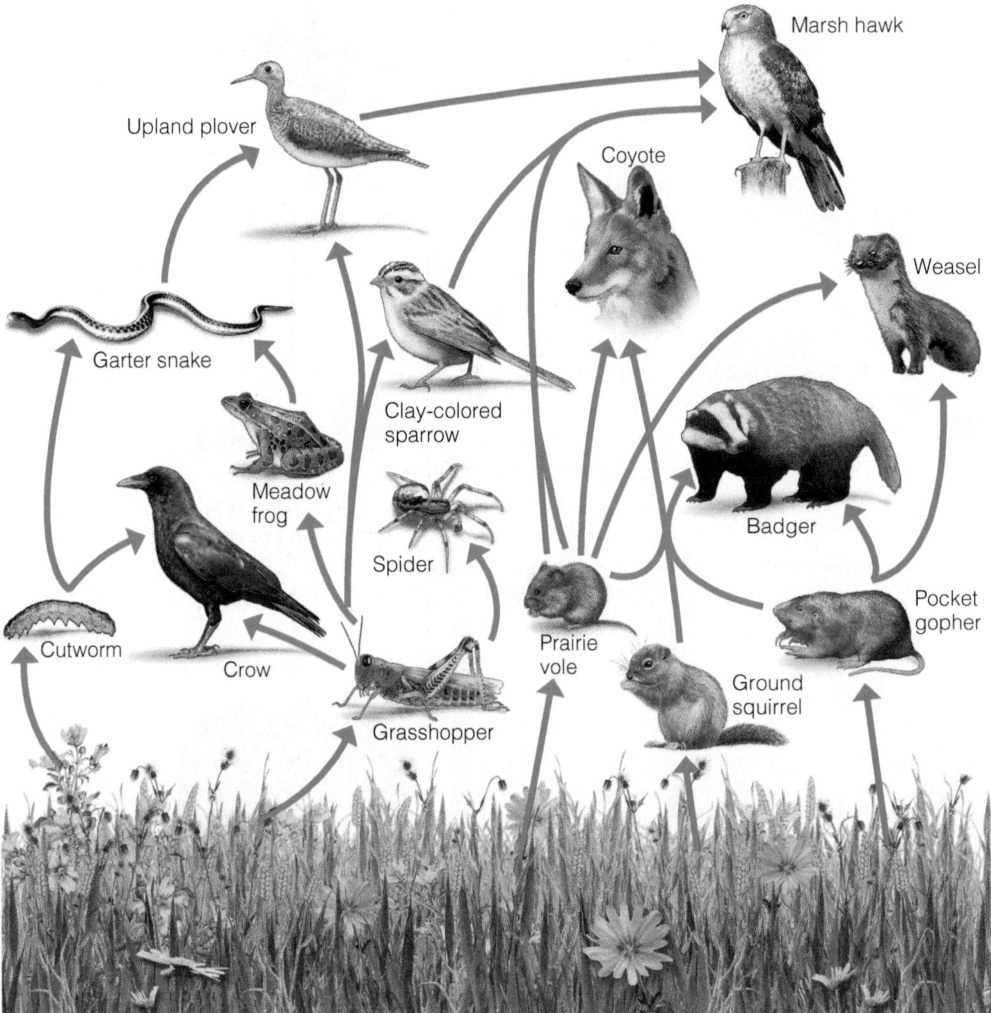

Figure 16.7 A food web for a prairie grassland community in the midwestern United States. Arrows flow from prey (consumed) to predator (consumer).

interact in the process of acquiring the resources necessary for metabolism, growth, and reproduction.

An abstract representation of feeding relationships within a community is the **food chain**. A food chain is a descriptive diagram—a series of arrows, each pointing from one species to another, representing the flow of food energy from prey (the consumed) to predator (the consumer). For example, grasshoppers eat grass, clay-colored sparrows eat grasshoppers, and marsh hawks prey on the sparrows. We write this relationship as follows:

$$\text{grass} \longrightarrow \text{grasshopper} \longrightarrow \text{sparrow} \longrightarrow \text{hawk}$$

Feeding relationships in nature, however, are not simple, straight-line food chains. Rather, they involve many food chains meshed into a complex food web with links leading from primary producers through an array of consumers (**Figure 16.7**). Such **food webs** are highly interwoven, with linkages representing the complex interactions of predator and prey.

A simple hypothetical food web is presented in **Figure 16.8** to illustrate the basic terminology used to describe the structure of food webs. Each circle represents a

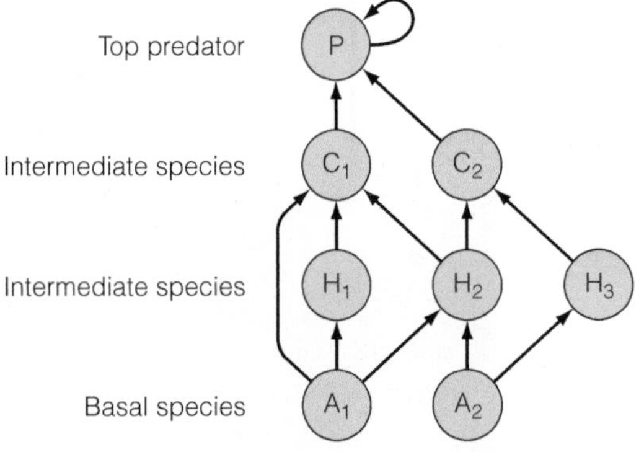

Figure 16.8 Hypothetical food web illustrating the various categories of species. A_1 and A_2 feed on no other species in the food web and are referred to as basal species (typically autotrophs). H_1, H_2, and H_3 are herbivores. C_2 is a carnivore, and C_1 is defined as an omnivore because it feeds at more than one trophic level. Species designated as H and C are all intermediate species because they function as both predators and prey within the food web. P is a top predator because it is eaten by no other species within the food web. P also exhibits cannibalism because this species feeds on itself.

species, and the arrows from the consumed to the consumer are termed **links**. The species in the webs are distinguished by whether they are basal species, intermediate species, or top predators. **Basal species** feed on no other species but are fed on by others. **Intermediate species** feed on other species and they are prey of other species. **Top predators** are not subject to predators; they prey on intermediate and basal species. These terms refer to the structure of the web rather than to strict biological reality.

Food webs can provide a useful tool for analyzing the structure of communities and a number of measures have been developed to quantify food web structure. As stated previously, each arrow linking predator (consumer) and prey (consumed) is referred to as a link or linkage. The maximum number of links in a food web is a direct function of the species richness, *S*. For a food web consisting of *S* species—assuming that each species may link to every other species including itself—the maximum number of links is S^2. The actual number of observed links in a food web (*L*) expressed as a proportion of the maximum possible number of links (S^2) provides a measure of food web **connectance (*C*)**:

$$C = L/S^2$$

An alternative measure of food web connectance considers only the number of possible unidirectional links (the link between any two species flows in only one direction). In this case, the maximum number of links is: $S(S - 1)/2$. It is important to note which approach is being used when reporting results.

Linkage density (*LD*) is a measure of the average number of links per species in the food web. It is calculated as the total number of observed links in the food web (*L*) divided by the total number of species (*S*):

$$LD = L/S$$

The length of any given food chain within the food web is measured as the number of links between a top predator (see Figure 16.8) and the base of the web (basal species). The **mean chain length (*ChLen*)** is the arithmetic average of the lengths of all chains in a food web. Examples of each of these measures (connectance, linkage density, and mean chain length) using a hypothetical food web is presented in **Figure 16.9a**.

It is apparent from the measures of food web structure presented that the number of possible species interactions (links) in a community increases with species richness (*S*), but how does species richness actually influence the complexity of food webs? Jennifer Dunne of the Santa Fe Institute (New Mexico) and colleagues examine the food web structure of a wide variety of terrestrial, freshwater, and marine ecosystems. Results of the analysis indicate that connectance decreases with species richness, whereas both linkage density and mean chain length increase as the number of species in the community increases (**Figure 16.10**).

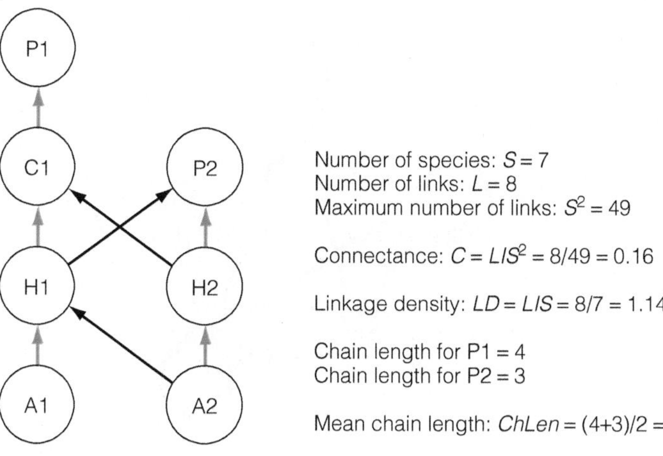

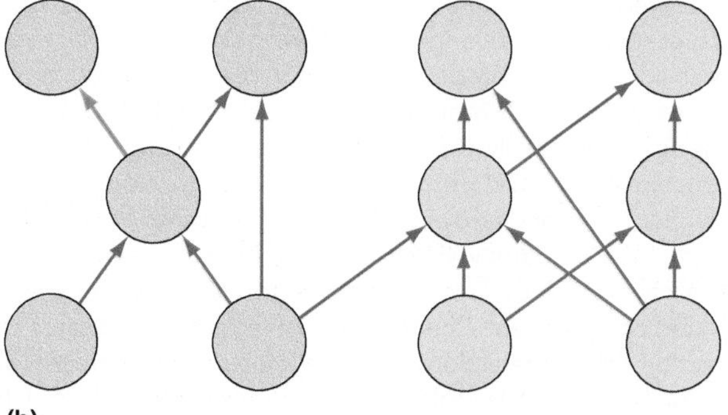

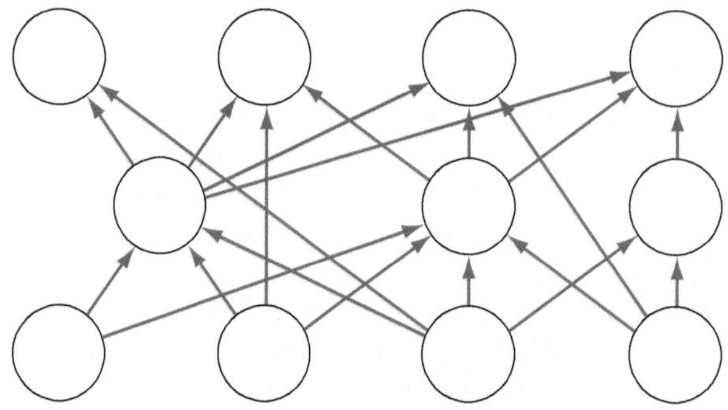

Figure 16.9 (a) Hypothetical food web composed of seven species illustrating the properties of food web connectance, linkage density, chain length (links shown in red), and mean chain length. (b) Two food webs composed of eleven species. The one on the left is compartmentalized, whereas the one on the right is not. The two compartments of the compartmentalized food chain are identified by different colors for the component species. Note that the two compartments are linked only by a single link.

As species richness increases, the structure of food webs become more complex and often food webs become compartmentalized. Species within the same compartment (group of species) interact frequently among themselves but show fewer interactions with species from other compartments (**Figure 16.9b**). For example, Enrico Rezende of the Universitat Autònoma de Barcelona (Spain) and colleagues analyzed a Caribbean marine food web depicting a total of 3313 trophic interactions between 249 species (**Figure 16.11a**). Their analyses indicated a division of the food web into five distinct compartments (**Figure 16.11b**). The researchers determined that the compartments were associated with differences in body size, the range of prey sizes selected, use of shore versus off-shore habitats, and their associated predators.

Although any two species are linked by only a single arrow representing the relationship between predator (the consumer) and prey (the consumed), the dynamics of

(a)

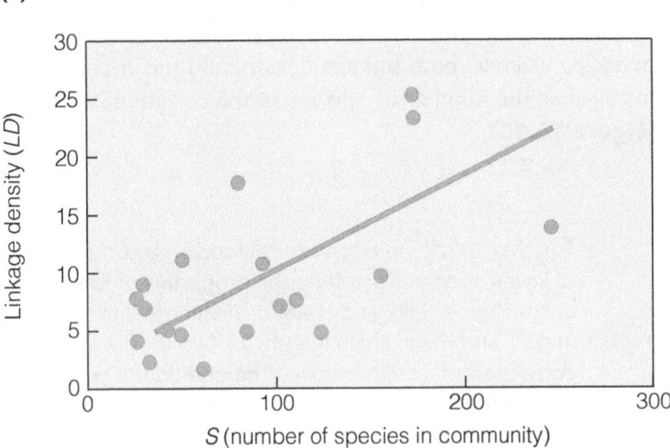

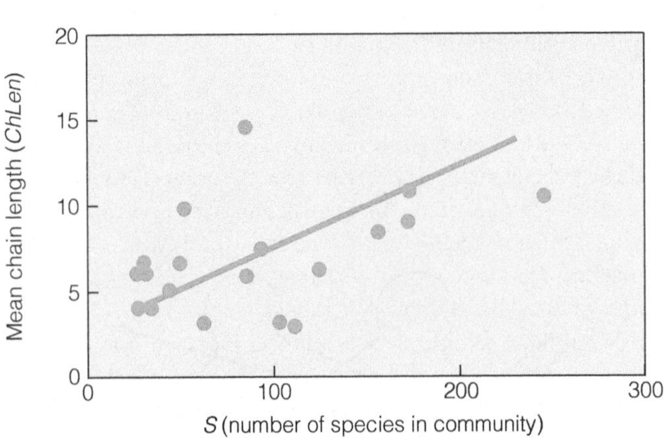

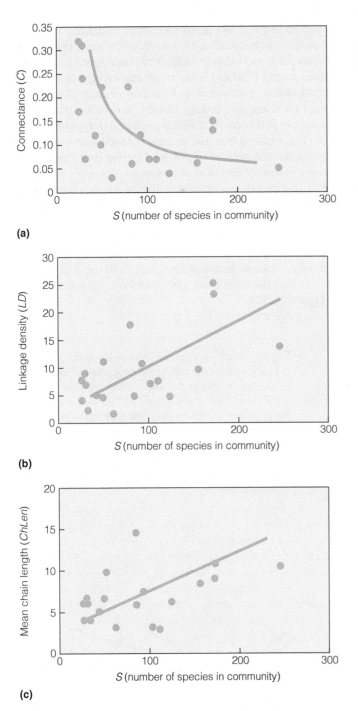

Figure 16.10 Relationship between species richness (*S*) and (a) connectance (*C*), (b) linkage density (*LD*), and (c) mean chain length (*ChLen*) for 19 food webs from a variety of marine, freshwater, and terrestrial communities.
(Data from Dunne et al. 2004.)

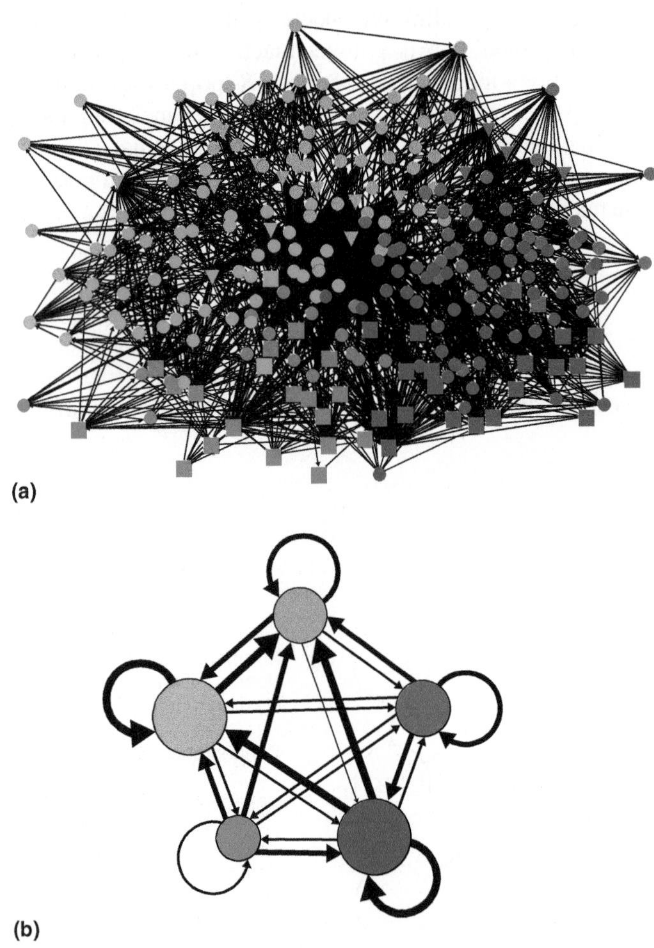

Figure 16.11 Compartmentalized structure of a Caribbean food web. (a) The entire food web. Symbols of different colors represent species belonging to different compartments, whereas each link (arrow) represents a predator–prey interaction. Squares, circles and triangles represent non-fish, bony fish, and shark species, respectively. (b) Diagram of the compartmentalized structure of the food web in (a). Each circle represents a compartment, and arrows indicate the flow of biomass from the prey to the predator within (loops) and between compartments. The size of each circle is proportional to the number of species in that compartment. The thickness of the arrows indicates the fraction of the interactions between the two compartments in relation to the total number of interactions in the entire food web.
(From Rezende et al. 2009.)

communities cannot be understood solely in terms of direct interactions between species. For example, a predator may reduce competition between two prey species by controlling their population sizes below their respective carrying capacities. An analysis of the mechanisms controlling community structure must include these "indirect" effects represented by the structure of the food web; we will explore this topic in more detail later (Chapter 17).

The simple designation of feeding relationships using the graphical approach of food webs can become incredibly complex in communities of even moderate diversity. For this reason, ecologists often simplify the representation of food webs by lumping species into broader categories that represent general feeding groups based on the source from which they derive energy. Earlier, we defined organisms that derive energy from sunlight as autotrophs, or primary producers (Part Two). Organisms that derive energy from consuming plant and animal tissue are called heterotrophs, or secondary producers and are further subdivided into herbivores, carnivores, and omnivores based on their consumption of plant tissues, animal tissues, or both. These feeding groups are referred to as **trophic levels**, after the Greek word *trophikos*, meaning "nourishment."

16.6 Species within a Community Can Be Classified into Functional Groups

The grouping of species into trophic levels is a functional classification; it defines groups of species that derive their energy (food) in a similar manner. Another approach is to subdivide each trophic level into groups of species that exploit a common resource in a similar fashion; these groups are termed **guilds**. The concept of guilds was first introduced by the ecologist Richard Root of Cornell University to describe groups of functionally similar species in a community. For example, hummingbirds and other nectar-feeding birds form a guild of species that exploits the common resource of flowering plants in a similar fashion. Likewise, seed-eating birds could be grouped into another feeding guild within the broader community. Because species within a guild draw on a shared resource, there is potential for strong interactions, particularly interspecific competition, between the members, but weaker interactions with the remainder of their community.

Classifying species into guilds can simplify the study of communities, which allows researchers to focus on more manageable subsets of the community. Yet by classifying species into guilds based on their functional similarity, ecologists can also explore questions about the very organization of communities. Just as we can use the framework of guilds to explore the interactions of the component species within a guild, we can also use this framework to pose questions about the interactions between the various guilds that compose the larger community. At one level, a community can be a complex assembly of component guilds interacting with each other and producing the structure and dynamics that we observe.

In recent years, ecologists have expanded the concept of guilds to develop a more broadly defined approach of classifying species based on function rather than taxonomy. The term **functional type** is now commonly used to define a group of species based on their common response to the environment, life history characteristics, or role within the community. For example, plants may be classified into functional types based on their photosynthetic pathway (C_3, C_4, and CAM), which, as we have seen earlier, relates to their ability to photosynthesize and grow under different thermal and moisture environments (Chapter 6). Similarly, plant ecologists use the functional classification of shade-tolerant and shade-intolerant to reflect basic differences in the physiology and morphology of plant species in response to the light environment (Section 6.8). Grouping plants or animals into the categories of iteroparous and semelparous also represents a functional classification based on the timing of reproductive effort (Chapter 10, Section 10.8).

As with the organization and classification of species into guilds, using functional groups allows ecologists to simplify the structure of communities into manageable units for study and to ask basic questions about the factors that structure communities, as we shall see later in the discussion of community dynamics (Chapter 18).

16.7 Communities Have a Characteristic Physical Structure

Communities are characterized not only by the mix of species and by the interactions among them—their biological structure—but also by their physical features. The physical structure of the community reflects abiotic factors, such as the depth and flow of water in aquatic environments. It also reflects biotic factors, such as the spatial arrangement of the resident organisms. For example, the size and height of the trees and the density and spatial distribution of their populations help define the physical attributes of the forest community.

The forms and structures of terrestrial communities are defined primarily in terms of their vegetation. Plants may be tall or short, evergreen or deciduous, herbaceous or woody. Such characteristics can describe growth forms. Thus, we might speak of shrubs, trees, and herbs and further subdivide the categories into needle-leaf evergreens, broadleaf evergreens, broadleaf deciduous trees, thorn trees and shrubs, dwarf shrubs, ferns, grasses, forbs, mosses, and lichens. Ecologists often classify and name terrestrial communities based on the dominant plant growth forms and their associated physical structure: forests, woodlands, shrublands, or grassland communities (see Chapter 23).

In aquatic environments, communities are also classified and named in terms of the dominant organisms. Kelp forests, seagrass meadows, and coral reefs are examples of such dominant species. However, the physical structure of aquatic communities is more often defined by features of the abiotic environment, such as water depth, flow rate, or salinity (see Chapter 24).

Every community has an associated vertical structure (**Figure 16.12**), a stratification of often distinct vertical layers. On land, the growth form of the plants largely determines this vertical structure—their size, branching, and leaves—and this vertical structure in turn influences, and is influenced by, the vertical gradient of light (see Section 4.2). A well-developed forest ecosystem (Figure 16.12a), for example, has multiple layers of vegetation. From top to bottom, they are the canopy, the understory, the shrub layer, the herb or ground layer, and the forest floor.

The upper layer, the **canopy**, is the primary site of energy fixation through photosynthesis. The canopy structure has a major influence on the rest of the forest. If the canopy is fairly open, considerable sunlight will reach the lower layers. If ample water and nutrients are available, a well-developed **understory** and shrub strata will form. If the canopy is dense and closed, light levels are low, and the understory and shrub layers will be poorly developed.

In the forests of the eastern United States, the understory consists of tall shrubs such as witch hobble (*Viburnum alnifolium*), understory trees such as dogwood (*Cornus* spp.) and hornbeam (*Carpinus caroliniana*), and younger trees, some of which are the same species as those in the canopy. The nature of the **herb layer** depends on the soil moisture and nutrient conditions, slope position, density of the canopy and understory, and exposure of the slope, all of which vary from place to place throughout the forest. The final layer, the **forest floor**, is where the important process of decomposition takes place and

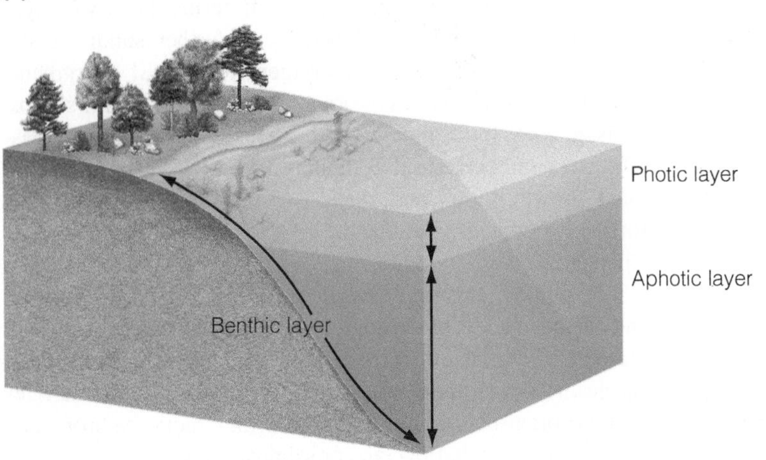

Figure 16.12 Vertical stratification of different communities: (a) temperate deciduous forest, (b) tropical savanna, and (c) lake. Stratification in terrestrial communities is largely biological. Dominant vegetation affects the physical structure of the community and the microclimatic conditions of temperature, moisture, and light. Stratification in aquatic communities is largely physical, influenced by gradients of oxygen, temperature, and light.

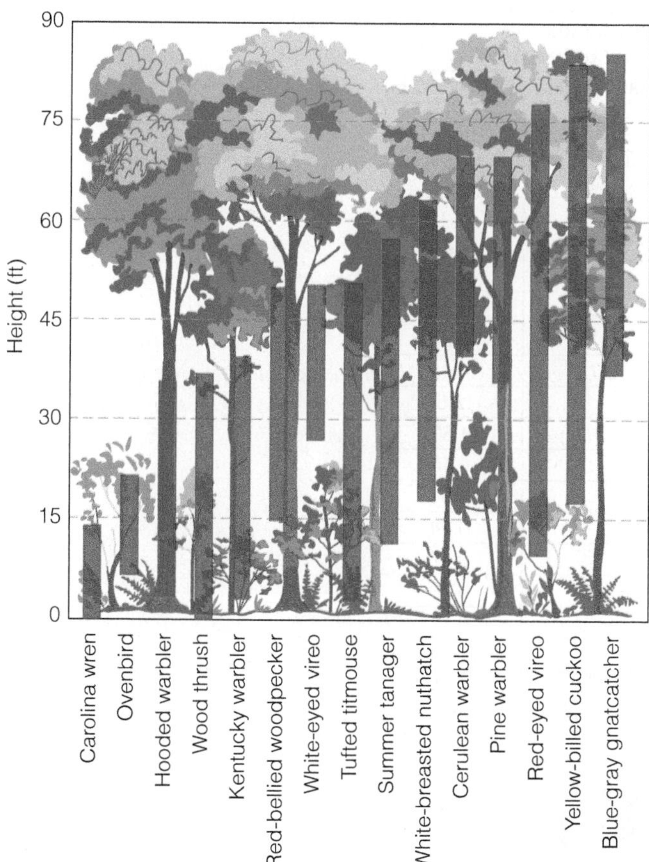

Figure 16.13 Vertical distribution of bird species within the forest community on Walker Branch watershed, Oak Ridge, Tennessee. Height range represented by colored bars is based on total observations of birds during the breeding season regardless of activity. (Data from Anderson and Shugart 1974.)

where microbial organisms feeding on decaying organic matter release mineral nutrients for reuse by the forest plants (see Chapter 21).

In the savanna communities found in the semi-arid regions of Africa, the vertical structure of the vegetation is largely defined by two distinct layers: an herbaceous layer typically dominated by grasses and a woody plant layer dominated by shrubs or trees of varying stature and density dependent on rainfall (Figure 16.12b; also see Chapter 23, Section 23.3).

The strata of aquatic ecosystems such as lakes and oceans are determined largely by the physical characteristics of the water column. As we discussed in Chapter 3, open bodies of water (lakes and oceans) have distinctive profiles of temperature and oxygen (see Sections 3.4 and 3.6). In the summer, well-stratified lakes have a surface layer of warm, well-mixed water high in oxygen, the epilimnion; a second layer, the metalimnion, which is characterized by a thermocline (a steep and rapid decline in temperature relative to the waters above and below); and the hypolimnion, a deep, cold layer of dense water at about 4°C (39°F), often low in oxygen (see Figures 3.8, 3.9, 3.12, and 3.13). Two distinct vertical layers are also recognized based on light penetration through the water column (Figure 16.12c; also see Section 3.3, Figure 3.7): an upper layer, the **photic layer**, where the availability of light supports photosynthesis, and a deeper layer of waters, the **aphotic layer**, an area without light. The bottom layer of sediments, where decomposition is most active, is referred to as the **benthic layer**.

Characteristic organisms inhabit each available vertical layer, or stratum, in a community. In addition to the vertical distribution of plant life already described, various types of consumers and decomposers occupy all levels of the community (although decomposers are typically found in greater abundance in the forest floor [soil surface] and sediment [benthic] layers). Considerable interchange takes place among the vertical strata, but many highly mobile animals are confined to only a few layers (**Figure 16.13**). Which species occupies a given vertical layer may change during the day or season. Such changes reflect daily and seasonal variations in the physical environment such as humidity, temperature, light, and oxygen concentrations in the water, shifts in the abundance of essential resources such as food, or different requirements of organisms for the completion of their life cycles. For example, zooplankton migrate vertically in the water column during the course of the day in response to varying light and predation (**Figure 16.14**).

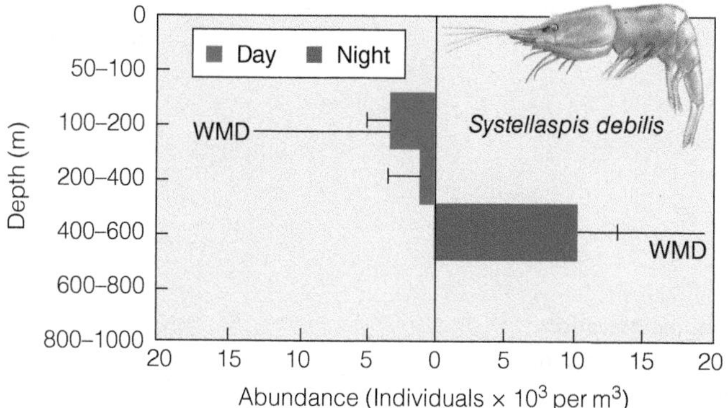

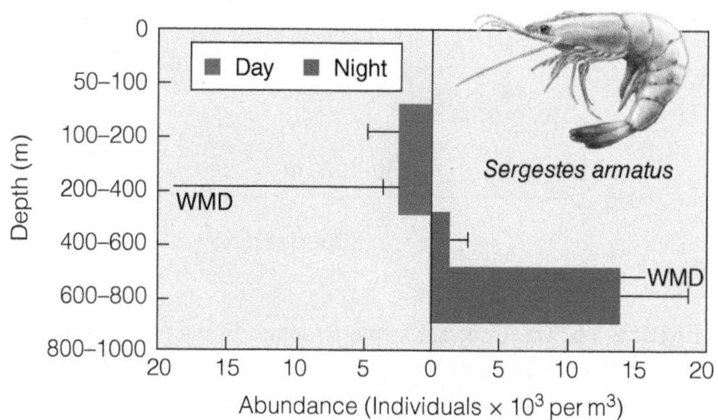

Figure 16.14 Daily vertical migration patterns of two decapod species (marine zooplankton) off the coast of Namibia (Africa). Mean (+1 standard deviation) of daytime and nighttime distribution. Weighted mean depths (WMD) are shown as solid line. (Adapted from Schukat et al. 2013.)

Species	Common name	Relative abundance (% Total number of individuals) 1920–2140 m	1680–1920 m	1370–1680 m
Abies nobilis	Noble Fir	64.15	40.10	2.41
Tsuga mertensiana	Mountain hemlock	35.77	18.97	0.07
Pinus monticola	Western white pine	0.08	0.03	
Pseudotsuga menziesii	Douglas fir		0.16	14.75
Libocedrus decurrens	Incense-cedar		0.35	2.44
Abies concolor	White fir		39.50	54.32
Pinus lambertiana	Sugar pine		0.03	0.98
Corylus rostrata	Beaked hazelnut		0.13	2.64
Acer glabrum	Rocky Mountain maple		0.73	6.44
Chamaecyparis lawsoniana	Lawson cypress			12.55
Taxus brevifolia	Pacific yew			1.83
Pinus ponderosa	Ponderosa pine			0.10
Acer macrophyllum	Oregon maple			0.51
Salix spp.	Willow			0.14
Alnus spp.	Alder			0.10
Amelanchier florida	Florida juneberry			0.10
Sobrum americana	American Mountain-ash			0.17
Castanopsis chrysophylla	Giant chinkapin			0.44

(a)

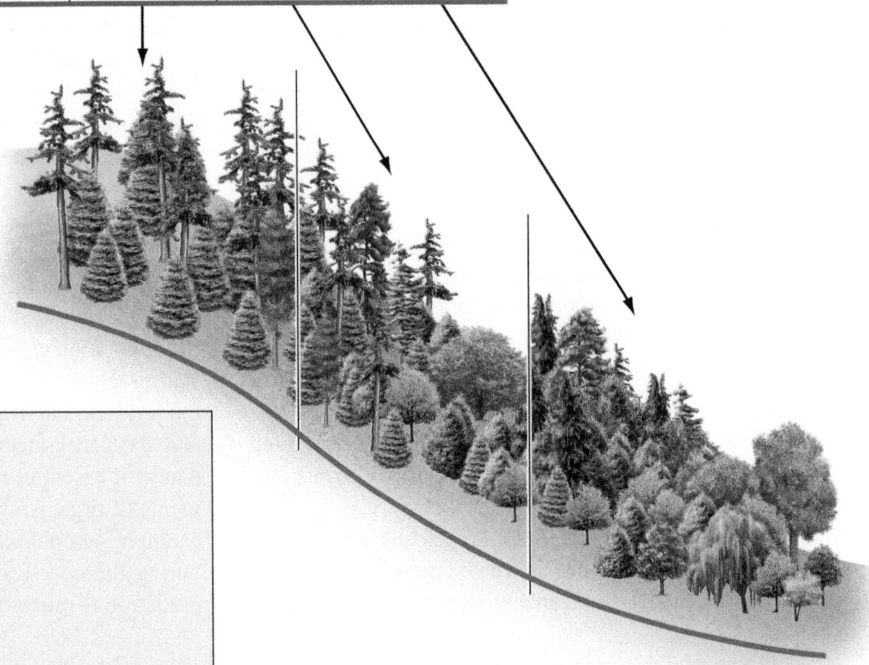

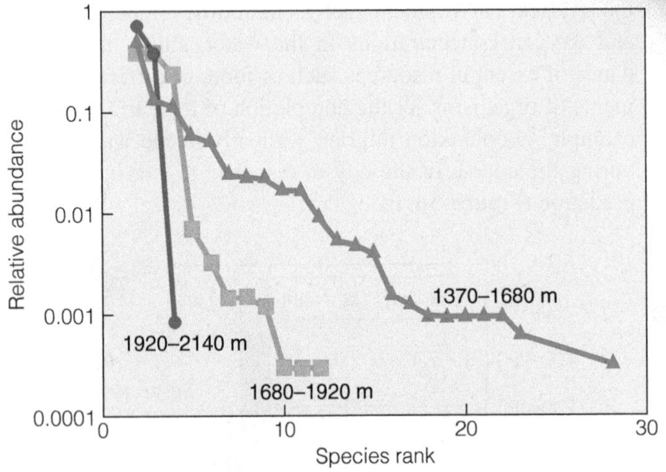

(b)

Figure 16.15 Changes in the structure of the forest communities along an elevation gradient in the Siskiyou Mountains of northwestern California and southwestern Oregon. (a) Changes in the relative abundance of tree species (percentage of total number of individuals) for three segments of the elevation gradient: 1370–1680, 1680–2140, and 1920–2140 m. (b) Rank-abundance curves for the three forest communities corresponding to the three segments of the elevation gradient.

(Data from Whittaker, R. 1960. Vegetation of the Siskiyou Mountains, Oregon and California. Ecological Monographs 30: 279–338. Table 12, pg. 294.)

16.8 Zonation Is Spatial Change in Community Structure

As we move across the landscape, the biological and physical structure of the community changes. Often these changes are small, subtle ones in the species composition or height of the vegetation. However, as we travel farther, these changes often become more pronounced. For example, in a study of the vegetation of the Siskiyou Mountains of northwestern California and southwestern Oregon, the eminent plant ecologist Robert Whittaker of Cornell University provides a description of changes in the structure of the forest communities along an elevation gradient from the base of the mountains to the summit. A description of changes in the forest community along part of this elevation gradient is presented in **Figure 16.15**. At mid-elevations (1370–1680 meters [m]) the forest community is dominated by white fir (*Abies concolor*) with a diverse array of conifer and deciduous species. As you move up in elevation (1680–1920 m), white fir remains a dominant component of the community; however, a second species, noble fir (*Abies nobilis*), a minor species at lower elevations, emerges as co-dominant. In addition to the shift in species composition, there is a decline in species richness from 17 tree species to 9 species. As you move further up the slope (1920–2140 m), there is once again both a shift in species composition and a further decline in species richness. At this elevation the community is effectively limited to only two species: Noble fir and mountain hemlock (*Tsuga mertensiana*). What Whittaker observed was a gradual change in the species composition and decline in species diversity in the forest community as one moves up in elevation. Besides changes in the vegetation, the animal species—insects, birds, and small mammals—that occupy the forest also change. These changes in the physical and biological structures of communities as one moves across the landscape are referred to as **zonation**.

Patterns of spatial variation in community structure or zonation are common to all environments—aquatic and terrestrial. **Figure 16.16** provides an example of zonation in a salt marsh along the northeastern coastline of North America. In moving from the shore and through the marsh to the upland, notice the variations in the physical and biological structures of the communities. The dominant plant growth forms in the marsh are grasses and sedges. These growth forms give way to shrubs and trees as we move to dry land and the depth of the water table increases. In the zone dominated by grasses and sedges, the dominant species change as we move back from the tidal areas. These differences result from various environmental changes across a spatial gradient, including microtopography, water depth, sediment oxygenation, and salinity. The changes are marked by distinct plant communities that are defined by changes in dominant plants as well as in structural features such as height, density, and spatial distribution of individuals.

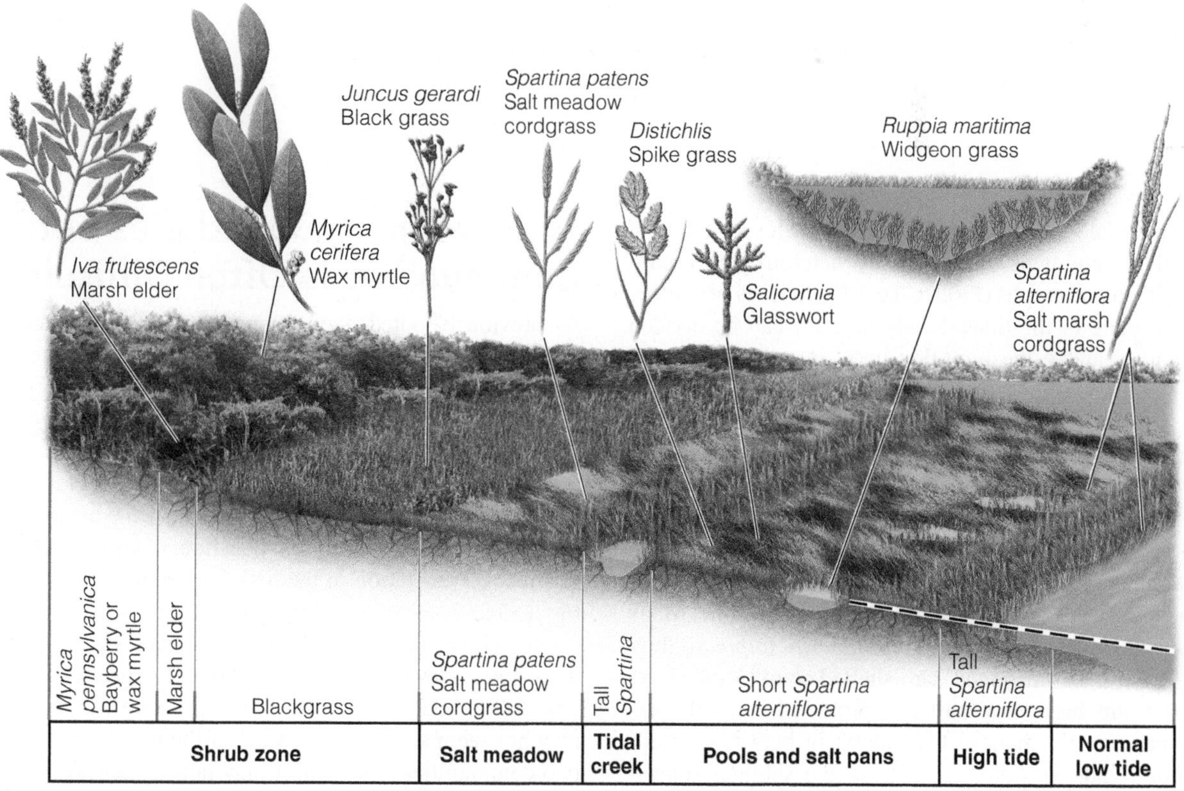

Figure 16.16 Patterns of zonation in an idealized New England salt marsh, showing the relationship of plant distribution to microtopography and tidal submergence.

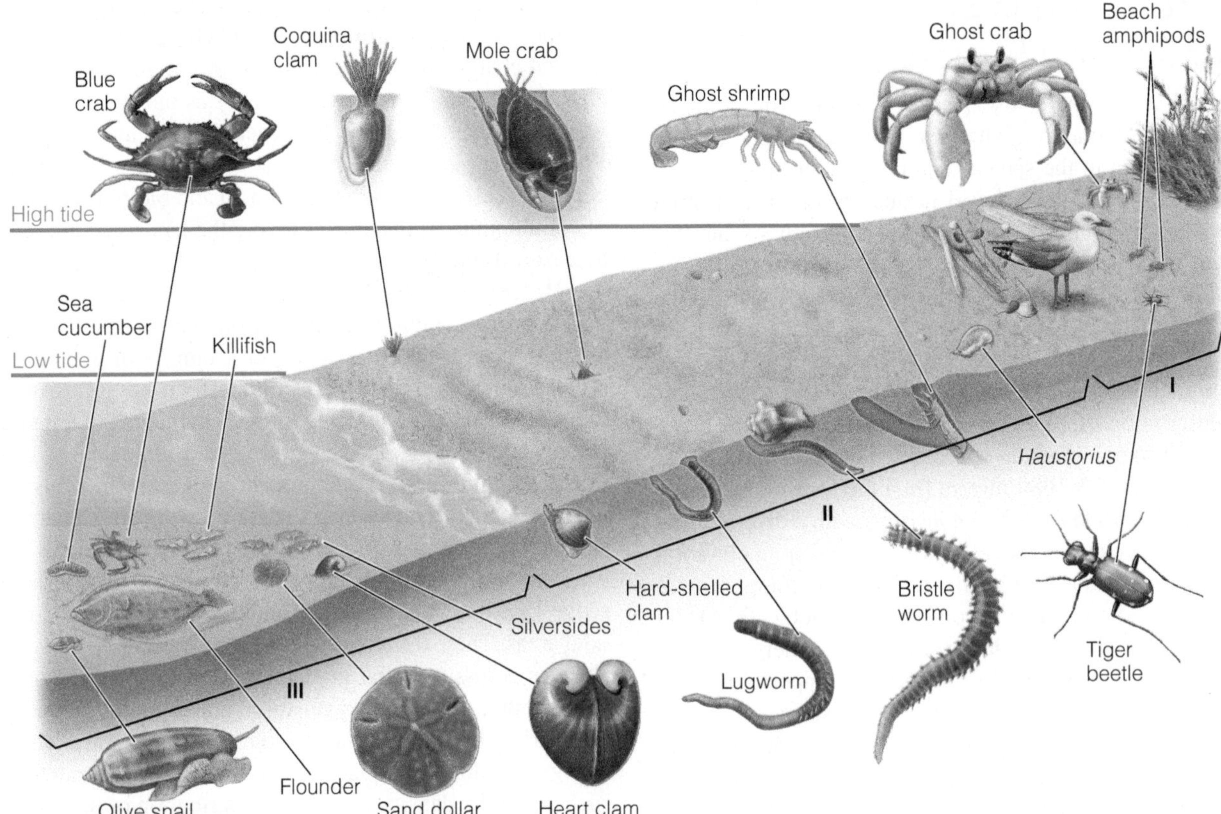

Figure 16.17 Life on a sandy ocean beach along the mid-Atlantic coast is an example of zonation dominated by changes in the fauna. The distribution of organisms changes along a gradient from land to sea as a function of the degree and duration of inundation during the tidal cycle. I—supratidal zone (above high-tide line): ghost crabs and Beach amphipods. II—intertidal zone (between high- and low-tide lines): ghost shrimp, bristle worms, clams, lugworms, mole crabs. III—subtidal zone (below low-tide line): flounder, blue crab, sea cucumber. The blue lines indicate high and low tides.

The intertidal zone of a sandy beach provides an example in which the zonation is dominated by heterotrophic organisms rather than autotrophs (**Figure 16.17**). Patterns of species distribution relate to the tides. Sandy beaches can be divided into supratidal (above the high-tide line), intertidal (between the high- and low-tide lines), and subtidal (below the low-tide line; continuously inundated) zones; each are home to a unique group of animal organisms. Pale, sand-colored ghost crabs (*Ocypode quadrata*) and beach fleas (*Talorchestia* and *Orchestia* spp.) occupy the upper beach, or supratidal zone. The intertidal beach is the zone where true marine life begins. An array of animal species adapted to the regular periods of inundation and exposure to the air are found within this zone. Many of these species, such as the mole crab (*Emerita talpoida*), lugworm (*Arenicola cristata*), and hard-shelled clam (*Mercenaria mercenaria*), are burrowing animals, protected from the extreme temperature fluctuations that can occur between periods of inundation and exposure. In contrast, the subtidal zone is home to a variety of vertebrate and invertebrate species that migrate into and out of the intertidal zone with the changing tides.

16.9 Defining Boundaries between Communities Is Often Difficult

As previously noted, the community is a spatial concept involving the species that occupy a given area. Ecologists typically distinguish between adjacent communities or community types based on observable differences in their physical and biological structures: the different species assemblages characteristic of different physical environments. How different must two adjacent areas be before we call them separate communities? This is not a simple question. Consider the elevation gradient of vegetation in the Siskiyou Mountains illustrated in Figure 16.15. Given the difference in species composition that occurs with changes in elevation, most ecologists would define these three elevation zones as different vegetation communities. As we hike up the mountainside, however, the distinction may not seem so straightforward. If the transition between the two communities is abrupt, it may not be hard to define community boundaries. But if the species composition and patterns of dominance shift gradually, the boundary is not as clear.

Ecologists use various sampling and statistical techniques to delineate and classify communities. Generally, all employ some measure of community similarity or difference (see **Quantifying Ecology 16.1**). Although it is easy to describe the similarities and differences between two areas in terms of species composition and structure, actually classifying areas into distinct groups of communities involves a degree of subjectivity that often depends on the study objectives and the spatial scale at which vegetation is being described.

The example of forest zonation presented in Figure 16.15 occurs over a relatively short distance moving up the mountainside. As we consider ever-larger areas, differences in community structure—both physical and biological—increase. An example is the pattern of forest zonation in Great Smoky Mountains National Park (**Figure 16.18**). The zonation is a complex pattern related to elevation, slope position, and exposure. Note that the description of the forest communities in the park contains few species names. Names like *hemlock forest* are not meant to suggest a lack of species diversity; they are just a shorthand method of naming communities for the dominant tree species. Each community could be described by a complete list of species, their population sizes, and their contributions to the total biomass (as with the communities in Table 16.1 or Figure 16.15). However, such lengthy descriptions are unnecessary to communicate the major changes in the structure of communities across the landscape. In fact, as we expand the area of interest to include the entire eastern United States, the nomenclature for classifying forest communities becomes even broader. In **Figure 16.19**, which is

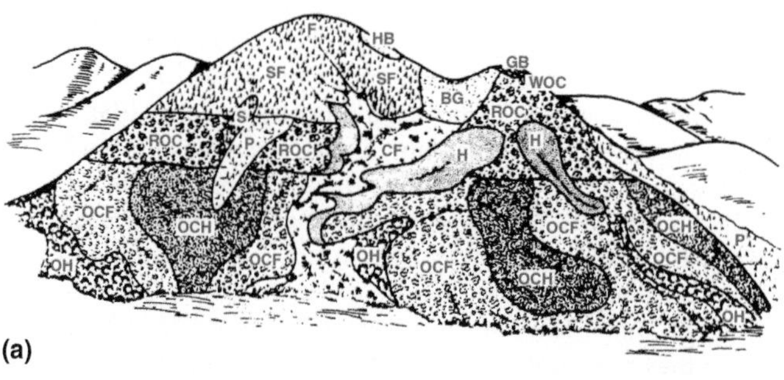

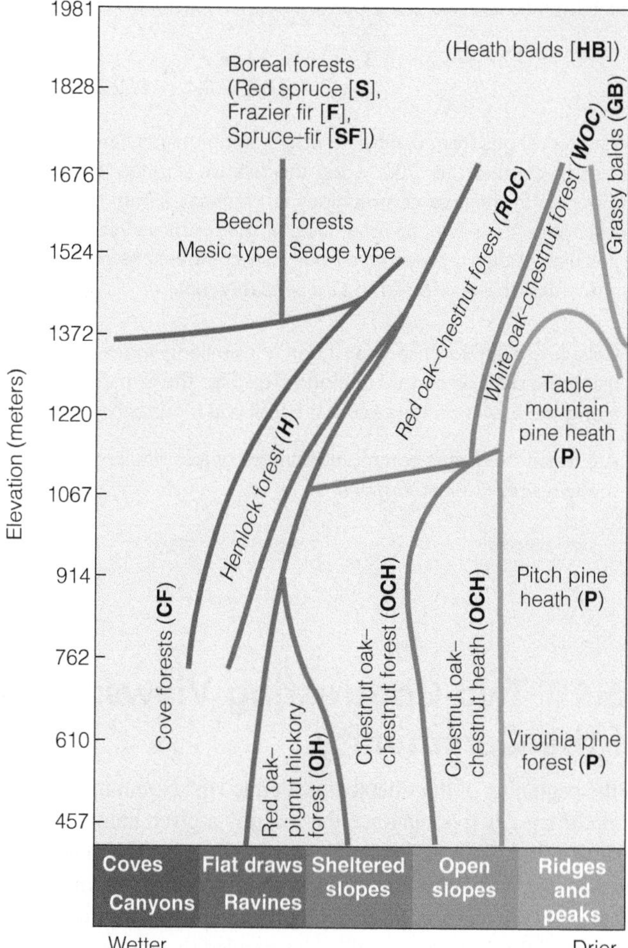

Figure 16.18 Two descriptions of forest communities in Great Smoky Mountains National Park. (a) Topographic distribution of vegetation types on an idealized west-facing mountain and valley. (b) Idealized arrangement of community types according to elevation and aspect.
(Adapted from Whittaker 1954.)

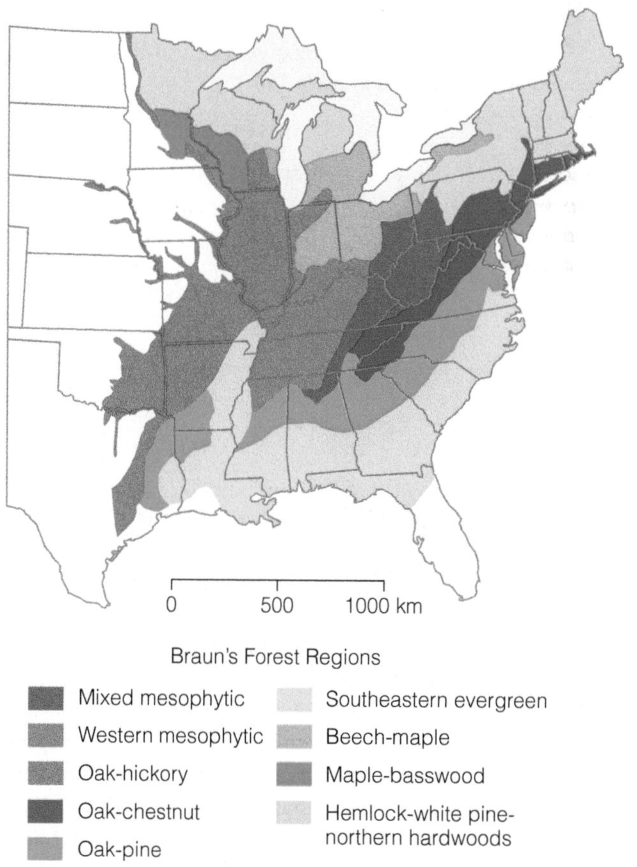

Figure 16.19 Large-scale distribution of deciduous forest communities in the eastern United States is defined by nine regions.
(Adapted from Dyer 2006.)

QUANTIFYING ECOLOGY 16.1 Community Similarity

When we say that a community's structure changes as we move across the landscape, we imply that the set of species that define the community differ from one place to another. But how do we quantify this change? How do ecologists determine where one community ends and another begins? Distinguishing between communities based on differences in species composition is important in understanding the processes that control community structure as well as in conservation efforts to preserve natural communities.

Various indexes have been developed that measure the similarity between two areas or sample plots based on species composition. Perhaps the most widely used is Sorensen's coefficient of community (*CC*). The index is based on species presence or absence. Using a list of species compiled for the two sites or sample plots that are to be compared, the index is calculated as:

Number of species common to both communities

$$CC = 2c/(s_1 + s_2)$$

Number of species in community 1 — Number of species in community 2

As an example of this index, we can use the two forest communities presented in Table 16.1:

$$s_1 = 24 \text{ species}$$

$$s_2 = 10 \text{ species}$$

$$c = 9 \text{ species}$$

$$CC = \frac{(2 \times 9)}{(24 + 10)} = \frac{18}{24} = 0.529$$

The value of the index ranges from 0, when the two communities share no species in common, to 1, in which the species composition of the two communities is identical (all species in common).

The *CC* does not consider the relative abundance of species. It is most useful when the intended focus is the presence or absence of species. Another index of community similarity that is based on the relative abundance of species within the communities being compared is the **percent similarity (*PS*)**.

To calculate *PS*, first tabulate species abundance in each community as a percentage (as was done for the two communities in Table 16.1). Then add the lowest percentage for each species that the communities have in common. For the two forest communities, 16 species are exclusive to one community or the other. The lowest percentage of those 16 species is 0, so they need not be included in the summation. For the remaining nine species, the index is calculated as follows:

$$PS = 29.7 + 4.7 + 4.3 + 0.8 + 3.6 + 2.9 + 0.4 + 0.4 + 0.4 = 47.2$$

This index ranges from 0, when the two communities have no species in common, to 100, when the relative abundance of the species in the two communities is identical. When comparing more than two communities, a matrix of values can be calculated that represents all pairwise comparisons of the communities; this is referred to as a *similarity matrix*.

1. Calculate both Sorensen's and percent similarity indexes using the data presented in Figure 16.15 for the forests along the elevation transect in the Siskiyou Mountains.
2. Are these two forest communities more or less similar than the two sites in West Virginia?

a broad-scale description of forest zonation in the eastern United States developed by E. Lucy Braun, all of Great Smoky Mountains National Park shown in Figure 16.18 (located in southeastern Tennessee and northwestern North Carolina) is described as a single forest community type: Oak-chestnut, a type that extends from New York to Georgia.

These large-scale examples of zonation make an important point that we return to when examining the processes responsible for spatial changes in community structure: our very definition of community is a spatial concept. Like the biological definition of population, the definition of community refers to a spatial unit that occupies a given area (see Chapter 8). In a sense, the distinction among communities is arbitrary, based on the criteria for classification. As we shall see, the methods used in delineating communities as discrete spatial units have led to problems in understanding the processes responsible for patterns of zonation (see Chapter 17).

16.10 Two Contrasting Views of the Community

At the beginning of this chapter, we defined the community as the group of species (populations) that occupy a given area, interacting either directly or indirectly. Interactions can have both positive and negative influences on species populations. How important are these interactions in determining community structure? In the first half of the 20th century, this question led to a major debate in ecology that still influences our views of the community.

When we walk through most forests, we see a variety of plant and animal species—a community. If we walk far enough, the dominant plant and animal species change (see Figure 16.15). As we move from hilltop to valley, the structure of the community differs. But what if we continue our walk over the next hilltop and into the adjacent valley? We would most likely notice that although the communities on the

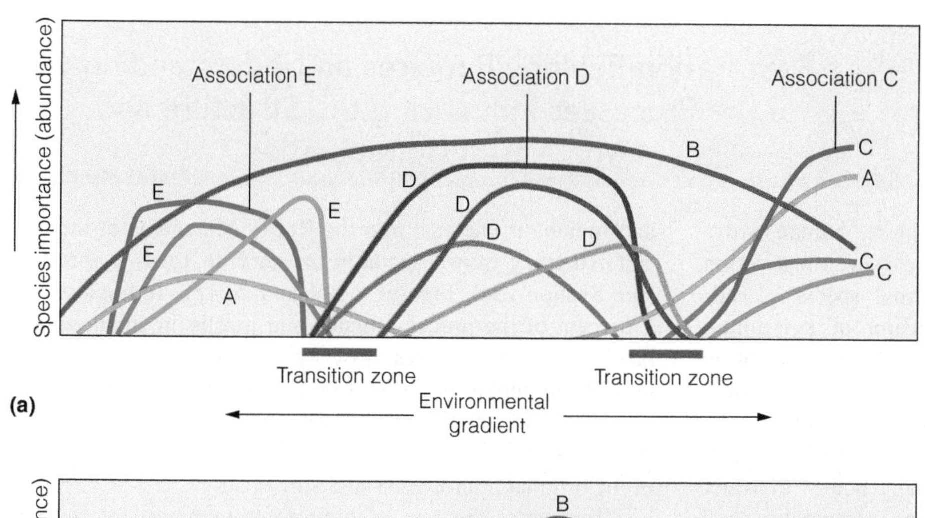

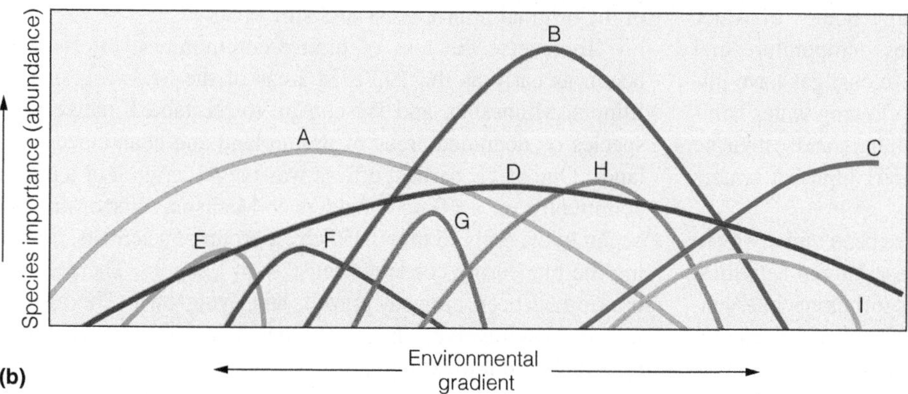

Figure 16.20 Two models of community. (a) The organismal, or discrete, view of communities proposed by Clements. Clusters of species (Cs, Ds, and Es) show similar distribution limits and peaks in abundance. Each cluster defines an association. A few species (e.g., A) have sufficiently broad ranges of tolerance that they occur in adjacent associations but in low numbers. A few other species (e.g., B) are ubiquitous. (b) The individualistic, or continuum, view of communities proposed by Gleason. Clusters of species do not exist. Peaks of abundance of dominant species, such as A, B, and C, are merely arbitrary segments along a continuum.

hilltop and valley are quite distinct, the communities on the two hilltops or valleys are quite similar. As a botanist might put it, they exhibit relatively consistent floristic composition. At the International Botanical Congress of 1910, botanists adopted the term *association* to describe this phenomenon. An association is a type of community with (1) relatively consistent species composition, (2) a uniform, general appearance (physiognomy), and (3) a distribution that is characteristic of a particular habitat, such as the hilltop or valley. Whenever the particular habitat or set of environmental conditions repeats itself in a given region, the same group of species occurs.

Some scientists of the early 20th century thought that association implied processes that might be responsible for structuring communities. The logic was that the existence of clusters or groups of species that repeatedly associate was indirect evidence for either positive or neutral interactions among them. Such evidence favors a view of communities as integrated units. A leading proponent of this thinking was the Nebraskan botanist Frederic Clements. Clements developed what has become known as the **organismic concept of communities**. Clements likened associations to organisms, with each species representing an interacting, integrated component of the whole. Development of the community through time (a process termed *succession*) was viewed as development of the organism (see Chapter 18).

As depicted in **Figure 16.20a**, the species in an association have similar distributional limits along the environmental gradient in Clements's view, and many of them rise to maximum abundance at the same point. Transitions between adjacent communities (or associations) are narrow, with few species in common. This view of the community suggests a common evolutionary history and similar fundamental responses and tolerances for the component species (see Chapter 5 and Section 12.6). Mutualism and coevolution play an important role in the evolution of species that make up the association. The community has evolved as an integrated whole; species interactions are the "glue" holding it together.

In contrast to Clements's organismal view of communities was botanist H. A. Gleason's view of community. Gleason stressed the individualistic nature of species distribution. His view became known as the **individualistic**, or **continuum concept**. The continuum concept states that the relationship among coexisting species (species within a community) is a result of similarities in their requirements and tolerances, not to strong interactions or common evolutionary history. In fact, Gleason concluded that changes in species abundance along environmental gradients occur so gradually that it is not practical to divide the vegetation (species) into associations. Unlike Clements, Gleason asserted that species distributions along environmental gradients do not form clusters but rather represent the independent responses of species. Transitions are gradual and difficult to identify (**Figure 16.20b**). What we refer to as the community is merely the group of species found to coexist under any particular set of environmental conditions. The major difference between these two views is the importance of interactions—evolutionary and current—in the structuring of communities. It is tempting to choose between these views, but as we will see, current thinking involves elements of both perspectives.

ECOLOGICAL Issues & Applications

Restoration Ecology Requires an Understanding of the Processes Influencing the Structure and Dynamics of Communities

As we have discussed in previous chapters, human activities have led to population declines and even extinction of a growing number of plant and animal species. Land-use changes associated with the expansion of agriculture (Chapter 9, *Ecological Issues & Applications*) and urbanization (Chapter 12, *Ecological Issues & Applications*) have resulted in dramatic declines in biological diversity associated with the loss of essential habitats. Likewise, dams have removed sections of turbulent river and created standing bodies of water (lakes and reservoirs), affecting flow rates, temperature and oxygen levels, and sediment transport. These changes have impacted not only the species that depend on flowing water habitats (see Figure 9.15) but also coastal wetlands and estuarine environments that depend on the continuous input of waters from river courses (see Chapter 25).

In recent years, considerable efforts have been under way to restore natural communities affected by these human activities. This work has stimulated a new approach to human intervention that is termed **restoration ecology**. The goal of restoration ecology is to return a community or ecosystem to a close approximation of its condition before disturbance by applying ecological principles. Restoration ecology involves a continuum of approaches ranging from reintroducing species and restoring habitats to attempting to reestablish whole communities.

The least intensive restoration effort involves the rejuvenation of existing communities by eliminating invasive species (Chapter 8, *Ecological Issues & Applications*), replanting native species, and reintroducing natural disturbances such as short-term periodic fires in grasslands and low-intensity ground fires in pine forests. Lake restoration involves reducing inputs of nutrients, especially phosphorus, from the surrounding land that stimulate growth of algae, restoring aquatic plants, and reintroducing fish species native to the lake. Wetland restoration may involve reestablishing the hydrological conditions, so that the wetland is flooded at the appropriate time of year, and the replanting of aquatic plants (**Figure 16.21**).

More intensive restoration involves recreating the community from scratch. This kind of restoration involves preparing the site, introducing an array of appropriate native species over time, and employing appropriate management to maintain the community, especially against the invasion of nonnative species from adjacent surrounding areas. A classic example of this type of restoration is the ongoing effort to reestablish the tallgrass prairie communities of North America.

When European settlers to North America first explored the region west of the Mississippi River, they encountered a landscape on a scale unlike any they had known in Europe. The forested landscape of the east gave way to a vast expanse of grass and wildflowers. The prairies of North America once covered a large portion of the continent, ranging from Illinois and Indiana in the east into the Rocky Mountains of the west and extending from Canada in the north to Texas in the south (see Section 23.4, Figures 23.14 and 23.15). Today less than 1 percent of the prairie remains and mostly in small isolated patches, which is the result of a continental-scale transformation of this region to agriculture (see Figure 9.17). For example, in the state of Illinois, tallgrass prairie once covered more than 90,000 km^2, whereas today estimates are that only 8 km^2 of the original prairie grassland still exists.

To reverse the loss of prairie communities, efforts were begun as early as the 1930s in areas of the Midwest, such as Illinois, Minnesota, and Wisconsin, to reestablish native plant species on degraded areas of pastureland and abandoned croplands. One of the earliest efforts was the re-creation of a prairie community on a 60-acre field near Madison, Wisconsin, that began in the early to mid-1930s by a group of scientists, including the pioneering conservationist Aldo Leopold. The previous prairie had been plowed, grazed, and overgrown. The restoration process involved destroying occupying weeds and brush, reseeding and replanting native prairie species, and burning the site once every two to three years to approximate a natural fire regime (**Figure 16.22**). After nearly 80 years, the plant community now resembles the original native prairie (**Figure 16.23**).

These early efforts were in effect an attempt to reconstruct native prairie communities—the set of plant and animal species that once occupied these areas. But how does one start to rebuild an ecological community? Can a community be constructed by merely bringing together a collection of species in one place?

Figure 16.21 Volunteers help National Oceanic and Atmospheric Administration (NOAA) scientists prepare sea-grass shoots for planting in the Florida Keys. The plantings help enhance recovery of areas where sea-grass communities have been damaged or large-scale die-off has occurred.

(a)

(b)

Figure 16.22 Photographs of early efforts in the restoration of a prairie community at the University of Wisconsin Arboretum (now the John T. Curtis Prairie). (a) In 1935, a Civilian Conservation Corps camp was established and work began on the restoration effort. (b) Early experiments established the critical importance of fire in maintaining the structure and diversity of the prairie community.

Many early reconstruction efforts met with failure. They involved planting whatever native plant species might be available in the form of seeds, often on small plots surrounded by agricultural lands. The native plant species grew, but their populations often declined over time. Early efforts failed to appreciate the role of natural disturbances in maintaining these communities. Fire has historically been an important feature of the prairie, and many of the species were adapted to periodic burning. In the absence of fire, native species were quickly displaced by nonnative plant species from adjacent pastures.

Prairie communities are characterized by a diverse array of plant species that differ in the timing of germination, growth, and reproduction over the course of the growing season. The result is a shifting pattern of plant populations through time that provides a consistent resource base for the array of animal species throughout the year. Attempts at restoration that do not include this full complement of plant species typically cannot attract and support the animal species that characterize native prairie communities.

The size of restoration projects was often a key factor in their failure. Small, isolated fragments tend to support species at low population levels and are thus prone to local extinction. These isolated patches were too distant from other patches of native grassland for the natural dispersal of other species, both plant and animal. Isolated patches of prairie often lacked the appropriate pollinator species required for successful plant reproduction.

Much has been learned from early attempts at restoring natural communities, and many restoration efforts have since succeeded. Restored prairie sites at Fermi National Accelerator Laboratory in northern Illinois are the product of more than 40 years of effort and now contain approximately 1000 acres; it is currently the largest restored prairie habitat in the world.

Attempts at reconstructing communities raise countless questions about the structure and dynamics of ecological communities, questions that in one form or another had been central to the study of ecological communities for more than a century. What controls the relative abundance of species within the community? Are all species equally important to the functioning and persistence of the community? How do the component species interact with each other? Do these interactions restrict or enhance the presence of other species? How do communities change through time? How does the community's size influence the number of species it can support? How do different communities on the larger landscape interact?

As we shall see in the chapters that follow, ecological communities are more than an assemblage of species whose geographic distributions overlap. Ecological communities represent a complex web of interactions whose nature changes as environmental conditions vary in space and time.

Figure 16.23 Curtis Prairie at the University of Wisconsin Arboretum. Native prairie vegetation has been restored on this 60-acre tract of land that was once used for agriculture.

SUMMARY

Biological Structure 16.1

A community is the group of species (populations) that occupy a given area and interact either directly or indirectly. The biological structure of a community is defined by its species composition, that is, the set of species present and their relative abundances.

Diversity 16.2

The number of species in the community defines species richness. Species diversity involves two components: species richness and species evenness, which reflect how individuals are apportioned among the species (relative abundances).

Dominance 16.3

When a single or a few species predominate within a community, they are referred to as *dominants*. The dominants are often defined as the most numerically abundant; however, in populations or among species in which individuals can vary widely in size, abundance alone is not always a sufficient indicator of dominance.

Keystone Species 16.4

Keystone species are species that function in a unique and significant manner, and their effect on the community is disproportionate to their numerical abundance. Their removal initiates changes in community structure and often results in a significant loss of diversity. Their role in the community may be to create or modify habitats or to influence the interactions among other species.

Food Webs 16.5

Feeding relationships can be graphically represented as a food chain: a series of arrows, each pointing from one species to another that is a source of food. Within a community, many food chains mesh into a complex food web with links leading from primary producers to an array of consumers. Species that are fed on but that do not feed on others are termed *basal species*. Species that feed on others but are not prey for other species are termed *top predators*. Species that are both predators and prey are termed *intermediate species*.

Functional Groups 16.6

Groups of species that exploit a common resource in a similar fashion are termed *guilds*. Functional group or functional type is a more general term used to define a group of species based on their common response to the environment, life history characteristics, or role within the community.

Physical Structure 16.7

Communities are characterized by physical structure. In terrestrial communities, structure is largely defined by the vegetation. Vertical structure on land reflects the life-forms of plants. In aquatic environments, communities are largely defined by physical features such as light, temperature, and oxygen profiles. All communities have an autotrophic and a heterotrophic layer. The autotrophic layer carries out photosynthesis. The heterotrophic layer uses carbon stored by the autotrophs as a food source. Vertical layering provides the physical structure in which many forms of animal life live.

Zonation 16.8

Changes in the physical structure and biological communities across a landscape result in zonation. Zonation is common to all environments, both aquatic and terrestrial. Zonation is most pronounced where sharp changes occur in the physical environment, as in aquatic communities.

Community Boundaries 16.9

In most cases, transitions between communities are gradual, and defining the boundary between communities is difficult. The way we classify a community depends on the scale we use.

Concept of the Community 16.10

Historically, there have been two contrasting concepts of the community. The organismal concept views the community as a unit, an association of species, in which each species is a component of the integrated whole. The individualistic concept views the co-occurrence of species as a result of similarities in requirements and tolerances.

Restoration Ecology Ecological Issues & Applications

The goal of restoration ecology is to return a community or ecosystem to a close approximation of its condition before disturbance by applying ecological principles. Restoration ecology requires an understanding of the basic processes influencing the structure and dynamics of ecological communities.

STUDY QUESTIONS

1. Distinguish between species richness and species evenness.
2. Distinguish between a dominant and a keystone species.
3. Why are plants classified as basal species in the structure of a food web?
4. Are all carnivores top predators? What distinguishes a top predator in the structure of a food chain?
5. Contrast the vertical stratification of an aquatic community with that of a terrestrial community.
6. Define zonation.
7. In Figure 16.18, the vegetation of Great Smoky Mountains National Park is classified into distinct community types. Does this approach suggest the organismal or individualistic concept of communities? Why?

FURTHER READINGS

Classic Studies

Pimm, S. L. 1982. *Food webs.* Chicago: University of Chicago Press.

Although first published more than 30 years ago, this book remains the most complete and clearest introduction to the study of food webs. New edition published in 2002.

Recent Research

Brown, J. H. 1995. *Macroecology.* Chicago: University of Chicago Press.

In this book, Brown presents a broad perspective for viewing ecological communities over large geographic regions and long timescales.

Estes, J., M. Tinker, T. Williams, and D. Doak. 1998. "Killer whale predation on sea otters linking oceanic and near-shore ecosystems." *Science* 282:473–476.

An excellent example of the role of keystone species in the coastal marine communities of western Alaska.

Falk, D. A., M. Palmer, and J. Zedler. 2006. *Foundations of restoration ecology*. Washington, D. C.: Island Press.

Overview of ecological principles and approaches to restoring natural communities and ecosystems.

Mittelbach, G. G. 2012. *Community Ecology*. Sunderland, MA: Sinauer Associates.

An excellent text that provides an overview of the study of ecological communities.

Morin, Peter J. 1999. *Community Ecology.* Oxford: Blackwell Science.

This text provides a good overview of the field of community ecology.

Pimm, S. L. 1991. *The balance of nature.* Chicago: University of Chicago Press.

An excellent example of the application of theoretical studies on food webs and the structure of ecological communities to current issues in conservation ecology.

Power, M. E., D. Tilman, J. Estes, B. Menge, W. Bond, L. Mills, G. Daily, J. Castilla, J. Lubchenco, and R. Paine. 1996. "Challenges in the quest for keystones." *Bioscience* 46:609–620.

This article reviews the concept of keystone species as presented by many of the current leaders in the field of community ecology.

Ricklefs, R. E., and D. Schluter, eds. 1993. *Ecological communities: Historical and geographic perspectives.* Chicago: University of Chicago Press.

This pioneering work examines biodiversity in its broadest geographical and historical contexts, exploring questions relating to global patterns of species richness and the historical events that shape both regional and local communities.

CHAPTER 17 Factors Influencing the Structure of Communities

Douglas-fir and western hemlock with an abundance of dead wood and decomposing logs—a setting characteristic of old-growth forests.

CHAPTER GUIDE

THE COMMUNITY IS A GROUP of plant and animal species that inhabit a given area. As such, understanding the biological structure of the community depends on understanding the distribution and abundance of species. Thus far we have examined a wide variety of topics addressing this broad question, including the adaptation of organisms to the physical environment, the evolution of life history characteristics and their influence on population demography, and the interactions among different species. Previously, we examined characteristics that define both the biological and physical structure of communities and described the structure of community change as one moves across the landscape (Chapter 16). However, the role of science is to go beyond description and to answer fundamental questions about the processes that give rise to these observed patterns. What processes shape these patterns of community structure? How will communities respond to the addition or removal of a species? Why are communities in some environments more or less diverse than others? Here, we integrate our discussion of the adaptation of organisms to the physical environment presented previously with the discussion of species interactions to explain the processes that control community structure in a wide variety of communities (Parts Two and Four).

17.1 Community Structure Is an Expression of the Species' Ecological Niche

As we discussed in Chapter 16, the biological structure of a community is defined by its species composition, that is, the species present and their relative abundances. For a species to be a component of an ecological community at a given location, it must first and foremost be able to survive. The environmental conditions must fall within the range under which the species can persist—its range of environmental tolerances. The range of conditions under which individuals of a species can function are the consequences of a wide variety of physiological, morphological, and behavioral adaptations. As well as allowing an organism to function under a specific range of environmental conditions, these same adaptations also limit its ability to do equally well under different conditions. As a result, species differ in their environmental tolerances and performance (ability to survive, grow, and reproduce) along environmental gradients. We have explored many examples of this premise. Plants adapted to high-light environments exhibit characteristics that preclude them from being equally successful under low-light conditions (Chapter 6). Animals that regulate body temperature through ectothermy (poikilotherms) are able to reduce energy requirements during periods of resource shortage. Dependence on external sources of energy, however, limits diurnal and seasonal periods of activity and the geographic distribution of poikilotherms (Chapter 7). Each set of adaptations enable a species to succeed (survive, grow, and reproduce) under a given set of environmental conditions, and conversely, restricts or precludes success under different environmental conditions. These adaptations determine the fundamental niche of a species (Section 12.6).

The concept of the species' fundamental niche provides a starting point to examine the factors that influence the structure of communities. We can represent the fundamental niches of various species with bell-shaped curves along an environmental gradient, such as mean annual temperature or elevation (**Figure 17.1a**). The response of each species along the gradient is defined in terms of its population abundance.

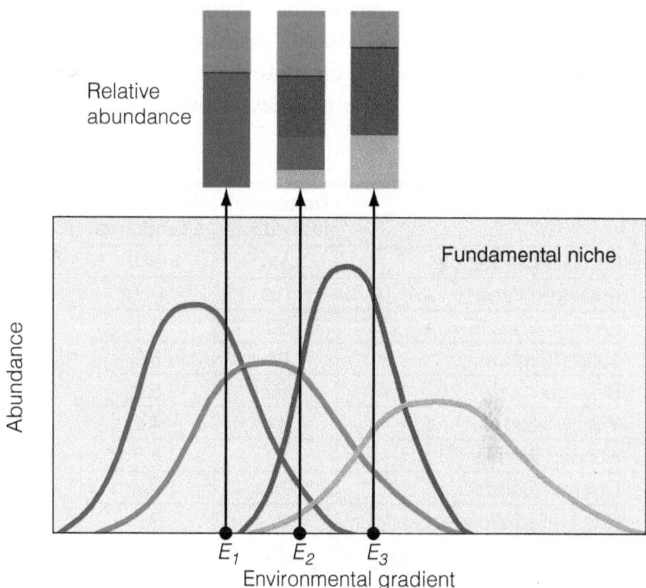

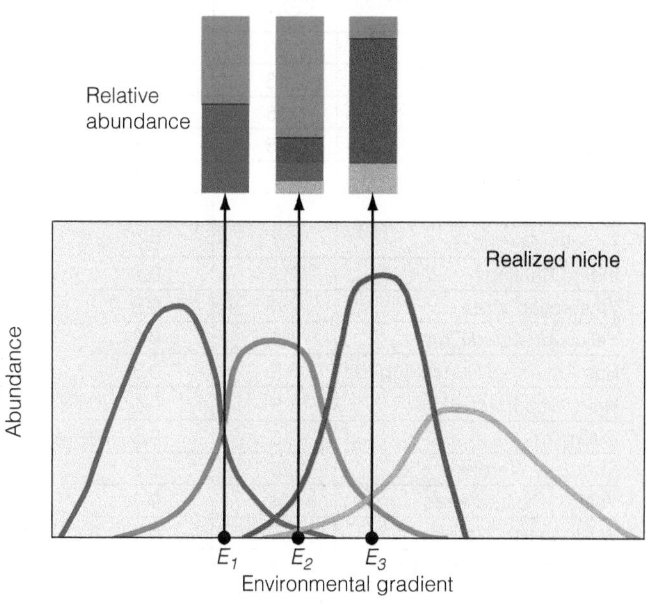

Figure 17.1 (a) Hypothetical example of the fundamental niches (potential responses in the absence of species interactions) of four species represented by their distributions and abundances along an environmental gradient. Their relative abundances at any point along the gradient (E_1, E_2, and E_3) provide a first estimate of community structure. (b) The actual community structure at any point along the gradient is a function of the species' realized niches—the species' potential responses as modified by their interaction with other species present.

Although the fundamental niches overlap, each species has limits beyond which it cannot survive. The distribution of fundamental niches along the environmental gradient represents a primary constraint on the structure of communities. For a location that corresponds to a given point along the environmental gradient, only a subset of species will be potentially present in the community, and their relative abundances at that point provide a first approximation of the expected community structure (Figure 17.1a). As environmental conditions change from location to location, the possible distribution and abundance of species changes, which changes the community structure. For example, **Figure 17.2** is a description of the biological structure of the breeding bird community on the Walker Branch Watershed in east Tennessee (species present and their relative abundances). The figure shows the maps of geographic range and population abundance of four of the bird species that are components of the bird community on the watershed. As we discussed previously, these geographic distributions reflect the occurrence of suitable environmental conditions (within the range of environmental tolerances; Chapter 8). Note that the geographic distributions of the four species are quite distinct, and the Walker Branch Watershed in east Tennessee represents a relatively small geographic region where the distributions of these four species overlap. As we move from this site in east

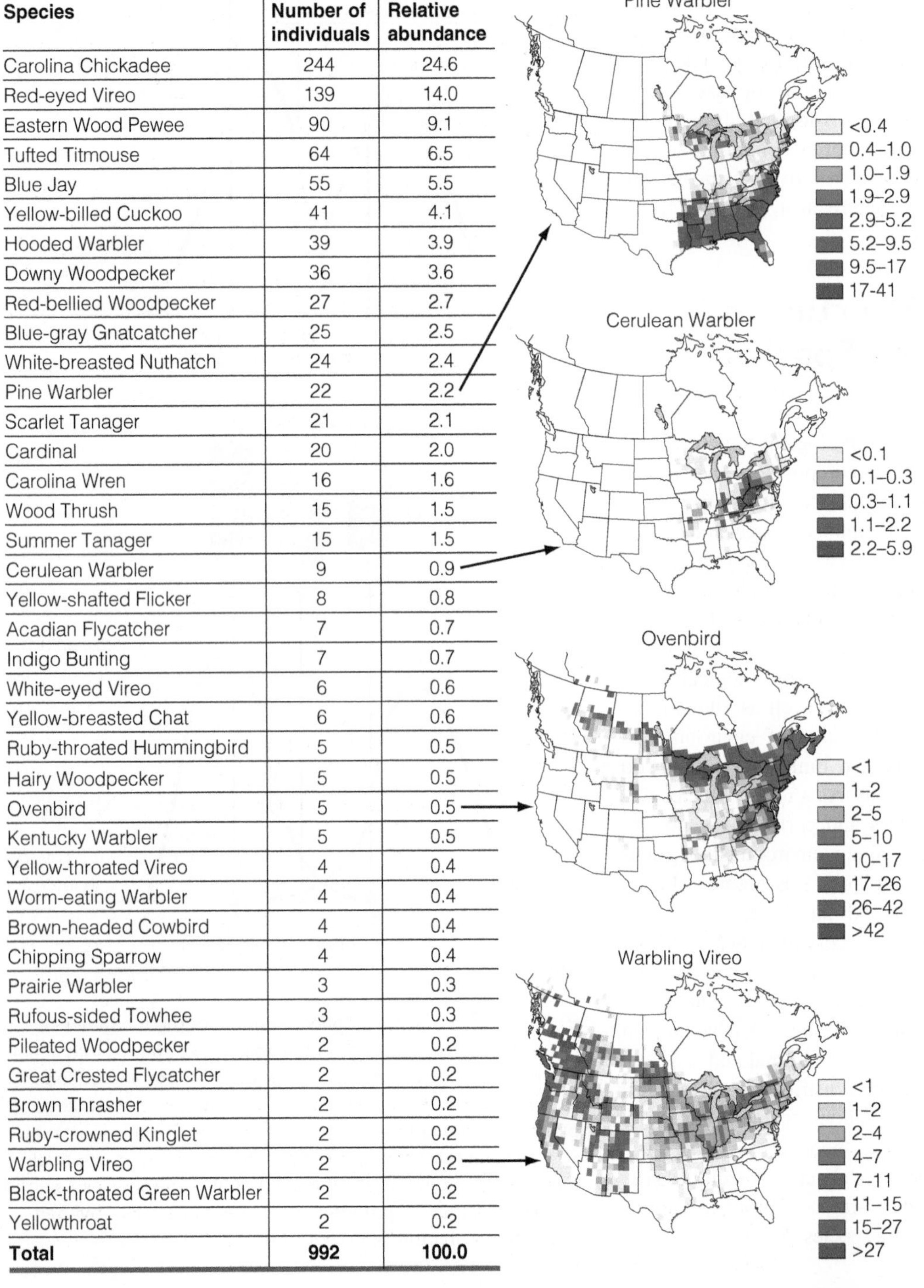

Species	Number of individuals	Relative abundance
Carolina Chickadee	244	24.6
Red-eyed Vireo	139	14.0
Eastern Wood Pewee	90	9.1
Tufted Titmouse	64	6.5
Blue Jay	55	5.5
Yellow-billed Cuckoo	41	4.1
Hooded Warbler	39	3.9
Downy Woodpecker	36	3.6
Red-bellied Woodpecker	27	2.7
Blue-gray Gnatcatcher	25	2.5
White-breasted Nuthatch	24	2.4
Pine Warbler	22	2.2
Scarlet Tanager	21	2.1
Cardinal	20	2.0
Carolina Wren	16	1.6
Wood Thrush	15	1.5
Summer Tanager	15	1.5
Cerulean Warbler	9	0.9
Yellow-shafted Flicker	8	0.8
Acadian Flycatcher	7	0.7
Indigo Bunting	7	0.7
White-eyed Vireo	6	0.6
Yellow-breasted Chat	6	0.6
Ruby-throated Hummingbird	5	0.5
Hairy Woodpecker	5	0.5
Ovenbird	5	0.5
Kentucky Warbler	5	0.5
Yellow-throated Vireo	4	0.4
Worm-eating Warbler	4	0.4
Brown-headed Cowbird	4	0.4
Chipping Sparrow	4	0.4
Prairie Warbler	3	0.3
Rufous-sided Towhee	3	0.3
Pileated Woodpecker	2	0.2
Great Crested Flycatcher	2	0.2
Brown Thrasher	2	0.2
Ruby-crowned Kinglet	2	0.2
Warbling Vireo	2	0.2
Black-throated Green Warbler	2	0.2
Yellowthroat	2	0.2
Total	**992**	**100.0**

Figure 17.2 The structure of the breeding bird community on the Walker Branch Watershed in Oak Ridge, Tennessee (United States) expressed in terms of the species' relative abundances (percentage of total individuals). Maps of the geographic range and abundance of four of the species are shown. The four species have distinct geographic distributions and the Walker Branch Watershed (location shown as yellow dot on maps) falls within a very limited region where the distributions of the four species overlap. As you move across eastern North America, the set of bird species whose distribution overlaps changes, and so does the species composition of the bird communities.

([a] Data from Anderson and Shugart 1974. [b] North American Breeding Bird Survey, U.S. Geological Survey.)

Tennessee to other regions of eastern North America, the set of bird species whose distributions overlap and their corresponding relative abundances change, and subsequently, so does the biological structure of the bird community.

This view of community represents what ecologists refer to as a **null model**. It assumes that the presence and abundance of the individual species found in a given community are solely a result of the independent responses of each individual species to the prevailing abiotic environment. Interactions among species have no significant influence on community structure. Considering the examples of species interactions that we have reviewed in the chapters of Part Four, this assumption must seem somewhat odd. However, it is helpful as a framework for comparing the actual patterns observed within the community. For example, this particular null model is the basis for experiments in which the interactions between two species (competition, predation, parasitism, and mutualism) are explored by physically removing one species and examining the population response of the other (Part Four). If the population of the remaining species does not differ from that observed previously in the presence of the removed species, we can assume that the apparent interspecific interaction has no influence on the remaining species' abundance within the community.

A great deal of evidence, however, indicates that species interactions do influence both the presence and abundance of species within communities. As we have seen in the examples presented in the chapters of Part Four, species interactions modify the fundamental niche of species involved, influencing their relative abundance, and in some cases, their distribution along environmental gradients. The resulting shifts in species' responses as a result of interactions with other species determine their realized niche (Section 12.6). The process of interspecific competition can reduce the abundance of or even exclude some species from a community, and positive interactions such as facilitation and mutualism can enhance the presence of a species or even extend a species' distribution beyond that defined by its fundamental niche (see this chapter, **Field Studies: Sally D. Hacker**). In contrast to our null model developed previously, in which our first approximation of community structure was based on the species' fundamental niche (Figure 17.1a), as we shall see in the following sections, the biological structure of the community is an expression of the species' realized niches (**Figure 17.1b**).

17.2 Zonation Is a Result of Differences in Species' Tolerance and Interactions along Environmental Gradients

We have now seen that the biological structure of a community is first constrained by the species' environmental tolerances—its fundamental niche. In turn, the fundamental niche is modified through interactions with other species (realized niche). Competitors and predators, for example, can restrict a species from a community; conversely, mutualists can facilitate a species' presence and abundance within the community. As we move across the landscape, variations in the abiotic environment alter these constraints on species' distribution and abundance. Differences in environmental tolerances among species and changes in the nature of species interactions (see Sections 12.4 and 13.9) result in shifts in the species present and their relative abundance (see Figure 17.1). These spatial changes in community structure are referred to as *zonation* (Section 16.8).

The rocky intertidal environment of coastal marine ecosystems (Section 25.2, Figures 25.1, 25.2 and 25.3) provides an excellent example of how the interaction of environmental tolerance and species interactions can give rise to a distinctive pattern of zonation. The intertidal zone of rocky shorelines is

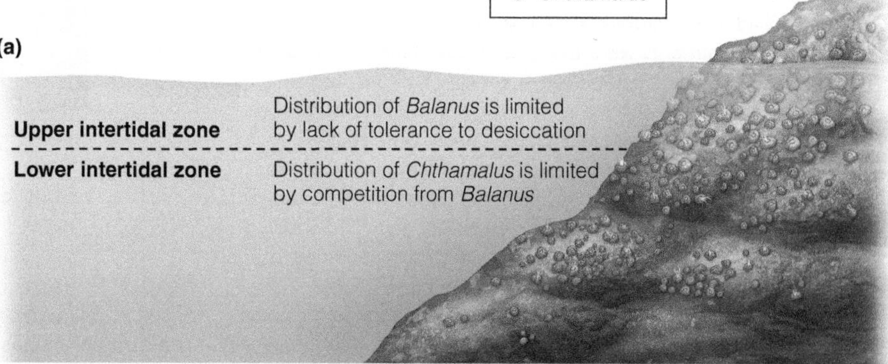

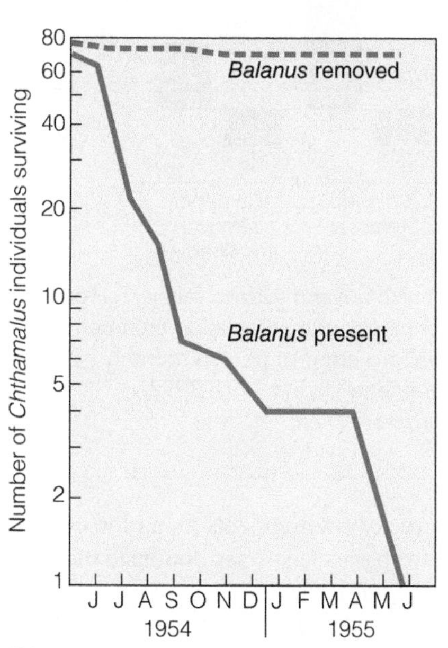

Figure 17.3 (a) Vertical zonation of the two dominant barnacle species, *Chthamalus stellatus* and *Balanus balanoides*, in the rocky intertidal environment of the Scottish coast. *Chthamalus* dominates the upper intertidal zone, and *Balanus* dominates the lower zone. Experiments show that *Balanus* is excluded from the upper zone as a result of its inability to tolerate desiccation when the upper zone is exposed during low tide. In contrast, when *Balanus* is experimentally removed from sites in the lower intertidal zone *Chthamalus* is able to survive and grow. (b) Differences in the survival (number of individuals) of *Chthamalus* over the period of 1954–1955 in lower intertidal zone in areas where *Balanus* is present and areas where *Balanus* has been experimentally removed. (Adapted from Connell 1961.)

FIELD STUDIES Sally D. Hacker

Department of Zoology, Oregon State University, Corvallis, Oregon

Salt marsh plant communities are ideal for examining the forces that structure natural communities. They are typically dominated by a small number of plant species that form distinct zonation patterns (see Figure 16.16). Seaward distribution of marsh plant species is set by harsh physical conditions such as waterlogged soils and high soil salinities, whereas terrestrial borders are generally set by competitive interactions (see Figure 13.11). Yet marsh plants also have strong ameliorating effects on these harsh physical conditions. Shading by marsh plants limits surface evaporation and the accumulation of soil salts. In addition, the transport of oxygen to the rooting zone (rhizosphere) by marsh plants can alleviate anaerobic substrate conditions. How might these modifying effects of marsh plants on the physical environment influence the structure of salt marsh communities? This question has been central to the research of ecologist Sally Hacker of Oregon State University.

To examine the role of plant–physical environment interactions on salt marsh plant zonation, Hacker focused on the terrestrial border of New England marshes. In southern New England, terrestrial marsh borders are dominated by the perennial shrub *Iva frutescens* (marsh elder) mixed with the rhizomatous perennial rush *Juncus gerardi* (black grass), which also dominates the lower marsh elevations (see Figure 17.6). The seaward border of the *Iva* zone is often characterized by low densities of stunted (35–50 cm) adult plants, whereas at higher elevations *Iva* are taller (up to 150 cm), more productive, and reach higher densities.

Previous studies suggested that *Iva* is relatively intolerant of high soil salinities and waterlogged soil conditions. Given the potential role of marsh plants to modify the local environment, Hacker hypothesized that the modifying effects of *Juncus* on soil environment function to extend the seaward distribution of *Iva*. To test this hypothesis, Hacker and colleague Mark Bertness of Brown University applied one of three treatments to randomly selected adult *Iva* shrubs on the seaward border of the *Iva* zone. The three treatments were designed to examine the effects of *Juncus* neighbors on established adult *Iva*: (1) all *Juncus* within a 0.5-m radius of each *Iva* plant were regularly clipped to ground level (neighbor removal, or NR), (2) all *Juncus* were clipped (as in NR) and then the soil was covered with a water-permeable fabric (shaded neighbor removal, or SNR), and (3) control. The use of fabric in the SNR treatment mimics the effect of *Juncus* shading on soil salinities without the effects of *Juncus* transporting oxygen to the rhizosphere (increased soil oxygen).

Soil physical conditions (soil salinity and redox [a measure of oxygen content of soils]) and *Iva* performance (photosynthetic rates and leaf production) were monitored in all treatments for a two-year period.

Removing *Juncus* plants in the neighborhood of *Iva* shrubs strongly affected local physical conditions (**Figure 1**). Removing *Juncus* neighbors more than doubled soil salinities in contrast to other treatments and led to more than an order-of-magnitude drop in soil redox, suggesting that the presence of *Juncus* neighbors increases soil oxygen levels. Because shading plots without *Juncus* (SNR treatment) prevented salinity increases but did not influence soil redox, the NR and SNR treatments separated the effects caused by both salt buffering and soil oxidation from those caused only by soil oxidation.

Photosynthetic rate and leaf production of *Iva* individuals in the treatment where *Juncus* neighbors were removed (NR) declined significantly in comparison to either the shaded neighbor removal (SNR) or control (C) treatments (**Figure 2**). Fourteen months after the experimental treatments were established, all *Iva* in the NR treatment were dead. These results show that soil salinity is the primary factor influencing the performance of *Iva* across the gradient. They also show that the presence of

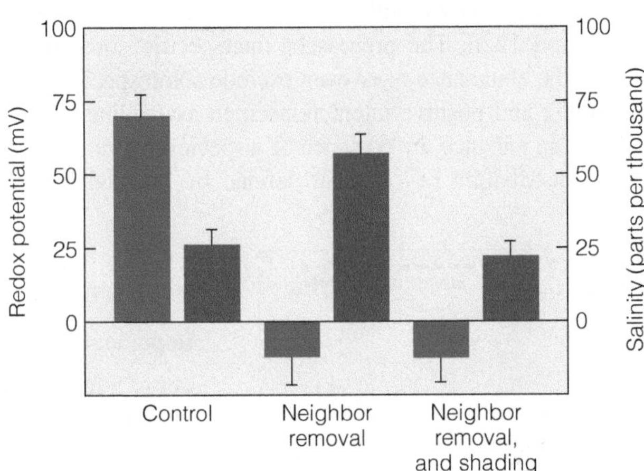

Figure 1 Redox potential (blue) and surface salinity (green) of soil in the *Iva frutescens* neighbor manipulation treatments. The data are means (±standard error) of pooled monthly (June–September) measurements during 1991–1992. (Adapted from Bertness and Hacker 1994.)

characterized by dramatic changes in environmental conditions over the tidal cycle (see Section 3.9). At high tide it is submerged, and at low tide it is exposed and subject to extreme changes in temperature, moisture, and solar radiation (see Sections 25.1 and 25.2 for a detailed discussion). As a result, the upper and lower limits of dominant species are often very sharply defined within this environment. In the rocky environments along the coast of northwestern Scotland, two barnacle species dominate the intertidal zone. *Chthamalus stellatus* is the dominant barnacle species in the upper intertidal zone, and *Balanus balanoides* dominates the lower intertidal zone (**Figure 17.3a**). In one of the earliest field studies aimed at examining the role of species interactions

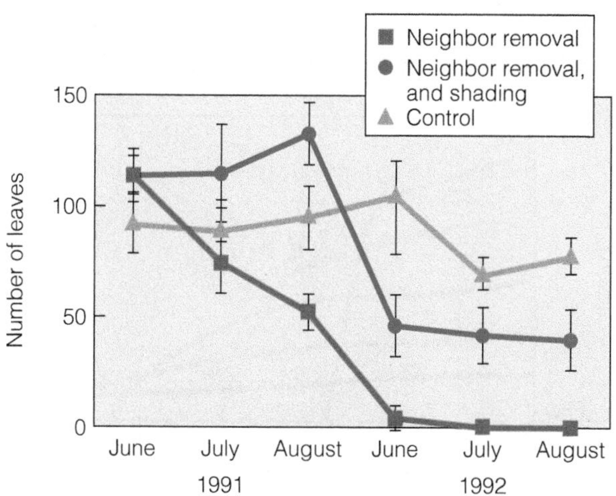

Figure 2 Total number of leaves on adult *Iva frutescens* under experimentally manipulated conditions, with and without neighbors. All data are means (±standard error).

Juncus, with its superior ability to withstand waterlogging and salt stress, modifies physical conditions in such a manner as to create a hospitable environment for *Iva*, and allow this species' distribution to extend to lower intertidal habitats.

The *Iva–Juncus* interaction has interesting consequences for higher trophic levels in the marsh. The most common insects living on *Iva* are aphids (*Uroleucon ambrosiae*) and their predators, ladybird beetles (*Hippodamia convergens* and *Adalia bipunctata*). Interestingly, aphids are most abundant on short, stunted *Iva* in the lower intertidal zone, despite having far higher growth rates on the taller *Iva* shrubs in the upper intertidal zone. This reduced growth rate occurs because ladybird beetles prefer tall structures, and increased predation on tall plants restricts aphids to the poorer-quality *Iva* plants in the lower marsh. These findings prompted the investigators to hypothesize that the *Iva–Juncus* interaction is critical in maintaining aphid populations in the marsh.

To explore this hypothesis, Hacker examined the abundance of aphids and ladybird beetles on the *Iva* individuals with (treatment C) and without (treatment NR) *Juncus* neighbors. To determine how the aphid population growth rates were affected by the absence of *Juncus*, Hacker calculated aphid population growth rates as the per capita rate of increase per day.

The overall percentage of plants with aphids and ladybird beetle predators was significantly higher for *Iva* plants without *Juncus* (NR) than for control (C) plants. This result suggests that *Juncus* neighbors influence stunted *Iva* by making the plants less noticeable to potentially colonizing aphids as well as the aphid's predators (ladybird beetles). Although the removal of neighboring *Juncus* increased the proportion of *Iva* individuals colonized by aphids, growth rates for aphid populations (**Figure 3**) were lower on the stunted *Iva* without *Juncus* (NR) than for control individuals (C). Even though aphids are better at finding stunted *Iva* host plants when *Juncus* is removed, by late summer (August), population growth rates were negative on *Iva* individuals without neighbors—indicating that food quality of *Iva* host plants decreases such that aphids were unable to produce enough offspring to replace themselves.

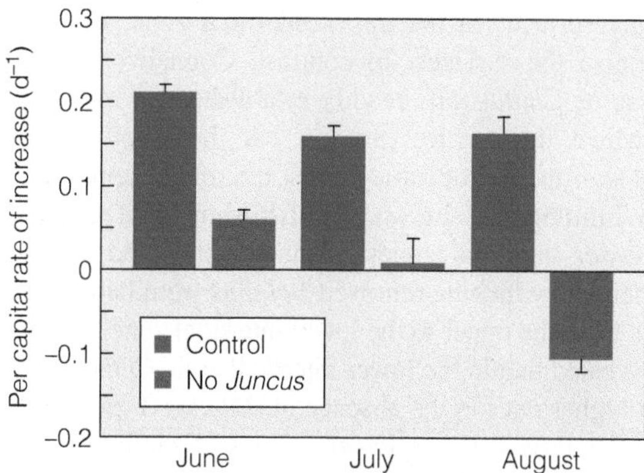

Figure 3 The per capita rate of population increase for the period June–August (1993) on control and no *Juncus* plants (neighbors removed).

Bibliography

Bertness, M. D., and S. D. Hacker. 1994. "Physical stress and positive associations among marsh plants." *The American Naturalist* 144:363–372.

Hacker, S. D., and M. D. Bertness. 1996. "Trophic consequences of a positive plant interaction." *The American Naturalist* 148:559–575.

Hacker S. D., and M. D. Bertness. 1999. "Experimental evidence for the factors maintaining plant species diversity in a New England salt marsh." *Ecology* 80: 2064–2073.

1. How do the results presented in Figures 1 and 2 suggest that salinity is the factor limiting seaward distribution of *Iva* in the marsh?

2. How does the presence of *Juncus* function to maintain the aphid populations in the marsh?

on community structure, the ecologist Joseph Connell of the University of California–Santa Barbara, performed a series of now-classic experiments on these two species of barnacles at a site along the Scottish coast. Connell established a series of plots from the upper to the lower intertidal zone in which he conducted a periodic census of populations by mapping the location of every barnacle. In this way, he was able to monitor interactions and determine the fate of individuals.

Results of the study show that the vertical distribution of newly settled larvae of the two species overlap broadly within the intertidal zone, so dispersal was not an important factor determining the pattern of species distribution. Rather, the distribution of the two species is a function in differences in the physiological tolerance and competitive ability along the vertical environmental gradient within the intertidal zone. *Balanus* is limited to the lower intertidal zone because it cannot tolerate the desiccation that results from prolonged exposure to the air in the upper intertidal zone. Even when *Chthamalus* was removed from rock surfaces in the upper intertidal zone, *Balanus* did not colonize the surfaces. In contrast, Connell observed that the larvae of *Chthamalus* readily established on rock surfaces below where the species persists, but the colonists die out within a short period of time. To test the role of competition as a factor limiting the successful establishment of *Chthamalus* in the lower intertidal zones, Connell conducted a series of experiments in which he removed *Balanus* from half of each of the plots from the upper to the lower intertidal zone. The experiments revealed that in the lower intertidal zone *Chthamalus* survived at higher rates in the absence of *Balanus* (**Figure 17.3b**). *Chthamalus* thrived in the lower regions of the intertidal zone where it does not naturally occur, which indicated that increased time of submergence (tolerance) is not the factor limiting the distribution of the species to higher positions on the shoreline. Observations showed that the two barnacle species compete for space. *Balanus* has a heavier shell and a much faster growth rate, allowing individuals to smother, undercut, or crush establishing *Chthamalus*. In addition, crowding caused reduced size and reproduction by surviving *Chthamalus* individuals, further adding to the population effects of increased mortality.

Connell's experiments clearly show that the spatial changes in species distribution—community zonation—along the intertidal gradient are a result of the trade-off between tolerance to environmental stress (dessication) and competitive ability. *Balanus* is limited to middle and lower intertidal zones as a function of restrictions on its physiological tolerance to desiccation (fundamental niche), whereas *Chthamalus* is restricted to the upper intertidal zone as a result of interspecific competition (realized niche). The asymmetry of competition between these two species leads to the competitive exclusion of *Chthamalus* from intertidal environments in which it is physiologically capable of flourishing (within its fundamental niche). What accounts for these differences in tolerance and competitive ability? Is there a relationship between tolerance to environmental stress and competitive ability that underlies this pattern of trade-offs that give rise to the pattern of zonation? Often superior competitive ability for resources is associated with a higher metabolic or growth rate, which often restricts (or is physiologically incompatible with) the ability to tolerate environmental stress.

Our previous discussion of plant adaptations to resource availability provides some insight into the trade-off between competitive ability and tolerance to environment stress (Chapter 6) and how this trade-off influences the relative competitive abilities of plant species across environmental gradients (Chapter 13). Adaptations of plants to variations in the availability of light, water, and nutrients result in a general pattern of trade-offs between the characteristics that enable a species to survive and grow under low resource availability and those that allow for high rates of photosynthesis and growth under high resource availability (**Figure 17.4a**). Competitive success in plants is often linked to their growth rate and the acquisition of resources (see Chapter 13). Species that have the highest growth rate and acquire most of the resources at any given point on the resource gradient often have the competitive advantage there. The differences in adaptations to resource availability among the species in Figure 17.4a result in a competitive advantage for each species over the range of resource conditions under which they have the greatest growth rate relative to the other plant

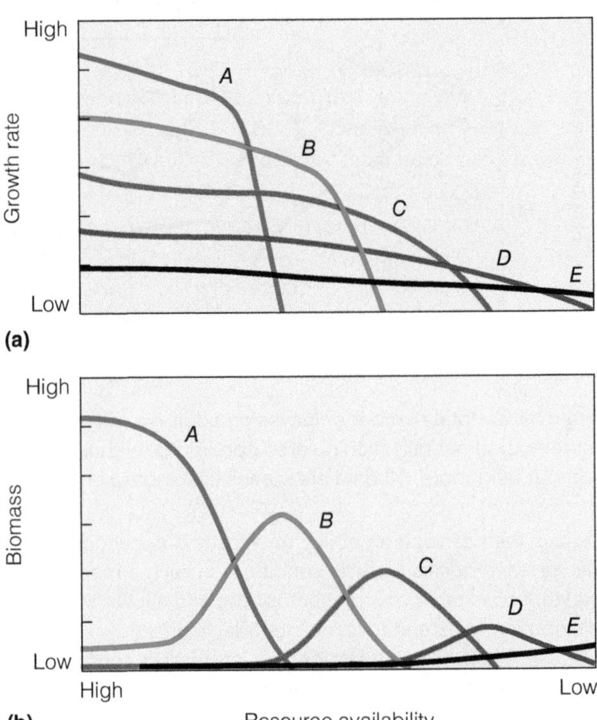

Figure 17.4 (a) General pattern of trade-off between a species' ability to survive and grow under low resource availability and the maximum growth rate achieved under high resource availability (such as light, water, and soil nutrients). The result is an inverse relationship between physiological maximum growth rate and the minimum resource requirement for hypothetical plant species *A–E*. Assuming that the superior competitor at any point along the resource gradient (x-axis) is the species with the highest growth rate, the species' relative competitive ability changes with resource availability. (b) The outcome of competition is a pattern of zonation in which lower boundaries (value of resource availability) for the species are the result of differences in their tolerances for low resource availability and upper boundaries are a product of competition.
(Adapted from Smith and Huston 1989.)

Interpreting Ecological Data

Q1. *In the hypothetical example in graph (a), under what resource conditions (availability) is the growth rate of each plant species assumed to be optimal?*

Q2. *What is the rank order of the hypothetical plant species in graph (a), moving from most to least tolerant of resource limitation?*

Q3. *If species B was removed from the hypothetical community, how would the predicted distribution of species A along the resource gradient (x-axis) in graph (b) change? How would you expect the distribution of species C to change? Why?*

species present (see discussion of changing competitive ability along resource gradients in Section 13.9). The result is a pattern of zonation along the gradient (**Figure 17.4b**) that reflects the changing relative competitive abilities. The lower boundary of each species along the gradient is defined by its ability to tolerate resource limitation (survive and maintain a positive carbon balance), whereas the upper boundary is defined by competition. Such a trade-off in tolerance and competitive ability can be seen in the examples presented previously of interspecific competition and the distribution of cattail species with water depth (Figure 12.13), zonation in New England salt marshes (Figure 13.11), and in the distribution of grass species in the semi-arid regions of southeastern Arizona (Figure 13.13).

Competition among plant species rarely involves a single resource, however. The greenhouse experiments of R. H. Groves and J. D. Williams examining competition between populations of subterranean clover and skeletonweed (Figure 13.7) and the field experiments of James Cahill (see Section 13.8) clearly show that there is an interaction between competition for both aboveground (light) and belowground resources (water and nutrients). The differences in adaptations relating to the acquisition of above- and belowground resources when they are in short supply can result in changing patterns of competitive ability along gradients where these two classes of resources co-vary. Allocating carbon to the production of leaves and stems provides increased access to the resources of light but at the expense of allocating carbon to the production of roots. Likewise, allocating carbon to the production of roots increases access to water and soil nutrients but limits the production of leaves, and therefore, the future rate of carbon gain through photosynthesis. As the availability of water (or nutrients) increases along a supply gradient, the relative importance of water and light as limiting resources shift. As a result, the competitive advantage shifts from those species adapted to low availability of water (high root production) to those species that allocate carbon to leaf production and height growth but that require higher water availability to survive (**Figure 17.5**).

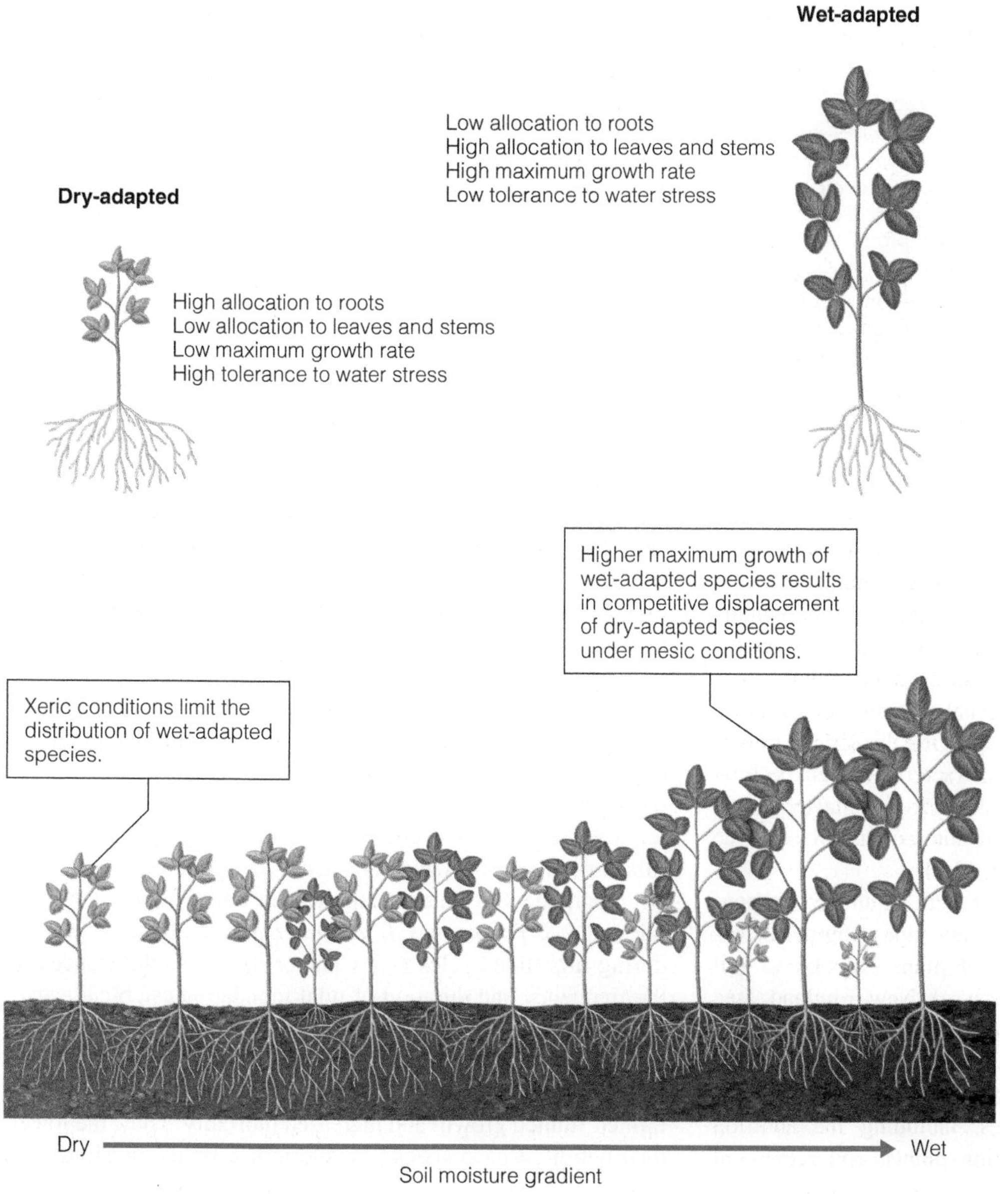

Figure 17.5 General trends in plant adaptations (characteristics) that increase fitness along a soil moisture gradient. For species adapted to low soil moisture (dry adapted), allocation to the production of roots at the expense of leaves aids in acquiring water and reducing transpiration, allowing the plant to survive under xeric conditions (tolerance). For species adapted to high soil moisture, allocation to leaves and stems at the expense of roots aids in achieving high rates of growth when water is readily available. As water availability increases, the overall increase in plant growth results in competition for light as some individuals overtop others. A shift in allocation to height growth (stems) and the production of leaves increases a plant's growth rate and competitive ability.

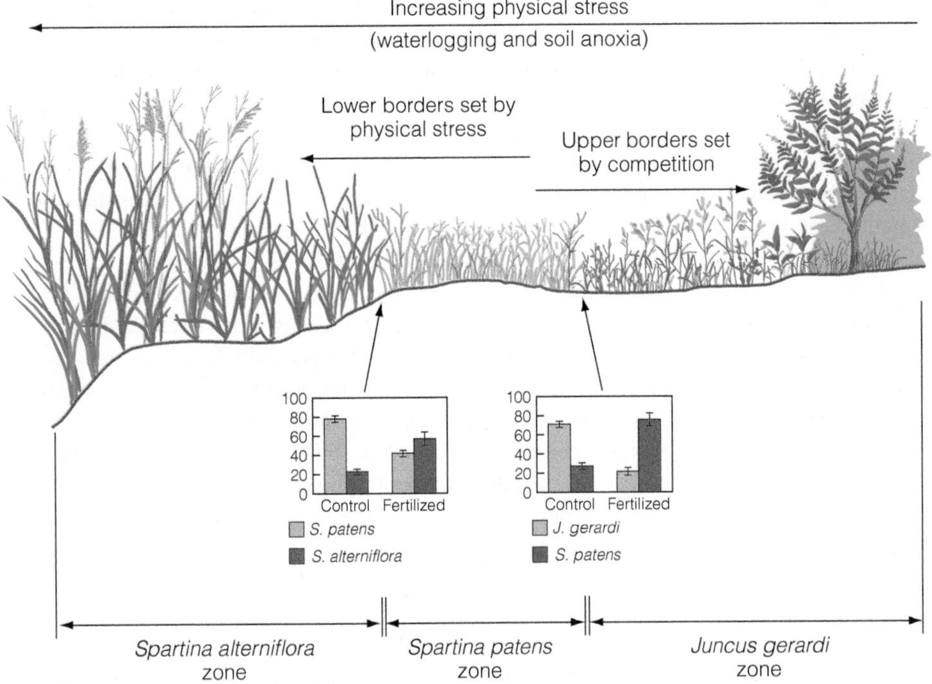

Figure 17.6 Patterns of plant zonation and physical stress along an elevation gradient in a tidal salt marsh. The lower marsh experiences daily flooding by the tides, but the upper zones are inundated only during the high-tide cycles. The higher water levels of the lower marsh result in lower oxygen levels in the sediments and higher salinities. The lower (elevation) boundary of each species is determined by its tolerance to physical stress, and the upper boundaries are limited by competition. Bar graphs show shifts in percentage of cover by the two adjoining species within the border zones under normal (control) conditions and when fertilizer was added to increase nutrient availability. Note that the increase in nutrient availability resulted in the subordinate competitor under ambient conditions (control) becoming the dominant (superior competitor).
(Adapted from Emery et al. 2001.)

Interpreting Ecological Data

Q1. *In the transition zone between the areas dominated by Spartina alterniflora and Spartina patens, which of the two species was the superior competitor (dominated) in the experimental plots under control conditions? Was the competitive outcome altered when nutrient availability was enhanced in the experimental plots (fertilized)? How so?*

Q2. *In the transition zone between the areas dominated by Spartina patens and Juncus gerardi, which of the two species was the superior competitor (dominated) in the experimental plots under control conditions? Was the competitive outcome altered when nutrient availability was enhanced in the experimental plots (fertilized)? How so?*

Q3. *What do the results of these experiments suggest about the role of nutrients in limiting the distribution of plant species along the gradient from low (sea side) to high (land side) marsh?*

This framework of trade-offs between the set of characteristics that enable individuals of a plant species to survive and grow under low resource conditions as compared to the set of characteristics that would enable those same individuals to maximize growth and competitive ability under higher resource conditions is a powerful tool for understanding changes in the structure and dynamics of plant community structure along resource gradients. However, applying this simple framework of trade-offs in phenotypic characteristics can become more complicated when dealing with environmental gradients and communities in which there are interactions of resource and nonresource factors (see Section 13.6).

The complex nature of competition along an environmental gradient involving both resource and nonresource factors is nicely illustrated in the pattern of plant zonation in salt marsh communities along the coast of New England (see Figure 16.16). Nancy Emery and her colleagues at Brown University conducted a number of field experiments to identify the factors responsible for patterns of species distribution in these coastal communities, including the addition of nutrients, removal of neighboring plants, and reciprocal transplants (i.e., planting species in areas where they are not naturally found to occur along the gradient). Experiment results indicate that the patterns of zonation reflect an interaction between the relative competitive abilities of species in terms of acquiring nutrients and the ability of plant species to tolerate increasing physical stress. The low marsh is dominated by *Spartina alterniflora* (smooth cordgrass), which is a large perennial grass with extensive rhizomes. The upper edge of *S. alterniflora* is bordered by *Spartina patens* (saltmeadow cordgrass), which is a perennial turf grass, and is replaced at higher elevations in the marsh by *Juncus gerardi* (black needle rush), which is a dense turf grass (**Figure 17.6**). Although the low marsh experiences daily flooding by the tides, the *S. patens* and *J. gerardi* zones are inundated only during high-tide cycles (see Chapter 3). These differences in the frequency and duration of tidal inundation establish a spatial gradient of increasing salinity, waterlogging, and reduced oxygen levels across the marsh. Individuals of *S. patens* and *J. gerardi* that were transplanted to lower marsh positions exhibited stunted growth and increased mortality. Thus, the lower distribution of each species is determined by its physiological tolerance to the physical stress imposed by tidal inundation (its fundamental niche). In contrast, individuals of *S. alterniflora* and *S. patens* exhibited increased growth when transplanted to higher marsh positions where the neighboring plants had been removed. They were excluded by competition from higher marsh positions when neighboring plants were present (not removed). These results indicate that the upper distribution of each species in the marsh was limited by competition.

At first, this example would seem to be a clear case of the trade-off between adaptations for stress tolerance and competitive ability (high growth rate and resource use), as suggested in Figure 17.4 (also see competition experiments between *S. alterniflora* and *S. patens* in Figure 13.11). However, such was not the case. The experimental addition of nutrients to the marsh indeed changed the outcome of competition but not in the manner that might be predicted. The addition of nutrients completely reversed the relative competitive abilities of the species, allowing the distributions of *S. alterniflora* and *S. patens* to shift to higher marsh positions (see Figure 17.6).

J. gerardi, dominant under ambient (low) nutrient conditions, allocates more carbon to root biomass than either species of *Spartina* does. That renders *Juncus* more competitive under conditions of nutrient limitation but limits its tolerance of the higher water levels of the lower marsh. In contrast, *S. alterniflora* allocates a greater proportion of carbon to aboveground tissues, producing taller tillers (stems and leaves), which is an advantage in the high water levels of the lower marsh. The trade-off in allocation to belowground and aboveground tissues results in the competitive hierarchy, and thus, the patterns of zonation observed under ambient conditions. When nutrients are not limiting (nutrient addition experiments), competition for light dictates the competitive outcome among marsh plants. The greater allocation of carbon to height growth by the *Spartina* species increased its competitive ability in the upper marsh.

In the salt marsh plant community, a trade-off between competitive ability belowground and the ability to tolerate the physical stress associated with the low oxygen and high salinity levels of the lower marsh appears to drive zonation patterns across the salt marsh landscape. In this environment, the stress gradient does not correspond to the resource gradient as in Figure 17.4, which allows the characteristics for stress tolerance to enhance competitive ability under high resource availability.

17.3 Species Interactions Are Often Diffuse

As we have seen in the previous chapters and sections, most studies that examine the role of species interactions on community structure typically focus on the direct interaction between two, or at best, a small subset of the species found within a community. As a result, such studies most likely underestimate the importance of species interactions on the structure and dynamics of communities because interactions are often diffuse and involve a number of species (see Section 12.5).

The late ecologist Robert MacArthur of Princeton University first coined the term *diffuse competition* to describe the total competitive effects of a number of interspecific competitors. If the relative abundance of a species in the community is a function of competitive interactions with a single competitor (**Figure 17.7a**), then an experiment that removes that competitor may be able to assess the importance

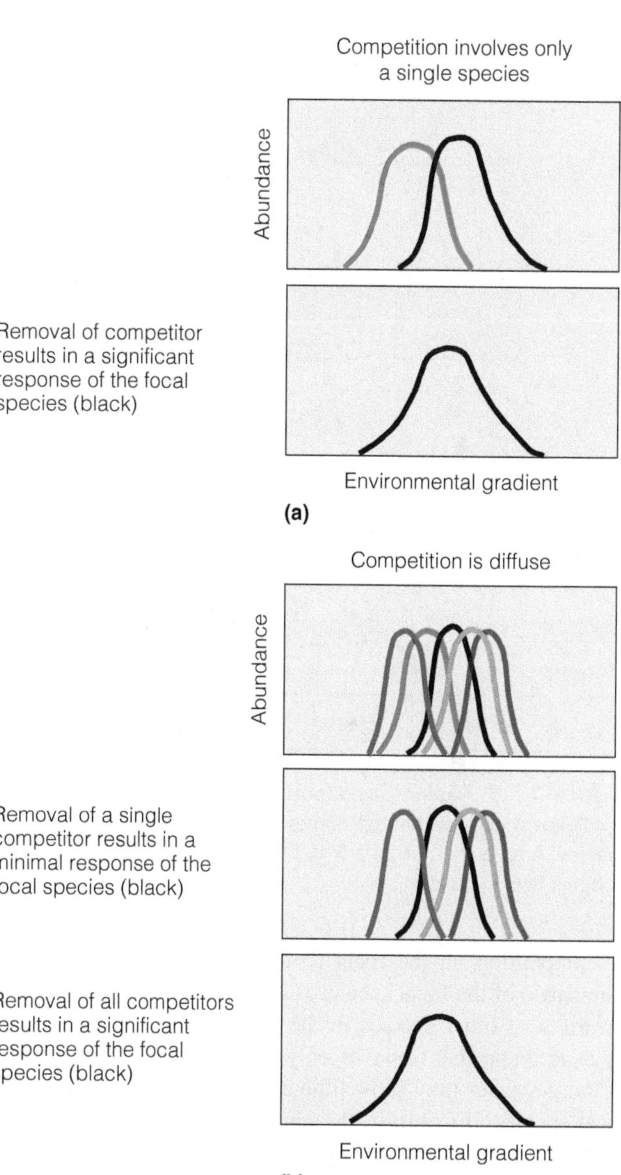

Figure 17.7 Illustration of diffuse competition. (a) When the abundance of a species (black curve) is influenced by competition with only a single species (red curve), the removal of that species can have a significant effect on the response of the species along the environmental gradient. (b) When the abundance of the species is influenced by a number of competing species (diffuse competition), the removal of a single species (red) may have a minimal (insignificant) effect on the response of the focal species (black). However, if all competitors are removed a significant response is observed.

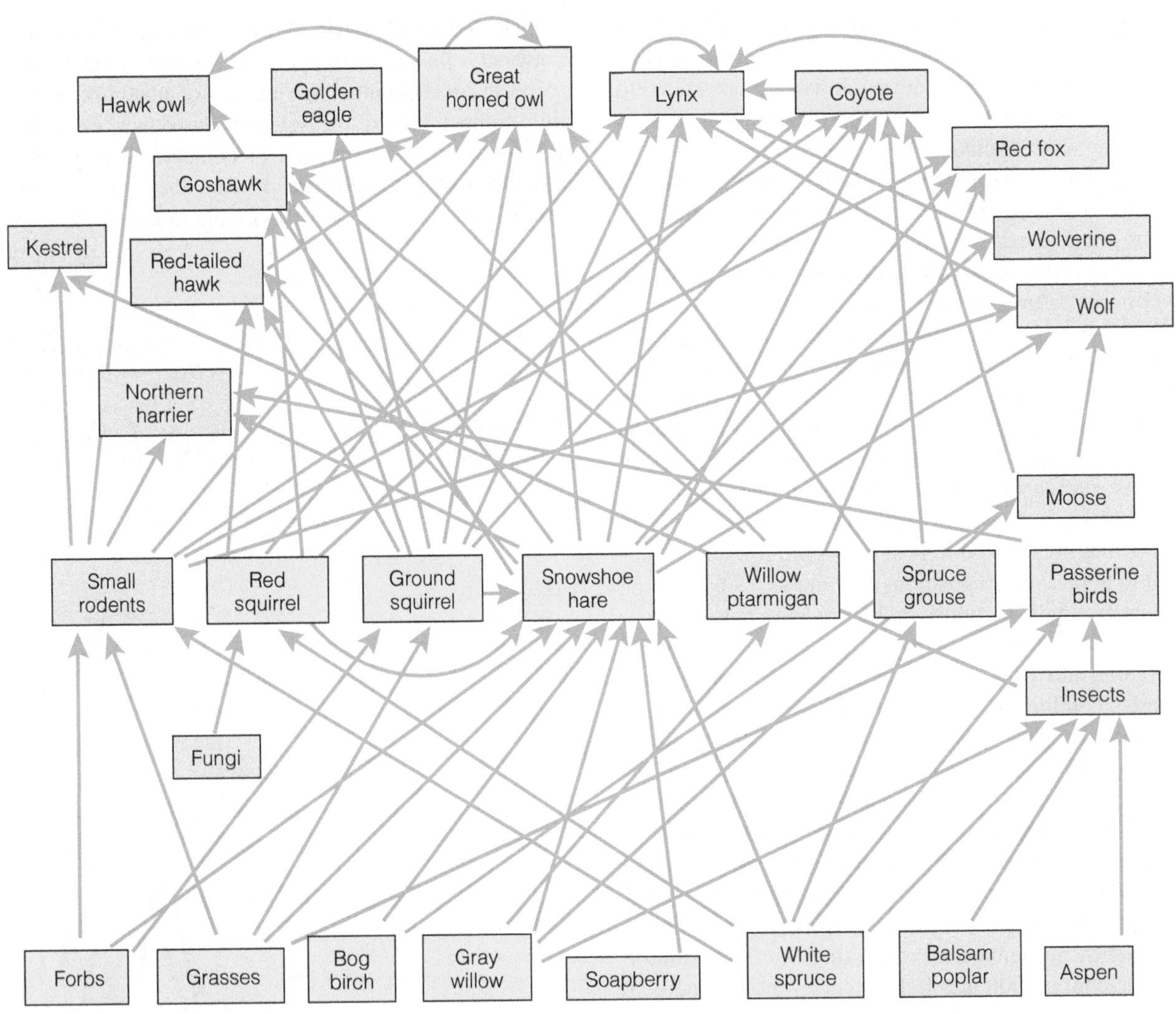

Figure 17.8 A generalized food web for the boreal forests of northwestern Canada. Dominant species within the community are shown in green. Arrows link predator with prey species. Arrows that loop back to the same species (box) represent cannibalism.
(Adapted from Krebs 2001.)

of competition on the focal species. However, if the relative abundance of the focal species is impacted by competition with a variety of other species in the community (**Figure 17.7b**), an experiment that removes only one or even a small number of those species may show little effect on the abundance of the focal species. In contrast, the removal or population reduction of the suite of competing species may result in a significant positive impact on the focal species. The work of ecologist Norma Fowler at the University of Texas provides an example. She examined competitive interactions within an old-field community by selectively removing species of plants from experimental plots and assessing the growth responses of remaining species. Her results showed that competitive interactions within the community tended to be rather weak and diffuse because removing a single species had relatively little effect. The response to removing groups of species, however, tended to be much stronger, suggesting that individual species compete with several other species for essential resources within the community.

Diffuse interactions in which one species may be influenced by interactions with many different species is not limited to competition. In the example of predator–prey cycles in Chapter 14, a variety of predator species (including the lynx, coyote, and horned owl) are responsible for periodic cycles observed in the snowshoe hare population (Section 14.14). Examples of diffuse mutualisms relating to both pollination and seed dispersal were presented previously in our discussion, where a single plant species may depend on a variety of animal species for successful reproduction (Chapter 15, Sections 15.13 and 15.14). Although food webs present only a limited view of species interactions within a community, they are an excellent means of illustrating the diffuse nature of species interactions. Charles J. Krebs of the University of British Columbia developed a generalized food web for the boreal forest communities of northwestern Canada (**Figure 17.8**). This food web contains the plant–snowshoe hare–carnivore system discussed previously (Chapter 14, Figure 14.24). The arrows point from prey to predator, and an arrow that circles back to the same box (species) represents cannibalism (e.g., great horned owl and lynx). Although this food web shows only the direct links between predator and prey, it also implies the potential for competition among predators for a shared prey resource, and it illustrates the diffuse nature of species interactions within this community. For example, 11 of the 12 predators present within the community prey on snowshoe hares. Any single predator species may have a limited effect on the snowshoe hare population, but the combined impact of multiple predators can regulate the snowshoe hare population. This same example illustrates the diffuse nature of competition within this community. Although the 12 predator species feed on a wide variety of prey species, snowshoe hares represent an important shared food resource for the three dominant predator species: lynx, great horned owl, and coyote.

17.4 Food Webs Illustrate Indirect Interactions

Food webs also illustrate a second important feature of species interactions within the community: indirect effects. Indirect interactions occur when one species does not interact with a second species directly but instead influences a third species that does directly interact with the second. For example, in the food web presented in Figure 17.8, lynx do not directly interact with white spruce; however, by reducing snowshoe hare and other herbivore populations that feed on white spruce, lynx's predation can positively affect the white spruce population (survival of seedlings and saplings). The key feature of indirect interactions is that they may arise throughout the entire community because of a single direct interaction between only two component species.

By affecting the outcome of competitive interactions among prey species, predation provides another example of indirect effects within food webs. Robert Paine of the University of Washington was one of the first ecologists to demonstrate this point. The intertidal zone along the rocky coastline of the Pacific Northwest is home to a variety of mussels, barnacles, limpets, and chitons, which are all invertebrate herbivores. All of these species are preyed on by the starfish (*Pisaster*; **Figure 17.9**). Paine conducted an experiment in which he removed the starfish from some areas (experimental plots) while leaving other areas undisturbed for purposes of comparison (controls). After he removed the starfish, the number of prey species in the experimental plots dropped from 15 at the beginning of the experiment to 8. In the absence of predation, several of the mussel and barnacle species that were superior competitors excluded the other species and reduced overall diversity in the community. This type of indirect interaction is called **keystone predation**, in which the predator enhances one or more less competitive species by reducing the abundance of the more competitive species (see discussion of keystone species in Section 16.4).

Ecologist Robert Holt of the University of Florida first described the conditions that might promote a type of indirect

(a)

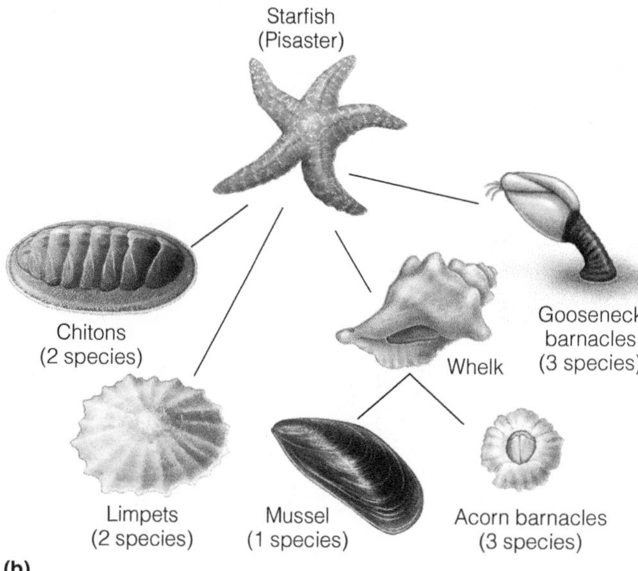

(b)

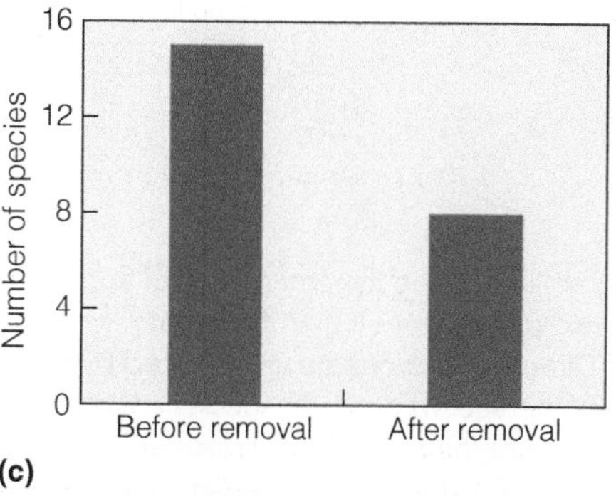

(c)

Figure 17.9 (a) The rocky intertidal zone of the Pacific Northwest coast is inhabited by a variety of species including starfish, barnacles, limpets, chitons, and mussels. (b) A food web of this community shows that the starfish (*Pisaster*) preys on a variety of invertebrate species. (c) The experimental removal of starfish from the community reduced the diversity of prey species as a result of increased competition.
(Adapted from Paine 1969.)

interaction he referred to as **apparent competition**. Apparent competition occurs when two species that do not compete with each other for limited resources affect each other indirectly by being prey for the same predator (**Figure 17.10**). Consider the example of the red squirrel and snowshoe hare in the food web of the boreal forest presented in Figure 17.8. These two species do not interact directly and draw on different food resources. The red squirrel is primarily a granivore (feeding on seeds), and the snowshoe hare is a browser, feeding on buds, braches, and twigs of low lying woody vegetation. Both species, however, are prey for the goshawk (predatory species of hawk). An increase in the red squirrel population might result in an increase in the goshawk population (numerical response; see Section 14.6), which in turn would negatively affect the population of snowshoe hare as a result of increased predation. The decline in the population of snowshoe hare in response to the increase in population density of red squirrel, which at first might be seen as a result of competition, is in fact a result of an indirect interaction mediated by the numerical increase of a third species, their common predator the goshawk.

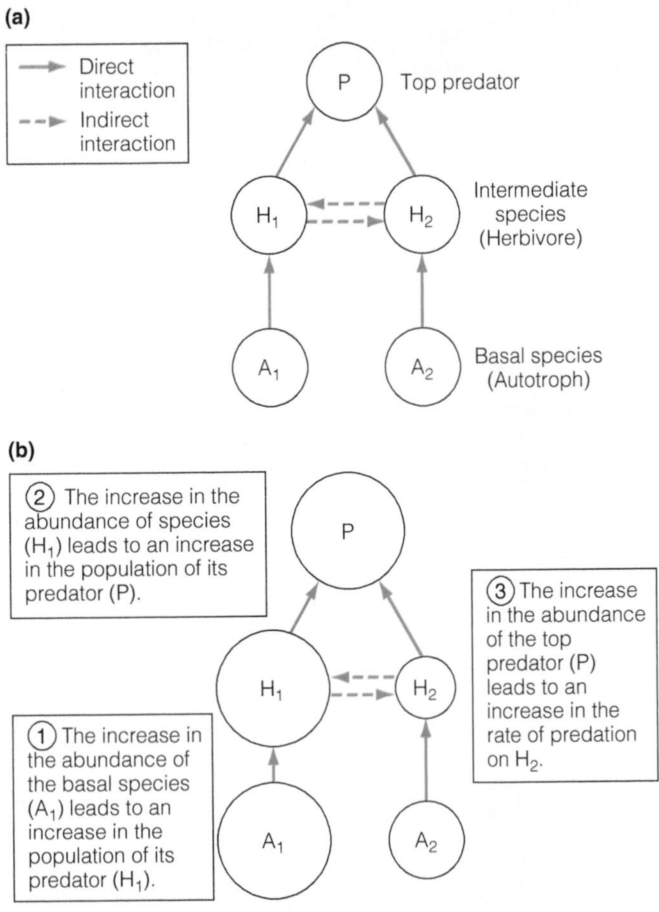
(a)
Direct interaction
Indirect interaction
P
Top predator
H1
H2
Intermediate species (Herbivore)
A1
A2
Basal species (Autotroph)
(b)
② The increase in the abundance of species (H1) leads to an increase in the population of its predator (P).
P
③ The increase in the abundance of the top predator (P) leads to an increase in the rate of predation on H2.
H1
H2
① The increase in the abundance of the basal species (A1) leads to an increase in the population of its predator (H1).
A1
A2

Figure 17.10 Diagram illustrating the emergence of apparent competition between two prey species (H_1 and H_2) that have a common predator (P). Direct interactions are represented by a solid arrow, and indirect interactions between species are indicated by a dashed arrow. The size of the populations is indicated by the size of the respective circles. (a) The two intermediate species (H_1 and H_2) feed on different basal species (A_1 and A_2) and therefore do not interact directly (no competition). Both species, however share a common predator (P). (b) An increase in the population of H_1 (resulting from an increase in its prey, A_1) results in an increase in the top predator population (P). In turn the increase in P results in an increase in the rate of predation on H_2. The corresponding decline in H_2 with the increase in H_1 has the appearance of a competitive interaction (apparent competition) even though the two species do not interact directly.

Apparent competition is an interesting concept that is illustrated by the structure of food webs. But does it really occur in nature? Many studies have identified community patterns that are consistent with apparent competition, and there is convincing experimental evidence of apparent competition in intertidal, freshwater, and terrestrial communities. Ecologists Christine Müller and H. C. J. (Charles) Godfray of Imperial College (Berkshire, England) conducted one such study. Müller and Godfray examined the role of apparent competition between two species of aphids that do not interact directly, yet share a common predator. The nettle aphid (*Microlophium carnosum*) feeds only on nettle plants (*Urtica* spp.), whereas the grass aphid (*Rhopalosiphum padi*) feeds on a variety of grass species. Although these two aphid species use different plant resources within the field community, they share a common predator: the ladybug beetle (Coccinellidae). In their study, the researchers placed potted nettle plants containing colonies of nettle aphids in plots of grass within the field community that contained natural populations of grass aphids (**Figure 17.11**). On a subset of the grass plots, they applied fertilizer that led to rapid grass growth and an increase in the local population of grass aphids. Nettle aphid colonies adjacent to the fertilized plots suffered a subsequent decline in population density when compared to colonies that were adjacent to unfertilized plots (control plots with low grass aphid populations). The reduced population of nettle aphids in the vicinity of high population densities of grass aphids (fertilized plots) was the result of increased predation by ladybug beetles, attracted to the area by the high concentrations of grass aphids; it was not as a result of direct resource competition between the two aphid species.

Some indirect interactions have negative consequences for the affected species, as in the preceding case of apparent competition. In other cases, however, indirect interactions between species can be positive. An example comes from a study of subalpine ponds in Colorado by Stanley Dodson of the University of Wisconsin. It involves the relationships between two herbivorous species of *Daphnia* and their predators, a midge larva (*Chaoborus*) and a larval salamander (*Ambystoma*). The salamander larvae prey on the larger of the two *Daphnia* species, whereas the midge larvae prey on the small species (**Figure 17.12**). In a study of 24 pond communities in the mountains of Colorado, Dodson found that where salamander larvae were present, the number of large *Daphnia* was low and the number of small *Daphnia*, high. However, in ponds where salamander larvae were absent, small *Daphnia*

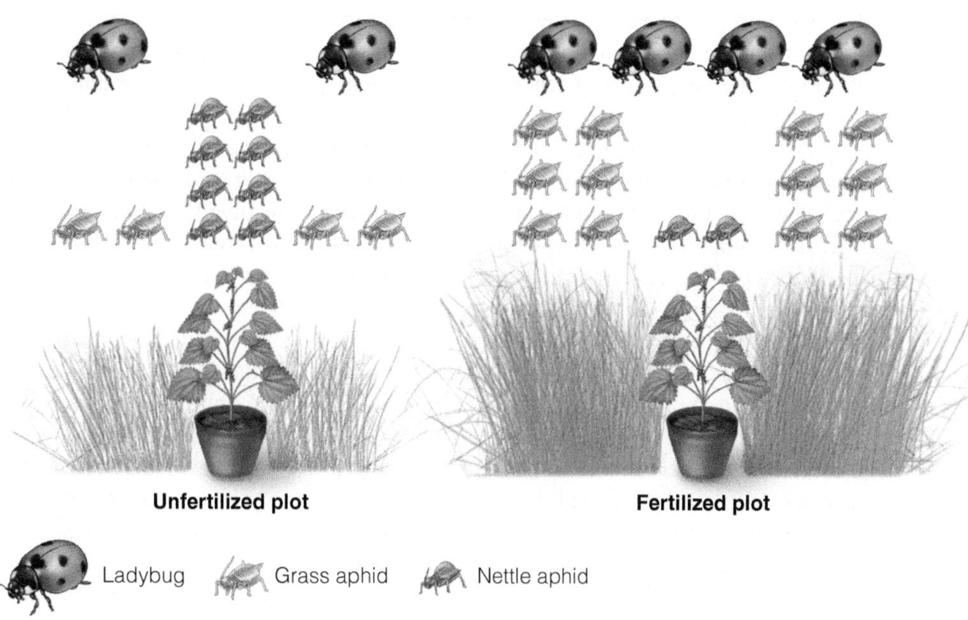

Figure 17.11 Example of a field experiment illustrating apparent competition. The grass aphid and the nettle aphid use different plant species as food and habitat in grassland communities. The grass aphid inhabits and feeds on grass plants, and the nettle aphid inhabits and feeds on nettle plants. The two species, however, share a common predator, the ladybug. Potted nettle plants containing nettle aphids were placed in each of two experimental grass plots: fertilized plots and unfertilized plots (control). Grass productivity increased on fertilized plots resulting in an increase in the population of grass aphids and attracted a greater number of predators (lady bugs). The result was an increased rate of predation on nettle aphids and a subsequent decline in their numbers on nettle plants in the fertilized plots.
(Based on Muller and Godfray 1997.)

were absent and midges could not survive. The two species of *Daphnia* apparently compete for the same resources. When the salamander larvae are not present, the larger of the two *Daphnia* species can outcompete the smaller. With the salamander larvae present, however, predation reduces the population growth rate of the larger *Daphnia*, allowing the two species to coexist. In this example, two indirect positive interactions arise. The salamander larvae indirectly benefit the smaller species of *Daphnia* by reducing the population size of its competitor. Subsequently, the midge apparently depends on the presence of salamander larvae for its survival in the pond. The indirect interaction between the midge and the larval salamander is referred to as **indirect commensalism** because the interaction is beneficial to the midge but neutral to the larval salamander. When the indirect interaction is beneficial to both species, the indirect interaction is termed **indirect mutualism**.

This role of indirect interactions can be demonstrated only in controlled experiments involving manipulations of the species populations involved. The importance of indirect interactions remains highly speculative, but experiments such as those just presented strongly suggest that indirect interactions among species—both positive and negative—can be an integrating force in structuring natural communities. There is a growing appreciation within ecology for the role of indirect effects in shaping community structure, and understanding these complex interactions is more than an academic exercise; it has direct implications for conservation and management of natural communities.

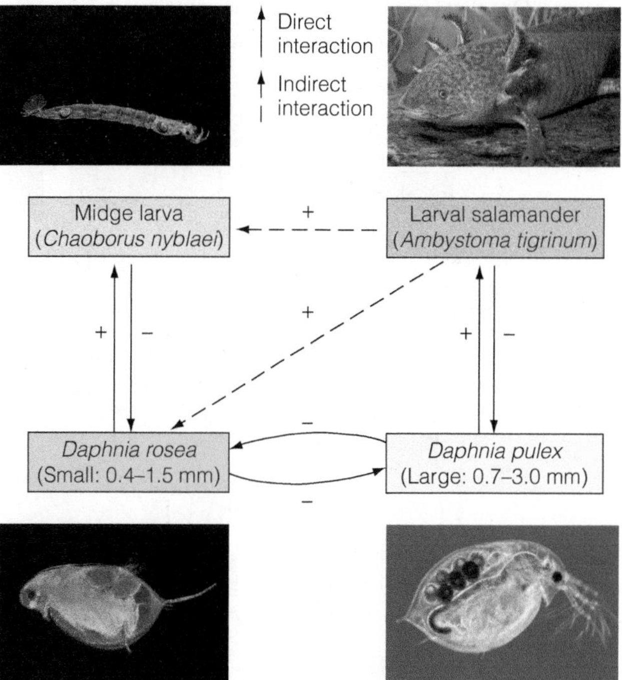

Figure 17.12 Diagram showing the relationship among the midge larva (*Chaoborus*), larval salamander (*Ambystoma*), and two species of *Daphnia* (*Daphnia rosea* and the larger *Daphnia pulex*) that inhabit pond communities in the mountains of Colorado. Removing salamander larvae from some ponds resulted in the competitive exclusion of *D. rosea* by *D. pulex* and the local extinction of the midge population that preyed on it.
(Data from Dodson, 1974.)

As with the example of starfish in the intertidal zone, removing a species from the community can have many unforeseen consequences. For example, Joel Berger of the University of Nevada and colleagues have examined how the local extinctions of grizzly bears (*Ursus arctos*) and wolves (*Canis lupus*) from the southern Greater Yellowstone ecosystem, resulting from decades of active predator control, have affected the larger ecological community (see Chapters 17, *Ecological Issues & Applications*). One unforeseen consequence of losing these large predators is the decline of bird populations that use the vegetation along rivers (riverine habitat) within the region. The elimination of large predators from the community resulted in an increase in the moose population (prey species). Moose selectively feed on willow (*Salix* spp.) and other woody species that flourish along the river shorelines. The increase in moose populations dramatically affected the vegetation in riverine areas that provide habitat for a wide variety of bird species and led to the local extinction of some populations.

17.5 Food Webs Suggest Controls of Community Structure

The wealth of experimental evidence illustrates the importance of both direct and indirect interactions on community structure. On that basis, rejecting the null model as presented in Section 17.1 would be justified. However, given the complexity of direct and indirect interactions suggested by food webs, how can we begin to understand which interactions are important in controlling community structure and which are not? Are all species interactions important? Does some smaller subset of interactions exert a dominant effect, whereas most have little impact beyond those species directly involved (see discussion of food web compartmentalization in Section 16.5)? The hypothesis that all species interactions are important in maintaining community structure suggests that the community is like a house of cards—that is, removing any one species may have a cascading effect on all others. The hypothesis that only a smaller subset of species interactions control community structure suggests a more loosely connected assemblage of species.

These questions are at the forefront of conservation ecology because of the dramatic decline in biological diversity that is a result of human activity (see Chapters 9 and 12, *Ecological Issues & Applications*). Certain species within the community can exert a dominant influence on its structure, such as the predatory starfish that inhabits the rocky intertidal communities. However, the relative importance of most species in the functioning of communities is largely a mystery. One approach being used to understand the influence of species diversity on the structure and dynamics of communities is grouping species into functional categories based on criteria relating to their function within the community. For example, the concept of guilds is a functional grouping of species based on sharing

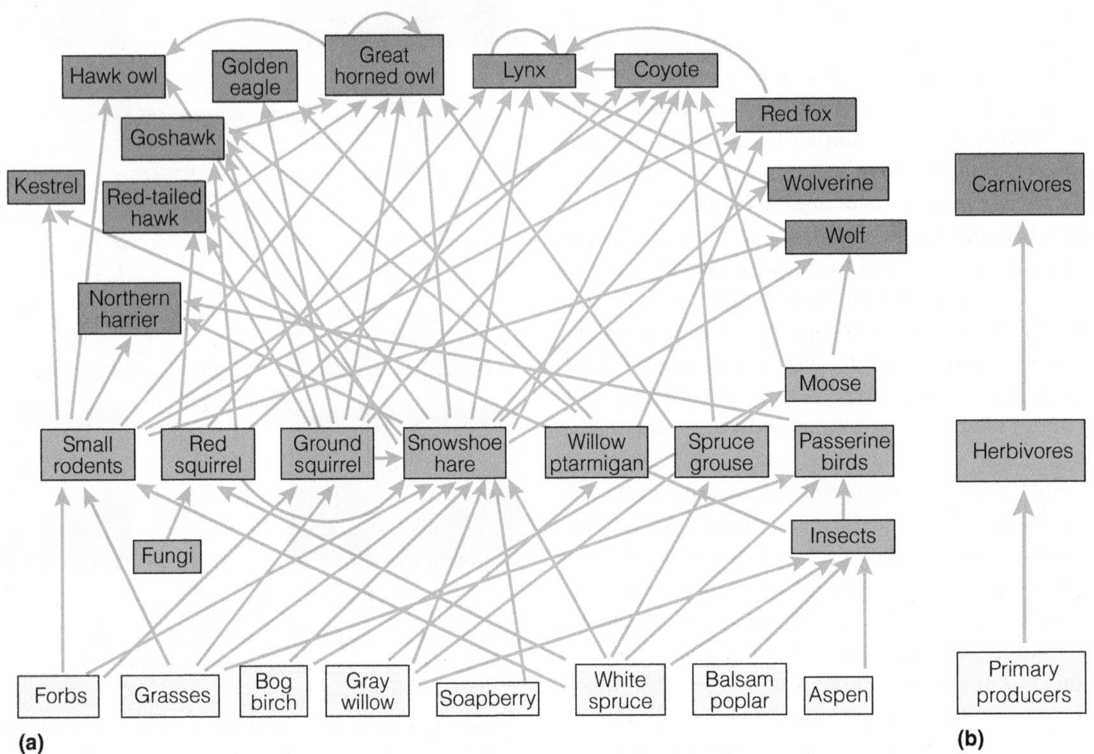

Figure 17.13 (a) Aggregation of species forming the food web for Canadian boreal forests presented in Figure 18.3 into trophic levels (generalized feeding groups). (b) As with the food web, arrows point from prey to predator.

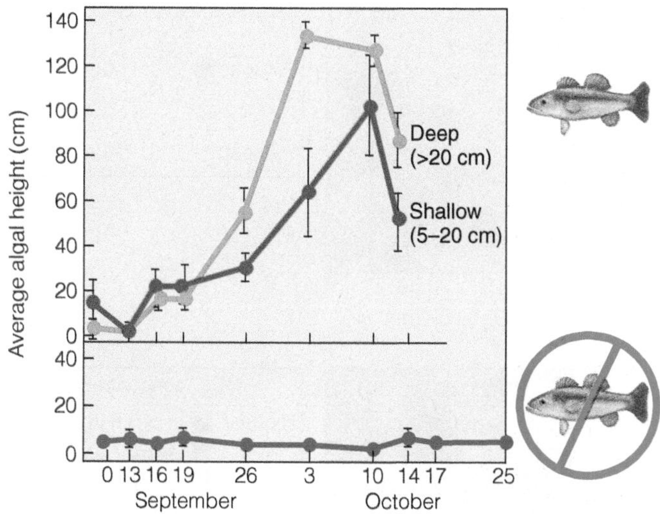

Figure 17.14 Changes in the abundance of algae in sections of the river (pools) in which the top predator, largemouth bass, was either absent or present (added to pool). Algal abundance measure as average algal height (cm) in both shallow and deep waters of the experimental pools. The observed increase in algae in the presence of bass is a result of increased predation on *Campostoma* (minnows), the dominant herbivore in the community. Vertical bars represent ±2 standard error.
(Adapted from Powers et al 1987.)

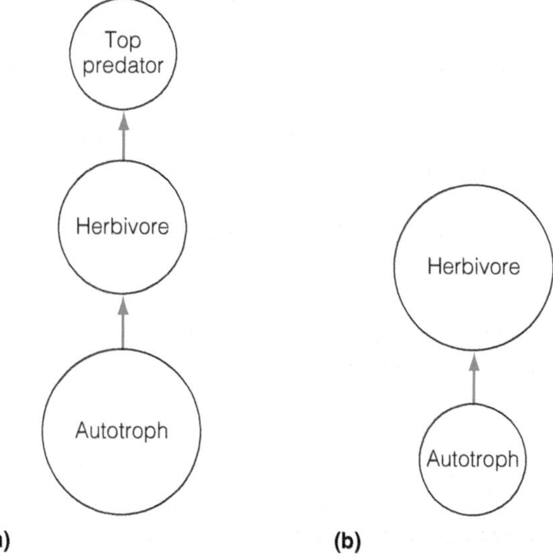

Figure 17.15 Simple example of a food web trophic cascade. (a) The food web consists of a single basal species (autotroph), an intermediate species (herbivore) that feeds on the basal species, and a top predator. (b) The removal of the top predator results in an increase in the intermediate species on which it preys. In turn, the increase in the intermediate species results in a decrease in the basal species on which it preys. The reverse occurs if the top predator is added to a community in which it is absent (from b to a).

similar functions within the community or exploiting the same resource (e.g., grazing herbivores, pollinators, cavity-nesting birds; see Section 16.6). By aggregating species into a smaller number of functional groups, researchers can explore the processes controlling community structure in more general terms. For example, what is the role of mammalian predators in boreal forest communities? This functional grouping of species can be seen in the food web presented in Figure 17.8, in which the categories (boxes) of forbs, grasses, small rodents, insects, and passerine birds represent groups of functionally similar species.

One way to simplify food webs is to aggregate species into trophic levels (Section 16.5). The food web presented in Figure 17.8 has been aggregated into three trophic levels: primary producers, herbivores, and carnivores (Figure 17.13). Although this is an obvious oversimplification, using this approach raises some fundamental questions concerning the processes that control community structure.

As with food webs, the arrows in a simple food chain based on trophic levels point in the direction of energy flow—from autotrophs to herbivores and from herbivores to carnivores. The structure of food chains suggests that the productivity and abundance of populations at any given trophic level are controlled (limited) by the productivity and abundance of populations in the trophic level below them. This phenomenon is called **bottom-up control**. Plant population densities control the abundance of herbivore populations, which in turn control the densities of carnivore populations in the next trophic level. However, as we have seen from the previous discussion of predation and food webs, **top-down control** also occurs when predator populations control the abundance of prey species.

Work by Mary Power and her colleagues at the University of Oklahoma Biological Station suggests that the role of top predators (carnivores) on community structure can extend to lower trophic levels, influencing primary producers (autotrophs) as well as herbivore populations. Power and colleagues showed that a top predator, the largemouth bass (*Micropterus salmoides*), had strong indirect effects that cascaded through the food web to influence the abundance of benthic algae in stream communities of the midwestern United States. In these stream communities, herbaceous minnows (primarily *Campostoma anomalum*) graze on algae, and in turn, largemouth bass feed on the minnows. During periods of low flow, isolated pools form in the streams. As part of the experiment, bass were removed from some pools and the populations of algae and minnows were monitored. Pools with bass had low minnow populations and a luxuriant growth of algae (Figure 17.14). In contrast, pools from which the bass were removed had high minnow populations and low populations (biomass) of algae. In this example, top predators (carnivores) were shown to control the abundance of plant populations (primary producers) indirectly through their direct control on herbivores (also see Chapter 20, Field Studies: Brian Silliman). This type of top-down control is referred to as a **trophic cascade**. A trophic cascade occurs when a predator in a food web suppresses the abundance of their prey (intermediate species) such that it increases the abundance of the next lower trophic level (basal species) on which the intermediate species feeds (Figure 17.15).

A now-famous article written by Nelson Hairston, Fred Smith, and Larry Slobodkin first introduced the concept of top-down control with the frequently quoted "the world is green" proposition. These three ecologists proposed that the world is green (plant biomass accumulates) because predators keep herbivore populations in check. Although this proposition is supported by a growing body of experimental studies such as those by Power and her colleagues, experimental data required to test this hypothesis are still limited, particularly in terrestrial ecosystems. However, the proposition continues to cause great debate within the field of community ecology. We will return to the topic when discussing factors that control primary productivity (Chapter 20).

17.6 Environmental Heterogeneity Influences Community Diversity

As we have seen thus far, the biological structure (species composition) of a community reflects both the direct response (survival, growth, and reproduction) of the component species to the prevailing abiotic environmental conditions, as well as their interactions (direct and indirect). In turn, as environmental conditions change from location to location, so will the set of species that can potentially occupy the area and the manner in which they interact. This framework has helped us understand why the biological structure of a community changes as we move across a landscape from hilltop to valley or from the shoreline into the open waters of a lake or pond. However, environmental conditions are typically not homogeneous even within a given community. For example, ecologist Philip Robertson and colleagues quantified spatial variation in soil nitrogen and moisture across an abandoned agricultural field in southeastern Michigan. Once used for agriculture, the site was abandoned in the late 1920s and reverted to an old-field community composed of a variety of forb, grass, and shrub species. Detailed sampling of a 0.5-hectare (ha) plot within the old field revealed considerable spatial variation (more than an order of magnitude) in soil moisture and nitrate at this spatial scale (**Figure 17.16**). Studies similar to that of Robertson and his colleagues have shown comparable patterns of fine-scale environmental variation within forest, intertidal, and benthic communities.

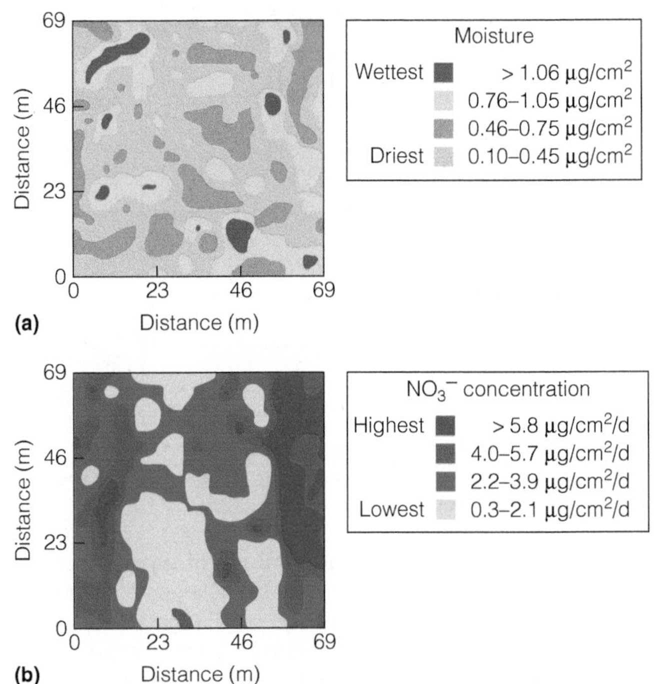

Figure 17.16 Variations in (a) soil moisture and (b) nitrogen (nitrate; NO_3^-) production in an old-field community in Michigan (abandoned agricultural field).
(Adapted from Robertson et al. 1998.)

But how does environmental heterogeneity within a community influence patterns of diversity? Do variations in environmental conditions translate into an area's ability to support more species? Some examples that we have considered in previous chapters provide an answer to this question in plant communities. Heterogeneity in the light environment of the forest floor caused by the death of canopy trees (gap formation) has been shown to increase tree species diversity in forest ecosystems. The increase in available light below canopy gaps allows for the survival and growth of shade-intolerant species that otherwise would be excluded from the community (see Section 6.8, Figure 6.10).

A good example of the influence of environmental heterogeneity comes from the link between vegetation structure and bird species diversity. The structural features of the vegetation that influence habitat suitability for a given bird species are related to a variety of species-specific needs relating to food, cover, and nesting sites. Because these needs vary among species, the structure of vegetation has a pronounced influence on the diversity of bird life within the community. Increased vertical structure means more resources and living space and a greater diversity of potential habitats (see Section 16.7, Figure 16.13). Grasslands, with their two strata, support six or seven species of birds, all of which nest on the ground. A deciduous forest in eastern North America may support 30 or more species occupying different strata. The scarlet tanager (*Piranga olivacea*) and wood pewee (*Contopus virens*) occupy the canopy, the hooded warbler (*Wilsonia citrina*) is a forest shrub species, and the ovenbird (*Seiurus aurocapillus*) forages and nests on the forest floor.

The late Robert MacArthur of Princeton University was the first ecologist to quantify the relationship between the structural heterogeneity of vegetation and the diversity of animal species that depend on the vegetation as habitat. He measured bird species diversity and the structural heterogeneity of vegetation in 13 communities in the northeastern United States. The communities represented a variety of structures, from grassland to deciduous forest. Bird species diversity in each

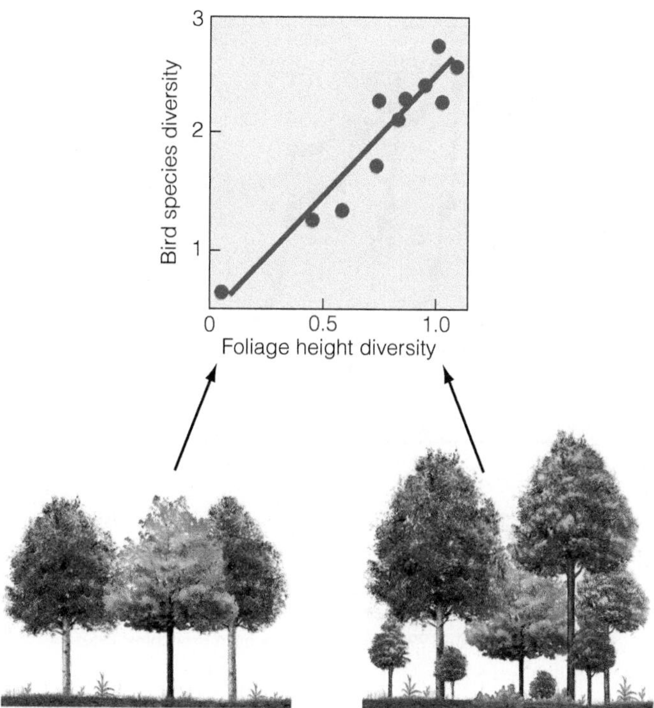

Figure 17.17 Relationship between bird species diversity and foliage height diversity for deciduous forest communities in eastern North America. Foliage height diversity is a measure of the vertical structure of the forest. The greater the number of vertical layers of vegetation, the greater the diversity of bird species present in the forest.
(Adapted from MacArthur and MacArthur 1961.)

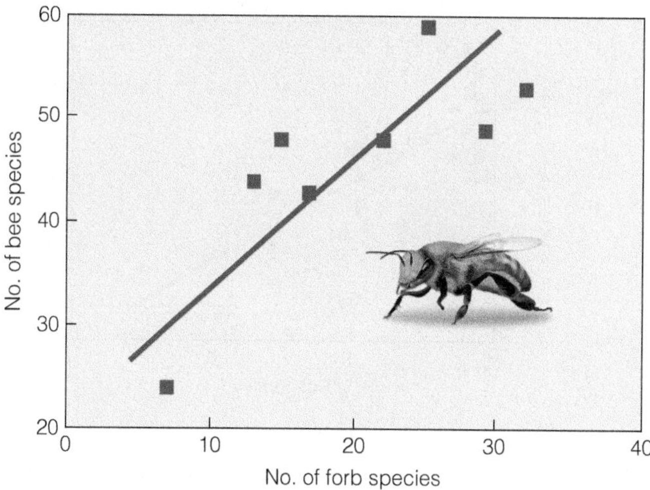

Figure 17.18 The relationship between forb species richness and bee species richness for eight prairie grassland communities.
(Adapted from Reed 1993.)

community was measured using an index of species diversity (see Section 16.2). To quantify the structural heterogeneity of the vegetation, MacArthur developed an index of foliage height diversity. The value of the index increased with the number of vertical layers, the maximum height of the vegetation, and the relative abundance of vegetation (biomass) within the vertical layers. By comparing the two indexes, MacArthur found a strong relationship between bird species diversity and the index of foliage height diversity for the various communities (**Figure 17.17**). Since the publication of this pioneering work by MacArthur in the early 1960s, similar relationships between the structural diversity of habitats and the diversity of animal species within a community have been reported for a wide variety of taxonomic groups in both terrestrial and aquatic environments.

Just as an increase in the diversity of potential habitats within a community results in an increase in the number of animal species that can be supported, an increase in the diversity of food resources within a community can likewise potentially increase the diversity of consumers that depend on those food resources. This appears to be the case with nectar-feeding insects and the flowering plant species on which they feed. Ecologist Catherine Reed of the University of Minnesota quantified patterns of species diversity for both flowering plants and their insect pollinators in eight prairie grassland communities in south central Minnesota. Reed found a significant positive correlation between bee species richness and forb species richness (**Figure 17.18**). Prairie grasslands that supported a greater number of forb species likewise supported a greater number of bee species that depend on nectar from these flowering plants as their food resources (see Section 12.3 and 15.13). The positive correlation between the number of bee and forb species for the study sites appears to be based on two factors: a greater number of food resources for bee species that specialize on feeding from a single plant species (flower type) and a more consistent food source throughout the season for bee species that are generalist feeders (i.e., feed on many plant species). The latter results because the various species of forbs flower at different times throughout the growing season.

17.7 Resource Availability Can Influence Plant Diversity within a Community

We examined the role of nutrient availability on plant processes (see Section 6.11). In general, increased availability of nutrients can support higher rates of photosynthesis, plant growth, and a higher density of plants per unit area. It might seem somewhat odd, therefore, that a variety of studies have shown an inverse relationship between nutrient availability and plant diversity in communities.

Michael Huston, an ecologist at Texas State University, examined the relationship between the availability of soil nutrients and species richness at 46 tropical rain forest sites in Costa Rica. Huston found an inverse relationship between species

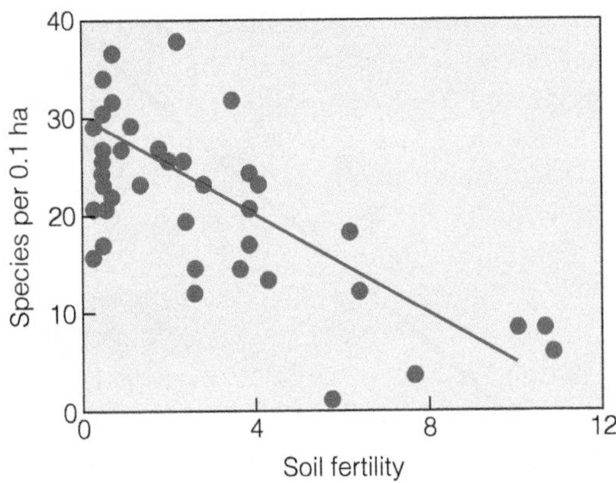

Figure 17.19 Relationship of tree species richness (species per 0.1 ha) to a simple index of soil fertility for 46 forest communities in Costa Rica. The fertility index is the sum of the percentage values of phosphorus, potassium, and calcium obtained by dividing the value of each nutrient by the mean value of that nutrient for all 46 sites.
(Adapted from Huston 1980.)

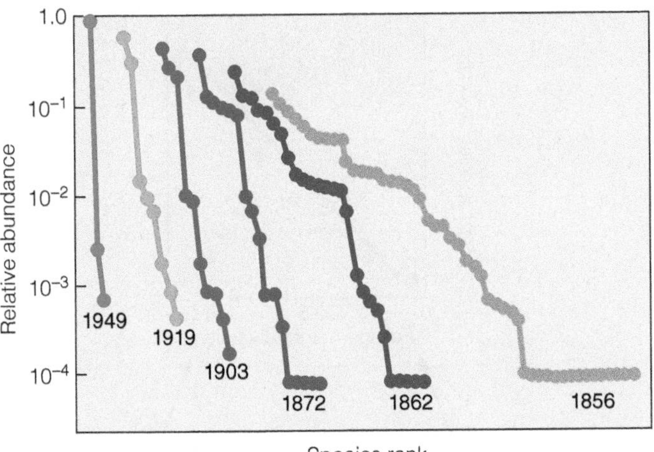

Figure 17.20 Rank abundance curves illustrating changes in plant species richness and evenness for the Rothamsted Park Grass Experiment (England) over time. The grassland has been subjected to the continuous application of nitrogen fertilizer since 1856.
(Adapted from Tokeshi 1993. Based on data from Brenchley 1958.)

richness and a composite index of soil fertility (**Figure 17.19**). Tropical forest communities on soils with lower nutrient availability supported a greater number of tree species (species richness) than did communities on more fertile soils. Huston hypothesized that the inverse relationship results from reduced competitive displacement under low nutrient availability. Low nutrient availability reduces growth rates and supports a lower density and biomass of vegetation. Species that might dominate under higher nutrient availability cannot realize their potential growth rates and biomass and thus are unable to displace slower-growing, less competitive species.

A wide variety of field and laboratory experiments have supported the hypothesis put forward by Huston. In a series of competition experiments under controlled greenhouse conditions, Fakhri Bazzaz of Harvard University and the British ecologist John Harper found that two herbaceous plant species—white mustard (*Sinapis alba*) and cress (*Lepidium sativum*)—coexisted on less fertile soil, whereas *Lepidium* was driven to extinction by *Sinapis* under conditions of higher soil fertility.

The Park Grass experiment was begun at Rothamsted Experimental Station in Great Britain in 1859 to examine the effects of fertilizers on yield and quality of hay from permanently maintained grasslands. This experiment has continued for more than 140 years. Beginning with a uniform mixture of grass and other herbaceous species, various types, quantities, and schedules of fertilization have been applied to experimental plots within the field. Changes in species composition began as early as the second year and increased through time until a relatively stable community structure was achieved. The unfertilized plots are the only ones retaining the original diversity of species that were planted. In all cases, the number of species was reduced by fertilization, and the most heavily fertilized plots became dominated by only a few grass species (**Figure 17.20**). Nearly identical results to those of the Park Grass experiment have been obtained in other fertilization experiments in both agricultural and natural grassland communities.

Greenhouse and field experiments both leave little doubt that changes in nutrient availability can greatly alter the composition and structure of plant species in communities. In all experimental studies to date, the effect of increasing nutrient availability has been to decrease diversity. But what processes cause this decrease in diversity, allowing some plant species to displace others under conditions of high nutrient availability? In a series of field experiments, ecologist James Cahill of the University of Alberta (Canada) examined how competition in grassland communities shifts along a gradient of nutrient availability. The experiments revealed a shift in the importance of below- and aboveground competition and the nature of their interaction under varying levels of nutrient availability. Cahill's work indicates that competition for below- and aboveground resources differs in an important way. Competition for belowground resources is *size symmetric* because nutrient uptake is proportional to the size of the plant's root system. Symmetric competition results when individuals compete in proportion to their size, so that larger plants cause a large decrease in the growth of smaller plants, and small plants cause a small (but proportionate to their size) decrease in the growth of larger plants. In contrast, competition for light (aboveground) is generally one-sided or *size asymmetric*; larger plants have a disproportionate advantage in competition for light by shading smaller ones, resulting in initial size differences being compounded over time. Any factor that reduces the growth rate of a plant initiates a positive feedback loop that decreases the plant's likelihood of obtaining a dominant position in the developing size hierarchy.

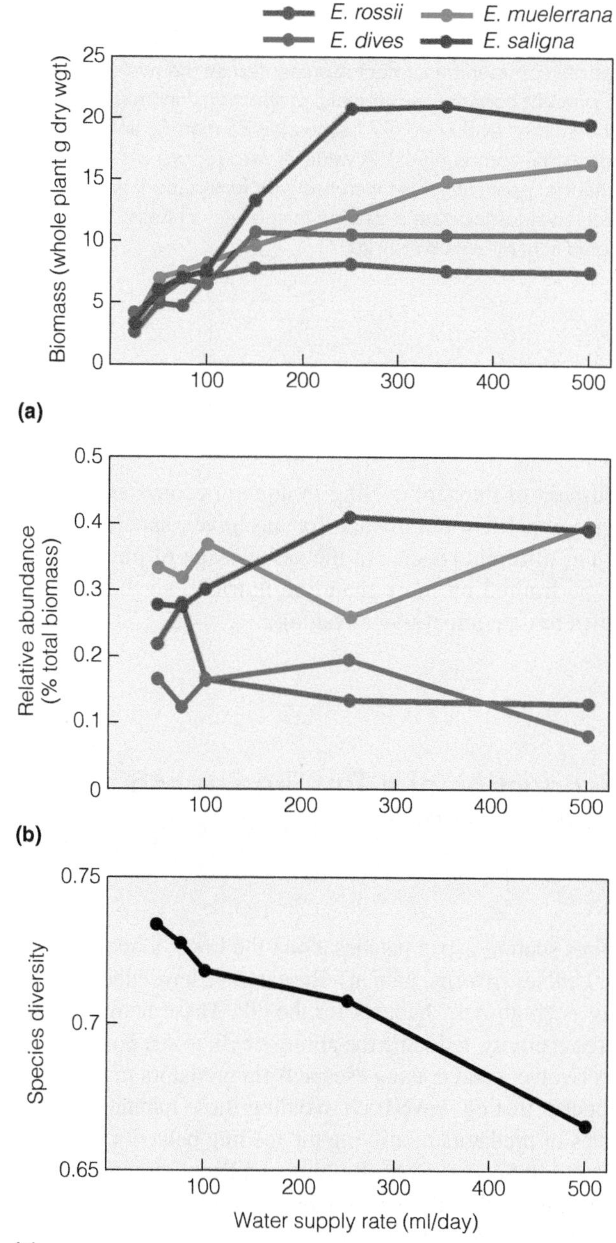

Figure 17.21 Changes in species evenness along an experimental gradient of belowground resource availability. (a) Response of seedlings of four *Eucalyptus* species when grown as single individuals along an experimental gradient of water treatments. Standard nutrient solution was used for water treatments, so water treatments represent a combined increase in both water and nutrient availability. (b) Relative abundances (percentage of total biomass) of the four species when grown in mixed cultures along the same experimental gradient. Under low resource availability, the distribution of biomass is more equitable, whereas increasing resource availability results in competitive dominance by the species having the highest potential growth rate under high resource availability (see a). (c) The result is a decline in species evenness (Simpson index) with increasing availability of belowground resources.
(Data from Austin et al. 2009 and unpublished data T. M. Smith.)

Under low nutrient availability, plant growth rate, size, and density are low for all species. Competition primarily occurs belowground and therefore is symmetric. Competitive displacement is low, and diversity is maintained. As nutrient availability increases, growth rate, size, and density increase. Species that maintain higher rates of photosynthesis and growth exhibit a disproportionate increase in size. As faster-growing species overtop the others, creating a disparity in light availability, competition becomes strongly asymmetric. Species that achieve high rates of growth and stature under conditions of high soil fertility eventually outcompete and displace the slower-growing, smaller-stature species, thus reducing the species richness of the community.

This shift in the nature of competition for above- and belowground resources along a gradient of belowground resource availability is well illustrated by the work of Mike Austin (Commonwealth Scientific and Industrial Research Organisation) and colleagues at the Australian National University. The researchers conducted a series of greenhouse experiments in which seedlings of four species of *Eucalyptus* were grown under different water treatments (belowground resource availability). Seedlings were grown in pots either individually (no competition) or in mixtures containing equal numbers of the four species (interspecific competition). The water provided in the treatments was a standard nutrient solution, so the treatments represent a combined gradient of water and nutrient availability. Results of the competition experiments (**Figure 17.21b**) show that under low belowground resource availability, the growth of all four species is depressed, and the biomasses of the four species are relatively equal. As the availability of belowground resources increases, however, those species with the higher potential growth rates under high resource availability (**Figure 17.21a**) outcompete the slower-growing species, dominating total biomass of the pot. The result is a decline in species evenness (as measured by the Simpson index of evenness; see Section 16.2) with increasing availability of belowground resources (**Figure 17.21c**).

In contrast to the inverse relationship between soil fertility (nutrient availability) and plant species richness observed in terrestrial plant communities, the pattern between nutrient availability and the species richness of autotrophic organisms in aquatic communities is quite different. Ecologist Helmut Hillebrand of the University of Cologne, Germany, and colleagues reviewed the results of 97 published studies in which nutrient availability was experimentally manipulated in both terrestrial (53) and aquatic communities (19 freshwater and 23 marine). Their analyses reveal that, unlike the pattern of decreasing species diversity observed in terrestrial plant communities, fertilization results in an increase in the species richness of autotrophs in both freshwater and marine communities (**Figure 17.22**). Although there is no consensus as to what causes the observed differences between terrestrial and aquatic communities, several factors have been suggested, including differences in the role of competition in terrestrial and marine environments. Unlike the competitive

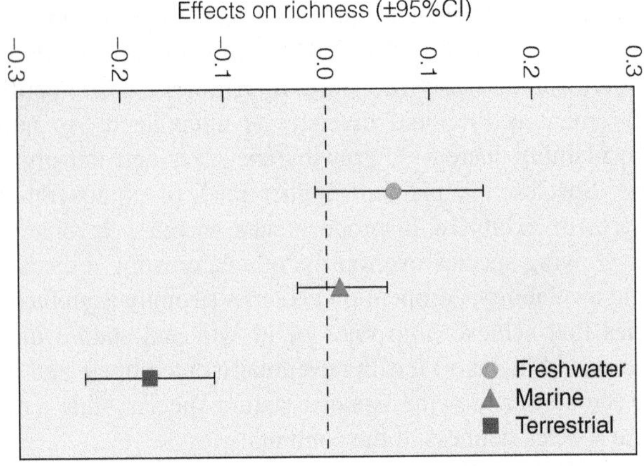

Figure 17.22 Average effect of fertilization on autotroph species richness in freshwater, marine, and terrestrial communities. Effect represented as the average (±95 percent confidence intervals) proportional increase for the studies evaluated (19 freshwater, 23 marine, and 59 terrestrial communities). A value of zero represents no change, positive values increase with fertilization, and negative values indicate a decline in species richness. (Adapted from Hillebrand et al. 2007.)

displacement of species under high nutrient availability discussed previously for terrestrial plant communities, competitive exclusion among phytoplankton species in pelagic waters is less common. Reduced competition may result from rapid changes in nutrient levels that occur in the water column (see discussion of nutrient cycling in aquatic ecosystems) and restrict competitive dominance for any given species (Chapter 21). Or, different species in the assemblage of phytoplankton may be limited by more than one nutrient, so that no single species has a competitive advantage.

ECOLOGICAL Issues & Applications

The Reintroduction of a Top Predator to Yellowstone National Park Led to a Complex Trophic Cascade

A trophic cascade occurs when predators suppress the population of their prey (herbivores), thereby releasing the next lower trophic level (autotrophs) from predation (Section 17.5, Figure 17.15). One of the most dramatic and well-documented examples of a trophic cascade is the recovery of vegetation in regions of the Yellowstone National Park following the reintroduction of wolves in the winter of 1995–1996. The reintroduction of wolves has led to a decline in their primary prey, elk, which in turn feed on the stems and shoots of shrubs and smaller deciduous trees, such as aspen, cottonwood, and willow.

Despite attempts at active management of the elk herd during the period from the 1930s to the 1960s, the population of elk in the in the Yellowstone Park dramatically increased over the seven decades following the extermination of wolves in the 1920s (see Chapter 13, *Ecological Issues & Applications*). As a result of increased browsing by elk, the populations of deciduous vegetation in the park declined, with little or no regeneration of species such as aspen in areas heavily used by the elk herd. The decline in the elk herd since the reintroduction of wolves has led to significant changes in the vegetation of the park. Ecologists Robert Beschta and William Ripple of Oregon State University have been monitoring changes in the vegetation in Yellowstone Park since the reintroduction of wolves in the mid-1990s. Their research shows a clear pattern of increasing populations of the major deciduous tree species (aspen, cottonwood, and willow) associated with a decline in browsing by elk populations (**Figure 17.23**). Interestingly, this recovery is not occurring in all areas of the park, but it is associated with specific habitats such as aspen patches along the lower slopes and along river courses (riverine habitat). Researchers have identified these areas as "high-risk" habitats for the elk. These habitats provide poorer visibility, reducing the ability of elk to see potential predators (wolves) and making escape from predators more difficult. It appears that elk have been avoiding these habitats (other examples of predators modifying the feeding behavior of prey can be found in Section 14.8). Beschta and Ripple found significant differences in the levels of browsing and the recovery of trees between high- and low-risk sites (**Figure 17.24**).

In addition to the increases in recruitment and growth of deciduous tree species, reduced browsing pressure by elk has resulted in an increase in the populations of berry-producing shrubs within the recovering stands of aspen trees. In a study of aspen stands in an area of the park where foraging by elk has declined since the reintroduction of wolves, Beschta and Ripple found a significant increase in the stature of aspen trees and an associated development of fruit-bearing shrub species in the developing understory. These berry-producing shubs provide an important food source for a variety of invertebrates, birds, and mammals. For example, the researchers found a significant increase in the percentage of fruits from these shrub species in the diets of grizzly bears compared to the period prior to reintroduction of wolves. The result is an inverse relationship between fruit consumption by grizzly bear and the elk population (**Figure 17.25a**), which is an example of an indirect positive effect of wolves on grizzly bear in the Yellowstone food chain (**Figure 17.25b**).

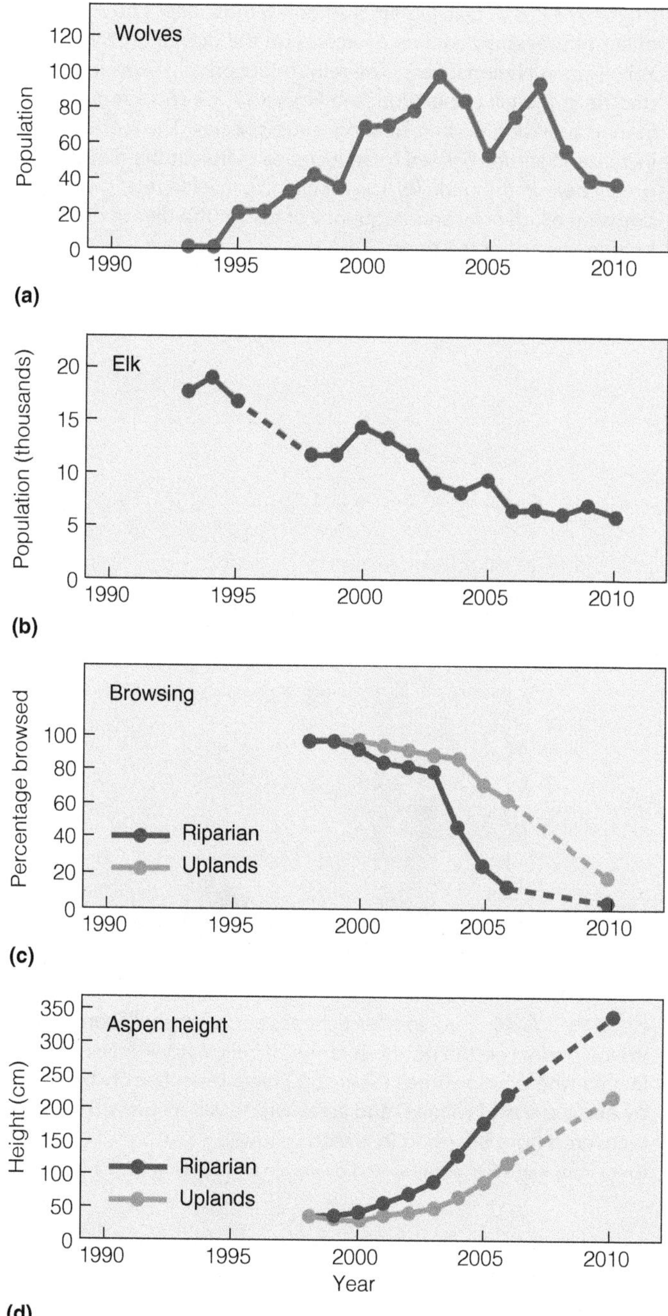

Figure 17.23 Trends in (a) wolf populations, (b) minimum elk populations from annual counts, (c) percentage of aspen leaders browsed, and (d) mean aspen heights (early springtime heights after winter browsing but before summer growth) for the northern range of the Yellowstone National Park since the reintroduction of wolves in the winter of 1995–1996.
(Data from Ripple, W.J. and R.L. Beschta. 2012. Trophic cascades in Yellowstone: The first 15 years after wolf reintroduction. Biological Conservation 145:205–213. Fig. 1, pg. 207.)

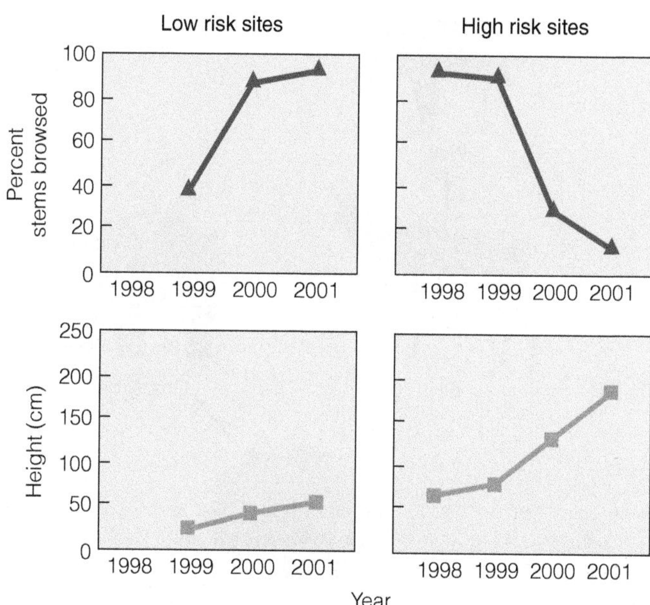

Figure 17.24 Percentage of stems browsed and mean plant height for young cottonwood plants on both low predation risk sites (left graphs) and high predation risk sites (right graphs) at Soda Butte in Yellowstone National Park. High-risk sites show less elk browsing and taller cottonwoods than low-risk sites.
(Data from Ripple, W.J. and R.L. Beschta. 2003. Wolf reintroduction, predation risk, and cottonwood recovery in Yellowstone National Park. Forest Ecology and Management 184:299–313. Fig. 5, pg. 308.)

The example of increased food resources from berry-producing shrubs illustrates that the positive effects of an increase in woody plant species because of decreased browsing by elk are not limited to the plant community. The changes in the plant community have the potential to impact the availability of habitat and food resources to a wide variety of vertebrate and invertebrate species within the Yellowstone community. Small herbivores such as rodents, hares, and rabbits may already be benefiting with additional cover and food because of decreases in browsing by elk and changes in the structure of the plant communities. A study of the avian community within the park by ecologist Lisa Baril of Montana State University shows that populations of passerine birds have been positively affected by the regrowth of willow trees resulting from decreased browsing pressure by elk. Baril found that the increased willow growth results in more structurally complex habitat that subsequently allowed for greater songbird richness and diversity (**Figure 17.26**).

Beaver have also increased since wolf reintroduction. The increase in beaver is likely a result, at least in part, of the resurgence of willow communities. In areas of the park where populations have risen, beaver feed almost exclusively on the newly growing willow trees. Increases in beaver populations have tremendous implications for both the hydrology and biodiversity of the riparian environments. Modifications of streams by beaver activity can decrease the erosion of stream banks and create wetland habitats ultimately influencing plant, vertebrate, and invertebrate diversity and abundance. A study of beaver activity on streams in Wyoming by Mark McKinstry and colleagues at the Wyoming Fish and Wildlife Cooperative Research Unit (Laramie, Wyoming) found that streams with beaver ponds have 75 times more abundant waterfowl than those without beaver ponds. Beaver ponds also provide habitat for a wide variety of invertebrates, amphibians, reptiles, and fish.

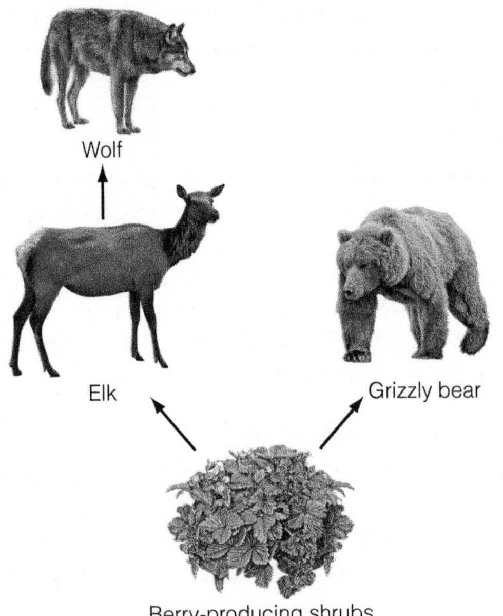

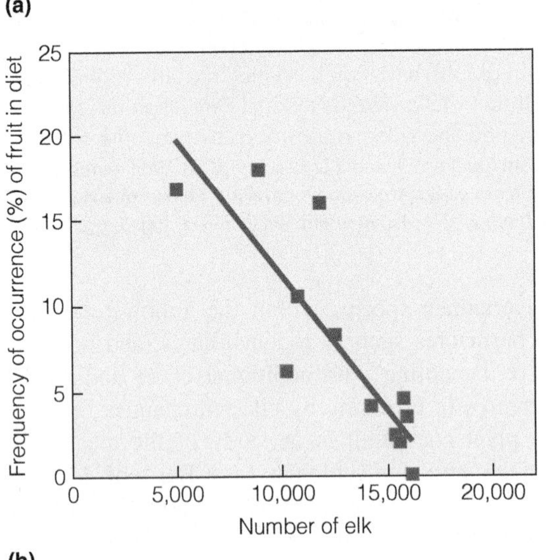

Figure 17.25 (a) Diagram illustrating the indirect positive effect of the reintroduction of wolves on the diet of grizzly bear in Yellowstone National Park. The reintroduction of wolves led to a decline in the elk population (see Figure 17.23) and a reduction in their browsing on fruit-bearing shrub species. The subsequent increase in shrubs has led to an increase in fruit in the diet of grizzly bear in the park. (b) The result is an inverse relationship between elk density and frequency of fruit in the diet of grizzly bear in regions of the park.
(Adapted from Ripple et al. 2013.)

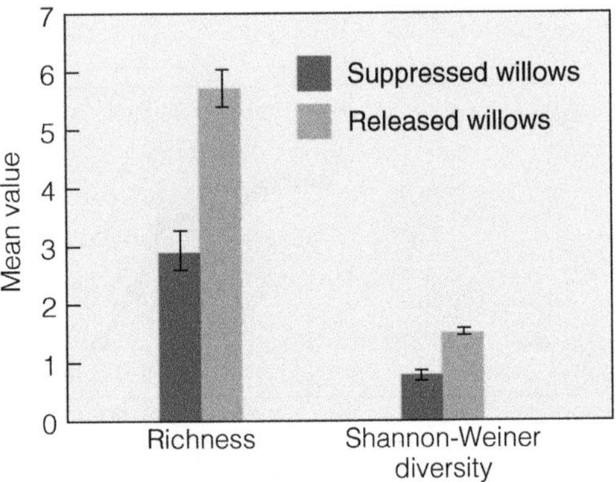

Figure 17.26 Bird species richness and diversity (Shannon–Weiner index) on the northern range of Yellowstone National Park in area where willow trees (*Salix* spp.) have been heavily browsed by elk (supressed willows) and areas where willow growth has recovered from browsing (released willows).
(Data from Bari 2009 as presented by Ripple and Beschta 2012.)

SUMMARY

Niche and Community Structure 17.1

The range of environmental conditions tolerated by a species defines its fundamental niche. These constraints on the ability of species to survive and flourish will limit their distribution and abundance to a certain range of environmental conditions. Species differ in the range of conditions they tolerate. As environmental conditions change in both time and space, the possible distribution and abundance of species also changes. This framework provides a null model against which to compare observed community patterns. Community structure reflects the species' realized niches, which are the fundamental niches modified by species interactions.

Zonation 17.2

The biological structure of a community is first constrained by the environmental tolerances of the species (fundamental niche), which are then modified through direct and indirect interactions with other species (realized niche). As we move across the landscape, variations in the physical environment alter the nature of both these constraints on species distribution and abundance, giving rise to patterns of zonation. There is a general trade-off between a species' stress tolerance and its competitive ability along gradients of resource availability. This trade-off can result in patterns of zonation across the landscape where variations in resource availability exist. The relationship between stress tolerance and competitive ability is more complex along gradients that include both resource and nonresource factors, such as temperature, salinity, or water depth.

Diffuse Interactions 17.3

Experiments that examine only two potentially interacting species tend to underestimate the importance of species interactions in communities because interactions are often diffuse and involve a number of species. In diffuse competition, direct interaction between any two species may be weak, making it difficult to determine the effect of any given species on another. Collectively, however, competition may be an important factor limiting the abundance of all species involved. Diffuse interactions involving predation and competition can be seen in the structure of food webs.

Indirect Interactions 17.4

Food webs also illustrate indirect interactions among species within the community. Indirect interactions occur when one species does not interact with a second species directly, but influences a third species that does interact directly with the second. For example, a predator may increase the population density of one or more inferior competitors by reducing the abundance of the superior competitor it preys on. Indirect positive interactions result when one species benefits another indirectly through its interactions with others, reducing either competition or predation.

Controls on Community Structure 17.5

To understand the role of species interactions in structuring communities, food webs are often simplified by placing species into functional groups based on their similarity in using resources or their role within the community. One such functional classification divides species into trophic levels based on general feeding groups (primary producers, herbivores, carnivores, etc.). The resulting food chains suggest the possibility of either bottom-up (primary producer) or top-down (top carnivore) control on community structure and function.

Environmental Heterogeneity 17.6

Environmental conditions are not homogeneous within a given community, and spatial variations in environmental conditions within the community can function to increase diversity by supporting a wider array of species. The structure of vegetation has a pronounced influence on the diversity of animal life within the community. Increased vertical structure means more resources and living space and a greater diversity of habitats.

Resource Availability 17.7

A variety of studies have shown an inverse relationship between nutrient availability and plant diversity in communities. By reducing growth rates, low nutrient availability functions to reduce competitive displacement. As nutrient availability in terrestrial communities increases, competition shifts from belowground (symmetrical competition) to aboveground (asymmetrical competition). The net result is an increase in competitive displacement and a reduction in plant species diversity as the faster-growing, taller plant species dominate the light resource.

Unlike the pattern observed in terrestrial plant communities, fertilization results in an increase in the species richness of autotrophs in both freshwater and marine communities.

Top Predator and Trophic Cascade Ecological Issues & Applications

The reintroduction of a top predator, the wolf, to the Yellowstone National Park has led to a trophic cascade in which the wolves have reduced the populations of elk, and the reduction in elk populations has resulted in dramatic changes in the vegetation of the community. Vegetation changes have influenced the food web structure and diversity of the community.

STUDY QUESTIONS

1. How does the species' fundamental niche function to constrain community structure?
2. The number of tree species that occur in a hectare of equatorial rain forest in eastern Africa can exceed 250. In contrast, the number of tree species occurring in a hectare of tropical woodland in southern Africa rarely exceeds three. In which forest community (rain forest or woodland) do you think diffuse competition would be the most prevalent? Why?
3. Define indirect interaction in the context of food webs.
4. Give an example of how predation can result in indirect positive interactions between species.
5. Contrast bottom-up and top-down control in the structure of a food web.
6. What is a trophic cascade?
7. In the ecologist Mary Power's work presented in Figure 17.14, the top predators appear to control plant productivity by controlling the abundance of herbivores (their prey). Now suppose we were to conduct a second experiment and reduce plant productivity by using some chemical that had no direct effect on the consumer organisms. If the results show that reduced plant productivity reduces herbivore populations, in turn leading to the decline of the top predator, what type of control would this imply? How might you reconcile the findings of these two experiments?
8. Using Chapter 6 (Sections 6.11 and 6.12) as your resource, what characteristics might enable a plant species (*species 1*) to tolerate low soil nutrient availability? How might these characteristics limit the maximum growth

rates under high soil nutrient conditions? Conversely, what characteristics might enable a plant species (*species 2*) to maintain high growth rates under high soil nutrient availability? How might these characteristics limit the plant species' ability to tolerate low soil nutrient conditions? Now predict the outcome of competition between *species 1* and *2* in two plant communities, one with low soil nutrients and the other with abundant nutrients. Discuss in terms of tolerance and competition.

9. How does the structure of vegetation within a community influence the diversity of animal life?
10. Contrast symmetric and asymmetric competition. How does the availability of soil nutrients shift the nature of competition from symmetric to asymmetric?

FURTHER READINGS

Classic Studies

Connell, J. 1961. "The influence of interspecific competition and other factors on the distribution of the barnacle *Chthamalus stellatus*." *Ecology* 42:710–723.
This article presents one of the early field experiments examining the role of interspecific competition in patterns of community zonation.

Paine, R. T. 1966. "Food web complexity and species diversity." *The American Naturalist* 100:65–75.
Classic field study that reveals the role of indirect interaction in community structure.

Current Research

Brown, J., T. Whithham, S. Ernest, and C. Gehring. 2001. "Complex species interactions and the dynamics of ecological systems: Long-term experiments." *Science* 93:643–650.
Review of long-term experiments that have revealed some of the complex interactions occurring within ecological communities.

Huston, M. 1994. *Biological diversity.* Cambridge, United Kingdom: Cambridge University Press.
An essential resource for those interested in community ecology. Huston provides an extensive review of geographic patterns of species diversity and presents a framework for understanding the distribution of species in both space and time.

McPeek, M. 1998. "The consequences of changing the top predator in a food web: A comparative experimental approach." *Ecological Monographs* 68:1–23.
This article presents a series of experiments directed at understanding the role of top predators in structuring communities. An excellent example of the application of experimental manipulations to understanding species interactions in ecological communities.

Pimm, S. L. 1991. *The balance of nature.* Chicago: University of Chicago Press.
An excellent example of the application of theoretical studies on food webs and the structure of ecological communities to current issues in conservation ecology.

Power, M. E. 1992. "Top-down and bottom-up forces in food webs: Do plants have primacy?" *Ecology* 73:733–746.
This is an excellent discussion of the role of predation in structuring communities, including a contrast between bottom-up and top-down controls on the structure of food webs.

Ripple, W. J., and R. L. Beschta. 2012. "Trophic cascades in Yellowstone: The first 15 years after wolf reintroduction." *Biological Conservation* 145:205–213.
Review article providing an excellent overview of research that has examined the impact of wolf reintroduction to the Yellowstone ecosystem.

Rosenzweig, M. 1995. *Species diversity in space and time.* Cambridge, UK: Cambridge University Press.
This book combines empirical studies and ecological theory to address various topics relating to patterns of species diversity. An excellent discussion of large-scale patterns of biological diversity over geological time.

Terborgh, J., and J. A. Estes (eds.). 2010. *Trophic cascades: Predators, prey, and the changing dynamics of nature.* Washington, D.C.: Island Press.
This edited volume provides a diversity of examples of trophic cascades in both terrestrial and aquatic ecosystems.

A diverse grassland community occupies a site once used for agriculture. The study of processes involved in the colonization of abandoned agricultural lands has given ecologists important insights into the dynamics of plant communities.

CHAPTER GUIDE

WE HAVE EXAMINED the variety of processes that interact to influence the structure of communities (Chapter 17). The changing nature of community structure across the landscape (zonation) reflects the shifting distribution of populations in response to changing environmental conditions, as modified by the interactions (direct and indirect) among the component species. Yet the structure (physical and biological) of the community also changes through time; it is dynamic, reflecting the population dynamics of the component species. The vertical structure of the community changes with time as plants establish themselves, mature, and die. The birthrates and death rates of species change in response to environmental conditions, resulting in a shifting pattern of species dominance and diversity through time. This changing pattern of community structure through time—community dynamics—is the topic of this chapter.

18.1 Community Structure Changes through Time

Community structure varies in time as well as in space. Suppose that rather than moving across the landscape, as with the examples of zonation presented earlier (see Section 16.8, Figures 16.15–16.19), we stand in one position and observe the area as time passes. For example, abandoned cropland and pastureland are common sights in agricultural regions in once forested areas of eastern North America (see Chapter 18, *Ecological Issues & Applications*). No longer tended, the land quickly grows up in grasses, goldenrod (*Solidago* spp.), and weedy herbaceous plants. In a few years, these same weedy fields are invaded by shrubby growth—blackberries (*Rubus* spp.), sumac (*Rhus* spp.), and hawthorn (*Crataegus* spp.). These shrubs eventually are replaced by pine trees (*Pinus* spp.), which with time form a closed canopy forest. As time passes, deciduous hardwood species develop in the understory. Many years later, this abandoned land supports a forest of maple (*Acer* spp.), oak (*Quercus* spp.), and other hardwood species (Figure 18.1). The process you would have observed, the gradual and seemingly directional change in community structure through time from field to forest, is called **succession**. Succession, in its most general definition, is the temporal change in community structure. Unlike zonation, which is the spatial change in community structure across the landscape, succession refers to changes in community structure at a given location on the landscape through time.

The sequence of communities from grass to shrub to forest historically has been called a **sere** (from the word *series*), and each of the changes is a seral stage. Although each **seral stage** is a point on a continuum of vegetation through time, it is often recognizable as a distinct community. Each stage has its characteristic structure and species composition. A seral stage may last only one or two years, or it may last several decades. Some stages may be missed completely or may appear only in abbreviated or altered form. For example, when forest trees immediately colonize an abandoned field (as in Figure 18.1), the shrub stage appears to be bypassed; however, structurally, the role of shrubs is replaced by the incoming young trees.

Figure 18.1 Generalized representation of succession on an abandoned agricultural field in eastern North America.

Like zonation, the process of succession is generally common to all environments, both terrestrial and aquatic. The ecologist Wayne Sousa of the University of California–Berkeley carried out a series of experiments designed to examine the process of succession in a rocky intertidal algal community in southern California. A major form of natural disturbance in these communities is the overturning of rocks by the action of waves. Algal populations then recolonize these cleared surfaces. To examine this process, Sousa placed concrete blocks in the water to provide new surfaces for colonization. Over time, the study results show a pattern of colonization and replacement, with other species displacing populations that initially colonized the concrete blocks (Figure 18.2). This is the process of succession. The initial, or **early successional species** (often referred to as **pioneer species**), are usually characterized by high growth rates, smaller size, high degree of dispersal, and high rates of per capita population growth. In contrast, **late successional species** generally have lower rates of dispersal and colonization, slower per capita growth rates, and are larger and longer-lived. As the terms *early* and *late succession* imply, the patterns of species replacement with time are not random. In fact, if Sousa's experiment were to be repeated tomorrow, we would expect the resulting patterns of colonization and local extinction (the successional sequence) to be similar to those presented in Figure 18.2.

A similar pattern of succession occurs in terrestrial plant communities. Figure 18.3 depicts the patterns of woody plant species replacement after forest clearing (clear-cutting) at the Hubbard Brook Experimental Forest in New Hampshire. Before forest clearing in the late 1960s, seedlings and saplings of beech (*Fagus grandifolia*) and sugar maple (*Acer*

(a)

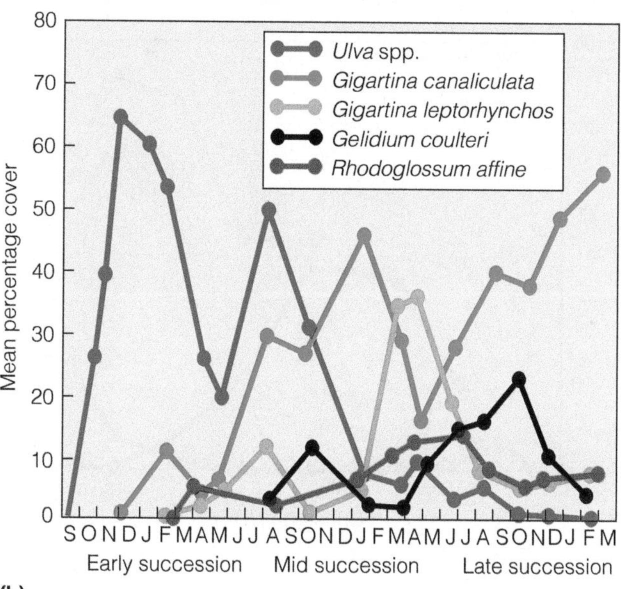

(b)

Figure 18.2 (a) Rocky intertidal zone along the California coast. (b) Mean percentage of five algal species that colonized concrete blocks introduced into the rocky intertidal zone in September 1974. Note the change in species dominance over time.
(Adapted from Sousa 1979.)

Interpreting Ecological Data

Q1. At what time does the species Gigartina canaliculata first appear in the experiments (month and year)? At what period during the experiment does this species of algae dominate the community?

Q2. Which algal species dominates the community during the first year of succession? Which species dominates during the last year of the experiment?

Q3. Which algal species never dominate the community (greatest relative abundance)?

Q4. During which period of the observed succession (early, mid, or late) is overall species diversity the highest?

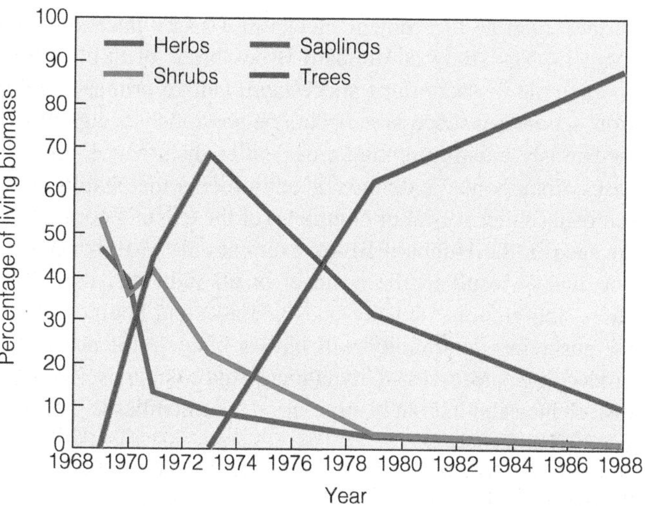

(a)

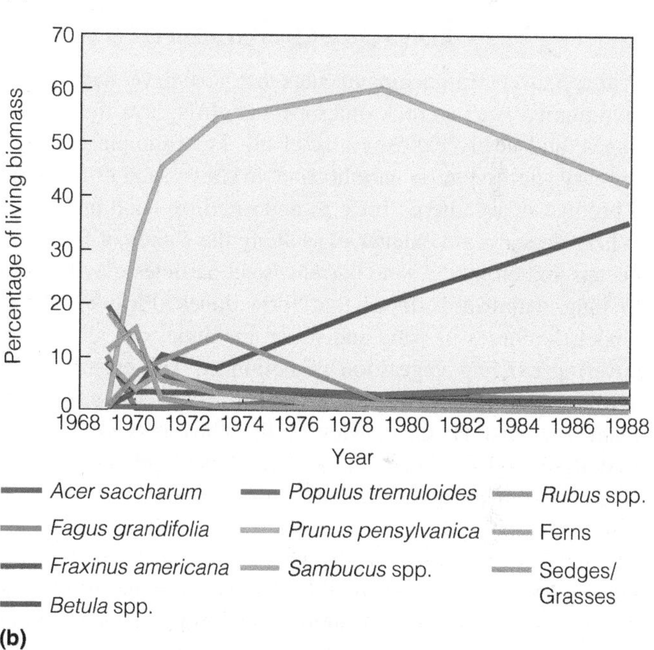

(b)

Figure 18.3 (a) Changes in the relative abundance (percentage of living biomass) of different plant growth forms in the Hubbard Brook Experimental Forest (Watershed 2) after a clear-cut.
(b) Changes in the woody species that make up the sapling and tree categories in (a) over the same period.
(Data from Reiners 1992.)

saccharum) dominated the understory. Large individuals of these two tree species dominated the canopy, and the seedlings represent successful reproduction of the parent trees. After the larger trees were removed by timber harvest, the numbers of beech and maple seedlings declined and were soon replaced by herbaceous species (ferns, sedges, and grasses), raspberry thickets, and seedlings of sun-adapted (shade-intolerant), fast-growing, early successional tree species such as pin cherry (*Prunus pennsylvanica*) and yellow birch (*Betula alleghaniensis*). Over the next 20 years these early successional species came to dominate the site, and now after half a century these species are currently being replaced by the late successional species of beech and sugar maple that previously dominated the site.

The two studies just presented point out the similar nature of successional dynamics in two different environments. They also present examples of two different types of succession: primary and secondary. **Primary succession** occurs on a site previously unoccupied by a community—a newly exposed surface such as the cement blocks in a rocky intertidal environment. The study at Hubbard Brook after forest clearing is an example of **secondary succession**. Unlike primary succession, secondary succession occurs on previously occupied sites (previously existing communities) after disturbance. In these cases, disturbance is defined as any process that results in the removal (either partial or complete) of the existing community. As seen in the Hubbard Brook example, the disturbance does not always result in the removal of all individuals. In these cases, the amount (density and biomass) and composition of the surviving community will have a major influence on the proceeding successional dynamics. Additional discussion of disturbance and its role in structuring communities is presented later (Chapter 19).

18.2 Primary Succession Occurs on Newly Exposed Substrates

Primary succession begins on sites that have never supported a community, such as rock outcrops and cliffs, lava fields, sand dunes, and newly exposed glacial till. For example, consider primary succession on an inhospitable site: a sand dune. Sand, a product of weathered rock, is deposited by wind and water. Where deposits are extensive, as along the shores of lakes and oceans and on inland sand barrens, sand particles may be piled in long, windward slopes that form dunes (**Figure 18.4a**). Under the forces of wind and water, the dunes can shift, often covering existing vegetation or buildings. The establishment and growth of plant cover acts to stabilize the dunes. The late plant ecologist H. C. Cowles of the University of Chicago first described colonization of sand dunes and the progressive development of vegetation in his pioneering classic study (published in 1899) of plant succession on the dunes of Lake Michigan. Later work by the ecologist John Lichter of the University of Minnesota would quantify the patterns first described by Cowles by examining a chronosequence of dunes (determined by radiocarbon dating) on the northern border of Lake Michigan (**Figure 18.4b**). A **chronosequence** (or chronosere) is a series of sites within an area that are at different stages of succession (seral stages). Because it is not always possible to monitor a site for the decades or centuries over which the process of succession occurs, it is often necessary to identify sites of different ages that represent the various stages of succession. In effect, the use of a chronosequence substitutes space for time.

In the process of primary succession on the newly formed dunes (Figure 18.4b), grasses, especially beach grass (*Ammophila breviligulata*), are the most successful pioneering plants and function to stabilize the dunes with their extensive system of rhizomes (see Section 8.1). Once the dunes are stabilized, mat-forming shrubs invade the area. Subsequently, the vegetation shifts to dominance by trees—first pines and then oak. Because of low moisture reserves in the sand, oak is rarely replaced by more moisture-demanding (mesophytic) trees. Only on the more favorable leeward slopes and in depressions, where microclimate is more moderate and where moisture can accumulate, does succession proceed to more mesophytic trees such as sugar maple, basswood, and red oak. Because these trees shade the soil and accumulate litter on the soil surface, they act to improve nutrients and moisture conditions. On such sites, a mesophytic forest may become established without going through the oak and pine stages. This example emphasizes one aspect of primary succession: the colonizing species ameliorate the environment, paving the way for invasion of other species.

(a)

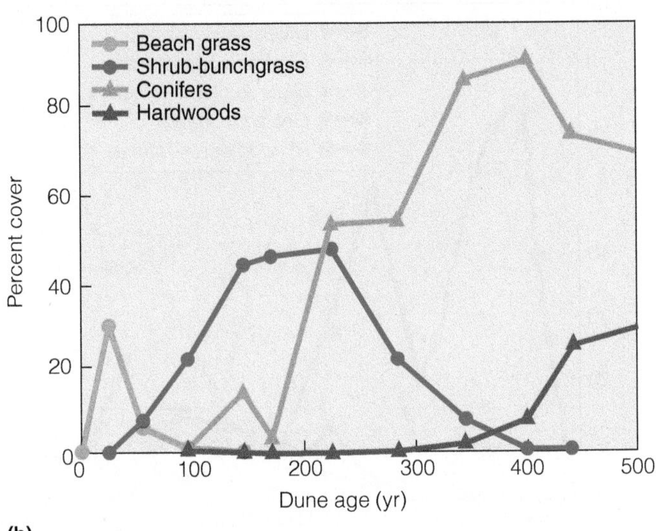

(b)

Figure 18.4 (a) Coastal sand dunes along Lake Michigan. (b) Changes in percentage of cover of the dominant plant groups during primary succession of the dune systems along Lake Michigan. Data represent changes observed along a chronosequence of sites ranging in age from 25 to 440 years. (Adapted from Lichter 1998.)

Newly deposited alluvial soil on a floodplain represents another example of primary succession. Over the past 200 years, the glacier that once covered the entire region of Glacier Bay National Park, Alaska, has been retreating (melting; **Figures 18.5**. As the glacier retreats, a variety of species such as alder (*Alnus* spp.) and cottonwood (*Populus* spp.) initially colonize the newly exposed landscape. Eventually, the later successional tree species of spruce (*Picea* spp.) and hemlock (*Tsuga* spp.) replace these early successional species, and the resulting forest (Figure 18.5c) resembles the forest communities in the surrounding landscape.

(a)

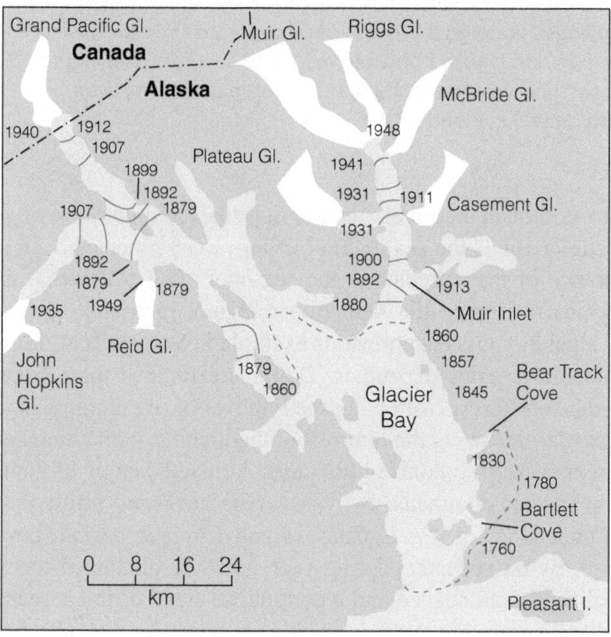

(b)

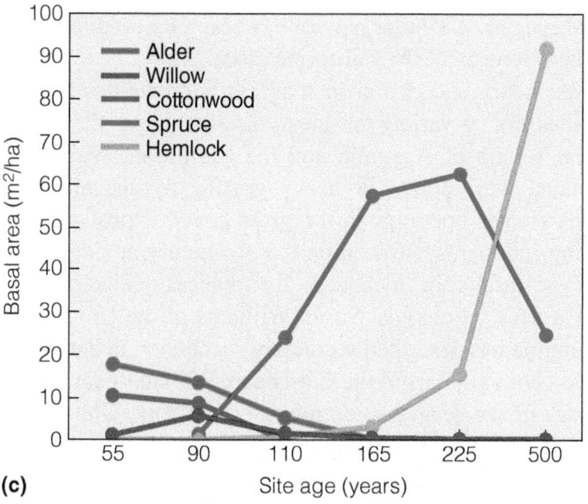

(c)

Figure 18.5 (a) Primary succession along riverine environments of Glacier Bay National Park, Alaska. (b) The Glacier Bay fjord complex in southeastern Alaska, showing the rate of ice recession since 1760. As the ice retreats, it leaves moraines along the edge of the bay in which primary succession occurs. (c) Changes in community composition (basal area of woody plant species) with age for sites (time since sediments first exposed) at Glacier Bay. (Adapted from Hobbie 1994.)

18.3 Secondary Succession Occurs after Disturbances

A classic example of secondary succession in terrestrial environments is the study of old-field succession in the Piedmont region of North Carolina by the eminent plant ecologist Dwight Billings (Duke University) in the late 1930s (see Figure 18.1). During the first year after a crop field has been abandoned, the ground is claimed by annual crabgrass (*Digitaria sanguinalis*); its seeds, lying dormant in the soil, respond to light and moisture and germinate. However, the crabgrass's claim to the ground is short-lived. In late summer, the seeds of horseweed (*Lactuca canadensis*), a winter annual, ripen. Carried by the wind, the seeds settle on the old field, germinate, and by early winter have produced rosettes. The following spring, horseweed, off to a head start over crabgrass, quickly claims the field. During the second summer, other plants invade the field: white aster (*Aster ericoides*) and ragweed (*Ambrosia artemisiifolia*).

By the third summer, broomsedge (*Andropogon virginicus*), a perennial bunchgrass, colonizes the field. Abundant organic matter and the ability to exploit soil moisture efficiently permits broomsedge to dominate the field. About this time, pine seedlings, finding room to grow in open places among the clumps of broomsedge, invade the field. Within 5 to 10 years, the pines are tall enough to shade the broomsedge. Eventually, hardwood species such as oaks and ash grow up through the pines, and as the pines die, they take over the field (**Figure 18.6**). Development of the hardwood forest continues as shade-tolerant trees and shrubs—dogwood, redbud, sourwood, hydrangea, and others—fill the understory.

Figure 18.6 Decline in the abundance of shortleaf pine (*Pinus echinata*) and increase in the density of hardwood species oak (*Quercus*) and hickory (*Carya*) during secondary succession on abandoned farmland in the Piedmont region of North Carolina. (Data from Billings, William D. "The structure and development of old field shortleaf pine stands and certain associated physical properties of soil," Ecological Monographs, July 1938.)

Similarly, studies of physical disturbance in marine environments have demonstrated secondary succession in seaweed, salt marsh, mangrove, seagrass, and coral reef communities. Ecologist David Duggins of the University of Washington examined the process of secondary succession after disturbance in the subtidal kelp forests of Torch Bay, Alaska (**Figure 18.7**).

Figure 18.7 (a) Idealized representation of kelp succession following the removal of the dominant herbivore (sea urchin) in Torch Bay, Alaska. During the first year, annual species dominate the site. *Nereocystis luetkeana* forms a canopy, and an understory of *Alaria fistulosa* and *Costaria costata* developed. By the second and third year, however, all annual species had declined in abundance and a continuous stand of the perennial species *Laminaria groenlandica* developed. (b) A kelp forest off the Aleutian Islands of Alaska dominated by *Laminaria groenlandica* (foreground).
(Based on Duggins 1980.)

(a)

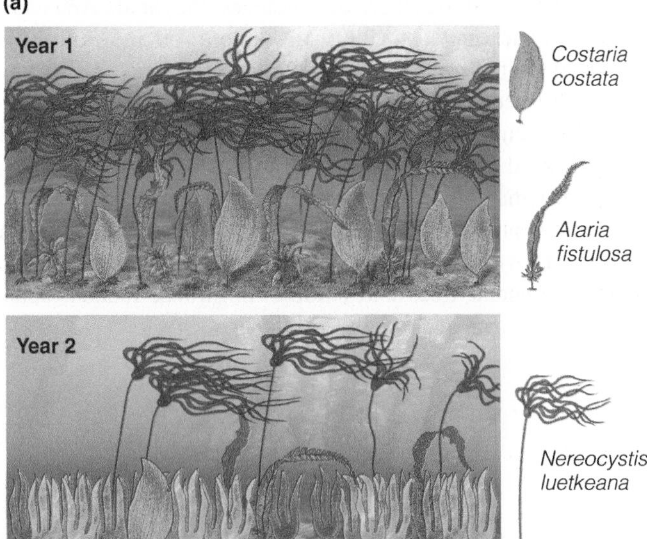

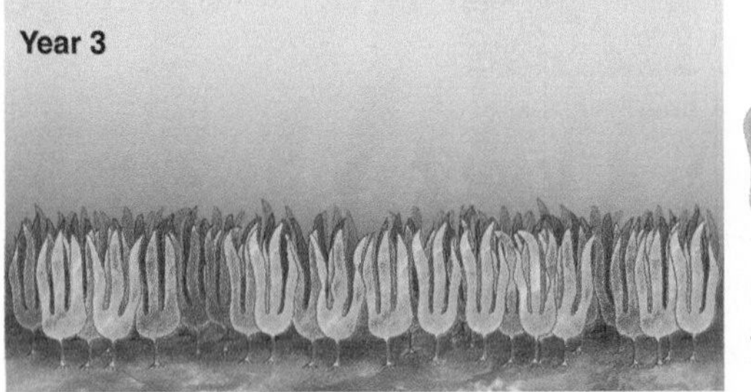

(b)

Figure 18.8 Disturbed areas (light-colored areas) within seagrass communities in Florida Bay. These areas, called *blowouts*, undergo a process of recovery that involves a shift in species dominance from macroalgae to seagrass (see Figure 18.9). Insert shows an area dominated by turtle grass (*Thalassia testudinum*) in Florida Bay.

The dominant herbivore in the subtidal communities of the north Pacific is the sea urchin (*Strongylocentrotus* spp.). In the absence of their predators, the sea otter (*Enhydra lutris*), sea urchins overgraze the kelp (macroalgae), removing virtually all algal biomass (see previous example in Section 16.4, Figure 16.6). In a series of studies, Duggins examined the recovery of the kelp forests following the removal of sea urchins. In the first year following the removal of the urchins, both annual and perennial kelps colonized the plots. A mixed canopy of annual kelp species dominated by *Nereocystis luetkeana* formed, and an understory of *Alaria fistulosa* and *Costaria costata* developed. By the second and third year, however, all annual species declined in abundance and a continuous stand of the perennial species *Laminaria groenlandica* developed. As a result of the dense canopy formed by *Laminaria*, shading and abrasion of the substrate suppressed the further recruitment and growth of annual species. A similar pattern has been observed in the subtidal kelp forests off the California coast.

Secondary succession in seagrass communities has been described for a variety of locations, including the shallow tropical waters of Australia and the Caribbean. Wave action associated with storms or heavy grazing by sea turtles and urchins creates openings in the grass cover, exposing the underlying sediments. Erosion on the down-current side of these openings results in localized disturbances called *blowouts* (**Figure 18.8**). Ecologist Susan Williams of the University of Washington has described secondary succession in detail in the seagrass communities of the Caribbean. Williams examined the recovery of the seagrass community (St. Croix, United States Virgin Islands) on a number of experimental plots following the removal of vegetation.

Rhizophytic macroalgae, comprised mostly of species of *Halimeda* and *Penicillus* (**Figure 18.9**), initially colonized the disturbed sites. These algae have some sediment-binding capability, but their ability to stabilize the sediments is minimal, and their major function in the early successional stage seems to be the contribution of sedimentary particles as they die and decompose. After the first year, algal densities begin to decline. There was no evidence that rhizophytic algae inhibited recolonization of the seagrasses, which invaded the plots during the first few months following disturbance. The density of the early successional species of seagrass, manatee grass (*Syringodium filiforme*), increased linearly during the first 15 months, eventually declining as the slower-growing, later-successional species, turtle grass (*Thalassia testudinum*) colonized the plots. The leaves and extensive rhizome and root systems of the sea grasses effectively trap and retain particles, increasing the organic matter of the sediment, and the once-disturbed area again resembles the surrounding seagrass community.

18.4 The Study of Succession Has a Rich History

The study of succession has been a focus of ecological research for more than a century. Early in the 20th century, botanists E. Warming in Denmark and Henry Cowles in the United States largely developed the concept of ecological succession. The intervening years have seen a variety of hypotheses attempting to address the processes that drive succession, that is, the seemingly consistent directional change in species composition through time.

Frederic Clements (1916, 1936) developed a theory of plant succession and community dynamics known as the *monoclimax hypothesis*. The community is viewed as a highly integrated superorganism and the process of succession represents the gradual and progressive development of the community to the ultimate, or climax stage (see Section 16.10 for further discussion). The process was seen as analogous to the development of an individual organism.

In 1954, Frank Egler proposed a hypothesis he termed *initial floristic composition*. In Egler's view, the process of succession at any site is dependent on which species get there first. Species replacement is not an orderly process because some species suppress or exclude others from colonizing the site. No species is competitively superior to another. The colonizing species that arrive first inhibit any further establishment of newcomers. Once the original colonizers eventually die, the site then becomes accessible to other species. Succession is therefore individualistic and dependent on the particular species that colonize the site and the order in which they arrive.

In 1977, ecologists Joseph Connell of University of California–Santa Barbara and Ralph Slatyer of Australian National University proposed a generalized framework for viewing succession that considers a range of species interactions and responses through succession. They offered three models.

The *facilitation model* states that early successional species modify the environment so that it becomes more suitable for later successional species to invade and grow to maturity. In effect, early-stage species prepare the way for late-stage species, facilitating their success (see Chapter 15 for discussion of facilitation).

The *inhibition model* involves strong competitive interactions. No one species is completely superior to another. The first species to arrive holds the site against all invaders. It makes the site less suitable for both early and late successional

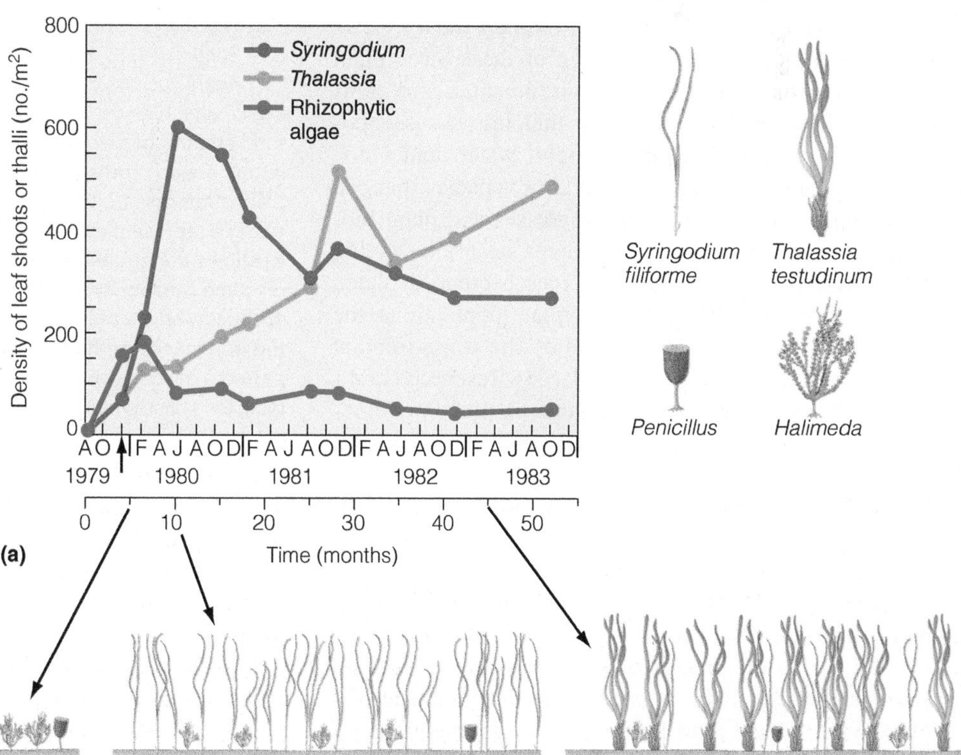

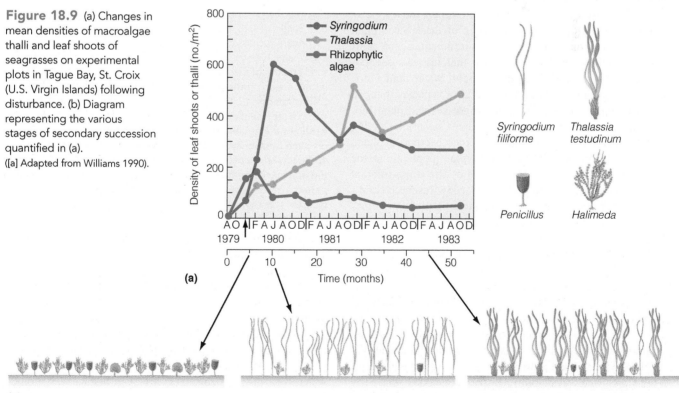

Figure 18.9 (a) Changes in mean densities of macroalgae thalli and leaf shoots of seagrasses on experimental plots in Tague Bay, St. Croix (U.S. Virgin Islands) following disturbance. (b) Diagram representing the various stages of secondary succession quantified in (a).
([a] Adapted from Williams 1990).

species. As long as it lives and reproduces, the first species maintains its position. The species relinquishes it only when it is damaged or dies, releasing space to another species. Gradually, however, species composition shifts as short-lived species give way to long-lived ones.

The *tolerance model* holds that later successional species are neither inhibited nor aided by species of previous stages. Later-stage species can invade a newly exposed site, establish themselves, and grow to maturity independently of the species that precede or follow them. They can do so because they tolerate a lower level of some resources. Such interactions lead to communities composed of those species most efficient in exploiting available resources. An example might be a highly shade-tolerant species that could invade, persist, and grow beneath the canopy because it is able to exist at a lower level of one resource: light. Ultimately, through time, one species would prevail.

Since the work of Connell and Slatyer, the search for a general model of plant succession has continued among ecologists. The life history classification of plants put forward by ecologist J. Phillip Grime of the University of Sheffield is based on three primary plant strategies (see Section 10.13, Figures 10.25 and 10.26). Species exhibiting the *R*, or ruderal, strategy rapidly colonize disturbed sites but are small in stature and short-lived. Allocation of resources is primarily to reproduction, with characteristics allowing for a wide dispersal of propagules to newly disturbed sites. Predictable habitats with abundant resources favor species that allocate resources to growth, favoring resource acquisition and competitive ability (*C* strategy). Habitats in which resources are limited favor stress-tolerant species (*S* strategy) that allocate resources to maintenance. Grime's theory views succession as a shift in the dominance of these three plant strategies in response to changing environmental conditions (habitats). Following the disturbance that initiates secondary succession, essential resources (light, water, and nutrients) are abundant, selecting for ruderal (*R*) species that can quickly colonize the site. As time progresses and plant biomass increases, competition for resources occurs, selecting for competitive (*C*) species. As resources become depleted as a result of high demand by growing plant populations, the (*C*) species will eventually be replaced by the stress-tolerant (*S*) species that are able to persist under low resource conditions. The pattern of changing dominance in plant strategies in response to changing environmental conditions is shown in **Figure 18.10**.

Plant ecologist Fakhri Bazzaz of Harvard University approached developing an understanding of plant succession by examining the nature of successional environments and the eco-physiological characteristics of different functional groups of plants involved in the process of colonization and replacement during the process of succession. He focused on characteristics of seed dispersal, storage, germination, and species photosynthetic and growth response to resource gradients of light and water availability. Bazzaz concluded that early and late successional plants have contrasting physiological characteristics that enable them to flourish in the contrasting environmental conditions presented by early and late successional habitats (**Table 18.1**).

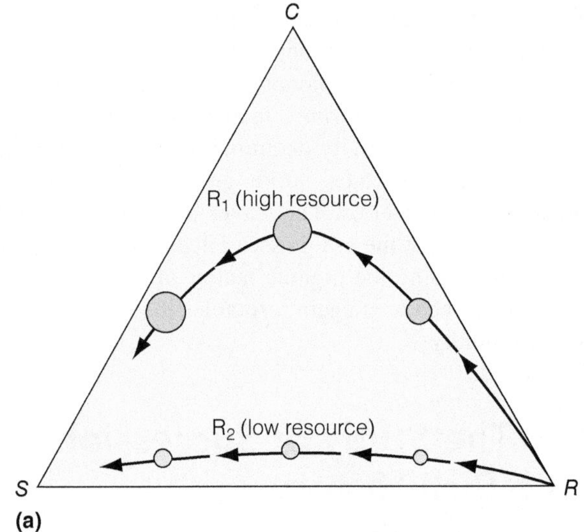

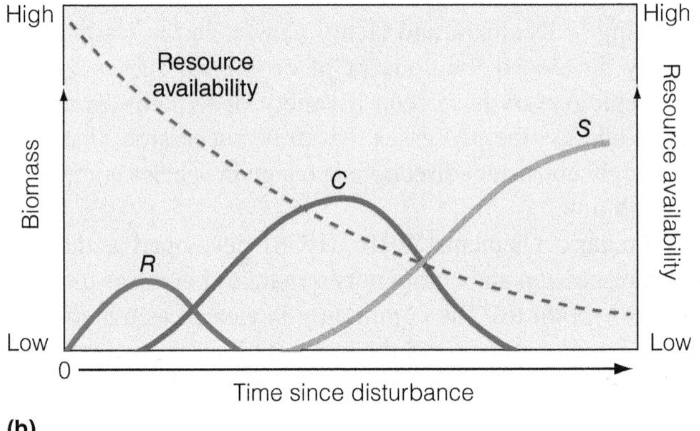

Figure 18.10 (a) Grime's triangular model of plant strategies (see Section 10.13, Figure 10.26) showing the pathway of secondary succession under conditions of low and high resource availability (high vs. low nutrient sites). The circles indicate the total amount of plant biomass at each stage of succession. The letters at each corner of the triangle represent the three primary plant strategies (*R*, *C*, and *S*). Points at any location within the triangle represent intermediate strategies. The trajectory for R_1 shows the greater importance of competitive species when resource availability is high. The role of *C* (competitive) species under low resources (curve R_2) would be low or absent. (b) Shift in the dominance of the three primary plant strategies and changing patterns of resource availability as succession proceeds. (Adapted from Grime 1979.)

Michael Huston of Texas State University and Thomas Smith of the University of Virginia proposed a model of community dynamics based on plant adaptations to environmental gradients. Their model is based on the cost-benefit concept that plant adaptations for the simultaneous use of two or more resources are limited by physiological and life history constraints. Their model focuses on the resources of light and water. The plants themselves largely influence variations in light levels within the community, whereas the availability of water is largely a function of climate and soils. Succession is

Table 18.1 Physiological characteristics of early and late successional plants

Attribute	Early Successional	Late Successional
Seeds		
dispersal in time	long	short
secondary (induced) dormancy	common	uncommon
Seeds		
germination enhanced by light	yes	no
fluctuating temperatures	yes	no
high nitrogen concentrations	yes	no
Light saturation intensity	high	low
Light compensation point	high	low
Efficiency at low light	low	high
Photosynthetic rates	high	low
Respiration rates	high	low
Transpiration rates	high	low
Stomatal conductance	high	low
Acclimation potential	high	low
Recovery from resource limitation	fast	slow
Resource acquisition rates	fast	slow

(Adapted from Bazzaz 1979.)

interpreted as a temporal shift in species dominance, primarily in response to changes in light availability. There is an inverse relationship between the ability to survive and grow under low light conditions and the ability to photosynthesize and grow at high rates when the availability of light is high (see Section 6.8, Figure 6.8). The pattern of species dominance shifts from fast-growing, shade-intolerant species in early succession to slower-growing, shade-tolerant species later in succession.

The ecologist David Tilman of the University of Minnesota proposed a model of succession based on a trade-off in characteristics that enable plants to compete for the essential resources of nitrogen and light, referred to as the *resource-ratio hypothesis*. The ability to effectively compete for light is associated with the allocation of carbon to the production of aboveground tissues (leaves and stems). Conversely, the ability to effectively compete for nitrogen is associated with the production of root tissue. This pattern of changing allocation of carbon under varying nutrient and light resources is discussed in Chapter 6. Succession comes about as the relative availability of nitrogen and light change through time. In Tilman's model, the availability of these two essential resources is inversely related. Environmental conditions range from habitats with soils poor in nitrogen but with a high availability of light at the soil surface to habitats with nitrogen-rich soils and low availability of light. Community composition changes along this gradient as the ratio of nitrogen to light changes (**Figure 18.11**).

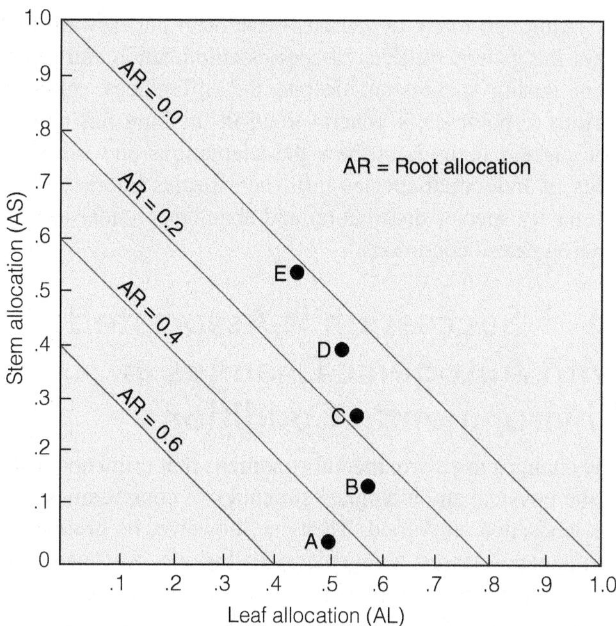

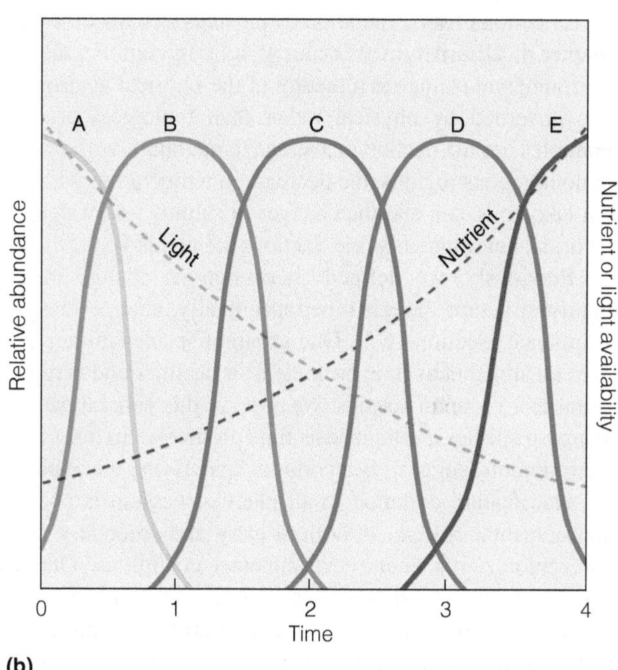

Figure 18.11 (a) Allocation triangle showing possible patterns of carbon allocation to leaves (AL), stems (AS), and roots (AR) by plants. Note the AL + AS + AR = 1. Each diagonal line is an equal allocation to roots. The allocation patterns of five species (A–E) are shown. These allocation patterns reflect a shift in plant strategies from adaptation to low nutrient and high light (species A) levels, to adaptation to high nutrient and low light (species E) levels.
(Adapted from Tilman 1994.)

(b) Tilman's resource ratio hypothesis of succession. During the initial stages of succession, light availability is high and nitrogen is low. As time progresses, light availability declines and nitrogen availability increases. The shift in species dominance over this period reflects the changing relative competitiveness of the five hypothetical plant species defined.
(Adapted from Tilman 1988.)

Although many hypotheses have been put forward to explain the general patterns of species colonization and replacement during succession, despite the differences among the various hypotheses, a general trend in thinking has emerged. The current focus is on how the adaptations and life history traits of individual species influence species interactions and ultimately species distribution and abundance under changing environmental conditions.

18.5 Succession Is Associated with Autogenic Changes in Environmental Conditions

The changes in environmental conditions that bring about shifts in the physical and biological structures of communities across the landscape are varied. They can, however, be grouped into two general classes: autogenic and allogenic. **Autogenic** environmental change is a direct result of the presence and activities of organisms within the community. For example, the vertical profile of light in a forest is a direct result of the interception and reflection of solar radiation by the trees (see Section 4.2 and Chapter 4, Quantifying Ecology 4.1). In contrast, **allogenic** environmental change is a feature of the physical environment; it is governed by physical rather than biological processes. Examples are the decline in average temperature with elevation in mountainous regions, the decrease in temperature with depth in a lake or ocean, and the changes in salinity and water depth in coastal environments (see Sections 2.8 and 3.4).

Previously, we defined succession as change in community structure though time, specifically, change in species dominance (Section 16.3). One group of species initially colonizes an area, but as time progresses, it declines and is replaced by another group of species. We observe this general pattern of changing species dominance as time progresses in most natural environments, suggesting a common underlying mechanism.

One feature common to all plant succession is autogenic environmental change. In both primary and secondary succession, colonization alters environmental conditions. One clear example is the alteration of the light environment. Leaves reflecting and intercepting solar radiation create a vertical profile of light within a plant community. In moving from the canopy to ground level, less light is available to drive the processes of photosynthesis (see Chapter 4, Quantifying Ecology 4.1). During the initial period of colonization, few if any plants are present. In the case of primary succession, the newly exposed site has never been occupied. In the case of secondary succession, plants have been killed or removed by some disturbance. Under these circumstances, the availability of light at the ground level is high, and seedlings are able to establish themselves. As plants grow, their leaves intercept sunlight, reducing the availability of light to shorter stature plants (Figure 18.12). This reduction in available light decreases rates of photosynthesis, slowing the growth of the shaded individuals. Assuming that not all plant species can photosynthesize and grow at the same rate,

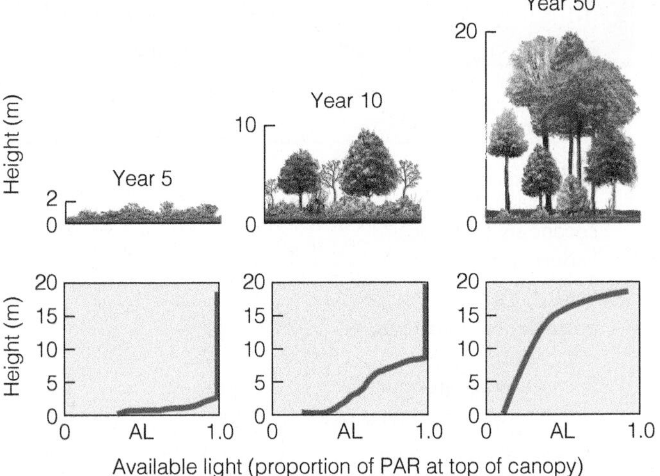

Figure 18.12 Changing vertical light profile through succession. Following disturbance, the site is dominated by low-stature herbaceous vegetation. As time progresses, the stature of the vegetation increases, and as the height of the canopy increases, the vertical profile of light changes, reflecting the increased leaf area and range of height over which the leaf area is distributed (see Chapter 4, Quantifying Ecology 4.1).

plant species that can grow tall the fastest have greater access to the light resource. They subsequently reduce the availability of light to the slower-growing species, enabling the fast-growing species to outcompete the other species and dominate the site. However, in changing the availability of light below the canopy, the dominant species create an environment that is more suitable for the species that will later displace them as dominants.

Recall that not all plant species respond in the same way to variation in available light. There is a fundamental physiological trade-off between the adaptations that enable high rates of growth under high light conditions and the ability to continue growth and survival under shaded conditions (Section 6.8, Figure 6.8). In the early stages of plant succession, shade-intolerant species can dominate because of their high growth rates. Shade-intolerant species grow above and shade the slower-growing, shade-tolerant species. As time progresses and light levels decline below the canopy, however, seedlings of the shade-intolerant species cannot grow and survive in the shaded conditions (see Section 6.8, Figure 6.10). At that time, although shade-intolerant species dominate the canopy, no new individuals are recruited into their populations. In contrast, shade-tolerant species are able to germinate and grow under the canopy. As the shade-intolerant plants that form the canopy die, shade-tolerant species in the understory replace them.

Figure 18.13 shows this pattern of changing population recruitment, mortality, and species composition through time in the forest community in the Piedmont region of North Carolina presented in Figure 18.6. Fast-growing, shade-intolerant pine species dominate in early succession. Over time, the number of new pine seedlings declines as the light decreases at the

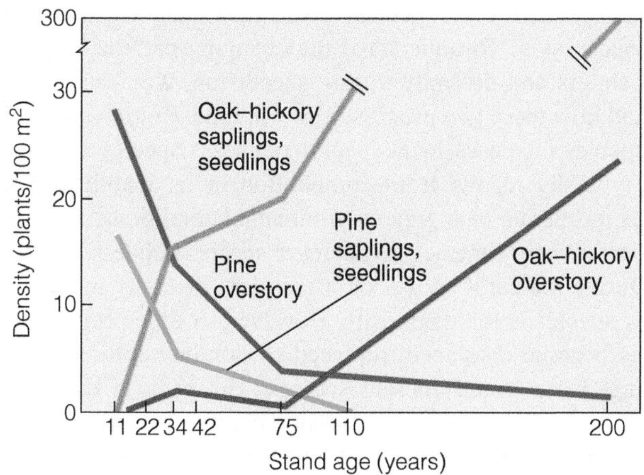

Figure 18.13 Dominance shift of overstory and understory (seedlings and saplings) of pines, oaks, and hickories during secondary succession in the Piedmont region of North Carolina (see Figure 18.6). Early successional pine species initially dominate the site. Pine seedling regeneration declines as the light decreases in the understory. Shade-tolerant oak and hickory seedlings establish themselves under the reduced light conditions. As pine trees in the overstory die, oak and hickory replace them as the dominant species in the canopy.
(Data from Billings 1938.)

forest floor. Shade-tolerant oak and hickory species, however, are able to establish seedlings in the shaded conditions of the understory. As the pine trees in the canopy die, the community shifts from a forest dominated by pine species to one dominated by oaks and hickories.

In this example, succession results from changes in the relative environmental tolerances and competitive abilities of the species under autogenically changing environmental conditions (light availability). Shade-intolerant species are able to dominate the early stages of succession because of their ability to grow quickly in the high light environment. However, as autogenic changes in the light environment occur, the ability to tolerate and grow under shaded conditions enables shade-tolerant species to rise to dominance.

Light is not the only environmental factor that changes during the course of succession, however. Other autogenic changes in environmental conditions can influence patterns of succession. The seeds of some plant species cannot germinate on the surface of mineral soil; these seeds require the buildup of organic matter on the soil surface before they can germinate and become established. In the examples of secondary succession on sand dunes (Figure 18.4) and in seagrass communities (Figure 18.9), early colonizing species function to stabilize the sediments and add organic matter, allowing for later colonization by other plant species.

Consider the example of primary succession on newly deposited glacial sediments (see Figure 18.5). Because of the absence of a well-developed soil, little nitrogen is present in these newly exposed surfaces, thus restricting the establishment, growth, and survival of most plant species. However, those terrestrial plant species that have the mutualistic association with nitrogen-fixing *Rhizobium* bacteria are able to grow and dominate the site (see Section 15.11, Figure 15.11). These plants provide a source of carbon (food) to the bacteria that inhabit their root systems. In return, the plants have access to the atmospheric nitrogen fixed by the bacteria. Alder, which colonizes the newly exposed glacial sediments in Glacier Bay, is one such plant species (see Figure 18.5c).

As individual alder shrubs shed their leaves or die, the nitrogen they contain is released to the soil through the processes of decomposition and mineralization (**Figure 18.14**; see also Chapter 21). Now other plant species can colonize the site. As nitrogen becomes increasingly available in the soil, species that do not have the added cost of mutualistic association and that exhibit faster rates of growth and recruitment come to dominate the site. As in the Piedmont forest example, succession is a result of autogenic change in the environment and the relative competitive abilities of the species colonizing the site.

The exact nature of succession varies from one community to another, and it involves a variety of species interactions and responses that include facilitation, competition, inhibition, and differences in environmental tolerances. However, in all cases, the role of temporal, autogenic changes in environmental conditions and the differential response of species to those changes are key features of community dynamics.

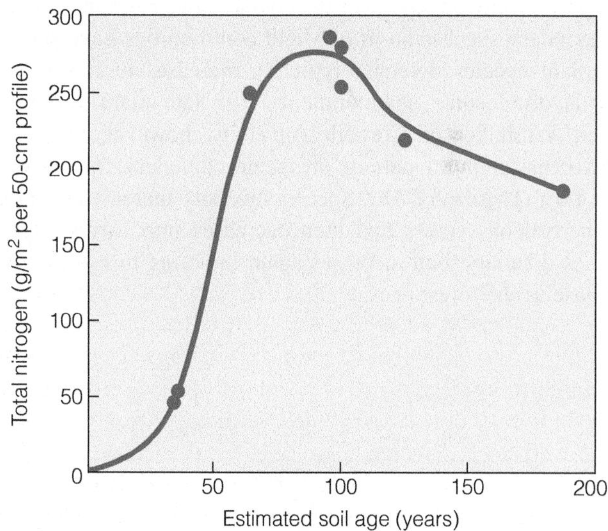

Figure 18.14 Changes in soil nitrogen during primary succession in Glacier Bay National Park since the retreating glacier exposed the surface for colonization by plants. Initially, the virtual absence of nitrogen limits site colonization to alder, which has a mutualistic association with *Rhizobium* bacteria, allowing it access to atmospheric nitrogen. As plant litter decomposes, nitrogen is released to the soil through mineralization. With the buildup of soil organic matter and nitrogen levels, other plant species are able to colonize the site and displace alder (see Figure 18.5).
(Adapted from Crocker and Major 1955.)

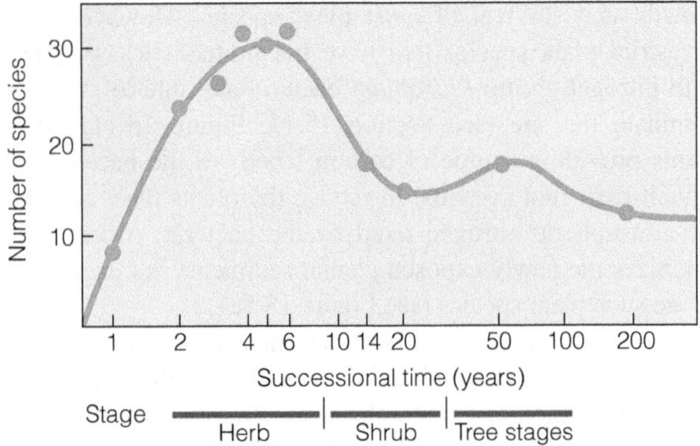

Figure 18.15 Changes in plant diversity during secondary succession of an oak–pine forest in Brookhaven, New York. Diversity is reported as species richness in 0.3-ha samples. Species richness increases into the late herbaceous stages, declines into the shrub stage, increases once again into the early forest stages, and declines thereafter. The peaks in diversity correspond to periods of transition between these stages, where species from both stages are present at the site.
(Data from Whittaker 1975, Whittaker and Woodman 1968.)

18.6 Species Diversity Changes during Succession

In addition to shifts in species dominance, patterns of plant species diversity change over the course of succession. Studies of secondary succession in old-field communities have shown that plant species diversity typically increases with site age (that is, time since abandonment). The late plant ecologist Robert Whittaker of Cornell University, however, observed a different temporal pattern of species diversity for sites in New York (**Figure 18.15**). Species diversity increases into the late herbaceous stages and then decreases into shrub stages. Species diversity then increases again in young forest, only to decrease as the forest ages.

The processes of species colonization and replacement drive succession. To understand the changing patterns of species richness and diversity during succession, we need to understand how these two processes vary in time. Colonization by new species increases local species richness. Species replacement typically results from competition or an inability of a species to tolerate changing environmental conditions. Species replacement over time acts to decrease species richness.

During the early phases of succession, diversity increases as new species colonize the site. However, as time progresses, species become displaced, replaced as dominants by slower-growing, more shade-tolerant species. The peak in diversity during the middle stages of succession corresponds to the transition period, after the arrival of later successional species but before the decline (replacement) of early successional species. The two peaks in diversity seen in Figure 18.15 correspond to the transition between the herbaceous- and shrub-dominated phases, when both groups of plants are present, and the transition between early and later stages of woody plant succession. Species diversity declines as shade-intolerant tree species displace the earlier successional trees and shrubs.

The rate of displacement is influenced by the growth rates of species involved in the succession. If growth rates are slow, the displacement process moves slowly; if growth rates are fast, displacement occurs more quickly. This observation led Michael Huston, an ecologist at Texas State University, to conclude that patterns of diversity through succession vary with environmental conditions (particularly resource availability) that directly influence the rates of plant growth. By slowing the population growth rate of competitors that eventually displace earlier successional species, the period of coexistence is extended, and species diversity can remain high (**Figure 18.16**). This hypothesis predicts the highest diversity at low to intermediate levels of resource availability by extending the period of coexistence.

Disturbance can have an effect similar to that of reduced growth rates by extending the period during which species coexist. In the simplest sense, disturbance acts to reset the clock in succession (**Figure 18.17**). By reducing or eliminating

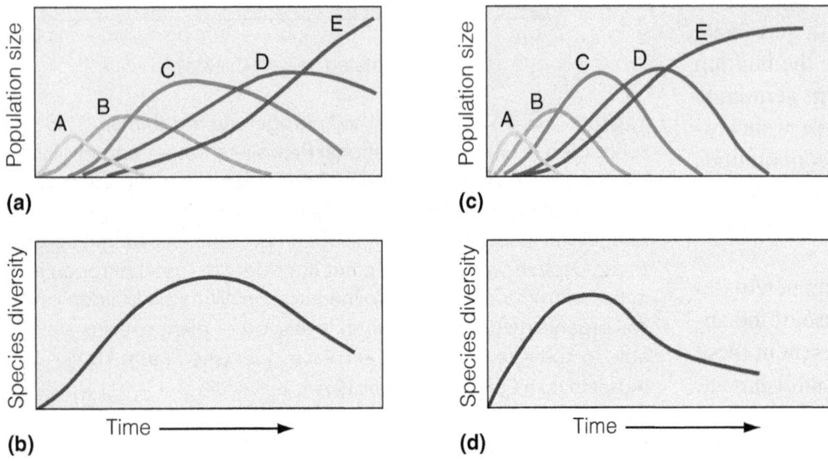

Figure 18.16 (a) Hypothetical succession involving five plant species (A–E), and (b) the associated temporal pattern of species diversity. Note that species diversity increases initially as new species colonize the site. However, diversity declines as autogenically changing environmental conditions and competition result in the displacement of early successional species. (c) When the growth rates of the five species are doubled, the succession progresses more quickly, and (d) the pattern of species diversity reflects the earlier onset of competition and more rapid displacement of early successional species. As a result, the period over which species diversity is at its maximum is reduced.
(Data from Huston, Michael A., Biological Diversity: the Coexistence of Species on Changing Landscapes [Cambridge University Press, 1994].)

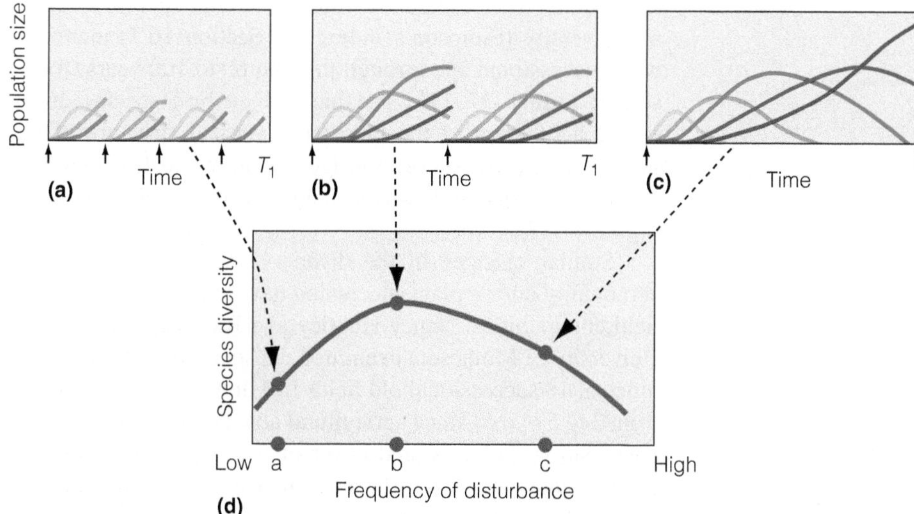

Figure 18.17 Patterns of succession for five hypothetical plant species under three levels of disturbance frequency: frequent, intermediate, and none. Time of the disturbance is shown as an arrow on the x-axis. Note the differences in species composition and diversity at time T_1 on the x-axis. (a) Under high frequency of disturbance, the absence of later successional species reduces overall diversity. (b) At intermediate frequency of disturbance, all species coexist, and diversity is at a maximum. (c) When disturbance is absent, later successional species eventually displace the earlier ones, and again diversity is low. (d) The general form of the relationship between species diversity and frequency of disturbance is a curve with maximum diversity at intermediate frequency and magnitude of disturbance. (Data from Huston, Michael A., Biological Diversity: the Coexistence of Species on Changing Landscapes [Cambridge University Press, 1994].)

plant populations, the site is once again colonized by early successional species, and the process of colonization and species replacement begins again. If the frequency of disturbance (defined by the time interval between disturbances) is high, then later successional species will never have the opportunity to colonize the site. Under this scenario, diversity remains low. In the absence of disturbance, later successional species displace earlier ones and species diversity declines. At an intermediate frequency of disturbance, colonization can occur, but competitive displacement is held to a minimum. The pattern of high diversity at intermediate frequencies of disturbance was proposed independently by Michael Huston and by Joseph Connell of the University of California–Santa Barbara and is referred to as the **intermediate disturbance hypothesis**.

18.7 Succession Involves Heterotrophic Species

Although our discussion and examples of succession have thus far focused on temporal changes in the autotrophic component of the community (plant succession), associated changes in the heterotrophic component also occur. As plant succession advances, changes in the structure and composition of the vegetation result in changes in the animal life that depends on the vegetation as habitat. Each successional stage has its own distinctive fauna. Because animal life is often influenced more by structural characteristics than by species composition, successional stages of animal life may not correspond to the stages identified by plant ecologists.

During plant succession, animals can quickly lose their habitat as species composition and structure of the vegetation changes. For example, we can return to the patterns of secondary succession that occurs in abandoned agricultural lands in eastern North America presented in Figure 18.1 (**Figure 18.18**). In the early stages of succession following abandonment, grasslands and old fields support meadowlarks, meadow mice, and grasshoppers. When woody plants—both

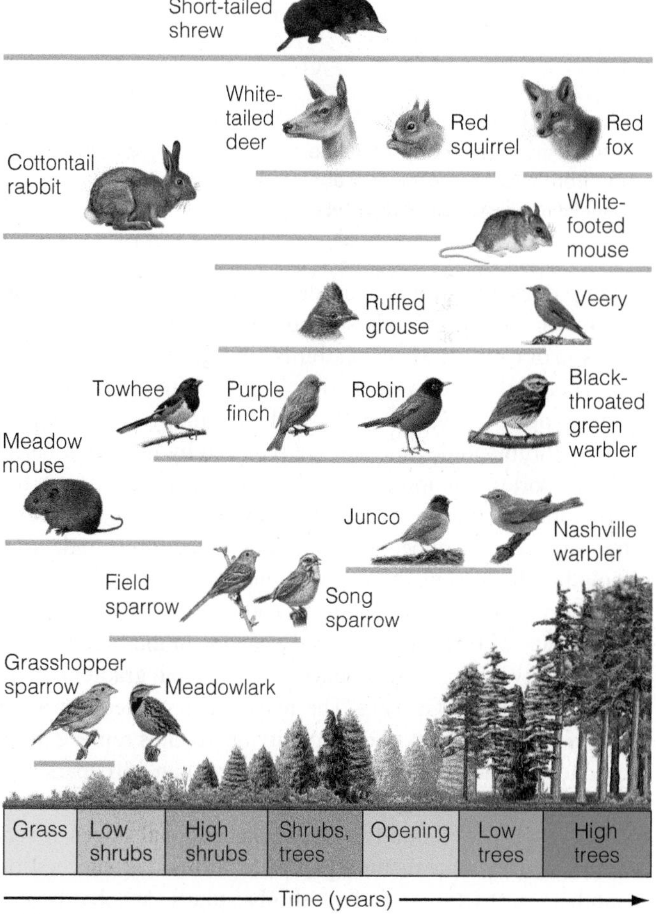

Figure 18.18 Changes in the composition of animal species inhabiting various stages of plant succession, from old-field to conifer forest, in central New York. Species appear or disappear as vegetation density and height change. Brown lines represent the range of vegetation (stages) inhabited by the associated species.

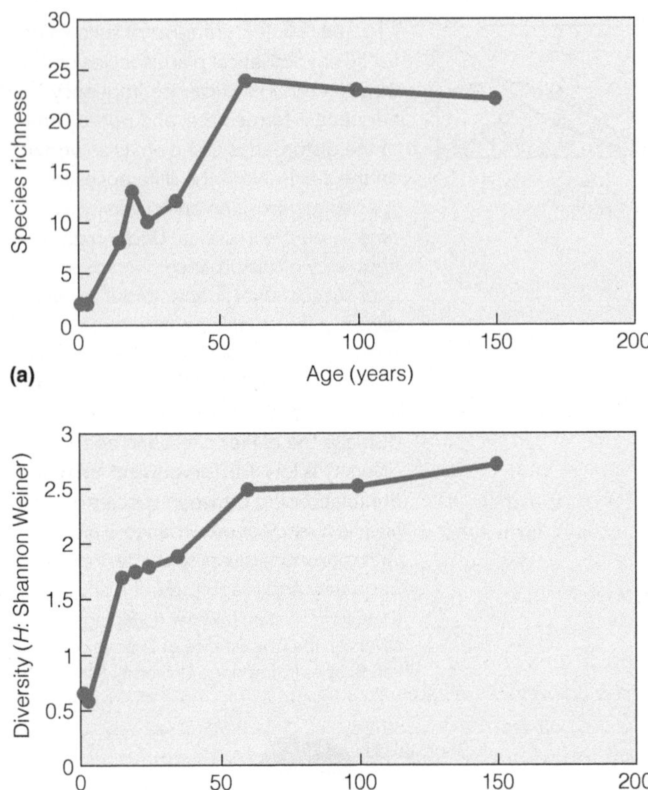

Figure 18.19 Species (a) richness and (b) diversity of the bird community on a chronosequence of sites in Georgia ranging in age from 1 to 150 years after abandonment from agriculture.
(Data from Johnston and Odum 1956.)

young trees and shrubs—invade, a new structural element appears. Grassland animals soon disappear, and shrubland animals take over. Towhees, catbirds, and goldfinches claim the thickets, and meadow mice give way to white-footed mice. As woody plant growth proceeds and the canopy closes, species of the shrubland decline and are replaced by birds and insects of the forest canopy. As succession proceeds, the vertical structure becomes more complex. New species appear, such as tree squirrels, woodpeckers, and birds of the forest understory, including hooded warblers and ovenbirds.

Ecologists Davis Johnson and Eugene Odum of the University of Georgia examined changes in the breeding bird community along a secondary successional gradient in the coastal region of Georgia. The researchers carried out a census of the breeding bird community on 10 sites representing a chronosequence ranging from 1 to more than 150 years since abandonment following agriculture. They found that the sites could be classified into four broad successional stages or seres dominated by four distinct plant life forms that succeed one another: herbs (grass forbs), shrubs, pines, and hardwoods. The occurrence of most bird species is limited to a given seral stage, but some species persist through many stages. As a result, each seral stage was characterized by a unique bird community. The researchers found that both species richness and diversity (Simpson's index; see Section 16.2) increased with successional age through the first 60 to 100 years (forest stages; **Figure 18.19**). The increase in bird species diversity is a function of the increasing vertical structure of the vegetation during succession (see relationship between bird species diversity and foliage height diversity, Section 17.6, Figure 17.17).

Similar changes in the diversity of the small mammal community during plant succession have been observed in old-field communities. Nancy Huntley and Richard Inouye of the University of Minnesota examined the small mammal communities of 18 successional old fields in Minnesota ranging in age from 2 to 57 years since agricultural abandonment. The species composition, biomass, and cover of the vegetation in the old fields changed along the chronosequence, and the species richness of the small mammal community likewise increased with field age (**Figure 18.20**). The researchers found that increase in both abundance and diversity of small mammals was related to increases in plant cover and nitrogen content (increased nitrogen content of food plants).

Maria Barberaena-Arias and T. Mitchell Aide of the University of Puerto Rico examined species composition of insects inhabiting the forest floor during secondary succession of tropical forests in Puerto Rico. The researchers documented chronosequences of secondary forest succession at four locations (different regions) on the island. At each of the four locations, sites representing approximately 5, 30, and 60 years following agricultural abandonment were sampled throughout the year to determine the species composition of the insect community. At each of the four locations, the researchers observed associated changes in the forest floor insect community during the process of plant succession. Species diversity of the insect community increased with forest age (**Figure 18.21a**). The increase in species diversity of insects on the forest floor was associated with the increased accumulation of leaf litter (dead and decomposing leaves on

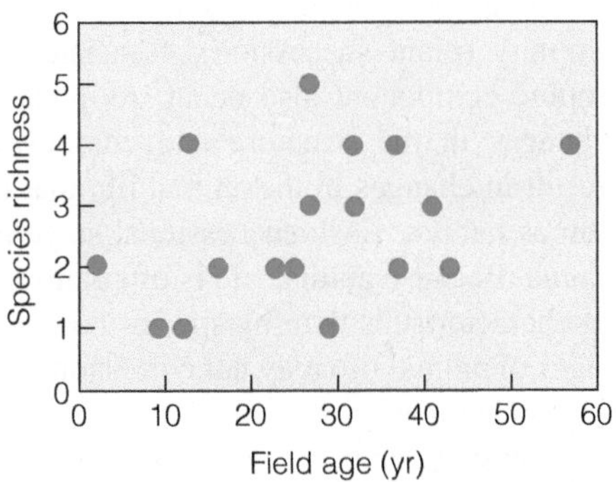

Figure 18.20 Total number of small mammal species captured for each field as a function of field age (years since abandonment from agriculture).
(Adapted from Huntley and Inouye 1987.)

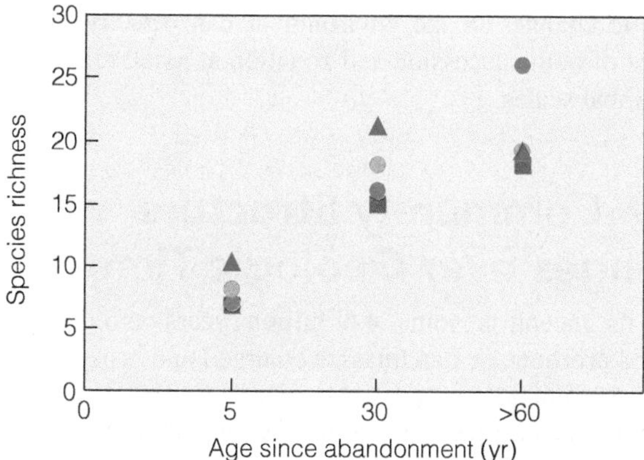

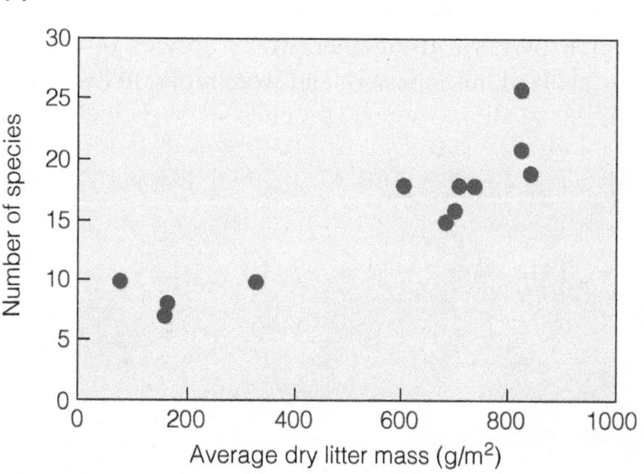

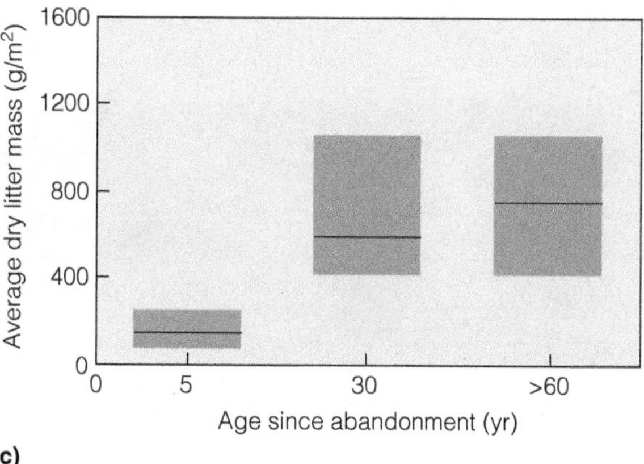

Figure 18.21 (a) Species richness of forest floor insects sampled at successional sites on the island Puerto Rico representing 5, 30, and 60 years following abandonment (agricultural field). Symbols represent the mean value for each of the four locations included in the study. (b) Species richness during succession is positively correlated with quantity of dead leaves on the forest floor (average dry litter mass; data are for each of the 12 study sites: 3 sites at each of 4 locations), (c) which increased with forest age. Graph shows dry litter mass in the 5-, 30-, and 60-year-old sites. Data were pooled for locations within each forest age. Boxes represent 5–95 percentiles; lines within the boxes represent the median value.
(Adapted from Barberaena-Arias and Aide 2003.)

the forest floor), providing a more complex array of habitats and resources (**Figures 18.21b** and **18.21c**).

In each of the examples presented, the changes in species composition and diversity of the heterotrophic communities over time are a product of changes in the associated vegetation during the process of plant succession. As illustrated in Figure 18.18, changes in the composition and structure of the vegetation through time alter the availability of habitats and resources available, shifting the array of animal species that can survive, grow, and reproduce within the community. In other cases, however, heterotrophic succession can be a product of autogenic changes in the environmental conditions brought about by the heterotrophic organisms themselves. A well-studied example of this type of succession is provided by the observed changes in the heterotrophic communities involved in decomposition. Dead plant tissues, animal carcasses, and droppings form substrates on which communities of organisms involved in decomposition exist. Within these communities, groups of fungi and animals succeed one another in a process of colonization and replacement that relates to changes in the physical and chemical properties of the substrate through time. These changes in the substrate are a direct function of the feeding activities of the decomposer organisms. We will examine the process of decomposition and associated changes in the decomposer (heterotrophic) community in detail in Chapter 21.

18.8 Systematic Changes in Community Structure Are a Result of Allogenic Environmental Change at a Variety of Timescales

The focus on succession thus far has been on shifting patterns of community structure in response to autogenic changes in environmental conditions. Such changes occur at timescales relating to the establishment and growth of the organisms that make up the community. However, purely abiotic environmental (allogenic) change can produce patterns of succession over timescales ranging from days to millennia or longer. Environmental fluctuations that occur repeatedly during an organism's lifetime are unlikely to influence patterns of succession among species with that general life span. For example, annual fluctuations in temperature and precipitation influence the relative growth responses of different species in a forest or grassland community, but they have little influence on the general patterns of secondary succession outlined in Figures 18.1 and 18.3. In contrast, shifts in environmental conditions that occur at periods as long as or longer than the organisms' life span are likely to result in successional shifts in species dominance. For example, seasonal changes in temperature, photoperiod, and light intensity produce a well-known succession of dominant phytoplankton in freshwater lakes that is repeated with very little variation each year. Seasonal succession of phytoplankton in Lawrence Lake, a small temperate lake in

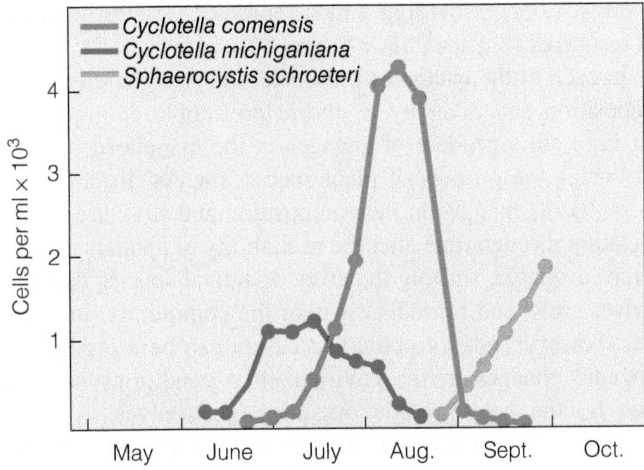

Figure 18.22 Temporal changes in the abundance of dominant phytoplankton species during the period May through October in Lawrence Lake, Michigan (1979). Mean generation times of species range from 1 to 10 days.
(Adapted from Crumpton and Wetzel 1982.)

Michigan, is presented in **Figure 18.22**. Periods of dominance are correlated with species' optimal temperature, nutrient, and light requirements, all of which systematically change over the growing season. Competition and seasonal patterns of predation by herbivorous zooplankton also interact to influence the temporal patterns of species composition.

Over a much longer timescale of decades to centuries, patterns of sediment deposition can have a major influence on the successional dynamics of coastal and freshwater communities. Ponds and small lakes act as a settling basin for inputs of sediment from the surrounding watershed (**Figure 18.23**). These sediments form an oozy layer that provides a substrate for rooted aquatics such as the branching green algae, *Chara*, and pondweeds. These plants bind the loose matrix of bottom sediments and add materially to the accumulation of organic matter. Rapid addition of organic matter and sediments reduces water depth and increases the colonization of the basin by submerged and emergent vegetation. That, in turn, enriches the water with nutrients and organic matter. This enrichment further stimulates plant growth and sedimentation and expands the surface area available for colonization by larger species of plants that root in the sediments. Eventually, the substrate, supporting emergent vegetation such as sedges and cattails, develops into a marsh. As drainage improves and the land builds higher, emergent plants disappear. Meadow grasses invade to form a marsh meadow in forested regions and wet prairie in grass country. Depending on the region, the area may pass into grassland, swamp woodland of hardwoods or conifers, or peat bog.

Over an even longer timescale, changes in regional climate directly influence the temporal dynamics of communities. The shifting distribution of tree species and forest communities during the 18000 years that followed the last glacial maximum in eastern North America is an example of how long-term allogenic changes in the environment can directly influence patterns of both succession and zonation at local, regional, and even global scales.

18.9 Community Structure Changes over Geologic Time

Since its inception some 4.6 billion years ago, Earth has changed profoundly. Landmasses emerged and broke into continents. Mountains formed and eroded, seas rose and fell, and ice sheets advanced to cover large expanses of the Northern and Southern Hemispheres and then retreated. All these changes affected the climate and other environmental conditions from one region of Earth to another. Many species of plants and animals evolved, disappeared, and were replaced by others. As

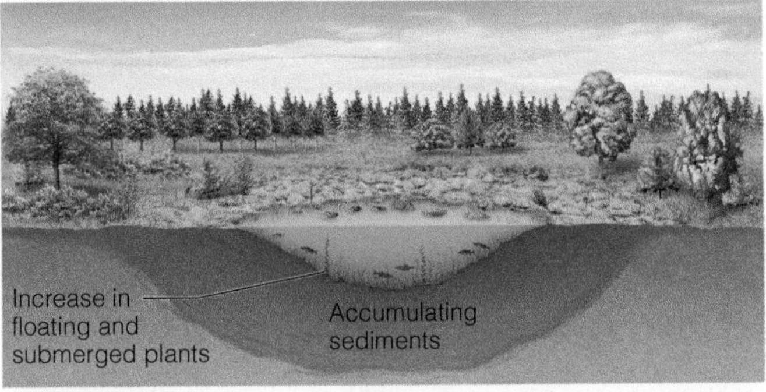

Figure 18.23 Generalized diagram of lake/pond succession in which the lake gradually fills with sediments (organic and inorganic), shrinking the area as time progresses. The area progresses from a marshy wetland to a meadow as the process reaches its final stage.

environmental conditions changed, so did the distribution and abundance of plant and animal species.

Records of plants and animals composing past communities lie buried as fossils: bones, insect exoskeletons, plant parts, and pollen grains. The study of the distribution and abundance of ancient organisms and their relationship to the environment is **paleoecology**. The key to explaining present-day distributions of animals and plants can often be found in paleoecological studies. For example, paleoecologists have reconstructed the distribution of plants in eastern North America after the last glacial maximum of the Pleistocene.

The Pleistocene (approximately 2.6 million to 11,700 B.P.) was an epoch of great climatic fluctuations throughout the world. Some 20 glacial cycles occurred during which ice sheets advanced and retreated. At maximum glacial extent, up to 30 percent of Earth's surface was covered by ice. The last great ice sheet, the Laurentian, reached its maximum advance about 20,000 to 18,000 B.P. during the Wisconsin glaciation stage in North America (**Figure 18.24**). Canada was under ice. A narrow belt of tundra about 60 to 100 km wide bordered the edge of the ice sheet and probably extended southward into the high Appalachians. Boreal forest, dominated by spruce and jack pine (*Pinus banksiana*), covered most of the eastern and central United States as far as western Kansas.

As the climate warmed and the ice sheet retreated northward, plant species invaded the glaciated areas. The maps in **Figure 18.25** reflect the advances of four major tree genera in eastern North America after the retreat of the ice sheet. Margaret Davis of the University of Minnesota developed these maps from patterns of pollen deposition in sediment cores taken from lakes in eastern North America. By examining the presence and quantity of pollen deposited in sediment layers and radiocarbon dating the sediments, she was able to obtain a picture of the spatial and temporal dynamics of tree communities over the past 18,000 years.

These analyses identify plants at the level of genus rather than species because, in many cases, we cannot identify species from pollen grains. Note that different genera, and associated species, expanded their distribution northward with the retreat of the glacier at markedly different rates. Differences in the rates of range expansion are most likely the result of the differences in temperature responses of the species, distances and rates at which seeds can disperse, and interactions among species. The implication is that, during the past 18,000 years, the distribution and abundance of species and the subsequent structure of forest communities in eastern North America have changed dramatically (**Figure 18.26**).

18.10 The Concept of Community Revisited

Our initial discussion of the processes influencing community structure and dynamics contrasted two views of the community (see Section 16.10). Through his organismal concept, Frederic Clements viewed the community as a quasi-organism made up of interdependent species. By contrast, in his individualistic or continuum concept, H. A. Gleason saw the community as an arbitrary concept and stated that each species responds independently to the underlying features of the environment. Research reveals that, as with most polarized debates, the

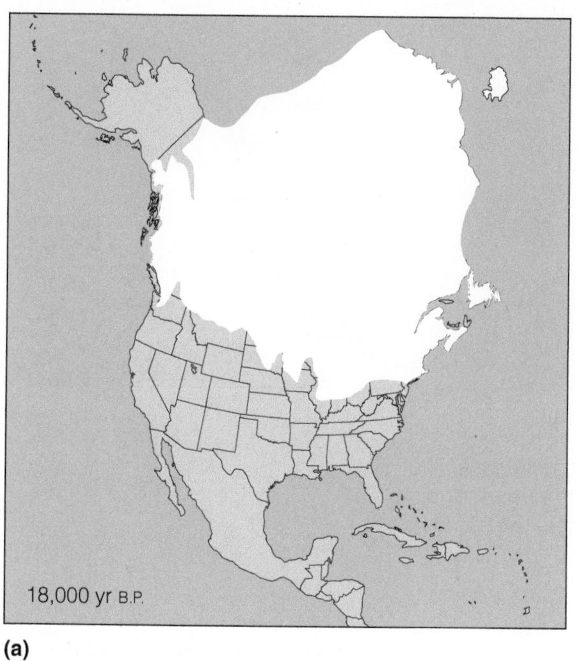

(a)

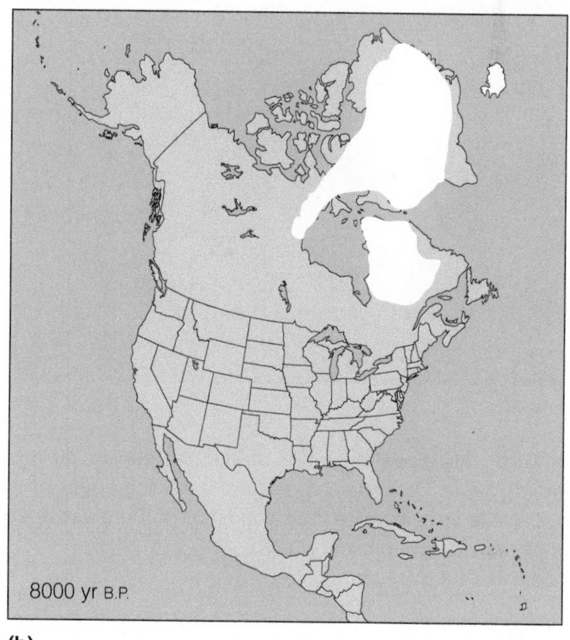

(b)

Figure 18.24 Distribution of the glacial ice sheet on the North American continent at (a) 18,000 and (b) 8000 years before present (B.P.).

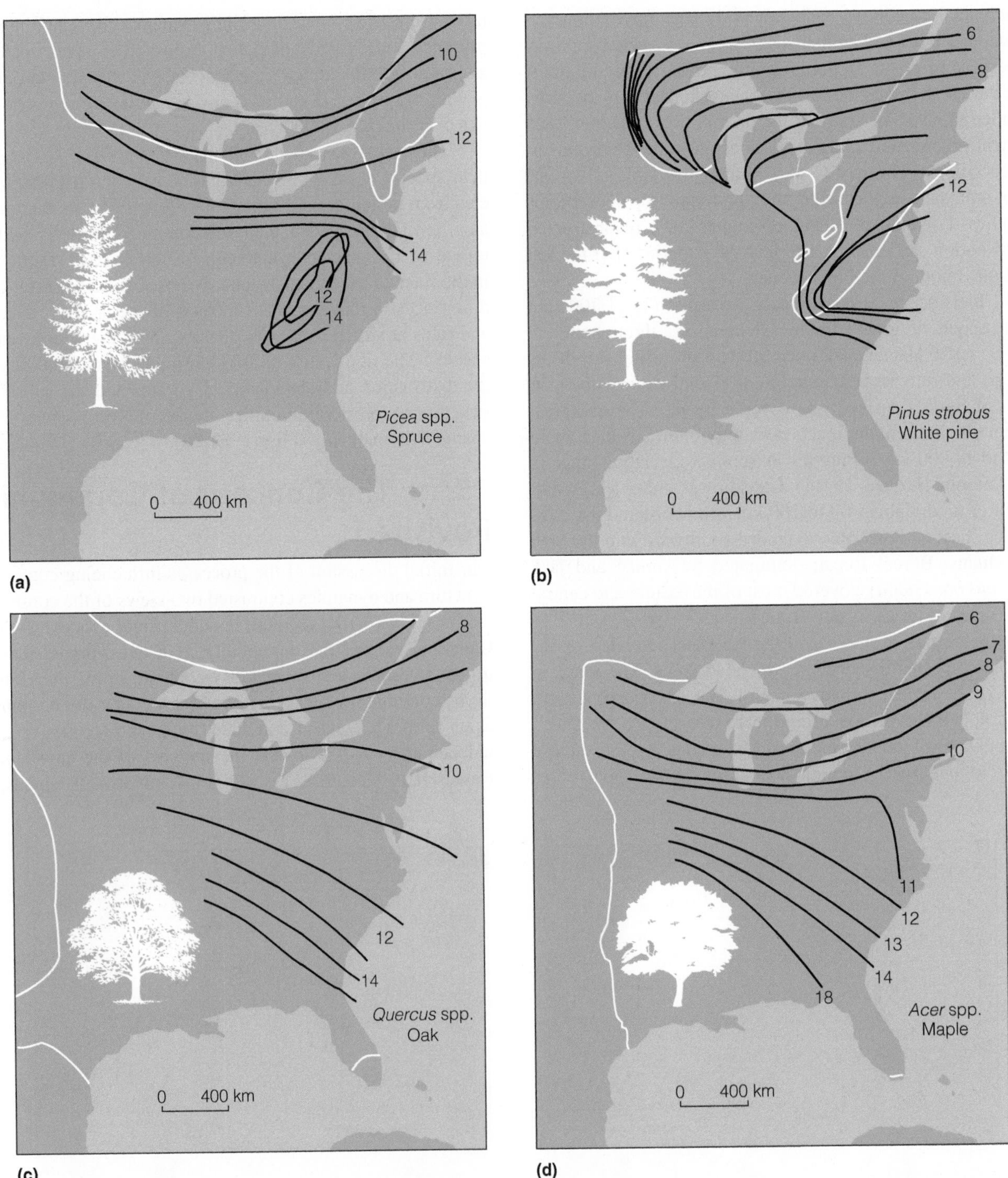

Figure 18.25 Postglacial migration of four tree genera: (a) spruce, (b) white pine, (c) oak, and (d) maple. Dark lines represent the leading edges of the northward-expanding populations. White lines indicate the boundaries of the present-day ranges. The numbers are thousands of years before present (B.P.).
(Adapted from Davis 1981.)

reality lies somewhere in the middle, and our viewpoints are often colored by our perspective. The organismal community is a spatial concept. As we stand in the forest, we see a variety of plant and animal species interacting and influencing the overall structure of the forest. The continuum view is a population concept, focusing on the response of the component species to the underlying features of the environment.

A simple example of the continuum concept is presented in **Figure 18.27**, which represents a transect up a mountain in an area with four plant species present. The distributions of the

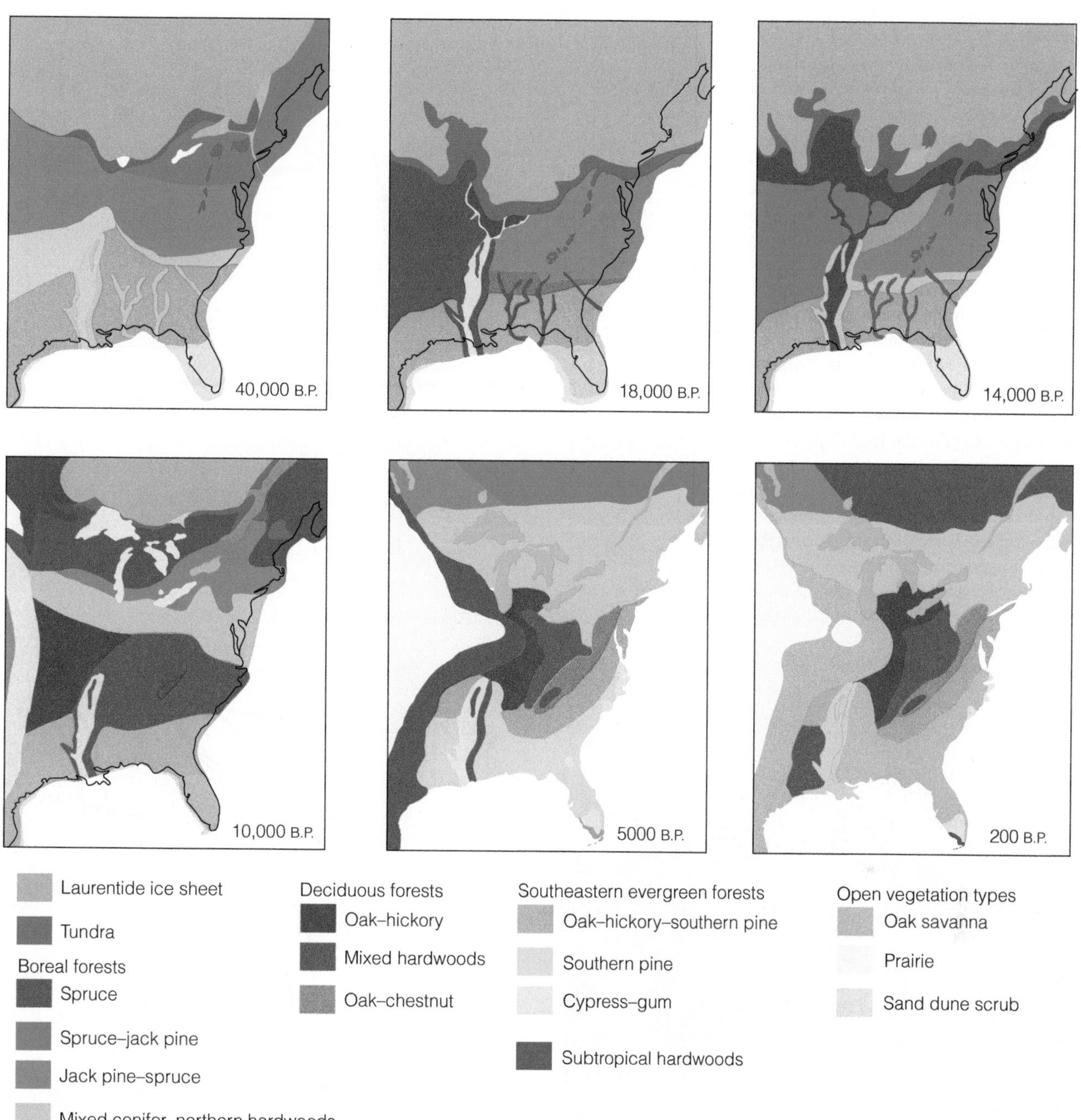

Figure 18.26 Changes in the distribution of plant communities during and after the retreat of the Wisconsin ice sheet; the changes are reconstructed from pollen analysis at sites throughout eastern North America.
(Adapted from Delcourt and Delcourt 1981.)

four plant species are presented in two ways. In the first view, the species distributions are plotted as a function of altitude or elevation. Note that the four species exhibit a continuum of species regularly replacing one another in a sequence of A, B, C, and D with increasing altitude—similar to the individualistic view of communities. The second view of species distributions is their spatial location along the transect. As we move up the mountainside, the distributions of the four species are not continuous. As a result, we might recognize several species associations as we walk along the transect. These associations are identified in Figure 18.27 by different symbols representing the combination of species. Communities composed of coexisting species are a consequence of the spatial pattern of the landscape.

The two views are quite different yet consistent. Each species has a continuous response to the environmental variable of elevation. Yet it is the spatial distribution of that environmental variable across the landscape that determines the overlapping

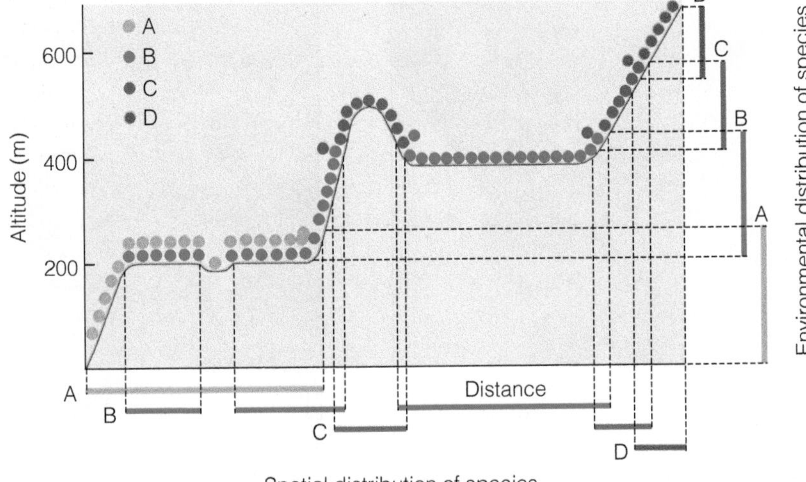

Figure 18.27 Patterns of co-occurrence for four plant species on a landscape along a gradient of altitude. Environmental distributions of the four species are presented in two ways: (1) as the spatial distributions of the species along a transect on the mountainside and (2) as a function of their response to altitude. Note that the species' responses to elevation are continuous, but their spatial distributions along the transect are discontinuous. Patterns of species composition along the mountain gradient result from the spatial pattern of environmental conditions (altitude) and the individual responses of the species. Responses of the species to elevation are consistent with the individualistic or continuum view of communities proposed by H. A. Gleason (see Figure 16.20b). However, consistent patterns of species co-occurrence across the landscape are a function of the spatial distribution of environmental conditions (topographic pattern). Repeatable patterns of species co-occurrence in similar habitats are consistent with the idea of plant associations supported by Frederic Clements (see Figure 16.20a).
(Adapted from Austin and Smith 1989.)

(a)

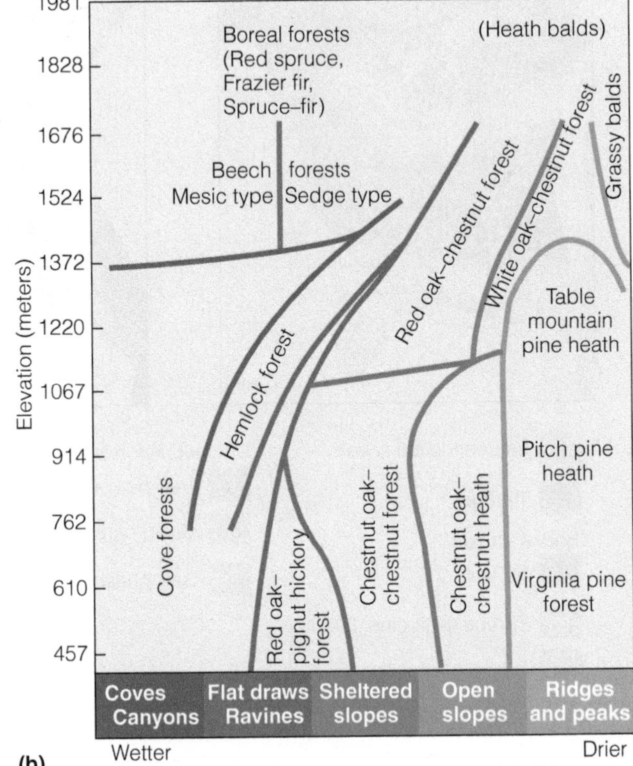

(b)

White pine
Pitch pine
Hemlock
Beech
Red maple
White oak
Dogwood
Virginia pine
American basswood
Tulip poplar
Chesnut oak
Scarlet oak
Percentage of stand
0 10 20 30 40
1 2 3 4 5 6 7 8 9 10 11 12 13
Mesic
Moisture gradient
Xeric
(c)

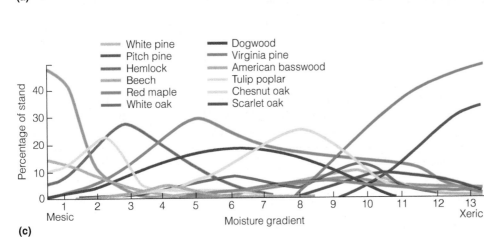

Figure 18.28 Two perspectives of plant communities in (a) Great Smoky Mountains National Park. (b) Distribution of plant communities (associations) in the park in relation to elevation (*y*-axis) and slope position and aspect (*x*-axis). Communities are classified based on the dominant tree species. (c) Distribution and abundance of major tree species that make up vegetation communities in the park, plotted along the gradient of moisture availability (a function of slope position and aspect).
([b]; Adapted from Whittaker 1954.) ([c]; Adapted from Whittaker 1956.)

patterns of species distributions, that is, the composition of the community.

This same approach can be applied to the patterns of forest communities in Great Smoky Mountains National Park (**Figure 18.28a**). Different elevations and slope positions are characterized by unique tree communities, identified by and named for the dominant tree species (**Figure 18.28b**). When presented in this fashion, the distributions of plant communities appear to support the organismal model of communities put forward by Clements. Yet if we plot the distributions of major tree species along a direct environmental gradient, such as soil moisture availability (**Figure 18.28c**), the species appear to be distributed independently of one another, thus supporting Gleason's view of the community.

The simple example in Figure 18.27 examines only one feature of the environment (elevation), yet the structure of communities is the product of a complex interaction of pattern and process. Species respond to a wide array of environmental factors that vary spatially and temporally across the landscape, and the interactions among organisms influence the nature of those responses. The product is a dynamic mosaic of communities that occupy the larger landscape.

ECOLOGICAL Issues & Applications

Community Dynamics in Eastern North America over the Past Two Centuries Are a Result of Changing Patterns of Land Use

Old-field communities, such as the one shown in Figure 18.1, are a common sight in the eastern portion of the United States. These fields represent the early stages in the process of secondary succession, a process that began with the abandonment of agricultural lands (cropland or pasture) and that will eventually lead to forest (see Figure 18.1). Although more than 50 percent of the United States land area east of the Mississippi River is currently covered by forest, the vast majority of these forest communities are less than 100 years old, the product of a continental-scale shift in land use that has occurred over the past 200 years.

When colonists first arrived on the eastern shores of North America in the 17th century, the landscape was dominated by forest. American Indians historically used fire to clear areas for planting crops or to create habitat for game species. However, their impact on the landscape was minor compared to what was to come as a result of the westward expansion of European colonization. The clearing of forest was driven by the need for agricultural lands and forest products, and as the population of colonists grew and expanded westward, so did the clearing of land (**Figure 18.29**). By the 19th

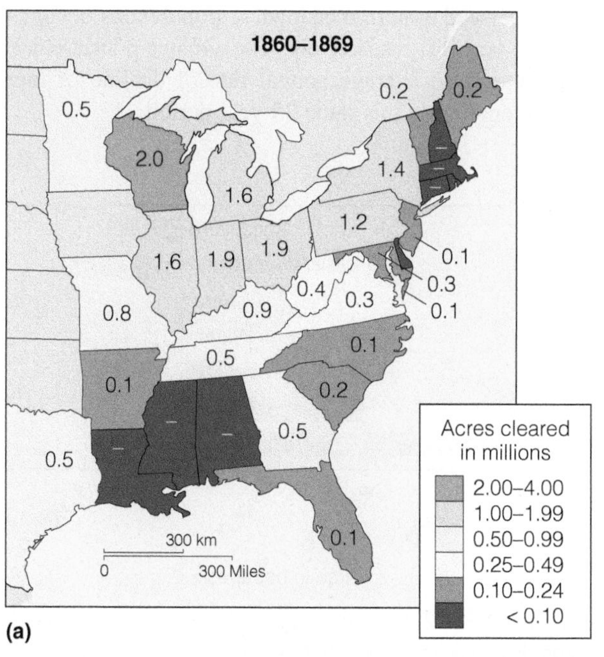

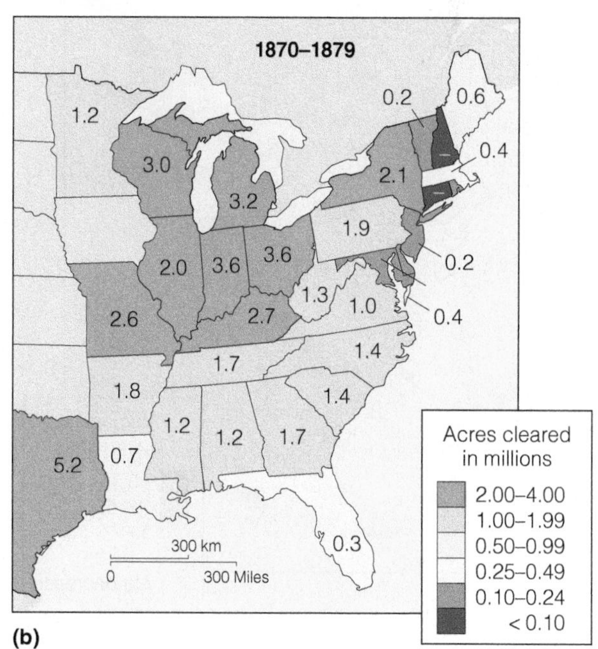

Figure 18.29 Millions of acres of forest lands were cleared for agriculture during the decades of (a) 1860–1869 and (b) 1870–1879 in the eastern United States. Note the westward expansion and accelerating rate of clearing during this period.
(Adapted from Williams 1989.)

century, most of the forests in eastern North America had been felled for agriculture. But by the early part of the 20th century, this trend was reversed.

The Dust Bowl period in the 1930s saw the beginning of the decline in small family farms in the agricultural regions west of the Mississippi (see discussion of Dust Bowl in Chapter 4, *Ecological Issues & Applications*). With the mechanization of agriculture and the large-scale production of chemical fertilizers by the late 1940s (see Chapter 21, *Ecological Issues & Applications*), agriculture in the West underwent a major transition, moving from small, family-owned farms to large commercial farms. The rise of large-scale commercial agriculture hastened the decline in agriculture east of the Mississippi River—a decline that began in the 1800s with the end of the large plantation farms in the Southern states. By the 1930s, the amount of agricultural land in the east had peaked, and it has been declining ever since (**Figure 18.30**). Since 1972 alone, more than 4500 square miles of farmland in the eastern United States has been abandoned and is currently reverting to forest.

The result of these trends is a regional-scale pattern of secondary succession in the eastern United States that has seen a shift from agricultural lands to old-field communities and the eventual reestablishment of forests. As we have seen in the various examples presented throughout the chapter, this transition has major implications for the species composition and patterns of species diversity. The reforestation of eastern North America enhanced the population and diversity of plant and animal species that depend on forested habitats (**Figure 18.31**) but have led to the decline of species dependent on the grassland habitats that were maintained by agricultural land-use practices. There are many examples of population decline in grassland birds in the eastern United States, most notably the extinction of the heath hen from the northeast. Likewise, over the 25-year period from 1966 to 1991, New England upland sandpiper and eastern meadowlark populations declined by 84 and 97 percent, respectively. The greater prairie-chicken has experienced an average annual rate of decline of more than 10 percent during this same 25-year period.

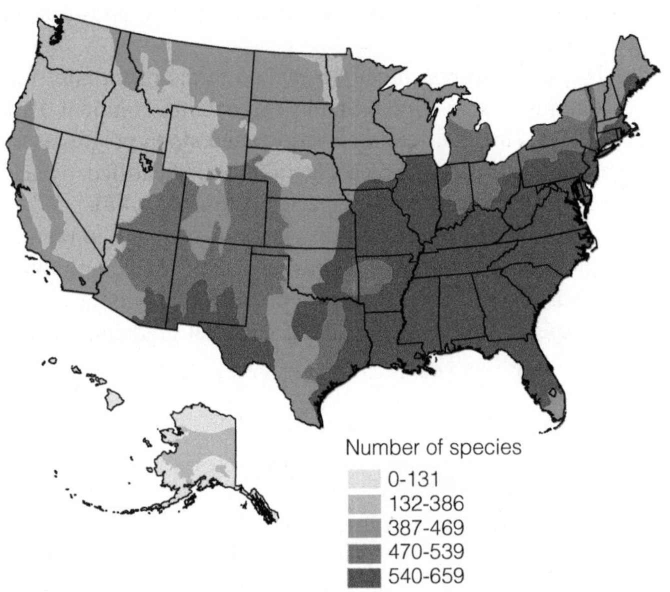

Figure 18.31 Geographic variation in the number of forest-associated species (all taxa).
(Adapted from Flather et al. 2003.)

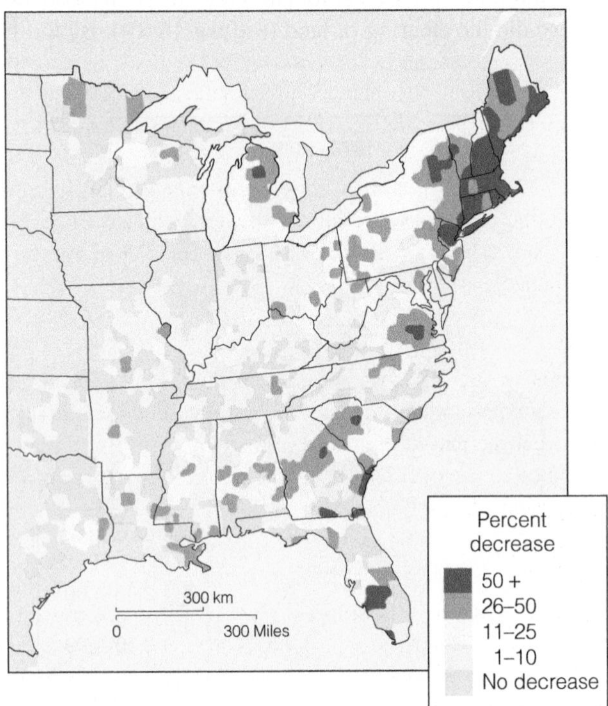

Figure 18.30 Percentage of decrease in agricultural land in the eastern United States since its peak in 1930.
(Data from Williams 1989.)

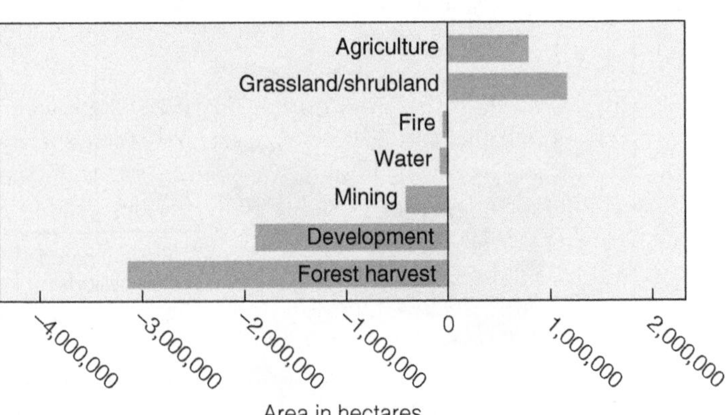

Figure 18.32 Net loss of gain of eastern forest by land cover type. Net gains from agriculture and grassland/forest represent reforestation.
(Data from Drummond and Loveland 2010.)

Despite the long-term trend of reforestation in the eastern regions of North America over the past century, recent studies indicate that this trend may be starting to reverse. Mark Drummond and Thomas Loveland examined land-use changes in the eastern United States from 1973 to 2000 as part of the United States Geological Survey's (USGS) Land Cover Trends project, using satellite data, survey data, and ground photographs. Over the observation period (1973–2000) the researchers document a 4.1 percent decline in total forest area, a net loss equivalent to more than 3.7 million hectares. The researchers found considerable regional variation, with net loss being particularly marked in the coastal plains of southeast. The major sources of forest decline are forestry activities, including both the harvesting of current forested lands and their conversion to pine plantations and development from urban expansion (**Figure 18.32**).

The future dynamics of the landscapes of eastern North America will certainly depend on the changing patterns of land use and the growing demand for land for urban development. However, the potential for future changes to the global climate system as a result of human activities (see Chapter 2, *Ecological Issues & Applications*) may well be the primary determinate of the future of the forests in the United States, a topic we will explore in detail in Chapter 27.

SUMMARY

Succession 18.1

With time, natural communities change. This gradual sequential change in the relative abundance of species in a community is succession. Opportunistic, early successional species yield to late successional species. Succession occurs in all environments. The similarity of successional patterns in different environments suggests a common set of processes.

Primary Succession 18.2

Primary succession begins on sites devoid of or unchanged by organisms. Examples include newly formed sand dunes, lava flows, or newly exposed glacial sediments.

Secondary Succession 18.3

Secondary succession begins after disturbance on sites where organisms are already present. Terrestrial examples include abandoned agricultural lands or the reestablishment of vegetation after logging or fire. In aquatic ecosystems, disturbances caused by storms, wave action, or herbivory can initiate the process of secondary succession.

History 18.4

The study of succession has been a focus of ecological research for more than a century. The intervening years have seen a variety of hypotheses attempting to address the processes that drive succession. These hypotheses include a variety of processes related to colonization, facilitation, competition, inhibition, and differences in environmental tolerances.

Autogenic Environmental Change 18.5

Environmental changes can be autogenic or allogenic. Autogenic changes are a direct result of the activities of organisms in the community. Changes in environmental conditions independent of organisms are allogenic. Succession is the progressive change in community composition through time in response to autogenic changes in environmental conditions. One example is the changing light environment and the shift in dominance from fast-growing, shade-intolerant plants to slow-growing, shade-tolerant plants observed in terrestrial plant succession. Autogenic changes in nutrient availability, soil organic matter, and stabilization of sediments can likewise have a major influence on succession.

Species Diversity and Succession 18.6

Patterns of species diversity change during the course of succession. Species colonization increases species richness, whereas species replacement acts to decrease the number of species present. Species diversity increases during the initial stages of succession as the site is colonized by new species. As early successional species are displaced by later arrivals, species diversity tends to decline. Peaks in diversity tend to occur during succession stages that correspond to the transition period, after the arrival of later successional species but before the decline of early successional species. Patterns of diversity during succession are influenced by resource availability and disturbance.

Heterotrophic Species 18.7

Changes in the heterotrophic component of the community also occur during succession. Successional changes in vegetation affect the nature and diversity of animal life. Certain sets of species are associated with the structure of vegetation found during each successional stage.

Allogenic Environmental Change 18.8

Fluctuations in the environment that occur repeatedly during an organism's lifetime are unlikely to influence patterns of succession among species with that general life span. Allogenic, abiotic environmental changes that occur over timescales greater than the longevity of the dominant organisms can produce patterns of succession over timescales ranging from days to millennia or longer.

Long-Term Changes 18.9

The current pattern of vegetational distribution reflects the glacial events of the Pleistocene. Plants retreated and advanced with the movements of the ice sheets. The rates and distances of their advances are reflected in the present-day ranges of species and the distribution of plant communities.

Community Revisited 18.10

The community is a spatial concept; the individual continuum is a population concept. Each species has a continuous response to an environmental gradient, such as elevation or moisture. Yet the spatial distribution of that environmental variable across the landscape determines the overlapping patterns of distribution, that is, the composition of the community.

Reforestation Ecological Issues & Applications

Although more than 50 percent of the U.S. land area east of the Mississippi River is currently covered by forest, the vast majority of these forest communities are less than 100 years old, the product of a continental-scale shift in land use that has occurred over the past 200 years. In recent decades, the trend of net forest gain has begun to reverse as demands for urban development and forest products have increased.

STUDY QUESTIONS

1. Distinguish between zonation and succession.
2. Distinguish between primary and secondary succession.
3. Defoliation of oak trees by gypsy moth larvae caused the death of extensive forest stands in the Blue Ridge Mountains of central Virginia. The recovery of these forest communities after defoliation includes the growth of existing trees and shrubs that have escaped defoliation, as well as colonization of the site by tree species outside the community. Is this an example of primary or secondary succession? Why?
4. Use the discussion of secondary succession in an abandoned agricultural field presented in Section 18.3 to differentiate among the models of facilitation, competition, and tolerance as they apply to the process of succession.
5. Classification of plant species into three primary life history strategies (*R*, *C*, and *S*) proposed by the plant ecologist Philip Grime was discussed in Chapter 10, Section 10.13. Which of these three plant strategies is most likely to characterize the colonization of a newly disturbed site (early successional species)? Which of the three plant strategies is most likely to characterize later successional species?
6. Why is the ability to tolerate low resource availability often associated with plant species that dominate during the later stages of succession?
7. In Section 17.7, we discussed the difference between size-symmetrical and size-asymmetrical competition for resources. How might the nature of these two types of competition shift during the process of plant succession?
8. If the vertical structure of the vegetation increases during the process of terrestrial plant succession, how might the pattern of animal species diversity respond? (*Hint:* See Section 17.6.)

FURTHER READINGS

Classic Studies

Bazzaz, F. A. 1979. "The physiological ecology of plant succession." *Annual Review of Ecology and Systematics* 10:351–371.

This now-classic article contrasts the physiology of plant species characteristic of different stages of succession. Bazzaz provides a framework for understanding the process of succession as the result of varying plants' adaptations to changing environmental conditions through time.

Clements, F. E. 1936. "Nature and structure of the climax." *Journal of Ecology* 24:252–284.

Clement's initial article in which he introduces the monoclimax hypothesis of plant succession.

Egler, F. E. 1954. "Vegetation science concepts I. Initial floristics composition, a factor in old-field vegetation development." *Vegetation* 4:412–417.

The original articles in which Egler introduces the initial floristics hypothesis of plant succession.

Golley, F., ed. 1978. *Ecological succession.* Benchmark Papers. Stroudsburg, PA: Dowden, Hutchinson, and Ross.

This edited volume presents a historical view of ecological theory relating to the process of succession. Golley presents a variety of original manuscripts with commentaries that trace the development of theoretical thinking on the subject.

Recent Research

Bazzaz, F. A. 1996. *Plants in changing environments: Linking physiological, population, and community ecology.* New York: Cambridge University Press.

This book integrates a variety of laboratory and field studies aimed at understanding the dynamics of plant communities.

Chapin, F. S., III, L. Walker, C. Fastie, and L. Sharman. 1994. "Mechanisms of primary succession following deglaciation at Glacier Bay, Alaska." *Ecological Monographs* 64:149–175.

A detailed study of the process of primary succession in Glacier Bay, Alaska.

Huston, M., and T. M. Smith. 1987. "Plant succession: Life history and competition." *American Naturalist* 130:168–198.

This article presents a framework for understanding the process of plant succession as a result of the trade-offs in the evolution of plant life history characteristics and the shifting nature of competition as environmental conditions change through time.

Smith, T. M., and M. Huston. 1987. "A theory of spatial and temporal dynamics of plant communities." *Vegetation* 3:49–69.

Expands the framework first developed by Huston and Smith (see preceding article) to understand patterns of zonation and succession in plant communities from the perspective of evolutionary trade-offs in plant life history characteristics.

Tilman, D. 1985. "The resource-ratio hypothesis of succession." *American Naturalist* 125:827–852.

This article presents the framework for the resource-ratio hypothesis and its role in successional dynamics.

Tilman, D. 1988. *Plant strategies and the dynamics and structure of plant communities.* Princeton, NJ: Princeton University Press.

In this book, Tilman draws upon a wide range of field and laboratory studies to develop a theoretical framework for understanding pattern and process in plant communities.

West, D. C., H. H. Shugart, and D. B. Botkin, eds. 1981. *Forest succession: Concepts and application.* New York: Springer-Verlag.

An excellent reference on patterns and processes relating to succession in forest communities.

Arial photograph of an agricultural landscape in Prima County, Arizona. Note the contrast in patterns between areas of managed lands and areas shaped by natural processes.

CHAPTER GUIDE

WE HAVE DEFINED a community as a group of potentially interacting species occupying a given area (Chapter 16). Although by definition ecological communities have a spatial boundary, they likewise have a spatial context within the larger landscape. Consider the view of the Virginia countryside in Figure 19.1. It is a patchwork of forest, fields, golf course, hedgerows, pine plantation, pond, and human habitations. This patchwork of different types of land cover is called a **mosaic**, using the analogy of mosaic art in which an artist combines many small pieces of variously colored material to create a larger pattern or image (Figure 19.2). The artist creates the emerging pattern, defining boundaries by using different shapes and colors of materials to construct the mosaic. In a similar fashion, the landscape mosaic is a product of the boundaries defined by changes in the physical and biological structure of the distinct communities that form its elements. In the artist's mosaic, the elements interact only visually to present the emerging image. By contrast, the landscape mosaic is dynamic. Patches and their boundaries—the structural and functional components of the landscape—interact in a variety of ways depending on their size and spatial arrangement, and they change through time. The study of the composition, structure, and function of landscapes is called **landscape ecology**.

Landscape ecology differs in a number of ways from the other areas of ecological study that we have examined thus far. First, the focus of landscape ecology is on spatial heterogeneity, which characterizes the spatial pattern of the elements that make up the landscape, determines the process that gives rise to the patterns, and examines how those patterns change through time. Landscape ecology is the study of linking pattern and process. Second, landscape ecology is distinguished by its focus on broader spatial extents than those traditionally studied in ecology. Third, landscape ecology often focuses on the role of humans in creating and affecting landscape patterns and processes.

Figure 19.1 A view of a Virginia landscape showing a mosaic of patches consisting of different types of land cover: natural forest, plantations, fields, water, and rural development.

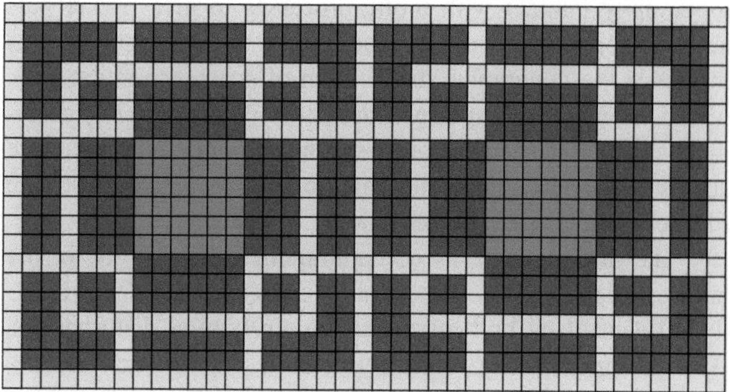

Figure 19.2 An artist's mosaic creating an image with an assemblage of pieces of colored glass, stone, or other materials.

19.1 A Variety of Processes Gives Rise to Landscape Patterns

When we view a landscape we notice the composition, that is, the types of land cover or communities (patches) and their spatial arrangement (as in Figure 19.1). How does the mosaic of patches develop on the landscape? The spatial structure of each landscape is unique and results from the interactions of a variety of factors, including abiotic (physical), biotic, natural disturbances, and human activities.

Climate and geology interact to define the landform, determining patterns of drainage, erosion, and deposition (Figure 19.3a). Climate, geology (rocks and minerals), and

Figure 19.3 (a) Landscape in which topographic variations are a product of surface water flow and erosion. (b) Slight variation in elevation and associated water table allow patches of forest to grow in the landscape of the Everglades National Park in south Florida.

(a)

(b)

topography interact with biotic processes to give rise to patterns of soil formation (see Chapter 4). These abiotic factors influence the geographic distribution of species (Section 8.2), which together with species interactions (Section 17.2) determine changes in species composition and community structure across the landscape (zonation; see **Figure 19.3b**, also see Figures 16.15–16.18 for examples).

The distinct patterns of communities that we see in the landscape, as well as the species inhabiting them, are heavily influenced by both past and present natural disturbances. A disturbance is any relatively discrete event—such as fire, windstorm, flood, extremely cold temperatures, or drought—that disrupts community structure and function. Disturbances both create and are influenced by patterns on the landscape (**Figure 19.4**). For example, communities on ridge tops are more susceptible to damage from wind and ice storms, and bottomland communities along streams and rivers are more susceptible to flooding. In turn, these disturbances result in new patterns of patches on the landscape.

Natural disturbances are not limited to physical processes. Biological processes can also function as disturbances that structure landscapes. Grazing by domestic animals is a common cause of disturbance. For example, in the southwestern United States, domestic cattle disperse seeds of mesquite (*Prosopis* spp.) and other shrubs through their droppings, enabling those species to invade already overgrazed grassland. This type of disturbance is not caused solely by domestic species. In many parts of eastern North America, overpopulations of white-tailed deer (*Odocoileus virginianus*) have decimated herbaceous plants and shrubs in the forest understory, eliminated forest regeneration, and destroyed habitat for forest understory wildlife. The African elephant (*Loxodonta africana*) has long been considered a major influence on the development of savanna communities. The combination of high density and restricted movement that result from populations being limited to the confines of national parks and conservation areas, can result in the large-scale destruction of woodlands (see Figure 16.4).

Figure 19.4 Examples of landscape patterns created by natural disturbances. (a) Mosaic pattern of the burn following the Rodeo-Chediski Forest Fire in the Apache-Sitgreaves National Forest, Arizona. (b) Trees blown down by a line of severe thunderstorms that moved through Pine County, Minnesota, and northwest Wisconsin, producing widespread wind damage.

(a)

(b)

Beavers (*Castor canadensis*) modify many forested areas in North America and Europe. By damming streams, beavers alter the structure and dynamics of flowing water ecosystems. When dammed streams flood lowland areas, beavers convert forested stands into wetlands. Pools behind dams become catchments for sediments. By feeding on aspen, willow, and birch, beavers maintain stands of these trees, which otherwise would be replaced by later successional species. Thus, the action of beavers creates a diversity of patches—pools, open meadows, and thickets of willow and aspen—within the larger landscape (**Figure 19.5a**).

Birds may seem an unlikely cause of major vegetation changes, but in the lowlands along the west coast of Hudson Bay, large numbers of the lesser snow goose (*Chen caerulescens caerulescens*) have affected the brackish and freshwater marshes. Snow geese grub for roots and rhizomes of graminoid plants in early spring and graze intensively on leaves of grasses and sedges in summer. The dramatic increase in the number of geese has stripped large areas of their vegetation, resulting in the erosion of peat and the exposure of underlying glacial gravels (**Figure 19.5b**). Plant species colonizing these patches differ from the surrounding marsh vegetation, giving rise to a mosaic of patches on the landscape.

Outbreaks of insects such as gypsy moths, spruce budworms, and pine bark beetle defoliate large areas of forest and cause the death or reduced growth of affected trees. The result of mortality of canopy trees is to create patches of open canopy, altering patterns of plant succession (**Figure 19.5c**).

Some of the most dominant factors influencing the structure of landscapes are human activities. Because human activity is ongoing and involves continuous management of an ecosystem, it affects ecosystems more profoundly than natural disturbances do. One of the more permanent and radical changes in vegetation communities occurs when humans remove natural communities and replace them with cultivated cropland and pastures. Ongoing fragmentation of larger tracts of land by development activities (both rural and urban) often reflect early land-survey methods set on straight lines with no attempts to follow topography and natural boundaries (**Figure 19.6a**).

Another large-scale disturbance is timber harvesting (Figure 19.6b). Disturbance to the forest depends on the logging methods employed, which range from removing only selected trees to clearing entire blocks of timber.

The result of human activities is often the fragmentation of larger continuous tracts of habitats, such as forest, shrubland, or grassland into a mosaic of smaller, often isolated patches (Figure 19.7). This process, referred to as **habitat fragmentation**, not only functions to reduce the availability of habitat for species—which is the leading cause of population decline and species extinction (see discussion in Chapter 9, *Ecological Issues & Applications*)—but it also fragments populations, the consequences of which we will explore in the following sections.

Figure 19.5 Examples of landscape pattern created by the activities of animals. (a) A small beaver dam about 2 m high along a stream in the Rocky Mountains. The reservoir of water behind the dam alters the stream flow. (b) The bare patches on this tundra landscape along the Hudson Bay coastline were created by the feeding activity of snow geese (see green plant growth in fenced exclosure). (c) Patches of living trees create a mosaic in a matrix of dead trees which have been killed by the pine bark beetle.

(a)

(b)

(c)

19.2 Landscape Pattern Is Defined by the Spatial Arrangement and Connectivity of Patches

A landscape is a spatially heterogeneous area, a mosaic of elements referred to as patches. **Patches** are areas that are more or less homogeneous compared to their surroundings.

Figure 19.6 Examples of landscape patterns created by human land use. (a) Block clear-cutting in a coniferous forest in the western United States. Such cutting fragments the forest. (b) Matrix of agricultural fields interspersed with patches of forest and other land cover types. Note the contrast between straight lines defining boundaries created by surveyors as compared to natural boundaries created by river course.

(a)

(b)

The communities that surround a patch constitute its **matrix**. For example in the photograph presented in Figure 19.1, the open fields represent patches of grassland vegetation embedded in the broader landscape. These patches differ from their surroundings. They vary in size and shape and are embedded within the matrix of surrounding patches from which they are often separated by distinct boundaries. **Boundaries** are the place where the edge of one patch meets the edge of another adjacent patch (or surrounding matrix). There is one additional feature that defines the structure of the landscape, the spatial arrangement of the patches: their context within the broader landscape. The proximity of patches to each other influences the ability for interactions among patches to occur: this is called their **connectivity**. For example, the proximity of patches influences the ability of individuals or propagules to disperse between patches (a topic we will examine in detail in following sections). One feature of the landscape that can increase connectivity is the presence of **corridors**, which are routes that facilitate movement between patches. Often these corridors are strips of habitat or cover that are similar to the patches they connect. Landscapes are therefore comprised of three main elements: patches, boundaries, and corridors and all are embedded in a matrix (Figure 19.8).

Different landscapes differ in the configuration of these elements. Recall that landscape ecology is the study of the effect of pattern on process. Therefore, it is important to describe landscape structure in terms of its pattern of patches, corridors, and boundaries. How we define these elements of the landscape, however, is dependent on the organism or process that is being examined. For example consider a forested landscape (forest cover defines the matrix) in eastern North America in which patches of grassland habitat are embedded (Figure 19.9). To the meadow vole (*Microtus pennsylvanicus*), which is a species of small mammal that inhabits these old-field (successional) environments, the landscape represents patches of potential grassland habitat separated by forest (unsuitable habitat) that present a barrier to movement between patches. Each patch might support a local population,

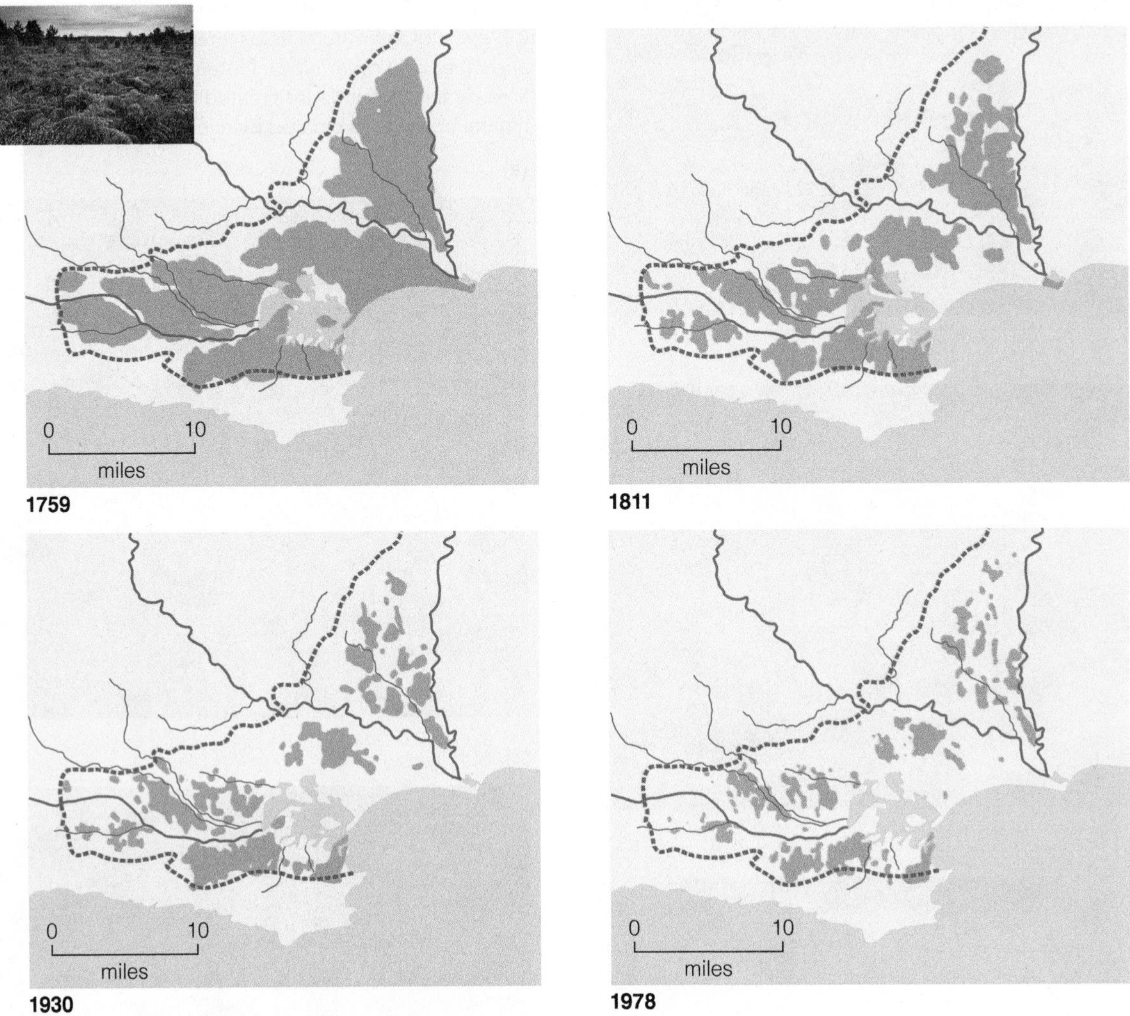

Figure 19.7 Fragmentation and isolation of Poole Basin, Dorset, England. Between 1759 and 1978, the area lost 86 percent of its heathland (40,000 ha to 6000 ha), changing from 10 large blocks separated by rivers to 1084 pieces—nearly half of these sites are less than 1 ha, and only 14 sites are larger than 100 ha.
(Adapted from Webb and Haskins 1980.)

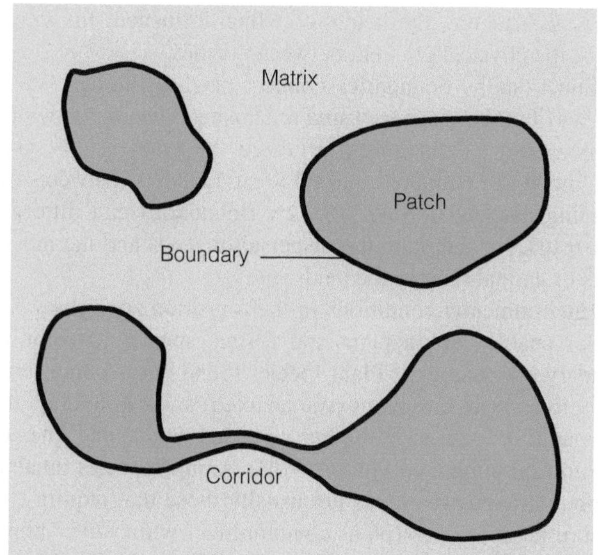

Figure 19.8 Elements of landscape pattern. Patches are areas of more or less homogeneous cover (habitat) in relation to the surrounding environment, referred to as the matrix. Boundaries are the place where the edge of one patch meets the edge of another adjacent patch or matrix. Corridors are strips of habitat or cover that is similar to the patches which they connect and increase the connectivity of the landscape.

with dispersal among populations influenced by the proximity of patches and the existence of corridors that facilitate movement (connectivity). To the milkweed bug (*Oncopeltus fasciatus*), which lives, feeds, and reproduces on individuals of the common milkweed plant (*Asclepias syriaca*), each patch of grassland habitat might function as a landscape. Each milkweed plant or patch of milkweed plants might support a local population of milkweed bugs, and the spatial distribution of milkweed within the field now defines the spatial distribution of patches within the grassland matrix. This example illustrates an important point. A landscape is not defined by size, but it is defined by the scale relative to the organism or process of interest. Thus, the spatial extent of a landscape may be a few square meters (m) or many square kilometers (km) depending on the specific process or organism being studied.

19.3 Boundaries Are Transition Zones that Offer Diverse Conditions and Habitats

The place where the edge of one patch meets the edge of another is called a boundary, which is an area of contact, separation, or transition between patches. Some boundaries

Figure 19.9 Example of defining landscape based on the organism or process being examined. (a) For the meadow vole (*Microtus pennsylvanicus*), the landscape is defined as a mosaic of grassland patches embedded in a matrix of forest. (b) For the milkweed bug (*Oncopeltus fasciatus* shown in insert on milkweed pod), which lives, feeds, and reproduces on individuals of the common milkweed plant (*Asclepias syriaca* shown in insert), each grassland area functions as a landscape of milkweed habitat patches embedded in a matrix of grass.

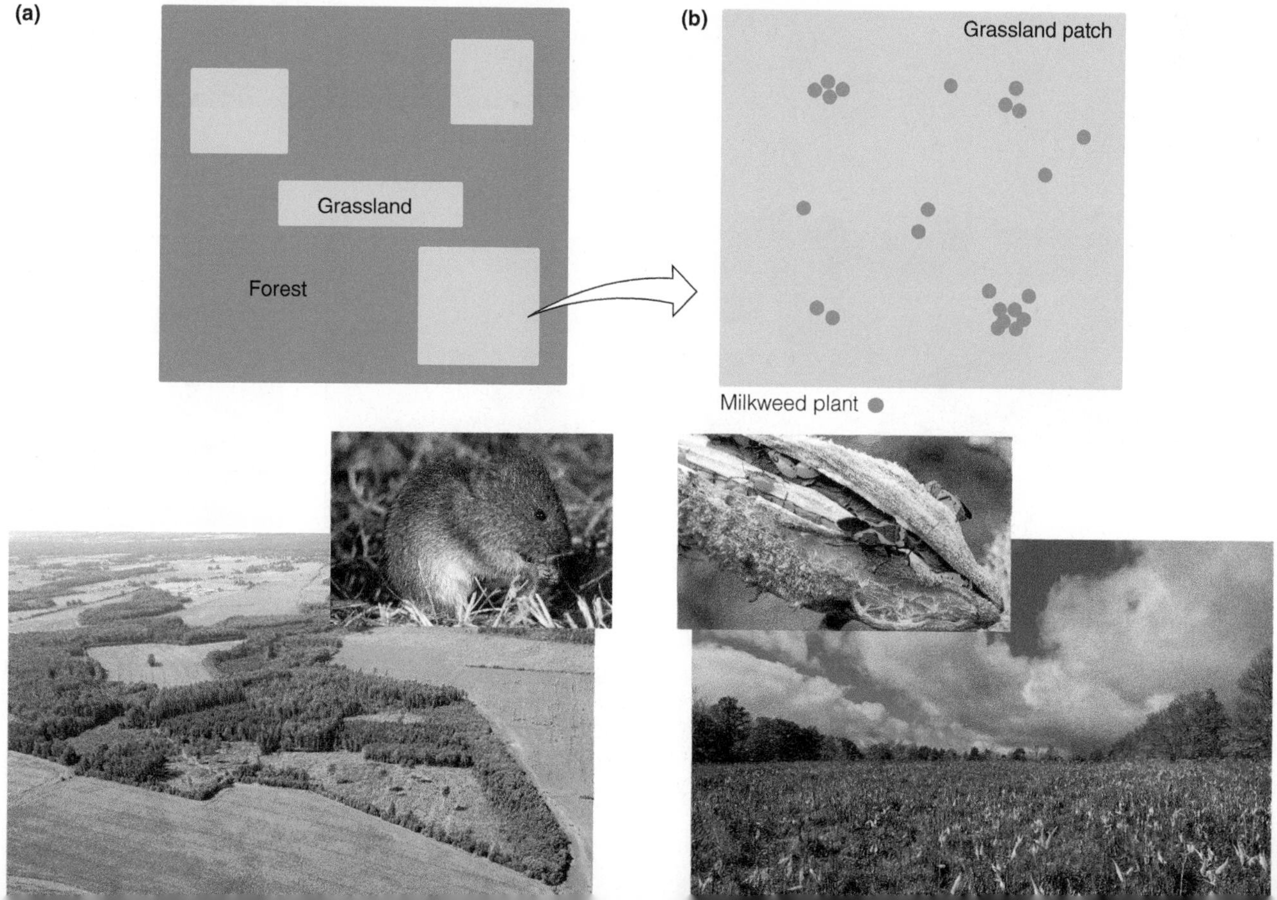

(a)
Grassland
Forest

(b)
Grassland patch
Milkweed plant ●

indicate an abrupt change in the physical conditions—topography, substrate, soil type, or microclimate—between communities. For example, the boundary between the terrestrial and aquatic environments at a pond's edge is defined by the intersection of the water table with the land surface (**Figure 19.10a**). Where abrupt changes in the physical environment give rise to associated changes in community structure, these boundaries are usually stable and often permanent features of the landscape. Other boundaries, however, result from natural disturbances such as fires, storms, and flood or from human-related activities such as timber harvesting, agriculture, and housing developments (**Figure 19.10b**). Such boundaries are typically transient, subject to successional changes over time.

Some boundaries between landscape patches are narrow and abrupt with a sharp contrast between the adjoining patches, such as between a forest and an adjacent agricultural field (see Figure 19.10b). Others are much broader, forming a transition zone called an **ecotone** between the adjoining patches (**Figure 19.11b**). Boundaries may vary in length and may be straight (**Figure 19.11a**), convoluted (**Figure 19.11c**), or perforated (**Figure 19.11d**). In addition to width, boundaries have a vertical structure, the height of which influences the steepness of the physical gradient between patches.

Functionally, boundaries connect patches through fluxes or flows of material, energy, and organisms. The height, width, and porosity of boundaries influence the gradients of wind flow, moisture, temperature, and solar radiation between the adjoining patches (**Figure 19.12**). Boundaries can differentially restrict or facilitate the dispersal of seeds and the movements of animals across the landscape.

Environmental conditions in the transition zones between patches enable certain plant and animal species to colonize boundary environments. Plant species found in such areas tend to be more shade-intolerant (sun adapted) and can tolerate the dry conditions caused by higher air temperatures and rates of evapotranspiration (see Figure 19.12). Animal species inhabiting boundary environments are usually those that require two or more habitat types (plant communities) within their home range or territory. For example, the ruffed grouse (*Bonasa umbellatus*) requires new forest openings with an abundance of herbaceous plants and low shrubs for feeding its young, a dense stand of sapling trees for nesting cover, and mature forests for winter food and cover. Because the home range size

Figure 19.10 Examples of boundaries. (a) The boundary between the aquatic environment of the pond and the terrestrial environment surrounding it is defined by the intersection of the water table and the land surface. (b) The boundary between the agricultural field and adjacent forest is maintained by continued mowing and tilling of the ground in the agricultural field.

(a)

(b)

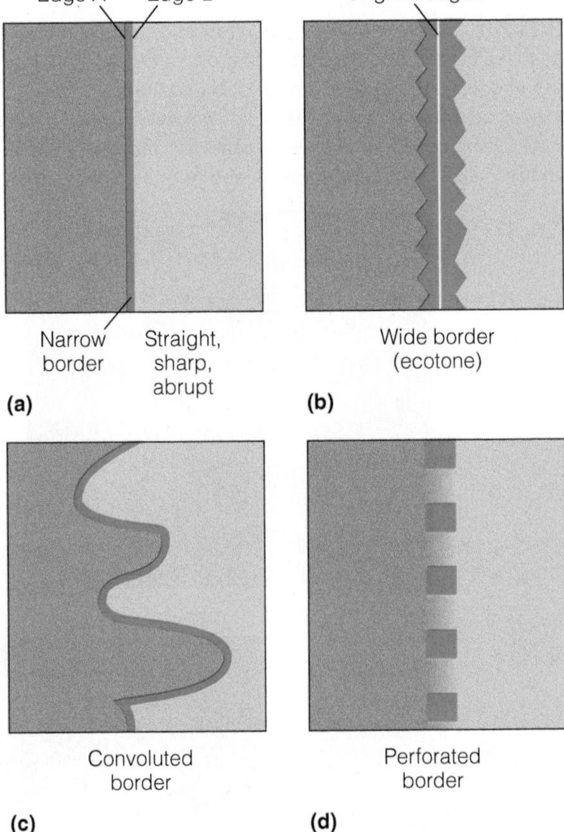

Figure 19.11 Types of boundaries. (a) Narrow, sharp, abrupt boundary created where the edges of two patches meet; (b) wide boundary creating an ecotone between two adjacent patches; (c) convoluted boundary; and (d) perforated boundary.

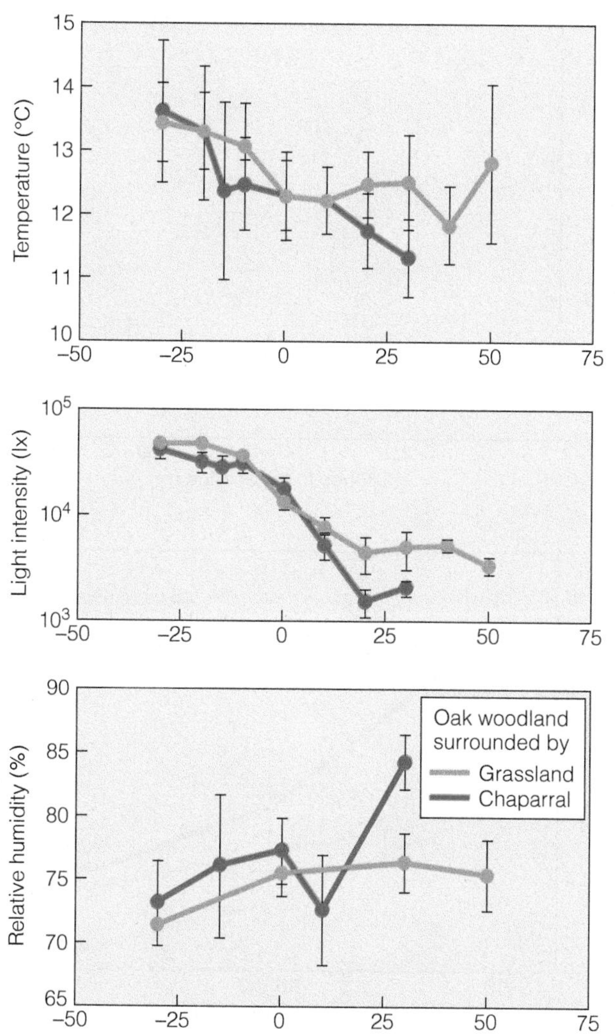

Figure 19.12 Microhabitat variation (temperature, light intensity, and relative humidity) across edges between oak woodland patches and two different matrix habitats—chaparral and grassland—in the Santa Cruz Mountains of central California. For the *x*-axis, positive values indicate the distance from an edge into the woodland habitat; negative values indicate the distance into the matrix habitat. Bars represent standard errors of the means. Temperature and relative humidity were measured at a height of 1 m above ground level. Light intensity was measured at a height of 2 m. Light intensity: lx × 0.0185 = PAR(μmol/m^2/s). (Adapted from Sisk et al. 1997.)

Interpreting Ecological Data

Q1. *How does air temperature (top graph) change as you move out of an oak woodland patch into a surrounding chaparral habitat?*

Q2. *In general, how does relative humidity (bottom graph) change as you move out of an oak woodland patch into a surrounding chaparral habitat?*

Q3. *Why are the observed changes in relative humidity opposite those observed for temperature? (Hint: Section 2.5, Figure 2.15.)*

of a ruffed grouse is from 4 to 8 hectares (ha), the whole range of habitats must be contained within this single area. Other species, such as the indigo bunting (*Passerina cyanea*), are restricted exclusively to the edge environment (**Figure 19.13**) and are referred to as **edge species**.

Because boundaries often blend elements from the adjacent patches (particularly in the case of ecotones), their structure and composition are often very different from the adjacent patches. Thus, boundaries offer unique habitats with relatively easy access to the adjacent communities. These diverse conditions enable these boundary zones to support plant and animal species from adjacent patches, as well as those species adapted to the edge environment (**Figure 19.14**). As a result, boundaries are often populated by a rich diversity of life. This phenomenon, called the **edge effect**, is influenced by the area of boundary available (length and width) and by the degree of contrast between adjoining plant communities. In general, the greater the contrast between adjoining patches, the greater the diversity of species. Therefore, a boundary between forest and grassland should support more species than would a boundary between a young and a mature forest.

Although edge effect may increase species diversity, it can also create problems. Narrow, abrupt boundaries appear to be attractive to predators. Predators often use edges as travel lanes, which increases rates of predation in edge habitats as compared to the interior of habitats. Ecologist Maiken Winter of the University of Missouri and colleagues examined the nesting success of grassland birds in patches of tall-grass prairie in southwestern Missouri. The researchers examined patterns of nesting success as a function of distance from

Figure 19.13 Map of territories (blue areas) of a true edge species, the indigo bunting (*Passerina cyanea*), which inhabits woodland edges, hedgerows, roadside thickets, and large gaps in forests that create edge conditions. The male requires tall, open song perches, and the female, a dense thicket in which to build a nest. (Adapted from Whitcomb et al. 1976.)

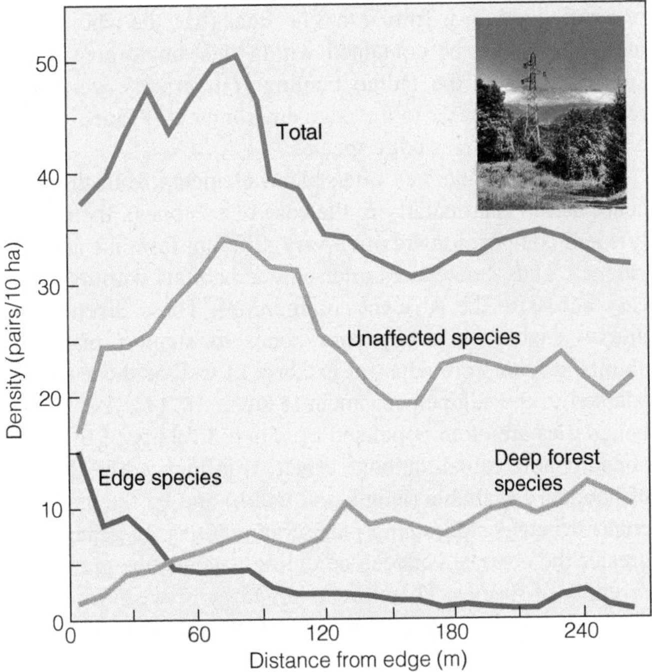

Figure 19.14 Example of edge effect. Trends in the density (pairs/10 ha) of breeding pairs of birds from the power-line corridor edge to deeper forest. The total density has been partitioned into three functional groups of species: edge species, species typically found to occupy forest interior environments (deep forest species), and those species whose distribution is unaffected by edge and interior environment. Note the increased density of birds in the region adjacent to the forest edge. (Adapted from Kroodsma 1982.)

the boundary between grassland patches and adjacent forest. Winter and colleagues found that the frequency of nest predation was highest in the edge environments and declined with distance from the forest edge (**Figure 19.15a**). The decline in observed predation rate was a result of a decrease in mammalian predator activity with distance from the edge of forest patches (**Figure 19.15b**).

Boundaries can further alter interactions among species by either restricting or facilitating dispersal across the landscape. For example, dense, thorny shrubs growing in the boundary of forest and field can block some animals from passing through.

Boundaries are dynamic, changing in space and time. In the example illustrated in **Figure 19.16**, the boundary between forest and field is characterized by an abrupt difference in the vertical structure of the vegetation. As the vegetation within the boundary grows taller, the vertical gap between the forest crown and boundary diminishes, which forms a continuous vertical profile of vegetation. Barring disturbance, the boundary expands horizontally as vegetation characteristic of the boundary encroaches into the patch. Plant species change as size and environmental conditions within the boundary change and competitive interactions among boundary species increase.

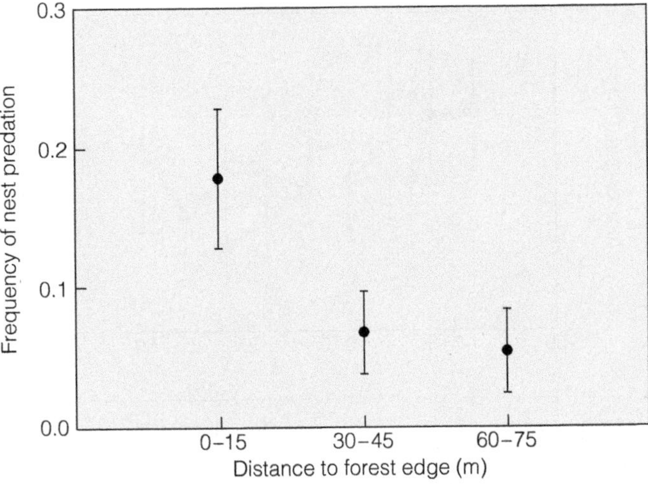

(a)

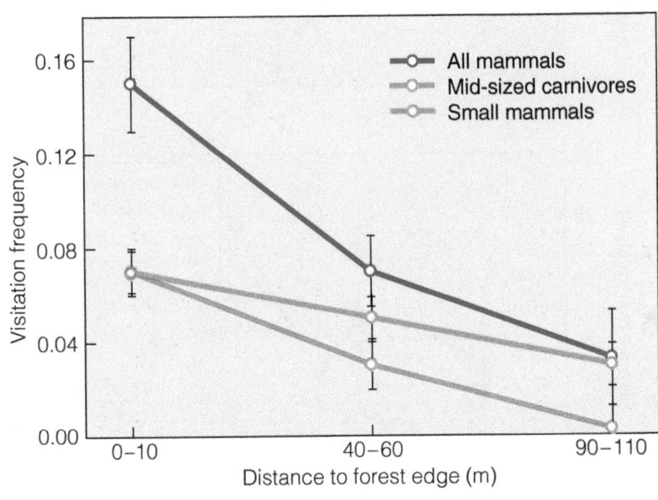

(b)

Figure 19.15 (a) Predation of nests by carnivores at three distance intervals from the forest edge in southwestern Missouri prairie fragments (circles represent mean values, and bars indicate standard errors). (b) Visitation frequency of mammalian predators measured at three distance intervals from a forested edge in the same fragments of native tallgrass prairie shown in (a). (Adapted from Winter et al. 2000.)

19.4 Patch Size and Shape Influence Community Structure

Patch size has a crucial influence on community structure, species diversity, and the presence and absence of species. As a general rule, large patches of habitat contain a greater number of individuals (population size) and species (species richness) than do small patches. The increase in population size for a given species with increasing area is simply a function of increasing the carrying capacity for the species (see Chapter 11). The greater the area, the greater the number of potential home ranges (or territories) that can be supported within the patch. Within the animal community, there is a general relationship between body size (weight) and the size of an animal's home range (see

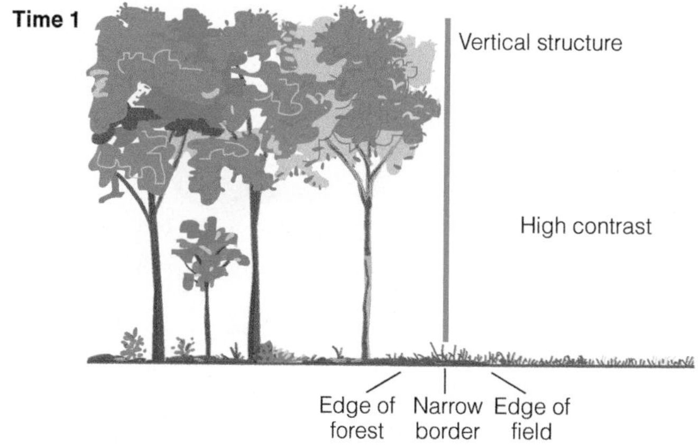

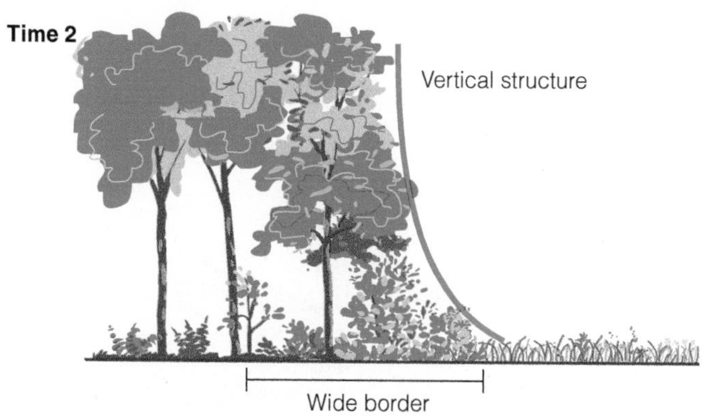

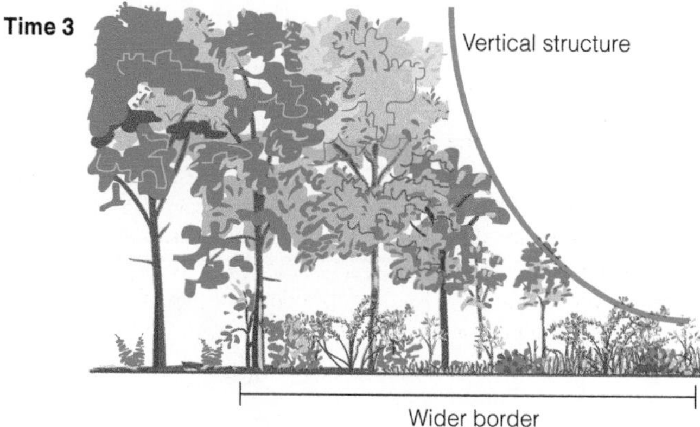

Figure 19.16 Changes in vertical and horizontal structure of a boundary through time.

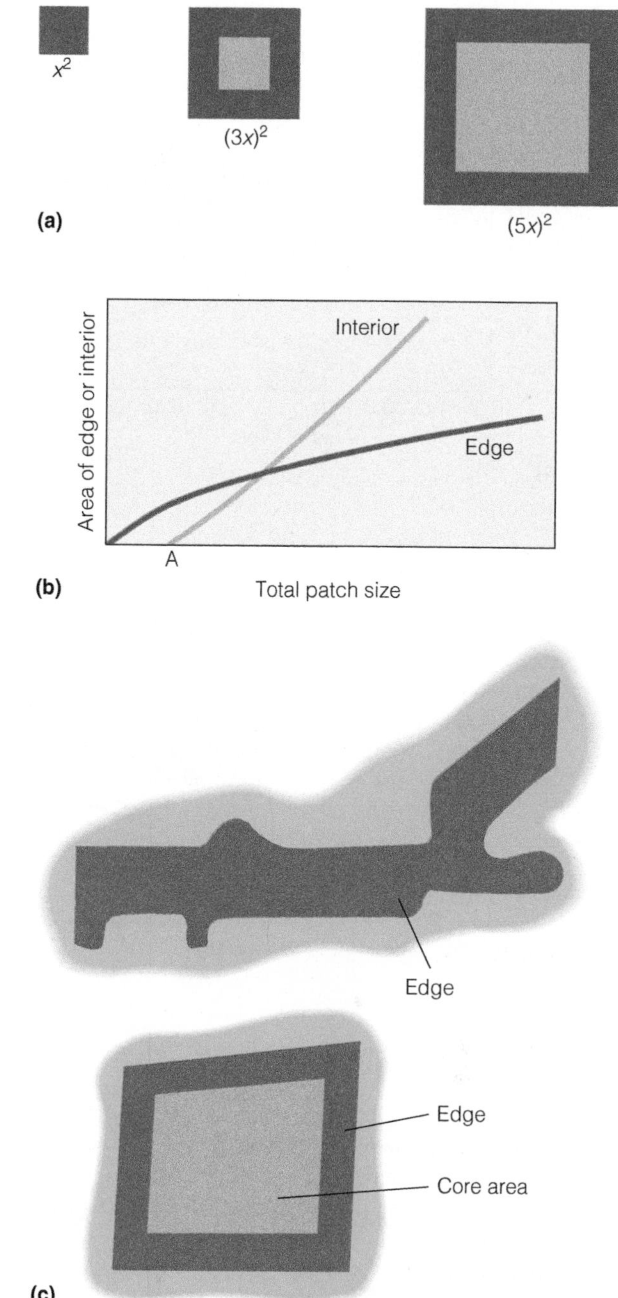

Figure 19.17 Relationship of habitat patch size to edge and interior conditions. All habitat patches are surrounded by edge. (a) Assuming that the depth of the edge remains constant, the ratio of edge to interior decreases as the habitat size increases. When the patch is large enough to maintain shaded, moist conditions, an interior begins to develop. (b) The general relationship between patch size and area of edge and interior. Below point A, the habitat is all edge. As size increases, interior area increases, and the ratio of edge to interior decreases. (c) This relationship of size to edge holds for a square or circular habitat patch. Long, narrow woodland islands whose widths do not exceed the depth of the edge are edge communities, even though the area may be the same as that of square or circular ones.

Figure 11.16). In addition, for a given body size, the home range of carnivores is greater than that of herbivores. Thus, large predators such as grizzly bears, wolves, and mountain lions will be limited to much larger, contiguous patches of habitat.

The relationship between patch size and species richness is more complex. Larger patches are more likely to contain variations in topography and soils that give rise to a greater diversity of plant life (both taxonomic and structural), which in turn create a wide array of habitats for animal species (see Section 17.6).

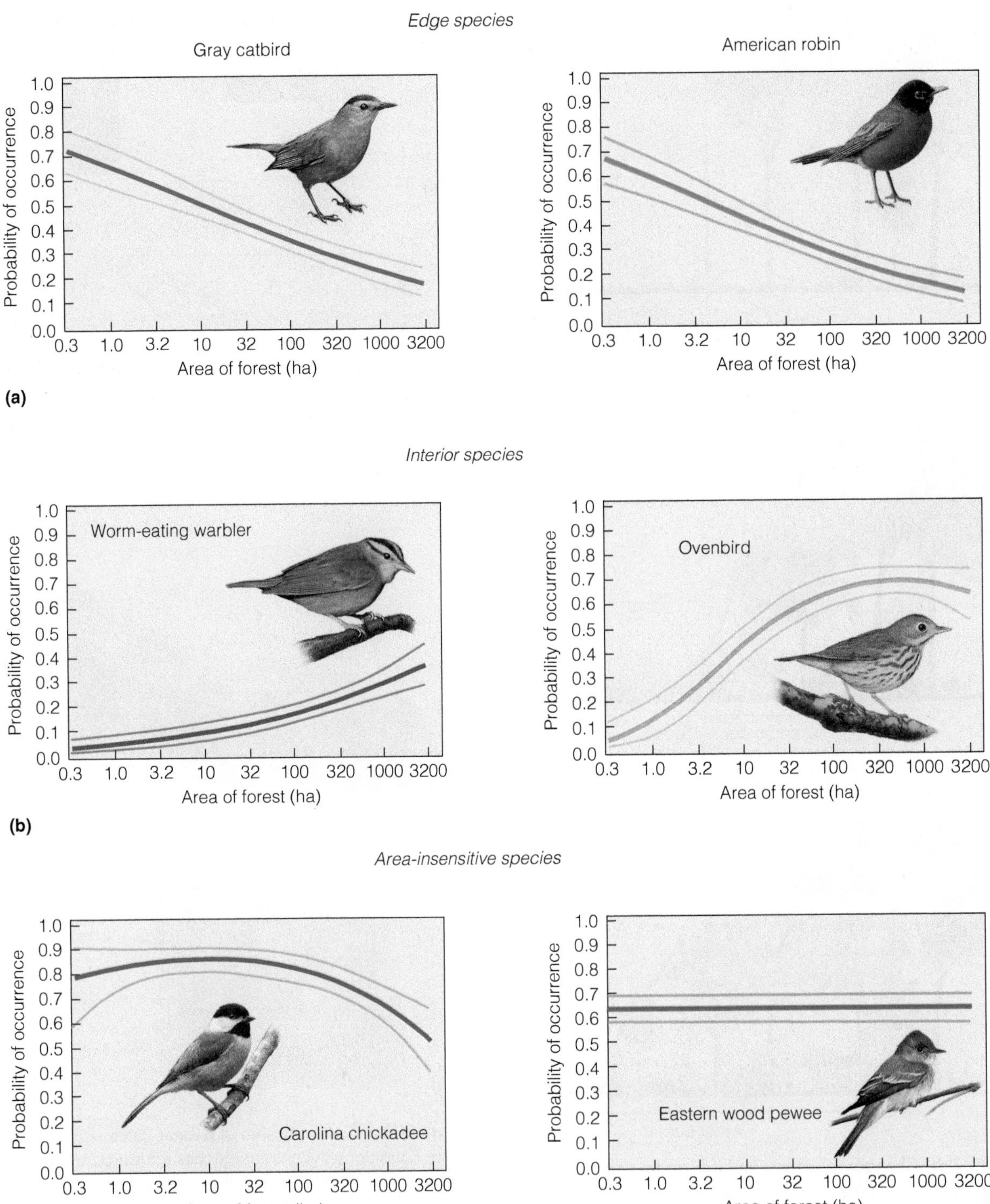

Figure 19.18 Difference in habitat responses between edge species and area-sensitive or interior species. The graphs indicate the probability of detecting these species from a random point in patches of various sizes. Light colored lines indicate the 95 percent confidence intervals for the predicted probabilities. (a) The catbird and the robin are familiar edge species. As patch size increases, the probability of finding them decreases. (b) The worm-eating warbler and the ovenbird are ground-nesting birds of the forest interior. The probability of finding them in small patches is low. (c) In contrast to edge and interior species, other species—such as the Carolina chickadee and Eastern wood pewee—are insensitive to patch area (probability of encountering species is independent of patch size).
(Adapted from Robbins et al. 1989.)

Another feature of patch size relates to the difference between the habitats provided by edge and interior environments. The size and shape of patches affect the relative abundance of edge (or perimeter) and interior environments. Only when a patch becomes large enough to be deeper than its boundary can it develop interior conditions (**Figure 19.17a**). For example, at one extreme, a very small patch is all boundary or edge habitat; but as patch size increases, the ratio of boundary to interior decreases (**Figure 19.17b**). Altering the shape of a patch can change this boundary-to-interior relationship. For example, the long, thin habitat in **Figure 19.17c** is all boundary or edge. Such long, narrow patches of woodland, whose depth does not exceed the width of the boundary, are boundary (or edge) communities regardless of the total patch area.

In contrast to edge species (see Section 19.3 and Figure 19.13), other species, termed **interior species**, require environmental conditions characteristic of interior habitats and stay away from the abrupt changes associated with boundary environments. **Figure 19.18** shows the relationship between forest area (patch size) and the probability of occurrence of six different bird species of eastern North America. Edge species, such as the gray catbird (*Dumetella carolinensis*) and American robin (*Turdus migratorius*), have a high probability of occurring in small forest patches dominated by edge environments. As patch size increases, the probability of finding these birds decreases. In contrast, the ovenbird (*Seiurus aurocapillus*) and the worm-eating warbler (*Helmitheros vermivorus*) are species adapted to the interior of older forest stands, which are characterized by large trees and sparse shrub cover in the understory layer. Accordingly, the probability is low that they will occur in small patches. Intermediate to these two groups are **area-insensitive species**, such as the Carolina chickadee (*Parus carolinensis*) and the Eastern wood pewee (*Contopus virens*).

The minimum size of habitat needed to maintain interior species differs between plants and animals. For plants, patch size per se is less important to the persistence of a species than are environmental conditions. For many shade-tolerant plant species found in the forest interior, the minimum size is the area needed to sustain moisture and light conditions typical of the interior. That area depends in part on the ratio of edge to interior and on the nature of the surrounding boundary habitat. If the stand is too small or too open, sunlight and wind will penetrate and dry the interior environment, eliminating herbaceous and woody species that require moister soil conditions. For example, in the northeastern United States, forest fragmentation can result in the decline of moisture-requiring (mesic) species such as sugar maple and beech while encouraging the growth of more xeric species such as oaks. In a study of the impacts of forest fragmentation on species composition of forest patches in southern Chile, Christian Echeverria of the Universidad Austral de Chile (Valdivia, Chile) and colleagues examined the changes in the abundance of edge and interior shrub and tree species as a function of patch size. Plant species were classified into interior and edge functional groups based on relative shade tolerance and habitat requirements described by previous studies. The researchers found that the species richness of edge species decreased with increasing patch size, while that of interior species increased (**Figure 19.19**; compare Figure 19.19 with pattern for birds in Figure 19.18).

Numerous studies that have examined bird species diversity in both forest and grassland patches reveal a pattern of increasing species richness with patch size (**Figure 19.20**), but only up to a point. The ecologist R. F. Whitcomb and colleagues studied patterns of species diversity in forest patches in western New Jersey. Their findings suggest that maximum bird diversity is achieved with woodlands 24 ha in size. Similar patterns were observed in studies investigating the species composition of bird communities occupying forest patches in agricultural regions in Illinois and Ontario, Canada. With patches of intermediate size, a general pattern of maximum species diversity results from the negative correlation between edge species and the size of habitat patches, combined with the positive correlation between interior species and increased area (see Figure 19.18).

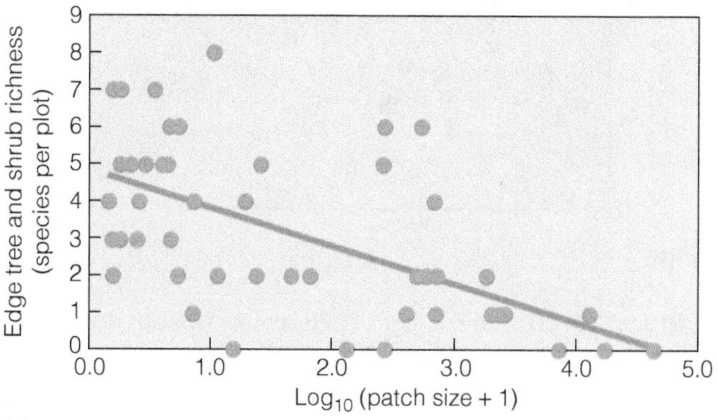

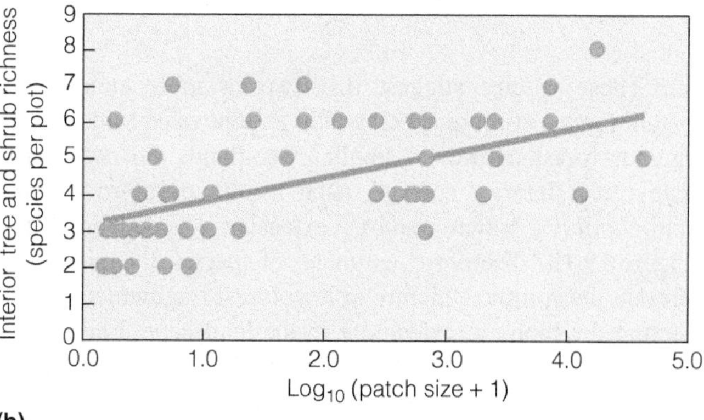

Figure 19.19 Relationships between $\log_{10}$ (patch size +1, in hectares) and (a) edge tree and shrub species richness and (b) interior tree and shrub species richness for temperate rain forest patches located in on a fragmented landscape in southern Chile. As forest fragments increased in size, the richness of interior tree and shrub species (shade-tolerant) increased, whereas the richness of edge tree and shrub species (shade-intolerant) declined. (Adapted from Echeverría et al. 2007.)

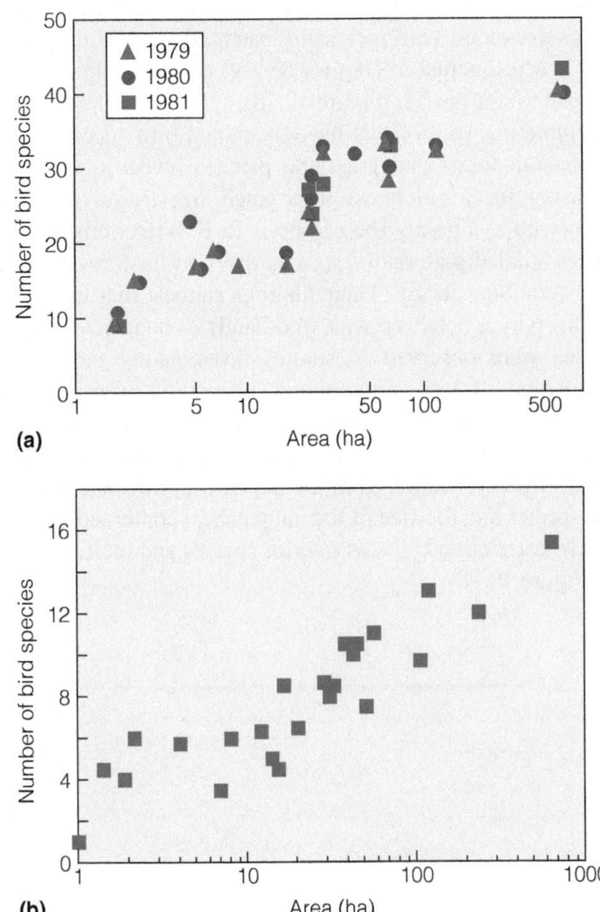

Figure 19.20 The number of bird species (species richness) plotted as a function of the area of (a) woodland or (b) grassland habitat. Area (x-axis) in both graphs is presented on a log scale. Different symbols in (a) refer to results from surveys conducted during three different time periods.
(Whitcomb et al. 1981, Herkert 1994.)

These studies suggest that two or more small forest patches support more species than an equivalent area of contiguous forest. However, smaller woodlands did not support true forest interior species, such as the ovenbird (*Seiurus aurocapillus*), which requires extensive wooded areas (see Figure 19.18). Therefore, estimates of species diversity do not present the complete picture of how forest fragmentation is affecting the biological diversity in the landscape. Large forest tracts with a high degree of heterogeneity are required to support the range of bird species characteristic of both edge and interior habitats.

Although the relation of patch size to species diversity focuses strongly on forests, the same concept applies to other landscapes—such as grasslands (see Figure 19.20b), shrublands, and marshes—that are all highly fragmented by cropland, grazing, sagebrush eradication, and housing developments. Many grassland species such as grasshopper sparrows (*Ammodramus savannarum*), western meadowlarks (*Sturnella neglecta*), and prairie grouse (*Tympanuchus* spp.), and shrubland species such as sage grouse (*Centrocercus urophasianus*)—all interior species—are experiencing serious decline as these landscapes become fragmented and patch size decreases (see Chapter 9, *Ecological Issues & Applications*).

19.5 Landscape Connectivity Permits Movement between Patches

It is the type, number, and spatial arrangement of patches on the landscape that influence movement of organisms among patches, and ultimately the population dynamics of species and community structure. The degree to which the landscape facilitates or impedes the movement of organisms among patches is referred to as **landscape connectivity**.

There are two components to landscape connectivity: structural and functional. **Structural connectivity** relates to the physical arrangement of habitat patches on the landscape; that is, the degree to which patches are contiguous or physically linked to one another. **Functional connectivity** describes the degree to which the landscape facilitates the movement of organisms and is a function of both the physical structure of the landscape as well as the behavioral responses of organisms to the structure. Functional connectivity is therefore both landscape- and species-specific. Distinguishing between these two types of connectivity is important because structural connectivity does not necessarily represent functional connectivity. Depending on its mode of and limits to dispersal, the physical structure (as measured by its structural connectivity) of a given landscape may facilitate the movement of one species, while presenting a barrier to effective dispersal for another. Conversely, patches do not necessarily need to be structurally connected to be functionally connected. Species capable of flight, such as butterflies and birds, are often able to move among patches across considerable distances. Therefore, assessing landscape connectivity requires a species-centered approach that considers factors such as a species' mode and range of dispersal and its ability to move through different types of habitat. For example, Oliver Honnay and colleagues at the University of Leuven (Belgium) examined patterns of colonization of forest plant species on a fragmented landscape in central Belgium. The researchers found that for all species the probability of successful colonization of a forest patch declined with distance from the nearest source population (forest patch occupied by a given species; **Figure 19.21a**); however, the relationship varied among species based on their modes of dispersal (**Figure 19.21b**). Colonization success was greater for species with seeds dispersed by birds and mammals as compared to dispersal by wind or ants. In effect, for the same structural connectivity (the spatial arrangement of forest patches on the landscape) the functional connectivity of the landscape differed among species based on their mode of seed dispersal. The net effect of reduced colonization rates with increasing isolation, however, represents a reduction in species richness with increasing distance to nearest neighboring patches (**Figure 19.21c**).

One structural feature of the landscape that can facilitate an organism's movement between patches of suitable habitat is

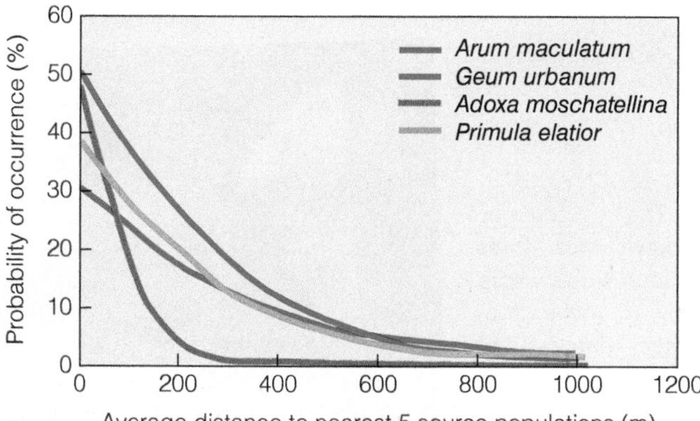

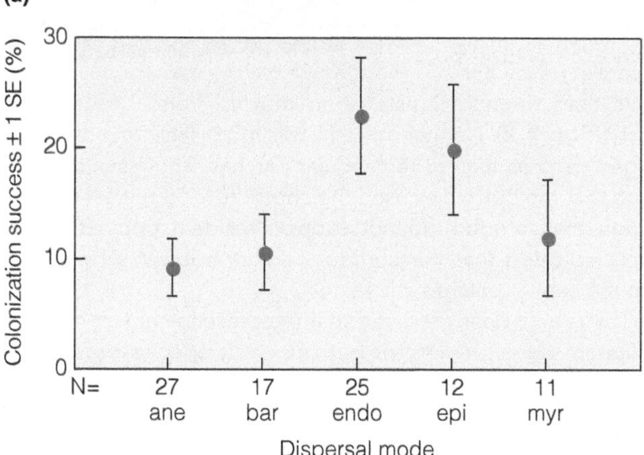

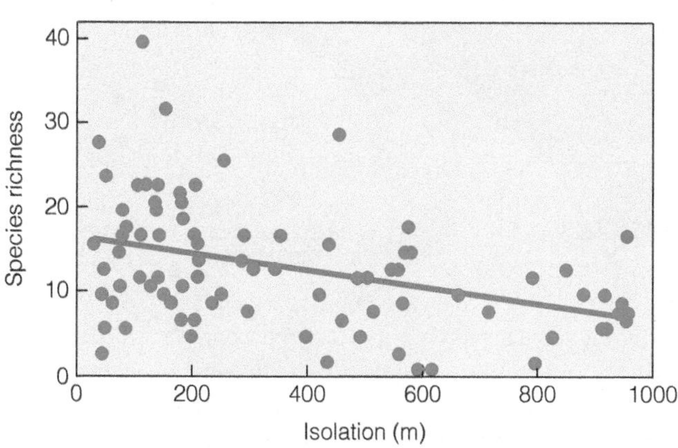

Figure 19.21 Influence of patch isolation on plant species colonization and species richness in a fragmented forest landscape in northern Belgium. (a) Estimated probability of occurrence of four herbaceous plant species in a target forest patch as a function of the average distance to the five nearest source populations of that species (occupied patches). (b) Relation between species dispersal mode and colonization success in the fragmented landscape. Ane, anemochore (dispersed by wind); bar, barochore (no dispersal adaptations); endo, endozoochore (dispersed by ingestion by birds and mammals); epi, epizoochore (dispersed by adherence to birds and mammals); myr, myrmecochore (dispersed by ants). The number of plant species (N=) representing each mode of dispersal is noted on the x-axis. (c) The relationship between species richness in early successional forest patches (age: 11 to 35 years) and patch isolation (calculated as the mean shortest [edge-to-edge] distance to the five nearest older forest patches). (Adapted from Honnay et al. 2002 and Jacquemyn et al. 2001.)

a corridor. Typically, corridors are strips of vegetation similar to the patches they connect but different from the surrounding matrix in which they are set (see this chapter, **Field Studies: Nick A. Haddad**). Many corridors are human made. Some may be narrow-line corridors, such as hedgerows and lines of trees planted as windbreaks, bridges over fast-flowing streams, highway median strips, and drainage ditches (**Figure 19.22a**). Wider bands of vegetation—called *strip corridors*—can consist of both interior and edge environments. Such corridors may be broad strips of woodlands left between housing developments, power-line rights-of-way, and belts of vegetation along streams and rivers (**Figure 19.22b**).

Corridors probably function best as travel lanes for individuals moving within the bounds of their home range, but when corridors interconnect to form networks, they offer dispersal routes for species traveling between habitat patches. They enhance the movement of organisms beyond what is possible through the adjacent matrix. By facilitating the movement among different patches, corridors can encourage gene flow between subpopulations and help reestablish species in habitats that have experienced local extinction. Corridors also act as filters, providing dispersal routes for some species but not others. Different-sized gaps in corridors allow certain organisms to cross while restricting others; this is the **filter effect**. For example, Gary Fry of the Norwegian Institute of Nature Research found that some butterfly species (Lepidoptera), despite their apparent high dispersal ability, are impeded by hedgerows

Figure 19.22 Examples of corridors: (a) hedgerow and (b) riverine vegetation.

(a)

(b)

FIELD STUDIES Nick A. Haddad

Department of Zoology, North Carolina State University, Raleigh, North Carolina

Corridors are thought to facilitate movement between connected patches of habitat, thus increasing gene flow, promoting reestablishment of locally extinct populations, and increasing species diversity within otherwise isolated areas. The potential utility of corridors in the conservation of biological diversity has attracted much attention, but often, the proposed value of corridors was based more on intuition than on empirical evidence. However, recent studies by ecologists, such as Nick Haddad of North Carolina State University, are providing valuable data on the role of corridors in facilitating species dispersal among otherwise isolated habitat patches on the larger landscape.

Haddad studies the influence of corridors on the dispersal and population dynamics of butterfly species. He has chosen butterflies as a focal organism because their life histories are well defined, they are generally associated with a narrow range of resources, and they are relatively easy to study at large spatial scales.

In cooperation with the United States Forest Service, Haddad and colleagues established eight 50-ha landscapes on the 1240 km^2 Savannah River Site, a National Environmental Research Park in South Carolina, to examine the effects of corridors on the dispersal and population dynamics of butterfly species. All landscapes are composed of mature (40- to 50-year-old) forest dominated by loblolly pine (*Pinus taeda*) and longleaf pine (*Pinus palustris*). Within each landscape, five early successional habitat patches were created by cutting and removing all trees and then burning the cleared areas.

Haddad compared movement rates from a 1-ha central patch—created at the center of each landscape—to four surrounding peripheral patches—created at the same time—each 150 m from the central patch (**Figure 1**). A 25-m wide corridor connected the central patch to one of the peripheral patches ("connected patches"). All other peripheral patches ("unconnected patches") were equal in size to the area of the connected patch plus the area of the corridor (1.375 ha); this controlled for effects of increased patch area in the connected patches. In unconnected patches, the corridor's area was added either as 75-m "wings" projecting from the sides of patches ("winged patches") or as additional habitat added to the backs of patches ("rectangular" patches). By comparing the rates of movement from the central patch into the connected patches versus movement into unconnected patches, the biologists were able to test the hypothesis that corridors function as conduits of movement.

To examine the effects of corridors on individual butterflies, Haddad tracked the movement of two species—the common buckeye (*Junonia coenia*) and the variegated fritillary (*Euptoieta claudia*). Both are common in early successional habitats and rare in mature forest habitats on the Savannah River Site. He marked naturally occurring butterflies in the central patches and recaptured marked individuals in the peripheral patches. All captured butterflies were marked, and the locations of initial and subsequent recaptures were recorded.

The common buckeye was three to four times more likely to move from center patches to connected patches than to unconnected patches, and the variegated fritillary was twice as likely to move down corridors than through forests when moving from the center patch (**Figure 2**). Neither butterfly was more likely to move to winged patches than to rectangular patches. The researchers found that the butterflies used the corridors for dispersal even though the corridors do not support resident populations, which indicated that the corridor habitat is lower in quality than the habitat patches it links.

The results clearly showed that the presence of a corridor facilitated the movement of both butterfly species between connected patches, even after controlling for patch size and shape. But in this experiment, the length of the corridor was

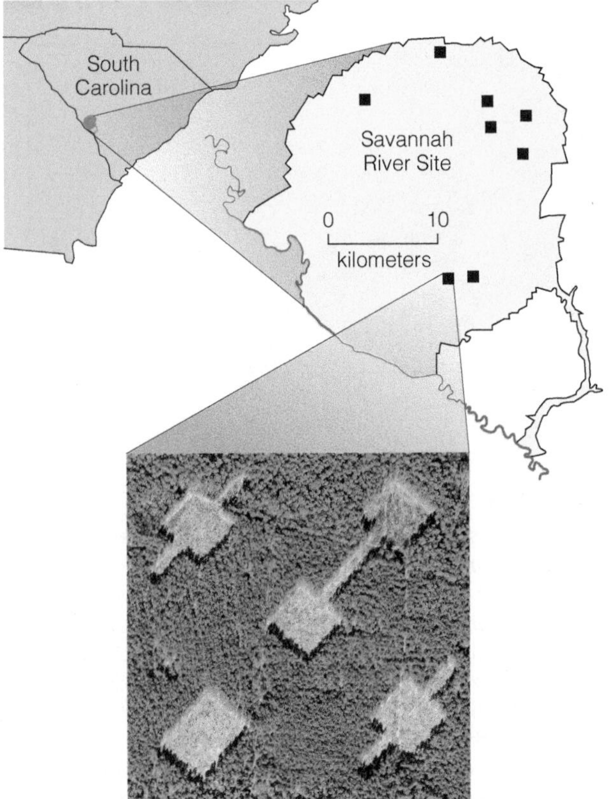

Figure 1 Map of experimental landscape locations and aerial photograph of one landscape, showing patch configuration.

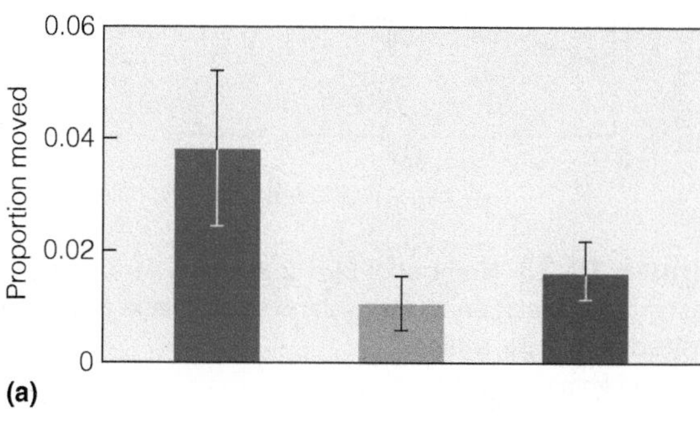

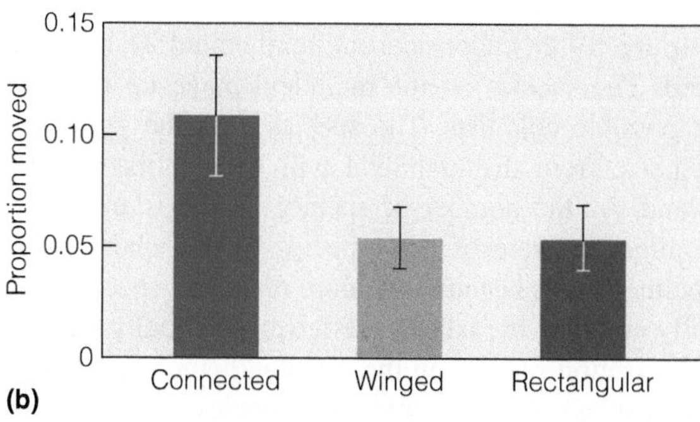

Figure 2 Movement rates of butterfly species between connected and isolated patches. *Junonia coenia* (a) and *Euptoieta claudia* (b) both moved between connected patches more often than between isolated patches. Data in both panels are means ±1 standard error for proportion of individuals marked in the central patch and recaptured in connected, winged, and rectangular peripheral patches. Go to www.masteringbiology.com to perform confidence intervals and t-tests.
(Haddad 1999a.)

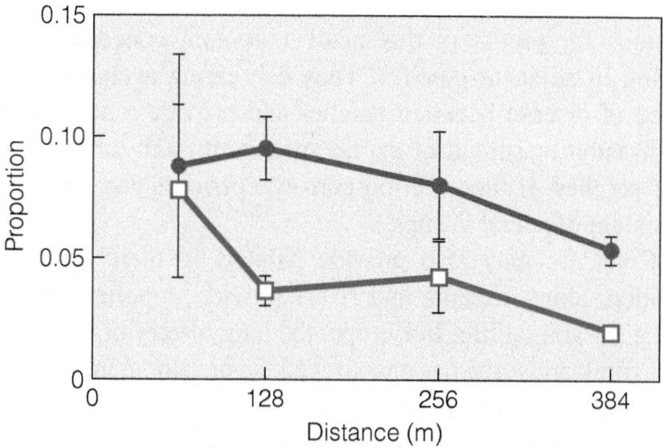

Figure 3 Mean proportion (±1 standard error) of individuals (males) *Junonia* marked in a patch who moved one of four distances to an adjacent patch. Circles indicate mean proportions moving between patches connected by a corridor; squares indicate mean proportions moving between unconnected patches.
(Based on Haddad, Nick M., "Corridor and Distance Effects on Interpatch Movement: A Landscape Experiment with Butterflies" Ecological Application, May 1999, Fig. 3A, p. 618.)

fixed at 150 m. How might the length of the corridor influence patterns of dispersal? To examine the influence of corridor length on patterns of butterfly dispersal, Haddad once again created experimental patches and corridors at the Savannah River Site. The experiment consisted of square patches of equal size (1.64 ha). The size of the square patches was fixed at 128 m on a side, and all patches were oriented in the same direction. Two characteristics of the patches were varied: interpatch distance and the presence or absence of a connecting corridor. Distances between patches were 64, 128, 256, or 384 m (half, one, two, or three times the width of the patch).

In this study, Haddad examined the movement patterns of the same two butterfly species, the common buckeye (*J. coenia*) and the variegated fritillary (*E. claudia*). Results of the mark-recapture studies showed that individuals of both species moved more frequently between patches connected by corridors than between unconnected patches, just as was observed in the previous study. However, interpatch movement was negatively related to interpatch distance (**Figure 3**), and the density of both butterfly species was significantly higher in connected patches.

The studies by Haddad and colleagues have expanded our understanding of the utility of corridors in patterns of dispersal, which is a key factor aiding the persistence of populations in a fragmented landscape. The unique design of these experiments has allowed us to test specific hypotheses regarding the influence of the size and shape of corridors on their utility in conservation efforts.

Bibliography

Haddad, N. 1999a. "Corridor and distance effects on interpatch movements: A landscape experiment with butterflies." *Ecological Applications* 9:612–622.

Haddad, N., and K. Baum. 1999b. "An experimental test of corridor effects on butterfly densities." *Ecological Applications* 9:623–633.

Haddad, N., and J. J. Tewksbury. 2005. "Low-quality habitat corridors as movement conduits for two butterfly species." *Ecological Applications* 15:250–257.

Tewksbury, J., D. Levey, N. Haddad, S. Sargent, J. Orrock, A. Weldon, B. Danielson, J. Brinkerhoff, E. Damschen, and P. Townsend. 2002. "Corridors affect plants, animals, and their interactions in fragmented landscapes." *Proceedings of the National Academy of Sciences of the United States of America* 99:12923–12926.

1. What life history characteristics might you expect for a species that benefits from the presence of corridors?

(see Figure 19.22a). Fry found that butterflies followed linear boundary features and crossed them only when there were gaps more than 1 meter wide. Even 1 meter high hedges without gaps were found to significantly reduce butterfly movement.

Corridors can have negative effects other that presenting barriers to dispersal. For example, corridors offer scouting positions for predators that need to remain concealed while hunting in adjacent patches. They can create avenues for the spread of disease between patches and provide a pathway for the invasion or spread of exotic species through the matrix to other patches. If they are too narrow, corridors can inhibit the movement of social groups.

Corridors may also provide habitats in their own right. Corridors along streams and rivers provide important riparian habitat for animal life. In Europe, the long history of hedgerows in the rural landscape has encouraged the development of typical hedgerow animal and plant communities (see Figure 19.22a). In suburban and urban settings, corridors provide habitat for edge species and act as stopover habitat for migrating birds.

Roads—corridors designed as dispersal routes for humans—have a negative impact by dissecting the landscape, impeding the movement of organisms by intersecting migratory and dispersal routes, and effectively dividing populations of many species (see this chapter, *Ecological Issues & Applications*). High-speed roads are a major source of mortality for wildlife ranging in size from large mammals to tiny insects.

19.6 The Theory of Island Biogeography Applies to Landscape Patches

As we have seen in the preceding sections, the community structure of a patch on the landscape is a function of its size, shape, and connectivity (position on the landscape relative to other patches). Yet how do these characteristics interact to determine community structure? The theory of island biogeography, first developed by Robert MacArthur (formerly of Princeton University) and Edward O. Wilson (Harvard University) in 1963, provides a framework to understand how size and connectivity can interact to influence patterns of species richness on patches within the landscape. The theory was first developed to explain patterns of species richness on islands. Early naturalist-explorers and biogeographers noted that large islands hold more species than small islands do (**Figure 19.23**). Johann Reinhold Forster, a naturalist on Captain Cook's second voyage to the Southern Hemisphere (1772–1775), noted that the number of different species found on islands depended on the island's size. The zoogeographer P. Darlington offered a rule of thumb: On islands, a 10-fold increase in land area leads to a doubling of the number of species.

The theory of island biogeography is quite simple: the number of species established on an island represents a dynamic equilibrium between the immigration of new colonizing species and the extinction of previously established ones

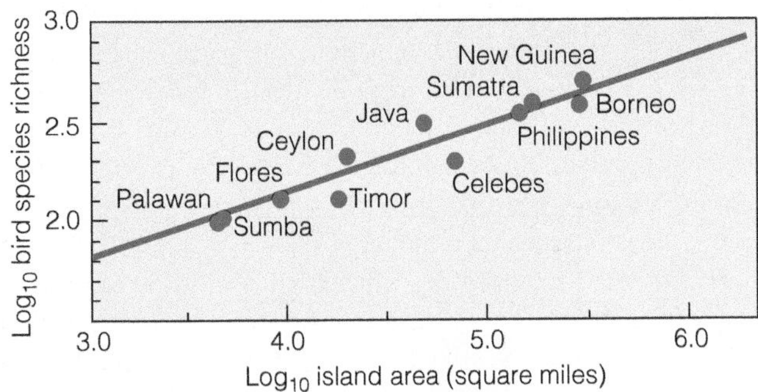

Figure 19.23 Number of bird species on various islands of the East Indies in relation to area (island size). The *x*- and *y*-axes are plotted on a $\log_{10}$ scale.
(Adapted from Preston 1962.)

(**Figure 19.24**). Consider an uninhabited island off the mainland. The species on the mainland make up the species pool of possible colonists. The species with the greatest ability to disperse from the mainland will be the first to colonize the island. As the number of species on the island increases, the immigration rate of new species to the island declines. The decline results because the more mainland species that successfully colonize the island, the fewer potentially new species for colonization remain on the mainland (the source of immigrating species). When all mainland species exist on the island, the rate of immigration will be zero.

If we assume that extinctions occur at random, the rate of species extinction on the island will increase with the number of species occupying the island based purely on chance. Other factors, however, will amplify this effect. Later immigrants may be unable to establish populations because earlier arrivals will already have used available habitats and resources. As the number

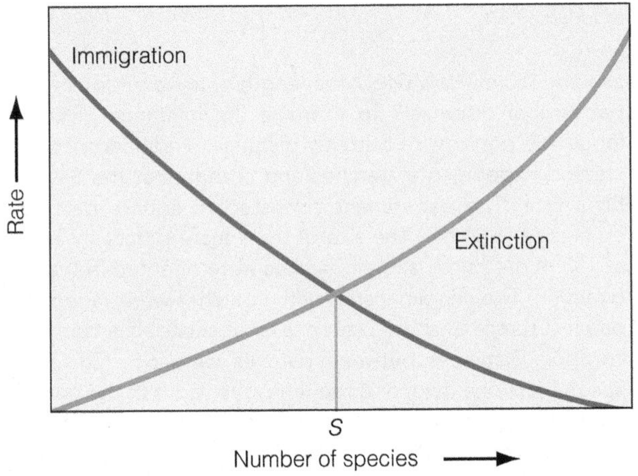

Figure 19.24 According to the theory of island biogeography, immigration rate declines with increasing species richness (*x*-axis) and extinction rate increases. The balance between rates of extinction and immigration (immigration rate = extinction rate) defines the equilibrium number of species (*S*) on the island.

of species increases, competition among the species will most likely increase, causing a progressive increase in the extinction rate. An equilibrium species richness (S) is achieved when the immigration rate equals the extinction rate (see Figure 19.24). If the number of species inhabiting the island exceeds this value, the extinction rate is greater than the immigration rate, resulting in a decline in species richness. If the number of species falls below this value, the immigration rate is greater than the extinction rate and the number of species increases. At equilibrium, the number of species residing on the island remains stable, although the composition of species may change. The rate at which one species is lost and a replacement species is gained is the **turnover rate**.

The distance of the island from the mainland and the size of the island both affect equilibrium species richness (**Figure 19.25**). The greater the distance between the island and the mainland, the less likely that many immigrants successfully complete the journey. The result is a decrease in the equilibrium number of species (Figure 19.25a). On larger islands, extinction rates, which vary with area, are lower because a greater area generally contains a wider array of resources and habitats. For this reason, large islands can support more individuals of each species as well as meet the needs of a wider variety of species. The lower rate of extinction on larger islands results in a higher equilibrium number of species as compared to smaller islands (Figure 19.25b).

Although the theory of island biogeography was applied initially to oceanic islands, there are many other types of "islands." Mountaintops, bogs, ponds, dunes, areas fragmented by human land use, and individual hosts of parasites are all essentially island habitats. As Daniel Simberloff of the University of Tennessee, one of the first ecologists to experimentally test the predictions of this theory, put it: "Any patch of habitat isolated from similar habitat by different, relatively inhospitable terrain traversed only with difficulty by organisms of the habitat patch may be considered an island."

So does island biogeography apply to landscape patches? As we have seen from the studies presented thus far in the chapter, the general patterns of increasing species richness with patch size (see Figure 19.20) and patch connectivity (see Figure 19.21) are in agreement with the predictions of the theory, but are these patterns a function of the dynamic equilibrium between colonization and extinction of species on habitat patches? Numerous studies have shown that the probability of colonization of a habitat patch on the landscape is influenced by its proximity to neighboring patches that function as a source of colonizers (as in Figure 19.21), but there are some fundamental differences between landscape patches and oceanic islands in presenting barriers to dispersal and colonization. Oceanic islands are terrestrial environments surrounded by an aquatic barrier to dispersal. They are inhabited by organisms of various species that arrived there by chance dispersal over a long period of time or represent remnant populations that existed in the area long before isolation. By contrast, the organisms that inhabit landscape patches are samples of populations that extend over a much wider area. These patch communities contain fewer species than are found in the larger area, and only a few individuals may represent each species. However, unlike oceanic islands, terrestrial landscape patches are associated with other terrestrial environments, which often present fewer barriers to movement and dispersal among patches. The matrix habitat in which the patches are embedded is often the determining factor in which species are successfully retained in isolated patches.

Although the theory of island biogeography has been influential in our understanding of the role of colonization and local extinction (extinction of population on an individual patch), the emergence of metapopulation theory has proven to provide a better framework for understanding patterns of community structure on landscapes.

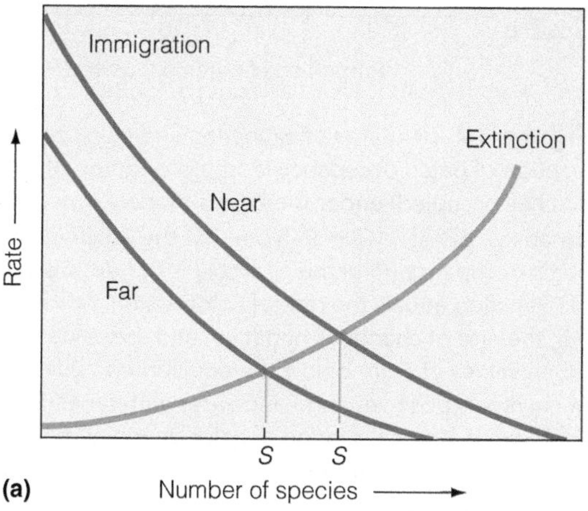

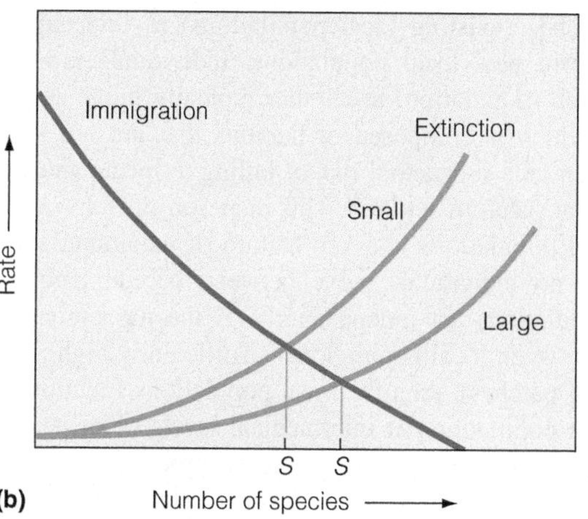

Figure 19.25 (a) Immigration rates are distance related. Islands near a mainland have a higher immigration rate and associated equilibrium species richness (S) than do islands distant from a mainland. (b) Extinction rates relate to area and are higher on small islands than on large ones. The equilibrium number of species varies according to island size, and larger islands have greater equilibrium species richness than do smaller islands.

19.7 Metapopulation Theory Is a Central Concept in the Study of Landscape Dynamics

In contrast to island biogeography, which examines the species richness of a single habitat patch as a function of the colonization by new species and local extinction of current resident species, metapopulation theory examines the colonization and local extinction of local populations of a given species on the array of patches (potential habitats) embedded on the broader landscape.

For a given species, we can view the landscape as patches of potential habitat varying in size, quality, and degree of isolation from one another. Each has the potential to support a distinct, partially isolated subpopulation (local population) possessing its own population dynamics, and links to varying degrees to other patches on the landscape by individuals moving among patches (dispersal). This is the concept of the metapopulation introduced in Chapter 8 (Section 8.2, Figure 8.8). The concept of the metapopulation provides a framework for examining the dynamics of a species distributed as discrete populations on the larger landscape.

Metapopulation dynamics differs from our discussion of population dynamics thus far in that it is governed by two sets of processes operating at two distinct spatial scales. At the scale of the individual patch, individuals move and interact with one another in the course of their routine activities. Population growth and regulation at the local scale are governed by the demographic processes of birth and death discussed in Chapter 9.

At the scale of the landscape—the metapopulation scale—dynamics are governed by the interaction of local populations, namely the process of dispersal and colonization. In theory, all local populations (population on a single patch) have a probability of extinction, so the long-term persistence of the metapopulation depends on the process of (re)colonization.

Colonization involves the movement of individuals from occupied patches (existing local populations) to unoccupied patches to form new local populations. Individuals moving from one patch (population) to another typically move across the matrix, which is composed of habitats that are not suitable, and often face substantial risk of failing to locate another suitable habitat patch to settle in. This dispersal of individuals between local populations is a key feature of metapopulation dynamics. If no individuals move between habitat patches, the local populations act independently. If the movement of individuals between local populations is sufficiently high (free access among patches), then the local populations function as a single large population. At intermediate levels of dispersal, however, a dynamic emerges in which the processes of local extinction and recolonization achieve some balance where the metapopulation exists as a shifting mosaic of occupied and unoccupied habitat patches. The metapopulation concept is therefore closely linked with the processes of population turnover—extinction and establishment of new populations—and the study of metapopulation dynamics is essentially the study of the conditions under which these two processes are in balance.

The fundamental idea of metapopulation persistence is a dynamic balance between the extinction of local populations and recolonization of empty habitat patches. In 1970, Richard Levins of Harvard University proposed a simple model of metapopulation dynamics, in which metapopulation size is defined by the fraction of (discrete) habitat patches (P) on the landscape occupied at any given time (t). The change in the fraction of habitat patches occupied by local populations through time ($\Delta P/\Delta t$) can therefore be defined as the difference between the rates of colonization (C) and extinction.

$$\Delta P/\Delta t = C - E$$

We can think of metapopulation growth ($\Delta P/\Delta t$) in a manner analogous to our discussion of population growth in Chapter 9, in which the change in the population (ΔN) over a given time interval (Δt) can be expressed as the difference between the rates of birth and death ($\Delta N/\Delta t = b - d$). In the case of the metapopulation, the processes of birth and death are replaced by the rates of colonization (C) and extinction (E).

In Levins's original model (see **Quantifying Ecology 19.1**), each subpopulation occupying a habitat patch has a fixed probability of extinction, and therefore the rate of at which occupied patches go extinct (E) increases linearly with P (**Figure 19.26**). The rate of colonization is a function of

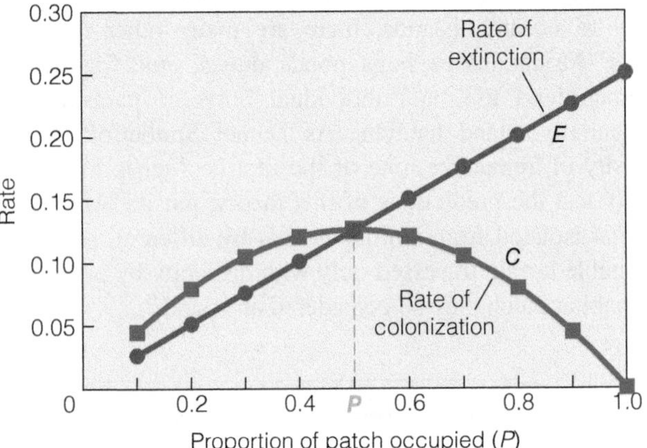

Figure 19.26 Rates of extinction and colonization as a function of patch occupancy (P, the proportion of available habitat patches occupied) under the Levins model of metapopulation dynamics: $\Delta P/\Delta t = C - E$. Note that the equilibrium value of patch occupancy (**P**) in this example is 0.5. At this value, the rate of extinction equals the rate of colonization. At values of P above 0.5, the rate of change is negative, and the value of P declines. When values of P are below the equilibrium value (<0.5), the rate of change is positive, and P increases with time. (See Quantifying Ecology 19.1 for calculation of colonization and extinction curves and equilibrium value of P.)

Interpreting Ecological Data

Q1. *Why is the rate of change ($\Delta P/\Delta t$) negative for values of P above the equilibrium (**P**)? Answer in terms of the influence on the rates of extinction and colonization.*

Q2. *Conversely, why is the rate of change ($\Delta P/\Delta t$) positive for values of P below the equilibrium (**P**)?*

QUANTIFYING ECOLOGY 19.1 Model of Metapopulation Dynamics

In 1970, Richard Levins proposed a simple model of metapopulation dynamics in which metapopulation size is defined by the fraction of (discrete) habitat patches (P) occupied at any given time (t). Within a given time interval, each subpopulation occupying a habitat patch has a probability of going extinct (e). Therefore, if P is the fraction of patches that is occupied during the time interval, the rate at which subpopulations go extinct (E) is defined as

$$E = eP$$

The rate of colonization of empty patches (C) depends on the fraction of empty patches ($1 - P$) available for colonization and the fraction of occupied patches providing colonists (P), multiplied by the probability of colonization (m), a constant that reflects the rate of movement (dispersal) of individuals between habitat patches. Therefore, the colonization rate is

$$C = [mP(1 - P)]$$

The change in metapopulation, defined as the fraction of habitat patches occupied by local populations through time ($\Delta P/\Delta t$) can therefore be defined as the difference between the rates of colonization (C) and extinction (E):

$$\Delta P/\Delta t = C - E$$

or

$$\Delta P/\Delta t = [mP(1 - P)] - eP$$

For any given values of e and m, we can plot the rates of extinction (E) and colonization (C) as a function of the proportion of habitat patches occupied (P; see Figure 19.26). The extinction rate increases linearly with P, and the colonization rate forms a convex curve, initially rising with the proportion of patches occupied and then declining as the proportion approaches 1.0 (all patches are occupied). The value of P where the lines cross represents the equilibrium value, $\mathbf{P}$. At this value of P, the extinction and colonization rates are equal ($E = C$) and the metapopulation growth rate is zero ($\Delta P/\Delta t = 0$). It is an equilibrium value because when the fraction of patches occupied (P) is below this value ($\mathbf{P}$), the rate of colonization exceeds the rate of extinction and the number of occupied habitat patches increases. Conversely, if the value of P exceeds $\mathbf{P}$, the rate of extinction exceeds the rate of colonization and the size of the metapopulation (number of occupied patches) declines.

By setting $\Delta P/\Delta t$ equal to zero, we can solve for the equilibrium value $\mathbf{P}$:

$$\frac{\Delta P}{\Delta t} = 0$$

$$[mP(1 - P)] - eP = 0$$

Using simple algebraic substitution, we can solve for P (the equilibrium value $\mathbf{P}$)

$$mP(1 - P) = eP$$

by dividing both sides of the equation by P

$$m(1 - P) = e$$

then dividing both sides of the equation by m

$$1 - P = \frac{e}{m}$$

and subtracting 1 from both sides

$$-P = -1 + \frac{e}{m}$$

then multiplying both sides of the equation by -1

$$\mathbf{P} = 1 - \frac{e}{m}$$

For the metapopulation to persist, the equilibrium value $\mathbf{P}$ must be greater than zero, so the probability of extinction (e) does not exceed the probability of colonization m.

the proportion of patches already occupied, initially rising with the proportion of patches occupied (greater abundance of populations as source of colonists) and then declining as the proportion approaches 1.0 (all patches occupied; see Figure 19.26). The value of P where the lines cross represents the equilibrium value, $\mathbf{P}$. At this value of P, the extinction and colonization rates are equal ($E = C$), and the metapopulation growth rate is zero ($\Delta P/\Delta t = 0$). It is an equilibrium value because when the fraction of patches occupied (P) is below this value ($\mathbf{P}$), the rate of colonization exceeds the rate of extinction and the number of occupied habitat patches increases. Conversely, if the value of P exceeds $\mathbf{P}$, the rate of extinction exceeds the rate of colonization and the size of the metapopulation (number of occupied patches) declines. So just as in the logistic model—in which the population density (N) tends to the equilibrium population size represented by the carrying capacity (K)—in the metapopulation model, the metapopulation density, P (proportion of patches occupied), tends to the equilibrium metapopulation size represented by $\mathbf{P}$ (see Quantifying Ecology 19.1).

The Levins model makes several assumptions for the sake of mathematical simplicity. It assumes that all patches are equal in size and quality as habitat. It also assumes that each unoccupied patch has an equal probability of being colonized, and the local population inhabiting each occupied patch has an equal probability of going extinct. In reality, each of these assumptions may be (and most likely is) unrealistic for naturally occurring metapopulations. Local populations often differ in their susceptibility to extinction. Habitat patches differ in size and spatial position relative to each other, and these variations influence the probability of colonization.

The work of Oskar Kindvall and Ingemar Ahlen of the Swedish University of Agricultural Sciences illustrates the importance of the location and size of habitat patches on metapopulation dynamics. Kindvall and Ahlen conducted a study of the metapopulation dynamics of the bush cricket (*Metrioptera bicolor*) in the Vomb valley of Sweden. The bush cricket, a medium-sized (12–19 millimeter [mm]), flightless katydid (**Figure 19.27a**), inhabits grass and heathland patches of varying size and isolation within a landscape dominated by pine forest. The metapopulation was surveyed in 1986, 1989, and 1990. **Figure 19.27b** shows the distribution of potential habitat patches within the valley. During the study period, the proportion of available patches occupied varied between 72 percent and 79 percent. Patterns of occupancy were directly related to characteristics of the habitat patches and their influence on patterns of extinction and colonization.

Patches that were colonized during the observation period were less isolated than patches that were not colonized (**Figure 19.27c**). The chance of colonization decreases dramatically when the interpatch distance exceeds about 100 m. The longest interpatch distance recorded for colonization by the crickets was 250 m over agricultural fields. Colonized and uncolonized patches did not differ significantly with

(a)

(b)
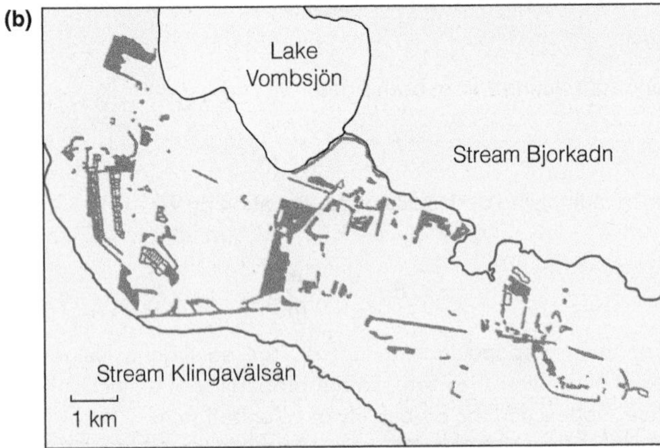

(c)
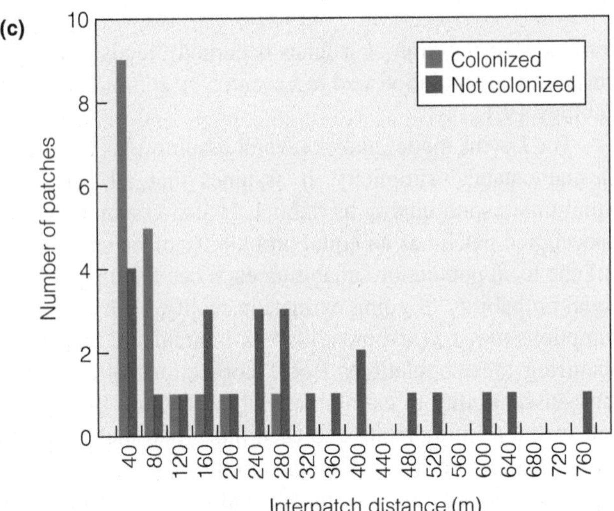

(d)
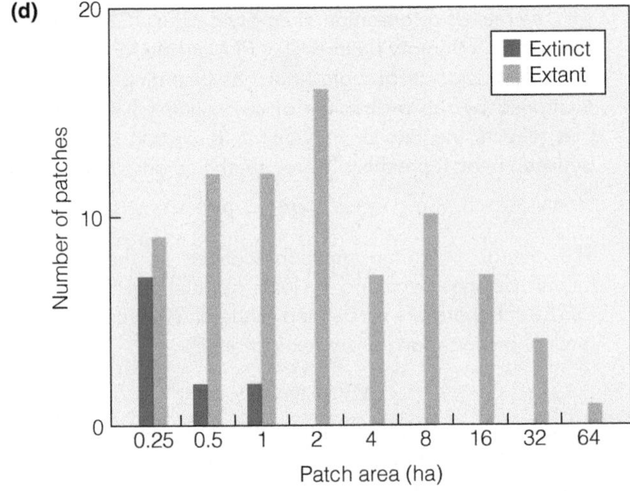

Figure 19.27 The Swedish metapopulation of bush cricket (*Metrioptera bicolor*) (a) is located south of Lake Vombsjön, which is surrounded by two streams (b). Patches of grassland with local populations that did not become extinct during the period of study are shown in brown. Other areas in tan were occupied either occasionally or not at all during the survey period. (c) The frequency distribution of interpatch distance between colonized and uncolonized patches during the survey period. (d) The frequency distribution of areas of patches with existing (extant) populations and patches where the local population became extinct during the survey period.
(Adapted from Kindvall 1992.)

Interpreting Ecological Data

Q1. *In Figure 19.27c, interpatch distance is a measure of the spatial isolation of the habitat patches from other patches of suitable habitat on the landscape. What is the maximum interpatch distance beyond which dispersal and colonization of bush crickets does not appear to be possible?*

Q2. *In Figure 19.27d, the total number of habitat patches in each size category (x-axis) that were occupied by bush cricket populations during the study period can be calculated by adding the values of occupied (extant) and extinct. Which size category of habitat patch has the highest value for occupancy during the study period? Which size category has the largest existing value for occupancy? What patch size appears to represent a threshold for the long-term persistence of local cricket populations?*

respect to patch size; however, patch size did influence the probability of extinction.

From 1986 to 1990, a total of 18 local populations became extinct. Habitat destruction or alteration, such as grazing, house building, or the application of pesticides, caused 6 of these extinctions, however the other 12 extinctions occurred on patches displaying no noticeable change in habitat. These 12 patches were significantly smaller than those with persisting populations (**Figure 19.27d**). The risk of local extinction apparently increases with decreasing patch size. The probability of local population extinction, however, appeared not to be influenced by patch isolation.

The influence of patch size on the persistence of local populations of bush cricket was found to be indirect, through the influence of patch area on the size of local populations. The researchers found a significant positive relationship between patch size and local population size (**Figure 19.28**). Data revealed that the risk of local extinction increases for a patch size of less than one-half ha (see Figure 19.27d), which corresponds to a critical population size of about 12 males (see Chapter 11, *Ecological Issues & Applications* for discussion of the minimum viable population concept). In 1990, only 67 percent of the suitable patches were larger than one-half ha; however, 79 percent of the habitat patches were occupied. The fraction of habitat patches occupied (*P*) was maintained at a higher value than would be expected under isolation of patches as a result of recolonization after local extinction. The fraction of occupied patches (*P*) was fairly constant during the survey period (five years); however, different patches were occupied each year.

As seen in the preceding example, both patch size and isolation influence local population dynamics. That is to say, metapopulation persistence depends simultaneously on patch size and isolation. The interaction between patch area and isolation is illustrated in the work of C. D. Thomas and T. M. Jones of Imperial College, England. Thomas and Jones examined patterns of extinction and colonization of patches of grassland habitat in the North and South Downs of southern England by the skipper butterfly (*Hesperia comma*; **Figure 19.29a**). The probability of local extinction declined with increasing patch area and increased with isolation. Conversely, the probability of colonization increased with patch area and declined with isolation from other local populations. The result, shown in **Figure 19.29b**, is a clear pattern of increasing probability of patch occupancy with declining isolation and increasing patch area, with increasing patch size compensating for increasing degree of isolation from neighboring populations.

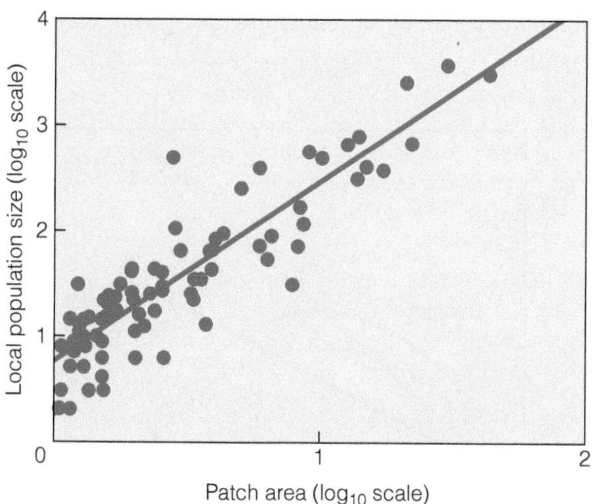

Figure 19.28 Relationship between local population size and patch area in the metapopulation of bush crickets, *Metrioptera bicolor* (see Figure 19.27 for map of site). Values for both population size and patch have been log transformed. (Adapted from Kindvall 1992.)

The influence of patch size and isolation on metapopulation dynamics is similar to the patterns observed for island size and isolation in island biogeography theory (see Figure 19.25). Isolation (distance from neighboring patches) decreases the probability of colonization and therefore decreases the rate of patch colonization (*C*). Increasing patch size decreases the probability of extinction and therefore the rate of extinction (*E*). Shifts in the rates of colonization and extinction function to change the equilibrium value of *P* (the proportion of patches occupied; **Figure 19.30**).

19.8 Local Communities Occupying Patches on the Landscape Define the Metacommunity

Some researchers have extended the framework of metapopulations to examine the dynamics involving interactions among local communities. Each habitat patch on the landscape is composed of a set of species that define the local community. The set of local communities (sets of species that interact with one another), occurring in discrete patches, linked by dispersal define the **metacommunity**. The crucial elements are multiple potentially interacting species, multiple patches at which interaction may occur, and dispersal by at least some of the species to link interactions among the sites. It is this focus on the community that creates the distinction between the concepts of metapopulation and metacommunity. The basic focus in metapopulation theory is to examine what determines the persistence of the metapopulation in a system of connected habitat patches, whereas the basic focus in metacommunity theory is to examine what regulates the co-existence of multiple species in that same system of connected habitat patches.

Just as with metapopulations, the analysis of metacommunity dynamics involves examining processes that occur at the landscape scale. The interaction among communities is influenced by size, shape, and spatial arrangement of the habitat patches and the matrix in which they are embedded. In general, small fragmented communities support few species at lower trophic levels, whereas large communities support greater species diversity and higher trophic levels. The spatial

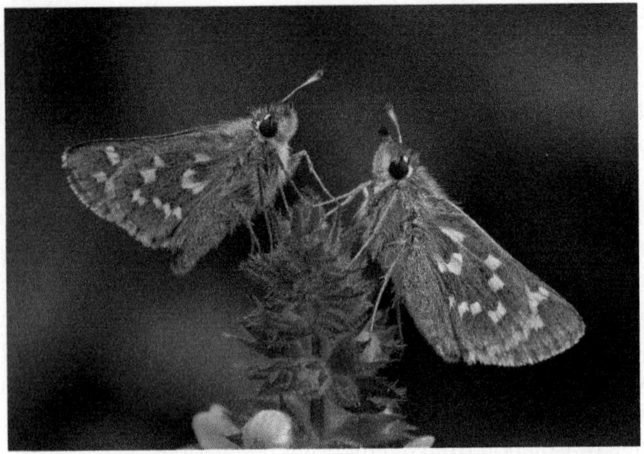

(a)

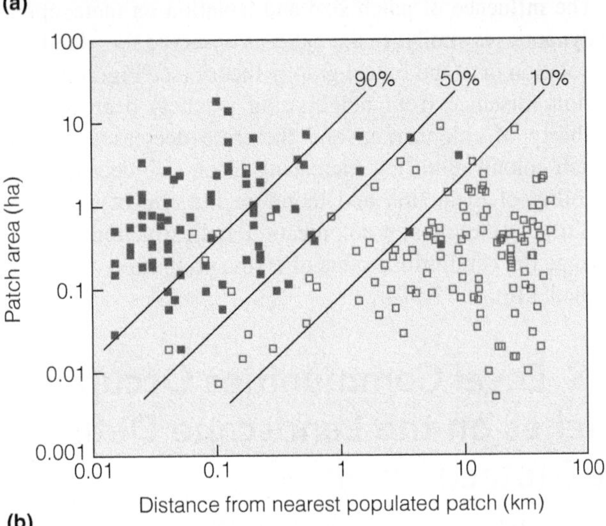

(b)

Figure 19.29 (a) Distribution of occupied (solid) and vacant (open) habitat patches of the skipper butterfly (*Hesperia comma*) in southern Britain (1991) in relation to patch area and isolation from the nearest populated patch. (b) Lines give the 90 percent, 50 percent, and 10 percent probabilities of occupancy. Note the compensatory effect of isolation and patch area on occupancy. (Adapted from Thomas 1993.)

Interpreting Ecological Data

The three lines in Figure 19.29b define the combined values of patch isolation (distance from nearest populated patch) and patch size (area) that include 90 percent, 50 percent, and 10 percent of the occupied habitat patches (solid squares) on the landscape during the study period. Use these lines to answer the following questions.

Q1. *What is the approximate probability that a 10-ha habitat patch 1 km from the nearest populated patch will be occupied by skipper butterflies?*

Q2. *What is the approximate area necessary to provide a 50 percent probability that a habitat patch 1 km from any neighboring populated patches will be colonized and occupied by a local population of skipper butterflies?*

configuration and species composition of patches influence dispersal within the metacommunity, and therefore the processes of colonization among the local communities. In turn, the species composition of patches will determine the nature of species interactions, which influences the potential for new species to colonize a patch as well as the ability of current species to persist. The presence of competitors in a habitat patch may limit the ability of a species to successfully colonize or increase its probability of extinction should it be able to initially become established in the patch. The absence of mutualistic species may have a similar effect. For example, should a species of flowering plant successfully disperse to a habitat patch, and if its pollination and successful reproduction is dependent on the presence of certain pollinating species (such as insects or birds), the absence of those species in the patch will limit its successful colonization. Successional changes will also influence species composition of some patches relative to others as environmental conditions of a patch change through time (see Chapter 18).

The model of metapopulation dynamics views the equilibrium number of patches occupied by a given species as the balance between colonization and extinction of local populations. The model of metacommunity dynamics examines how species interactions influence the metapopulation dynamics of the component species that make up the local communities. By providing a framework for examining species distributed as discrete populations within the larger landscape, both the concepts of metapopulation and metacommunity are central to the study of landscape dynamics.

19.9 The Landscape Represents a Shifting Mosaic of Changing Communities

Unlike the artist's mosaic shown in Figure 19.2 with its fixed pattern, the mosaic of communities defining the landscape is ever changing. Disturbances—large and small, frequent and infrequent—alter the biological and physical structures of communities making up the landscape, giving way to the

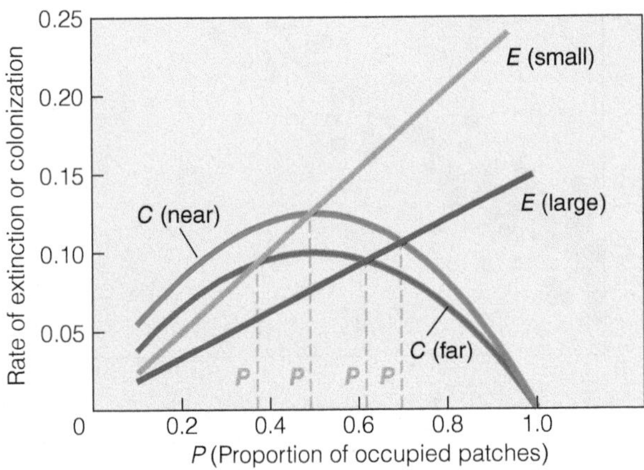

Figure 19.30 Changes in the equilibrium proportion of habitat patches occupied (***P***) for habitat patches of differing size (small and large) and isolation (near and far). Increasing patch size functions to decrease the rate of extinction (*E* large), and increasing distance from neighboring patches functions to decrease the rate of colonization (*C* far).

process of succession. This view of the landscape suggests a **shifting mosaic** composed of patches, each in a phase of successional development (**Figure 19.31**). The ecologists F. Herbert Borman and Gene Likens applied this concept to describe the process of succession in forested landscapes, using the term *shifting-mosaic steady state*. The term *steady state* is a statistical description of the collection of patches and thus refers to the average state of the forest. In other words, the mosaic of patches shown in Figure 19.31 is not static. Each patch is continuously changing. Disturbance causes the patches in the mosaic that are currently classified as late successional to revert to early successional. Patches currently classified as early successional undergo shifts in species composition, and later successional species come to dominate. Although the overall mosaic is continuously changing, the average composition of the landscape (average overall patches) may remain fairly constant—in a steady state. This example of a continuously changing population of patches that remains fairly constant when viewed collectively rather than individually is similar to the concept of a stable age distribution presented in Chapter 9 (Section 9.7). In a population with a stable age distribution, the proportion of individuals in each age class remains constant even though individuals are continuously entering and leaving the population through births and deaths.

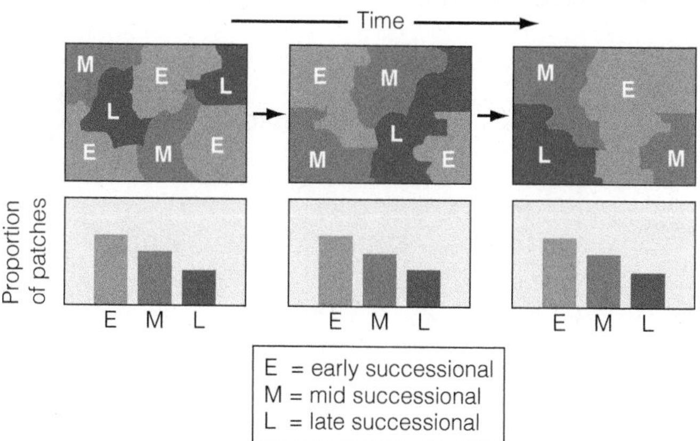

Figure 19.31 Representation of the concept of shifting mosaic steady state in a forested landscape where patches are in varying stages of succession (early, mid, and late successional stages; see Chapter 18). The characteristics of the individual forest patches change through time, but the collective properties of the patches remain relatively constant—in a steady state.

Returning to the Virginia countryside of Figure 19.1, we can now see the mosaic of patches—forest, fields, golf course, hedgerows, pine plantation, pond, and human habitations—not as a static image but as a dynamic landscape. Most of the forested lands were once fields and pastureland. During the late 19th and early 20th centuries, when agriculture in the region declined, farmers abandoned their fields and the land reverted to forest (see Chapter 18, *Ecological Issues & Applications*). The current mosaic of land cover is maintained by active processes, many of which are forms of human-induced disturbance. Within these patches, the communities function as islands, some bridged by corridors. Their populations are part of larger metapopulations, which in turn are components of the metacommunity linked by the dispersal of individuals. As time passes, the landscape will continue to change. Patterns of land use will shift boundaries, succession will alter the structure of communities, and natural disturbances such as fire and storms will form new dynamic patches within the mosaic.

ECOLOGICAL Issues & Applications

Corridors Are Playing a Growing Role in Conservation Efforts

Given the ever-growing pressures placed on lands by the human population, and fragmentation of existing habitats, preservation of biological diversity depends more and more on the establishment of designated protected areas. According to the International Union for the Conservation of Nature (IUCN) there are currently more than 11,000 strictly protected areas (categories Ia, Ib, and II of the IUCN classification that includes nature reserves, wilderness areas, and national parks) that have been designated worldwide, covering some 6.16 million km^2. Additionally, more than 55,000 partially protected areas combine to cover an additional 5.67 million km^2. That may seem like a large area of land, but most of the protected areas are relatively small. More than half of current nature reserves cover an area of 100 km^2 or less (**Figure 19.32**). Although these smaller protected areas may be adequate for maintaining populations of smaller species, larger species of herbivores and carnivores require much larger land areas to maintain viable populations (see discussion of minimum viable populations in Chapter 11, *Ecological Issues & Applications*). The combined total of protected lands accounts for only 11.9 percent of Earth's total land surface, and current marine preserves account for less than 1 percent of marine environments.

With few exceptions, most large tracts of land that will function as future protected areas have already been established and fall within the current system of nature reserves. However, new and smaller reserves are being established throughout the world, and lands under limited protection (such as the United States National Forests) are continuously being reclassified into categories of increased protection. For example, in 2002 more than 10,000 acres of the George Washington National Forest (Virginia) were designated as wilderness areas (Priest and Three Ridges), and in 2011 the Elkhorn Ridge Wilderness in California became the newest area within the Bureau of Land Management system that was designated as a wilderness area, which is the highest designation of protection for federal lands within the United States. However, many

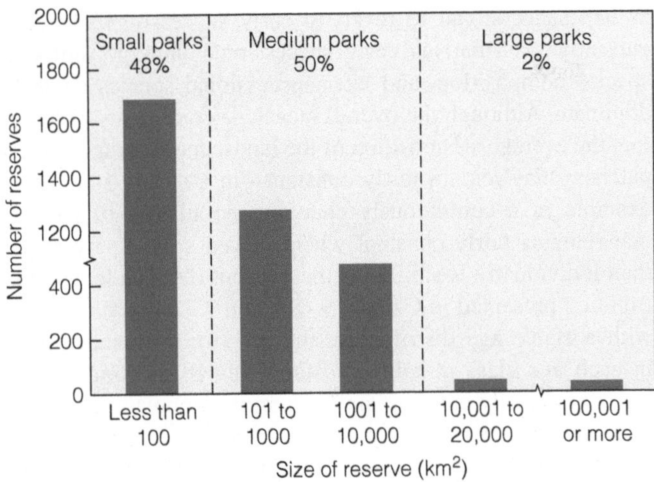

Figure 19.32 Size distribution of nature reserves around the world, as listed by the International Union for the Conservation of Nature (IUCN).

current conservation efforts are drawing on basic principles of landscape ecology and focusing efforts on working with the existing protected lands, providing buffer zones and corridors that enhance their conservation value.

Strategies exist for aggregating small nature reserves and other protected areas into larger conservation blocks. Nature reserves are often embedded in a larger matrix of habitat managed for resource extraction, such as timber harvest, grazing, or farmland. If protecting biological diversity can be incorporated as a secondary priority into the management plan of these lands, a greater representation of species and habitats can be protected.

An approach that has gained growing attention in conservation efforts is to view nature reserves in an area or region as habitat patches on the broader landscape and focus benefits that can be gained by increasing connectivity. The approach is to link isolated protected areas into one large system through the use of habitat corridors—which are areas of protected land running between the reserves. Such corridors can facilitate the dispersal of plants and animals from one reserve to another. Corridors may assist species that migrate seasonally to different habitats to obtain food or breed.

This principle was put into practice in Costa Rica to link two wildlife reserves, the Braulio Carrillo National Park and La Selva Biological Station (**Figure 19.33**). A 7700-ha corridor of forest several kilometers wide, known as the La Selva Protection Zone, was set aside to provide a link that allows at least 75 species of birds to seasonally migrate between the two conservation areas. Subsequently, the corridor has been incorporated into the National Park.

Another example of the conservation benefits of establishing landscape corridors comes from the southeastern United States. In 1988, the Nature Conservancy together with the United States Forest Service acquired Pinhook Swamp, a tract of approximately 24,000 ha that links the Okefenokee National Wildlife Refuge and the Osceola National Forest (**Figure 19.34**). The Okefenokee National Wildlife Refuge spans some 162,000 ha across the borders of the states of Georgia and Florida and is the largest wildlife refuge in the eastern United States. Sixteen kilometers to the south, the 65,000-ha Osceola National Forest is an area of wetlands, swamps, and upland pine forests. On its own, the Osceola National Forest is not large enough to support viable populations of red-cockaded woodpecker, black bear, and other species requiring large home ranges. Consolidation of these lands through the establishment of a corridor provides a contiguous area of habitat in excess of 250,000 ha and better sustains populations of these larger species.

In some rare cases, actions are being taken to link established protected areas that go beyond the limited use of

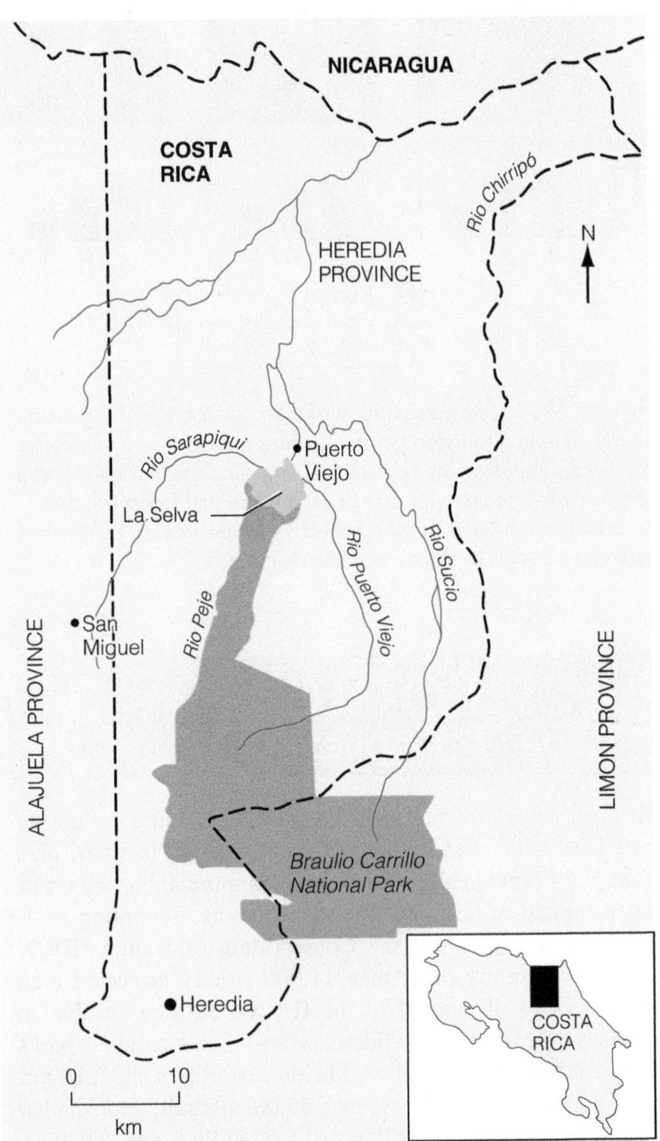

Figure 19.33 The corridor linking La Selva Biological Station and Braulio Carrillo National Park in Costa Rica. Originally known as the La Selva Protection Zone, the area has since been incorporated into the Braulio Carrillo National Park.
(R.M. Timm et al. 1989, "Mammals of the La Selva-Braulio Carrillo Complex, Costa Rica" United States Department of the Interior Fish and Wildlife Service, North American Fauna 75 Fig 1, pg. 5.)

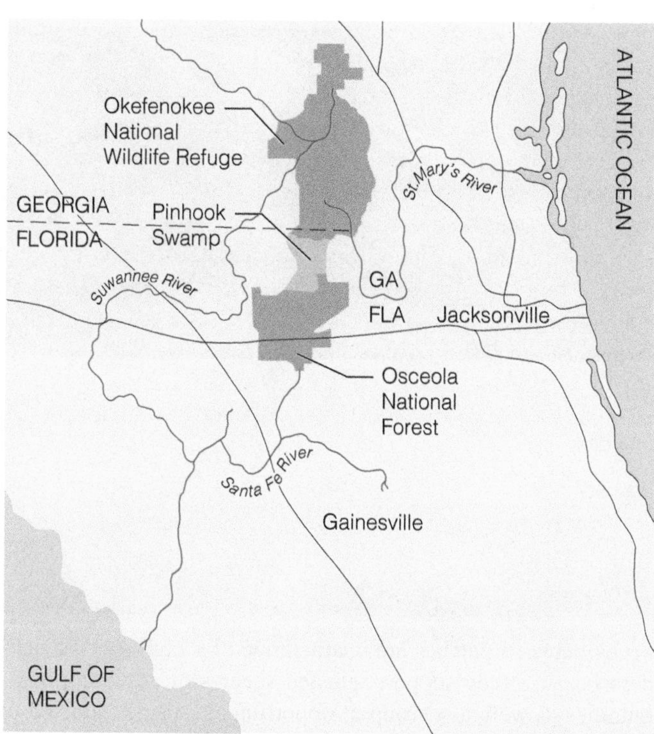

Figure 19.34 The acquisition and protection of Pinhook Swamp in Florida maintains a natural corridor between the Okefenokee National Wildlife Refuge and the Osceola National Forest.
(Based on A.F. Bennett, 2003. "Linkages in the landscape" IUCN Forest Conservation Programme. Conserving Forest Ecosystems Series 1. Fig. 9-3, pg 183.)

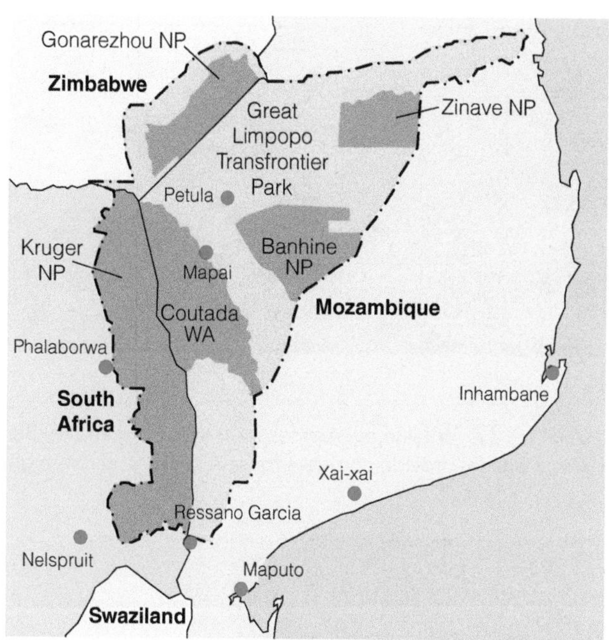

Figure 19.35 Map of the Great Limpopo Transfrontier Park, an international conservation effort that involves linking three existing national parks: Kruger National Park (South Africa), Gonarezhou National Park, (Zimbabwe), and Coutada Wildlife Area, Banhine and Zinave National Parks (Mozambique).

corridors. Efforts are under way in southern Africa to establish the Great Limpopo Transfrontier Park (**Figure 19.35**), an international conservation effort that involves linking three existing national parks: Kruger National Park (South Africa), Gonarezhou National Park, (Zimbabwe), and Coutada Wildlife Area, Banhine and Zinave National Parks (Mozambique). Establishing the transfrontier park would create a conservation area measuring 100,000 square kilometers, making it one of the world's largest contiguous nature reserves.

The establishment of corridors to increase landscape connectivity does not always require the acquisition of large tracts of land or international agreements. In many cases, connectivity can be increased by removing or bypassing existing barriers to movement that have been constructed to facilitate human activities. For example, seasonal movement of pronghorn antelope (*Antilocapra americana*) between Grand Teton National Park and the Upper Green River Valley in northwestern Wyoming is the longest remaining migration of any land mammal in North America. During this migration to their winter range, the pronghorn face numerous natural obstacles including rivers and a high mountain pass. But some of the greatest obstacles are human made. Unlike deer and elk, pronghorn antelope will not readily jump over fences, and as a result, the fences that border ranches along the migration route limit movement and are a source of mortality (**Figure 19.36**). Combined efforts by U.S. federal agencies, nongovernmental conservation organizations, and private citizens have focused on removing fences or at least establishing openings to allow for movement of the herds along the migration route.

Some of the most dangerous places along the migratory path are highways, where hundreds of animals are killed each year while attempting to cross these human-made barriers. Newly established wildlife overpasses and underpasses are now enabling pronghorn and other animals to safely cross these roadways (**Figure 19.37**). The practice of establishing wildlife overpasses is an example of a simple solution to breaking down barriers to wildlife movement that is being widely applied within nature reserves and public highways.

Figure 19.36 Unlike deer and elk, pronghorn antelope do not readily jump over fences. Fences represent a significant barrier to seasonal migration, and the removal of fences along the migration route has greatly aided in opening a migration route between summer and winter ranges.

(a)

(b)

Figure 19.37 Wildlife overpasses in (a) Wyoming and (b) Banff National Park (Alberta, Canada) provide corridors for safe passage across highways.

SUMMARY

Creating Landscapes 19.1

A variety of factors give rise to landscapes including abiotic, biotic, natural disturbances, and human activities. Climate and geology interact to define the land. Abiotic and biotic factors interact to determine species composition. Disturbance, a discrete event that disrupts communities and populations, also initiates succession and creates diversity. Natural disturbances can be a powerful force for change in the landscape. Fire is a natural large-scale disturbance that has both beneficial and adverse effects. Other major natural disturbance regimes include wind, floods, storms (especially hurricanes), and animals. Through their activities they can transform landscape mosaics. Major human-induced disturbances include logging, mining, agriculture, and development function to fragment the landscape.

Landscape Pattern 19.2

A landscape is a spatially heterogeneous area composed of patches embedded within a matrix. Boundaries are the place where the edge of one patch meets the edge of another. The proximity of patches to each other influences landscape connectivity, which is the ability for interactions to occur among patches. A landscape is not defined by its size, but by its scale relative to the organism or process of interest.

Boundaries and Edges 19.3

The place where the edges of two different patches meet is a boundary. A boundary may be produced by a sharp environmental change, such as a topographical feature or a shift in soil type—or created by some form of disturbance that is limited in extent and changes through time. Some boundaries are narrow and abrupt; others are wide and form a transition zone, or ecotone, between adjoining patches. A boundary also has a vertical structure that influences physical gradients between patches. Functionally, a boundary connects patches through fluxes or flows of material, energy, and organisms. Typically, transition zones between patches are characterized by high species richness because they support selected species of adjoining communities as well as a group of opportunistic species adapted to edges—a phenomenon called *edge effect*.

Patch Size, Shape, and Diversity 19.4

A positive relationship exists between area and species diversity. Generally, large areas support more species than small areas do. The increase in species diversity with increasing patch size is related to several factors. Many species are area sensitive; they require large, unbroken blocks of habitat. Larger areas typically encompass a greater number of microhabitats and thus support a greater array of animal species. Another feature of patch size relates to differences between the habitats provided by boundary and interior environments. In contrast to edge species, interior species require environmental conditions found in the interior of large habitat patches, away from the abrupt changes in environmental conditions associated with edge environments.

Connectivity 19.5

The degree to which the landscape facilitates or impedes the movement of organisms among patches is referred to as landscape connectivity. Structural connectivity relates to the physical arrangement of habitat patches on the landscape, and functional connectivity describes the degree to which the landscape facilitates the movement of organisms. Linking one patch to another are corridors, which are the strips of habitat similar to a patch but unlike the surrounding matrix. Corridors act as conduits or travel lanes, function as filters or barriers, and provide dispersal routes among patches.

Island Biogeography 19.6

The theory of island biogeography proposes that the number of species an island holds represents a balance between immigration and extinction. The island's distance from a mainland or source of potential immigrants influences immigration rates. Thus, islands farther from a mainland would receive fewer immigrants than would islands closer to the mainland. The area of an island influences extinction rates. Because they hold fewer individuals of a species and their habitat varies less, small islands have higher extinction rates than large islands do. In habitat patches, as in islands, large areas support more species than do small areas.

Metapopulations 19.7

Habitat fragmentation and human exploitation have reduced many species to isolated or semi-isolated populations. These subpopulations inhabiting fragmented habitats form metapopulations. Metapopulation persistence is a dynamic balance between the extinction and recolonization of empty habitat patches. Colonization involves the movement of individuals from occupied patches to unoccupied patches to form new local populations. The ability of individuals to disperse between habitat patches is directly related to their spatial arrangement on the landscape. The rate of colonization declines with increasing isolation (distance from adjacent patches). The risk of local extinction increases with decreasing patch size.

Metacommunities 19.8

Communities occupying habitat patches connected by movements of the constituent species make up the metacommunity. The interaction among communities is influenced by size, shape, spatial arrangement of the habitat patches and the matrix in which they are embedded. The spatial configuration and species composition of patches influence dispersal and the processes of colonization among the local communities. In turn, the species composition of patches determines the nature of species interactions, which influences the potential for new species to colonize a patch as well as the ability of current species to persist.

Landscape as a Shifting Mosaic 19.9

The landscape is dynamic, and patches are in various stages of development and disturbance. This process suggests a landscape pattern that represents a shifting mosaic. The term *steady state* is a statistical description of the collection of patches; it describes the average state of the landscape.

Corridors and Conservation Ecological Issues & Applications

An approach that is being adopted in conservation efforts is linking existing isolated protected areas into one large system through the use of habitat corridors, which are areas of protected land running between the reserves. Such corridors can facilitate the dispersal of plants and animals from one reserve to another. Corridors also assist species that migrate seasonally to different habitats to obtain food or breed.

STUDY QUESTIONS

1. In what way is the study of spatial patterns the defining characteristic of landscape ecology?
2. How do variations in the physical environment (geology, topography, soils, and climate) give rise to the landscape patterns in your region? Contrast these patterns to the influence of human activities in defining the mosaic of patches that forms the surrounding landscape.
3. Why do edges and ecotones often support a greater diversity of species than do the adjoining communities?
4. How do corridors aid in maintaining the diversity of habitat patches?
5. Species comprising a community differ in their dispersal behavior. How might the landscape connectivity—both structural and functional—influence species composition and diversity of the various patches making up the metacommunity?
6. How does the proportion of edge to interior habitat change with increasing patch size?
7. How does species diversity change with increasing patch size?
8. Contrast edge and interior species.
9. The theory of island biogeography envisions the species richness of island communities as a balance between the processes of colonization and local extinction. In what way can this theory be applied to isolated habitat patches in terrestrial environments?
10. How do island size and isolation (distance to neighboring islands) influence patterns of species richness?
11. Define *metapopulation*. How does this concept relate to the spatial distribution (dispersion) of the population over a geographic region?
12. How do the size and spatial arrangement of habitat patches (local populations) influence metapopulation dynamics? Discuss in terms of probabilities of extinction and colonization.
13. How does the theory of island biogeography relate to the model of metapopulation dynamics?
14. How has the landscape around your home or neighborhood changed since you were a young child? What processes have been responsible for the changes?

FURTHER READINGS

Classic Studies

Den Boer, P. 1981. "On the survival of populations in a heterogeneous and variable environment." *Oecologia* 50:39–53.
An early and important article that examines the importance of the metapopulation concept in understanding the persistence of populations in variable environments.

Forman, R. T. T., and M. Gordon, M. 1981. "Patches and structural components for a landscape ecology." *BioScience* 31:733–740.
One of the earliest articles describing the study of landscapes and defining the emerging field of landscape ecology.

Levins, R. 1969. "Some demographic and genetic consequences of environmental heterogeneity for biological control." *Bulletin of the Entomological Society of America* 15:237–240.

Levins, R. 1970. Extinction. In: *Some Mathematical Problems in Biology* (M. Gesternhaber, ed.), 77–107. Providence, Rhode Island: American Mathematical Society.
In these two articles Levins first introduces the concept of metapopulation.

Turner, M. 1989. "Landscape ecology: The effects of pattern and process." *Annual Review of Ecology and Systematics* 20:171–197.
One of the earliest reviews of the emerging field of landscape ecology. This article provides an overview of terms and concepts that are central to the theory of ecological landscapes.

Urban, D., R. V. O'Neill, and H. H. Shugart. 1987. "Landscape ecology." *BioScience* 37:119–127.
An early and important article in the development of landscape ecology. This well-written and illustrated article is a good introduction to many of the central concepts in landscape ecology.

Current Research

Cadenasso, M. L., S. T. A. Pickett, K. E. Weathers, and C. D. Jones. 2003. "A framework for a theory of ecological boundaries." *BioScience* 53:750–758.
This article provides a framework for the study of ecological boundaries, focusing on flows of organisms, materials, and energy in heterogeneous landscape mosaics.

Collinge, S. K. 2009. *Ecology of fragmented landscapes.* Baltimore: John Hopkins University Press.
Synthesizes body of research on landscape fragmentation and discusses restoration, conservation, and planning.

Forman, R. T. T. 1995. *Land mosaics: The ecology of landscapes and regions.* New York: Cambridge University Press.
A synthesis exploring the ecology of heterogeneous land areas, where natural processes and human activities spatially interact to produce a continually changing mosaic.

Hanski, I. 1999. *Metapopulation ecology.* New York: Oxford University Press.
The definitive reference text written by a leading scientist in the study of metapopulations. Provides a wealth of illustrated examples relating to topics covered in this chapter.

Hilty, J. A., W. Z. Lidicker, Jr., and A. M. Merenlender. 2006. *Corridor ecology: The science and practice of linking landscapes for biodiversity conservation.* Washington, DC: Island Press.
An excellent review of corridor ecology and its application in creating, enhancing, and maintaining connectivity between natural areas.

Holyoak, M., M. A. Leibold, and R. D Holt (eds). 2005. *Metacommunities: Spatial dynamics and ecological communities.* Chicago: University of Chicago Press.
Collection of empirical, theoretical, and synthetic chapters focusing on how communities work in fragmented landscapes.

Lindenmayer, D. B., and J. Fischer. 2006. *Habitat fragmentation and landscape change: An ecological and conservation synthesis.* Washington, DC: Island Press.
Reviews ecological problems caused by landscape change and discusses the relationships among landscape change, habitat fragmentation, and biodiversity.

Rosenberg, D. K., B. R. Noon, and E. C. Meslow. 1997. "Biological corridors: Forms, function and efficiency." *BioScience* 47:677–687.
An excellent introduction to the topic of landscape corridors. This review article presents numerous examples of field research addressing the role of corridors in conservation ecology.

Turner, M., R. H. Gardner, and R. V. O'Neill. 2001. *Landscape ecology in theory and practice.* New Yory, N.Y.: Springer.
This text provides an excellent overview of the field of landscape ecology.

MasteringBiology®

Students Go to **www.masteringbiology.com** for assignments, the eText, and the Study Area with practice tests, animations, and activities.

Instructors Go to **www.masteringbiology.com** for automatically graded tutorials and questions that you can assign to your students, plus Instructor Resources.

A Rocky Mountain alpine tundra ecosystem carpeted with cotton grass (*Eriophorum angustifolium*) in full bloom.

CHAPTER GUIDE

20.1 The Laws of Thermodynamics Govern Energy Flow
20.2 Energy Fixed in the Process of Photosynthesis Is Primary Production
20.3 Climate and Nutrient Availability Are the Primary Controls on Net Primary Productivity in Terrestrial Ecosystems
20.4 Light and Nutrient Availability Are the Primary Controls on Net Primary Productivity in Aquatic Ecosystems
20.5 External Inputs of Organic Carbon Can Be Important to Aquatic Ecosystems
20.6 Energy Allocation and Plant Life-Form Influence Primary Production
20.7 Primary Production Varies with Time
20.8 Primary Productivity Limits Secondary Production
20.9 Consumers Vary in Efficiency of Production
20.10 Ecosystems Have Two Major Food Chains
20.11 Energy Flows through Trophic Levels Can Be Quantified
20.12 Consumption Efficiency Determines the Pathway of Energy Flow through the Ecosystem
20.13 Energy Decreases in Each Successive Trophic Level
ECOLOGICAL Issues & Applications Human Consumption

THE SUNLIGHT THAT FLOODS EARTH is the ultimate source of energy that drives the dynamics of the planet. Solar energy arrives at the top of the atmosphere as electromagnetic radiation (including light) in particles of energy known as *photons*. When the photons reach the atmosphere, land, and water, some are transformed into another form of energy—heat—that warms the atmosphere, land surface, and waters of the oceans, drives the water cycle, and causes the currents of air (winds) and water (see Chapters 2 and 3) to form. Some of the photons that reach plants are transformed into photochemical energy used in photosynthesis (see Chapter 6). That energy, stored in the chemical bonds of carbohydrates and other carbon-based compounds, becomes the source of energy for other living organisms. In this way, the story of energy within an ecosystem is in large part a story of carbon in the form of organic matter—the living and dead tissues of plants and animals.

All ecological processes involve the transfer of energy, and ecosystems are no different from physical systems such as the atmosphere because they are subject to the same physical laws. Here, we will explore the pathways, efficiencies, and constraints that characterize energy flow through the ecosystem. But first, we must examine the physical laws governing the flow of energy.

20.1 The Laws of Thermodynamics Govern Energy Flow

Energy exists in two forms: potential and kinetic. **Potential energy** is stored energy; it is capable of and available for performing work. **Kinetic energy** is energy in motion. It performs work at the expense of potential energy. Work occurs when a force acts on an object causing its displacement.

Two laws of thermodynamics govern the expenditure and storage of energy. The **first law of thermodynamics** relates to the conservation of energy and states that energy is neither created nor destroyed. It may change form, pass from one place to another, or act on matter in various ways. Regardless of what transfers and transformations take place, however, no gain or loss in total energy occurs. Energy is simply transferred from one form or place to another. When wood burns, the potential energy released from the molecular bonds of the wood equals the kinetic energy generated as heat (thermal energy). When a chemical reaction results in the loss of energy from the system (in this case the wood), it is called an **exothermic reaction**.

On the other hand, some chemical reactions must absorb energy in order to proceed. These are **endothermic reactions**. Here, too, the first law of thermodynamics holds true. In photosynthesis, for example, the molecules of the products (simple sugars) store more energy than do the reactants that combined to form the products. The extra energy stored in the products is acquired from outside the system—from the sunlight harnessed by the chlorophyll within the leaf (see Chapter 6). Again, there is no gain or loss in total energy. Although the total amount of energy in any reaction, such as burning wood, does not increase or decrease, much of the potential energy degrades into a form incapable of doing further work. It is transferred from the system to the surrounding environment as heat. This reduction in potential energy is commonly referred to as **entropy**. The transfer of energy involves the **second law of thermodynamics**. This law states that when energy is transferred or transformed, part of the energy assumes a form that cannot pass on any further. Entropy increases. When coal is burned in a boiler to produce steam, some of the energy creates steam and some of the energy is dispersed as heat to the surrounding air. The same thing happens to energy in the ecosystem. As energy is transferred from one organism to another in the form of food, a portion is stored as energy in living tissue, whereas a large part of that energy is dissipated as heat and entropy increases.

At first, biological systems do not seem to conform to the second law of thermodynamics. The tendency of life is to produce order out of disorder, to decrease rather than increase entropy. The second law theoretically applies to **closed systems** in which no energy or matter is exchanged with the surrounding environment. With the passage of time, closed systems tend toward maximum entropy; eventually, no energy is available to do work. Living systems, however, are **open systems** with a constant input of energy in the form of solar radiation, providing the means to counteract entropy.

21.2 Energy Fixed in the Process of Photosynthesis Is Primary Production

The rate at which carbon dioxide in the atmosphere or water is converted into organic compounds by autotrophs is referred to as *primary productivity* because it is the first and basic form of energy storage in ecosystems. With the limited exception of chemotrophs (see Sections 6.1 and 24.11 for discussion of chemotrophs), which obtain energy through the oxidation of electron donating molecules in their environment, the flow of energy through an ecosystem starts with the harnessing of sunlight in the process of photosynthesis.

Gross primary productivity is the total rate of photosynthesis, or energy assimilated by the autotrophs. Like all other organisms, autotrophs must expend energy in the process of respiration (see Chapter 6). The rate of energy storage as organic matter after respiration is **net primary productivity (NPP)**. NPP can be described by the following equation:

Net primary productivity (NPP)	=	Gross primary productivity (GPP)	−	Respiration by autotrophs (R)

Productivity is usually expressed in units of energy per unit area per unit time: kilocalories per square meter per year (kcal/m^2/yr). However, productivity may also be expressed in units of dry organic matter: (g/m^2/yr). As pointed out by the late ecologist Eugene Odum, in all these definitions, the term *productivity* and the phrase *rate of production* may be used interchangeably. Even when the word *production* is used, a time element is always assumed or understood, so one should always state the time interval.

The amount of accumulated organic matter found in an area at a given time is the **standing crop biomass**. Biomass is usually expressed as grams of organic matter per square meter (g/m^2) or some other appropriate unit of area. Standing crop biomass differs from productivity. Productivity is the rate at which organic matter is created by photosynthesis. Biomass is the amount of organic matter present at any given time.

The simplest and most common method of measuring net primary production in terrestrial ecosystems is to estimate the change in standing crop biomass (SCB) over a given time interval.

$(t_2 - t_1)$:

$$\Delta SCB = SCB(t_2) - SCB(t_1)$$

Two possible losses of biomass over the time period must also be recognized: loss of biomass as a result of the death of plants (D), and loss of biomass resulting from consumption by consumer organisms (C). The estimate of net primary productivity is then:

$$NPP = (\Delta SCB) + D + C.$$

In aquatic ecosystems, the most common method of estimating NPP is the light/dark bottle method (**Figure 20.1**). Because oxygen is one of the most easily measured products of both photosynthesis and respiration, a good way to gauge primary productivity in an aquatic ecosystem is to measure the concentration of dissolved oxygen (see Section 6.1). In one set of clear glass "light bottles," a water sample from the aquatic ecosystem (and associated autotrophic organisms—phytoplankton) is allowed to incubate in the sealed bottle for a defined time period. If photosynthesis is greater than respiration, oxygen will accumulate in the water, providing an estimate of NPP. Water is also incubated over the same time period in another set of "dark bottles" (painted dark to prevent light from reaching the water). Because the lack of light will prevent photosynthesis, the oxygen content of the water will decline as a function of respiration. The difference between the values of oxygen in the light (photosynthesis + respiration) and dark (respiration) bottles at the end of the time period therefore provides an estimate of total photosynthesis, or gross primary productivity.

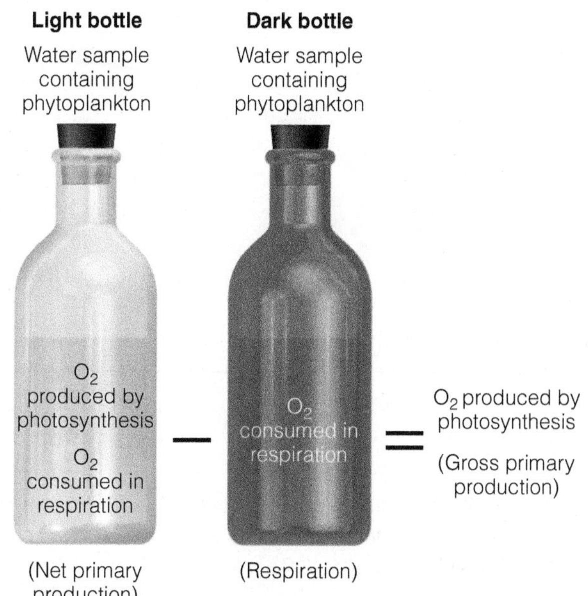

Figure 20.1 Paired light and dark bottles are used to measure photosynthesis (gross production), respiration, and net primary production by phytoplankton in aquatic ecosystems. A sample of water containing phytoplankton (primary producers) is placed in both bottles and allowed to incubate for a period of time. In the light (clear) bottle, O_2 is produced in photosynthesis and consumed in respiration. The resulting change (increase) in O_2 concentration represents the difference in the rates at which these two processes occur: net primary productivity. Lacking light to drive the process of photosynthesis, only respiration occurs in the dark bottle. As a result, O_2 concentration declines. The difference between the O_2 concentrations of the water from the light and dark bottles at the end of the incubation period represents the rate of O_2 produced in photosynthesis: gross primary productivity.

20.3 Climate and Nutrient Availability Are the Primary Controls on Net Primary Productivity in Terrestrial Ecosystems

An array of environmental factors, including climate, influence the productivity of terrestrial ecosystems. Measured estimates of NPP for various terrestrial ecosystems around the world are plotted in **Figure 20.2** as a function of the mean annual precipitation (Figure 20.2a) and mean annual temperature (Figure 20.2b) for each site. NPP increases with increasing mean annual temperature and rainfall. Increasing mean annual temperature is directly related to the annual intercepted solar radiation at the site, reflecting both an increase in mean daily temperature and the length of the growing season (see Chapter 2, Section 2.2). The length of the growing season is defined as the period (number of days) during which temperatures are warm enough to support photosynthesis. As a result, sites with a higher mean annual temperature typically support higher rates of photosynthesis and are associated with a longer time period over which photosynthesis can occur (**Figure 20.3**).

As we have seen, for photosynthesis and productivity to occur, plants' stomata must open to take in carbon dioxide (see Chapter 6). When the stomata are open, water is lost from the leaf to the surrounding air (transpiration). For plants' stomata to remain open, roots must replace the lost water. The higher the rainfall, the more water is available for transpiration. The amount of water available to the plant will therefore

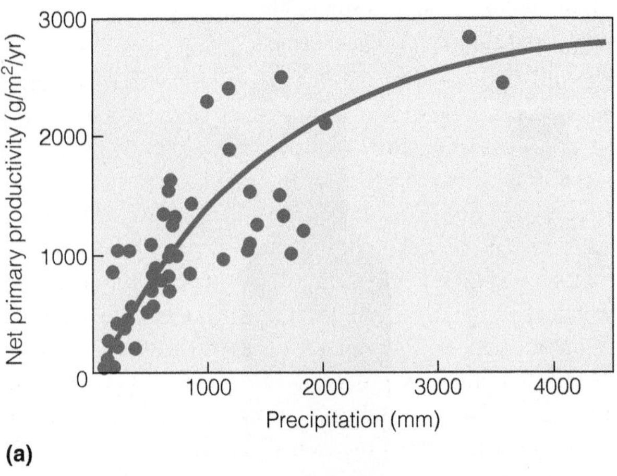

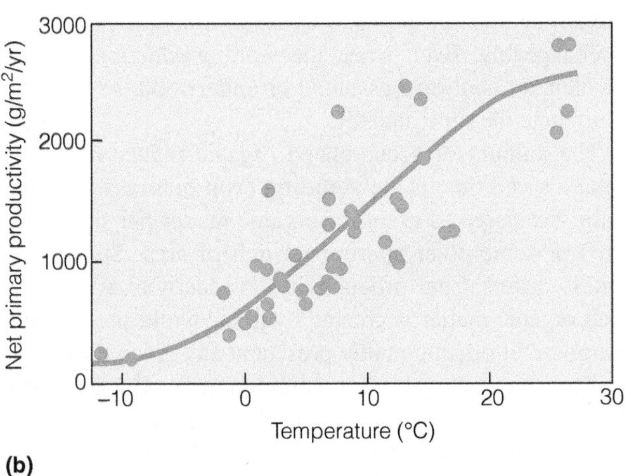

Figure 20.2 Net primary productivity for a variety of terrestrial ecosystems (a) as a function of mean annual precipitation and (b) as a function of mean annual temperature. (Adapted from Lieth 1973.)

limit both the rate of photosynthesis and the amount of leaves (surface area that is transpiring) that can be supported (see Section 6.9). The combination of these factors determines the rate of primary productivity.

Although the two graphs in Figure 20.2 show independent effects of temperature and precipitation on primary productivity, in reality the influence of these two factors is closely related. Warm air temperatures increase the potential for evaporation and therefore increase rates of transpiration and plant water demand (see Section 2.5). If temperatures are warm but water availability is low, productivity will also be low. Conversely, if temperatures are low, rates of photosynthesis and productivity will be low regardless of the availability of water. This interaction between temperature and water on the process of NPP explains the high degree of variation in NPP observed in Figures 20.2a and 20.2b with increasing values of mean annual temperature and precipitation. For example, in Figure 20.2b, values of NPP for sites having a mean annual temperature of approximately 12°C range from a low of 900 to a high of more than 2500 g/m^2/yr. This range of values reflects differences in the corresponding mean annual precipitation at these sites, with low values of productivity associated with low rainfall sites and high values associated with high precipitation sites. Similarly, in Figure 20.2a, variation in values of NPP for sites receiving approximately the same annual precipitation reflects differences in the mean annual temperature.

It is the combination of warm temperatures and an adequate water supply for transpiration that gives the highest primary productivity. This pattern is reflected in **Figure 20.4**, which relates the NPP of various ecosystems to estimates of actual evapotranspiration (AET). AET is the combined value of surface evaporation and transpiration (see Section 3.1). It reflects both the demand and the supply of water to the

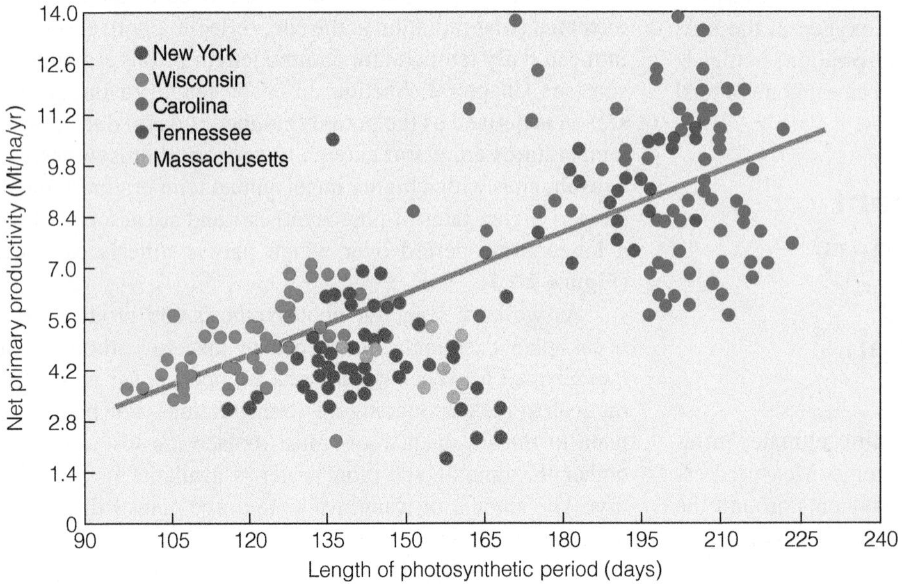

Figure 20.3 Relationship between net primary productivity and the length of the growing season for deciduous forest stands in eastern North America. Each point represents a single forest site. The (regression) line represents the general trend of increasing productivity with increasing length of the growing season (largely a function of latitude). (After Leith 1975.)

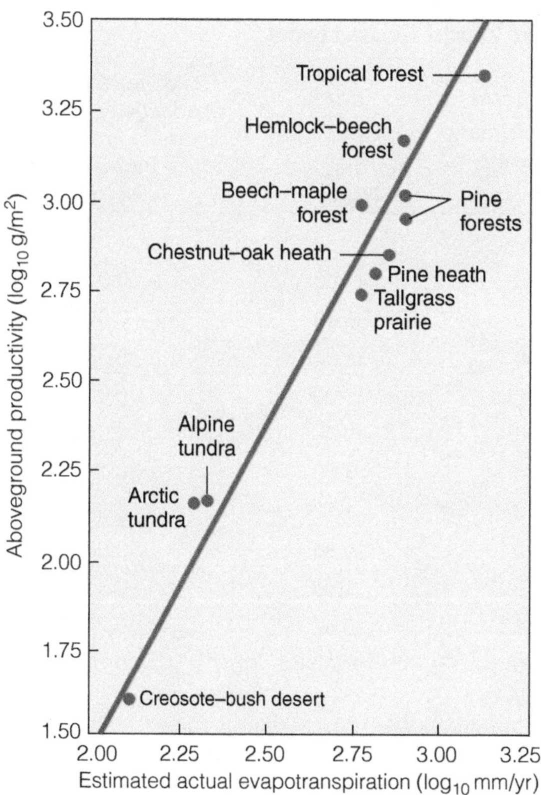

Figure 20.4 Relationship between aboveground net primary productivity and (actual) evapotranspiration for a range of terrestrial ecosystems. Evapotranspiration—the combination of evaporation and transpiration at a site—depends on both precipitation and temperature (see Chapter 3).
(Adapted from MacArthur and Connell 1966.)

ecosystem. The demand is a function of incoming radiation and temperature, whereas the supply is a function of precipitation.

The influence of climate on primary productivity in terrestrial ecosystems is reflected in the global patterns presented in **Figure 20.5**. These patterns of primary productivity reflect the global patterns of temperature and precipitation presented earlier (Figures 2.6 and 2.16). In addition, measured estimates of NPP for a variety of ecosystems are summarized in **Table 20.1**. The regions of highest NPP are located in the equatorial zone where the combination of year-round warm temperatures and precipitation supports high rates of photosynthesis and leaf area (tropical rain forest). Moving north and south from the Equator, the seasonality of precipitation increases (see discussion of intertropical convergence zone, Section 2.6). This decreases the growing season, and subsequently, the values of NPP. Continuing into the temperate regions (midlatitudes), an increasing seasonality of temperature functions to reduce the mean annual temperature and restrict the length of the growing season (see Figure 20.3). In addition, as one moves from the coast to the interior of the continents, both mean annual temperature and precipitation decline, reducing values of NPP (Section 2.7).

In addition to climate, the availability of essential nutrients required for plant growth directly affects ecosystem productivity. The availability of nutrients in the soil influences the rate of nutrient uptake, photosynthesis, and plant growth; the net result is a general pattern of increasing NPP with increasing soil nutrient availability (discussed in Chapter 6, Section 6.11).

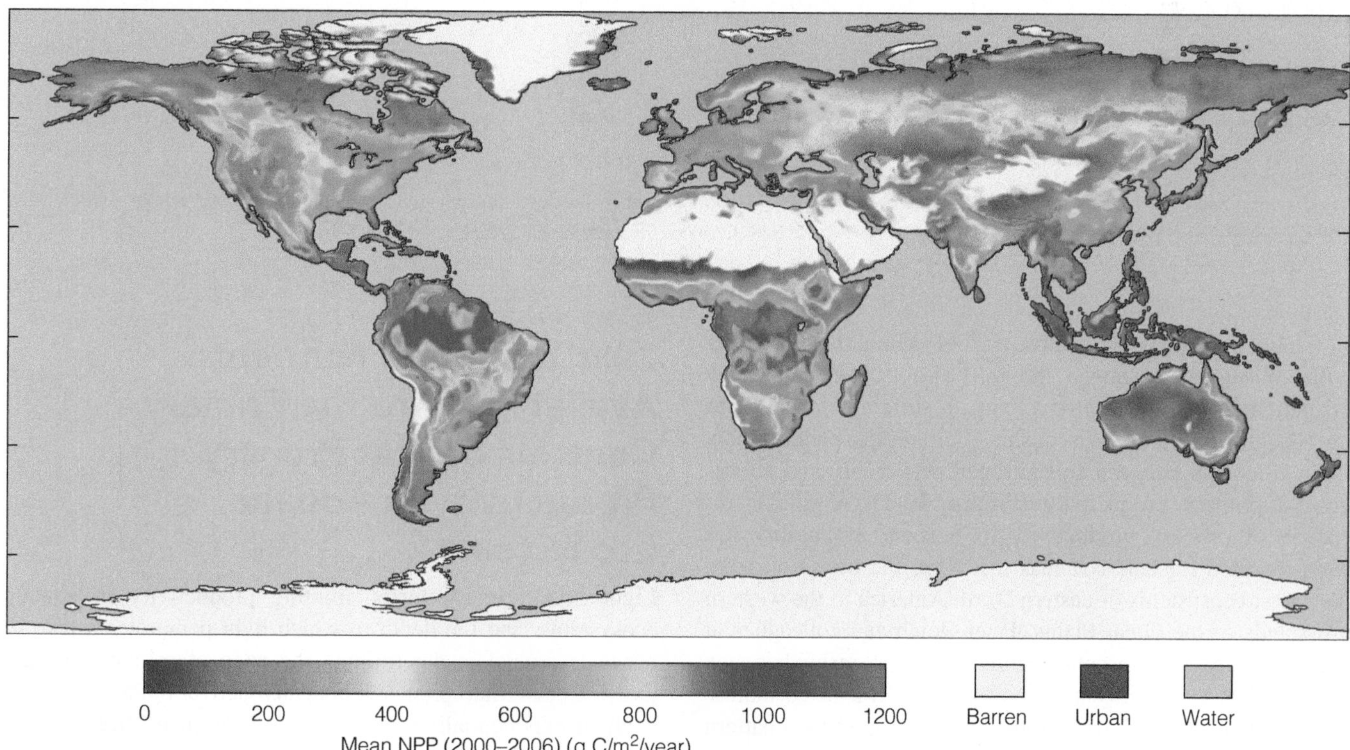

Figure 20.5 A global map of net primary productivity for the terrestrial surface (ecosystems).
(NASA/MODIS.)

Table 20.1 **Net Primary Productivity and Plant Biomass of World Ecosystems**

Ecosystems (in order of productivity)	Area (10^6km^2)	Mean Net Primary Productivity per Unit Area (g/m^2/yr)	World Net Primary Productivity (10^9 Mt/yr)	Mean Biomass per Unit Area (kg/m^2)	Relative Net Primary Productivity (Mean Net Primary Productivity/ Mean Biomass per Unit Area (g/g/yr)
Continental					
Tropical rain forest	17.0	2000.0	34.00	44.00	0.045
Tropical seasonal forest	7.5	1500.0	11.30	36.00	0.042
Temperate evergreen forest	5.0	1300.0	6.40	36.00	0.036
Temperate deciduous forest	7.0	1200.0	8.40	30.00	0.040
Boreal forest	12.0	800.0	9.50	20.00	0.040
Savanna	15.0	700.0	10.40	4.00	0.175
Cultivated land	14.0	644.0	9.10	1.10	0.585
Woodland and shrubland	8.0	600.0	4.90	6.80	0.088
Temperate grassland	9.0	500.0	4.40	1.60	0.313
Tundra and alpine meadow	8.0	144.0	1.10	0.67	0.215
Desert shrub	18.0	71.0	1.30	0.67	0.106
Rock, ice, sand	24.0	3.3	0.09	0.02	–
Swamp and marsh	2.0	2500.0	4.90	15.00	0.167
Lake and stream	2.5	500.0	1.30	0.02	25.0
Total continental	149.0	720.0	107.09	12.30	
Marine					
Algal beds and reefs	0.6	2000.0	1.10	2.00	1.0
Estuaries	1.4	1800.0	2.40	1.00	1.8
Upwelling zones	0.4	500.0	0.22	0.02	25.0
Continental shelf	26.6	360.0	9.60	0.01	36.0
Open ocean	332.0	127.0	42.00	0.003	42.3
Total marine	361.0	153.0	55.32	0.01	15.3
World total	**510.0**	**320.0**	**162.41**	**3.62**	

Source: Adapted from Whittaker 1975.

Relative net primary productivity (RNPP) is calculated by dividing the value of net primary productivity (column 3) by the corresponding value of mean biomass (column 5). Values of mean biomass must first be converted to units of g/m^2. The resulting units for RNPP are g/g/yr.

John Pastor of the University of Minnesota, together with colleagues, examined the role of nitrogen availability on patterns of primary productivity in different forest types on Blackhawk Island, Wisconsin. Their results clearly show the relationship between soil nitrogen availability and aboveground primary productivity (**Figure 20.6**). A similar response of primary productivity to nutrient availability has been reported for oak savannas that form the transition from the forest ecosystems of eastern North America to the western grasslands of the Great Plains. Peter Reich and colleagues at the University of Minnesota examined the relationship between soil nitrogen availability and aboveground NPP in 20 mature oak savanna stands in Minnesota. Their results show a pattern of increasing primary productivity with available nitrogen in these mixed tree–grass ecosystems (**Figure 20.7**).

20.4 Light and Nutrient Availability Are the Primary Controls on Net Primary Productivity in Aquatic Ecosystems

Light is a primary factor limiting productivity in aquatic ecosystems, and the depth to which light penetrates a lake or ocean is crucial in determining the zone of primary productivity. Recall that photosynthetically active radiation (PAR) declines exponentially with water depth (**Figure 20.8**, see also Figure 3.7). The photosynthetic rate and subsequently the gross productivity of phytoplankton are highest at intermediate levels

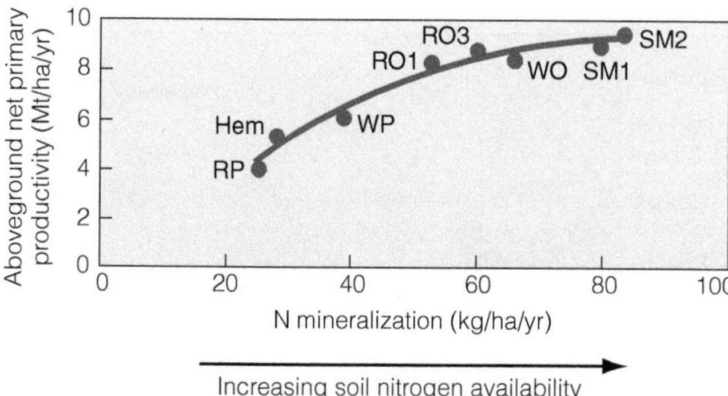

Figure 20.6 Relationship between net primary production and nutrient availability. Aboveground productivity increases with increasing nitrogen availability (N mineralization rate) for a variety of forest ecosystems on Blackhawk Island, Wisconsin. Abbreviations refer to the dominant trees in each stand: Hem, hemlock; RP, red pine; RO, red oak; WO, white oak; SM, sugar maple; WP, white pine.
(Adapted from Pastor et al. 1984.)

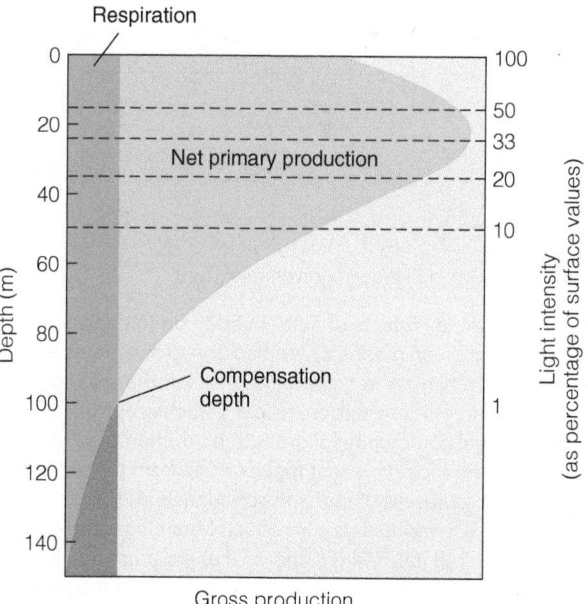

Figure 20.8 Changes in available light, gross productivity, respiration, and net primary productivity (gross productivity − respiration) with water depth. Respiration rate is relatively constant with depth, whereas gross productivity (photosynthesis) declines with depth as a function of declining available light. The depth at which gross productivity is equal to respiration (net photosynthesis equal to zero) is called the *compensation depth*.

of PAR (see Figure 6.2). On the other hand, the respiration rate does not change significantly with depth. This means that as the phytoplankton go deeper in the water column, the photosynthetic rate declines as the light intensity decreases until at some point the rate of photosynthesis (gross production) is equal to the rate of respiration, and NPP is zero (see Figure 20.8). This zone is referred to as the **compensation depth** and corresponds to the depth at which the availability of light is equal to the light compensation point (discussed in Chapter 6; see Figure 6.2).

In the oceans, nutrients in the deeper waters must be transported to the surface waters, where light (PAR) is sufficient to support photosynthesis. As a result, nutrients—particularly nitrogen, phosphorus, and iron—are a major limitation on primary productivity in the oceans (see Sections 21.10 and 24.13). John Downing, an ecologist at Iowa State University, together with his colleagues Craig Osenburg and Orlando Sarnelle, examined results from more than 300 nutrient-enrichment experiments conducted in marine habitats around the world (**Figure 20.9a**). The authors found that nitrogen (N) addition stimulated phytoplankton growth the most, followed closely by the addition of iron (Fe). In contrast, the addition of phosphorus (P) did not, on average, stimulate phytoplankton growth. These results confirm the prevailing view among marine ecologists that nitrogen and iron are the two most limiting nutrients in marine environments. These results, however, represent average responses and do not account for differences among habitats. The magnitude of response to nutrient enrichment varied significantly among marine environments, particularly for phosphorus (**Figure 20.9b**). Growth response to phosphorus addition in the more polluted waters of the nearshore environments (bays, estuaries, and harbors) was largely negative. In contrast, in the more pristine coastal and oceanic environments, the positive response was nearly as great as that observed for nitrogen.

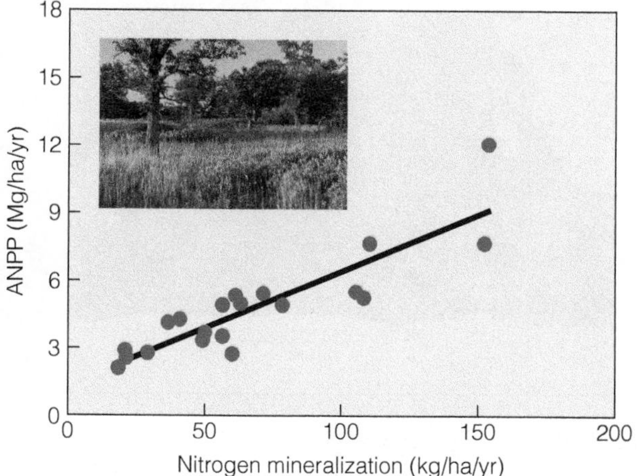

Figure 20.7 Relationship between aboveground net primary productivity (ANPP) and nitrogen availability (nitrogen mineralization rate) for 20 oak savanna sites in Minnesota.
(Adapted from Reich et al. 2001.)

The role of environmental constraints on the primary productivity of the world's oceans can be seen in the map presented in **Figure 20.10** (also see Table 20.1). For two reasons, the most productive waters of the oceans are the shallow waters of the coastal environments. First, shallow waters allow for a greater transport of nutrients from the bottom sediments to the surface waters, aided by wave action and the changing tides. Second, coastal waters receive a large input of nutrients carried from terrestrial ecosystems by rivers and streams (see Section 22.2).

The constraints on NPP in freshwater ecosystems are not always as easy to interpret as those operating in marine ecosystems. Solar radiation limits primary productivity in

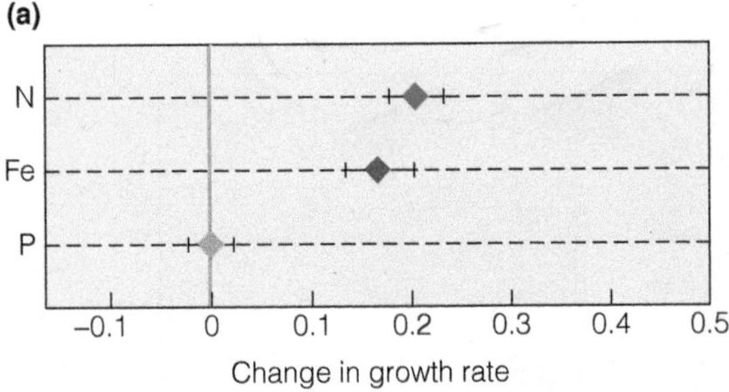

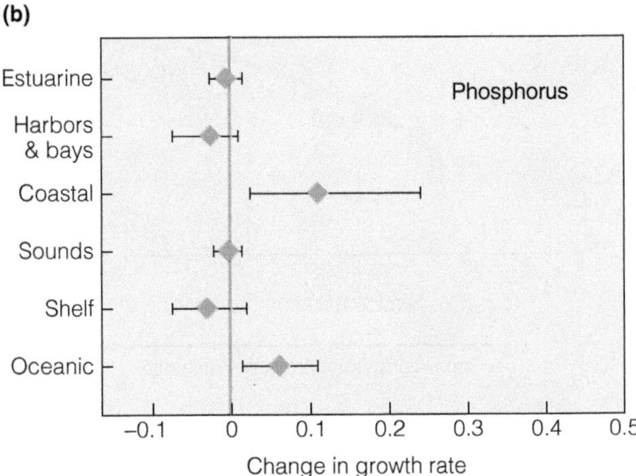

Figure 20.9 (a) Effects of nutrient addition (nitrogen, iron, and phosphorus) on marine phytoplankton growth rates in 303 experiments. Changes in growth rate measured as change in per unit (per gram, per unit carbon, or per unit chlorophyll), a growth rate of an algal community following the addition of surplus nutrients. Diamonds represent mean values from the experiments, and the bars represent 95 percent confidence intervals. The solid orange line represents zero effect. Mean response values are based on 148 (N), 114 (P), and 35 (Fe) experiments. (b) The change in phytoplankton growth rate as a result of the phosphorus addition varied among different marine environments.
(Adapted from Downing et al. 1999.)

lake ecosystems, but the close link between light intensity and temperature makes it difficult to evaluate these two factors independently (see Section 3.4). The role of nutrient availability on primary productivity in lake ecosystems, however, has been well established. Ecologists P. J. Dillon and F. H. Rigler of the University of Toronto examined the relationship between summer chlorophyll and spring total phosphorus concentration for 19 lakes in southern Ontario. Water chlorophyll concentration provides a simple and accurate estimate of phytoplankton standing biomass and productivity. The researchers combined their results with data reported in the literature for other North American lakes. The results of their analysis, presented in **Figure 20.11**, show a clear pattern of increasing primary productivity with phosphorus concentration. Similar results have been reported for studies in which the nutrient concentrations of lake water have been manipulated experimentally through fertilization.

As with lakes, both light and nutrient availability have been shown to constrain NPP in stream ecosystems. Amy Rosemond of Vanderbilt University and colleagues conducted

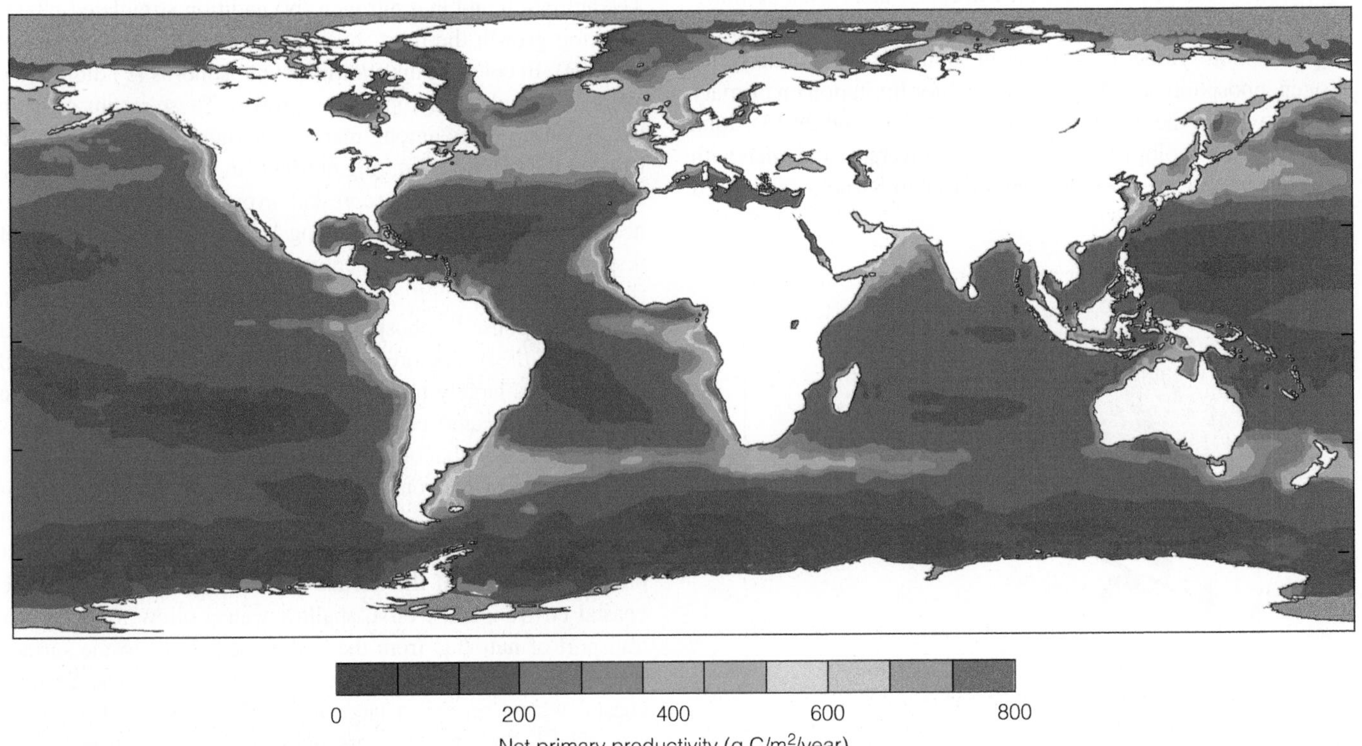

Figure 20.10 Geographic variation in primary productivity of the world's oceans. Note that the highest productivity is along the coastal regions, and areas of lowest productivity are in the open ocean (see Chapter 24 for detailed discussion).

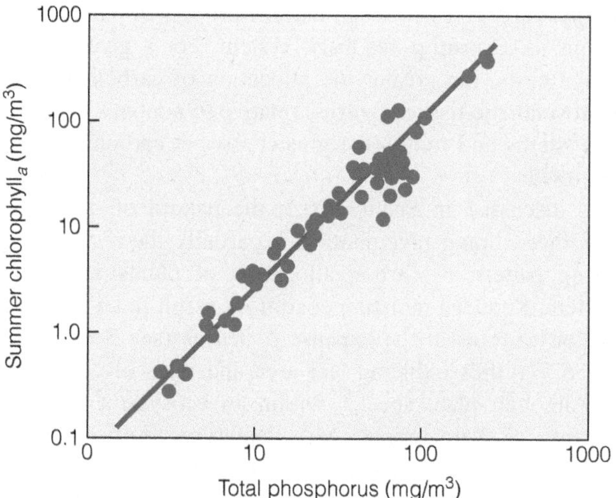

Figure 20.11 Relationship between summer average chlorophyll (an estimate of phytoplankton net primary productivity) content (y-axis) and spring total phosphorus concentration (x-axis) for northern temperate lake ecosystems (each point represents a single lake).
(Adapted from Dillon and Rigler 1974.)

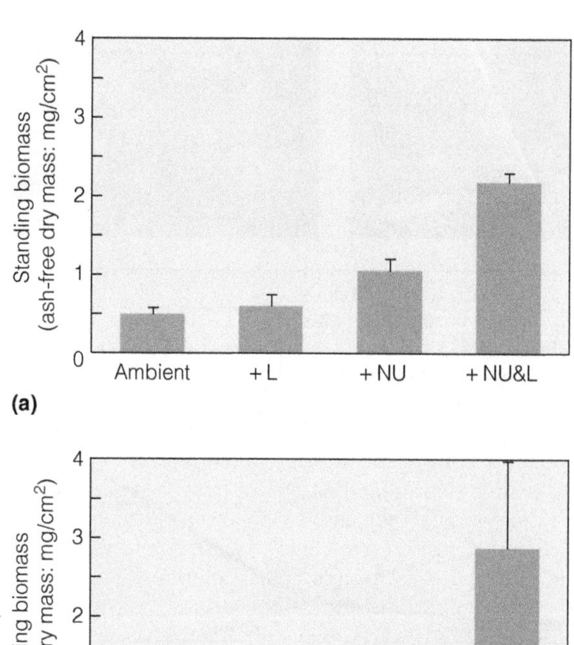

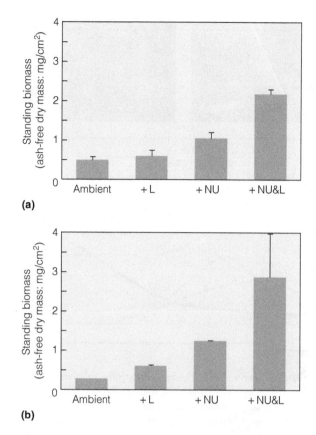

Figure 20.12 Standing biomass (periphyton) in experimental stream channels under different treatments during the last week of (a) fall experiment (11/3 – 12/20) and (b) spring experiment (3/7 – 4/26). Ambient: shaded, low nutrient treatment; +L, high light, low nutrient treatment; +NU, shaded, high nutrient treatment; +NU&L, high light, high nutrient treatment.
(Adapted from Rosemond et al. 2000.)

a series of experiments in which irradiance and nutrient concentrations were manipulated in experimental stream channels located on the Walker Branch Watershed in eastern Tennessee (United States). The researchers found that irradiance limited productivity and standing biomass during the summer months when the canopy leaf cover was at its greatest and light levels in the stream at their lowest on the forested watershed. In contrast, nutrients were more limiting during the fall and spring, periods during which the canopy is more open and light levels reaching the forest floor and stream are higher (**Figure 20.12**). During all seasons (spring, summer, and fall), however, the greatest increases in productivity and standing biomass were observed when both light and nutrient levels were enhanced.

Although numerous experimental manipulations of water nutrient concentrations in stream and river (lotic) ecosystems have shown an increase in NPP with nutrient concentration, primary productivity in these ecosystems is typically low in comparison to standing water (lentic) ecosystems. An important source of organic carbon for these ecosystems is inputs of dead organic matter from adjacent terrestrial ecosystems.

20.5 External Inputs of Organic Carbon Can Be Important to Aquatic Ecosystems

Organic carbon produced within an ecosystem is described as **autochthonous**, whereas inputs from outside the ecosystem are described as **allochthonous**. In aquatic ecosystems, the autochthonous input is provided through photosynthesis by aquatic plants, attached algae in shallow waters, and by phytoplankton in the open waters. However, a substantial proportion of organic carbon in these ecosystems is allochthonous, derived from dead organic matter from adjacent terrestrial ecosystems, entering the water as both dissolved organic matter (DOM) and particulate organic matter (POM).

The relative importance of allochthonous sources of carbon varies widely among different aquatic ecosystems. In most marine ecosystems, autochthonous inputs of organic carbon dominate because of the high NPP of resident populations of phytoplankton. In contrast, small stream ecosystems flowing through forested catchments derive most of their organic carbon from dead plant materials deposited into the water from the surrounding forest ecosystem. Shading from trees limits available light in the waters, preventing any significant growth of phytoplankton or attached algae (see light treatments in Figure 20.12). In these streams, allochthonous carbon provides energy resources that support levels of consumer abundance beyond those that could be supported from autochthonous sources (NPP within the ecosystem).

As the stream widens farther downstream, the shading by adjacent trees becomes limited to the stream margins, and the increase in available light in the water column allows for an increase in NPP by submerged aquatic plants, algae, and phytoplankton. The result is a continuum of the relative importance of allochthonous and autochthonous sources of carbon in the energy balance of flowing water ecosystems from the smaller

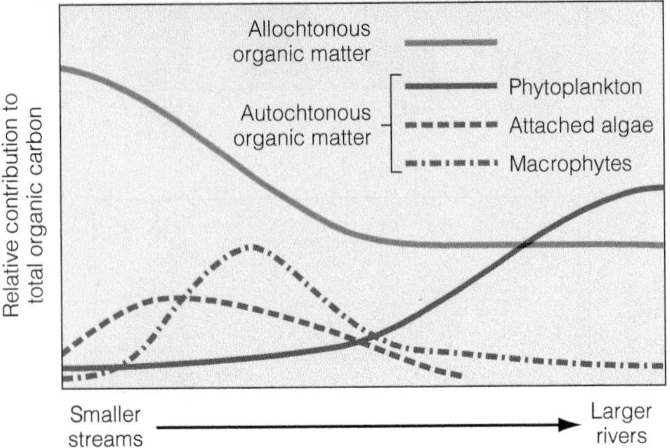

Figure 20.13 Generalized scheme showing the relative contributions of allochthonous organic matter and autocthonous production by aquatic plants (macrophytes), attached algae, and phytoplankton in the transition of a stream to a river ecosystem. (From Wetzel 1975.)

headwater streams to larger river systems (**Figure 20.13**) referred to as the **river continuum concept** (see Section 24.7 for broader discussion of this concept).

As with flowing water ecosystems, the relative importance of internal and external sources of organic carbon varies among lake ecosystems as a function of their morphology (size and shape) and the nature of the surrounding catchment. In larger lake ecosystems, autochthonous inputs generally dominate. Allochthonous inputs of both DOM and POM vary seasonally with the volume of water from streams and rivers that flows into the lake. In contrast, in smaller lakes allochthonous inputs can be significant. In a study of two small lake ecosystems (1.1 and 2.5 hectares [ha]) occupying a forested catchment in Michigan, ecologist Michael Pace of the University of Virginia and colleagues found that upward of 50 percent of organic carbon that moved through the food chain was from adjacent terrestrial sources (allochthonous). We will further examine these linkages between terrestrial and aquatic ecosystems later in Chapter 21.

20.6 Energy Allocation and Plant Life-Form Influence Primary Production

Previously, we examined the implications of how plants allocate the carbon fixed in photosynthesis (carbon allocation) to the process of plant growth (see Chapter 6, **Quantifying Ecology 6.1**). You will recall that the process of plant growth functions as a positive feedback system. For a given rate of photosynthesis, the greater the allocation of carbon (energy) to photosynthetic tissues (leaves) relative to nonphotosynthetic tissues (stems and roots), the greater the net carbon gain and plant growth.

As discussed in Section 20.3, the pattern of decreasing NPP with declining precipitation is partially the result of the changing pattern of carbon allocation of plants within the ecosystem. Reduced moisture conditions result in an increased allocation to roots at the expense of leaves (see Section 6.9, Figure 6.17), thus reducing leaf area and rates of net carbon gain. Although plant species within an ecosystem exhibit a wide range of characteristics and adaptations to microclimatic conditions (such as in the case of shade-tolerant and shade-intolerant species), the average patterns of carbon allocation for different ecosystems reflect the general pattern of carbon allocation exhibited by individual plants in response to such environmental gradients as moisture availability (see Figures 6.17 and 6.18). In a recent review of more than 700 studies that have examined patterns of carbon allocation in different ecosystems, Karel Mokany of Australian National University and colleagues found that estimates of the ratio of belowground to aboveground biomass (root-to-shoot ratio [R:S]) range from a low of 0.20 in tropical rain forest ecosystems to 1.2 for arid shrublands and to a high of 4.5 in desert ecosystems. These differences reflect the greater allocation to roots relative to aboveground tissues (leaves and stem) with decreasing annual precipitation (see Figure 6.17).

The decline in NPP from mesic to more xeric environments shown in Figure 20.2 is paralleled by a reduction in standing crop biomass and the accumulation of NPP over time (**Figure 20.14**). The ecosystems with greater NPPs are those with the greater standing biomass. The relationship of increasing standing biomass to increasing NPP is seen in both terrestrial and marine environments.

Recall from our discussion of the growth of individual plants that larger plants generally have a greater net carbon gain or absolute growth rate (grams per unit time) than do smaller plants (see Chapter 6, Quantifying Ecology 6.1). This situation often changes, however, when we examine the relative growth rate, or biomass gain per unit of plant mass (grams per gram plant mass per unit time). The same holds true for the collective growth of plants within the ecosystem. The ratio of NPP to standing biomass from Table 20.1—relative net primary productivity (RNPP [units of g/g/yr])—represents an index similar to that of relative growth rate for individual plants: the rate of biomass accumulation per unit of plant biomass present. A comparison of RNPP with the average standing biomass in each of the terrestrial ecosystems shows a pattern inverse to that discussed for Figure 20.14. For example, although the NPP of a temperate forest (0.04 g/g/yr) is more than twofold greater than that of a temperate grassland, if we calculate the productivity per unit of standing biomass, the grassland ecosystem's NPP is almost an order of magnitude greater than that of the forest (0.31 g/g/yr). This reflects the general pattern of higher relative growth rate for grasses when compared to trees.

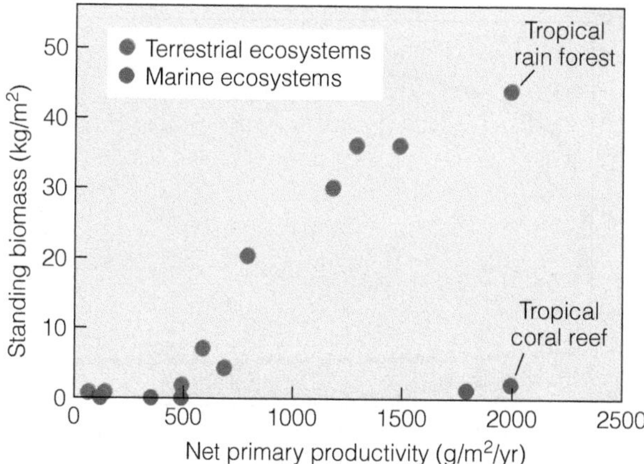

Figure 20.14 Relationship between standing biomass and net primary productivity for ecosystems in Table 20.1. Note the difference in the standing biomass per unit of net primary productivity between terrestrial and aquatic ecosystems (see text for discussion).

The same inverse relationship between standing biomass and RNPP observed for terrestrial ecosystems also occurs in marine ecosystems. The interpretation, however, is quite different. In the terrestrial ecosystems represented in Table 20.1, the longevity of dominant plant species is typically much greater than the period over which NPP is measured (annual NPP). This is not, however, the case for most marine ecosystems. Phytoplankton (microscopic algae) are the dominant net primary producers in open-water ecosystems. These species are short-lived (weeks) with high rates of reproduction. As a result, there is a constant turnover of the populations, with many generations occurring during the period over which annual NPP is measured. As a result of the fast turnover of the populations, the standing biomass at any time interval is low as compared to the accumulated NPP over the course of the year. This accounts for the extremely high value of RNPP for the open ocean (42.3 g/g/yr) as compared to all terrestrial ecosystems (see Table 20.1).

20.7 Primary Production Varies with Time

Primary production also varies within an ecosystem with time and age (**Figure 20.15**). Both photosynthesis and plant growth are directly influenced by seasonal variations in environmental conditions (Figure 20.15a). Regions with cold winters or distinct wet and dry seasons have a period of plant dormancy when primary productivity ceases. In the wet regions of the tropics, where conditions are favorable for plant growth year-round, there is little seasonal variation in primary productivity.

Year-to-year variations in primary productivity within an area can occur as a result of climatic variation (Figure 20.15b). In a long-term experiment, known as Park Grass, at the Rothamsted Experimental Station in Hertfordshire (England), grass yields have been recorded since 1856. Yields have been recorded, with constant treatments using standard measurement methods, since 1965 (**Figure 20.16**). Late-summer yields are depressed in hot, dry summers. The three lowest-yielding years were 1976, 1990, and 1995—all years with hot, dry summers.

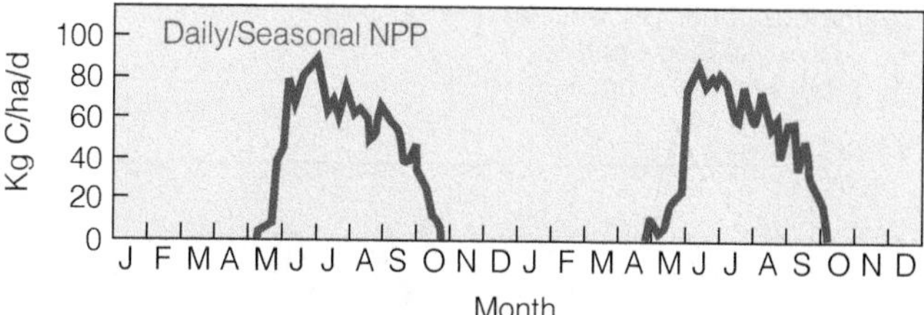

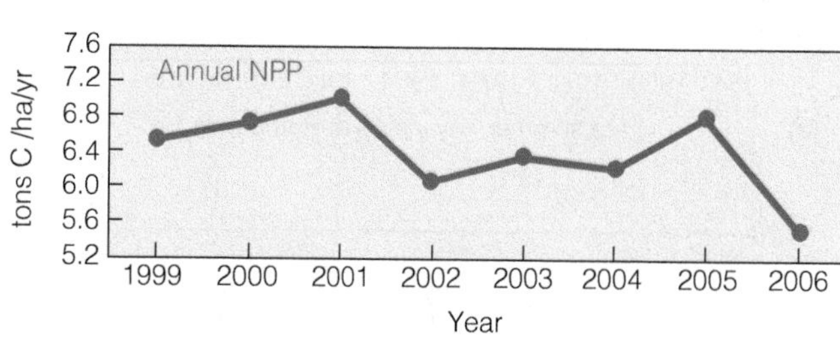

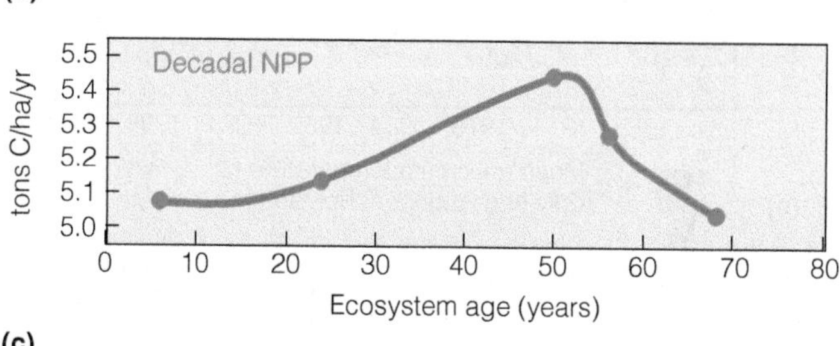

Figure 20.15 Patterns of net primary productivity (NPP) at different timescales in a temperate forest ecosystem. (a) Daily net primary productivity changes during the growing season in response to climate variables including solar radiation and precipitation, while the duration of NPP during the growing season (spring green-up and autumn leaf fall) is largely a function of photoperiod. (b) Annual NPP changes from one year to the next in response to longer-term trends in climate, including shifts in solar radiation caused by differences in cloud cover from year to year. (c) Decadal patterns of NPP track changes in the structure of the forest through the process of succession (Chapter 18).
(From Grough 2011.)

Disturbances such as herbivory and fire can also lead to year-to-year variations in NPP at a site (see Section 19.1). Overgrazing of grasslands by cattle and sheep or defoliation of forests by such insects as the gypsy moth can significantly reduce NPP. Fire in grasslands may increase productivity in wet years but reduce it in dry years.

NPP also varies with stand age (Figure 20.15c), particularly in ecosystems that are dominated by woody vegetation. Trees and woody shrubs can survive for a long time, which greatly influences how they allocate energy. Early in life, leaves make up more than one-half of their biomass (dry weight), but as trees age, they accumulate more woody growth. Trunks and stems become thicker and heavier, and the ratio of leaves to woody tissue changes (**Figure 20.17**). Although the

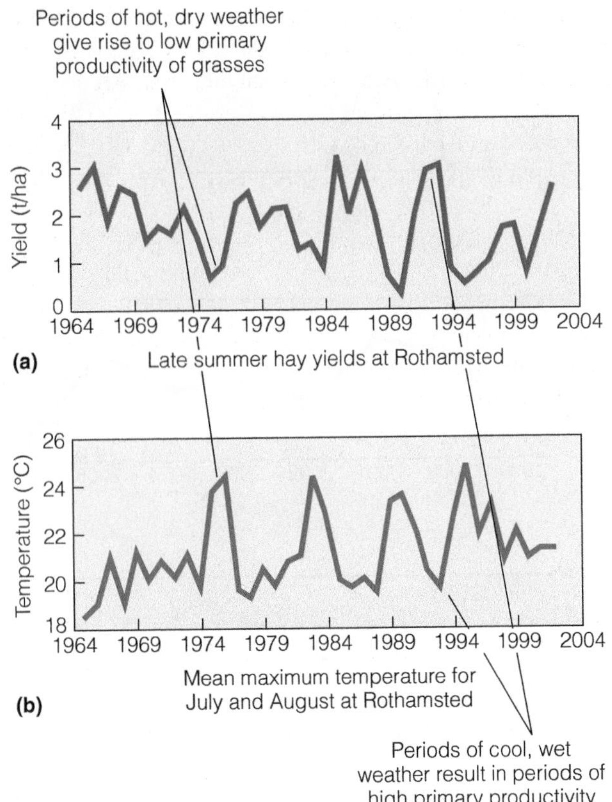

Figure 20.16 Seasonal variations in (a) grass productivity and (b) mean maximum temperature for the period of July and August at the Rothamsted Experimental Station in Hertfordshire (England) for the period 1965–2004.
(Adapted from Sparks and Potts 2004.)

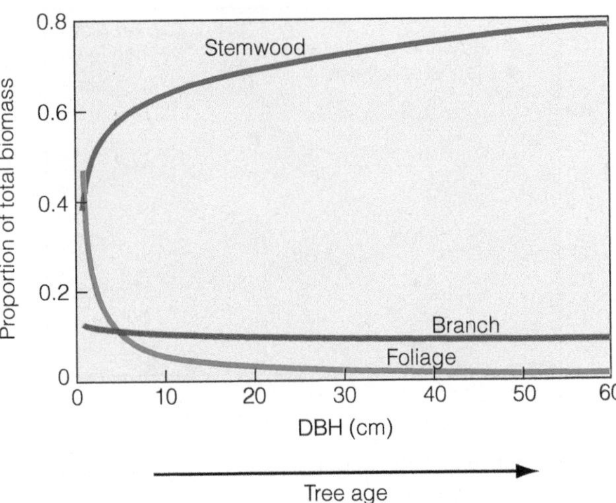

Figure 20.17 Changes in the proportion of biomass in foliage (leaves), branch, and stemwood (bole) for white oak trees (*Quercus alba*) as a function of tree size (diameter of the trunk at 1.5 m above the ground, or DBH).

Interpreting Ecological Data

Q1. *Go to Quantifying Ecology 6.1 on page 106. Assuming that leaf area (cm²) increases linearly with leaf mass (g), how does the leaf area ratio of the oak tree change with age?*

Q2. *Based on the discussion of relative growth rate presented in Quantifying Ecology 6.1, hypothesize how relative growth rate might change as the tree increases in size (DBH) and age.*

annual growth and turnover of leaves and fine roots may account for the majority of annual NPP, in a forest stand leaves account for only 1 to 5 percent of the total standing crop biomass (**Figure 20.18**). The production system (leaf mass) that supplies the energy is considerably less than the biomass it supports. Thus, as the woody plant grows, much of the energy goes into support and maintenance (respiration), which increase as the plant ages. This pattern of growth and energy allocation has implications for the pattern of NPP of forests through the process of stand development (see Figure 20.15c).

Stith Gower, a forest ecologist at the University of Wisconsin, examined the potential causes of declining productivity with stand age with colleagues Ross McMurtrie and Danuse Murty of Australian National University. The authors found that as the age of a forest stand increases, more and more of the living biomass occurs as woody tissue whereas the leaf area remains relatively constant or declines (**Figure 20.19a**). As the stand ages, rates of both photosynthesis and respiration decline. In addition, more of the gross production (photosynthates) is used for maintenance (respiration of woody tissues) and less remains for growth. The result is a pattern of increasing primary productivity during the early stages of stand development followed by a decline as the forest ages and the standing biomass increases (**Figure 20.19b**).

20.8 Primary Productivity Limits Secondary Production

Net primary production is the energy available to the heterotrophic component of the ecosystem. Either herbivores or decomposers eventually consume virtually all primary productivity, but often it is not all used within the same ecosystem. Humans or other agents, such as wind or water currents, may disperse the NPP of any given ecosystem to another food chain outside the ecosystem (see this chapter, *Ecological Issues & Applications*). For example, about 45 percent of the net production of a salt marsh is lost to adjacent estuarine waters (see Chapter 25).

Some energy in the form of plant material, once consumed, passes from the body as waste products (such as feces and urine). Of the energy assimilated, part is used as heat for metabolism (see Section 7.8). The remainder is available for maintenance—capturing or harvesting food, performing muscular work, and keeping up with wear and tear on the animal's body. The energy used for maintenance is eventually lost to the surrounding environment as heat. The energy left over from maintenance and respiration goes into production, including the growth of new tissues and the production of young (allocation to reproduction; see Chapter 10). This net energy allocated to production is called **secondary production** (or **secondary productivity**). Secondary production represents the formation of living mass of heterotrophic organisms over some period of time (grams per unit area per unit time). It is the heterotrophic

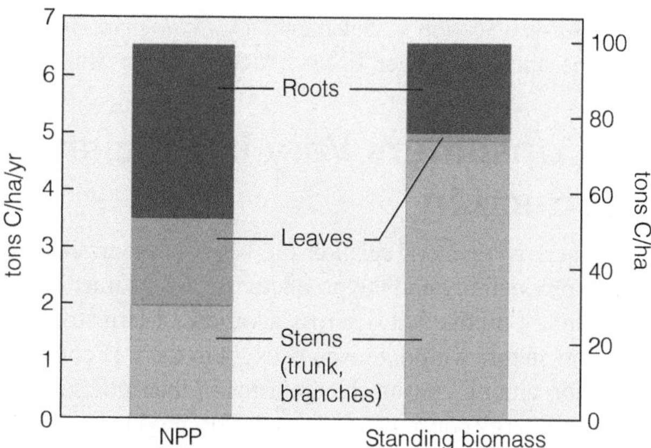

Figure 20.18 Net primary productivity (NPP) and standing biomass allocation for a 90-year old deciduous forest in Michigan estimated from forest inventories in which growth is quantified over time.
(From Gough 2011.)

equivalent of net primary production by autotrophs. Secondary productivity is greatest when the birthrate of the population and the growth rate of individuals are highest.

Secondary production depends on primary production for energy, and therefore primary productivity should function as a constraint on secondary productivity within the ecosystem. Ecologist Sam McNaughton of Syracuse University compiled data from 69 studies that reported both NPP and secondary productivity for terrestrial ecosystems ranging from Arctic tundra to tropical forests. A number of general patterns emerge. As expected, both herbivore biomass (**Figure 20.20a**) and consumption of primary productivity by herbivores (**Figure 20.20b**) increased with primary productivity. Likewise, secondary productivity of herbivores increased with primary productivity (**Figure 20.20c**). Inspection of the relationship between consumption and NPP (Figure 20.20b) revealed that forests exhibit less consumption per unit of primary productivity than do grasslands. Restricting the estimates of NPP to foliage only (leaves), however, reduced these differences.

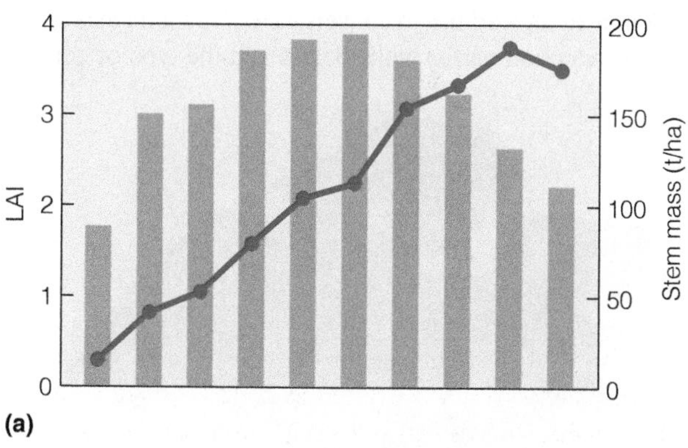

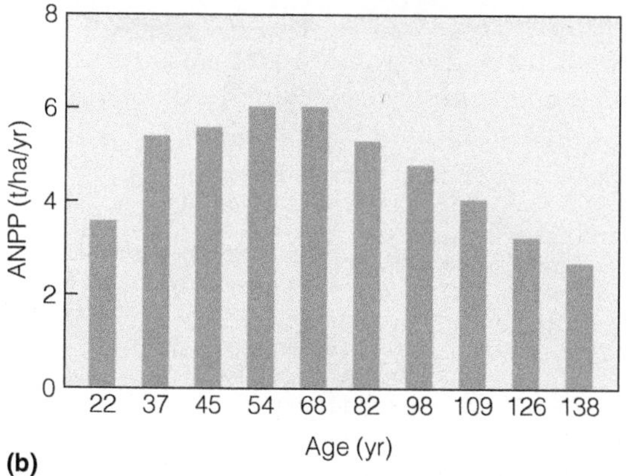

Figure 20.19 (a) Aboveground stem mass (filled circles), leaf area index (LAI; green bars), and (b) aboveground net primary productivity (ANPP) for stands of the boreal needle-leaf evergreen conifer (*Picea abies*) of differing age.
(Adapted from Gower et al. 1996.)

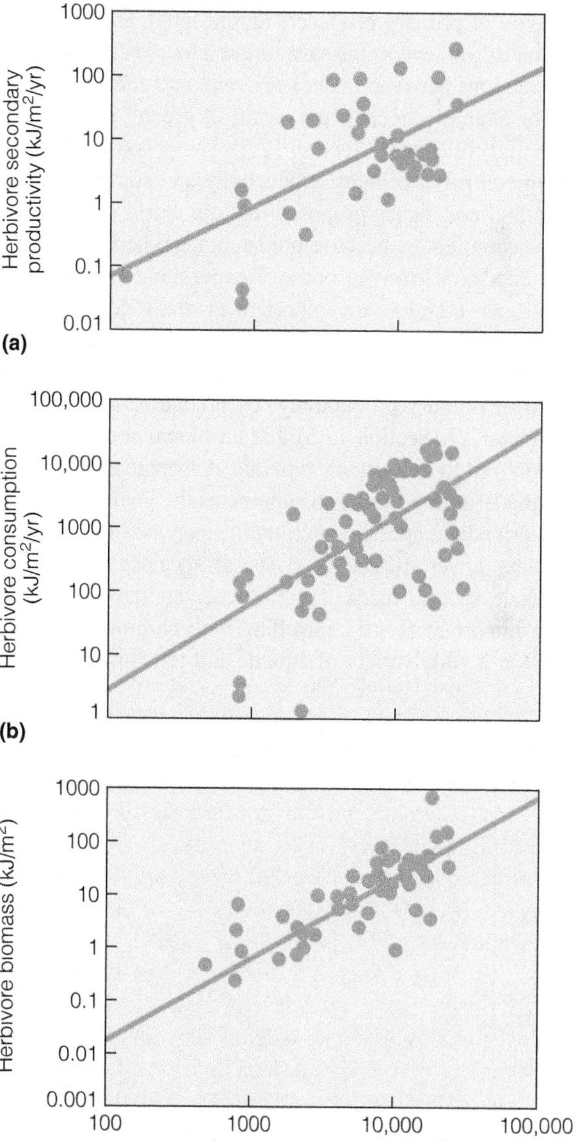

Figure 20.20 Relationship between aboveground net primary productivity and (a) net secondary productivity of herbivores, (b) consumption, and (c) herbivore biomass. Units are KJ/m^2/yr except for biomass, which is KJ/m^2. Each point represents a different terrestrial ecosystem.
(Adapted from McNaughton et al. 1989; Nature Publishing Group.)

A similar relationship to that observed by McNaughton for terrestrial ecosystems has been observed between phytoplankton production (primary productivity) and zooplankton production (secondary productivity) in lake ecosystems. Michael Brylinsky and K. H. Mann of Dalhousie University (Canada) examined the relationship between phytoplankton and zooplankton productivity in 43 lakes and 12 reservoirs distributed from the tropics to the Arctic. The researchers found a significant positive relationship between phytoplankton productivity and the productivity of both herbivorous and carnivorous (**Figure 20.21**) zooplankton.

The relationships presented in Figures 20.20 and 20.21 suggest bottom-up control of the flow of energy through the ecosystem, wherein populations and productivity of secondary producers (herbivores) are controlled by the populations and productivity of primary producers (autotrophs). However, in our discussion of food webs, we saw a more complex picture of species interactions between plant, herbivore, and carnivore populations. For example, recall "the world is green" proposition of Hairston and colleagues (Section 17.5). The proposition suggests top-down control of primary productivity and standing biomass. Hairston and colleagues proposed that the world is green (plant biomass accumulates) because predators keep herbivore populations in check. A growing body of experimental data suggests that top-down controls are important in many ecosystems and that patterns of NPP are influenced not only by abiotic conditions but also by controls on herbivore populations (and rates of consumption of primary productivity) by predators. You will recall from Chapter 17 (Section 17.5) that top-down control in a food web is referred to as a *trophic cascade*. A trophic cascade occurs when a predator in a food web suppresses the abundance of their prey (intermediate species) such that it increases the abundance of the next lower trophic level (basal species) on which the intermediate species feeds. Trophic cascades have been shown to be an important factor controlling both community structure and NPP in a wide variety of aquatic and terrestrial ecosystems (for examples see Section 17.5, Chapter 17, *Ecological Issues & Applications*, and this chapter, **Field Studies: Brian Silliman**).

20.9 Consumers Vary in Efficiency of Production

Although there is a general relationship between the availability of primary productivity and the productivity of consumer organisms (secondary productivity) across a variety of terrestrial and aquatic ecosystems, within a given ecosystem there is considerable variation among consumer organisms in their efficiency in transforming energy consumed into secondary production. These differences can be illustrated using the following simple model of energy flow through a consumer organism (**Figure 20.22**).

Of the food ingested by a consumer (*I*), a portion is assimilated across the gut wall (*A*), and the remainder is expelled from the body as waste products (*W*). Of the energy that is assimilated, some is used in respiration (*R*) and the remainder goes to production (*P*), which includes production of new tissues as well as reproduction. The ratio of assimilation to ingestion (*A*/*I*), the **assimilation efficiency**, is a measure of the efficiency with which the consumer extracts energy from food. The ratio of production to assimilation (*P*/*A*), the **production efficiency**, is a measure of how efficiently the consumer incorporates assimilated energy into secondary production.

A consumer's ability to convert the energy it ingests into secondary production varies with species and the type of consumer.

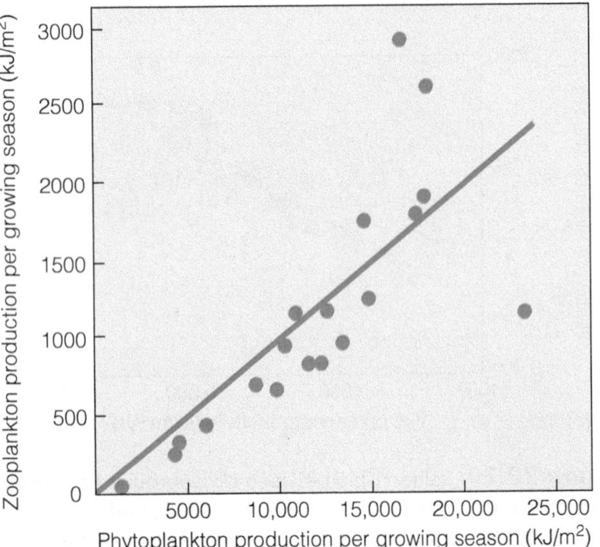

Figure 20.21 Relationship between phytoplankton (primary) and zooplankton (secondary) productivity in lake ecosystems. (Adapted from Brylinsky and Mann 1973.)

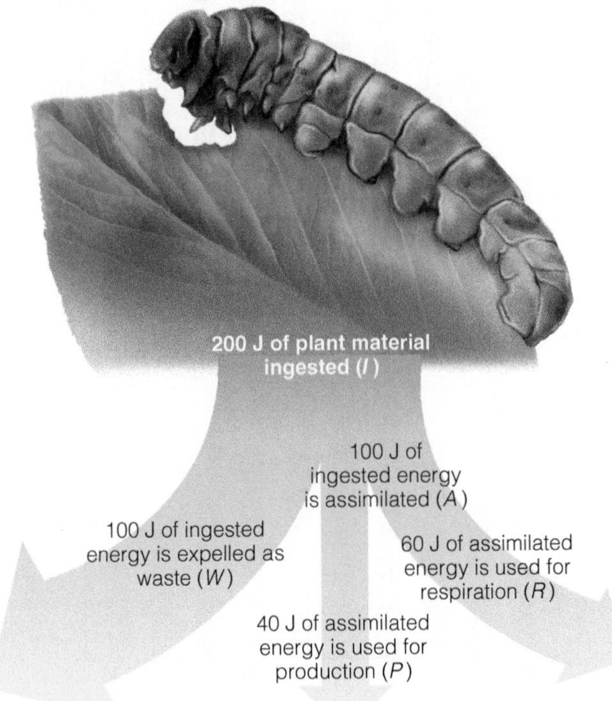

Figure 20.22 Simple model of energy flow through a consumer. Of the 200 Joules of plant material ingested by a consumer (*I*), a 100 J is assimilated across the gut wall (*A*), and the remaining 100 J is expelled from the body as waste products (*W*). Of the 100 J of energy that is assimilated, 60 J is used in respiration (*R*) and the remaining 40 J goes to production (*P*).

Assimilation efficiencies vary widely among ectotherms and endotherms. Endotherms are much more efficient than ectotherms. However, carnivores, even ectothermic ones, have a higher assimilation efficiency (approximately 80 percent) than herbivores (20 percent to 50 percent). Predatory spiders feeding on invertebrates have assimilation efficiencies of more than 90 percent.

Production efficiency varies mainly according to taxonomic class (**Table 20.2**). Invertebrates in general have high efficiencies (30–40 percent), losing relatively little energy in respiratory heat and converting more assimilated energy into production. Among the vertebrates, ectotherms have intermediate values of production efficiency (approximately 10 percent). In contrast, endotherms, with their high level of energy expenditure associated with maintaining a constant body temperature, convert only 1 to 2 percent of their assimilated energy into production (see Section 7.11).

For endotherms, body size also influences production efficiency. You may recall that the mass-specific metabolic rate (kcal/g body weight/hr) increases exponentially with decreasing body mass (Section 7.10, Figure 17.21). An increase in mass-specific metabolic rate lowers production efficiency.

20.10 Ecosystems Have Two Major Food Chains

Organic compounds fixed by autotrophs are the source of energy on which the rest of life on Earth depends. This energy stored by autotrophs is passed along through the ecosystem in a series of steps of eating and being eaten—known as a *food chain* (see Section 16.5). Feeding relationships within a food chain are defined in terms of trophic or consumer levels. From a functional rather than a species viewpoint, all organisms that obtain their energy in the same number of steps from the autotrophs or primary producers belong to the same trophic level in the ecosystem. The first trophic level belongs to the primary producers, the second level to the herbivores (first-level consumers), and the higher levels to the carnivores (second-level consumers). Some consumers occupy a single trophic level, but many others, such as omnivores, occupy more than one trophic level (see Section 7.2).

Food chains are descriptive. They represent a more abstract expression of the food webs presented earlier (Chapters 16 and 17). Major feeding groups are defined based on a common source of energy, such as autotrophs, herbivores, and carnivores. Each feeding group is then linked to others in a manner that represents the flow of energy (see Figure 17.13 for a simple food chain). Boxes represent the three feeding groups: autotrophs, herbivores, and carnivores. The arrows linking the boxes represent the direction of energy flow.

Within any ecosystem there are two major food chains: the **grazing food chain** and the **detrital food chain** (**Figure 20.23**). The distinction between these two food chains is the source of energy for the first-level consumers, the herbivores. In the grazing food chain, the source of energy is living plant biomass or net primary production. In the detrital food chain, the source of energy is dead organic matter or detritus. In turn, the herbivores in each food chain are the source of energy for the carnivores, and so on. Cattle grazing on pastureland, deer browsing in the forest, insects feeding on leaves in the forest canopy, or zooplankton feeding on phytoplankton in the water column all represent first-level consumers of the grazing food chain. In contrast, a variety of invertebrates—such as snails, beetles, millipedes, and earthworms, as well as fungi and bacteria—represent first-level consumers of the detrital food chain (see Chapter 21).

Table 20.2 Production Efficiency ($P/A \times 100$) of Various Animal Groups

Group	*P/A* (%)
Mice	4.10
Voles	2.63
Other mammals	2.92
Birds	1.26
Fish	9.74
Social insects	8.31
Orthoptera	41.67
Hemiptera	41.90
All other insects	41.23
Mollusca	21.59
Crustacea	24.96
All other noninsect invertebrates	27.68
Noninsect invertebrates	
Herbivores	18.81
Carnivores	25.05

Source: Data from Humphreys 1979.

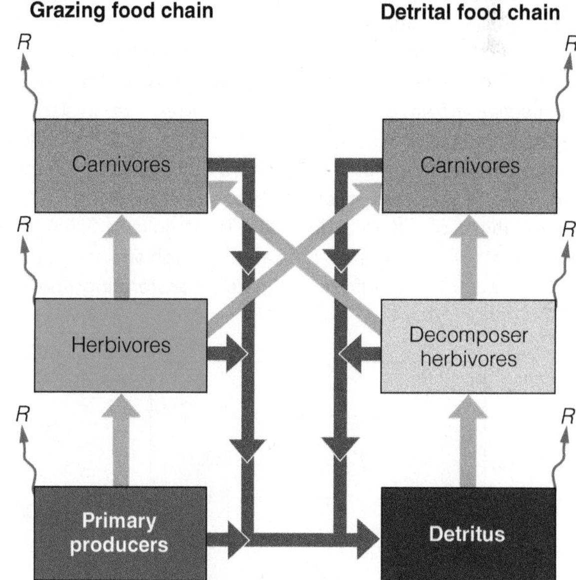

Figure 20.23 Two parts of any ecosystem: a grazing food chain and a detrital food chain. Orange arrows linking trophic levels represent the flow of energy associated with ingestion. The blue arrows from each trophic level represent the loss of energy through respiration. The brown arrows represent a combination of dead organic matter (unconsumed biomass) and waste products (feces and urine).

FIELD STUDIES *Brian Silliman*

Division of Marine Science & Conservation, Nicholas School of the Environment, Duke University, Durham, North Carolina

The salt marshes that fringe the coastline of eastern North America are among the most productive ecosystems in the world. For the past half-century, the prevailing theory found in the ecological literature and textbooks has been that productivity in these coastal ecosystems is controlled by physical conditions (water depth, frequency of inundation, salinity, etc.) and nutrient availability, referred to as "bottom-up" control (see Sections 17.5 and 20.4). But as a result of research conducted by ecologist Brian Silliman of Duke University, this long-held view is being brought into question. The focus of Silliman's research is the role of consumer organisms in the salt marshes of the southeastern United States. What has emerged from his work is a rich, complex picture of salt marshes that involves the interactions of marsh plants, "fungus-farming" snails, and an array of predators, including the blue crab.

The salt-marsh tidal zones of eastern North America are dominated by salt-marsh cordgrass (*Spartina alterniflora*). The most abundant and widespread grazer in these communities is the marsh periwinkle (*Littoraria irrorata*; see insert, **Figure 3**). The marsh periwinkle is a small snail, reaching 2.5 cm in length, with population densities upward of 500 individuals per square meter within the tidal zone dominated by *Spartina*.

While Silliman was a graduate student working in the salt marshes along the eastern shore of Virginia, he explored the role of the herbivory on patterns of net primary productivity (NPP) and standing biomass. Before Silliman's work, it was assumed that the grazers had little influence on the growth of *Spartina* plants and the overall productivity of the marsh. Snails were believed to function largely as part of the decomposer food chain, feeding on dead and dying plant tissues. Silliman designed an experiment to assess the influence of grazing snails on *Spartina* growth (individual plants) and productivity (collective growth of plants). He used cages, 1 m^2 in size, made from a fine-mesh wire fencing material, to establish different experimental treatments (see Figure 3). In some of the cages, snails were excluded; in others, snails were added to establish populations of differing densities. The results of Silliman's experiments were dramatic. Moderate to high snail densities led to runaway grazing effects, ultimately transforming one of the most productive ecosystems in the world into a barren mudflat (**Figure 1**). The effect of snails on plant growth and productivity is not through the direct consumption of green plant tissues, but by preparing the leaf tissue for colonization by their preferred food: fungus. As the snail crawls along the leaf surface, it scrapes the surface with its band of sawlike teeth called *radulae*, creating wounds (radulations) that run lengthwise on the leaf surface and kill the surrounding tissues (Figure 1). While it travels, the snail also deposits feces containing fungal spores and nutrients, effectively stimulating the establishment and growth of fungus.

In a series of follow-up experiments with colleague Steven Newell of the University of Georgia Marine Institute, Silliman demonstrated a mutualistic relationship between snails and fungi at the expense of the *Spartina* plants, the resource on which both depend. The snails employ a low-level food production strategy whereby they prepare a favorable environment for fungal growth, provide substrate to promote growth, add supplemental nutrients and propagules, and consume fungus. Although this type of facilitation, known as fungus farming, has been reported for some beetle, termite, and ant species, the work by Silliman and his colleague was the first reported case of this type of cultivation behavior in organisms other than insects.

Given the potential of the snails to so dramatically reduce plant growth and marsh primary productivity at even moderate population densities, how do the salt marshes remain so productive? To answer this question, Silliman turned his attention to the structure of the marsh's food chain. The periwinkle has several potential predators, including the terrapin, mud crab, and blue crab. Silliman hypothesized that predation maintains snail populations below the densities at which they have a devastating impact on *Spartina* plants. This type of control on

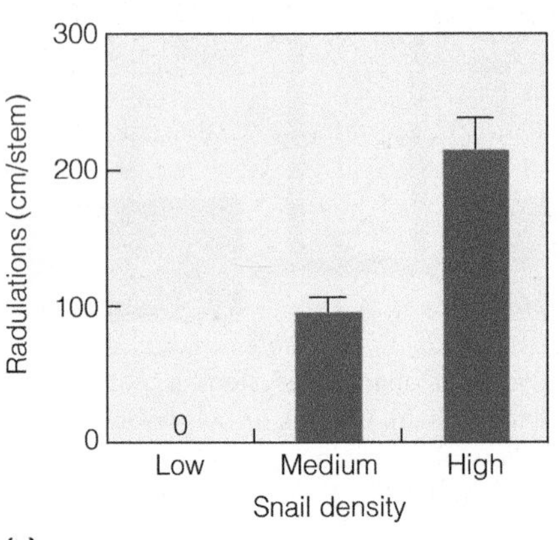

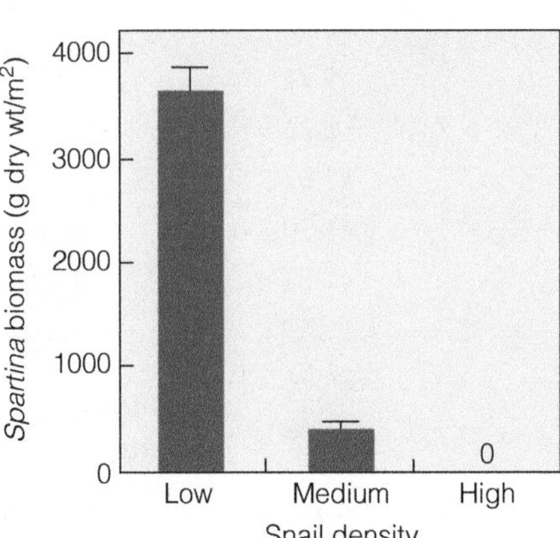

Figure 1 Effects of snail density on (a) grazing intensity (radulations) and (b) *Spartina* standing crop after eight months for experimental treatments in the tall zone of the marsh. Bars represent mean values, with ± 1 standard error of the mean indicated by vertical lines. (Adapted from Silliman and Bertness 2002.)

plant productivity, in which predators limit populations of herbivores, is called *top-down* control (see Section 17.5).

To establish the role of predators in determining the distribution of snails within the marsh, Silliman once again used wire mesh cages—this time not to exclude the snails, but to exclude their predators. Other plots (areas of marsh) without cages were monitored for comparison (control plots). The experiment ran for a year, after which the snail populations were estimated in both the exclosure and control plots. The experiment results supported the hypothesis that predators control the distribution and abundance of snail populations. Once again, the results were dramatic. The density of juvenile snails was 100 times greater in the plots where predators were excluded as compared to the control plots. Predators such as the blue crab do indeed control snail populations in the marsh.

Thanks to Silliman's research, a new, more complex picture of the salt marshes of the eastern U.S. shorelines emerged. Rather than the classical view of a bottom-up control on NPP, in which plant growth and productivity are controlled primarily by the physical conditions and nutrient availability, a new top-down model of control on plant growth and productivity emerged in which consumers exert a strong control on plant structure and productivity within the ecosystem. This phenomenon is referred to as a *trophic cascade* (see Section 17.5) because the effects of predators cascade down the food web to influence primary productivity (**Figure 2**).

How important are these top-down controls on the structure and productivity of salt marsh ecosystems? Since the mid-1990s, large expanses of southeastern salt marsh—hundreds of square kilometers—have experienced unprecedented dieback.

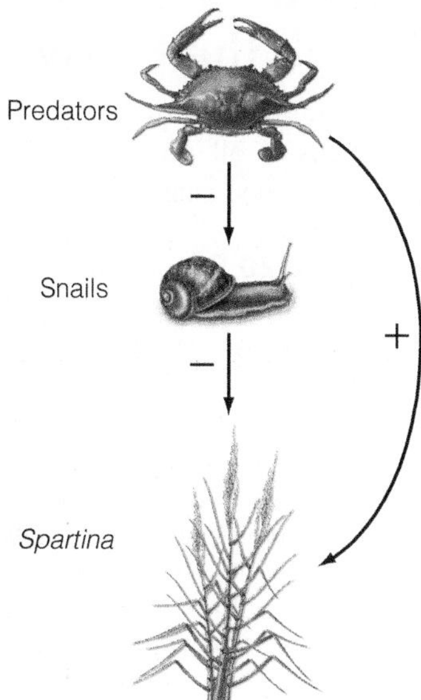

Figure 2 The proposed mechanism of the marsh *trophic cascade*. Grazing by snails has a negative impact on *Spartina* populations. Likewise, predators have a negative impact on snail populations. The net effect is that predators have a positive indirect effect on *Spartina* populations by reducing rates of herbivory by snails.

Figure 3 The marsh periwinkle on Spartina leaf shown in insert. The effect of snail exclusion cage on *Spartina* biomass on the die-off border at the Georgia marsh site.
(From Silliman et al. 2005.)

The discovery of the trophic cascade presented in Figure 2 implies that over-harvesting of snail predators, such as blue crabs, may be an important factor contributing to this regional die-off of salt marshes. Across the southeastern United States, densities of blue crabs have declined by 40 to 80 percent over the last two decades largely as a function of harvesting by humans. To examine the possible role of grazing herbivores and the concept of trophic cascade on the observed marsh die-off, Silliman and colleagues examined the extent and intensity of grazing impacts at marsh die-off sites that spanned 1200 km of shoreline. The study revealed extreme densities of plant-grazing snails, commonly 500 to 2000 individuals/m^2, aggregated in extensive fronts on die-off borders (**Figure 3**). Experimental manipulation of snail populations on the high-density borders and in remnant *Spartina* patches revealed that snail grazing contributes to the expanding dieback in these marsh ecosystems.

Bibliography

Silliman, B. R., and J. C. Zieman. 2001. "Top-down control on *Spartina alterniflora* production by periwinkle grazing in a Virginia salt marsh." *Ecology* 82:2830–2845.

Silliman, B. R., and M. D. Bertness. 2002. "A trophic cascade regulates salt marsh primary production." *Proceedings of the National Academy of Sciences USA* 99:10500–10505.

Silliman, B. R., and S. Y. Newell. 2003. "Fungal farming in a snail." *Proceedings of the National Academy of Sciences USA* 100:15643–15648.

Silliman, B. R., J. van de Koppel, M. D. Bertness, L. E. Stanton and I. A. Mendelssohn. 2005. "Drought, snails, and large-scale die-off of southern U.S. salt marshes." *Science* 310:1803–1806.

1. How does the model of trophic cascade presented in Figure 3 relate to the concept of indirect mutualism developed in Chapter 17 (Section 17.3)?

2. Suppose we were to introduce a top predator to this ecosystem that feeds on blue crabs. How might this action change the nature of the trophic cascade outlined in Figure 3?

Figure 20.23 combines the two food chains to produce a generalized model of trophic structure and energy flow through an ecosystem. The two food chains are linked. The initial source of energy for the detrital food chain is the input of dead organic matter and waste materials from the grazing food chain. This linkage appears as a series of brown arrows from each of the trophic levels in the grazing food chain, leading to the box designated as detritus or dead organic matter. There is one notable difference in the flow of energy between trophic levels in the grazing and decomposer food chains. In the grazing food chain, the flow is unidirectional, with net primary production providing the energy source for herbivores, herbivores providing the energy for carnivores, and so on. In the decomposer food chain, the flow of energy is not unidirectional. The waste materials and dead organic matter (organisms) in each of the consumer trophic levels are "recycled," returning as an input to the dead organic matter box at the base of the detrital food chain. In addition, the distinction between the grazer and consumer food chains is often blurred at the higher trophic levels (carnivores) because predators rarely distinguish whether prey draw their resources from primary producers or detritus. For example, the diet of an insectivorous bird might include beetle species that feed on detritus, as well as species that feed on green leaf tissues.

20.11 Energy Flows through Trophic Levels Can Be Quantified

To quantify the flux of energy through the ecosystem using the conceptual model of the food chains just discussed, we need to return to the processes involved in secondary production discussed in Section 20.9: consumption, ingestion, assimilation, respiration, and production. We will examine a single trophic compartment (**Figure 20.24a**). The energy available to a given trophic level (designated as n) is the production of the next-lower level ($n - 1$); for example, net primary production (P_1) is the available energy for grazing herbivores (trophic level 2). Following the simple model of energy flow through a consumer organism presented in Section 20.9, some proportion of that productivity is consumed or ingested (I); the remainder makes its way to the dead organic matter of the detrital food chain. Some portion of the energy consumed is assimilated by the organisms (A), and the remainder is lost as waste materials (W) to the detrital food chain. Of the energy assimilated, some is lost to respiration, shown as the arrow labeled R that is leaving the upper left corner of the box, and the remainder goes to production (P). We quantify this flow with the energetic efficiencies defined in Section 20.9: the assimilation efficiency (A/I) and the production efficiency (P/A). One additional index of energetic efficiency, however, must be introduced: consumption efficiency. The **consumption efficiency**, the ratio of ingestion to production at the next-lower trophic level (I_n/P_{n-1}) defines the amount of available energy being consumed. Sample values of these efficiencies for an invertebrate herbivore in the grazing food chain are provided in **Figure 20.24b**. Using these efficiency values, we can track the fate of a given amount of energy (1000 kcal) available to herbivores in the form of NPP through the herbivore trophic level.

If we apply efficiency values for each trophic level in the grazing and detrital food chains, we can calculate the flow of energy through the whole ecosystem. The production from each trophic level provides the input to the next-higher level, and unconsumed production (dead individuals) and waste products from each trophic level provide input into the dead organic matter compartment. The entire flow of energy through the ecosystem is a function of the initial transformation of solar energy into NPP. All energy entering the ecosystem as NPP is eventually lost through respiration.

20.12 Consumption Efficiency Determines the Pathway of Energy Flow through the Ecosystem

Although the general model of energy flow presented in Figure 20.23 pertains to all ecosystems, the relative importance of the two major food chains and the rate of energy flow

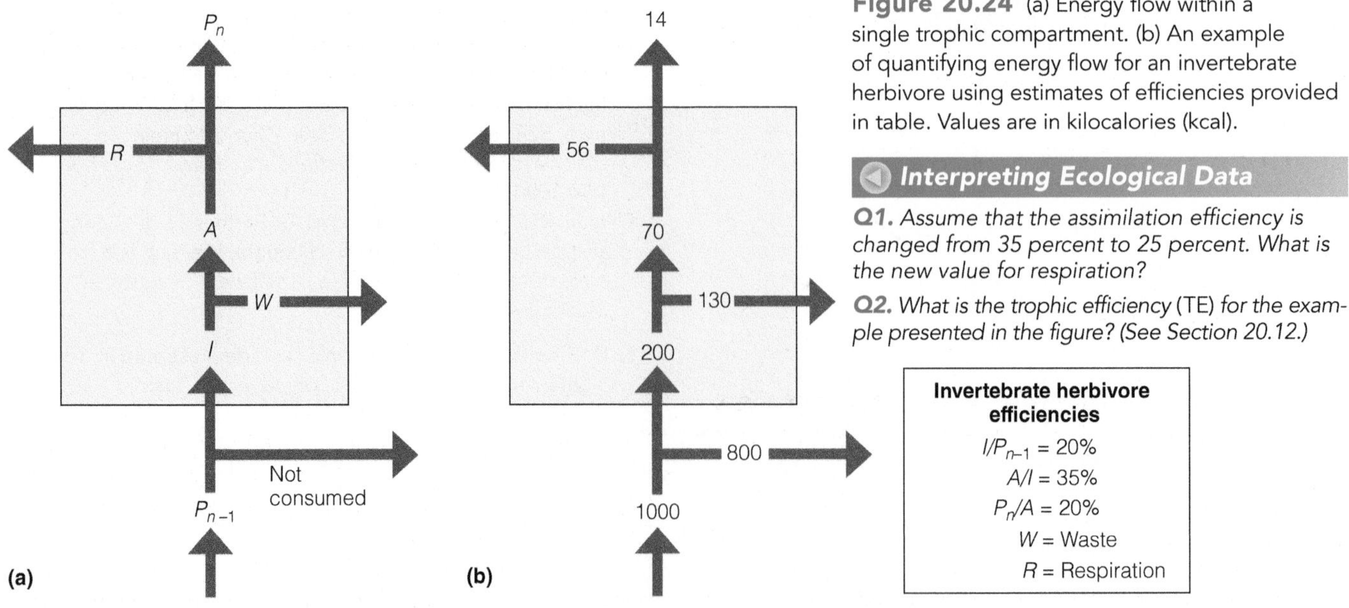

Figure 20.24 (a) Energy flow within a single trophic compartment. (b) An example of quantifying energy flow for an invertebrate herbivore using estimates of efficiencies provided in table. Values are in kilocalories (kcal).

Interpreting Ecological Data

Q1. *Assume that the assimilation efficiency is changed from 35 percent to 25 percent. What is the new value for respiration?*

Q2. *What is the trophic efficiency (TE) for the example presented in the figure? (See Section 20.12.)*

Invertebrate herbivore efficiencies
I/P_{n-1} = 20%
A/I = 35%
P_n/A = 20%
W = Waste
R = Respiration

through the various trophic levels can vary widely among different types of ecosystems. The consumption efficiency (I_n/P_{n-1}) defines the amount of available energy produced by any given trophic level (P_{n-1}) that is consumed by the next-higher level (I_n). Values of consumption efficiency for the various consumer trophic levels therefore determine the pathway of energy flow through the food chain, providing a basis for comparison of energy flow through different ecosystems.

Despite its conspicuousness, the grazing food chain is not the major one in most terrestrial and many aquatic ecosystems. Only in some open-water aquatic ecosystems do the grazing herbivores play the dominant role in energy flow. Ecologists Helene Cyr of the University of Toronto (Canada) and Michael Pace of the University of Virginia compiled published measurements of herbivore consumption rates (herbivore consumption efficiency), herbivore biomass, and primary productivity for a wide range of aquatic and terrestrial ecosystems (**Figure 20.25**). Although there is considerable variation in both environments, some generalizations do emerge from their analysis. Aquatic ecosystems dominated by phytoplankton have higher rates of herbivory (median value of 79 percent) than do those in which vascular plants (submerged and emergent) dominate (median value of 30 percent). In contrast, only 17 percent of primary productivity (median value) is removed by herbivores in terrestrial ecosystems. Therefore, in most terrestrial and shallow-water ecosystems, with their high standing biomass and relatively low harvest of primary production by herbivores, the detrital food chain is dominant. In deep-water aquatic ecosystems, with their low standing biomass, rapid turnover of organisms, and high rate of harvest, the grazing food chain may be dominant.

In terrestrial ecosystems, distinct differences in consumption efficiency and energy flow exist between forest and grassland ecosystems. Nelson Hairston of Cornell University reviewed a wide range of studies that examined patterns of energy flow through terrestrial ecosystems, providing a comparison of consumption efficiencies for herbivores (primary producer → herbivore) and their predators (herbivore → carnivore). The author found an average consumption efficiency of 3.7 percent for herbivores inhabiting deciduous forest ecosystems, whereas herbivores inhabiting grassland ecosystems had a value of 9.3 percent (both values lower than the average for terrestrial ecosystems reported by Cyr and Pace). Much smaller differences were observed for the consumption efficiency of predators inhabiting the two ecosystem types. Predators inhabiting forests had a value of 89.9 percent, whereas predators inhabiting grassland ecosystems had an average value of 77 percent.

Patterns of energy flow through flowing-water ecosystems (streams and rivers) differ markedly from both terrestrial and standing-water ecosystems (lakes and oceans). By comparison, stream and river ecosystems have extremely low NPP, and the grazing food chain is minor (see Chapter 24). The detrital food chain dominates and depends on inputs of dead organic matter from adjacent terrestrial ecosystems (see Section 20.4).

Figure 20.26 graphically represents the different patterns of energy transfer in the four different ecosystems just discussed: forest, grassland, standing water, and flowing water.

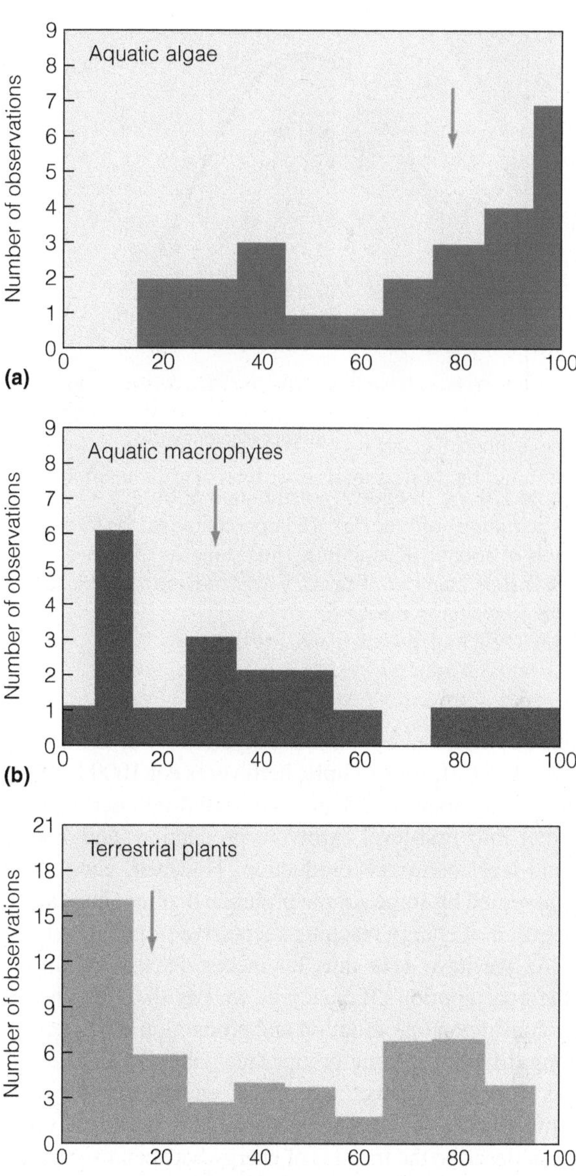

Figure 20.25 Results from a review of studies that examined rates of herbivory in different ecosystems. Histograms represent the percentage of net primary productivity consumed by herbivores in ecosystems dominated by (a) algae (phytoplankton), (b) rooted aquatic plants, and (c) terrestrial plants. Number of observations refers to the number of experiments having a given level of consumption. Red arrows indicate the median value. Note that herbivores consume a significantly greater proportion of phytoplankton productivity than do either aquatic or terrestrial plants. (Adapted from Cyr and Pace 1993.)

20.13 Energy Decreases in Each Successive Trophic Level

Based on the preceding discussion and the analysis presented in Figure 20.24, we can conclude that the quantity of energy flowing into a trophic level decreases with each successive trophic level in the food chain. This pattern occurs because not all energy is used for production. An ecological rule of thumb is that only 10 percent of the energy stored as biomass in a

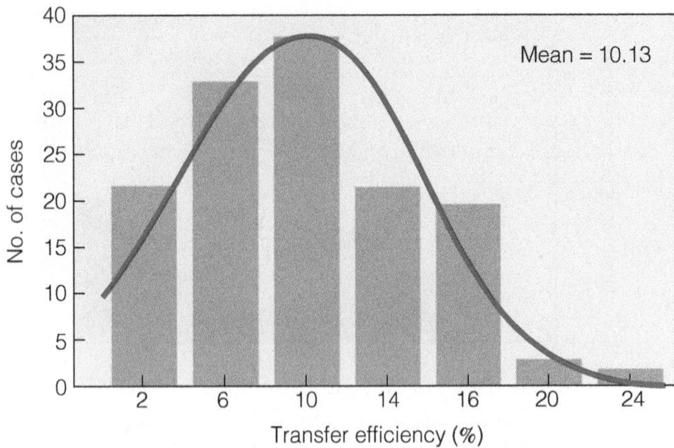

Figure 20.26 Frequency distribution of trophic efficiencies (energy transfer efficiencies: *TE* in percentages) for 48 tropic models of aquatic ecosystems. The estimates of *TE* are based on 140 transfers (number of cases: *y*-axis) between trophic levels in the 48 ecosystems modeled.
(Adapted from Pauly and Christensen 1995.)

given trophic level is converted to biomass at the next-higher trophic level. If, for example, herbivores eat 1000 kcal of plant energy, only about 100 kcal is converted into herbivore tissue, 10 kcal into first-level carnivore production, and 1 kcal into second-level carnivore production. However, ecosystems are not governed by some simple principle that regulates a constant proportion of energy reaching successive trophic levels.

As we have seen thus far in our discussion, differences in the consumption efficiency as well as the efficiency of energy conversion (assimilation and production efficiencies) exist among different feeding groups (see Table 20.2). These differences directly influence the rate of energy transfer from one trophic level to the next-higher level. A measure of efficiency used to describe the transfer of energy between trophic levels is called the *trophic efficiency*. The **trophic efficiency** (*TE*) is the ratio of productivity in a given trophic level (P_n) to the trophic level its organisms feed on (P_{n-1}):

$$TE = P_n / P_{n-1}$$

Daniel Pauly and Villy Christensen of the University of British Columbia examined the energy transfer efficiency reported in 48 different studies of aquatic ecosystems (**Figure 20.26**). There is considerable variation among studies and trophic levels, but the mean value of 10.13 percent is close to the general rule of 10 percent transfer between trophic levels.

An important consequence of decreasing energy transfers through the food web is a corresponding decrease in the standing biomass of organisms within each successive trophic level. If we sum all of the biomass or energy contained in each trophic level, we can construct pyramids for the ecosystem (**Figure 20.27**). The pyramid of biomass indicates by weight, or other means of measuring living material, the total bulk of organisms or fixed energy present at any one time—the standing crop. Because some energy or material is lost at each successive trophic level, the total mass supported at each level is limited by the rate at which energy is being stored at the next-lower level. In general, the biomass of producers must be greater than that of the herbivores they support, and the biomass of herbivores must be greater than that of carnivores. That circumstance results in a narrowing pyramid for most ecosystems (Figure 20.27a).

This arrangement does not hold for all ecosystems. In such ecosystems as lakes and open seas, primary production is concentrated in the phytoplankton. These microscopic organisms have a short life cycle and rapid reproduction. They are heavily grazed by herbivorous zooplankton that are larger and longer-lived. Thus, despite the high productivity of algae, their biomass is low compared to that of zooplankton herbivores (Figure 20.27b). The result is an inverted pyramid, with a lower standing biomass of primary producers (phytoplankton) and herbivores (zooplankton).

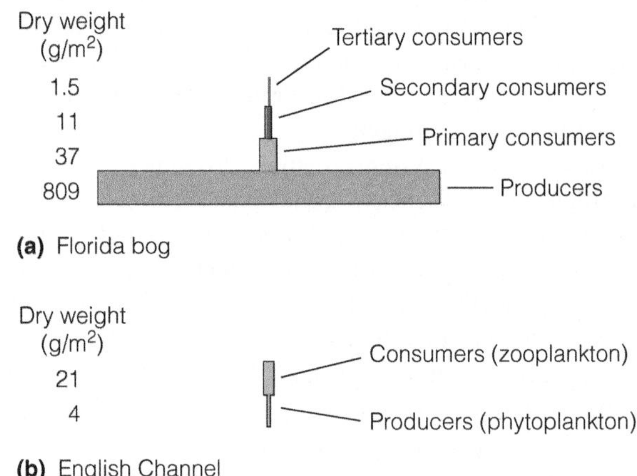

Figure 20.27 Biomass pyramids for the consumer food chain of (a) a bog ecosystem in Florida and (b) the marine ecosystem of the English Channel. The pyramid for the marine ecosystem is inverted due to the high productivity but fast turnover of phytoplankton populations (short life span and high rate of consumption by zooplankton).

ECOLOGICAL Issues & Applications: Humans Appropriate a Disproportionate Amount of Earth's Net Primary Productivity

Although we represent only 1 of more than 1.4 million known species inhabiting our planet, humans use a vastly disproportionate share of Earth's resources. This pattern of resource use is most evident in the human appropriation of NPP. The basic question of how much of Earth's primary productivity is used by humans was first posed in the 1970s by the ecologists Robert Whittaker and Gene Likens at Cornell University, yet it would take more than a decade before the first comprehensive estimate would emerge. In a paper published in 1986, ecologist Peter Vitousek and colleagues at Stanford University used three different approaches to estimate the fraction of NPP that humans have appropriated. A low estimate was derived by calculating the amount of NPP people use directly—as food, fuel, fiber, or timber. An intermediate estimate was also calculated that includes all the productivity of lands devoted entirely to human activities (such as the NPP of croplands, as opposed to the portion of crops actually eaten). This estimate also included the energy human activity consumes, such as in setting fires to clear land. A third approach provided a high estimate, which further included productive capacity lost as a result of converting open land to cities and forests to pastures or because of desertification or overuse (overgrazing, excessive erosion). Results of these three approaches (low, intermediate, and high) yielded a wide range of estimates: 3 percent, 19 percent, and 40 percent, respectively. These estimates are a remarkable level of use for a species that represents approximately 0.5 percent of the total heterotrophic biomass on Earth.

Since the initial analyses conducted by Vitousek and colleagues in the mid-1980s, advances in satellite technology have greatly enhanced scientists' ability to monitor patterns of land use and primary productivity at a continental to global scale. In 2004, Marc Imhoff and Lahouari Bounoua, researchers at NASA's Goddard Space Flight Center (GSFC; Greenbelt, Maryland), and colleagues undertook an analysis of patterns of human use of NPP at a global scale using satellite-derived data (see **Quantifying Ecology 20.1**).

The researchers defined human appropriation of terrestrial NPP (HANPP) as the amount of terrestrial NPP required to derive food and fiber products consumed by humans, including the organic matter that is lost during harvesting and processing of whole plants into end products. This definition most closely approximates the intermediate scenario defined by Vitousek and colleagues. Using data compiled by the Food and Agricultural Organization (FAO) of the United Nations on products consumed in 1995 for 230 countries in seven categories (plant foods, meat, milk, eggs, wood, paper, and fiber), a per capita value of HANPP was calculated for each country. The per capita estimate of HANPP was then applied to a gridded database of the human population at a spatial resolution of 0.25° (latitude and longitude). This spatial scale was chosen to match the spatial resolution of NPP derived from the satellite-derived vegetation index (NDVI; see Quantifying Ecology 20.1). The resulting map depicts the spatial pattern of HANPP, showing where the products of terrestrial photosynthesis are consumed (**Figure 20.28a**). By combining the maps of global NPP and HANPP, the researchers were able to map HANPP as a percentage of NPP, providing a spatially explicit balance of NPP "supply" and "demand" (**Figure 20.28b**). The resulting map reveals a great deal of regional variation in the appropriation of NPP.

Summing for the globe, the researchers estimate annual HANPP to be 24.2 Pg organic matter (1 Pg = 10^{15} g), or approximately 20 percent of terrestrial annual NPP (with high and low estimates of 14–26 percent). Some regions, however, such as Western Europe and south central Asia, consume more than 70 percent of their NPP. In contrast, HANPP in other regions (typically in the wet tropics) is less than 15 percent. The lowest value (approximately 6 percent) is found in South America.

Both population and per capita consumption interact to determine the human ecological impact at a regional scale. From Figure 20.29b, the role of population is clear despite vast differences in consumption among nations. For example, east and south central Asia, with almost half of the world's population, appropriates 72 percent of its regional NPP despite having the lowest per capita consumption of any region. Affluence also plays an important role. The average annual per capita HANPP for industrialized countries is almost double that of developing

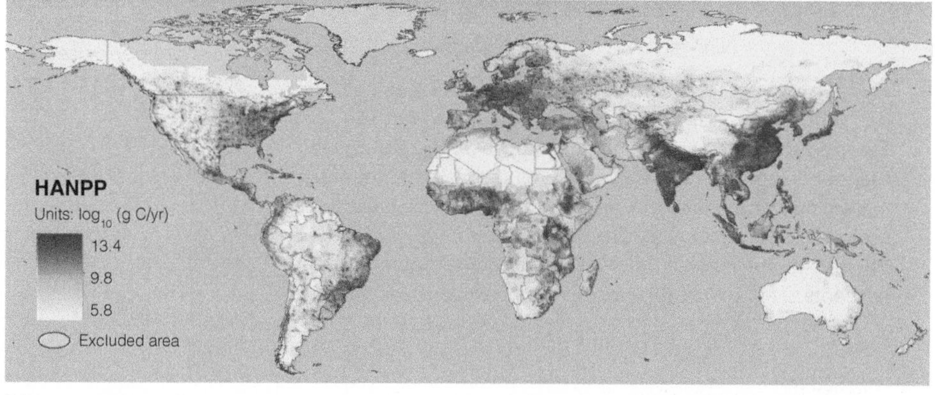

(a)

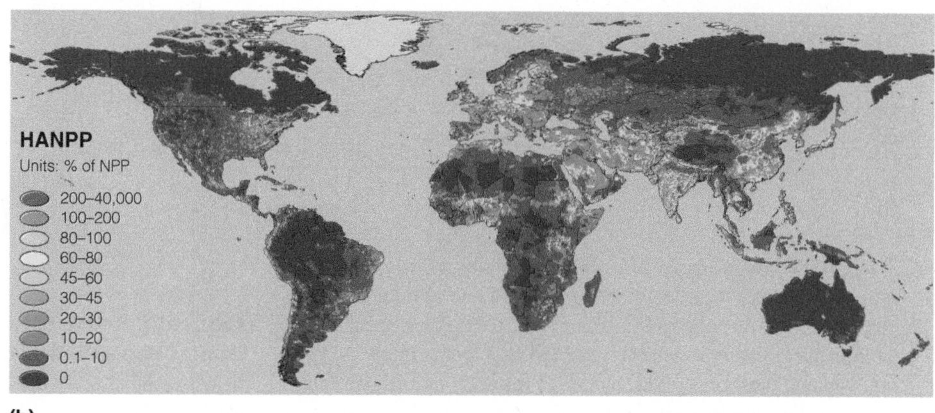

(b)

Figure 20.28 Spatial distribution of the annual NPP resources required by the human population, as measured by (a) HANPP, and (b) HANPP as a percentage of local NPP. (Imhoff et al. 2004.)

QUANTIFYING ECOLOGY 20.1 Estimating Net Primary Productivity Using Satellite Data

Terrestrial ecologists use a variety of techniques to estimate net primary productivity (NPP) at a site. Methods range from clipping sample plots in grasslands to estimating seasonal grass production, measuring the diameter growth of individual trees within a forest stand, to continually measuring the exchange of carbon dioxide between the vegetation and the atmosphere. Estimating primary productivity over a much larger area, however, requires a level of sampling that is generally not feasible. One way that is being used by ecologists to estimate NPP over large areas is a simple model based on the maximum efficiency of carbon (carbon dioxide) uptake by plants for each unit of sunlight absorbed by the leaves (grams C per unit of solar energy per second; see Figure 6.2). Many researchers have estimated this maximum efficiency (called e_{max}) for a variety of species and ecosystems by relating field-based estimates of net primary productivity with measures of absorbed photosynthetically active radiation (PAR; see Section 6.2; **Figure 1**). Results indicate that the maximum amount of carbon fixed into plant tissues per unit of sunlight absorbed is roughly 0.4–0.8 g C per megajoule of solar energy. These estimates are for natural ecosystems. Highly productive agricultural systems that are irrigated and fertilized can have e_{max} values of more than 2.0.

Using this approach, NPP can be estimated using the following equation:

$$NPP = APAR \times e_{max} \times Efact$$

where APAR is the absorbed PAR (megajoules per unit area per year), e_{max} is the maximum efficiency of light use, and *Efact* is a factor that accounts for the processes that prevent the plant from reaching e_{max} (such as resource limitations or extreme temperatures). A variety of techniques have been proposed for calculating *Efact*, but in general, *Efact* is expressed as some function of precipitation, temperature, and nutrients. The key to this approach is having an estimate of the amount of PAR absorbed by the plant canopy over a given area, and for this ecologists turn to satellite technology. A National Oceanic and Atmospheric Administration (NOAA) meteorological satellite has an onboard sensor that records spectral reflectance in the visible and infrared portions of the electromagnetic spectrum (see Figure 2.1). The advanced, very high resolution radiometer (AVHRR) on the NOAA satellite provides global coverage at a spatial resolution of 1.1 km at least once per day during the daylight hours.

Because green plants use PAR (400–700 nm) in the process of photosynthesis, they display a unique spectral reflectance pattern in the visible and near-infrared wavelengths. This unique spectral signature has been used to develop a vegetation index that discriminates living plants from the surrounding rock, soil, and water: the normalized difference vegetation index (NDVI) is calculated as follows:

$$NDVI = \frac{(NIR + VIS)}{(NIR - VIS)}$$

where NIR is canopy reflectance in the near-infrared (700–1200 nm), and VIS is canopy reflectance in the visible part of the spectrum (400–700 nm).

Because vegetation has a uniquely high reflectance in the NIR and uniquely low reflectance in the VIS, the NDVI goes up with increasing absorption of PAR. The NDVI index has therefore been used to predict APAR at a regional and global scale for use in the simple model of NPP presented previously (**Figure 2**). An example of the application of this approach is presented in this chapter, *Ecological Issues & Applications*.

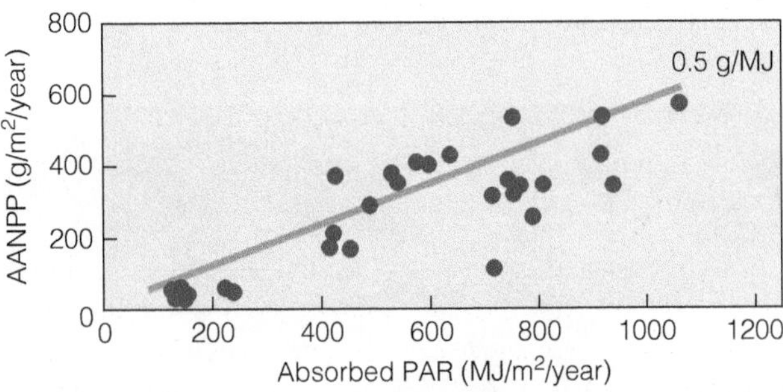

Figure 1 Relationship between annual absorbed photosynthetically active radiation (APAR) and annual aboveground net primary productivity (AANPP) for spruce stands in northern Minnesota. Annual e_{max} (maximum efficiency of carbon uptake by plants for each unit of sunlight absorbed by the leaves) is estimated by the simple linear regress of AANPP as a function of APAR: 0.5 g/MJ.
(Data from Goetz et al. 1996.)

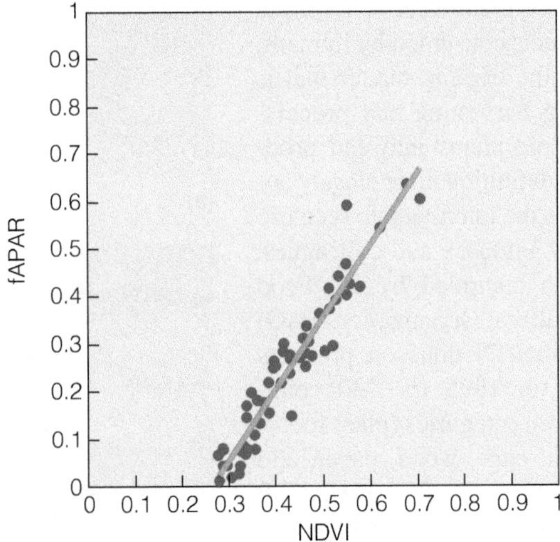

Figure 2 Relationship between the fraction of photosynthetically active radiation absorbed (fAPAR) and NDVI for 68 sites in the Sahel region of North Africa. fAPAR = APAR/PAR, where PAR is the total photosynthetically active radiation at the site over the year.
(Adapted from Fensholt et al. 2004.)

nations, which are home to 83 percent of the global population. If the per capita HANPP of the developing nations increased to match that of the industrialized countries, global HANPP would increase by 75 percent to a value of 35 percent of the current global NPP.

As technologies improve our ability to monitor patterns of land-use change at a global scale, estimates of the impacts of human activities continue to be updated. In 2007, Helmut Haberl, Fridolin Krausmann, and colleagues at the Institute of Social Ecology, Klagenfurt University (Vienna, Austria) and Potsdam Institute for Climate Impact Research (Potsdam, Germany) published an updated analysis of the human appropriation of NPP using improved global databases and integrating them into a high-resolution geographic information systems (GIS) data set. These data, in combination with a dynamic global vegetation model (DGVM), were used to derive a comprehensive assessment of global HANPP. Their analyses provide an estimate of human-induced changes in ecosystems on a global grid with 5-minute geographical resolution (approximately 10 × 10 km at the equator) for the year 2000. The authors define HANPP as the combined effect of harvest and productivity changes induced by land use on the availability of NPP in ecosystems. That is, HANPP is calculated as the difference between the NPP of the potential plant cover that would prevail in the absence of human activities and the fraction of NPP remaining in ecosystems after their harvest by humans.

Although Haberl and colleagues's definition of HANPP differs slightly from that of the previous analysis by Imhoff and Bounoua, their overall results are similar. The researchers found an aggregate global HANPP value of 15.6 Pg C/yr or 23.8 percent of potential NPP (as compared to approximately 20 percent in the previous analysis), of which 53 percent was contributed by harvest, 40 percent by land-use-induced productivity changes, and 7 percent by human-induced fires. The regional patterns were likewise similar, with the highest observed values in southern Asia (63 percent) and Europe (40–52 percent), whereas the topical regions of Central and South America and sub-Saharan Africa had some of the lowest values of HANPP (16–18 percent).

Given the inherent variation (definition of HANPP, data sources, spatial resolution, time period of analysis) in the three studies presented, they are remarkably similar in their general findings (HANPP of approximately 20 percent). Yet growth of the human population, economic development, and technological advances have brought about remarkable changes in both the supply and demand for basic natural resources over this period. How have these changes influenced HANPP over the past century, and how might they influence demand into the future?

In 2013, Fridolin Krausmann, Helmut Haberl, and their colleagues published a follow-up analysis examining the trends in HANPP over the period from 1910 to 2005. Over this period the human population has increased more than 4-fold and economic output by 17-fold; however, the researchers estimate the HNAPP has only increased by approximately 2-fold, increasing from 13 to 25 percent over this same period. How was the increase in demand resulting from the exponential increases in human population and economic growth met with only a doubling of HANPP over the 20th century? The answer is the dramatic increase in the primary productivity of lands under active management that has occurred over this period of time; namely dramatic increases in agricultural yields (approximately 4-fold for global grain yields, see Chapter 21, *Ecological Issues & Applications*). This increase in efficiency, however, has come at serious environmental costs (see this chapter, *Ecological Issues & Applications* for discussion and example).

So what does the future hold? Over the next four decades, the human population is expected to grow by some 40 percent and the world economy by 3-fold. On the other hand, agricultural production is predicted to grow by 60 to 100 percent over the same period. The researchers predict that if we continue with the rate of gain in efficiency realized over the past century, then HANPP might be limited to 27 to 29 percent by the year 2050. However, providing for the growing energy demand through the use of biofuels could increase global HANPP to as high as 44 percent, further increasing the human appropriation of Earth's natural resources.

SUMMARY

Laws of Thermodynamics 20.1

Energy flow in ecosystems supports life. Energy is governed by the laws of thermodynamics. The first law states that although energy can be transferred, it cannot be created or destroyed. The second law states that as energy is transferred, a portion ceases to be usable. As energy moves through an ecosystem, much of it is lost as heat of respiration. Energy degrades from a more organized to a less organized state (entropy). However, a continuous flux of energy from the Sun prevents ecosystems from running down.

Primary Production 20.2

The flow of energy through an ecosystem starts with the harnessing of sunlight by green plants through a process referred to as *primary production*. The total amount of energy fixed by plants is *gross primary production*. The amount of energy remaining after plants have met their respiratory need is net primary production in the form of plant biomass. The rate of primary production is *net primary productivity*, which is measured in units of weight per unit area per unit time.

Terrestrial Ecosystems 20.3

Productivity of terrestrial ecosystems is influenced by climate, especially temperature and precipitation. Temperature influences the photosynthetic rate, and the amount of available water limits photosynthesis and the amount of leaves that can be supported. Warm, wet conditions make the tropical rain forest the most productive terrestrial ecosystem. Nutrient availability also directly influences rates of primary productivity.

Aquatic Ecosystems 20.4

Light is a primary factor limiting productivity in aquatic ecosystems, and the depth to which light penetrates is crucial to determining the zone of primary productivity. Nutrient availability is the most pervasive influence on the productivity of oceans. The most productive ecosystems are shallow coastal waters, coral reefs, and estuaries, where nutrients are more available. Nutrient availability is also a dominant factor limiting net primary productivity in lake ecosystems. In rivers and streams, net primary productivity is low, with inputs of dead organic matter from adjacent terrestrial ecosystems being an important source of energy input.

External Inputs 20.5

In many aquatic ecosystems, a substantial proportion of organic carbon is derived from dead organic matter from adjacent terrestrial ecosystems. The relative importance of external sources of organic carbon varies widely among different aquatic ecosystems. In large rivers, lakes, and most marine systems, the majority of organic carbon is derived internally from photosynthesis by autotrophs. In contrast, in smaller streams and lakes the dominant source is often external sources of organic carbon.

Energy Allocation 20.6

Energy fixed by plants is allocated to different parts of the plant and to reproduction. How much is allocated to each component is a function of the plant life-form as well as the environmental conditions. The pattern of allocation directly influences standing biomass and productivity rate.

Temporal Variation 20.7

Primary production in an ecosystem varies with time. Seasonal and yearly variations in moisture and temperature directly influence primary production. In ecosystems dominated by woody vegetation, net primary production declines with age. As the ratio of woody biomass to foliage increases, more of gross production goes into maintenance.

Secondary Production 20.8

Net primary production is available to consumers directly as plant tissue or indirectly through animal tissue. Change in biomass of heterotrophs, including weight change and reproduction, is secondary production. Secondary production depends on primary production. Any environmental constraint on primary production will constrain secondary production in the ecosystem.

Efficiency of Energy Use 20.9

Efficiency of production varies. Endotherms have high assimilation efficiency but low production efficiency because they must expend so much energy on respiration. Ectotherms have low assimilation efficiency but high production efficiency because lower rates of respiration enable them to allocate more energy toward growth.

Food Chains and Energy Flow 20.10

The various members of a food web can be grouped into categories called *trophic* or *feeding levels*. Autotrophs occupy the first trophic level. Herbivores that feed on autotrophs make up the next trophic level. Carnivores that feed on herbivores make up the third and higher trophic levels.

Energy flow in ecosystems takes two routes: one through the grazing food chain, the other through the detrital food chain. The bulk of production is used by organisms that feed on dead organic matter. The two food chains are linked by the input of dead organic matter and wastes from the consumer food chain functioning as input into the detrital food chain.

Quantifying Energy Flow 20.11

At each trophic level, estimates of the efficiency of energy exchange are defined as *consumption efficiency*, which is the proportion of available energy being consumed; *assimilation efficiency*, which is the portion of energy ingested that is assimilated and not lost as waste material; and *production efficiency*, which is the portion of assimilated energy that goes to growth rather than respiration. These estimates of efficiency can be used to quantify the flow of energy through the food chain.

Consumption Efficiency 20.12

Consumption efficiency determines the flow of energy through the ecosystem. The detrital food chain dominates in terrestrial ecosystems, with only a small proportion of net primary productivity being consumed by herbivores. In open-water ecosystems, such as lakes and oceans, a greater proportion of primary productivity is consumed by herbivores. Consumption efficiency of predators is more similar among these ecosystems.

Energy Pyramids 20.13

The quantity of energy flowing into a trophic level decreases with each successive trophic level in the food chain. This pattern occurs because not all energy is used for production. An ecological rule of thumb is that only 10 percent of the energy stored as biomass in a given trophic level is converted to biomass at the next-higher trophic level. A plot of the total weight of individuals at each successive level produces a tapering pyramid. In aquatic ecosystems, however, where there is a rapid turnover of small aquatic producers, the pyramid of biomass becomes inverted.

Human Consumption Ecological Issues & Applications

Although humans represent only approximately 0.5 percent of the total heterotrophic biomass on Earth, our species accounts for the appropriation of approximately 20 percent of the planet's net primary productivity.

STUDY QUESTIONS

1. How do the concepts of community and ecosystem differ?
2. Relate the following terms: gross primary productivity, autotrophic respiration, and net primary productivity.
3. Contrast net primary productivity and standing biomass for an ecosystem.
4. How do temperature and precipitation interact to influence net primary productivity in terrestrial ecosystems?
5. How does net primary productivity vary with water depth in standing-water ecosystems (lakes and oceans)? What is the basis for the vertical profile of net primary productivity in these ecosystems?
6. What environmental factors might influence the (light) compensation depth of a lake ecosystem?
7. How does primary productivity function as a constraint on secondary productivity in ecosystems?
8. What does the top-down model of food chain structure imply about the role of secondary producers in controlling net primary productivity and standing biomass within ecosystems?
9. How do assimilation efficiency and production efficiency relate to the flow of energy through a trophic level?
10. What is the difference in energy allocation and production efficiency between endotherms and ectotherms?
11. What are the two major food chains, and how are they related?
12. How does consumption efficiency differ between terrestrial and marine ecosystems?

FURTHER READINGS

Classic Studies

Gates, D. M. 1985. *Energy and ecology.* Sunderland, MA: Sinauer Associates.
This text shows the function of ecological systems in terms of energy flow and covers a range of topics from photosynthesis to ecosystem productivity.

Gosz, J. R., R. T. Holmes, G. E. Likens, and F. H. Bormann. 1978. "The flow of energy in a forest ecosystem." *Scientific American* 238:92–102.
This article summarizes one of the most detailed analyses of energy flow through an ecosystem and provides a contrast to the more aggregated view based on food chains presented in this chapter.

Leith, H., and R. H. Whittaker, eds. 1975. *Primary productivity in the biosphere.* Ecological Studies Vol. 14. New York: Springer-Verlag.
A classic volume exploring patterns of primary productivity at a global scale. An excellent resource for comparing primary productivity patterns in the various terrestrial and aquatic ecosystems on Earth.

Lindeman, R. L. 1942. "The trophic-dynamics aspect of ecology." *Ecology* 23:399–418.
The classic article that developed the concept of viewing ecosystems in terms of productivity and energy flow.

National Academy of Science. 1975. *Productivity of world ecosystems.* Washington, DC: National Academy of Science.
This volume provides an excellent summary of world primary production.

Current Research

Aber, J. D., and J. M. Melillo. 2001. 2nd edition. *Terrestrial ecosystems*. Philadelphia: Saunders College Publishing.
This well-written and illustrated text gives readers an excellent introduction to ecosystem ecology.

Chapin, F. S., P. A. Matson, and H. A. Mooney. 2002. *Principles of terrestrial ecosystem ecology.* New York: Springer-Verlag.
This well-written and illustrated text is aimed at a more advanced readership but provides an excellent reference for exploring topics covered in Part Six in more depth.

Golley, F. B. 1994. *A history of the ecosystem concept in ecology.* New Haven, CT: Yale University Press.
This book presents a historical overview of the development of the study of ecosystems in the broader framework of ecology.

Howarth, R. W. 1988. "Nutrient limitation of net primary production in marine ecosystems." *Annual Review of Ecology and Systematics* 19:89–110.
This article reviews research on the role of nutrients in limiting net primary productivity of the world's oceans.

Colorful decomposers such as honey mushrooms (*Armillaria mellea*) reside on the forest floor throughout much of continental North America.

CHAPTER GUIDE

THE FLOW OF ENERGY THROUGH ECOSYSTEMS is the story of the element carbon (Chapter 20). Beginning with the fixing of carbon dioxide into simple organic carbon compounds in the process of primary productivity, carbon moves through the food chain, ultimately returning to the atmosphere through the process of cellular respiration. Primary productivity, however, depends on the uptake of an array of essential mineral (inorganic) nutrients by plants and other autotrophs and their incorporation into living tissues (Chapter 6). The source of carbon is carbon dioxide in the atmosphere, but what is the source of the array of other essential elements life depends on? Each element has its own story of origin and movement through the ecosystem, and we shall explore these pathways, known as biogeochemical cycles, later (Chapter 22). In general, however, the source of these essential nutrients is either the atmosphere, as in the case of carbon, or rocks and minerals (via weathering; a topic discussed in Chapter 4). Once in the soil or water, the nutrients are taken up by plants and then move through the food chain. In fact, nutrients in organic form, stored in living tissues, represent a significant proportion of nutrients within ecosystems. So what is the fate of these nutrients once they make their way into the food chain? As these living tissues senesce, the nutrients are returned to the soil or sediments in the form of dead organic matter, where they make their way through the decomposer (detrital) food chain. But unlike carbon, most of the nutrients are recycled within the ecosystem. Various microbial decomposers transform the organic nutrients into a mineral form, and the nutrients are once again available to the plants for uptake and incorporation into new tissues. This process, called **nutrient cycling**, is an essential feature of all ecosystems. It represents a recycling of nutrients within the ecosystem.

In this chapter, we will examine the processes involved in the recycling of nutrients within the ecosystem. A central focus of our discussion will be the processes of decomposition and nutrient mineralization and the environmental factors that control the rate at which these processes proceed. We also will explore how this general process, common to all ecosystems, varies between terrestrial and aquatic ecosystems, thus setting the stage for discussing specific biogeochemical cycles presented in Chapter 22.

21.1 Most Essential Nutrients Are Recycled within the Ecosystem

The cycling of nutrients within a terrestrial ecosystem is represented in **Figure 21.1**. We use the essential element nitrogen as an example to follow the pathway taken by nutrients from the soil to the vegetation and back again to the soil.

As with all essential nutrients, plants require nitrogen in an inorganic or mineral form. Nitrogen in the soil solution (in the form of ammonium and nitrate) is taken up through plant roots and used to produce proteins, enzymes, and a variety of other nitrogen-based compounds. This step represents a transformation of nitrogen from inorganic to organic form—namely, nitrogen contained in living plant tissues. The availability of nitrogen and other nutrients in the soil solution will limit the rate of uptake by plants and subsequently the rate of net primary productivity (NPP; see Sections 6.11 and 20.3). In the case of nitrogen, the rate of plant uptake will directly influence the rate of photosynthesis (see Figure 6.23).

As plant tissues senesce, nutrients are returned to the soil surface in the form of dead organic matter. Before senescence occurs, however, plants absorb some of the nutrients from senescing tissues into the perennial parts of the plant to be stored and used in producing new tissues. This process of recycling nutrients within the plant is called **retranslocation** or **resorption**. For example, as the days of autumn become shorter in the deciduous forests of temperate regions, the synthesis of chlorophyll (the light-harvesting pigment that gives leaves their green color) by plants begins to decline (Section 6.1). The yellow

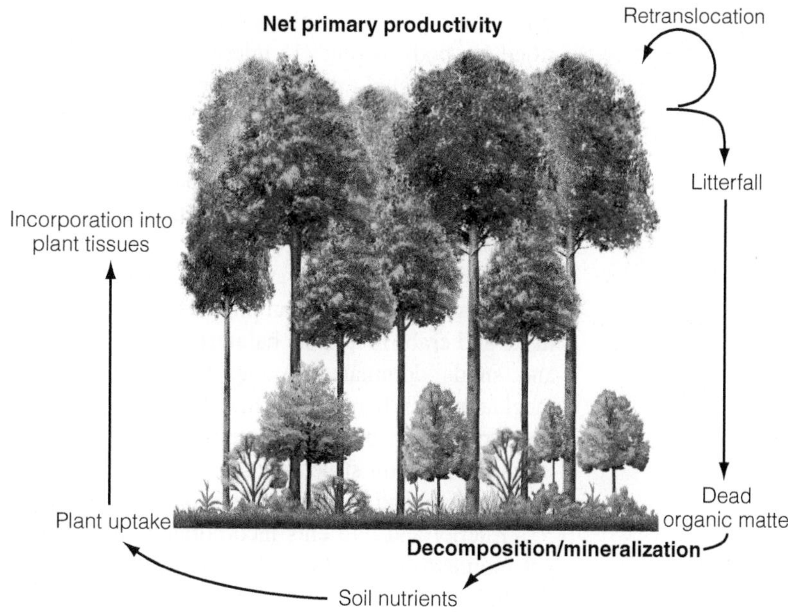

Figure 21.1 A generalized model of nutrient cycling in a terrestrial ecosystem.

and orange pigments (carotinoids and xanthophylls), which are always present within the leaf, begin to show. The senescing cells of the leaves also produce other chemicals, particularly anthocyanins, which are responsible for red and purple colors. The leaves of some species, particularly the oaks, contain high quantities of tannins that produce brown colors. As senescence occurs, water and nutrients are being drawn into the stems and away from the leaves. The plant can recover as much as 70 percent of the nitrogen from the green leaves before they senesce and fall to the forest floor (**Table 21.1**), thus reducing the amount of nutrients returned to the soil as dead organic matter.

Once on the forest floor, various decomposer organisms break down and consume the dead plant tissues, transforming the organic nutrients into a mineral form through the process of mineralization (discussed in detail in Section 21.5). The cycle is now complete, and the nutrients are once again available to the plants for uptake and incorporation into plant tissues.

21.2 Decomposition Is a Complex Process Involving a Variety of Organisms

The key process in the recycling of nutrients within ecosystems is decomposition. Decomposition is the breakdown of chemical bonds formed during the construction of plant and animal tissues. Whereas photosynthesis involves the incorporation of solar energy, carbon dioxide, water, and inorganic nutrients into organic compounds (living matter), decomposition involves the release of energy originally fixed by photosynthesis, carbon dioxide, and water, and ultimately the conversion of organic compounds into inorganic nutrients. Decomposition is a complex of many processes, including leaching, fragmentation, changes in physical and chemical structure, ingestion, and excretion of waste products. These processes are accomplished by a variety of decomposer organisms. All heterotrophs function to some degree as decomposers. As they digest food, they break down organic matter, alter it structurally and chemically, and release it partially in the form of waste products. However, what we typically refer to as decomposers are organisms that feed on dead organic matter or detritus. These organisms include microbial decomposers, a group made up primarily of bacteria and fungi, and detritivores, which are animals that feed on dead material including dung (**Figure 21.2**).

Table 21.1 **Nitrogen Content of Living and Senescent Leaves for Nine Tree Species Found in Central Virginia***

Species	Green Leaf % N	Senescent Leaf % N	Retranslocation % N
White oak	2.08	0.82	60.6
Scarlet oak	2.14	0.85	60.3
Southern red oak	1.88	0.60	68.1
Red maple	1.96	0.76	61.2
Tulip poplar	2.55	0.90	64.7
Virginia pine	1.62	0.54	66.7
American hornbeam	2.20	1.16	47.3
Sweetgum	1.90	0.59	68.9
Sycamore	2.10	0.90	57.1

*All values are expressed as a percentage of dry weight. Percent nitrogen retranslocation is the difference between the nitrogen content of green and senescent leaves expressed as a percentage of green leaf nitrogen content.

The innumerable organisms involved in decomposition are categorized into several major groups based on their size and function. Organisms most commonly associated with the process of decomposition are the microflora, composed of the bacteria and fungi. Bacteria may be aerobic, requiring oxygen for metabolism (respiration), or they may be anaerobic, able to carry on their metabolic functions without oxygen by using inorganic compounds. This type of respiration by anaerobic bacteria, commonly found in the mud and sediments of aquatic habitats and in the rumen of ungulate herbivores, is fermentation.

Bacteria are the dominant decomposers of dead animal matter, whereas fungi are the major decomposers of plant material. Fungi extend their hyphae into the organic material to withdraw nutrients. Fungi range in type from species that feed on highly soluble, organic compounds, such as glucose, to the more complex hyphal fungi that invade tissues with their hyphae (Figure 21.2a).

Bacteria and fungi secrete enzymes into plant and animal tissues to break down the complex organic compounds. Some of the resulting products are then absorbed as food. After one group has exploited the material to the extent of its ability, a different group of bacteria or fungi able to use the remaining material moves in. Thus, a succession of microflora are involved in decomposing the organic matter until it is finally reduced to inorganic nutrients.

Decomposition is aided by the fragmentation of leaves, twigs, and other dead organic matter (detritus) by invertebrate detritivores. These organisms fall into four major groups, classified by body width: (1) microfauna and microflora (<100 μm) which include protozoans and nematodes inhabiting the water in soil pores, (2) mesofauna (between 100 μm and 2 millimeters [mm]) which include mites (Figure 22.2b), potworms, and springtails that live in soil air spaces, (3) macrofauna (2–20 mm), and (4) megafauna ($>$ 20 mm). The last two categories are represented by millipedes, earthworms (Figure 22.2c), and snails in terrestrial habitats and by annelid worms, smaller crustaceans—especially amphipods and isopods—and mollusks and crabs in aquatic habitats (Figure 21.2d). Earthworms and snails dominate the megafauna. The macrofauna and megafauna can burrow into the soil or substrate to create their own space, and megafauna, such as earthworms, have major influences on soil structure (see Chapter 4). These detritivores feed on plant and animal remains and on fecal material.

Energy and nutrients incorporated into bacterial and fungal biomass do not go unexploited in the decomposer world. Feeding on bacteria and fungi are the microbivores. Making

Figure 21.2 (a) Fungi and bacteria are major decomposers of plant and animal tissues. (b) Mites and springtails are among the most abundant of small detritivores. (c) Earthworms and millipedes are large detritivores in terrestrial ecosystems, and (d) mollusks and crabs play a similar role in aquatic ecosystems.

up this group are protozoans such as amoebas, springtails (Collembola), nematodes, larval forms of beetles (Coleoptera), and mites (Acari). Smaller forms feed only on bacterial and fungal hyphae. Because larger forms feed on both microflora and detritus, members of this group are often difficult to separate from detritivores.

21.3 Studying Decomposition Involves Following the Fate of Dead Organic Matter

Decomposers, like all heterotrophs, derive the energy and most of the nutrients they require by consuming organic compounds. Energy is obtained through the oxidation of carbon compounds—carbohydrates (such as glucose)—in the process of respiration (see Section 6.1). Ecologists study the process of decomposition by designing experiments that follow the decay of dead plant and animal tissues through time. The most widely used approach is the use of litterbags to examine the decomposition of dead plant tissues (called *plant litter*). Litterbags are mesh bags constructed of synthetic material that does not readily decompose (**Figure 21.3**). The holes in the bag must be large enough to allow decomposer organisms to enter and feed on the litter but small enough to prevent decomposing plant material from falling out of the bag (holes typically 1–2 mm), although this compromise in mesh size typically restricts access to the larger decomposer organisms.

A fixed amount of litter material is placed in each bag. In the experiments presented in **Figure 21.4**, 30 litterbags filled with 5 grams (g) of leaf litter for each of the three tree species were buried in the litter layer of the forest. At six intervals during the course of a year, five bags were collected for each species and their contents were dried and weighed in the laboratory. From these data, the decomposition rate can be determined for each of the species (see **Quantifying Ecology 21.1**).

Figure 21.3 Litterbag experiment. In this example, a known quantity of senescent leaves is placed in mesh bags on the forest floor. Bags are retrieved at various intervals, and the mass loss as a result of consumption by decomposers is tracked through time.

Note in Figure 21.4 that the mass of litter remaining in the bags decreases continuously as time progresses. As decomposer organisms consume the litter, using the carbon as a source of energy, the carbon is eventually lost to the atmosphere as carbon dioxide in the process of respiration. It is important to note, however, that in this approach of studying decomposition, as time passes, the mass of organic matter remaining in the litterbag includes original plant material as well as the bacteria and fungi (microbial decomposers) that have colonized and grown on the plant litter. Because of the difficulty in doing so, few studies have quantified the changing contribution of primary (original plant material) and secondary (microbial biomass) organic matter in the decomposing litter (remaining mass).

Microbial ecologist Martin Swift of the University of Zimbabwe was able to estimate the growth of fungi during decomposition of wood sawdust by measuring the change in chitin content, an organic compound that is the main component of fungal cell walls. By the end of the experiment, a 39 percent weight loss was recorded for the sawdust, but the biomass estimate showed that 58 percent of the remaining mass was composed of living and dead fungal biomass. The apparent decomposition rate (k) of the sawdust was 0.04/week, but the rate more than doubled—to 0.09/week—when recalculated to exclude the fungal biomass (for a discussion of k, see Quantifying Ecology 21.1). This shift in the proportion of remaining organic matter in plant and decomposer biomass is crucial to understanding the dynamics of other nutrients, such

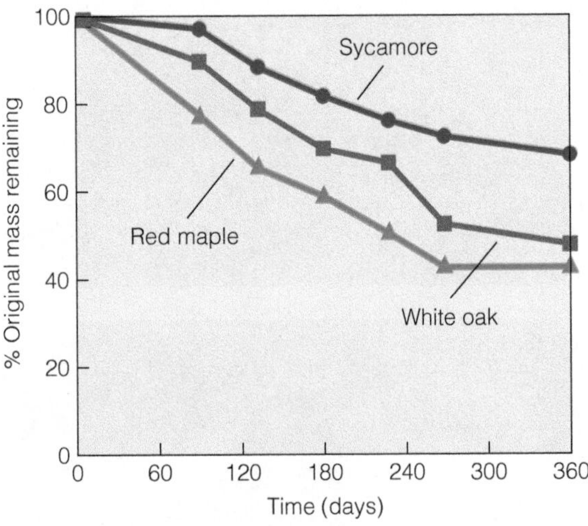

Figure 21.4 Results of a litterbag experiment in central Virginia designed to examine the decomposition of fallen leaves from red maple, white oak, and sycamore trees. Decomposition is expressed as the percentage of the original mass remaining at different times during the first year of the experiment.

Interpreting Ecological Data

Q1. *What proportion of the original leaf mass has been lost (consumed) for each of the three species at day 180 in the experiment?*

Q2. *Which of the three species has the slowest rate of decomposition?*

QUANTIFYING ECOLOGY 21.1 Estimating the Rate of Decomposition

Litterbag experiments such as the ones presented in Figures 21.4 and 21.6 are the primary means by which ecologists study the process of decomposition. By collecting multiple (replicate) litterbags at regular intervals during the process of decay, researchers can plot the proportion of mass loss (or proportion of mass remaining) through time. These data, in turn, can be used to estimate the rate of decomposition. Each point in **Figure 1** represents the average of five replicate litterbags collected on the given day during the experiment. The *x*-axis shows time (weeks), and the *y*-axis shows the percentage of the original mass remaining.

The mass loss through time is generally expressed as a negative exponential function:

$$\text{Original mass remaining} = e^{-kt}$$

Here, *e* is the natural logarithm, *t* is the time unit used (years, months, weeks, or days), and *k* is the decomposition coefficient, which defines the slope of the negative exponential curve. The decomposition coefficient can be estimated using regression techniques (fitting an exponential regression model to the data set: $y = e^{-kx}$, where *y* is the original mass remaining and *x* is the corresponding time in units of weeks). This technique was used for the data presented in Figure 1, and the resulting equations are presented. The estimates of *k* for red maple and Virginia pine are 0.0167 and 0.0097, respectively, in units of proportion mass loss per week. In the current example, the values are multiplied by 100 to convert them from proportion to percentage. The higher the value of *k*, the faster the rate of decomposition.

Data from the same litterbag study for a third tree species, sycamore (*Platanus occidentalis*), is presented in **Table 1**. Each value represents the remaining dry mass of the original 7 grams dry weight of leaf litter that was placed in the bag at the beginning of the experiment (day 0). The percentage of original mass remaining is therefore calculated by dividing each value by 7 and multiplying by 100 to convert the proportion to a percentage.

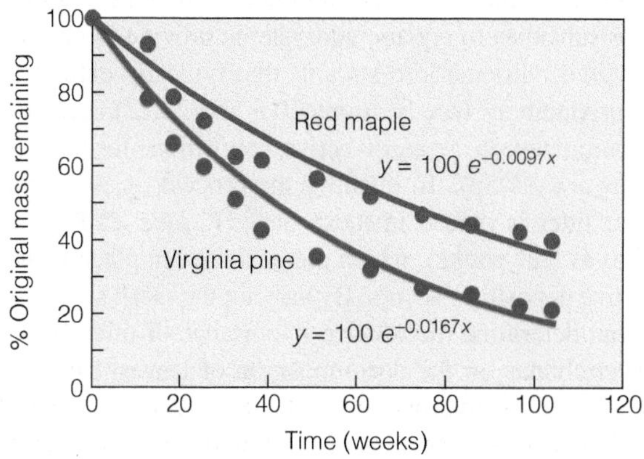

Figure 1 Data from two litterbag experiments that examined the rate of decomposition for red maple (*Acer rubra*) and Virginia pine (*Pinus virginiana*) leaf litter over a period of two years. Each point represents the average mass remaining in five replicate litterbags sampled during that period. The lines represent the predicted relationship based on the solution for the decomposition coefficient *k* using the negative exponential model discussed in the text. Values of *k* were estimated using a nonlinear regression technique.

1. Using the data in Table 1, calculate the mean value for the five replicate samples at each time period, and then convert the mean value to percent original mass remaining (%OMR). Now plot the resulting mean values of %OMR (*y*-axis) as a function of time (*x*-axis), as shown in Figure 1. For purposes of comparison, you must convert days to weeks.
2. How does the general pattern of decomposition (%OMR) through time compare to that of red maple and Virginia pine presented in Figure 1?
3. How do you think the lignin concentration of the sycamore leaves compares to that of red maple or Virginia pine? Why?

Table 1 Litterbag Study after the Decomposition of Sycamore Leaf Litter*

	Day											
Litterbag	0	90	131	179	228	269	360	445	525	603	680	730
Replicate1	7	6.92	5.82	5.69	5.29	4.89	4.66	4.33	4.06	3.78	3.84	3.71
Replicate2	7	6.84	6.08	5.59	5.28	4.99	4.87	4.24	3.99	4.09	3.71	3.83
Replicate3	7	6.91	5.98	5.83	5.38	5.18	4.75	4.38	4.26	4.01	3.60	3.78
Replicate4	7	6.75	5.74	5.88	5.44	5.13	4.92	4.50	4.21	3.88	3.92	3.58
Replicate5	7	6.72	5.88	5.78	5.41	5.21	4.72	4.19	4.08	3.87	3.82	3.65

*All values in grams dry weight. Initial weight of leaf litter in each bag was 7.0 g.

as nitrogen, during the decomposition process; that topic is examined in Section 21.5.

A similar approach to litterbag experiments is used to evaluate the decomposition of plant litter in stream ecosystems. Leaf litter subsidies to aquatic ecosystems provide large inputs of energy and nutrients into streams that typically exhibit low primary productivity (see Sections 20.5 and 20.12). Leaf litter often accumulates in areas of active deposition, forming **leaf packs** (Figure 21.5a). To quantify the process of decomposition, plant litter is placed in mesh bags (Figure 21.5b); also referred to as **leaf packs**), which are anchored in place in areas where active deposition occurs. By varying the mesh size, investigators can determine the relative importance of microbes and macroinvertebrates on the decomposition of leaves. Fine-mesh screens allow only microbes to colonize and decompose the leaves, whereas coarse screens also allow the larger organisms to enter and feed on the material (Figure 21.5c). As with terrestrial litterbags, the leaf packs are placed in the stream for several weeks to measure the weight loss and chemical changes of the leaves during the process of decomposition (Figure 21.6).

Figure 21.5 (a) Inputs of plant litter from the surrounding terrestrial environment can form areas of deposition, known as *leaf packs*, in stream ecosystems. (b) Much like the use of litterbag experiments by terrestrial ecologists, stream ecologists use mesh bags to simulate natural leaf packs and examine the processes of decomposition. (c) Smaller mesh bags allow access to only microbial decomposers, whereas larger mesh bags allow access to the diversity of invertebrate decomposers that inhabit streams.

Unlike terrestrial litterbag experiments, leaf pack studies typically quantify the diversity of micro- and macro-decomposers during decomposition. (A detailed discussion of decomposition in stream ecosystems is presented in Chapter 24, Section 24.6.)

Litterbags and leaf packs are also used to examine the decomposition of plant litter in wetland and marsh ecosystems. The mesh bags are tethered to stakes to prevent them from being displaced by the tidal currents.

21.4 Several Factors Influence the Rate of Decomposition

Not all organic matter decomposes at the same rate. For example, note the differences in decomposition rate for the leaf litter of the three tree species reported in Figure 21.4. From the wealth of studies completed over the past 50 years, some generalizations concerning factors that influence decomposition rate have emerged. The rate of decay (mass loss) is related to (1) the quality of plant litter as a substrate (food source) for microorganisms and soil fauna active in the decomposition process and (2) features of the physical environment that directly influence decomposer populations, namely soil properties (e.g., texture and pH) and climate (temperature and precipitation).

Characteristics that influence the quality of plant litter as an energy source relate directly to the types and quantities of carbon compounds present; that is, the types of chemical bonds present and the size and three-dimensional structure of the molecules in which these bonds are formed. Carbon is plentiful in plant remains, typically making up 45–60 percent of the total

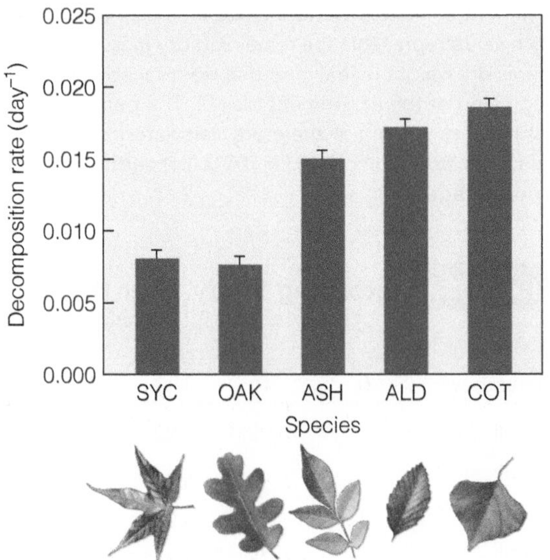

Figure 21.6 Decomposition rates for leaf litter from five tree species submerged in streams. Experiments used submerged litterbags (leaf packs) that were sampled at five intervals over a period of 83 days. Decomposition rates expressed as mass loss per day (see Quantifying Ecology 21.1). SYC: *Platanus wrightii*; OAK: *Quercus gambelii*; ASH: *Fraxinus velutina*; ALD: *Alnus oblongifolia*; COT: *Populus fremontii*.
(Adapted from Leroy and Marks 2006.)

dry weight of plant tissues, but not all carbon compounds are of equal quality as an energy source for microbial decomposers. Glucose and other simple sugars, the first products of photosynthesis, are high-quality sources of carbon. These molecules are physically small. The breakage of their chemical bonds yields much more energy than required to synthesize the enzymes needed to break them. Cellulose and hemicellulose are the main constituents of cell walls. These compounds are more complex in structure and therefore more difficult to decompose than simple carbohydrates. They are of moderate quality as a substrate for microbial decay. The much larger lignin molecules are among the most complex and variable carbon compounds in nature. There is no precise chemical description of lignin; rather, it represents a class of compounds. These compounds possess very large molecules, intricately folded into complex three-dimensional structures that effectively shield much of the internal structure from attack by enzyme systems. As such, lignins, major components of wood, are among the slowest components of plant tissue to decompose. Lignin compounds are of such low quality as a source of energy that they yield almost no net gain of energy to microbes during decomposition. Bacteria do not degrade lignins; they are broken down by only a single group of fungi, the Basidiomycetes (which includes the mushrooms; see Figure 21.2a).

Variation in the consumption rates of different carbon compounds is revealed in an experiment examining the rate at which carbon was consumed during the decomposition of straw that was placed on a soil surface (**Figure 21.7**). The total carbon content of the straw, expressed as a percentage of the original mass, declined exponentially during the period of the 80-day study. However, when the total carbon was partitioned into various classes of carbon compounds, the decomposition rates of these compounds varied widely. Proteins, simple sugars, and other soluble compounds made up some 15 percent of the original total carbon content. These compounds decomposed very quickly, disappearing completely within the first few days of the experiment. Cellulose and hemicellulose made up some 60 percent of the original carbon content. Although these compounds decomposed more slowly than the proteins and simple sugars, by three weeks into the experiment they had been completely broken down. The third category of carbon compounds examined, the lignins, made up some 25 percent of the total original carbon. These compounds were broken down very slowly during the course of the experiment, and the vast majority of lignins remained intact by day 80. As decomposition proceeded during the experiment, the quality of the carbon resource declined. High- and intermediate-quality

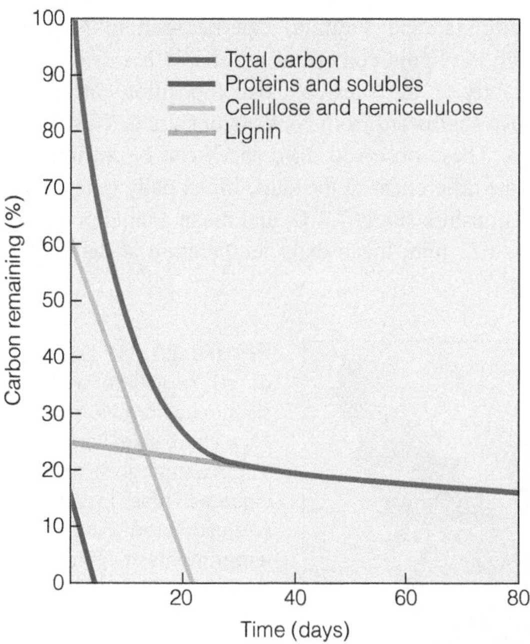

Figure 21.7 Variation in the rates of decay (mass loss) of different classes of carbon compounds in an experiment examining the decomposition of straw on the soil surface. At any time, the sum of the three classes of carbon compounds is equal to the value for total carbon.
(Adapted from Swift, Heal, and Anderson 1979.)

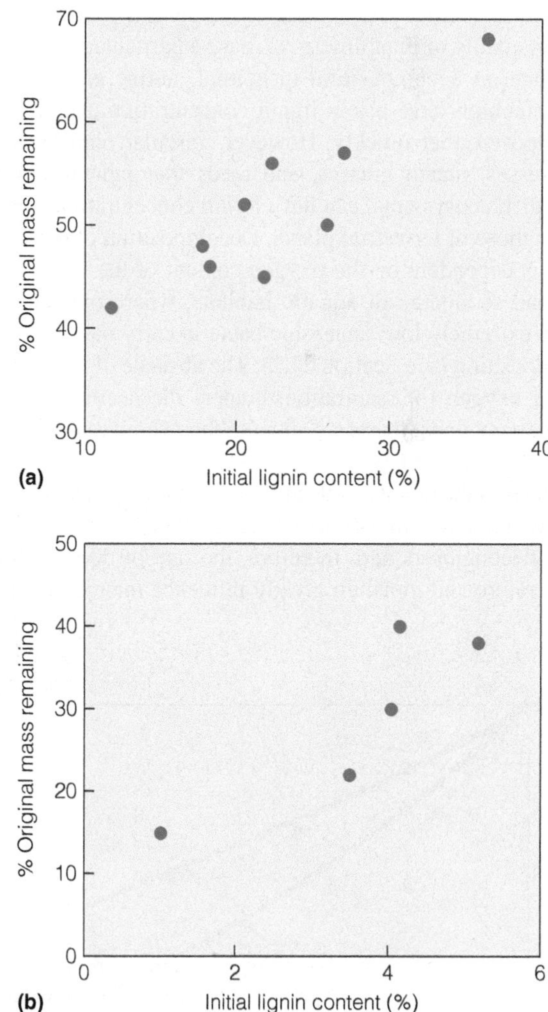

Figure 21.8 Relationship between initial lignin content of litter material and rate of decomposition for a variety of plant litters in (a) terrestrial and (b) aquatic environments. Each point on the graphs represents an individual plant species.
(Adapted from Klap et al. 1999.)

carbon compounds declined at a relatively rapid rate. Thus, the proportion of total carbon remaining as lignin compounds continually increased with time. The increasing component of lignin lowered the overall quality of the remaining litter as an energy source for microbial decomposers, therefore slowing the decomposition rate.

Because of its low quality as an energy source, the proportion of carbon contained in lignin-based compounds is used as an index of litter quality for decomposer organisms. The difference in the decomposition rates of the three species shown in Figure 21.4 is a direct result of their initial lignin content. The freshly fallen leaves of red maple (*Acer rubrum*) have a lignin content of 11.7 percent as compared to 17.7 percent for white oak (*Quercus alba*) and 36.4 percent for sycamore (*Platanus occidentalis*) leaves. In general, there is an inverse relationship between the decomposition rate for plant litter and its lignin content at the start of decomposition. This inverse relationship between decomposition rate and the carbon quality of plant litter as a source of energy for decomposers has been reported for a wide range of plant species inhabiting both terrestrial and aquatic environments (**Figure 21.8**). Carbon quality of plant litters can have a particularly important influence on decomposition in coastal marine environments. Phytoplankton have a low lignin concentration and therefore decompose rather quickly. However, vascular plants, such as sea grasses, marsh grasses, and reeds that inhabit estuarine and marsh ecosystems, can have lignin concentrations that approach those of terrestrial plants. Decomposition of these plant litters is dependent on the oxygen content of the water. In the mud and sediments of aquatic habitats, where oxygen levels can be extremely low, anaerobic bacteria carry out most of the decomposition (see Section 21.2). The absence of fungi, which require oxygen for respiration, hinders the decomposition of lignin compounds, therefore slowing the overall rate of decomposition (**Figure 21.9**).

Besides the quality of dead organic matter as a food source, the physical environment also directly affects both macro- and micro-decomposers and, therefore, the rate of decomposition. Temperature and moisture greatly influence microbial activity.

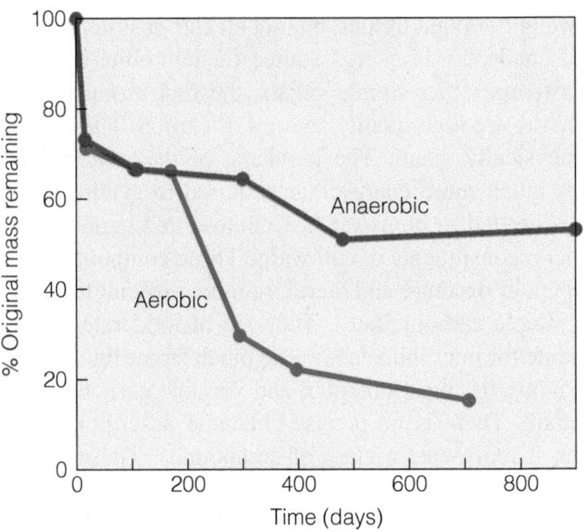

Figure 21.9 Results of a litterbag experiment designed to examine the decomposition of *Spartina alternifolia* litter exposed to aerobic (litterbags on the marsh surface) and anaerobic (buried 5–10 cm below the marsh surface) conditions.
(Adapted from Valiela 1984.)

Low temperatures reduce or inhibit microbial activity, as do dry conditions. The optimum environment for microbes is warm and moist. As a result, decomposition rates are highest in warm, wet climates (see this chapter, **Field Studies: Edward [Ted] A. G. Schuur**). Alternate wetting and drying and continuous dry spells tend to reduce microflora activity and populations.

This effect of climate on the decomposition of red maple leaves at three sites in eastern North America (New Hampshire, West Virginia, and Virginia) can be seen in **Figure 21.10**. Although the lignin content of red maple leaves does not differ significantly at the sites, the decomposition rate increases as you move southward from New Hampshire to West Virginia and Virginia. These observed differences can be attributed directly to climate differences at the sites. Mean daily temperature at the New Hampshire site is 7.2°C, and mean annual potential evaporation is 621 mm; mean daily temperature at the West Virginia

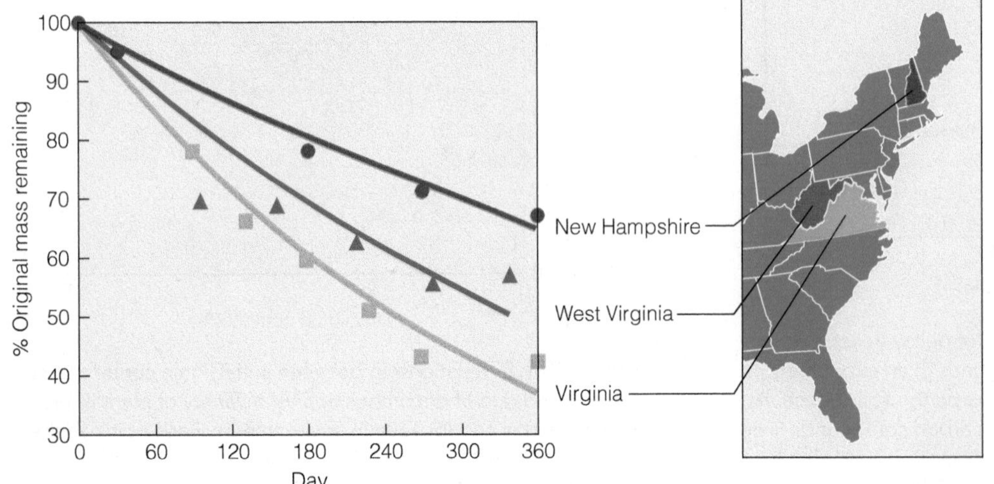

Figure 21.10 Decomposition of red maple litter at three sites in eastern North America: New Hampshire (circles), West Virginia (triangles), and Virginia (squares). Mass loss through time was estimated using litterbag experiments at each site. The decline in decomposition rate from north to south is a direct result of changes in climate, primarily temperature.
(Adapted from Melillo et al. 1982, New Hampshire; Mudrick et al. 1994, West Virginia; Smith 2002, Virginia.)

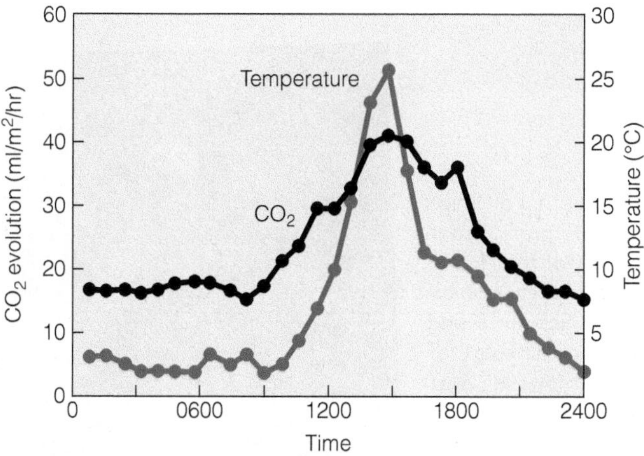

Figure 21.11 Diurnal changes in air temperature and decomposition in a temperate deciduous forest. Decomposition rate is measured indirectly as the release (evolution) of carbon dioxide (CO_2) from decomposing litter on the forest floor. The release of CO_2 is a measure of the respiration of decomposer organisms.

(Adapted from Whitkamp and Frank 1969.)

site is 12.2°C, and mean annual potential evaporation is 720 mm; mean daily temperature at the Virginia site is 14.4°C, and mean potential evaporation is 806 mm.

The direct influence of temperature on decomposers results in a distinct diurnal pattern of microbial activity as measured by microbial respiration from the soil (**Figure 21.11**). The daily temperature pattern is closely paralleled by the release of carbon dioxide from the soil as a result of the respiration of microbial decomposers.

21.5 Nutrients in Organic Matter Are Mineralized During Decomposition

Just as dead organic matter varies in the quality of carbon compounds as an energy source for microbial decomposers, likewise the nutrient quality of dead organic material varies greatly. The macronutrient nitrogen can serve as an example. Most dead leaf material has a nitrogen content in the range of 0.5–1.5 percent (see Table 21.1). The higher the nitrogen content of the dead leaf, the higher the nutrient value for the microbes and fungi that feed on the leaf.

As the dead organic matter is consumed, the microbial decomposers—bacteria and fungi—transform nitrogen and other elements contained in organic compounds into inorganic (or mineral) forms (**Figure 21.12**). This process is called **mineralization**. For example, the inorganic form of nitrogen, ammonia, is a waste product of microbial metabolism. The same decomposers that are responsible for mineralization also require nitrogen for their own growth and reproduction. Therefore, whenever mineralization occurs, **immobilization**—the uptake and assimilation of mineral nitrogen by microbial decomposers—runs counter to it (see Figure 21.12). Because both of these

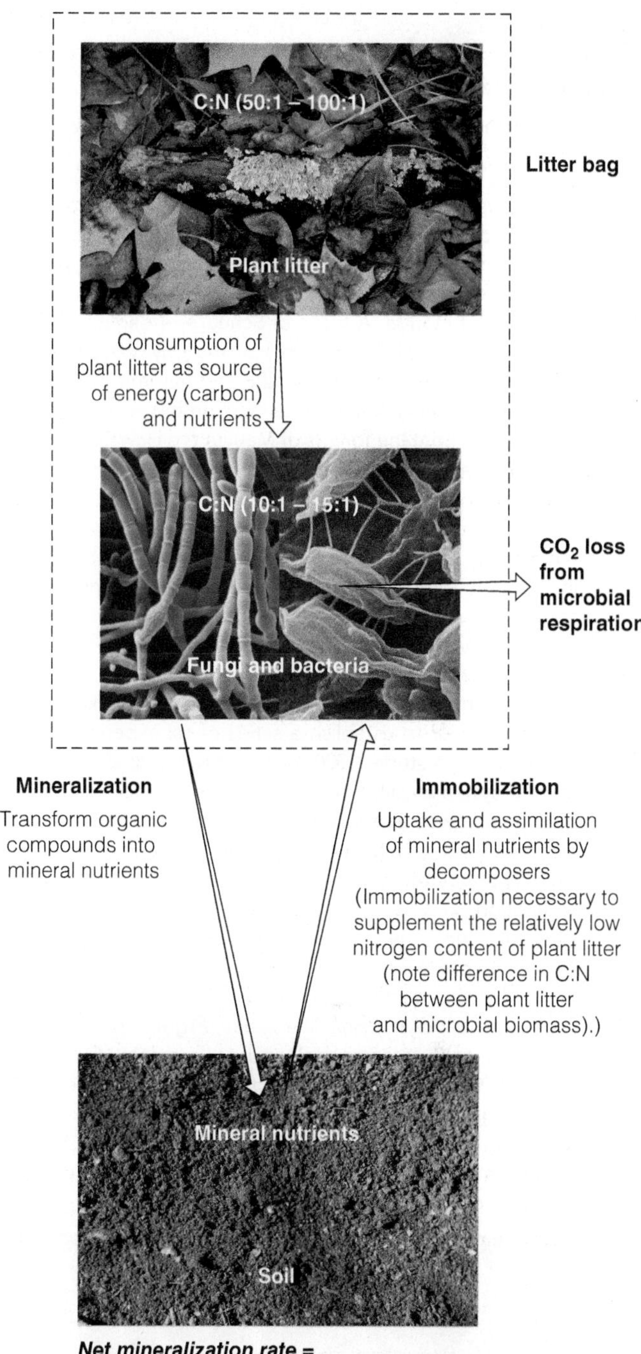

Figure 21.12 Diagram illustrating the exchanges between litterbag (delineated by the dashed line) and soil in a standard litterbag experiment used to quantify the process of decomposition. Decomposer organisms (bacteria and fungi) colonize the plant litter. As litter is consumed, a significant proportion of carbon is respired, and nutrients bound in organic matter are mineralized and released to the soil. To convert plant carbon to microbial biomass, mineral nutrients are taken up in the process of immobilization. The difference in the rates of mineral nutrient release (mineralization) and immobilization is the rate of net mineralization. Note that as time progresses, the residual organic matter in the litterbag is composed of a growing proportion of microbial biomass as the original plant material is consumed, respired, and converted into microbial biomass.

FIELD STUDIES Edward (Ted) A. G. Schuur

Department of Botany, University of Florida, Gainesville, Florida

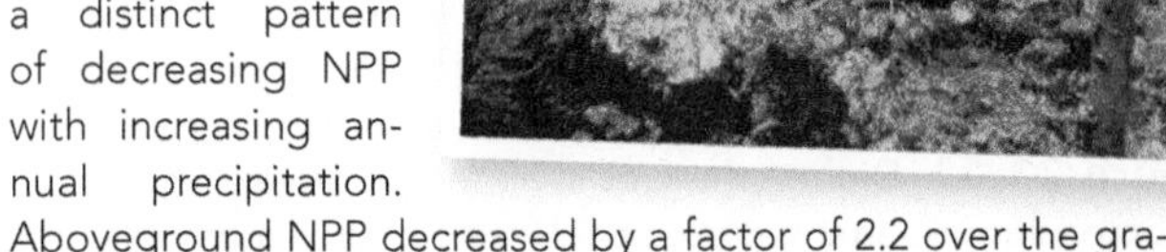

The warm, wet environments of the tropical rain forest support the highest rates of net primary productivity (NPP) and decomposition of any terrestrial ecosystem on Earth. Or so we thought before the work of ecologist Ted Schuur of the University of Florida. A focus of Schuur's research is how species characteristics and features of the physical environment interact to control patterns of NPP and nutrient cycling in terrestrial ecosystems. Much of his work on this subject has been done in the montane forests of Maui in the Hawaiian Islands.

The island of Maui provides a unique environment in which to explore the interactions between biotic and abiotic controls on ecosystem processes. Ecologists are drawn to it for two reasons. First, the interaction of topography and prevailing easterly trade winds results in a diversity of microclimatic conditions on the island. Second, the Hawaiian Islands flora is relatively species poor; thus, a few species and genera occupy the broad range of environmental conditions.

The rain shadow created by the 3055-m Haleakala volcano allowed Schuur to establish a series of six experimental plots at a constant altitude (1300 m) and mean annual precipitation ranged systematically from 2200 mm/yr (mesic) to more than 5000 mm/yr (wet) as a function of aspect relative to the prevailing trade winds (see Section 2.8). Other environmental factors—such as temperature regime, parent material, ecosystem age, vegetation, and topographic relief—were similar among the sites, allowing Schuur to focus on the role of precipitation on ecosystem processes. The forest canopy at all sites on this gradient was consistently dominated by the native evergreen tree *Metrosideros polymorpha* (Myrtaceae; **Figure 1**).

Initial experiments designed to estimate the aboveground NPP at the sites revealed an unexpected result. The prevailing conceptual (and empirical) model of how NPP relates to precipitation is shown in Chapter 20 (Figure 20.2). NPP increases with increasing precipitation, leveling off under the conditions of high annual precipitation typical of that found in the wet tropics. What Schuur found in the forests of Maui, however, was a distinct pattern of decreasing NPP with increasing annual precipitation. Aboveground NPP decreased by a factor of 2.2 over the gradient (1000 g/m^2 to $<$ 500 g/m^2).

Figure 1 Typical forest of the Makawao and Koolau Forest Reserves.

What mechanism could be at play? How could increasing precipitation result in a decrease in forest productivity? Measurements of the chemical composition of leaves collected from trees at the six sites along the mountainside provided some clues. Two characteristics varied systematically across the gradient of rainfall: Leaf nitrogen concentration decreased and the concentration of lignin increased with increasing annual precipitation at the sites. Both of these characteristics are known to influence the processes of decomposition and mineralization (see Figures 21.8 and 21.15). This evidence pointed to the possibility that systematic changes in plant characteristics along this gradient may reduce rates of decomposition and nutrient cycling and subsequently limit nutrient availability for NPP.

To test this hypothesis, Schuur conducted a set of experiments designed to examine how both plant and site (physical environment) characteristics influenced the process of nutrient cycling across the six study sites. The dynamics of carbon and nitrogen during decomposition were examined using litterbag experiments (see Sections 21.3 and 21.5). In each litterbag experiment, replicate samples (bags) were collected at each of five time intervals during a 15-month period (1, 3, 6, 9, and 15 months).

The litterbag experiment results revealed that the decomposition rate of leaves decreased by a factor of 6.4 across the precipitation gradient, whereas the rate of nitrogen release from decomposing litter declined by a factor of 2.2 across the gradient (**Figure 2**). The litterbag experiments, therefore, presented a consistent picture of declining decomposition and nitrogen cycling across the gradient of mean annual rainfall.

Differences in the rate of nitrogen release from decomposing litter were reflected in the availability of nitrogen in the soil. Available (inorganic) soil nitrogen declined with increasing annual precipitation at the sites, as did oxygen availability in the soil.

The results clearly showed that reduced nitrogen availability was responsible for the decline in NPP across the rainfall gradient. But what factor (or factors) was controlling the rate of nutrient cycling in these forest ecosystems?

In these forests, the litter decomposition rates and nutrient release slow with increasing rainfall, apparently as a result of decreased soil oxygen availability and the production of low-quality litter at wetter sites. These two factors are

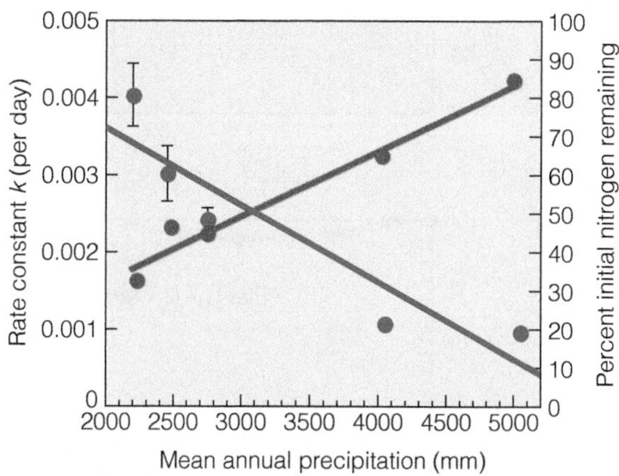

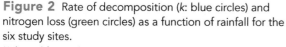

Figure 2 Rate of decomposition (*k*: blue circles) and nitrogen loss (green circles) as a function of rainfall for the six study sites.
(Adapted from Schuur 2001.)

interrelated, however. Because of high precipitation, low soil oxygen reduces rates of decomposition and nitrogen release (mineralization). In turn, low soil nitrogen potentially leads to lower leaf concentrations and litter quality, which in turn further decreases decomposition rates (see Figure 21.21).

Schuur's work in the forests of Maui has important implications for the current debate over how climate change will influence terrestrial ecosystems (see Chapter 27). According to current empirical models, precipitation has little effect on NPP and decomposition in the humid tropical ecosystems (see Figure 20.2). When this relationship is extended to wetter environments within the tropics using Schuur's data, however, an inverse relationship emerges (**Figure 3**). Because tropical forest is the largest terrestrial biome and accounts for one-third of the potential terrestrial NPP, the nature of this relationship

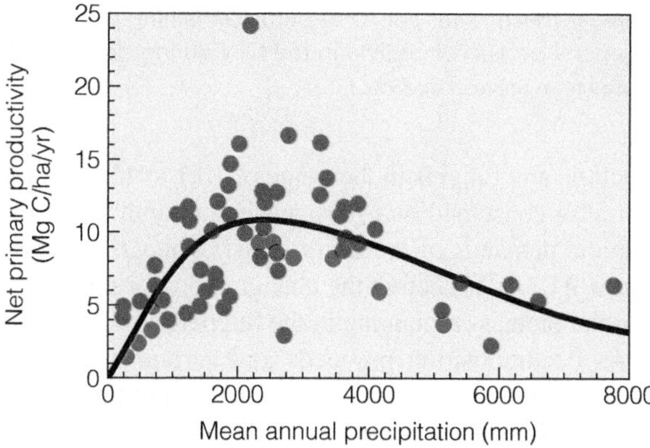

Figure 3 Relationship between net primary productivity and mean annual precipitation.
(Adapted from Schuur 2003.)

is crucial to understanding how projected climate change will influence the global carbon balance.

Although the previous work by Schuur clearly shows a reduction in the rates of decomposition and nutrient cycling with increasing precipitation in the montane forests of the wet tropics, the Hawaiian precipitation gradient sites are dominated by a single tree species and on the same parent material of similar age. Although the studies provide valuable insights into the effect of changes in mean annual precipitation on forest ecosystem dynamics, it is not clear if high rainfall has the same effect in species-rich lowland tropical forests that dominate much of the wet tropics. To determine if the results from the montane forest of Hawaii represent a more general relationship between precipitation and decomposition in the wet tropics, Schuur and colleague Juan Posada of the Universidad del Rosario (Bogota, Colombia) examined seven neotropical lowland forests with a mean annual precipitation ranging between 2650 and 9510 mm to examine the relationships between decomposition rate, nutrient availability, and rainfall. The results of the study are in agreement with the previous studies by Schuur that show a negative effect of increasing precipitation on nutrient availability. In the lowland rain forest studies, phosphorus (P) rather than nitrogen appears to be the limiting factor on NPP. Schuur and Posada found that retranslocation (resorption) of phosphorus increases with increasing mean annual precipitation, resulting in a decline in the phosphorus content of senescent leaves. The lower phosphorus availability results in a slower turnover rate of carbon (decomposition rate), resulting in an indirect control of precipitation on the processes of decomposition and nutrient cycling in these lowland tropical forests through the direct effect on litter quality.

Bibliography

Schuur, E. A. G. 2001. "The effect of water on decomposition dynamics in mesic to wet Hawaiian montane forests." *Ecosystems* 4:259–273.

Schuur, E. A. G. 2003. "Productivity and global climate revisited: The sensitivity of tropical forest growth to precipitation." *Ecology* 84:1165–1170.

Schuur, E. A. G., and P. A. Matson. 2001. "Net primary productivity and nutrient cycling across a mesic to wet precipitation gradient in Hawaiian montane forest." *Oecologia* 128:431–442.

Posada, J. M., and E. A. G. Schuur. 2011. "High precipitation regime reduces nutrient availability and carbon cycling in tropical lowland wet forest." *Oecologia* 165:783–795.

1. What type of experiment could Schuur undertake at his research site in Hawaii to determine if nitrogen availability in the soil is directly limiting the nutrient concentration of leaves and subsequently the rates of net primary productivity?

2. Why would reduced oxygen concentrations in the soil function to reduce rates of decomposition?

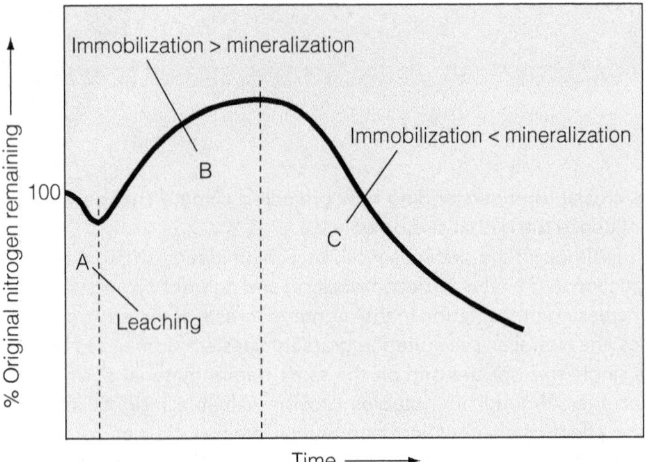

Figure 21.13 Idealized graph showing the change in nitrogen content of plant litter during decomposition. The initial phase (A) corresponds to the leaching of soluble compounds. Nitrogen content then increases above initial concentrations (phase B) as the rate of immobilization exceeds the rate of mineralization. As decomposition proceeds, the rate of nitrogen mineralization exceeds that of immobilization, and there is a net release of nitrogen from the litter (phase C).

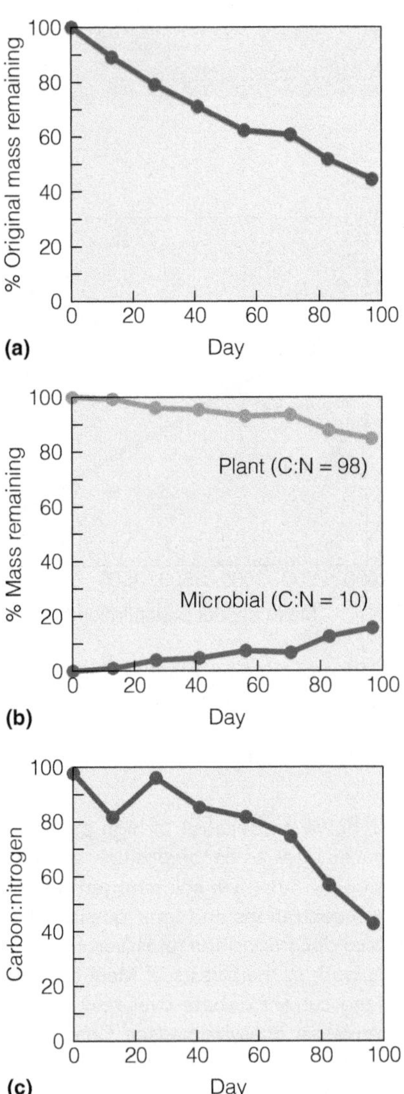

Figure 21.14 Results from a litterbag experiment designed to examine the changing composition of decomposing winter rye in an agricultural field. (a) Mass loss occurred throughout the 100 days of the experiment. (b) The proportion of the remaining mass in plant and microbial (fungal) biomass (living and dead). (c) Because the ratio of carbon to nitrogen (C/N) of the microbial biomass is much lower than that of the remaining plant litter, there is a general pattern of decline in the C/N during decomposition. (Adapted from Beare et al. 1992.)

processes—mineralization and immobilization—are taking place as decomposer organisms are consuming the litter, the supply rate of mineral nutrients to the soil during the process of decomposition—the **net mineralization rate**—is the difference between the rates of mineralization and immobilization.

Changes in the nutrient content of litter during decomposition are typically examined concurrently with changes in mass and carbon content in the litterbag experiments described in Section 21.3. As with litter mass (see Figure 21.4) and carbon content (see Figure 21.7), the nitrogen content of the remaining litter can be expressed as a percentage of the nitrogen content of the original litter mass. Changes in the nitrogen content of plant litter during decomposition typically conform to three stages (**Figure 21.13**). Initially, the amount of nitrogen in the litter declines as water-soluble compounds are leached from the litter. This stage can be very short and, in terrestrial environments, depends on soil moisture levels.

After the initial period of leaching, nitrogen content typically increases as microbial decomposers immobilize nitrogen from outside the litter. For this reason, nitrogen concentrations can rise to more than 100 percent, actually exceeding the initial nitrogen content of the litter material. To understand how this can occur, first recall from our discussion that the remaining organic matter in the litterbags includes the original dead leaf material as well as the living and dead microorganisms (see Figures 21.12 and 21.14). Second, the nitrogen content of the fungi and bacteria is considerably higher than that of the plant material they are feeding on. The ratio of carbon to nitrogen (C/N; grams of carbon per gram of nitrogen) is a widely used index to characterize the nitrogen content of different tissues. The C/N of plant litter is typically in the range of 50:1 to 100:1, whereas the C/N of bacteria and fungi is in the range of 10:1 to 15:1. As the plant material is consumed and nitrogen is immobilized to meet the metabolic demands of the decomposers, the C/N ratio declines (**Figure 21.14**), reflecting the changing proportion of plant and microbial biomass remaining in the litterbag.

As decomposition proceeds and carbon quality declines (as a result of a higher proportional fraction of lignin), the mineralization rate exceeds the immobilization rate. The result is a net release of nitrogen to the soil (or water).

The pattern presented in Figure 22.13 is idealized. The actual pattern of nitrogen dynamics during decomposition

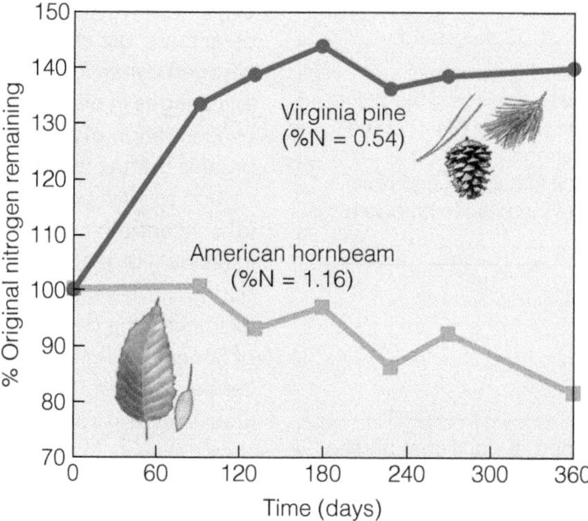

Figure 21.15 Change in the nitrogen content of leaf litter from two tree species inhabiting the forests of central Virginia: American hornbeam and Virginia pine. Nitrogen content is expressed as the percentage of the original mass of nitrogen remaining at different times during the first year of the experiment. All data are from litterbag experiments. Note the difference between the two species in the initial nitrogen content of the leaf litter and the subsequent rates of immobilization.

depends on the initial nutrient content of the litter material. If the nitrogen content of the litter material is high, then mineralization may exceed the rate of immobilization from the onset of decomposition and nitrogen concentration of the litter will not increase above the initial concentration (**Figure 21.15**).

Although the discussion and examples just presented have focused on nitrogen, the same pattern of immobilization and mineralization as a function of litter nutrient content applies to all essential nutrients (**Figure 21.16**). As with nitrogen, the exact pattern of dynamics during decomposition is a function of the nutrient content of the litter and the demand for the nutrient by the microbial population.

21.6 Decomposition Proceeds as Plant Litter Is Converted into Soil Organic Matter

We have thus far discussed a process of decomposition in which decomposer organisms consume plant litter. In doing so, they mineralize nutrients and alter the chemical composition of residual organic matter. This process continues as the litter degrades into a dark brown or black homogeneous organic matter known as *humus*. As this organic matter becomes embedded in the soil matrix, it is referred to as *soil organic matter* (see Section 4.7). Most field studies, including those presented previously, are relatively short term, examining the initial stages of decomposition over a period of one or two years at most. The work of Bjorn Berg and colleagues at the Swedish University of Agricultural Sciences, however, provides one of the most complete pictures of long-term decomposition of plant litter.

Berg and colleagues examined the decomposition of leaf (needle) litter in a Scots pine (*Pinus sylvestris*) forest in central Sweden over a period of five years. Mass loss continued over the five-year study, and carbon continued to decline as carbon dioxide was lost to the atmosphere through microbial respiration (**Figure 21.17**). Simultaneously, the need for decomposers to convert plant organic matter with an intial C/N ratio of 134:1 into microbial biomass (C/N of approximately 10:1) results in a prolonged period of immobilization (Figure 21.17a) and increasing nitrogen content of the residual organic matter (Figure 21.17b). The net result is an increasing nitrogen concentration and a declining C/N of the residual organic matter (Figure 21.17c).

As the litter decomposition progresses, the rate of mass loss (decomposition rate) declines. This decline is a result of the preferential consumption of easily digested, high-energy-yielding carbon compounds (termed *labile organic matter*), leaving increasingly recalcitrant (hard to break down) compounds in the remaining litter, illustrated by the increasing fraction of lignin-based compounds in the residual organic

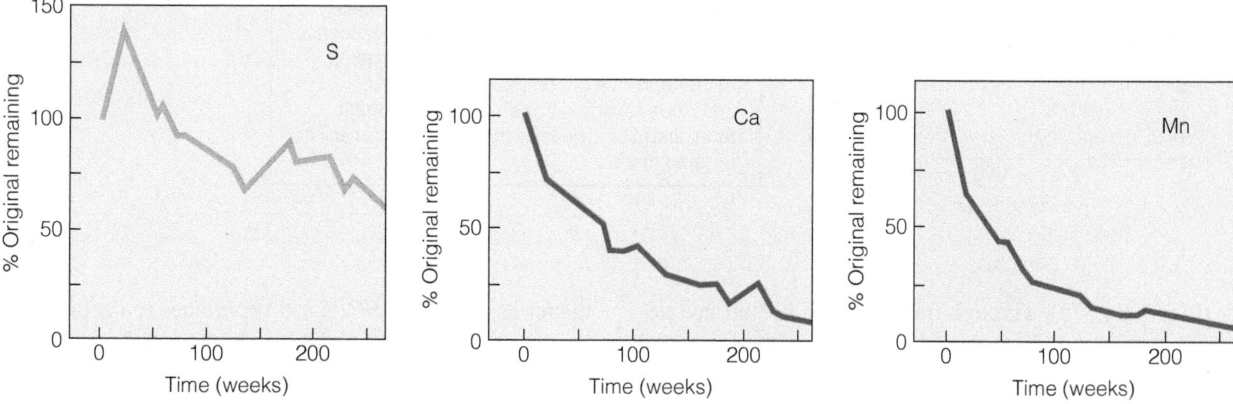

Figure 21.16 Patterns of immobilization and mineralization for sulfur (S), calcium (Ca), and manganese (Mn) in decomposing needles of Scots pine. Results are from a litterbag experiment continued over a period of five years.
(Adapted from Staaf and Berg 1982.)

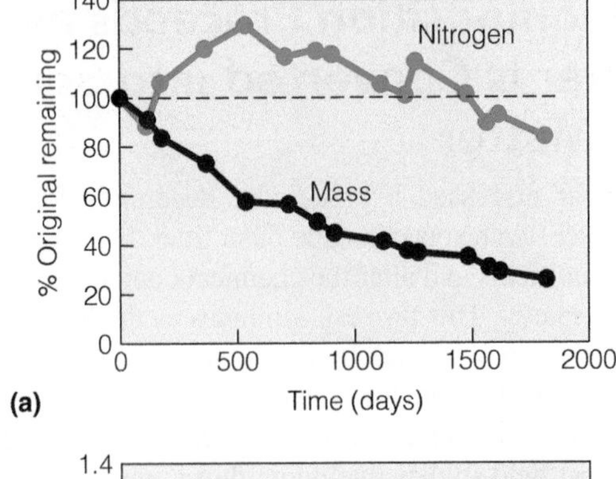

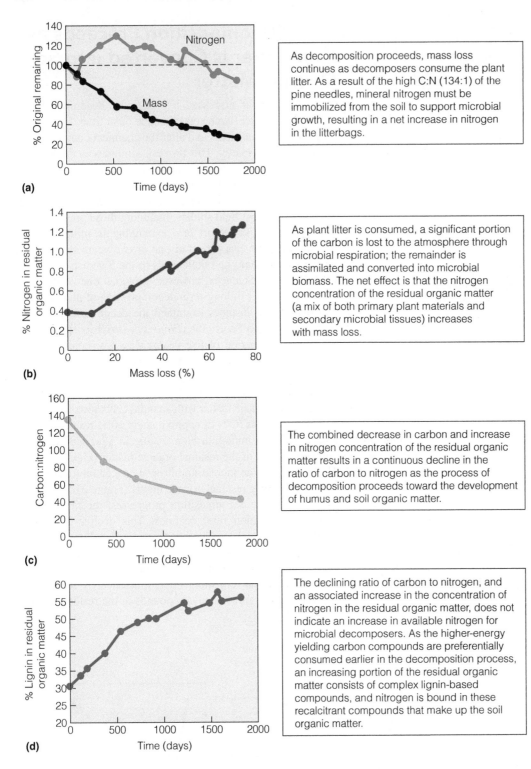

As decomposition proceeds, mass loss continues as decomposers consume the plant litter. As a result of the high C:N (134:1) of the pine needles, mineral nitrogen must be immobilized from the soil to support microbial growth, resulting in a net increase in nitrogen in the litterbags.

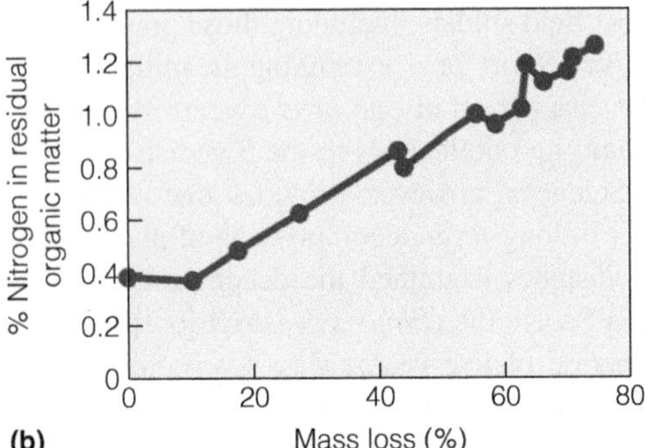

As plant litter is consumed, a significant portion of the carbon is lost to the atmosphere through microbial respiration; the remainder is assimilated and converted into microbial biomass. The net effect is that the nitrogen concentration of the residual organic matter (a mix of both primary plant materials and secondary microbial tissues) increases with mass loss.

(c)

The combined decrease in carbon and increase in nitrogen concentration of the residual organic matter results in a continuous decline in the ratio of carbon to nitrogen as the process of decomposition proceeds toward the development of humus and soil organic matter.

(d)

The declining ratio of carbon to nitrogen, and an associated increase in the concentration of nitrogen in the residual organic matter, does not indicate an increase in available nitrogen for microbial decomposers. As the higher-energy yielding carbon compounds are preferentially consumed earlier in the decomposition process, an increasing portion of the residual organic matter consists of complex lignin-based compounds, and nitrogen is bound in these recalcitrant compounds that make up the soil organic matter.

Figure 21.17 Patterns of (a) mass loss and nitrogen dynamics, (b) changes in nitrogen concentration of residual organic matter, (c) ratio of carbon to nitrogen, and (d) concentration of lignin in residual organic matter during a five-year experiment examining the decomposition of Scots pine leaf litter in central Sweden.
(Data from Berg et al. 1982.)

matter (Figure 21.17d). Through fragmentation by soil invertebrates and chemical alterations, the litter is converted into soil organic matter. As microbes die, chitin and other recalcitrant components of their cell walls comprise an increasing proportion of the residual organic matter (litter plus microbial mass), leading to the production of humus. All of these processes contribute to a gradual reduction in the quality of soil organic matter as it ages. The C/N ratio continues to decline as decomposition proceeds; the carbon is respired, and some of the mineralized nitrogen is incorporated into humus and combines with mineral particles in the soil to form colloids.

The decline in C/N ratio is not, however, an indicator of increased nitrogen availability because the nitrogen becomes incorporated into chemical structures that are recalcitrant, that

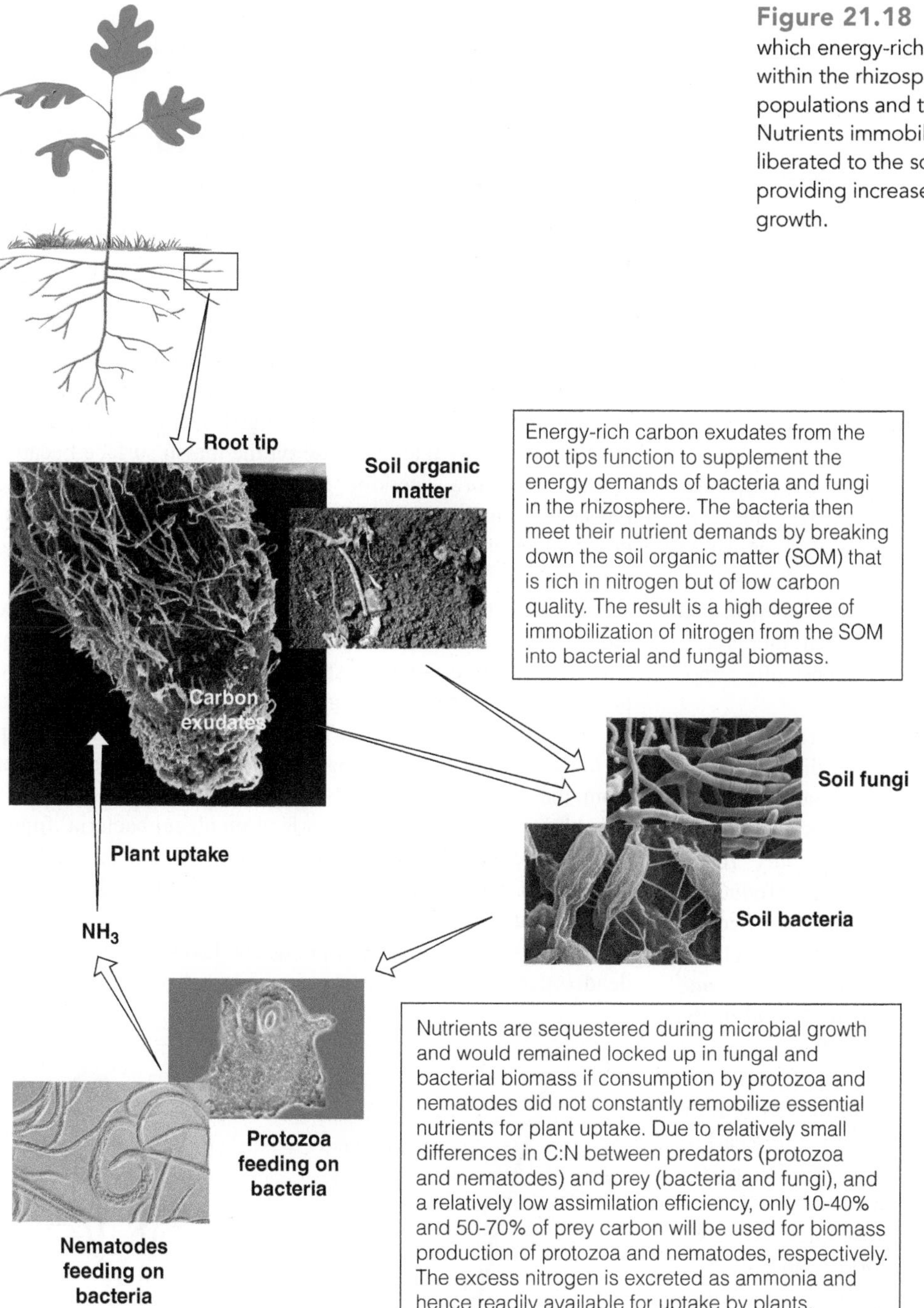
Root tip
Soil organic matter
Energy-rich carbon exudates from the root tips function to supplement the energy demands of bacteria and fungi in the rhizosphere. The bacteria then meet their nutrient demands by breaking down the soil organic matter (SOM) that is rich in nitrogen but of low carbon quality. The result is a high degree of immobilization of nitrogen from the SOM into bacterial and fungal biomass.
Carbon exudates
Soil fungi
Plant uptake
Soil bacteria
NH3
Nutrients are sequestered during microbial growth and would remained locked up in fungal and bacterial biomass if consumption by protozoa and nematodes did not constantly remobilize essential nutrients for plant uptake. Due to relatively small differences in C:N between predators (protozoa and nematodes) and prey (bacteria and fungi), and a relatively low assimilation efficiency, only 10-40% and 50-70% of prey carbon will be used for biomass production of protozoa and nematodes, respectively. The excess nitrogen is excreted as ammonia and hence readily available for uptake by plants.
Protozoa feeding on bacteria
Nematodes feeding on bacteria

Figure 21.18 Illustration of the soil microbial loop in which energy-rich carbon exudates from the plant roots within the rhizosphere enhance the growth of microbial populations and the breakdown of soil organic matter. Nutrients immobilized in microbial biomass are then liberated to the soil through predation by microbivores, providing increased mineral nutrients to support plant growth.

is, carbon compounds of very low quality that yield little energy to decomposers. At that point, the original plant litter has been degraded to a form where decomposition proceeds very slowly; the rate is largely a function of carbon quality. Soil organic matter typically has a residence time of 20 to 50 years, although it can range from 1 or 2 years in a cultivated field to thousands of years in environments that support slow rates of decomposition (dry or cold). Humus decomposes very slowly, but because of its abundance it represents a significant portion of the carbon and nutrients released from soils.

21.7 Plant Processes Enhance the Decomposition of Soil Organic Matter in the Rhizosphere

More than a century ago, the soil scientist Lorenz Hiltner put forward the term **rhizosphere** to describe a region of the soil where plant roots function. It is an active zone of root growth and death, characterized by intense microbial and fungal activity. Decomposition in the rhizosphere is more rapid than in the surrounding soil. The rhizosphere makes up virtually all of the soil in fine-rooted grasslands, where the average distance between roots is about 1 mm. Root density is far less per unit of soil volume in forested ecosystems (often 10 mm between roots). Roots alter the chemistry of the rhizosphere by secreting carbohydrates into the soil. These carbohydrates may account for as much as 40 percent of dry matter production of plants. This large expenditure of carbon must be of fundamental importance for plants to afford the significant trade-off in carbon allocation.

The growth of bacteria in the rhizosphere is supported by the abundant source of high-quality, energy-rich carbon of the root exudates (exuded matter). Bacterial growth is limited most strongly by nutrient availability because the exudates are energy rich, but very low in nitrogen and other essential nutrients for microbial growth. Bacteria acquire nutrients for their growth by breaking down soil organic matter. In other words, plant roots use carbon-rich exudates to supplement the decomposition process of low carbon quality organic matter in the rhizosphere. Nutrient immobilization occurs during microbial growth, and nutrients would remain sequestered in bacterial biomass if predation by protozoa and nematodes did not constantly remobilize essential nutrients for plant uptake.

Unlike the significant difference in the C/N ratio between bacteria and the plant litter they feed on (see discussions in Sections 21.4 and 21.5), there is a relatively small difference in C/N ratio between predators and bacterial prey. Because of their relatively low assimilation efficiency, only 10–40 percent of prey carbon will be used for biomass production of protozoa and 50–70 percent for nematodes (see Section 20.9). The remainder is lost as carbon dioxide in cellular respiration. Excess nitrogen is excreted as ammonia and hence readily available for uptake by plant roots within the rhizosphere. The interplay between microbial decomposers and microbivores determines the rate of nutrient cycling in the rhizosphere and strongly enhances the availability of mineral nutrients to plants. This process of supplementing carbon to microbial decomposers in the rhizosphere, enhancing the decomposition of soil organic matter, and subsequently releasing mineral nutrients for plant uptake by microbial grazers is referred to as the **soil microbial loop** (**Figure 21.18**). It is similar in structure to the concept of the microbial loop in aquatic ecosystems discussed in Section 21.8 (and Chapter 24, Section 24.10).

Populations of protozoa and nematodes fluctuate. As populations decline, their readily decomposable tissues also enter the detrital food chain. Protozoan biomass can equal that of all other soil animal groups except for earthworms. As much as 70 percent of soil respiration can be the result of protozoa and an additional 15 percent of nematodes. Production rates of microbivores can reach 10–12 times their standing biomass with a minimum generation time of 2–4 hours, suggesting a significant effect on nutrient mineralization within the soil.

Some estimates conclude that at the global scale, rhizosphere processes use approximately 50 percent of the energy fixed by photosynthesis in terrestrial ecosystems, contribute roughly 50 percent of the total carbon dioxide emitted from terrestrial ecosystems, and mediate virtually all aspects of nutrient cycling.

21.8 Decomposition Occurs in Aquatic Environments

Decomposition in aquatic ecosystems follows a pattern similar to that in terrestrial ecosystems but with some major differences influenced by the watery environment. As in terrestrial environments, decomposition involves leaching, fragmentation, colonization of detrital particles by bacteria and fungi, and consumption by detritivores and microbivores. In coastal environments, permanently submerged plant litters decompose more rapidly than do those on the marsh surface because (1) they are more accessible to detritivores, and (2) the stable physical environment is more favorable to microbial decomposers than are the alternative periods of wetting and drying that characterize tidal environments (**Figure 21.19**).

In flowing-water ecosystems, aquatic fungi colonize leaves, twigs, and other particulate matter. One group of aquatic arthropods—called shredders—fragments the organic particles, in the process eating bacteria and fungi on the surface of the litter (see Section 24.10). Downstream, another group of invertebrates—the filtering and gathering collectors—filter from the water fine particles and fecal material left by the shredders. Grazers and scrapers feed on algae, bacteria, fungi, and organic matter collected on rocks and large debris (all discussed in Chapter 24). Algae take up nutrients and dissolved organic matter from the water.

In the still, open water of ponds and lakes and in the ocean, dead organisms and other organic material, which are called

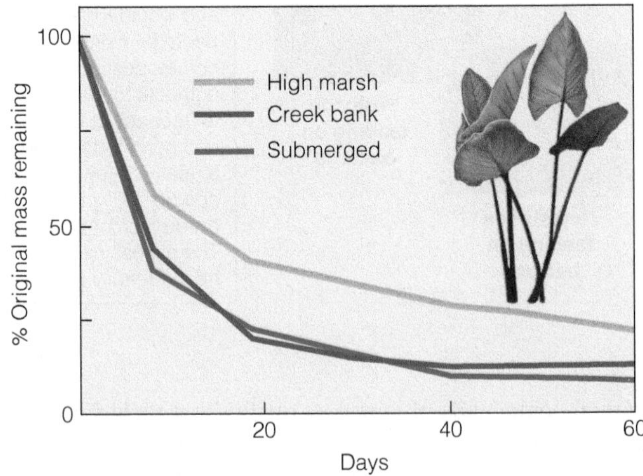

Figure 21.19 Decomposition of leaves of arrow arum (*Peltandra virginica*) in a tidal freshwater marsh. Decomposition is a measure of the percent of original mass remaining in litterbags under three conditions: irregularly flooded high marsh exposed to alternate periods of wetting and drying, creek bed flooded two times daily (tidal), and permanently submerged. The litter that is consistently wet (high marsh) has the highest rate of decomposition.
(Adapted from Odum and Heywood 1978.)

particulate organic matter (POM), drift toward the bottom. On its way, POM is constantly ingested, digested, and mineralized until much of the organic matter settles on the bottom in the form of humic compounds (see discussion of humus formation in Chapter 4 and Section 21.6). How much depends partly on the depth of the water the particulate matter falls through. In shallow water, much of it may arrive in relatively large fragments to be further fragmented and digested by bottom-dwelling detritivores such as crabs, snails, and mollusks.

Bacteria work on the bottom, or benthic, organic matter. Bacteria living on the surface can carry on aerobic respiration, but the oxygen supply is exhausted within a few centimeters below the surface of the sediment. Under this anoxic condition, a variety of bacteria capable of anaerobic respiration take over decomposition, which proceeds at a much slower rate than in the aerobic environment of the surface and shallow sediments (see Figure 21.9).

Aerobic and anaerobic decomposition in the benthic environment form only a part of the decomposition process. Dissolved organic matter (DOM) in the water column also provides a source of fixed carbon for decomposition. Major sources of DOM are the free-floating macroalgae, phytoplankton, and zooplankton inhabiting the open water. On death, their bodies break up and dissolve within 15 to 30 minutes, too rapidly for any bacterial action to occur. Phytoplankton and other algae also excrete quantities of organic matter at certain stages of their life cycles, particularly during rapid growth and reproduction. For example, during photosynthesis the marine alga *Fucus vesiculosus* produces an exudate high in carbon content. This DOM then becomes a substrate for bacterial growth.

Ciliates and zooplankton eat bacteria and in turn excrete nutrients in the form of exudates and fecal pellets in the water. Zooplankton, in the presence of abundant food, consume more than they need and excrete half or more of the ingested material as fecal pellets. These pellets make up a significant fraction of the suspended material, providing a substrate for further bacterial decomposition. An important component of the aquatic nutrient cycle is the microbial loop. The microbial loop is the trophic pathway through which DOM is reintroduced to the food web by being incorporated into bacteria, which in turn are consumed by ciliates and zooplankton (see Section 24.10 for a detailed discussion of the microbial loop in aquatic food chains). The lighter nature of DOM (in contrast to particulate) allows it to remain longer in the upper waters, so that nutrients entering the microbial loop have a greater likelihood of remaining in the upper waters of the photic zone to support further primary productivity.

21.9 Key Ecosystem Processes Influence the Rate of Nutrient Cycling

You can see from Figure 21.1 that the internal cycling of nutrients through the ecosystem depends on the processes of primary production and decomposition. Primary productivity determines the rate of nutrient transfer from inorganic to organic form (nutrient uptake), and decomposition determines the rate of transformation of organic nutrients into inorganic form (net mineralization rate). Therefore, the rates at which these two processes occur directly influence the rates at which nutrients cycle through the ecosystem. But how do these two key processes interact to limit the rate of the internal cycling of nutrients through the ecosystem? The answer lies in their interdependence.

For example, consider the cycling of nitrogen, an essential nutrient for plant growth. The direct link between soil nitrogen availability, rate of nitrogen uptake by plant roots, and the resulting leaf nitrogen concentrations was discussed previously (Chapter 6, Figure 6.23). The maximum rate of photosynthesis is strongly correlated with nitrogen concentrations in the leaves because certain compounds directly involved in photosynthesis (e.g., rubisco and chlorophyll) contain a large portion of leaf nitrogen (see Figure 6.23c). Thus, availability of nitrogen in the soil (sediments or water in the case of aquatic ecosystems) directly affects rates of ecosystem primary productivity via the influence of nitrogen on photosynthesis and carbon uptake.

A low availability of soil nitrogen reduces net primary production (the total production of plant tissues) as well as the nitrogen concentration of the plant tissues produced (again, see Figure 6.23b). Thus, the reduced availability of soil nitrogen influences the input of dead organic matter to the decomposer food chain by reducing both the total quantity of dead organic matter produced and its nutrient concentration. The net effect is a lower return of nitrogen in the form of dead organic matter.

Both the quantity and quality of organic matter as a food source for decomposers directly influence the rates of decomposition and nitrogen mineralization (nutrient release). Lower nutrient concentrations in the dead organic matter promote immobilization of nutrients from the soil and water to meet the nutrient demands of decomposer populations (see Figure 21.15). This immobilization effectively reduces nutrient availability to the plants (reduces the rate of net mineralization), which adversely affects primary productivity (see Figures 20.6 and 20.7).

One can now appreciate the feedback system that exists in the internal cycling of nutrients within an ecosystem (**Figure 21.20**). Reduced nutrient availability can have the effect of reducing both the nutrient concentration of plant tissues (primarily leaf tissues) and NPP. This reduction lowers the total amount of nutrients returned to the soil in dead organic matter. The reduced quantity and quality (nutrient concentration) of organic matter entering the decomposer food chain increase immobilization and reduce the availability of nutrients for uptake by plants. In effect, low nutrient availability begets low nutrient availability. Conversely, high nutrient availability encourages high plant tissue concentrations and high NPP. In turn, the associated high quantity and quality of dead organic matter encourage high rates of net mineralization and nutrient supply in the soil.

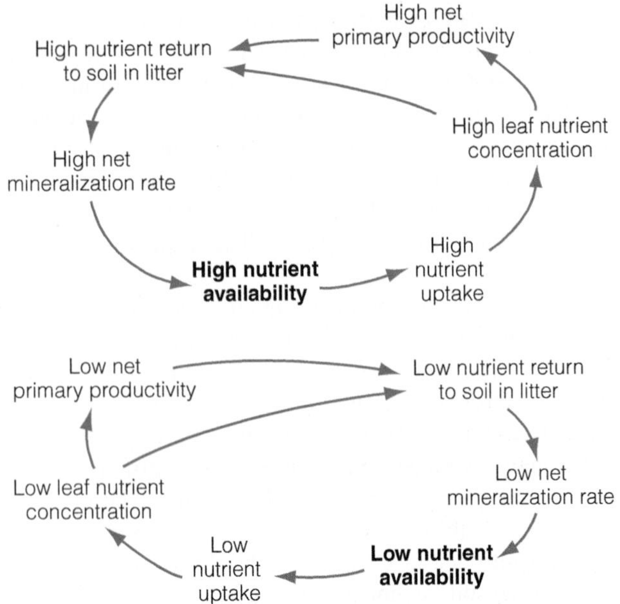

Figure 21.20 Feedback that occurs between nutrient availability, net primary productivity, and nutrient release in decomposition for initial conditions of low and high nutrient availability.
(Adapted from Chapin 1980.)

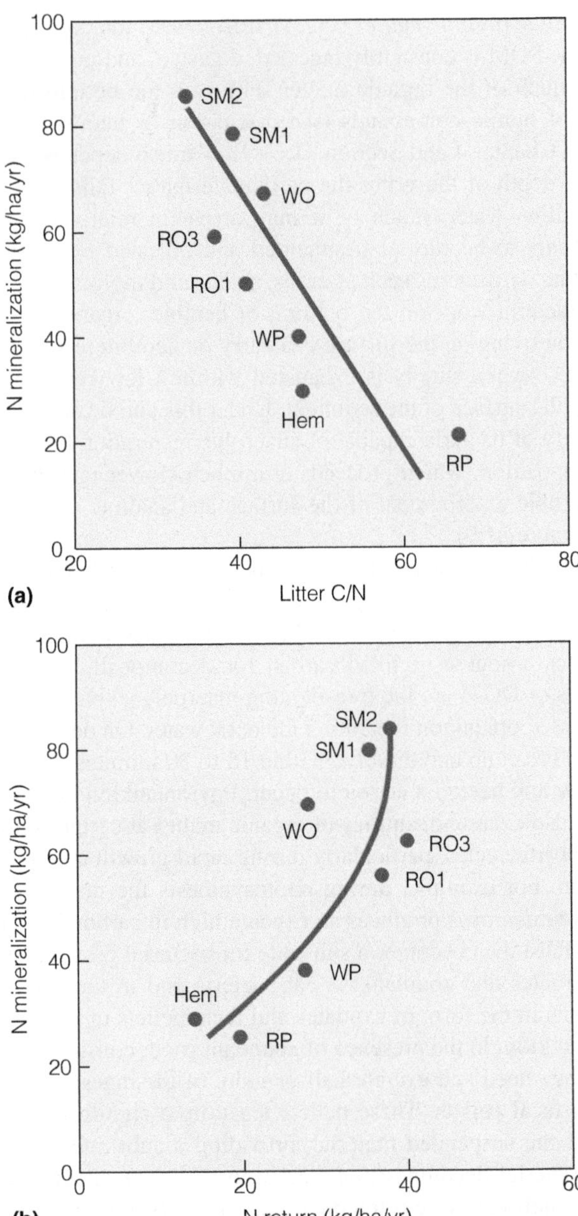

Figure 21.21 Relationship between (a) litter quality (carbon to nitrogen [C/N]) and nitrogen mineralization rate (nitrogen availability) and (b) nitrogen mineralization rate and nitrogen returned in annual litterfall for a variety of forest ecosystems on Blackhawk Island, Wisconsin. Abbreviations refer to the dominant trees in each stand: Hem, hemlock; RP, red pine; RO, red oak; WO, white oak; SM, sugar maple; WP, white pine.
(Adapted from Pastor et al. 1984.)

The feedback between litter quality, nutrient cycling, and NPP is illustrated in the work of John Pastor of the University of Minnesota. Pastor and colleagues examined aboveground production and nutrient cycling (nitrogen and phosphorus) in a series of forest stands along a gradient of soil texture on Blackhawk Island, Wisconsin. Tree species producing higher-quality litter (lower C/N ratio) dominated sites with a progressively finer soil texture (silt and clay content; see Section 4.6). The higher-quality litter resulted in a higher rate of nutrient mineralization (release; **Figure 21.21a**). Higher rates of nutrient availability in turn resulted in a higher rate of primary productivity (see Figure 20.6) and nutrient return in litterfall (**Figure 21.21b**). The net effect was to increase the rate at which nitrogen and phosphorus cycle through the forest stands. The changes in species composition and litter quality along the soil gradient were directly related to the influence of soil texture on plant available moisture (see Section 4.8 and Figure 4.10).

21.10 Nutrient Cycling Differs between Terrestrial and Open-Water Aquatic Ecosystems

Nutrient cycling is an essential process in all ecosystems and represents a direct (cyclic) link between NPP and decomposition. However, the nature of this link varies among ecosystems, particularly between terrestrial and aquatic ecosystems.

In virtually all ecosystems, there is a vertical separation between the zones of production (photosynthesis) and decomposition (**Figure 21.22**). In terrestrial ecosystems, the plants themselves bridge this physical separation between the decomposition zone at the soil surface and the productivity zone in the plant canopy; the plants physically exist in both zones. The root systems provide access to the nutrients made available in the soil through decomposition, and the vascular system within the plant transports these nutrients to the sites of production (canopy).

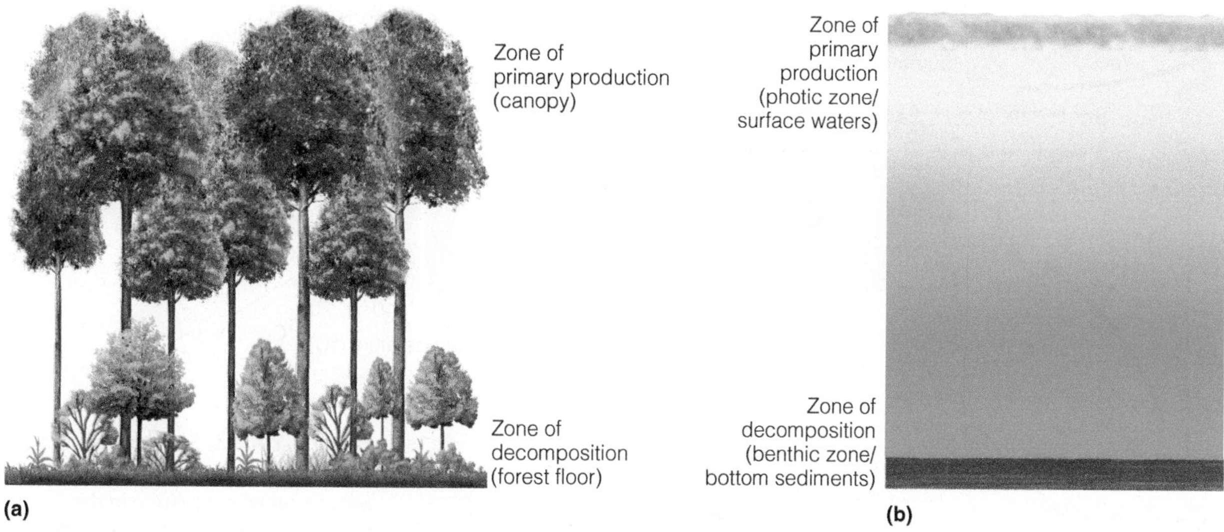

Figure 21.22 Comparison of the vertical zones of production and decomposition in (a) a terrestrial (forest) and (b) an open-water (lake) ecosystem. In the terrestrial ecosystem, the two zones are linked by the vegetation (trees). However, this is not the case in the lake ecosystem.

In aquatic ecosystems, however, plants do not always function to bridge the zones of production and decomposition. In shallow-water environments of the shoreline, emergent vegetation such as cattails, cordgrasses (*Spartina* spp.), and sedges are rooted in the sediments. Here, as in terrestrial ecosystems, the zones of decomposition and production are linked directly by the plants. Likewise, submerged vegetation, such as seagrass (see Figure 18.8), is rooted in the sediments. The plants extend up the water column into the photic zone, the shallower waters where light levels support higher productivity (see Figures 16.12 and 24.3). However, as water depths increase, primary production is dominated by free-floating phytoplankton within the upper waters (photic zone). In deeper water, the zones of decomposition in the bottom sediments and waters (the benthic zone) are physically separated from the surface waters, where temperatures and light availability support primary productivity. This physical separation between the zones where nutrients become available through decomposition and the zone of productivity where nutrients are needed to support photosynthesis and plant growth is a major factor controlling the productivity of open-water ecosystems (e.g., lakes and oceans).

To understand how nutrients are transported vertically from deeper waters to the surface, where temperature and light conditions can support primary productivity, we must examine the vertical structure of the physical environment in open-water ecosystems first presented in Chapter 3 (Section 3.4). The vertical structure of open-water ecosystems, such as lakes or oceans, can be divided into three rather distinct zones (**Figure 21.23**; also see Chapter 16). The epilimnion, or surface water, is relatively warm because of the interception of solar radiation. The oxygen content is also relatively high because of the diffusion of oxygen from the atmosphere into the surface waters (see Section 3.6). In contrast, the hypolimnion, or deep water, is cold and relatively low in oxygen. The transition zone between the surface and deep waters is characterized by a steep temperature gradient called the *thermocline*. In effect, the vertical structure can be represented as a warm, low-density surface layer of water on top of a denser, colder layer of deep water; these layers are separated by the rather thin zone of the thermocline. This vertical structure and physical separation of the epilimnion and hypolimnion have an important influence on the distribution of nutrients and subsequent patterns of primary productivity in aquatic ecosystems. The colder, deeper waters where decomposition occurs are relatively nutrient rich, but temperature and light conditions cannot support high productivity. In contrast, the surface waters are relatively nutrient poor; however, this is the zone where temperatures and light support high productivity.

Although winds blowing over the water surface cause turbulence that mixes the waters of the epilimnion, this mixing does not extend into the colder, deeper waters. As autumn and winter approach in the temperate and polar zones, the amount of solar radiation reaching the water surface decreases, and the temperature of the surface water declines. As the water temperature of the epilimnion approaches that of the hypolimnion, the thermocline breaks down, and mixing throughout the profile can take place (Figure 21.23). If surface waters become cooler than the deeper waters, they will begin to sink, displacing deep waters to the surface. This process is called *turnover* (see Section 3.4). With the breakdown of the thermocline and mixing of the water column, nutrients are brought up from the bottom to the surface waters. With the onset of spring, increasing temperatures and light in the epilimnion give rise to a peak in productivity because

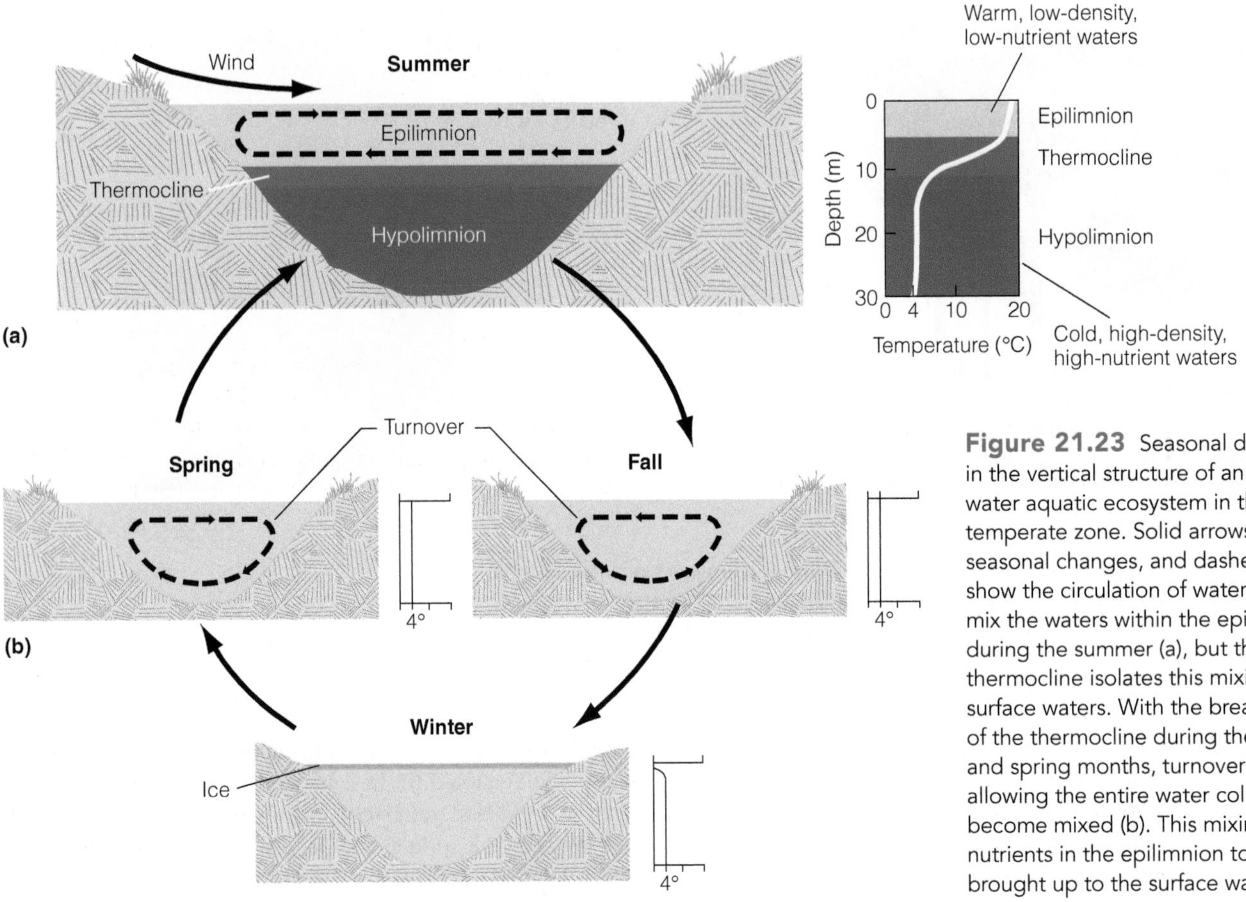

Wind
Summer
Epilimnion
Thermocline
Hypolimnion
(a)
Warm, low-density, low-nutrient waters
Epilimnion
Thermocline
Hypolimnion
Cold, high-density, high-nutrient waters
Depth (m)
Temperature (°C)
Turnover
Spring
Fall
4°
(b)
Winter
Ice
4°

Figure 21.23 Seasonal dynamics in the vertical structure of an open-water aquatic ecosystem in the temperate zone. Solid arrows track seasonal changes, and dashed arrows show the circulation of waters. Winds mix the waters within the epilimnion during the summer (a), but the thermocline isolates this mixing to the surface waters. With the breakdown of the thermocline during the fall and spring months, turnover occurs, allowing the entire water column to become mixed (b). This mixing allows nutrients in the epilimnion to be brought up to the surface waters.

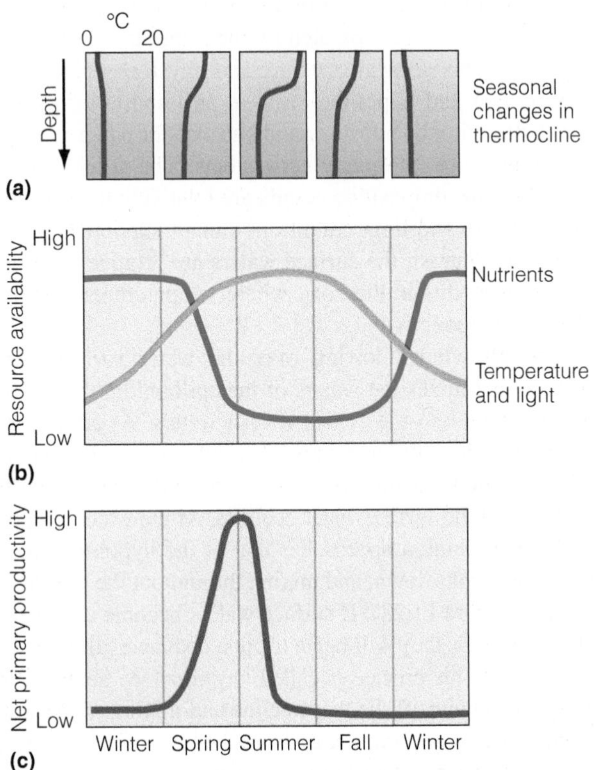

Figure 21.24 Seasonal dynamics of (a) the thermocline and associated changes in (b) the availability of light and nutrients, and (c) net primary productivity of the surface waters.

of the increased availability of nutrients in the surface waters. As the spring and summer progress, the nutrients in the surface water are used by autotrophs, reducing the nutrient content of the water, and a subsequent decline in productivity occurs. The annual cycle of productivity in these ecosystems (**Figure 21.24**) is a direct result of thermocline dynamics and the consequent behavior of the vertical distribution of nutrients.

21.11 Water Flow Influences Nutrient Cycling in Streams and Rivers

Inputs in the form of dead organic matter from adjacent terrestrial ecosystems (leaves and woody debris), rainwater, and subsurface seepage bring nutrients into streams. Although the internal cycling of nutrients follows the same general pathway as that discussed for terrestrial and open-water ecosystems (see Figure 21.1), the continuous, directional movement of water affects nutrient cycling in stream ecosystems. Jack Webster of the University of Georgia was the first to note that because nutrients are continuously being transported downstream, a spiral rather than a cycle better represents the cycling of nutrients. He coined the term **nutrient spiraling** to describe this process.

Nutrients in terrestrial and open-water ecosystems are recycled more or less in place. An atom of nutrients passes

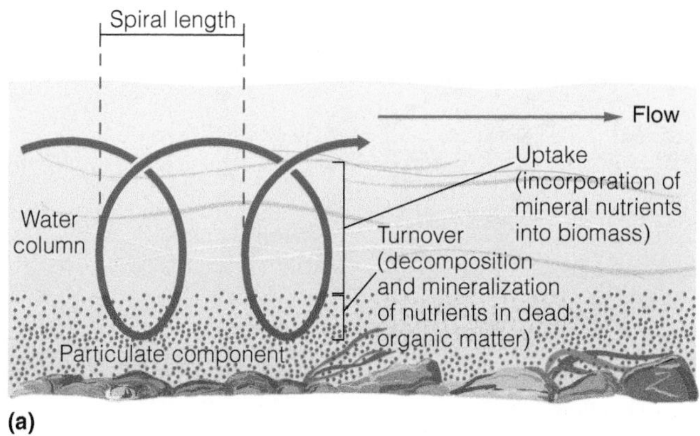

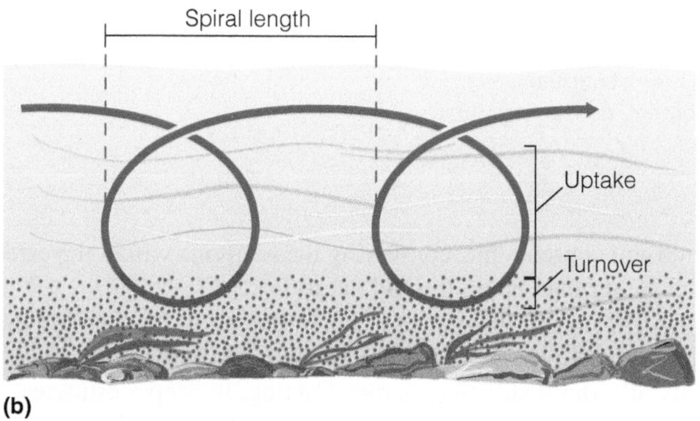

Figure 21.25 Nutrient spiraling between organic matter and the water column in a stream ecosystem. Uptake and turnover take place as nutrients flow downstream. The tighter the spiraling, the longer the nutrients remain in place. (a) Tight spiraling; (b) open spiraling.
(Adapted from Newbold et al. 1982.)

from the soil or water column to plants and consumers and passes back to the soil or water in the form of dead organic matter. Then it is recycled within the same location within the ecosystem, although losses do occur. Cycling essentially involves time. Flowing water has an added element: a spatial cycle. Nutrients in the form of organic matter are constantly being carried downstream. How quickly these materials are carried downstream depends on how fast the water moves and the physical and biological factors that hold nutrients in place. Physical retention involves storage in wood detritus such as logs and snags, in debris caught in pools behind logs and boulders, in sediments, and in patches of aquatic vegetation. Biological retention occurs through the uptake and storage of nutrients in animal and plant tissue.

The processes of recycling, retention, and downstream transport may be pictured as a spiral lying horizontally (longitudinally) in a stream (**Figure 21.25**). One cycle in the spiral is the uptake of an atom of nutrient, its passage through the food chain, and its return to water, where it is available for reuse. Spiraling is measured as the distance needed to complete one cycle. The longer the distance required, the more open the spiral; the shorter the distance, the tighter the spiral. If dead leaves and other debris are physically held in place long enough to allow organisms to process the organic matter, the spiral is tight. This type of physical retention is especially important in fast headwater streams, which can rapidly lose organic matter downstream. Organisms can function to both open and tighten the spiral. Organisms that shred and fragment the organic matter can open the spiral by facilitating the transport of organic materials downstream. Other organisms tighten the spiral by physically storing dead organic matter.

J. D. Newbold and colleagues at Oak Ridge National Laboratory in Tennessee experimentally determined how quickly phosphorus moved downstream in a small woodland brook on the Walker Branch watershed. They determined that phosphorus moved downstream at the rate of 10.4 m a day and cycled once every 18.4 days. The average downstream distance of one cycle (spiral) was 190 m. In other words, 1 atom of phosphorus on the average completed one cycle from the water compartment and back again for every 190 m of downstream travel.

21.12 Land and Marine Environments Influence Nutrient Cycling in Coastal Ecosystems

Coastal ecosystems are among the most productive environments. Water from most streams and rivers eventually drains into the oceans, and the place where this freshwater joins saltwater is called an *estuary* (see Section 3.10). Estuaries are semi-enclosed parts of the coastal ocean where seawater is diluted and partially mixed with water coming from the land. As the rivers meet the ocean, the current velocity drops and sediments are deposited within a short distance (referred to as a *sediment trap*; **Figure 21.26**). The buildup of sediments creates alluvial plains about the estuary, giving rise to mudflats and salt marshes that are dominated by grasses and small shrubs rooted in the mud and sediments (see Chapter 25 for detailed descriptions of these ecosystems). Nutrient cycling in these ecosystems differs from that of terrestrial, open-water, and stream ecosystems discussed thus far. In a way, it combines some features of each. As in terrestrial ecosystems, the dominant plants are rooted in the sediments and therefore function to link the zones of decomposition and primary production. Submerged plants take up nutrients from the sediments as well as directly from the water column. As in streams and rivers, the directional (horizontal) flow of water transports organic matter (energy) and nutrients both into (inputs) and out of (outputs) the ecosystem.

Nutrients are carried into the coastal marshes by precipitation, surface water (streams and rivers), and groundwater. In addition, the rise and fall of water depth with the tidal cycle serves to flush out salts and other toxins from the marshes and brings in nutrients from the coastal waters by a process referred to as the **tidal subsidy**. The tidal cycle also serves to replace oxygen-depleted waters within the surface sediments with oxygenated water.

The salt marsh is a detrital system with only a small portion of its primary production consumed by herbivores. Almost three-quarters of the detritus (dead organic matter) produced in the salt-marsh ecosystem is broken down by bacteria and

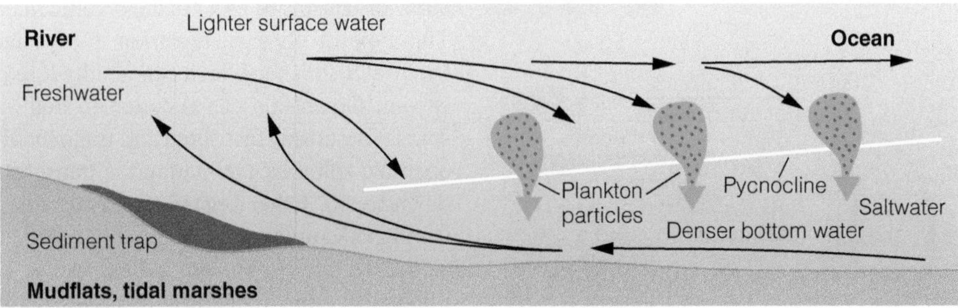

Figure 21.26 Circulation of freshwater and saltwater in an estuary functions to trap nutrients. A salty wedge of intruding seawater on the bottom produces a surface flow of lighter freshwater and a counterflow of heavier brackish water. These layers are physically separated by variations in water density arising from both salt concentration and temperature differences. The zone of maximum vertical difference in density—the pycnocline—functions much like the thermocline in lake ecosystems. Living and dead particles settle through the pycnocline into the countercurrent and are carried up-estuary along with their nutrient content, conserving the nutrients within the estuary rather than being flushed out to sea.
(Adapted from Correll 1978.)

fungi. Nearly 50 percent of the total NPP is lost through respiration by microbial decomposers. The low oxygen content of the sediments favors anaerobic bacteria. They can carry on their metabolic functions without oxygen by using inorganic compounds such as sulfates rather than oxygen in the process of fermentation (see Section 6.1).

A substantial portion (usually 20–45 percent) of the NPP of a salt marsh is exported to adjacent estuaries. The exact nature of this exchange depends on the geomorphology of the basin (its shape and nature of the opening to the sea) and the magnitude of tidal and freshwater flows (fluxes). Each salt marsh differs in the way carbon and other nutrients move through the food web as well as in the route taken and amount of nutrients exported. Some salt marshes are dependent on tidal exchanges and import more than they export, whereas others export more than they import. Some material may be exported to the estuary as mineral nutrients or physically as detritus, as bacteria, or as fish, crabs, and intertidal organisms within the food web. Although inflowing water from rivers and coastal marshes carries mineral nutrients into the estuary, primary production is regulated more by internal nutrient cycling than by external sources. This internal cycling involves the release of nutrients through decomposition and mineralization within the bottom sediments, as well as excretion of mineralized nutrients by herbivorous zooplankton.

As in the tidal marshes, nutrients and oxygen are carried into the estuary by the tides. Typical estuaries maintain a "salt wedge" of intruding seawater on the bottom, producing a surface flow of freshwater and a counterflow of more brackish, heavier water (see Figure 21.26). These layers are physically separated by variations in water density arising from both salt concentration and temperature differences. The zone of maximum vertical difference in density, the **pycnocline**, functions much like the thermocline in lake ecosystems. Living and dead particles settle through the pycnocline into the countercurrent and are carried up the estuary along with their nutrient content, thus conserving the nutrients within the estuary rather than being flushed out to sea.

The regular movement of freshwater and saltwater into the estuary, coupled with the shallowness and turbulence, generally allows for sufficient vertical mixing. In deeper estuaries, a thermocline may form during the summer months. In this case, the seasonal pattern of vertical mixing and nutrient cycling will be similar to the pattern discussed for open-water ecosystems in Section 21.10.

21.13 Surface Ocean Currents Bring about Vertical Transport of Nutrients

The global pattern of ocean surface currents influences patterns of surface-water temperature, productivity, and nutrient cycling (Chapter 2, Figure 2.13). The Coriolis effect drives the patterns of surface currents. But how deep does this lateral movement of water extend vertically into the water column? In general, the lateral flow is limited to the upper 100 m, but in certain regions the lateral movement can bring about a vertical circulation or upwelling of water. Along the western margins of the continents, the surface currents flow along the coastline and toward the equator (see Figure 2.13 and Section 3.9). At the same time, these surface waters are pushed offshore by the Coriolis effect. The movement of surface waters offshore results in deeper, more nutrient-rich waters being transported vertically to the surface (Figure 3.16a).

Surface currents give rise to a similar pattern of upwelling in the equatorial waters. As the two equatorial currents flow west, they are deflected to the right north of the equator and to the left south of the equator. Where this occurs, subsurface water is transported vertically, bringing cold waters, rich in nutrients, to the surface (see Figure 3.16b). These regions of nutrient-rich waters are highly productive and support some of the world's most important fisheries (see Figure 20.10).

ECOLOGICAL Issues & Applications

Agriculture Disrupts the Process of Nutrient Cycling

As we discussed in Section 21.9 (Figure 21.20), in natural ecosystems there is a tight link between the processes of NPP and decomposition. NPP determines the quantity and quality of organic matter available to decomposer populations. In turn, the process of decomposition determines the rate of nutrient release to the soil (net mineralization rate), constraining plant uptake and NPP. In the practice of agriculture, however, this balance is disrupted. Plants, and the nutrients they contain, are harvested as crops, and as a result, the organic matter does not return to the soil surface to undergo the process of decomposition and mineralization. The inevitable result is a decline in soil fertility and future plant productivity. For this reason, in all agricultural systems it is necessary to either allow for a period of time for natural processes to reestablish soil nutrient concentrations between harvests, or to add nutrients to the soil from outside sources. The manner in which the nutrient status of the soil is maintained to allow for plant growth and crop production is a defining characteristic of different traditional and modern agricultural practices.

Figure 21.27 (a) In swidden agriculture, a plot of forest is cleared and burned to allow crops to be planted. The ashes are an important source of nutrients. (b) Sequence of areas used for swidden agriculture in the Yucatán region of Mexico. The area in the foreground has recently been cleared for planting. The area directly behind it has been abandoned (fallow) for one year, and the forest in the background has been in the process of regeneration (secondary succession) for approximately 10 years.

A method of traditional subsistence farming that is practiced primarily in the tropical regions is shifting cultivation, or **swidden agriculture**. This method of traditional agriculture involves a rotating cultivation technique in which trees (and other vegetation) are first cut down and burned to clear land for planting (Figure 21.27). The burning of felled trees and brush serves two purposes. First, it removes debris, thus clearing the land for planting and ensuring that the plot is relatively free of weeds. Second, the resulting ash is high in mineral nutrients, promoting plant growth. The plot is then cultivated and crops are harvested. A characteristic of this type of agriculture is a decline in productivity with each successive harvest because each time crops are harvested, nutrients in the form of plant tissues are being removed from the plot (Figure 21.28). Eventually the site is abandoned as farmland for a period of time (3 to 30 years) referred to as *fallow*, and allowed to revert to natural vegetation through secondary succession (see Chapter 18, Section 18.3). If left undisturbed for a sufficient time, the nutrient status of the site recovers to precultivation levels (Figure 21.29; see Section 18.5 for discussion of nutrient dynamics during succession). At that point, the site can once again be cleared and planted. In the meantime, other areas have been cleared, burned, and planted. So in effect, this type of agriculture represents a shifting, highly heterogeneous patchwork of plots in various stages of cultivation and secondary succession.

The swidden cultivation system is currently practiced by an estimated 500 million people or 7 percent of the world population, primarily in the regions of the topics. It represents a sustainable form of agriculture when time is permitted for regrowth of natural vegetation and the recovery of soil nutrients. But it also requires sufficient land area to allow for the appropriate rotation period. The problem currently being faced in many parts of the tropics is that growing populations are placing ever-increasing demands on the land, and sufficient recovery periods are not always possible. In these cases, the land is quickly degraded and yield continuously declines.

Unlike swidden agriculture, all other agricultural systems depend on supplementing soil nutrient supplies between harvests through the use of fertilizers. **Fertilizers** are any substance added to the soil that contains chemical elements that improve soil fertility and enhance the growth and productivity

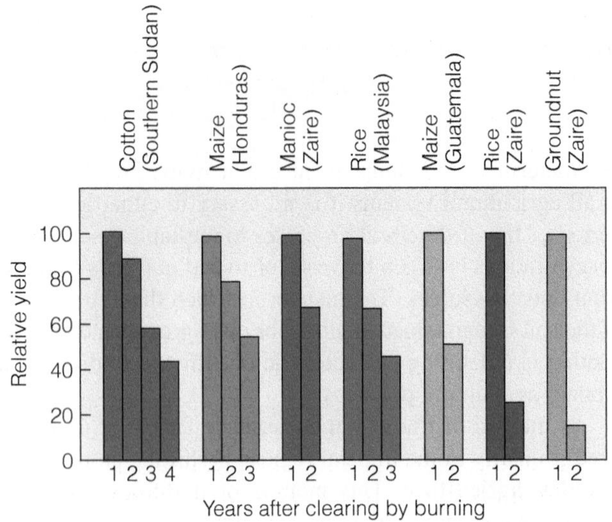

Figure 21.28 Patterns of declining productivity in successive years in swidden agricultural systems for various crops in several regions of the tropics.

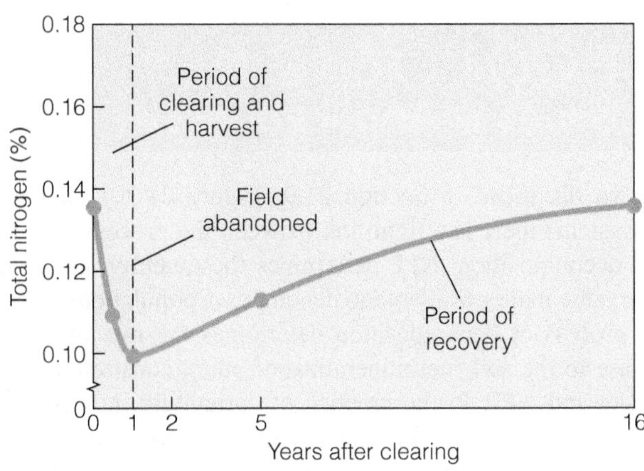

Figure 21.29 Changes in total soil nitrogen during a cycle of clearing, harvest, abandonment, and recovery for a swidden agricultural system in Costa Rica (Central America). Note that nitrogen levels decline until the plot is abandoned after the first year. Soil nitrogen recovers to original levels by year 16. (Adapted from Ewel et al. 1981.)

of plants (**Figure 21.30**). Fertilizers fall into two broad categories: organic and inorganic (mineral).

Organic fertilizers are composed of organic matter derived from plants or animals. Organic fertilizers include naturally occurring organic matters such as manure, guano, peat, algae, and dead plant materials (including crop residues and cover crops that are plowed into the soil, referred to as "green manure"), and processed organic matters such as compost, bone and blood meal, fish meal, and various plant and algal extracts. Recall from our previous discussions (Sections 6.11 and 21.1) that plants require nutrients in a mineral or inorganic form. Therefore, organic fertilizers do not directly supply plants with available nutrients; rather, these organic compounds provide a substrate (food source) for decomposers. The organic matter must undergo the process of decomposition and mineralization before nutrients are available for uptake by plant roots. As a result, the chemical composition of the organic matter being supplied as a fertilizer is critical to determining the rates of decomposition and exchange of nutrients with the soil (net mineralization rate). For example, in Section 21.5 we discussed that the net release of nitrogen during decomposition (net mineralization rate) is directly influenced by the nitrogen content of the dead organic matter as represented by the ratio of carbon to nitrogen (C/N). When the C/N of the organic matter is high, the nitrogen content relative to the carbon (energy source) is insufficient to meet the demands of the decomposers and the rate of immobilization exceeds the rate of mineralization (see Figure 21.13). The

(a)

(b)

Figure 21.30 (a) The primary source of fertilizers before World War II was organic matter as seen with this farmer hauling a trailer of manure to be spread on the crop fields. (b) Modern agriculture depends on the application of inorganic-chemical fertilizers.

result is a net decrease in the amount of mineral (inorganic) nitrogen in the soil (net mineralization rate < 0). It is critical for an organic material used as fertilizer to have a C/N (nitrogen content) that provides the necessary balance of carbon and nitrogen for the decomposers such that the rate of immobilization is less than the rate of mineralization from the start of decomposition following application, and therefore there is an immediate net release of mineral nitrogen to the soil (net mineralization rate > 0). The C/N at which this occurs is generally referred to as the "critical C/N" (see Figure 21.17 for dynamics of carbon and nitrogen during decomposition). In general, organic materials having a C/N less than 20 result in positive net mineralization (net mineralization rate > 0), whereas a C/N greater than 30 generally results in net immobilization (net mineralization rate < 0), effectively removing mineral nutrients from the soil over the short term (until the C/N declines to 20:1). A sample of typical C/N for a variety of organic materials (including those commonly used as organic fertilizer) is presented in **Table 21.2**.

The second category of fertilizers is **inorganic** or **mineral fertilizers** (referred to as chemical fertilizers when it is of a synthetic origin). Unlike organic fertilizers, inorganic fertilizers provide nutrients to the soil which are immediately available for uptake by the plant roots. However, they are also readily leached from the soil and a significant proportion of inorganic fertilizers applied to agricultural fields leach into groundwater or surface waters where they can result in serious environmental problems (see topic of dead zones in Chapter 24, *Ecological Issues & Applications*).

As early as the 18th century, inorganic fertilizers were mined from natural geologic mineral deposits; however, their availability was limited and organic fertilizers were the dominant nutrient supplements used in agricultural systems. As the world population increased, the growing demand for food was accompanied by an increasing need for fertilizers, and by the late 1800s there was growing concern about depletion of the limited sources of nitrogen used as mineral fertilizer (potassium nitrate [KNO_3] sodium nitrate [$NaNO_3$]). In the early 20th century, however, the German physical chemist Fritz Haber developed the synthetic ammonia process, which made the manufacture of ammonia (NH_3) from atmospheric N_2 economically feasible. Carl Bosch, an industrial chemist, soon translated the process developed by Haber for synthesizing liquid ammonia into a large-scale process using a catalyst and high-pressure methods. By 1913, a chemical plant was operating in Germany, producing ammonia using what has now become known as the Haber–Bosch process.

By the 1920s ammonia-producing plants, based on the Haber–Bosch process, were being built in the United States and in Europe outside of Germany; however, ammonia from these plants was still not economically feasible for wide-scale agricultural use as compared to the traditional source of nitrogen from organic fertilizers. That would change, however, with the advent of World War II. The war effort in the United States saw a dramatic increase in the industrial production of ammonia for the manufacturing of explosives, leading to cheaper and more efficient production methods. Following the end of World War II, the focus of the industrial production of ammonia shifted from the war effort to the agricultural field. Beginning in the late 1940s, the use of inorganic nitrogen fertilizers has increased exponentially leading to dramatic increases in agricultural productivity over the same period (**Figure 21.31**). In an analysis of the sources of nitrogen in global crop production, Vaclav Smil of the University of Manitoba (Canada) estimates that of the approximately 110 Tg (Tg $= 10^{12}$ g) of nitrogen fertilizers added to agricultural croplands each year, 71 percent is in the form of inorganic fertilizers, with the remaining 29 percent from organic materials (organic fertilizers). It is estimated that the nitrogen content of more than one-third of the protein consumed by the world's population is a product of the Haber–Bosch process.

Table 21.2 Sample of Typical C:N for a Variety of Organic Materials (Including Those Commonly Used as Organic Fertilizer)

Organic Matter	C:N Ratio
Fresh grass	< 10:1
Poultry manure	10:1
Dairy manure	17:1
Clover and alfalfa	13:1
Compost	15:1
Wheat, oat or rye straw	80:1
Oak leaves	90:1
Fresh sawdust	400:1

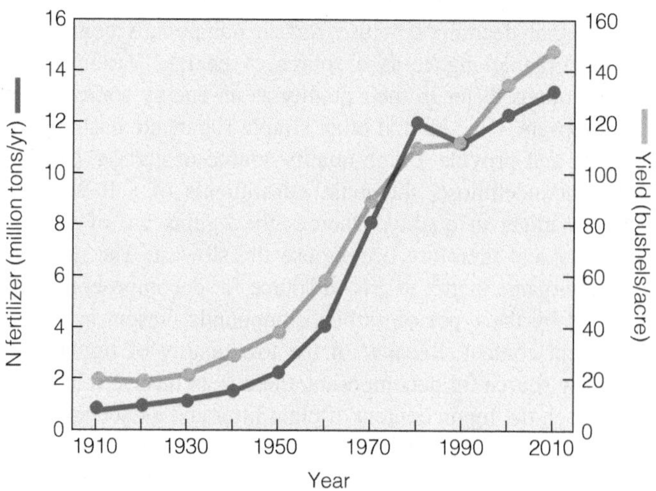

Figure 21.31 Changes in the application of inorganic nitrogen fertilizer and corresponding changes in the yield of corn per acre in the United States over the period of 1910–2010.
(Data from United States Department of Agriculture.)

SUMMARY

Nutrient Cycling 21.1

As plants take up nutrients from the soil or water, they become incorporated into living tissues: organic matter. As the tissues senesce, the dead organic matter returns to the soil or sediment surface. Various decomposers transform the organic nutrients into mineral form, and they are once again available for uptake by plants. This process is called *internal cycling*.

Decomposition Processes 21.2

Decomposition is the breakdown of chemical bonds formed during the construction of plant and animal tissues. Decomposition involves an array of processes including leaching, fragmentation, digestion, and excretion. An essential function of decomposers is the release of organically bound nutrients into an inorganic form. The wide variety of organisms involved in decomposition is classified into groups based on both function and size. The microflora, comprising the bacteria and fungi, are the group most commonly associated with decomposition. Bacteria can be grouped as either aerobic or anaerobic based on their requirement of oxygen to carry out respiration. Invertebrate detritivores are classified based on body size. Microbivores feed on bacteria and fungi.

Litterbag Studies 21.3

Decomposers derive their energy and most of their nutrients by consuming organic compounds. Ecologists study the process of decomposition by designing experiments that follow the decay of dead plant and animal tissues through time. The most widely used approach is litterbags. A fixed amount of dead organic matter is placed in mesh bags, and the rate of loss is followed through time.

Factors Influencing Decomposition 21.4

Microbial decomposers use carbon compounds contained in dead organic matter as a source of energy. Various carbon compounds differ in their quality as an energy source for decomposers. Glucose and other simple sugars are easily broken down and provide a high-quality source of carbon. Cellulose and hemicellulose, the main constituents of cell walls, are intermediate in quality, whereas the lignins are of very low quality and therefore decompose the slowest. The quality of dead organic matter as a food source for decomposers is influenced by the types of carbon compounds present and by the nutrient content. Because of the low quality of lignins as an energy source for decomposers, there is an inverse relationship between the lignin content of plant litter and its decomposition rate. The physical environment directly influences both macro- and micro-decomposers. Temperature and moisture greatly influence microbial activity. Low temperatures and moisture inhibit microbial activity. As a result, the highest decomposition rates occur during warm, wet conditions. The influence of temperature and moisture on decomposer activity results in geographic variation in rates of decomposition that relate directly to climate.

Nutrient Mineralization 21.5

The nutrient quality of dead organic matter depends on its nutrient content. As microbial decomposers break down dead organic matter, they transform nutrients tied up in organic compounds into an inorganic form. This process is called *nutrient mineralization*. The same organisms responsible for mineralization reuse some of the nutrients they have produced, incorporating the inorganic nutrients into an organic form. This process is called *nutrient immobilization*. The difference between the rates of mineralization and immobilization is the net mineralization rate, which represents the net release of nutrients to the soil or water during decomposition. The rates of mineralization and immobilization during decomposition are related to the nutrient content of the dead organic matter being consumed.

Soil Organic Matter 21.6

The process of decomposition continues as the litter degrades into a dark brown or black homogeneous organic matter known as *humus*. As this organic matter becomes embedded in the soil matrix, it is referred to as *soil organic matter*. Although high in nitrogen, humus decomposes very slowly because of the low quality of carbon compounds. Despite the slow rate of decomposition—because of its sheer abundance—the decomposition of humus represents a significant portion of the carbon and nutrients released from soils.

Rhizosphere 21.7

The rhizosphere is an active zone of root growth and death, characterized by intense microbial and fungal activity. Decomposition in the rhizosphere is more rapid than in the bulk soil because plant roots use carbon-rich exudates to supplement the decomposition process of low carbon quality organic matter. Nutrients immobilized in bacteria biomass are then released to the soil as microbivores (protozoa and nematodes) that feed on the bacteria.

Decomposition in Aquatic Environments 21.8

Decomposition in aquatic ecosystems varies as a function of water depth and flow rate. In flowing waters (streams and rivers), various specialized detritivores are involved in the breakdown of plant litter imported from adjacent terrestrial ecosystems. In open-water environments, dead organisms and other organic matter, called *particulate organic matter* (POM), drift downward to the bottom. On its way, POM is constantly being ingested, digested, and mineralized until much of the organic matter is in the form of humic compounds by the time it reaches the bottom sediments. Bacteria decompose organic matter on the bottom sediments, using aerobic or anaerobic respiration depending on the supply of oxygen. Dissolved organic matter (DOM) in the water column also provides a source of carbon for decomposers.

Rate of Nutrient Cycling 21.9

The rate at which nutrients cycle through the ecosystem is directly related to the rates of primary productivity (nutrient uptake) and decomposition (nutrient release). Environmental factors that influence these two processes affect the rate at which nutrients cycle through the ecosystem.

Comparison of Terrestrial and Aquatic Ecosystems 21.10

There is typically a vertical separation between the zones of primary production and decomposition. In terrestrial and shallow-water ecosystems, plants function to bridge this gap. In open-water ecosystems, there is a physical separation between these zones that limits nutrient availability in the surface waters. The thermocline functions to limit the movement of nutrients from the bottom (benthic) zone (cold) to the surface (warm) waters. During the winter season, the thermocline breaks down, allowing for a mixing of the water column and the movement of nutrients into the surface waters. This seasonality of the thermocline and mixing of the water column controls seasonal patterns of productivity in these ecosystems.

Stream Ecosystems 21.11

The continuous, directional movement of water affects nutrient cycling in stream ecosystems. Because nutrients are continuously being transported downstream, a spiral rather than a cycle better represents the cycling of nutrients. One cycle in the spiral is the uptake of an atom of nutrient, its passage through the food chain, and its return to water, where it is available for reuse. The cycle length involved is related to the flow rate of the stream and to the physical and biological mechanisms available for nutrient retention.

Coastal Ecosystems 21.12

Water from most streams and rivers eventually drains into the oceans, giving rise to estuary and salt-marsh ecosystems along the coastal environment. As in streams and rivers, the directional flow of water functions to transport both organic matter and nutrients into and out of the ecosystem. The rise and fall of water depth with the tidal cycle serves to flush out salts and other toxins from the marshes and brings in nutrients from the coastal waters. The combined effect of the inward (toward the coast) movement of saltwater and the outward flow of freshwater develops a countercurrent that carries both living and dead particles and the nutrients they contain back toward the coastline. This mechanism conserves nutrients within the estuary and salt-marsh ecosystems.

Vertical Transport in Oceans 21.13

The global pattern of surface currents brings about the transport of deep, nutrient-rich waters to the surface in the coastal regions. As surface currents move waters away from the western coastal margins, deep water moves to the surface, carrying nutrients with it. A similar pattern of upwelling occurs in the equatorial regions of the oceans, where surface currents move to the north and south.

Agriculture Ecological Issues & Applications

In the practice of agriculture and harvesting of crops, organic matter does not return to the soil surface to undergo the process of decomposition and mineralization. The result is a decline in soil fertility and future plant productivity. In swidden agriculture, farmland is abandoned for a period of time allowing the reestablishment of soil nutrients during secondary succession. All other forms of agriculture require the use of fertilizers, either organic or inorganic. Organic fertilizers must be decomposed by soil microbes to release their nutrients in a plant-available form. Inorganic fertilizers provide nutrients in a form that is immediately available for uptake by plants.

STUDY QUESTIONS

1. Define decomposition. What are the major microbial decomposers of plant and animal material?
2. (a) How does the type of carbon compounds present in dead organic matter influence its quality as an energy source for decomposers? (b) How does lignin concentration influence the decomposition of plant litter?
3. How does the concentration of lignin-based compounds in the residual (remaining) organic matter change during the decomposition of plant litter?
4. (a) Contrast the processes of mineralization and immobilization. (b) What does an increase in the nitrogen content of decomposing plant tissues imply about the relative rates of mineralization and immobilization?
5. How does the initial ratio of carbon to nitrogen (C/N) of plant litter influence the relative rates of nitrogen mineralization and immobilization during the initial stages of decomposition?
6. What is the difference in the ratio of carbon to nitrogen (C/N) of plant litter and the tissues of microbial decomposers, and how does the difference influence the nutrient dynamics during decomposition?
7. Tree species that inhabit the forests of the far north (boreal forests) are characterized by low concentrations of nitrogen in their leaves. How might this factor influence nitrogen cycling in these forests?
8. Contrast nutrient cycling in terrestrial and open-water aquatic ecosystems. What is the outstanding difference?
9. How do changes in the thermocline influence seasonal patterns of net primary productivity in lake ecosystems?
10. Go back to the global map of primary productivity of marine ecosystems in Figure 20.10. Identify areas of high productivity in the equatorial region that might be related to the process of upwelling.
11. How does the continuous, directional flow of water influence the cycling of nutrients in stream ecosystems?
12. What mechanism functions to conserve nutrients in estuary ecosystems?

FURTHER READINGS

Classic Studies

Bormann, F. H., and G. E. Likens. 1967. "Nutrient cycling." *Science* 155:424–429.
Early study of nutrient cycling in forested watersheds.

Likens, G. E., and F. H. Bormann. 1995. *Biogeochemistry of a forest ecosystem.* 2nd ed. New York: Springer-Verlag.
This now-classic text provides a 32-year record of the structure, dynamics, and nutrient cycling of the Hubbard Brook ecosystem.

Newbold, J. D., J. W. Elwood, R. V. O'Neill, and A. L. Sheldon. 1983. "Phosphorus dynamics in a woodland stream ecosystem: A study of nutrient spiraling." *Ecology* 64:1249–1265.
This early study helped change ecologists' view of nutrient cycling in flowing-water ecosystems.

Olson, J. S., 1963. "Energy storage and the balance of producers and decomposers in ecological systems." *Ecology* 44: 322–331.
This classic article presents the first quantitative model of decomposition in natural ecosystems.

Swift, M. J., O. W. Heal, and J. M. Anderson. 1979. *Decomposition in terrestrial ecosystems.* Berkeley: University of California Press.
An older, but extensive reference on decomposition processes in terrestrial ecosystems.

Current Research

Aber, J. D., and J. M. Melillo. 2001. *Terrestrial ecosystems.* 2nd ed. Philadelphia: Saunders College Publishing.
This text provides an excellent introduction to the topic of nutrient cycling in terrestrial ecosystems.

Bertness, M. D. 1992. "The ecology of a New England salt marsh." *American Scientist* 80:260–268.
This article presents a fine overview of nutrient cycling in these important coastal ecosystems.

Hooper, D. U., and P. M. Vitousek. 1998. "Effects of plant composition and diversity on nutrient cycling." *Ecological Monographs* 68:121–149.
Although technical, this article presents a good example of the influences of plant species characteristics on the collective process of internal nutrient cycling within terrestrial ecosystems.

Pastor, J., and S. D. Bridgham. 1999. "Nutrient efficiency along nutrient availability gradients." *Oecologia* 118:50–58.
This article provides an excellent discussion of variation in the processes involved in nutrient cycling along an environmental gradient of nutrient availability.

Wagener, S. M., M. W. Oswood, and J. P. Schimel. 1998. "Rivers and soils: Parallels in carbon and nutrient processing." *BioScience* 48:104–108.
An excellent discussion of the similarities and differences in the processes of decomposition and nutrient cycling in terrestrial and stream ecosystems.

CHAPTER

22 Biogeochemical Cycles

Oxygen bubbles on the surface of an aquatic plant produced as a by-product of the process of photosynthesis. Modern atmospheric concentrations of oxygen are a result of the accumulation of oxygen produced in the process of photosynthesis.

CHAPTER GUIDE

WE HAVE EXAMINED the internal cycling of nutrients within the ecosystem, driven by the processes of net primary productivity and decomposition (Chapter 21). The internal cycling of nutrients within the ecosystem is a story of biological processes. But not every transformation of elements in the ecosystem is biologically mediated. Many chemical reactions take place in abiotic components of the ecosystem: the atmosphere, water, soil, and parent material. The weathering of rocks and minerals releases certain elements into the soil and water, making them available for uptake by plants. The energy from lightning produces small amounts of ammonia (NH_3) from molecular nitrogen and water in the atmosphere, providing an input of nitrogen to aquatic and terrestrial ecosystems. Other processes, such as the sedimentation of calcium carbonate in marine environments, remove elements from the active process of internal cycling (see Section 3.5).

Each element has its own story, but all nutrients flow from the nonliving to the living and back to the nonliving components of the ecosystem in a more or less cyclic path known as the **biogeochemical cycle** (from *bio*, "living"; *geo* for the rocks and soil; and *chemical* for the processes involved). We will expand our view of nutrient flow through the ecosystem from our previous discussion of internal cycling, which was dominated by the biological processes of uptake and decomposition, to include a wider array of both biotic and abiotic processes. We will also examine in detail some of the major biogeochemical cycles.

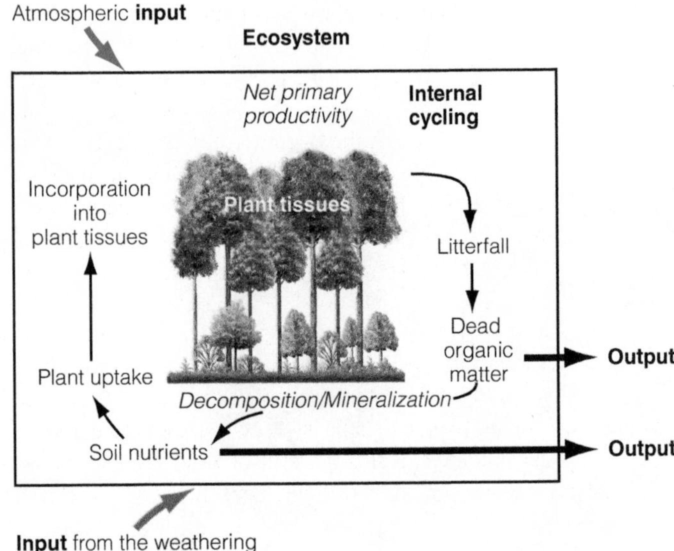

Figure 22.1 A generalized representation of the biogeochemical cycle of an ecosystem. The three common components—inputs, internal cycling, and outputs—are shown in bold. Key processes involved in the internal cycling of nutrients within ecosystems—net primary productivity and decomposition—are shown in italics.

22.1 There Are Two Major Types of Biogeochemical Cycles

There are two basic types of biogeochemical cycles: gaseous and sedimentary. This classification is based on the primary source of nutrient input into the ecosystem. In gaseous cycles, the main pools of nutrients are the atmosphere and the oceans. For this reason, gaseous cycles are distinctly global. The gases most important for life are nitrogen, oxygen, and carbon dioxide. These three gases—in stable quantities of 78 percent, 21 percent, and 0.03 percent, respectively—are the dominant components of Earth's atmosphere.

In sedimentary cycles, the main pool is the soil, rocks, and minerals. The mineral elements required by living organisms come initially from inorganic sources. Available forms occur as salts dissolved in soil water or in lakes, streams, and seas. The mineral cycle varies from one element to another, but essentially it consists of two phases: the rock phase and the salt solution phase. Mineral salts come directly from Earth's crust through weathering (see Section 4.4). The soluble salts then enter the water cycle. With water, the salts move through the soil to streams and lakes and eventually reach the seas, where they remain indefinitely. Other salts return to Earth's crust through sedimentation. They become incorporated into salt beds, silts, and limestone. After weathering, they enter the cycle again.

There are many different kinds of sedimentary cycles. Cycles such as the sulfur cycle are a hybrid of the gaseous and the sedimentary cycles because they have major pools in Earth's crust as well as in the atmosphere. Other cycles, such as the phosphorus cycle, have no significant gaseous pool; the element is released from rock and deposited in both the shallow and deep sediments of the sea.

Gaseous and sedimentary cycles involve biological and nonbiological processes. Both cycles are driven by the flow of energy through the ecosystem, and both are tied to the water cycle (see Section 3.1, Figure 3.2). Water is the medium that moves elements and other materials through the ecosystem. Without the cycling of water, biogeochemical cycles would cease.

Although the biogeochemical cycles of the various essential nutrients required by autotrophs and heterotrophs differ in detail, from the perspective of the ecosystem all biogeochemical cycles have a common structure. They share three basic components: inputs, internal cycling, and outputs (**Figure 22.1**).

22.2 Nutrients Enter the Ecosystem via Inputs

The input of nutrients to the ecosystem depends on the type of biogeochemical cycle. Nutrients with a gaseous cycle, such as carbon and nitrogen, enter the ecosystem via the atmosphere. In contrast, nutrients such as calcium and phosphorus have sedimentary cycles, with inputs dependent on the weathering of rocks and minerals (see Section 4.4). The process of soil formation and the resulting soil characteristics have a major influence on processes involved in nutrient release and retention (see Chapter 4).

Supplementing nutrients in the soil are nutrients carried by rain, snow, air currents, and animals. Precipitation brings appreciable quantities of nutrients, called **wetfall**. Some of these nutrients, such as tiny dust particles of calcium and sea salt, form the nuclei of raindrops; others wash out of the atmosphere as the rain falls. Some nutrients are brought in by airborne particles and aerosols, collectively called **dryfall**. Between 70 and 90 percent of rainfall striking the forest canopy reaches the forest floor. As it drips through the canopy (throughfall) and runs down the stems (stemflow), rainwater picks up and carries with it nutrients deposited as dust on leaves and stems together with nutrients leached from them. Therefore, rainfall reaching the forest floor is richer in calcium, sodium, potassium, and other nutrients than rain falling in the open at the same time.

The major sources of nutrients for aquatic life are inputs from the surrounding land in the form of drainage water, detritus, and sediment and from the atmosphere in the form of precipitation. Flowing-water aquatic systems (streams and rivers) are highly dependent on a steady input of dead organic matter from the watersheds they flow through (see Chapters 21 and 24).

22.3 Outputs Represent a Loss of Nutrients from the Ecosystem

The export of nutrients from the ecosystem represents a loss that must be offset by inputs if a net decline is not to occur. Export can occur in a variety of ways, depending on the specific biogeochemical cycle. Carbon is exported to the atmosphere in the form of carbon dioxide via the process of respiration by all living organisms (see Section 6.1). Likewise, various microbial and plant processes result in the transformation of nutrients to a gaseous phase that can subsequently be transported from the ecosystem to the atmosphere. Examples of these processes are provided in the following sections, which examine specific biogeochemical cycles.

Transport of nutrients from the ecosystem can also occur in the form of organic matter. Organic matter from a forested watershed can be carried from the ecosystem through surface flow of water in streams and rivers. The input of organic carbon from terrestrial ecosystems constitutes most of the energy input into stream ecosystems (see Chapter 24). Organic matter can also be transferred between ecosystems by herbivores. Moose feeding on aquatic plants can transport and deposit nutrients to adjacent terrestrial ecosystems in the form of feces. Conversely, the hippopotamus (*Hippopotamus amphibius*) feeds at night on herbaceous vegetation near the body of water where it lives. Large quantities of nutrients are then transported in the form of feces and other wastes to the water.

Although the transport of organic matter can be a significant source of nutrient loss from an ecosystem, organic matter plays a key role in recycling nutrients because it prevents rapid losses from the system. Large quantities of nutrients are bound tightly in organic matter structure; they are not readily available until released by activities of decomposers.

Some nutrients are leached from the soil and carried out of the ecosystem by underground water flow to streams. These losses may be balanced by inputs to the ecosystem, such as the weathering of rocks and minerals (see Chapter 4).

Considerable quantities of nutrients are withdrawn permanently from ecosystems by harvesting, especially in farming and logging, because biomass is directly removed from the ecosystem. In such ecosystems, these losses must be replaced by applying fertilizer; otherwise, the ecosystem becomes impoverished (see Chapter 21, *Ecological Issues & Applications*). In addition to the nutrients lost directly through biomass removal, both farming and logging can result in the leaching of nutrients from the ecosystem by altering processes involved in internal cycling, such as reduction in the uptake of soil nutrients following the removal of plants.

Depending on its intensity, fire kills vegetation and converts varying proportions of the biomass and soil organic matter to ash (see Chapter 19 for a discussion of fire as disturbance). Besides the loss of nutrients through volatilization and airborne particles, the addition of ash changes the soil's chemical and biological properties. Many nutrients become readily available, and phosphorus in ash is subject to rapid mineralization, which is a process known as *pyromineralization*. If not taken up by vegetation during recovery, nutrients may be lost from the ecosystem through leaching and erosion. Stream-water runoff is often greatest after fire because of reduced water demand for transpiration. High nutrient availability in the soil, coupled with high runoff, can lead to large nutrient losses from the ecosystem.

22.4 Biogeochemical Cycles Can Be Viewed from a Global Perspective

The cycling of nutrients and energy occurs within all ecosystems, and it is most often studied as a local process, that is, the internal cycling of nutrients within the ecosystem and the identification of exchanges both to (inputs) and from (outputs) the ecosystem. Through these processes of exchange, the biogeochemical cycles of differing ecosystems are linked.

Often, the output from one ecosystem represents an input to another, as in the case of exporting nutrients from terrestrial to aquatic ecosystems. The processes of exchanging nutrients among ecosystems require viewing the biogeochemical cycles from a much broader spatial framework than that of a single ecosystem. This is particularly true of nutrients that go through a gaseous cycle. Because the main pools of these nutrients are the atmosphere and the ocean, they have distinctly global circulation patterns. In the following sections, we explore the cycling of carbon, nitrogen, phosphorus, sulfur, and oxygen and examine the specific processes involved in their movement through the ecosystem. We will then expand our model of biogeochemical cycling to provide a framework for understanding the global cycling of these elements, which are crucial to life.

22.5 The Carbon Cycle Is Closely Tied to Energy Flow

Carbon, a basic constituent of all organic compounds, is involved in the fixation of energy by photosynthesis (see Chapter 6). Carbon is so closely tied to energy flow that the two are inseparable. In fact, we often express ecosystem productivity in terms of grams (g) of carbon fixed per square meter per year (see Section 20.2).

The source of all carbon, both in living organisms and fossil deposits, is carbon dioxide in the atmosphere and the waters of Earth. Photosynthesis draws carbon dioxide from the air and water into the living component of the ecosystem. Just as energy flows through the grazing food chain, carbon passes to herbivores and then to carnivores. Primary producers and consumers release carbon back to the atmosphere in the form of carbon dioxide by respiration. The carbon in plant and animal tissues eventually goes to the reservoir of dead organic matter. Decomposers release it to the atmosphere through respiration.

Figure 22.2 shows the cycling of carbon through a terrestrial ecosystem. The difference between the rate of carbon uptake by plants in photosynthesis and release by respiration is the net primary productivity (in units of carbon). The difference between the rate of carbon uptake in photosynthesis and the rate of carbon loss as a result of autotrophic and heterotrophic respiration is the **net ecosystem productivity**.

Several processes, particularly the rates of primary productivity and decomposition, determine the rate at which carbon cycles through the ecosystem. Both processes are influenced strongly by environmental conditions such as temperature and precipitation (see Section 20.3). In warm, wet ecosystems such as a tropical rain forest, production and decomposition rates are high, and carbon cycles through the ecosystem quickly. In cool, dry ecosystems, the process is slower. In ecosystems in which temperatures are very low, decomposition is slow, and dead organic matter accumulates (see Chapter 23). In swamps and marshes, where dead material falls into the water, organic

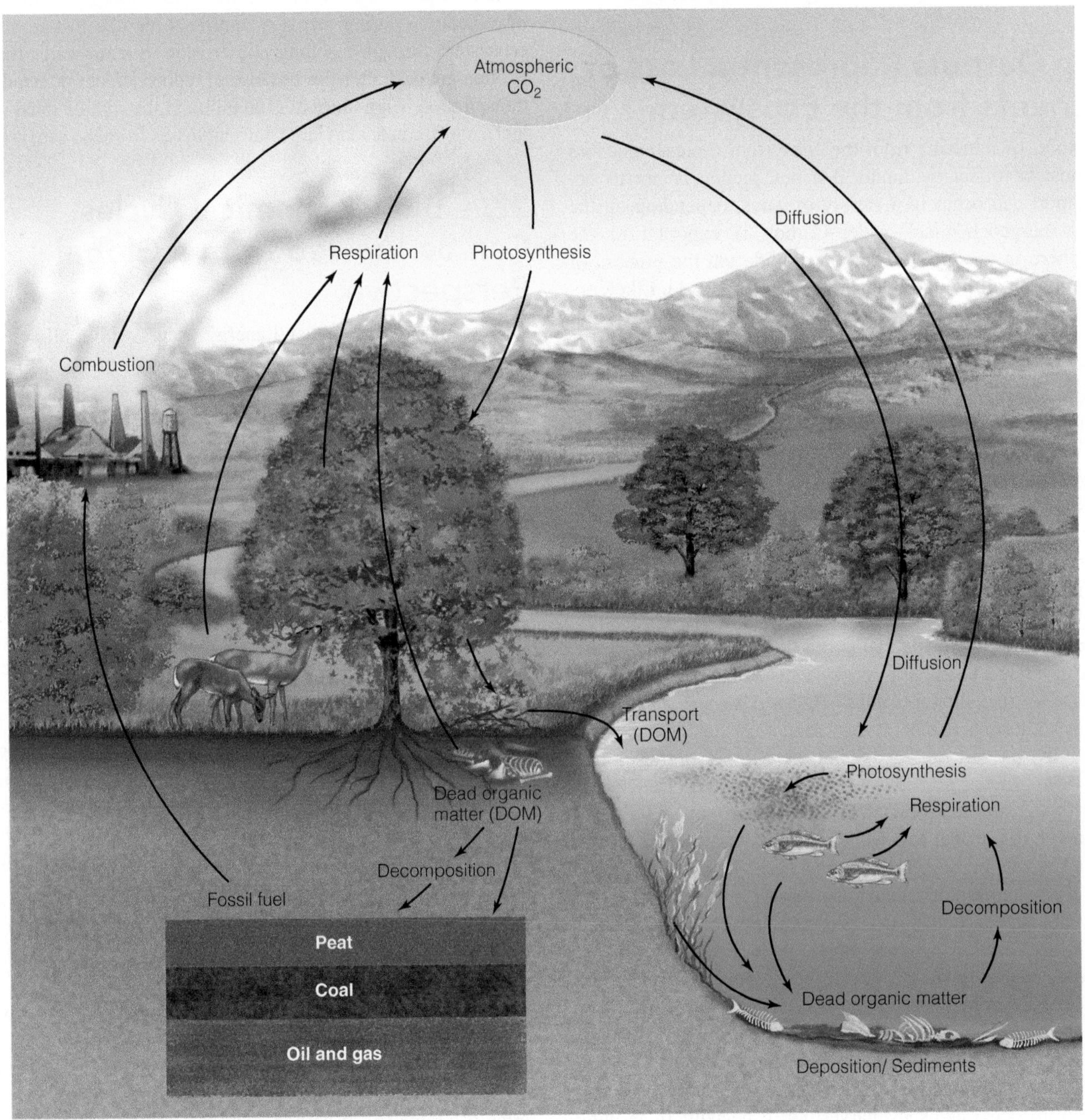

Figure 22.2 The carbon cycle as it occurs in both terrestrial and aquatic ecosystems.

material does not completely decompose. When stored as raw humus or peat, carbon circulates slowly (see Chapter 25). Over geologic time, this buildup of partially decomposed organic matter in swamps and marshes has formed fossil fuels (oil, coal, and natural gas).

Similar cycling takes place in freshwater and marine environments (see Figure 22.2). Phytoplankton use the carbon dioxide that diffuses into the upper layers of water or is present as carbonates and convert it to plant tissue. The carbon then passes from the primary producers through the aquatic food chain. Carbon dioxide produced through respiration is either reused or reintroduced to the atmosphere by diffusion from the water surface to the surrounding air (see Section 3.7 for a discussion of the carbon dioxide–carbonate system in aquatic ecosystems).

Significant amounts of carbon can be bound as carbonates in the bodies of mollusks and foraminifers and incorporated into their exoskeletons (shells, etc.). Some of these carbonates dissolve back into solution, and some become buried in the bottom mud at varying depths when the organisms die. Because it is isolated from biotic activity, this carbon is removed from cycling. On incorporation into bottom sediments over geologic time, it may appear in coral reefs and limestone rocks.

22.6 Carbon Cycling Varies Daily and Seasonally

If you were to measure the concentration of carbon dioxide in the atmosphere above and within a forest on a summer day, you would discover that it fluctuates throughout the day (**Figure 22.3**). At daylight when photosynthesis begins, plants start to withdraw carbon dioxide from the air, and the concentration declines sharply. By afternoon when the temperature is increasing and relative humidity is decreasing, the rate of photosynthesis declines, and the concentration of carbon dioxide in the air surrounding the canopy increases. By sunset, photosynthesis ceases—carbon dioxide is no longer being withdrawn from the atmosphere—respiration increases, and the atmospheric concentration of carbon dioxide increases sharply. A similar diurnal fluctuation occurs in aquatic ecosystems.

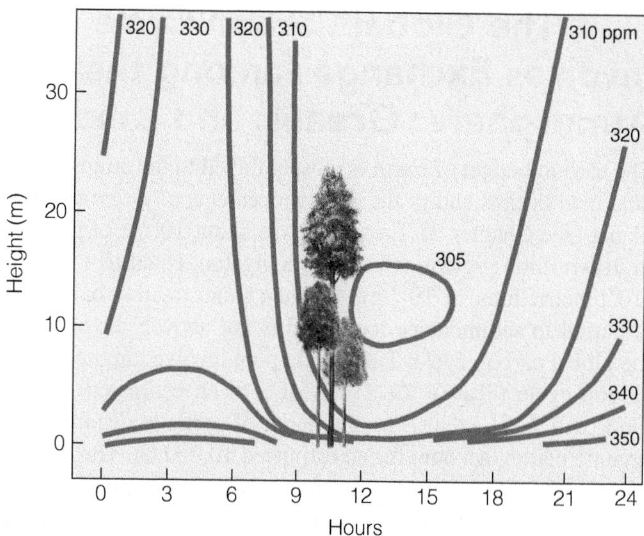

Figure 22.3 Daily flux of carbon dioxide (CO_2) in a forest. Isopleths (lines) define concentration gradients. Note the consistently high level of CO_2 on the forest floor—the site of microbial respiration. Atmospheric CO_2 in the forest is lowest from mid-morning to late afternoon. CO_2 levels are highest at night, when photosynthesis shuts down and respiration pumps CO_2 into the atmosphere.
(Adapted from Baumgartner 1968.)

Likewise, the production and use of carbon dioxide undergoes a seasonal fluctuation relating both to the temperature and the timing of the growing and dormant seasons (**Figure 22.4**). With the onset of the growing season when the landscape is greening, the atmospheric concentration begins to drop as plants withdraw carbon dioxide through photosynthesis. As the growing season reaches its end, photosynthesis declines or ceases, respiration is the dominant process, and atmospheric concentrations of carbon dioxide rise. Although these patterns of seasonal rise and decline occur in both aquatic and terrestrial ecosystems, the fluctuations are much greater in terrestrial environments. As a result, these fluctuations in atmospheric concentrations of carbon dioxide are more pronounced in the Northern Hemisphere with its much larger land area.

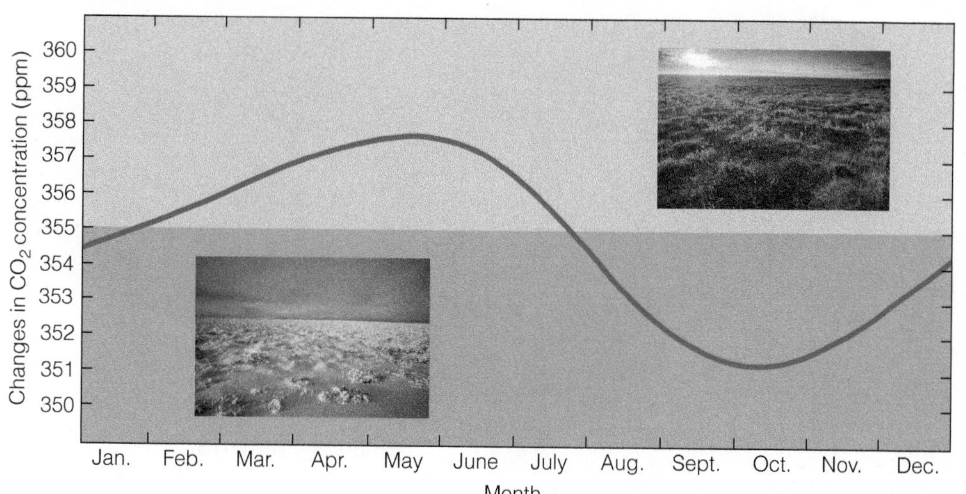

Figure 22.4 Variation in atmospheric concentration of carbon dioxide (CO_2) during a typical year in Barrow, Alaska. Concentrations increase during the winter months, declining with the onset of photosynthesis during the growing season (May–June).
(Adapted from Pearman and Hyson 1981.)

22.7 The Global Carbon Cycle Involves Exchanges among the Atmosphere, Oceans, and Land

The carbon budget of Earth is closely linked to the atmosphere, land, and oceans and to the mass movements of air around the planet (see Chapter 2). Earth contains about 10^{23} g of carbon, or 100 million gigaton (Gt) (Gt is a gigaton, equal to 1 billion [10^9] metric tons, or 10^{15} g). All but a small fraction of carbon is buried in sedimentary rocks and is not actively involved in the global carbon cycle. The carbon pool involved in the global carbon cycle (**Figure 22.5**) amounts to an estimated 55,000 Gt. Fossil fuels, created by the burial of partially decomposed organic matter, account for an estimated 10,000 Gt. The oceans contain the vast majority of the active carbon pool—about 38,000 Gt—mostly as bicarbonate and carbonate ions (see Section 3.7). Dead organic matter in the oceans accounts for 1650 Gt of carbon; living matter, mostly phytoplankton (primary producers), accounts for 3 Gt. The terrestrial biosphere (all terrestrial ecosystems) contains an estimated 1500 Gt of carbon as dead organic matter and 560 Gt as living matter (biomass). The atmosphere holds about 750 Gt of carbon.

In the ocean, the surface water acts as the site of the main exchange of carbon between atmosphere and ocean. The ability of the surface waters to take up carbon dioxide is governed largely by the reaction of carbon dioxide with the carbonate ion to form bicarbonates (see Section 3.7). In the surface water, carbon circulates physically by means of currents and biologically through photosynthesis by phytoplankton and movement through the food chain. The net exchange of carbon dioxide between the oceans and atmosphere as a result of both physical and biological processes results in a net uptake of 1 Gt per year by the oceans, and burial in sediments accounts for a net loss of 0.5 Gt of carbon per year.

Uptake of carbon dioxide from the atmosphere by terrestrial ecosystems is governed by gross production (photosynthesis). Losses are a function of autotrophic and heterotrophic respiration—the latter being dominated by microbial decomposers. Until recently, exchanges of carbon dioxide between the landmass and the atmosphere (uptake in photosynthesis and release by respiration and decomposition) were believed to be nearly in equilibrium (Figure 22.5). However, more recent research suggests that the terrestrial surface is acting as a carbon sink, with a net uptake of carbon dioxide from the atmosphere (see Chapter 27).

Of considerable importance in the terrestrial carbon cycle are the relative proportions of carbon stored in soils and in living vegetation (biomass). Carbon stored in soils includes dead organic matter on the soil surface and in the underlying mineral soil. Estimates place the amount of soil carbon at 1500 Gt compared with 560 Gt in living biomass.

The average amount of carbon per unit volume of soil increases from the tropical regions poleward to the boreal forest and tundra (see Chapter 23). Low values for the tropical forest reflect high rates of decomposition, which compensate for high productivity and litterfall. Frozen tundra soil and waterlogged soils of swamps and marshes have the greatest accumulation of dead organic matter because factors such as low temperature, saturated soils, and anaerobic conditions function to inhibit decay.

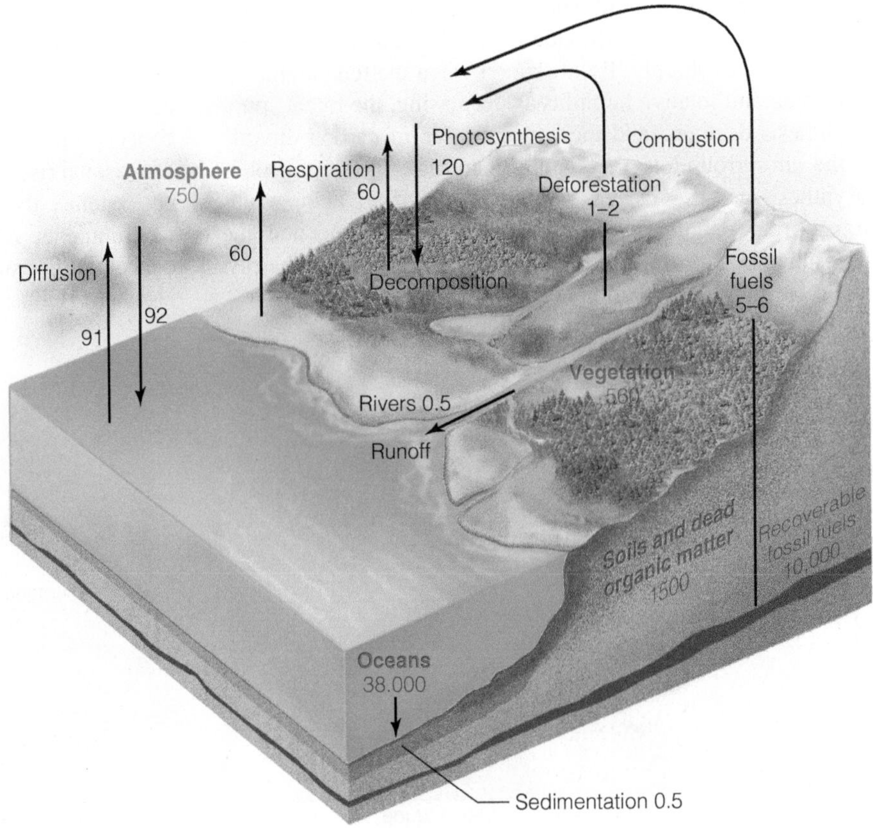

Figure 22.5 The global carbon cycle. The sizes of the major pools of carbon are labelled in red, and arrows indicate the major exchanges (fluxes) among them. All values are in gigatons (Gt) of carbon, and exchanges are on an annual timescale. The largest pool of carbon—geologic—is not included because of the slow rates (geologic timescale) of transfer with other active pools. (Adapted from Edmonds 1992.)

Interpreting Ecological Data

Q1. *Which two fluxes in Figure 22.5 provide an estimate of global net primary productivity (NPP)? What is the estimate of global NPP in the figure?*

Q2. *Which of the major pools identified in the figure represent an estimate of global standing crop biomass?*

Q3. *Which fluxes can be used to provide an estimate of global net ecosystem productivity? What is the estimate of global net ecosystem productivity in the figure?*

22.8 The Nitrogen Cycle Begins with Fixing Atmospheric Nitrogen

Nitrogen is an essential constituent of protein, which is a building block of all living tissue. Nitrogen is generally available to plants in only two chemical forms: ammonium (NH_4^+) and nitrate (NO_3^-). Thus, although Earth's atmosphere is almost 80 percent nitrogen gas, it is in a form (N_2) that is not available for uptake (assimilation) by plants. Nitrogen enters the ecosystem via two pathways, and the relative importance of each varies greatly among ecosystems (**Figure 22.6**). The first pathway is atmospheric deposition. This can be in wetfall—such as rain, snow, or even cloud and fog droplets—and in dryfall, such as aerosols and particulates. Regardless of the form of atmospheric deposition, nitrogen in this pathway is supplied in a form that is already available for uptake by plants.

The second pathway for nitrogen to enter ecosystems is via **nitrogen fixation**. This fixation comes about in two ways. One is high-energy fixation. Cosmic radiation, meteorite trails, and lightning provide the high energy needed to combine nitrogen with the oxygen and hydrogen of water. The resulting ammonia and nitrates are carried to Earth's surface in rainwater. Estimates suggest that less than 0.4 kg N/ha comes to Earth annually in this manner. About two-thirds of this amount is deposited as ammonia and one-third as nitric acid (HNO_3).

The second method of fixation is biological. This method produces approximately 10 kilograms (kg) N/yr for each hectare (ha) of Earth's land surface, or roughly 90 percent of the fixed nitrogen contributed each year. This fixation is accomplished by symbiotic bacteria living in mutualistic association with plants, by free-living aerobic bacteria, and by cyanobacteria (blue-green algae; see Section 15.11 and Figure 15.11). Fixation splits molecular nitrogen (N_2) into two atoms of free nitrogen. The free nitrogen atoms then combine with hydrogen to form two molecules of ammonia (NH_3). The process of fixation requires considerable energy. To fix 1 g of nitrogen, nitrogen-fixing bacteria associated with the root system of a plant must expend about 10 g of glucose, which is a simple sugar produced by the plant in photosynthesis.

In agricultural ecosystems, *Rhizobium* bacteria associated with approximately 200 species of leguminous plants are the preeminent nitrogen fixers (see Figure 15.11). In nonagricultural systems, some 12,000 species—from cyanobacteria to

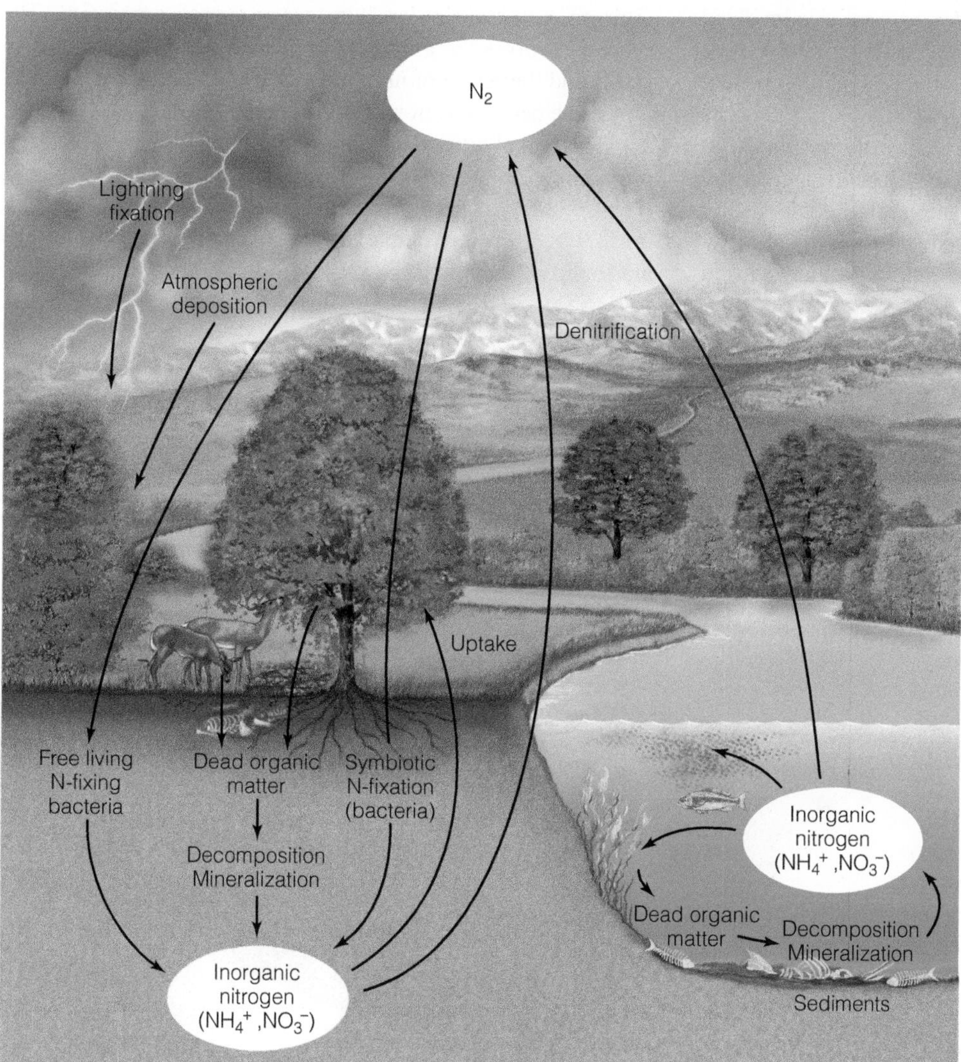

Figure 22.6 The nitrogen cycle in terrestrial and aquatic ecosystems.

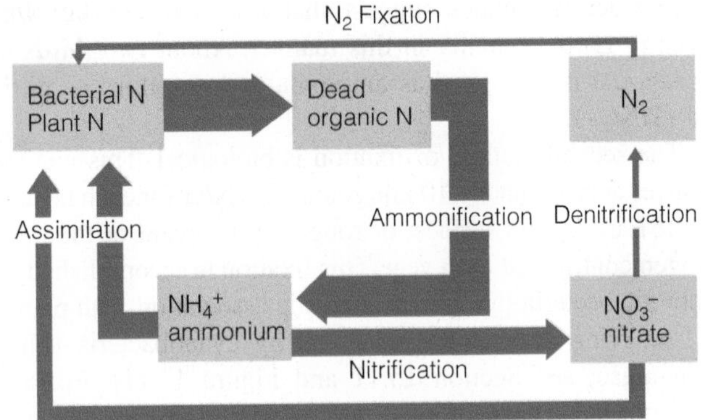

Figure 22.7 Bacterial processes involved in nitrogen cycling.

nodule-bearing plants—are responsible for nitrogen fixation. Also contributing to the fixation of nitrogen are free-living soil bacteria; the most prominent of the 15 known genera are the aerobic *Azotobacter* and the anaerobic *Clostridium*. Cyanobacteria are another important group of largely non-symbiotic nitrogen fixers. Of some 40 known species, the most common are in the genera *Nostoc* and *Calothrix*, which are found in soil as well as in aquatic habitats. Certain lichens (*Collema tunaeforme* and *Peltigera rufescens*) are also involved in nitrogen fixation. Lichens with nitrogen-fixing ability possess nitrogen-fixing cyanobacteria as their algal component (see Section 15.10, Figure 15.10).

Ammonium in the soil can be used directly by plants. In addition to atmospheric deposition, NH_4^+ occurs in the soil as a product of microbial decomposition of organic matter wherein NH_3 is released as a waste product of microbial activity (see Section 21.5). This process is called **ammonification** (**Figure 22.7**). Most soils have an excess of H^+ (slightly acidic) and the NH_3 is rapidly converted to ammonium (NH_4^+) (see Section 3.7). Interestingly, because NH_3 is a gas, the transfer of nitrogen back to the atmosphere (volatilization) can occur in soils with a pH close to 7 (neutral)—a low concentration of H^+ ions to convert ammonia to ammonium. Volatilization can be especially pronounced in agricultural areas where both nitrogen fertilizers and lime (to decrease soil acidity) are used extensively.

In some ecosystems, plant roots must compete for NH_4^+ with two groups of aerobic bacteria, which use it as part of their metabolism (see Figure 22.7). The first group, *Nitrosomonas*, oxidizes NH_4^+ to NO_2^-, and a second group, *Nitrobacter*, oxidizes NO_2^- to NO_3^-. This process is called **nitrification**. Once nitrate is produced, several things may happen to it. First, plant roots may take it up. Second, **denitrification** may occur under anaerobic conditions, when another group of bacteria (*Pseudomonas*) chemically reduces NO_3^- to N_2O and N_2. These gases are then returned to the atmosphere. The anaerobic conditions necessary for denitrification are generally rare in most terrestrial ecosystems (but can occur seasonally). These conditions, however, are common in wetland ecosystems and in the bottom sediments of open-water aquatic ecosystems (see Chapters 24 and 25).

Finally, nitrate is the most common form of nitrogen exported from terrestrial ecosystems in stream water, although in undisturbed ecosystems the amount is usually quite small because of the great demand for nitrogen. Indeed, the amount of nitrogen recycled within the ecosystem is usually much greater than the amount either entering or leaving the ecosystem through inputs and outputs.

Because both nitrogen fixation and nitrification are processes mediated by bacteria, they are influenced by various environmental conditions, such as temperature and moisture. However, one of the more important factors is soil pH. Both processes are usually greatly limited in extremely acidic soils because of the inhibition of bacteria under those conditions.

Inputs of nitrogen can vary, but the internal cycling of nitrogen is fairly similar from ecosystem to ecosystem. The process involves the assimilation of ammonium and nitrate by plants and the return of nitrogen to the soil, sediments, and water via the decomposition of dead organic matter.

The global nitrogen cycle follows the pathway of the local nitrogen cycle presented previously, only on a grander scale (**Figure 22.8**). The atmosphere is the largest pool, containing 3.9×10^{21} g. Comparatively small amounts of nitrogen are found in the biomass (3.9×10^{15} g) and soils (95×10^{15}

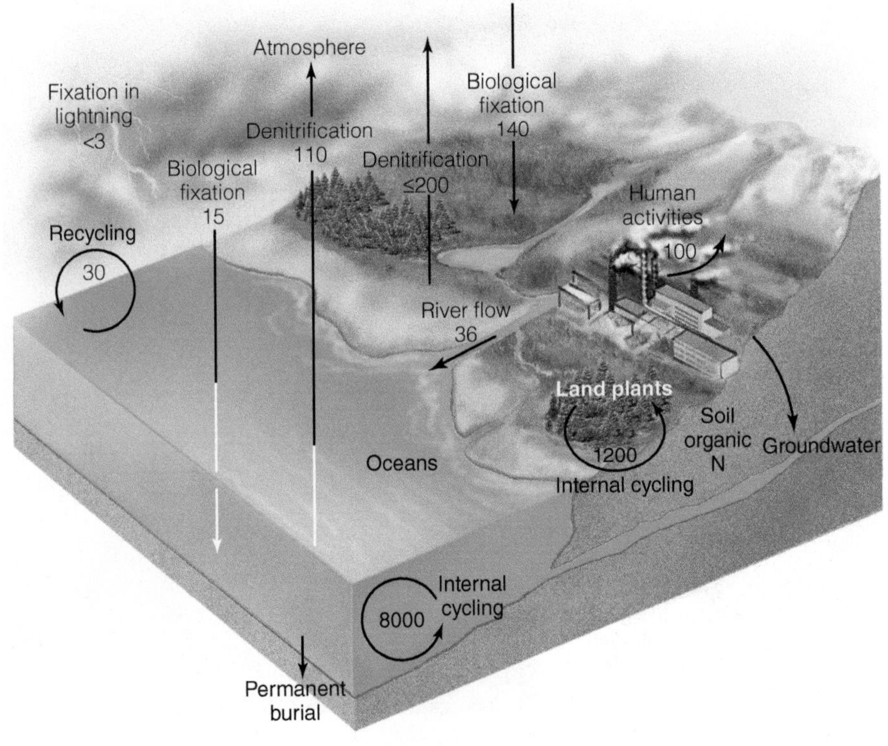

Figure 22.8 The global nitrogen cycle. Each flux is shown in units of 10^{12} g N/yr. (Adapted from Schlesinger 2013.)

to 140 × 10^{15} g) of terrestrial ecosystems. Global estimates of denitrification in terrestrial ecosystems vary widely but are of the order 200 × 10^{12} g/yr, and more than half of that total occurs in wetland ecosystems. Major sources of nitrogen to the world's oceans are dissolved forms in the freshwater drainage from rivers (36 × 10^{12} g/yr) and inputs in precipitation (30 × 10^{12} g/yr). Biological fixation accounts for another 15 × 10^{12} g/yr. Denitrification accounts for an estimated flux of 110 × 10^{21} g N/yr from the world's oceans to the atmosphere.

There also are small but steady losses from the biosphere to the deep sediments of the ocean and to sedimentary rocks. In return, there is a small addition of new nitrogen from the weathering of igneous rocks and from volcanic activity.

Human activity has significantly influenced the global nitrogen cycle. Major human sources of nitrogen input are agriculture, industry, and automobiles; in recent decades, anthropogenic inputs of nitrogen into aquatic and terrestrial ecosystems have been a growing cause of concern (see this chapter, *Ecological Issues & Applications*). The first major intrusion probably came from agriculture when people began burning forests and clearing land for crops and pasture. Heavy application of chemical fertilizers to croplands disturbs the natural balance between nitrogen fixation and denitrification, and a considerable portion of nitrogen fertilizers is lost as nitrates to groundwater and runoff that find their way into aquatic ecosystems (see Chapter 24, *Ecological Issues & Applications*).

Automobile exhaust and industrial high-temperature combustion add nitrous oxide (N_2O), nitric oxide (NO), and nitrogen dioxide (NO_2) to the atmosphere. These oxides can reside in the atmosphere for up to 20 years, drifting slowly up to the stratosphere. There, ultraviolet light reduces nitrous oxide to nitric oxide and atomic oxygen (O). Atomic oxygen reacts with oxygen (O_2) to form ozone (O_3; see Section 22.12).

22.9 The Phosphorus Cycle Has No Atmospheric Pool

Phosphorus occurs in only minute amounts in the atmosphere. Therefore, the phosphorus cycle can follow the water (hydrological) cycle only part of the way—from land to sea (**Figure 22.9**). Because phosphorus lost from the ecosystem in this way is not returned via the biogeochemical cycle, phosphorus is in short supply under undisturbed natural conditions. The natural scarcity of phosphorus in aquatic ecosystems is emphasized by the explosive growth of algae in water receiving heavy discharges of phosphorus-rich wastes.

The main reservoirs of phosphorus are rock and natural phosphate deposits. Phosphorus is released from these rocks and minerals by weathering, leaching, erosion, and mining for use as agricultural fertilizers. Nearly all of the phosphorus in terrestrial ecosystems comes from the weathering of calcium phosphate minerals. In most soils, only a small fraction of the total phosphorus is available to plants. The major process regulating phosphorus availability for net primary production is the internal cycling of phosphorus from organic to inorganic forms (nutrient cycling; see Chapter 21). Some of the available phosphorus in terrestrial ecosystems escapes and is exported to lakes and seas.

In marine and freshwater ecosystems, the phosphorus cycle moves through three states: particulate organic phosphorus, dissolved organic phosphates, and inorganic phosphates. Organic phosphates are taken up quickly by all forms of phytoplankton, which in turn are eaten by zooplankton and detritus-feeding organisms. Zooplankton may excrete as much

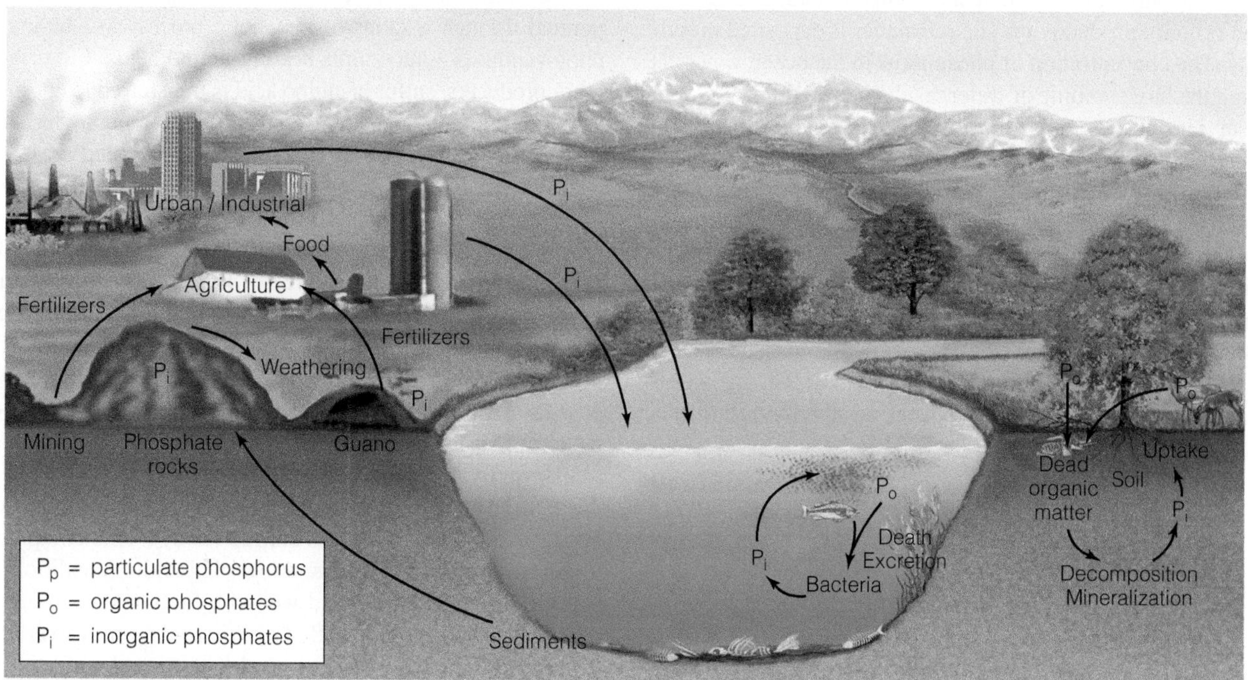

Figure 22.9 The phosphorus cycle in aquatic and terrestrial ecosystems.

phosphorus daily as they store in their biomass, returning it to the cycle. More than half of the phosphorus zooplankton excrete is inorganic phosphate, which is taken up by phytoplankton. The remaining phosphorus in aquatic ecosystems exists in organic compounds that may be used by bacteria, which fail to regenerate much dissolved inorganic phosphate. Bacteria are consumed by the microbial grazers, which then excrete the phosphate they ingest. Part of the phosphate is deposited in shallow sediments and part in deep water. In the process of ocean upwelling, the movement of deep waters to the surface brings some phosphates from the dark depths to shallow waters, where light is available to drive photosynthesis (see Sections 3.8 and 21.13). These phosphates are taken up by phytoplankton. Part of the phosphorus contained in the bodies of plants and animals sinks to the bottom and is deposited in the sediments. As a result, surface waters may become depleted of phosphorus, and the deep waters become saturated. Much of this phosphorus becomes locked up for long periods of time in the hypolimnion and bottom sediments, and some is returned to the surface waters by upwelling.

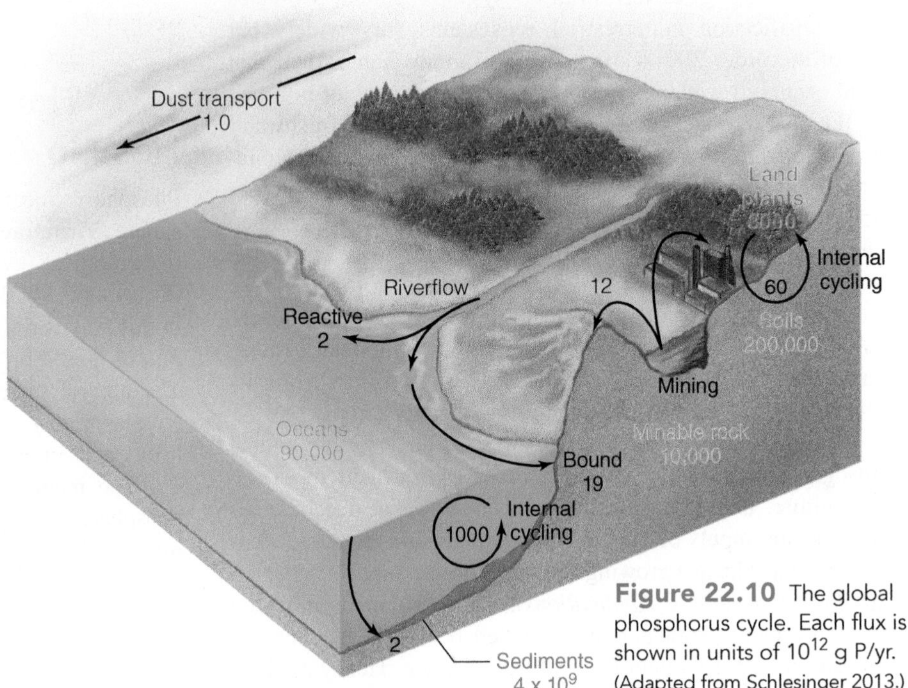

Figure 22.10 The global phosphorus cycle. Each flux is shown in units of 10^{12} g P/yr. (Adapted from Schlesinger 2013.)

The global phosphorus cycle (**Figure 22.10**) is unique among the major biogeochemical cycles in having no significant atmospheric component, although airborne transport of phosphorus in soil dust and sea spray is of the order 1×10^{12} g P/yr.

Rivers transport approximately 21×10^{12} g P/yr to the oceans, but only about 10 percent of this amount is available for net primary productivity. The remainder is deposited in sediments. The concentration of phosphorus in the ocean waters is low, but the large volume of water results in a significant global pool of phosphorus. The turnover of organic phosphorus in the surface waters occurs within days, and the vast majority of phosphorus taken up in primary production is decomposed and mineralized (internally cycled) in the surface waters. However, approximately 2×10^{12} g P/yr is deposited in the ocean sediments or transported to the deep waters. In the deep waters, organic phosphorus converted into inorganic, soluble forms remains unavailable to phytoplankton in the surface waters until transported by upwelling (see Section 3.8). On a geological timescale, uplifting and subsequent weathering return this phosphorus to the active cycle.

22.10 The Sulfur Cycle Is Both Sedimentary and Gaseous

The sulfur cycle has both sedimentary and gaseous phases (**Figure 22.11**). In the long-term sedimentary phase, sulfur is tied up in organic and inorganic deposits, released by weathering and decomposition, and carried to terrestrial ecosystems in salt solution. The gaseous phase of the cycle permits sulfur circulation on a global scale.

Sulfur enters the atmosphere from several sources: the combustion of fossil fuels, volcanic eruptions, exchange at the surface of the oceans, and gases released by decomposition. It enters the atmosphere initially as hydrogen sulfide (H_2S), which quickly interacts with oxygen to form sulfur dioxide (SO_2). Atmospheric sulfur dioxide, which is soluble in water, is carried back to the surface in rainwater as weak sulfuric acid (H_2SO_4). Whatever the source, sulfur in a soluble form is taken up by plants and incorporated through a series of metabolic processes—starting with photosynthesis—into sulfur-bearing amino acids. From the primary producers, sulfur in amino acid is transferred to consumers.

Excretion and death carry sulfur from living material back to the soil and to the bottom of ponds, lakes, and seas, where bacteria release it as hydrogen sulfide or sulfate. One group, the colorless sulfur bacteria, reduces hydrogen sulfide to elemental sulfur and then oxidizes it to sulfuric acid. Green and purple bacteria, in the presence of light, use hydrogen sulfide during photosynthesis. Best known are the purple bacteria found in salt marshes and in the mudflats of estuaries. These organisms can transform hydrogen sulfide into sulfate, which is then recirculated and taken up by producers or used by bacteria that further transform the sulfates. Green bacteria can transform hydrogen sulfide into elemental sulfur.

Sulfur, in the presence of iron and under anaerobic conditions, precipitates as ferrous sulfide (FeS_2). This compound is highly insoluble in neutral and low pH (acidic) conditions, and it is firmly held in mud and wet soil. Sedimentary rocks containing ferrous sulfide—called *pyritic rocks*—may overlie coal deposits. When exposed to air during deep and surface mining for coal, the ferrous sulfide reacts with oxygen. In the presence of water, it produces ferrous sulfate ($FeSO_4$) and sulfuric acid.

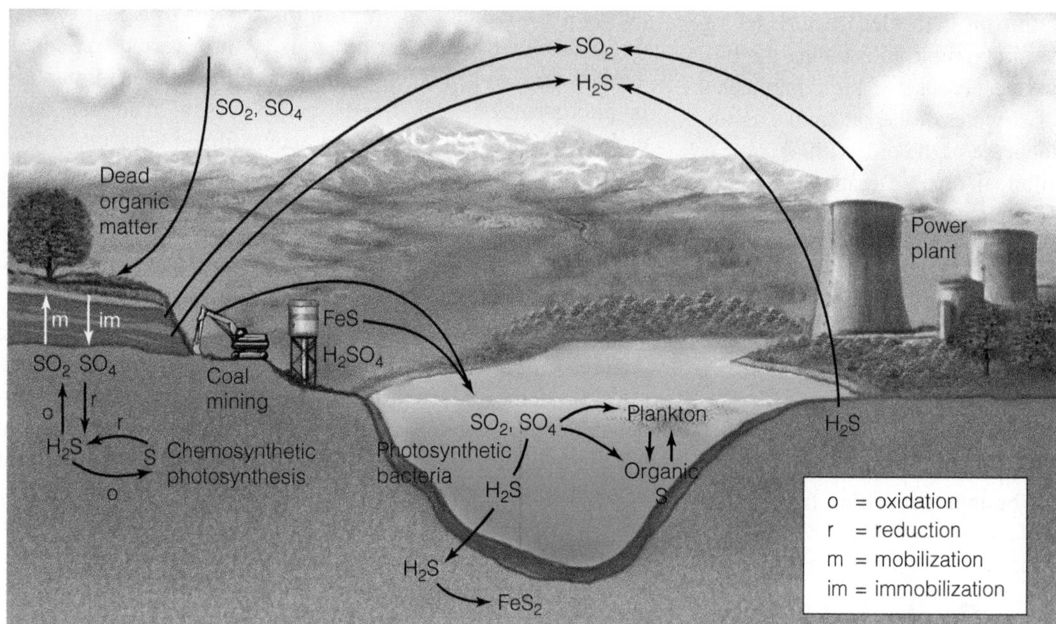

Figure 22.11 The sulfur cycle. Note the two components: sedimentary and gaseous. Major sources from human activity are the burning of fossil fuels and acidic drainage from coal mines.

In this manner, sulfur in pyritic rocks, suddenly exposed to weathering by human activities, discharges sulfuric acid, ferrous sulfate, and other sulfur compounds into aquatic ecosystems. These compounds destroy aquatic life. They have converted hundreds of kilometers of streams in the eastern United States to lifeless, highly acidic water.

22.11 The Global Sulfur Cycle Is Poorly Understood

The global sulfur cycle is presented in **Figure 22.12**. Although a great deal of research now focuses on the sulfur cycle—particularly the role of human inputs—our understanding of the global sulfur cycle is primitive.

The gaseous phase of the sulfur cycle permits circulation on a global scale. The annual flux of sulfur compounds through the atmosphere is of the order 30×10^{12} g. The atmosphere contains not only sulfur dioxide and hydrogen sulfide but also sulfate particles. The sulfate particles become part of dry deposition (dryfall); the gaseous forms combine with moisture and are transported in precipitation (wetfall).

The oceans are a large source of aerosols that contain sulfate (SO_4); however, most are redeposited in the oceans as precipitation and dryfall (see Figure 22.12). Dimethylsulfide ($[CH_3]_2S$) is the major gas emitted from the oceans that is generated by biological processes. Estimates of 16×10^{12} g S/yr make it the largest natural source of sulfur gases released to the atmosphere.

Various biological sources of sulfur emissions from terrestrial ecosystems exist, but collectively they cause only a minor flux to the atmosphere. The dominant sulfur gas emitted from freshwater wetlands and anoxic (oxygen-depleted) soils is hydrogen sulfide (H_2S). Emissions from plants are poorly

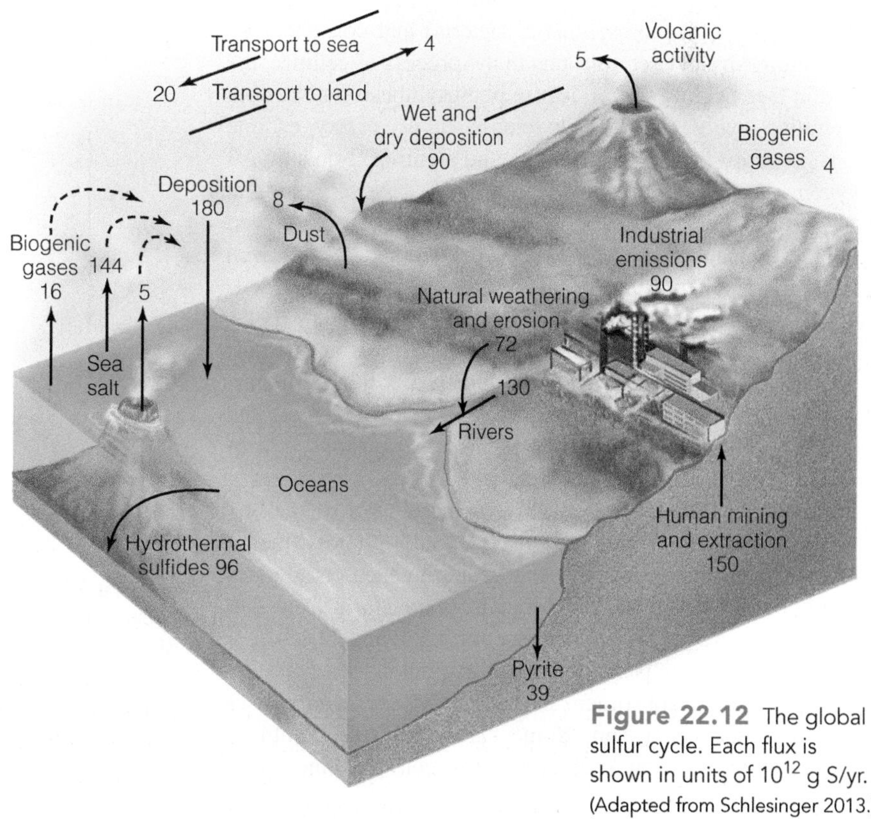

Figure 22.12 The global sulfur cycle. Each flux is shown in units of 10^{12} g S/yr. (Adapted from Schlesinger 2013.)

understood, but forest fires emit on the order of 3×10^{12} g of sulfur annually. It is almost impossible to estimate the biological turnover of sulfur dioxide because of the complicated cycling within the biosphere. Estimates of the net annual assimilation of sulfur by marine plants are of the order 130×10^{12} g. Adding the anaerobic oxidation of organic matter brings the total to an estimated 200×10^{12} g (see Section 21.4).

Volcanic activity also contributes to the global biogeochemical cycle of sulfur. Major events, such as the eruption of Mt. Pinatubo in 1991, released on the order of 5×10^{12} to 10×10^{12} g of sulfur. When volcanic activity is averaged over a long period, the annual global flux is of the order 10×10^{12} g of sulfur.

Human activity plays a dominant role in the biogeochemical cycle of sulfur. Thus, to complete the picture of the global sulfur cycle, we must include the inputs resulting from industrial activity.

22.12 The Oxygen Cycle Is Largely under Biological Control

The major source of free oxygen that supports life is the atmosphere. There are two significant sources of atmospheric oxygen. One is the breakup of water vapor through a process driven by sunlight. In this reaction, the water molecules (H_2O) are disassociated to produce hydrogen and oxygen. Most of the hydrogen escapes into space. If the hydrogen did not escape, it would recombine with the oxygen to form water vapor again.

The second source of oxygen is photosynthesis, which has been active only since life began on Earth (**Figure 22.13**). Oxygen is produced by photosynthetic autotrophs (green plants, algae, and photosynthetic bacteria) and consumed by both autotrophs and heterotrophs in the process of cellular respiration (see Section 6.1). Because photosynthesis and aerobic respiration involve the alternate release and use of oxygen, one would seem to balance the other, so no significant quantity of oxygen would accumulate in the atmosphere. Nevertheless, at some time in Earth's history, the amount of oxygen introduced into the atmosphere had to exceed the amount taken up in respiration (including the decay of organic matter) and geological processes, such as the oxidation of sedimentary rocks. Part of the oxygen present in the atmosphere is from the past imbalance between photosynthesis and respiration in plants. Undecomposed organic matter in the form of fossil fuels and carbon in sedimentary rocks represents a net positive flux of oxygen to the atmosphere (see Section 22.7).

The other main reservoirs of oxygen are water and carbon dioxide. All the reservoirs are linked through photosynthesis. Oxygen is also biologically exchangeable in such compounds as nitrates and sulfates, which organisms transform from ammonia and hydrogen sulfide (see Sections 22.8 and 22.10).

Because oxygen is so reactive, its cycling in the ecosystem is complex. As a constituent of carbon dioxide, it circulates throughout the ecosystem. Some carbon dioxide combines with calcium to form carbonates. Oxygen combines with nitrogen compounds to form nitrates, with iron to form ferric oxides, and with other minerals to form various oxides. In these states, oxygen is temporarily withdrawn from circulation. In photosynthesis, the freed oxygen is split from the water molecule. The oxygen is then reconstituted into water during cellular respiration in both plants and animals. Part of the atmospheric oxygen is reduced to ozone by high-energy ultraviolet radiation.

Ozone is an ambivalent atmospheric gas. In the stratosphere, 10 to 40 km above Earth, it shields the planet from biologically harmful ultraviolet radiation. Close to the ground, ozone is a damaging pollutant, cutting visibility, irritating eyes and respiratory systems, and injuring or killing plant life. In the stratosphere, ozone is diminished by its reaction with human-caused pollutants. In the troposphere, ozone is born from the union of nitrogen oxides with oxygen in the presence of sunlight.

A cycling reaction requiring sunlight maintains ozone in the stratosphere. Solar radiation breaks the O–O bond in O_2. Freed oxygen atoms rapidly combine with O_2 to form O_3. At the same time, a reverse reaction consumes ozone to form O and O_2. Under natural conditions in the stratosphere, a balance exists between the rates of ozone formation and destruction. In recent times, however, catalysts—some human caused and some biologically derived—injected into the stratosphere have been reactive enough to reduce stratospheric ozone. Among them are chlorofluorocarbons (CFCs), methane (CH_4), both natural and human caused, and nitrous oxide from denitrification and synthetic nitrogen fertilizer. Of particular concern is chlorine

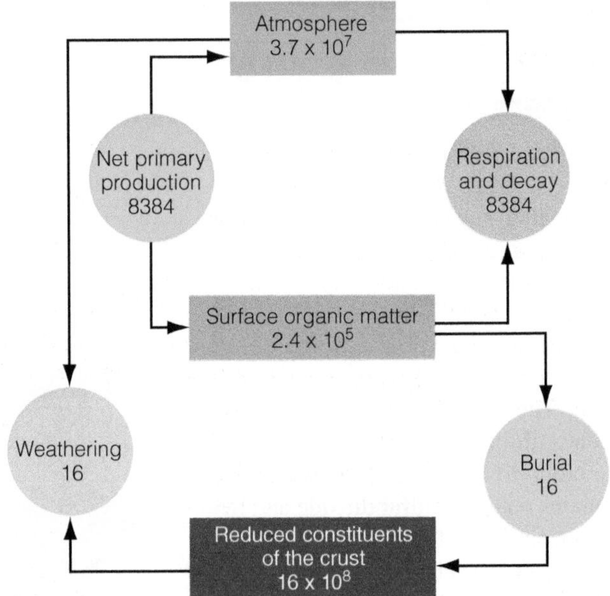

Figure 22.13 A simple model for the global biogeochemical cycle of free oxygen (O_2). Data are expressed in units of 10^{12} moles of O_2 per year or the equivalent amount of reduced compounds. Note that a small imbalance in the ratio of photosynthesis to respiration can result in a net storage of reduced organic materials in the crust and an accumulation of O_2 in the atmosphere. (Adapted from Schlesinger 2013.)

monoxide (ClO) derived from CFCs and used in aerosol spray propellants (banned in the United States), refrigerants, solvents, and other sources. This form of chlorine can break down ozone.

22.13 The Various Biogeochemical Cycles Are Linked

Although we have introduced each of the major biogeochemical cycles independently, they are all linked in various ways. In specific cases, they are linked through their common membership in compounds that form an important component of their cycles. Examples are the links between calcium and phosphorus in the mineral apatite—a phosphate of calcium—and the link between nitrogen and oxygen in nitrate. In general, cycled nutrients are all components of living organisms and constituents of organic matter. As a result, they travel together in their odyssey through internal cycling.

Autotrophs and heterotrophs require nutrients in different proportions for different processes. For example, photosynthesis uses six moles of water (H_2O) and produces six moles of oxygen (O_2) for every six moles of carbon dioxide (CO_2) that is transformed into one mole of sugar ($C_6H_{12}O_6$). (See equation for photosynthesis, Chapter 6.) The proportions of hydrogen, oxygen, and carbon involved in photosynthesis are fixed. Likewise, a fixed quantity of nitrogen is required to produce a mole of rubisco, the enzyme that catalyzes the fixation of carbon dioxide in photosynthesis (see Section 6.1). Therefore, the nitrogen content of a rubisco molecule is the same in every plant, independent of species or environment. The same is true for the variety of amino acids, proteins, and other nitrogenous compounds that are essential for the synthesis of plant cells and tissues. The branch of chemistry dealing with the quantitative relationships of elements in combination is called **stoichiometry**. The stoichiometric relationships among various elements involved in processes related to carbon uptake and plant growth have an important influence on the cycling of nutrients in ecosystems.

Because of similar relationships among the variety of macronutrients and micronutrients required by plants, the limitation of one nutrient can affect the cycling of all the others. As an example, consider the link between carbon and nitrogen (Chapter 21). Although the nitrogen content of a rubisco molecule is the same in every plant independent of species or environment, plants can differ in the concentration of rubisco found in their leaves and, therefore, in their concentration of nitrogen (grams N per gram dry weight). Plants growing under low nitrogen availability will have a lower rate of nitrogen uptake and less nitrogen for the production of rubisco and other essential nitrogen-based compounds. In turn, the lower concentrations of rubisco result in lower rates of photosynthesis and carbon gain (see Section 6.11). In turn, the lower concentrations of nitrogen in the leaf litter influence the relative rates of immobilization and mineralization and subsequent nitrogen release to the soil in decomposition (see Section 21.5). In this way, nitrogen availability and uptake by plants influence the rate at which carbon and other essential plant nutrients cycle through the ecosystem.

Conversely, the variety of other essential nutrients and environmental factors that directly influence primary productivity, and thus the demand for nitrogen, influence the rate of nitrogen cycling through the ecosystem. In fact, the cycles of all essential nutrients for plant and animal growth are linked because of the stoichiometric relationships defining the mixture of chemicals that make up all living matter.

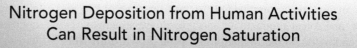

ECOLOGICAL Issues & Applications

Nitrogen Deposition from Human Activities Can Result in Nitrogen Saturation

Net primary productivity in most terrestrial forest ecosystems is limited by the availability of soil nitrogen, yet in recent decades, human activities have caused a dramatic increase in the rates of nitrogen deposition. In North America, anthropogenic activities such as fossil fuel combustion and high-intensity agriculture have increased the inputs of nitrogen oxides in the atmosphere far above natural inputs. Nitrogen oxides quickly undergo a variety of chemical reactions in the atmosphere and therefore do not reside in the atmosphere for long periods but rather are deposited in the region where the emissions originated. The result is that deposition rates vary widely among geographic regions (**Figure 22.14**).

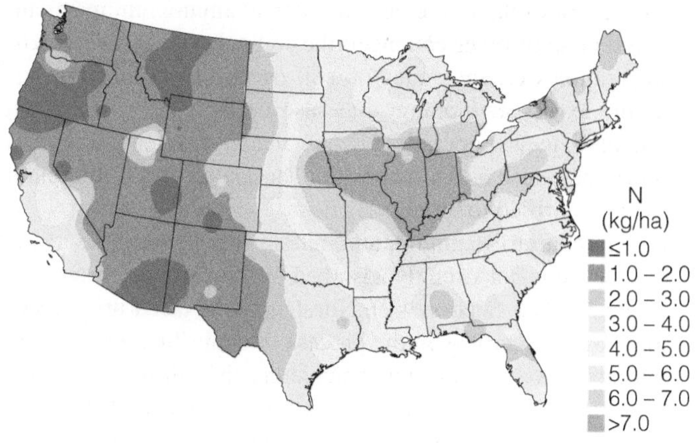

Figure 22.14 Estimated inorganic nitrogen deposition from nitrate and ammonium in 1998.
(National Atmospheric Deposition Program.)

The concentration of nitrogen in the soil solution influences the rate of plant uptake and plant tissue concentration (see Chapter 6, Section 6.11). In turn, there is a strong relationship between photosynthetic capacity and leaf nitrogen because of a greater concentration of enzymes and pigments used in

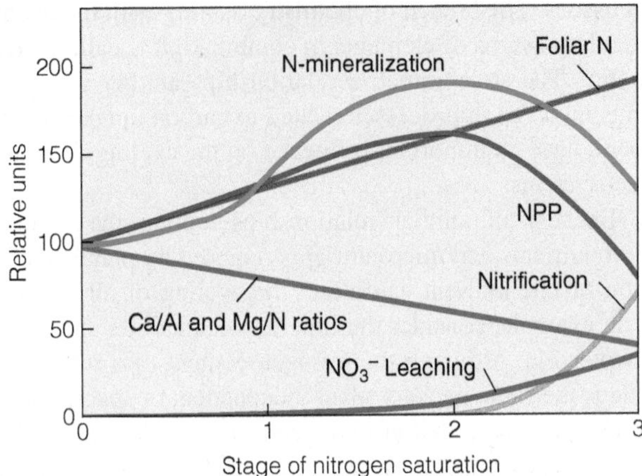

Figure 22.15 Hypothesized response of temperate forest ecosystems to long-term nitrogen (N) additions. In stage 1, N-mineralization increases, which results in increased net primary productivity (NPP). In stage 2, NPP and N-mineralization decline because of decreasing calcium-to-aluminum (Ca/Al) and magnesium-to-nitrogen (Mg/N) ratios and to soil acidification. Nitrification also increases as excess ammonium is available. Finally, in stage 3, nitrate leaching increases dramatically. (Adapted from Aber et al. 1998.)

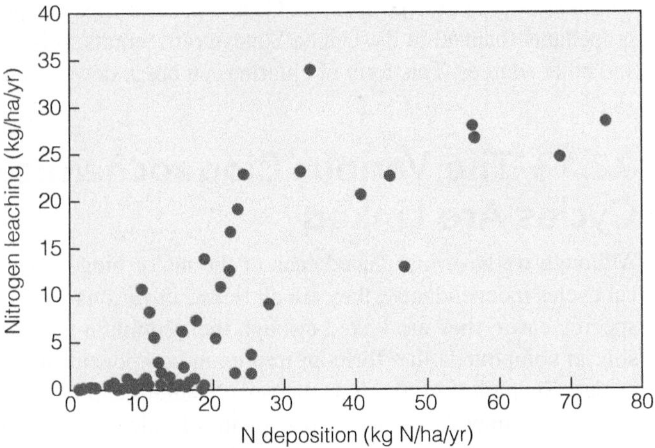

Figure 22.16 Relationship between atmospheric nitrogen deposition and leaching of nitrogen in European forests. (Adapted from Dise and Knight 1995.)

photosynthesis (see Figure 6.23). The net result is that initially nitrogen deposition acts as a fertilizer, increasing rates of net primary productivity. However, as water and other nutrients become more limiting relative to nitrogen, these ecosystems may approach "nitrogen saturation." If the nitrogen supply continues to increase, a complex series of changes to soil and plant processes may ultimately lead to soil acidification and forest decline (**Figure 22.15**).

Most nitrogen deposited is in the form of either nitrate or ammonium, though ammonium is dominant. In the early stages of nitrogen saturation, plants take up most of the nitrogen, and forest productivity and growth are stimulated by nitrogen inputs. As the limitations on plant growth shift from nitrogen to other resources, more ammonium becomes available in the soil. High concentrations of ammonium may cause the release of other cations to the soil solution by overwhelming cation exchange sites on soil particles (see Section 4.9). Some excess ammonium may also be used by microbes and increase their populations, thus increasing other microbial processes, such as decomposition, but much of the ammonium will be nitrified to nitrate.

In contrast to ammonium, nitrate is highly mobile in soils because it is not strongly adsorbed to soil particles through ion exchange (see Section 4.9). Nitrate may be taken up by plants or used by microbes in the process of denitrification, reducing nitrate to N_2 gas and thus completing the nitrogen cycle (see Section 22.8, Figure 22.7). Nitrate that is not used by plants or microbes may be leached to groundwater and surface water. As a result, increased leaching of inorganic nitrogen (primarily nitrates) is associated with regions of high atmospheric deposition (**Figure 22.16**). The result is nutrient enrichment of streams and lakes (see discussion of problems associated with nutrient enrichment in aquatic ecosystems in Chapter 24, *Ecological Issues & Applications*).

When nitrate is leached from the soil, a loss of base cations and soil acidification may result. As the pH of the soil decreases, the buffering ability of cation exchange may be exhausted, resulting in the release of aluminum (Al) ions to the soil solution (see Section 4.9). Aluminum can be leached to aquatic ecosystems, where it is toxic. In addition, high aluminum concentrations in the soil solution can have detrimental effects on forest ecosystems.

In the later stages of nitrogen saturation, productivity is expected to decrease and plant mortality may increase. This is partly because of nutrient imbalances resulting from the overwhelming availability of nitrogen. The health of plants is affected primarily by the relative concentrations of nutrients as opposed to their absolute abundances. As aluminum concentration in the soil solution increases because of soil acidification, the calcium-to-aluminum (Ca/Al) and magnesium-to-aluminum (Mg/Al) ratios of the foliage (leaf tissues) decrease. This decrease is partly the result of the higher affinity of aluminum during the passive uptake of nutrients in soil solution by roots. These relative nutrient proportions have been correlated with declining spruce (*Picea*) populations in both the northeastern United States and Europe. For example, Steve McNulty of the Coweeta Hydrologic Laboratory (North Carolina) and colleagues John Aber and Steven Newman of the University of New Hampshire examined the response of foliar nutrient concentration and growth of trees in a high elevation spruce–fir forest in southeastern Vermont to nitrogen addition treatments. The researchers found that increasing nitrogen inputs resulted in a decline in the foliar calcium-to-aluminum ratio, which in turn resulted in reduced tree growth (**Figure 22.17**). The decline in the calcium-to-aluminum ratio was the result of both a decline in calcium and an increase in aluminum concentration in leaves with increased nitrogen additions to the experimental plots.

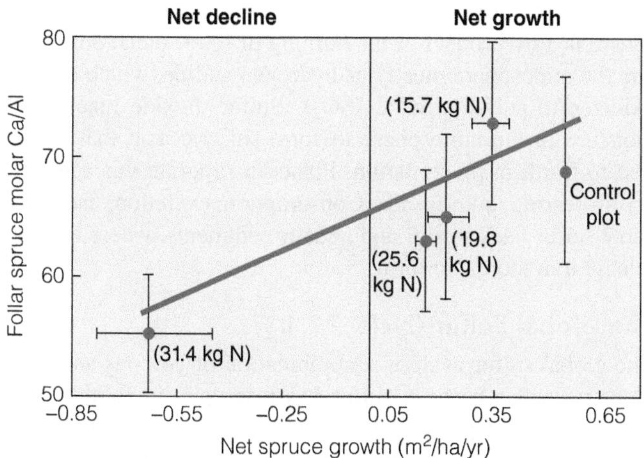

Figure 22.17 Relationship between the foliar calcium-to-aluminum ratio (Ca/Al) and annual basal area growth of spruce trees on 10 spruce–fir plots. Values in parentheses are the nitrogen (N) addition treatments (four paired N addition treatments, and one paired control) on Mt. Ascutney, Vermont. Circles represent mean values, and bars represent the standard error.
(Adapted from McNulty et al. 1996.)

The reduced uptake of calcium by trees can influence growth rates in a number of ways. Calcium has several important roles in plant functioning, including the production of new sapwood—the outer wood on the bole of the tree that contains the active vascular tissues for water transport. A reduction in calcium can therefore result in a reduction in sapwood production and water transport, which in turn would result in decreasing the leaf area that the tree can support. The combined effect is reduced net uptake of carbon by the plant (see Figure 22.17).

In addition to calcium, reduced magnesium concentrations (associated with low foliar magnesium-to-aluminum and magnesium-to-nitrogen ratios) can also influence tree growth rates and overall health. Magnesium is an important element in a variety of plant enzymes, particularly chlorophyll. Observed decreases in the magnesium-to-nitrogen ratio with nitrogen deposition are associated with higher foliar nitrogen concentrations. New growth stimulated by high nitrogen levels results in the retranslocation of magnesium from mature needles, resulting in the decline in the magnesium-to-nitrogen ratio, nutrient deficiency, and yellowing.

SUMMARY

Biogeochemical Cycles 22.1

Nutrients flow from the living to the nonliving components of the ecosystem and back in a perpetual cycle. Through these cycles, plants and animals obtain nutrients necessary for their survival and growth. There are two basic types of biogeochemical cycles: the gaseous, represented by oxygen, carbon, and nitrogen cycles, whose major pools are in the atmosphere, and the sedimentary, represented by the sulfur and phosphorus cycles, whose major pools are in Earth's crust. The sedimentary cycles involve two phases: salt solution and rock. Minerals become available through the weathering of Earth's crust, enter the water cycle as a salt solution, take diverse pathways through the ecosystem, and ultimately return to Earth's crust through sedimentation. All nutrient cycles have a common structure and share three basic components: inputs, internal cycling, and outputs.

Inputs 22.2

The input of nutrients to the ecosystem depends on whether the biogeochemical cycle is gaseous or sedimentary. The availability of essential nutrients in terrestrial ecosystems depends heavily on the nature of the soil. Supplementing the nutrients in soil are the nutrients carried by rain, snow, air currents, and animals. The major sources of nutrients for aquatic life are inputs from the surrounding land (in the form of drainage water, detritus, and sediments).

Outputs 22.3

Nutrient levels within an ecosystem can decline as a result of nutrient export. Export can occur in a variety of ways, depending on the specific biogeochemical cycle. A major means of transportation is in the form of organic matter carried by surface flow of water in streams and rivers. Leaching of dissolved nutrients from soils into surface water and groundwater also represents a significant export in some ecosystems. Harvesting of biomass in forestry and agriculture represents a permanent withdrawal from the ecosystem. Fire is also a major source of nutrient export in some terrestrial ecosystems.

Global Cycles 22.4

The biogeochemical cycles of various ecosystems are linked. As such, it is important to view the biogeochemical cycles of many elements from a global perspective.

The Carbon Cycle 22.5

The carbon cycle is inseparable from energy flow. Carbon is assimilated as carbon dioxide by plants, consumed in the form of plant and animal tissue by heterotrophs, released through respiration, mineralized by decomposers, accumulated in standing biomass, and withdrawn into long-term reserves. Carbon cycles through the ecosystem at a rate that depends on the rates of primary productivity and decomposition. Both processes are faster in warm, wet ecosystems. In swamps and marshes, organic material stored as raw humus or peat circulates slowly, forming oil, coal, and natural gas. Similar cycling takes place in freshwater and marine environments.

Temporal Variation in the Carbon Cycle 22.6

Cycling of carbon exhibits daily and seasonal fluctuations. Carbon dioxide builds up at night, when respiration increases. During the day, plants withdraw carbon dioxide from the air and carbon dioxide concentration drops sharply. During the growing season, atmospheric concentration drops.

The Global Carbon Cycle 22.7

Earth's carbon budget is closely linked to the atmosphere, land, and oceans and to the mass movements of air around the planet. In the ocean, surface water acts as the main site of carbon exchange between the atmosphere and ocean. The ability of the surface waters to take up carbon dioxide is governed largely by the reaction of carbon dioxide with the carbonate ion to form bicarbonates. The uptake of carbon dioxide from the atmosphere by terrestrial ecosystems is governed by gross production (photosynthesis). Losses are a function of autotrophic and heterotrophic respiration; the latter are dominated by microbial decomposers.

The Nitrogen Cycle 22.8

The nitrogen cycle is characterized by the fixation of atmospheric nitrogen by mutualistic nitrogen-fixing bacteria associated with the roots of many plants, largely legumes, and cyanobacteria. Other processes are ammonification, the breakdown of amino acids by decomposer organisms to produce ammonia, nitrification, the bacterial oxidation of ammonia to nitrate and nitrates, and denitrification, the reduction of nitrates to gaseous nitrogen. The global nitrogen cycle follows the pathway of the local nitrogen cycle just described, only on a grander scale. The atmosphere is the largest pool, with comparatively small amounts of nitrogen found in the biomass and soils of terrestrial ecosystems. Major sources of nitrogen to the world's oceans are dissolved forms in the freshwater drainage from rivers and inputs in precipitation.

The Phosphorus Cycle 22.9

The phosphorus cycle has no significant atmospheric pool. The main pools of phosphorus are rock and natural phosphate deposits. The terrestrial phosphorus cycle follows the typical biogeochemical pathways. In marine and freshwater ecosystems, however, the phosphorus cycle moves through three states: particulate organic phosphorus, dissolved organic phosphates, and inorganic phosphates. Involved in the cycling are phytoplankton, zooplankton, bacteria, and microbial grazers. The global phosphorus cycle is unique among the major biogeochemical cycles in having no significant atmospheric component, although airborne transport of phosphorus occurs in the form of soil dust and sea spray. Nearly all of the phosphorus in terrestrial ecosystems is derived from the weathering of calcium phosphate minerals. The transfer of phosphorus from terrestrial to aquatic ecosystems is low under natural conditions; however, the large-scale application of phosphate fertilizers and the disposal of sewage and wastewater to aquatic ecosystems result in a large input of phosphorus to aquatic ecosystems.

The Sulfur Cycle 22.10

Sulfur has both gaseous and sedimentary phases. Sedimentary sulfur comes from the weathering of rocks, runoff, and decomposition of organic matter. Sources of gaseous sulfur are decomposition of organic matter, evaporation of oceans, and volcanic eruptions. Much of the sulfur released to the atmosphere is a by-product of the burning of fossil fuels. Sulfur enters the atmosphere mostly as hydrogen sulfide, which quickly oxidizes to sulfur dioxide (SO_2). Sulfur dioxide reacts with moisture in the atmosphere to form sulfuric acid that is carried to Earth in precipitation. Plants incorporate this acid into sulfur-bearing amino acids. Consumption, excretion, and death carry sulfur back to soil and aquatic sediments, where bacteria release it in inorganic form.

The Global Sulfur Cycle 22.11

The global sulfur cycle is a combination of gaseous and sedimentary cycles because sulfur has reservoirs in Earth's crust and in the atmosphere. The sulfur cycle involves a long-term sedimentary phase in which sulfur is tied up in organic and inorganic deposits, is released by weathering and decomposition, and is carried to terrestrial and aquatic ecosystems in salt solution. The bulk of sulfur first appears in the gaseous phase as a volatile gas, hydrogen sulfide (H_2S), in the atmosphere, which quickly oxidizes to form sulfur dioxide. Once in soluble form, sulfur is taken up by plants and incorporated into organic compounds. Excretion and death carry sulfur in living material back to the soil and to the bottoms of ponds, lakes, and seas, where sulfate-reducing bacteria release it as hydrogen sulfide or as a sulfate.

The Oxygen Cycle 22.12

Oxygen, the by-product of photosynthesis, is very active chemically. It combines with a wide range of chemicals in Earth's crust, and it reacts spontaneously with organic compounds and reduced substances. It is involved in oxidizing carbohydrates in the process of respiration to release energy, carbon dioxide, and water. The current atmospheric pool of oxygen is maintained in a dynamic equilibrium between the production of oxygen in photosynthesis and its consumption in respiration. An important constituent of the atmospheric reservoir of oxygen is ozone (O_3).

Biogeochemical Cycles Are Linked 22.13

All of the major biogeochemical cycles are linked; the nutrients that cycle are all components of living organisms, constituents of organic matter. Stoichiometric relationships among the various elements involved in plant processes related to carbon uptake and plant growth have an important influence on the cycling of nutrients in ecosystems.

Nitrogen Saturation Ecological Issues & Applications

Human activities are resulting in an ever-increasing atmospheric deposition of nitrogen to aquatic and terrestrial ecosystems. In terrestrial ecosystems, nitrogen deposition initially acts as a fertilizer, increasing rates of net primary productivity. However, as water and other nutrients become more limiting relative to nitrogen, these ecosystems may approach nitrogen saturation. If the nitrogen supply continues to increase, a complex series of changes to soil and plant processes may ultimately lead to soil acidification and forest decline.

STUDY QUESTIONS

1. How are the processes of photosynthesis and decomposition involved in the carbon cycle?
2. In the temperate zone, is the atmospheric concentration of carbon dioxide higher during the day or night? Why?
3. Characterize the following processes in the nitrogen cycle: fixation, ammonification, nitrification, and denitrification.
4. What biological and nonbiological mechanisms are responsible for nitrogen fixation?
5. What is the source of sulfur in the sulfur cycle? Why does the sulfur cycle have characteristics of both sedimentary and gaseous cycles?
6. What is the source of phosphorus in the phosphorus cycle?
7. What is the major source of phosphorus input into aquatic ecosystems?
8. What is the role of photosynthesis and decomposition in the oxygen cycle?
9. How would the development of large deposits of fossil fuels have influenced the oxygen concentration of the atmosphere over geologic time?

FURTHER READINGS

Classic Studies

Bolin, B., and R. Cook, (Eds.). 1983. *The major biogeochemical cycles and their interactions*. Wiley, New York.
Edited volume providing some of the earlier research efforts as quantifying major biogeochemical cycles at a global scale.

Current Research

Aber, J. D. 1992. "Nitrogen cycling and nitrogen saturation in temperate forest ecosystems." *Trends in Ecology and Evolution* 7:220–223.

Aber, J., W. McDowell, K. Nadelhoffer, A. Magill, G. Berntson, M. Kamakea, S. McNulty, W. Currie, L. Rustad, and I. Fernandez. 1998. "Nitrogen saturation in temperate forest ecosystems: Hypotheses revisited." *BioScience* 48:921–934.
These two articles by John Aber and colleagues provide an excellent introduction to the concept and consequences of nitrogen saturation in terrestrial ecosystems.

Galloway, J. N., A. R. Townsend, J. W. Erisman, M. Bekunda, Z. Cai, J. R. Freney, L. A. Martinelli, S. P. Seitzinger, M. A. Sutton. 2008. "Transformation of the nitrogen cycle: Recent trends, questions and potential solutions." *Science* 320:889–892.

Gorham, E. 1991. "Biogeochemistry: Its origins and development." *Biogeochemistry* 13.3:199–239.
The author presents a thorough and scholarly review of the history of biogeochemistry.

Post, W. M., T. H. Peng, W. R. Emanuel, A. W. King, V. H. Dale, and D. L. DeAngelis. 1990. "The global carbon cycle." *American Scientist* 78:310–326.
An excellent updated review of the global carbon cycle—what we know and do not know about it.

Sarmiento, J. L., and N. Gruber, 2006. *Ocean biogeochemical dynamics.* Princeton, NJ: Princeton University Press.
This text is an excellent overview of the circulation and interactions of the major elements in the oceans.

Schlesinger, W. H. 2013. *Biogeochemistry: An analysis of global change.* 3rd ed. London: Academic Press.
A superb text on the subject of biogeochemical cycles. Provides an excellent discussion of nutrient cycling in both terrestrial and aquatic environments.

Sprent, J. I. 1988. *The ecology of the nitrogen cycle.* New York: Cambridge University Press.
This text provides an ecological perspective on the processes involved in the nitrogen cycle.

Vitousek, P. M., J. Aber, R. W. Howarth, G. E. Likens, P. A. Matson, D. W. Schindler, W. H. Schlesinger, and G. D. Tilman. 1997. "Human alteration of the global nitrogen cycle: Sources and consequences." *Ecological Applications* 7:737–750.
An excellent discussion of how human activities have changed the nitrogen cycle.

CHAPTER 23 Terrestrial Ecosystems

Spectacular fall color is a hallmark of the eastern deciduous mixed hardwood forest.

CHAPTER GUIDE

23.1 Terrestrial Ecosystems Reflect Adaptations of the Dominant Plant Life-Forms

23.2 Tropical Forests Characterize the Equatorial Zone

23.3 Tropical Savannas Are Characteristic of Semiarid Regions with Seasonal Rainfall

23.4 Grassland Ecosystems of the Temperate Zone Vary with Climate and Geography

23.5 Deserts Represent a Diverse Group of Ecosystems

23.6 Mediterranean Climates Support Temperate Shrublands

23.7 Forest Ecosystems Dominate the Wetter Regions of the Temperate Zone

23.8 Conifer Forests Dominate the Cool Temperate and Boreal Zones

23.9 Low Precipitation and Cold Temperatures Define the Arctic Tundra

ECOLOGICAL Issues & Applications Forest Management

IN 1939, ECOLOGISTS F. E. Clements (Carnegie Institution of Washington) and V. E. Shelford (University of Illinois) introduced an approach for combining the broad-scale distribution of plants and associated animals into a single classification system. In their book *Bio-ecology*, Clements and Shelford referred to these biotic units as **biomes**. Biomes are classified according to the predominant plant types. There are at least eight major terrestrial biome types, but there may be more, depending on how finely the biomes are classified. These include tropical forest, temperate forest, conifer forest (taiga or boreal forest), tropical savanna, temperate grasslands, chaparral (shrublands), tundra, and desert (**Figure 23.1**). These broad categories reflect the relative contribution of three general plant life-forms: trees, shrubs, and grasses. A closed canopy of trees characterizes forest ecosystems. Woodland and savanna ecosystems are characterized by the codominance of grasses and trees (or shrubs). As the names imply, shrubs are the dominant plant form in shrublands, and grasses dominate in grasslands. Desert is a general category used to refer to areas with scarce plant cover.

When the plant ecologist Robert Whittaker of Cornell University plotted these biome types on gradients of mean annual temperature and mean annual precipitation, he found they formed a distinctive climatic pattern, as graphed in **Figure 23.2**. As the graph indicates, boundaries between biomes are broad and often indistinct as they blend into one another. Besides climate, other factors such as topography, soils,

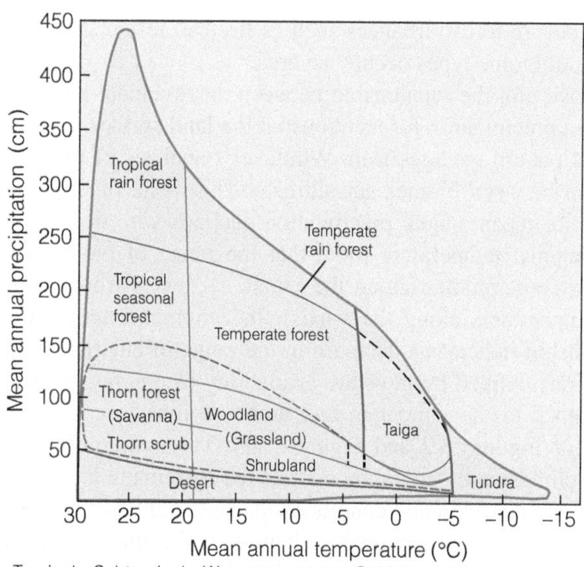

Figure 23.2 The pattern of terrestrial biomes in relation to temperature and moisture. Where the climate varies, soil can shift the balance between types. The dashed line encloses environments in which either grassland or one of the types dominated by woody plants may prevail.
(Adapted from Whittaker 1970.)

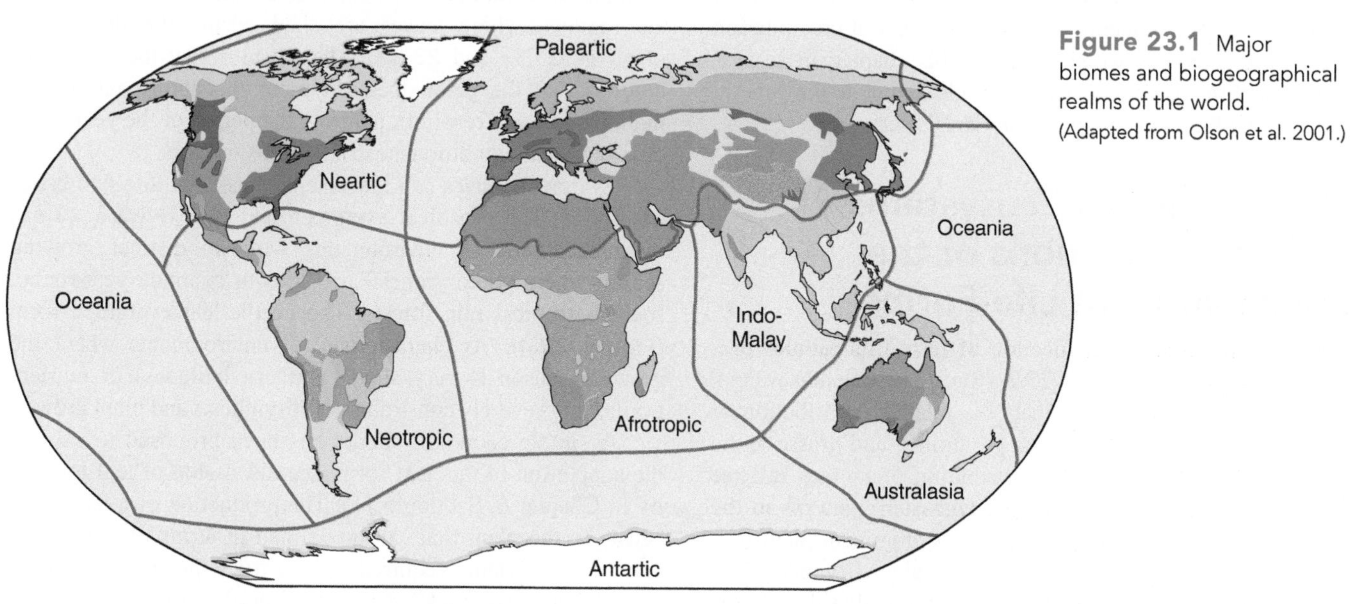

Figure 23.1 Major biomes and biogeographical realms of the world.
(Adapted from Olson et al. 2001.)

and exposure to disturbances such as fire can influence which of several biome types occupy an area.

If we plot the relationship between mean annual temperature and precipitation for locations on the land surface, another general pattern emerges from Whittaker's analysis of the relationship between biomes and climate. The range of observed values for mean annual precipitation declines with decreasing mean annual temperature (note that the range of biomes defined by precipitation along the *y*-axis decreases with declining temperatures along the *x*-axis). In geographic terms, this relationship indicates a decrease in the range of environmental conditions defined by moisture availability as one moves from the tropics to the temperate and arctic regions (see label on *x*-axis of Figure 23.2 and Figure 2.17). This relationship between climate and geography reflects the systematic latitudinal pattern of environmental conditions discussed in Chapter 2 that result directly from seasonal variations in the influx of solar radiation to Earth's surface. Mean annual temperature decreases from the equator to the poles, whereas seasonal variation in temperatures (and day length) increases (see Figures 2.5 and 2.8). The result is a decline in the growing season (period over which photosynthesis and plant growth can be maintained).

The systematic variation in climate with latitude is not limited to temperature. Average annual precipitation decreases with increasing latitude as a result of the interaction of humidity and temperature (see Section 2.6 and Figure 2.17). With declining temperatures, the amount of moisture that can be held in the air declines, reducing the overall amount of precipitation (see Figure 2.15). As we shall see in this chapter, these systematic patterns of climate across the globe dictate the general distribution of terrestrial biomes on Earth's surface.

23.1 Terrestrial Ecosystems Reflect Adaptations of the Dominant Plant Life-Forms

Given that the broad classification of terrestrial biomes presented in Figures 23.1 and 23.2 (forest, woodland/savanna, shrubland, and grassland) reflects the relative contribution of three general plant life-forms (trees, shrubs, and grasses), the question of what controls the distribution of biomes relative to climate becomes: Why are there consistent patterns in the distribution and abundance of these three dominant plant life-forms that relate to climate and the physical environment? The answer to this question lies in the adaptations that these three very different plant life-forms possess, as well as the advantages and constraints arising from these adaptations under different environmental conditions.

Although the broad categories of grasses, shrubs, and trees each represent a diverse range of species and characteristics, they have fundamentally different patterns of carbon allocation and morphology (see Chapter 6). Grasses allocate less carbon to the production of supportive tissues (stems) than do woody plants (shrubs and trees), enabling grasses to maintain a higher proportion of their biomass in photosynthetic tissues (leaves). For woody plants, shrubs allocate a lower percentage of their resources to stems than do trees. The production of woody tissue gives the advantage of height and access to light, but it also has the associated cost of maintenance and respiration. If this cost cannot be offset by carbon gain through photosynthesis, the plant is unable to maintain a positive carbon balance and dies (see Chapter 6). As a result, as environmental conditions become adverse for photosynthesis (dry, low nutrient concentrations, or short growing season and cold temperatures), trees decline in both stature and density until they can no longer persist as part of the plant community.

Within the broad classes of forest and woodland ecosystems in which trees are dominant or codominant, leaf form is another plant characteristic that ecologists use to classify ecosystems. Leaves can be classified into two broad categories based on their longevity. Leaves that live for only a single year or growing season are classified as **deciduous**, whereas those that live beyond a year are called **evergreen**. The deciduous leaf is characteristic of environments with a distinct growing season. Leaves are typically shed at the end of the growing season and then regrown at the beginning of the next. Deciduous leaf type is further divided into two categories based on dormancy period. Winter-deciduous leaves are characteristic of temperate regions, where the period of dormancy corresponds to low (below freezing) temperatures (**Figures 23.3a** and **23.3b**; also see discussion of plant adaptations in Chapter 6). Drought-deciduous leaves are characteristic of environments with seasonal rainfall, especially in the subtropical and tropical regions, where leaves are shed during the dry period (**Figures 23.3c** and **23.3d**). The advantage of the deciduous habit is that the plant does not incur the additional cost of maintenance and respiration during the period of the year when environmental conditions restrict photosynthesis.

Evergreen leaves can likewise be classified into two broad categories. The broadleaf evergreen leaf type (**Figure 23.4a**) is characteristic of environments with no distinct growing season where photosynthesis and growth continue year-round, such as tropical rain forests. The needle-leaf evergreen form (**Figure 23.4b**) is characteristic of environments where the growing season is very short (northern latitudes) or nutrient availability severely constrains photosynthesis and plant growth.

A simple economic model has been proposed to explain the adaptation of this leaf form (see discussion of leaf longevity in Chapter 6, Section 6.11). The production of a leaf has a "cost" to the plant that can be defined in terms of the carbon and other nutrients required to construct the leaf. The time required to "pay back" the cost of production (carbon) will be a function of the rate of net photosynthesis (carbon gain). If environmental conditions result in low rates of net photosynthesis, the period of time required to pay back the cost of production will be longer. If the rate of photosynthesis is low enough, it may not be possible to pay back the cost over the period of a single growing season. A plant adapted to such environmental conditions cannot "afford" a deciduous leaf form, which requires producing new leaves every year. The leaves of evergreens, however, may survive for several years. So under this model, we can view the needle-leaf evergreen as a plant adapted for survival in an environment with a distinct growing season, in which conditions limit the plant's ability to produce enough carbon through photosynthesis during the growing season to pay for the cost of producing the leaves.

On combining the simple classification of plant life-forms and leaf type with the large-scale patterns of climate presented previously, we can begin to understand the distribution of biome types relative to the axes of temperature and precipitation shown in Figure 23.2. Ecosystems characteristic of warm, wet climates with no distinct seasonality are dominated by broadleaf evergreen trees and are called tropical (and subtropical) rain forest. As conditions become drier, with a distinct dry season, the broadleaf evergreen habit gives way to drought-deciduous trees that characterize the seasonal tropical forests. As precipitation declines further, the stature and density of these trees declines, giving rise to the woodlands and savannas that are characterized by the coexistence of trees (shrubs) and grasses. As precipitation further declines, trees can no longer be supported, giving rise to the arid shrublands (thorn scrub) and desert.

The temperature axis represents the latitudinal gradient from the equator to the poles (see geographical labels on x-axis of Figure 23.2). Moving from the broadleaf evergreen forests of the wet tropics into the cooler, seasonal environments of the temperate regions, the dominant trees are winter-deciduous. These are the regions of temperate deciduous forest. In areas of the temperate region where precipitation is insufficient to support trees, grasses dominate and give rise to the prairies

(a)

(b)

(c)

(d)

Figure 23.3 Examples of winter- and drought-deciduous trees. Temperate deciduous forest in central Virginia during (a) summer and (b) winter seasons. Semiarid savanna/woodland in Zimbabwe, Africa, during (c) rainy and (d) dry seasons.

Figure 23.4 Examples of evergreen trees. (a) Broadleaf evergreen trees dominate the canopy of this tropical rain forest in Queensland, Australia. (b) Needle-leaf evergreen trees (foxtail pine) inhabit the high-altitude zone of the Sierra Nevada in western North America.

(a)

(b)

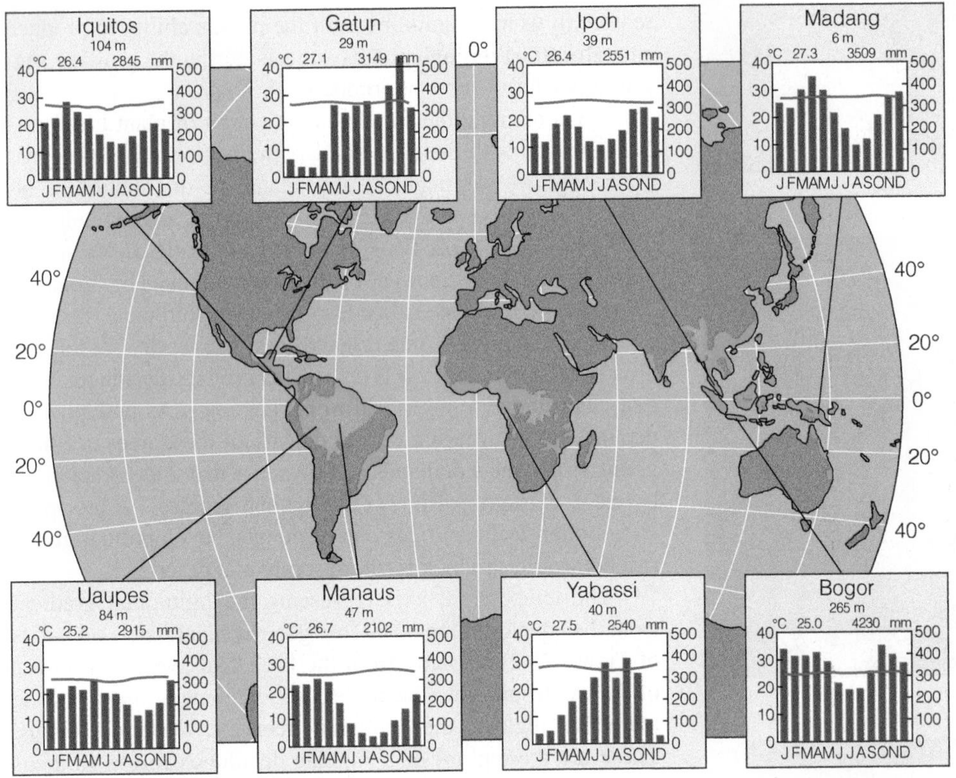

Figure 23.5 Geographic distribution of Earth's tropical forest ecosystems and associated climate diagrams showing long-term patterns of monthly temperature and precipitation for selected locations. Note the lack of seasonality in mean monthly temperatures, which are above 20°C. Although the rainfall in some regions is seasonal, note that minimum monthly precipitation is typically above 60 mm, and total annual precipitation above 2000 mm.
(Adapted from Archibold 1995.)

of North America, the steppes of Eurasia, and the pampas of Argentina. Moving poleward, the temperate-deciduous forests give way to the needle-leaf–dominated forests of the boreal region (conifer forest or taiga). As temperatures become more extreme and the growing season shorter, trees can no longer be supported, and the short-stature shrubs and sedges (grasslike plants of the family Cyperaceae) characteristic of the tundra dominate the landscape ecosystems of the arctic region.

In the following sections, we will examine the eight major categories of terrestrial biomes outlined in Figure 23.2. We begin each section by relating their geographic distribution to the broad-scale constraints of regional climate, as outlined in Figure 23.2, as well as to associated patterns of seasonality in temperature and precipitation (see **Quantifying Ecology 23.1**) that function as constraints on the dominant plant life-forms and patterns of primary and secondary productivity. In our discussion, we emphasize the unique physical and biological characteristics defining these broad categories of terrestrial ecosystems (biomes).

23.2 Tropical Forests Characterize the Equatorial Zone

The tropical rain forests are restricted primarily to the equatorial zone between latitudes 10° N and 10° S (**Figure 23.5**), where the temperatures are warm throughout the year and rainfall occurs almost daily. The largest and most continuous region of rain forest in the world is in the Amazon basin of South America (**Figure 23.6**). The second largest is located in Southeast Asia, and the third largest is in West Africa around the Gulf of Guinea and in the Congo basin. Smaller rain forests occur along the northeastern coast of Australia, the windward

Figure 23.6 Tropical rain forests in (a) Amazon Basin (South America) and (b) Malaysia (Southeast Asia). Despite being taxonomically distinct, these two tropical rain forest regions are dominated by broadleaf evergreen trees and support vigorous plant growth year-round. Tropical rain forests represent the most diverse and productive terrestrial ecosystems on our planet.

(a)

(b)

QUANTIFYING ECOLOGY 23.1 Climate Diagrams

As illustrated in Figure 23.2, the distribution of terrestrial biomes is closely related to climate. The measures of regional climate used in Whittaker's graph (Figure 23.2) are mean annual temperature and precipitation. Yet, as we will see in the discussion of the various biome types, the distribution of terrestrial ecosystems is influenced by other aspects of climate as well, namely seasonality of both temperature and precipitation. Topographic features such as mountains and valleys also influence the climate of a region.

To help understand the relationship between regional climate and the distribution of terrestrial ecosystems, for each biome discussed in this chapter (tropical forest, savanna, etc.), we present a map showing its global distribution. Accompanying the map is a series of climate diagrams. The diagrams describe the local climate at representative locations around the world where a particular biome type is found. See **Figure 1** for a representative climate diagram, which we have labeled to help you interpret the information it presents. As you study the diagram, take particular note of the patterns of seasonality.

1. In Figure 23.12, what is the distinctive feature of the climate diagrams for these tropical savanna ecosystems? How do the patterns differ between sites in the Northern and Southern Hemispheres? What feature of Earth's climate system discussed in Chapter 2 is responsible for these distinctive patterns?
2. In Figure 23.23, what feature of the climate is common to all mediterranean ecosystems?

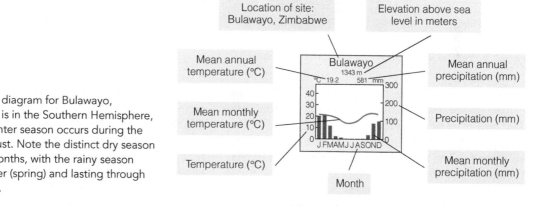

Figure 1 Climate diagram for Bulawayo, Zimbabwe. This city is in the Southern Hemisphere, where the cooler winter season occurs during the period of May–August. Note the distinct dry season during the winter months, with the rainy season beginning in October (spring) and lasting through the summer months.

side of the Hawaiian Islands, the South Pacific Islands, the east coast of Madagascar, northern South America, and southern Central America.

The climate of tropical rain forest regions varies geographically but is typically characterized by a mean temperature of all months exceeding 18°C and minimum monthly precipitation above 60 millimeters (mm; see climate diagrams for representative tropical rain forest sites in Figure 23.5). Within the lowland forest zone, mean annual temperatures typically exceed 25°C with an annual range less than 5°C.

Tropical rain forests have a high diversity of plant and animal life. Covering only 6 percent of the land surface, tropical

(a)

(b)

Figure 23.7 Examples of primate species that inhabit the tropical rain forests of the world: (a) the chimpanzee (*Pan troglodytes*) inhabits the tropical rain forests of Central Africa, and (b) the orangutan (*Pongo pygmaeus*) inhabits the tropical rain forests of Borneo (Southeast Asia).

Figure 23.8 Vertical stratification of a tropical rain forest.

rain forests account for more than 50 percent of all known plant and animal species. Tree species number in the thousands. A 10-km^2 area of tropical rain forest may contain 1500 species of flowering plants and up to 750 species of trees. The richest area is the lowland tropical forest of peninsular Malaysia, which contains some 7900 species.

Nearly 90 percent of all nonhuman primate specie live in the tropical rain forests of the world (**Figure 23.7**). Sixty-four species of New World primates—small mammals with prehensile tails—live in the trees. The Indo-Malaysian forests are inhabited by a number of primates, many with a limited distribution within the region. The orangutan, an arboreal ape, is confined to the island of Borneo. Peninsular Malaysia has seven species of primates, including three gibbons, two langurs, and two macaques. The long-tailed macaque is common in disturbed or secondary forests, and the pig-tailed macaque is a terrestrial species adaptable to human settlements. The tropical rain forest of Africa is home to mountain gorillas and chimpanzees. The diminished rain forest of Madagascar holds 39 species of lemurs (see Chapter 9, *Ecological Issues & Applications* and Figure 9.20).

Tropical rain forests may be divided into five vertical layers (**Figure 23.8**): emergent trees, upper canopy, lower canopy, shrub understory, and a ground layer of herbs and ferns. Conspicuous in the rain forest are lianas—climbing vines—growing upward into the canopy, epiphytes growing on the trunks and branches, and strangler figs (*Ficus* spp.) that grow downward from the canopy to the ground. Many large trees develop plank-like outgrowths called *buttresses* (**Figure 23.9**). They function as prop roots to support trees rooted in shallow soil that offers poor anchorage. The floor of a tropical rain forest is thickly laced with roots, both large and small, forming a dense mat on the ground.

The continually warm, moist conditions in rain forests promote strong chemical weathering and rapid leaching of soluble materials. The characteristic soils are oxisols, which are deeply weathered with no distinct horizons (see Chapter 4 for discussion and classification of soils). Ultisols may develop in areas with more seasonal precipitation regimes and are typically associated with forested regions that exhibit seasonal soil moisture deficits. Areas of volcanic activity in parts of Central and Southeast Asia, where recent ash deposits quickly weather, are characterized by andosols (see Figure 4.12).

The warmer, wetter conditions of the tropical rain forest result in high rates of net primary productivity and subsequent high annual rates of litter input to the forest floor. Little litter accumulates, however, because decomposers consume the dead

Figure 23.9 Plank-like buttresses help to support tall rain forest trees.

(a)

(b)

Figure 23.10 A tropical dry forest in Costa Rica during the (a) rainy and (b) dry season. Most of the tropical dry forests in Central America have disappeared from land clearing for agriculture.

organic matter almost as rapidly as it falls to the forest floor. Most of the nutrients available for uptake by plants are a result of the rapidly decomposed organic matter that is continuously falling to the soil surface. Growing plants, however, rapidly absorb these nutrients. The average time for leaf litter to decompose is 24 weeks.

Moving from the equatorial zone to the regions of the tropics that are characterized by greater seasonality in precipitation, the broadleaf evergreen forests are replaced by the dry tropical forests (**Figure 23.10**). Dry tropical forests undergo a dry season whose length is based on latitude. The more distant the forest is from the equator, the longer is the dry season—in some areas, up to eight months. During the dry season, the drought-deciduous trees and shrubs drop their leaves. Before the start of the rainy season, which may be much wetter than the wettest time in the rain forest, the trees begin to leaf. During the wet season, the landscape becomes uniformly green.

The largest proportion of tropical dry forest is found in Africa and South America, to the south of the zones dominated by rain forest. These regions are influenced by the seasonal migration of the Intertropical Convergence Zone (see Section 2.6, Figure 2.18). In addition, areas of Central America, northern Australia, India, and Southeast Asia are also classified as dry tropical forest. Much of the original forest, especially in Central America and India, has been converted to agricultural and grazing land.

23.3 Tropical Savannas Are Characteristic of Semiarid Regions with Seasonal Rainfall

The term *savanna* was originally used to describe the treeless areas of South America. Now it is generally applied to a range of vegetation types in the drier tropics and subtropics characterized by a ground cover of grasses with scattered shrubs or trees. Savanna includes an array of vegetation types representing a continuum of increasing cover of woody vegetation, from open grassland to widely spaced shrubs or trees and to woodland (**Figure 23.11**). In South America, the more densely wooded areas are referred to as cerrado. The campos and llano are characterized by a more open appearance (lower density of trees), and thorn scrub is the dominant cover of the caatinga. In Africa, the miombo, mopane, and *Acacia* woodlands can be distinguished from the more open and park-like bushveld. Scattered individuals of *Acacia* and *Eucalyptus* dominate the mulga and brigalow of Australia.

The physiognomic diversity of the savanna vegetation reflects the different climate conditions occurring throughout this

Figure 23.11 Savanna ecosystems, such as the (a) cerrano of South America and (b) mulga woodlands of central Australia are characterized by a ground cover of grasses with scattered shrubs or trees.

(a)

(b)

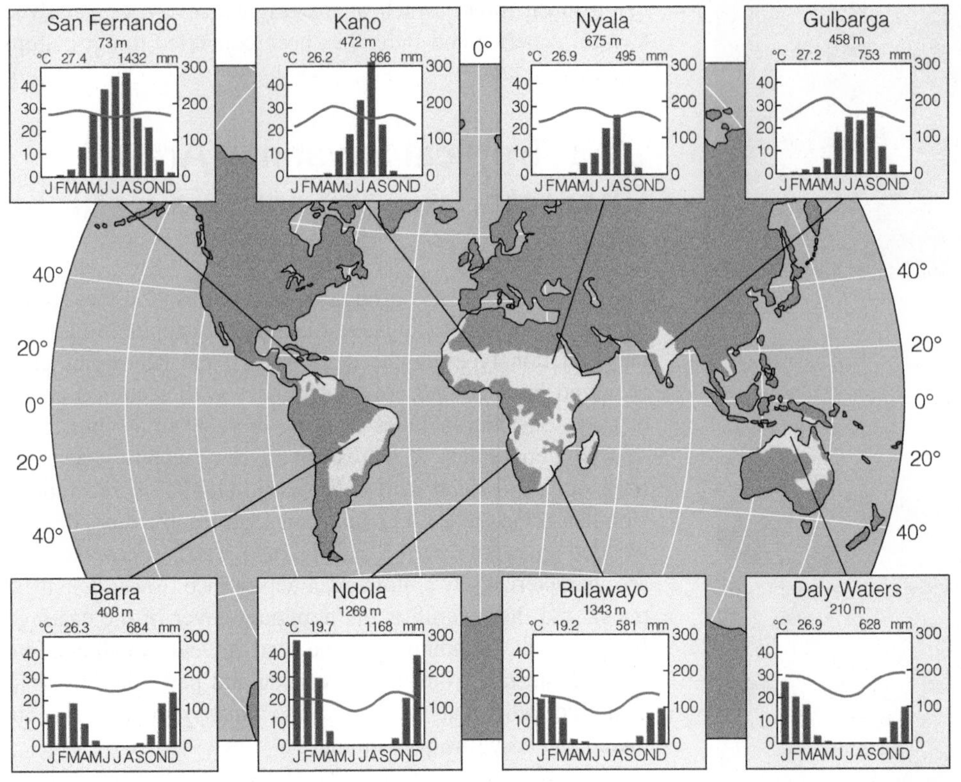

Figure 23.12 Geographic distribution of Earth's tropical savanna ecosystems and associated climate diagrams showing long-term patterns of monthly temperature and precipitation for selected locations. Seasonal patterns of temperature are similar to those of the tropical forest sites presented in Figure 24.5, reflecting the tropical and subtropical climates at these sites. Note, however, the distinct seasonality of precipitation that characterizes these ecosystems (also note the shift in timing of rainy season for Northern and Southern Hemisphere locations, reflecting the seasonal shift of the Intertropical Convergence Zone).
(Adapted from Archibold 1995.)

widely distributed ecosystem (**Figure 23.12**). Moisture appears to control the density of woody vegetation, a function of both rainfall (amount and distribution) and soil—its texture, structure, and water-holding capacity (**Figure 23.13**; also see Chapter 4).

Savannas are associated with a warm continental climate with distinct seasonality in precipitation and a large interannual (year to year) variation in total precipitation (see climate diagrams for representative savanna sites in Figure 23.12). Mean monthly temperatures typically do not fall below 18°C, although during the coldest months in highland areas, temperatures can be considerably lower. There is seasonality in temperatures, and maximum temperatures occur at the end of

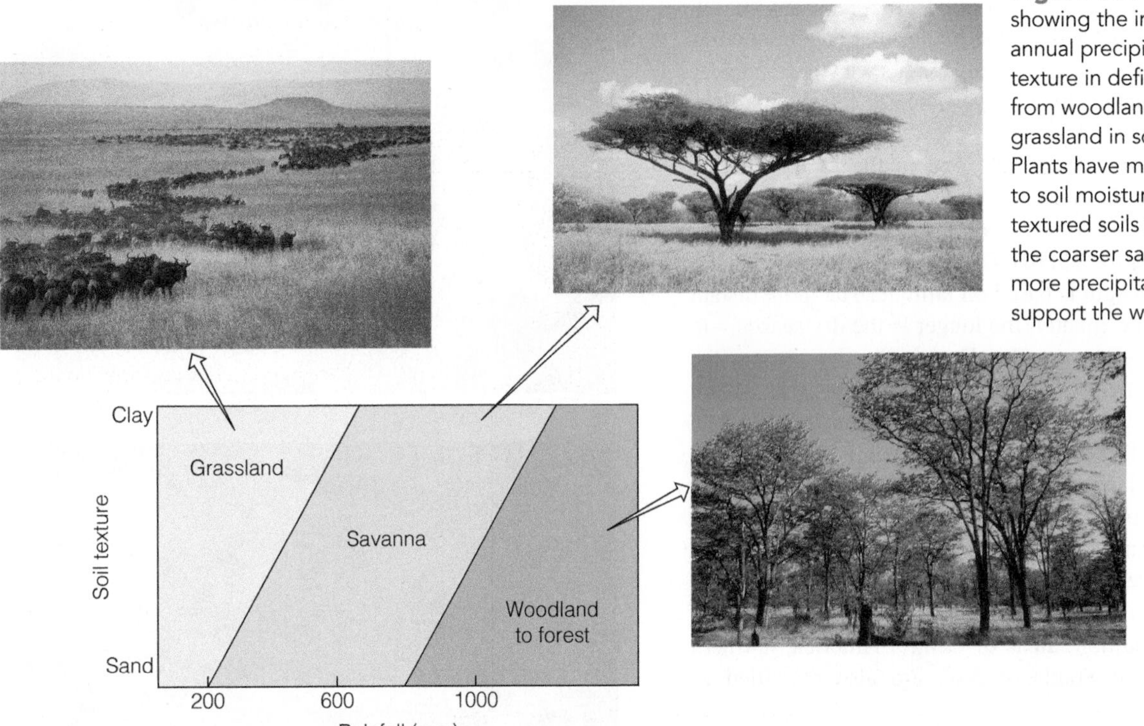

Figure 23.13 Diagram showing the interaction between annual precipitation and soil texture in defining the transition from woodland to savanna and grassland in southern Africa. Plants have more limited access to soil moisture on the heavily textured soils (clays) than on the coarser sands, so annually more precipitation is needed to support the woody plants.

the wet season. The nature of the vegetation cover, however, is more closely determined by the amount and seasonality of precipitation than by temperature.

Savannas, despite their differences in vegetation, exhibit a certain set of characteristics. Savannas occur on land surfaces of little relief—often on old plateaus, interrupted by escarpments and dissected by rivers. Continuous weathering in these regions has produced nutrient-poor oxisols, which are particularly deficient in phosphorus. Alfisols are common in the drier savannas, whereas entisols are associated with the driest savannas (see Figure 4.12). Subject to recurrent fires, the dominant vegetation is fire adapted. Grass cover with or without woody vegetation is always present, and the woody component is short-lived—individuals seldom survive for more than several decades. Savannas are characterized by a two-layer vertical structure because of the ground cover of grasses and the presence of shrubs or trees (see Figure 16.12b).

The yearly cycle of plant activity and subsequent productivity in tropical savannas is largely controlled by the markedly seasonal precipitation and corresponding changes in available soil moisture. Most leaf litter is decomposed during the wet season, and most woody debris is consumed by termites during the dry season.

The microenvironments associated with tree canopies can influence species distribution, productivity, and soil characteristics. Stem flow and associated litter accumulation result in higher soil nutrients and moisture under tree canopies, often encouraging increased productivity and the establishment of species adapted to the more shaded environments.

Savannas can support a large and varied assemblage of herbivores—invertebrate and vertebrate, grazing and browsing. The African savanna, visually at least, is dominated by a large and diverse ungulate fauna of at least 60 species that share the vegetative resources. Some species, such as the wildebeest and zebra, are migratory during the dry season (see Figure 7.10 for example).

Savanna vegetation supports an incredible number of insects: flies, grasshoppers, locusts, crickets, carabid beetles, ants, and detritus-feeding dung beetles and termites. Mound-building termites excavate and move tons of soil, mixing mineral soil with organic matter. Some species construct extensive subterranean galleries and others accumulate organic matter.

Preying on the ungulate fauna is an array of carnivores including the lion, leopard, cheetah, hyena, and wild dog. Scavengers, including vultures and jackals, subsist on the remains of prey killed by carnivores.

23.4 Grassland Ecosystems of the Temperate Zone Vary with Climate and Geography

Natural grasslands occupy regions where rainfall is between 25 and 80 centimeters (cm) a year, but they are not exclusively climatic. Many exist through the intervention of fire and human activity. Conversions of forests into agricultural lands and the planting of hay and pasturelands extended grasslands into once forested regions. Formerly covering about 42 percent of the land surface of Earth, natural grasslands have shrunk to less than 12 percent of their original size because of conversion to cropland and grazing lands.

The natural grasslands of the world occur in the midlatitudes in midcontinental regions, where annual precipitation declines as air masses move inward from the coastal environments (**Figure 23.14**; see Section 2.7 for discussion of continental

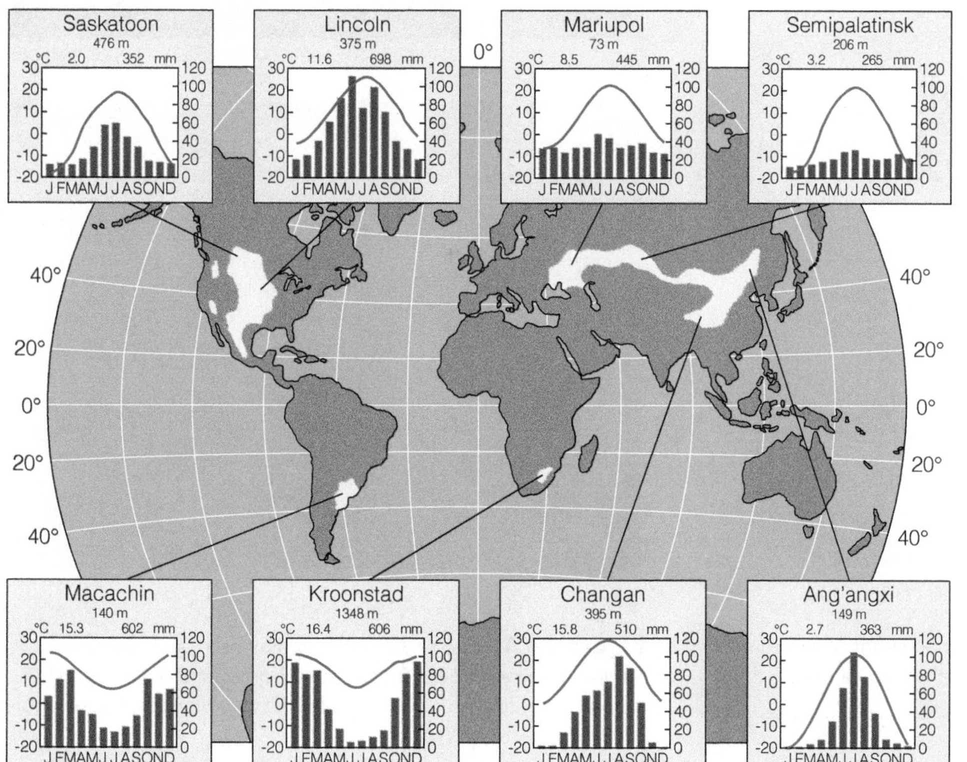

Figure 23.14 Geographic distribution of Earth's temperate grassland ecosystems and associated climate diagrams showing long-term patterns of monthly temperature and precipitation for selected locations. Most of the major grassland regions are mid-continental with a distinct seasonality in temperature. Mean annual precipitation is typically well below 1000 mm, comparable in range to that observed for tropical and subtropical savannas but less than that of temperate deciduous forests (compare with Figure 24.27). (Adapted from Archibold 1995.)

patterns of precipitation). In the Northern Hemisphere, these regions include the prairies of North America and the steppes of central Eurasia. In the Southern Hemisphere, grasslands are represented by the pampas of Argentina and the veld of the high plateaus of southern Africa. Smaller areas occur in southeastern Australia and the drier parts of New Zealand.

The temperate grassland climate is one of recurring drought, and much of the diversity of vegetation cover reflects differences in the amount and reliability of precipitation. Grasslands do the least well where precipitation is lowest and the temperatures are high. They are tallest in stature and the most productive where mean annual precipitation is greater than 800 mm and mean annual temperature is above 15°C. Thus, native grasslands of North America, influenced by declining precipitation from east to west, consist of three main types distinguished by the height of the dominant species: tallgrass, mixed-grass, and shortgrass prairie (**Figure 23.15**). **Tallgrass prairie** (**Figure 23.16a**) is dominated by big bluestem (*Andropogon gerardi*), growing 1 meter (m) tall with flowering stalks 1 to 3.5 m tall. **Mixed-grass prairie** (**Figure 23.16b**), typical of the Great Plains, is composed largely of needlegrass–grama grass (*Bouteloua–Stipa*). South and west of the mixed prairie and grading into the desert regions is the **shortgrass prairie** (**Figure 23.16c**), dominated by sod-forming blue grama (*Bouteloua gracilis*) and buffalo grass (*Buchloe dactyloides*), which has remained somewhat intact, and desert grasslands. From southeastern Texas to southern Arizona and south into Mexico lies the **desert grassland**,

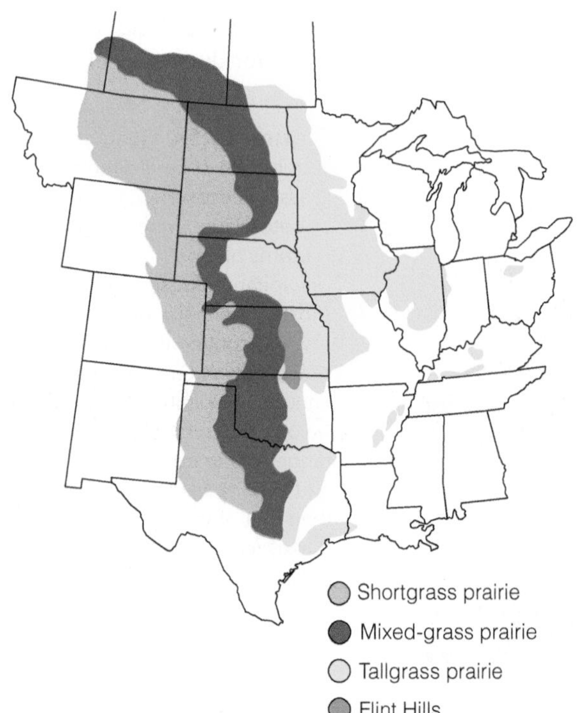

Figure 23.15 Map showing the original extent of shortgrass, mixed-grass, and tallgrass prairies in North America before the arrival of Europeans.
(After Reichman 1987.)

(a)

(b)

(c)

Figure 23.16 North American grasslands. (a) A remnant tallgrass prairie in Iowa; (b) the mixed-grass prairie has been called "daisyland" for the diversity of its wildflowers; (c) shortgrass steppe in western Wyoming.

similar in many respects to the shortgrass plains, except that three-awn grass (*Aristida* spp.) replaces buffalo grass. Confined largely to the Central Valley of California is **annual grassland**. It is associated with a mediterranean climate (see Section 23.6) characterized by rainy winters and hot, dry summers. Growth occurs during early spring, and most plants are dormant in summer, turning the hills a dry tan color accented by the deep green foliage of scattered California oaks.

At one time, the great grasslands of the Eurasian continent extended from eastern Europe to western Siberia south to Kazakhstan. These **steppes**, treeless, except for ribbons and patches of forest, are divided into four belts of latitude, from the mesic meadow steppes in the north to semiarid grasslands in the south.

In the Southern Hemisphere, the major grasslands exist in southern Africa and southern South America. Known as **pampas**, the South American grasslands extend westward in a large semicircle from Buenos Aires and cover about 15 percent of Argentina. These pampas have been modified by the introduction of European forage grasses and alfalfa (*Medicago sativa*), and the eastern tallgrass pampas have been converted to wheat and corn. In Patagonia, where annual rainfall averages about 25 cm, the pampas change to open steppe.

The **velds** of southern Africa (not to be confused with savanna) occupy the eastern part of a high plateau 1500 to 2000 m above sea level in the Transvaal and the Orange Free State.

Australia has four types of grasslands: arid tussock grassland in the northern part of the continent, where the rainfall averages between 20 and 50 cm, mostly in the summer; arid hummock grasslands in areas with less than 20 cm rainfall; coastal grasslands in the tropical summer rainfall region; and subhumid grasslands along coastal areas where annual rainfall is between 50 and 100 cm. However, the introduction of fertilizers, nonnative grasses, legumes, and sheep grazing have changed most of these grasslands.

Grasslands support a diversity of animal life dominated by herbivorous species, both invertebrate and vertebrate. Large grazing ungulates and burrowing mammals are the most conspicuous vertebrates (**Figure 23.17**). The North American grasslands were once dominated by huge migratory herds of millions of bison (*Bison bison*) and the forb-consuming pronghorn antelope (*Antilocarpa americana*). The most common burrowing rodent was the prairie dog (*Cynomys* spp.), which along with gophers (*Thomomys* and *Geomys* spp.) and the mound-building harvester ants (*Pogonomyrex* spp.), appeared to be instrumental in developing and maintaining the ecological structure of the shortgrass prairie.

The Eurasian steppes and the Argentine pampas lack herds of large ungulates. On the pampas, the two major large herbivores are the pampas deer (*Ozotoceros bezoarticus*), and, farther south, the guanaco (*Lama guanicoe*), a small relative of the camel. These species, however, are greatly reduced in number compared with the past.

The African grassveld once supported great migratory herds of wildebeest (*Connochaetes taurinus*) and zebra (*Equus* spp.) along with the associated carnivores, the lion (*Panthera leo*), leopard (*Panthera pardus*), and hyena (*Crocuta crocuta*). The great ungulate herds have been destroyed and replaced with sheep, cattle, and horses.

Many forms of Australian marsupial mammals evolved that are the ecological equivalents of placental grassland mammals. The dominant grazing animals are several kangaroo species, especially the red kangaroo (*Macropus rufus*) and the gray kangaroo (*Macropus giganteus*).

Grasslands evolved under the selective pressure of grazing. Thus, up to a point, grazing stimulates primary production. Although the most conspicuous grazers are large herbivores, the major consumers in grassland ecosystems are invertebrates. The heaviest consumption takes place belowground, where the dominant herbivores are nematodes.

The most visible feature of grassland is the tall, green, ephemeral herbaceous growth that develops in spring and dies back in autumn. One of the three strata in the grassland, it arises from the crowns, nodes, and rosettes of plants hugging the soil. The ground layer and the belowground root layer are the other two major strata of grasslands. The highly developed root layer can make up more than half the total plant biomass and typically extends fairly deep into the soil.

Depending on their history of fire and degree of grazing and mowing, grasslands accumulate a layer of mulch that retains moisture and, with continuous turnover of fine roots, adds organic matter to the mineral soil. Dominant soils of the grasslands are mollisols with a relatively thick, dark-brown to black surface horizon that is rich in organic matter (see Figure 4.12). Soils typically become thinner and paler in the drier regions because less organic material is incorporated into the surface horizon.

Figure 23.17 North American grasslands were once dominated by (a) large grazing ungulates such as bison and (b) burrowing mammals such as the prairie dog.

(a)

(b)

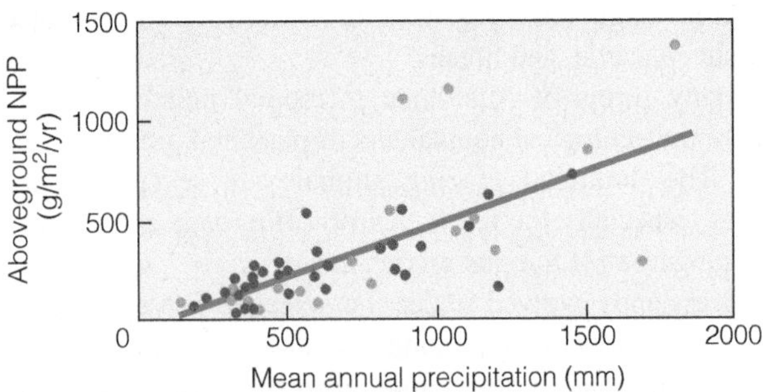

Figure 23.18 Relationship between aboveground net primary production (NPP) and mean annual precipitation for 52 grassland sites around the world. Each point represents a different grassland site. North American grasslands are indicated by dark-green dots. (Adapted from Lauenroth 1979.)

The productivity of temperate grassland ecosystems is primarily related to annual precipitation (**Figure 23.18**), yet temperature can complicate this relationship. Increasing temperatures have a positive effect on photosynthesis but can actually reduce productivity by increasing the demand for water.

23.5 Deserts Represent a Diverse Group of Ecosystems

The arid regions of the world occupy from 25 to 35 percent of Earth's landmass (**Figure 23.19**). The wide range reflects the various approaches used to define desert ecosystems based on climate conditions and vegetation types. Much of this land lies between 15° and 30° latitude, where the air that is carried aloft along the Intertropical Convergence Zone subsides to form the semipermanent high-pressure cells that dominate the climate of tropical deserts (see Figure 2.17). The warming of the air as it descends in addition to cloudless skies result in intense radiation heat during the summer months.

Temperate deserts lie in the rain shadow of mountain barriers or are located far inland, where moist maritime air rarely penetrates. Here, temperatures are high during the summer but can drop to below freezing during the winter months. Thus, the lack of precipitation, rather than continually high temperature, is the distinctive characteristic of all deserts.

Most of the arid environments are found in the Northern Hemisphere. The Sahara, the world's largest desert, covers approximately 9 million km^2 of North Africa. It extends the breadth of the African continent to the deserts of the Arabian Peninsula, continuing eastward to Afghanistan and Pakistan and finally terminating in the Thar Desert of northwest India. The temperate deserts of Central Asia lie to the north. The most westerly of these is the Kara Kum desert region of Turkmenistan. Eastward lie the high-elevation deserts of western China and the high plateau of the Gobi Desert.

A similar transition to temperate desert occurs in western North America. Here, the Sierra Nevada effectively blocks the passage of moist air into the interior of the Southwest. Mountain ranges run parallel to the Sierras throughout the northern part of this region, and desert basins occur on the eastern sides of these ranges.

Apart from the drier parts of southern Argentina, the deserts of the Southern Hemisphere all lie within the subtropical high-pressure belt that mirrors that of the Northern Hemisphere (see preceding discussion). Cold ocean currents also contribute

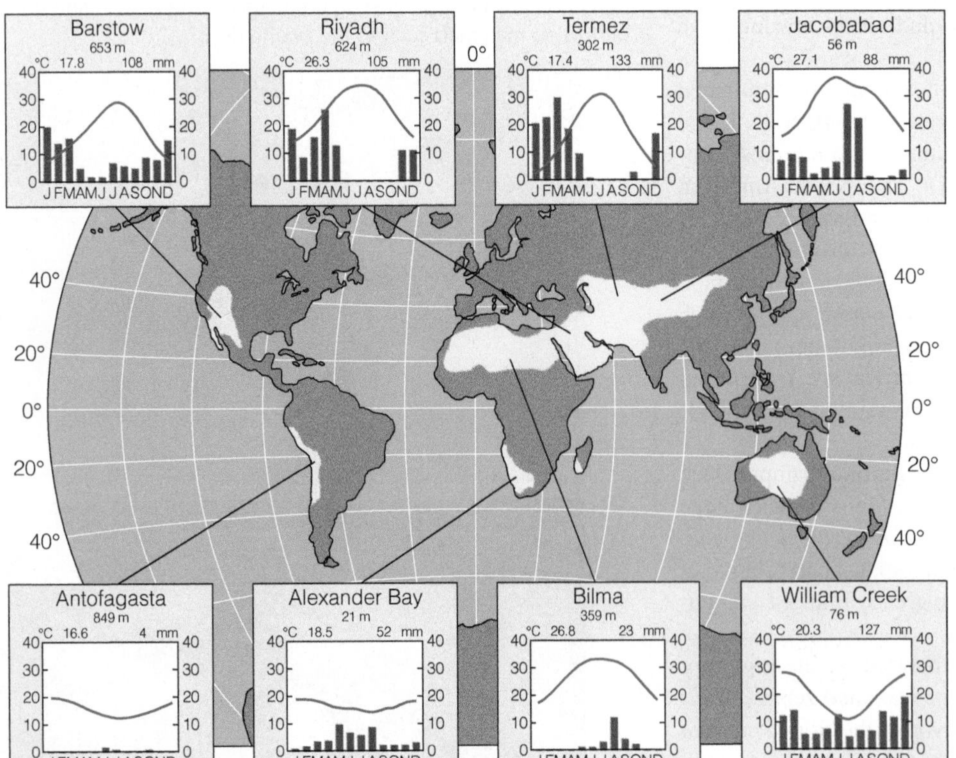

Figure 23.19 Geographic distribution of Earth's desert (arid) ecosystems and associated climate diagrams showing long-term patterns of monthly temperature and precipitation for selected locations. Seasonal variations in temperature differ among these representative sites; but at all sites, annual precipitation is well below the potential evaporative demand, resulting in a low level of soil moisture available to provide for primary productivity. (Adapted from Archibold 1995.)

(a)

(b)

Figure 23.20 Two examples of desert scrub. (a) Northern desert shrubland in Wyoming is dominated by sagebrush (*Artemisia*). Although classified as a cool desert plant, sagebrush forms one of the most important shrub types in North America. (b) Saltbrush shrubland in Victoria, Australia, is dominated by *Atriplex* and is an ecological equivalent of the Great Basin shrublands in North America.

to the development of arid coastal regions (see Section 2.4). Drought conditions are severe along a narrow strip of the coast that includes Chile and Peru. The drier parts of Argentina lie in the rain shadow of the Andes.

The deserts of southern Africa include three regions. The Namib Desert occupies a narrow strip of land that runs along the west coast of Africa from southern Angola to the border of the cape region of South Africa. This region continues south and east across South Africa as the Karoo, which merges with the Kalahari Desert to the north in Botswana. The most extensive region of arid land in the Southern Hemisphere is found in Australia, where more than 40 percent of the land is classified as desert.

Deserts are not the same everywhere. Differences in moisture, temperature, soil drainage, topography, alkalinity, and salinity create variations in vegetation cover, dominant plants, and groups of associated species. There are hot deserts and cold deserts, extreme deserts and semideserts, ones with enough moisture to verge on being grasslands or shrublands, and gradations between those extremes within continental deserts.

Cool deserts—including the Great Basin of North America, the Gobi, Takla Makan, and Turkestan deserts of Asia—and high elevations of hot deserts are dominated by *Artemisia* and chenopod shrubs (**Figure 23.20**). They may be considered shrub steppes or desert scrub. In the Great Basin of North America, the northern, cool, arid region lying west of the Rocky Mountains is the northern desert scrub. The climate is continental, with warm summers and prolonged cold winters. The vegetation falls into two main associations: one is sagebrush, dominated by *Artemisia tridentata*, which often forms pure stands; the other is shadscale (*Atriplex confertifolia*), a C_4 species, and other chenopods (halophytes—tolerant of saline soils).

A similar type of desert scrub exists in the semiarid inland of southwestern Australia. Many chenopod species, particularly the saltbushes of the genera *Atriplex* and *Maireana*, form extensive low shrublands on low riverine plains.

The hot deserts range from those lacking vegetation to ones with some combination of chenopods, dwarf shrubs, and succulents (**Figure 23.21**). Creosote bush (*Larrea divaricata*) and bur sage (*Franseria* spp.) dominate the deserts of southwestern North America—the Mojave, the Sonoran, and the Chihuahuan. Areas of favorable moisture support tall growths of *Acacia* spp., saguaro (*Cereus giganteus*), palo verde (*Cercidium* spp.), ocotillo (*Fouquieria* spp.), yucca (*Yucca* spp.), and ephemeral plants.

Figure 23.21 Two examples of hot deserts. (a) The Chihuahuan Desert in Nuevo Leon, Mexico. The substrate of this desert is sand-sized particles of gypsum. (b) Dunes in the Saudi Arabian desert near Riyadh. Note the extreme sparseness of vegetation.

(a)

(b)

Figure 23.22 A spadefoot toad, named for the black, sharp-edged "spades" on its hind feet, emerges from its desert burrow to breed when the rains come.

Both plants and animals adapt to the scarcity of water by either drought evasion or drought resistance. Drought-evading plants flower only when moisture is present. They persist as seeds during drought periods, ready to sprout, flower, and produce seeds when moisture and temperature are favorable. If no rains come, these ephemeral species do not germinate and grow.

Drought-evading animals, like their plant counterparts, adopt an annual cycle of activities or go into estivation or some other dormant stage during the dry season. For example, the spadefoot toad (*Scaphiopus*; **Figure 23.22**) remains underground in a gel-lined underground cell, making brief reproductive appearances during periods of winter and summer rains. If extreme drought develops during the breeding season, many animals such as lizards and birds do not reproduce.

Desert plants may be deep-rooted woody shrubs, such as mesquite (*Prosopis* spp.) and *Tamarix*, whose taproots reach the water table, rendering them independent of water supplied by rainfall. Some plants, such as *Larrea* and *Atriplex*, are deep-rooted perennials with superficial laterals that extend as far as 15 to 30 m from the stems. Other perennials, such as the various species of cactus, have shallow roots that often extend no more than a few centimeters below the surface.

Despite their aridity, desert ecosystems support a surprising diversity of animal life, including a wide assortment of beetles, ants, locusts, lizards, snakes, birds, and mammals. The mammals are mostly herbivorous species. Grazing herbivores of the desert tend to be generalists and opportunists in their mode of feeding. They consume a wide range of species, plant types, and parts. Desert rodents—particularly the family Heteromyidae—and ants feed largely on seeds and are important in the dynamics of desert ecosystems. Seed-eating herbivores can eat up to 90 percent of the available seeds. That consumption can distinctly affect plant composition and plant populations. Desert carnivores, such as foxes and coyotes, have mixed diets that include leaves and fruits; even insectivorous birds and rodents eat some plant material. Omnivory, rather than carnivory and complex food webs, seems to be the rule in desert ecosystems.

The infrequent rainfall coupled with high rates of evaporation limit the availability of water to plants, so primary productivity is low. Most desert soils are poorly developed aridisols and entisols, and the sparse cover of arid lands limits the ability of vegetation to heavily modify the soil environment (see Figure 4.12). Underneath established plants, however, "islands of fertility" can develop because of higher litter input and the enrichment by wastes from animals that seek shade, particularly under shrubs.

23.6 Mediterranean Climates Support Temperate Shrublands

Shrublands—plant communities where the shrub growth form is either dominant or codominant—are difficult types of ecosystems to categorize, largely because of the difficulty in characterizing the term shrub itself. In general, a shrub is a plant with multiple woody, persistent stems but no central trunk and a height from 4.5 to 8 m. However, under severe environmental conditions, even many trees do not exceed that size. Some trees—particularly individuals that coppice (resprout from the stump) after destruction of the aboveground tissues by fire, browsing, or cutting—are multistemmed, and some shrubs can have large, single stems. In addition, the shrub growth form can be a dominant component of a variety of tropical and temperate ecosystems, including the tropical savannas and scrub desert communities (see Section 23.3, respectively). However, in five widely disjunct regions along the western margins of the continents, between 30° and 40° latitude, are found the mediterranean ecosystems dominated by evergreen shrubs and sclerophyllous trees that have adapted to the distinctive climate of summer drought and cool, moist winters.

The five regions of mediterranean ecosystems include the semiarid regions of western North America, the regions bordering the Mediterranean Sea, central Chile, the cape region of South Africa, and southwestern and southern Australia (**Figure 23.23**). The mediterranean climate has hot, dry summers, with at least one month of protracted drought, and cool, moist winters (see representative climate diagrams in Figure 23.23). About 65 percent of the annual precipitation falls during the winter months. Winter temperatures typically average 10–12°C with a risk of frost. The hot, dry summer climates of the mediterranean regions arise from the seasonal change in the semipermanent high-pressure zones that are centered over the tropical deserts at about 20° N and 20° S (see discussion in Section 23.5). The persistent flow of dry air out of these regions during the summer brings several months of hot, dry weather. Fire is a frequent hazard during these periods.

All five regions support similar-looking communities of xeric broadleaf evergreen shrubs and dwarf trees known as sclerophyllous (*scleros,* "hard"; *phyll,* "leaf") vegetation with a herbaceous understory. Sclerophyllous vegetation possesses small leaves, thickened cuticles, glandular hairs, and sunken stomata—all characteristics that function to reduce water loss during the hot, dry summer period (**Figure 23.24**). Vegetation in each of the mediterranean systems also shares adaptations to fire and to low nutrient levels in the soil.

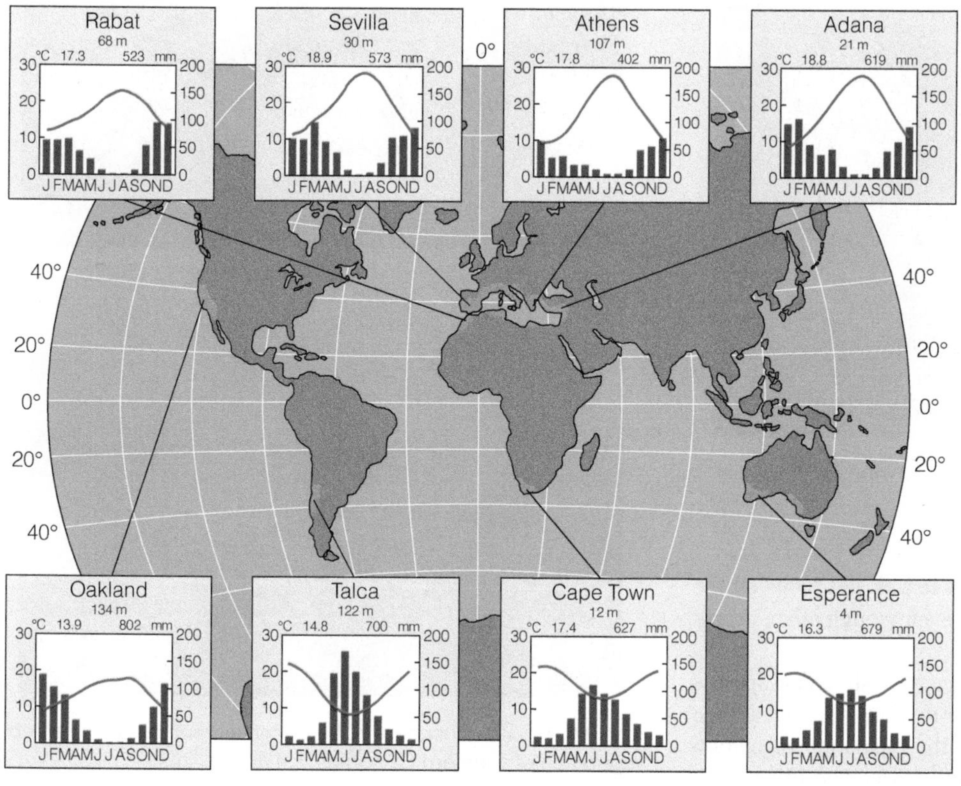

Figure 23.23 Geographic distribution of Earth's mediterranean ecosystems and associated climate diagrams showing long-term patterns of monthly temperature and precipitation for selected locations. Note that these ecosystems are characterized by a winter rainy season and dry summers.
(Adapted from Archibold 1995.)

The largest area of mediterranean ecosystem forms a discontinuous belt around the Mediterranean Sea in southern Europe and North Africa. Much of the area is currently or was once dominated by mixed evergreen woodland supporting species such as holm oak (*Quercus ilex*) and cork oak (*Quercus suber*). Often, these two species grow in mixed stands in association with strawberry tree (*Arbutus unedo*) and various species of shrubs. The easternmost limit of these ecosystems is in the coastal areas of Syria, Lebanon, and Israel, where they grade into the arid lands of the Middle East. Here, deciduous oak species are more abundant. Desert vegetation extends across North Africa as far as Tunisia, with mediterranean shrub and woodland extending through the northern coastal areas of Algeria and Morocco.

The mediterranean zone in southern Africa is restricted to the mountainous region of the Cape Province, where the vegetation is known as *fynbos*. The vegetation is composed primarily of broadleaf proteoid (Proteaceae) and ericoid (Ericaceae) shrubs that grow to a height of 1.5 to 2.5 m (**Figure 23.25**). In southwest Australia, the mediterranean shrub community known as mallee is dominated by low-growing *Eucalyptus*, 5 to 8 m in height, with broad sclerophyllous leaves.

In North America, the sclerophyllous shrub community is known as chaparral, a word of Spanish origin meaning a thicket of shrubby evergreen oaks (**Figure 23.26**). California chaparral, dominated by scrub oak (*Quercus berberidifolia*) and chamise (*Adenostoma fasciculatum*), is evergreen, winter-active, and summer-dormant. Another shrub type, also designated as chaparral, is found in the Rocky Mountain foothills. Dominated by Gambel oak (*Quercus gambelii*), it is winter-deciduous.

The matorral shrub communities of central Chile occur in the coastal lowlands and on the west-facing slopes of the Andes. Most of the matorral species are evergreen shrubs 1 to 3 m in height with small sclerophyllous leaves, although drought-deciduous shrubs are also found.

For the most part, mediterranean shrublands lack an understory and ground litter and are highly flammable. Many species have seeds that require the heat and scarring action of fire to induce germination. Without fire, chaparral grows taller and

(a)

(b)

(c)

Figure 23.24 Sclerophyllous leaves of some tree and shrub species inhabiting mediterranean shrublands (chaparral) of California: (a) chamise (*Adenostoma fasciculatum*), (b) scrub oak (*Quercus dumosa*), and (c) chinquapin (*Chrysolepis sempervirens*).

Figure 23.25 Mediterranean vegetation (fynbos) of the Western Cape region of South Africa.

Figure 23.26 Chaparral is the dominant mediterranean shrub vegetation of southern California.

denser, building up large fuel loads of leaves and twigs on the ground. In the dry season the shrubs, even though alive, nearly explode when ignited.

After fire, the land returns either to lush green sprouts coming up from buried root crowns or to grass if a seed source is nearby. As the regrowth matures, the chaparral vegetation once again becomes dense, the canopy closes, the litter accumulates, and the stage is set for another fire.

Shrub communities have a complex of animal life that varies with the region. In the mediterranean shrublands, similarity in habitat structure has resulted in pronounced parallel and convergent evolution among bird species and some lizard species, especially between the Chilean mattoral and the California chaparral. In North America, chaparral and sagebrush communities support mule deer (*Odocoileus hemionus*), coyotes (*Canis latrans*), a variety of rodents, jackrabbits (*Lepus* spp.), and sage grouse (*Centrocercus urophasianus*). The Australian mallee is rich in birds, including the endemic mallee fowl (*Leipoa ocellata*), which incubates its eggs in a large mound. Among the mammalian life are the gray kangaroo (*Macropus giganteus*) and various species of wallaby (Macropodidae).

The diverse topography and geology of the mediterranean environments give rise to a diversity of soil conditions, but soils are typically classified as alfisols (see Figure 4.12). The soils of the regions are generally deficient in nutrients, and litter decomposition is limited by low temperatures during the winter and low soil moisture during the summer months. These ecosystems vary in productivity depending on the annual precipitation and the severity of summer drought.

23.7 Forest Ecosystems Dominate the Wetter Regions of the Temperate Zone

Climatic conditions in the humid midlatitude regions give rise to the development of forests dominated by broadleaf deciduous trees (**Figure 23.27**). But in the mild, moist climates of the Southern Hemisphere, temperate evergreen forests become predominant. Deciduous forest once covered large areas of Europe and China, parts of North and South America, and the highlands of Central America. The deciduous forests of Europe and Asia, however, have largely disappeared, cleared over the centuries for agriculture. In eastern North America, the deciduous forest consists of several forest types or associations (**Figure 23.28**), including the mixed mesophytic forest of the unglaciated Appalachian plateau, the beech–maple and northern hardwood forests (with pine and hemlock) in northern regions that eventually grade into the boreal forest (see Section 24.8), the maple–basswood forests of the Great Lakes states, the oak–chestnut (now oak since the die-off of the American chestnut) or central hardwood forests, which cover most of the Appalachian Mountains, the magnolia–oak forests of the Gulf Coast states, and the oak–hickory forests of the Ozarks. In North America, temperate deciduous forests reach their greatest development in the mesic forests of the central Appalachians, where the number of tree species is unsurpassed by any other temperate area in the world.

The Asiatic broadleaf forest, found in eastern China, Japan, Taiwan, and Korea, is similar to the North American deciduous forest and contains several plant species of the same genera as those found in North America and western Europe. However, broadleaf evergreen species become increasingly present in Japan, South Korea, and southern China and in the wet foothills of the Himalayas. In southern Europe, their presence reflects the transition into the mediterranean region. Evergreen oaks and pines are also widely distributed in the

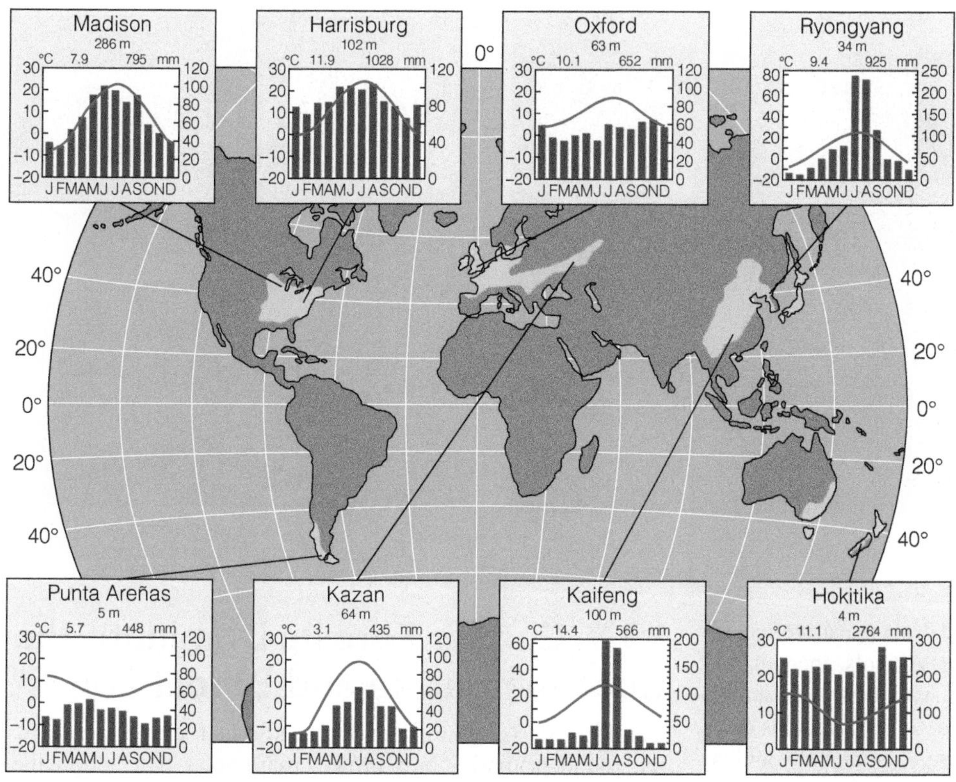

Figure 23.27 Geographic distribution of Earth's temperate forest ecosystems and associated climate diagrams showing long-term patterns of monthly temperature and precipitation for selected locations. These ecosystems are characterized by distinct seasonality in temperature with adequate precipitation during the growing season to support a closed canopy of trees.
(Adapted from Archibold 1995.)

southeastern United States, where they are usually associated with poorly developed sandy or swampy soils.

In the Southern Hemisphere, temperate deciduous forests are found only in the drier parts of the southern Andes. In southern Chile, broadleaf evergreen rain forests have developed in an oceanic climate that is virtually frost-free. Evergreen forests are also found in New Zealand, Tasmania, and parts of southeastern Australia where the winter temperatures are moderated by the coastal environment. Climate regions in these areas are similar to those of the Pacific Northwest of North America, but here the predominant species are conifers.

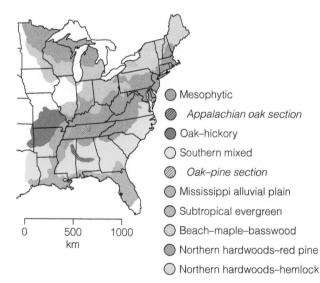

Figure 23.28 Large-scale distribution of temperate forest communities in the eastern United States, derived from contemporary data. Compare with Figure 17.11, depicting the original forest pattern.
(Adapted from Dyer 2006.)

In the broadleaf deciduous forests of the temperate region, the end of the growing season is marked by the autumn colors of foliage shortly before the trees enter into their leafless winter period (**Figure 23.29**). The trees resume growth in the spring in response to increasing temperatures and longer day lengths. Many herbaceous species flower at this time before the developing canopy casts a heavy shade on the forest floor.

Highly developed, unevenly aged deciduous forests usually have four vertical layers or strata (see Figure 16.12a). The upper canopy consists of the dominant tree species, below which is the lower tree canopy, or understory. Next is the shrub layer, followed finally by the ground layer of herbs, ferns, and mosses. The diversity of animal life is associated with this vertical stratification and the growth forms of plants (see Figure 16.13). Some animals, particularly forest arthropods, spend most of their lives in a single stratum; others range over two or more strata. The greatest concentration and variety of life in the forest occurs on and just below the ground layer. Many animals—the soil and litter invertebrates in particular—remain in the subterranean stratum. Others, such as mice, shrews, ground squirrels, and forest salamanders, burrow into the soil or litter for shelter and food. Larger mammals live on the ground layer and feed on herbs, shrubs, and low trees. Birds move rather freely among several strata but typically favor one layer over another (see Figure 16.13).

Figure 23.29 A temperate forest of the Appalachian region: (a) the canopy during autumn, and (b) interior of the forest during spring. The forest is dominated by oaks (*Quercus* spp.) and yellow poplar (*Liriodendron tulipifera*), with an understory of redbud (*Cercis canadensis*) in bloom.

Differences in climate, bedrock, and drainage are reflected in the variety of soil conditions present. Alfisols, inceptisols, and ultisols are the dominant soil types with alfisols typically associated with glacial materials in more northern regions (see Figure 4.12). Primary productivity varies geographically and is influenced largely by temperatures and the length of the growing season (see Section 20.3). Leaf fall in deciduous forests occurs over a short period in autumn, and the availability of nutrients is related to rates of decomposition and mineralization (see Chapter 21).

23.8 Conifer Forests Dominate the Cool Temperate and Boreal Zones

Conifer forests, dominated by needle-leaf evergreen trees, are found primarily in a broad circumpolar belt across the Northern Hemisphere and on mountain ranges, where low temperatures limit the growing season to a few months each year (**Figure 23.30**). The variable composition and structure of these forests reflect the wide range of climatic conditions in which they grow. In central Europe, extensive coniferous

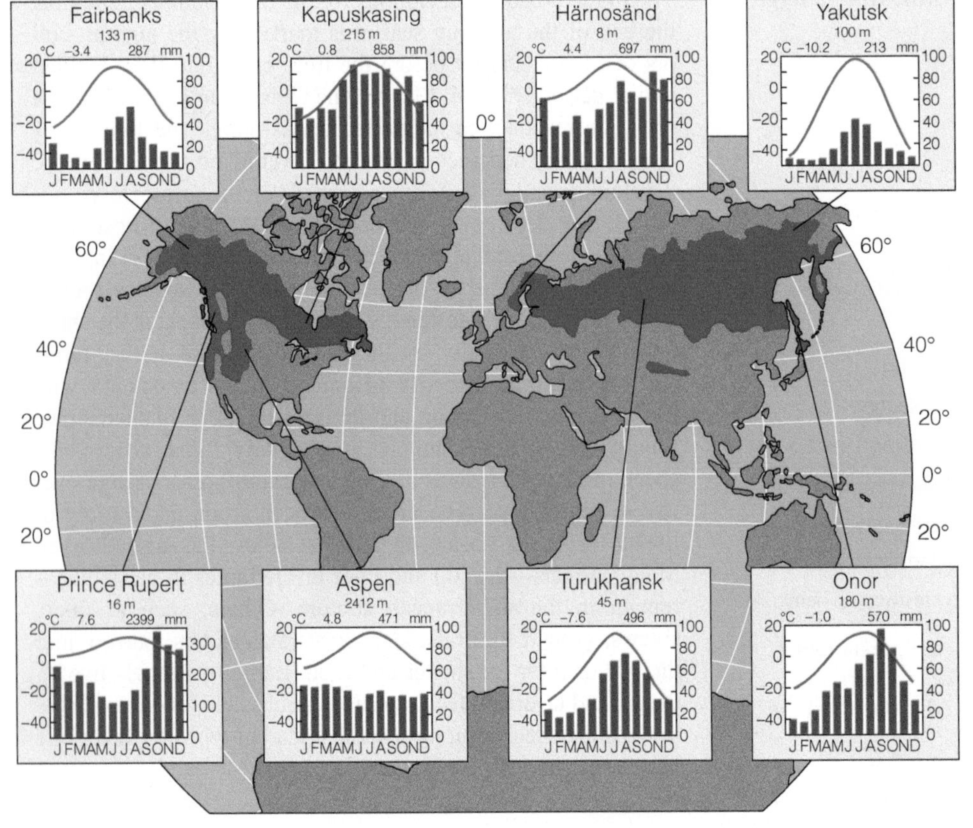

Figure 23.30 Geographic distribution of Earth's conifer forest ecosystems and associated climate diagrams showing long-term patterns of monthly temperature and precipitation for selected locations. Regions supporting conifer forest ecosystems have a lower mean annual temperature and shorter growing season than areas that support temperate deciduous forest (see Figure 23.27).
(Adapted from Archibold 1995.)

Figure 23.31 Two coniferous forest types. (a) A Norway spruce in the Tarvisio region of Italy. (b) A montane coniferous forest in the Rocky Mountains. The dry, lower slopes support ponderosa pine; the upper slopes are cloaked with Douglas fir.

forests, dominated by Norway spruce (*Picea abies*), cover the slopes up to the subalpine zone in the Carpathian Mountains and the Alps (**Figure 23.31a**). In North America, several coniferous forests blanket the Rocky, Wasatch, Sierra Nevada, and Cascade mountains. At high elevations in the Rocky Mountains grows a subalpine forest dominated by Engelmann spruce (*Picea engelmannii*) and subalpine fir (*Abies lasiocarpa*). Middle elevations have stands of Douglas fir, and lower elevations are dominated by open stands of ponderosa pine (*Pinus ponderosa*; **Figure 23.31b**) and dense stands of the early successional conifer, lodgepole pine (*Pinus contorta*). The largest tree of all, the giant sequoia (*Sequoiadendron giganteum*), grows in scattered groves on the western slopes of the California Sierra. In addition, the mild, moist climate of the Pacific Northwest supports a highly productive coastal forest extending along the coastal strip from Alaska to northern California.

The largest expanse of conifer forest—in fact, the largest vegetation formation on Earth—is the boreal forest, or taiga (Russian for "land of little sticks"). This belt of coniferous forest, encompassing the high latitudes of the Northern Hemisphere, covers about 11 percent of Earth's terrestrial surface (see Figure 23.30). In North America, the boreal forest covers much of Alaska and Canada and spills into northern New England, with fingers extending down the western mountain ranges and into the Appalachians. In Eurasia, the boreal forest begins in Scotland and Scandinavia and extends across the continent, covering much of Siberia, to northern Japan.

Three major vegetation zones make up the taiga (**Figure 23.32**): (1) the forest–tundra ecotone with open stands

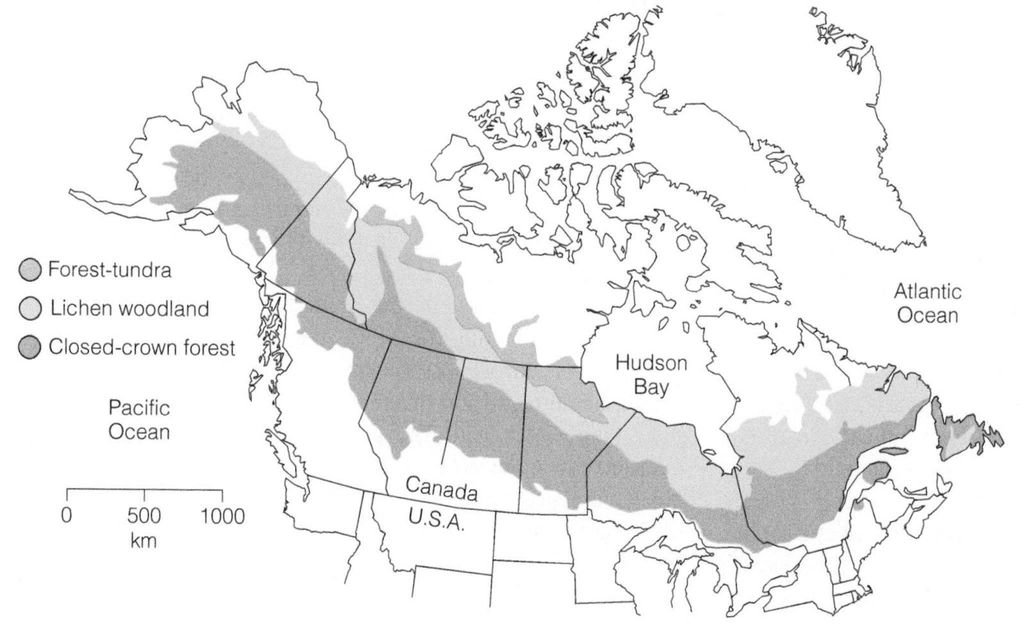

Figure 23.32 Major subdivisions of the North American boreal forest. (Adapted from Payette et al. 2001.)

Figure 23.33 Black spruce is a dominant conifer in the North American taiga.

of stunted spruce, lichens, and moss; (2) the open lichen woodland with stands of lichens and black spruce; and (3) the main boreal forest (**Figure 23.33**) with continuous stands of spruce and pine broken by poplar and birch on disturbed areas. This boreal–mixed forest grades into the temperate forest of southern Canada and the northern United States. Primarily occupying formerly glaciated land, the taiga is also a region of cold lakes, bogs, rivers, and alder thickets.

A cold continental climate with strong seasonal variation dominates the taiga. The summers are short, cool, and moist; the winters are long, harsh, and dry, with a prolonged period of snowfall. The driest winters and the greatest seasonal fluctuations are in interior Alaska and central Siberia, which experience seasonal temperature extremes (differences between minimum and maximum annual temperatures) of as much as 100°C.

Much of the taiga is under the controlling influence of permafrost, which impedes infiltration and maintains high soil moisture. Permafrost is the perennially frozen subsurface that may be hundreds of meters deep. It develops where the ground temperatures remain below 0°C for extended periods of time. Its upper layers may thaw in summer and refreeze in winter. Because the permafrost is impervious to water, it forces all water to remain and move above it. Thus, the ground stays soggy even though precipitation is low, enabling plants to exist in the driest parts of the Arctic.

Fires are recurring events in the taiga. During periods of drought, fires can sweep over hundreds of thousands of hectares. All of the boreal species, both broadleaf trees and conifers, are well adapted to fire. Unless too severe, fire provides a seedbed for regeneration of trees. Light surface burns favor early successional hardwood species. More severe fires eliminate hardwood competition and favor spruce and jack pine regeneration.

Because of the global demand for timber and pulp, vast areas of the boreal forest across North America and Siberia are being clear-cut with little concern for their future, see this chapter, *Ecological Issues & Applications*. This exploitation can alter the nature and threaten the survival of the boreal forest.

The boreal forest has a unique animal community. Caribou (*Rangifer tarandus*), wide-ranging and feeding on grasses, sedges, and especially lichens, inhabit open spruce–lichen woodlands. Joining the caribou is the moose (*Alces alces*), called elk in Eurasia, the largest of all deer. It is a lowland mammal feeding on aquatic and emergent vegetation as well as alder and willow. Competing with moose for browse is the cyclic snowshoe hare (*Lepus americanus*). The arboreal red squirrel (*Sciurus hudsonicus*) inhabits the conifers and feeds on young pollen-bearing cones and seeds of spruce and fir, and the quill-bearing porcupine (*Erethizon dorsatum*) feeds on leaves, twigs, and the inner bark of trees. Preying on these is an assortment of predators including the wolf, lynx (*Lynx canadensis* and *L. lynx*), pine martin (*Martes americana*), and owls. The taiga is also the nesting ground of migratory neotropical birds and the habitat of northern seed-eating birds such as crossbills (*Loxia* spp.), grosbeaks (*Coccothraustes* spp.), and siskins (*Carduelis* spp.).

Of great ecological and economic importance are major herbivorous insects such as the spruce budworm (*Choristoneura fumiferana*). Although they are major food items for insectivorous summer birds, these insects experience periodic outbreaks during which they defoliate and kill large expanses of forest.

Compared to more temperate forests, boreal forests have generally low net primary productivity; they are limited by low nutrients, cooler temperatures, and the short growing season. Likewise, inputs of plant litter are low compared to the forests of the warmer temperate zone. However, rates of decomposition are slow under the cold, wet conditions, resulting in the accumulation of organic matter. Soils are primarily spodosols characterized by a thick organic layer (see Figure 4.12). The mineral soils beneath mature coniferous forests are comparatively infertile, and growth is often limited by the rate at which mineral nutrients are recycled through the ecosystem.

23.9 Low Precipitation and Cold Temperatures Define the Arctic Tundra

Encircling the top of the Northern Hemisphere is a frozen plain, clothed in sedges, heaths, and willows, dotted with lakes, and crossed by streams (**Figure 23.34**). Called *tundra*, its name comes from the Finnish *tunturi*, meaning "a treeless plain." The arctic tundra falls into two broad types: tundra with up to 100 percent plant cover and wet to moist soil (**Figure 23.35**), and polar desert with less than 5 percent plant cover and dry soil.

Conditions unique to the Arctic tundra are a product of at least three interacting forces: (1) the permanently frozen deep layer of permafrost; (2) the overlying active layer of organic matter and mineral soil that thaws each summer and freezes the following winter; and (3) vegetation that reduces warming and retards thawing in summer. Permafrost chills the soil, retarding the general growth of plant parts both above- and belowground, limiting the activity of soil microorganisms, and diminishing the aeration and nutrient content of the soil.

Alternate freezing and thawing of the upper layer of soil creates the unique, symmetrically patterned landforms

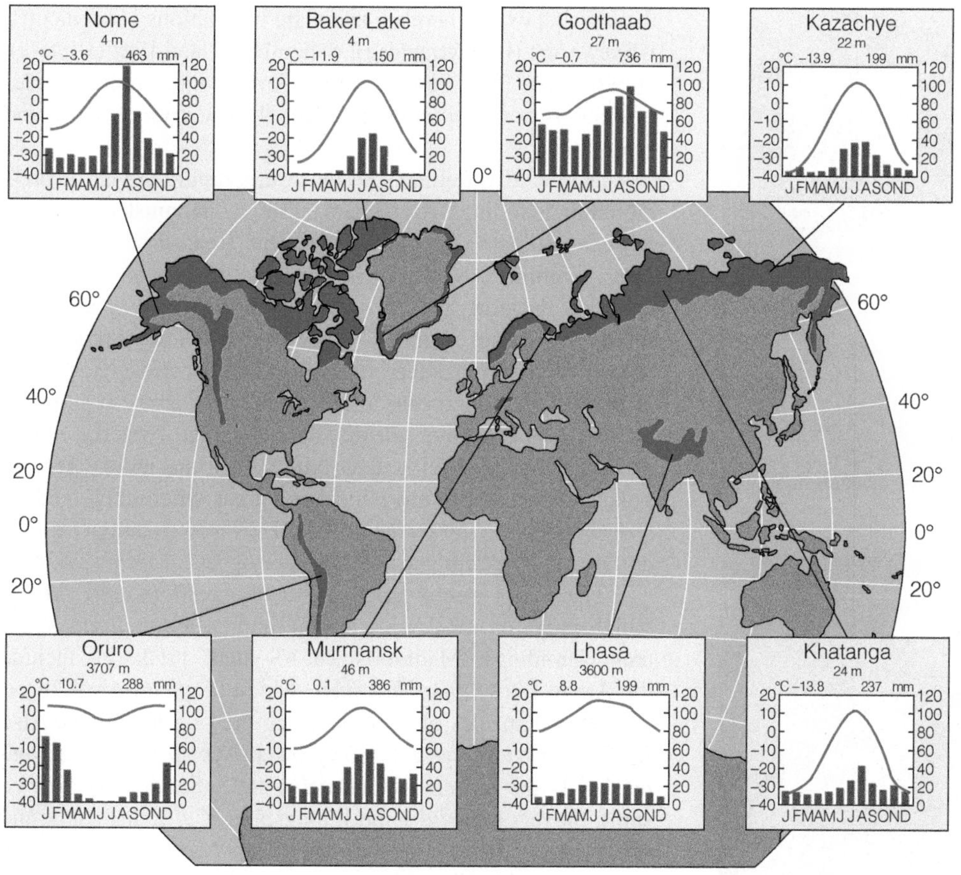

Figure 23.34 Geographic distribution of Earth's tundra ecosystems and associated climate diagrams showing long-term patterns of monthly temperature and precipitation for selected locations. Tundra ecosystems are characterized by lower mean temperatures, a shorter growing season, and lower annual precipitation than regions supporting conifer forest (see Figure 24.30 for comparison). (Adapted from Archibold 1995.)

typical of the tundra (**Figure 23.36**). The frost pushes stones and other material upward and outward from the mass to form a patterned surface of frost hummocks, frost boils, earth stripes, and stone polygons. On sloping ground, creep, frost thrusting, and downward flow of supersaturated soil over the permafrost form solifluction terraces, or "flowing soil." This gradual downward creep of soils and rocks eventually rounds off ridges and other irregularities in topography. Such molding of the landscape by frost action, called cryoplanation, is far more important than erosion in wearing down the Arctic landscape.

Structurally, the vegetation of the tundra is simple. The number of species tends to be low, and growth is slow. Only those species able to withstand constant disturbance of the soil, buffeting by the wind, and abrasion from wind-carried particles of soil and ice can survive. Low ground is covered with a complex of cotton grasses, sedges, and *Sphagnum*. Well-drained sites support heath shrubs, dwarf willows and birches, herbs, mosses, and lichens. The driest and most exposed sites support scattered heaths and crustose and foliose lichens growing on the rock. Arctic plants propagate themselves almost entirely by vegetative means, although viable seeds many hundreds of years old exist in the soil.

Plants are photosynthetically active on the Arctic tundra about three months out of the year. As snow cover disappears, plants commence photosynthetic activity. They maximize use of the growing season and light by photosynthesizing during the 24-hour daylight period, even at midnight when light is

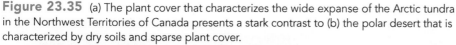

Figure 23.35 (a) The plant cover that characterizes the wide expanse of the Arctic tundra in the Northwest Territories of Canada presents a stark contrast to (b) the polar desert that is characterized by dry soils and sparse plant cover.

(a)

(b)

(a)

(b)

Figure 23.36 Patterned landforms typical of the tundra region: (a) frost hummocks, and (b) polygons. Alternate freezing and thawing of the upper layer of soil creates the symmetrically patterned landforms.

one-tenth that of noon. The nearly erect leaves of some Arctic plants permit almost complete interception of the low angle of the Arctic sun.

Much of the photosynthate goes into the production of new growth, but about one month before the growing season ends, plants cease to allocate photosynthate to aboveground biomass. They withdraw nutrients from the leaves and move them to roots and belowground biomass, sequestering 10 times the amount stored by temperate grasslands.

Structurally, most of the tundra vegetation is underground. Root-to-shoot ratios of vascular plants range from 3:1 to 10:1. Roots are concentrated in the upper soil that thaws during the summer, and aboveground parts seldom grow taller than 30 cm. It is not surprising, then, that the belowground net annual production is typically three times that of the aboveground productivity.

The tundra hosts fascinating animal life, even though the diversity of species is low. Invertebrates are concentrated near the surface, where there are abundant populations of segmented whiteworms (Enchytraeidae), collembolas, and flies (Diptera), chiefly crane flies. Summer in the Arctic tundra brings hordes of black flies (*Simulium* spp.), deer flies (*Chrysops* spp.), and mosquitoes.

Dominant vertebrates on the Arctic tundra are herbivores, including lemmings, Arctic hare, caribou, and musk ox (*Ovibos moschatus*). Although caribou have the greatest herbivore biomass, lemmings, which breed throughout the year, may reach densities as great as 125 to 250 per hectare; they consume three to six times as much forage as caribou do. Arctic hares (*Lepus arcticus*) that feed on willows disperse over the range in winter and congregate in more restricted areas in summer. Caribou are extensive grazers, spreading out over the tundra in summer to feed on sedges. Musk oxen are more intensive grazers, restricted to more localized areas where they feed on sedges, grasses, and dwarf willow. Herbivorous birds are few, dominated by ptarmigan and migratory geese.

The major Arctic carnivore is the wolf (*Canus lupus*), which preys on musk ox, caribou, and, when they are abundant, lemmings. Medium-sized to small predators include the Arctic fox (*Alopex lagopus*), which preys on Arctic hare, and several species of weasel, which prey on lemmings. Also feeding on lemmings are snowy owls (*Nyctea scandiaca*) and the hawk-like jaegers (*Stercorarius* spp.). Sandpipers (*Tringa* spp.), plovers (*Pluvialis* spp.), longspurs (*Calcarius* spp.), and waterfowl, which nest on the wide expanse of ponds and boggy ground, feed heavily on insects.

At lower latitudes, alpine tundra occurs in the higher mountains of the world. The alpine tundra is a severe environment of rock-strewn slopes, bogs, meadows, and shrubby thickets (**Figure 23.37**). It is a land of strong winds, snow, cold, and widely fluctuating temperatures. During summer, the temperature on the surface of the soil ranges from 40 to 0°C. The atmosphere is thin so light intensity, especially ultraviolet, is high on clear days. Alpine tundras have little permafrost, and it is confined mostly to very high elevations. Lacking permafrost, soils are drier. Only in alpine wet meadows and bogs do soil moisture conditions compare with those of the Arctic. Precipitation, especially snowfall and humidity, is higher in the alpine regions than in the Arctic tundra, but steep topography induces a rapid runoff of water.

Figure 23.37 Rocky Mountains alpine tundra.

ECOLOGICAL Issues & Applications

The Extraction of Resources from Forest Ecosystems Involves an Array of Management Practices

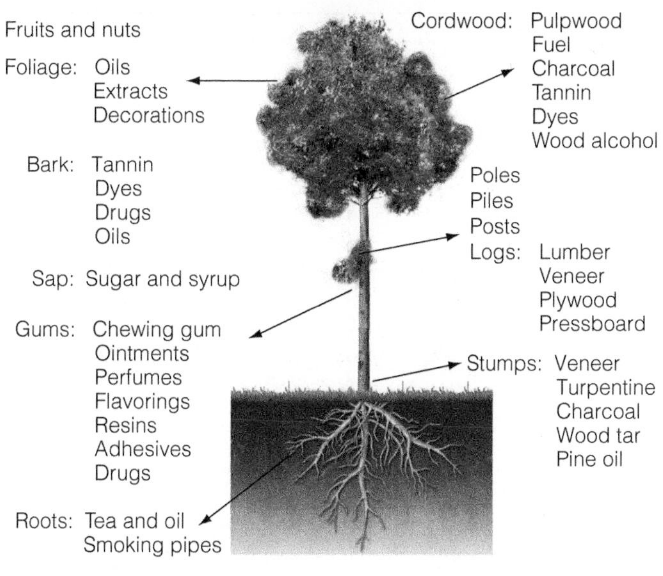

Figure 23.38 A variety of products derived from forests.

Global distribution of original and remaining forests

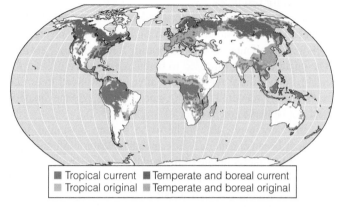

Tropical current | Temperate and boreal current
Tropical original | Temperate and boreal original

Figure 23.39 Globally, about half of the forest that was present under modern (post-Pleistocene) climatic conditions, and before the spread of human influence, has disappeared—largely because of human activities.
(Adapted from United Nations, FAO.)

Forest ecosystems cover approximately 35 percent of Earth's surface and provide a wealth of resources, including fuel, building materials, and food (**Figure 23.38**). Although plantations provide a growing percentage of forest resources, more than 90 percent of global forest resources are still harvested from native forests.

Globally, about half of the forest that was present under modern (post-Pleistocene) climatic conditions—and before the spread of human influence—has disappeared largely through the impact of human activities (**Figure 23.39**). The spread of agriculture and animal husbandry, the harvesting of forests for timber and fuel, and the expansion of populated areas have all taken their toll on forests. The causes and timing of forest loss differ among regions and forest types, as do the current trends in change in forest cover. In the face of increasing demand and declining forest cover, the sustainable production of forest resources requires achieving a balance between net growth and harvest. To achieve this end, foresters have an array of silvicultural and harvesting techniques from clear-cutting to selection cutting.

Clear-cutting involves removing the forest and reverting it to an early stage of succession (**Figure 23.40a**). The area harvested can range from thousands of hectares to small patch cuts of a few hectares designed to create habitat for wildlife species that require an opening within the forest (see Section 19.4). Postharvest management varies widely for clear-cut areas. When natural forest stands are clear-cut, there is generally no follow-up management. Stands are left to regenerate naturally from existing seed and sprouts on the site and the input of seeds from adjacent forest stands. With no follow-up management, clear-cut areas can be badly disturbed by erosion that affects subsequent recovery of the site as well as adjacent aquatic communities.

Figure 23.40 Examples of (a) clear-cut and (b) shelterwood (seed-tree) forest harvest.

(a)

(b)

Harvest by clear-cutting is the typical practice on forest plantations, but here intensive site management follows clearing. Plant materials that are not harvested (branches, leaves, and needles) are typically burned to clear the site for planting. After clearing, seedlings are planted and fertilizer applied to encourage plant growth. Herbicides are often used to discourage the growth of weedy plants that would compete with the seedlings for resources.

The **seed-tree**, or **shelterwood**, system is a method of regenerating a new stand by removing all trees from an area except for a small number of seed-bearing trees (**Figure 23.40b**). The uncut trees are intended to be the main source of seed for establishing natural regeneration after harvest. Seed trees can be uniformly scattered or left in small clumps, and they may or may not be harvested later.

In many ways, the shelterwood system is similar to a clear-cut because generally not enough trees are left standing to affect the microclimate of the harvested area. The advantage of the shelterwood approach is that the seed source for natural regeneration is not limited to adjacent stands. This can result in improved distribution (or stocking) of seedlings as well as a more desirable mix of species.

Like any silviculture system, shelterwood harvesting requires careful planning to be effective. Trees left on the site must be strong enough to withstand winds and capable of producing adequate seed, seedbed conditions must be conducive to seedling establishment (this may require a preparatory treatment during or after harvest), and follow-up management may be required to fully establish the regeneration.

In **selection cutting**, mature single trees or groups of trees scattered through the forest are removed. Selection cutting produces only small openings or gaps in the forest canopy. Although this form of timber harvest can minimize the scale of disturbance within the forest caused by direct removal of trees, the network of trails and roads necessary to provide access can be a major source of disturbance (to both plants and soils). Selective cutting also can cause changes in species composition and diversity because only certain species are selectively removed.

Regardless of the differences in approach, some general principles apply if the harvesting of resources is to be sustainable. Forest trees function in the manner discussed for competition in plant populations (Chapter 11, Section 11.3). Whether a forest is planted as seedlings or grown by natural regeneration, its establishment begins with a population of small individuals (seedlings) that grow and compete for the essential resources of light, water, and nutrients. As biomass in the forest increases, the density of trees decreases and the average tree size increases as a result of self-thinning (**Figure 23.41**; also see Section 11.5, Figure 11.9). For a stand to be considered economically available for harvest (referred to as being in an *operative state*), minimum thresholds must be satisfied for the harvestable volume of timber per hectare and average tree size (see Figure 23.41); these thresholds vary depending on the species. In plantation forestry, for a given set of thresholds (timber volume and average tree size), the initial stand density (planting density) can be controlled to influence the timing of the stand's availability for harvest (**Figure 23.42**).

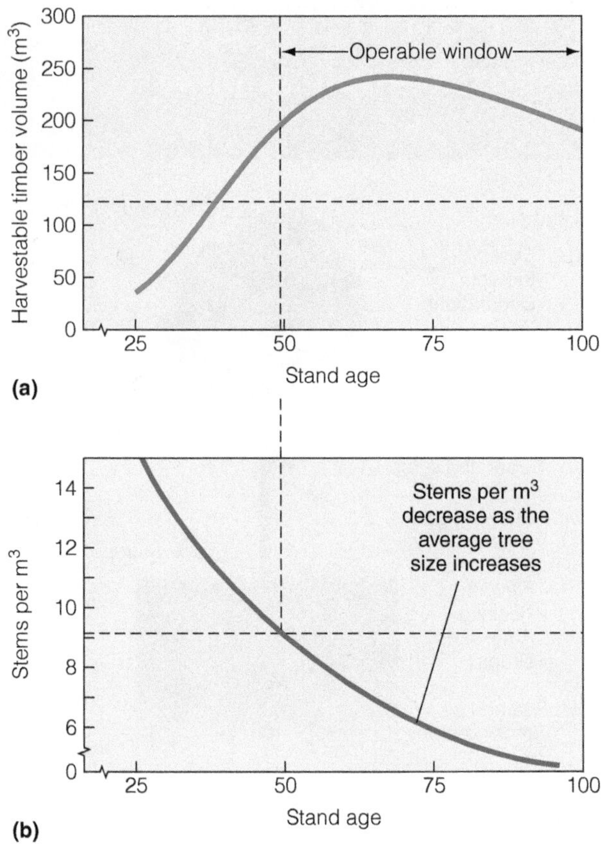

Figure 23.41 Two criteria are used to determine when a stand of trees is suitable for harvest (referred to as the *operable window*): (1) salable volume of wood per hectare (m^3/ha), and (2) average tree size as measured by the stems per cubic meter (the number of trees required to make a cubic meter of wood volume). In the example shown, dashed horizontal lines represent the criteria of (a) minimum salable wood volume of 100 m^3/ha, and (b) minimum average tree size of 9 stems/m^3. Dashed vertical lines indicate the earliest stage at which both criteria are met. (Adapted from Oriens et al. 1986.)

Interpreting Ecological Data

Q1. What does graph (a) imply about the change in average tree size (diameter or height) with stand age?

Q2. Assume that a decision is made to harvest the stand when it reaches the minimum salable wood volume (100 m^3/ha). How many stems (trees) per m^3 would be harvested?

After trees are harvested, a sufficient time must pass for the forest to regenerate. For sustained yield, the time between harvests must be sufficient for the forest to regain the level of biomass it had reached at the time of the previous harvest. Rotation time depends on a variety of factors related to the tree species, site conditions, type of management, and intended use of the trees being harvested. Wood for paper products (pulpwood), fence posts, and poles are harvested from fast-growing

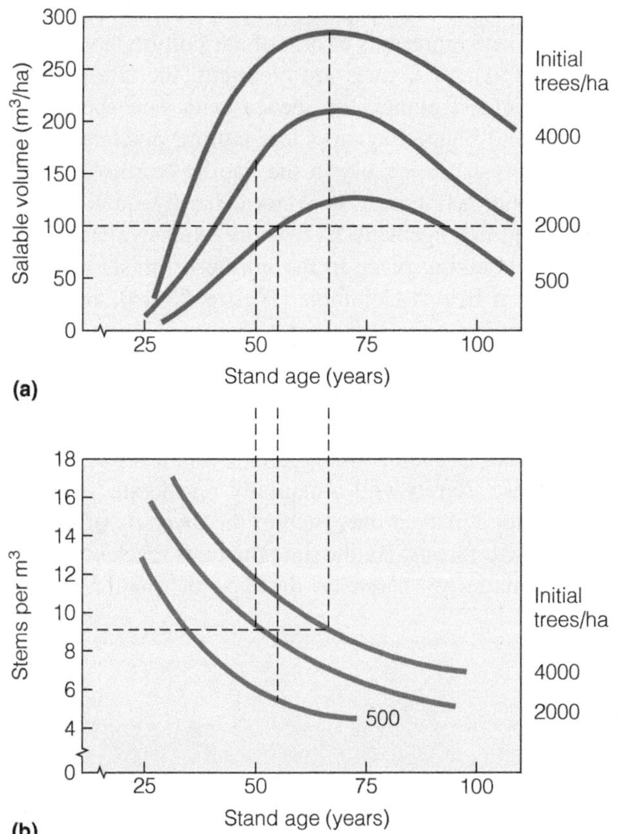

Figure 23.42 Effects of initial stand density on timing of stand availability for harvest (operable window). As in Figure 23.41, dashed horizontal lines represent criteria of (a) minimum marketable wood volume of 100 m^3/ha and (b) a minimum average tree size of 9 stems/m^3. Dashed vertical lines indicate the earliest stage at which both criteria are met. Note that the intermediate planting density of 2000 trees per hectare provides the earliest operable window.
(Adapted from Oriens et al. 1983.)

Interpreting Ecological Data?

Q1. The analysis includes three initial planting densities: 500, 2000, and 4000 trees/ha. At which stand age does each of the three initial planting densities achieve the minimum constraint for average tree size?

Q2. Given the requirements of minimum wood volume and average tree size as defined, which of the initial planting densities meets these requirements at the earliest stand age (earliest operable window)?

species, allowing a short rotation period (15–40 years). These species are often grown in highly managed plantations where trees can be spaced to reduce competition and fertilized to maximize growth rates. Trees harvested for timber (saw logs) require a much longer rotation period. Hardwood species used for furniture and cabinetry are typically slower growing and may have a rotation time of 80 to 120 years. Sustained forestry of these species works best in extensive areas where blocks of land can be maintained in different age classes.

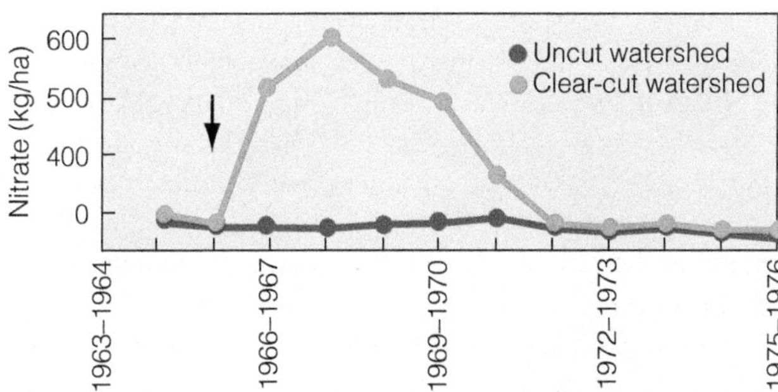

Figure 23.43 Temporal changes in the nitrate concentration of stream water for two forested watersheds in Hubbard Brook, New Hampshire. The forest on one watershed was clear-cut (noted by arrow), and the other forest was undisturbed. Note the large increase in concentrations of nitrate in the stream on the clear-cut watershed. This increase is a result of increased decomposition and nitrogen mineralization after clear-cutting. The nitrogen then leached into the surface water and groundwater.
(Adapted from Likens and Borman 1995.)

As with agricultural crops, a significant amount of nutrients are lost from the forest when trees are harvested and removed (see Chapter 22, *Ecological Issues & Applications*). The loss of nutrients in plant biomass is often compounded by further losses from soil erosion and various postharvest management practices—particularly the use of fire. The reduction of nutrients reduces plant growth, requiring a longer rotation period for subsequent harvests or causing reduced forest yield if the rotation period is maintained. Forest managers often counter the loss of nutrients by using chemical fertilizers, which create other environmental problems for adjacent aquatic ecosystems (see Chapter 24, *Ecological Issues & Applications*).

In addition to the nutrients lost directly through biomass removal, logging can also result in the transport of nutrients from the ecosystem by altering processes involved in internal cycling. The removal of trees in clear-cutting and other forest management practices increases the amount of radiation (including direct sunlight) reaching the soil surface. The resulting increase in soil temperatures promotes decomposition of remaining soil organic matter and causes an increase in net mineralization rates (see Sections 21.4 and 21.5). This increase in nutrient availability in the soil occurs at the same time that demand for nutrients is low because plants have been removed and net primary productivity is low. As a result, there is a dramatic increase in the leaching of nutrients from the soil into ground and surface waters (**Figure 23.43**). This export of nutrients from the ecosystem results from decoupling the two processes of nutrient release in decomposition and nutrient uptake in net primary productivity.

Sustained yield is a key concept in forestry and is practiced to some degree by large timber companies and federal and state forestry agencies. But all too often, industrial forestry's approach to sustained yield is to grow trees as a crop

Figure 23.44 Large-scale clear-cut in British Columbia, Canada.

rather than maintaining a forest ecosystem. Their management approach represents a form of agriculture in which trees are grown as crops: trees are clear-cut, the site is sprayed with herbicides, planted or seeded with one species, then clear-cut and planted again. Clear-cutting practices in some national forests, especially in the Pacific Northwest and the Tongass National Forest in Alaska, hardly qualify as sustained-yield management. Even more extensive clear-cutting of forests is taking place in the northern forests of Canada, especially in British Columbia (**Figure 23.44**), and in large areas of Siberia.

The problem of sustained-yield forestry is its economic focus on the resource with little concern for the forest as a biological community. A carefully managed stand of trees, often reduced to one or two species, is not a forest in an ecological sense. Rarely will a naturally regenerated forest, and certainly not a planted one, support the diversity of life found in old-growth forests. By the time the trees reach economic or financial maturity—based on the type of rotation—they are cut again.

SUMMARY

Ecosystem Distribution and Plant Adaptations 23.1

Terrestrial ecosystems can be grouped into broad categories called *biomes*. Biomes are classified according to the predominant plant types. There are at least eight major terrestrial biome types: tropical forest, temperate forest, conifer forest (taiga or boreal forest), tropical savanna, temperate grasslands, chaparral (shrublands), tundra, and desert. These broad categories reflect the relative contribution of three general plant life-forms: trees, shrubs, and grasses. Interaction between moisture and temperature is the primary factor limiting the nature and geographic distribution of terrestrial ecosystems.

Tropical Forests 23.2

Seasonality of rainfall determines the types of tropical forests. Rain forests, associated with high seasonal rainfall, are dominated by broadleaf evergreen trees. They are noted for their enormous diversity of plant and animal life. The vertical structure of the forest is divided into five general layers: emergent trees, high upper canopy, low tree stratum, shrub understory, and a ground layer of herbs and ferns. Conspicuous in the rain forest are the lianas or climbing vines, epiphytes growing up in the trees, and stranglers growing downward from the canopy to the ground. Many large trees develop buttresses for support. Nearly 90 percent of nonhuman primate species live in tropical rain forests.

Tropical rain forests support high levels of primary productivity. The high rainfall and consistently warm temperatures also result in high rates of decomposition and nutrient cycling.

Dry tropical forests undergo varying lengths of dry season, during which trees and shrubs drop their leaves (drought-deciduous). New leaves are grown at the onset of the rainy season. Most dry tropical forests have been lost to agriculture and grazing and other disturbances.

Tropical Savannas 23.3

Savannas are characterized by a codominance of grasses and woody plants. Such vegetation is characteristic of regions with alternating wet and dry seasons. Savannas range from grass with occasional trees to shrubs to communities where trees form an almost continuous canopy as a function of precipitation and soil texture. Productivity and decomposition in savanna ecosystems are closely tied to the seasonality of precipitation.

Savannas support a large and varied assemblage of both invertebrate and vertebrate herbivores. The African savanna is dominated by a large, diverse population of ungulate fauna and associated carnivores.

Temperate Grasslands 23.4

Natural grasslands occupy regions where rainfall is between 250 and 800 mm a year. Once covering extensive areas of the globe, natural grasslands have shrunk to a fraction of their original size because of conversion to cropland and grazing lands.

Grasslands vary with climate and geography. Native grasslands of North America, influenced by declining precipitation from east to west, consist of tallgrass prairie, mixed-grass prairie, shortgrass prairie, and desert grasslands. Eurasia has steppes; South America, the pampas; and southern Africa, the veld. Grassland consists of an ephemeral herbaceous layer that arises from crowns, nodes, and rosettes of plants hugging the ground. It also has a ground layer and a highly developed root layer. Depending on the history of fire and degree of grazing and mowing, grasslands accumulate a layer of mulch.

Grasslands support a diversity of animal life dominated by herbivorous species, both invertebrate and vertebrate. Grasslands once supported herds of large grazing ungulates such as bison in North America, migratory herds of wildebeest in Africa, and marsupial kangaroos in Australia. Grasslands evolved under the selective pressure of grazing. Although the most conspicuous grazers are large herbivores, the major consumers are invertebrates. The heaviest consumption takes place belowground, where the dominant herbivores are nematodes.

Deserts 23.5

Deserts occupy about one-seventh of Earth's land surface and are largely confined to two worldwide belts between 15° N and 30° S latitude. Deserts result from dry descending air masses within these regions, the rain shadows of coastal mountain ranges, and remoteness from oceanic moisture. Two broad types of deserts exist: cool deserts, exemplified by the Great Basin of North America, and hot deserts, like the Sahara. Deserts are structurally simple—scattered shrubs, ephemeral plants, and open, stark topography. In this harsh environment, ways of circumventing aridity and high temperatures by either evading or resisting drought have evolved in plants and animals. Despite their aridity, deserts support a diversity of animal life, notably opportunistic herbivorous species and carnivores.

Shrubland 23.6

Shrubs have a densely branched, woody structure and low height. Shrublands are difficult to classify because of the variety of climates in which shrubs can be a dominant or codominant component of the plant community. But in five widely disjunct regions along the western margins of the continents between 30° and 40° latitude are found the mediterranean ecosystems. Dominated by evergreen shrubs and sclerophyll trees, these biomes have adapted to the distinctive climate of summer drought and cool, moist winters. These shrublands are fire adapted and highly flammable.

Temperate Forests 23.7

Broadleaf deciduous forests are found in the wetter environments of the warm temperate region. They once covered large areas of Europe and China, but their distribution has been reduced by human activity. In North America, deciduous forests are still widespread. They include various types such as beech–maple and oak–hickory forest; the greatest development is in the mixed mesophytic forest of the unglaciated Appalachians. Well-developed deciduous forests have four strata: upper canopy, lower canopy, shrub layer, and ground layer. Vertical structure influences the diversity and distribution of life in the forest. Certain species are associated with each stratum.

Conifer Forests 23.8

Coniferous forests of temperate regions include the montane pine forests and lower-elevation pine forests of Eurasia and North America and the temperate rain forests of the Pacific Northwest.

North of the temperate coniferous forest is the circumpolar taiga, or boreal forest, the largest biome on Earth. Characterized by a cold continental climate, the taiga consists of four major zones: the forest ecotone, open boreal woodland, main boreal forest, and boreal–mixed forest ecotone.

Permafrost, the maintenance of which is influenced by tree and ground cover, strongly influences the pattern of vegetation, as do recurring fires. Spruces and pines dominate boreal forest with successional communities of birch and poplar. Ground cover below spruce is mostly moss; in open spruce and pine stands, the cover is mostly lichen.

Major herbivores of the boreal region include caribou, moose, and snowshoe hare. Predators include the wolf, lynx, and pine martin.

Tundra 23.9

The Arctic tundra extends beyond the tree line at the far north of the Northern Hemisphere. It is characterized by low temperature, low precipitation, a short growing season, a perpetually frozen subsurface (the permafrost), and a frost-molded landscape. Plant species are few, growth forms are low, and growth rates are slow. Over much of the Arctic, the dominant vegetation is cotton grass, sedge, and dwarf heaths. These plants exploit the long days of summer by photosynthesizing during the 24-hour daylight period. Most plant growth occurs underground. The animal community is low in diversity but unique. Summer in the Arctic brings hordes of insects, providing a rich food source for shorebirds. Dominant vertebrates are lemming, Arctic hare, caribou, and musk ox. Major carnivores are the wolf, Arctic fox, and snowy owl.

Alpine tundras occur in the mountains of the world. They are characterized by widely fluctuating temperatures, strong winds, snow, and a thin atmosphere.

Forest Management Ecological Issues & Applications

More than 90 percent of global forest resources, which include fuel, building materials, and food, are harvested from native forests. Sustainable production of forest resources requires achieving a balance between net growth and harvest. To achieve this end, foresters have an array of silvicutural and harvesting techniques.

STUDY QUESTIONS

1. What are the main differences between tropical rain forest and tropical dry forest?
2. What are the major strata in the tropical rain forest?
3. How does the warm, wet environment of tropical rain forests influence rates of net primary productivity and decomposition?
4. What types of trees characterize tropical rain forest (leaf type)?
5. What distinguishes savannas from grassland ecosystems?
6. Under what climate conditions do you find tropical savanna ecosystems?

7. How does seasonality influence rates of net primary productivity and decomposition in savanna ecosystems?
8. What features of regional climate lead to the formation of desert ecosystems?
9. What climate characterizes mediterranean ecosystems?
10. What type of leaves characterize mediterranean plants?
11. What types of leaves characterize the trees of temperate forest ecosystems?
12. How does seasonality of temperature influence the structure and productivity of temperate forest ecosystems?
13. What climate is characteristic of temperate grasslands?
14. How does annual precipitation influence the structure and productivity of grassland ecosystems?
15. What type of trees characterizes boreal forest?
16. What is permafrost, and how does it influence the structure and productivity of boreal forest ecosystems?
17. What physical and biological features characterize Arctic tundra?
18. How does alpine tundra differ from Arctic tundra?

FURTHER READINGS

Archibold, O. W. 1995. *Ecology of world vegetation*. London: Chapman & Hall.
An outstanding reference for those interested in the geography and ecology of terrestrial ecosystems.

Bliss, L. C., O. H. Heal, and J. J. Moore, eds. 1981. *Tundra ecosystems: A comparative analysis*. New York: Cambridge University Press.
A major reference book on the geography, structure, and function of these high-latitude ecosystems.

Bonan, G. B., and H. H. Shugart. 1989. "Environmental factors and ecological processes in boreal forests." *Annual Review of Ecology and Systematics* 20:1–18.
An excellent review article providing a good introduction to boreal forest ecosystems.

Bruijnzeel, L. A., F. N. Scatena, and L. S. Hamilton 2011. *Tropical montane cloud forests*. New York: Cambridge University Press.
Comprehensive overview of high elevation cloud forests, examining geographic distribution, climate, soils, hydrological processes, and cloud forest conservation and management.

Evenardi, M., I. Noy-Meir, and D. Goodall, eds. 1986. *Hot deserts and arid shrublands of the world. Ecosystems of the World* 12A and 12B. Amsterdam: Elsevier Scientific.
A major reference work on the geography and ecology of the world's deserts.

French, N., ed. 1979. *Perspectives on grassland ecology*. New York: Springer-Verlag.
This text provides a good summary of grassland ecology.

Murphy, P. G., and A. E. Lugo. 1986. "Ecology of tropical dry forests." *Annual Review of Ecology and Systematics* 17:67–88.
This article provides an overview of the distribution and ecology of tropical dry forests, one of the most endangered terrestrial ecosystems.

Quinn, R. D., and S. C. Keeley. 2006. *Introduction to California chaparral*. Berkeley: University of California Press.
An overview of the chaparral ecosystem with an excellent discussion of the role and impact of fire.

Reichle, D. E., ed. 1981. *Dynamic properties of forest ecosystems*. Cambridge, UK: Cambridge University Press.
The major reference source on the ecology and function of forest ecosystems throughout the world.

Reichman, O. J. 1987. *Konza prairie: A tallgrass natural history*. Lawrence: University Press of Kansas.
An excellent introduction to the tallgrass prairie ecosystem; emphasizes the interrelationships of plants, animals, and landscapes.

Richards, P. W. 1996. *The tropical rain forest: An ecological study*, 2nd ed. New York: Cambridge University Press.
A thoroughly revised edition of a classic book on the ecology of tropical rain forests.

Sinclair, A. R. E., and P. Arcese, eds. 1995. *Serengeti II: Dynamics, management, and conservation of an ecosystem*. Chicago: University of Chicago Press.
A masterful study of an ecosystem. A valuable reference on the ecology of tropical savanna ecosystems.

Sinclair, A., C. Packer, and S. Mduma, eds. 2008. *Serengeti III. Human Impacts on Ecosystem Dynamics*. Chicago: University Chicago Press.
This book is a follow up to *Serengeti II*. It updates information on ecosystem dynamics, conservation, and changes brought about by the growing population of pastoralists and agriculturalists surrounding the Serengeti.

MasteringBiology®

Students Go to **www.masteringbiology.com** for assignments, the eText, and the Study Area with practice tests, animations, and activities.

Instructors Go to **www.masteringbiology.com** for automatically graded tutorials and questions that you can assign to your students, plus Instructor Resources.

CHAPTER 24 Aquatic Ecosystems

Waves heavily influence the pattern of life on rocky shores along the California coastline.

CHAPTER GUIDE

WHEREAS SCIENTISTS CLASSIFY terrestrial ecosystems according to their dominant plant life-forms, classification of aquatic ecosystems is largely based on features of the physical environment. A major feature influencing the adaptations of organisms to the aquatic environment is water salinity (see Section 3.5). For this reason, aquatic ecosystems fall into two major categories: freshwater and saltwater (or marine). These categories are further divided into several ecosystem types based on substrate, depth and flow of water, and type of dominant organisms (typically autotrophs).

Ecologists subdivide marine ecosystems into two broad categories of coastal and open-water systems. Freshwater ecosystems are classified on the basis of water depth and flow. Flowing-water, or **lotic**, ecosystems include rivers and streams. Non-flowing water, or **lentic** ecosystems, include ponds, lakes, and inland wetlands.

All aquatic ecosystems, both freshwater and marine, are linked directly or indirectly as components of the hydrological cycle (**Figure 24.1**; also see Section 3.1). Water that evaporated from oceans and terrestrial environments falls as precipitation. The precipitation that remains on the land surface (i.e., does not infiltrate into the soil or evaporate) follows a path determined by gravity and topography—more specifically, geomorphology. Flowing-water ecosystems begin as streams. These streams, in turn, coalesce into rivers as they follow the topography of the landscape, or they collect in basins and floodplains to form standing-water ecosystems such as ponds, lakes, and inland wetlands. Rivers eventually flow to the coast and form estuaries, which represent the transition from freshwater to marine environments.

In this chapter, we will examine the basic characteristics of aquatic ecosystems, both freshwater and marine. Beginning first with freshwater ecosystems, we will examine lentic ecosystems (lakes and ponds). We then turn our attention to lotic ecosystems, following the path and changing characteristics of streams as they coalesce to form rivers, eventually flowing to coastal environments. After examining estuarine environments, we conclude by examining the marine environments that cover more than 70 percent of Earth's surface.

Figure 24.1 Idealized landscape/seascape showing the linkages among the various types of aquatic ecosystems via the water cycle.

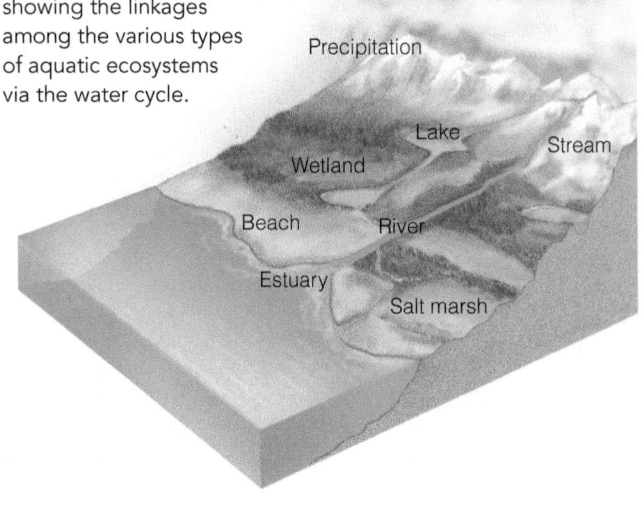

24.1 Lakes Have Many Origins

Lakes and ponds are inland depressions containing standing water (**Figure 24.2**). They vary in depth from 1 meter (m) to more than 2000 m and they range in size from small ponds of less than a hectare (ha) to large lakes covering thousands of square kilometers. Ponds are small bodies of water so shallow that rooted plants can grow over much of the bottom. Some lakes are so large that they mimic marine environments. Most ponds and lakes have outlet streams, and both may be more or less temporary features on the landscape, geologically speaking (see Section 18.8, Figure 18.23).

Some lakes have formed through glacial erosion and deposition. Abrading slopes in high mountain valleys, glaciers carved basins that filled with water from rain and melting snow to form tarns. Retreating valley glaciers left behind crescent-shaped ridges of rock debris, called moraines, which dammed up water behind them. Many shallow kettle lakes and potholes were left behind by the glaciers that covered much of northern North America and northern Eurasia.

Lakes also form when silt, driftwood, and other debris deposited in beds of slow-moving streams dam up water behind them. Loops of streams that meander over flat valleys and floodplains often become cut off by sediments, forming crescent-shaped oxbow lakes.

Shifts in Earth's crust, uplifting mountains or displacing rock strata, sometimes develop water-filled depressions. Craters of some extinct volcanoes have also become lakes. Landslides may block streams and valleys to form new lakes and ponds.

Many lakes and ponds form through nongeological activity. Beavers dam streams to make shallow but often extensive ponds (see Figure 19.5a). Humans create huge lakes by damming rivers and streams for power, irrigation, or water storage and construct smaller ponds and marshes for recreation, fishing, and wildlife. Quarries and surface mines can also form ponds.

24.2 Lakes Have Well-Defined Physical Characteristics

All lentic ecosystems share certain characteristics. Life in still-water ecosystems depends on light. The amount of light penetrating the water is influenced by natural attenuation, by silt and other material carried into the lake, and by the growth of phytoplankton (see Chapter 4, **Quantifying Ecology 4.1** and Figure 20.8). Temperatures vary seasonally and with depth (Figure 3.9). Oxygen can be limiting, especially in summer, because only a small proportion of the water is in direct contact with air, and respiration by decomposers

Figure 24.2 Lakes and ponds fill basins or depressions in the land. (a) A rock basin glacial lake, or tarn, in the Rocky Mountains. (b) Swampy tundra in Siberia is dotted with numerous ponds and lakes. (c) An oxbow lake formed when a bend in the river was cut off from the main channel. (d) A human-constructed, old New England millpond. Note the floating vegetation.

(a)

(b)

(c)

(d)

on the bottom consumes large quantities (see Figure 3.12). Thus, variation in oxygen, temperature, and light strongly influences the distribution and adaptations of life in lakes and ponds (see Chapter 3 for more detailed discussion).

Ponds and lakes may be divided into both vertical and horizontal strata based on penetration of light and photosynthetic activity (**Figure 24.3**). The horizontal zones are obvious to the eye; the vertical ones, influenced by depth of light penetration, are not. Surrounding most lakes and ponds and engulfing some ponds completely is the **littoral zone**, or shallow-water zone, in which light reaches the bottom, stimulating the growth of rooted plants. Beyond the littoral is open water, the **limnetic zone**, which extends to the depth of light penetration. Inhabiting this zone are microscopic phytoplankton (autotrophs) and zooplankton (heterotrophs) as well as **nekton**, free-swimming organisms such as fish. Beyond the depth of effective light penetration is the **profundal zone**. Its beginning is marked by the compensation depth of light, the point at which respiration balances photosynthesis (see Figure 20.8). The profundal zone depends on a rain of organic material from the limnetic zone for energy. Common to both the littoral and profundal zones is the third vertical stratum—the **benthic zone**, or bottom region, which is the primary place of decomposition. Although these zones are named and often described separately, all are closely dependent on one another in the dynamics of lake ecosystems.

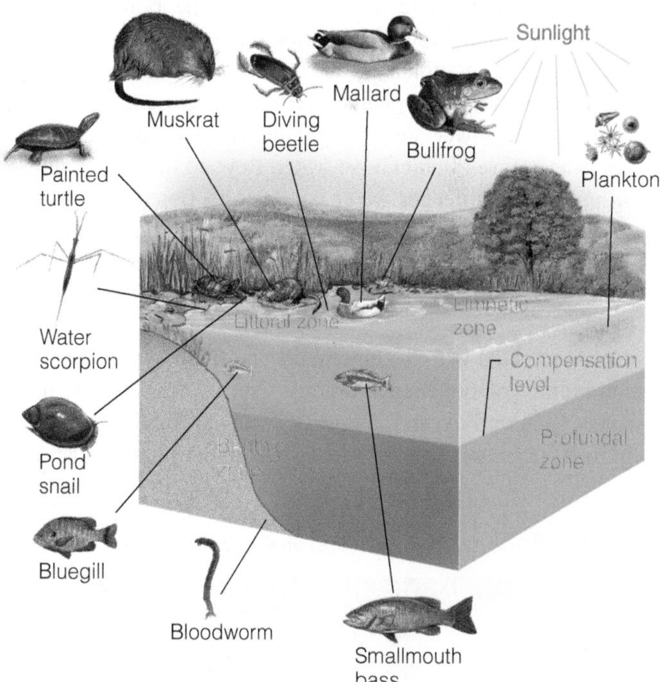

Figure 24.3 Major zones of a lake in midsummer: littoral, limnetic, profundal, and benthic. The compensation level is the depth where light levels are such that gross production in photosynthesis is equal to respiration, so net production (primary) is zero. The organisms shown are typical of the various zones in a lake community.

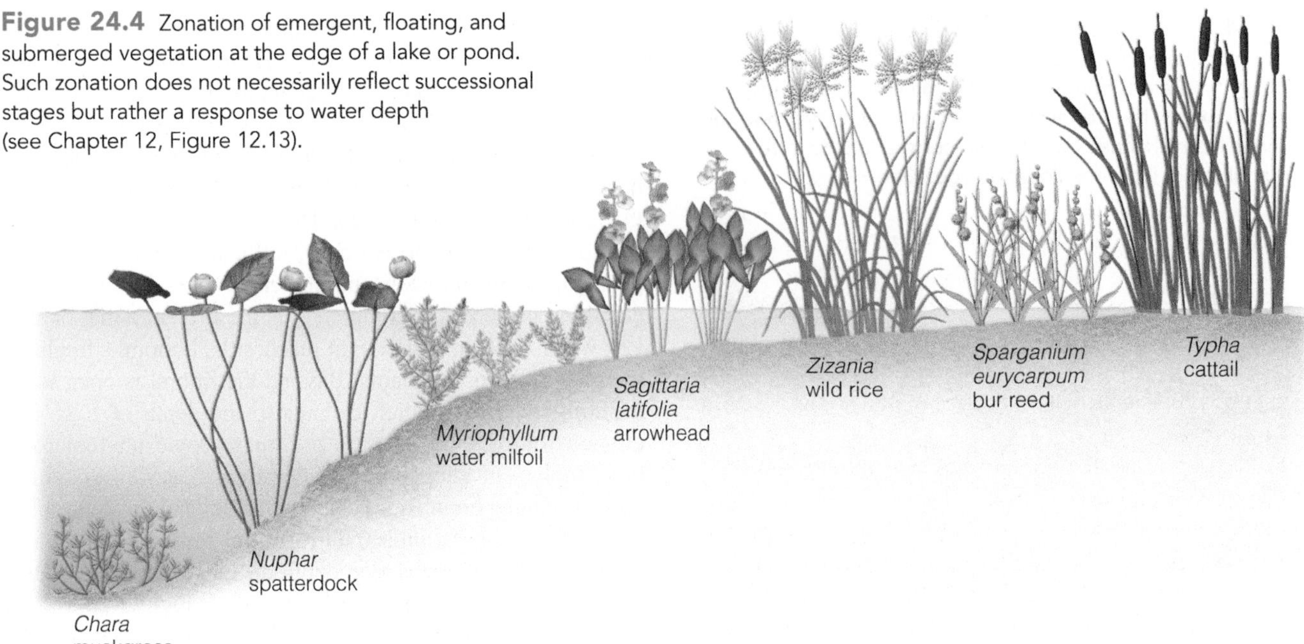

Figure 24.4 Zonation of emergent, floating, and submerged vegetation at the edge of a lake or pond. Such zonation does not necessarily reflect successional stages but rather a response to water depth (see Chapter 12, Figure 12.13).

24.3 The Nature of Life Varies in the Different Zones

Aquatic life is richest and most abundant in the shallow water about the edges of lakes and ponds as well as in other places where sediments have accumulated on the bottom and decreased the water depth (**Figure 24.4**). Dominating these areas is emergent vegetation such as cattails (*Typha* spp.) and sedges (Cyperaceae), that is, plants whose roots are anchored in the bottom mud, lower stems are immersed in water, and upper stems and leaves stand above water. Beyond the emergents and occupying even deeper water is a zone of floating plants such as pondweed (*Potamogeton*) and pond lily (*Nuphar* spp.). In depths too great for floating plants live submerged plants, such as species of pondweed with their finely dissected or ribbon-like leaves.

Associated with the emergents and floating plants is a rich community of organisms, among them hydras, snails, protozoans, dragonflies and diving insects, pickerel (*Esox* spp.), sunfish (*Lepomis* spp.), herons (Ardeidae), and blackbirds (*Agelaius* spp. and *Xanthocephalus xanthocephalus*). Many species of pond fish have compressed bodies, permitting them to move easily through the masses of aquatic plants. The littoral zone contributes heavily to the large input of organic matter into the system.

The main forms of life in the limnetic zone are phytoplankton and zooplankton (**Figure 24.5**). Phytoplankton, including desmids, diatoms, and filamentous algae, are the primary producers in open-water ecosystems and form the base on which the rest of life in open water depends. Also suspended in the water column are small grazing animals, mostly tiny crustaceans that feed on the phytoplankton. These animals form an important link in energy flow in the limnetic zone.

During the spring and fall turnovers, plankton are carried downward, but at the same time nutrients released by decomposition on the bottom are carried upward to the impoverished surface layers (see Section 21.10 and Figure 21.24). In spring when surface waters warm and stratification develops, phytoplankton have access to both nutrients and light. A spring bloom develops, followed by rapid depletion of

Figure 24.5 (a) Phytoplankton and (b) zooplankton.

(a)

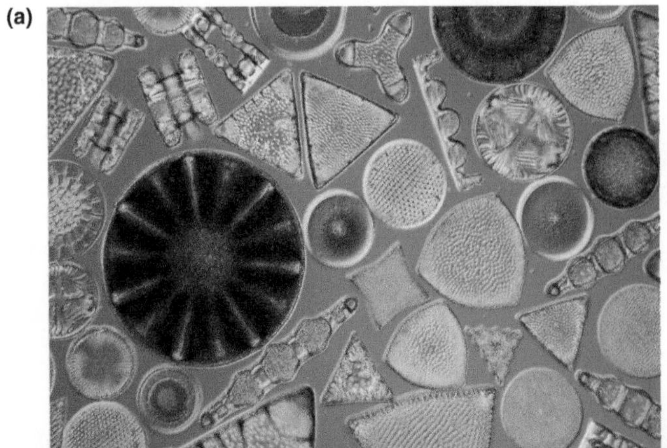

(b)

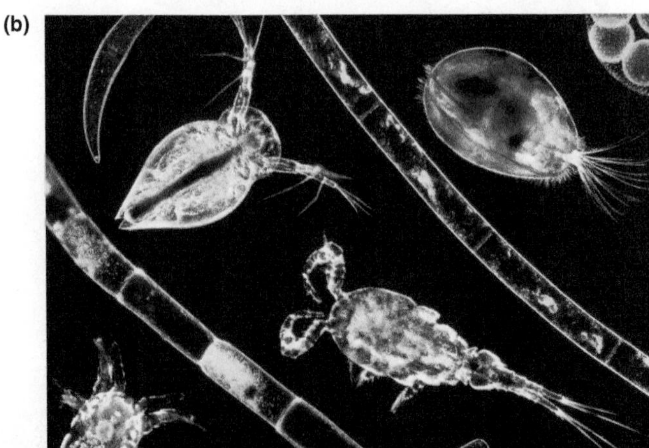

nutrients and a reduction in planktonic populations, especially in shallow water.

Fish make up most of the nekton in the limnetic zone. Their distribution is influenced primarily by food supply, oxygen, and temperature. During the summer, large predatory fish such as largemouth bass (*Micropterus salmoides*), pike (*Esox lucius*), and muskellunge (*Esox masquinongy*), inhabit the warmer epilimnion waters, where food is abundant. In winter, these species retreat to deeper water. In contrast, lake trout (*Salvelinus namaycush*) require colder water temperatures and move to greater depths as summer advances. During the spring and fall turnover, when oxygen and temperature are fairly uniform throughout, both warm-water and cold-water species occupy all levels.

Life in the profundal zone depends not only on the supply of energy and nutrients from the limnetic zone above but also on the temperature and availability of oxygen. In highly productive waters, oxygen may be limiting because the decomposer organisms so deplete it that little aerobic life can survive. Only during spring and fall turnovers, when organisms from the upper layers enter this zone, is life abundant in profundal waters.

Easily decomposed substances drifting down through the profundal zone are partly mineralized while sinking. The remaining organic debris—dead bodies of plants and animals of the open water, and decomposing plant matter from shallow-water areas—settles on the bottom. Together with quantities of material washed in, these substances make up the bottom sediments, the habitat of benthic organisms.

The bottom ooze is a region of great biological activity—so great, in fact, that the oxygen curves for lakes and ponds show a sharp drop in the profundal water just above the bottom. Because the organic muck is so low in oxygen, the dominant organisms there are anaerobic bacteria. Under anaerobic conditions, however, decomposition cannot proceed to create inorganic end products. When the amounts of organic matter reaching the bottom are greater than can be used by bottom fauna, they form a muck that is rich in hydrogen sulfide and methane.

As the water becomes shallower, the benthos changes. The action of water, plant growth, and recent organic deposits modifies the bottom material, typically consisting of stones, rubble, gravel, marl, and clay. Increased oxygen, light, and food encourage a richness of life not found on the bottom of the profundal zone.

Closely associated with the benthic community are organisms collectively called **periphyton** or **aufwuchs**. They are attached to or move on a submerged substrate but do not penetrate it. Small aufwuchs communities colonize the leaves of submerged aquatic plants, sticks, rocks, and debris. Periphyton—mostly algae and diatoms living on plants—are fast growing and lightly attached. Aufwuchs on stones, wood, and debris form a more crustlike growth of cyanobacteria, diatoms, water moss, and sponges.

24.4 The Character of a Lake Reflects Its Surrounding Landscape

Because of the close relationship between land and water ecosystems, lakes reflect the character of the landscape in which they occur. Water that falls on land flows over the surface or moves through the soil to enter springs, streams, and lakes. The water transports with it silt and nutrients in solution. Human activities including road construction, logging, mining, construction, and agriculture add another heavy load of silt and nutrients—especially nitrogen, phosphorus, and organic matter. These inputs enrich aquatic systems by a process called **eutrophication**. The term **eutrophy** (from the Greek *eutrophos*, "well nourished") means a condition of being rich in nutrients.

A typical eutrophic lake (**Figure 24.6a**) has a high surface-to-volume ratio; that is, the surface area is large relative to depth. Nutrient-rich deciduous forest and farmland often surround it. An abundance of nutrients, especially nitrogen and phosphorus, flowing into the lake stimulates a heavy growth of algae and other aquatic plants. Increased photosynthetic production leads to increased recycling of nutrients and organic compounds, stimulating even further growth.

Phytoplankton concentrate in the warm upper layer of the water, giving it a murky green cast. Algae, inflowing organic debris and sediment, and remains of rooted plants drift to the bottom, where bacteria feed on this dead organic matter. Their activities deplete the oxygen supply of the bottom sediments

Figure 24.6 (a) A eutrophic lake. Note the floating algal mats on the water surface. (b) An oligotrophic lake in Montana.

(a)

(b)

Figure 24.7 River delta formed by the deposition of sediments.

and deep water until this region of the lake cannot support aerobic life. The number of bottom species declines, although the biomass and numbers of organisms remain high. In extreme cases, oxygen depletion (anoxic conditions) can result in the die-off of invertebrate and fish populations (see this chapter, *Ecological Issues & Applications*).

In contrast to eutrophic lakes and ponds are oligotrophic bodies of water (**Figure 24.6b**). **Oligotrophy** is the condition of being poor in nutrients. Oligotrophic lakes have a low surface-to-volume ratio. The water is clear and appears blue to blue-green in the sunlight. The nutrient content of the water is low; nitrogen may be abundant, but phosphorus is highly limited. A low input of nutrients from surrounding terrestrial ecosystems and other external sources is mostly responsible for this condition. Low availability of nutrients causes low production of organic matter that leaves little for decomposers, so oxygen concentration remains high in the hypolimnion. The bottom sediments are largely inorganic. Although the numbers of organisms in oligotrophic lakes and ponds may be low, species diversity is often high.

Lakes that receive large amounts of organic matter from surrounding land, particularly in the form of humic materials that stain the water brown, are called **dystrophic** (from *dystrophos*, "ill-nourished"). These bodies of water occur generally on peaty substrates, or in contact with peaty substrates in bogs or heathlands that are usually highly acidic (see Section 25.6). Dystrophic lakes generally have highly productive littoral zones. This littoral vegetation dominates the lake's metabolism, providing a source of both dissolved and particulate organic matter.

24.5 Flowing-Water Ecosystems Vary in Structure and Types of Habitats

Even the largest rivers begin somewhere back in the hinterlands as springs or seepage areas that become headwater streams, or they arise as outlets of ponds or lakes. A few rivers emerge fully formed from glaciers. As a stream drains away from its source, it flows in a direction and manner dictated by the lay of the land and the underlying rock formations. Joining the new stream are other small streams, spring seeps, and surface water.

Just below its source, the stream may be small, straight, and swift, with waterfalls and rapids. Farther downstream, where the gradient is less steep, velocity decreases and the stream begins to meander, depositing its load of sediment as silt, sand, or mud. At flood time, a stream drops its load of sediment on surrounding level land, over which floodwaters spread to form floodplain deposits.

Where a stream flows into a lake or a river into the sea, the velocity of water is suddenly checked. The river then is forced to deposit its load of sediment in a fan-shaped area about its mouth to form a delta (**Figure 24.7**). Here, its course is carved into several channels, which are blocked or opened with subsequent deposits. As a result, the delta becomes an area of small lakes, swamps, and marshy islands. Material the river fails to deposit in the delta is carried out to open water and deposited on the bottom.

Because streams become larger on their course to rivers and are joined along the way by many others, we can classify them according to order (**Figure 24.8**). A small headwater stream

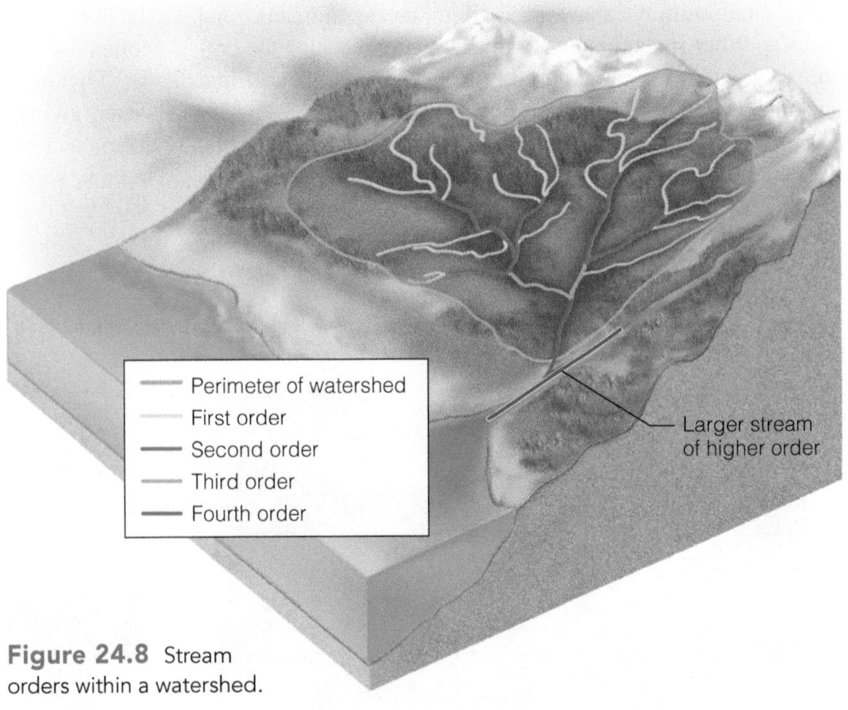

Figure 24.8 Stream orders within a watershed.

with no tributaries is a first-order stream. Where two streams of the same order join, the stream becomes one of higher order. If two first-order streams unite, the resulting stream becomes a second-order one; and when two second-order streams unite, the stream becomes a third-order one. The order of a stream increases only when a stream of the same order joins it. It does not increase with the entry of a lower-order stream. In general, headwater streams are orders first to third; medium-sized streams, fourth to sixth; and rivers, greater than sixth.

The velocity of a current molds the character and structure of a stream (see **Quantifying Ecology 24.1**). The shape and steepness of the stream channel, its width, depth, and roughness of the bottom, and the intensity of rainfall and rapidity of snowmelt all affect velocity. Fast streams are those whose velocity of flow is 50 centimeters (cm)/second (s) or higher. At this velocity, the current removes all particles less than 5 millimeters (mm) in diameter and leaves behind a stony bottom.

(a)

(b)

Figure 24.9 (a) A fast mountain stream. The gradient is steep, and the bottom is largely bedrock. (b) A slow stream is deeper and has a lower slope gradient.

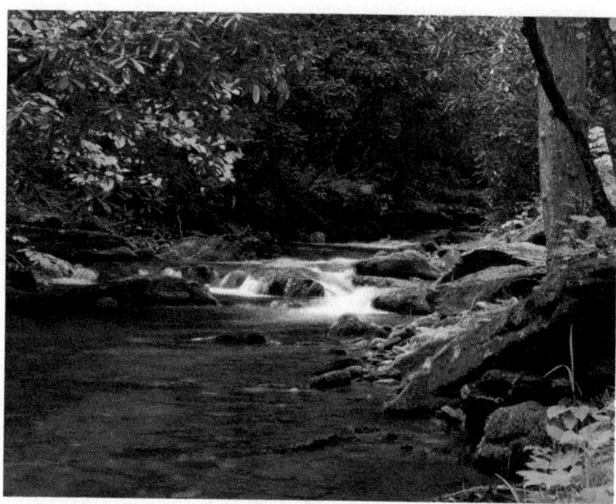

Figure 24.10 Two different but related habitats in a stream: a riffle (background) and a pool (foreground).

High water increases the velocity; it moves bottom stones and rubble, scours the streambed, and cuts new banks and channels. As the gradient decreases and the width, depth, and volume of water increase, silt and decaying organic matter accumulate on the bottom. The character of the stream changes from fast water to slow (**Figure 24.9**), with an associated change in species composition.

Flowing-water ecosystems often alternate between two different but related habitats: the turbulent riffle and the quiet pool (**Figure 24.10**). Processes occurring in the rapids influence the waters of the pool below; in turn, the waters of the rapids are influenced by events in the pools upstream.

Riffles are the sites of primary production in the stream. Here the periphyton or aufwuchs, organisms that are attached to or move on submerged rocks and logs, assume dominance. Periphyton, which occupy a position of the same importance as phytoplankton in lakes and ponds, consist chiefly of diatoms, cyanobacteria, and water moss.

Above and below the riffles are the pools. Here, the environment differs in chemistry, intensity of current, and depth. Just as the riffles are the sites of organic production, so the pools are the sites of decomposition. Here, the velocity of the current slows enough for organic matter to settle. Pools, the major sites of carbon dioxide production during the summer and fall, are necessary for maintaining a constant supply of bicarbonate in solution (see Section 3.7). Without pools, photosynthesis in the riffles would deplete the bicarbonates and result in smaller and smaller quantities of available carbon dioxide downstream.

24.6 Life Is Highly Adapted to Flowing Water

Living in moving water, inhabitants of streams and rivers face the challenge of remaining in place without being swept downstream. Unique adaptations have evolved among these organisms

QUANTIFYING ECOLOGY 24.1 Streamflow

The ecology of a stream ecosystem is determined largely by its streamflow, which is the water discharge occurring within the natural streambed or channel. The rate at which water flows through the stream channel influences the water temperature, oxygen content, rate of nutrient spiraling, physical structure of the benthic environment, and subsequently the types of organisms inhabiting the stream. As such, streamflow is an important parameter used by ecologists to characterize the stream environment.

Flow is defined simply as the volume of water moving past a given point in the stream per unit time. As such, it can be estimated from the cross-sectional area of the stream channel and the velocity of the flow as follows:

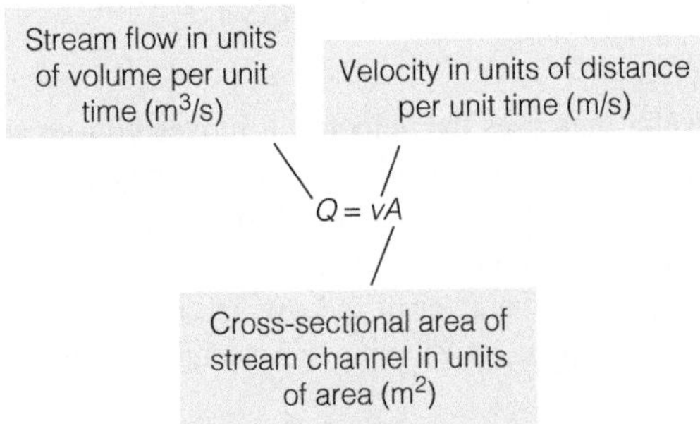

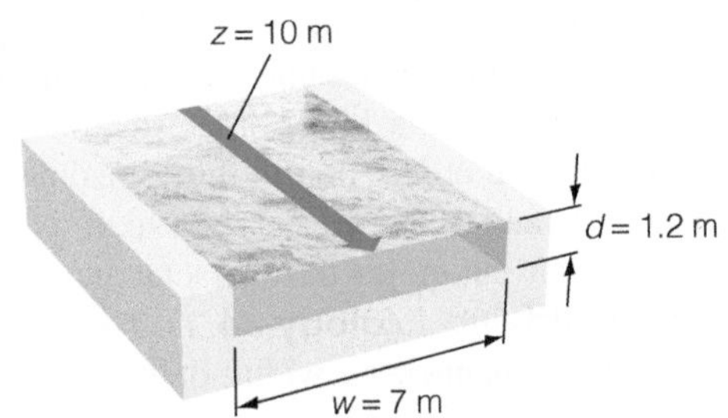

Figure 1 Idealized representation of a stream channel (water shown in blue, land surface in yellow). The cross-sectional area (*A*) of the stream can be calculated as the stream width (*w*) times the depth (*d*). The velocity of water flow is calculated as the distance traveled (*z*) per unit time. For example, if a position on the water surface travels downstream 10 m over a period (*t*) of 20 s, the velocity of the water is z/t or 10 m/20 s = 0.5 m/s.

As shown in **Figure 1**, the cross-sectional area (*A*) can be calculated by measuring the depth (*d*) and width (*w*) of the stream and multiplying the two ($A = w \times d$). Estimates of depth and width can be easily made for a point along the stream channel by using a tape measure and meter stick. Velocity (*v*) can generally be thought of as distance (*z*) traveled over time (*t*); see Figure 1. The velocity can be estimated using a "current" or "flow" meter. One flow meter commonly used in streams and rivers is an apparatus with rotating cups (**Figure 2a**). The flow causes the cups to rotate (**Figure 2b**), and the number of rotations is monitored electronically. Number of rotations (distance traveled) per unit time provides a measure of velocity.

For example, in the simple representation of a stream channel depicted in Figure 1, let us assume that the stream depth (*d*) is 1.2 m and the stream width (*w*) is 7 m. The cross-sectional area (*A*) of the stream is then 8.4 m^2 (1.2 m $\times$ 7 m = 8.4 m^2). If the measured velocity (*v*) is 0.5 m/s, then the streamflow (*Q*) is 4.2 m^3/s (8.4 m^2 $\times$ 0.5 m/s = 4.2 m^3/s). In reality, the profile of the stream channel is never as simple as the rectangular profile presented in Figure 1, and the water velocity varies as a function of depth and position relative to the stream bank. For this reason, multiple measurements of depth and velocity are taken across the stream profile. For example, the stream profile in **Figure 3** is 6 m wide; however, the depth varies across the width of the stream channel. A simple approach to estimating

that help them deal with life in the current (**Figure 24.11a**). A streamlined form, which offers less resistance to water current, is typical of many animals found in fast water. Larval forms of many insect species have extremely flattened bodies and broad, flat limbs that enable them to cling to the undersurfaces of stones where the current is weak. The larvae of certain species of caddisflies (Trichoptera) construct protective cases of sand or small pebbles and cement them to the bottoms of stones. Sticky undersurfaces help snails and planarians cling tightly and move about on stones and rubble in the current. Among the plants, water moss (*Fontinalis* spp.) and heavily branched, filamentous algae cling to rocks by strong holdfasts. Other algae grow in cushion-like colonies or form sheets—covered with a slippery, gelatinous coating—that lie flat against the surfaces of stones and rocks.

All animal inhabitants of fast-water streams require high, near-saturation concentrations of oxygen and moving water to keep their absorbing and respiratory surfaces in continuous contact with oxygenated water. Otherwise, a closely adhering film of liquid, impoverished of oxygen, forms a cloak about their bodies.

In slow-flowing streams where current is at a minimum, streamlined forms of fish give way to fish species such as smallmouth bass (*Micropterus dolomieu*), whose compressed bodies enable them to move through beds of aquatic vegetation. Pulmonate snails (Lymnaeacea) and burrowing mayflies (Ephemeroptera) replace rubble-dwelling insect larvae. Bottom-feeding fish, such as catfish (Akysidae), feed on life on the silty bottom, and back-swimmers and water striders inhabit sluggish stretches and still backwaters (**Figure 24.11b**).

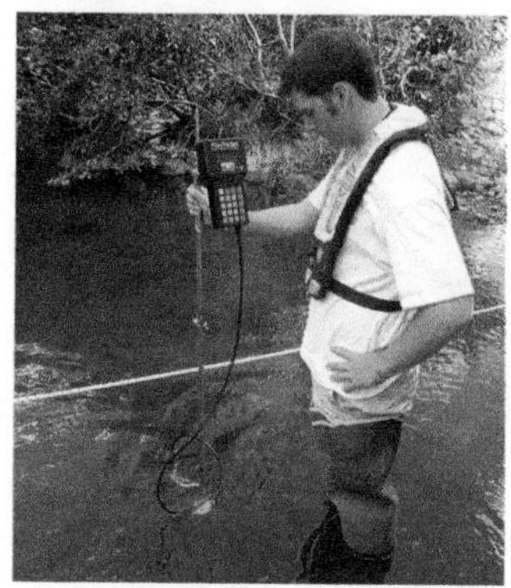

(a)

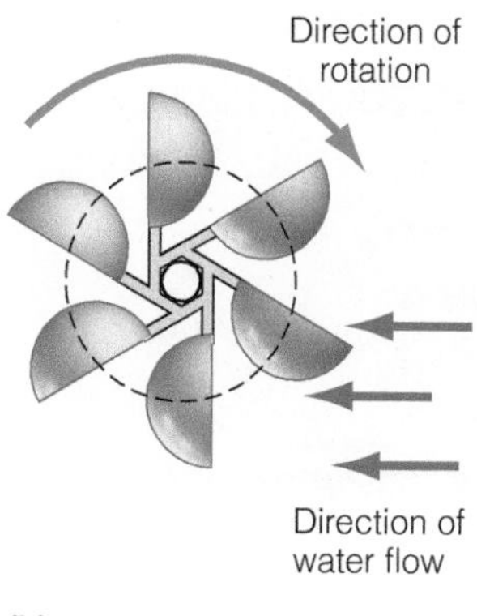

(b)

Figure 2 (a) Stream velocity can be easily measured using a cup-type flow meter. (b) As the water flows, the cups rotate about a fixed point on the instrument. The number of rotations per unit time provides an estimate of water velocity.

flow for this stream would be to sample water depth and velocity at several locations along the width of the stream, using the average values of water depth and velocity to calculate the value of streamflow. In most cases, however, stream ecologists will use a much more elaborate sampling scheme, estimating water depth at regular intervals along the stream profile and water velocity at several depths at each interval.

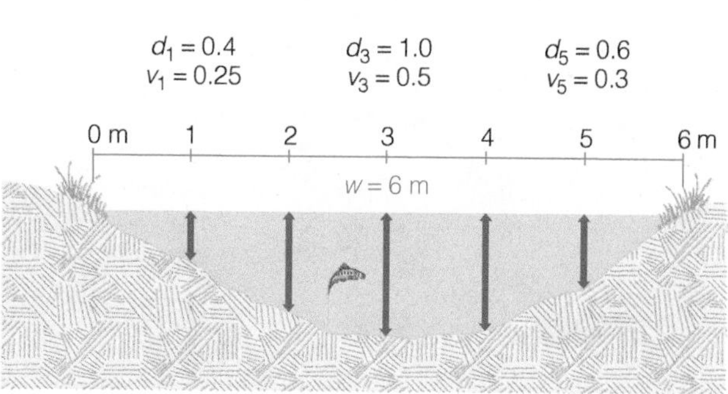

Figure 3 Cross-sectional view of a stream channel. To estimate the cross-sectional area (A), measurements of depth (d) and velocity (v) are taken at several locations along the width of the stream. By averaging these values of depth and velocity, a more accurate estimate of cross-sectional and velocity can be obtained for calculating streamflow (Q).

The cross-sectional area and velocity of a stream varies through time based on the amount of water being discharged from the surrounding watershed. In turn, the amount of water discharged reflects the input of water to the surrounding watershed from precipitation. As a result, an accurate picture of streamflow requires a systematic sampling of the stream morphology (width and depth) and velocity through time.

1. In Figure 3, the water depths at 1, 3, and 5 m from the left stream bank are given as 0.4, 1.0, and 0.6 m, respectively. Estimates of velocity for the three locations are 0.25, 0.5, and 0.3 m/s, respectively. Estimate streamflow by representing depth and velocity as the simple mean of the three samples.
2. Why might the water velocity (v) decrease from the center of the stream channel to the banks?

Invertebrate inhabitants are classified into four major groups based on their feeding habits (**Figure 24.12**). **Shredders**, such as caddisflies (Trichoptera) and stoneflies (Plecoptera), make up a large group of insect larvae. They feed on coarse particulate organic matter (CPOM) that is $>$ 1 mm in diameter—mostly leaves that fall into the stream. The shredders break down the CPOM, feeding on the material not so much for the energy it contains as for the bacteria and fungi growing on it. Shredders assimilate about 40 percent of the material they ingest and pass off 60 percent as feces.

When broken up by the shredders and partially decomposed by microbes, the leaves, along with invertebrate feces, become part of the fine particulate organic matter (FPOM), that is $<$ 1 mm but $>$ 0.45 micrometers (μm) in diameter. Drifting downstream and settling on the stream bottom, FPOM is picked up by another feeding group of stream invertebrates, the **filtering collectors** and **gathering collectors**. The filtering collectors include, among others, the larvae of black flies (Simuliidae) with filtering fans and the net-spinning caddisflies. Gathering collectors, such as the larvae of midges, pick up particles from stream-bottom sediments. Collectors obtain much of their nutrition from bacteria associated with the fine detrital particles.

While shredders and collectors feed on detrital material, another group, the **grazers**, feeds on the algal coating of stones and rubble. This group includes the beetle larvae, water penny (*Psephenus* spp.), and a number of mobile caddisfly larvae.

Figure 24.11 Life in (a) a fast stream and (b) a slow stream. Fast stream: (1) black-fly larva (Simuliidae); (2) net-spinning caddisfly (*Hydropsyche* spp.); (3) stone case of caddisfly; (4) water moss (*Fontinalis*); (5) algae (*Ulothrix*); (6) mayfly nymph (*Isonychia*); (7) stonefly nymph (*Perla* spp.); (8) water penny (*Psephenus*); (9) hellgrammite (dobsonfly larva, *Corydalis cornuta*); (10) diatoms (Diatoma); (11) diatoms (Gomphonema); (12) crane-fly larva (Tipulidae). Slow stream: (13) dragonfly nymph (Odonata, *Anisoptera*); (14) water strider (*Gerris*); (15) damselfly larva (Odonata, *Zygoptera*); (16) water boatman (Corixidae); (17) fingernail clam (*Sphaerium*); (18) burrowing mayfly nymph (*Hexegenia*); (19) bloodworm (Oligochaeta, *Tubifex* spp.); (20) crayfish (*Cambarus* spp.). The fish in the fast stream is a brook trout (*Salvelinus fontinalis*). The fish in the slow stream are, from left to right: northern pike (*Esox lucius*), bullhead (*Ameiurus melas*), and smallmouth bass (*Micropterus dolomieu*).

Much of the material they scrape loose enters the drift as FPOM. Another group, associated with woody debris, is composed of the **gougers**, which are invertebrates that burrow into waterlogged limbs and trunks of fallen trees.

Feeding on the detrital feeders and grazers are predaceous insect larvae and fish such as the sculpin (*Cottus*) and trout (Salmoninae). Even these predators do not depend solely on aquatic insects; they also feed heavily on terrestrial invertebrates that fall or wash into the stream.

Because of the current, quantities of CPOM, FPOM, and invertebrates tend to drift downstream to form a traveling benthos. This drift is a normal process in streams, even in the absence of high water and abnormally fast currents. Drift is so characteristic of streams that a mean rate of drift can serve as an index of a stream's production rate.

24.7 The Flowing-Water Ecosystem Is a Continuum of Changing Environments

From its headwaters to its mouth, the flowing-water ecosystem is a continuum of changing environmental conditions (**Figure 24.13**). Headwater streams (orders first to third) are usually swift, cold, and in shaded forested regions. Primary productivity in these streams is typically low, and they depend heavily on the input of detritus from terrestrial streamside vegetation, which contributes more than 90 percent of the organic input. Even when headwater streams are exposed to sunlight and autotrophic production exceeds inputs from adjacent terrestrial ecosystems, organic matter produced invariably enters

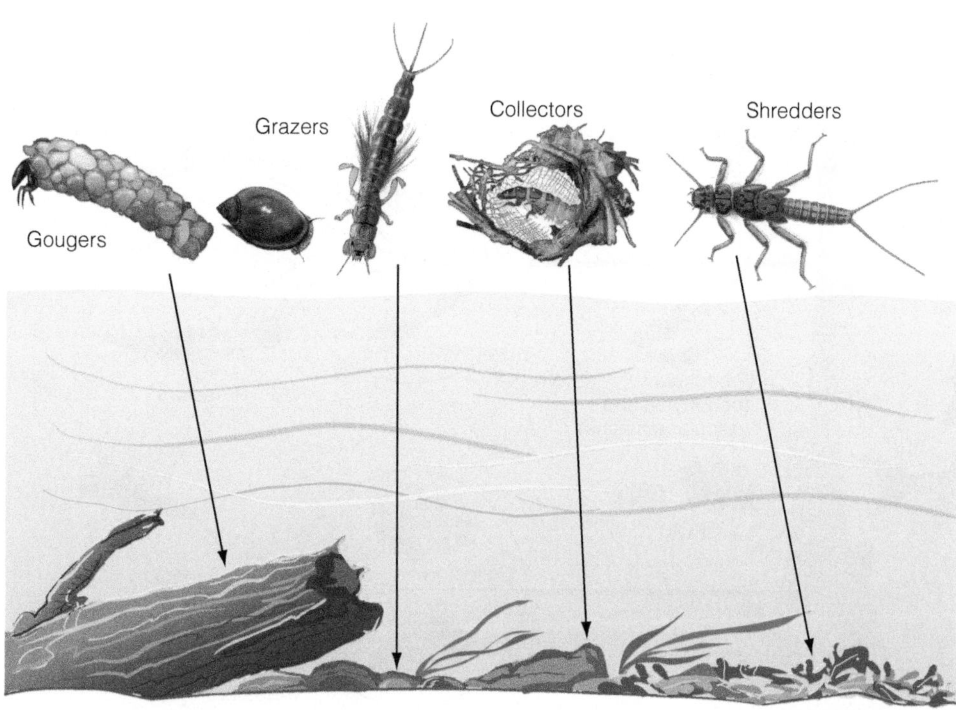

Figure 24.12 The four major feeding groups within the stream community: gougers, grazers, collectors, and shredders. Processing the leaves and other particulate matter are bacteria and fungi.

the detrital food chain. Dominant organisms are shredders, processing large-sized litter and feeding on CPOM, and collectors, processors of FPOM. Populations of grazers are minimal, reflecting the small amount of autotrophic production. Predators are mostly small fish—sculpins, darters, and trout. Headwater streams, then, are accumulators, processors, and transporters of particulate organic matter of terrestrial origin. As a result, the ratio of gross primary production to respiration is less than 1.

As streams increase in width to medium-sized creeks and rivers (orders fourth to sixth), the importance of riparian vegetation and its detrital input decreases. With more surface water exposed to the sun, water temperature increases, and as the elevation gradient declines, the current slows. These changes bring about a shift from dependence on terrestrial input of particulate organic matter to primary production by algae and rooted aquatic plants. Gross primary production now exceeds community respiration. Because of the lack of CPOM, shredders disappear. Collectors, feeding on FPOM transported downstream, and grazers, feeding on autotrophic production, become the dominant consumers. Predators show little increase in biomass but shift from cold-water species to warm-water species, including bottom-feeding fish such as suckers (Catostomidae) and catfish.

As the stream order increases from sixth to tenth and higher, riverine conditions develop. The channel is wider and deeper. The flow volume increases, and the current becomes slower. Sediments accumulate on the bottom. Both riparian and autotrophic production decrease. A basic energy source is FPOM, used by bottom-dwelling collectors that are now the dominant consumers. However, slow, deep-water and dissolved organic matter (DOM), which is $< 0.45\ \mu m$ in diameter, support a minimal phytoplankton and associated zooplankton population.

Throughout the downstream continuum, the community capitalizes on upstream feeding inefficiency. Downstream adjustments in production and the physical environment are reflected in changes in consumer groups.

24.8 Rivers Flow into the Sea, Forming Estuaries

Waters of most streams and rivers eventually drain into the sea. The place where freshwater joins saltwater is called an *estuary*. Estuaries are semi-enclosed parts of the coastal ocean where seawater is diluted and partially mixed with freshwater coming from the land (**Figure 24.14**). Here, the one-way flow of freshwater streams and rivers into an estuary meets the inflowing and outflowing saltwater tides. This meeting sets up a complex of currents that vary with the structure of the estuary (size, shape, and volume), season, tidal oscillations, and winds. Mixing waters of different salinities and temperatures create a counterflow that works as a nutrient trap (see Figure 21.26). Inflowing river waters most often impoverish rather than fertilize the estuary, with the possible exception of phosphorous. Instead, nutrients and oxygen are carried into the estuary by the tides. If vertical mixing occurs, these nutrients are not swept back out to sea but circulate up and down among organisms, water, and bottom sediments (see Figure 21.26).

Organisms inhabiting the estuary face two problems: maintaining their position and adjusting to changing salinity. Most estuarine organisms are benthic. They attach to the bottom, bury themselves in the mud, or occupy crevices and crannies. Mobile inhabitants are mostly crustaceans and fish, largely the young of species that spawn offshore in high-salinity water.

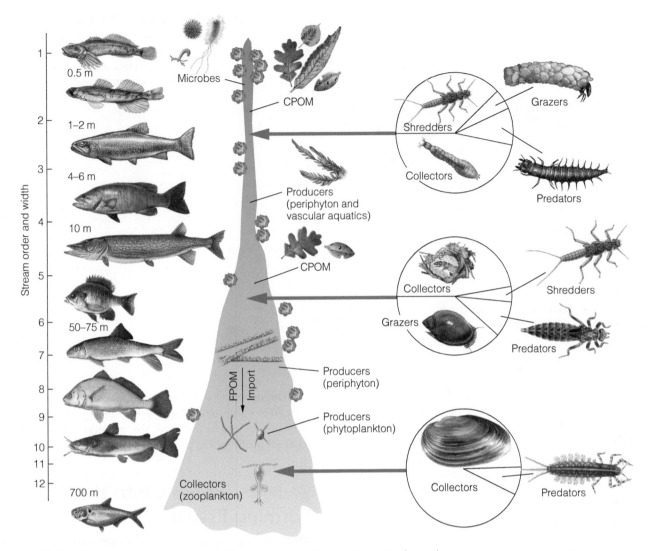

Figure 24.13 Changes in consumer groups along the river continuum. Stream order and width (m) are shown on the axis to the left of the figure. The headwater stream is strongly heterotrophic, dependent on terrestrial input of detritus. The dominant consumers are shredders and collectors. As stream size increases, the input of organic matter shifts to primary production by algae and rooted vascular plants. The major consumers are now collectors and grazers. As the stream grows into a river, the lotic system shifts back to heterotrophy. A phytoplankton population may develop. The consumers are mostly bottom-dwelling collectors. The fish community likewise changes as one moves downstream (from top to bottom as shown: sculpin, darter, brook trout, smallmouth bass, pickerel, sunfish, sucker, freshwater drum, catfish, and shad).

Figure 24.14 Estuary on the east coast of Australia.

Planktonic organisms are wholly at the mercy of the currents. Seaward movements of streamflow and ebb tide transport plankton out to sea, and the rate of water movement determines the size of the plankton population.

Salinity dictates the distribution of life in the estuary. The vast majority of the organisms inhabiting an estuary are marine, able to withstand full seawater. Some estuarine inhabitants cannot withstand lowered salinities, and these species decline along a salinity gradient from the ocean to the river's mouth. Sessile and slightly motile organisms have an optimum salinity range within which they grow best. When salinities exceed this range in either direction, populations decline.

Figure 24.15 Oyster reef in estuarine environment.

Anadromous fish are those that live most of their lives in saltwater and return to freshwater to spawn. These fish are highly specialized to endure the changes in salinity. Some species of fish, such as the striped bass (*Morone saxatilis*), spawn near the interface of fresh and low-salinity water. The larvae and young fish move downstream to more saline waters as they mature. Thus, for the striped bass, an estuary serves as both a nursery and as a feeding ground for the young. Anadromous species such as the shad (*Alosa* spp.) spawn in freshwater, but the young fish spend their first summer in an estuary and then move out to the open sea. Species such as the croaker (Sciaenidae) spawn at the mouth of the estuary, but the larvae are transported upstream to feed in plankton-rich, low-salinity areas.

The oyster bed and oyster reef are the outstanding communities of the estuary (**Figure 24.15**). The oyster (Ostreidae) is the dominant organism about which life revolves. Oysters may be attached to every hard object in the intertidal zone, or they may form reefs—areas where clusters of living organisms grow cemented to the almost buried shells of past generations. Oyster reefs usually lie at right angles to tidal currents, which bring planktonic food, carry away wastes, and sweep the oysters clean of sediment and debris. Closely associated with oysters are encrusting organisms such as sponges, barnacles, and bryozoans, which attach themselves to oyster shells and depend on the oysters or algae for food.

In shallow estuarine waters, rooted aquatics such as the sea grasses widgeongrass (*Ruppia maritima*) and eelgrass (*Zostera marina*) assume major importance (**Figure 24.16**; also see Figure 18.8). These aquatic plants represent complex systems supporting many epiphytic organisms. Such communities are important to certain vertebrate grazers, such as geese, swans, and sea turtles, and they provide a nursery ground for shrimp and bay scallops.

Figure 24.16 Sea-grass meadow in the Chesapeake Bay dominated by eelgrass (*Zostera marina*).

24.9 Oceans Exhibit Zonation and Stratification

The marine environment is marked by several differences compared to the freshwater world. It is large, occupying 70 percent of Earth's surface, and it is deep, in places more than 10 km. The surface area lit by the sun is small compared to the total volume of water. This small volume of sunlit water and the dilute solution of nutrients limit primary production. All of the seas are interconnected by currents, influenced by wave actions and tides, and characterized by salinity (see Chapter 3).

Just as lakes exhibit stratification and zonation, so do the seas. The ocean itself has two main divisions: the **pelagic**, or whole body of water, and the benthic zone, or bottom region. The pelagic is further divided into two provinces: the **neritic** province, which is water that overlies the continental shelf, and the **oceanic** province. Because conditions change with depth, the pelagic is divided into several distinct vertical layers or zones (**Figure 24.17**). From the surface to about 200 m is the **epipelagic zone**, or **photic zone**, in which there are sharp gradients in illumination, temperature, and salinity. From 200 to 1000 m is the **mesopelagic zone**, where little light penetrates and the temperature gradient is more even and gradual, without much seasonal variation. This zone contains an oxygen-minimum layer and often the maximum concentration of nutrients (nitrate and phosphate). Below the mesopelagic is the **bathypelagic zone**, where darkness is virtually complete, except for bioluminescent organisms, temperature is low, and water pressure is great. The **abyssopelagic zone** (Greek meaning "no bottom") extends from about 4000 m to the sea floor. The only zone deeper than this is the **hadalpelagic zone**, which includes areas found in deep-sea trenches and canyons.

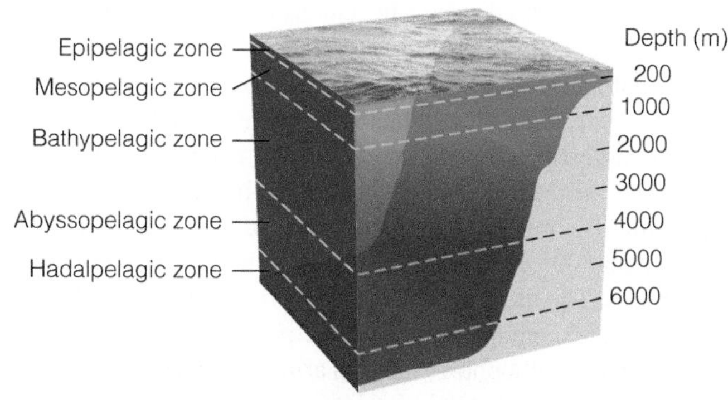

Figure 24.17 Major regions of the ocean.

24.10 Pelagic Communities Vary among the Vertical Zones

When viewed from the deck of a ship or from an airplane, the open sea appears to be monotonously the same. Nowhere can you detect as strong a pattern of life or well-defined communities, as you can over land. The reason is that pelagic ecosystems lack the supporting structures and framework provided by large, dominant plant life. The dominant autotrophs are phytoplankton, and their major herbivores are tiny zooplankton.

There is a reason for the smallness of phytoplankton. Surrounded by a chemical medium that contains the nutrients necessary for life in varying quantities, they absorb nutrients directly from the water. The smaller the organism, the greater is the surface-to-volume ratio (see Section 7.1). More surface area is exposed for the absorption of nutrients and solar energy. Seawater is so dense that there is little need for supporting structures (see Section 3.2).

Because they require light, autotrophs are restricted to the upper surface waters where light penetration varies from tens to hundreds of meters (see Section 20.4, Figure 20.8). In shallow coastal waters, the dominant marine autotrophs are attached algae—restricted by light requirements to a maximum depth of about 120 m. The brown algae (Phaeophyceae) are the most abundant, associated with the rocky shoreline. Included in this group are the large kelps—such as species of *Macrocystis*, which grows to a length of 50 m and forms dense subtidal forests in the tropical and subtropical regions (see Figure 4.1). The red algae (Rhodophyceae) are the most widely distributed of the larger marine plants. They occur most abundantly in the tropical oceans, where some species grow to depths of 120 m.

The dominant autotrophs of the open water are phytoplankton (see Figure 24.5a). Each ocean or region within an ocean appears to have its own dominant forms. Littoral and neritic waters and regions of upwelling are richer in plankton than are the mid-oceans. In regions of downwelling, the dinoflagellates—a large, diverse group characterized by two whiplike flagella—concentrate near the surface in areas of low turbulence. These organisms attain greatest abundance in warmer waters. In summer, they may concentrate in the surface waters in such numbers that they color it red or brown. Often toxic to vertebrates, such concentrations of dinoflagellates are responsible for "red tides." In regions of upwelling, the dominant forms of phytoplankton are diatoms. Enclosed in a silica case, diatoms are particularly abundant in Arctic waters.

The **nanoplankton**, which are smaller than diatoms, make up the largest biomass in temperate and tropical waters. Most abundant are the tiny cyanobacteria. The haptophytes—a group of primarily unicellular, photosynthetic algae that includes more than 500 species—are distributed in all waters except the polar seas. The most important members of this group, the coccolithophores, are a major source of primary production in the oceans. Coccolithophores have an armored appearance because of the calcium carbonate platelets, called coccoliths, covering the exterior of the cell (**Figure 24.18**).

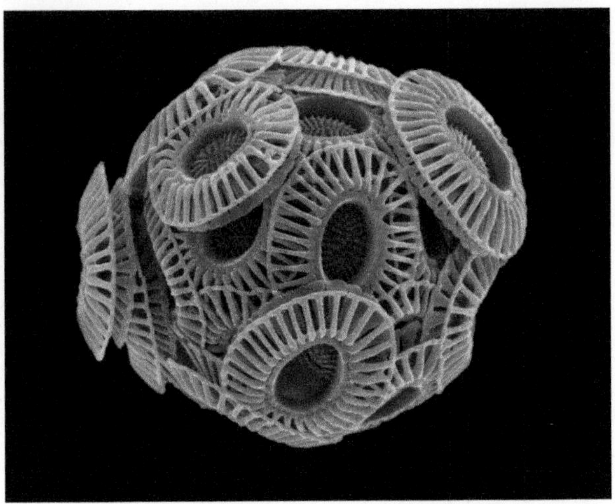

Figure 24.18 Coccolithophores have an armored appearance because of the calcium carbonate platelets, called coccoliths, covering the exterior of the cell.

Converting primary production into animal tissue is the task of herbivorous zooplankton, the most important of which are the copepods (see Figure 24.5b). To feed on the minute phytoplankton, most of the grazing herbivores must also be small—between 0.5 and 5 mm. Most grazing herbivores in the ocean are copepods, which are probably the most abundant animals in the world. In the Antarctic, the shrimplike euphausiids, or krill (**Figure 24.19**), fed on by baleen whales and penguins, are the dominant herbivores. Feeding on the herbivorous zooplankton are the carnivorous zooplankton, which include such organisms as the larval forms of comb jellies (Ctenophora) and arrow worms (Chaetognatha).

However, part of the food chain begins not with the phytoplankton, but with organisms even smaller. Bacteria and protists—both heterotrophic and photosynthetic—make up one-half of the biomass of the sea and are responsible for the largest part of energy flow in pelagic systems. Photosynthetic nanoflagellates (2–20 m) and cyanobacteria (1–2 m), responsible for a large part of photosynthesis in the sea, excrete a substantial fraction of their photosynthate in the form of dissolved organic material that heterotrophic bacteria use. In turn, heterotrophic nanoflagellates consume heterotrophic bacteria. This interaction introduces a feeding loop, the **microbial loop** (**Figure 24.20**), and adds several trophic levels to the plankton food chain.

Figure 24.19 Small euphausiid shrimps, called krill, are eaten by baleen whales and are essential to the marine food chain.

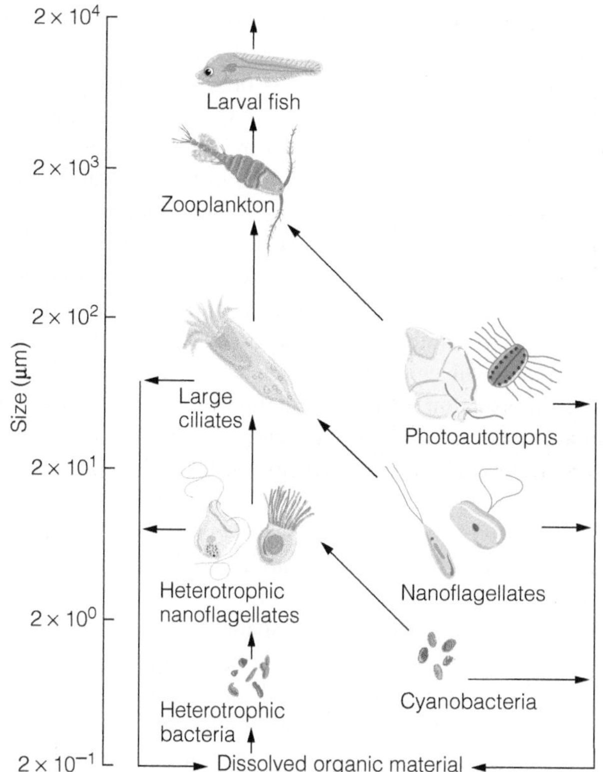

Figure 24.20 A representation of the microbial loop and its relationship to the plankton food web. Autotrophs are on the right side of the diagram, and heterotrophs are on the left.

Like phytoplankton, zooplankton live mainly at the mercy of the currents, but many forms of zooplankton have enough swimming power to exercise some control. Some species migrate vertically each day to arrive at a preferred level of light intensity. As darkness falls, zooplankton rise rapidly to the surface to feed on phytoplankton. At dawn, they move back down to preferred depths.

Feeding on zooplankton and passing energy along to higher trophic levels are the nekton, which are swimming organisms that can move at will in the water column. They range in size from small fish to large predatory sharks and whales, seals, and marine birds such as penguins. Some predatory fish, such as tuna, are more or less restricted to the photic zone. Others are found in deeper mesopelagic and bathypelagic zones or move between them, as the sperm whale does. Although the ratio in size of predator to prey falls within limits, some of the largest nekton organisms in the sea—the baleen whales (Mysticeti)—feed on disproportionately small prey, euphausiids, or krill (see Figure 24.19). By contrast, the sperm whale attacks very large prey such as the giant squid.

Residents of the deep have special adaptations for securing food. Darkly pigmented and weak bodied, many deep-sea fish depend on luminescent lures, mimicry of prey, extensible jaws, and expandable abdomens (enabling them to consume large items of food). In the mesopelagic region, bioluminescence reaches its greatest development—two-thirds of the species produce light. Fish have rows of luminous organs along their sides and lighted lures that enable them to bait prey and recognize other individuals of the same species. Bioluminescence is not restricted to fish. Squid and euphausiid shrimp possess searchlight-like structures complete with lens and iris, and some discharge luminous clouds to escape predators.

24.11 Benthos Is a World of Its Own

The term *benthic* refers to the floor of the sea, and **benthos** refers to plants and animals that live there. In a world of darkness, no photosynthesis takes place, so the bottom community is strictly heterotrophic (except in vent areas), depending entirely on the rain of organic matter drifting to the bottom. Patches of dead phytoplankton as well as the bodies of dead whales, seals, birds, fish, and invertebrates all provide an array of foods for different feeding groups and species. Despite the darkness and depth, benthic communities support a high diversity of species. In shallow benthic regions, the polychaete worms may be represented by more than 250 species, and the pericarid crustaceans by well more than 100. But the deep-sea benthos supports a surprisingly greater diversity. The number of species collected in more than 500 samples—of which the total surface area sampled was only 50 m^2—was 707 species of polychaetes and 426 species of pericarid crustaceans.

Important organisms in the benthic food chain are the bacteria of the sediments. Commonly found where large quantities of organic matter are present, bacteria may reach several tenths of a gram (g) per square meter in the topmost layer of silt. Bacteria synthesize protein from dissolved nutrients and in turn become a source of protein, fat, and oils for other organisms.

In 1977, oceanographers first discovered high-temperature, deep-sea hydrothermal vents along volcanic ridges in the ocean floor of the Pacific near the Galápagos Islands. These vents spew jets of superheated fluids that heat the surrounding water to 8 to 16°C, considerably higher than the 2°C ambient water. Since then, oceanographers have discovered similar vents on other volcanic ridges along fast-spreading centers of the ocean floor, particularly in the mid-Atlantic and eastern Pacific.

Vents form when cold seawater flows down through fissures and cracks in the basaltic lava floor deep into the underlying crust. The waters react chemically with the hot basalt, giving up some minerals but becoming enriched with others such as copper, iron, sulfur, and zinc. The water, heated to a high temperature, reemerges through mineralized chimneys rising up to 13 m above the sea floor. Among the chimneys are white smokers and black smokers (**Figure 24.21**). White-smoker chimneys rich in zinc sulfides issue a milky fluid with a temperature of less than 300°C. Black smokers, narrower chimneys rich in copper sulfides, issue jets of clear water from 300°C to more than 450°C that are soon blackened by precipitation of fine-grained sulfur–mineral particles.

Associated with these vents is a rich diversity of unique deep-sea life, confined to within a few meters of the vent

Figure 24.21 Black smoker. Deep under the oceans, next to hydrothermal vents issuing from cracked volcanic rocks (black smokers), live some of the most unusual animals ever seen. The photograph shows a black smoker in full flow, some 2250 m down on the ocean floor west of Vancouver Island, on the Juan de Fuca Ridge. The water temperatures in the chimneys exceed 400°C. Surrounding the vents are tubeworms, a bizarre group of animals found only in conditions where high levels of hydrogen sulfide can fuel their internal symbionts—bacteria that manufacture food for them. Distribution of this fauna is related to the history of the tectonic development of the ridges, thus linking the geology and biology of this peculiar ecosystem.

system. The primary producers are chemosynthetic bacteria that oxidize reduced sulfur compounds such as hydrogen sulfide (H_2S) to release energy used to form organic matter from carbon dioxide. Primary consumers include giant clams, mussels, and polychaete worms that filter bacteria from water and graze on the bacterial film on rocks (**Figure 24.22**).

24.12 Coral Reefs Are Complex Ecosystems Built by Colonies of Coral Animals

Lying in the warm, shallow waters about tropical islands and continental landmasses are coral reefs—colorful, rich oases within the nutrient-poor seas (**Figure 24.23**). They are a unique accumulation of dead skeletal material built up by carbonate-secreting organisms, mostly living coral (Cnidaria, Anthozoa) but also coralline red algae (Rhodophyta, Corallinaceae), green calcerous algae (*Halimeda*), foraminifera, and mollusks. Although various types of corals can be found from the water's surface to depths of 6000 m, reef-building corals are generally found at depths of less than 45 m. Because reef-building corals have a symbiotic relationship with algal cells, their distribution is limited to depths where sufficient solar radiation (photosynthetically active radiation) is available to support photosynthesis (zooxanthellae; see Section 15.10 and Figure 15.9). Precipitation of calcium from the water is necessary to form the coral skeleton. This precipitation occurs when water temperature and salinity are high and carbon dioxide concentrations are low. These requirements limit the distribution of reef-building corals to the shallow, warm tropical waters (20–28°C).

Coral reefs are of three basic types: (1) Fringing reefs grow seaward from the rocky shores of islands and continents. (2) Barrier reefs parallel shorelines of continents and islands and are separated from land by shallow lagoons. (3) Atolls are rings of coral reefs and islands surrounding a lagoon, formed when a volcanic mountain subsides beneath the surface. Such lagoons are about 40 m deep, usually connect to the open sea by breaks in the reef, and may have small islands of patch reefs. Reefs build up to sea level.

Coral reefs are complex ecosystems that begin with the complexity of the corals themselves. Corals are modular animals, anemone-like cylindrical polyps, with prey-capturing tentacles surrounding the opening or mouth. Most corals form sessile colonies supported on the tops of dead ancestors and cease growth when they reach the surface of the water. In the tissues of the gastrodermal layer live zooxanthellae, which are symbiotic, photosynthetically active, endozoic dinoflagellate algae that corals depend on for most efficient growth (see Chapter 15). On the calcareous skeletons live still other kinds of algae—the encrusting red and green coralline species and the filamentous species, including turf algae—and a large bacterial population. Also associated with coral growth are mollusks, such as giant clams (*Tridacna*, *Hippopus*), echinoderms, crustaceans, polychaete worms, sponges, and a diverse array of fishes, both herbivorous and predatory.

Figure 24.22 Pictured is a colony of giant tubeworms with vent fish and crabs, all highly specialized for and found only in the extreme environment of the hydrothermal vent ecosystem. (Courtesy of Richard Lutz, Rutgers University, Stephen Low Productions, and Woods Hole Oceanographic Institution.)

Figure 24.23 A rich diversity of coral species, algae, and colorful fish occupy this reef in Fiji (South Pacific Ocean).

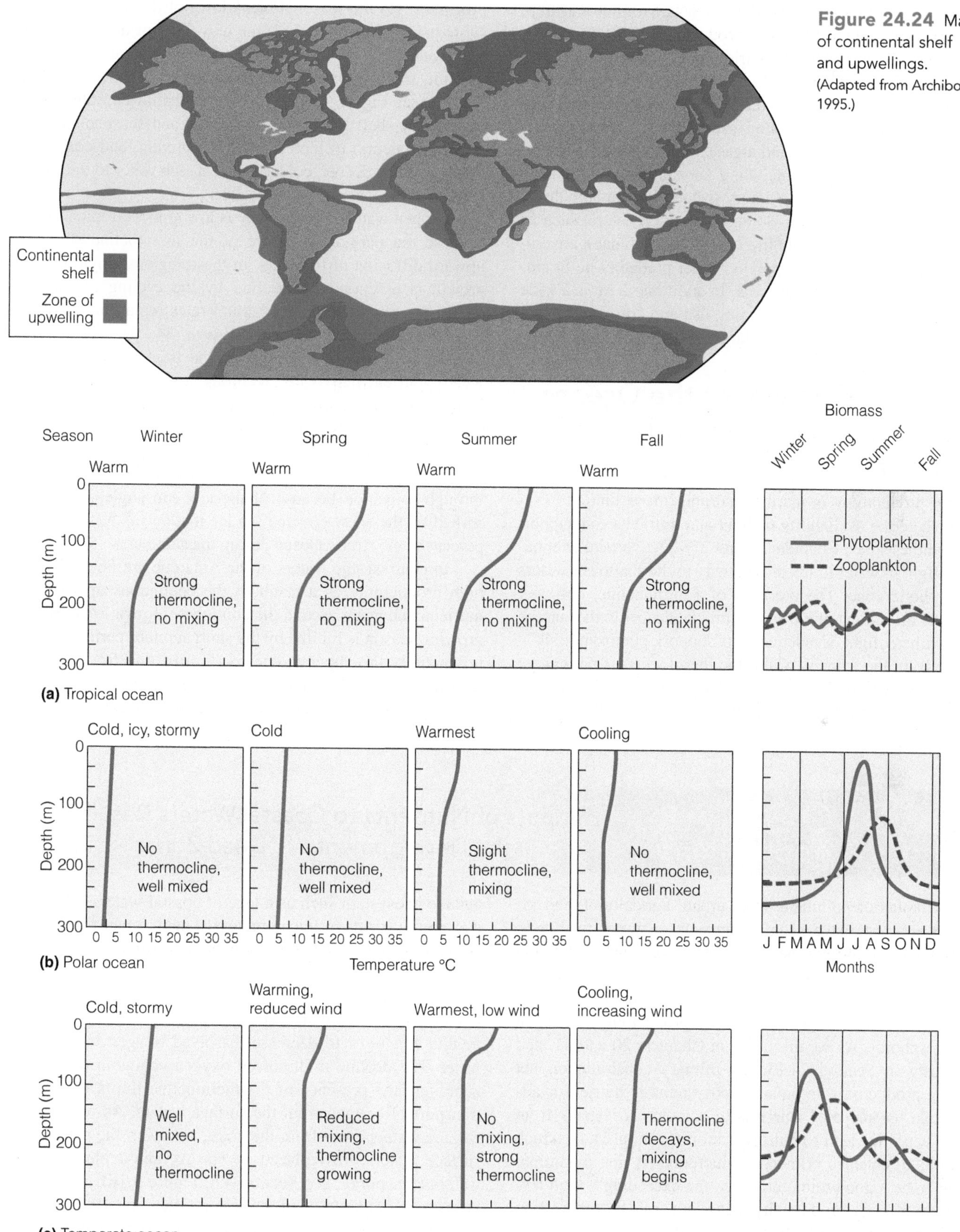

Figure 24.24 Map of continental shelf and upwellings.
(Adapted from Archibold 1995.)

Figure 24.25 Thermal profiles, extent of vertical mixing, and associated patterns of productivity in (a) tropical, (b) polar, and (c) temperate oceans during the four seasons of the year.
(From Nybakken 1997.)

Because the coralline community acts as a nutrient trap, offshore coral reefs are oases of productivity (1500 to 5000 g C/m^2/yr) within the relatively nutrient-poor, lower-productivity sea (15 to 50 g C/m^2/yr; see Figure 20.10). This productivity and the varied habitats within the reef support a great diversity of life—thousands of species of invertebrates (such as sea urchins, which feed on coral and algae). Many kinds of herbivorous fish graze on algae, including zooxanthellae within the coral tissues; hundreds of predatory species feed on both invertebrate and vertebrate prey. Some of these predators, such as puffers (Tetraodontidae) and filefish (Monacanthidae), are corallivores that feed on coral polyps. Other predators lie in ambush for prey in coralline caverns. In addition, there is a wide array of symbionts, such as cleaning fish and crustaceans, that pick parasites and detritus from larger fish and invertebrates.

24.13 Productivity of the Oceans Is Governed by Light and Nutrients

Primary productivity in marine environments is limited to regions where the availability of light and nutrients can support photosynthesis (see Chapters 20 and 21). The vertical attenuation of light in water limits productivity to the shallower waters of the photic zone. The presence of a thermocline, however, limits the movement of nutrients from the deeper to the surface waters where light is adequate to support photosynthesis—especially in the tropics, where the thermocline is permanent (see Section 21.10). The rate at which nutrients are returned to the surface, and therefore productivity, is controlled by two processes: (1) the seasonal breakdown of the thermocline and subsequent turnover, and (2) the upwelling of deeper nutrient-rich waters to the surface (see Sections 21.10 and 21.13). As a result, the highest primary productivity is found in coastal regions (see Figure 20.10). There, the shallower waters of the continental shelf allow for turbulence and seasonal turnover (where it occurs) to increase vertical mixing, and coastal upwelling brings deeper, colder, nutrient-rich water to the surface (**Figure 24.24**; see also Figure 3.16).

In open waters, productivity is low in most tropical oceans because the permanent nature of the thermocline slows the upward diffusion of nutrients. In these regions, phytoplankton growth is essentially controlled by the cycling of nutrients within the photic zone. Production rates remain more or less constant throughout the year (**Figure 24.25a**). The highest production in the open waters of the tropical oceans occurs where nutrient-rich water is brought to the surface in the equatorial regions, where upwelling occurs as surface currents diverge (see Figures 3.16 and 24.24).

Productivity is also low in the Arctic, mainly because of light limitations. A considerable amount of light energy is lost through reflection because of the low sun angle, or it is absorbed by the snow-covered sea ice that covers as much as 60 percent of the Arctic Ocean during the summer.

In contrast, the waters of the Antarctic are noted for their high productivity as a result of the continuous upwelling of nutrient-rich water around the continent (**Figure 24.25b**). The growing season is limited by the short summer period. Primary productivity in temperate oceans (**Figure 24.25c**) is strongly related to seasonal variation in nutrient supply, driven by the seasonal dynamics of the thermocline (see Section 21.10).

ECOLOGICAL Issues & Applications

Inputs of Nutrients to Coastal Waters Result in the Development of "Dead Zones"

Human inputs of pollutants from urban, agricultural, and industrial activities have negative impacts on water quality in both freshwater and marine ecosystems (see Chapter 22), and although it may at first seem counterintuitive, one of the major classes of pollutants are the essential mineral nutrients that support plant growth and net primary productivity—nitrogen and phosphorus. As we discussed in Chapters 20 and 21, and previously in Section 24.13, the primary constraint on net primary productivity in aquatic ecosystems is nutrient availability, so the input of nutrients to aquatic ecosystems from human activities lead to eutrophication (Section 24.4), which functions to enhance primary productivity. At first this might appear to be a good thing, however, the exceedingly high levels of net primary productivity that result from eutrophication can result in anoxia (depletion of oxygen) and the development of "dead zones."

The development of dead zones begins with the input of "unnaturally" high levels of nitrogen and phosphorus to an aquatic ecosystem such as a lake or coastal waters. The abundance of nutrients results in increased net primary productivity of phytoplankton and other aquatic autotrophs. As the autotrophs senesce they sink from the surface waters (photic zone) to the benthic zone, where bacteria feed on the dead organic matter. The increase in respiration by the decomposer community functions to decrease dissolved oxygen levels in the water. The decline in dissolved oxygen begins in the benthic zone, and the presence of the thermocline limits the vertical transport of oxygen from the surface waters. As winds move surface waters away from the coast, however, the oxygenated surface waters are replaced by the oxygen-depleted deeper waters (upwelling, see Section 3.8, Figure 3.16). This leaves no oxygen in any layer of the water. Often aquatic life moves toward the shoreline in an attempt to find oxygenated waters. Under extreme conditions, dissolved oxygen levels drop to such low levels that anoxia results in the death of marine organisms, hence the term *dead zone* (**Figure 24.26**).

Dead zones can be found worldwide (**Figure 24.27**). Marine dead zones can be found in the Baltic Sea, Black Sea, off the coast of Oregon, and in the Chesapeake Bay. Dead zones may also be found in lakes, such as Lake Erie. One of the largest dead zones in the world is in the Gulf of Mexico. The Gulf of Mexico dead zone is an area of hypoxic (less than 2 ppm dissolved oxygen) waters at the mouth of the Mississippi River. Its area varies in size, but cover an area of up to 7000 square miles. The zone occurs between the inner and mid-continental shelf region of the northern Gulf of Mexico, beginning at the Mississippi River delta and extending westward to the upper Texas coast (**Figure 24.28**).

The zone of anoxic water in the Gulf of Mexico is caused by nutrient enrichment (eutrophication) from the Mississippi River. Watersheds within the Mississippi River Basin drain much of the United States, from Montana to Pennsylvania and extend southward along the Mississippi River. Most of the nutrient input comes from major farming states in the Mississippi River Valley, including Minnesota, Iowa, Illinois, Wisconsin, Missouri, Tennessee, Arkansas, Mississippi, and Louisiana (**Figure 24.29a**). Nitrogen and phosphorous enter the river through upstream runoff of fertilizers, soil erosion, animal wastes, and sewage. An estimated 66 percent of nitrogen entering the Gulf of Mexico is derived from crop fields (largely corn and soybean) within the Mississippi River Basin (**Figure 24.29b**). The anoxic zone develops each spring as the rains leach chemical fertilizers from farm fields (see Chapter 21, *Ecological Issues & Applications*), and the size of the input and resulting dead zone is heavily influenced by weather conditions. For example, drought conditions in 2012 resulted in the fourth-smallest dead zones on record, measuring only 2889 square miles, an area slightly larger than Delaware.

The dead zone that forms in the northern Gulf of Mexico each summer threatens valuable commercial and recreational Gulf fisheries. Ongoing research indicates that long-term changes in species diversity and the structure of food webs are occurring, and numerous areas of the Gulf experience large-scale fish kills on an annual basis. The Gulf of Mexico currently supplies 72 percent of U.S. harvested shrimp, 66 percent of harvested oysters, and 16 percent of commercially harvested fish.

Figure 24.26 Dead fish from anoxic conditions in the Bayou Chaland area of Louisiana.

The future of the northern Gulf will require a concerted effort at the national level because the Mississippi River Basin encompasses such a vast area of the continent (see Figure 24.29a). The key to minimizing the Gulf dead zone is to address it at the source. There is a need to manage nutrients more efficiently in farm fields by using fewer fertilizers and adjusting the timing of fertilizer applications to limit runoff of excess nutrients into adjacent aquatic ecosystems. In addition, the restoration of wetlands and riparian ecosystems (see Chapter 25) can help to capture nutrients and reduce runoff. The federal government is also funding efforts to restore wetlands along the Gulf coast to naturally filter the water before it enters the Gulf.

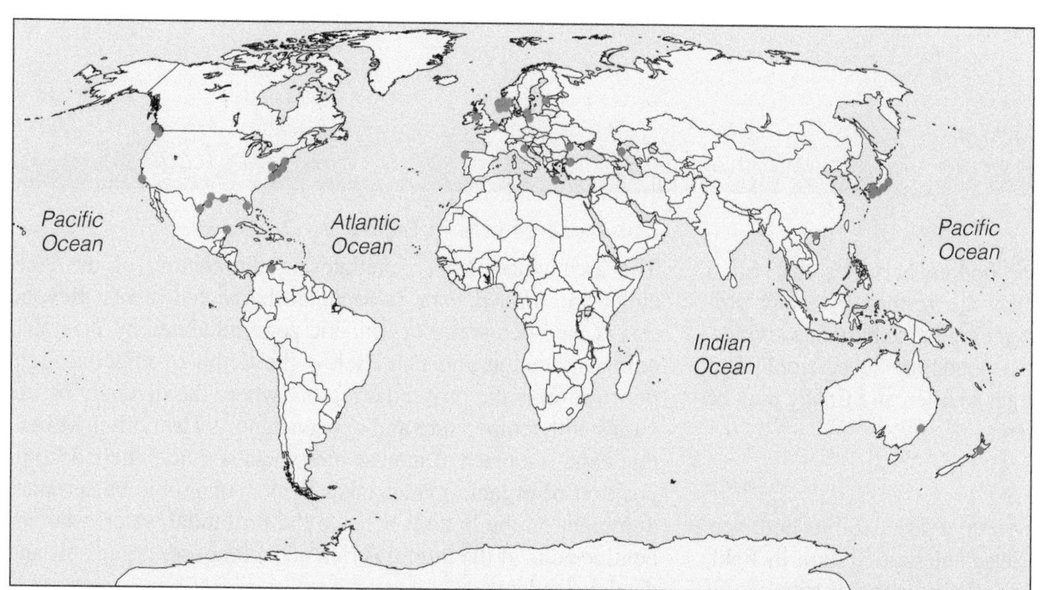

Figure 24.27 Dead zones occur in coastal waters and estuaries all over the globe, particularly near densely populated or intensively farmed areas. Each red dot on this map indicates a body of water prone to oxygen depletion and the formation of a dead zone.
(Map by Robert Simmon, based on data from Robert Diaz, Virginia Institute of Marine Science.)

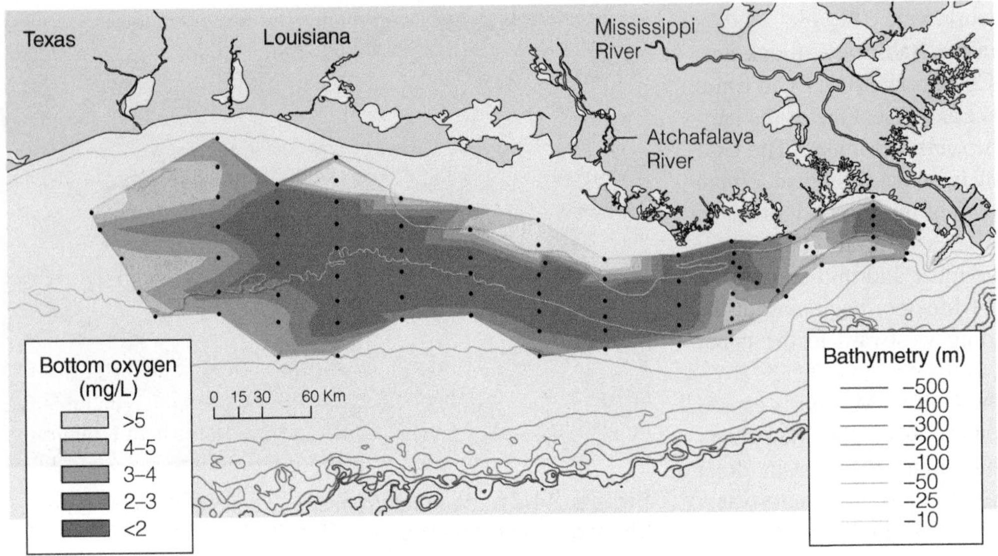

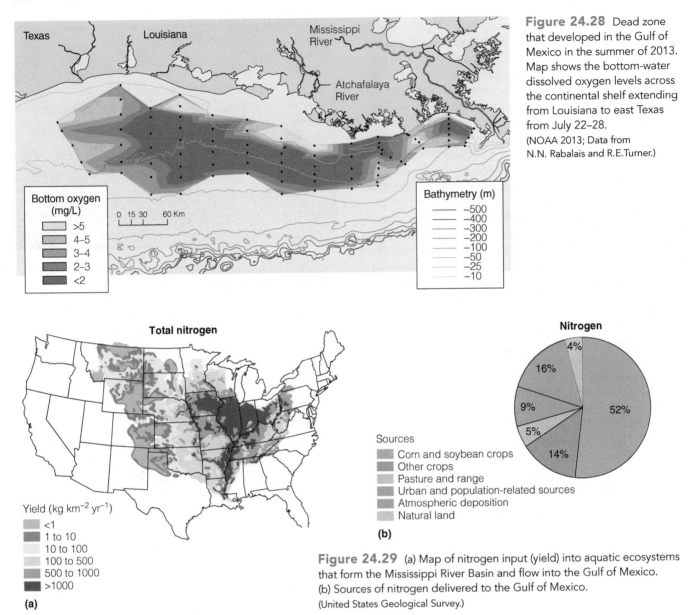

Figure 24.28 Dead zone that developed in the Gulf of Mexico in the summer of 2013. Map shows the bottom-water dissolved oxygen levels across the continental shelf extending from Louisiana to east Texas from July 22–28. (NOAA 2013; Data from N.N. Rabalais and R.E.Turner.)

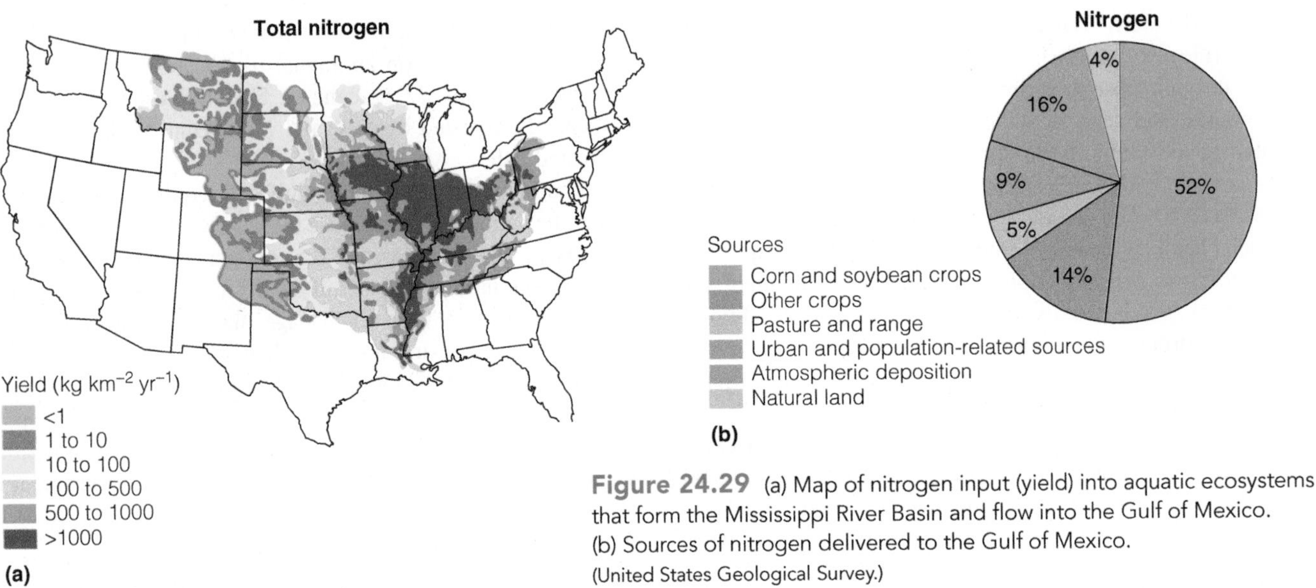

Figure 24.29 (a) Map of nitrogen input (yield) into aquatic ecosystems that form the Mississippi River Basin and flow into the Gulf of Mexico. (b) Sources of nitrogen delivered to the Gulf of Mexico. (United States Geological Survey.)

SUMMARY

Lakes Defined 24.1

Lake and pond ecosystems are bodies of water that fill a depression in the landscape. They are formed by many processes, ranging from glacial and geological to human activities. Geologically speaking, lakes and ponds are successional features. In time, most of them fill, get smaller, and finally may be replaced by a terrestrial ecosystem.

Lake Stratification 24.2

As a nearly self-contained ecosystem, a lake exhibits both vertical and horizontal gradients. Seasonal stratification in light, temperature, and dissolved gases influences the distribution of life in the lake.

Zonation of Life in Lakes 24.3

The area where light penetrates to the bottom of the lake, called the *littoral zone*, is occupied by rooted plants. Beyond this is the open-water or limnetic zone inhabited by plant and animal plankton and fish. Below the depth of effective light penetration is the profundal region, where the diversity of life varies with temperature and oxygen supply. The bottom or benthic zone is a place of intense biological activity where decomposition of organic matter takes place. Anaerobic bacteria are dominant on the bottom beneath the profundal water, whereas benthic zone of the littoral is rich in decomposer organisms and detritus feeders.

Nutrient Input into Lakes 24.4

Lakes are strongly influenced by their surrounding landscape. They may be classified as eutrophic (nutrient rich), oligotrophic (nutrient poor), or dystrophic (acidic and rich in humic material). Most lakes are subject to cultural eutrophication, which is the rapid addition of nutrients—especially nitrogen and phosphorus—from sewage, industrial wastes, and agricultural runoff.

Flowing-Water Habitat 24.5

Currents and their dependence on detrital material from surrounding terrestrial ecosystems set flowing-water ecosystems apart from other aquatic systems. Currents shape the life in streams and rivers and carry nutrients and other materials downstream. Flowing-water ecosystems change longitudinally in flow and size from headwater streams to rivers. They may be fast or slow, characterized by a series of riffles and pools.

Adaptations to Flowing Water 24.6

Organisms inhabiting fast-water streams are well-adapted to living in the current. They may be streamlined in shape, flattened to conceal themselves in crevices and underneath rocks, or attached to rocks and other substrates. In slow-flowing streams where current is minimal, streamlined forms of fish tend to be replaced by those with compressed bodies that enable them to move through aquatic vegetation. Burrowing invertebrates inhabit the silty bottom. Stream invertebrates fall into four major groups that feed on detrital material: shredders, collectors, grazers, and gougers.

River Continuum 24.7

Life in flowing water reflects a continuum of changing environmental conditions from headwater streams to the river mouth. Headwater streams depend on inputs of detrital material. As stream size increases, algae and rooted plants become important energy sources as reflected in the changing species composition of fish and invertebrate life. Large rivers depend on fine particulate matter and dissolved organic matter as sources of energy and nutrients. River life is dominated by filter feeders and bottom-feeding fish.

Estuaries 24.8

Rivers eventually reach the sea. The place where the one-way flow of freshwater meets the incoming and outgoing tidal water is an estuary. The intermingling of freshwater and tides creates a nutrient trap exploited by estuarine life. Salinity determines the nature and distribution of estuarine life. As salinity declines from the estuary up through the river, so do marine species. An estuary serves as a nursery for many marine organisms, particularly some of the commercially important finfish and shellfish because here the young are protected from predators and competing species unable to tolerate lower salinity.

Open Ocean 24.9

The marine environment is characterized by salinity, waves, tides, depth, and vastness. Like lakes, oceans are characterized by both stratification of temperature (and other physical parameters) and stratification of the organisms that inhabit the differing vertical strata. The open sea can be divided into several vertical zones. The hadalpelagic zone includes areas found in the deep-sea trenches and canyons. The abyssopelagic zone extends from the sea floor to a depth of about 4000 m. Above is the bathypelagic zone, void of sunlight and inhabited by darkly pigmented, bioluminescent animals. Above that lies the dimly lit mesopelagic zone, inhabited by characteristic species, such as certain sharks and squid. The bathypelagic and mesopelagic zones depend on a rain of detrital material from the upper lighted zone, the epipelagic zone, for their energy.

Ocean Life 24.10

Phytoplankton dominate the surface waters. The littoral and neritic zones are richer in plankton than the open ocean. Tiny nanoplankton, which make up the largest biomass in temperate and tropical waters, are the major source of primary production. Feeding on phytoplankton are herbivorous zooplankters, especially copepods. They are preyed on by carnivorous zooplankton. The greatest diversity of zooplankton, including larval forms of fish, occurs in the water over coastal shelves and upwellings; the least diversity occurs in the open ocean. Making up the larger life-forms are free-swimming nekton, ranging from small fish to sharks and whales. Benthic organisms (those living on the floor of the deep ocean) vary with depth and substrate. They are strictly heterotrophic and depend on organic matter that drifts to the bottom. They include filter feeders, collectors, deposit feeders, and predators.

Hydrothermal Vents 24.11

Along volcanic ridges are hydrothermal vents inhabited by unique and newly discovered life-forms, including crabs, clams, and worms. Chemosynthetic bacteria that use sulfates as an energy source account for primary production in these hydrothermal vent communities.

Coral Reefs 24.12

Coral reefs are nutrient-rich oases in nutrient-poor tropical waters. They are complex ecosystems based on anthozoan coral and coralline algae. Their productive and varied habitats support a high diversity of invertebrate and vertebrate life.

Ocean Productivity 24.13

Primary productivity in marine environments is limited to regions where the availability of light and nutrients can support photosynthesis and plant growth. The areas of highest productivity are coastal regions and areas of upwelling. In open oceans, especially in tropical areas, productivity is low because the permanent nature of the thermocline slows the upward diffusion of nutrients. Primary productivity in temperate oceans is strongly related to seasonal variation in nutrient supply, driven by the seasonal dynamics of the thermocline.

Dead Zones Ecological Issues & Applications

Inputs of nutrients to aquatic ecosystems from human activities result in eutrophication. The result is increased net primary productivity. With the death of autotrophs, increased decomposition can function to decrease the concentrations of dissolved oxygen, leading to anoxia and the development of dead zones in which low oxygen levels result in the death of marine organisms.

STUDY QUESTIONS

1. What distinguishes the littoral zone from the limnetic zone in a lake? What distinguishes the limnetic zone from the profundal zone?
2. What conditions distinguish the benthic zone from the other strata in lake ecosystems? What is the dominant role of the benthic zone?
3. Distinguish between oligotrophy, eutrophy, and dystrophy.
4. What physical characteristics are unique to flowing-water ecosystems? Contrast these conditions in a fast- and slow-flowing stream.
5. How do environmental conditions change along a river continuum?
6. What is an estuary?
7. Characterize the major life zones of the ocean—both vertical and horizontal.
8. How does temperature stratification in tropical seas differ from that of temperate regions of the ocean? How do these differences influence patterns of primary productivity?
9. What are hydrothermal vents, and what makes life around them unique?
10. What are coral reefs, and how do they form?
11. How can increased primary productivity in coastal waters resulting from the input of nutrients from human activities lead to anoxic conditions (low dissolved oxygen)?

FURTHER READINGS

Allan, J. D. 1995. *Stream ecology: Structure and functioning of running waters.* Dordrecht: Kluwer Academic Press.
An extensive reference on the ecology of stream ecosystems.

Grassle, J. F. 1991. "Deep-sea benthic diversity." *Bioscience* 41:464–469.
An excellent and well-written introduction for those interested in the strange and wonderful world of the deep ocean.

Gross, M. G., and E. Gross. 1995. *Oceanography: A view of Earth,* 7th ed. Englewood Cliffs, NJ: Prentice-Hall.
An excellent reference on the physical aspects of the sea.

Jackson, J. B. C. 1991. "Adaptation and diversity of reef corals." *Bioscience* 41:475–482.
This excellent review article relates patterns of species distribution to life history and disturbance.

Kaiser, M. J., M. Attrill, S. Jennings, D. N. Thomas, D. Barnes, A. S. Brierley, J. G. Hiddink, H. Kaartokallio, N. Polunin, and D. Raffaelli. 2011. *Marine ecology: Processes, systems, and impacts,* 2nd ed. New York, Oxford University Press.
Summarizes marine ecology from key processes involved in the marine environment to issues and challenges surrounding its future.

Nybakken, J. W., and M. D. Bertness. 2005. *Marine biology: An ecological approach,* 6th ed. Menlo Park, CA: Benjamin Cummings.
An excellent reference on marine life and ecosystems. Well written and illustrated.

Reddy, P. A. 2011. *Wetland ecology: Principles and conservation,* 2nd ed. New York: Cambridge University Press.
A well-organized text that covers the basic concepts of wetland ecology, drawing on examples from every major continent and wetland type.

Snelgrove, P. V. R. 2010. *Discoveries of the census of marine life: Making ocean life count.* New York: Cambridge University Press.
Filled with outstanding photographs, this book highlights some of the most dramatic findings from the recently completed 10-year Census of Marine Life. Based on the 10-year Census of Marine Life, involving a global network of 71 scientists in 80 nations assessing the diversity and abundance of life in the ocean.

Wetzel, R. G. 2001. *Limnology: Lake and river ecosystems,* 3rd ed. San Diego: Academic Press.
An outstanding reference that provides an introduction to the ecology of freshwater ecosystems.

Pond pine (*Pinus serotina*) and a diversity of floating and emergent aquatic plants dominate this area of cypress swamp in spring (Charleston, South Carolina).

CHAPTER GUIDE

WHEREVER LAND AND WATER MEET, there is a transitional zone that gives rise to a diverse array of unique ecosystems. In coastal environments, this zone lies between the terrestrial and marine environments. These environments are classified based on their underlying geology and substrate—sediment type, size, and shape. The product of marine erosion—the rocky shore—is the most primitive type of coast because it has been altered the least. Sandy beaches, found in wave-dominated, depositional settings, are highly dynamic environments subjected to continuous and often extreme change. Associated with estuarine environments or in the protected regions of coastal dunes are tidal mudflats, salt marshes, and mangrove forests.

The transitional zones between freshwater and land are characterized by terrestrial wetlands dominated by specialized plants that occur where the soil conditions remain saturated for most or all of the year. These are the marshes, swamps, bogs, and zones of emergent vegetation along rivers and lakes.

In this chapter text, we will examine these unique environments and the organisms that inhabit them.

(a)

25.1 The Intertidal Zone Is the Transition between Terrestrial and Marine Environments

Rocky, sandy, muddy, and either protected from or pounded by incoming swells, all intertidal shores have one feature in common: they are alternately exposed and submerged by the tides. Roughly, the region of the seashore is bounded on one side by the height of extreme high tide and on the other by the height of extreme low tide. Within these confines, conditions change hourly with the ebb and flow of the tides (see Section 3.9). At high tide, the seashore is a water world; at low tide, it belongs to the terrestrial environment, with its extremes in temperature, moisture, and solar radiation. Despite these changes, seashore inhabitants are essentially marine organisms adapted to withstand some degree of exposure to the air for varying periods of time.

At low tide, the uppermost layers of intertidal life are exposed to air, wide temperature fluctuations, intense solar radiation, and desiccation for a considerable period, whereas the lowest fringes on the intertidal shore may be exposed only briefly before the rising tide submerges them again. These varying conditions result in one of the most striking features of the coastal shoreline: the zonation of life.

25.2 Rocky Shorelines Have a Distinct Pattern of Zonation

All rocky shores have three basic zones (**Figure 25.1**), and each is characterized by dominant organisms (**Figure 25.2**). The approach to a rocky shore from the landward side is marked by a gradual transition from lichens and land plants to marine life dependent at least partly on the tidal waters. Moving from the terrestrial or **supralittoral** or **supratidal zone**, the first major change from the adjacent terrestrial environment appears at the **supralittoral fringe**, where saltwater comes only once every

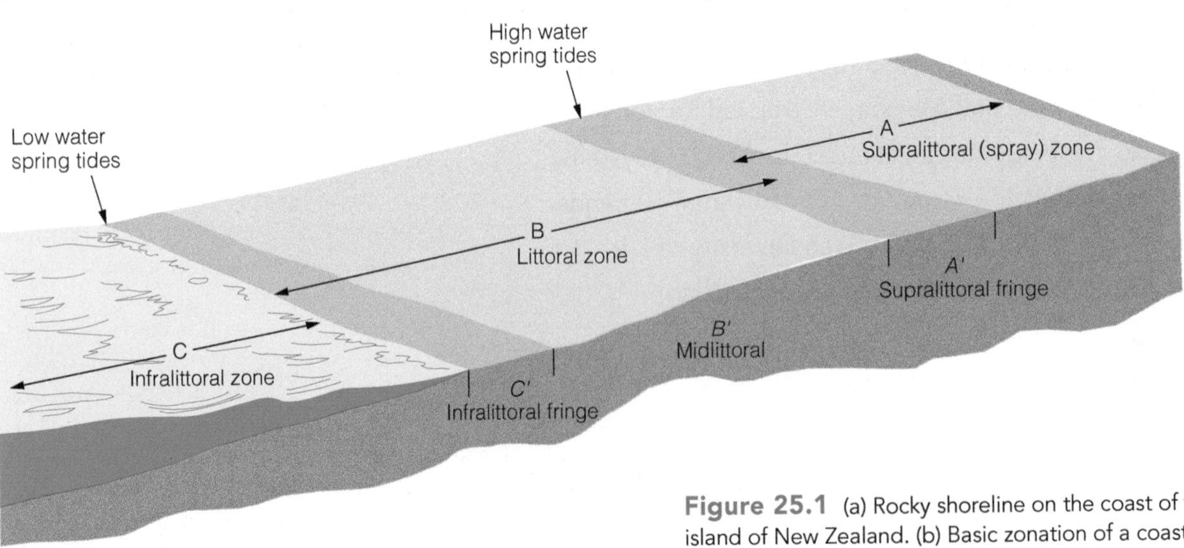

(b)

Figure 25.1 (a) Rocky shoreline on the coast of the south island of New Zealand. (b) Basic zonation of a coastal shoreline. Refer to this diagram while studying Figure 25.2.

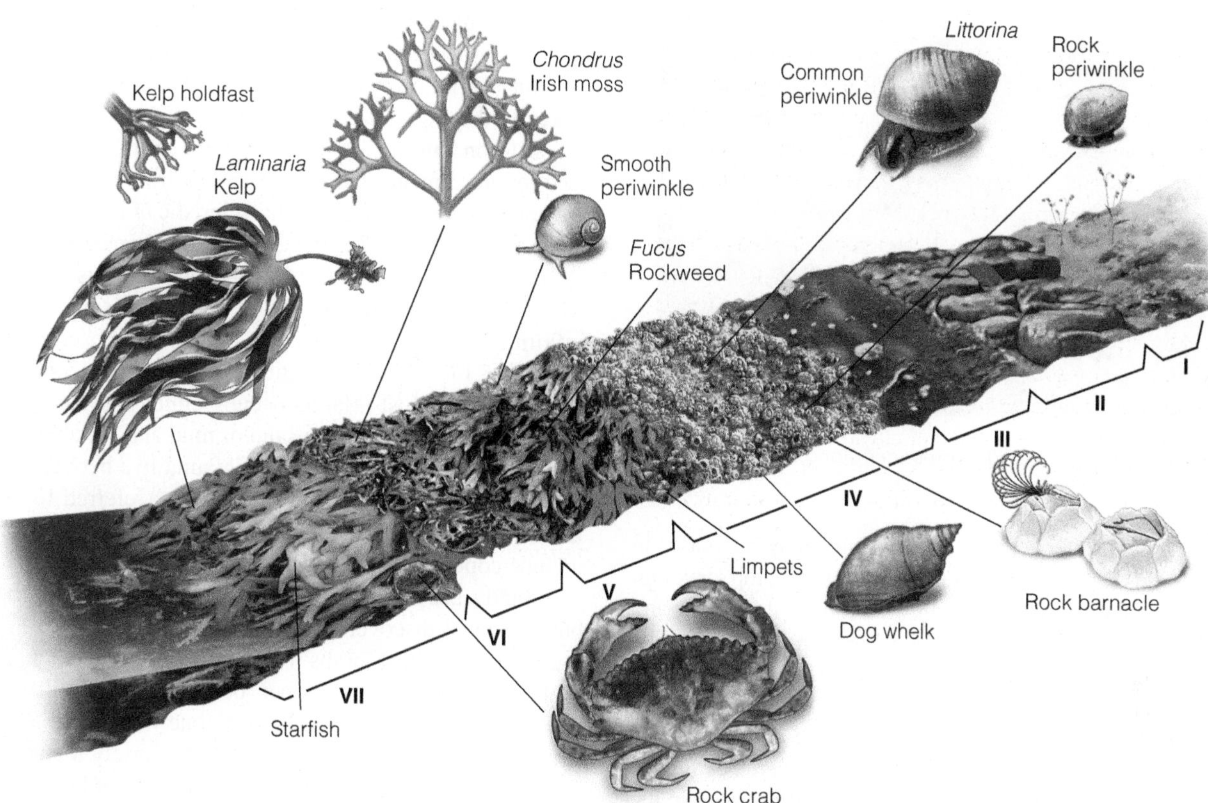

Figure 25.2 Zonation on a rocky shore along the North Atlantic. Compare with Figure 25.1b. I, land: lichens, herbs, grasses; II, bare rock; III, black algae and rock periwinkle (*Littorina*) zone; IV, barnacle (*Balanus*) zone: barnacles, dog whelks, common periwinkles, mussels, limpets; V, fucoid zone: rockweed (*Fucus*) and smooth periwinkles; VI, Irish moss (*Chondrus*) zone; VII, kelp (*Laminaria*) zone.

two weeks on the spring tides. It is marked by the black zone, named for the thin black layer of cyanobacteria (*Calothrix*) growing on the rock together with lichens (*Verrucaria*) and green alga (*Entophysalis*) above the high-tide waterline. Living under conditions that few plants could survive, these organisms represent an essentially nonmarine community. Common to this black zone are periwinkles, which are small snails of the genus *Littorina* (from which the term *littoral zone* is derived) that graze on wet algae covering the rocks.

Below the black zone lies the **littoral** or **intertidal zone**, which is covered and uncovered daily by the tides. In the upper reaches, barnacles are most abundant. Oysters, blue mussels, and limpets appear in the middle and lower portions of the littoral, as does the common periwinkle. Occupying the lower half of the littoral zone (midlittoral) of colder climates, and in places overlying the barnacles, is an ancient group of plants—brown algae, which are commonly known as rockweeds (*Fucus* spp.) and wrack (*Ascophyllum nodosum*). On the hard-surfaced shores that have been covered partly by sand and mud, blue mussels may replace the brown algae.

The lowest part of the littoral zone, uncovered only at the spring tides and not even then if wave action is strong, is the **infralittoral fringe**. This zone, exposed for short periods of time, consists of forests of large brown alga—*Laminaria* (one of the kelps)—with a rich undergrowth of smaller plants and animals among the holdfasts. Below the infralittoral fringe is the **infralittoral** or **subtidal zone**.

Grazing, predation, competition, larval settlement, and action of waves heavily influence the pattern of life on rocky shores. Waves bring in a steady supply of nutrients and carry away organic material. They keep the fronds of seaweeds in constant motion, moving them in and out of shadow and sunlight, allowing more even distribution of incident light and thus more efficient photosynthesis. By dislodging both plants and invertebrates from the rocky substrate, waves open up space for colonization by algae and invertebrates and reduce strong interspecific competition. Heavy wave action can reduce the activity of such predators as starfish and sea urchins that feed on sessile intertidal invertebrates. In effect, disturbance influences community structure.

The ebbing tide leaves behind pools of water in rock crevices, rocky basins, and depressions (**Figure 25.3**). These pools represent distinct habitats, which differ considerably from exposed rock and the open sea and even differ among themselves. At low tide, all pools are subject to wide and sudden fluctuations in temperature and salinity. Changes are most marked in shallow pools. Under the summer sun, temperatures may rise above the maximum that many organisms can tolerate. As water evaporates, especially in the shallower pools, salt crystals may appear around the edges. When rain or drainage from the surrounding

Figure 25.3 Tidal pools fill depressions along this length of rocky coastline in Maine.

land brings freshwater to the pools, salinity may decrease. In deep pools, this freshwater tends to form a layer on top, developing a strong salinity stratification in which the bottom layer and its inhabitants are little affected. If algal growth is considerable, oxygen will be high during the daylight hours but low at night, a situation that rarely occurs at sea. The rise of carbon dioxide at night lowers the pH (see Section 3.7). Most pools suddenly return to sea conditions on the rising tide and experience drastic and sudden changes in temperature, salinity, and pH. Life in the tidal pools must be able to withstand these extreme fluctuations.

25.3 Sandy and Muddy Shores Are Harsh Environments

Sandy and muddy shores often appear devoid of marine life at low tide in sharp contrast to the life-filled rocky shore (**Figure 25.4**), but sand and black mud are not as barren as they seem. Beneath them life lurks, waiting for the next high tide.

The sandy shore is a product of the harsh and relentless weathering of rock—both inland and along the shore. Rivers and waves carry the products of rock weathering and deposit them as sand along the edge of the sea. The size of the sand particles deposited influences the nature of the sandy beach, water retention during low tide, and the ability of animals to burrow through it. In sheltered areas of the coast, the slope of the beach may be so gradual that the surface appears to be flat. Because of the flatness, the outgoing tidal currents are slow, leaving behind a residue of organic material settled from the water. In these situations, mudflats develop.

Figure 25.4 A stretch of sandy beach washed by waves on the southern New Zealand coast. Although the beach appears barren, life is abundant beneath the sand.

Life on sand is almost impossible. Sand provides no surface for attachment of seaweeds and their associated fauna, and the crabs, worms, and snails characteristic of rocky crevices find no protection there. Life, then, is forced to live beneath the sand.

Life on sandy and muddy beaches consists of **epifauna**, which are organisms living on the sediment surface, and **infauna**, which are organisms living in the sediments (see Figure 16.17 for an illustration of patterns of zonation on a sandy beach). Most infauna occupy either permanent or semipermanent tubes within the sand or mud and are able to burrow rapidly into the substrate. Other infauna live between particles of sand and mud. These tiny organisms, referred to as **meiofauna**, range in size from 0.05 to 0.5 millimeters (mm) and include copepods, ostracods, nematodes, and gastrotrichs.

Sandy beaches also exhibit zonation related to the tides, but you must discover it by digging (see Figure 16.17). Pale, sand-colored ghost crabs (Ocypodinae) and beach hoppers (Talitridae) occupy the upper beach—the supralittoral. The intertidal beach—the littoral—is a zone where true marine life appears. Although sandy shores lack the variety found on rocky shores, the populations of individual species of largely burrowing animals often are enormous. An array of animals—among them starfish and the related sand dollar—can be found above the low-tide line in the littoral zone.

Organisms living within the sand and mud do not experience the same extreme fluctuations in temperature as do those on rocky shores. Although the surface temperature of the sand at midday may be 10°C (or more) higher than the returning seawater, the temperature a few centimeters below the sand remains almost constant throughout the year. Nor is there a great fluctuation in salinity, even when freshwater runs over the surface of the sand. Below 25 centimeters (cm), salinity is little affected.

Near and below the low-tide line live predatory gastropods, which prey on bivalves beneath the sand. In the same area lurk predatory portunid crabs (Portunidae) such as the blue crab (*Callinectes sapidus*) and green crab (*Carcinus maenas*) that feed on mole crabs (*Emerita*), clams, and other organisms. These species move back and forth with the tides. The incoming tides also bring small predatory fish, such as killifish and silversides. As the tide recedes, gulls and shorebirds scurry across the sand and mudflats to hunt for food.

The energy base for life on the sandy shore is an accumulation of organic matter. Most sandy beaches contain a certain amount of detritus from seaweeds, dead animals, and feces brought in by the tides. This organic matter accumulates within the sand in sheltered areas. It is subject to bacterial decomposition, which is most rapid at low tide. Some detrital-feeding organisms ingest organic matter largely as a means of obtaining bacteria. Prominent among them are many nematodes and copepods (Harpacticoida), polychaete worms (*Nereis*), gastropod mollusks, and lugworms (*Arenicola*), which are responsible for the conspicuous coiled and cone-shaped casts on the beach. Other sandy beach animals are filter feeders that obtain their food by sorting particles of organic matter from tidal water. Two of these—alternately advancing and retreating with the tide—are the mole crab and the coquina clam (*Donax*).

25.4 Tides and Salinity Dictate the Structure of Salt Marshes

Salt or **tidal marshes** occur in temperate latitudes where coastlines are protected from the action of waves within estuaries, deltas, and by barrier islands and dunes (**Figure 25.5**). The structure of a salt marsh is dictated by tides and salinity, which create a complex of distinctive and clearly demarked plant communities.

From the edge of the sea to the high land, zones of vegetation distinctive in form and color develop, reflecting a microtopography that lifts the plants to various heights within and above high tide (see Figure 16.16 for an illustration of vegetation zonation in a coastal salt marsh). Commonly found on the seaward edge of marshes and along tidal creeks of the Eastern coastline in North America are tall, deep green growths of salt-marsh cordgrass, *Spartina alterniflora*. Cordgrass forms a marginal strip between the open mud to the front and the high marsh behind. It has a high tolerance for salinity and is able to live in a semi-submerged state. To get air to its roots, which are buried in anaerobic mud, cordgrass has hollow tubes leading from the leaf to the root through which oxygen diffuses.

Above and behind the low marsh is the high marsh, standing at the level of mean high water. At this level, tall salt-marsh cordgrass gives way rather abruptly to a short form. This shorter form of *Spartina* is yellowish, in contrast to the tall, dark green form. This short form is an example of phenotypic plasticity in response to environmental conditions of the high marsh (see Chapter 5). The high marsh has a higher salinity and a decreased input of nutrients that result from a lower tidal exchange rate than in the low marsh. Here also grow the fleshy, translucent glassworts (*Salicornia* spp.; **Figure 25.6**) that turn bright red in fall, sea lavender (*Limonium carolinianum*), spearscale (*Atriplex patula*), and sea blite (*Suaeda maritima*).

Where the microelevation is about 5 cm above mean high water, short *Spartina alterniflora* and its associates are replaced by salt meadow cordgrass (*Spartina patens*) and an associate, spikegrass or saltgrass (*Distichlis spicata*). As the microelevation rises several more centimeters above mean high tide, and if there is some intrusion of freshwater, *Spartina* and *Distichlis* may be replaced by two species of black needlerush or black grass (*Juncus roemerianus* and *Juncus gerardi*)—so called because their dark green color becomes almost black in the fall. Beyond the black grass and often replacing it is a shrubby growth of marsh elder (*Iva frutescens*) and groundsel (*Baccharis halimifolia*). On the upland fringe grow bayberry (*Myrica pensylvanica*) and the pink-flowering sea holly (*Hibiscus palustris*).

Two conspicuous features of a salt marsh are the salt pans interspersed among meandering creeks. The creeks form an intricate system of drainage channels that carry tidal waters back out to sea (**Figure 25.7**). Their exposed banks support

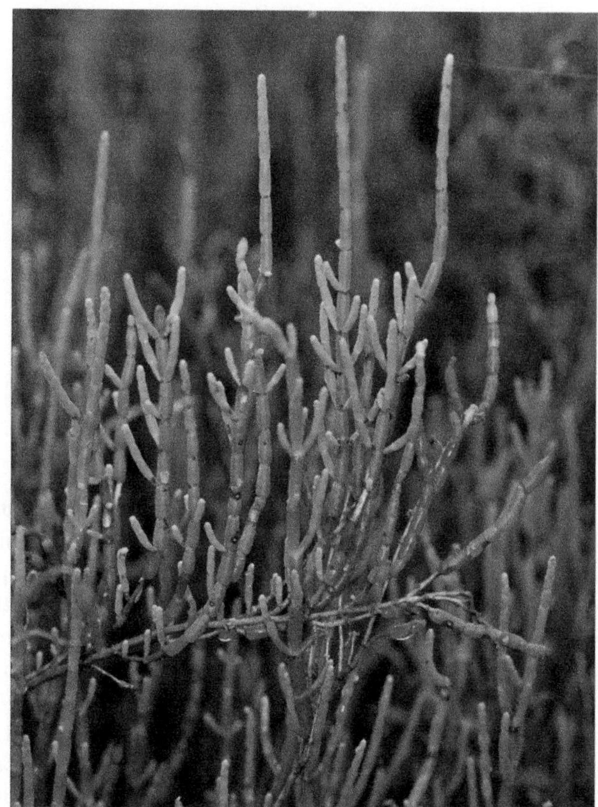

Figure 25.6 Glasswort dominates highly saline areas on the salt marsh. The plant, which turns red in fall, is a major food of overwintering geese.

Figure 25.5 Coastal salt marsh.

Figure 25.7 A tidal creek at high tide on the high marsh. Tall *Spartina* grows along the banks.

Figure 25.8 A salt pan or pool in the high marsh.

a dense population of mud algae, diatoms, and dinoflagellates that are photosynthetically active all year. Salt pans are circular to elliptical depressions flooded at high tide. At low tide, they remain filled with saltwater. If the pans are shallow enough, the water may evaporate completely, leaving an accumulating concentration of salt on the mud. The edges of these salt flats may be invaded by glasswort and spikegrass (**Figure 25.8**).

Although the salt marsh is not noted for its diversity, it is home to several interesting organisms. Some of the inhabitants are permanent residents in the sand and mud, others are seasonal visitors, and most are transients coming to feed at high and low tide.

Three dominant animals of the low marsh are ribbed mussels (*Modiolus demissus*), buried halfway in the mud, fiddler crabs (*Uca* spp.), running across the marsh at low tide, and marsh periwinkles (*Littorina* spp.) that move up and down the stems of *Spartina* and onto the mud to feed on alga. Three conspicuous vertebrate residents of the low marsh of eastern North America are the diamond-backed terrapin (*Malaclemys terrapin*), clapper rail (*Rallus longirostris*), and seaside sparrow (*Ammospiza maritima*).

In the high marsh, animal life changes as abruptly as the vegetation. The small, coffee-colored pulmonate snail (*Melampus*)—found by the thousands under the low grass—replaces the marsh periwinkle. The willet (*Catoptrophorus semipalmatus*) and seaside sharp-tailed sparrow (*Ammospiza caudacuta*) replace the clapper rail and seaside sparrow.

Low tide brings a host of predators into the marsh to feed. Herons, egrets, gulls, terns, willets, ibis, raccoons, and others spread over the exposed marsh floor and muddy banks of tidal creeks. At high tide, the food web changes as the tide waters flood the marsh. Small predatory fish such as the silversides (*Menidia menidia*), killifish (*Fundulus heteroclitus*), and four-spined stickleback (*Apeltes quadracus*), which are restricted to channel waters at low tide, spread over the marsh at high tide, as does the blue crab.

25.5 Mangroves Replace Salt Marshes in Tropical Regions

Replacing salt marshes on tidal flats in tropical regions are **mangrove forests** or **mangals** (**Figure 25.9**), which cover 60 to 75 percent of the coastline of the tropical regions. Mangrove

Figure 25.9 Mangroves replace tidal marshes in tropical regions.

forests develop where wave action is absent, sediments accumulate, and the muds are anoxic (without oxygen). They extend landward to the highest vertical tidal range, where they may be only periodically flooded. The dominant plants are mangroves, which include 8 families and 12 genera dominated by *Rhizophora*, *Avicennia*, *Bruguiera*, and *Sonneratia*. Growing with them are other salt-tolerant plants, mostly shrubs.

In growth form, mangroves range from short, prostrate forms to timber-size trees 30 meters (m) high. All mangroves have shallow, widely spreading roots, and many have prop roots coming from trunk and limbs (**Figure 25.10**). Many species have root extensions called *pneumatophores* that take in oxygen for the roots. The tangle of prop roots and pneumatophores slows the movement of tidal waters, allowing sediments to settle out. Land begins to move seaward, followed by colonizing mangroves.

Mangrove forests support a rich fauna, with a unique mix of terrestrial and marine life. Living and nesting in the upper branches are many species of birds, particularly herons and egrets. *Littorina* snails live on the prop roots and trunks of mangrove trees. Also attached to the stems and prop roots are oysters and barnacles, and on the mud at the base of the roots are detritus-feeding snails. Fiddler crabs and tropical land crabs burrow into the mud during low tide and live on prop roots and high ground during high tide. In the Indo-Malaysian mangrove forests live mudskippers, fish of the genus *Periophthalmus*, with modified eyes set high on the head. They live in burrows

Figure 25.10 Interior of a red mangrove stand. Note the prop roots.

in the mud and crawl about on the top of it. In many ways they act more like amphibians than fish. The sheltered waters about the roots provide a nursery and haven for the larvae and young of crabs, shrimp, and fish.

25.6 Freshwater Wetlands Are a Diverse Group of Ecosystems

The transitional zones between freshwater and land are characterized by terrestrial wetlands. These unique environments form ecotones between terrestrial and adjacent aquatic ecosystems, sharing characteristics of both. Wetlands cover 6 percent of Earth's surface. They are found in every climatic zone but are local in occurrence. Only a few—such as the Everglades in Florida, the Pantanal in Brazil, the Okavango in southern Africa (**Figure 25.11**), and the Fens of England—cover extensive areas of the landscape (see this chapter, *Ecological Issues & Applications*).

Wetlands range along a gradient from permanently flooded to periodically saturated soil (**Figure 25.12**) and support specialized plants that occur where the soil conditions remain saturated for most or all of the year. These hydrophytic (water-adapted) plants are adapted to grow in water or on soil that is periodically anaerobic (lacking oxygen) because of excess water (see Chapter 6). **Hydrophytic plants** are typically classified into one of three groups: (1) obligate wetland plants that require saturated soils, which include the submerged pondweeds, floating pond lily, emergent cattails and bulrushes, and trees such as bald cypress (*Taxodium distichum*); (2) facultative wetland plants that can grow in either saturated or upland soil and rarely grow elsewhere, such as certain sedges and alders, and trees such as red maple (*Acer rubrum*) and cottonwoods (*Populus* spp.); and (3) occasional wetland plants that are usually found out of wetland environments but can tolerate wetlands. The third group of plants is critical in determining the upper limit of a wetland along a gradient of soil moisture.

Wetlands most commonly occur in three topographic situations (**Figure 25.13**). Basin wetlands develop in shallow basins, ranging from upland depressions to filled-in lakes and ponds. Riverine wetlands develop along shallow and periodically flooded banks of rivers and streams. Fringe wetlands occur along the coasts of large lakes. The three types are partially separated because of the direction of water flow (see Figure 25.13). Water flow in basin wetlands is vertical as a result of precipitation and the downward infiltration of water into the soil. In riverine wetlands, water flow is unidirectional. In fringe wetlands, flow is bidirectional because it involves rising lake levels or tidal action. These flows transport nutrients and sediments into and out of the wetland.

Wetlands dominated by emergent herbaceous vegetation are **marshes** (**Figure 25.14**). With their reeds, sedges, grasses, and cattails, marshes are essentially wet grasslands. Forested wetlands are commonly called **swamps** (**Figure 25.15**). They may be deep-water swamps dominated by cypress (*Taxodium* spp.), tupelo (*Nyssa* spp.), and swamp oaks (*Quercus* spp.); or they may be shrub swamps dominated by alder (*Alnus* spp.)

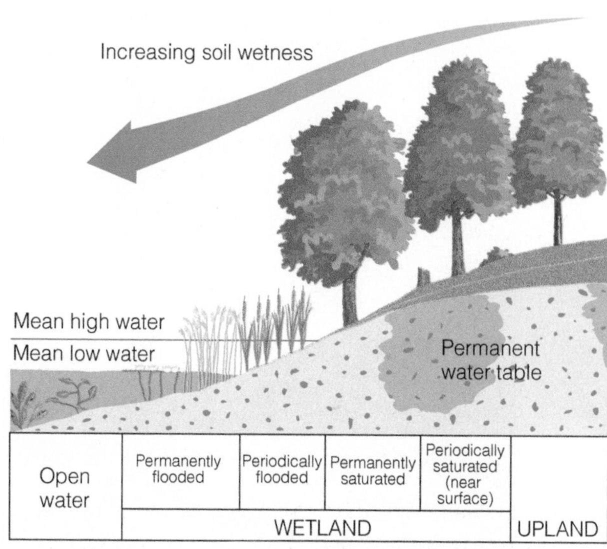

Figure 25.12 Location of wetlands along a soil-moisture gradient.

Figure 25.11 (a) Okavango Delta as photographed from the space shuttle. (b) The dry northwest corner of Botswana is the starting point for the annual summer floods of the Okavango River, which spills down through a vast network of narrow waterways, lagoons, oxbow lakes, and floodplains that covers some 22,000 km^2 of Botswana.

(a)

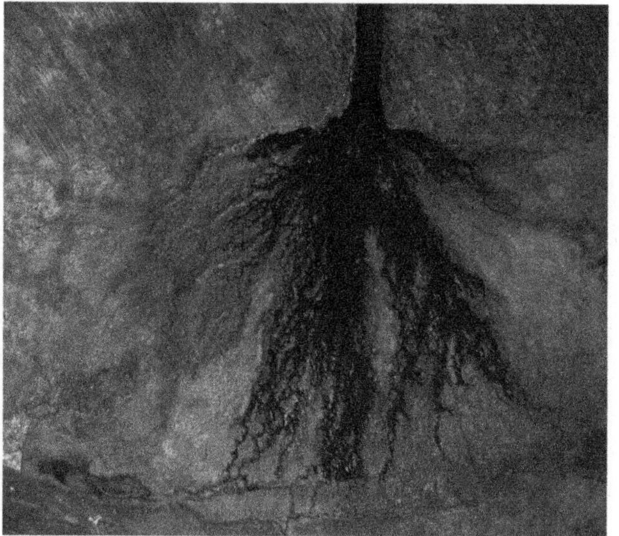

(b)

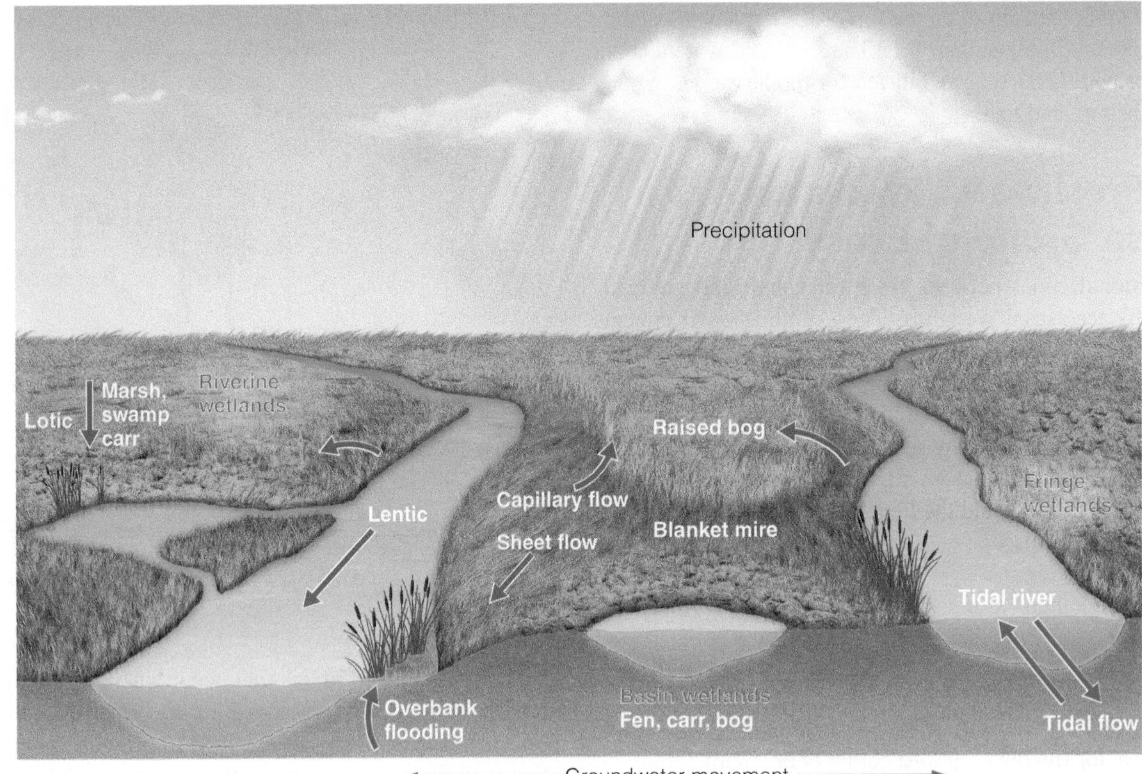

Figure 25.13 Water flow in various types of freshwater wetlands.

and willows (*Salix* spp.). Along many large river systems are extensive tracts of **bottomland** or **riparian woodlands** (**Figure 25.16**), which are occasionally or seasonally flooded by river waters but are dry for most of the growing season.

Wetlands that are characterized by an accumulation of partially decayed organic matter with time are called **peatlands** or **mires** (**Figure 25.17**). Organic matter accumulates because it is produced faster than it can decompose. The water table is at or near the soil surface, which creates anaerobic conditions that slow microbial activity. Mires that are fed by groundwater moving through mineral soil, from which they obtain most of their nutrients, and are dominated

Figure 25.14 The Horicon marsh in Wisconsin is an outstanding example of a northern marsh with well-developed emergent vegetation and patches of open water—an ideal environment for wildlife.

Figure 25.15 (a) A cypress deep-water swamp in the southern United States. (b) An alder (*Alnus*) shrub swamp with an herbaceous understory of skunk cabbage (*Symplocarpus foetidus*).

(a)

(b)

Figure 25.16 A riparian forest in Alabama.

Figure 25.17 An upland black spruce–tamarack bog in the Adirondack Mountains of New York.

(a)

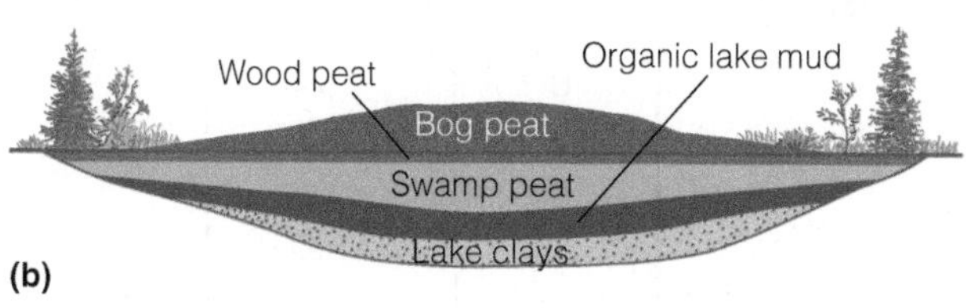

(b)

Figure 25.18 A raised bog (a) develops when an accumulation of peat (b) rises above the surrounding landscape.

by sedges are known as **fens**. Mires dependent largely on precipitation for their water supply and nutrients and that are dominated by *Sphagnum* are **bogs**. Mires that develop in upland situations—where decomposed, compressed peat forms a barrier to the downward movement of water, resulting in a perched water table (zone of saturation above an impermeable horizon) above mineral soil—are **blanket mires** and **raised bogs** (**Figure 25.18**). Raised bogs are popularly known as **moors**. Because bogs depend on precipitation for nutrient inputs, they are highly deficient in mineral salts and low in pH. Bogs also develop when a lake basin fills with sediments and organic matter carried by inflowing water. These sediments divert water around the lake basin and raise the surface of the mire above the influence of groundwater. Other bogs form when a lake basin fills in from above rather than from below, creating a floating mat of peat over open water. Such bogs are often termed *quaking bogs* (**Figure 25.19**).

25.7 Hydrology Defines the Structure of Freshwater Wetlands

Wetland structure is influenced by the phenomenon that creates it: its hydrology. Hydrology has two components. One involves the physical aspects of water and its movement: precipitation, surface and subsurface flow, direction and kinetic energy of water, and chemistry of the water. The other component is the **hydroperiod**, which involves duration, frequency, depth, and season of flooding. The length of the hydroperiod varies among types of wetlands. Basin wetlands have a longer hydroperiod. They usually flood during periods of high rainfall and draw down during dry periods. Both phenomena appear to be essential to the long-term existence of wetlands. Riverine wetlands have a short period of flooding associated with peak stream flow. The hydroperiod of fringe wetlands, influenced by wind and lake waves, may be short and regular and does not undergo the seasonal fluctuation characteristic of many basin marshes.

Hydroperiod influences plant composition because it affects germination, survival, and mortality at various stages of the plants' life cycles. The effect of hydroperiod is most pronounced in basin wetlands, especially those in the prairie regions of North America. In basins (called *potholes* in the prairie region) deep enough to have standing water throughout periods of drought, the dominant plants are submergents (**Figure 25.20**). If the wetland goes dry annually or during a period of drought, tall or midheight emergent species such as cattails dominate the marsh. If the pothole is shallow and flooded only briefly in the spring, then grasses, sedges, and forbs will make up a wet-meadow community.

If the basin is deep enough toward its center as well as large enough, then zones of vegetation may develop, ranging from submerged plants to deep-water emergents such as cattails and bulrushes, shallow-water emergents, and wet-ground species such as spike rush. Zonation reflects the response of plants to the hydroperiod. Those areas of wetland subject to a long hydroperiod support submerged

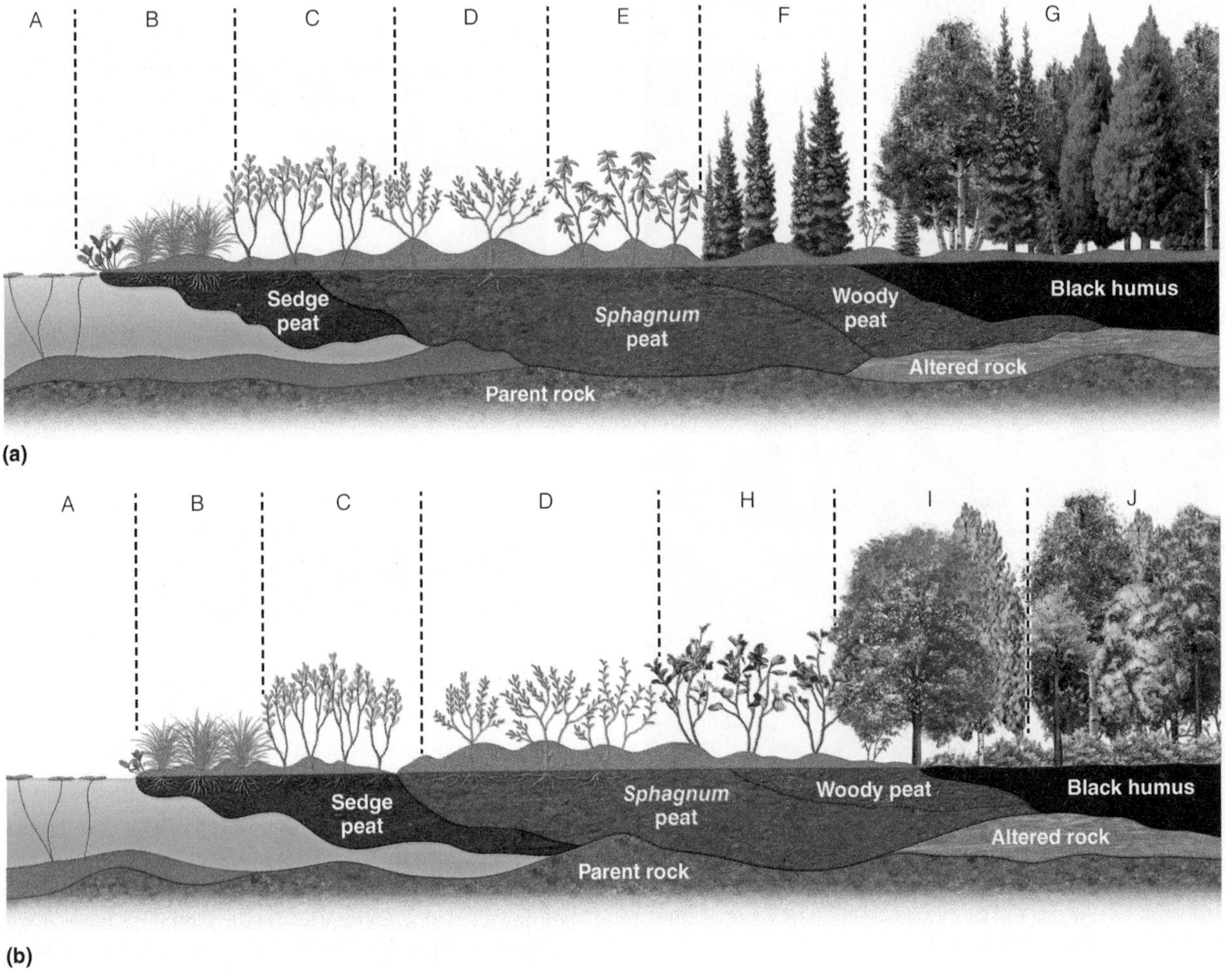

Figure 25.19 (a) Transect through a quaking bog, showing zones of vegetation, *Sphagnum* mounds, peat deposits, and floating mats. A, pond lily in open water; B, buckbean (*Menyanthes trifoliata*) and sedge; C, sweetgale (*Myrica gale*); D, leatherleaf (*Chamaedaphne calyculata*); E, Labrador tea (*Ledum groenlandicum*); F, black spruce; G, birch–black spruce–balsam fir forest. (b) An alternative vegetational sequence. H, alder; I, aspen, red maple; J, mixed deciduous forest.

and deep-water emergents; those with a short hydroperiod and shallow water are occupied by shallow-water emergents and wet-ground plants.

Periods of drought and wetness can induce vegetation cycles associated with changes in water levels. Periods of above-normal precipitation can raise the water level and drown the emergents to create a lake marsh dominated by submerged plants. During a drought, the marsh bottom is exposed by receding water, stimulating seed germination in the emergents and annuals characteristic of mudflats. When water levels rise again, the mudflat species drown, and the emergents survive and spread vegetatively.

Figure 25.20 Prairie potholes.

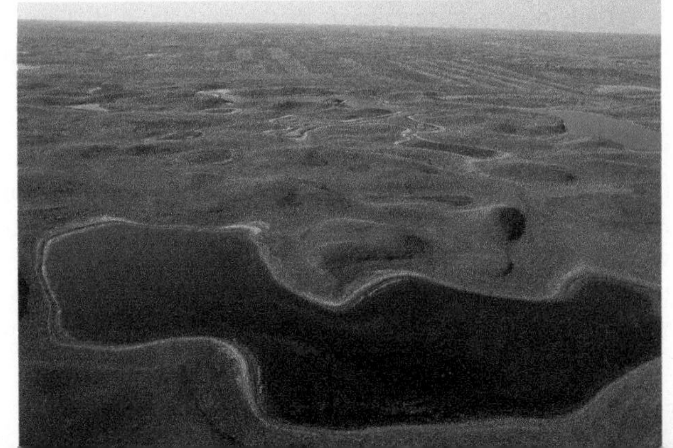

Peatlands differ from other freshwater wetlands in the accumulation of peat that results because organic matter is produced faster than it can be decomposed. In northern regions, acid-forming, water-holding *Sphagnum* add new growth on top of the accumulating remains of past moss generations, and their spongelike ability to hold water increases water retention on the site. As the peat blanket thickens, the water-saturated mat of moss and associated vegetation is raised above and insulated from mineral soil. The peat mat then becomes its own reservoir of water, creating a perched water table.

Peat bogs and mires generally form under oligotrophic and dystrophic conditions (see Section 24.4). Although usually associated with and most abundant in boreal regions of the Northern Hemisphere, peatlands also exist in tropical and subtropical regions. They develop in mountainous and coastal regions where hydrological conditions encourage an accumulation of partly decayed organic matter. Examples in coastal regions are the Everglades in Florida and the pocosins on the coastal plains of the southeastern United States.

25.8 Freshwater Wetlands Support a Rich Diversity of Life

Biologically, freshwater wetlands are among the richest and most interesting ecosystems. They support a diverse community of benthic, limnetic, and littoral invertebrates, especially crustaceans and insects. These invertebrates, along with small fishes, provide a food base for waterfowl, herons, gulls, and other birds, and supply the fat-rich nutrients ducks need for egg production and the growth of young. Amphibians and reptiles, notably frogs, toads, and turtles, inhabit the emergent growth, soft mud, and open water of marshes and swamps.

Herbivores are a conspicuous component of animal life. The dominant herbivore in prairie marshes is the muskrat (*Ondatra zibethicus*). Muskrats are the major prey for mink (*Mustela vison*), the dominant carnivore on the marshes. Other predators, including the raccoon, fox, weasel, and skunk, can seriously reduce the reproductive success of waterfowl on small marshes.

ECOLOGICAL Issues & Applications

Wetland Ecosystems Continue to Decline as a Result of Land Use

For centuries, we have looked at wetlands as forbidding, mysterious places: sources of pestilence, home to dangerous and pestiferous insects, and the abode of slimy, sinister creatures that rise out of swamp waters. They have been looked upon as places that should be drained for more productive uses by human standards: agricultural land, solid waste dumps, housing, industrial developments, and roads. The Romans drained the great marshes around the Tiber to make room for the city of Rome. William Byrd described the Great Dismal Swamp on the Virginia–North Carolina border as a "horrible desert, the foul damps ascend without ceasing." Despite the enormous amount of vacant dry land available in 1763, a corporation called the Dismal Swamp Land Company, owned in part by George Washington, failed in an attempt to drain the western end of the swamp for farmland. Although severely affected over the past 200 years, much of the swamp remains as a wildlife refuge.

Rationales for draining wetlands are many. The most persuasive relates to agriculture. Drainage of wetlands opens many hectares of rich organic soil for crop production. In prairie country, agriculturalists viewed the innumerable potholes (see Figure 25.20) as an impediment to efficient farming. Draining them tidies up fields and allows unhindered use of large agricultural machinery. There are other reasons, too. Landowners and local governments view wetlands as an economic liability that produces no economic return and provides little tax revenue. Many regard the wildlife that wetlands support as threats to grain crops. Elsewhere, wetlands are considered valueless lands, at best filled in and used for development. Some major wetlands have been in the way of dam development projects. For example, the large Pymatuning Lake in the states of Pennsylvania and Ohio covers what was once a 4200-hectare *Sphagnum*–tamarack bog (see Figure 25.19). Peat bogs in the northern United States, Canada, Ireland, and northern Europe are excavated for fuel, horticultural peat, and organic soil (**Figure 25.21**). In some areas, such exploitation threatens to wipe out peatland ecosystems.

Many remaining wetlands, especially in the north-central and southwestern United States, are contaminated and degraded by the pesticides and heavy metals carried into them by surface and subsurface drainage and sediments from surrounding croplands. Although inputs of nitrogen and phosphorus increase the productivity of wetlands, a concentration of herbicides, pesticides, and heavy metals poisons the water, destroys invertebrate life, and has debilitating effects on wildlife (including deformities, lowered reproduction, and death). Waterfowl in wetlands scattered throughout agricultural lands are also more exposed to predation, and without access to natural upland vegetation they breed less successfully.

Fifty-one percent of the human population of the contiguous United States, and globally 70 percent of all people, live within 80 kilometers (km) of the coastlines. With so much

(a)

(b)

Figure 25.21 Harvesting peat using (a) traditional and (b) mechanized methods.

humanity clustered near the coasts, it is obvious why coastal wetlands are threatened and disappearing rapidly. During colonial times, the area now embraced by the 50 United States contained some 160 million hectares of wetlands. Over the past 200 years, that area has decreased to 110 million hectares (**Figure 25.22**), and many of these remnants are degraded. Coastal Europe has lost 65 percent of its original tidal marshes, and 75 percent of the remaining are heavily managed. Since the 1980s, 35 percent of tropical mangrove forests have been diked for aquaculture—pond rearing of shrimp and fish—and cut for wood chips and charcoal. In a satellite-based assessment of the extent and changes in the global distribution of wetlands, Catherine Prigent and colleagues at Observatoire de Paris found that wetlands declined by 6 percent between 1993 and 2007 as a result of conversion for agriculture, drainage, and water diversion. The study found wetland loss was the greatest in the tropics and subtropics, where population growth and agricultural expansion is the greatest. Other research has shown that wetland loss, especially in Malaysia and Indonesia, has continued since 2007. Plantation development for palm oil and pulp and paper production are key drivers of wetland loss in Southeast Asia.

Commonly regarded as economic wastelands, salt marshes have been and are still being ditched, drained, and filled for real estate development (everyone likes to live at the water's edge), industrial development, and agriculture. Reclamation of marshes for agriculture is most extensive in Europe, where the high marsh is enclosed within a sea wall and drained. Most of the marshland and tideland in Holland has been reclaimed in this fashion. Many coastal cities such as Boston, Amsterdam, and much of London have been built on filled-in marshes. Salt marshes close to urban and industrial developments occasionally become polluted with spillages of oil, which becomes easily trapped within the vegetation.

Losses of coastal wetlands have a pronounced effect on the salt marsh and associated estuarine ecosystems. They are the nursery grounds for commercial and recreational fisheries. There is, for example, a positive correlation between the expanse of coastal marsh and shrimp production in the coastal waters of the Gulf of Mexico. Oysters and blue crabs are marsh dependent, and the decline of these important species relates to loss of salt marshes. Coastal marshes are major wintering grounds for waterfowl. One-half of the migratory waterfowl of the Mississippi Flyway depend on Gulf Coast wetlands, and the bulk of the snow goose population winters on coastal marshes from the Chesapeake Bay to North Carolina. Through grazing or uprooting, these geese may remove nearly 60 percent of the belowground production of marsh vegetation. Forced concentration of these wintering migratory birds into shrinking salt-marsh habitats could jeopardize marsh vegetation and the future of these salt-marsh ecosystems.

The loss of wetlands has reached a point where both environmental and socioeconomic values—including waterfowl habitat, groundwater supply and quality, floodwater storage, and sediment trapping—are in jeopardy. In the United States,

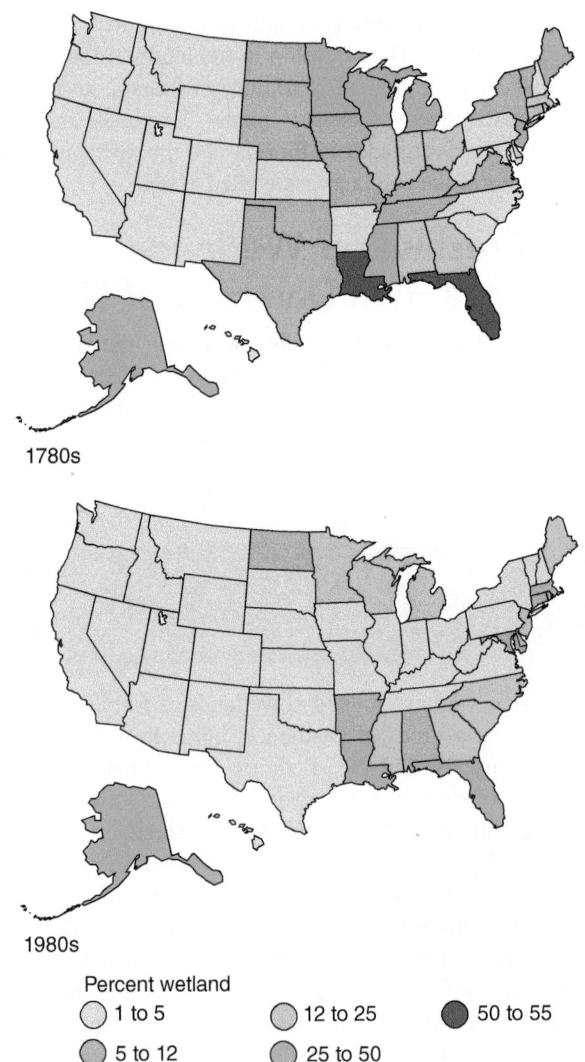

Figure 25.22 The loss of wetlands in the United States over the past 200 years.
(Adapted from United States Department of Agriculture.)

however, some progress has been made in recent decades to reduce the trend of continued loss. In a report developed for Congress, Thomas Dahl of the United States Department of Interior examined the status and trends in wetland ecosystems in the conterminous United States over the period from 1998 to 2004. The study indicated that there were an estimated 110.1 million acres (44.6 million ha) of wetlands in the conterminous United States in 2009, of which an estimated 95 percent of all wetlands were freshwater and 5 percent were in the marine or estuarine (saltwater) systems. Salt marsh made up an estimated 66.7 percent of all estuarine and marine wetland area, and forested wetlands made up the single largest category (49.5 percent) of wetland in the freshwater system. Dahl found that between 2004 and 2009, wetland area declined by an estimated 62,300 acres (25,200 ha). The reasons for this are complex and potentially reflect economic conditions, land-use trends, and changing wetland regulation and enforcement measures. Certain types of wetland exhibited decline whereas others increased in area. Collectively, marine and estuarine intertidal wetlands declined by an estimated 84,100 acres (34,050 ha) or an estimated 1.4 percent. In contrast, freshwater wetlands realized a slight increase in area between 2004 and 2009.

Although the United States has made some progress toward preserving the remaining wetlands through legislative action and land purchase, the future of freshwater wetlands is not secure. Apathy, hostility toward wetland preservation, political maneuvering, court decisions, and arguments over what constitutes a wetland allow the continued destruction of wetlands.

The Everglades National Park in South Florida is one of the largest natural wetlands in the world. Over the past century, the draining of lands and the diversion of water to meet growing residential and agricultural needs in this region have threatened this wetland ecosystem. Efforts are currently underway to restore the flow of water that is critical to preserving this unique ecosystem. Information on the history of the Everglades ecosystem and the Comprehensive Everglades Restoration Plan are available online at www.evergladesplan.org/.

SUMMARY

Intertidal Zone 25.1

Sandy shores and rocky coasts occur where the sea meets the land. The drift line marks the farthest advance of the tide on sandy shores. On rocky shores, a zone of black algal growth marks the tide line.

Rocky Coasts 25.2

The most striking feature of the rocky shore—the zonation of life—results from alternate exposure and submergence by the tides. The black zone marks the supralittoral fringe, the upper part of which is flooded only once every two weeks by spring tides. Submerged daily by the tides is the littoral zone, characterized by barnacles, periwinkles, mussels, and fucoid seaweeds. Uncovered only at spring tides is the infralittoral, which is dominated by large brown laminarian seaweeds, Irish moss, and starfish. Distribution and diversity of life across rocky shores are also influenced by wave action, competition herbivory, and predation. Left behind by outgoing tides are tidal pools, distinct habitats subject to wide fluctuations in temperature and salinity over a 24-hour period and inhabited by varying numbers of organisms, depending on the amount of emergence and exposure.

Sandy Beaches 25.3

Sandy beaches are a product of weathering of rock. Exposed to wave action, the beaches are subject to deposition and wearing away of the sandy substrate. Sandy and muddy shores appear barren of life at low tide; but beneath the sand and mud, conditions are more amenable to life than on the rocky shore. Zonation of life is hidden beneath the surface. The energy base for sandy and muddy shores is organic matter carried in by tides and made available by bacterial decomposition. Basic consumers are bacteria, which in turn are a major source of food for both deposit-feeding and filter-feeding organisms.

Salt Marshes 25.4

The interaction of salinity, tidal flow, and height produces a distinctive zonation of vegetation in salt marshes. Salt-marsh cordgrass dominates marshes flooded by daily tides. Higher microelevations that are shallow, flooded only by spring tides, and subject to higher salinity support salt meadow cordgrass and spikegrass. Salt-marsh animals are adapted to tidal rhythms. Detrital feeders such as fiddler crabs and their predators are active at low tide; filter-feeding ribbed mussels are active at high tide.

Mangrove Forests 25.5

In tropical regions, mangrove forests or mangals replace salt marshes and cover up to 70 percent of coastlines. Uniquely adapted to a tidal environment, many mangrove tree species have supporting prop roots that carry oxygen to the roots, and their seeds grow into seedlings on the tree and drop into the water to take root in the mud. Mangroves support a unique mix of terrestrial and marine life. The sheltered water about the prop roots provides a nursery for the larvae and young of crabs, shrimp, and fish.

Freshwater Wetlands 25.6

Wetlands can be defined as a community of hydrophytic plants occupying a gradient of soil wetness from permanently flooded to periodically saturated during the growing season. Hydrophytic plants are adapted to grow in water or on soil periodically deficient in oxygen. Wetlands dominated by grasses and herbaceous hydrophytes are marshes. Those dominated by wooded vegetation are forested wetlands (riparian forests) or shrub swamps. Wetlands characterized by an accumulation of peat are mires. Mires fed by water moving through the mineral soil and dominated by sedges are fens; those dominated by *Sphagnum* and dependent largely on precipitation for moisture and nutrients are bogs. Bogs are characterized by blocked drainage, an accumulation of peat, and low productivity.

Hydrology Structures Wetlands 25.7

The structure and function of wetlands are strongly influenced by their hydrology—both the physical movement of water and hydroperiod. Hydroperiod is the depth, frequency, and duration of flooding. Hydroperiod influence on vegetation is most evident in basin wetlands that exhibit zonation from deep-water submerged vegetation to wet-ground emergents.

Diversity of Wetland Life 25.8

Wetlands support a diversity of wildlife. Freshwater wetlands provide essential habitats for frogs, toads, turtles, and a diversity of invertebrate life. Nesting, migrant, and wintering waterfowl depend on these critical habitats.

Wetlands Decline Ecological Issues & Applications

Wetland ecosystems continue to be drained and converted to other land uses such as agriculture. The extent of both freshwater and coastal wetlands continues to decline the greatest losses occurring in the tropics and subtropics, where population growth and agricultural expansion is the greatest.

STUDY QUESTIONS

1. Describe the three major zones of the rocky shore.
2. What environmental stresses are organisms living in the rocky intertidal zone subject to?
3. How does life on sandy shores differ from that in the rocky intertidal zone? How is it the same?
4. What influences the major structural features (zonation) of a salt marsh?
5. Compare salt marshes with mangrove forests.
6. What is a freshwater wetland?
7. What are the three major types of wetlands in terms of hydrology (water flow)?
8. What is hydroperiod, and how does it relate to the structure of wetlands?
9. Characterize the types of wetlands by their vegetation.

FURTHER READINGS

Bertness, M. D. 1999. *The ecology of Atlantic shorelines.* Sunderland, MA: Sinauer Associates.
A well-written and illustrated introduction to coastal ecology.

Lugo, A. E. 1990. *The forested wetlands.* Amsterdam: Elsevier.
Excellent overview and discussion of the structure and function of these freshwater wetland ecosystems.

Lugo, A. E., and S. C. Snedaker. 1974. "The ecology of mangroves." *Annual Review of Ecology and Systematics,* 5:39–64.
This review article provides an excellent introduction to mangrove ecosystems.

Mathieson, A. C., and P. H. Nienhuis, eds. 1991. "Intertidal and littoral ecosystems." *Ecosystems of the world,* 24. Amsterdam: Elsevier.
A detailed, broad survey of the intertidal and littoral zones of the world.

Mitsch, W. J., and J. C. Gosslink. 2007. *Wetlands,* 4th ed. New York: John Wiley & Sons.
A pioneering text and major reference on the ecology of wetland ecosystems.

Moore, P. G., and R. Seed, eds. 1986. *The ecology of rocky shores.* New York: Columbia University Press.
A comprehensive, worldwide review of the rocky intertidal zone.

Niering, W. A., and B. Littlehales. 1991. *Wetlands of North America.* Charlottesville, VA: Thomasson-Grant.
An exceptionally illustrated survey of North American wetlands, both freshwater and salt.

Teal, J., and M. Teal. 1969. *Life and death of the salt marsh.* Boston: Little, Brown.
Although out of print, this beautifully written classic is an excellent introduction to salt-marsh ecology.

Valiela, I., J. L. Bowen, and J. K. York. 2001. "Mangrove forests: One of the world's threatened major tropical environments." *BioScience,* 51:807–815.
An overview of the impact of mariculture on mangroves.

Van der Valk, A., ed. 1989. *Northern prairie wetlands.* Ames, IA: Iowa State University Press.
Includes detailed studies of this unique wetland ecosystem that is rapidly disappearing.

Williams, M., ed. 1990. *Wetlands: A threatened landscape.* Cambridge, MA: Blackwell Publishers.
This comprehensive appraisal of the world's wetlands covers occurrence and composition, physical and biological dynamics, human impact, and management and preservation.

CHAPTER 26 Large-Scale Patterns of Biological Diversity

Although covering only 6 percent of the land surface, tropical rain forests contain more than 50 percent of all known terrestrial species.

CHAPTER GUIDE

EARTH'S ECOSYSTEMS SUPPORT an amazing diversity of species. Scientists have identified and named approximately 1.7 million species (**Figure 26.1**), and the task is not complete. Scientists are continuously discovering new species, and quantifying the actual number of species inhabiting Earth is an ongoing exercise (**Figure 26.2**). Some scientists, such as Harvard biologist E. O. Wilson, believe that the actual number of species may be close to 10 million.

Regardless of whether the actual number of species is 1.7 million or 10 million, the diversity of our planet is not static. Over evolutionary time, new species evolve while existing species fade away and become extinct. The diversity of our planet is a story of constant change.

Nor is biological diversity the same everywhere on Earth's surface. Distinct geographic patterns of diversity relate to environmental conditions that have influenced the evolution of species diversity as well as the ability of local environments to support a diverse community. In this chapter, we will examine these regional and global patterns of Earth's biological diversity, both in time and space.

26.1 Earth's Biological Diversity Has Changed through Geologic Time

In Chapter 18, we examined the temporal dynamics of species diversity on a successional timescale (year to centuries). In those examples, temporal changes in the local patterns of diversity reflect changes in the local distribution and abundance of species in response to changes in environmental conditions through time (see Section 18.5). On a longer timescale (1000s to 10,000s years) temporal patterns of diversity are influenced by the immigration and emigration of species as geographic

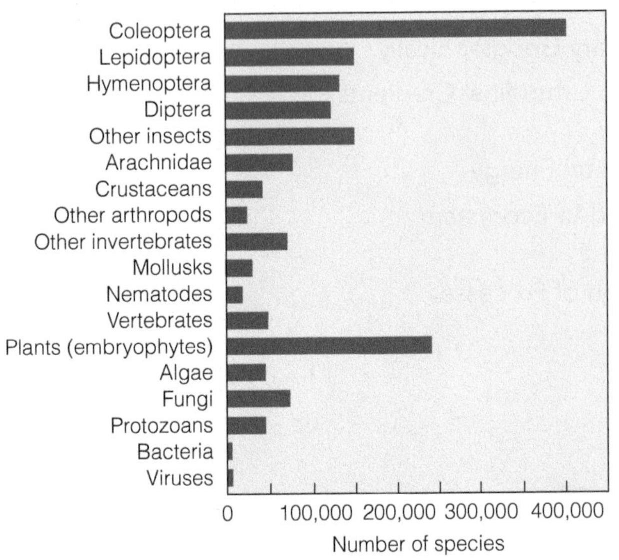

Figure 26.1 Number of living species of all kinds of organisms currently known. Species are classified into major taxonomic groups. Insects and plants dominate the diversity of living organisms. (Data from Wilson 1999.)

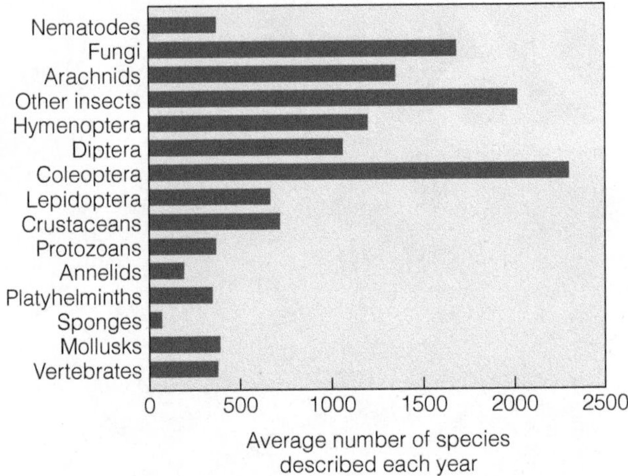

Figure 26.2 Many new (previously unknown) species are discovered each year. Most of these species belong to taxonomic groups that are relatively small in size.

distributions respond to regional and global changes in climate (see Section 18.9). Over geologic time, however, there have been dramatic long-term evolutionary changes in patterns of global diversity, largely as a result of speciation and global extinction.

Over the past 600 million years, the number of different types of organisms has been increasing. Among most groups of organisms for which data exist from the fossil record, the number of species has increased almost continuously since the taxonomic group first appeared in the fossil record. **Figure 26.3** represents the estimated species richness of fossilized invertebrates over geologic time. Species richness of this taxonomic group has increased over the past 600 million years with slight decreases during the late Devonian and Permian periods.

The evolution of diversity among vascular land plants presents a particularly interesting pattern (**Figure 26.4**). Since the appearance of these plants more than 400 million years ago, the number of species has increased almost continuously, but the groups that dominate the land flora have shifted dramatically through time. The early vascular plants—the rootless and

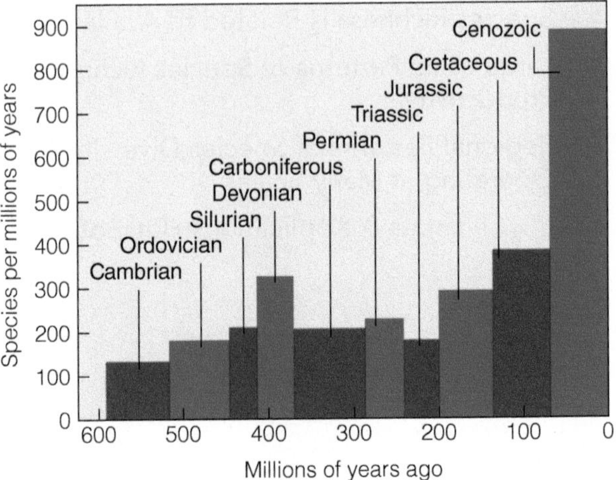

Figure 26.3 Estimated species richness of fossilized invertebrates over geologic time.

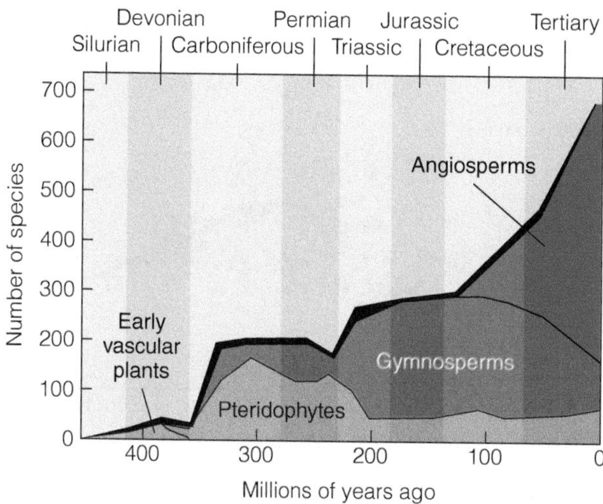

Figure 26.4 Pattern of expansion and reduction of major terrestrial plant groups during 400 million years of plant evolution. (Adapted from Niklas et al. 1983.)

leafless psilopsids—went extinct by the end of the Devonian and were replaced by pteridophytes (ferns), which flourished during the Carboniferous period. This group then decreased in abundance by the early Triassic. The decline of this group of plants coincided with the diversification of the gymnosperms (includes ginkgos, cycads, conifers), which in turn declined in abundance and diversity over the past 100 million years as angiosperms (flowering plants) diversified.

26.2 Past Extinctions Have Been Clustered in Time

Although the history of Earth's biological diversity is generally a story of increasing species richness, it has experienced periods of decline. The general pattern of increasing diversity through geologic time has also been accompanied by extinctions, which were not timed evenly through Earth's history (**Figure 26.5**). Most extinctions are clustered in geologically brief periods of time. One mass extinction occurred at the end

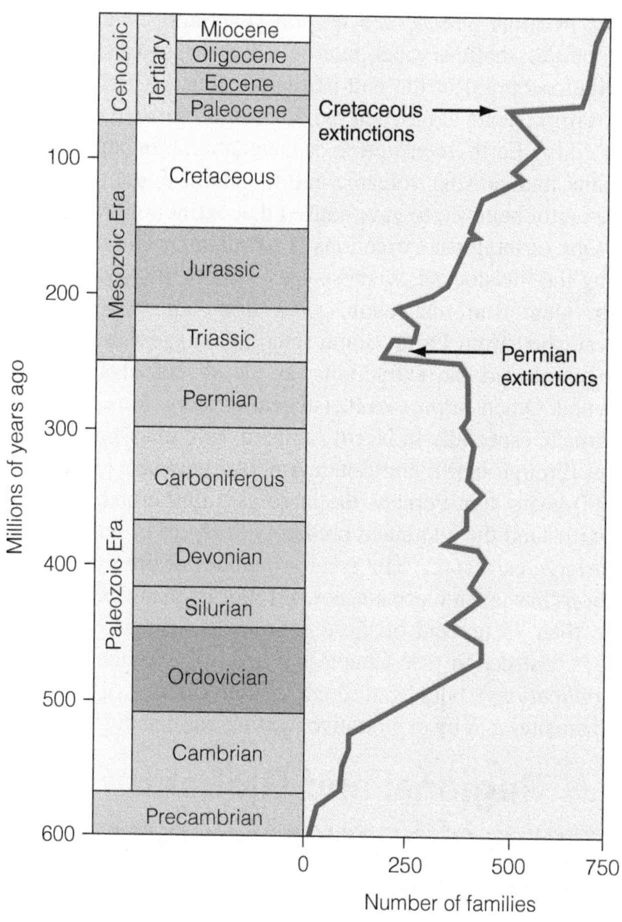

Figure 26.5 The geologic timescale and mass extinctions in the history of life. The fossil record profiles mass extinctions during geological times. The most recent mass extinction occurred during the Cretaceous, which wiped out more than half of all species, including the dinosaurs. The mass extinction event at the end of the Permian resulted in the loss of 96 percent of all marine species and perhaps as many as 50 percent of the total species on Earth.

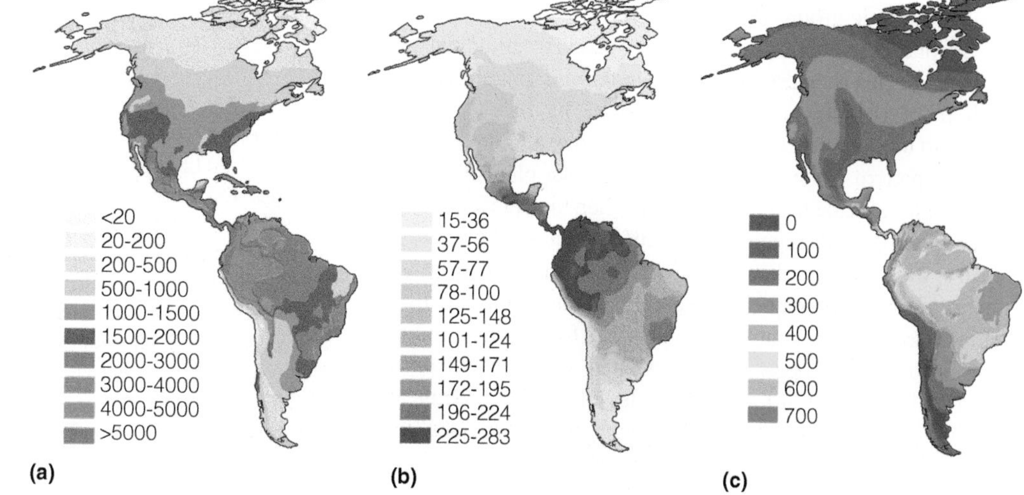

Figure 26.6 Geographic patterns of species richness for (a) vascular plants, (b) mammals, and (c) birds in the New World (North and South America). (From [a] Mutke, J. and Barthlott, W. 2005; [b] Kaufman and Willig 1998; [c] Hawkins et al. 2006.)

of the Permian period, 225 million years ago, when 90 percent of the shallow-water marine invertebrates disappeared. Another occurred at the end of the Cretaceous period, 65 to 125 million years ago, when the dinosaurs vanished. An asteroid striking Earth, interrupting oceanic circulation, altering the climate, and causing volcanic and mountain-building activity is currently believed to have caused that extinction event.

One of the great extinctions of mammalian life took place during the Pleistocene, when species such as the woolly mammoth, giant deer, mastodon, giant sloth, and saber-toothed cat vanished from Earth. Some scientists suggest that climate changes caused the extinctions, as ice sheets advanced and retreated. Others argue that Pleistocene hunters overkilled large mammals, especially in North America, as human populations swept through North and South America between 11,550 and 10,000 years ago. Perhaps the large grazing herbivores could not withstand the combined predatory pressure of humans and other large carnivores. The greatest number of present-day extinctions has taken place since A.D. 1600. Humans have caused more than 75 percent of these extinctions, primarily through habitat destruction (see Chapters 9 and 20, *Ecological Issues & Applications*), but also through the introduction of predators and parasites and by exploitative hunting and fishing.

26.3 Regional and Global Patterns of Species Diversity Vary Geographically

The 1.7 million species that have been identified are not distributed equally across Earth's surface. There are distinct geographic patterns of species richness (number of species [*S*]). In general, the number of terrestrial species decreases as one moves away from the equator toward the poles. The three maps in **Figure 26.6** illustrate distinct geographic patterns of species richness for vascular plants, mammals, and birds. Note the overall decrease in species richness as one moves from the tropics toward the poles. This pattern is more obvious if we plot species richness as a function of latitude (**Figure 26.7**).

The general pattern of a decline in species diversity moving from the tropics to the poles is also observed in aquatic environments—both freshwater and marine—for a wide variety of taxonomic groups (**Figure 26.8**).

Despite the general nature of the observed latitudinal gradients in species richness, the actual underlying pattern is often much more complex and often disrupted by other variables relating to topography or climate. For example, although there is a distinct pattern in species richness of vascular plants with latitude at a global scale (see Figure 26.7), when the data are viewed for longitudinal belts relating to the different continental masses (**Figure 26.9**), a number of significant deviations from the general trend can be seen. For example, in the longitudinal belt that includes the continents of Europe and Africa, two latitudinal peaks in diversity appear, one associated with the equatorial region and the other in the midlatitudes of the Mediterranean, a region of high species diversity and endemism (see this chapter, *Ecological Issues & Applications*, and Section 23.6).

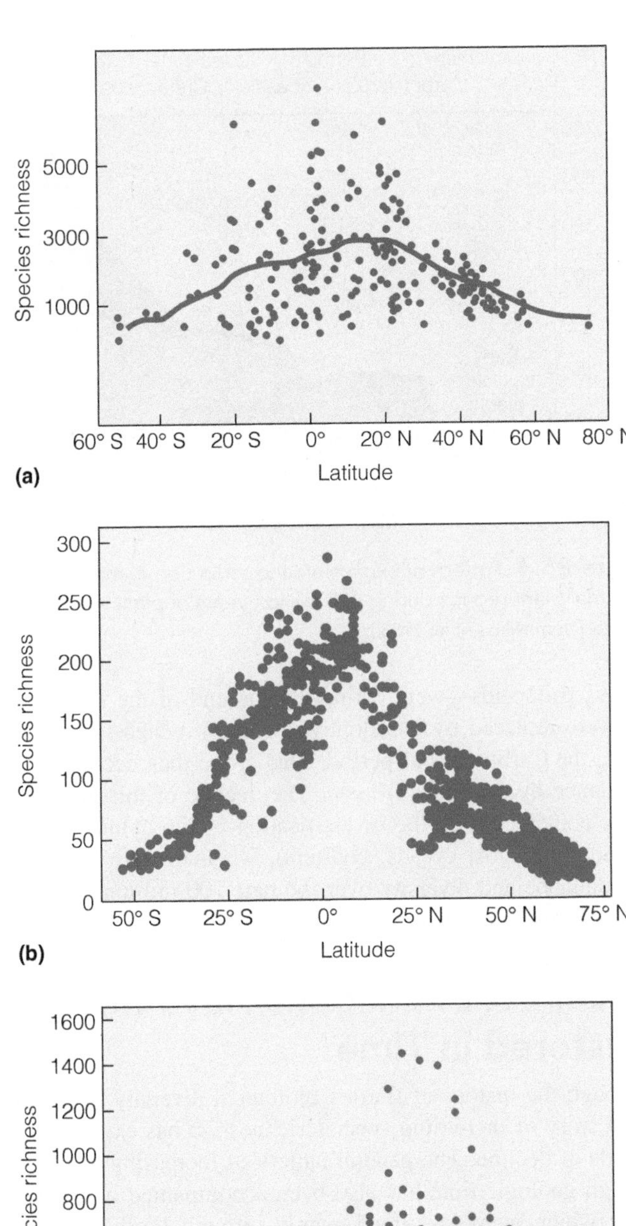

Figure 26.7 Examples of patterns of species richness as a function of latitude for the New World (North and South America). (a) Latitudinal gradients of vascular plant species richness. Each dot represents mean species number per 10,000 km^2 grid plotted against latitude of the geographic midpoint of the grid (Adapted from Mutke and Barthlott 2005). (b) Mammalian species richness in the continental New World as a function of latitude based on 2.5° cells (Data from Kaufman and Willig 1998). (c) Avian species richness in the continental New World in grid cells of approximately 611,000 km^2 (WORLDMAP grid).

(Adapted from Blackburn and Gaston 1996.)

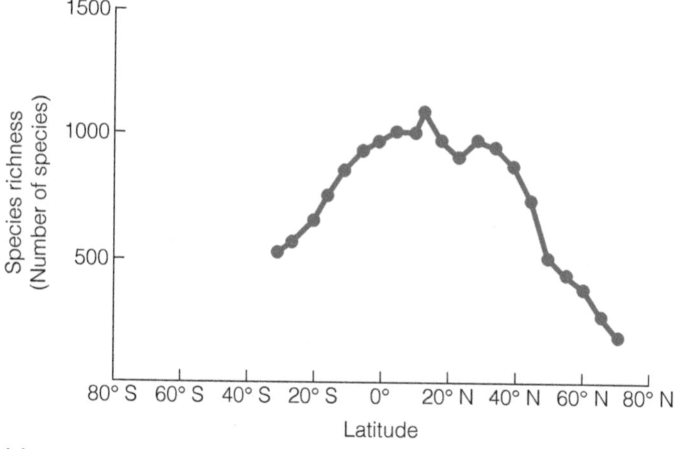

In addition, areas of lower diversity are observed in the subtropical desert regions of both Northern (Sahara Desert) and Southern Africa (Namib, Kalahari, and Karoo Deserts).

Deviations from the general latitudinal gradient of species richness are also observed for certain taxonomic groups, particularly for groups that are associated with specific environments or resources. For example, although the latitudinal pattern of species richness for vascular plants in the New World (North and South America) exhibits the general pattern of peak diversity in the equatorial region (see Figures 26.6 and 26.9), species richness for members of the plant family Cactaceae (cacti; **Figure 26.10**) is low in the equatorial region (mesic conditions) and reaches its maximum in the midlatitude zones dominated by a permanent zone of high pressure and low precipitation (arid desert regions; see Sections 2.6 and 23.5).

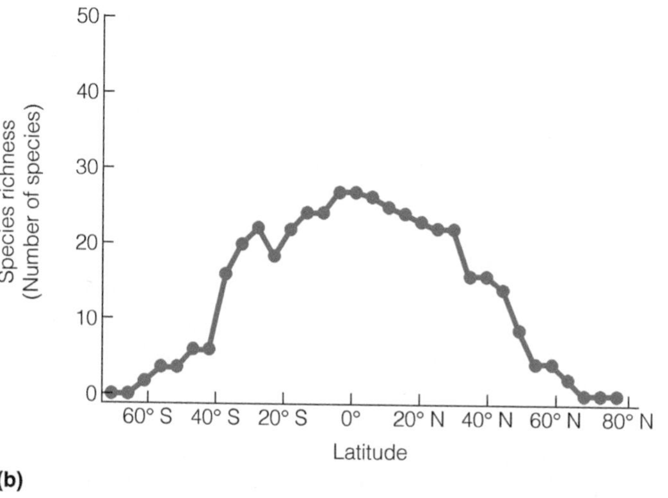

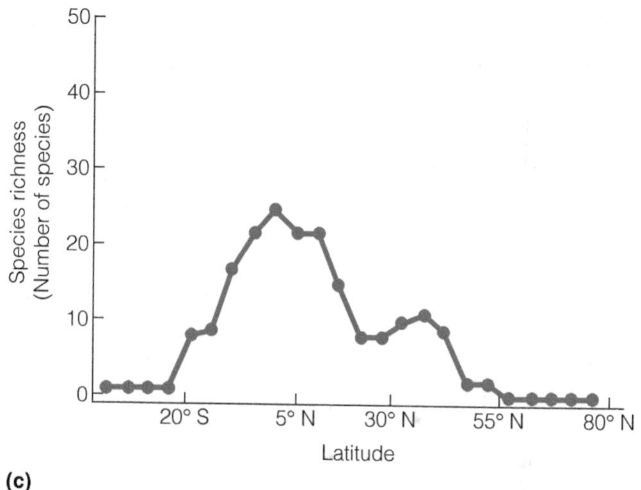

Figure 26.8 Latitudinal patterns of species richness in the eastern Atlantic Ocean: (a) fish (Data from Macpherson and Duarte 1994), (b) pelagic Salpida (minute floating marine tunicates), and (c) benthic (continental shelf) cephalopods (Cephalopoda), which are marine mollusks that includes nautilus, squid, cuttlefish, and octopus. Data expressed in 5° latitude bands.
(Adapted from Macpherson 2002.)

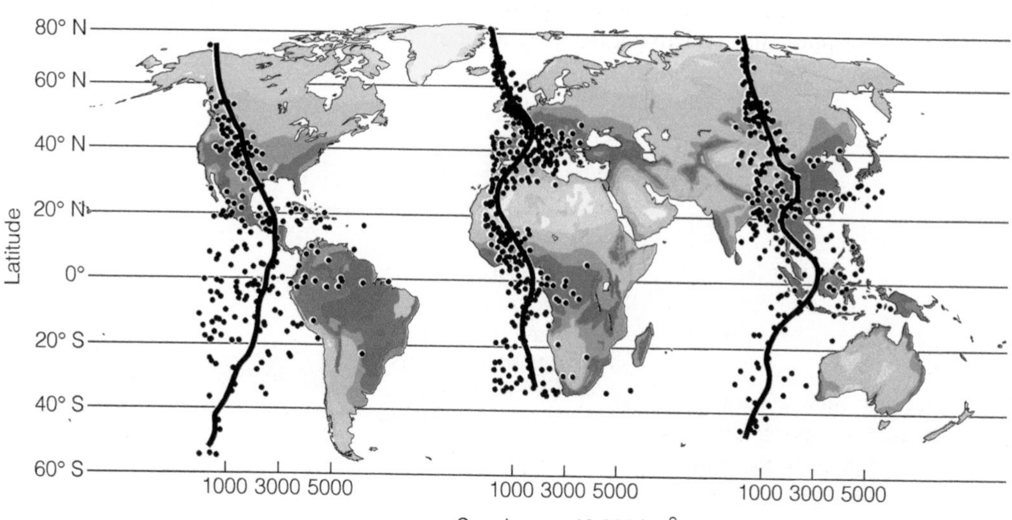

Figure 26.9 Latitudinal gradients of vascular plant species richness for three different longitudinal bands associated with the major continental regions (North and South America, Europe and Africa, and Asia and Australia). Each dot represents mean species richness plotted against latitude of the geographic midpoint of the grid cell (approximately 10,000 km^2). Data for North and South America is the same as that presented in Figure 26.7a.
(From Mutke and Barthlott 2005.)

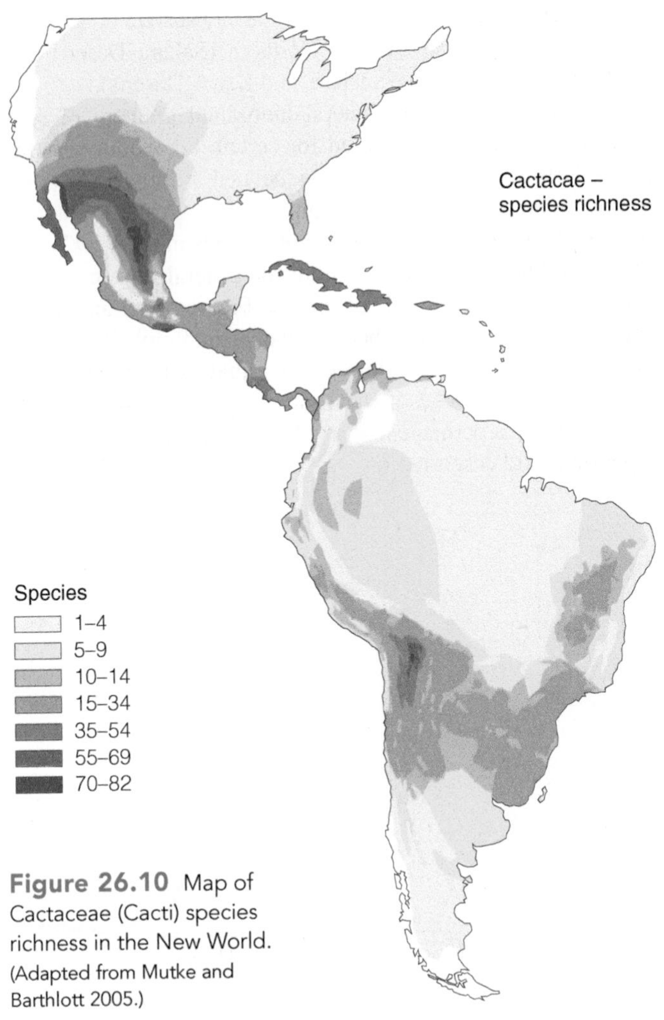

Figure 26.10 Map of Cactaceae (Cacti) species richness in the New World. (Adapted from Mutke and Barthlott 2005.)

26.4 Various Hypotheses Have Been Proposed to Explain Latitudinal Gradients of Diversity

What factors could possibly be responsible for the observed latitudinal gradient of species richness? Although scientists do not know the exact mechanisms underlying the geographic pattern of species diversity, more than 25 different mechanisms have been proposed, including the age of the community, stability of the climate over time, spatial heterogeneity of the environment, and ecosystem productivity. An example of the potential role of age of community and stability of climate can be found in the discussion of the role of periods of glacial expansion and retreat on the distribution of tree species and plant communities in North America presented in Chapter 18 (Section 18.9, Figures 18.24–18.26). During periods of glacial expansion in the Northern Hemisphere, the distribution of tree species shifted southward, moving northward again as the climate warmed and the glaciers retreated. In contrast, tropical regions, although undergoing changes in climate, did not experience the displacement and possible extinction of species on the same scale. For example, Jonathan Adams of University College of North Wales and Ian Woodward of the University of Sheffield examined the potential role of past patterns of glacial dynamics on current patterns of global tree species richness. Europe has far fewer tree species and genera than either North America or eastern Asia. Fossil evidence shows that west-central Europe had a much richer flora during the Upper-Tertiary (25–2 million years [Myr] B.P.) with many genera that now only survive in temperate regions of North America and Asia. These tree species seem to have been eliminated (regional extinction) from Europe during cold, dry periods of the Pleistocene Era (2–0.001 Myr B.P.). Adams and Woodward hypothesize that their failure to recolonize following glacial retreat may be a result of the failure of species to reach potential refugia (areas of suitable climate and edaphic conditions) during the glacial phase.

Another mechanism that has been proposed relates to the relationship between area (spatial extent) and species diversity discussed in Chapter 19 (Section 19.4, Figures 19.20 and 19.23). Recall that in general, species richness increases with increasing area. Could the latitudinal gradient of species richness at a continental and global scale be related to differences in geographic (spatial) area? Being spherical in shape, it is true that the surface area of the planet for a given latitudinal band (such as 5°) will decrease as you move from the equator to the poles (**Figure 26.11**). However, if we compare latitudinal patterns of species richness for terrestrial taxa with the corresponding latitudinal distribution of the global

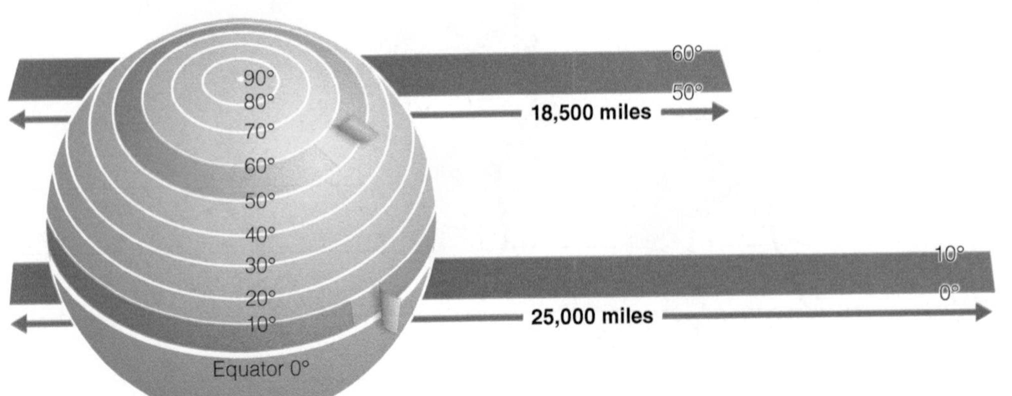

Figure 26.11 There is a greater surface between 0 and 10° north latitude than there is between 50 and 60° north. This is because Earth's circumference is largest at the equator. The circumference of Earth at the equator is 25,000 miles (40,000 km), whereas at 50° latitude it is 18,500 miles (30,000 km).

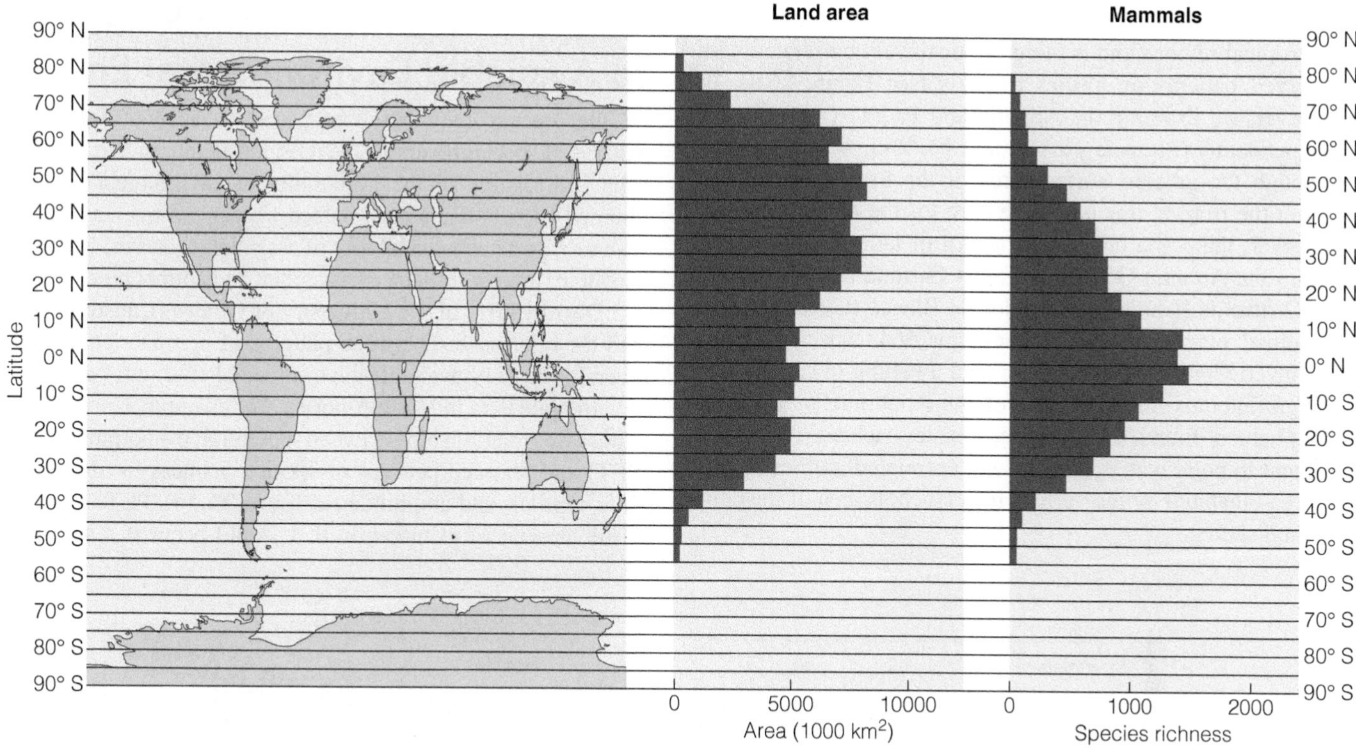

Figure 26.12 Comparison of variation in land area and species richness of mammals across 5° bands of latitude.

landmass, it becomes obvious that the gradients do not align (**Figure 26.12**). Differences in land area alone do not account for the current observed latitudinal gradients of terrestrial diversity. It has been hypothesized, however, that most of the land surface of the Earth was tropical or subtropical (climate zones) during the Tertiary, which could partly explain the greater diversity in the tropics today as an outcome of historical evolutionary processes (see Chapter 5).

When we examine the spatial distribution of the oceans as a function of latitude (**Figure 26.13a**), we see that the

Figure 26.13 (a) Percentage of total ocean surface area by latitude. (b) Map of Pacific and (c) Atlantic Ocean basins.

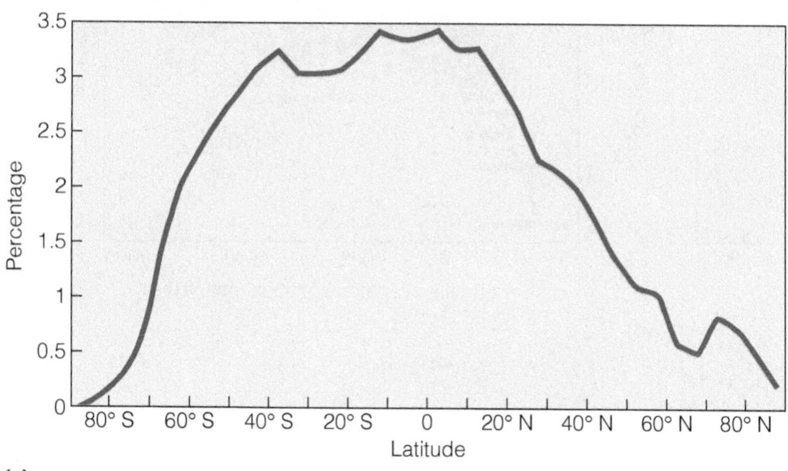

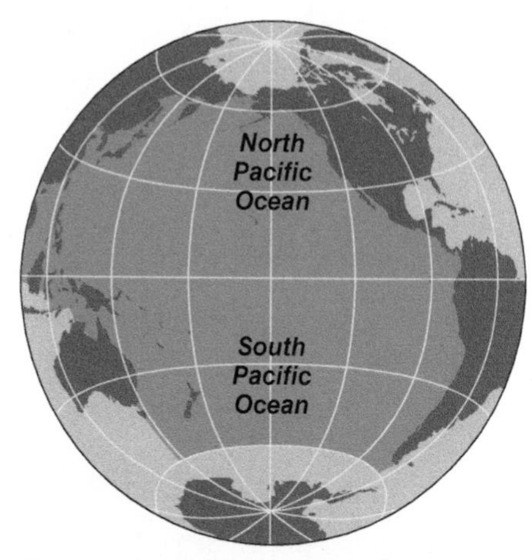

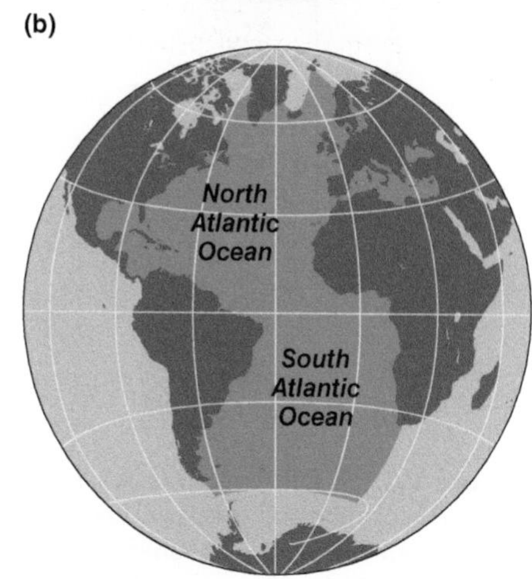

surface area of the oceans (as well as volume) is greatest in the tropical regions and is therefore positively correlated with observed patterns of species richness (see Figure 26.8). If, however, we examine the Atlantic and Pacific Ocean basins independently (**Figures 26.13b** and **26.13c**), we can see that although the greatest surface area of the Pacific Ocean lies within the tropics, this is not the case for the Atlantic Ocean. However, there is a consistent pattern of latitudinal diversity within the Atlantic Ocean. Ecologist Enrique Macpherson of the Centro de Estudios Avanzados de Blanes (Girona, Spain) examined marine species richness for 6643 species (fishes and invertebrates) in 10 different taxa dwelling in benthic and pelagic habitats on both sides of the Atlantic. For all taxa, Macpherson found a decline in species richness from the tropical to polar waters that was not correlated with estimates of area inhabited by the various taxa (such as area of the continental shelf).

26.5 Species Richness Is Related to Available Environmental Energy

Of the various hypotheses proposed to account for global patterns of species diversity, the most easily interpreted are those explicitly relating to the availability of environmental energy (thermal energy) that are associated with environmental features such as climate and availability of essential resources, which are known to directly influence basic plant and animal processes.

David Currie of the University of Ottawa (Canada) examined the relationship between patterns of species richness and several indices of available environmental energy for a number of vertebrate taxa in North America. For example, Currie found a positive relationship between potential evapotranspiration (PET) and regional patterns of species richness for mammals, birds, reptiles, and amphibians (**Figure 26.14**). PET is defined as the amount of evaporation that would occur if a sufficient

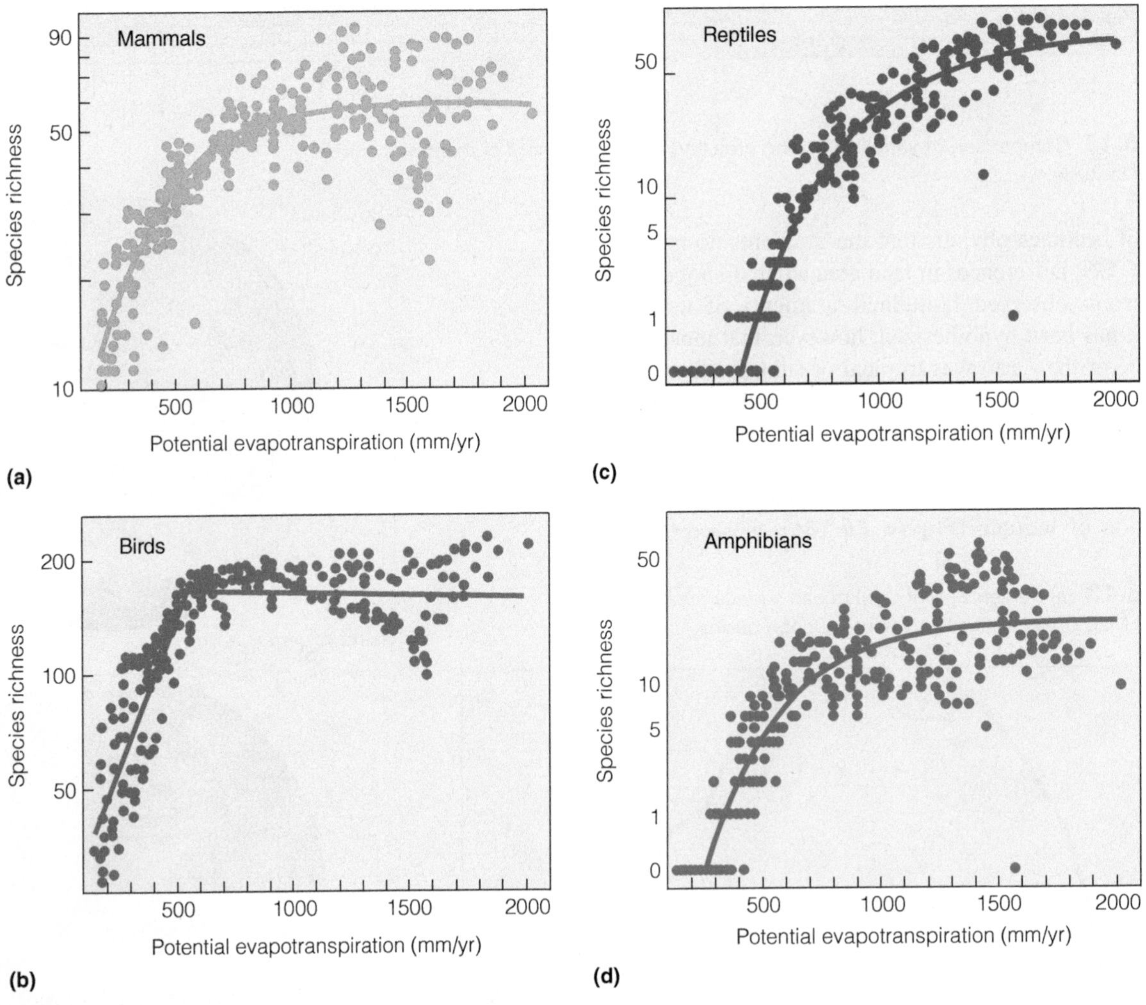

Figure 26.14 Relationship between annual estimate of potential evapotranspiration and species richness of (a) mammals, (b) birds, (c) reptiles, and (d) amphibians in North America.
(From Currie 1991.)

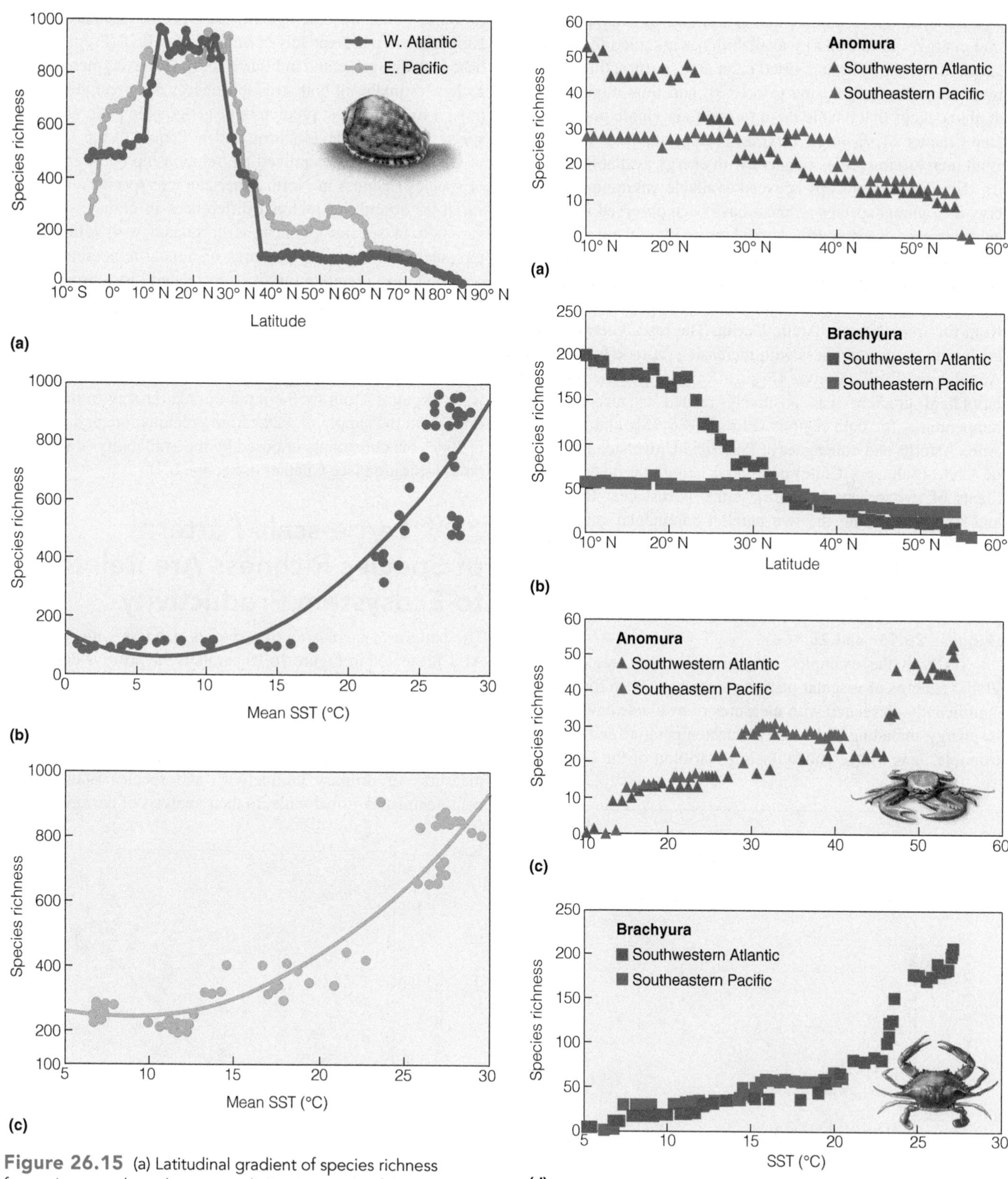

Figure 26.15 (a) Latitudinal gradient of species richness for marine prosobranch gastropods (marine snails of the subclass Prosobranchia, class Gastropoda) in eastern Pacific and western Atlantic Ocean. Species richness of (b) eastern Pacific and (c) western Atlantic as a function of mean sea surface temperature (SST).
(Adapted from Roy et al 1998.)

Figure 26.16 Latitudinal patterns of species richness for (a) Anomura and (b) Brachyura marine crabs along the eastern and western coasts of South America. Patterns of species richness for (c) Anomura and (d) Brachyura crabs plotted as a function of mean sea surface temperature (SST).
(Adapted from Astorga et al 2003.)

water source were available and is an index of integrated thermal energy. Greater energy availability is assumed to enable a greater biomass to be supported in an area. In turn, this enables more individual organisms to coexist, and thus more species at abundances that enable them to maintain viable populations (see Chapter 11, *Ecological Issues & Applications*). The result is an increase in species richness with energy availability.

Similar relationships between available environmental energy and animal species richness have been observed in marine environments. Kaustav Roy of the University of California–San Diego examined patterns of species richness of marine prosobranch gastropods (marine snails) living on the continental shelves of the western Atlantic and eastern Pacific Oceans, from the tropics to the Arctic Ocean. The researchers found a decline in species richness with increasing latitude for both the Atlantic and Pacific shelves (**Figure 26.15a**) and the observed latitudinal gradient was positively related to mean surface temperatures for both regions (**Figures 26.15b** and **26.15c**). Anna Astorga and colleagues at Pontificia Universidad Católica de Chile (Santiago, Chile) compared latitudinal diversity gradients of two groups of marine crabs (Crustacea: Brachyura and Anomura) along the two parallel continental coasts (east and west) of South America. The researchers found a decline in species richness with increasing latitude for both groups of crabs along both coasts (**Figures 26.16a** and **26.16b**) that was positively correlated with annual mean sea surface temperatures (**Figures 26.16c** and **26.16d**).

As with the examples of animal diversity, regional and global patterns of vascular plant diversity have been found to be significantly correlated with measures of available environmental energy, including both mean annual temperature and PET. For example, Jens Mutke and Wilhelme Barthlott of the University of Bonn found that the species richness of vascular plants in North America is positively correlated with PET (**Figure 26.17**; based on data presented in Figure 26.6). However, measures that include estimates of both available energy and precipitation have proven to be the best predictors of geographic patterns of plant species richness. David Currie and V. Paquin of the University of Ottawa, Canada, examined the relationship between patterns of species richness in North American tree species and several variables describing regional differences in climate. Although variation in species richness is correlated with climatic factors such as integrated measures of annual temperature, solar radiation, and precipitation, it is correlated most strongly with estimates of actual evapotranspiration (AET; **Figure 26.18**). AET is the flux of water from the terrestrial surface to the atmosphere through both evaporation and transpiration (see Chapters 3 and 20). It is a function of both the atmospheric demand for water brought about by the input of solar energy to the surface (PET) and the supply of water from precipitation, and as such it includes the constraints imposed by the availability of soil water on transpiration (see Chapter 6, Section 6.9).

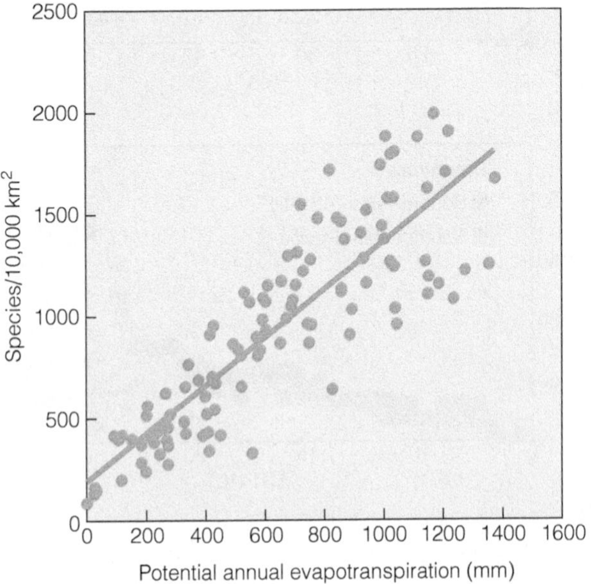

Figure 26.17 Relationship between vascular plant species richness and mean potential evapotranspiration for the region of North America. Species richness expressed as number of species per grid cell of 10,000 km^2.
(Adapted from Mutke and Barthlott 2005.)

26.6 Large-scale Patterns of Species Richness Are Related to Ecosystem Productivity

The pattern of increasing tree species richness with increasing AET presented in Figure 26.18 parallels the positive correlation between AET and net primary productivity (NPP) presented in Chapter 20 (see Figure 20.4), suggesting a relationship between plant diversity and primary productivity. In fact, a number of researchers have found a significant relationship between measures of primary productivity and species richness at a continental and global scale. In their analysis of current patterns

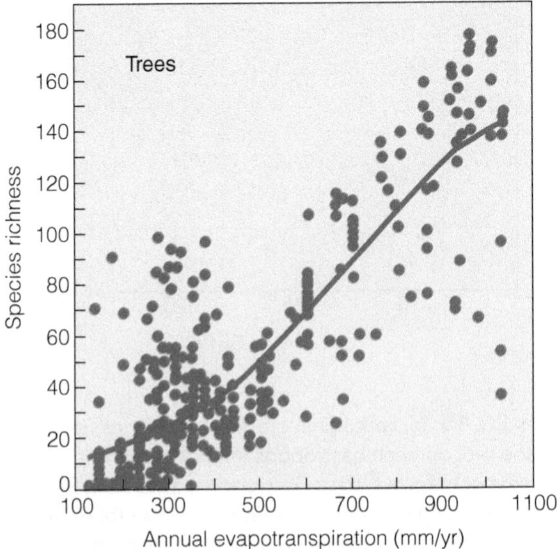

Figure 26.18 Relationship between annual measure of actual evapotranspiration (AET) and tree species richness for North America.
(From Currie and Paquin 1987.)

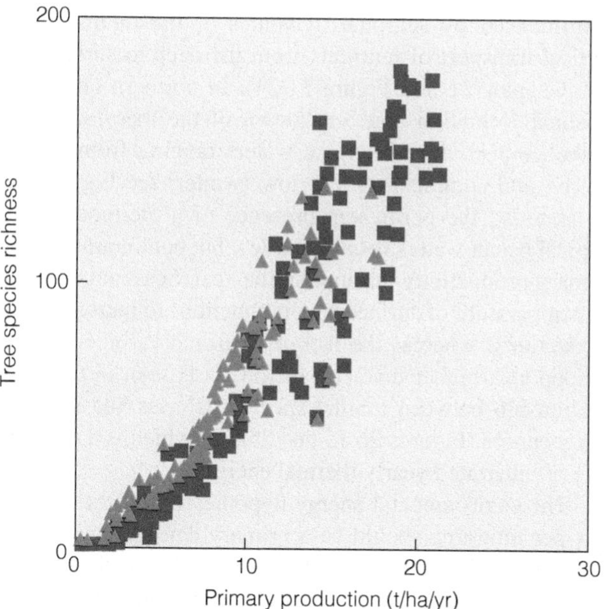

Figure 26.19 Relationship between tree species richness and net primary productivity for North America (squares) and Europe (triangles). Data represent mean values for 2.5° (latitude and longitude) grid cells.
(Adapted from Adams and Woodward 1989.)

of global tree species richness, Jonathan Adams of University College of North Wales and Ian Woodward of the University of Sheffield (see Section 26.4) found that species richness is positively correlated with estimates of NPP (**Figure 26.19**). This relationship indicates that environmental conditions favorable for photosynthesis and plant growth are capable of supporting a greater number of tree species. In addition, more productive environments may well give rise to increased plant diversity over evolutionary time.

The relationship between primary productivity and species richness is not limited to autotrophs. In his analysis of latitudinal patterns of vertebrate species richness in North America (see Section 26.5, Figure 26.14), Currie found that species richness of birds, mammals, reptiles, and amphibians were all positively related to estimates of NPP (**Figure 26.20a**). In addition, Currie found a positive correlation between plant (tree) and animal (bird, mammal, reptile, and amphibian) species richness (**Figure 26.20b**). This observed relationship between broad-scale patterns of animal diversity and plant productivity and diversity are consistent with patterns at the level of the community and ecosystem. In general, there is a positive correlation between primary and secondary productivity in both terrestrial (Figure 20.20) and aquatic (Figure 20.21) ecosystems. In addition, both the structural (Figure 17.17) and species diversity (Figure 17.18) of the plant community has been shown to be positively correlated with patterns of animal species diversity. These observed relationships are not surprising given the role of autotrophs as both food and habitat for heterotrophic organisms.

Although there is much observational evidence to support the hypothesis that the latitudinal gradient of species richness

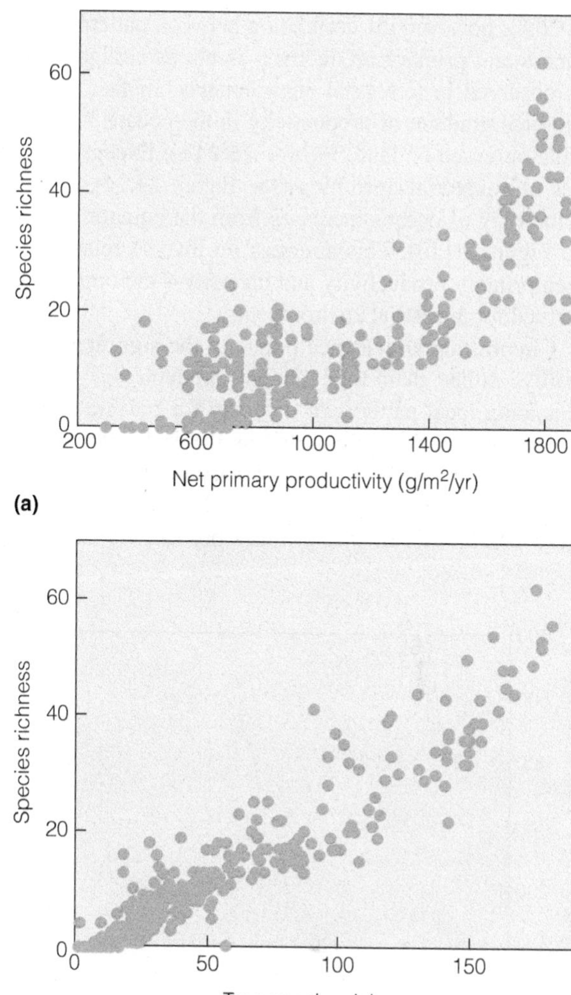

Figure 26.20 Species richness of North American amphibians (see Figure 26.14d) expressed as a function of (a) net primary productivity, and (b) tree species diversity.
(Adapted from Currie 1991.)

at a global scale are related to the availability of environmental energy, evaluating which aspect(s) of environmental energy may provide a mechanistic understanding is complicated by the fact that most of the indices discussed thus far are strongly correlated at a regional to global scale. The input of solar radiation to Earth's surface that determines both mean surface temperature and PET declines from the equator to the poles (see Figures 2.4, 2.6, and 2.9). Likewise, there is a general decline in mean annual precipitation with latitude (Figure 2.17), which together with temperatures defines patterns of AET. Given the strong relationship between AET and terrestrial NPP, the result is a distinct pattern of declining terrestrial NPP with latitude (**Figure 26.21a**) that parallels the observed patterns of terrestrial species diversity. For marine environments, however, the pattern is quite different.

The latitudinal gradients of species richness for marine organisms are similar to those observed for terrestrial organisms (decreasing with increasing latitude; see Figures 26.7

and 26.8); however, the correlation between patterns of species richness and primary productivity is not as straightforward as that observed in terrestrial environments. In fact, the general latitudinal gradient of productivity in the oceans is the reverse of that observed on land (**Figure 26.21b**). Except in localized areas of equatorial upwelling (see Figure 24.24), the primary productivity of oceans increases from the equator to the poles (see Figure 20.10). This suggests an inverse relationship between primary productivity and diversity—the opposite of that observed for terrestrial environments.

Circumstantial evidence points to the significance of seasonality, rather than total annual productivity, as a factor influencing local patterns of diversity for pelagic and benthic species. Observations show that as the influence of seasonal fluctuations in temperature on primary productivity increases, species richness declines and species dominance increases (see Figure 24.25). Recall that primary productivity in the ocean is influenced by seasonal dynamics of the thermocline and vertical transport of nutrients from the deep to surface waters (see Chapter 21 and Figure 21.24). In northern latitudes, the seasonal formation and breakdown of the thermocline result in the productivity of surface waters ranging from very high (spring and summer) to very low (winter; see Figure 24.25). By contrast, the permanent presence of a thermocline in the tropical ocean waters results in a low but continuous pattern of primary productivity throughout the year. Seasonal variation in the temperature of surface waters functions to increase primary productivity, whereas the lack of seasonal variation functions to support a higher diversity of life, as is seen in the positive relationship between marine species richness and sea surface temperatures (Figures 26.15 and 26.16), which is a direct measure of integrated yearly thermal energy.

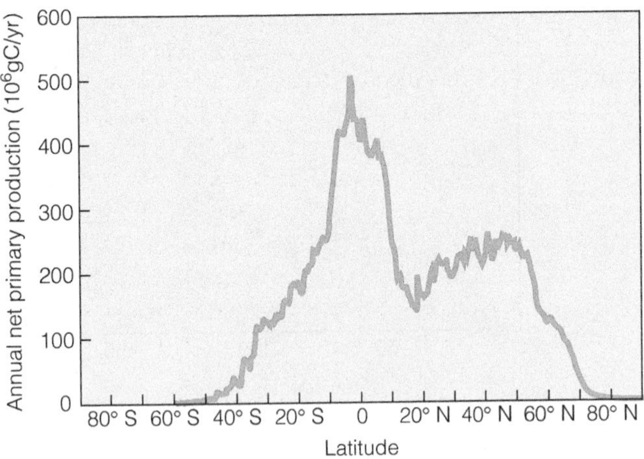

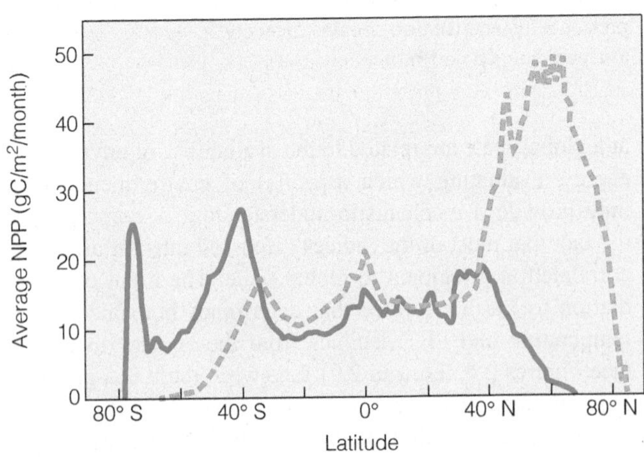

Figure 26.21 Patterns of global (a) terrestrial and (b) marine net primary productivity (NPP) as a function of latitude. For marine NPP, the sold line is the average Austral summer (December–February) NPP; the dashed line is the average Boreal summer (June–August) NPP.
(Adapted from [a] Xiao et al. 1997 and [b] Behrenfeld et al 2001.)

The environmental energy hypothesis predicts that energy flux per unit area should be a primary determinant of species richness and that the latitudinal gradient of species richness is a function of the corresponding gradient of energy flux to Earth's surface (Chapter 2). In the case of terrestrial plants, primary productivity represents a measure of the actual energy captured (absorbed) and the best correlate of primary productivity is AET, which is an index that includes the combined constraints of solar energy and available water on photosynthesis and plant growth in terrestrial environments. For terrestrial animals, the relationship appears to be more complicated. Terrestrial vertebrate richness appears to be less closely related to primary productivity than direct measures of energy flux, such as temperature and PET. One possible explanation is that energy available to vertebrates is better represented by thermal energy than by food energy. You will recall from Chapter 7 that poikilotherms (ectothermic) regulate their body temperature by directly absorbing thermal energy from their environment, and the metabolic costs incurred by homeotherms decrease with increasing ambient temperature.

Further evidence for the relationship between environmental energy and species richness in terrestrial environments can be found in the parallel relationship between elevational and latitudinal gradients of species richness. In mountainous regions, researchers have observed a pattern of decreasing species richness with increasing elevation. Patterns of species richness with increasing elevation for bird species in New Guinea and for mammals and vascular plants in the Himalayan Mountains are shown in **Figure 26.22**. The mechanisms underlying a decline in species richness with a rise in elevation may be similar to those involved with changing latitude. Variations in temperature, PET, AET, and primary productivity with increasing elevation parallel those observed with increasing latitude. However, the negative correlation between species richness and elevation are confounded because high-altitude communities generally occupy a smaller spatial area than do corresponding lowland communities in ecosystems located at equivalent latitudes. These high-elevation communities are also likely to be isolated from similar communities, suggesting the potential role of geographic isolation on local species immigration and extinction (see Chapter 19).

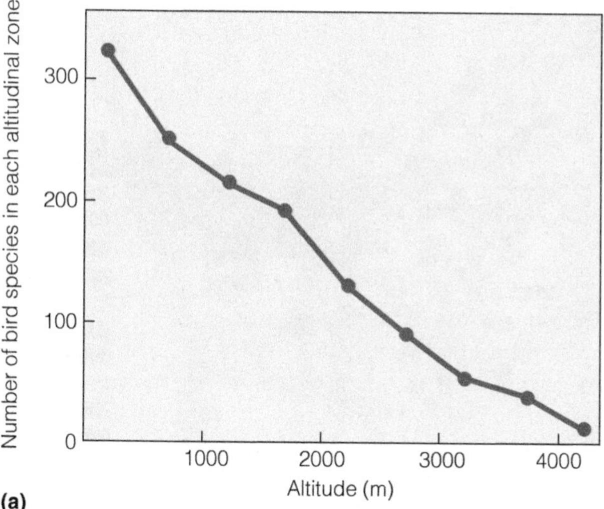

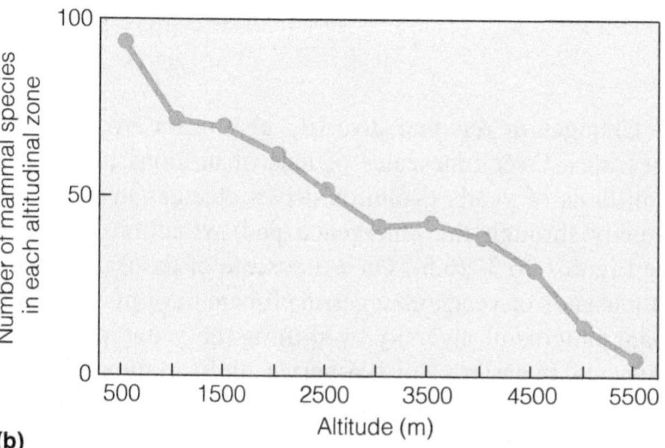

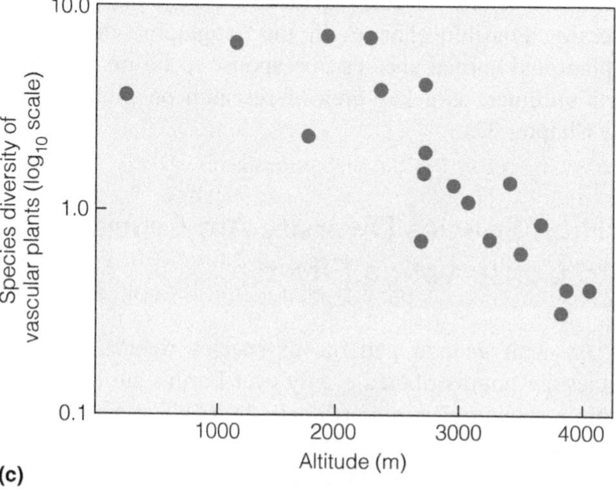

Figure 26.22 Relationship between species richness and altitude for (a) bird species in New Guinea, (b) mammal species in the Himalayas, and (c) vascular plants in the Himalayas.
(From Kikkawa and Williams 1971 [a]; from Hunter and Yonzon 1992 [b]; adapted from Whittaker 1977 [c].)

In contrast to terrestrial environments, the primary control on NPP in marine environments is nutrient availability in the surface waters (see Section 20.4), which is positively related to seasonal variations in temperature. The result is an inverse relationship between species richness and primary productivity. Yet in marine environments, species richness is positively related to sea surface temperatures, a direct measure of the energy flux per unit area. In the case of both terrestrial and marine environments, there is a consistent relationship between the direct flux of energy to the surface and patterns of species richness.

26.7 Regional Patterns of Species Diversity Are a Function of Processes Operating at Many Scales

The discussion of species diversity, even at the broad geographic scale that has been the focus of this chapter, is complicated by factors that relate directly to topics presented previously. For example, we discussed species diversity of individual communities in Chapter 17. Ecologists define species diversity at the spatial scale of individual communities as local or **alpha diversity**. Quantifying local patterns of diversity is hindered by the often difficult task of defining community boundaries. In addition, the relationship between species diversity and area makes it difficult to compare patterns of species diversity among communities and ecosystems that differ in size, such as lake ecosystems of varying size (see Chapter 19). Local patterns of plant and animal diversity also change over time during succession further compounding the difficulty of quantifying and comparing patterns of diversity among communities (see Figures 18.15, 18.19, and 18.20).

Within a given region, patterns of diversity are also influenced by environmental heterogeneity. The late ecologist Robert H. Whittaker of Cornell University defined **beta diversity** as the variation in species composition among sites (communities) in a geographic area (see Chapter 17). Increasing environmental heterogeneity within a region typically results in a corresponding increase in beta diversity because spatial differences in environmental constraints across the landscape function to restrict the distribution of species. For example, the diverse topography of mountainous regions generally supports a greater diversity of habitats and communities that does the consistent terrain of flatlands. The diversity of habitats and communities results in an increase in regional patterns of diversity. From east to west in North America, the number of species of trees, breeding birds, and mammals (see Figure 26.6) increases. This increased diversity on an east–west (longitude) gradient relates to an increased diversity of the environment both horizontally and altitudinally (and corresponding increases in beta diversity). Miquel De Cáceres of the University of Montreal (Quebec, Canada) and colleagues examined variation in beta diversity of tree species across a global network of forest plots. The researchers found that tree beta diversity increased with topographic variability within the forested site. As we observed

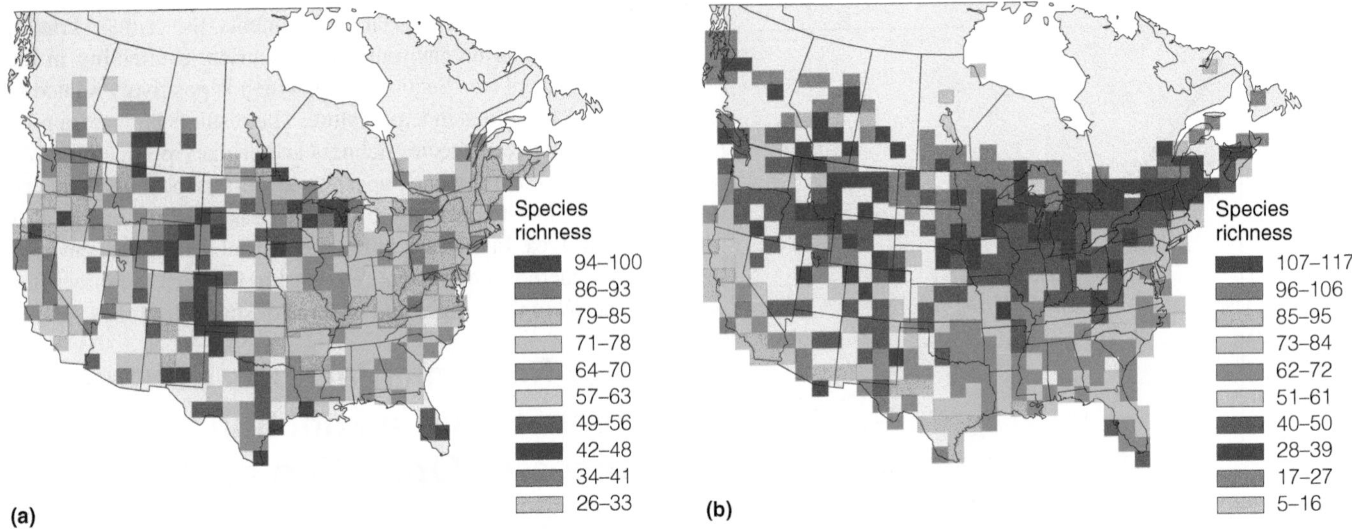

Figure 26.23 (a) Summer and (b) winter patterns of bird species richness across North America (20,000 km^2 grid cells).
(From Hurlbert and Haskell 2003.)

in Figure 26.9, variation in regional environmental conditions can result in deviations from the general trend in species richness with latitude observed at a continental to global scale.

The total species diversity (or species richness) across all communities within a geographic area is called regional or **gamma diversity**. Diversity at this scale corresponds to the patterns depicted in Figures 26.6. Comparison of even these broad-scale patterns of diversity at a continental or global scale can be confounded by time. For example, latitudinal patterns of bird species diversity presented in Figure 26.7 vary seasonally (**Figure 26.23**). More than 50 percent of the bird species that nest and breed in North America during the spring and summer months are migratory; they reside farther south in North, Central, and South America during the fall and winter months. Patterns of species migration alter seasonal patterns of regional diversity for a wide range of taxonomic groups.

Changes in regional diversity also occur over geologic timescales. Over timescales of tens of millions to hundreds of millions of years, evolution drives changes in patterns of diversity through the emergence and extinction of species (see Figures 26.3–26.5). On a timescale of thousands to tens of thousands of years, changes in climate have influenced regional patterns of diversity by shifting the geographic ranges of species. In eastern North America, shifts in the distribution of tree species after the last glacial maximum around 20000 years B.P. represent one such example (see Figures 18.25 and 18.26). Geographic ranges of many species in North America continue to shift, influencing local and regional patterns of diversity. Possible changes in the geographic distributions of plant and animal species in response to future changes in Earth's climate is a key area of research on global change (see Chapter 27).

ECOLOGICAL Issues & Applications — Regions of High Species Diversity Are Crucial to Conservation Efforts

Complicating the interpretation of broad-scale patterns of diversity is the fact that most of Earth's species are endemic—they have small, restricted geographic ranges (see Section 8.2). For example, of the world's approximately 10,000 bird species, more than 2500 are endemic, being restricted to a range smaller than 50,000 kilometers squared (km^2). Species of flora endemic to a single country represent 46–62 percent of the world flora. Of the thousands of new species being identified each year, virtually all of them are endemic to restricted regions within the tropics. It is the restricted distribution of these species that makes them extremely vulnerable to human activities that degrade or destroy their habitats. Of the species classified as threatened by the International Union for Conservation of Nature (IUCN), 91 percent are endemic.

As with general patterns of species richness, endemic species are not distributed evenly over Earth's surface or even within a geographic region (**Figure 26.24**). Certain regions of the world exhibit both high species richness and endemism. British ecologist Norman Myers defined these regions of unusually high diversity as **hotspots**. Myers developed the concept of biodiversity hotspots in 1988 to address the dilemma that conservationists face: What areas are the most important for preserving species?

The designation of a region as a hotspot of biological diversity is based on two factors: overall diversity of the region and significance of impact from human activities. Plant diversity is the biological basis for hotspot designation; to qualify as a hotspot, a region must support 1500 or more endemic plant

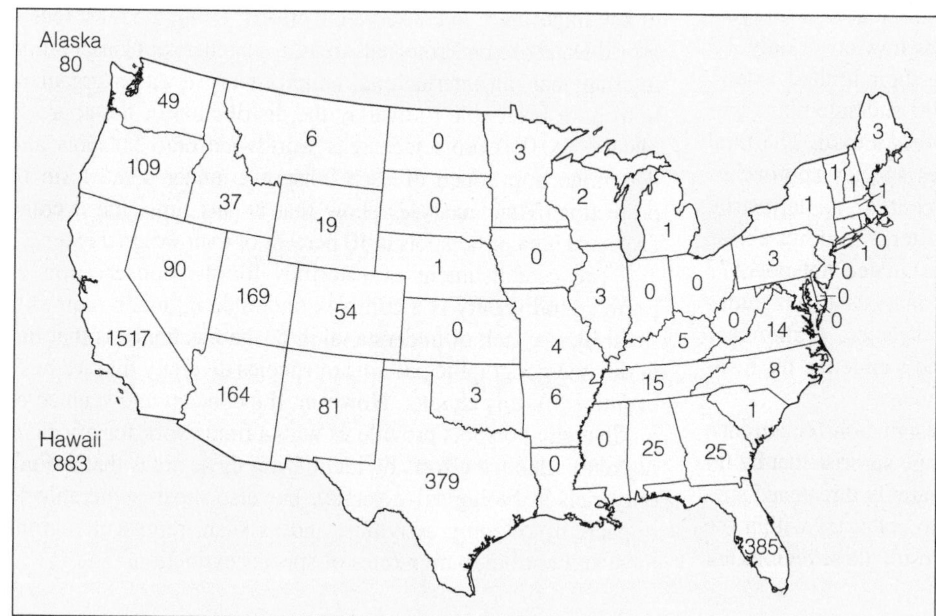

Figure 26.24 The number of plant species endemic to the different states of the United States.
(Adapted from Gentry 1986.)

species (0.5 percent of the global total), and the region must have lost more than 70 percent of its original habitat. Plants have been used as qualifiers because they are easy to identify and count as well as providing the basis of diversity in other taxonomic groups.

The original 25 biodiversity regions of Earth that were designated as hotspots by the IUCN in 2000 contain 44 percent of all plant species and 35 percent of all terrestrial vertebrate species in only 1.4 percent of the planet's land area. Several of these hotspots are tropical island archipelagos, such as the Caribbean and the Philippines, or relatively large islands, such as New Caledonia. However, other hotspots are continental islands. They are effectively isolated, being surrounded by deserts, mountain ranges, and seas. The Cape Floristic Province in South Africa is isolated by the aridity of the Kalahari, Karoo, and Namib deserts. Other landlocked islands are high mountains and mountain ridges. For communities inhabiting mountain ridges in the Tropical Andes (South America) and the Caucasus (central Asia), the lowlands are insurmountable barriers to dispersal.

An updated analysis of global biodiversity by Myers and colleagues at IUCN in 2004 expanded the number of biodiversity hotspots to 35 (**Figure 26.25**), which includes six previously overlooked areas. These are the Madrean Pine-Oak Woodlands of northern Mexico and the southwestern United States, southern Africa's Maputaland-Pondoland-Albany region, the Horn of Africa, the Irano-Anatolian region, the Mountains of Central Asia, and Japan. In addition, two hotspots have been subdivided because data are now sufficient to show that they contain quite distinctive biotas.

The 35 designated hotspots shown in Figure 26.25 once covered 15.7 percent of Earth's land surface; however,

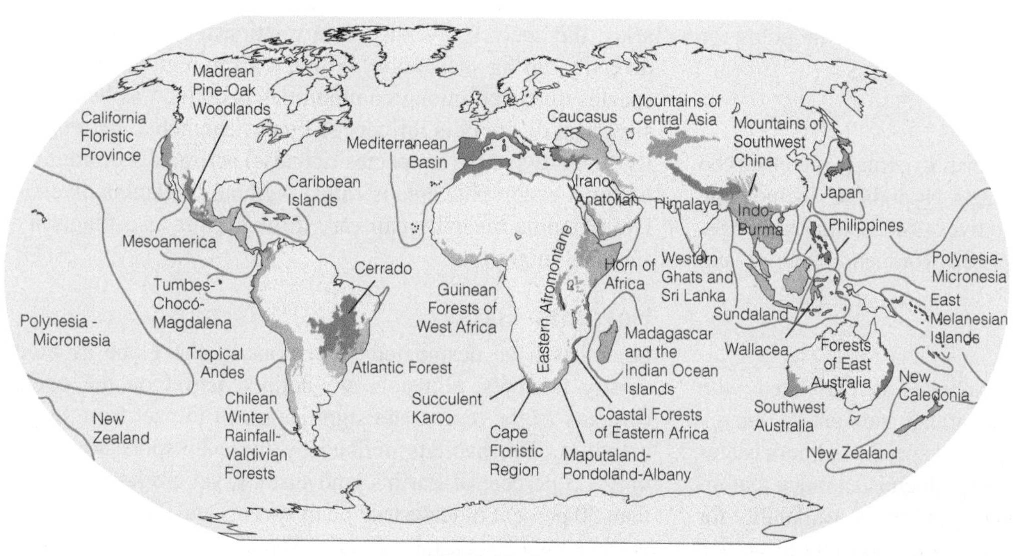

Figure 26.25 Thirty four areas of the world that the International Union for Conservation of Nature has designated as biodiversity hotspots.
(Adapted from Myers et al. 2004.)

86 percent of the hotspots' habitat has already been destroyed. As a result, intact remnants of the hotspots now cover only 2.3 percent of Earth's land surface. Despite their limited extent, these habitats are home to at least 150,000 endemic plant species, approximately 50 percent of the world's total. The total number of endemic terrestrial vertebrates species approaches 12,000, representing more than 40 percent of all terrestrial vertebrate species. In total, over 22,000 terrestrial vertebrate species are found to occur in the 35 designated hotspots, 77 percent of the world's total. The updated analysis also includes the first assessment of freshwater fish species. Preliminary analyses show that 29 percent of the world's endemic freshwater fish species occur within the hotspot zones.

Given the criteria used in hotspot designation (occurrence of endemics and high human impact), it is no surprise that by far the largest proportion of the 121,000 potentially threatened species in the tropical regions are endemic to countries within the 35 designated biodiversity hotspots. As a result, these regions are of key importance in conservation efforts. Using the most recent World Database on Protected Areas, researchers at Conservation International (an international nonprofit conservation organization) have been able to overlay the distribution of the approximately 100,000 protected areas worldwide onto hotspots and determine how much of each hotspot is under some form of protection. Their analyses show that at this time, the average protected area of hotspots is 10 percent of their original extent.

The establishment of priorities for the conservation of Earth's biodiversity is a complex undertaking, made more difficult by our lack of understanding of the mechanisms that underlay the geographic patterns of species diversity that we have examined in this chapter. However, the concept and science of biodiversity hotspots provide us with a framework for prioritizing our collective efforts by identifying those areas that are not only rich in biological diversity, but also most vulnerable to impacts by ongoing activities, and as such, represent regions most susceptible to high rates of species extinction.

SUMMARY

Temporal Patterns of Species Diversity 26.1

Earth's biological diversity has changed through evolutionary time. The fossil record suggests a pattern of increasing diversity over the past 600 million years.

Extinctions 26.2

Despite the overall pattern of increasing diversity through time, Earth's history is marked by periods of large-scale or mass extinctions. Two notable periods are the end of the Permian, when more than 90 percent of marine invertebrates disappeared from the fossil record, and the Cretaceous period, which saw the extinction of dinosaurs.

Geographic Patterns of Species Richness 26.3

The approximately 1.4 million species identified by scientists are not distributed evenly over Earth's surface. In general, species richness decreases from the equator toward the poles for both aquatic and terrestrial organisms.

Hypotheses 26.4

Various hypotheses have been put forth to explain the observed latitudinal patterns of species richness. Hypotheses include age of community, stability of climate over time, spatial heterogeneity of the environment, available environmental energy, and ecosystem productivity.

Environmental Energy 26.5

Species richness of terrestrial vertebrates is correlated with potential evapotranspiration, an integrated measure of energy input to the ecosystem. Regionally, plant species richness is correlated with actual evapotranspiration, which is an index that includes both the flux of thermal energy and water availability for evaporation. Geographic patterns of species diversity in marine environments are correlated with mean sea surface temperatures.

Species Richness and Primary Productivity 26.6

Geographic patterns of both terrestrial plants and animals are positively related to net primary productivity. Like patterns of species richness, terrestrial net primary productivity decreases with increasing latitude. The general pattern for marine environments appears opposite that observed for terrestrial environments. Patterns of species diversity for marine organisms are inversely related to ocean primary productivity. Latitudinal patterns of diversity appear to be influenced by the seasonality of primary productivity rather than by total productivity per se.

Local and Regional Diversity 26.7

Ecologists define diversity within a community or ecosystem as local or alpha diversity. Quantifying local patterns of diversity is hindered by difficulties in defining community boundaries, the species diversity–area relationship, and changes in diversity during succession. Within a given region, variation in species diversity among communities is defined as beta diversity. Beta diversity is influenced by environmental heterogeneity. Total diversity (or species richness) across all communities within a geographic area is called regional or gamma diversity. Even gamma diversity can vary through time as a function of seasonal migration.

Hotspots Ecological Issues & Applications

Scientists have designated 35 regions of the globe as biodiversity hotspots. Hotspots are defined based on the overall diversity of the region and significance of impact from human activities. The habitats defined by these hotspots now cover only 2.3 percent of Earth's land surface, yet are home to more than 50 percent of terrestrial plant species and 77 percent of the world's total terrestrial vertebrate species, making them critical areas for conservation efforts.

STUDY QUESTIONS

1. How has Earth's biological diversity changed over the past 600 million years?
2. What is a mass extinction?
3. How does species richness vary with latitude?
4. How are regional patterns of diversity in terrestrial environments related to measures of environmental energy such as potential evapotranspiration (PET)?
5. Why might the index of actual evapotranspiration be a better predictor of terrestrial plant species richness than PET?
6. What does the relationship between tree species richness and AET presented in Figure 26.18 imply about the relationship between tree species diversity and net primary productivity?
7. How do patterns of species richness in marine environments relate to sea surface temperatures?
8. How is primary productivity related to species richness in the oceans?
9. What is the difference between terrestrial and marine environments in the latitudinal patterns of NPP? Why does this difference occur?
10. How does environmental heterogeneity influence patterns of species diversity?
11. Contrast alpha, beta, and gamma diversity.
12. How are biodiversity hotspots related to patterns of endemism?

FURTHER READINGS

Classic Studies

Wilson, E. O. 1992. *The diversity of life.* Cambridge, MA: The Belknap Press of Harvard University Press.
A modern-day classic providing an overview of biodiversity over geological and evolutionary time—and human impacts on biodiversity.

Recent Research

Brown, J. H. 1995. *Macroecology.* Chicago: University of Chicago Press.
This book provides an excellent discussion of large-scale patterns of biological diversity over geologic time.

Cox, C. B., and P. D. Moore. 2000. *Biogeography: An ecological and evolutionary approach,* 6th ed. Oxford, UK: Blackwell Publishing.
An excellent text; provides an introduction to the geography and ecology of Earth's biological diversity.

Gaston, K. J. 2001. "Global patterns in biodiversity." *Nature* 405:220–227.
This review article provides an excellent introduction to the topic of global patterns of biodiversity.

Gaston, K. J., and T. M. Blackburn. 2000. *Pattern and process in macroecology.* Oxford, UK: Blackwell Publishing.
An excellent introduction to the topic of broad-scale patterns of species diversity; easy to read with an excellent array of research examples.

Huston, M. 1994. *Biological diversity: The coexistence of species on changing landscapes.* New York: Cambridge University Press.
An outstanding review of our current understanding of processes governing patterns of biological diversity, from a local to global scale.

Irigoien, X., J. Huisman, and R. P. Harris. 2004. "Global biodiversity patterns of marine phytoplantkton and zooplankton." *Nature,* 429:863–867.
This research article examines global patterns of phytoplankton and zooplankton diversity as they relate to patterns of primary productivity.

Myers, N., R. A. Mittermeier, C. G. Mittermeier, G. A. B. da Fonseca, and J. Kent. 2000. Biodiversity hotspots for conservation priorities. *Nature* 403:853–858.
This article provides an introduction to the concept of biodiversity hotspots and their importance to global conservation efforts.

Rex, M., and R. J. Etter. 2010. *Deep sea biodiversity, pattern, and scale.* Cambridge MA: Harvard University Press.
Relates the great discoveries by the Census of Marine Life of biodiversity patterns on the deep ocean floor to the physical and biological dynamics of the global ocean regions.

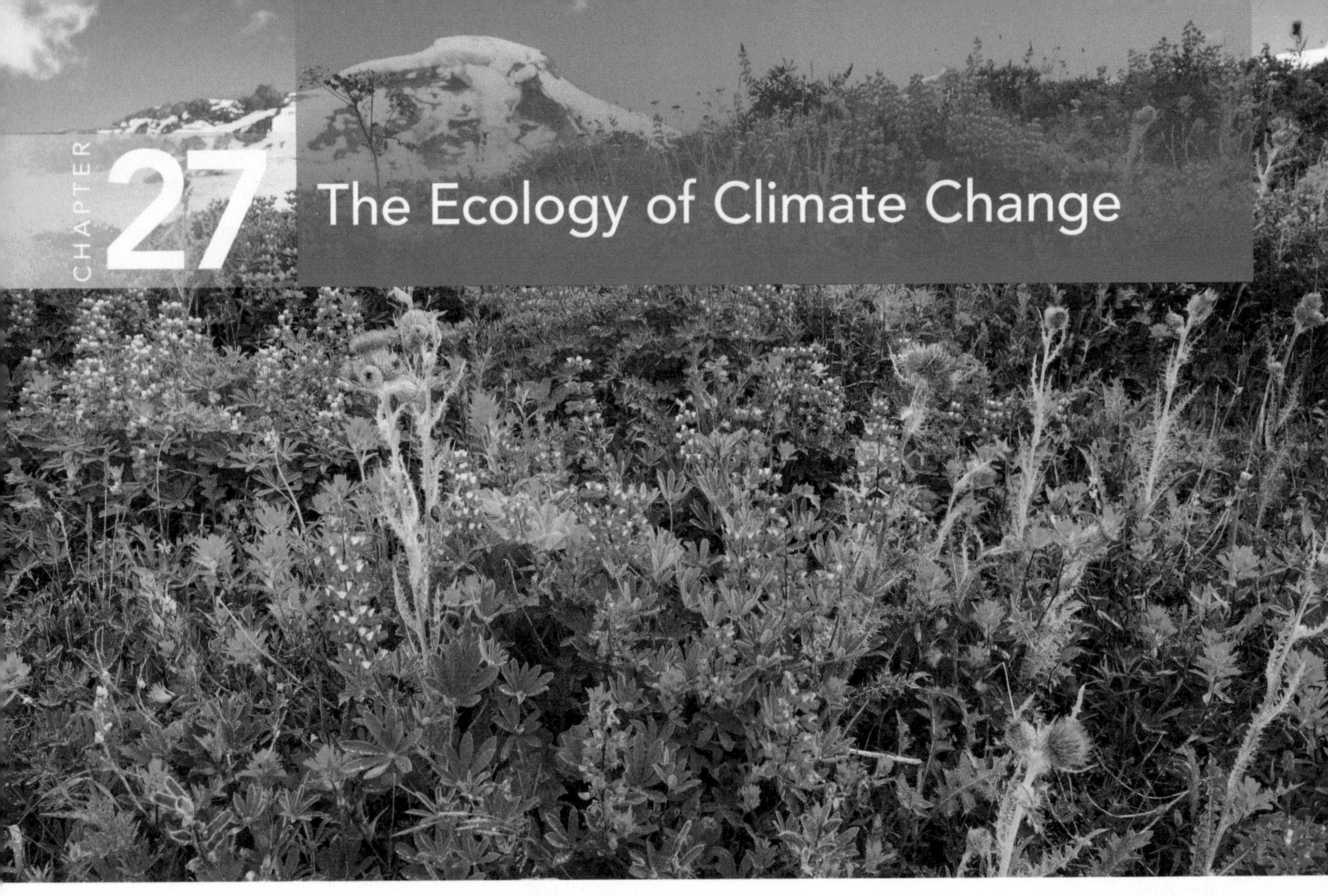

CHAPTER 27 The Ecology of Climate Change

Rising average temperatures over the past century have impacted the timing of ecological activities such as the onset of reproduction in plant and animal populations.

CHAPTER GUIDE

THE TERM *GLOBAL CLIMATE CHANGE* is redundant. Change is inherent in Earth's climate system. For example, although the tilt of Earth's axis relative to the Sun is 23.5°, giving rise to the seasons, Earth is actually wobbly (see Chapter 2). In fact, the tilt of Earth's axis varies from 22.1° to 24.5°. The amount of tilt in Earth's rotation affects the amount of sunlight striking the different parts of the globe, influencing patterns of global climate. This variation in the tilt of Earth takes place during a cycle of 41,000 years and is largely responsible for the ice ages—periods of glacial expansion and retreat (see Section 18.9, Figure 18.24).

In turn, variations in climate profoundly affect life on Earth. Paleoecology has recorded the responses of populations, communities, and ecosystems to climate changes during periods of glacial expansion and retreat over the past 100,000 years (see Figures 18.25 and 18.26). On an even longer timescale, the fossil record recounts a story of evolutionary change resulting from the dynamics of Earth's climate over geologic time. Throughout this text, we have seen countless examples of how climate influences the function of natural ecosystems—from the uptake of carbon dioxide by leaves in the process of photosynthesis to the distribution and productivity of Earth's ecosystems. But now we have entered a new era in the history of life on our planet, one in which a single species—humans—may have the ability to alter Earth's climate.

In Chapter 2 we first presented evidence that since the onset of the Industrial Revolution, the burning of fossil fuels has resulted in an exponential increase in the atmospheric concentrations of carbon dioxide and other greenhouse gases, and these changes to the chemistry of Earth's atmosphere have resulted in a general pattern of warming over the past century (Chapter 2, *Ecological Issues & Applications*). In addition, as the emissions of greenhouse gases continue into the future, human activities have the potential to further alter the global climate system.

In this chapter, we examine how human activities are altering Earth's climate and explore how these changes have impacted ecological systems by shifting the phenology and distribution of species, altering their interactions, and ultimately influencing the distribution and productivity of ecosystems. We will then explore the methods being used to predict how Earth's climate might continue to change into the future, and the implication of future climate changes to ecological systems.

27.1 Earth's Climate Has Warmed over the Past Century

As we first discussed in Chapter 2 (*Ecological Issues & Applications*), over the past century Earth's climate has warmed by an estimated 0.74°C (±0.2°C). The rate of warming during the latter period of the century has been approximately double that of the first (**Figure 27.1**), and it is believed to be greater than at any other time during the last 1000 years. These changes in temperature have not been distributed equally across Earth's surface; rather, regional changes have been highly heterogeneous (see Figure 27.1). The greatest warming has occurred in the polar regions, particularly the Arctic, which is warming twice as fast as other parts of the world.

Warming has also not occurred equally throughout the seasons. The greatest observed warming over the last half-century has been during the winter months (see Chapter 2, *Ecological Issues & Applications* and Figure 2.29). In addition, diurnal temperature ranges have decreased because minimum temperatures are increasing at about twice the rate of maximum temperatures. As a consequence, the freeze-free periods in most mid- and high-latitude regions are lengthening. Since 1900, the maximum area covered by seasonally frozen ground has decreased by about 7 percent in the Northern Hemisphere, with a decrease in spring of up to 15 percent.

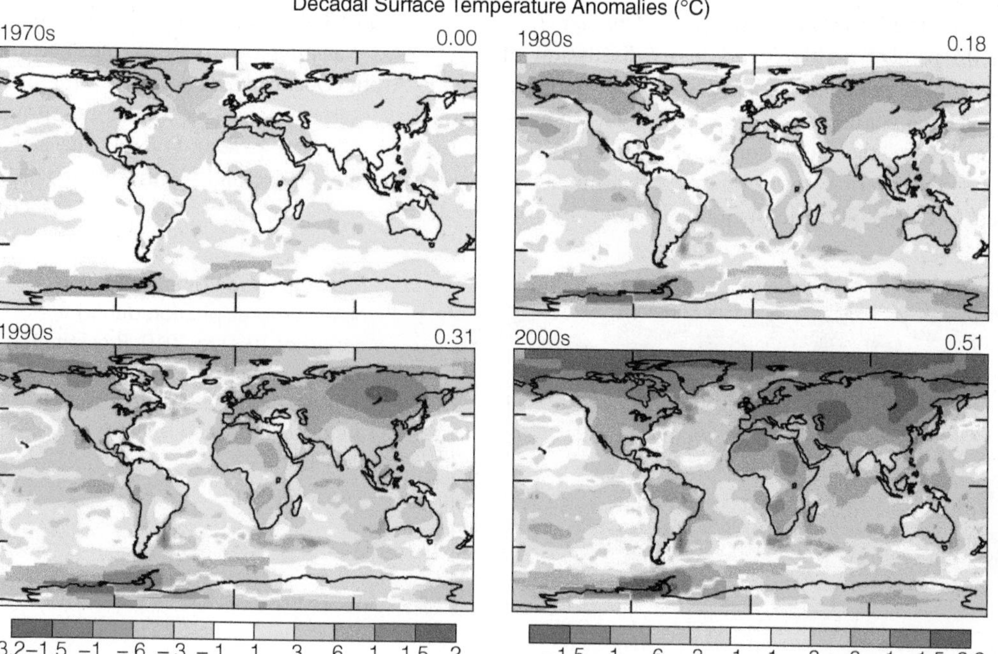

Figure 27.1 Decadal changes (°C) in mean annual surface temperatures for the period of 1970 to 2010. Changes (anomalies) were calculated as the difference between the current year mean surface temperature and the average surface temperature for the base period of 1951 to 1980 (a period over which temperatures were relatively stable; see Figure 2.26). The value reported in the upper right corner of each map is the global average temperature change for that decade. (From Hansen et al. 2010.)

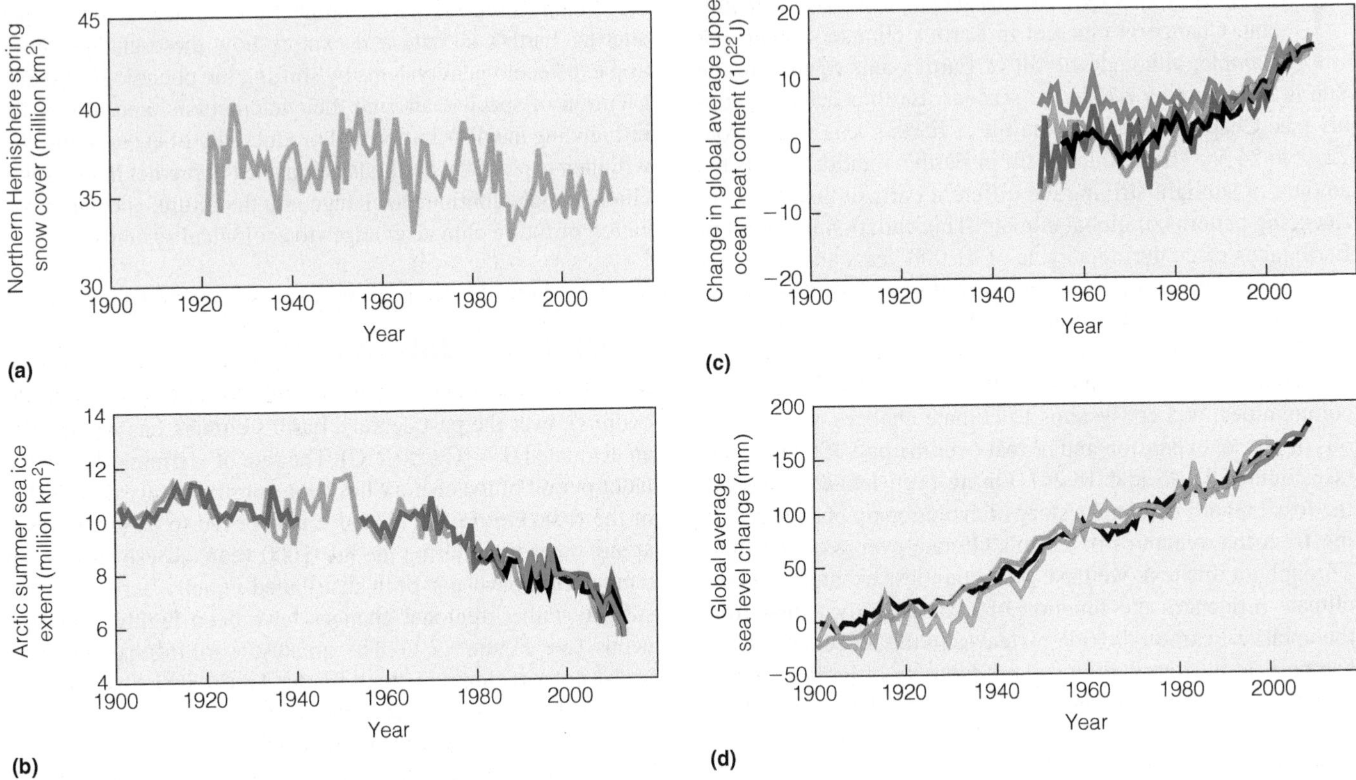

Figure 27.2 Changes in (a) Northern Hemisphere spring snow cover, (b) Arctic summer sea ice extent, (c) global average heat content of upper ocean waters, and (d) global average sea level over the past century.
(From Intergovernmental Panel on Climate Change 2013.)

The changes in Earth's surface energy balance (see Chapter 2) and associated warming have influenced features of the global climate system in addition to surface temperatures. Satellite data reveal a 10 percent decrease in snow cover and ice extent since the late 1960s (**Figure 27.2a**). Changes in the precipitation have also been neither spatially nor temporally uniform (**Figure 27.3**). In the mid- and high latitudes of the Northern Hemisphere a decadal increase of 0.5 percent mostly occurs in autumn and winter, whereas, in the tropics, precipitation has decreased by about approximately 0.3 percent per decade.

Changes have not been limited to the land surface. Observations since 1961 show that the average temperature of the global ocean has increased to depths of at least 3000 meters (m) and that the ocean has been absorbing more than 80 percent of the heat added to the climate system (**Figure 27.2b**). The warming of the surface waters of the ocean over the past 50 years has been slower than that observed for the land surface, however, because the ocean responds more slowly than the land as a result of its large thermal inertia (see Chapter 2, Section 2.7 and Chapter 3, Section 3.2 for further discussion of the topic). Warming of the ocean surface has been largest over the Arctic Ocean where satellite data since 1978 show the annual average Arctic sea ice extent has shrunk by 2.7 percent per decade (**Figure 27.2c**). Overall, the ocean warming has caused seawater to expand, contributing to sea level rise (**Figure 27.2d**).

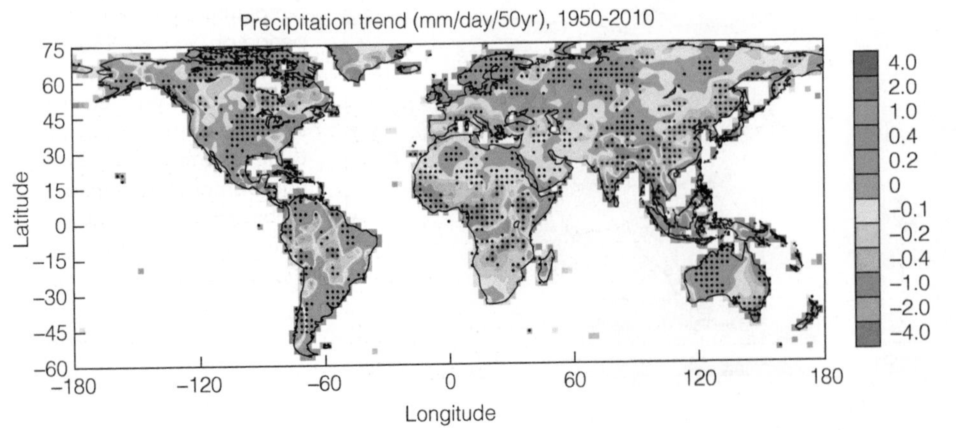

Figure 27.3 Linear trends in observed precipitation during the period of 1950 to 2010. Precipitation trend is in mm per day over the period of observation. Stippling indicates the observed trend is statistically significant.
(From Dai 2011.)

The observed changes in Earth's climate over the past century outlined here have had a broad and diverse impact on ecological systems, both terrestrial and aquatic.

27.2 Climate Change Has a Direct Influence on the Physiology and Development of Organisms

As we have discussed in Chapters 6 and 7, temperature has a direct effect on the basic metabolic and developmental processes of both plants and animals. How global patterns of warming over the past century have impacted organisms has been the intense focus of research.

Endothermic animals maintain body temperatures at a relatively constant level as a result of internal heat production through metabolic processes (see Chapter 7, Section 7.10). Therefore, environmental temperatures influence metabolic processes largely through determining heat exchange with the surrounding environment. Within a species, larger body size functions to reduce the dissipation of heat by reducing the surface area relative of body mass (see Figure 7.2), and therefore, mass-specific metabolic rate tends to decreases with increasing body size (see Figure 7.21). As a result, in endothermic species, there is a geographic trend for average body size to increase with decreasing mean annual temperature (or increasing latitude; see Section 7.10). Numerous studies have documented recent changes in animal body size for local populations over the timescale of decades to a century that are correlated to changes in local temperature (see Chapter 7, *Ecological Issues & Applications* for detailed discussion and examples). In general, results have shown a recent decrease in average body size for a variety of endothermic animal species (both bird and mammal) correlated with increasing temperatures, a pattern consistent with the hypothesis of that smaller body size is more energetically efficient under a warmer climate.

In contrast to endotherms, environmental temperatures have a direct effect on body temperature and metabolic rates in ecothermic animals (see Section 7.9). Species have a limited range of temperatures over which they can maintain basic metabolic processes and activities, and beyond which they cannot survive (thermal tolerance limits; see Figure 7.14). To examine the potential impacts of climate warming on ectotherms, researchers have focused on how changes in temperature directly affect metabolic rates within the species range of thermal tolerance, as well as how changes in temperature may detrimentally affect species currently living closest to their upper thermal tolerance limits (warming may exceed thermal tolerances).

Because temperatures have risen most rapidly in the mid- to high latitudes of the Northern Hemisphere (see Figure 27.1 and Figure 2.29 for changes in temperature by latitude), the majority of research on the impacts of warming has focused on these regions. In ectotherms, however, metabolic activity increases exponentially rather than linearly with temperature and, therefore, climate warming may have a greater proportional effect on the metabolic rates of ectotherms living in warmer rather than the cooler climates.

Michael Dillon of the University of Wyoming and colleagues used published data on a diverse array of ecothermic animals to derive a general (averaged) metabolic model for estimating mass-specific metabolic rates from global temperature data (high-frequency temperature data for the period of 1961 to 2009 from 3186 weather stations across the world). Predicted absolute changes in metabolic rates show a markedly different pattern from the observed changes in regional temperature (**Figure 27.4**). Metabolic rates increased most quickly in the tropics and north temperate zones, and less so in the Arctic. The predicted increase in metabolism in the tropics was large, despite the small rise in temperature because tropical warming took place in an environment that was initially warm. Such increases will have physiological and ecological impacts, such as an increased need for food and increased vulnerability to starvation unless food resources increase, possible reduced energy allocation for reproduction (see Chapter 10), increased rates of evaporative water loss in dry environments, and altered rate of birth and death.

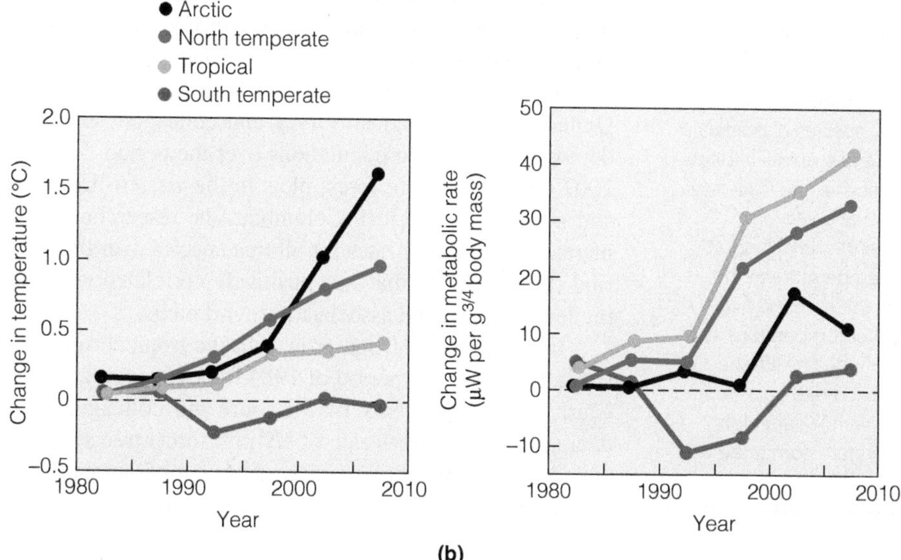

Figure 27.4 Global changes in temperature and ectotherm metabolic rates since 1980. (a) Changes in mean temperature (5-year averages) for Arctic (100 sites), northern temperate (356 sites), southern temperate (51 sites) and tropical (169 sites) regions. (b) Predicted absolute changes in mass normalized metabolic rates ($\mu W/g^{3/4}$ body mass) by geographic region. Both temperate and metabolic rates are expressed as the difference from the standard reference period of 1961–1990 calculated on a per site basis. Mass-specific metabolic rates were calculated using parameters estimated for an average ectotherm (see text).
(Data from Dillon et al. 2010.)

Recent studies using diverse physiological and biophysical approaches indicate that tropical ectotherms may be particularly vulnerable to climate warming even though observed and predicted tropical warming is relatively small. In Chapter 7, we saw that the thermal tolerance limits of species are closely aligned with the prevailing temperatures of the environments which they inhabit. For example, in a comparison between habitat temperature and acute thermal tolerance limits for porcelain crabs (genus *Petrolisthes*) inhabiting the intertidal environments of the eastern Pacific, Jonathon Stillmand and George Somero of Stanford University found that the upper thermal tolerance limits of the different species (temperature at which 50 percent mortality occurs [LT_{50}]) was positively correlated with the maximum temperatures recorded in the microhabitats (maximum habitat temperature [MHT]) inhabited by the crabs (**Figure 27.5**). Adaptive variation is clear. Tropical species are uniformly more heat tolerant than temperate species. Although the tropical species exhibit a higher temperature tolerance than species from temperate habitats, how might the species differ in their response to the threats posed by rising temperatures? The line of equality (LT_{50} = MHT) shown in Figure 27.5 provides an estimate of risk of heat death under current habitat conditions. In general, tropical species are most threatened by further increases in habitat temperature because current MHTs may reach or exceed lethal temperatures (LT_{50}). Furthermore, the researchers found that these species are further disadvantaged by possessing a relatively small ability to increase LT_{50} through acclimation (see Figures 5.7 and 7.18 for examples).

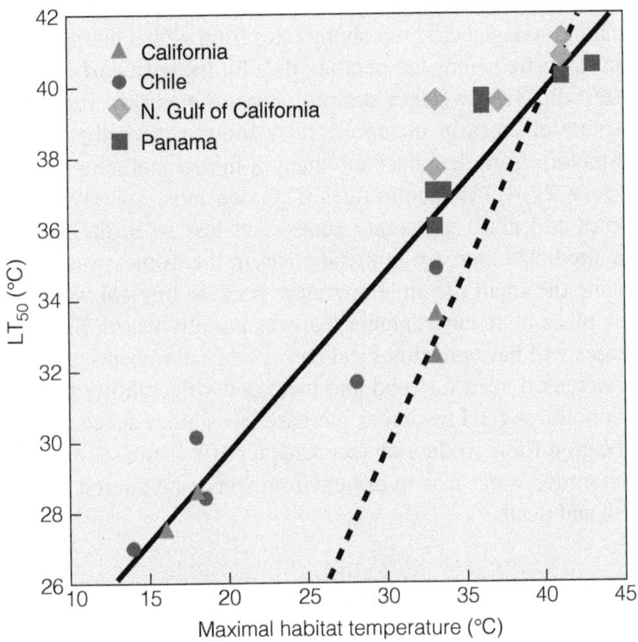

Figure 27.5 Relationship between maximum habitat temperature and upper thermal tolerance limits (T_{max}) for 20 species of porcelain crabs of the genus *Petrolisthes*, from intertidal and subtidal habitats throughout the eastern Pacific (temperate: California and Chile; tropical/subtropical: Panama and Northern Gulf of California). Each symbol represents a different crab species. Maximal habitat temperature is the maximum recorded water temperature at the geographic location and microhabitat used by the species. T_{max} represents the LT_{50} (lethal temperature; the water temperature at which 50 percent of test crabs died). The line of equality (Dashed line: LT_{50} = MHT) allows estimation of risk of heat death under current habitat conditions. Species with current MHT near LT_{50} are highly susceptible to further increases in water temperature compared to species where MHTs are far below values of LT_{50}.
(Modified from Stillman and Somero 2000 as presented in Somero 2010.)

The pattern seen among species of porcelain crabs has been observed for other groups of marine invertebrates that inhabit the intertidal zone. The hypothesis that the most warm-adapted species of marine intertidal invertebrates are likely to be the most threatened by climate warming agrees with broad conclusions reached in recent studies of terrestrial ectotherms from different latitudes. These analyses show that tropical ectotherms are more threatened by climate change than species from midlatitudes because tropical species live closer to their upper temperature tolerance limits, and in some cases, live at temperatures above those at which physiological processes operate at their thermal optima (see Figure 7.14).

Examining the impacts of recent climate change on terrestrial plant species is more complicated than for animal species because the response of plants to changes in both temperature and precipitation (soil water availability) are influenced by atmospheric concentrations of carbon dioxide (see Chapter 6, *Ecological Issues & Applications* for discussion of the direct response of plants to rising atmospheric carbon dioxide). As the primary cause of recent climate change, the atmospheric concentration of carbon dioxide has risen in parallel with global temperatures over the past century. In addition, responses to regional changes in climate can be modified by local edaphic and microhabitat conditions (soils, slope, aspect, etc.). Regardless of these complicating factors, a number of general, regional trends have been reported.

In the more mesic regions of eastern North America, the analysis of tree rings has shown an increase in growth rates over the past century that are correlated to an increase in the length of the growing season (defined by the dates of the first and last frost of the year). In regions of western North America, however, increasing water deficits associated with warming have led to an increase in tree mortality. Phillip van Mantgem of the United States Geological Survey and colleagues examined the demographics of tree populations over the period from 1955 to 2007 on 76 long-term forest plots in the western United States and southwestern British Columbia. The researchers found an increase in mortality rates for all tree species over the study period (**Figure 27.6**) that was positively correlated with increasing temperatures and associated water deficits.

In an analysis of tree growth in the tropical rainforests of Costa Rica over the period of 1984 to 2000, ecologist Deborah Clark of the University of Missouri and colleagues found a decrease in the growth rate of six rain forest tree species associated with increasing minimum daily temperatures (nighttime temperatures) over the study period. The observed decreases in

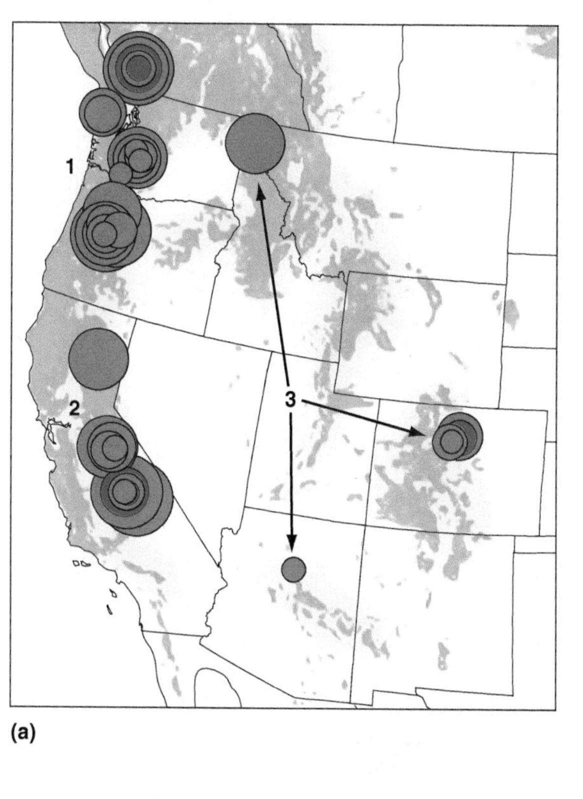

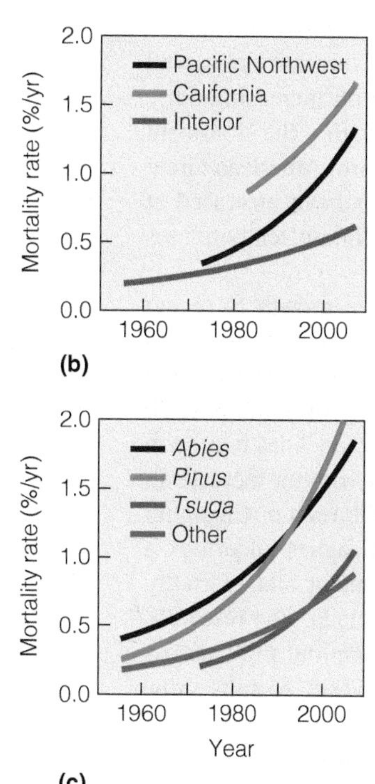

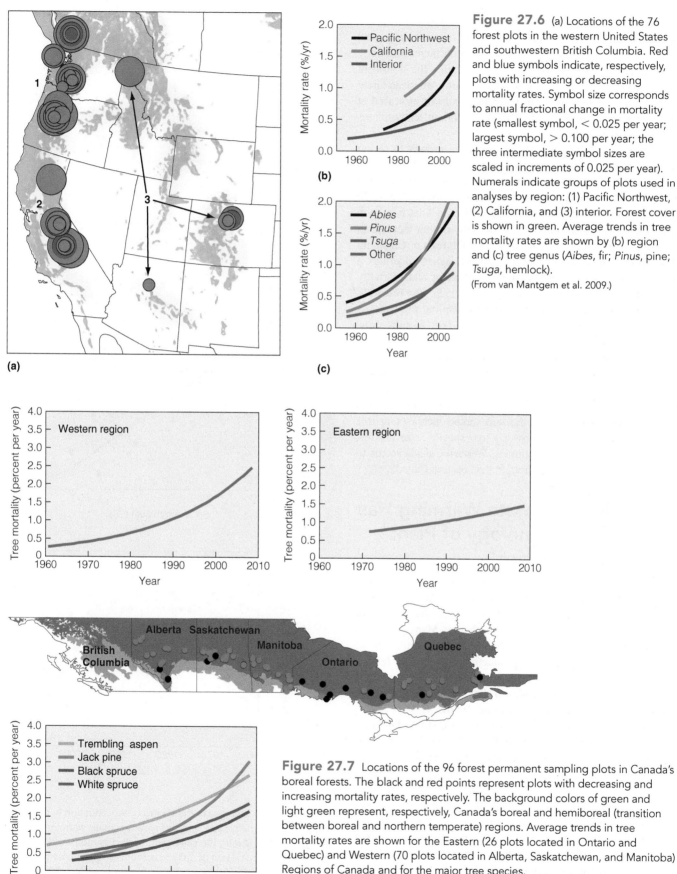

Figure 27.6 (a) Locations of the 76 forest plots in the western United States and southwestern British Columbia. Red and blue symbols indicate, respectively, plots with increasing or decreasing mortality rates. Symbol size corresponds to annual fractional change in mortality rate (smallest symbol, < 0.025 per year; largest symbol, > 0.100 per year; the three intermediate symbol sizes are scaled in increments of 0.025 per year). Numerals indicate groups of plots used in analyses by region: (1) Pacific Northwest, (2) California, and (3) interior. Forest cover is shown in green. Average trends in tree mortality rates are shown by (b) region and (c) tree genus (*Aibes*, fir; *Pinus*, pine; *Tsuga*, hemlock).
(From van Mantgem et al. 2009.)

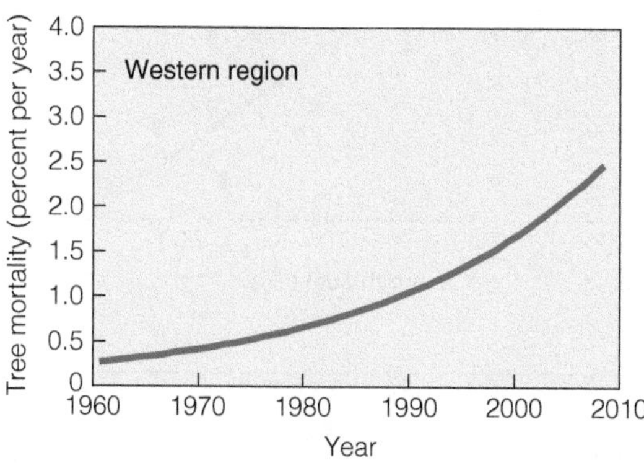

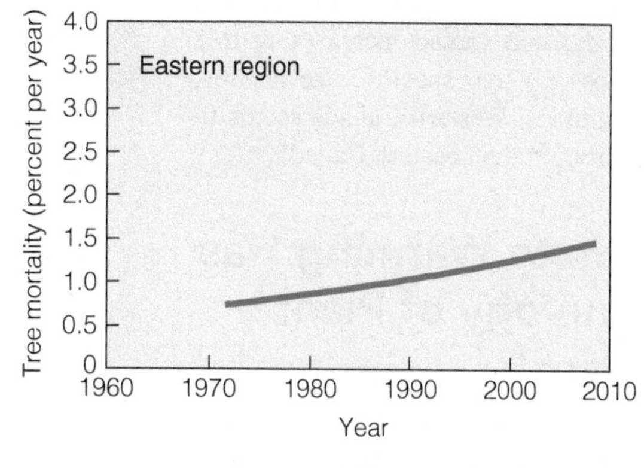

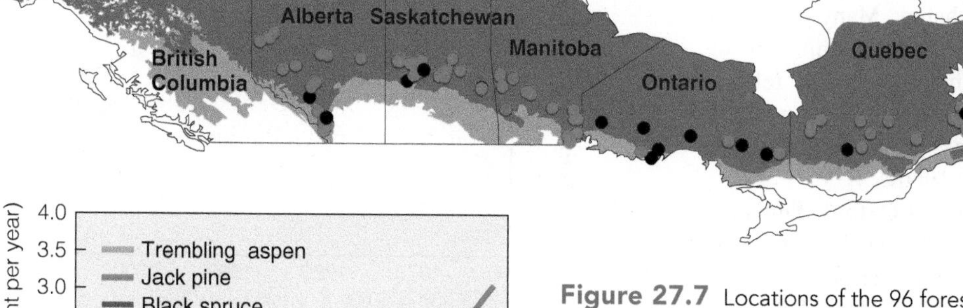

Figure 27.7 Locations of the 96 forest permanent sampling plots in Canada's boreal forests. The black and red points represent plots with decreasing and increasing mortality rates, respectively. The background colors of green and light green represent, respectively, Canada's boreal and hemiboreal (transition between boreal and northern temperate) regions. Average trends in tree mortality rates are shown for the Eastern (26 plots located in Ontario and Quebec) and Western (70 plots located in Alberta, Saskatchewan, and Manitoba) Regions of Canada and for the major tree species.
(Data from Peng, C. et al. 2011. A drought-induced pervasive increase in tree mortality across Canada's boreal forests. Nature Climate Change 1: 467–471. Fig. 1 and 2, pg. 468.)

growth rate are a result of higher rates of respiration. Richard Alward and colleagues at Colorado State University observed a similar trend of deceasing growth rate with increasing daily minimum temperatures for *Bouteloua gracilis*, the dominant C_4 grass of the shortgrass prairies of the North American Great Plains, where daily minimum temperatures have increased at approximately twice the rate of daily maximum temperatures over the past 50 years.

Studies examining the response of tree growth to recent climate change in the Arctic region have revealed mixed results. Although a number of studies have measured localized increases in tree growth rates with warming over the past 50 years, the majority of analyses have revealed declines in growth rates and increases in mortality associated with an increase in warming-related water stress throughout the region. Changhui Peng of the University of Quebec and colleagues undertook a detailed analysis of long-term tree recruitment and mortality across the entire boreal forest region of Canada. The researchers used data from 96 long-term forest sampling plots across Canada covering the period from 1963 to 2008. Results show that tree mortality rates increased by an overall average of 4.7 percent per year over the study period (**Figure 27.7**), with greater increases in mortality rates in western regions than in eastern regions (about 4.9 and 1.9 percent per year, respectively). The water stress associated with regional drought was the dominant contributor to the widespread increases in tree mortality rates observed across all tree species, size classes, elevations, longitudes, and latitudes. Western Canada seems to have been more sensitive to drought than eastern Canada.

27.3 Recent Climate Warming Has Altered the Phenology of Plant and Animal Species

Phenology, the timing of seasonal activities of plants and animals, is one of the most widely studied phenomena on which researchers have been able to track changes in the ecology of species in response to recent climate change. Many processes and activities such as migratory behavior, the termination of dormancy, or the onset of reproductive activity are related to seasonal changes in climate conditions, and long-term phenological data sets have shown that spring activities have occurred progressively earlier since the 1960s.

Anne Charmantier of the University of Oxford (England) and colleagues used data from a 47-year population study of the great tit (*Parus major*), a species of passerine bird common to the woodland of England, to examine potential changes in the onset of reproduction in response to recent patterns of regional warming. The population, located at Wytham Woods near Oxford, United Kingdom, shows a marked change over time in the date at which breeding occurs. Over the past 47 years (1961–2007), the mean egg-laying date of females has advanced (occurs earlier) by approximately 14 days (**Figure 27.8a**)—a change that began in the mid-1970s. Previous work with this population suggests that the onset of reproduction is a function of temperatures in the period

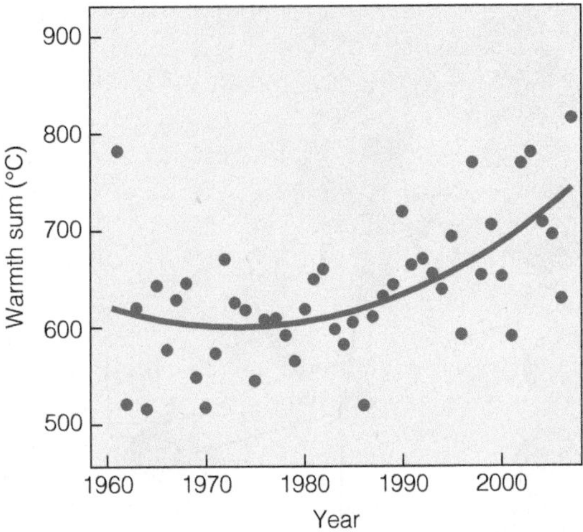

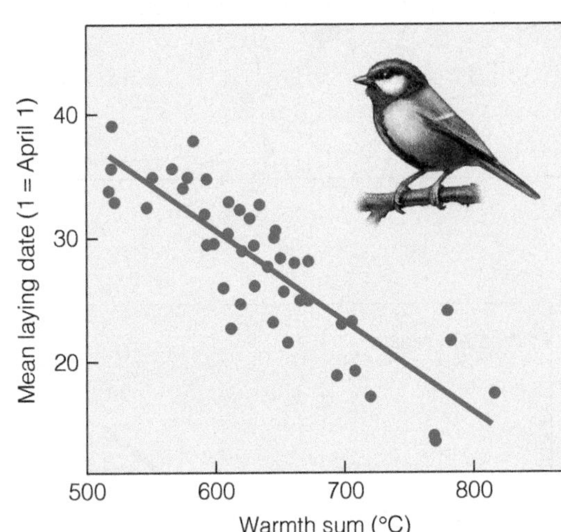

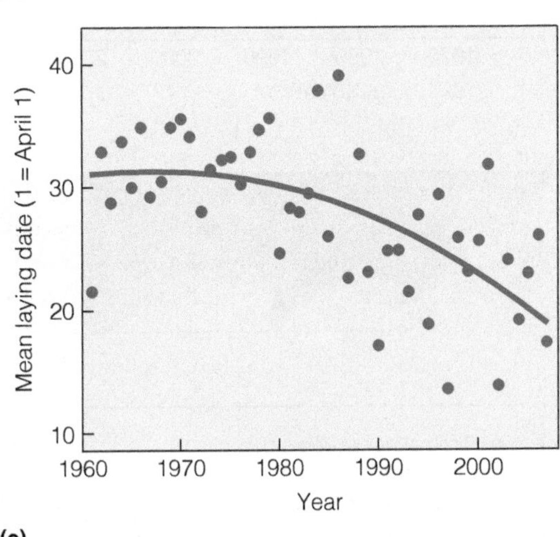

Figure 27.8 Changes in the timing of reproduction in the great tit (*Parus major*) as a result of recent climate warming. (a) Temperature during the pre-laying period as measured by spring warmth sum (sum of daily maximum temperatures between March 1 and April 25) increased over the study period. (b) Mean laying date is inversely related to warmth sum index. As a result, (c) mean laying date declined over the study period. (From Charmantier et al. 2008.)

preceding egg laying. In the current study, the researchers found a significant correlation between mean laying date and an index of spring "warmth," measured as the sum of daily maximum temperatures between March 1st and April 25th (**Figure 27.8b**). There has been a marked change in this index of spring temperatures over the study period, with a linear increase since the mid-1970s (**Figure 27.8c**).

For migratory birds, the timing of arrival on breeding territories and over-wintering grounds is a key determinant of survival and reproductive success. Peter Cotton of the University of Plymouth (England) examined the arrival and departure dates of 20 migratory bird species that spend the spring and summer breeding season in Oxfordshire, England, and migrate to sub-Saharan Africa for the winter. Using a data set that covered a period of 30 years, Cotton found that for 17 of the 20 species studied, the arrival date in Oxfordshire has advanced in response to increased temperature trends in their African over-wintering grounds (**Figure 27.9**). The departure date of migrant birds from Oxfordshire has also advanced in parallel with the change in arrival date. Cotton found that the timing of departure of migrant birds from the England site is correlated with increased summer minimum temperatures. Overall, the duration of stay of migrant birds in Oxfordshire has remained unchanged over the study period; however, the period of residency has shifted to an earlier date by an average of 8 days over the last 30 years.

Some of the most extensive data sets for changes in phenology in response to recent climate change come from plant populations, particularly at northern latitudes where the greatest amount of warming has occurred over the past century (see Figure 27.1). Elisabeth Beaubien and Andreas Hamann of the University of Alberta evaluated climate trends and the corresponding changes in flowering times for seven plant species in the central parklands of Alberta, Canada. Over the 71-year study period extending from 1936 to 2006, the researchers found substantial warming, which ranged from a 5.3°C in the mean monthly temperatures for February to an increase of 1.5°C for the month of May (**Figure 27.10a**). Flowering dates for the early blooming species (*Populus tremuloides* and *Anemone patens*) advanced by an average of two weeks over the study period. In contrast, the late-flowering species advanced between zero and six days (**Figure 27.10b**).

In a meta-analysis of studies that have examined the impacts of recent climate change on the phenology of plant and animal species, Camille Parmesan of the University of Texas examined published studies that have reported on 203 plant and animal species inhabiting the Northern Hemisphere (**Figure 27.11**). In her analysis, Parmesan found that amphibians had a significantly greater shift toward earlier breeding than all other taxonomic groups examined, advancing more than twice as fast as trees, birds, and butterflies. In turn, butterfly emergence or migratory arrival showed three times greater advancement than the first flowering of herbaceous plants, suggesting a possible impact on insect–plant interactions (see Chapter 15, Section 15.13). When analyzed across taxonomic groups, overall shifts in phenology are significantly greater at higher latitudes where observed warming has been greatest.

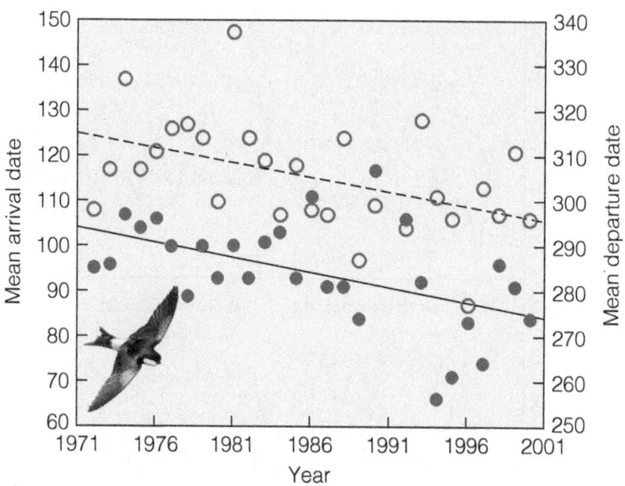

Figure 27.9 Trends in the arrival date (filled circles and solid line) and departure date (open circles and dashed line) of Northern House Martins (*Delichon urbica*) in Oxfordshire, England, from 1971 to 2000.
(From Cotton 2003.)

27.4 Changes in Climate Have Shifted the Geographic Distribution of Species

As we discussed in Chapter 8, climate has a direct influence on species' distributions, often because of species-specific physiological tolerances to temperature (see Chapters 6 and 7 for examples). In many cases, the northern (or upper elevation) boundary of a species' geographic distribution reflects constraints imposed by minimum temperatures, and with general warming trends, the latitudes and elevations at which these minimum temperature constraints occur has become shifted toward the poles or higher elevations. To the extent that dispersal and other environmental factors allow, species are expected to track the shifting climate and likewise shift their distributions poleward in latitude and upward in elevation.

In a study of the latitudinal distribution ranges of two closely related limpet species—*Lottia digitalis* and *Lottia austrodigitalis*—inhabiting the intertidal environments of the California coast, Lisa Crummett and Douglas Eernisse of California State University (Fullerton) found that over the 20-year period between 1978 and 1998, the southern range limit of the northern species (*L. digitalis*) is contracting northward, whereas the northern range limit of the southern species (*L. austrodigitalis*) is expanding northward (**Figure 27.12a**). This shift in the latitudinal range of the two limpet species is consistent with increases in shoreline ocean temperatures recorded for the region over this period (**Figure 27.12b**), and previous studies by the researchers that have shown that *L. austrodigitalis* (the southern species) has a higher thermal tolerance than *L. digitalis* (the northern species).

Camille Parmesan of the University of Texas and colleagues analyzed distributional changes over the past century for nonmigratory species of butterfly whose northern

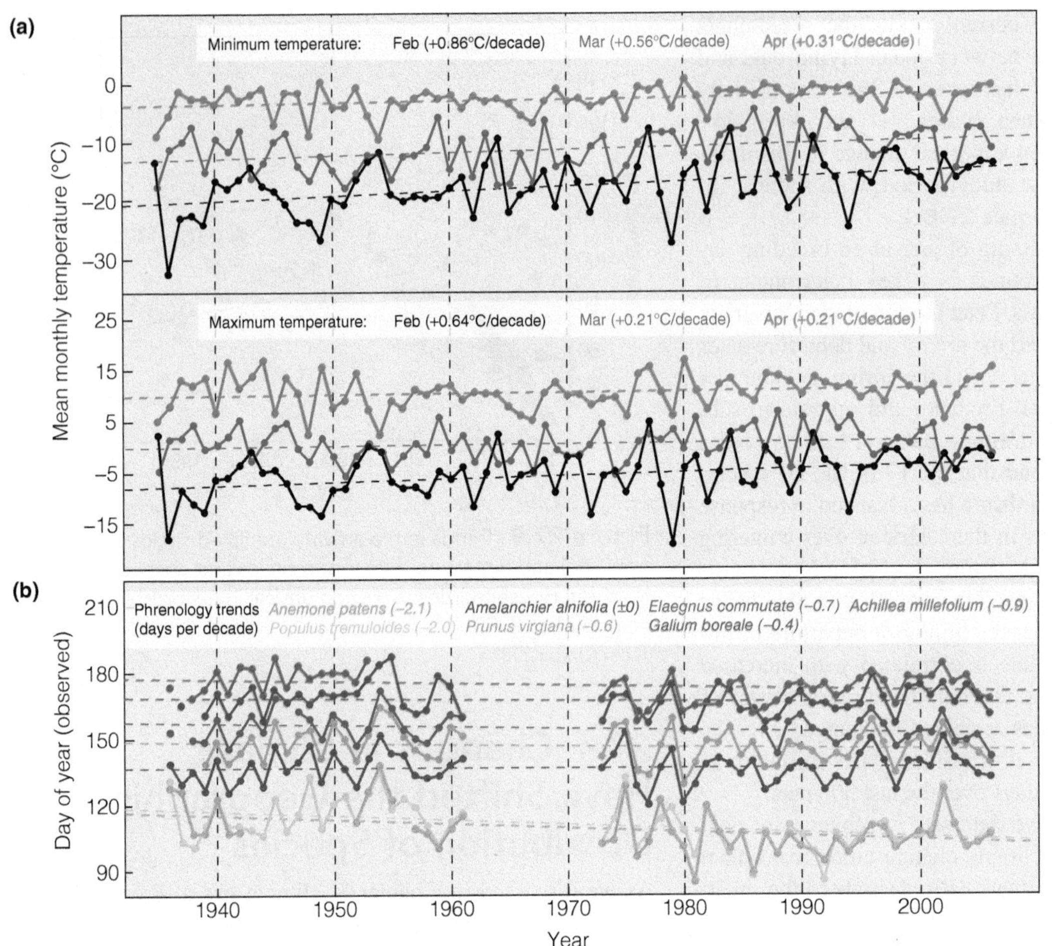

Figure 27.10 Temperature trends for the central Alberta, Canada study area for (a) the change in the mean monthly minimum temperature and the change in the mean monthly maximum temperature. (b) Corresponding trends in observations of first bloom for seven plant species in the study area species. (Adapted from Beaubien and Hamann 2011).

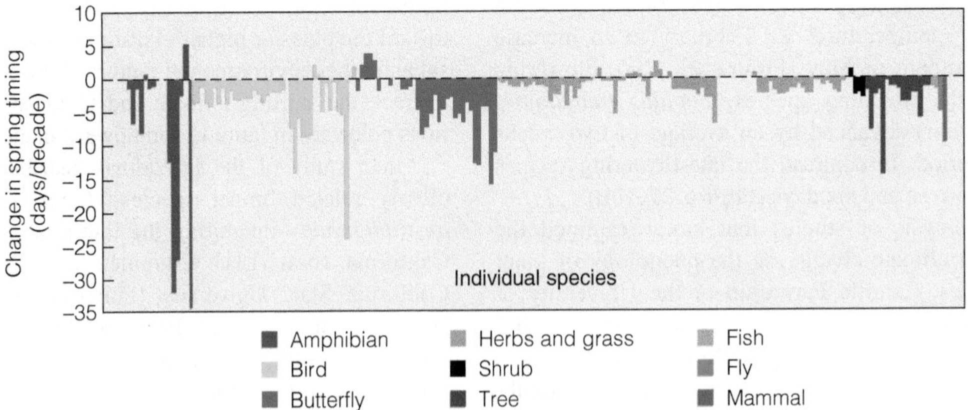

Figure 27.11 Summary of results from published studies that have examined the phenological responses of 203 different plant and animal species to recent changes in climate. Observed changes in the timing of spring events are expressed in days per decade for individual species grouped by taxonomy or functional type for the combined data set. Each bar represents a separate, independent species. Negative values indicate advancement (earlier phenology through time), whereas positive values indicate delay (later phenology through time). Note that most species experience an advance in the timing of spring activities. (From Parmesan 2007.)

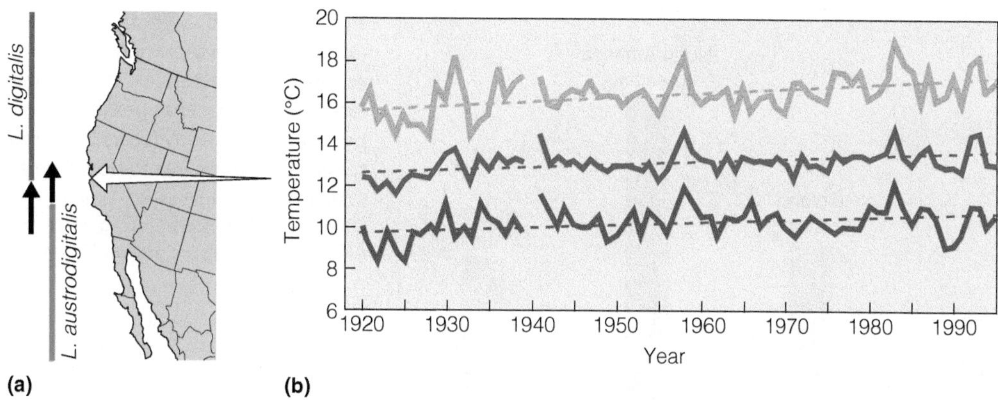

Figure 27.12 (a) Latitudinal distribution ranges of two closely related limpet species, *Lottia digitalis* and *Lottia austrodigitalis*, inhabiting the intertidal environments of the California coast as measured in 1998. Black arrows show contraction of the southern range for *L. digitalis* and expansion of the northern range for *L. austrodigitalis* between 1978 and 1998 in response to (b) increasing shoreline water temperatures (annual maximum, minimum and mean).
(Data from Crummett and Eernisse 2007 as modified by Somero 2010.)

boundaries were in northern Europe and whose southern boundaries were in southern Europe or northern Africa. For 65 percent of the 52 species included in the analysis, the northern boundaries have extended northward in the past 30–100 years. Nearly all northward shifts involved extensions at the northern boundary with the southern boundary remaining stable (**Figure 27.13**), or retracting northward.

Numerous studies of the geographic distribution of nonmigratory bird species in North America and Europe have likewise shown a consistent pattern of northward range expansion since the 1960s (**Figure 27.14**).

In contrast to the examples of range shift in response to climate warming by mobile animal species, observations of shifts in the geographic distribution of plant species in response to recent climate change are more limited. Kai Zhu, Christopher Woodall, and James Clark of Duke University used large-scale forest inventory data (United States Department of Agriculture Forest Service's Forest Inventory and Analysis data) to examine changes in the distribution of seedlings and trees for 92 species across the eastern United States. The researchers compared observed changes in species distribution over the period of 1999 to 2009 with patterns of 20th-century temperature and precipitation change. Results suggest that 54 of the 92 tree species examined show the pattern expected for a population undergoing range contraction, rather than expansion, at both northern and southern boundaries. Fewer species show a pattern consistent with a northward shift (19 species) and fewer still with a southward shift (15 species). Only 4 species are consistent with expansion at both range limits. When compared with the 20th-century climate changes that have occurred at the range boundaries themselves, there is no consistent evidence that shifts in the populations is greatest in areas where climate (temperature and precipitation) has changed most.

In the Arctic region, where increases in temperature over the past century have been greatest, numerous studies have

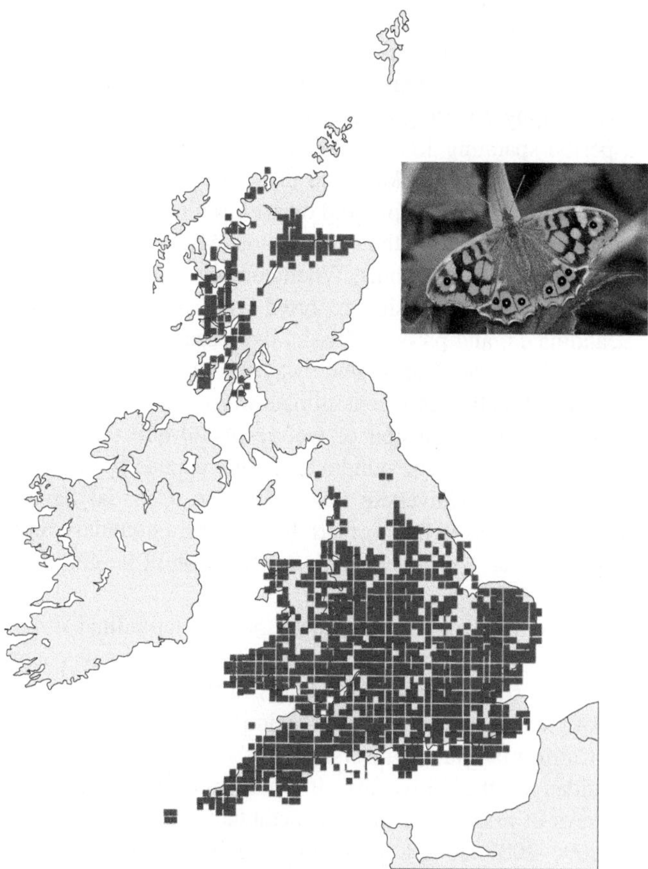

Figure 27.13 Changes in the range of Speckled Wood butterfly (*Pararge aegeria*) in Great Britain since 1940. Data are plotted on a 10 km × 10 km grid. A colored grid cell indicates more than one population observed during 1940–1969 (blue) or 1970–1997 (red).
(Data from Parmesan, C. et al. 1999. Poleward shifts in geographical ranges of butterfly species associated with regional warming. Nature 399: 579–583. Fig. 1-4, pg. 581–582.)

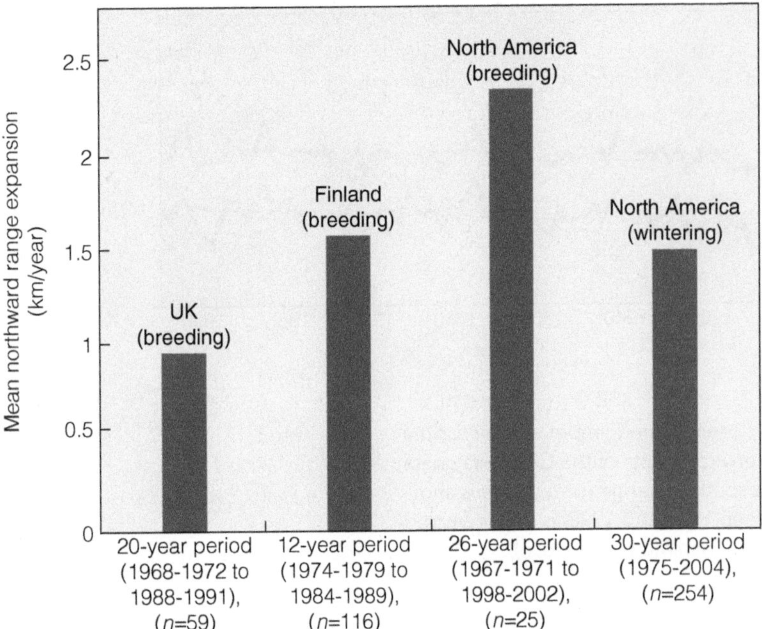

Figure 27.14 Summary of the results from four regional studies reporting the mean northward range expansion of bird species during the past 50 years as a result of regional changes in climate. Location, season, period of each study (years), and the number of bird species observed (*n*) are noted on each histogram. (Data from: UK, Thomas and Lennon 1999; Finland, Brommer 2004; North America/breeding, Hitch and Leberg 2007; North America/wintering, LaSorte and Thompson 2007.)

observed a northern expansion in the abundance and extent of shrub species into areas of tundra. For example, using repeat photography covering extensive areas of northern Alaska for a period spanning 1942 to 2002, Ken Tape and colleagues at the University of Alaska found that the cover of alder (*Alnus* spp.), willow (*Salix* spp.), and dwarf birch (*Betula nana*) have been increasing, with the change most easily detected on hill slopes and valley bottoms. When combined with similar patterns of shrub expansion reported from studies in Canada, Scandinavia, and parts of Russia, the researchers suggest that a vegetation transition is underway throughout the Arctic region as a function of recent climate warming. Consistent with the northward expansion of woody plants into the southern regions of the Arctic tundra, Isabelle Gamache and Serge Payette of the Université Laval (Québec, Canada) have reported that trees in the northern forest-tundra boundary region of Canada have experienced an acceleration of height growth since the 1970s.

In contrast to the limited data showing latitudinal shifts in the range of plant species in response to recent climate change, numerous studies have shown shifts in the elevation boundaries of plant species in montane regions. Ecologists Anne Kelly of California State University (Los Angeles) and Michael Goulden of the University of California (Irvine) compared surveys of plant cover that were conducted in 1977 and again in 2006–2007 along a 2314-m elevation gradient in Southern California's Santa Rosa Mountains. During the 30-year period between surveys, the region's climate warmed (0.4°C increase in mean temperature), the precipitation variability increased, and the amount of snow decreased. In a comparison of the two vegetation surveys, the researchers found that the average elevation of the dominant plant species increased by 65 m (**Figure 27.15**), with all but one species' distribution shifting up the mountain range in response to warming.

27.5 Recent Climate Change Has Altered Species Interactions

Recent climate change is apparently influencing interactions among species within existing ecological communities through its differential effects on the component species. Often the nature of species interactions are altered as a result of the species exhibiting different phonological responses to changes in key climate variables.

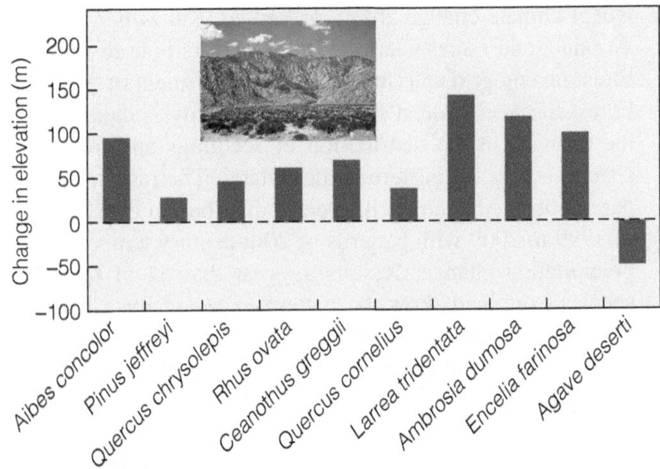

Figure 27.15 Change in elevational range (meters) of the 10 most widely distributed plant species along the Deep Canyon transect in Southern California's Santa Rosa Mountains between 1997 and 2007. Positive values represent a movement of the species' range up the mountain range (increase in elevation), whereas a negative value represents a contraction of the species' range (Data from Kelly and Goulden 2008.)

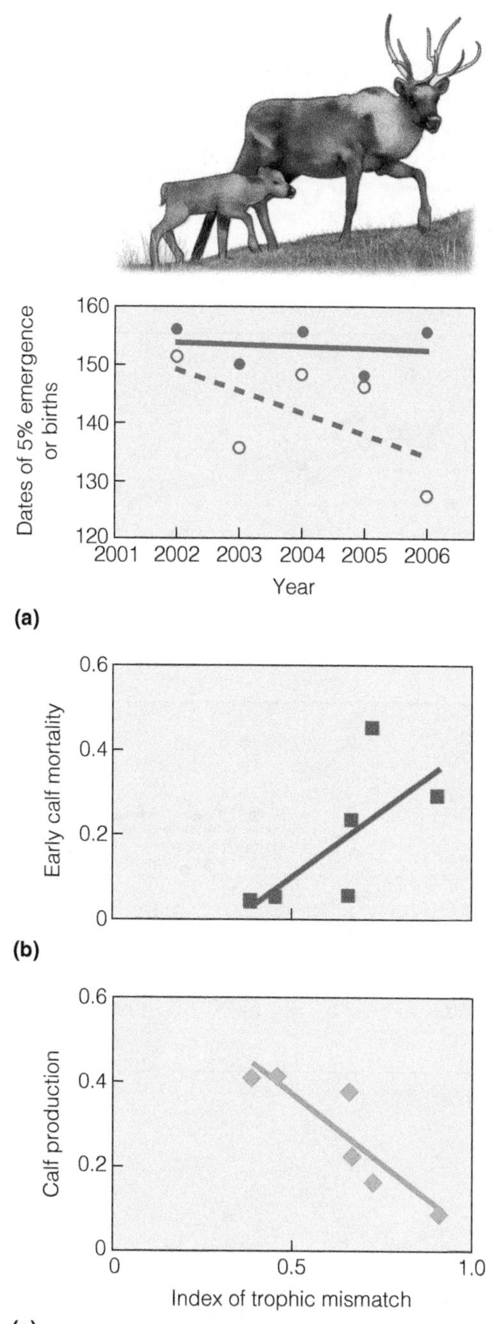

Figure 27.16 (a) Dates (in day of year) of emergence of 5 percent of forage species (open circles, dashed line) and of 5 percent of caribou births (filled circles, solid line) at the study site in Kangerlussuaq, West Greenland, during the period of continuous annual data collection from 2002 to 2006. Relation between the magnitude of trophic mismatch between caribou calving and plant phenology (as shown in (a)) and (b) early calf mortality, and (c) calf production. Calf production is estimated as the final proportion of calves observed each year. The index of the degree of trophic mismatch each year is based on the percentage of forage species emergent on the date at which 50 percent of caribou births have occurred. This index quantifies the temporal state of the forage resource midway through the season of caribou births.
(From Post and Forchhammer 2008.)

In seasonal environments, the production of offspring is timed to coincide with the annual peak of resource availability. For herbivores, this resource peak is represented by the annual onset and progression of the plant growing season. As plant phenology advances in response to climatic warming (see Figure 27.10), there is potential for development of a mismatch between the peak of resource demands by reproducing herbivores and the peak of resource availability (the emergence and growth of plants).

Colleagues Eric Post of Pennsylvania State University and Mads Forchhammer of the University of Aarhus (Denmark) collected data during six summers (1993, 2002–2006) on the annual timing and progression of the calving season in the Kangerlussuaq population of caribou in West Greenland and monitored plant phenology over the same period, recording the timing and progression of plant species emergence. Successful reproduction by the caribou population depends on synchronizing offspring production with the time of year when resources are most abundant or of highest quality. In the far north, nutritional content and digestibility of plants reach a peak soon after emergence and decline rapidly thereafter. Successful reproduction therefore depends on the timing of calving being closely linked to the onset of the plant growing season.

Over the course of their study, Post and Forchhammer found that the onset of the plant growing season (estimated by the data of emergence of plant species) advanced by 14.8 days. In contrast, the onset of calving advanced by only 1.28 days. The result is a rapidly developing mismatch between caribou reproduction and the timing of availability of their food supply (**Figure 27.16a**). The growing mismatch results from the difference in the environmental cues that trigger the beginning of the breeding season in the caribou population and the onset of plant growth in the region. Changes in day length (which effectively does not vary from year to year) cue the timing of seasonal migration of caribou to their summer ranges, where calves are born. In contrast, the onset of the plant growing season is highly correlated with mean spring temperature, which increased by 4.6°C over the study period.

The researchers found that the rate of calf mortality increased as the degree of mismatch between the timing of the calving season and onset of plant growth increased (**Figure 27.16b**). As a result, annual calf production declined with increasing mismatch, varying fourfold between the lowest and highest levels of mismatch observed (**Figure 27.16c**).

A similar development of a mismatch between the phenology of herbivores and their food resources has been observed in lake ecosystems. Monika Winder and Daniel Schindler of the University of Washington examined changes in the phenology of phytoplankton (autotroph) and zooplankton (heterotrophs) populations in Lake Washington (United States) over the period from 1962 to 2002 in response to climate warming over the same period. Spring water temperatures in Lake Washington exhibit significant warming trends and values in the upper 10-m water layer from March to June, increasing on average 1.4°C since 1962 (**Figure 27.17a**). In response to this warming, the spring phytoplankton bloom has advanced by 27 days over the study period (**Figure 27.17b**). A significant trend toward earlier timing of peak densities of the herbivorous rotifer *Keratella* (zooplankton) has likewise occurred, advancing by 21 days. In contrast, the timing of the annual spring peaks of the herbivorous zooplankton *Daphnia* exhibited no significant advance over the study period (**Figure 27.17c**). The offset in timing between the peak of spring phytoplankton bloom and peak in the spring *Daphnia* bloom has increased significantly over the study period resulting in a corresponding long-term decline in spring/summer *Daphnia* densities (**Figure 27.17d**).

In addition to changes in species interactions that occur as a result of divergence in phenological responses to climate warming, changes in climate alter the nature of interactions of species within the community by increasing the fitness of one species at the expense of another. The mountain pine beetle (*Dendroctonus ponderosae*), native to western North America, attacks most trees of the genus *Pinus*, and periodically erupts in epidemics. Over the past decade, epidemics of mountain pine beetle have been on an order of magnitude larger than previously recorded as a direct result of recent climate change. The increase in epidemics has resulted from a number of climate-related responses. Increased temperatures in the region have led to an increase in the frequency of drought, which has decreased tree health (see Section 27.2 and Figure 27.6) and increased their susceptibility to attack by beetles. Regional warming has also led to range expansion of the mountain pine beetle, particularly into higher elevation forests. Perhaps most important, however, has been the direct effects of warming on the life history of the beetle.

Research by ecologists Jeffry Mitton and Scott Ferrenberg of the University of Colorado has shown that after two decades of temperature increase in the Colorado Front Range, the mountain pine beetle flight season (period when adults emerge from brood trees and fly to attack new host trees) begins more than one month earlier and is approximately twice as long as indicated by the historical record. As a result, the life cycle in some populations has increased from one to two generations per year (**Figure 27.18**). Because their development is controlled by temperature, the beetles have responded to regional patterns of climate change with faster development, which exacerbates the current epidemic.

Recent warming in the Arctic has also altered the interaction between red (*Vulpes vulpes*) and Arctic fox (*Vulpes lagopus*). Numbers of Arctic fox (*Alopex lagopus*) are declining in Arctic regions of both North America and Europe. Research has shown that the northern limit of the red fox's geographic range is determined by temperature limitation (both directly and indirectly through resource abundance), whereas the southern limit of the Arctic fox's range is determined by interspecific competition with the red fox. Warming in the Arctic over the past four decades has led to a northward expansion of red fox populations, resulting in a decline in Arctic fox populations in areas of their southern distribution, a region in which both species have co-existed for centuries.

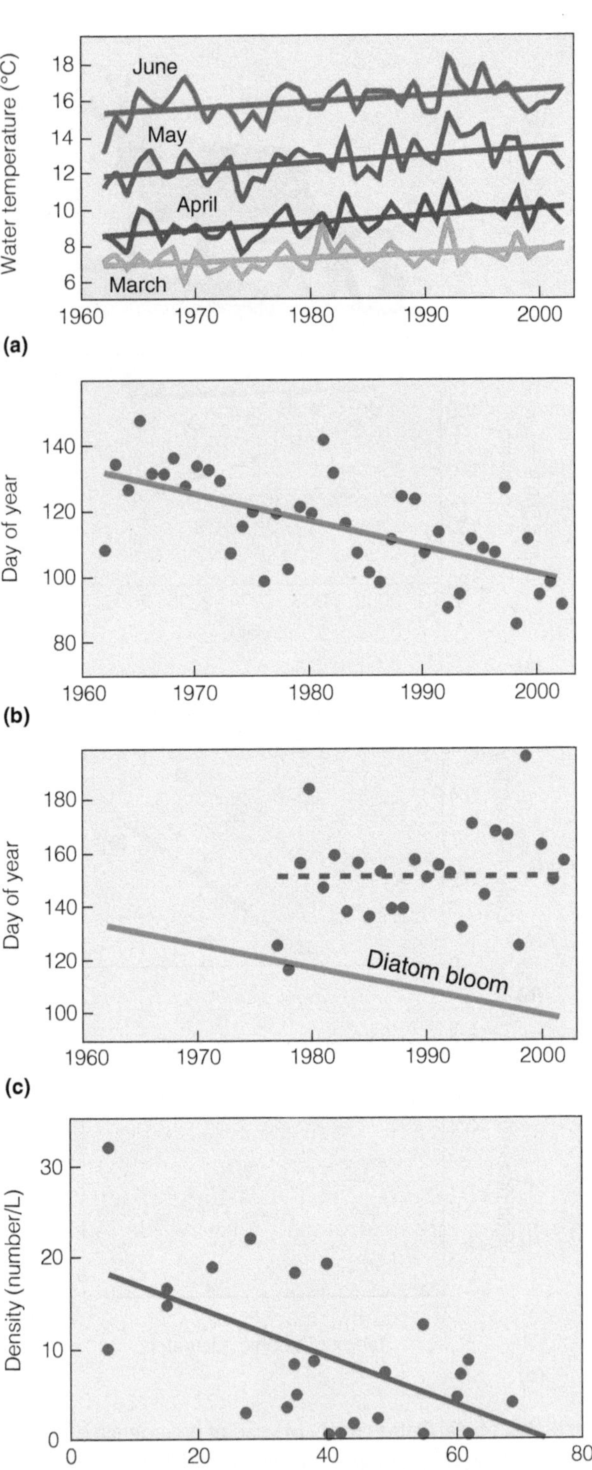

Figure 27.17 Physical and biological trends in Lake Washington, United States: (a) average monthly volume-weighted temperature of the upper 10-m water layer during spring (March through June), (b) timing of diatom bloom, (c) timing of diatom bloom (solid line) relative to annual timing of *Daphnia* peaks (circles, dashed line). (d) Relation of *Daphnia* densities in May to the mismatch (in days) between the timing of diatom bloom and *Daphnia* peak for the period of 1977–2002.
(Adapted from Winder and Schindler 2004.)

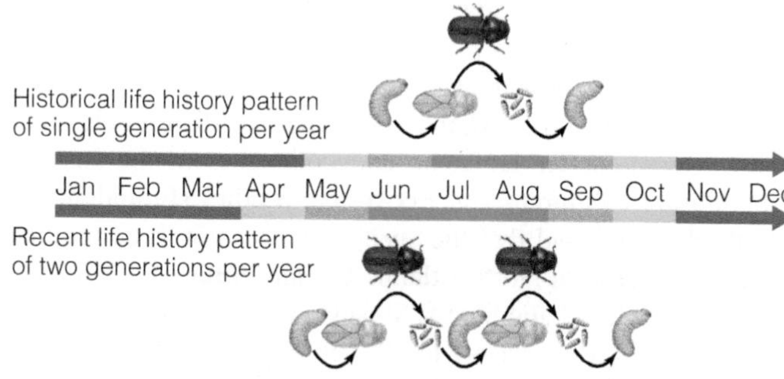

Figure 27.18 The historical mountain pine beetle life cycle of a single generation per year and the recently observed life cycle of two generations per year. Calendar arrow colors represent corresponding monthly temperature regimes: blue for < 0°C, yellow for 0–4.99°C, orange for 5–9.99°C, and red for 10°C and higher. (Adapted from Mitton and Ferrenberg 2012.)

27.6 Community Structure and Regional Patterns of Diversity Have Responded to Recent Climate Change

Recent changes in global climate have resulted in shifts in the geographic ranges of species and changes in the nature of species interactions that, in turn, have led to shifts in the species composition and diversity of both terrestrial and marine communities.

Richard Brusca of the University of Arizona and colleagues were able to provide a detailed analysis of the impacts of recent climate warming on patterns of plant community structure in the Santa Catalina Mountains by reexamining an earlier transect study undertaken by the plant ecologists Robert Whittaker and William Niering 50 years earlier (1963). During the intervening period, the mean annual rainfall for the region of the Santa Catalina Mountains has decreased, but the mean annual temperature has increased by 0.25°C per decade between 1949 and 2011 (**Figure 27.19a**).

Brusca and colleagues resampled the elevation transect originally reported by Whittaker and Niering and focused on the 27 most abundant species identified in the original study. The researchers discovered that the range of three quarters of these species has shifted significantly upslope—in some cases as much as 1000 feet (over 300 m)—or now grow in a narrower elevation range compared to their distribution reported in 1963 by Whittaker and Niering (**Figure 27.19b**). The main finding of the study is that the species compositions of the plant communities along the elevation gradient (the pattern of zonation; see Section 16.8 for discussion of community zonation) have changed significantly from those observed 50 years earlier. The observed changes in the patterns of zonation in the Santa Catalina Mountains are a result of individual species' elevation ranges shifting independently in response to regional changes in climate, leading to a reorganization of plant communities (see Chapter 17).

Some of the best examples of shifts in community structure in response to recent climate change have been reported for marine ecosystems. Gregory Beaugrand and colleagues at the Sir Alister Hardy Foundation for Ocean Science (Plymouth, England) reported large-scale changes in the species composition and diversity of calanoid copepod (Calanoida) crustaceans (zooplankton) in the Eastern North Atlantic Ocean and European shelf seas. The researchers compared the distribution of established associations of calanoid copepods over the period of 1958 to 1999 in response to changes in sea surface temperatures. The species associations reflect groups of calanoid copepod species that co-occur in communities associated with environmental conditions found in specific regions of the North Atlantic Ocean (defined as warm-temperate oceanic species association, southern shelf

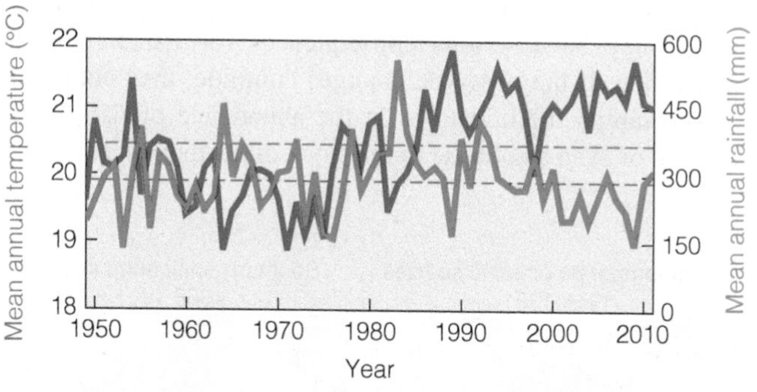

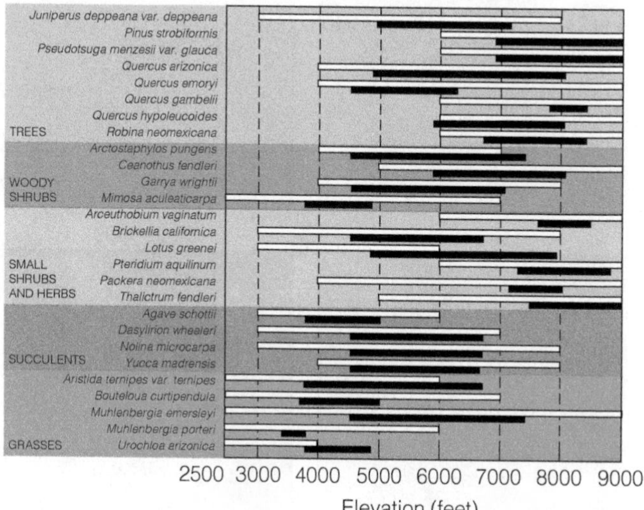

Figure 27.19 (a) Mean annual air temperature and rainfall for region of Tucson, Arizona, for the period from 1949 to 2011 (Data from United States National Weather Service). (b) Summary of elevation range of the 27 most common upland montane plants along the original elevational transect in the Santa Catalina Mountains (area of Tucson, AZ) reported by Whittaker and Niering in 1964. White bars are 1963 elevational range data from Whittaker and Niering study, whereas the black bars represent 2011 elevation data from the current study. (Brusca et al. 2013.)

edge association, cold temperate species association, subarctic species association, and Arctic species association). The copepod assemblages were based on approximately 177,000 samples collected by the Continuous Plankton Recorder survey, which has monitored plankton on a monthly basis in the North Atlantic since 1946.

Increases in regional sea surface temperatures (see Figure 27.1) have triggered a major reorganization of calanoid copepod species composition and biodiversity over the whole North Atlantic Basin (**Figure 27.20**) with all the species associations showing consistent long-term changes. During the last 40 years, there has been a northerly movement of warmer-water plankton species by 10° latitude in the northeast Atlantic and a similar retreat of colder-water plankton to the north. This geographical movement is much more pronounced than any documented for terrestrial organisms, most likely aided by the patterns of ocean circulation.

The researchers noted that the observed geographical shifts may have serious consequences for fisheries in the North Sea. If the observed changes continue, they could lead to substantial modifications in the abundance of fish, with a decline or even a collapse in the stock of northern species such as cod, which is already weakened by overexploitation (see Chapter 14, *Ecological Issues & Applications*).

In a study aimed at examining the effects of regional warming in the North Atlantic on patterns of fish species richness, Remment ter Hofstede and colleagues at the Institute for Marine Resources and Ecosystem Studies (Ijmuiden, The Netherlands) found that the species richness in three regional seas in the eastern North Atlantic Ocean (**Figure 27.21a**; NS: North Sea; CS: Celtic Sea; WS: west of Scotland) has changed over the period from 1997 to 2008 in response to higher water temperatures (**Figure 27.21b**).

For purposes of the analysis, species were categorized based on their geographic affinity: Atlantic (widespread), Boreal (northern, cold favoring), and Lusitanian (southern, warm favoring). Boreal fishes are those considered to be northerly taxa that extend northward to the Norwegian Sea and Icelandic waters. These fishes often have their southern limits of distribution around the British Isles. Lusitanian fishes are those that tend to be abundant from the Iberian Peninsula to as far north as the British Isles. Many of these species have distributions extending into the Mediterranean Sea and off the coast of northwest Africa. Atlantic species are those (often pelagic or

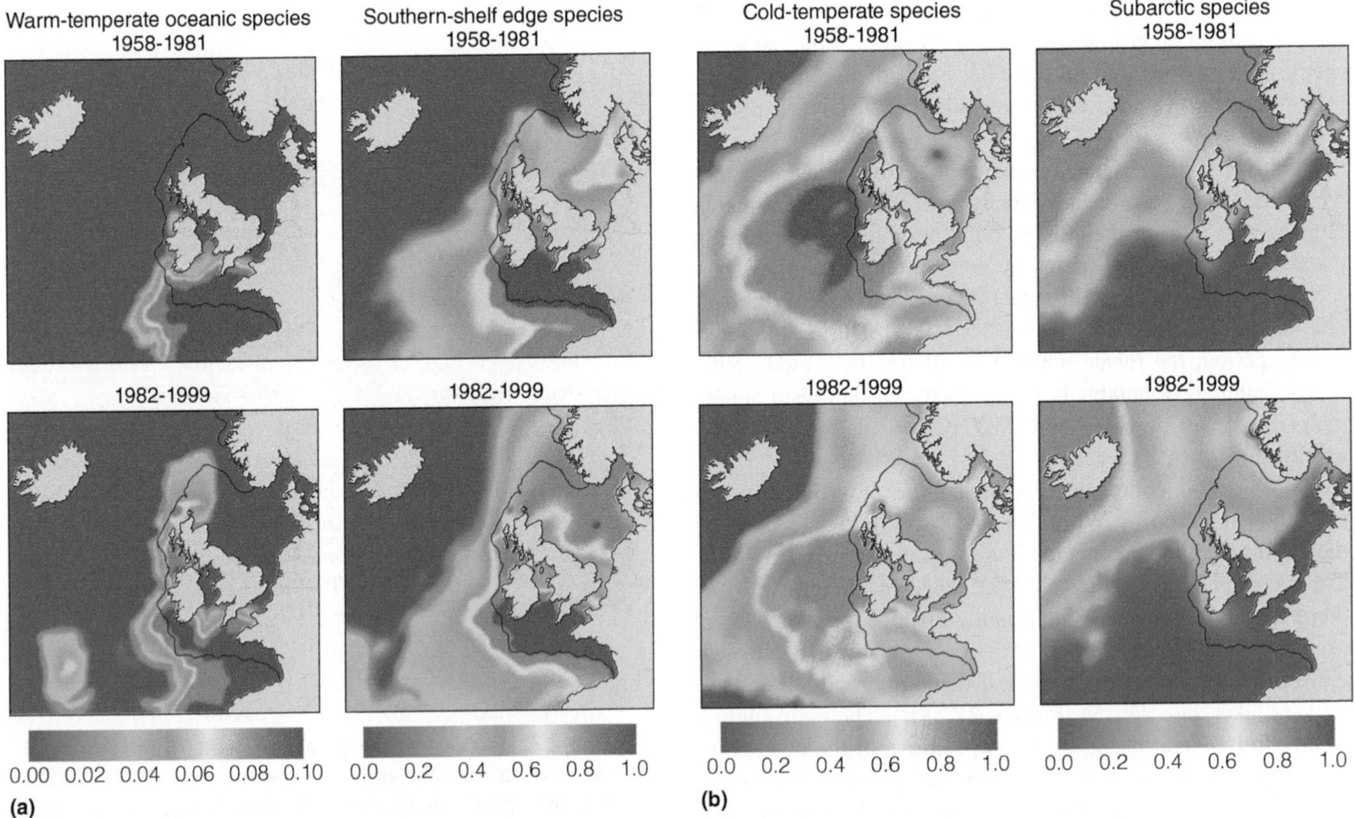

Figure 27.20 Observed northerly shift of zooplankton assemblages (functional categories of species) in the Northeast Atlantic over two periods (1958–1981 and 1982–1999). See text for description of species assemblages. There has been (a) a northerly shift of approximately 1000 km for warmer-water species during the past 40 years, whereas colder-water species (b) have contracted their range. Scale is the mean number of species per assemblage, which provides an index of abundance. For example, in a given region, a decrease in the number of warm-temperate species, associated with an increase in the number of cold-temperate and subarctic species, would suggest a community shift from a warm to a colder environment species.
(From Beaugrand et al. 2001 as presented by Hayes et al. 2005.)

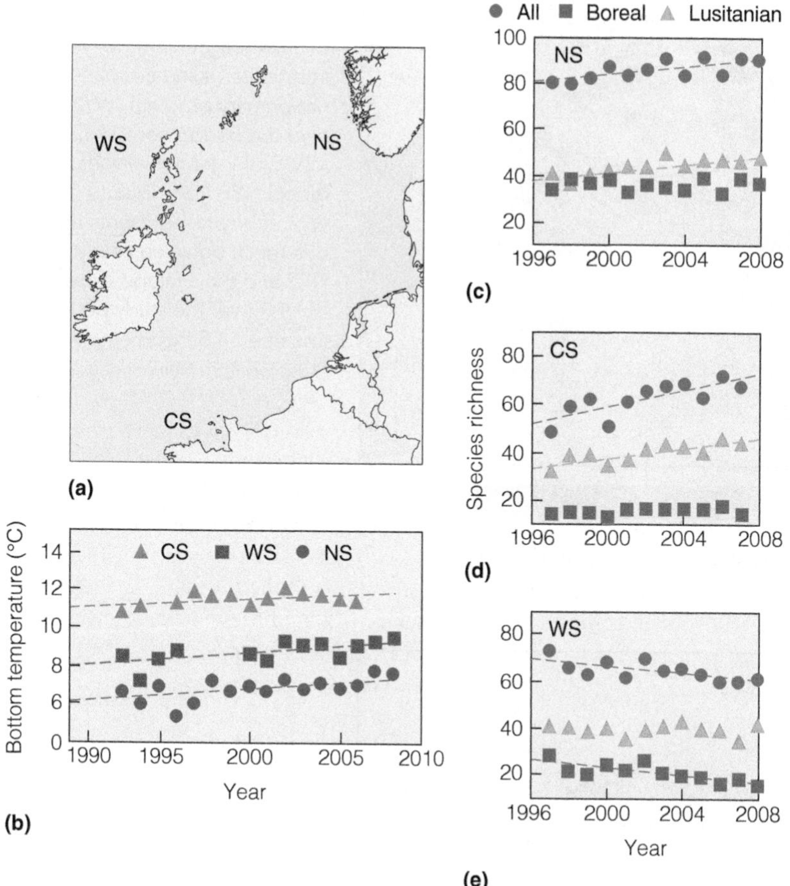

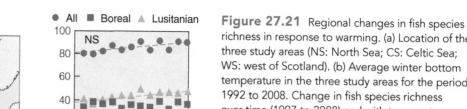

Figure 27.21 Regional changes in fish species richness in response to warming. (a) Location of the three study areas (NS: North Sea; CS: Celtic Sea; WS: west of Scotland). (b) Average winter bottom temperature in the three study areas for the period 1992 to 2008. Change in fish species richness over time (1997 to 2008) and with temperature. Species richness per year for All, Boreal (cold-water), and Lusitanian (warm-water) fish species in the (c) North Sea, (d) Celtic Sea, and (e) west of Scotland. Dashed lines indicate significant trends. See text for description of fish species groups. (Adapted from Hofstede et al. 2010.)

deep water) species that are widespread in the North Atlantic and include many of the deeper-water or mesopelagic species that may be widely distributed along the continental slope.

The researchers found that regional warming caused changes in local communities as a result of both species extinctions and latitudinal range shifts. Results indicate that species richness increased in both the North and Celtic Seas because of increases in the number of warmer Lusitanian species (**Figures 27.21c** and **27.21d**). In the area west of Scotland, species richness decreased because the number of colder Boreal species decreased (**Figure 27.21e**). In general, the increase in bottom water temperatures in the region has resulted in poleward advances of warmer Lusitanian species and a retreat in the colder Boreal species.

27.7 Climate Change Has Impacted Ecosystem Processes

Climate has a direct influence on the key ecosystem processes of net primary productivity (NPP; see Chapter 20) and decomposition (see Chapter 21), therefore controlling the rates at which nutrients cycle through the ecosystem.

As with field-based observations of the response of individual plant species to climate change (see Section 27.2), site-based studies of the response of local ecosystems are often difficult to interpret because of the confounding effects of edaphic factors (such as soils and topography), and the direct effects of increasing atmospheric concentrations of carbon dioxide on plant growth (Chapter 6, *Ecological Issues & Applications*). As a result, most studies that have examined changes in NPP and carbon storage resulting from climate change over the past several decades have used satellite-based estimates to evaluate changes at a regional to global scale.

As presented in Chapter 20 (**Quantifying Ecology 20.1** and Chapter 20, *Ecological Issues & Applications*), satellite-based measures of absorbed photosynthetically active radiation (APAR) can be used to estimate terrestrial NPP over large land areas. Ramakrishna Nemani of the University of Montana and NASA's Goddard Space Flight Center together with colleagues at a number of other institutions in the United States and Japan, used satellite data from the National Oceanic and Atmospheric Administration (NOAA) Pathfinder Advanced Very High Resolution Radiometer (AVHRR) to estimate global patterns of terrestrial NPP from 1982 to 1999. Changes in NPP over the study period are presented in **Figure 27.22**.

Results of the analysis indicate that global changes in climate (increased global temperatures, growing season, and precipitation) have eased critical climatic constraints on plant growth, such that net primary production increased 6 percent (3.4 petagrams of carbon) globally over the 18-year period. Ecosystems in all tropical regions and those in the high latitudes of the Northern Hemisphere accounted for 80 percent of the increase. The largest increase was in tropical ecosystems.

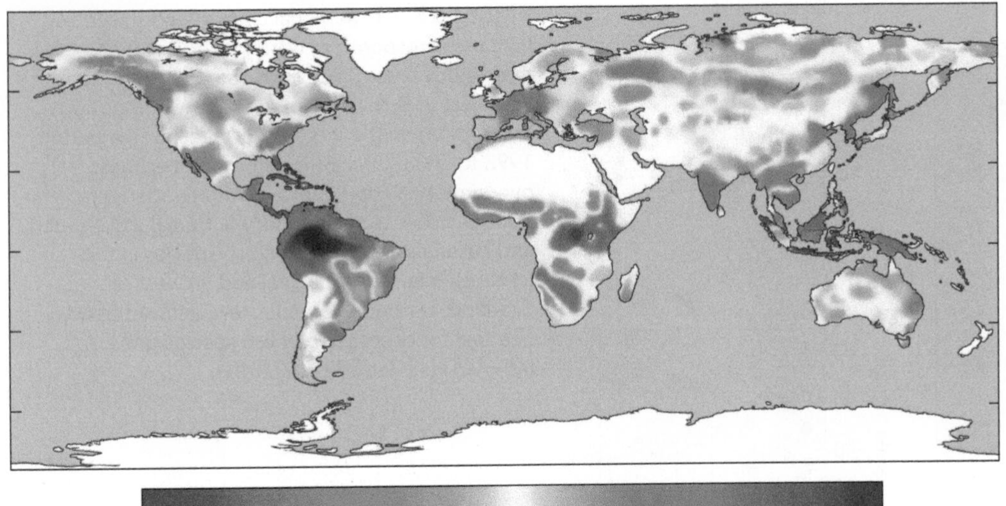

(a)

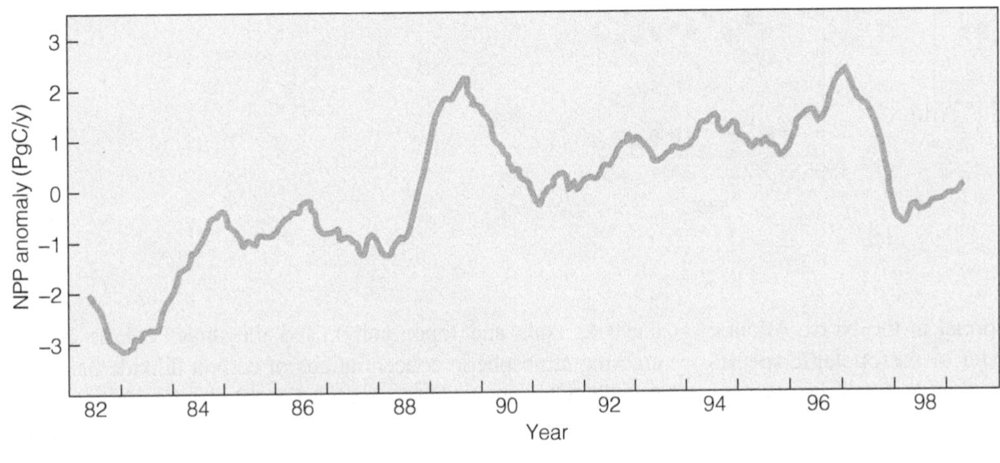

(b)

Figure 27.22 (a) Global map of changes (percent per year) in satellite-based estimates of net primary productivity (NPP) over the period from 1982 to 1999. (b) Interannual variation in global NPP from 1982 to 1999. NPP anomaly represents the difference between yearly annual NPP and the average value of NPP over the study period (mean = 54.5 PgC per year) (Adapted from Nemani et al. 2003.)

Amazon rain forests accounted for 42 percent of the global increase in net primary production, which is a product of decreased cloud cover and the resulting increase in solar radiation.

A more recent analysis by Maosheng Zhao and Steven W. Running of the University of Montana using the same methodology as Nemani and colleagues (including Steven Running) extended the analysis to cover the period of 2002 to 2009. The decade between 2000 and 2009 is the warmest since instrumental measurements began, which might at first suggest continuation of increases in NPP observed prior to 2000. Results of their analysis, however, suggest a reduction in the global NPP of 0.55 petagrams of carbon as a result of regional drying.

Figure 27.23 shows the spatial pattern of NPP trends over the past decade. NPP increased over large areas in the Northern Hemisphere but decreased in the Southern Hemisphere. Over the Northern Hemisphere, 65 percent of vegetated land area had experienced an increase in NPP, including large areas of North America, Western Europe, India, China, and the Sahel. Regions with decreased NPP include Eastern Europe, central Asia, and high latitudes of west Asia. In the Southern Hemisphere, 70 percent of vegetated land areas had decreased NPP, including large parts of South America, Africa, and Australia. The decline in NPP was a result of higher temperatures leading to increased evaporation and reduced water availability to support plant growth. Overall, the drying trend in the Southern Hemisphere and associated decreases in NPP counteracted the increased NPP over the Northern Hemisphere, resulting in the average global decline in NPP.

27.8 Continued Increase in Atmospheric Concentrations of Greenhouse Gases Is Predicted to Cause Future Climate Change

As human activities continue to increase the atmospheric concentration of carbon dioxide, how will rising concentrations of greenhouse gases influence the future global climate?

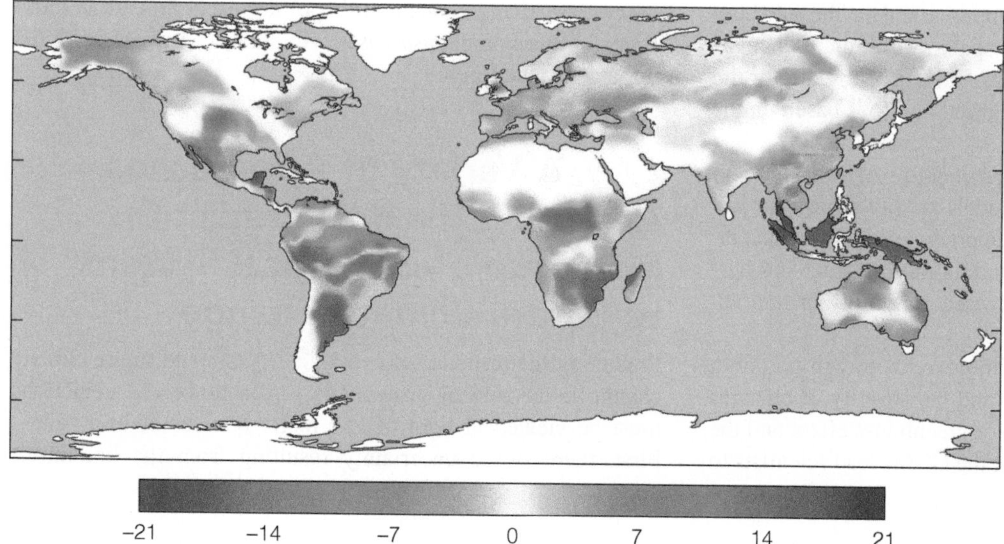

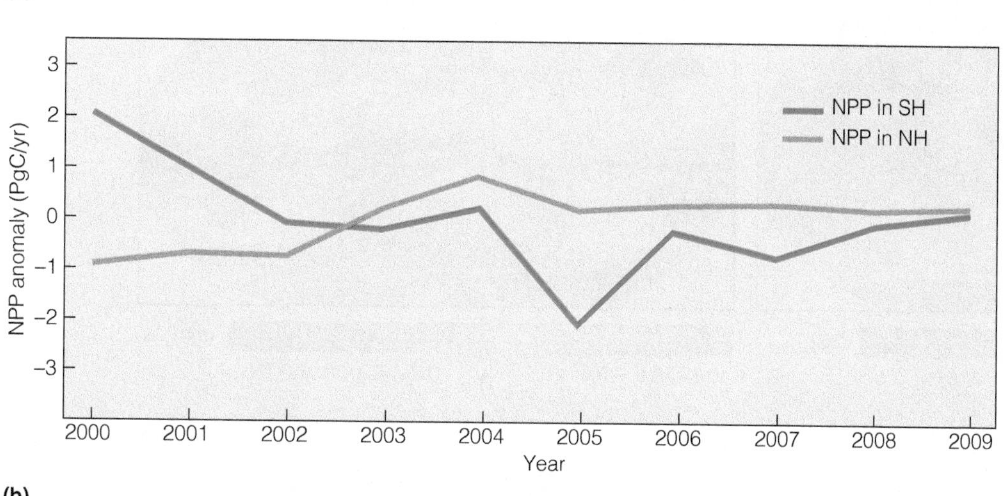

Figure 27.23 (a) Global map of changes (gC/m^2/yr) in satellite-based estimates of net primary productivity (NPP) from 2000 through 2009. (b) Interannual variation in total NPP over the Northern and Southern Hemispheres. NPP anomaly represents the difference between yearly annual NPP and the average value of NPP over the study period (mean = 53.5 PgC per year). (Adapted from Zhao and Running 2010.)

Scientists estimate that at current rates of emission, the preindustrial level of 280 parts per million (ppm) of carbon dioxide in the atmosphere will double sometime this century (398 ppm as of January 2014). Moreover, carbon dioxide is not the only greenhouse gas increasing because of human activities. Other significant components of the total greenhouse gas emissions include methane (CH_4), chlorofluorocarbons (CFCs), hydrogenated chlorofluorocarbons (HCFCs), and nitrous oxide (N_2O). Although much lower in concentration, some of these gases are much more effective at trapping heat than is carbon dioxide.

The role of greenhouse gases in warming Earth's surface is well established (see discussion in Chapter 2, *Ecological Issues & Applications*), but the specific influence that doubling the carbon dioxide concentration of the atmosphere will exert on the global climate system is much more uncertain. Atmospheric scientists have developed complex computer models of Earth's climate system—called **general circulation models (GCMs)**—to help determine how increasing concentrations of greenhouse gases may influence large-scale patterns of global climate. Although all use the same basic physical descriptions of climate processes, GCMs at different research institutions differ in their spatial resolution and in how they describe certain features of Earth's surface and atmosphere. As a result, the models differ in their predictions.

Despite these differences, certain patterns consistently emerge. All of the models predict an increase in the average global temperature as well as a corresponding increase in global precipitation. Findings published in 2013 by the Intergovernmental Panel on Climate Change (IPCC) suggest an increase in globally averaged surface temperature in the range of 1.1 to 6.4°C by the year 2100 (actual range depends on the specific scenario of greenhouse gas emission developed by the IPCC). These changes would not be evenly distributed over Earth's surface. As with observed patterns over the past century, predicted future warming is expected to be greatest during the winter months and in the northern latitudes. **Figure 27.24** shows the spatial variation in changes in mean temperature and precipitation at a global scale for the Northern Hemisphere winter and summer periods based on the most recent IPCC analyses.

Although in popular speech, *greenhouse effect* is synonymous with *global warming,* the models predict more than just hotter days. One of the most notable predictions foretells an increased variability of climate, including more storms and hurricanes, greater snowfall, and increased variability in rainfall, depending on the region.

As models of global climate improve, there will no doubt be further changes in the patterns and the severity of changes they predict. However, the physics of greenhouse effect and the consistent qualitative predictions of the GCMs lead scientists to believe that rising concentrations of atmospheric carbon dioxide and other greenhouse gases will significantly affect global climate into the future.

27.9 A Variety of Approaches Are Being Used to Predict the Response of Ecological Systems to Future Climate Change

Predicting the response of ecological systems to future climate change is an area of growing research; however, predictions must be viewed in light of two major sources of uncertainty. First, there is the uncertainty resulting from the limitations

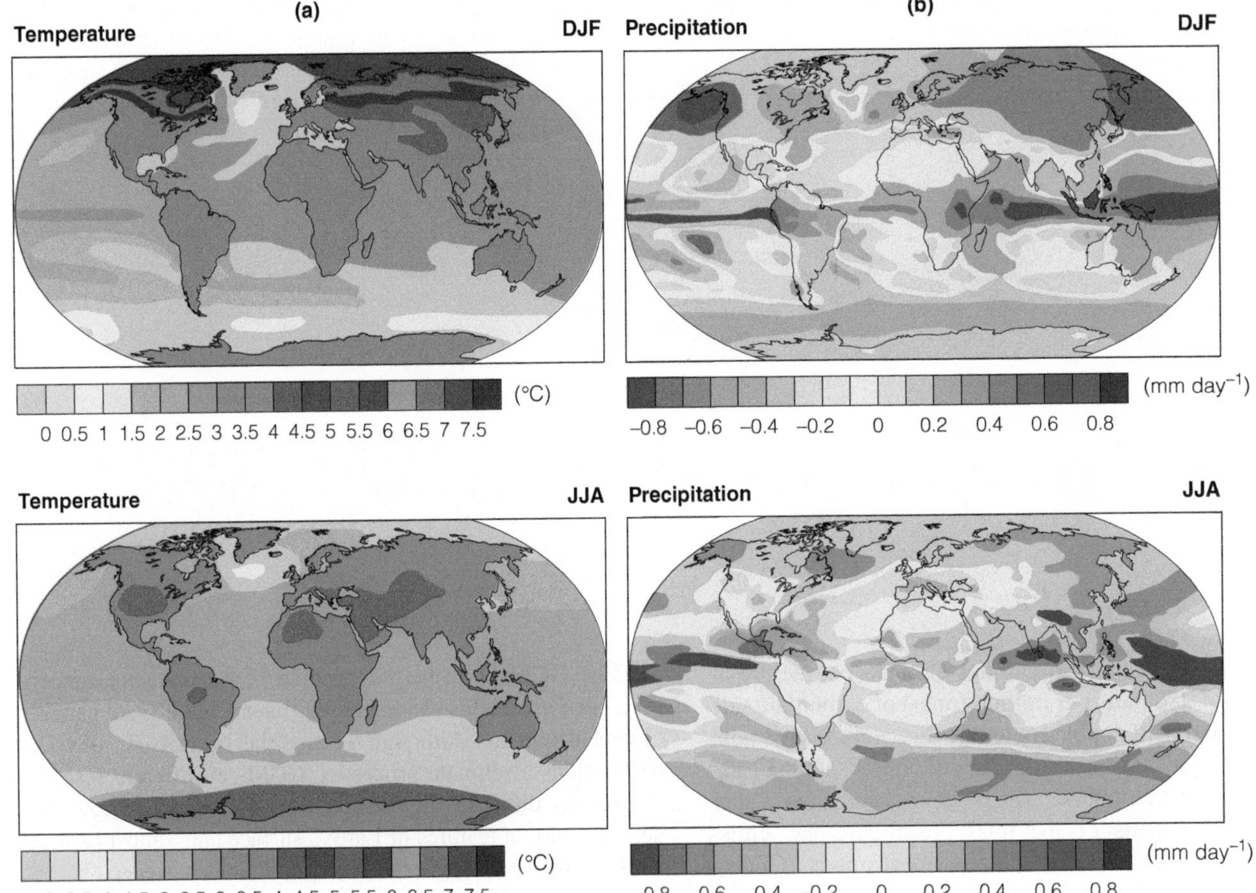

Figure 27.24 Mean changes in (a) surface air temperature (°C), and (b) precipitation (mm per day) for Northern Hemisphere winter (DJF—December, January, and February, top) and summer (JJA—June, July, and August, bottom) under a scenario of rising atmospheric concentrations of greenhouse gases developed by the Intergovernmental Panel on Climate Change. Results represent an average of the patterns predicted by the various global circulation models used in the fourth assessment. Changes are for the period 2080 to 2099 relative to 1980 to 1999. Note that most of the warming is predicted to occur in the more northern latitudes and during the winter months.
(Intergovernmental Panel on Climate Change 2007.)

in our understanding of processes that control the current distribution and abundance of species. Secondly, there is the uncertainty associated with the specific predictions of how the climate in a given region will change in response to elevated greenhouse gases. As such, our intent in the following discussion is not to present examples of current research as actual predictions of what will occur in the future, but rather as an overview of the methods being used by ecologists to evaluate the climate change scenarios being developed by climate scientists, such as those presented in the previous section (see Figure 27.24).

Research examining the possible impacts of future climate change on ecological systems is being undertaken at all levels of organization, from the potential response of individual species to patterns of ecosystem productivity and carbon exchange with the atmosphere at a global scale. Studies, however, can be classified into two broad categories: those that examine the response of ecological systems to experimental warming (and associated environmental factors), and those that use models of ecological systems to evaluate the response to future climate scenarios. In the following discussion, we will present examples of both.

Numerous experimental studies in a wide variety of environments (from polar to tropical) have shown that communities and ecosystems respond strongly to warming (see this chapter, **Field Studies: Erika Zavaleta**); most studies, however, are limited to a single location and are of short duration. In addition, comparisons among studies are difficult because of the variety of techniques used to achieve experimental warming, differences in experimental designs, and in the measurements used to assess responses. However, a number of coordinated international efforts have been undertaken that enable a comparison across locations using standardized methodologies. One such effort is by the International Tundra Experiment (ITEX).

ITEX is a network of Arctic and alpine research sites throughout the world where experimental and observational studies have been established by using standardized protocols to measure responses of tundra plants and plant communities to increased temperature. Investigators from 13 countries participate in the network of 11 sites (**Figure 27.25**). The passive warming treatments used in these studies increased plant-level air temperature by 1–3°C, which is in the range of predicted and observed warming for tundra regions. Responses were rapid and detected in whole plant communities after only two growing seasons. Overall, warming increased height and cover of deciduous shrubs and graminoid species (grasses, sedges, and rushes), decreased cover of mosses and lichens, and decreased species diversity and evenness. These results predict that warming will cause a decline in biodiversity across a wide variety of tundra, at least in the short term. They also provide experimental evidence confirming that recently observed increases in shrub cover in many tundra regions is a result of recent warming in the region (see Section 27.4 for example and discussion).

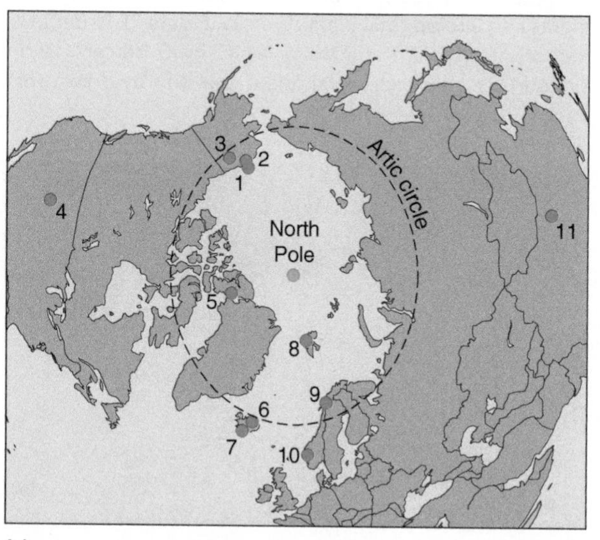

(a)

(b)

Figure 27.25 (a) Location of research sites making up the International Tundra Experiment (ITEX): (1) Barrow, United States; (2) Atqasuk, United States; (3) Toolik Lake, United States; (4) Niwot Ridge, United States; (5) Alexandra Fiord, Canada; (6) Audkuluheidi, Iceland; (7) Thingvellir, Iceland; (8) Svalbard, Norway; (9) Latnjajaure, Sweden; (10) Finse, Norway; and (11) Tibetan Plateau, China. (b) ITEX researchers use small, passive, clear plastic, open-top chambers to warm the tundra and extend the growing season. The chambers raise the daily temperature of the tundra plant canopy by 1.5 to 1.7°C, which is in the range predicted by global climate simulations.

FIELD STUDIES *Erika Zavaleta*

Environmental Studies Department, University of California, Santa Cruz

Human activities are altering Earth's atmosphere and climate in a variety of ways. Increases in atmospheric concentrations of carbon dioxide (CO_2) are contributing to rising global temperatures as well as changes in annual patterns of precipitation (see Sections 27.1 and Chapter 2, *Ecological Issues & Applications*). Global anthropogenic nitrogen (N) fixation (see Chapter 22, *Ecological Issues & Applications*) now exceeds all natural sources of N fixation, and its products include greenhouse gases such as nitrous oxide (N_2O).

How these global changes in climate and atmospheric chemistry may alter the diversity of plant communities by changing resource availability and affecting individual species performances is a question central to the research of ecologist Erika Zavaleta of the University of California, Santa Cruz. Since the mid-1990s, Zavaleta and her colleagues at Stanford University have been studying the response of California's grassland ecosystems to changes in climate, atmospheric CO_2, and N-deposition based on future scenarios developed for the region.

Zavaleta's studies were conducted in the California grassland at the Jasper Ridge Biological Preserve in the San Francisco Bay Area. The grassland community is composed of annual grasses, annual and biennial forbs, and occasional perennial bunchgrasses, forbs, and shrubs. Annual grasses are the community dominants, contributing the majority of plant biomass during the period of peak primary productivity in the growing season.

In this region's mediterranean-type climate, annual plants (both grasses and forbs) germinate with the onset of the fall–winter rains (see Section 23.6). Plants then set seed and senesce as water limitation becomes severe with the cessation of rain in March–May. The small stature and short life span of the annual plants that dominate these communities make this site an excellent experimental system to examine community response to altered environmental conditions over a period of multiple generations.

In 1997, Zavaleta and colleagues established 32 circular plots 2 m in diameter and surrounded each with a solid belowground partition to 50 cm depth. Each plot was then divided into four 0.78 m^2 quadrats, using solid partitions belowground and mesh partitions aboveground. Four global change treatments were applied to the experimental plots: (1) elevated CO_2 (ambient plus 300 ppm), (2) warming (80 watts [W]/m^2 of thermal radiation resulting in soil-surface warming of 0.8–1.0°C), (3) elevated precipitation (increased by 50 percent, including a growing-season extension of 20 days), and (4) N-deposition (increased by 7g/m^2/day).

The experimental design of the plots and application of treatments can be seen in the photograph in **Figure 1**. CO_2 was elevated by using a free-air system with emitters surrounding each plot and delivering pure CO_2 at the canopy level. Warming was applied with infrared lamps suspended over the center of each plot. Extra precipitation was delivered with overhead sprinklers and drip lines. The growing season extension was delivered in two applications, at 10 and 20 days after the last natural rainfall event. N-deposition was administered with liquid (in autumn) and slow-release (in winter) $Ca(NO_3)_2$ applications each year.

The four treatments were applied in one- to four-way treatment combinations (see **Figure 2** caption for description of treatment combinations), each replicated eight times. Treatments were begun in November 1998 and continued for a period of three years.

To evaluate the influence of the global change treatments on the grassland community within the experimental plots, a census was conducted in May of each year to determine plant species diversity on each plot. Plant diversity was quantified using species richness (total number of species).

After three years, three of the four global change treatments had altered total plant diversity (Figure 2). N-deposition reduced total plant species diversity by 5 percent, and elevated CO_2 reduced overall plant diversity by 8 percent. In

Figure 1 An experimental study plot at the Jasper Ridge Biological Reserve.

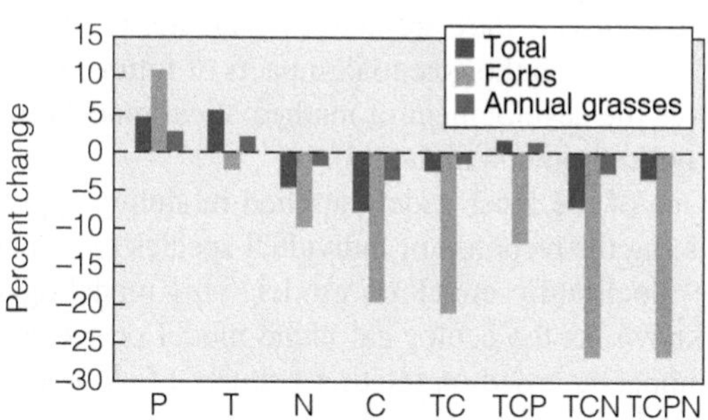

Figure 2 Changes in the total, forb, and annual grass diversity under single and combined global change treatments. Values are percent difference between controls and elevated levels for each treatment, based on values of mean species richness for each treatment. Treatments: C, CO_2; T, warming; P, precipitation; N, nitrogen; TC, warming and CO_2; TCP, warming, CO_2 and precipitation; TCN, warming, CO_2, and nitrogen; TCPN, warming, CO_2, precipitation, and nitrogen.

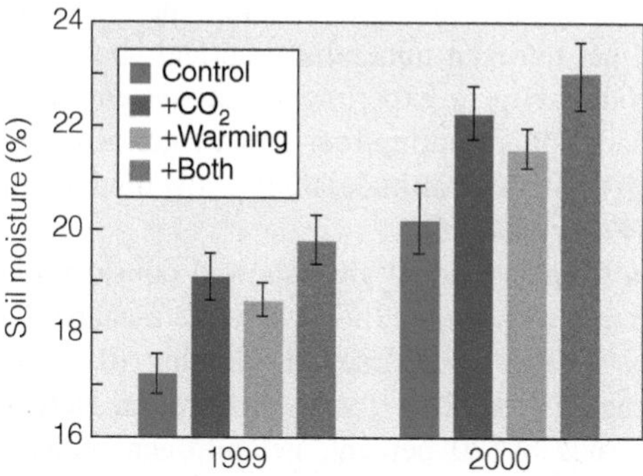

Figure 3 Warming and elevated CO_2 effects on spring soil moisture for 1999–2000. Values represent mean soil moisture from January to July for each year.

contrast, elevated precipitation increased plant diversity by 5 percent. The fourth treatment, elevated temperature, had no significant effect on plant species diversity in the experimental plots. The effects of elevated CO_2, N-deposition, and precipitation on total diversity were driven mainly by significant gains and losses of forb species (see Figure 2), which make up most of the native plant diversity in California grasslands. Reduced diversity in the N-deposition treatment was partly a function of the loss of all three N-fixing forb species at the site. In contrast to forb species, annual grass diversity was relatively unresponsive to all individual global change treatments.

All four treatment-combination scenarios produced mean declines in forb diversity of greater than 10 percent (see Figure 2). Diversity of this functional group, which includes many of the remaining native and rare species in California grasslands, seems susceptible to decline regardless of whether N-deposition and precipitation increase. The effects of these four treatment combinations on total plant diversity were not significant, however, because increases in perennial grass diversity partially offset losses of forb species.

Perhaps the most interesting and unexpected result from Zavaleta's experiments emerged from a comparison using only a subset of the treatments: elevated CO_2 (C), warming (T), and CO_2 plus warming (C+T). It is generally believed that global warming may increase aridity in water-limited ecosystems, such as the California grasslands, by accelerating evapotranspiration. However, the experiments conducted by Zavaleta and colleagues produced the reverse effect. Simulated warming increased spring soil moisture by 5–10 percent under both ambient and elevated CO_2 (**Figure 3**). This effect was not caused by decreasing leaf area or plant production under elevated temperatures, but rather by earlier plant senescence (in later May and early June) in the elevated temperature treatments. Lower transpirational water losses resulting from earlier senescence provide a mechanism for the unexpected rise in soil moisture, and this biotic link between warming and water balance may well prove to be an important influence on the response of grassland and savanna communities to climate change.

Bibliography

Zavaleta, E. S., B. D. Thomas, N. R. Chiariello, G. P. Asner, and C. B. Field. 2003. "Plants reverse warming effect on ecosystem water balance." *Proceedings of the National Academy of Sciences* USA 100:9892–9893.

Zavaleta, E. S., M. R. Shaw, N. R. Chiariello, H. A. Mooney, and C. B. Field. 2003. "Additive effects of simulated climate changes, elevated CO_2, and nitrogen deposition on grassland diversity." *Proceedings of the National Academy of Sciences* USA 100:7650–7654.

1. In the results of the combined treatments presented in Figure 2, how does increased precipitation influence the effects of elevated CO_2 and temperature on the diversity of forb species?

2. Based on the discussion of plant response to elevated CO_2 presented in Section 6.13, how might changes in stomatal conductance (and transpiration) influence soil moisture over the growing season?

The Network of Ecosystem Warming Studies, an activity of the Global Change and Terrestrial Ecosystems core project of the International Geosphere-Biosphere Program, provides another example of a coordinated international effort to examine the response of ecosystems to rising temperatures. The focus of these experiments is to examine the response of soil respiration, net nitrogen mineralization, and aboveground net primary productivity to experimental ecosystem warming at 32 research sites representing four broadly defined biomes, including high (latitude or altitude) tundra, low tundra, grassland, and forest (**Figure 27.26a**).

Results from individual sites showed considerable variation in response to warming, however a meta-analysis showed that, across all sites, experimental warming (0.3–6.0°C) for a period ranging from 2 to 9 years in duration increased soil respiration rates by 20 percent, net nitrogen mineralization rates by 46 percent, and NPP by 19 percent (**Figures 27.27b, 27.27c**, and **27.27d**).

The response of soil respiration to warming was generally larger in forested ecosystems compared to low tundra and grassland ecosystems, and the response of plant productivity was generally larger in low tundra ecosystems than in forest and grassland ecosystems. Overall, the ecosystem-warming experiments provide valuable insights on the response of terrestrial ecosystems to elevated temperature.

Although experiments allow for a direct assessment of the influence of warming (and associated increases in carbon dioxide) on the structure and function of communities and ecosystems, even the coordinated international programs presented are limited in their geographic scope and the range of climate change scenarios they are able to evaluate. For these reasons, most research on the potential impacts of future climate change involves the development of mathematical models (both statistical and computer simulation).

One of the most widely applied modeling approaches for evaluating the response of individual species to climate change is the **bioclimatic envelope model**. This modeling approach, also known as the ecological niche model or species distribution model, is based on relating features of climate (as well as edaphic factors) to geographic patterns of species occurrence (**Figure 27.27**). By establishing a quantitative relationship between climate and a species' current geographic distribution, the relationship (typically represented by a statistical model such as regression analysis) can then be used to map the potential geographic distribution of the species under changed climate conditions (see **Figure 27.28**).

Ecologists Louis Iverson, Anantha Prasad, Steve Matthews, and Matt Peters of the United States Department of Agriculture's Forest Service have used the basic approach of bioclimatic envelope models to explore the possible impacts

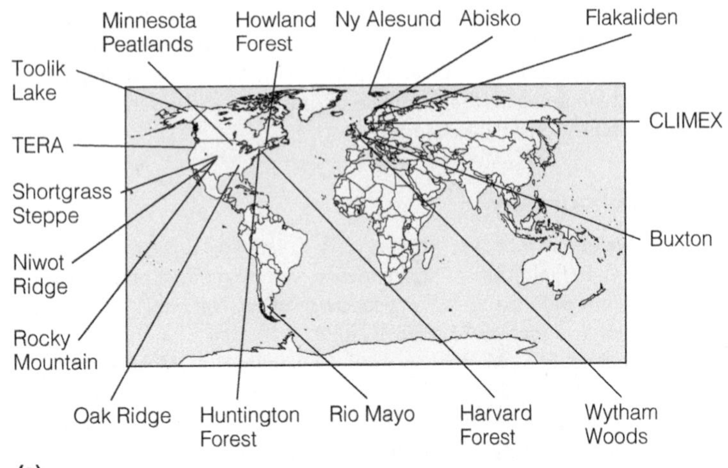

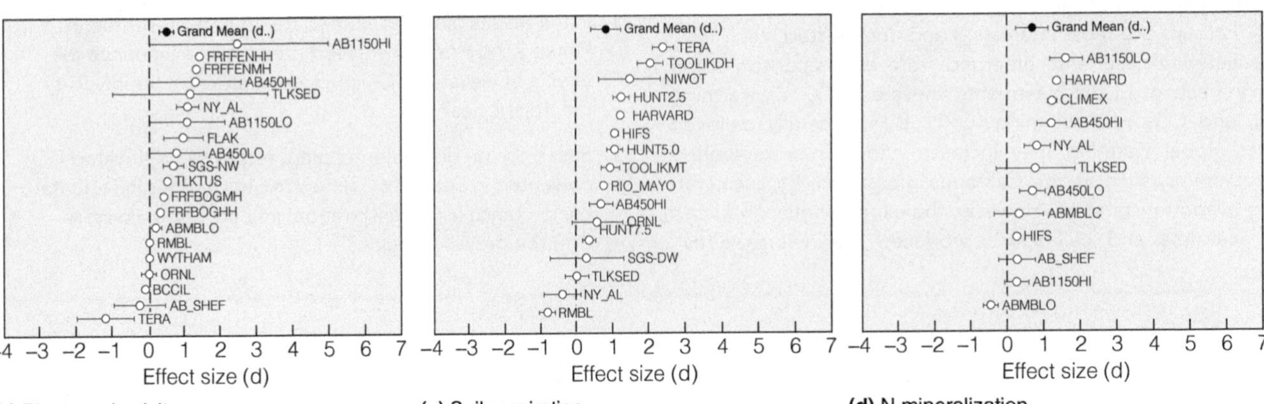

Figure 27.26 (a) The Global Change and Terrestrial Ecosystems Network of Ecosystem Warming Studies research sites. Mean effect sizes (open circles) and 95 percent confidence intervals for individual sites included in the meta-analysis for (b) plant productivity, (c) soil respiration, (d) nitrogen (N) mineralization. Grand mean effect sizes for experimental warming (closed circle) and 95 percent confidence intervals for each response variable are given at the top of each panel.
(Adapted from Rustad et al. 2001.)

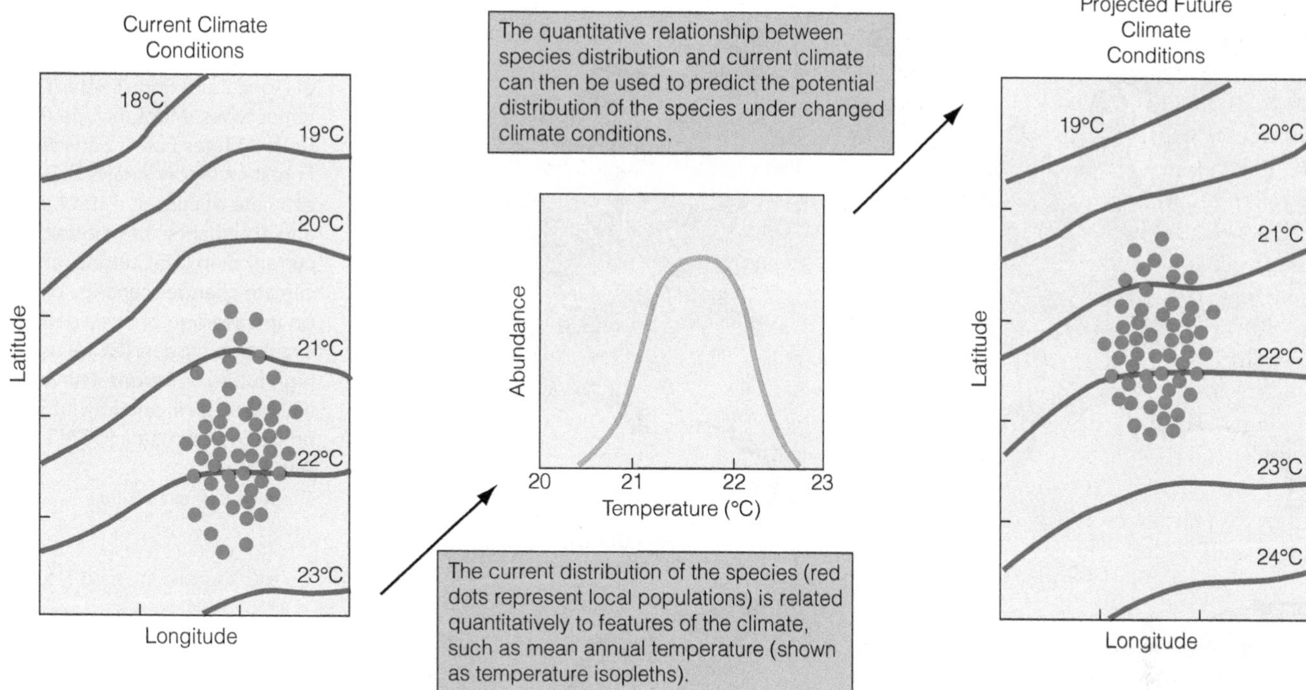

Figure 27.27 General procedure used in the development of a bioclimatic envelope model.

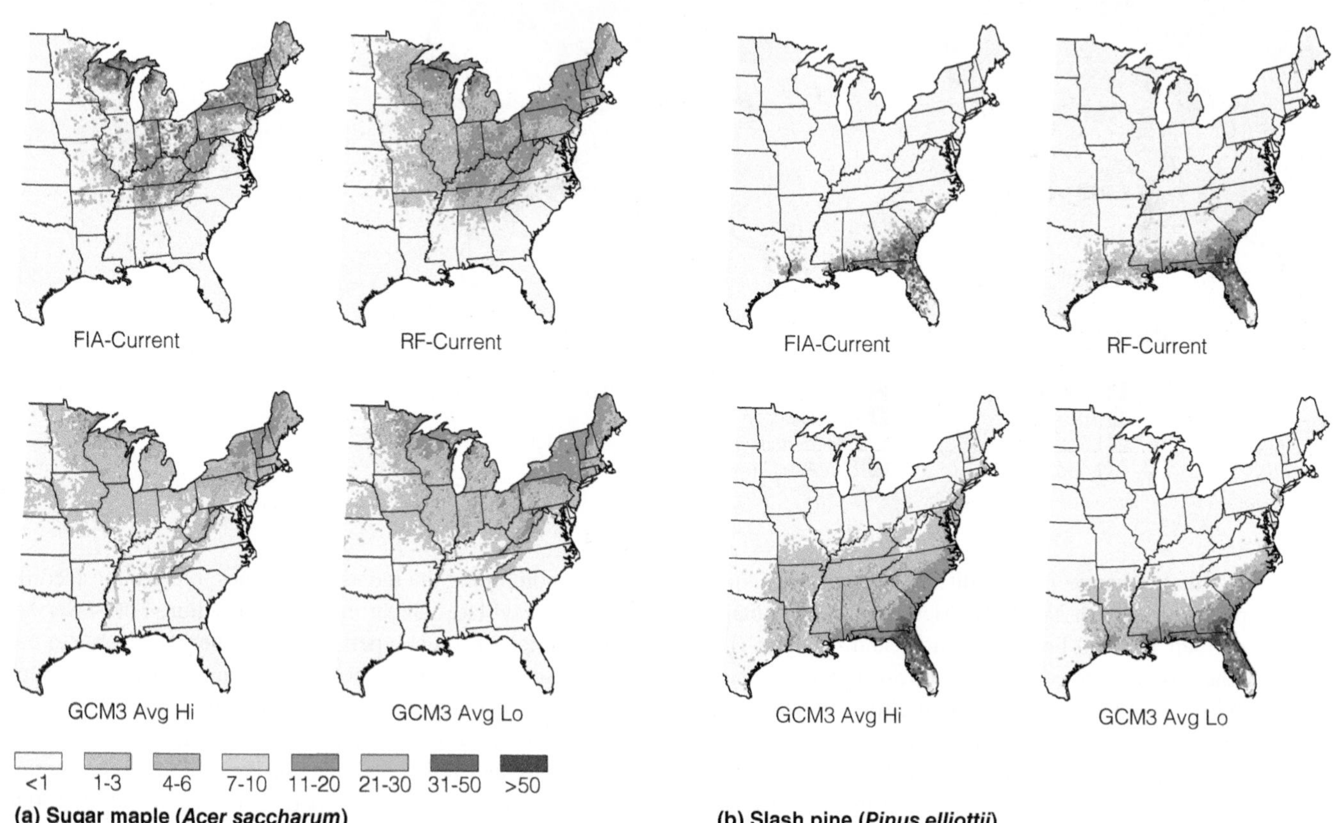

Figure 27.28 Maps of current and future potential distributions for (a) sugar maple (*Acer saccharum*) and (b) slash pine (*Pinus elliottii*). Maps include the United States Forest Service-Forest Inventory Analysis (FIA) estimate of current distribution and abundance, the modeled current map (RF-Current), and climate change scenarios based on the average of three general circulation models (GCM) using high future emissions estimates (GCM3 Avg Hi), and low future emissions estimates (GCM3 Avg Lo).
(From Iverson et al. 2008.)

Figure 27.29 Maps of current and future potential distributions for United States Department of Agriculture Forest Service forest types. Maps include the United States Forest Service-Forest Inventory Analysis (FIA) estimate of current distribution and abundance, the modeled current map (RF-Current), and climate change scenarios based on the average of three general circulation models (GCM) using high future emissions estimates (GCM3 Avg Hi) and low future emissions estimates (GCM3 Avg Lo).
(From Iverson et al. 2008.)

of future climate change on the distribution and abundance of tree species in eastern North America. Using data obtained from more than 100,000 Forest Inventory and Analysis (FIA) plots for the eastern United States, the researchers correlated the distribution of 134 tree species with a total of 36 environmental variables relating to climate, elevation, soils, and land use. Each species was modeled individually to show current and potential future distributions according to climate change scenarios for the region that were developed from three climate models (GCMs; see Section 27.8). Their analyses show that climate change could have large impacts on suitable habitat for tree species in the eastern United States (Figure 27.28).

By combining the individual species into groups according to the United States Department of Agriculture Forest Service's classification of forest types, the researchers were able to make some assessments of the potential changes in the geographic distribution forest types (forest communities; **Figure 27.29**).

Daniel McKenney and colleagues at the Canadian Forest Service have undertaken a similar analysis for the entire continent of North America, using bioclimatic envelope models of 130 tree species to examine changes in patterns of species diversity (richness) under a range of climate change scenarios (**Figure 27.30**).

The use of bioclimate models to evaluate the potential impacts of future climate change has not been limited to plant

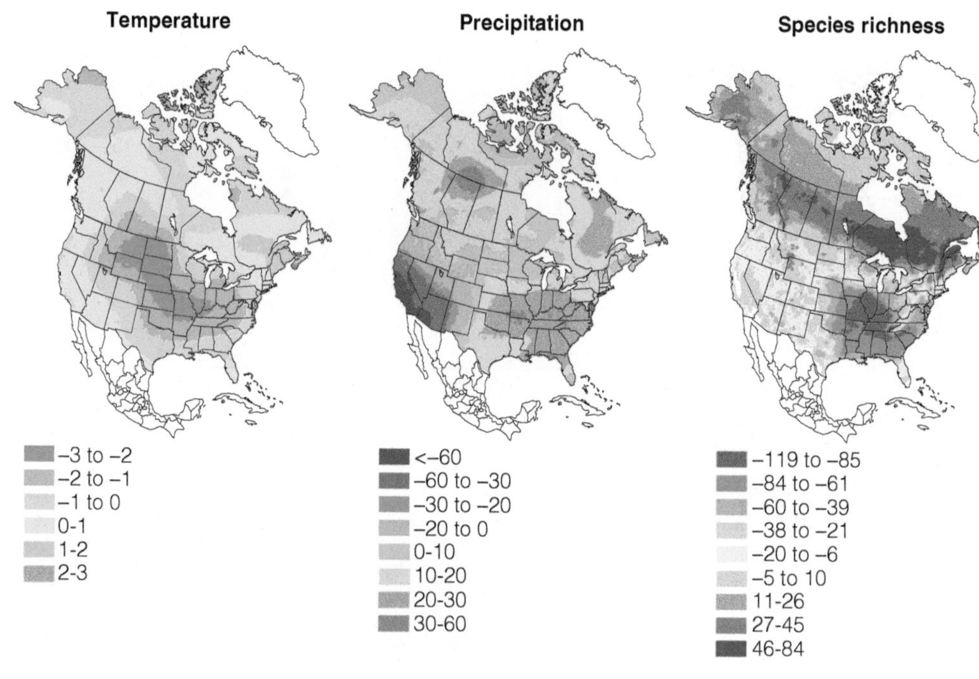

Figure 27.30 Projected changes in climate and tree species richness for North America under future climate change scenario of CGCM2 (Canadian Center for Climate Change Modeling and Analysis, Canada: CCCMA). Differences between current (1971–2000) and future (2071–2100) mean annual temperature (°C), annual precipitation (expressed as a percentage of current values), and tree species richness. (From McKenney et al. 2011.)

species. This basic approach of modeling has been applied to a wide variety of animal taxa, including birds, mammals, reptiles, and amphibians. The same research group from the U.S. Department of Agriculture's Forest Service that undertook the analysis of tree species in eastern North America presented previously (Iverson, Prasad, Matthews, and Peters) developed statistical models of the distributions of 147 bird species in the eastern United States using data from the North American Breeding Bird Survey (BBS). The researchers used climate, elevation, and the distributions of tree species to predict contemporary bird distributions. The resulting models were then used to predict the distributions of the bird species under three scenarios of future climate change (**Figure 27.31**).

For most taxonomic groups, however, we do not have sufficient information on the environmental factors controlling the distribution of individual species to allow for an analysis such as those presented. For other groups of organisms, we must depend on more general relationships between features of the environment and overall patterns of diversity. For example, we examined the work of ecologist David Currie of the University of Ottawa in correlating the broad-scale patterns of species diversity at the continental scale to features of the physical environment (see Chapter 26, Section 26.5). Currie found that the richness of most terrestrial animal groups, including vertebrates, covaries with features of the physical environment related to the energy and water balance of organisms: temperature, evapotranspiration, and incident solar radiation. Currie has used these relationships between climate (specifically, mean January and July temperature and precipitation) and species richness at a regional scale to predict changes in bird and mammal diversity for the conterminous United States under scenarios of future climate change (**Figure 27.32**). His analyses predict a northward shift in the regions of highest diversity, with species richness declining in the southern areas of the United States while increasing in New England, the Pacific Northwest, and in the Rocky Mountains and the Sierra Nevada.

27.10 Predicting Future Climate Change Requires an Understanding of the Interactions between the Biosphere and the Other Components of Earth's System

Increasing atmospheric concentrations of carbon dioxide and other greenhouse gases, and the potential changes in global climate patterns that may result, present a new class of ecological problems. To understand the effect of rising carbon dioxide emissions from fossil fuel burning and land clearing (see Chapter 2, *Ecological Issues & Applications*), we have to examine the carbon cycle linking the atmosphere, hydrosphere, biosphere, and geosphere on a global scale (see discussion of global carbon cycle in Section 22.7 and Figure 22.5). Of the carbon that is emitted to the atmosphere through human activities, only a fraction remains in the atmosphere, the reminder flows into the other components of Earth's system represented in the global carbon cycle presented in Figure 22.5. Current estimates show that approximately 50 percent of the carbon dioxide emitted to the atmosphere through the clearing of forests and burning of fossil fuels remains in the atmosphere, contributing to future warming. The remaining 50 percent is being taken up by terrestrial ecosystems (26 percent) and the

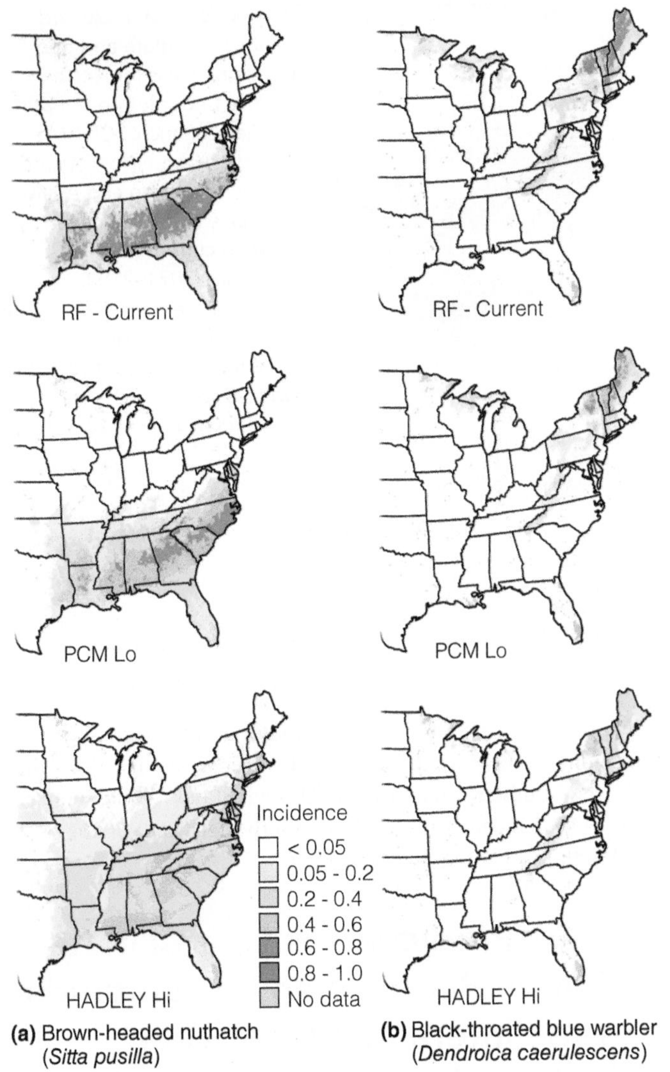

Figure 27.31 The current modeled distribution (RF-current) and projected suitable habitat under two climate change scenarios for the (a) brown-headed nuthatch (*Sitta pusilla*) and (b) black-throated blue warbler (*Dendroica caerulescens*). Climate change scenarios based on the Hadley Centre for Climate Change (England) general circulation model (GCM) representing high levels of future emissions (HADLEY Hi), and the National Center for Atmospheric Research (United States) GCM representing lower levels of future emissions (PCM Lo).
(From Matthews et al. 2011.)

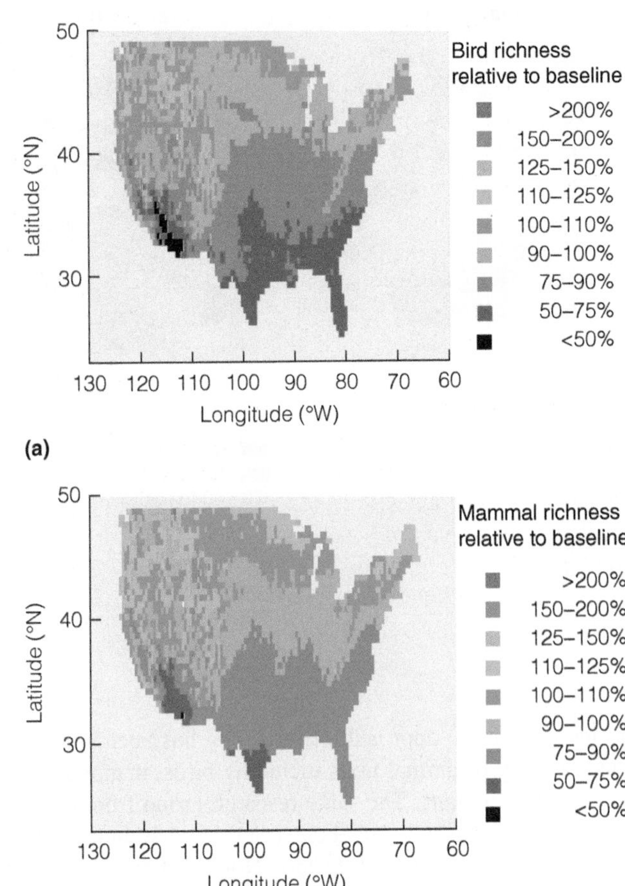

Figure 27.32 Changes in (a) bird, and (b) mammal species richness, relative to current species richness resulting from the climatic changes associated with doubling of atmospheric carbon dioxide. Richness was projected using five general circulation models (GCMs).
(Adapted from Currie 2001.)

oceans (24 percent). Carbon dioxide dissolved in the oceans forms carbonic acid, which leads to higher acidity at the surface and inhibits shell growth in marine animals and bleaches coral (see Chapter 3, *Ecological Issues & Applications*).

As we learned in Chapters 3 (Section 3.7) and 22 (Section 22.7), the primary factor controlling the exchange of carbon between the atmosphere and surface waters of the oceans is the physical process of diffusion. In contrast, the exchange of carbon between the atmosphere and land surface is controlled by two ecosystem processes: the net uptake of carbon by terrestrial ecosystems as NPP (photosynthesis minus autotrophic respiration; Chapter 20) and the loss of carbon from terrestrial ecosystems in decomposition (heterotrophic respiration; Chapter 21). The difference in the rate between these two processes (NPP – decomposition) represents net ecosystem productivity (NEP), which is the net uptake or loss of carbon by the ecosystem.

A key question in predicting the fate of future carbon dioxide emissions is how rising atmospheric concentrations of carbon dioxide and associated climate change will influence the exchange of carbon between the atmosphere and terrestrial ecosystems. If climate change results in an increase in NPP, the effect is a net removal of carbon dioxide from the atmosphere, therefore functioning as a negative feedback on the rising atmospheric concentrations of carbon dioxide. In contrast, if climate change results in a reduction in NPP or increases in rates of decomposition, the result would be a positive feedback, increasing atmospheric concentrations of carbon dioxide and the potential for further warming. Research examining these possible feedbacks between terrestrial ecosystems and the atmosphere has become a key component in the development of global climate models such as those discussed in Section 27.8. Numerous research groups have developed models that examine the exchange of carbon between terrestrial ecosystems and the atmosphere at a global scale, and although the specific structure of these models varies, they all simulate the basic processes of net primary productivity and decomposition.

One example of this research effort is the terrestrial biosphere model developed by the research group at the Potsdam Institute for Climate Impact Research in Potsdam, Germany. The model simulates key ecosystem processes governing terrestrial biogeochemistry and biogeography. Terrestrial plants are represented by functional categories based on their physiological (C_3 or C_4 photosynthesis), physiognomic (woody or herbaceous), and phenological (deciduous or evergreen) attributes, and the processes of photosynthesis and respiration are modeled in response to atmospheric and soil conditions. In addition, rates of aboveground litter, belowground litter, and soil organic matter decomposition are simulated, which, when combined with estimates of NPP based on photosynthesis and autotrophic respiration, provide estimates of NEP.

The researchers at the Potsdam Institute have used their terrestrial biosphere model to examine the consequences of future climate change scenarios based on five different general circulation models (GCMs; see Section 27.8) on the exchange of carbon between the land surface and atmosphere (**Figure 27.33**). Their findings show that there is great uncertainty in the predicted patterns of carbon exchange between terrestrial ecosystems and the atmosphere, largely as a result of large differences in the climate projections. In three out of five climate projections the terrestrial biosphere was a source of carbon to the atmosphere by the year 2100. A forth climate change scenario shows the terrestrial biosphere to act as a sink for atmospheric carbon, whereas in the fifth scenario it is approximately neutral.

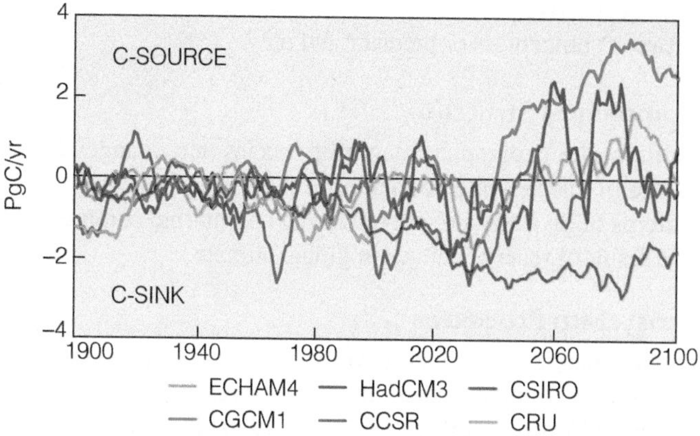

Figure 27.33 Variations in the exchange (flux) of carbon between the global land surface (terrestrial ecosystems) and the atmosphere for the period from 1900 to 2100 predicted by the Lund-Potsdam-Jena (LPJ) model under six different future climate change scenarios. Values are in petagrams (10^{12} g) of carbon per year. Positive values represent a net flux to the atmosphere (terrestrial ecosystems act as a carbon source to the atmosphere), whereas negative values represent a net uptake of carbon by terrestrial ecosystems (terrestrial ecosystems act as a carbon sink). Climate change scenarios: Canadian Centre for Climate Modelling and Analysis' CGCM1 model; Hamburg Max Planck Institute for Meteorology's ECHAM4 model; University of Tokyo Center for Climate System Research's and Japanese National Institute of Environmental Studies' CCSR model; Australian Commonwealth Scientific and Industrial Research Organisation's (CSIRO) Atmospheric Research Mark 2b model; United Kingdom Meteorological Office Hadley Centre's HadCM3 model. (From Schaphoff et al. 2006.)

SUMMARY

Recent Climate Warming 27.1

Over the past century Earth's climate has warmed by an estimated 0.74°C (±0.2°C), and regional changes have been highly heterogeneous. Associated with this general pattern of global warming has been a lengthening of the freeze-free periods in most mid- and high-latitude regions, a decrease in snow cover and ice extent, changes in regional precipitation patterns, and an increase in sea surface temperatures.

Physiology and Development 27.2

Temperature has a direct effect on the basic metabolic and developmental processes of both plants and animals. Warming in the northern latitudes has been associated with an increase in body size for local populations of endotherms over the timescale of decades to a century. For ectotherms, warming results in a direct increase in metabolic rates. The ability of tropical ectotherms to acclimate to warming might be limited where species currently live closest to their upper thermal tolerance limits. The impacts of climate change on terrestrial plant species is more complicated than for animal species because the response of plants to changes in both temperature and precipitation are influenced by atmospheric concentrations of carbon dioxide.

Phenology 27.3

Many processes and activities such as migratory behavior, the termination of dormancy, or the onset of reproductive activity are related to seasonal changes in climate conditions, and long-term phenological data sets have shown that, over the past century, spring activities have occurred progressively earlier.

Geographic Distribution 27.4

To the extent that dispersal and other environmental factors have allowed, recent warming has resulted in a shift in species' distributions poleward in latitude and upward in elevation.

Species Interactions 27.5

Recent climate change has been shown to influence interactions among species within existing ecological communities. Differences in the phenological responses of herbivores and their food resources have directly impacted population dynamics of herbivores. In other cases, changes in climate have favored one species over another, influencing the nature of interactions between competitors or predator and prey.

Community Structure 27.6

Shifts in the geographic ranges of species and changes in the nature of species interactions have led to regional changes in patterns of diversity in both terrestrial and marine communities as a result of recent changes in global climate.

Ecosystem Processes 27.7

Global changes in climate have eased several critical climatic constraints to plants such that net primary production increased globally between 1982 and 1999. Ecosystems in all tropical regions and those in the high latitudes of the Northern Hemisphere accounted for 80 percent of the observed increase. In contrast, the decade between 2000 and 2009 saw a reduction in the global net primary productivity, largely as a result of drying in the Southern Hemisphere.

Future Climate Change 27.8

Atmospheric scientists have developed complex computer models of Earth's climate system—called general circulation models (GCMs)—to help determine how increasing concentrations of greenhouse gases may influence large-scale patterns of global climate in the future. All of the models predict an increase in the average global temperature as well as a corresponding increase in global precipitation over the next century.

Impacts on Ecological Systems 27.9

Studies examining the possible impacts of future climate change on ecological systems can be classified into two broad categories: those that examine the response to experimental warming and those that use models of ecological systems to evaluate the response to future climate scenarios. Both approaches indicate the potential for major shifts in community structure and ecosystem processes under climate warming.

Feedback and Climate Change 27.10

To predict the effect of rising carbon dioxide emissions from fossil fuel burning and land clearing on future climate change, it is necessary to understand the influence of terrestrial ecosystems on the exchange of carbon between the land surface and atmosphere. To achieve this end, researchers are developing terrestrial biosphere models that can evaluate the exchange of carbon between terrestrial ecosystems and the atmosphere under various climate change scenarios.

STUDY QUESTIONS

1. How has the average surface temperature of the planet changed over the past century? Has this change occurred equally across all regions? If not, where have most of the changes occurred?
2. Contrast how increasing temperatures might affect ectotherms and endotherms.
3. Why might ectotherms in the tropics be more adversely affected by an increase in temperature than ectotherms in the temperate and Arctic regions?
4. What is phenology, and how does it relate to the timing of ecological processes?
5. How has regional warming influenced the timing of ecological processes?
6. Reproduction in herbivore species is generally timed to coincide with the peak growth period of plant populations (their food resources). If the timing of reproduction in herbivore species and the peak growth of plant populations respond differently to changing climate conditions, how might this influence the interaction between herbivores and their food resources?
7. How might a regional increase in temperatures influence the distribution of plant and animal species in mountainous areas?
8. If species within a region respond differently to changes in climate, how might these differences influence the structure of ecological communities?
9. In the example of plant populations responding to regional climate warming presented in Figure 27.19, the populations appear to respond independently rather than as a single community. Which of the two models of community structure presented in Chapter 16 (Section 16.10), the organismic concept of communities or the individualistic concept, do these observations support?
10. How have observed changes in global climate since 1982 influenced the global net primary productivity of terrestrial ecosystems?
11. What is a global circulation model?
12. How have warming experiments been used to examine the potential response of terrestrial communities and ecosystems in the Arctic regions to future climate change?
13. What is a bioclimatic envelope model, and how is it used to predict changes in the distribution of plant and animal species to future climate change?
14. How might an increase in the net ecosystem productivity of terrestrial ecosystems under climate warming influence the atmospheric concentrations of CO_2 (carbon dioxide)?

FURTHER READINGS

Anderegg, W. R. L., J. M. Kane, and L. D. L. Anderegg. 2012. "Consequences of widespread tree mortality triggered by drought and temperature stress." *Nature Climate Change* 3:30–36.

This article provides an excellent overview of global patterns of forest die-off as a consequence of recent climate change and the impacts on ecosystem services and land-climate interactions.

Hansen, J., R. Ruedy, M. Sato, and K. Lo. 2010. "Global surface temperature change." *Reviews of Geophysics* 48:RG4004.

An excellent overview of recent climate change.

IPCC. 2013: Summary for Policymakers. In: *Climate Change 2013: The physical science basis. Contribution of Working Group I to the Fifth Assessment Report of the Intergovernmental Panel on Climate Change*. Stocker,T.F., D. Qin, G. K. Plattner, M. Tignor, S.K. Allen, J. Boschung, A. Nauels, Y. Xia, V. Bex, and P.M. Midgley (eds.). Cambridge University Press, Cambridge, United Kingdom.

The summary of the 5th scientific assessment of climate change by the Intergovernmental Panel on Climate Change.

Parmesan, C., and G. Yohe. 2003. "A globally coherent fingerprint of climate change impacts across natural systems." *Nature* 421:37–42.

Pereira, H. M., P. W. Leadley, V. Proença, R. Alkemade, J. P. W. Scharlemann, J. F. Fernandez-Manjarrés, M. B. Araújo, P. Balvanera, R. Biggs, W. W. L. Cheung, L. Chini, H. D. Cooper, E. L. Gilman, S. Guénette, G. C. Hurtt, H. P. Huntington, G. M. Mace, T. Oberdorff, C. Revenga, P. Rodrigues, R. J. Scholes, U. R. Sumaila, and M. Walpole. 2010. "Scenarios for global biodiversity in the 21st century." *Science* 330:1496–1501.

This article provides an analysis of a variety of climate change scenarios on future biodiversity using a range of measures including extinctions, changes in species abundance, habitat loss, and distribution shifts.

Post, E., M. C. Forchhammer, M. S. Bret-Harte, T. V. Callaghan, T. R. Christensen, B. Elberling, A. D. Fox, O. Gilg, D. S. Hik, T. T. Høye, R. A. Ims, E. Jeppesen, D. R. Klein, J. Madsen, A. D. McGuire, S. Rysgaard, D. E. Schindler, I. Stirling, M. P. Tamstorf, N. J. C. Tyler, R. van der Wal, J. Welker, P. A. Wookey, N. M. Schmidt, and P. Aastrup. 2009. "Ecological dynamics across the Arctic associated with recent climate change." *Science* 325:1355–1358.

An excellent review of the ecological impacts of recent climate change in the Arctic.

Rustad, L. E, J. L. Campbell, G. M. Marion, R. J. Norby, M. J. Mitchell, A. E. Hartley, J. H. C. Cornelissen, and J. Gurevitch. 2001. "A meta-analysis of the response of soil respiration, net nitrogen mineralization, and aboveground plant growth to experimental ecosystem warming." *Oecologia* 126:543–562.

This review article provides an overview of the warming experiments conducted by the Global Change and Terrestrial Ecosystems Network of Ecosystem Warming Studies.

Walther, G. 2010. "Community and ecosystem responses to recent climate change." *Philosophical Transactions of the Royal Society B* 365: 2019–2024.

Excellent review of research quantifying the response of ecological systems to recent climate change.

Woodward, G., D. M. Perkins, and L. E. Brown. 2010. "Climate change and freshwater ecosystems: Impacts across multiple levels of organization." *Philosophical Transactions of the Royal Society B* 365: 2093–2106.

One of the few reviews of recent climate change on freshwater ecosystems.

MasteringBiology®

Students Go to **www.masteringbiology.com** for assignments, the eText, and the Study Area with practice tests, animations, and activities.

Instructors Go to **www.masteringbiology.com** for automatically graded tutorials and questions that you can assign to your students, plus Instructor Resources.

REFERENCES

Chapter 1 The Nature of Ecology

Bronowski, J. 1956. *Science and human values*. Harper & Row.

Brower, J. E., and J. H. Zar. 1984. *Field and laboratory methods for general ecology.* 2nd ed. Dubuque, IA: Wm. Brown.

Cox, G. W. 1985. *Laboratory manual for general ecology.* 5th ed. Dubuque, IA: Wm. Brown.

Egerton, F. N. (ed.). 1977. *History of American ecology.* New York: Arno Press.

Golley, F. B. 1993. *A history of the ecosystem concept in ecology: More than the sum of its parts.* New Haven, CT: Yale University Press.

Hagen, J. B. 1992. *An entangled bank: The origins of ecosystem ecology.* New Brunswick, NJ: Rutgers University Press.

Kingsland, S. 1995. *Modeling nature.* Chicago: University of Chicago Press.

McIntosh, R. P. 1985. *The background of ecology: Concept and theory.* New York: Cambridge University Press.

Sheall, J. (ed.). 1988. *Seventy-five years of ecology: The British Ecological Society.* Oxford: Blackwell Scientific.

Worster, D. 1977. *Nature's economy.* San Francisco: Sierra Club Books.

Chapter 2 Climate

Ahrens, C. D. 2003. *Meteorology today.* 7th ed. Pacific Grove, CA: Brooks/Cole.

Berry, R. G., and R. J. Corley. 1998. *Atmosphere, weather, and climate.* 7th ed. New York: Routledge.

Gates, D. M. 1962. *Energy exchange in the biosphere.* New York: Harper & Row.

Geiger, R. 1965. *Climate near the ground.* Cambridge, MA: Harvard University Press.

Graedel, T. E., and P. J. Crutzen. 1997. *Atmosphere, climate and change.* New York: Scientific American Library.

Intergovernmental Panel on Climate Change. 2007. Climate Change 2007: Fourth Assessment (AR4). Cambridge: Cambridge University Press. http://www.ipcc.ch/

Landsberg, H. E. 1970. Man-made climatic changes. *Science* 170:1265–1274.

Lee, R. 1978. *Forest microclimatology.* New York: Columbia University Press.

Chapter 3 The Aquatic Environment

Bainbridge, R., G. C. Evans, and O. Rackham (eds.). 1966. *Light as an ecological factor.* Oxford: Blackwell Scientific.

Clark, G. A. 1939. Utilization of solar energy by aquatic organisms. *Problems in Lake Biology* 10:27–38.

Doney, S.C., V.J. Fabry, R.A. Feely and J.A. Kleypas. 2009. Ocean acidification: The other CO_2 problem. *Annual Review of Marine Science* 1:169–192.

Garrison, T. 2005. *Essentials of oceanography.* 4th ed. St. Paul, MN: Brooks/Cole.

Hynes, H. B. N. 1970. *The ecology of running water.* Toronto: University of Toronto Press.

Likens, G. E. (ed.). 1985. *An ecosystem approach to aquatic ecology: Mirror Lake and environment.* New York: Springer-Verlag.

Nybakken, J. W., and M. Bertness. 2004. *Marine biology: An ecological approach.* 6th ed. San Francisco: Benjamin Cummings.

Wetzel, R. G. 1983. *Limnology.* 2nd ed. Philadelphia: Saunders College Publishing.

Chapter 4 The Terrestrial Environment

Boul, S. W., F. D. Hole, R. J. McCracken, and R. J. Southward. 1997. *Soil genesis and classification.* Ames: Iowa State University Press.

Brady, N. C., and R. W. Weil. 2001. *The nature and properties of soils.* 13th ed. Upper Saddle River, NJ: Prentice Hall.

Brown, L. R., and E. C. Wolf. 1984. *Soil erosion: Quiet crisis in the world economy.* Washington, DC: Worldwatch Institute.

Cernusca, A. 1976. Energy exchange within individual layers of a meadow. *Oecologica* 23:148.

Farb, P. 1959. *The living earth.* New York: Harper & Row.

Gliessman, S. R. (ed.) 1990. *Agroecology: Researching the ecological basis for sustainable agriculture.* Ecological Studies Series No. 78. New York: Springer-Verlag.

Hutchinson, B. A., and D. R. Matt. 1977. The distribution of solar radiation within a deciduous forest. *Ecological Monographs* 47:185–207.

Jenny, H. 1980. *The soil resource.* New York: Springer-Verlag.

Killham, K. 1994. *Soil ecology.* New York: Cambridge University Press.

Larcher, W. 1980. *Physiological Plant Ecology*. Springer-Verlag, NY.

Lutz, H. J., and R. F. Chandler. 1946. *Forest soils.* New York: Wiley.

Patton, T. R. 1996. *Soils: A new global view.* New Haven, CT: Yale University Press.

Pimentel, D. 1984. *Energy flow in agroecosystems.* Pages 121–132 in R. Lawrence, B. Stinner, and G. J. House, eds., *Agricultural ecosystems: Unifying concepts.* New York: John Wiley & Sons.

Pimentel, D. 2006. Soil erosion: A food and environmental threat. *Environment, Development and Sustainability* 8:119–137.

Reifsnyder, W. E., and H. W. Lull. 1965. *Radiant energy in relation to forests.* Technical Bulletin No. 1344. Washington, DC: U.S. Department of Agriculture.

Soil Survey Division Staff. 1993. *Soil survey manual.* Handbook 18. Washington, DC: U.S. Government Printing Office.

Way, D.A. and R.W. Pearcy. 2012. Sunflecks in trees and forests: from photosynthetic physiology to global change biology. *Tree Physiology* 32:1066–1081.

Yanga, W., W. Ni-Meistera, N.Y. Kiangb, P.R. Moorcroftc, A.H. Strahlerd and A. Oliphante. 2010. A clumped-foliage canopy radiative transfer model for a Global Dynamic Terrestrial Ecosystem Model II: Comparison to measurements. *Agricultural and Forest Meteorology* 150:895–907.

Chapter 5 Adaptation and Natural Selection

Angilletta, M.J., P. H. Niewiarowski, A. E. Dunham, A. D. Leache´, and W. P. Porter. 2004. Bergmann's Clines in Ectotherms: Illustrating a Life-History Perspective with Sceloporine Lizards. *The American Naturalist* 164:168–183.

Boag, P. T., and P. R. Grant. 1981. Intense natural selection in a population of Darwin's finches (*Geospiza*) in the Galapagos. *Science* 214:82–85.

————. 1984. The classical case of character release: Darwin's finches (*Geospiza*) on Isla Daphne Major, Galapagos. *Biological Journal of the Linnean Society* 22:243–287.

Brookes, G. and P. Barfoot. 2010. Global impact of biotech crops: Environmental effects, 1996–2008. *AgBioForum* 13:76–94.

Darwin, C. 1859. *The origin of species.* Philadelphia: McKay. [Reprinted from 6th London edition.]

Desmond, A., and J. Moore. 1991. *Darwin.* New York: Warner Books.

Fangue, N.A. and W. A. Bennett. 2003. Thermal tolerance responses of laboratory-acclimated and seasonally acclimatized Atlantic Stingray, *Dasyatis sabina. Copeia* 2003:315–325.Futuyma, D. J. 1984. *Evolutionary biology.* Sunderland, MA: Sinauer Associates.

Grant, P. 1999. *Ecology and evolution of Darwin's finches.* Princeton, NJ: Princeton University Press.

Grant, V. 1971. *Plant speciation.* New York: Columbia University Press.

————. 1985. *The evolutionary process: A critical review of evolutionary theory.* New York: Columbia University Press.

Hartl, D. 1988. *A primer of population genetics.* Sunderland, MA: Sinauer Associates.

Highton, R. 1995. Speciation in eastern North American salamanders of the genus *Plethodon. Annual Review of Ecology and Systematics* 26:579–600.

James, Clive. 2012. *Global Status of Commercialized Biotech/GM Crops: 2012*. ISAAA Brief No. 44. ISAAA: Ithaca, NY.

Jordan, N. 1992. Path Analysis of Local Adaptation in Two Ecotypes of the Annual Plant *Diodia teres* Walt.(Rubiaceae). *The American Naturalist* 140:149–165.

Lack, D. 1974. *Darwin's finches: An essay on the general biological theory of evolution.* London: Cambridge University Press.

Mayr, E. 1963. *Animal species and evolution.* Cambridge, MA: Harvard University Press.

————. 1991. *One long argument.* Cambridge, MA: Harvard University Press.

Montesinos-Navarro, A., J. Wig, F. X. Pico, and S. J. Tonsor. 2010. *Arabidopsis thaliana* populations show clinal variation in a climatic gradient associated with altitude. *New Phytologist* 189:282–294.

Reznick, D., F. H. Shaw, F. H. Rodd, and R. G. Shaw. 1997. Evaluation of the rate of evolution in natural populations of guppies (*Poecilia reticulata*). *Science* 275:1934–1937.

Robinson, B. W. 2000. Trade offs in habitat-specific foraging efficiency and the nascent adaptive divergence of sticklebacks in lakes. *Behaviour* 137:865–888.

Robinson, B. W., and S. Wardrop. 2002. Experimentally manipulated growth rate in threespine sticklebacks: Assessing trade-offs with developmental stability. *Environmental Biology of Fishes* 63:67–78.

Sultan, S.E. 2000. Phenotypic plasticity for plant development, function and life history. *Trends in Plant Science* 5:537–542.

Whitman, D.W. and A.A. Agrawal. 2009. What is phenotypic plasticity and why is it important? Pgs. 1–63, In: *Phenotypic Plasticity of Insects*. (D.W. Whitman and T.N. Ananthakrishnan, eds.). Science Publishing, Enfield, NJ.

Chapter 6 Plant Adaptations to the Environment

Abrams, M.D. and M.E. Kubiske. 1990. Leaf structural characteristics of 31 hardwood and conifer tree species in central Wisconsin: influence of light regime and shade-tolerance rank. *Forest Ecology and Management* 31:245–253.

Ainsworth, E.A. and A. Rogers. 2007. The response of photosynthesis and stomatal conductance to rising CO_2: mechanisms and environmental interactions. *Plant, Cell and the Environment* 30:258–270.

Augspurger, C. K. 1982. Light requirements of neotropical tree seedlings: A comparative study of growth and survival. *Journal of Ecology* 72: 777–795.

Austin, M. P., R. H. Groves, L. M. F. Fresco, and P. E. Laye. 1986. Relative growth of six thistle species along a nutrient gradient with multispecies competition. *Journal of Ecology* 73:667–684.

Austin, M.P., T.M. Smith, K.P. Van Niel, and A.B. Wellington. 2009. Physiological responses and statistical models of the environmental niche: a comparative study of two co-occurring Eucalyptus species. *Journal of Ecology* 97:496–507.

Bjorkman, O. 1973. Comparative studies on photosynthesis in higher plants. Page 53 in A. C. Geise, ed., *Photophysiology.* New York: Academic Press.

Bradshaw, A. D., M. J. Chadwick, D. Jowett, and R. W. Snaydon. 1964. Experimental investigations into the mineral nutrition of several grass species. IV. Nitrogen level. *Journal of Ecology* 52:665–676.

Cernusaka, L.A., L.B. Hutleya, J. Beringerc, J.A.M. Holtumd, and B.L. Turnere. 2011. *Agricultural and Forest Meteorology* 151:1462–1470

Chapin, F. S., III. 1980. The mineral nutrition of wild plants. *Annual Review of Ecology and Systematics* 11:233–260.

Chazdon, R. L., and R. W. Pearcy. 1991. The importance of sunflecks for forest understory plants. *BioScience* 41:760–765.

Dale, J. E. 1992. How do leaves grow? *BioScience* 42:423–432.

Davies, S. J. 1998. Photosynthesis of nine pioneer *Macaranga* species from Borneo in relation to life-history traits. *Ecology* 79:2292–2308.

Davies, S. J., P. A. Palmiotto, P. S. Ashton, H. S. Lee, and J. V. LaFrankie. 1998. Comparative ecology of 11 sympatric species of *Macaranga* in Borneo: Tree distribution in relation to horizontal and vertical resource heterogeneity. *Journal of Ecology* 86:662–673.

Feldman, L. J. 1988. The habits of roots. *BioScience* 38:612–618.

Field, C.B., R.B. Jackson and H.A. Mooney. 1995. Stomatal response to increased CO2: implications from plant to the global scale. *Plant, Cell and the Environment* 18:1214–1225

Field, C. B., and H. Mooney. 1986. The photosynthetic-nitrogen relationship in wild plants. Pages 25–55 in T. J. Givnish, ed., *On the economy of plant form and function.* Cambridge: Cambridge University Press.

Givnish TJ. 1984. Leaf and canopy adaptations in tropical forests. In: Medina E, Mooney HA, Vázquez-Yánes C, eds. *Physiological ecology of plants of the wet tropics*. The Hague, the Netherlands: Dr W Junk, pgs. 51–84.

Gomez-Aparicio, L., F. Valladares, and R. Zamora. 2006. Differential light responses of Mediterranean tree saplings: linking ecophysiology with regeneration niche in four co-occurring species. *Tree Physiology* 26: 947–958.

Grime, J. 1971. *Plant strategies and vegetative processes.* New York: Wiley.

Huston, M. and Smith, T.M. 1987. Plant succession: life history and competition. *American Naturalist* 130:168–198.

Iio, A., H. Fukasawa, Y. Nose, and Y. Kakubari. 2004. Stomatal closure induced by high vapor pressure deficit limited midday photosynthesis at the canopy top of *Fagus crenata* Blumeon on Naeba mountain in Japan. *Trees* 18:510–517.

Kirk, J. T. O. 1983. *Light and photosynthesis in aquatic systems.* New York: Cambridge University Press.

Kozlowski, T. T. 1982. Plant responses to flooding of soil. *BioScience* 34: 162–167.

Lambers, H., F. S. Chapin III, and T. L. Pons. 1998. *Plant physiological ecology.* New York: Springer.

Larcher, W. 1996. *Physiological plant ecology.* 3rd ed. New York: Springer-Verlag.

Mokany, K., R.J. Raison, and A.S. Prokushkin. 2006. Critical analysis of root : shoot ratios in terrestrial biomes. *Global Change Biology* 12:84–96.

Mooney, H. A. 1986. Photosynthesis. Chapter 11 in M. J. Crawley, ed., *Plant ecology.* New York: Blackwell Scientific.

Mooney, H. A., O. Bjorkman, J. Ehleringer, and J. Berry. 1976. Photosynthetic capacity of in situ Death Valley plants. *Carnegie Institute Yearbook* 75:310–413.

Parsons, T., J. R. Parsons, M. Takahashi, and B. Hargrave. 1984. *Biological oceanographic processes,* 3rd ed. New York: Pergamon Press.

Pearcy, R. W. 1977. Acclimation of photosynthetic and respiratory CO_2 to growth temperature in *Atriplex lentiformis. Plant Physiology* 61: 484–486.

Poorter, H., and M. Perez-Soba. 2002. Plant growth at elevated CO2. Pages 489–496 in H. A. Mooney and J. G. Canadell, eds., *Encyclopedia of global change.* Vol. 2: The Earth system: Biological and ecological dimensions of global environmental change. Chichester, England: John Wiley and Sons, Ltd.

Reich, P. B., D. S. Ellsworth, and M. B. Walters. 1998. Leaf structure (specific leaf area) regulates photosynthesis-nitrogen relations: Evidence from within and across species and functional groups. *Functional Ecology* 12:948–958.

Reich, P. B., M. G. Tjoelker, M. B. Walters, D. Vanderklein, and C. Buschena. 1998. Close association of RGR, leaf and root morphology, seed mass and shade tolerance in seedlings of nine boreal tree species grown in high and low light. *Functional Ecology* 12:327–338.

Schulze, E. D., R. H. Robichaux, J. Grace, P. W. Rundel, and J. R. Ehleringer. 1987. Plant water balance. *BioScience* 37:30–37.

Stokes, V.J., M.D. Morecrof and J.I.L. Morison. 2006. Boundary layer conductance for contrasting leaf shapes in a deciduous broadleaved forest canopy. *Agricultural and Forest Meteorology* 139:40–54

Teeri, J.A. and Stowe L.G. 1976. Climatic patterns and the distribution of C_4 grasses in North America. *Oecologia* 23: 1–12.

Turner, N. C., and P. J. Kramer (eds.). 1980. *Adaptation of plants to water and high temperature stress.* New York: Wiley.

Woodward, I., and T. Smith. 1993. Predictions and measurements of the maximum photosynthetic rate, A_{max} at a global scale. Pages 491–508 in E. D. Schultz and M. M. Caldwell, eds., *Ecophysiology of photosynthesis.* vol. 100. Berlin: Springer.

Chapter 7 Animal Adaptations to the Environment

Adkisson, P. L. 1966. Internal clocks and insect diapause. *Science* 154:234–241.

Bauwens, D., A.M. Castilla, and P. Mouton. 1999. Field body temperatures, activity levels and opportunities for thermoregulation in an extreme microhabitat specialist, the girdled lizard (*Cordylus macropholis*). *Journal of Zoology* 249:11–18.

Belovsky, G. E., and P. F. Jordan. 1981. Sodium dynamics and adaptations of a moose population. *Journal of Mammalogy* 63:613–621.

Blouin-Demers, G. and P.J. Weatherhead. 2001. Thermal ecology of black rate snakes (*Elaphe obsoleta*) in a thermally challenging environment. *Ecology* 82:3025–3043.

Cheatum, E. L., and C. W. Severinghaus. 1950. Variations in fertility of white-tailed deer related to range conditions. *Transactions of the North American Wildlife Conference* 15:170–189.

Chena, X., and X. Xub, X. Jia. 2003. Influence of body temperature on food assimilation and locomotor performance in white-striped grass lizards, *Takydromus wolteri* (Lacertidae). *Journal of Thermal Biology* 28: 385–391

Dodd, R. H. 1973. Insect nutrition: Current developments and metabolic implications. *Annual Review of Entomology* 18:381–420.

French, A. R. 1992. Mammalian dormancy. Pages 105–121 in T. E. Tomasi and T. H. Horton, eds., *Mammalian energetics.* Ithaca, NY: Comstock.

Gilles, R. (ed.). 1979. *The mechanisms of osmoregulation in animals.* New York: Wiley.

Glanville, E.J. and Seebacher, F.J. 2006. Compensation for environmental change by complementary shifts of thermal sensitivity and thermoregulatory behavior in an ectotherm. *Journal of Experimental Biology* 209:4869–4877

Heinrich, B. (ed.). 1981. *Insect thermoregulation.* New York: Wiley.

Hill, E. P., III. 1972. Litter size in Alabama cottontails as influenced by soil fertility. *Journal of Wildlife Management* 36:1199–1209.

Hill, R. W. 1992. The altricial/precocial contrast in the thermal relations and energetics of small mammals. Pages 122–159 in T. E. Tomasi and T. H. Horton, eds., *Mammalian energetics.* Ithaca, NY: Comstock.

Hill, R. W., and G. A. Wyse. 1989. *Animal physiology.* New York: Harper & Row.

James, F. C. 1971. Ordinations of habitat relationships among breeding birds. *Wilson*

Bulletin 83:215–236.

Jones, R. L., and H. P. Weeks. 1985. Ca, Mg, and P in the annual diet of deer in south-central Indiana. *Journal of Wildlife Management* 49:129–133.

Lee, R. E., Jr. 1989. Insect cold-hardiness: To freeze or not to freeze. *BioScience* 39:308–313.

Leith, H. (ed.). 1974. *Phenology and seasonality modeling.* New York: Springer-Verlag.

Lyman, C. P., A. Malan, J. S. Willis, and L. C. H. Wang. 1982. *Hibernation and torpor in mammals and birds.* New York: Academic Press.

Maloly, C. M. O. (ed.). 1979. *Comparative physiology of osmoregulation in animals.* New York: Academic Press.

McKechnie, A.E. and B.G. Lovegrove. 2001. Thermoregulation and the energetic significance of clustering behavior in the White-backed Mousebird (*Colius colius*). *Physiological and Biochemical Zoology* 74:238–249.

Schmidt-Neilsen, K. 1997. *Animal physiology: Adaptation and environment.* 5th ed. New York: Cambridge University Press.

Secord,R., J. I. Bloch, S. G. B. Chester, D. M. Boyer, A. R. Wood, S. L. Wing, M. J. Kraus, F. A. McInerney, and J. Krigbaum. 2012. Evolution of the Earliest Horses Driven by Climate Change in the Paleocene-Eocene Thermal Maximum. *Science* 335:959.

Stillman, J. and G.N. Somero. 2000. A comparative analysis of the upper thermal tolerance limits of eastern Pacific porcelain crabs, genus *Petrolisthes*: Influences of latitude, vertical zonation, acclimation, and phylogeny. *Physiological and Biochemical Zoology*. 73(2):200–208.

Storey, K. B., and J. M. Storey. 1996. Natural freezing survival in animals. *Annual Review of Ecology and Systematics* 27:365–386.

Taylor, C. R. 1972. The desert gazelle: A parody resolved. In G. M. O. Malory, ed., *Comparative physiology of desert animals. Symposium of the Zoological Society of London No. 31.* New York: Academic Press.

Teplitsky, C., J.A. Mills, J.S. Alho, J.W. Yarrall, and J. Merila. 2008. Bergman's rule and climate change revisited" disentangling environmental and genetic responses in a wild bird population. *Proceedings of the National Academy of Sciences* 105:13492–13496.

Weeks, H. P., Jr., and C. M. Kirkpatrick. 1978. Salt preferences and sodium drive phenology in fox squirrels and woodchuck. *Journal of Mammalogy* 59:531–542.

Weir, J. S. 1972. Spatial distribution of elephants in an African national park in relation to environmental sodium. *Oikos* 23:113.

Yom-Tova, Y., N. Leaderb, S. Yom-Tova, H. J. Baagøec. 2010. Temperature trends and recent decline in body size of the stone marten *Martes foina* in Denmark. *Mammalian Biology* 75: 146–150.

Yom-Tov, Y., A. Roos, P. Mortensen, Ø. Wiig, S. Yom-Tov, and T. M. Heggberget. 2010. Recent Changes in Body Size of the Eurasian Otter *Lutra lutra* in Sweden. *AMBIO* 39:496–503

Chapter 8 Properties of Populations

Alexander, M. M. 1958. The place of aging in wildlife management. *American Scientist* 6: 123–131.

Begon, M., J. L. Harper, and C. R. Townsend. 1996. *Ecology: Individuals, populations, and communities.* New York: Blackwell Scientific.

Bleich, V.C., J. D. Wehausen and S. A. HollSource. 1990. Desert-Dwelling Mountain Sheep: Conservation Implications of a Naturally Fragmented Distribution. *Conservation Biology* 4:383–390.

Blower, J. G., L. M. Cook, and J. A. Bishop. 1981. *Estimating the size of animal populations.* London: George Allen & Unwin.

Engle, L. G. 1960. Yellow-poplar seedfall pattern. *Central States Forest Experiment Station Note 143.*

Forman, R. T. T. 1964. Growth under controlled conditions to explain the hierarchical distribution of a moss, *Tetraphis pellucida. Ecological Monographs* 34:1–25.

Krebs, C. 2001. *Ecology: The experimental analysis of distribution and abundance.* 5th ed. San Francisco: Benjamin Cummings.

Krebs, C. J. 1989. *Ecological methodology.* 2nd ed. San Francisco: Benjamin Cummings.

Liebhold, A. M., J. A. Halverson, and G. A. Elmes. 1992. Gypsy moth invasion in North America: A quantitative analysis. *Journal of Biogeography* 19:513–520.

Loewenstein, E.F., P.S. Johnson and H.E. Garrett. 2000. Age and diameter structure of a managed uneven-aged oak forest. Canadian Journal of Forest Research 30: 1060–1070.

MacArthur, R. H., and J. H. Connell. 1966. *The biology of populations.* New York: John Wiley.

Mueller-Dombois, D., and H. Ellenberg. 1974. *Aims and methods of vegetation ecology.* New York: Wiley.

Root, T. 1988. Energy constraints on avian distributions and abundances. *Ecology* 69:330–339.

Southwood, T. R. E. 1978. *Ecological methods.* London: Chapman and Hall.

Valverde, T. and J. Silvertown. 1998. Variation in the Demography of a Woodland Understory Herb (*Primula vulgaris*) along the Forest Regeneration Cycle: Projection Matrix. *Journal of Ecology* 86:545–562

Wiens, J. 1973. Pattern and process in grassland bird communities. *Ecological Monographs* 43:247–270.

Chapter 9 Population Growth

Allee, W. C. 1945. *Animal aggregations. A study in general sociology,* Chicago: University of Chicago Press.

Begon, M., and M. Mortimer. 1995. *Population ecology: A unified study of plants and animals.* 2nd ed. Cambridge, MA: Blackwell Scientific.

Binkley, C. S., and R. S. Miller. 1983. Population characteristics of the whooping crane, *Grus americana. Canadian Journal of Zoology* 61:2768–2776.

Burkhead, N.M. 2012. Extinction Rates in North American Freshwater Fishes, 1900–2010. *BioScience* 62: 798–808.

Campbell, R. W. 1969. Studies on gypsy moth population dynamics. Pages 29–34 in *Forest insect population dynamics.* USDA Research Paper, NE-125.

Cannon, J. R. 1996. Whooping crane recovery: A case study in public and private cooperation in the conservation of an endangered species. *Conservation Biology* 10:813–821.

Caughley, G. 1977. *Analysis of vertebrate populations.* New York: Wiley.

Deevey, Jr., E.S. 1947. Life Tables for Natural Populations of Animals. *The Quarterly Review of Biology* 22:283–314.

Gotelli, N. J. 1995. *A primer of ecology.* Sunderland, MA: Sinauer Associates.

Hackney, E., and J. B. McGraw. 2001. Experimental demonstration of an allee effect in American ginseng. *Conservation Biology* 15:129–136.

Harper, J. L. 1977. *The population biology of plants.* London: Academic Press.

Johnston, R. F. 1956. Predation by short-eared owls in a *Salicornia* salt marsh. *Wilson Bulletin* 68:91–102.

Krebs, C. 2001. *Ecology: The experimental analysis of distribution and abundance.* 5th ed. San Francisco: Benjamin Cummings.

Lowe, V. P. W. 1969. Population dynamics of red deer (*Cervus elaphus* L.) on the Isle of Rhum. *Journal of Animal Ecology* 38:425–457.

MacLulich, D. A. 1937. Fluctuations in the numbers of varying hare (*Lepus americanus*). *University of Toronto Biological Series No 43.*

North American Bird Conservation Initiative Canada. 2012. *The State of Canada's Birds, 2012*. Environment Canada, Ottawa, Canada. 36 pages.

North American Bird Conservation Initiative, U.S. Committee, 2009. *The State of the Birds, United States of America, 2009.* U.S. Department of Interior: Washington, DC. 36 pages.Ostlie, W.R. and J.L. Haferman. 1999. Ecoregional conservation in the Great Plains. *Proceedings of the North American Prairie Conference* 16:136–148.

Pitelka, F. A. 1957. Some characteristics of microtine cycles in the Arctic. *Proceedings of 18th Biology Colloquium.* Salem: University of Oregon Press.

Scharitz, R. R., and J. R. McCormick. 1973. Population dynamics of two competing plant species. *Ecology* 54:723–740.

Scheffer, V. C. 1951. Rise and fall of a reindeer herd. *Scientific Monthly* 73:356–362.

Sherman, P.W. and M. L. Morton. 1984. Demography of Belding's Ground Squirrels. *Ecology* 65:1617–1628.

Stephens P. A., W. J. Sutherland, and R. P. Freckleton. 1999. What is the allee effect? *Oikos* 87:185–190.

Syvitski, J.P.M. and A. Kettner. 2011. Sediment flux and the Anthropocene. *Philosophical Transactions of the Royal Society A* 369:957–975.

Thornhill, N. W. (ed.). 1993. *The natural history of inbreeding and outbreeding: Theoretical and empirical perspectives.* Chicago: University of Chicago Press.

Tilley, S.G. 1980. Life Histories and Comparative Demography of Two Salamander Populations. *Copeia* 4:806–821.

Tinkle, D.W, J. D. Congdon and P. C. Rosen. 1981. Nesting Frequency and Success: Implications for the Demography of Painted Turtles. *Ecology* 62:1426–1432.

Vinegar, M.B. 1975. Demography of the Striped Plateau Lizard, *Sceloporus virgatus*. *Ecology* 56:172–182.

Chapter 10 Life History

Alcock, J. 1995. *Animal behavior: An evolutionary approach.* 4th ed. Sunderland, MA: Sinauer Associates.

Andersson, M., and Y. Iwasa. 1996. Sexual selection. *Trends in Ecology and Evolution* 11:53–58.

Bajema, C. J. (ed.). 1984. *Evolution by sexual selection theory. Benchmark Papers in Systemic and Evolutionary Biology.* New York: Scientific and Academic Editions.

Baker, M.C., T.K. Bjerke, H. Lampe and Y. Espmark. 1986. Sexual Response of Female Great Tits to Variation in Size of Males' Song Repertoires. *The American Naturalist* 128: 491–498.

Boyce, M. S. 1984. Restitution of *r* and *K* selection as modes of density-dependent natural selection. *Annual Review of Ecology and Systematics* 15:427–447.

Catchpole, C. K. 1987. Bird song, sexual selection, and female choice. *Trends in Ecology and Evolution* 2:94–97.

Clutton-Brock, T. H. 1984. Reproductive effort and terminal investment in iteroparous animals. *American Naturalist* 123(2):212–229.

Cody. M. L. 1966. A general theory of clutch size. *Evolution* 20:174–184.

Darwin, C. 1871. *The descent of man and selection in relation to sex.* London: Murray.

Dijkstra, C., A. Bult, S. Bijlsma, S. Daan, T. Meijer, and M. Zijlstra. 1990. Brood size manipulations in the kestrel *(Falco tinnunculus): effects on offspring and parent survival. Journal of Animal Ecology* 59:269–285.

Eis, S., E. H. Garman, and L. F. Ebel. 1965. Relation between cone production and diameter increment of Douglas fir (*Pseudotsuga*), grand fir (*Aibes grandis*), and western white pine (*Pinus monticola*). *Canadian Journal of Botany* 43:1553–1559.

Grime, J. P. 1977. Evidence for the existence of three primary strategies in plants and its relevance to ecological and evolutionary theory. *American Naturalist* 111:1169–1194.

———. 1979. *Plant strategies and vegetative processes.* New York: Wiley.

Gubernich, D. J., and P. H. Klopher (eds.). 1981. *Parental care in mammals.* New York: Plenum.

Hines, A.H. 1991. Fecundity and reproductive output in nine species of cancer crabs (Crustacea, Brachyura, Cancridae). *Canadian Journal Fisheries and Aquatic Science* 48:2767–275.

Johnsgard, P. A. 1994. *Arena birds: Sexual selection and behavior.* Washington, DC: Smithsonian Institution Press.

Jones, M. B. 1978. Aspects of the biology of the big handed crab, *Heterozius rohendifrons,* from Kaikoura, New Zealand. *New Zealand Journal of Zoology* 5:783–794.

Krebs, J. R., and N. D. Davies. 1981. *An introduction to behavioral ecology.* Oxford: Blackwell Scientific.

Lack, D. 1954. *The natural regulation of animal numbers.* Oxford: Oxford University Press (Clarendon Press).

MacArthur, R. H., and E. O. Wilson. 1967. *The theory of island biogeography.* Princeton, NJ: Princeton University Press.

McGregor, P. K., J. R. Krebs, and C. M. Perrins. 1981. Song repertoires and lifetime reproductive success in the great tit (*Parus major*). *American Naturalist* 118:49–59.

Mennill, D., L. M. Ratcliffe, and P. T. Boag. 2002. Female eavesdropping on male song contests in songbirds. *Science* 296:873.

Oksanen, T. A., E. Koskela and T. Mappes. 2002. Hormonal manipulation of offspring number: maternal effort and reproductive costs. *Evolution* 56:1530–1537.

Petrie, M. 1994. Improved growth and survival of offspring of peacocks with more elaborate trains. *Nature* 371:598–599.

Pianka, E. 1972. *r* and *K* selections or *b* and *d* selection? *American Naturalist* 100:65–75.

Primack, R. B. 1979. Reproductive effort in annual and perennial species of Plantago (*Plantaginaceae*). *American Naturalist* 114:51–62.

Primack, R.B. and P. Hall. 1990. Cost of reproduction in the pink lady slipper orchid: A four-year experimental study. *American Naturalist* 136: 638–656.

Reid, J.M., P. Arcese, A.L.E.V. Cassidy, S.M. Hiebert, J.N.M. Smith, P.K. Stoddard, A.B. Marr and L.F. Keller. 2004. Song repertoire size predicts initial mating success in male song sparrows, *Melospiza melodia. Animal Behaviour* 68:1055–1063.

Reid, J.M., P. Arcese, A.L.E.V. Cassidy, S.M. Hiebert, J.N.M. Smith, P.K. Stoddard, A.B. Marr and L.F. Keller. 2005. Fitness correlates of song repertoire size in free-living song sparrows (*Melospiza melodia*). *The American Naturalist* 165:299–310.

Reznick, D. A., H. Bryga, and J. A. Endler. 1990. Experimentally induced life-history evolution in a natural population. *Nature* 346:357–359.

Small, M. F. 1992. Female choice in mating. *American Scientist* 80:142–151.

Stearns, S. C. 1992. *The Evolution of Life Histories.* Oxford University Press.

Tamburi, N. E. and P. R. Martin. 2009. Reaction norms of size and age at maturity of *Pomacea canaliculata* under a gradient of food deprivation. *Journal of Molluscan Studies* 75:19–26.

Temeles, E.J. and W. J. Kress. 2010. Mate choice and mate competition by a tropical hummingbird at a floral resource. *Proceedings of the Royal Society B* 277:1607–1613.

Thornhill, R., and J. Alcock. 1983. *The evolution of insect mating systems.* Cambridge, MA: Harvard University Press.

Tsikliras, A. C., E. Antonopoulou and K. I. Stergiou. 2007. A phenotypic trade-off between previous growth and present fecundity in round sardinella *Sardinella aurita. Population Ecology* 49:221–227.

Warner, R. R. 1988. Sex change and size-advantage model. *Trends in Ecology and Evolution* 3:133–136.

Wasser, S. K. (ed.). 1983. *Social behavior of female vertebrates.* New York: Academic Press.

Wauters, L., and A. A. Dohondt. 1989. Body weight, longevity, and reproductive success in red squirrels (*Sciurus vulgaris*). *Journal of Animal Ecology* 58:637–651.

Werner, P. A., and W. J. Platt. 1976. Ecological relationships of co-occurring goldenrods (*Solidago:Compositae*). *American Naturalist* 110:959–971.

Whitham, T. G. 1980. The theory of habitat selection: Examined and extended using *Pemphigus aphids. American Naturalist* 115:449–466.

Willson, M. F. 1983. *Plant reproductive ecology.* New York: Wiley.

Wilson, E. O. 1980. *Sociobiology: The abridged edition.* Cambridge, MA: Harvard University Press.

Chapter 11 Intraspecific Population Regulation

Ball, R.A., Purcell, L.C. and Vories, E.D. 2000. Short-Season Soybean Yield Compensation in Response to Population and Water Regime. *Crop Science* 40: 1070–1078.

Berger, J. 1990. Persistence of different-sized populations: An empirical assessment of rapid ex-tinctions in bighorn sheep. *Conservation Biology* 4:91–98.

Berteaux, D., and S. Boutin. 2000. Breeding dispersal in female North American red squirrels. *Ecology* 81:1311–1326.

Both, C. and M.E. Visser. 2000. Breeding territory size affects fitness: an experimental

study on competition at the individual level. *Journal of Animal Ecology* 69:1021–1030.

Brashares, J.S., J. R. Werner and A. R. E. Sinclair. 2010. Social 'meltdown' in the demise of an island endemic: Allee effects and the Vancouver Island marmot. *Journal of Animal Ecology* 79:965–973.

Cahill, J. F. 2000. Investigating the relationship between neighbor root biomass and belowground competition: Field evidence for symmetric competition belowground. *Oikos* 90:311–320.

Chatworthy, J. N. 1960. *Studies on the nature of competition between closely related species.* D. Phil. diss., University of Oxford, Oxford, UK.

Chepko-Sade, B. D., and Z. T. Halpin. 1987. *Mammalian dispersal patterns.* Chicago: University of Chicago Press.

DeLoach, C.J. 1974. Rate of increase of populations of cabbage, green peach, and turnip aphids at constant temperatures. *Annals of the Entomological Society of America* 67:332–340.

Fery, R. L., and J. Janick. 1971. Response of corn (*Zea mays.* L.) to population pressure. *Crop Science* 11:220–224.

Flockhart, D.T.T., Martin, T.G., and Norris, D.R. 2012. Experimental Examination of Intraspecific Density-Dependent Competition during the Breeding Period in Monarch Butterflies (*Danaus plexippus*). *PLoS ONE* 7(9): e45080. doi:10.1371/journal.pone.0045080.

Fowler, C. W. 1981. Density dependence as related to life history strategy. *Ecology* 62:602–610.

Frazier, M.R., Huey, R.B. and D. Berrigan. 2006. Thermodynamic constrains the evolution of insect population growth rates: Warmer is better. *The American Naturalist.* 168: 512–520.

Gaines, M. S., and L. R. McClenaghan Jr. 1980. Dispersal in small mammals. *Annual Review of Ecology and Systematics* 11:163–169.

Greenwood, P. J., and P. H. Harvey. 1982. The natal and breeding dispersal in birds. *Annual Review of Ecology and Systematics* 13:1–21.

Grime, J. P. 1979. *Plant strategies and vegetative processes.* New York: Wiley.

Hackney, E. and J. B. McGraw. 2001. Experimental demonstration of an Allee effect in American ginseng. *Conservation Biology* 15:129–136.

Harestad, A. S., and F. L. Bunnell. 1979. Home range and body weight—a reevaluation. *Ecology* 60:389–402.

Harper, J. L. 1977. *Population biology of plants.* New York: Academic Press.

Hughes, R.N. and C. L. Griffiths. 1988. Self-Thinning in Barnacles and Mussels: The Geometry of Packing. *The American Naturalist* 132: 484–491.

Huston, M. and Smith, T.M. 1987. Plant succession: life history and competition. *American Naturalist* 130:168–198.

Jenkins, T. M., S. Diehl, K. W. Kratz, and S. D. Cooper. 1999. Effects of population density on individual growth of brown trout in streams. *Ecology* 80:941–956.

Jones, W. T. 1989. Dispersal distance and the range of nightly movements of Merriam's kangaroo rats. *Journal of Mammalogy* 20:29–34.

Keeley, E. R. 2001. Demographic responses to food and space competition by juvenile steelhead trout. *Ecology* 82:1247–1259.

King, A. A., and W. C. Schaffer. 2001. The geometry of a population cycles: A mechanistic model of snowshoe demography. *Ecology* 82:814–830.

Krebs, J., and N. B. Davies (eds.). 1991. *Behavioral ecology: An evolutionary approach.* 3rd ed. Oxford: Blackwell Scientific.

Krebs, J. R. 1971. Territory and breeding density in the great tit *Parus major. Ecology* 52:2–22.

Larsen, K. W., and S. Boutin. 1994. Movement, survival, and settlement of red squirrel (*Tamiasciurus hudsonicus*) offspring. *Ecology* 75:214–223.

Lett, P. F., R. K. Mohn, and D. F. Gray. 1981. Density-dependent processes and management strategy for the northwest Atlantic harp seal populations. Pages 135–158 in C. W. Fowler and T. D. Smith, eds., *Dynamics of large mammal populations.* New York: Wiley.

Massey, A., and J. D. Vandenbergh. 1980. Puberty delay by a urinary cue from female house mice in feral populations. *Science* 209:821–822.

McCullough, D. R. 1981. Population dynamics of the Yellowstone grizzly. Pages 173–196 in C. W. Fowler and T. D. Smith, eds., *Dynamics of large mammal populations.* New York: Wiley.

Mech, L. D., R. E. McRoberts, R. O. Peterson, and R. E. Page. 1978. Relationship of deer and moose populations to previous winter's snow. *Journal of Animal Ecology* 56:615–627.

Murdoch, W. W. 1994. Population regulation in theory and practice. *Ecology* 75:271–287.

Myers, K., C. S. Hale, R. Mykytowycz, and R. L. Hughes. 1971. The effects of varying density and space on sociality and health in animals. Pages 148–187 in A. E. Esser, ed., *Behavior and environment: The use of space by animals and men.* New York: Plenum.

Packard, J. M., and L. D. Mech. 1983. Population regulation in wolves. Pages 151–174 in F. L. Bunnell, D. S. Eastman, and J. M. Peak, eds., *Symposium on natural regulation of wildlife populations.* Moscow: University of Idaho Press.

Reed, D. H., J. J. O'Grady, B. W. Brook, J. D. Balloc, and R. Frankham. 2003. Estimates of minimum viable population sizes for vertebrates and factors influencing those estimates. *Biological Conservation* 113:23–34.

Relyea, R.A. 2002. Competitor-induced plasticity in tadpoles: consequences, cues and connections to predator-induced plasticity. *Ecological Monographs* 72:523–540.

Searcy, W. A. 1982. The evolutionary effects of mate selection. *Annual Review of Ecology and Systematics* 13:57–85.

Schonewald-Cox, C. M. 1983. Conclusions: Guidelines to management: A beginning attempt. Pages 414–445 in C. M. Schonewald-Cox, S. M. Chambers, B. MacBryde, and L. Thomas, eds., *Genetics and conservation: A reference for managing wild animal and plant populations.* Menlo Park, CA: Benjamin Cummings.

Schueller, A.M., Hansen, M.J., Newman, S.P. and Edwards, C.J. 2005. Density dependence of walleye maturity and fecundity in Big Crooked Lake, Wisconsin, 1997–2003. *North American Journal of Fisheries Management* 25: 841–847.

Sinclair, A. R. E. 1977. *The African buffalo: A study of resource limitations of populations.* Chicago: University of Chicago Press.

Smith, R. L. 1963. Some ecological notes on the grasshopper sparrow. *Wilson Bulletin* 75:159–165.

Smith, S. M. 1978. The underworld in a territorial adaptive strategy for floaters. *American Naturalist* 112:570–582.

Smith, T.M. and Goodman, P.S. 1986. The role of competition on the sructure and dynamics of *Acacia* (spp.) savannas in Southern Africa. *Journal of Ecology* 74:1031–1044.

Turchin, P. 1999. Population regulation: A synthetic view. *Oikos* 84:160–163.

Wang, L., A.M. Showalter and I.A. Ungar. 2005. Effects of intraspecific competition on growth and photosynthesis of *Atriplex prostrate. Aquatic Botany* 83:187–192.

Watkinson, A. R., and A. J. Davy. 1985. Population biology of salt marsh and dune annuals. *Vegetatio* 62:487–497.

Wolff, J. O. 1997. Population regulation in mammals: An evolutionary perspective. *Journal of Animal Ecology* 66:1–13.

Yoda, K., T. Kira, H. Ogawa, and K. Hozumi. 1963. Self-thinning in overcrowded pure stands under cultivated and natural conditions. *Journal Biology* 14:107–129.

Zimen, E. 1981. *The wolf: A species in danger.* New York: Delacourt Press.

Chapter 12 Species Interactions, Population Dynamics, and Natural Selection

Bascompte, J. 2010. Structure and dynamics of ecological networks. *Science* 329:765–766.

Brodie Jr., E.D., B.J. Ridenhour and E.D. Brodie. 2002. The evolutionary response of predators to dangerous prey: hotspots and coldspots in the geographic mosaic of coevolution between garter snakes and newts. *Evolution* 56:2067–2082.

Després, L. and M. Cherif. 2004. The Role of Competition in Adaptive Radiation: A Field Study on Sequentially Ovipositing Host-Specific Seed Predators. *Journal of Animal Ecology* 73:109–116.

Elton, C. S. (2001). *Animal Ecology*. University of Chicago Press.

Grace, J.B. and R.G. Wetzel. 1981. Habitat partitioning and competitive displacement in cattails (*Typha*): Experimental field studies. *The American Naturalist* 118: 463–474.

Grinnell, J. (1917). The niche-relationships of the California Thrasher. *Auk* 34: 427–433.

Hutchinson, G. E. 1957. Concluding remarks. *Cold Spring Harbor Symposia on Quantitative Biology* 22 (2): 415–427.

Lindberg, A. B., and J. M. Olesen. 2001. The fragility of extreme specialization: *Passiflora mixta* and its pollinating hummingbird *Ensifera ensifera. Journal of Tropical Ecology* 17:323–329.

Nosil, P., and B. J. Crespi. 2006. Experimental evidence that predation promotes divergence in adaptive radiation. *Proceedings of the National Academy of Sciences* 103 (24):9090–9095.

Nosil, P. 2007. Divergent host plant adaptation and reproductive isolation between ecotypes of *Timema cristinae* walking sticks. *American Naturalist* 169:151–162.

Shochat, E., S.B. Lerman, J. Anderies, P.S. Warren, S.H. Faeth and C.H. Nilon. 2010. Invasion, competition and biodiversity loss in urban ecosystems. *BioScience* 60:199–208.

Thompson, J. N. 1988. Variation in interspecific interactions. *Annual Review of Ecology and Systematics* 19:65–87

Travis, J. 1996. The significance of geographic variation in species interactions. *American Naturalist* 148 (Supplement):S1-S8.

Chapter 13 Interspecific Competition

American Naturalist. 1983. Interspecific competition papers. vol. 122, no. 5 (November).

Austin, M. P. 1999. A silent clash of paradigms: Some inconsistencies in community ecology. *Oikos* 86:170–178.

Austin, M. P., R. H. Groves, L. M. F. Fresco, and P. E. Laye. 1986. Relative growth of six thistle species along a nutrient gradient with multispecies competition. *Journal of Ecology* 73:667–684.

Bazzaz, F. A. 1996. *Plants in changing environments.* New York: Cambridge University Press.

Berger, K.M and E.M. Gese. 2007. Does interference competition with wolves limit the distribution and abundance of coyotes? *Journal of Animal Ecology* 76:1075–1085.

Bertness, M.D. 1991. Zonation of *Spartina patens* and *Spartina alterniflora* in New England Salt Marsh. *Ecology* 72:138–148.

Bolnick, D. I., T. Ingram, W. E. Stutz, L. K. Snowberg, O. L. Lau and J. S. Paull. 2010. Ecological release from interspecific competition leads to decoupled changes in population and individual niche width. *Proceedings of the Royal Society* B 277:1789–1797.

Brown, J. C., O. J. Reichman, and D. W. Davidson. 1979. Granivory in desert ecosystems. *Annual Review of Ecology and Systematics* 10:210–227.

Connell, J. H. 1975. Some mechanisms producing structure in natural communities: A model and evidence from field experiments. Pages 460–490 in M. L. Cody and J. Diamond, eds., *Ecology and evolution of communities.* Cambridge, MA: Harvard University Press.

———. 1983. On the prevalence and relative importance of interspecific competition: Evidence from field experiments. *American Naturalist* 122:661–696.

Darwin, C. 1859. *The origin of species.* Philadelphia: McKay. [Reprinted from 6th London edition.]

Dayan, T., D. Simberloff, E. Tchernov, and Y. Yom-Tov. 1990. Feline canines: Community-wide character displacement among the small cats of Israel. *American Naturalist* 136:39–60.

Dye, P. J., and P. T. Spear. 1982. The effects of bush clearing and rainfall variability on grass yield and composition in south-west Zimbabwe. *Zimbabwe Journal of Agricultural Research* 20:103–118.

Emery, N. C., P. J. Ewanchuk, and M. D. Bertness. 2001. Competition and salt marsh plant zonation: Stress tolerators may be dominant competitors. *Ecology* 82:2471–2485.

Fynn, R. W. S., Morris, C. D. and K. P. Kirkman. 2005. Plant strategies and trait trade-offs influence trends in competitive ability along gradients of soil fertility and disturbance. *Journal of Ecology* 93:384–394.

Gause, G. F. 1934. *The struggle for existence.* Baltimore: Williams and Wilkins.

Grace, J. B., and R. G. Wetzel. 1981. Habitat partitioning and competitive displacement in cattails (*Typha*): Experimental field studies. *American Naturalist* 118:463–474.

Grant, P. 1999. *Ecology and evolution of Darwin's finches.* Princeton, NJ: Princeton University Press.

Groves, R. H., and J. D. Williams. 1975. Growth of skeleton weed (*Chondrilla juncea*) as affected by growth of subterranean clover (*Trifolium subterraneum L.*) and infection by *Puccinia chondrilla. Australian Journal of Agricultural Research* 26:975–983.

Gurevitch, J. L. 1986. Competition and the local distribution of the grass *Stipa neomexicana. Ecology* 67:46–57.

Gurevitch, J. L., L. Morrow, A. Wallace, and J. J. Walch. 1992. Meta-analysis of competition in field experiments. *American Naturalist* 140:539–572.

Heller, H. C., and D. Gates. 1971. Altitudinal zonation of chipmunks (*Eutamias*): Energy budgets. *Ecology* 52:424–443.

Lotka, A. J. 1925. *Elements of physical biology.* Baltimore: Williams and Wilkins.

Park, T. 1954. Experimental studies of interspecies competition: 2. Temperature, humidity and competition in two species of *Trilobium. Physiological Zoology* 27:177–238.

Pianka, E. R. 1978. *Evolutionary ecology.* 3rd ed. New York: HarperCollins.

Pickett, S. T. A., and F. A. Bazzaz. 1978. Organization of an assemblage of early successional species on a soil moisture gradient. *Ecology* 59:1248–1255.

Rice, E. L. 1984. *Allelopathy.* 2nd ed. Orlando, FL: Academic Press.

Schoener, T. W. 1983. Field experiments on interspecific competition. *American Naturalist* 122:240–285.

Taniguchi, Y., F.J. Rahel, D.C. Novinger, and K.G. Gerow. 1998. Temperature mediation of competitive interactions among three fish species that replace each other along longitudinal stream gradients. *Canadian Journal of Fisheries and Aquatic Science* 55:1894–1901.

Tilman, D. M., M. Mattson, and S. Langer. 1981. Competition and nutrient kinetics along a temperature gradient: An experimental test of a mechanistic approach to niche theory. *Limnology and Oceanography* 26:1020–1033.

Volterra, V. 1926. Variation and fluctuations of the numbers of individuals in animal species living together. Reprinted on pages 409–448 in R. M. Chapman (1931), *Animal Ecology.* New York: McGraw-Hill.

Werner, E. E., and J. D. Hall. 1976. Niche shifts in sunfishes: Experimental evidence and significance. *Science* 191:404–406.

———. 1979. Foraging efficiency and habitat switching in competing sunfish. *Ecology* 60:256–264.

Wiens, J. A. 1977. On competition and variable environments. *American Scientist* 65:590–597.

Chapter 14 Predation

Agrawal, A. A. 2004. Resistance and susceptibility of milkweed to herbivore attack: Consequences of competition, root herbivory, and plant genetic variation. *Ecology* 85:2118–2133.

Bricklemeyer, E. C., Jr., S. Ludicello, and H. J. Hartmann. 1989. Discarded catch in U.S. commercial marine fisheries. Pages 259–295 in *Audubon wildlife report 1989/1990.* San Diego: Academic Press.

Cooper, S. M., and N. Owen-Smith. 1986. Effects of plant spinescence on large mammalian herbivores. *Oecologica* 68:446–455.

Davies, N. B. 1977. Prey selection and social behavior in wagtails (*Aves monticillidae*). *Journal of Animal Ecology* 46:37–57.

Fox, L. R. 1975. Cannibalism in natural populations. *Annual Review of Ecology and Systematics* 6:87–106.

Frost, B. W. 1972. Effects of size and concentration of food particles on the feeding behavior of the marine planktonic copepod *Calanus pacificus. Limnology and Oceanography* 17:805–815.

Gause, G.F. (1934). *The struggle for existence*. Baltimore, MD: Williams & Wilkins.

Godfray, H. C. J. 1994. *Parasitoids: Behavioral and evolutionary ecology.* Princeton, NJ: Princeton University Press.

Hassell, M. P., and R. M. May. 1974. Aggregation in predator sand insect parasites and its effect on stability. *Journal of Animal Ecology* 43:567–594.

Holling, C. S. 1959. The components of predation as revealed by a study of small mammal predation of the European sawfly. *Canadian Entomologist* 91:293–320.

————. 1966. The functional response of invertebrate predators to prey density. *Memoirs Entomological Society of Canada* 48:1–86.

Huffaker, C.B. 1958. Experimental Studies on Predation: Dispersion Factors and Predator-Prey Oscillations. *Hilgardia* 27:342–383.

Hughes, R. N. and M. I. Croy. 1993. An experimental analysis of frequency-dependent predation (switching) in the 15-spined stickleback, *Spinachia spinachia. Journal of Animal Ecology* 62:341–352.

Irons, D. B., R. G. Anthony, and J. A. Estes. 1986. Foraging strategies of glaucous-winged gulls in a rocky intertidal community. *Ecology* 67:1460–1474.

Jedrzejewski, W., B. Jedrzejewski, and L. Szymura. 1995. Weasel population response, home range, and predation on rodents in a deciduous forest in Poland. *Ecology* 76:179–195.

Keith, L. B., J. R. Cary, O. J. Rongstad, and M. C. Brittingham. 1984. Demography and ecology of a declining snowshoe hare population. *Journal of Wildlife Management* 90:1–43.

Krebs, C. J., R. Boonstra, S. Boutin, and A. R. E. Sinclair. 2001. What drives the 10-year cycle of snowshoe hares? *BioScience* 51:25–35.

Krebs, J. R., and N. B. Davies (eds.). 1984. *Behavioral ecology: An evolutionary approach.* 2nd ed. Oxford: Blackwell Scientific.

Leonard, G. H., M. D. Bertness, and P. O. Yund. 1999. Crab predation, water-borne cues, and inducible defenses in the blue mussel *Mytilus edulis. Ecology* 80:1–14.

Lima, S. L. 1998. Nonlethal effects in the ecology of predator-prey interactions. *BioScience* 48:25–34.

Mazncourt, de C., M. Loreau, and U. Dieckmann. 2001. Can the evolution of plant defense lead to plant-herbivore mutualism? *American Naturalist* 158:109–123.

Mooney, H. A., and Gulmon, S. L. 1982. Constraints on leaf structure and function in reference to herbivory. *BioScience* 32:198–206.

Nelson, E. H., C. E. Matthews, and J. A. Roenheim. 2004. Predators reduce prey population growth by inducing change in behavior. *Ecology* 85:1853–1858.

O'Donoghue, M., S. Boutin, C. J. Krebs, G. Zuleta, D. L. Murray, and E. J. Hofer. 1998. Functional response of coyotes and lynx to the snowshoe hare cycle. *Ecology* 79:1193–1208.

O'Donoghue, M., S. Boutin, C. J. Krebs and E. J. Hofer. 1997. Numerical responses of coyotes and lynx to the snowshoe hare cycle. *Oikos* 80: 150–162.

Polis, G. 1981. The evolution and dynamics of intraspecific predation. *Annual Review of Ecology and Systematics* 12:225–251.

Polis, G., and R. D. Holt. 1992. Intraguild predation: The dynamics of complex trophic interactions. *Trends in Ecology and Evolution* 7:- 151–154.

Relyea, R. A. 2001. The relationship between predation risk and antipredator responses in larval anurans. *Ecology* 82:541–554.

Salamolard, M., Butet, A., Leroux, A. & Bretagnolle, V. 2000. Responses of an avian predator to variations in prey density at a temperate latitude. *Ecology* 81:2428–2441.

Schnell, D. E. 1976. *Carnivorous plants of the United States and Canada.* Winston-Salem, NC: John F. Blair.

Scrimgeour, G.J. and J.M. Culp. 1994. Feeding while evading predators by a *lotic* mayfly: linking short-term foraging behaviours to long-term fitness consequences. *Oecologia* 100:128–134.

Stephens, D. W., and J. W. Krebs. 1987. *Foraging theory.* Princeton, NJ: Princeton University Press.

Taylor, R. J. 1984. *Predation.* New York: Chapman and Hall.

Van Valen, L. 1973. A new evolutionary law. *Evolutionary Theory* 1:1–30.

Wagner, J. D., and D. H. Wise. 1996. Cannibalism regulates densities of young wolf spiders: Evidence from field and laboratory experiments. *Ecology* 77:639–652.

Williams, K., K. G. Smith, and F. M. Stevens. 1993. Emergence of 13-year periodical cicada (*Cicadidae: Magicicada*): Phenology, mortality, and predator satiation. *Ecology* 74:1143–1152.

Chapter 15 Parasitism and Mutualism

Addicott, J. F. 1986. Variations in the costs and benefits of mutualism: The interaction between yucca and yucca moths. *Oeocologica* 70:486–494.

Allan, B.F., F. Keesing and R.S. Ostefeld. 2003. Effect of Forest Fragmentation on Lyme Disease Risk. *Conservation Biology* 17: 267–272. Fig.1, pg. 270

Anderson, R. C. 1963. The incidence, development, and experimental transmission of *Pneumostrongylus tenuis* Dougherty (*Metastrongyloidae: Protostrongyliidae*) of the menges of the white-tailed deer (*Odocoilus virginianus borealis*) in Ontario. *Canadian Journal of Zoology* 41:775–792.

Anderson, R. C., and A. K. Prestwood. 1981. Lungworms. In W. R. Davidson, ed., *Diseases and parasites of white-tailed deer.* Miscellaneous Publication No. 7. Tallahassee, FL: Tall Timbers Research Station.

Bacon, P. J. 1985. *Population dynamics of rabies in wildlife.* New York: Academic Press.

Ball, D. M., J. F. Pedersen, and G. D. Lacefield. 1993. The tall fescue endophyte. *American Scientist* 81:370–379.

Barbour, A. G., and D. Fish. 1993. The biological and social phenomenon of Lyme disease. *Science* 260:1610–1616.

Barrett, J. A. 1983. Plant-fungus symbioses. In D. Futuyma and M. Slakin, eds., *Coevolution.* Sunderland, MA: Sinauer Associates.

Barth, F. G. 1991. *Insects and flowers: The biology of a partnership.* Princeton, NJ: Princeton University Press.

Beattie, A. J., and D. C. Culver. 1981. The guild of myrmecohores in the herbaceous flora and West Virginia forests. *Ecology* 62:107–115.

Bentley, B. L. 1977. Extrafloral nectarines and protection by pugnacious bodyguards. *Annual Review of Ecology and Systematics* 8:407–427.

Blount, J. D., N. B. Metcalfe, T. R. Birkhead, and P. F. Surai. 2003. Carotenoid modulation of immune function and sexual attractiveness in zebra finches. *Science* 300:125–127.

Boucher, D. H. (ed.). 1985. *The biology of mutualism.* Oxford: Oxford University Press.

Boucher, D. H., S. James, and H. D. Keller. 1982. The ecology of mutualism. *Annual Review of Ecology and Systematics* 13:315–347.

Burdon, J. J. 1987. *Diseases and plant population biology.* Cambridge: Cambridge University Press.

Buskirk, J. V., and R. S. Ostfeld. 1995. Controlling Lyme disease by modifying the density and species composition of tick hosts. *Ecological Applications* 5:1133–1140.

Clay, K. 1988. Fungal endophytes of grasses: A defensive mutualism between plants and fungi. *Ecology* 60:10–16.

————. 1990. Fungal endophytes of grasses. *Annual Review of Ecology and Systematics* 21:275–297.

Croll, N. A. 1966. *Ecology of parasites.* Cambridge, MA: Harvard University Press.

Dobson, A. P., and E. R. Carper. 1996. Infectious diseases and human population history. *BioScience* 46:115–125.

Doebler, S. A. 2000. The rise and fall of the honeybee. *BioScience* 50: 738–732.

Edwards, M. A., and U. McDonald. 1982. *Animal diseases in relation to-animal conservation.* New York: Academic Press.

Feinsinger, P. 1983. Coevolution and pollination. Pages 282–310 in D. Futuyma and M. Slatkin, eds., *Coevolution.* Sunderland, MA: Sinauer Associates.

Fenner, F., and F. N. Ratcliffe. 1965. *Myxamatosis.* Cambridge: Cambridge University Press.

Futuyma, D. J., and M. Slatkin (eds.). 1983. *Coevolution.* Sunderland, MA: Sinauer Associates.

Garnett, G. P., and E. C. Holmes. 1996. The ecology of emergent infectious disease. *BioScience* 46:127–135.

Grenfell, B. T. 1988. Gastrointestinal nematode parasites and the stability and productivity of intensive ruminal grazing systems. Philosophical Transactions Royal Society, London. *British Biological Science* 321:541–563.

Hamilton, W. D., and M. Zuk. 1982. Heritable true fitness and bright birds: A role for parasites? *Science* 218:384–387.

Hanzawa, F. M., A. J. Beattie, and D. C. Culvar. 1988. Directed dispersal: Demographic analysis of an ant-seed mutualism. *American Naturalist* 131:1–13.

Harley, J. L., and S. E. Smith. *The biology of mycorrhizae.* London: Academic Press.

Heithaus, E. R. 1981. Seed predation by rodents on three ant-dispersed plants. *Ecology* 63:136–145.

Hocking, B. 1975. Ant-plant mutualism: Evolution and energy. Pages 78–90 in L. E. Gilbert and P. H. Raven, eds., *Coevolution of animals and plants.* Austin: University of Texas Press.

Holmes, J. 1983. Evolutionary relationships between parasitic helminths and their hosts. In D. J. Futuyma and M. Slakin, eds., *Coevolution.* Sunderland, MA: Sinauer Associates.

Hougen-Eitzman, D., and M. D. Rausher. 1994. Interactions between herbivorous insects and plant-insect coevolution. *American Naturalist* 143:677–697.

Howe, H. F. 1986. Seed dispersal by fruit-eating birds and mammals. Pages 123–190 in D. R. Murra, ed., *Seed dispersal.* Sydney: Academic Press.

Howe, H. F., and L. C. Westley. 1988. *Ecological relationships of plants and animals.* New York: Oxford University Press.

Hutchins, H. E. 1994. Role of various animals in the dispersal and establishment of white-barked pine in the Rocky Mountains, USA. Pages 163–171 in W. C. Schmidt and F-K. Holtmeier, comps., *Proceedings—International workshop on subalpine stone pines and their environment: The status of our knowledge*; 1992 September 5–11; St. Moritz, Switzerland. *General Technical Report INT-GTR-309.* Ogden, UT: USDA Forest Service, Intermountain Research Station.

Iowa, K., and M. D. Rausher. 1997. Evolution of plant resistance to multiple herbivores: Quantifying diffuse evolution. *American Naturalist* 149:316–335.

Janzen, D. H. 1966. Coevolution of mutualism between ants and acacias in Central America. *Evolution* 20:249–275.

Korstian, C. F., and P. W. Stickel. 1927. The natural replacement of blight-killed chestnut. *USDA Miscellaneous Circular 100.* Washington, DC: U.S. Government Printing Office.

Lafferty, K. D., and A. K. Morris. 1996. Altered behavior of parasitized killifish increases susceptibility to predation by bird final hosts. *Ecology* 77:1390–1397.

Lindstrom, E. R., et al. 1994. Disease reveals the predator: Sarcoptic mange, red fox predation, and prey populations. *Ecology* 74:1041–1049.

Margulis, L. 1982. *Early life.* Boston: Science Books International.

————. 1998. *Symbiotic planet.* New York: Basic Books.

Maser, C., J. M. Trappe, and R. A. Nussbaum. 1978. Fungal–small mammal interrelationship with emphasis on Oregon coniferous forests. *Ecology* 59:799–809.

Mattes, H. 1994. Coevolutional aspects of stone pines and nutcrackers. Pages 31–35 in W. C. Schmidt and F-K. Holtmeier, comps., *Proceedings–International workshop on subalpine stone pines and their environment: The status of our knowledge*; 1992 September 5–11; St. Moritz, Switzerland. *General Technical Report INT-GTR-309.* Ogden, UT: USDA Forest Service, Intermountain Research Station.

May, R. M. 1983. Parasitic infections as regulators of animal populations. *American Scientist* 71:36–45.

May, R. M., and R. M. Anderson. 1979. Population biology of infective diseases: Part II. *Nature* 280:455–461.

McNeill, W. H. 1976. *Plagues and peoples.* New York: Doubleday.

Moore, J. 1995. The behavior of parasitized animals. *BioScience* 45:89–96.

Muscatine, L., and J. W. Porter. 1977. Reef corals: Mutualistic symbioses adapted to nutrient-poor environments. *BioScience* 27:454–460.

Ostfeld, R. S. 1997. The ecology of Lyme disease risk. *American Scientist* 85:338–346.

Randolph, S. E. 1975. Patterns of the distribution of the tick *Ixodes trianguliceps birula* on its host. *Journal of Animal Ecology* 44:451–474.

Real, L. A. (ed.). 1983. *Pollination biology.* Orlando, FL: Academic Press.

————. 1996. Sustainability and the ecology of infectious disease. *BioScience* 46:88–97.

Salamolard, M., A. Butet, A. Leroux and V. Bretagnolle. 2000. Responses of an avian predator to variations in prey density at a temperate latitude. *Ecology* 81:2428–2441.

Schall, J. J. 1983. Lizard malaria: Cost to vertebrate host's reproductive success. *Parasitology* 87:1–6.

Seitz, R. D., R. N. Lipcuis, A. H. Hines and D. B. Eggleston. 2001. Density-dependent predation, habitat variation, and the persistence of marine bivalve prey. *Ecology* 82:2435–2451.

Sheldon, B. C., and S. Verhulst. 1996. Ecological immunity: Costly parasite defenses and trade-offs in evolutionary ecology. *Trends in Ecology and Evolution* 11:317–321.

Spielman, A., M. I. Wilson, J. F. Levine, and J. Piesman. 1985. Ecology of *Ixodes dammini*–borne human babesiosis and Lyme disease. *Annual Review of Entomology* 30:439–460.

Tomback, D. F., and Y. B. Linhart. 1990. The evolution of bird-dispersed pines. *Evolutionary Ecology* 4:185–219.

Vittor , A.Y., W. Pan, R. H. Gilman , J. Tielsch , G. Glass , T. Shields , W. Sánchez-Lozano, V. V. Pinedo , E. Salas-Cobos , S. Flores , and J. A. Patz. 2009. Linking Deforestation to Malaria in the Amazon: Characterization of the Breeding Habitat of the Principal Malaria Vector, *Anopheles darlingi. American Journal of Tropical Medicine and Hygene* 81:5–12.

Wakelin, D. 1997. Parasites and the immune system. *BioScience* 47:32–40.

Warner, R. E. 1968. The role of introduced diseases in the extinction of endemic Hawaiian avifauna. *Condor* 70:101–120.

Wilson, E. O. 1975. *Sociobiology: The new synthesis.* Cambridge, MA: Harvard University Press.

Woodard, T. N., R. J. Gutierrez, and W. H. Rutherford. 1974. Bighorn lamb production, survival and mortality in south-central Colorado. *Journal of Wildlife Management* 28:381–391.

Zuk, M. 1991. Parasites and bright birds: New data and new predictions. Pages 317–327 in J. E. Loge and M. Zuk, eds., *Bird–parasite interactions.* Oxford: Oxford University Press.

Chapter 16 Community Structure

Anderson, S., and H. H. Shugart. 1974. Avian community analyses of Walker Branch Watershed. *ORNL Publication No.* 623. Oak Ridge, TN: Environmental Sciences Division, Oak Ridge National Laboratory.

Braun, E. L. 1950. *Deciduous forests of eastern North America.* New York: McGraw-Hill.

Clements, F. E. 1916. Plant succession: An analysis of the development of vegetation. *Carnegie Inst. Wash. Publ. 242.*

Dunne, J.A., R.J. Williams, and N.D. Martinez. 2004. Network structure and robustness of marine food webs. *Marine Ecology Progress Series* 273:291–302.

Dyer, J.M. 2006. Revisiting the Deciduous Forests of Eastern North America. *BioScience* 56: 341–352.

Edgell, R.A. and C.C. Long. 2009. *Wabash River: Carroll, Cass, Huntington, Miami, Tippecanoe, and Wabash Counties 2008 Fish Management Report.* Fisheries Section, Indiana Department of Natural Resources, Division of Fish and Wildlife. Indianapolis, IN.

Estes, J., M. Tinker, T. Williams, and D. Doak. 1998. Killer whale predation on sea otters linking oceanic and nearshore ecosystems. *Science* 282:473–476.

Gleason, H. A. 1926. The individualistic concept of the plant association. *Bulletin Torrey Botanical Club* 53:1–20.

Magurran, A.E. 2004. *Measuring biological diversity.* Blackwell Scientific, Oxford.

Magurran, A.E. 2004. *Ecological diversity and its measurement.* Princeton University Press, Princeton, NJ.

Morin, P. J. 1999. *Community ecology.* Oxford: Blackwell Scientific.

Parker, D.M. 2008. *The effects of elephants at low densities and after short occupation time on ecosystems of the Eastern Cape Province, South Africa.* Ph.D.. Dissertation. Rhodes University, South Africa.

Phillip, D.A.T. 1998. *Biodiversity of freshwater fishes in Trinidad and Tobago.* Ph.D. dissertation, University of St. Andrews, St. Andrews, UK. pg. 99.

Pimm, S. L. 1982. *Food webs.* New York: Chapman & Hall.

Power, M. E., D. Tilman, J. Estes, B. Menge, W. Bond, L. Mills, G. Daily, J. Castilla, J. Lubchenco, and R. Paine. 1996. Challenges in the quest for keystones. *BioScience* 46:609–620.

Rezende, E.L., E.M. Albert, M. A. Fortuna, and J. Bascompte. 2009. Compartments in a marine food web associated with phylogeny, body mass, and habitat structure. *Ecology Letters* 12: 779–788.

Riginos, C. and J.B. Grace. 2008. Savanna tree density, herbivores and herbaceous community: bottom-up vs top-down effects. *Ecology* 89: 2228–2238.

Riginos, C., J.B. Grace, D.J. Augustine, and T.P. Young. 2009. Local versus landscape-scale effects of savanna trees on grasses. *Journal of Ecology* 97:1337–1345.

Schukat, A., M. Bode, H. Auel, R. Carballo, B. Martin, R. Koppelmann, and W. Hagen. 2013. Pelagic decapods in the northern Benguela upwelling system: Distribution, ecophysiology and contribution to active carbon flux. *Deep-Sea Research I* 75:146–156.

Shannon, C. E., and W. Wiener. 1963. *The mathematical theory of communications.* Urbana: University of Illinois Press.

Simpson, E. H. 1949. Measurement of diversity. *Nature* 163:688.

Whittaker, R. H. 1956. Vegetation of the Great Smoky Mountains. *Ecological Monographs* 26:1–80.

Whittaker, R. 1960. Vegetation of the Siskiyou Mountains, Oregon and California. *Ecological Monographs* 30: 279–338.

Chapter 17 Factors Influencing the Structure of Communities

Baril, L.M. et al. 2011. Songbird response to increased willow (Salix spp.) growth in Yellowstone's northern range. Ecological Applications 21:2283–2296

Baril, L. M. 2009. *Change in deciduous woody vegetation, implications of increased willow (Salix spp.) growth for bird species diversity, and willow species composition in and around Yellowstone National Park's northern range.* Thesis. Montana State University, Bozeman, Montana, USA.

Bazzaz, F. A., and J. L. Harper. 1976. Relationship between plant weight and numbers in mixed populations of *Sinapis arvensis* (L.) *Rabenh.* and *Lepidium sativum* L. *Journal of Applied Ecology* 13:211–216.

Berger, J., P. B. Stacey, L. Bellis, and M. P. Johnson. A mammalian predator-prey imbalance: Grizzly bear and wolf extinction affect avian neotropical migrants. *Ecological Applications* 11:947–960.

Cahill, J. F. 1999. Fertilization effects on interactions between above- and belowground competition in an old field. *Ecology* 80:466–480.

Cohen, J. E. 1989. Food webs and community structure. Pages 181–202 in J. Roughgarden, R. May, and S. Levin, eds., *Perspectives in ecological theory.* Princeton, NJ: Princeton University Press.

Connell, J. H. 1961. The influence of interspecific competition and other factors on the distribution of the barnacle *Chthamalus stellatus. Ecology* 42:710–723.

Dodd, M., J. Silvertown, K. McConway, J. Potts, and M. Crawley. (1994) Biomass stability in the plant communities of the Park Grass Experiment: The influence of species richness, soil pH and biomass. *Philosophical Transactions of the Royal Society* B.346:185–193.

Dodson, S. I. 1970. Complementary feeding niches sustained by size-selective predation. *Limnology and Oceanography* 15:131–137.

Ehrlich, P. R., and A. Ehrlich. 1981. *Extinction.* New York: Random House.

Emery, N. C., P. J. Ewanchuk, and M. D. Bertness. 2001. Competition and salt marsh plant zonation: Stress tolerators may be dominant competitors. *Ecology* 82:2471–2485.

Fowler, N. L. 1981. Competition and coexistence in a North Carolina grassland II: The effects of the experimental removal of species. *Journal of Ecology* 69:843–854.

Gotelli, N. J. 2001. Research frontiers in null model analysis. *Global Ecology and Biogeography* 10:337–343.

Gotelli, N. J., and G. R. Graves. 1996. *Null models in ecology.* Herndon, VA: Smithsonian Institution.

Hairston, N. G., F. E. Smith, and L. B. Slobodkin. 1960. Community structure, population control, and competition. *American Naturalist* 94:421–425.

Hillebrand, H. et al. 2007. Consumer versus resource control of producer diversity depends on ecosystem type and producer community structure. *Proceedings of the National Academy of Sciences* 104: 10904–10909.

Holt, R. D. 1977. Predation, apparent competition and the structure of prey communities. *Theoretical Population Biology* 12:197–229.

Huston, M. 1979. A general hypothesis of species diversity. *American Naturalist* 113:81–101.

Huston, M. A. 1980. Soil nutrients and tree species richness in Costa Rican forests. *Journal of Biogeography* 7:147–157.

Krebs, C. 2001. *Ecology.* 5th ed. San Francisco: Benjamin Cummings.

Krebs, C. J., S. Boutin, and R. Boonstra (eds.). 2001. *Vertebrate community dynamics in the Kluane Boreal Forest.* Oxford: Oxford University Press.

Krebs, C. J., S. Boutin, R. Boonstra, A. R. E. Sinclair, J. N. M. Smith, M. R. T. Dale, K. Martin, and R. Turkington. 1995. Impact of food and predation on the snowshoe hare cycle. *Science* 269:1112–1115.

MacArthur, R. H., and J. W. MacArthur. 1961. On bird species diversity. *Ecology* 42:594–598.

Muller, C.B. and H. C. J. Godfray. 1997. Apparent competition between two aphid species. *Journal of Animal Ecology* 66:57–64.

Paine, R. T. 1966. Food web complexity and species diversity. *American Naturalist* 100:65–75.

————. 1969. The Pisaster-Tegula interaction: Prey patches, predator food preference and intertidal community structure. *Ecology* 50:950–961.

Pimm, S. L. 1982. *Food webs.* New York: Chapman & Hall.

Power, M. E., M. J. Matthews, and A. J. Stewart. 1985. Grazing minnows, piscivorous bass, and stream algae: Dynamics of a strong interaction. *Ecology* 66:1448–1456.

Reed, C. 1993. *Reconstruction of Pollinator Communities on Restored Prairies in Eastern Minnesota.* Final Report to the Minnesota Department of Natural Resources Nongame Wildlife Program. Minnesota Department of Natural Resources.

Ricklefs, R. E., and D. Schluter (eds.). 1993. *Ecological communities: Historical and geographical perspectives.* Chicago: University of Chicago Press.

Ripple, W.J. and R.L. Beschta. 2012. Trophic cascades in Yellowstone: The first 15 years after wolf reintroduction. *Biological Conservation* 145:205–213.

Ripple, W.J. and R.L. Beschta. 2003. Wolf reintroduction, predation risk, and cottonwood recovery in Yellowstone National Park. *Forest Ecology and Management* 184:299–313.

Ripple, W.J. et al. 2013. Ripple, W.J., Beschta, R.L., Fortin, J.K. and C.T. Robbins. Trophic cascades from wolves to grizzly bears in Yellowstone *Journal of Animal Ecology* doi: 10.1111/1365–2656.12123.

Robertson, G. P., M. A. Huston, F. C. Evans, and J. M. Tiedje. 1988. Spatial variability in a successional plant community: Patterns of nitrogen availability. *Ecology* 69:1517–1524.

Silvertown, J., M. Dodd, K. McConway, J. Potts, and M. Crawley. 1994. Rainfall, biomass variation, and community composition in the Park Grass Experiment. *Ecology* 75:2430–2437.

Smith, T. M., and M. Huston. 1987. A theory of spatial and temporal dynamics of plant communities. *Vegetatio* 3:49–69.

Smith, T. M., F. I. Woodward, and H. H. Shugart (eds.). 1996. *Plant functional types: Their relevance to ecosystem properties and global change.* Cambridge: Cambridge University Press.

Tokeshi, M. 1993. Species abundance patterns and community structure. *Advances in Ecological Research* 24: 111–181.

Walker, B. H. 1992. Biodiversity and ecological redundancy. *Conservation Biology* 6:18–23.

Whittaker, R. H. 1960. Vegetation of the Siskiyou Mountains, Oregon and California. *Ecological Monographs* 23:41–78.

————. 1975. *Communities and ecosystems.* New York: Macmillan.

Chapter 18 Community Dynamics

Austin, M. P., and T. M. Smith. 1989. A new model of the continuum concept. *Vegetatio* 83:35–47.

Barberaena-Arias, M. and T.M. Aide 2003. Species diversity and trophic composition of litter insects during plant secondary succession. *Caribbean Journal of Science* 29:161–169.

Bazzaz, F. A. 1979. The physiological ecology of plant succession. *Annual Review of Ecology and Systematics* 10:351–371.

Billings, W. D. 1938. The structure and development of oldfield shortleaf pine stands and certain associated physical properties of the soil. *Ecological Monographs* 8:437–499.

Bormann, F., and G. E. Likens. 1979. *Patterns and process in forested ecosystems.* New York: Springer-Verlag.

Clements, F. E. 1916. Plant succession: An analysis of the development of vegetation. *Carnegie Inst. Wash. Publ. 242.*

————. 1936. Nature and structure of the climax. *Journal of Ecology* 24:252–284.

Connell, J. H. 1978. Diversity in tropical rain forests and coral reefs. *Science* 199:1302–1310.

Connell, J. H., and R. O. Slatyer. 1977. Mechanisms of succession in natural communities and their role in community stability and organization. *American Naturalist* 111:1119–1144.

Cowles, H. C. 1899. The ecological relations of the vegetation on the sand dunes of Lake Michigan. *Botanical Gazette* 27:95–117, 167–202, 281–308, 361–391.

Crocker, R. L., and J. Major. 1955. Soil development in relation to vegetation and surface age at Glacier Bay, Alaska. *Journal of Ecology* 43:427–488.

Davis, M. B. 1983. Holocene vegetational history of the eastern United States. Pages 166–188 in H. E. Wright Jr., ed., *Late quaternary environments of the United States. Vol. II: The Holocene.* Minneapolis: University of Minnesota Press.

Delcourt, P. A., and H. R. Delcourt. 1981. Vegetation maps for eastern North America, 40,000 yr bp to present. Pages 123–166 in R. Romans, ed., *Geobotany.* New York: Plenum Press.

Drummond, M.A. and T.R. Loveland. 2010. Land-use pressure and a transition to forest-cover loss in the Eastern United States. *BioScience* 60:286–298.

Duggins, D. O. 1980. Kelp beds and sea otters: An experimental approach. *Ecology* 61:447–453.

Egler, F. 1954. Vegetation science concepts. I. Initial floristic composition, a factor in old field vegetation development. *Vegetatio* 14:412–417.

Flather, C.H., T.H. Ricketts, C.H. Sieg, M.S. Knowles, J.P. Fay, and J. McNees. 2003. Criterion 1: Conservation of Biological Diversity. Indicator 6: The Number of Forest-Dependent Species. In Darr, D. (comp.), *Technical Document Supporting the 2003 National Report on Sustainable Forests.* Washington, DC: USDA Forest.

Golley, F. (ed.). 1978. *Ecological succession. Benchmark Papers.* Stroudsburg, PA: Dowden, Hutchinson, and Ross.

Grime, J. P. 1977. Evidence for the existence of three primary strategies in plants and its rele-vance to ecological and evolutionary theory. *American Naturalist* 111:1169–1194.

————. 1979. *Plant strategies and vegetative processes.* New York: Wiley.

Hobbie, E. A. 1994. *Nitrogen cycling during succession in Glacier Bay, Alaska.* Masters thesis. University of Virginia, Charlottesville, VA.

Huntley, N. and R. Inouye. 1987. Small mammal populations of an old-field chronosequence: Successional patterns and associations with vegetation. *Journal of Mammology* 68:739–745.

Huston, M. A. 1979. A general hypothesis of species diversity. *American Naturalist* 113:81–101.

————. 1994. *Biological diversity: The coexistence of species on changing landscapes.* New York: Cambridge University Press.

Huston, M., and T. M. Smith. 1987. Plant succession: Life history and competition. *American Naturalist* 130:168–198.

Johnston, D.W. and E.P Odum. 1956. Breeding bird populations in relation to plant succession on the Piedmont of Georgia. *Ecology* 37:50–62.

Lichter, J. 1998. Primary succession and forest development on coastal Lake Michigan sand dunes. *Ecological Monographs* 68:487–510.

Maser, C., and J. M. Trappe (eds.). 1984. The seen and unseen world of the fallen tree. *USDA Forest Service General Technical Report PNW-164.*

Olson, J. S. 1958. Rates of succession and soil changes in southern Lake Michigan sand dunes. *Botanical Gazette* 119:125–170.

Oosting, H. J. 1942. An ecological analysis of the plant communities of Piedmont, North Carolina. *American Midland Naturalist* 28:1–26.

Pastor, J., R. J. Naiman, B. Dewey, and P. McInnes. 1988. Moose, microbes, and boreal forest. *BioScience* 38:770–777.

Reiners, W.A. 1992. Twenty Years of Ecosystem Reorganization Following Experimental Deforestation and Regrowth Suppression. *Ecological Monographs* 62:503–523.

Smith, R. L. 1959. Conifer plantations as wildlife habitat. *New York Fish and Game Journal* 5:101–132.

Sousa, W. P. 1979. Disturbance in marine intertidal boulder fields: The non-equilibrium maintenance of species diversity. *Ecology* 60:1225–1239.

————. 1984. Intertidal mosaics: Patch size, propagule availability, and spatially variable patterns of succession. *Ecology* 65:1918–1935.

Sprugle, D. G. 1976. Dynamic structure of wave-generated *Abies balsamea* forests in northeastern United States. *Journal of Ecology* 64:889–911.

Tilman, D. 1988. *Plant Strategies and the Dynamics and Structure of Plant Communities.* Monographs in Population Biology 26. Princeton University Press, Princeton, NJ. 360 pp.

Warming, J. E. B. 1909. *Oecology of plants.* Oxford: Oxford University Press.

Watts, M. T. 1976. *Reading the American landscape.* New York: Macmillan.

West, D. C., H. H. Shugart, and D. B. Botkin (eds.). 1981. *Forest succession: Concepts and application.* New York: Springer-Verlag.

Whittaker, R. H. 1956. Vegetation of the Great Smoky Mountains. *Ecological Monographs* 26:1–80.

Whittaker, R. H., and G. M. Woodwell. 1968. Dimension and production relations of trees and shrubs in the Brookhaven forest, New York. *Ecology* 56:1–25.

————. 1969. Structure and function, production, and diversity of the oak-pine forest at Brookhaven, New York. *Journal of Ecology* 57: 155–174.

Williams, M. 1989. *Americans and their forests: A historical geography.* Oxford: Oxford University Press.

Williams, S.L. 1990. Experimental studies of Caribbean seagrass bed development. *Ecological Monographs* 60:449–469.

Zieman, J. C. 1982. The ecology of seagrasses in south Florida: A community profile. *U.S. Fish and Wildlife Service FWS/OBS-82/25.*

Chapter 19 Landscape Ecology

Bormann, F. H., and G. E. Likens. 1979. *Pattern and process in a forested ecosystem.* New York: Springer-Verlag.

Brown, J. H., and A. Kodrich-Brown. 1977. Turnover rates in insular biogeography: Effect of immigration on extinction. *Ecology* 58:445–449.

Chepko-Sade, B. D., and Z. T. Halpin (eds.). 1987. *Mammalian dispersal patterns: The effect of social structure on population genetics.* Chicago: University of Chicago Press.

Delcourt, H. R. 1987. The impact of prehistoric agriculture and land occupation on natural vegetation. *Trends in Ecology and Evolution* 2:39–44.

Echeverría, C., A.C. Newton, A. Lara, J.M.R. Benayas, and D.A. Coomes. 2007. Impacts of forest fragmentation on species composition and forest structure in the temperate landscape of southern Chile. *Global Ecology and Biogeography* 16:426–439.

Fahrig, L., and G. Merriam. 1994. Conservation of fragmented populations. *Conservation Biology* 8:50–59.

Forman, R. T. T. 1995. *Land mosaics: The ecology of landscapes and regions.* New York: Cambridge University Press.

Forman, R. T. T., and M. Gordon. 1981. Patches and structural components for a landscape ecology. *BioScience* 31:733–740.

Freemark, K. E., and H. G. Merriam. 1986. Importance of area and habitat heterogeneity to bird assemblages in temperate forest fragments. *Biological Conservation* 36:115–141.

Galli, A. E., C. F. Leck, and R. T. T. Forman. 1976. Avian distribution patterns in forest islands of different sizes in central New Jersey. *Auk* 93:356–364.

Gray, J. T., and W. H. Schlesinger. 1981. Nutrient cycling in Mediterranean-type ecosystems. In P. C. Miller, ed., *Resource use by chaparral and matorral.* New York: Springer-Verlag.

Guthery, F. S., and R. L. Bingham. 1992. On Leopold's principle of edge. *Wildlife Society Bulletin* 20:340–344.

Hanski, I. 1990. Density dependence, regulation and variability in animal populations. Philo-sophical Transactions of the Royal Society London 330:141–150.

————. 1991. Single-species metapopulation dynamics. Biological Journal of the Linnean Society 42:17–38.

————. 1994. Patch occupancy dynamics in fragmented landscapes. Trends in Ecology and Evolution 9:131–135.

————. 1999. Metapopulation ecology. Oxford: Oxford University Press.

Hanski, I., and M. Gilpin. 1991. Metapopulation dynamics: Brief history and conceptual domain. Biological Journal of the Linnean Society 42:3–16. Harris, L. D. 1984. *The fragmented forest.* Chicago: University of Chicago Press.

————. 1988. Edge effects and the conservation of biological diversity. *Conservation Biology* 2:212–215.

Hanski, I. A., and M. E. Gilpin (eds.). 1997. *Metapopulation biology: Ecology, genetics, and evolution.* Academic Press, San Diego, CA.

Herkert, J.R. 1994.The effects of habitat fragmentation on midwestern grassland bird communities. *Ecological Applications* 4:461–471.
Honnay, O., K. Verheyen, J. Butaye, H. Jacquemyn, B. Bossuyt and M. Hermy. 2002. Possible effects of habitat fragmentation and climate change on the range of forest plant species. *Ecology Letters* 5:525–530.
Jacquemyn, H., J. Butaye and M. 2001. Forest plant species richness in small, fragmented mixed deciduous forest patches: the role of area, time and dispersal limitation. *Biogeography* 28: 801–812.
Johnson, H. B. 1976. *Order upon the land.* Oxford: Oxford University Press.
Kerbes, R. H., P. M. Kotanen, and R. L. Jeffries. 1990. Destruction of wetland habitats by lesser snow geese: A keystone species on the west coast of Hudson Bay. *Journal of Applied Ecology* 27:242–258.
Kindvall, O. 1996. Habitat heterogeneity and survival in a bush cricket metapopulation. *Ecology* 77:207–214.
Kindvall, O., and I. Ahlen. 1992. Geometrical factors and metapopulation dynamics of the bush cricket, *Metrioptera bicolor. Conservation Biology* 6:520–529.
Kroodsma, R.L. 1982. Edge effect on breeding forest birds along a power-line corridor. *Journal of Applied Ecology* 19:361–370.
Leopold, A. 1933. *Game management.* New York: Scribner.
Levenson, J. B. 1981. Woodlots as biogeographic islands in southeastern Wisconsin. Pages 13–39 in R. L. Burgess and D. M. Sharpe, eds., *Forest island dynamics in man-dominated landscapes.* New York: Springer-Verlag.
Levins, R. 1969. Some demographic and genetic consequences of environmental heterogeneity for biological control. *Bulletin of the Entomological Society of America* 15: 237–240.
Levin S.A.1974. Dispersion and Population Interactions. *The American Naturalist* 108: 207–228.
MacArthur, R. H., and E. O. Wilson. 1967. *The theory of island biogeography.* Princeton, NJ: Princeton University Press.
Marquis, D. A. 1974. The impact of deer browsing on Allegheny hardwood regeneration. *USDA Forest Service Research Paper NE-308.*
————. 1981. Effect of deer browsing on timber production in Allegheny hardwood forests of northwestern Pennsylvania. *USDA Forest Service Research Paper NE-475.*
Mooney, H. A., et al. (eds.). 1981. Proceedings: Symposia on fire regimes and ecosystem properties. *General Technical Report WO-26.* Washington, DC: USDA Forest Service.
Naiman, R. J., J. M. Melillo, and J. E. Hobbie. 1986. Ecosystem alteration of boreal forest streams by beaver (*Castor canadensis*). *Ecology* 67: 1254–1269.
Preston, F. W. 1962. The canonical distribution of commonness and rarity. *Ecology* 43:185–215, 410–432
Ranney, J. W. 1977. Forest island edges—their structure, development, and importance to regional forest ecosystem dynamics. *EDFB/IBP Cont. No. 77/1*, Oak Ridge National Laboratory, Oak Ridge, TN.
Ranney, J. W., M. C. Brunner, and J. B. Levenson. 1981. The importance of edge in the structure and dynamics of forest lands. Pages 67–91 in R. L. Burgess and D. M. Sharpe, eds., *Forest island dynamics in man-dominated landscapes* (Ecological Studies No. 41). New York: Springer-Verlag.
Robbins, C. S., D. K. Dawson, and B. A. Dowell. 1989. Habitat area requirements of breeding forest birds of the Middle Atlantic States. *Wildlife Monographs* 103.
Romme, W. H., and D. H. Knight. 1982. Landscape diversity: The concept applied to Yellowstone Park. *BioScience* 32:664–670.
Rosenberg, D. K., B. R. Noon, and E. C. Meslow. 1997. Biological corridors: Forms, function and efficiency. *BioScience* 47:677–687.
Runkle, J. R. 1985. Disturbance regimes in temperate forests. Pages 17–34 in S. T. A. Pickett and P. S. White, eds., *Ecology of natural disturbance and patch dynamics.* Orlando, FL: Academic Press.
Schonewald-Cox, C. M., S. M. Chambers, B. MacBryde, and W. L. Thomas (eds.). 1983. *Genetics and conservation.* Menlo Park, CA: Benjamin Cummings.
Schwartz, M. W. 1997. *Conservation in highly fragmented landscapes.* New York: Chapman & Hall.
Simberloff, D. S. 1974. Equilibrium theory of island biogeography and ecology. *Annual Review of Ecology and Systematics* 5:161–182.
Simberloff, D. S., and E. O Wilson. 1969. Experimental zoogeography of islands. The colonization of empty islands. *Ecology* 50:278–296.
Sisk, T.D., Haddad, N.M., Ehrlich, P.R. 1997. Bird assemblages in patchy woodlands: modeling the effects of edge and matrix habitats. *Ecological Applications* 7:1170–1180.
Temple, S. A. 1986. Predicting impacts of habitat fragmentation on forest birds: A comparison of two models. Pages 301–304 in J. Verner, M. L. Morrison, and C. T. Ralph, eds., *Wildlife 2000: Modeling habitat relations of terrestrial vertebrates.* Madison: University of Wisconsin Press.
Thomas, C., M. Singer, and D. Boughton. 1996. Catastrophic extinction of population sources in a butterfly metapopulation. *American Naturalist* 148:957–975.
Thomas, C., and T. Jones. 1993. Partial recovery of a skipper butterfly (Hesperia comma) from population refuges: Lessons for conservation in a fragmented landscape. *Journal of Animal Ecology* 62:472–481.
Thomas, J. 1983. The ecology and conservation of *Lysandra bellargus* in Britain. *Journal of Applied Ecology* 20:59–83.
Thomas, J. W., R. G. Anderson, C. Master, and E. L. Bull. 1979. Snags. Pages 60–77 in J. W. Thomas, ed., *Wildlife habitats in managed forests (the Blue Mountains of Oregon and Washington), USDA Forest Service Agricultural Handbook No. 553.*
Turner, M. 1989. Landscape ecology: The effects of pattern and process. *Annual Review of Ecology and Systematics* 20:171–197.
————. 1998. Landscape ecology. In S. I. Dodson, T. F. H. Allen, S. R. Carpenter, A. R. Ives, R. L. Jeanne, J. F. Kitchell, N. E. Langston, and M. G. Turner, *Ecology.* Oxford: Oxford University Press.
Urban, D., R. V. O'Neill, and H. H. Shugart. 1987. Landscape ecology. *BioScience* 37:119–127.
Walker, B. 1989. Diversity and stability in ecosystem conservation. Pages 125–130 in D. Western and M. C. Pearl, eds., *Conservation for the twenty-first century.* New York: Oxford.
Watt, A. S. 1947. Pattern and process in the plant community. *Journal of Ecology* 35:1–22.
Webb, N.R. and Haskins, L.E. 1980. An ecological survey of heathland in the Poole Basin Dorset, England in 1978. *Biological Conservation* 17: 281–296.
Whitcomb, R. E., J. F. Lynch, M. K. Klimkiewicz, C. S. Robbins, B. L. Whitcomb, and D. Bystrak. 1981. Effects of forest fragmentation on avifauna of the eastern deciduous forest. Pages 125–205 in R. L. Burgess and D. M. Sharpe, eds., *Forest island dynamics in man-dominated landscapes.* New York: Springer-Verlag.
Winter, M., D.H. Johnson, and J. Faaborg. 2000. Evidence for edge effects on multiple levels in tallgrass prairie. *Condor* 102: 256–266.

Chapter 20 Ecosystem Energetics

Aber, J. D., and J. M. Melillo. 1991. *Terrestrial ecosystems.* Philadelphia: Saunders College Publishing.
Anderson, J. M., and A. MacFadyen (eds.). 1976. *The role of terrestrial and aquatic organisms in the decomposition process. Seventeenth Symposium, British Ecological Society.* Oxford: Blackwell Scientific.
Andrews, R. D., D. C. Coleman, J. E. Ellis, and J. S. Singh. 1975. Energy flow relationships in a shortgrass prairie ecosystem. Pages 22–28 in *Proceedings 1st International Congress of Ecology.* The Hague: W. Junk.
Begon, M., J. L. Harper, and C. R. Townsend. 1996. *Ecology: Individuals, populations, and communities.* New York: Blackwell Scientific.
Brylinsky, M. 1980. Estimating the productivity of lakes and reservoirs. Pages 411–418 in E. D. Le Cren and R. H. Lowe-McConnell, eds., *The functioning of freshwater ecosystems. International Biological Programme No. 22.* Cambridge: Cambridge University Press.
Brylinsky, M., and K. H. Mann. 1973. An analysis of factors governing productivity in lakes and reservoirs. *Limnology and Oceanography* 18:1–14.
Coe, M. J., D. H. Cummings, and J. Phillipson. 1976. Biomass and production of large African herbivores in relation to rainfall and primary production. *Oecologica* 22:341–354.
Cooper, J. P. 1975. *Photosynthesis and productivity in different environments.* New York: Cambridge University Press.
Cowan, R. L. 1962. Physiology of nutrition of deer. Pages 1–8 in *Proceedings 1st National White-tailed Deer Disease Symposium.* Maurice F. Baker, ed.. University of Georgia Continuing Education. Athens, GA.

Cry, H., and M. L. Pace. 1993. Magnitude and patterns of herbivory in aquatic and terrestrial ecosystems. *Nature* 361:148–150.

DeAngelis, D. L., R. H. Gardner, and H. H. Shugart. 1980. Productivity of forest ecosystems studies during the IBP: The woodlands data set. In D. E. Reichle, ed., *Dynamic properties of forest ecosystems. International Biological Programme 23.* Cambridge: Cambridge University Press.

Dillon, P. J., and F. H. Rigler. 1974. The phosphorus–chlorophyll relationship in lakes. *Limnology and Oceanography* 19:767–773.

Downing, J. A., C. W. Osenberg, and O. Sarnelle. 1999. Meta-analysis of marine nutrient-enrichment experiments: Systematic variation in the magnitude of nutrient limitation. *Ecology* 80:1157–1167.

Elton, C. 1927. *Animal ecology.* London: Sidgwick & Jackson.

Fensholt, R., I. Sandholt, and M.S Rasmussen. 2004. Evaluation of MODIS LAI, fAPAR and the relation between fAPAR and NDVI in a semi-arid environment using in situ measurements. *Remote Sensing of Environment* 91: 490–507.

Gates, D. M. 1985. *Energy and ecology.* Sunderland, MA: Sinauer Associates.

Goetz, S. J. and S. D. Prince (1996). Remote sensing of net primary production in boreal forest stands. *Agricultural and Forest Meteorology* 78(3):149–179.

Golley, F. B., and H. Leith. 1972. Basis of organic production in the tropics. Pages 1–26 in P. M. Golley and F. H. Golley, eds., *Tropical ecology with an emphasis on organic production.* Athens: University of Georgia Press.

Gough, C. M. (2012) *Terrestrial Primary Production: Fuel for Life.* Nature Education Knowledge 3(10):28.

Gower, S. T., R. E. McMurtie, and D. Murty. 1996. Aboveground net primary productivity declines with stand age: Potential causes. *Trends in Ecology and Evolution* 11:378–383.

Haberl, H., K.H. Erb , F. Krausmann , V. Gaube , A. Bondeau, C. Plutzar, S. Gingrich, W. Lucht, and M. Fischer-Kowalski. 2007. Quantifying and mapping the human appropriation of net primary production in earth's terrestrial ecosystems. *Proceedings of the National Academy of Sciences* 104:12942–12947.

Hairston, N. J., Jr., and N. G. Hairston Sr. 1993. Cause-effect relationships in energy flow, trophic structure, and interspecific interactions. *American Naturalist* 143:379–411.

Imhoff, M., L. Bounoua, T. Ricketts, C. Loucks, R. Harriss, and W. T. Lawrence. 2004. Global patterns in human consumption of net primary productivity. *Nature* 439:370–373.

Krausmanna, F., K. Erba, S. Gingricha, H. Haberla, A. Bondeaub, V. Gaubea, C. Lauka, C. Plutzara, and T.D. Searchingerd. 2013. Global human appropriation of net primary production doubled in the 20th century. *Proceedings of the National Academy of Sciences* 110: 10324–10329.

Leith, H. 1973. Primary production: Terrestrial ecosystems. *Human Ecology* 1:303–332.

————. 1975. Primary productivity in ecosystems: Comparative analysis of global patterns. Pages 67–88 in W. H. van Dobben and R. H. Lowe-McConnell, eds., *Unifying concepts in ecology.* The Hague: W. Junk.

Leith, H., and R. H. Whittaker (eds.). 1975. Primary productivity in the biosphere. *Ecological Studies.* vol. 14. New York: Springer-Verlag.

MacArthur, R. H., and J. H. Connell. 1966. *The biology of populations.* New York: Wiley.

McNaughton, S. J., M. Oesterheid, D. A. Frank, and K. J. Williams. 1989. Ecosystem-level patterns of primary productivity and herbivory in terrestrial habitats. *Nature* 341:142–144.

National Academy of Science. 1975. *Productivity of world ecosystems.* Washington, DC: National Academy of Science.

Odum, E. P. 1983. *Basic Ecology.* Philadelphia: Saunders College Publishing.

Pastor, J., J. D. Aber, C. A. McClaugherty, and J. M. Melillo. 1984. Aboveground production and N and P cycling along a nitrogen mineralization gradient on Blackhawk Island, Wisconsin. *Ecology* 65:256–268.

Pauly, D., and V. Christensen. 1995. Primary production required to sustain global fisheries. *Nature* 374:255–257.

Petrusewicz, K. (ed.). 1967. *Secondary productivity of terrestrial ecosystems.* Warsaw, Poland: Pantsworve Wydawnictwo Naukowe.

Phillipson, J. J. 1966. *Ecological energetics.* New York: St. Martin's Press.

Reich, P. B., D. A. Peterson, K. Wrage, and D. Wedin. 2001. Fire and vegetation effects on productivity and nitrogen cycling across a forest-grassland continuum. *Ecology* 82:1703–1719.

Reichle, D. E. 1970. *Analysis of temperate forest ecosystems.* New York: Springer-Verlag.

————. (ed.). 1981. *Dynamic properties of forest ecosystems.* Cambridge: Cambridge University Press.

Rosemond, A.D., P.J. Mulholland, and S.H. Brawley. 2000. Seasonally shifting limitation of stream periphyton: response of algal populations and assemblage biomass and productivity to variation in light, nutrients, and herbivores .*Canadian Journal of Fisheries and Aquatic Science* 57: 66–75.

Ryan, M. G., D. Binkley, and J. H. Fownes. 1997. Age-related decline in forest productivity: Pattern and process. *Advances in Ecological Research* 27:213–262.

Schindler, D. W. 1977. Evolution of phosphorus limitation in lakes. *Science* 195:260–262.

————. 1978. Factors regulating phytoplankton production and standing crop in the world's freshwaters. *Limnology and Oceanography* 23: 478–486.

Transeau, E. N. 1926. The accumulation of energy by plants. *Ohio Journal of Science* 26:1–10.

Vitousek, P.M., P.R. Ehrlich, A.H. Ehrlich, P.A. Matson. 1986. Human appropriation of the products of photosynthesis. *BioScience* 36:368–373.

Whittaker, R. H., and G. E. Likens. 1973. Carbon in the biota. In G. M. Woodwell and E. V. Pecan, eds., *Carbon and the biosphere conference 72501.* Springfield, VA: National Technical Information Service.

Wiegert, R. C. (ed.). 1976. *Ecological energetics. Benchmark Papers.* Stroudsburg, PA: Dowden, Hutchinson, and Ross. Collection of papers surveying the development of the concept.

Chapter 21 Decomposition and Nutrient Cycling

Aber, J., and J. Melillo. 1982. Nitrogen immobilization in decaying hardwood leaf litter as a function of initial nitrogen and lignin content. *Canadian Journal of Botany* 60:2263–2269.

————. 1991. *Terrestrial ecosystems.* Philadelphia: Saunders College Publishing.

Anderson, J. M., and A. MacFadyen (eds.). 1976. The role of terrestrial and aquatic organisms in the decomposition process. *Seventeenth Symposium, British Ecological Society.* Oxford: Blackwell Scientific.

Austin, A. T., and P. M. Vitousek. 2000. Precipitation, decomposition and litter decomposability of *Metrosideros polymorpha* in native forests in Hawaii. *Journal of Ecology* 88:129–138.

Beare, M. H., R. W. Parmelee, P. F. Hendrix, W. Cheng, D. C. Coleman, and D. A. Crossley. 1992. Microbial and fungal interactions and effects on litter nitrogen and decomposition in agroecosystems. *Ecological Monographs* 62:569–591.

Chapin, F. S. 1990. The mineral nutrition of wild plants. *Annual Review of Ecology and Systematics* 11:233–260.

Correll, D. L. 1978. Estuarine productivity. *BioScience* 28:646–650.

Ewel, J., C. Berish, B. Brown, N. Price, and J. Raich. 1981. Slash and burn impacts on a Costa Rican wet forest site. *Ecology* 62:816–829.

Fenchel, T. 1988. Marine plankton food chains. *Annual Review of Ecology and Systematics* 19:19–38.

Fenchel, T., and T. H. Blackburn. 1979. *Bacteria and mineral cycling.* London: Academic Press.

Fenchel, T., and P. Harrison. 1976. The significance of bacterial grazing and mineral cycling for the decomposition of particulate detritus. Pages 285–321 in J. M. Anderson and A. MacFayden, eds., *The role of terrestrial and aquatic organisms in the decomposition process.* New York: Blackwell Scientific.

Fletcher, M., G. R. Gray, and J. G. Jones. 1987. *Ecology of microbial communities.* New York: Cambridge University Press.

Gliessman, S. R. (ed.) 1990. *Agroecology: Researching the ecological basis for sustainable agriculture.* Ecological Studies Series No. 78. New York: Springer-Verlag.

Klap, V., P. Louchouarn, J. J. Boon, M. A. Hemminga, and J. van Soelen. 1999. Decomposition dynamics of six salt marsh halophytes as determined by cupric oxide oxidation and direct temperature-resolved mass spectrometry. *Limnology and Oceanography* 44:1458–1476.

Likens, G. E., and F. H. Bormann. 1974. Linkages between terrestrial and aquatic ecosystems. *BioScience* 24(8):447–456.

Lovelock, J. 1979. *GAIA: A new look at life on Earth.* Oxford: Oxford University Press.

Meentemeyer, V. 1978. Macroclimate and lignin control of decomposition. *Ecology* 59:465–472.

Melillo, J., J. Aber, and J. Muratore. 1982. Nitrogen and lignin control of hardwood leaf litter decomposition dynamics. *Ecology* 63:621–626.

Mudrick, D., M. Hoosein, R. Hicks, and E. Townsend. 1994. Decomposition of leaf litter in an Appalachian forest: Effects of leaf species, aspect, slope position and time. *Forest Ecology and Management* 68:231–250.

Newbold, J. D., R. V. O'Neill, J. W. Elwood, and W. Van Winkle. 1982. Nutrient spiraling in streams: Implications for nutrient and invertebrate activity. *American Naturalist* 20:628–652.

Nybakken, J. W. 1988. *Marine biology: An ecological approach.* 2nd ed. New York: Harper & Row.

Odum, W. E., and M. A. Haywood. 1978. Decomposition of intertidal freshwater marsh plants. Pages 89–97 in R. E. Good, D. F. Whigham, and R. L. Simpson, eds., *Freshwater wetlands.* New York: Academic Press.

Pastor, J., J. D. Aber, C. A. McClaugherty, and J. M. Melillo. 1984. Aboveground production and N and P cycling along a nitrogen mineralization gradient on Blackhawk Island, Wisconsin. *Ecology* 65:256–268.

Pesek, J. (ed.). 1989. *Alternative agriculture.* Washington, DC: National Academy Press.

Pimentel, D. 1984. Energy flow in agroecosystems. Pages 121–132 in R. Lawrence, B. Stinner, and G. J. House, eds., *Agricultural ecosystems: Unifying concepts.* New York: John Wiley & Sons.

Smil, V. 2011. Nitrogen cycle and world food production. *World Agriculture* 2:9–1.

Smil, V. 1999. Nitrogen in crop production: An account of global flows. *Global Biogeochemical Cycles* 13:647–662.

Smith, T. M. Spatial variation in leaf-litter production and decomposition within a temperate forest: The influence of species composition and diversity. *Journal of Ecology.* In review.

Staff, H., and B. Berg. 1982. Accumulation and release of plant nutrients in decomposing Scots pine needle litter. Long-term decomposition in a Scots pine forest. II. *Canadian Journal of Botany* 60:1516–1568.

Swift, M. J., O. W. Heal, and J. M. Anderson. 1979. *Decomposition in terrestrial ecosystems.* Berkeley: University of California Press.

Valiela, I. 1984. *Marine ecological processes.* New York: Springer-Verlag.

Vitousek, P. M., S. W. Andariese, P. A. Matson, L. Morris, and R. L. Sanford. 1992. Effects of harvest intensity, site preparation, and herbicide use on soil nitrogen -transformations in a young loblolly pine plantation. *Forest Ecology and Management* 49:277–292.

Webster, J. R. 1975. *Analysis of potassium and calcium dynamics in stream ecosystems on three Appalachian watersheds of contrasting vegetation.* Ph.D. diss. University of Georgia.

Webster, J. R., D. J. D'angelo, and G. T. Peters. 1991. Nitrate and phosphate uptake in streams at Coweeta Hydrological Laboratory. *Internationale Verein Limnologie* 24: 1681–1686.

Whitkamp, M., and M. L. Frank. 1969. Evolution of carbon dioxide from litter, humus and subsoil of a pine stand. *Pedobiologia* 9:358–365.

Witherspoon, J. P. 1964. Cycling of cesium-134 in white oak trees. *Ecological Monographs* 34:403–420.

Chapter 22 Biogeochemical Cycles

Aber, J. D., and A. H. Magill. 2004. Chronic nitrogen additions at the Harvard Forest: The first fifteen years of a nitrogen saturation experiment. *Forest Ecology and Management* 196:1–6.

Aber, J. D., K. N. Nadelhoffer, P. Steudler, and J. M. Melillo. 1989. Nitrogen saturation in northern forest ecosystems. *BioScience* 39:378–386.

Alexander, M. (ed.). 1980. *Biological nitrogen fixation.* New York: Plenum.

Binkley, D., C. T. Driscoll, H. L. Allen, P. Schoeneberger, and D. McAvoy. 1989. *Acidic deposition and forest soils: Context and case studies in southeastern United States.* New York: Springer-Verlag.

Bolin, B., E. T. Degens, S. Kempe, and P. Ketner (eds.). 1979. *The global carbon cycle. SCOPE 13.* New York: John Wiley.

Bormann, F. H., and G. E. Likens. 1979. *Pattern and process in a forested ecosystem.* New York: Springer-Verlag.

Dise, N.B. and R.F. Wright. 1995. Nitrogen leaching from European forests in relation to nitrogen deposition. *Forest Ecology and Management* 71: 153–161.

McNulty, S.G., J. D. Aber , and S. D. Newman. 1996. Nitrogen saturation in a high elevation New England spruce-fir stand. *Forest Ecology and Management* 84:109–121.

Pomeroy, L. R. 1974. Cycles of essential elements. *Benchmark Papers in Ecology.* Stroudsburg, PA: Dowden, Hutchinson, and Ross.

Post, W. M., T. H. Peng, W. R. Emanuel, A. W. King, V. H. Dale, and D. L. DeAngelis. 1990. The global carbon cycle. *American Scientist* 78:310–326.

Schlesinger, W. H. 1997. *Biogeochemistry: An analysis of global change.* 2nd ed. London: Academic Press.

Schulze, E. D., O. L. Lange, and O. Oren. 1989. *Forest decline and air pollution: A study of spruce* (Picea abies) *on acid soils.* New York: Springer-Verlag.

Smith, W. H. 1990. *Air pollution and forests: Interaction between air contaminants and forest ecosystems.* 2nd ed. New York: Springer-Verlag.

Sprent, J. I. 1988. *The ecology of the nitrogen cycle.* New York: Cambridge University Press.

Wellburn, A. 1988. *Air pollution and acid rain: The biological impact.* Essex, England: Longman Scientific and Technical; New York: Wiley.

Chapter 23 Terrestrial Ecosystems

Archibold, O. W. 1995. *Ecology of world vegetation.* London: Chapman and Hall.

Bailey, R. G. 1996. *Ecosystem geography.* New York: Springer.

Likens, G.E and F.H. Bormann. 1995. *Biogeochemistry of a Forested Ecosystem.* Springer-Verlag, N.Y. 162 pp.

Orians, G. H., J. Buckley, W. Clark, M. Gilpin, J. Lehman, R. May, G. Robilliard, and D. Simberloff. 1986. *Ecological knowledge and environmental problem-solving.* Washington, DC: National Academy Press.

Tropical Forests

Ashton, P. S. 1988. Dipterocarp biology as a window to understanding of tropical forest structure. *Annual Review of Ecology and Systematics* 19:347–370.

Bawa, K. S. 1990. Plant-pollinator interactions in tropical rain forests. *Annual Review of Ecology and Systematics* 21:399–422.

Collins, M. (ed.). 1990. *The last rain forests: A world conservation atlas.* New York: Oxford University Press.

Denslow, J. S. 1987. Tropical rainforest gaps and tree species diversity. *Annual Review of Ecology and Systematics* 18:431–451.

Erwin, T. L. 1988. The tropical forest canopy: The heart of biotic diversity. In E. O. Wilson and F. M. Peters, eds., *Biodiversity.* Washington, DC: National Academy Press.

Fleming, T. H., R. Breitwisch, and G. H. Whitesides. 1987. Patterns of tropical vertebrate frugivore diversity. *Annual Review of Ecology and Systematics* 18:71–90.

Golley, F. B. (ed.). 1983. *Tropical forest ecosystems: Structure and function. Ecosystems of the World No. 14A.* Amsterdam: Elsevier Scientific.

Jordan, C. F., and J. R. Kline. 1972. Mineral cycling: Some basic concepts and their application in a tropical rain forest. *Annual Review of Ecology and Systematics* 3:33–50.

Lathwell, D. J., and T. L. Grove. 1986. Soil-plant relations in the tropics. *Annual Review of Ecology and Systematics* 17:1–16.

Leigh, E. G., Jr. 1975. Structure and climate in tropical rain forest. *Annual Review of Ecology and Systematics* 6:67–86.

Murphy, P. G., and A. E. Lugo. 1986. Ecology of tropical dry forests. *Annual Review of Ecology and Systematics* 17:67–88.

Richards, P. W. 1996. *The tropical rain forest: An ecological study.* 2nd ed. New York: Cambridge University Press.

Simpson, B. B., and J. Haffer. 1978. Spatial patterns in the Amazonian rain forest. *Annual Review of Ecology and Systematics* 9:497–518.

Sutton, S. L., T. C. Whitmore, and A. C. Chadwick (eds.). 1983. *Tropical rain forests: Ecology and management.* Oxford: Blackwell Scientific.

Tomlinson, P. B. 1987. Architecture of tropical plants. *Annual Review of Ecology and Systematics* 18:1–21.

Walter, H. 1971. *Ecology of tropical and subtropical vegetation.* Edinburgh: Oliver & Boyd.

Whitmore, T. C. 1984. *Tropical rainforests of the Far East.* Oxford: Clarendon Press.

————. 1990. *An introduction to tropical rain forests.* New York: Oxford University Press.

Temperate Forests

Bormann, F. H., and G. E. Likens. 1979. *Pattern and process in a forest ecosystem.* New York: Springer-Verlag.

Brinson, M. M., B. L. Swift, R. C. Plantico, and J. S. Barclay. 1981. *Riparian ecosystems: Their ecology and status. FWS/OBS-81/17.* Kearneysville, WV: Eastern Energy and Land Use Team, U.S. Fish and Wildlife Service.

Curtis, J. T. 1959. *The vegetation of Wisconsin.* Madison: University of Wisconsin Press.

Davis, M. B. (ed.). 1996. *Eastern old-growth forests: Prospects for rediscovery and recovery.* Covelo, CA: Island Press.

Duvigneaud, P. (ed.). 1971. *Productivity of forest ecosystems.* Paris: UNESCO.

Lang, G. E., W. A. Reiners, and L. H. Pike. 1980. Structure and biomass dynamics of epiphytic lichen communities of balsam fir forests in New Hampshire. *Ecology* 63:541–550.

Peterken, G. F. 1996. *Natural woodlands: Ecology and conservation in north temperate regions.* New York: Cambridge University Press.

Polunin, O., and M. Walters. 1985. *A guide to the vegetation of Britain and Europe.* Oxford: Oxford University Press.

Reichle, D. E. (ed.). 1981. *Dynamic properties of forest ecosystems.* Cambridge: Cambridge University Press.

Conifer Forest

Bonan, G. B., and H. H. Shugart. 1989. Environmental factors and ecological processes in boreal forests. *Annual Review of Ecology and Systematics* 20:1–18.

Edmonds, R. L. (ed.). 1981. *Analysis of coniferous forest ecosystems in the western United States.* Stroudsburg, PA: Dowden, Hutchinson & Ross.

Franklin, J. F., and C. T. Dyrness. 1973. Natural vegetation of Oregon and Washington. *USDA Forest Service Gen. Tech. Rept. PNW8.* Corvallis, OR: USDA Forest Service.

Knystautas, A. 1987. *The natural history of the USSR.* New York: McGraw-Hill.

Larsen, J. A. 1980. *The boreal ecosystem.* New York: Academic Press.

————. 1989. *The northern forest border in Canada and Alaska. Ecological Studies 70.* New York: Springer-Verlag.

Polunin, O., and M. Walters. 1985. *A guide to the vegetation of Britain and Europe.* Oxford: Oxford University Press.

Woodland/Savanna

Bourliere, F. (ed.). 1983. *Tropical savannas: Ecosystems of the world 13.* Amsterdam: Elsevier Scientific.

Plumb, T. R. 1979. Proceedings of the symposium on the ecology, management, and utilization of California oaks. *General Technical Report PSW-44.* Berkeley, CA: Pacific Southwest Forest and Range Experiment Station, Forest Service, U.S. Dept. of Agriculture.

Sarimiento, G. 1984. *Ecology of neotropical savannas.* Cambridge, MA: Harvard University Press.

Sinclair, A. R. E., and P. Arcese (eds.). 1995. *Serengeti II: Dynamics, management, and conservation of an ecosystem.* Chicago: University of Chicago Press.

Sinclair, A. R. E., and M. Norton-Griffiths (eds.). 1979. *Serengeti: Dynamics of an ecosystem.*

Tothill, J. C., and J. J. Mott (eds.). 1985. *Ecology and management of the world's ecosystems.* Canberra: Australian Academy of Science.

Grasslands

Breymeyer, A., and G. Van Dyne (eds.). 1980. *Grasslands, systems analysis and man.* Cambridge: Cambridge University Press.

Callenback, E. 1996. *Bring back the buffalo: A sustainable future for America's Great Plains.* Covelo, CA: Island Press.

Coupland, R. T. (ed). 1979. *Grassland ecosystems of the world: Analysis of grasslands and their uses.* Cambridge: Cambridge University Press.

Duffy, E. 1974. *Grassland ecology and wildlife management.* London: Chapman and Hall.

French, N. (ed.). 1979. *Perspectives on grassland ecology.* New York: Springer-Verlag.

Hodgson, J., and A. W. Illius (eds.). 1966. *Ecology and management of grazing systems.* New York: CAB International.

Levin, S. A. (ed.). 1993. Forum: Grazing theory and rangeland management. *Ecological Applications* 3:1–38.

Manning, D. 1995. *Grassland: History, biology, politics, and promise of the American prairie.* New York: Penguin Books.

Reichman, O. J. 1987. *Konza prairie: Tallgrass natural history.* Lawrence: University Press of Kansas.

Risser, P. G., et al. 1981. *The true prairie ecosystem.* Stroudsburg, PA: Dowden, Hutchinson & Ross.

Weaver, J. E. 1954. *North American prairie.* Lincoln, NE: Johnson.

Shrublands

Castri, F. Di, D. W. Goodall, and R. L. Specht (eds.). 1981. *Mediterranean-type shrublands. Ecosystems of the World No. 11.* Amsterdam: Elsevier Scientific.

Castri, F. Di., and H. A. Mooney (eds.). 1973. *Mediterranean-type ecosystems. Ecological Studies No. 7.* New York: Springer-Verlag.

Groves, R. H. 1981. *Australian vegetation.* Cambridge: Cambridge University Press.

McKell, C. M. (ed.). 1983. *The biology and utilization of shrubs.* Orlando, FL: Academic Press.

Polunin, O., and M. Walters. 1985. *A guide to the vegetation of Britain and Europe.* Oxford: Oxford University Press.

Specht, R. L. (ed.). 1979, 1981. *Heathlands and related shrublands. Ecosystems of the World 9A and 9B.* Amsterdam: Elsevier Scientific.

Desert

Brown, G. W., Jr. (ed.). 1976–1977. *Desert biology,* 2 vols. New York: Academic Press.

Evenardi, M., I. Noy-Meir, and D. Goodall (eds.). 1985–1986. *Hot deserts and arid shrublands of the world. Ecosystems of the World 12A and 12B.* Amsterdam: Elsevier Scientific.

Wagner, F. H. 1980. *Wildlife of the deserts.* New York: Harry W. Abrams.

Tundra

Bliss, L. C., O. H. Heal, and J. J. Moore (eds.). 1981. *Tundra ecosystems: A comparative analysis.* New York: Cambridge University Press.

Brown, J., P. C. Miller, L. L. Tieszen, and F. L. Bunnell. 1980. *An arctic ecosystem: The coastal tundra at Barrow, Alaska.* Stroudsburg, PA: Dowden, Hutchinson & Ross.

Furley, P. A., and W. A. Newey. 1983. *Geography of the biosphere.* London: Butterworths.

Rosswall, T., and O. W. Heal (eds.). 1975. *Structure and function of tundra ecosystems. Ecological Bulletins 20.* Stockholm: Swedish Natural Sciences Research Council.

Smith, A. P., and T. P. Young. 1987. Tropical alpine plant ecology. *Annual Review of Ecology and Systematics* 18:137–158.

Sonesson, M. (ed.). 1980. *Ecology of a subarctic mire. Ecological Bulletins 30.* Stockholm: Swedish Natural Sciences Research Council.

Wielgolaski, F. E. (ed.). 1975. *Fennoscandian tundra ecosystems. Part I: Plants and microorganisms. Part II: Animals and systems analysis.* New York: Springer-Verlag.

Zwinger, A. H., and B. E. Willard. 1972. *Land above the trees.* New York: Harper & Row.

Chapter 24 Aquatic Ecosystems

Diaz, R.J. and R. Rosenberg. 2008. Spreading dead zones and consequences for marine ecosystems. *Science* 321: 926–929.

Freshwater and Estuarine Ecosystems

Barnes, R. S. K., and K. H. Mann. 1980. *Fundamentals of aquatic ecosystems.* New York: Blackwell Scientific.

Brock, T. D. 1985. *A eutrophic lake: Lake Mendota, Wisconsin.* New York: Springer-Verlag.

Carpenter, S. A. 1980. Enrichment of Lake Wingra, Wisconsin, by submerged macrophyte decay. *Ecology* 61:1145–1155.

Carpenter, S. A., and J. F. Kitchell. 1984. Plankton community structure and limnetic primary production. *American Naturalist* 124:159–172.

Carpenter, S. A., J. F. Kitchell, and J. Hodgson. 1985. Cascading trophic interactions and lake productivity. *BioScience* 35:634–639.

Cummins, K. W. 1974. Structure and function of stream ecosystems. *BioScience* 24:631–641.

————. 1979. Feeding ecology of stream invertebrates. *Annual Review of Ecology and Systematics* 10:147–172.

Gore, J. A., and G. E. Petts. 1989. *Alternatives in regulated river management.* Boca Raton, FL: CRC Press.

Hutchinson, G. E. 1957–1967. *A treatise on limnology. Vol. 1: Geography, physics, and chemistry. Vol. 2: Introduction to lake biology and limnoplankton.* New York: Wiley.

Hynes, H. B. N. 1970. *The ecology of running water.* Toronto: University of Toronto Press.

Ketchum, B. H. (ed.). 1983. *Estuaries and enclosed seas. Ecosystems of the World 26.* Amsterdam: Elsevier Scientific.

Likens, G. E. (ed.). 1985. *An ecosystem approach to aquatic ecology: Mirror Lake and environment.* New York: Springer-Verlag.

Macan, T. T. 1970. *Biological studies of English lakes.* New York: Elsevier Scientific.

————. 1973. *Ponds and lakes.* New York: Crane, Russak.

————. 1974. *Freshwater ecology.* 2nd ed. New York: John Wiley & Sons.

Maitland, P. S. 1978. *Biology of fresh waters.* New York: John Wiley & Sons.

McLusky, D. S. 1989. *The estuarine ecosystem.* 2nd ed. New York: Chapman & Hall.

Meyer, J. L. 1990. A blackwater perspective on riverine ecosystems. *BioScience* 40:643–651.

Petts, G. E. 1984. *Impounded rivers: Perspectives for ecological management.* New York: Wiley.

Stanford, J. A., and A. P. Covich (eds.). 1988. Community structure and function in temperate and tropical streams. *Journal of North American Benthological Society* 7:261–529.

Vannote, R. L., G. W. Minshall, K. W. Cummins, J. R. Sedell, and C. E. Cushing. 1980. The river continuum concept. *Canadian Journal of Fisheries and Aquatic Science* 37:130–137.

Wiley, M. 1976. *Estuarine processes.* New York: Academic Press.

Oceans

Grassle, J. F. 1985. Hydrothermal vent animals: Distribution and biology. *Science* 229:713–717.

————. 1989. Species diversity in deep-sea communities. *Trends in Ecology and Evolution* 4:12–15.

————. 1991. Deep-sea benthic diversity. *BioScience* 41:464–469.

Gross, M. G. 1982. *Oceanography: A view of Earth.* 3rd ed. Englewood Cliffs, NJ: Prentice-Hall.

Hardy, A. 1971. *The open sea: Its natural history.* Boston: Houghton Mifflin.

Hayman, R. M., and R. C. McDonald. 1985. The ecology of deep sea hot springs. *American Scientist* 73:441–449.

Kinne, O. (ed.). 1978. *Marine ecology.* 5 vols. New York: John Wiley and Sons Ltd.

Marshall, N. B. 1980. *Deep-sea biology: Developments and perspectives.* New York: Garland STMP Press.

Nybakken, J. W. 1997. *Marine biology: An ecological approach.* 4th ed. Menlo Park, CA: Benjamin Cummings.

Rex, M. A. 1981. Community structure in the deep-sea benthos. *Annual Review of Ecology and Systematics* 12:331–353.

Steele, J. 1974. *The structure of a marine ecosystem.* Cambridge, MA: Harvard University Press.

Hiatt, R. W., and D. W. Strasburg. 1960. Ecological relationships of the fish fauna on coral reefs of the Marshall Islands. *Ecological Monographs* 30:66–120.

Huston, M. 1985. Patterns of species diversity on coral reefs. *Annual Review of Ecology and Systematics* 16:149–177.

Jackson, J. B. C. 1991. Adaptation and diversity of reef corals. *BioScience* 41:475–482.

Jones, O. A., and R. Endean (eds.). 1973, 1976. *Biology and geology of coral reefs.* Vols. II, III. New York: Academic Press.

Nybakken, J. W. 1997. *Marine biology: An ecological approach.* 4th ed. Menlo Park, CA: Benjamin Cummings.

Pomeroy, L. R., and E. J. Kuenzler. 1969. Phosphorus turnover by coral reef animals. Pages 478–483 in D. J. Nelson and F. E. Evans, eds., *Symposium on radioecology conference.* 670503. Springfield, VA: National Technical Information Services.

Reaka, M. J. (ed.). 1985. *Ecology of coral reefs.* Symposia Series for Undersea Research, NOAA Underseas Research Program 3. Washington DC: U.S. Department of Commerce.

Sale, P. F. 1980. The ecology of fishes on coral reefs. *Annual Review of Oceanography and Marine Biology* 18:367–421.

Valiela, I. 1984. *Marine biology processes.* New York: Springer-Verlag.

Wellington, G. W. 1982. Depth zonation of corals in the Gulf of Panama: Control and facilitation by resident reef fishes. *Ecological Monographs* 52:223–241.

Wilson, R., and J. Q. Wilson. 1985. *Watching fishes: Life and behavior on coral reefs.* New York: Harper & Row.

Chapter 25 Coastal and Wetland Ecosystems

Dahl, T.E. 2011. Status and Trends of Wetlands in the Conterminous United States 2004 to 2009. U.S. Department of the Interior, Fish and Wildlife Service, Washington, D.C. 108 pp. http://www.fws.gov/wetlands/Status-And-Trends-2009/index.html.

Coastal Ecosystems

Bertness, M. D. 1999. *The ecology of Atlantic shorelines.* Sunderland, MA: Sinauer Associates.

Carson, R. 1955. *The edge of the sea.* Boston: Houghton Mifflin.

Eltringham, S. K. 1971. *Life in mud and sand.* New York: Crane, Russak.

Leigh, E. G., Jr. 1987. Wave energy and intertidal productivity. *Proceedings of the National Academy of Sciences* 84:1314.

Lubchenco, J. 1978. Algal zonation in the New England rocky intertidal community: An experimental analysis. *Ecology* 61:333–344.

Mathieson, A. C., and P. H. Nienhuis (eds.). 1991. *Intertidal and littoral ecosystems. Ecosystems of the world 24.* Amsterdam: Elsevier Scientific.

Moore, P. G., and R. Seed (eds.). 1986. *The ecology of rocky shores.* New York: Columbia University Press.

Newell, R. C. 1970. *Biology of intertidal animals.* New York: Elsevier Scientific.

Stephenson, T. A., and A. Stephenson. 1973. *Life between the tidemarks on rocky shores.* San Francisco: Freeman.

Underwood, A. J., E. J. Denley, and M. J. Moran. 1983. Experimental analyses of the structure and dynamics of midshore rocky intertidal communities in New South Wales. *Oecologica* 56:202–219.

Yonge, C. M. 1949. *The seashore.* London: Collins.

Wetlands

Bertness, M. D. 1984. Ribbed mussels and *Spartina alterniflora* production on a New England marsh. *Ecology* 65:1794–1807.

————. 1999. *The ecology of Atlantic shorelines.* Sunderland, MA: Sinauer Associates.

Bildstein, K. L., G. T. Bancroft, P. J. Dugan et al. 1991. Approaches to the conservation of coastal wetlands in the Western Hemisphere. *Wilson Bulletin* 103:218–254.

Chapman, V. J. 1976. *Mangrove vegetation.* Leutershausen, Germany: J. Cramer.

————. (ed.). 1977. *Wet coastal ecosystems.* Amsterdam: Elsevier Scientific.

Clark, J. 1974. *Coastal ecosystems: Ecological considerations for the management of the coastal zone.* Washington, DC: Conservation Foundation.

Cowardin, L. M., V. Carter, and E. C. Golet. 1979. *Classification of wetlands and deepwater habitats of the United States.* Washington, DC: U.S. Department of Interior, Fish and Wildlife Service FWS/OBS-79/31.

Dahl, T. E. 1990. *Wetland losses in the United States, 1780's to 1980's.* Washington, DC: U.S. Department of Interior, Fish and Wildlife Service.

Dugan, P. (ed.). 1993. *Wetlands in danger: A world conservation atlas.* New York: Oxford University Press.

Ewel, K. C. 1990. Multiple demands on wetlands. *BioScience* 40:660–666.
Ewel, K. C., and H. T. Odum (eds.). 1986. *Cypress swamps.* Gainesville: University Presses of Florida.
Gore, A. P. J. (ed.). 1983. *Mire, swamp, bog, fen, and moor. Ecosystems of the world 4A and 4B.* Amsterdam: Elsevier Scientific.
Haines, B. L., and E. L. Dunn. 1985. Coastal marshes. Pages 323–347 in B. F. Chabot and H. A. Mooney, eds., *Physiological ecology of North American plant communities.* New York: Chapman and Hall.
Hopkinson, C. S., and J. P. Schubauer. 1984. Static and dynamic aspects of nitrogen cycling in the salt marsh graminoid *Spartina alterniflora. Ecology* 65:961–969.
Howarth, R. W., and J. Teal. 1979. Sulfate reduction in a New England salt marsh. *Limnology and Oceanography* 24:999–1013.
Jefferies, R. L., and A. J. Davy (eds.). 1979. *Ecological processes in coastal environments.* Oxford: Blackwell Scientific.
Long, S. P., and C. F. Mason. 1983. *Salt marsh ecology.* New York: Chapman & Hall.
Lugo, A. E. 1990. *The forested wetlands.* Amsterdam: Elsevier Scientific.
Lugo, A. E., and S. C. Snedaker. 1974. The ecology of mangroves. *Annual Review of Ecology and Systematics* 5:39–64.
Mitsch, W. J., and J. C. Gosslink. 1993. *Wetlands.* 2nd ed. New York: Van Nostrand Reinhold.
Moore, P. D., and D. J. Bellemany. 1974. *Peatlands.* New York: Springer-Verlag.
Niering, W. A., and B. Hales. 1991. *Wetlands of North America.* Charlottesville, VA: Thomasson-Grant.
Nixon, S. W., and C. A. Oviatt. 1973. Ecology of a New England salt marsh. *Ecological Monographs* 43:463–498.
Nybakken, J. W. 1997. *Marine biology: An ecological approach.* 4th ed. Menlo Park, CA: Benjamin Cummings.
Pomeroy, L. R., and R. G. Wiegert (eds.). 1981. *The ecology of a salt marsh.* New York: Springer-Verlag.
Teal, J. 1962. Energy flow in a salt marsh ecosystem of Georgia. *Ecology* 43:614–624.
Tiner, R. W. 1991. The concept of a hydrophyte for wetland identification. *BioScience* 41:236–247.
Valiela, I., and J. M. Teal. 1979. The nitrogen budget of a salt marsh ecosystem. *Nature* 47:337–371.
Valiela, I., J. M. Teal, and W. G. Denser. 1978. The nature of growth forms in salt marsh grass *Spartina alternifolia. American Naturalist* 112:461–370.
Van der Valk, A. (ed.). 1989. *Northern prairie wetlands.* Ames: Iowa State University Press.
Weller, M. W. 1981. *Freshwater wetlands: Ecology and wildlife management.* Minneapolis: University of Minnesota Press.
Zedler, J., T. Winfield, and D. Mauriello. 1992. *The ecology of Southern California coastal marshes: A community profile.* U.S. Fish and Wildlife Service Office of Biological Services FWS/OBS 81/54.

Chapter 26 Large-Scale Patterns of Biological Diversity

Adams, J.M. and F.I. Woodward. 1989. Patterns in tree species richness as a test of the glacial extinction hypothesis. *Nature* 339: 699–701.
Angel, M. V. 1991. Variations in time and space: Is biogeography relevant to studies of long-time scale change? *Journal of the Marine Biological Association of the United Kingdom* 71:191–206.
Archibold, O. W. 1995. *Ecology of world vegetation.* London: Chapman and Hall.
Astorga, A., M. Fernández, E.E. Boschi and N. Lagos. 2003. Two oceans, two taxa and one mode of development: latitudinal diversity patterns of South American crabs and test for possible causal processes. *Ecology Letters* 6:420–427.
Bailey, R. G. 1996. *Ecosystem geography.* New York: Springer.
Bailey, R. G. 1978. *Description of the ecoregions of the United States.* Ogden, UT: USDA Forest Service Intermountain Region.
Bartlein, P. J., T. Webb, and E. C. Fleri. 1984. Climate response surfaces from pollen data for some eastern North American taxa. *Journal of Biogeography* 13:35–57.
Behrenfeld, M.J., R.O'Malley, D.Siegel, C. McClain, J. Sarmiento, G. Feldman, A. Milligan, P. Falkowski, R. Letelier, E. Boss. 2006. Climate-driven trends in contemporary ocean productivity. *Nature* 444: 752–755.
Blackburn, T.M. and K. J. Gaston. 1996. Spatial Patterns in the Species Richness of Birds in the New World. *Ecography* 19:369–376.
Clements, F. E., and V. E. Shelford. 1939. *Bio-ecology.* New York: McGraw-Hill.
Conservation International. 2004. *Biodiversity Hotspots Revisited.* Conservation Synthesis, Center for Applied Biodiversity Science at Conservation International. http://www.biodiversityhotspots.org/xp/Hotspots/resources/maps.xml
Currie, D. J. 1991. Energy and large-scale biogeographical patterns of animal and plant species richness. *American Naturalist* 137:27–49.
Currie, D. J., and V. Paquin. 1987. Large-scale biogeographical patterns of species richness in trees. *Nature* (London) 329:326–327.
DeCandolle, A. P. A. 1874. *Constitution dans le Regne Vegetal de Groupes Physiologiques Applicables a la Geographie Ancienne et Moderne.* Archives des Sciences Physiques et Naturelles. May, Geneva.
Gentry A.H. 1986. Endemism in tropical versus temperate communities. In: Soule´ M.E. (ed) *Conservation Biology: The Science of Scarcity and Diversity.* Sinauer Associates, Sunderland, Massachusetts, pp. 153–181.
Hawkins, B.A. et al. 2006. Post-Eocene climate change, niche conservatism, and the latitudinal diversity gradient of New World birds. *Journal of Biogeography* 33: 770–780.
Holdridge, L. R. 1947. Determination of wild plant formations from simple climatic data. *Science* 105:367–368.
————. 1967. Determination of world plant formation from simple climatic data. *Science* 130:572.
Hunter, M. L., and P. Yonzon. 1992. Altitudinal distributions of birds, mammals, people, forests and parks in Nepal. *Conservation Biology* 7:420–423.
Hurlbert, A.H. and J.P. Haskell. 2003. The Effect of Energy and Seasonality on Avian Species Richness and Community Composition. *American Naturalist* 161:83–97.
Huston, M. 1994. *Biological diversity: The coexistence of species on changing landscapes.* New York: Cambridge UniversityPress.
Kaufman, D.M. and M.R. Willig. 1998. Latitudinal patterns of mammalian species richness in the New World: the effects of sampling method and faunal group. *Journal of Biogeography* 25:795–805.
Kikkawa, J., and W. T. Williams. 1971. Altitudinal distribution of land birds in New Guinea. *Search* 2:64–69.
Koppen, W. 1918. Klassification der Klimate nach Temperatur, Niederschlag, und Jahres lauf. *Petermann's Mittelungen* 64:193–203, 243–248.
Macpherson, E. and Duarte, M. C. 1994. Patterns in species richness, size, and latitudinal range of East Atlantic fishes. *Ecography* 17: 242–248.
Macpherson, E. 2002. Large-Scale Species-Richness Gradients in the Atlantic Ocean. *Proceedings of the Royal Society of London B.* 269:1715–1720.
Merriam, C. H. 1890. Results of a biological survey of the San Francisco mountain region and the desert of the Little Colorado, Arizona. *North American Fauna* 1–136.
————. 1899. Zone temperatures. *Science* 9:116.
Mittermeier, R. A., Robles-Gil, P., Hoffmann, M., Pilgrim, J. D., Brooks, T. B., Mittermeier, C. G., Lamoreux, J. L. and Fonseca, G. A. B. 2004. *Hotspots Revisited: Earths Biologically Richest and Most Endangered Ecoregions.* CEMEX, Mexico City, Mexico 390pp.
Mutke, J. and Barthlott, W. 2005. Patterns of vascular plant diversity at continental to global scales. *Biologiske Skrifter* 55: 521–531.
Myers, N., R.A. Mittermeier, C.G. Mittermeier, G.A. B. da Fonseca and J. Kent. 2000. Biodiversity hotspots for conservation priorities. *Nature* 403:853–858.
Myers, N. 2003. Biodiversty hotspots revisited. *BioScience* 53:796–797.
Niklas, K. J., B. H. Tiffney, and A. H. Knoll. 1983. Patterns in land plant diversification. *Nature* 303:293–299.
Prentice, I. C., P. J. Bartlein, and T. Webb. 1991. Vegetation and climate change in eastern North America since the last glacial maximum: A response to continuous climate forcing. *Ecology* 72:2038–2056.
Rex, M. A., C. T. Stuart, R. R. Hessler, J. A. Allen, H. A. Sanders, and G. D. F. Wilson. 1993. Global-scale latitudinal patterns of species diversity in the deep-sea benthos. *Nature* 365:636–639.
Roy, K., D. Jablonski, J.W. Valentine, and G. Rosenberg. 1998. Marine latitude diversity gradients: Tests of causal hypotheses. *Proceedings of the National Academy of Science* 95: 3699–3702.

Schimper, A. F. W. 1903. *Plant-geography upon a physiological basis.* Oxford: Clarendon.

Whittaker, R. H. 1975. *Communities and ecosystems.* London: Collier-Macmillan.

————. 1977. Evolution of species diversity in land communities. *Evolutionary Biology* 10:1–67.

Wilson, E. O. 1999. *The diversity of life.* New York: Norton.

Xiao, X. et al. 1997. Linking a Global Terrestrial Biogeochemical Model with a 2-Dimensional Climate Model: Implications for Global Carbon Budget. *Tellus* 49B:18–37.

Chapter 27 The Ecology of Climate Change

Beaubien, E. and A. Hamann. 2011. Spring flowering response to climate change between 1936 and 2006 in Alberta, Canada. *BioScience* 61: 514–524.

Beaugrand, G. et al. 2002. Reorganization of North Atlantic Marine Copepod Biodiversity and Climate. *Science* 296:1692–1694.

Brommer, J. (2004) The range margins of northern birds shift polewards. *Ann. Zool. Fennici* 41: 391–397.

Brusca, R.C. et al. 2013. Dramatic response to climate change in the Southwest: Robert Whittaker's 1963 Arizona Mountain plant transect revisited. *Ecology and Evolution* 3: 3307–3319.

Charmantier, A. et al. 2008. Adaptive phenotypic plasticity in response to climate change in a wild bird population *Science* 320:800–803.

Cotton, P.A. 2003. Avian migration phenology and global climate change. *Proceedings of the National Academy of Sciences* 100: 12219–12222.

Currie D. J. 2001. Projected effects of climate change on patterns of vertebrate and tree species richness in the conterminous United States. *Ecosystems* 4:216–225.

Dai, A. 2013. Increasing drought under global warming in observations and models. *Nature Climate Change* 3:52–58.

Dillon, M.E. et al. 2010. Global metabolic impacts of recent climate warming. *Nature* 467:704–707.

Hansen, J., R. Ruedy, Mki. Sato, and K. Lo, 2010: Global surface temperature change. *Review of Geophysics* 48, RG4004, doi:10.1029/2010RG000345.

Hitch, A. T. and Leberg, P. L. (2007) Breeding distributions of north American bird species moving north as a result of climate change. *Conservation Biology* 21: 534–539.

Hofstede, R. et al. 2010. Regional warming changes fish species richness in the eastern North Atlantic Ocean. *Marine Ecology Progress Series* 414:1–9.

Intergovernmental Panel on Climate Change. 2013. *Working Group I Contribution to the IPCC Fifth Assessment Report Climate Change 2013: The Physical Science Basis Summary for Policymakers.* http://www.ipcc.ch/report/ar5/wg1/

Iverson, L.R., et al. 2008. Estimating potential habitat for 134 eastern US tree species under six climate scenarios. *Forest Ecology and Management* 254:390–406.

Kelly, A.E. and M.L. Goulden. 2008. Rapid shifts in plant distribution with recent climate change. *Proceedings of the National Academy of Sciences* 105: 11823–11826.

La Sorte, F. A. and Thompson, F. R. (2007) Poleward shifts in winter ranges of North American birds. *Ecology* 88: 1803–1812.

Matthews, S.N., et al. 2011. Changes in potential habitat of 147 North American breeding bird species in response to redistribution of trees and climate following predicted climate change. *Ecography* 34: 933–945.

McKenney, D.W., et al. 2011. Revisiting projected shifts in the climate envelopes of North American trees using updated general circulation models. *Global Change Biology* 17:2720–2730.

Mitton, J.B. and S.M. Ferrenberg. 2012. Mountain pine beetle develops an unprecedented summer generation in response to climate warming. *The American Naturalist* 179: E163-E171.

Nemani, R.R., et al. 2003. Climate-Driven Increases in Global Terrestrial Net Primary Production from 1982 to 1999. *Science* 300:1560–1563.

Parmesan, C. 2007. Influences of species, latitudes and methodologies on estimates of phenological response to global warming. *Global Change Biology* 13: 1860–1872.

Parmesan, C. et al. 1999. Poleward shifts in geographical ranges of butterfly species associated with regional warming. *Nature* 399: 579–583.

Peng, C. et al. 2011. A drought-induced pervasive increase in tree mortality across Canada's boreal forests. *Nature Climate Change* 1: 467–471.

Post, E. and M.C. Forchhammer. 2008. Climate change reduces reproductive success of an Arctic herbivore through trophic mismatch. *Philosophical Transactions of the Royal Society B* 363: 2369–2375.

Rustad, L.E. et al. 2001. A meta-analysis of the response of soil respiration, net nitrogen mineralization, and aboveground plant growth to experimental ecosystem warming. *Oecologia* 126:543–562.

Schaphoff, S., et al. 2006. Terrestrial biosphere carbon storage under alternative climate projections. *Climatic Change* 74: 97–122.

Somero, G.N. 2010. The physiology of climate change: how potentials for acclimatization and genetic adaptation will determine 'winners' and 'losers'. *The Journal of Experimental Biology* 213: 912–920.

Thomas, C. D. and Lennon, J. J. (1999) Birds extend their ranges northwards. *Nature* 399: 213.

Van Mantgem, P.J. et al. 2009. Widespread increase of tree mortality rates in the western United States. *Science* 323:521–524.

Winder, M. and D.E. Schindler. 2004. Climate change uncouples trophic interactions in an aquatic ecosystem. *Ecology* 85: 2100–2106.

Zhao, M. and S.W. Running. 2010. Drought-induced reduction in global terrestrial net primary production from 2000 through 2009. *Science* 329:940–943.

GLOSSARY

A horizon. Surface stratum of mineral soil, characterized by maximum accumulation of organic matter, maximum biological activity, and loss of such materials as iron, aluminum oxides, and clays.

Abiotic. Nonliving; the abiotic component of the environment includes soil, water, air, light, nutrients, and the like.

Abundance. The number of individuals of a species in a given area.

Abyssopelagic zone. Oceanic depth from 4000 m to ocean floor.

Acclimation. Reversible phenotypic changes in an individual organism in response to changing environmental conditions—a form of phenotypic plasticity.

Accumulation curve. Records the total number of species revealed during a survey, as additional sample units are added to the pool of all previously observed or collected samples.

Acidity. A measure of the abundance of hydrogen ions (H^+) in solution; the greater the number of hydrogen ions, the more acidic.

Adaptation. A genetically determined characteristic (behavioral, morphological, or physiological) that improves an organism's ability to survive and reproduce under prevailing environmental conditions.

Adaptive radiation. Evolution from a common ancestor of divergent forms adapted to distinct ways of life.

Age distribution. The proportion of individuals in various age classes for any one year. See also age structure.

Age-specific mortality rate. The proportion of deaths occurring per unit time in each age group within a population.

Age-specific birth rate. Average number of offspring produced per individual per unit time as a function of age class.

Aggregative response. Movement of predators into areas of high prey density.

Alkaline (alkalinity). A measure of the abundance of hydroxyl ions (OH^-) in solution; the greater the number of hydroxyl, the more alkaline.

Allee effect. Reduction in reproduction or survival under low population densities.

Allele. One of two or more alternative forms of a gene that occupies the same relative position or locus on homologous chromosomes.

Allele frequency. The proportion of a given allele among all the alleles present at the locus in the population.

Allelopathy. Effect of metabolic products of plants (excluding microorganisms) on the growth and development of other nearby plants.

Allochthonous. Organic matter originating in a place other than where it is found (produced outside the ecosystem).

Allogenic. Refers to successional change brought about by a change in the physical environment.

Alpha diversity. The variety of organisms occurring in a given place or habitat; compare beta diversity, gamma diversity.

Altricial. Condition among birds and mammals of being hatched or born usually blind and too weak to support their own weight.

Amensalism. Relationship between species that is negative to individuals of one species, but neutral to the other.

Ammonification. Breakdown of proteins and amino acids, especially by fungi and bacteria, with ammonia as the excretory by-product.

Anion. Ion carrying a negative charge.

Annual grassland. Grassland in California dominated by exotic annual grasses that reseed every year, replacing native perennial grasses.

Apoematism. See warning coloration.

Aphotic layer (zone). A deepwater area of marine ecosystems below the depth of effective light penetration.

Apparent competition. Occurs when a single species of predator feeds on two prey species supporting a higher predator density, increasing the rate at which prey are consumed. As a result, the density of both prey species is lowered, suggesting a competitive interaction.

Aqueous solution. Solution in which water is the solvent.

Area-insensitive species. Species that are at home in any size habitat patch.

Asexual reproduction. Any form of reproduction, such as budding, that does not involve the fusion of gametes.

Assimilation. Transformation or incorporation of a substance by organisms; absorption and conversion of energy and nutrients into constituents of an organism.

Assimilation efficiency. Ratio of assimilation to ingestion; a measure of efficiency with which a consumer extracts energy from food.

Assortative mating. When individuals choose mates nonrandomly with respect to their genotype, or more specifically, select mates based on some phenotypic trait.

Aufwucks. Small animals and plants that adhere to open surfaces in aquatic environments surface growth.

Autochthonous. Organic matter formed or originating from the place where it is found (produced within the ecosystem).

Autogamy. Pollination of the ovules of a flower by its own pollen (self-pollination).

Autogenic. Self-generated.

Autotrophs. Organisms that produce organic material from inorganic chemicals and some source of energy.

Available water capacity. Supply of water available to plants in a well-drained soil.

B horizon. Soil stratum beneath the A horizon, characterized by an accumulation of silica, clay, and iron and aluminum oxides, and possessing blocky or prismatic structure.

Basal species. Species that feed on no other species but are fed upon by others.

Batesian mimicry. Resemblance of a palatable or harmless species, the mimic, to an unpalatable or dangerous species, the model.

Bathypelagic zone. Lightless zone of the open ocean, lying above the abyssal or bottom water, usually above 4000 m.

Behavioral defenses. Aggressive and submissive postures or actions that threaten or deter enemies.

Behavioral ecology. The study of the behavior of an organism in its natural habitat.

Benthic layer (zone). The area of the sea or lake bottom.

Benthos. Animals and plants living on the bottom of a lake or sea, from the high-water mark to the greatest depth.

Beta diversity. Diversity (differences in species composition) among communities in a geographic area

Biogeochemical cycle. Movement of elements or compounds through living organisms.

Bioclimatic envelope model. Statistical models that correlate features of climate (as well as edaphic factors) to geographic patterns of species occurrence.

Biomass. Weight of living material, usually expressed as dry weight per unit area.

Biomes. Major regional ecological community of plants and animals; usually corresponds to plant ecologists' and European ecologists' classification of plant formations and life zones.

Biosphere. Thin layer about Earth in which all living organisms exist.

Biotic. Applied to the living component of an ecosystem.

Birth rate. The rate of births in a population (births per population size per unit time).

Blanket mire. Large area of upland dominated by sphagnum moss and dependent upon precipitation for a water supply; a moor.

Bog. Wetland ecosystem characterized by an accumulation of peat, acid conditions, and dominance of sphagnum moss.

Border. The place where the edge of one vegetation patch meets the edge of another.

Bottomland. Land bordering a river that floods when the river overflows its banks.

Bottom-up control. Influence of producers on the trophic levels above them in a food web.

Boundary (border). The place where the edge of one patch meets the edge of another adjacent patch (or surrounding matrix).

Boundary layer. A layer of still air close to or at the surface of an object.

Buoyancy. The power of a fluid to exert an upward force on a body placed in it.

Bundle sheath cells. Cells surrounding small vascular bundles in the leaves of vascular plants.

C horizon. Soil stratum beneath the solum (A and B horizons), little affected by biological activity or soil-forming processes.

C3 plant. Any plant that produces as its first step in photosynthesis the three-carbon compound phosphoglyceric acid.

C4 photosynthetic pathway. Photosynthetic pathway involving two distinct types of photosynthetic cells: the mesophyll cells and the bundle sheath cells. CO_2 initially reacts with PEP to form 4-carbon compounds in the mesophyll cells. These compounds are then transported into the bundle sheath cells where they are converted into CO_2 and undergo the C3 photosyntheic pathway (Calvin cycle).

Calcification. Process of soil formation characterized by accumulation of calcium in lower horizons.

CAM pathway (crassulacean acid metabolism). Photosynthetic pathway (using same processes involved in C4 pathway) that separates the processes of carbon dioxide uptake and fixation when growing under arid conditions; it takes up gaseous carbon dioxide at night, when stomata are open, and converts it into simple sugars during the day, when stomata are closed.

Canopy. Uppermost layer of vegetation formed by trees; also the uppermost layer of vegetation in shrub communities or in any terrestrial plant community where the upper layer forms a distinct habitat.

Capillary water. That portion of water in the soil held by capillary forces between soil particles.

Character displacement. Shift in species' morphology, behavior, or physiology as a result of natural selection resulting from interspecific competition.

Carnivore. Organism that feeds on animal tissue; taxonomically, a member of the order Carnivora (Mammalia).

Carrying capacity (*K*). Number of individual organisms the resources of a given area can support, usually through the most unfavorable period of the year.

Categorical data. Qualitative observations that fall into separate and distinct categories.

Cation. Part of a dissociated molecule carrying a positive electrical charge.

Cation exchange capacity (CEC). Ability of a soil particle to absorb positively charged ions.

Chaparral. Vegetation consisting of broadleaved evergreen shrubs, found in regions of mediterranean-type climate.

Character displacement. The principle that two species are more different where they occur together than where they are separated geographically.

Chemical defense. The use by organisms of bitter, distasteful, or toxic secretions that deter potential enemies.

Chemical weathering. The action of a set of chemical processes such as oxidation, hydrolysis, and reduction operating at the atomic and molecular level that break down and re-form rocks and minerals.

Chromosome. One of a group of threadlike structures of different lengths and sizes in the nuclei of cells of eukaryote organisms.

Chronosequences. Groups of sites within the same area that are in different stages of succession.

Clear-cutting. Forest harvesting procedure in which all trees on the site are cut and removed.

Climate. Long-term average pattern of local, regional, or global weather.

Cline. A measurable, gradual change over a geographic region in the average of some phenotypic character, such as size or coloration; or it can be a gradient in genotypic frequency.

Clone. A population of genetically identical individuals resulting from asexual reproduction.

Closed system. A system that neither receives inputs from nor contributes output to the external environment.

CO_2 fertilization effect. see Fertilization effect

Codominance. Occurs when the physical expression of the heterozygous individual is indeterminate between those of the homozygotes. Each allele has a specific proportional effect that it contributes to the phenotype.

Coevolution. Joint evolution of two or more non-interbreeding species that have a close ecological relationship; through reciprocal selective pressures, the evolution of one species in the relationship is partially dependent on the evolution of the other.

Cohesion. The ability of water molecules, because of hydrogen bonding, to stick firmly to each other, restricting external forces to break these bonds.

Cohort. A group of individuals of the same age.

Colloid. Negatively charged particles in the soil that provide surfaces with high cation exchange capacity.

Commensalism. Relationship between species that is beneficial to one, but neutral or of no benefit to the other.

Community. A group of interacting plants and animals inhabiting a given area.

Community ecology. Study of the living component of ecosystems; description and analysis of patterns and processes within the community.

Compensation depth. In aquatic ecosystems, the depth of the water column at which light intensity reaching plants is just sufficient for the rate of photosynthesis to balance the rate of respiration.

Competition. Any interaction that is mutually detrimental to both participants, occurring between species that share limited resources.

Competition coefficient. Coefficient used in Lokta-Volterra model of interspecific competition that relates individuals of one species to the carrying capacity of the species with which it is competing.

Competitive exclusion principle. Hypothesis that when two or more species coexist using the same resource, one must displace or exclude the other.

Competitive release. Niche expansion in response to reduced interspecific competition.

Condensation. The transformation of water vapor to a liquid state.

Conduction. Direct transfer of heat energy from one substance to another.

Conformer. Species for which changes in external environmental conditions induce internal changes in the body that parallel the external conditions

Connectance (connectivity). also see functional and structural connectivity)

Connectivity. The extent to which a species or a population can move among patches within the matrix.

Conservation ecology. A synthetic field that applies principles of ecology, biogeography, population genetics, economics, sociology, and other fields to the maintenance of biological diversity. Also called conservation biology.

Constitutive defense. Fixed feature of an organism, such as object resemblance, that deters predators.

Consumption efficiency. Ratio of ingestion to production or energy available; defines the amount of available energy being consumed.

Contest competition. Competition in which a limited resource is shared only by dominant individuals; a relatively constant number of individuals survive, regardless of initial density.

Cotinentality. Land areas farther from the coast (or other large bodies of water) experience a greater seasonal variation in temperature than do coastal areas.

Continuous data. Numerical data in which any value within an interval is possible, limited only by the ability of the measurement device.

Continuum concept. The view, first proposed by H. A. Gleason, that vegetation is a continuous variable in a continuously changing environment; therefore, no two vegetational communities are identical, and associations of species arise only from similarities in requirements. (Also see individualistic concept.)

Convection. Transfer of heat by the circulation of a liquid or gas.

Coriolis effect. Physical consequence of the law of conservation of angular momentum; as a result of Earth's rotation, a moving object veers to the right in the Northern Hemisphere and to the left in the Southern Hemisphere relative to Earth's surface.

Corridor. A strip of a particular type of vegetation that differs from the land on both sides.

Countercurrent heat exchange. An anatomical and physiological arrangement by which heat exchange takes place between outgoing warm arterial blood and cool venous blood returning to the body core; important in maintaining temperature homeostasis in many vertebrates.

Crude birthrate. The number of young produced per unit of population.

Crude density. The number of individuals per unit area; compare ecological density.

Cryptic coloration. Coloration of organisms that makes them resemble or blend into their habitat or background.

Currents. Water movements that result in the horizontal transport of water masses.

Deciduous. Of leaves, shed during a certain season (winter in temperate regions; dry season in the tropics); of trees, having deciduous parts.

Decomposer. Organism that obtains energy from the breakdown of dead organic matter to simpler substances; most precisely refers to bacteria and fungi.

Decomposition. Breakdown of complex organic substances into simpler ones.

Definitive host. Host in which a parasite becomes an adult and reaches maturity.

Demographic stochasticity. Random variations in birth and death rates that occur in a population from year to year.

Demographic transition. The transition from high birth and death rates to low birth and death rates.

Denitrification. Reduction of nitrates and nitrites to nitrogen by microorganisms.

Density dependence. Regulation of population growth by mechanisms controlled by the size of the population; effect increases as population size increases.

Density-dependent fecundity. Decline in fecundity (birth) rate with increasing population size.

Density-dependent growth. An inverse relationship between population density and individual growth.

Density-dependent mortality. Increase in mortality (death) rate with increasing population size.

Density independence. Being unaffected by population density; regulation of growth is not tied to population density.

Desert grassland. Grassland of hot, dry climates, with rainfall varying between 200 and 500 mm, dominated by bunchgrasses and widely interspersed with other desert vegetation.

Detrital food chain. Food chain in which detritivores consume detritus or litter, mostly from plants, with subsequent transfer of energy to various trophic levels; ties into the grazing food chain; compare grazing food chain.

Detritivore. Organism that feeds on dead organic matter; usually applies to detritus-feeding organisms other than bacteria and fungi.

Detritus. Fresh to partly decomposed plant and animal matter.

Developmental plasticity. Differences in phenotypic traits for a given genotype under different environmental conditions that reflect differences in the allocation of biomass to different tissues (leaves, stem, and roots) during the growth and development of the individual plant.

Dew point temperature. Temperature at which condensation of water in the atmosphere begins.

Diapause. A period of dormancy, usually seasonal, in the life cycle of an insect, in which growth and development cease and metabolism greatly decreases.

Diffuse competition. Competition in which a species experiences interference from many other species that deplete the same resources.

Diffusion. Spontaneous movement of particles of gases or liquids from an area of high concentration to an area of low concentration.

Dioecious. Having male and female reproductive organs on separate plants; compare monoecious.

Directional selection. Selection favoring individuals at one extreme of the phenotype in a population.

Discrete data. Numerical data in which only certain values are possible, such as with integer values or counts.

Disease. Any deviation from a normal state of health.

Dispersal. Leaving an area of birth or activity for another area.

Disruptive selection. Selection in which two extreme phenotypes in the population leave more offspring than the intermediate phenotype, which has lower fitness.

Distribution. The spatial location or area occupied by a population.

Disturbance. A discrete event in time that disrupts an ecosystem, community, or population, changing substrates and resource availability.

Diversity. Abundance of different species in a given location; species richness.

Diversity index. The mathematical expression of species richness of a given community or area.

Dominant. Population possessing ecological dominance in a given community and thereby governing type and abundance of other species in the community.

Dominant allele. An allele that is expressed in either the homozygous or heterozygous state.

Dryfall. Nutrients brought into an ecosystem by airborne particles and aerosols.

Dynamic composite life table. Pooled cohort of individuals born over several time periods instead of just one.

Dynamic life table. Fate of a group of individuals born at the same time and followed from birth to death.

Dystrophic. Term applied to a body of water with a high content of humic or organic matter, often with high littoral productivity and low plankton productivity.

E horizon. Mineral horizon characterized by the loss of clay, iron, or aluminum, and a concentration of quartz and other resistant minerals in sand and silt sizes; light in color.

Early successional species. Plant species characterized by high dispersal rates, ability to colonize disturbed sites, short life span, and shade intolerance.

Ecological density. Density measured in terms of the number of individuals per area of available living space; compare crude density.

Ecological niche. The range of physical and chemical conditions under which a species can persist (survive and reproduce) and the array of essential resources it utilizes.

Ecology. The study of relations between organisms and their natural environment, living and nonliving.

Ecosystem. The biotic community and its abiotic environment, functioning as a system.

Ecosystem ecology. The study of natural systems with emphasis on energy flow and nutrient cycling.

Ecotone. Transition zone between two structurally different communities; wide borders that form a transition zone between adjoining patches.

Ecotype. Subspecies or race adapted to a particular set of environmental conditions.

Ectoparasite. Parasite, such as a flea, that lives in the fur, feathers, or skin of the host.

Ectothermy. Determination of body temperature primarily by external thermal conditions.

Edge effect. Response of organisms, animals in particular, to environmental conditions created by the edge.

Edge species. Species that are restricted exclusively to the edge or border environment.

El Niño–Southern Oscillation (ENSO). Global event arising from large-scale interactions between ocean and atmosphere.

Elaiosome. Shiny, oil-containing, ant-attracting tissue on the seed coats of many plants.

Emigration. Movement of part of a population permanently out of an area.

Endemic. Restricted to a given region.

Endoparasite. Parasite that lives within the body of the host.

Endothermic reaction. Chemical reaction that gains energy from the environment.

Endothermy. Regulation of body temperature by internal heat production; allows maintenance of appreciable difference between body temperature and external temperature.

Entropy. Transformation of matter and energy to a more random, more disorganized state.

Environmental science. Scienfic discipline that examines the impact of humans on the natural environment and as such covers a wide range of topics including agronomy, soils, demography, agriculture, energy, and hydrology.

Environmental stochasticity. Random variations in the environment that directly affect birth and death rates.

Epifauna. Benthic animals that live on or move over the surface of a substrate.

Epiflora. Benthic plants that live on the surface of a substrate.

Epilimnion. Warm, oxygen-rich upper layer of water in a lake or other body of water, usually seasonal.

Epipelagic zone. The lighted area of the ocean; also called photic zone.

Equatorial low. A low air pressure zone near the surface of the equatorial zone.

Estivation. Dormancy in animals during a period of drought or a dry season.

Estuary. A partially enclosed embayment where freshwater and seawater meet and mix.

Ethology. Scientific study of animal behavior.

Eutrophic. Term applied to a body of water with high nutrient content and high productivity.

Eutrophication. Nutrient enrichment of a body of water; called cultural eutrophication when accelerated by introduction of massive amounts of nutrients from human activity.

Eutrophy. Condition of being nutrient rich.

Evaporation. Loss of water vapor from soil or open water or another exposed surface.

Evapotranspiration. Sum of the loss of moisture by evaporation from land and water surfaces and by transpiration from plants.

Evergreen. Applied to trees and shrubs for which there is no complete seasonal loss of leaves; two types, broadleaf and needle-leaf.

Evolution. Change in gene frequency through time resulting from natural selection and producing cumulative changes in characteristics of a population.

Evolutionary ecology. Integrated study of evolution, genetics, natural selection, and adaptations within an ecological context; evolutionary interpretation of population, community, and ecosystem ecology.

Exothermic reaction. Chemical reaction that releases heat to the environment.

Exploitation competition. Competition by a group or groups of organisms that reduces a resource to a point that adversely affects other organisms. (See also exploitation.)

Exponential population growth (r). Instantaneous rate of population growth, expressed as proportional increase per unit of time.

Externalities. When the actions of one individual or a group affect another individual's well-being, but relevant costs are not reflected in market prices.

Facilitation. (facultative interactions) A situation where one species benefits from the presence or action of another.

Fecundity. Potential ability of an organism to produce eggs or young; rate of production of young by a female.

Fecundity table. Shows the number of offspring produced per unit time; constructed by using the survivorship column from the life table and the age-specific birthrates; the mean number of females born in each age group of females.

Fen. Slightly acidic wetland, dominated by sedges, in which peat accumulates.

Fermentation. Breakdown of carbohydrates and other organic matter under anaerobic conditions.

Fertilization effect (CO_2 fertilization effect). The higher rates of diffusion and photosynthesis under elevated atmospheric concentrations of CO_2.

Fertilizer. Any substance added to the soil that contains chemical elements that improve soil fertility and enhance the growth and productivity of plants.

Field capacity. Amount of water held by soil against the force of gravity.

Filter effect. Corridors that provide dispersal routes for some species but restrict the movement of others.

Filtering collectors. Stream insect larvae that feed on fine particulate matter by filtering it from water flowing past them.

Finite multiplication rate. Expressed as lambda, λ, the geometric rate of increase by discrete time intervals; given a stable age distribution, lambda can be used as a multiplier to project population size.

First law of thermodynamics. Energy is neither created nor destroyed; in any transfer or transformation, no gain or loss of total energy occurs.

Fitness. Genetic contribution by an individual's descendants to future generations.

Flashing coloration. Hidden markings on animals that, when quickly exposed, startle or divert the attention of a potential predator.

Food chain. Movement of energy and nutrients from one feeding group of organisms to another in a series that begins with plants and ends with carnivores, detrital feeders, and decomposers.

Food web. Interlocking pattern formed by a series of interconnecting food chains.

Forest floor. Term describing the ground layer of leaves and detritus; site of decomposition.

Frequency distribution. A count of the number of observations (frequency) having a given score or value.

Functional connectivity. The degree to which the landscape facilitates the movement of organisms

Functional type (group). A collection of species that exploit the same array of resources or perform similar functions within the community.

Functional response. Change in rate of exploitation of a prey species by a predator in relation to changing prey density.

Fundamental niche. Total range of environmental conditions under which a species can survive and reproduce (also see ecological niche).

Gamma diversity. Differences among similar habitats in widely separated regions.

Gap. Opening made in a forest canopy by some small disturbance such as windfall; death of an individual tree or group of trees that influences the development of vegetation beneath.

Gathering collectors. Stream insect larvae that pick up and feed on particles from stream-bottom sediment.

Gene. Unit material of inheritance; more specifically, a small unit of a DNA molecule, coded for a specific protein to produce one of the many attributes of a species.

Gene expression. The process of creating proteins through the genetic code in DNA.

Gene flow. Exchange of genetic material between populations.

Gene frequency. Relative abundance of different alleles carried by an individual or a population.

Gene pool. The sum of all the genes of all individuals in a population.

General circulation models. Computer models that help determine how increasing concentrations of greenhouse gases may influence large-scale patterns of global climate.

Genet. A genetic individual that arises from a single fertilized egg.

Genetic differentiation. When genetic variation occurs among subpopulations of the same species.

Genetic drift. Random fluctuation in allele frequency over time, due to chance alone without any influence by natural selection; important in small populations.

Genetic engineering. The process of directly altering an organism's genome.

Genome. The collective term for all the DNA in a cell.

Genotype. Genetic constitution of an organism.

Genotypic frequency. The proportion of various genotypes in a population; compare gene frequency.

Geographic isolates. Groups of populations that are semi-isolated from one another by some extrinsic barrier; compare subspecies.

Geographic range. The distribution described by all the individuals of a species encompassed in a defined area.

Geometric population growth. Population growth described over discrete time intervals.

Gleization. A process in waterlogged soils in which iron, because of an inadequate supply of oxygen, is reduced to a ferrous compound, giving dull gray or bluish mottles and color to the horizons.

Global ecology. The study of ecological systems on a global scale.

Gougers. Stream insect larvae that burrow into waterlogged limbs and trunks of fallen trees.

Grazers. Stream invertebrates that feed on algal coating on rocks and other substrates.

Grazing food chain. Food chain in which primary producers (green plants) are eaten by grazing herbivores, with subsequent energy transfers to other trophic levels. Compare detrital food chain.

Greenhouse effect. Selective energy absorption by carbon dioxide in the atmosphere, which allows short wavelength energy to pass through but absorbs longer wavelengths and reflects heat back to Earth.

Greenhouse gas. A gas that absorbs longwave radiation and thus contributes to the greenhouse effect when present in the atmosphere; includes water vapor, carbon dioxide, methane, nitrous oxides, and ozone.

Gross primary production. Energy fixed per unit area by photosynthetic activity of plants before respiration; total energy flow at the secondary level is not gross production, but rather assimilation, because consumers use material already produced with respiratory losses.

Gross reproductive rate. Sum of the mean number of females born to each female age group.

Groundwater. Water that occurs below Earth's surface in pore spaces within bedrock and soil, free to move under the influence of gravity.

Guild. A group of populations that utilize a gradient of resources in a similar way.

Gyre. Circular motion of water in major ocean basins.

Habitat. Place where a plant or animal lives.

Habitat fragmentation. The fragmentation of larger continuous tracts of habitats, such as forest, shrubland, or grassland into a mosaic of smaller, often isolated patches.

Habitat selection. Behavioral responses of individuals of a species involving certain environmental cues used to choose a potentially suitable environment.

Hadalpelagic zone. That part of the ocean below 6000 m.

Hardy–Weinberg principle. The proposition that genotypic ratios resulting from random mating remain unchanged from one generation to another, provided natural selection, genetic drift, and mutation are absent.

Hemiparasite. A parasitic plant that contains chlorophyll and is therefore capable of photosynthesis.

Herb layer. Lichens, moss, ferns, herbaceous plants, and small woody seedlings growing on the forest floor.

Herbivore. Organism that feeds on plant tissue.

Herbivory. Feeding on plants.

Hermaphrodite (hermaphroditic). Organism possessing the reproductive organs of both sexes.

Heterotherm. An organism that during part of its life history becomes either endothermic or ectothermic; hibernating endotherms become ectothermic, and foraging insects such as bees become endothermic during periods of activity; they are characterized by rapid, drastic, repeated changes in body temperature.

Heterotrophs. Organisms that are unable to manufacture their own food from inorganic materials and thus rely on other organisms, living and dead, as a source of energy and nutrients.

Heterozygous. Containing two different alleles of a gene, one from each parent, at the corresponding loci of a pair of chromosomes.

Hibernation. Winter dormancy in animals, characterized by a great decrease in metabolism.

Histogram. Type of bar graph.

Holoparasite. A parasitic plant that contains no chlorophyll and is therefore not capable of photosynthesis, as a result it cannot exist without a host plant.

Home range. Area over which an animal ranges throughout the year.

Homeostasis. Maintenance of a nearly constant internal environment in the midst of a varying external environment; more generally, the tendency of a biological system to maintain itself in a state of stable equilibrium.

Homeotherm. Animal with a fairly constant body temperature; also spelled *homoiotherm* and *homotherm.*

Homologous chromosomes. Corresponding chromosomes from male and female parents that pair during meiosis.

Homozygous. Containing two identical alleles of a gene at the corresponding loci of a pair of chromosomes.

Horizon. Major zone or layer of soil, with its own particular structure and characteristics.

Host. Organism that provides food or other benefit to another organism of a different species; usually refers to an organism exploited by a parasite.

Hotspot. Regions of unusually high species diversity.

Humus. Organic material derived from partial decay of plant and animal matter.

Hydrogen bonding. A type of bond occurring between an atom of oxygen or nitrogen and a hydrogen atom joined to oxygen or nitrogen on another molecule; responsible for the properties of water.

Hydrological cycle. See water cycle.

Hydroperiod. In wetlands, the duration, frequency, depth, and season of flooding.

Hydrophytic plants. Plants adapted to grow in water or on soil that is periodically anaerobic because of excess water.

Hydrothermal vent. Place on ocean floor where water, heated by molten rock, issues from fissures; vent water contains sulfides oxidized by chemosynthetic bacteria, providing support for carnivores and detritivores.

Hyperosmotic. Having a higher concentration of salts in the body tissue than does the surrounding water.

Hypervolume. The multidimensional space of a species niche; compare niche.

Hypolimnion. Cold, oxygen-poor zone of a lake, below the thermocline.

Hypoosmotic. Having a lower concentration of salts in the body tissue than does the surrounding water.

Hypothesis. Proposed explanation for a phenomenon; we should be able to test it, accepting or rejecting it based on experimentation.

Immigration. Arrival of new individuals into a habitat or population.

Immobilization. Conversion of an element from inorganic to organic form in microbial or plant tissue, rendering the nutrient unavailable to other organisms.

Inbreeding. Mating among close relatives.

Inbreeding depression. Detrimental effects of inbreeding.

Indirect interaction. When one species does not interact with a second species directly, but instead influences a third species that does have a direct interaction with the second.

Indirect commensalism. Indirect interaction between two species that is classified as commensalism (benefits one species and is neutral to the other).

Indirect mutualism. Indirect interaction between two species that is classified as multualistic (benefits both species involved).

Individualistic concept. The view, first proposed by H. A. Gleason, that vegetation is a continuous variable in a continuously changing environment; therefore, no two vegetational communities are identical, and associations of species arise only from similarities in requirements.

Induced defense. Defense response brought about or induced by the presence or action of a predator; for example, alarm pheromones.

Infauna. Organisms living within a substrate.

Infection. Diseased condition arising when pathogenic microorganisms enter a body, become established, and multiply.

Infiltration. Downward movement of water into the soil.

Infralittoral fringe. Region below the littoral region of the sea.

Infralittoral or subtidal zone. Region below the littoral zone and infralittoral fringe. This zone is covered by water even at low tide and dominated by algae.

Inorganic fertilizer. Fertlizer that provides nutrients in an inorganic or mineral form (referred to as chemical fertilizers when it is of a synthetic origin).

Interception. The capture of rainwater by vegetation, from which the water evaporates and does not reach the ground.

Interference competition. Competition in which access to a resource is limited by the presence of a competitor.

Interior species. Organisms that require large areas of habitat, even though their home ranges may be small.

Intermediate disturbance hypothesis. The concept that species diversity is greatest in those habitats experiencing a moderate amount of disturbance, allowing the coexistence of early and late successional species.

Intermediate host. Host that harbors a developmental phase of a parasite; the infective stage or stages can develop only when the parasite is independent of its definitive host; compare definitive host.

Intermediate species. Species that feed on other species, and they themselves are prey of other species.

Internal cycling (also see nutrient cycling). Movement or cycling of nutrients through components of ecosystems.

Intersexual selection. Choice of a mate, usually by the female.

Interspecific competition. Competition between individuals of different species.

Intertidal zone. Area lying between the lines of high and low tide.

Intertropical convergence zone (ITCZ). The boundary zone separating the northeast trade winds of the Northern Hemisphere from the southeast trade winds of the Southern Hemisphere.

Intrasexual selection. Competition among members of the same sex for a mate; most common among males and characterized by fighting and display.

Intraspecific competition. Competition among individuals of the same species

Intrinsic rate of Increase. The per capita rate of growth of a population that has reached a stable age distribution and is free of competition and other growth restraints. (Also referred to as instantaneous per capita rate of growth.)

Invasive species. A species that is not native (indigenous) to that ecosystem (also referred to as non-native or alien species).

Ion. An atom that is electrically charged as a result of a loss of one or more electrons or a gain of electrons.

Ion exchange capacity. Total number of charged sites on soil particles within a volume of soil

Island biogeography. (see Theory of island biogeography)

Isosmotic. A characteristic describing an organism with body fluids that have the same osmotic pressure as seawater.

Iteroparous. Having multiple broods over a lifetime.

K-strategist. A competitive species with a stable population with a slow growth rate, long-lived individuals; they produce relatively few young or seeds, and provide parental care.

Keystone predation. Predation that is central to the organization of a community; the predator enhances one or more inferior competitors by reducing the abundance of the superior competitor.

Keystone species. A species whose activities have a significant role in determining community structure.

Kinetic energy. Energy associated with motion; performs work at the expense of potential energy.

Landscape. An area of land (or water) composed of a patchwork of communities and ecosystems.

Landscape connectivity. Degree to which the landscape facilitates or impedes the movement of organisms among patches; comprised of functional and structural connectivity.

Landscape ecology. Study of structure, function, and change in a heterogeneous landscape composed of interacting ecosystems.

La Niña. A global climate phenomenon characterized by strong winds and cool ocean currents flowing westward from the coastal waters of South America to the tropical Pacific Ocean.

Lapse rate. The rate at which atmospheric temperature decreases with an increase in altitude.

Late successional species. Long-lived, shade-tolerant plant species that supplant early successional species.

Latent heat. Amount of heat given up when a unit mass of a substance converts from a liquid to a solid state, or the amount of heat absorbed when a substance converts from the solid to liquid state.

Laterization. Soil-forming process in hot, humid climates, characterized by intense oxidation; results in loss of bases and in a deeply weathered soil composed of silica.

Leaching. Dissolving and washing of nutrients out of soil, litter, and organic matter.

Leaf area index. The total leaf area of a plant exposed to incoming light energy relative to the ground surface area beneath the plant.

Leaf area ratio (LAR). Total area of leaves per total plant weight.

Leaf pack. Areas of accumulated leaf litter in stream ecosystems; term also refers to litterbags used in stream decomposition experiments.

Leaf weight ratio (LWR). The total weight of leaves expressed as a proportion of total plant weight

Lek. Communal courtship area males use to attract and mate with females.

Lentic. Pertaining to standing water, such as lakes and ponds; a population is limited by the lowest amount needed of an essential nutrient.

Life expectancy. The average number of years to be lived in the future by members of a population.

Life table. Tabulation of mortality and survivorship of a population; static, time-specific, or vertical life tables are based on a cross section of a population at a given time; dynamic, cohort, or horizontal life tables are based on a cohort followed throughout life.

Light compensation point. Depth of water or level of light at which photosynthesis and respiration balance each other.

Light saturation point. Amount of light at which plants achieve the maximum rate of photosynthesis.

Limnetic. Pertaining to or living in the open water of a pond or lake.

Limnetic zone. Shallow-water zone of a lake or sea, in which light penetrates to the bottom.

Link. In a food web, the arrows leading from a consumer to a species being consumed.

Linkage density. A measure of the average number of links per species in the food web.

Littoral zone. Shallow water of a lake, in which light penetrates to the bottom, permitting submerged, floating, and emergent vegetative growth; also shore zone of tidal water between high-water and low-water marks.

Littoral or intertidal zone. Area of the shoreline between the high tide and low tide lines that is covered and uncovered daily by the tides

Local (alpha) diversity. Number of species in a small area of homogeneous habitat. Also called alpha diversity.

Locus. Site on a chromosome occupied by a specific gene.

Logistic model of population growth (logistic equation). Mathematical expression for the population growth curve in which rate of increase decreases linearly as population size increases.

Long-day organism. Plant or animal that requires long days—days with more than a certain minimum of daylight—to flower or come into reproductive condition.

Longwave radiation. Infrared radiation that occurs as wavelengths longer than 3 or 4 microns.

Lotic. Pertaining to flowing water.

Macronutrients. Essential nutrients plants and animals need in large amounts.

Macroparasite. Any of the parasitic worms, lice, fungi, and the like that have comparatively long generation times, spread by direct or indirect transmission, and may involve intermediate hosts or vectors.

Mangal. A mangrove swamp.

Mangrove forest. Tropical inshore communities dominated by several species of mangrove trees and shrubs capable of growth and reproduction in areas inundated daily by seawater.

Marsh. Wetland dominated by grassy vegetation such as cattails and sedges.

Mating system. Pattern of mating between individuals in a population.

Matric potential. Tendency of water to adhere to surfaces.

Matrix. The communities that surround a patch on the landscape.

Maximum sustainable yield. The maximum rate at which individuals can be harvested from a population without reducing its size; recruitment balances harvesting.

Mean chain length. The arithmetic average of the lengths of all chains in a food web, where length is defined as the number of links between the top predator and basal species.

Mechanical weathering. Breakdown of rocks and minerals in place by disintegration processes such as freezing, thawing, and pressure that do not involve chemical reactions. (Also referred to as physical weathering.)

Mediterranean-type climate. Semiarid climate characterized by a hot, dry summer and a wet, mild winter.

Meiofauna. Benthic organisms within the size range from 1 to 0.1 mm; interstitial fauna.

Mesic. Moderately moist.

Mesopelagic. Uppermost lightless pelagic zone.

Mesophyll. Specialized tissue located between the epidermal layers of a leaf; palisade mesophyll consists of cylindrical cells at right angles to upper epidermis and contains many chloroplasts; spongy mesophyll lies next to the lower epidermis and has interconnecting, irregularly shaped cells with large intercellular spaces.

Metacommunity. Set of local communities that are linked by the dispersal of multiple potentially interacting species

Metapopulation. A population broken into sets of subpopulations held together by dispersal or movements of individuals among them.

Microbial loop. Feeding loop in which bacteria take up dissolved organic matter produced by plankton and nanoplankton consume the bacteria; adds several trophic levels to the plankton food chain.

Microclimate. Climate on a very local scale, which differs from the general climate of the area; influences the presence and distribution of organisms.

Micronutrient. Essential nutrient needed in very small quantities by plants and animals.

Microparasite. Any of the viruses, bacteria, and protozoans, characterized by small size, short generation time, and rapid multiplication.

Migration. Intentional, directional, usually seasonal movement of animals between two regions or habitats; involves departure and return of the same individual; a round-trip movement.

Mimicry. Resemblance of one organism to another or to an object in the environment, evolved to deceive predators.

Mineral fertilizer. (see Inorganic fertilizer)

Mineralization. Microbial breakdown of humus and other organic matter in soil to inorganic substances.

Minimum dynamic area (MDA). Area of suitable habitat necessary for maintaining a minimum viable population.

Minimum viable population (MVP). Size of a population that, with a given probability, will ensure the population's existence for a stated period of time.

Mire. Wetland characterized by an accumulation of peat.

Mixed-grass prairie. Grassland in mid North America, characterized by great variation in precipitation and a mixture of largely cool-season shortgrass and tallgrass species.

Models. In theoretical and systems ecology, an abstraction or simplification of a natural phenomenon, developed to predict a new phenomenon or to provide insight into existing ones; in mimetic association, the organism mimicked by a different organism.

Modular organism. Organism that grows by repeated iteration of parts, such as branches or shoots of a plant; some parts may separate and become physically and physiologically independent.

Monoculture. Planting of a single plant species.

Monoecious. Having male and female reproductive organs separated in different floral structures on the same plant; compare hermaphrodite, dioecious.

Monogamy. In animals, mating and maintenance of a pair bond with only one member of the opposite sex at a time.

Moor. A blanket bog or peatland.

Mortality rate. The probability of dying; the ratio of number dying in a given time interval to the number alive at the beginning of the time interval.

Mosaic. A pattern of patches, corridors, and matrices in the landscape.

Mullerian mimicry. When many unpalatable or venomous species share a similar color pattern.

Mutation. Transmissible changes in the structure of a gene or chromosome.

Mutualism. Relationship between two species in which both benefit.

Mycorrhizae. Association of fungus with roots of higher plants, which improves the plants' uptake of nutrients from the soil.

Myrmecochores. Plants that possess ant-attracting substances on their seed coats.

Nanoplankton. Plankton with a size range from 2 to 20 mm.

Natural selection. Differential success (reproduction and survival) of individuals that results in elimination of maladaptive traits from a population.

Negative assortative mating. Occurs when mates are phenotypically less similar to each other than expected by chance.

Negative feedback. Homeostatic control in which an increase in some substance or activity ultimately inhibits or reverses the direction of the processes leading to the increase.

Nekton. Aquatic animals that are able to move at will through the water.

Neritic. Marine environment embracing the regions where landmasses extend outward as a continental shelf.

Net assimilation rate (NAR). Assimilation of new plant tissue per unit leaf area per unit time (g/cm^2/time); component of relative growth rate.

Net ecosystem productivity. Difference between the rates of net primary productivity and carbon lost through consumer and decomposer respiration.

Net mineralization rate. Difference between the rates of mineralization and immobilization.

Net photosynthesis. Difference between the rate of carbon uptake in photosynthesis and carbon loss in respiration.

Net primary productivity (NPP). The rate of energy storage as organic matter after respiration.

Net radiation. The difference between incoming (absorbed) and outgoing (emitted) electromagnemtic radiation (both shortwave and longwave) for an object.

Net reproductive rate. Average number of female offspring produced by an average female during her lifetime.

Niche. See ecological niche.

Nitrification. Breakdown of nitrogen-containing organic compounds into nitrates and nitrites.

Nitrogen fixation. Conversion of atmospheric nitrogen to forms usable by organisms.

Nominal data. Categorical data in which objects fall into unordered categories, such as hair color or sex.

Norm of reaction. The set of phenotypes expressed by a single genotype across a range of environmental conditions.

Null model. Assumes that the presence and abundance of the individual species found in a given community are solely a result of the independent responses of each individual species to the prevailing physical environment.

Numerical data. Data in which objects are measured based on some quantitative trait.

Numerical response. Change in size of a population of predators in response to change in density of its prey.

Nutrient cycle (cycling). Pathway of an element or nutrient through the ecosystem, from assimilation by organisms to release by decomposition (see also internal cycling).

Nutrient spiraling. In flowing-water ecosystems, the combined processes of nutrient cycling and downstream transport.

O horizon. Surface stratum of mineral soil, dominated by organic material and consisting of undecomposed or partially decomposed plant materials, such as dead leaves.

Object resemblance. A prey species assumes the appearance of some feature in the environment, such as a leaf, to avoid detection.

Oceanic. Referring to regions of the sea with depths greater than 200 m that lie beyond the continental shelf.

Oligotrophic. Term applied to a body of water low in nutrients and in productivity.

Oligotrophy. Nutrient-poor condition.

Omnivore. An animal (heterotroph) that feeds on both plant and animal matter.

Open system. System with exchanges of materials and energy to the surrounding environment.

Operative environmental temperature. body temperature that occurs when the snake occupies each of these environments.

Optimal foraging theory. Tendency of animals to harvest food efficiently, selecting food sizes or food patches that supply maximum food intake for energy expended.

Ordinal data. Categorical data in which order is important, such as reproductive status.

Organic fertilizer. Fertilizer composed of organic matter derived from plants or animals.

Organismic concept of community. Idea that species, especially plant species, are integrated into an internally interdependent unit; upon maturity and death of the community, another identical plant community replaces it.

Osmosis. Movement of water molecules across a differentially permeable membrane in response to a concentration or pressure gradient.

Osmotic potential. The attraction of water across a membrane; the more concentrated a solution, the lower is its osmotic potential.

Outcrossing. Reproduction in which pollen from one plant fertilises another plant.

Paleoecology. Study of ecology of past communities by means of the fossil record.

Pampas. Temperate South American grassland, dominated by bunchgrasses; much of the moister pampas are under cultivation.

Parasitism. Relationship between two species in which one benefits while the other is harmed (although not usually killed directly).

Parasitoidism. Relationship between insect larva that kills its host by consuming the host's soft tissues before pupation or metamorphosis into an adult.

Parent material. The material from which soil develops.

Parthenogenesis. Development of an individual from an egg that did not undergo fertilization.

Patch. An area of habitat that differs from its surroundings and has sufficient resources to allow a population to persist.

Peatland. Any ecosystem dominated by peat; compare bog, mire, and fen.

Pelagic. Referring to the open sea.

PEP carboxylase. The enzyme phosphoenolpyruvate carboxylase that catalyzes the fixation of CO_2 into four-carbon acids, malate, and aspartate.

Periphyton. In freshwater ecosystems, organisms that are attached to submerged plant stems and leaves; see aufwuchs.

Permafrost. Permanently frozen soil in arctic regions.

Phenotype. Physical expression of a characteristic of an organism, determined by both genetic constitution and environment.

Phenotypic plasticity. Ability to change form under different environmental conditions.

Pheromone. Chemical substance released by an animal that influences behavior of others of the same species.

Photic layer (zone). Lighted water column of a lake or ocean, inhabited by plankton.

Photoinhibition. The slowing or stopping of a plant process by light.

Photosynthesis. Use of light energy by plants to convert carbon dioxide and water into simple sugars.

Photosynthetically active radiation (PAR). That part of the light spectrum between wavelengths of 400 and 700 nm that is used by plants in photosynthesis.

Physiological ecology. Study of the physiological functioning of organisms in relation to their environment.

Phytoplankton. Small, floating plant life in aquatic ecosystems; planktonic plants.

Pioneer species. Plants that invade disturbed sites or appear in early stages of succession.

Plankton. Small, floating or weakly swimming plants and animals in freshwater and marine ecosystems.

Pleistocene. Geological epoch extending from about 2 million to 10,000 years ago, characterized by recurring glaciers; the Ice Age.

Pneumatophore. An erect respiratory root that protrudes above waterlogged soils; typical of bald cypress and mangroves.

Podzolization. Soil-forming process in which acid leaches the A horizon and iron, aluminum, silica, and clays accumulate in lower horizons.

Poikilotherm. An organism whose body temperature varies according to the temperature of its surroundings.

Polar easterlies. Easterly wind located at high latitudes poleward of the subpolar low.

Polyandry. Mating of one female with several males.

Polyculture. Planting of several plant species.

Polygamy. Acquisition by an individual of two or more mates, none of which is mated to other individuals.

Polygyny. Mating of one male with several females.

Population. A group of individuals of the same species living in a given area at a given time.

Population density. The number of individuals in a population per unit area.

Population ecology. Study of how populations grow, fluctuate, spread, and interact intraspecifically and interspecifically.

Population genetics. The study of changes in gene frequency and genotypes in populations.

Population projection table. Chart of growth of a population developed by calculating the births and mortality of each age group over time.

Population regulation. Mechanisms or factors within a population that cause it to decrease when density is high and increase when density is low.

Positive assortative mating. Occurs when mates are phenotypically more similar to each other than expected by chance.

Positive feedback. Control in a system that reinforces a process in the same direction.

Potential energy. Energy available to do work.

Potential evapotranspiration. Amount of water that would be transpired under constantly optimal conditions of soil moisture and plant cover.

Practical salinity unit (psu). The total amount of dissolved material in seawater, expressed as parts per thousand (‰).

Precipitation. All the forms of water that fall to Earth; includes rain, snow, hail, sleet, fog, mist, drizzle, and the measured amounts of each.

Precocial. In birds, hatched with down, open eyes, and ability to move about; in mammals, born with open eyes and ability to follow the mother, as fawns and calves can.

Predation. Relationship in which one living organism serves as a food source for another.

Predator defenses. Evolved characteristics that help prey avoid detection or capture.

Predator satiation. A predator defense mechanism involving the physiological timing of reproduction by a prey species, plant or animal, to produce a maximum number of seeds or young within a short period—more than predators can possibly consume—thus allowing a greater percentage of offspring to escape.

Primary producer. Green plant or chemosynthetic bacterium that converts light or chemical energy into organismal tissue.

Primary production. Production by green plants over time.

Primary productivity. Rate at which plants produce biomass per unit area per unit time.

Primary succession. Vegetational development starting on a new site never before colonized by life.

Production. Amount of energy formed by an individual, population, or community per unit time.

Production efficiency. The ratio of production to assimilation; a measure of the efficiency of a consumer to incorporate assimilated energy into secondary production.

Profundal zone. Deep zone in aquatic ecosystems, below the limnetic zone.

Promiscuity. Member of one sex mates with more than one member of the opposite sex, and the relationship terminates after mating.

Protective armor. Hard outer covering of an animal body, such as shells of turtles and spines of porcupines, that deters or makes the owner somewhat invulnerable to most enemies.

Pycnocline. Area in the water column where the highest rate of change in density occurs for a given change in depth.

Pyromineralization. Mineralization of nutrients bound in organic compounds by fire.

Qualitative inhibitors. The secondary compounds that function as defenses against herbivory that are present in small to minute quantities.

Qualitative traits. Phenotypic characteristics that fall into a limited number of discrete categories.

Quantitative inhibitors. The secondary compounds that are produced by the plant in large quantities.

Quantitative traits. Phenotypic characteristics that have a continuous distribution.

Quaking bog. Bog characterized by a floating mat of peat and vegetation over water.

***r*-strategist.** Short-lived species with a high reproductive rate at low population densities; characterized by relatively small body size, rapid growth rate, large number of offspring, and lack of parental care.

Rain forest. Permanently wet forest of the tropics; also the wet coniferous forest of the Pacific Northwest of the United States.

Rain shadow. A dry region on the leeward side of a mountain range resulting from a reduction in rainfall.

Raised bog. A bog in which the accumulation of peat has raised the surface above both the surrounding landscape and the water table; it develops its own perched water table.

Ramet. An individual member of a plant clone.

Rank-abundance diagram. Plots of relative abundance of each species against rank, defined as the order of species from the most to the least abundant.

Reabsorption. See retranslocation.

Realized niche. Portion of the fundamental niche that a species actually exploits as a result of interactions with other species.

Recessive allele. Applies to an allele whose phenotypic effect is expressed in the homozygous state and masked in the presence of an allele in organisms heterozygous for that gene.

Regolith. The unconsolidated debris layer composed of exposed rock overlaying the hard, unweathered rock in most places on the Earth's surface. It varies in depth from virtually nonexistent to tens of meters.

Regulator. Animals that use a variety of biochemical, physiological, morphological, and behavioral mechanisms to regulate their internal environments over a broad range of external environmental conditions.

Relative abundance. Proportional representation of a species in a community or sample of a community.

Relative growth rate (RGR). Weight gained during a specified period of time.

Relative humidity. Water vapor content of air at a given temperature, expressed as a percentage of the water vapor needed for saturation at that temperature.

Reproductive effort. Proportion of its resources an organism expends on reproduction.

Resorption. (see retranslocation)

Respiration. (also cellular respiration or aerobic respiration) Harvesting of energy from the chemical breakdown of simple sugars and other carbohydrates; the oxidation of carbohydrates to generate energy in the form of ATP that takes place exclusively in the mitochondria.

Restoration ecology. Applying principles of ecosystem development and function to the restoration and management of disturbed lands.

Rete. A large network or discrete vascular bundle of intermingling small blood vessels carrying arterial and venous blood that acts as a heat exchanger in mammals and certain fish and sharks.

Retranslocation. Recycling of nutrients within a plant.

Rhizosphere. Describes a region of the soil where plant roots function. It is an active zone of root growth and death, characterized by intense microbial and fungal activity.

Riparian woodland. Woodland along the bank of a river or stream; riverbank forests are often called gallery forests, see also bottomland.

River continuum concept. The relative importance of allochthonous sources of carbon in the energy balance of flowing water ecosystems (as compared to autochthonous sources) decreases from the smaller headwater streams to larger river systems.

Root-to-shoot ratio (R:S). Ratio of the weight of roots to the weight of shoots of a plant.

Rubisco. Enzyme in photosynthesis that catalyzes the initial transformation of CO_2 into sugar.

Ruminant. Ungulate with a three-chamber or four-chamber stomach; in the large first chamber, or rumen, bacteria ferment plant matter.

Salinity. A measure of the total quantity of dissolved substances in water in parts per thousand (0/00) by weight.

Salinization. The process of accumulation of soluble salts in soil, usually by upward capillary movement from a salty groundwater source. Also called salination.

Salt or tidal marshes. Communities of emergent vegetation rooted in a soil alternately flooded and drained by tidal action.

Sample. That part of a population that is selected for observation by an investigator (a subset of the total population).

Saprophyte. Plant that draws its nourishment from dead plant and animal matter, mostly the former.

Saturation. The state of a soil when the water content is greater than the pore space can hold.

Saturation vapor pressure. Water vapor content of a parcel of air when it is saturated; saturation is defined as condition when rate of evaporation equals the rate of condensation.

Savanna. Tropical grassland, usually with scattered trees or shrubs.

Scale. Level of resolution within the dimensions of time and space; spatial proportion as a ratio of length on a map to actual length.

Scaling. The process by which most morphological and physiological features change as a function of body size in a predictable way.

Scatter plot. A type of graph constructed by defining two axes (x and y), each representing one of the two variables being examined. The individuals are then plotted as points on the graph.

Scavenger. Animal that feeds on dead animals or on animal products, such as dung.

Scramble competition. Intraspecific competition in which limited resources are shared to the point that no individual survives.

Scrapers. Aquatic invertebrates that feed on algal coating on stones and rubble in streams; also called grazers.

Search image. Mental image formed in predators, enabling them to find prey more quickly and to concentrate on a common type of prey.

Second law of thermodynamics. In any energy transfer or transformation, part of the energy assumes a form that cannot be passed on any farther.

Secondary compounds. Chemicals that are not involved in the basic metabolism of plant cells.

Secondary producers. Organisms that derive energy from consuming plant or animal tissue and breaking down assimilated carbon compounds.

Secondary production. Production by consumer organisms over time.

Secondary productivity. Rate at which heterotrophs produce biomass per unit area per unit time.

Secondary succession. Development of vegetation after a disturbance.

Seed-tree. Method of forest harvest in which a small number of trees are left on the site (uncut) to provide a source of seeds for natural regeneration of the population.

Selection cutting. Method of forest harvesting in which only selected individual trees of high commercial value are removed from the forest stand.

Selective agent. The environmental cause of fitness differences among organisms within a population with difference phenotypes.

Selective breeding. Selecting individuals that exhibit a desired trait, and mating them with individuals exhibiting the same trait (or traits).

Self-thinning. Progressive decline in density of plants associated with the increasing size of individuals.

Semelparous. Having only a single reproductive effort in a lifetime, over one short period of time.

Sequential hermaphrodite. An individual organism that changes sex from female to male or male to female at some time in its life cycle.

Seral stage. Following a series of stages; a point in a continuum of vegetation through time.

Sere. The series of successional stages on a given site that lead to a terminal community.

Sessile. Not free to move about; permanently attached to a substrate.

Set point. In a homeostatic system, it refers to the normal or desired state

Sex ratio. The relative number of males to females in a population.

Sexual dimorphism. The occurrence of morphological differences other than primary sexual characteristics that distinguish males from females in a species.

Sexual reproduction. Two individuals produce haploid gametes (egg and sperm) that combine to form a diploid cell.

Shade-intolerant. Growing and reproducing best under high light conditions; growing poorly and failing to reproduce under low light conditions.

Shade-tolerant. Able to grow and reproduce under low light conditions.

Shannon index. An index of diversity that considers both species richness and evenness.

Shelterwood. See seed-tree cut.

Shifting mosaic. Constantly changing pattern of patches as each patch passes through successive stages of development.

Shortgrass prairie. Westernmost grasslands of the Great Plains, characterized by infrequent rainfall, low humidity, and high winds; dominated by shallow-rooted, sod-forming grasses.

Shortwave radiation. Infrared radiation that occurs as wavelengths shorter than 3 or 4 microns.

Shredders. Aquatic invertebrates that feed on coarse particulate matter in streams.

Siblicide. Killing of an offspring by another offspring of the same parents.

Simpson's index (*D*). A measure of the probability that two individuals randomly selected from a sample will belong to the same species.

Simultaneous hermaphrodite. An individual organism that possesses both male and female sex organs at the same time in its life cycle.

Social dominance. Physical dominance of one individual over another, usually maintained by some manifestation of aggressive behavior.

Soil horizon. (see horizon)

Soil microbial loop. Process of supplementing carbon to microbial decomposers in the rhizosphere, enhancing the decomposition of soil organic matter, and subsequently releasing mineral nutrients for plant uptake by microbial grazers.

Soil profile. Distinctive layering of horizons in the soil.

Soil texture. Relative proportions of the three particle sizes (sand, silt, and clay) in the soil.

Solute. A substance that is dissolved in solution.

Solution. Liquid that is a homogeneous mixture of two or more substances.

Solvent. Dissolving agent of a solution.

Sorensen's coefficient of community. Index of similarity between two stands or communities. The index ranges from 0 to indicate communities with no species in common to 100 to indicate communities with identical species composition.

Speciation. Separation of a population into two or more reproductively isolated populations.

Species diversity. Measurement that relates number and relative abundances of species within a community (see species evenness and species richness).

Species evenness. A component of species diversity index; a measure of the distribution of individuals among total species occupying a given area.

Species richness. Number of species in a given area.

Specific heat. Amount of energy that must be added or removed to raise or lower the temperature of a substance by a specific amount.

Specific leaf area. The area of leaf per gram of leaf weight (cm^2/g).

Stabilizing selection. Selection favoring the middle in the distribution of phenotypes.

Stable age distribution. Constant proportion of individuals of various age classes in a population through population changes.

Standing crop biomass. Total amount of biomass per unit area at a given time.

Static life table. See life table.

Statistical population. Set of objects about which inferences can be drawn.

Steppe. Name given to Eurasian grasslands that extend from eastern Europe to western Siberia and China.

Stoichiometry. Branch of chemistry dealing with the quantitative relationships of elements in combination.

Stomata. Pores in the leaf or stem of a plant that allow gaseous exchange between the internal tissues and the atmosphere.

Structural connectivity. Degree to which patches on a landscape are contiguous or physically linked to one another.

Subpolar low. A zone of low air pressure near the surface of the Earth.

Subspecies. Geographical unit of a species population, distinguishable by morphological, behavioral, or physiological characteristics.

Subtropical high. A semipermanent high-air-pressure belt at the surface encircling the Earth.

Succession. The gradual and seemingly directional change in community structure through time (temporal changes in community structure).

Supercooling. In ectotherms, lowering body temperature below freezing without freezing body tissue, by means of solutes (particularly glycerol).

Supralittoral fringe. The highest zone on the intertidal shore, bounded below by the upper limit of barnacles and above by the upper limit of *Littorina* snails.

Supralittoral or supratidal zone. The terrestrial zone adjacent to the shoreline (supralittoral fringe).

Surface runoff. The excess water that flows across the surface of the ground when the soil is saturated during heavy rains.

Surface tension. Elastic film across the surface of a liquid, caused by the attractive forces between molecules at the surface of the liquid.

Survivorship. In the analysis of life tables, it is the probability at birth of surviving to any given age (x).

Survivorship curve. A graph describing the survival (l_x) of a cohort of individuals in a population from birth to the maximum age reached by any one member of the cohort.

Sustained yield. Yield per unit time equal to production per unit time in an exploited population.

Swamp. Wooded wetland in which water is near or above ground level.

Swidden agriculture. Farming systems that alternate periods of annual cropping with extended fallow periods. Also referred to as shifting cultivation; fire is used to clear fallow areas for cropping.

Switching. Changing the diet from a less abundant to a more abundant prey species.

Symbiosis. Situation in which two dissimilar organisms live together in close association.

Systems ecology. The application of general system theory and methods to ecology.

Taiga. The northern circumpolar boreal forest.

Tallgrass prairie. A narrow belt of tall grasses dominated by big bluestem that once ran north and south adjacent to the deciduous forest of eastern North America; presence maintained by fire; largely destroyed by cultivation.

Target of selection. The phenotypic trait that natural selection acts directly upon.

Temperate rain forest. Forest in regions characterized by mild climate and heavy rainfall that produces lush vegetative growth; one example is the coniferous forest of the Pacific Northwest of North America.

Temperature. A measure of the average speed or kinetic energy of atoms and molecules in a substance.

Territory. Area defended by an animal; varies among animal species according to social behavior, social organization, and resource requirements.

Theory. An integrated set of hypotheses that together explain a broader set of observations than any single hypothesis.

Theory of island biogeography. Theory stating that the number of species established on an island represents a dynamic equilibrium between the immigration of new colonizing species and the extinction of previously established ones.

Thermal conductivity. Ability to conduct or transmit heat.

Thermocline. Layer in a thermally stratified body of water in which temperature changes rapidly relative to the remainder of the body.

Thermoneutral zone. Range of environmental temperatures within which the metabolic rates are minimal.

Throughfall. That part of precipitation that falls through vegetation to the ground.

Tidal overmixing. Mixing of freshwater and seawater when a tidal wedge of seawater moves upstream in a tidal river faster than freshwater moves seaward; seawater on the surface tends to sink as lighter freshwater rises to the surface.

Tidal subsidy. Nutrients carried to coastal ecosystems and wastes carried away by tidal cycles.

Time-specific life table. A population sampled in some manner to obtain the distribution of age classes during a single time period.

Top-down control. Influence of predators on the structure of lower trophic levels in a food web.

Top predator. Species not subjected to predators; they prey on intermediate and basal species.

Topography. Physical structure of the landscape.

Torpor. Temporary great reduction in an animal's respiration, with loss of motion and feeling; reduces energy expenditure in response to some unfavorable environmental condition, such as heat or cold.

Trade winds. Tropical easterly winds that blow in a steady direction from the subtropical high-pressure areas to the equatorial low-pressure areas between the latitudes 30° and 40° north and south; these winds are generally northeasterly in the Northern Hemisphere and southeasterly in the Southern Hemisphere.

Transpiration. Loss of water vapor from a plant to the outside atmosphere.

Trophic. Related to feeding.

Trophic cascade. Occurs when a predator in a food web suppresses the abundance of their prey (intermediate species) such that it increases the abundance of the next lower trophic level (basal species) on which the intermediate species feeds.

Trophic efficiency. Ratio of productivity in a given trophic level with the trophic level on which it feeds.

Trophic level. Functional classification of organisms in an ecosystem according to feeding relationships, ranging from first-level autotrophs through succeeding levels of herbivores and carnivores.

Trophic structure. Organization of a community based on the number of feeding or energy transfer levels.

Tundra. Area in an arctic or alpine (high mountain) region, characterized by bare ground, absence of trees, and growth of mosses, lichens, sedges, forbs, and low shrubs.

Turgor pressure. The state in a plant cell in which the protoplast is exerting pressure on the cell wall due to intake of water by osmosis.

Turnover. Vertical mixing of layers in a body of water, brought about by seasonal changes in temperature.

Turnover rate. In the theory of island biogeography, for an equilibrium species richness on an island (S), it is the rate at which one species is lost through extinction and a replacement species is gained through immigration.

Type I functional response. Rate of prey mortality due to predation is constant, a function of the efficiency of predators.

Type II functional response. Per capita rate of predation increases in a decelerating fashion only up to a maximum rate that is attained at some high prey density.

Type III functional response. Rate of prey consumed; slow at first and then increasing in an S-shaped (sigmoid) fashion as the rate of predation reaches a maximum.

Ubiquitous. Having widespread geographic distribution.

Understory. Growth of medium-height and small trees beneath the canopy of a forest; sometimes includes a shrub layer as well.

Unitary organism. An organism, such as an arthropod or vertebrate, whose growth to adult form follows a determinate pathway, unlike modular organisms whose growth involves indeterminate repetition of units of structure.

Upwelling. In oceans and large lakes, a water current or movement of surface waters produced by wind that brings nutrient-loaded colder water to the surface; in open oceans, regions where surface currents diverge deep waters, which rise to the surface to replace departing waters.

Urban ecology. The study of the ecology of organisms in the context of the urban environment.

Vapor pressure. The amount of pressure water vapor exerts independent of dry air.

Vapor pressure deficit. The difference between saturation vapor pressure and the actual vapor pressure at any given temperature.

Vector. Organism that transmits a pathogen from one organism to another.

Vegetative reproduction. Asexual reproduction in plants by means of specialized multicellular organs, such as bulbs, corms, rhizomes, stems, and the like.

Veld. Extensive grasslands in the east of the interior of South Africa, largely confined to high terrain.

Viscosity. Property of a fluid that resists the force that causes it to flow.

Visible light. Light comprising wavelengths of 3400 to 740 nanometers.

Warning coloration. Conspicuous color or markings on an animal that serve to discourage potential predators.

Water balance. Maintenance of the balance of water between organisms and their surrounding environment.

Water cycle. Movement of water between atmosphere and Earth by way of precipitation and evaporation (see hydrological cycle).

Water potential. Measure of energy needed to move water molecules across a semipermeable membrane; water tends to move from areas of high or less negative potential to areas of low or more negative potential.

Water-use efficiency. Ratio of net primary production to transpiration of water by a plant.

Watershed. Entire region drained by a waterway into a lake or reservoir; total area above a given point on a stream that contributes water to the flow at that point; the topographic dividing line from which surface streams flow in two different directions.

Weather. The combination of temperature, humidity, precipitation, wind, cloudiness, and other atmospheric conditions at a specific place and time.

Weathering. Physical and chemical breakdown of rock and its components at and below Earth's surface (also see mechanical and chemical weathering).

Westerlies. The dominant east-to-west motion of the winds centered over the middle latitudes of both hemispheres.

Wetfall. Component of acid deposition that reaches Earth by some form of precipitation; wet deposition.

Wetland. A general term applied to open-water habitats and seasonally or permanently waterlogged land areas; defining the extent of a wetland is controversial because of conflicting land-use demands.

Wilting point. Moisture content of soil at which plants wilt and fail to recover their turgidity when placed in a dark, humid atmosphere; measured by oven drying.

Xeric. Dry, especially in soil.

Yield. Individuals or biomass removed or harvested from a population per unit time.

Zero-growth isocline. An isocline along which the net population growth rate is zero.

Zonation. Changes in the physical and biological structures of communities as one moves across the landscape; spatial changes in community structure.

Zooplankton. Floating or weakly swimming animals in freshwater and marine ecosystems; planktonic animals.

CREDITS

Text/Illustration Credits

Chapter 1 p. 2 Gladwin Hill, The New York Times, November 30, 1969; p. 2 Ernst Haeckel, The History of Creation, 1866

Chapter 2 2.4 http://earthobservatory.nasa.gov/Features/EnergyBalance/page3.php; 2.8b UCAR/The COMET Program. Reprinted by permission.; 2.9 Based on The COMET Program; 2.26 http://data.giss.nasa.gov/gistemp/2011/; 2.27 IPCC Third Assessment Report- Climate Change 2001: Working Group I: The Scientific Basis. Summary for Policy Makers, Fig. 2a; 2.28a CDIAC: EPI 2009.; 2.29 http://data.giss.nasa.gov/gistemp/2005/; p. 31 Report of Working Group I, "The Physical Science Basis," IPCC, 2007

Chapter 3 3.12 Adapted from Likens, Gene, An Ecosystem Approach to Aquatic Ecology: Mirror Lake and Its environment (Springer-Verlag, 1985). Reprinted by permission of G.E. Likens.; 3.19 Adapted from Doney et al. 2009.; Table 3.1 Adapted from Doney, S.C, V.J. Farby, R.A. Feely and J.A. Kleypas. 2009. Ociean Acidification: The Other C)2 Problem. Annual Review of Marine Science 1:169–192 Figure 4, p. 176; p. 49 Doney, S.C., V.J. Farby, R.A. Feely and J.A. Kleypas (2009), "Ocean Acidification: The Other CO_2 problem" Annual Review of Marine Science 1:169–192

Chapter 4 4.2 Adapted from Larcher, Walter, Physiological Plant Ecology, 2nd Edition 1980. Springer Verlag.; 4.13 Adapted from USGS, Soil Conservation Service; 4.15 Map copyright by David Burns, Fasttrack Teaching Materials. Used by permission. www .fasttrackteaching.com/burns; p. 58 Jenny, Hans, The Soil Resource: Origin and Behavior. © Copyright 1980 Springer-Verlag.; p. 64 Bulletin 55, U.S. Bureau of Soils, 1909

Chapter 5 p. 70 Ernest Mayr, What Evolution Is Basic Books, 2001; p. 70 Charles Darwin, Origin of the Species, 1859; 5.7 Fangue, N.A. and W. A. Bennett 2003. Thermal tolerance responses of laboratory-acclimated and seasonally acclimatized Atlantic Stingray, Dasyatis sabina. Copeia 2003:315–325. Data from Table 2, pg. 320. Reprinted by permission.; p. 70 Charles Darwin, Origin of the Species, 1859; 5.8 Grant, Peter R.; Ecology and Evaluation of Darwin's Finches. © 1986 Princeton University Press. Reprinted by permission of Princeton University Press.; 5.9 Grant, Peter R.; Ecology and Evaluation of Darwin's Finches. © 1986 Princeton University Press. Reprinted by permission of Princeton University Press.; 5.10 Based on data in Table 6 (pg. 268) from Grant, Peter R. Ecology and Evolution of Darwin's Finches, Princeton University Press, 1986; 5.11 Adapted from Boag, Peter T. and Grsnt, Peter R., "Intense Natural Selection in a Population of Darwin's Finches (Geospizinae) in the Galapagos," Science, October 1981. Reprinted by permission of the author.; 5.12 Grant, Peter R.; Ecology and Evaluation of Darwin's Finches. © 1986 Princeton University Press. Reprinted by permission of Princeton University Press.; 5.13 Grant, Peter R.; Ecology and Evaluation of Darwin's Finches. © 1986 Princeton University Press. Reprinted by permission of Princeton University Press.; 5.15a Reprinted by permission of the Ecological Society of America from an illustration by Laura Nagel, "Adaptative radiation in sticklebacks: Size, shape and habitat use efficiency" Ecology, vol. 74, no. 3, 1993; permission conveyed through Copyright Clearance Center, Inc.; 5.15b Adapted from Robinson 2000.; 5.16 From Angilletta, M.J., P.H. Niewiarowski, A.E. Dunhma, A.D. Leache, and W.P. Porter, 2004. Bergman's Clines in Ectotherms: Illustrating a Life-history Perspective with Sceloporine Lizards. The American Naturalist 164:168–183., Figure 5a and b pg 176. Reprinted by permission of the University of Chicago Press.; 5.17 Based on Montesinos-Navarro, A., J. Wig, F. X. Pico, and S. J. Tonsor. 2010. Arabidopsis thaliana populations show clinal variation in a climatic gradient associated with altitude. New Phytologist 189:282–294. Figure 4.; 5.18 Based on data from Table 1, pg.154 of Jordan, N. 1992. Path Analysis of Local Adaptation in Two Ecotypes of the Annual Plant Diodia teres Walt.(Rubiaceae) The American Naturalist 140:149–165.; 5.20 Grant, Peter R.; Ecology and Evaluation of Darwin's Finches. © 1986 Princeton University Press. Reprinted by permission of Princeton University Press.; 5.21 Grant, Peter R.; Ecology and Evaluation of Darwin's Finches. © 1986 Princeton University Press. Reprinted by permission of Princeton University Press.; 5.22 Based on Patel, Nipam H., "Evolutionary biology: How to build a longer beak," Nature, August 3, 2006 pp 515–516; 5.24 Adapted from Boag, Peter T. and Grsnt, Peter R., "Intense Natural Selection in a Population of Darwin's Finches (Geospizinae) in the Galapagos," Science, October 1981. Reprinted by permission.

Chapter 6 6.8 Adapted from Gomez-Aparicio, L., F. Valladares, and R. Zamora. 2006. Differential light responses of Mediterranean tree saplings: linking ecophysiology with regeneration niche in four co-occurring species.Tree Physiology 26:947–958; 6.9 Adapted from P. B. Reich et al., "Photosynthesis and respiration rates depend on leaf and root morphology and nitrogen concentration in nine boreal tree species differing in relative growth rate," Functional Ecology 12 (3): 395–40 (June 1998). Copyright © 1998 British Ecological Society.; 6.10 Adapted from C. K. Augspurger, "Light requirements of neotropical tree seedlings," Journal of Ecology 72 (3): 777–795 (Nov. 1984). Copyright © 1984 British Ecological Society; 6.12a Based on data from Iio, A., H.Fukasawa, Y. Nose, and Y.Kakubari.2004. Stomatal closure induced by high vapor pressure deficit limited midday photosynthesis at the canopy top of Fagus crenata Blume on Naeba mountain in Japan. Trees 18:510–517. Figure 1, pg. 512.; 6.13 Data from Austin,M.P., T.M. Smith, K.P. Van Niel, and A.B. Wellington2009. Physiological responses and statistical models of the environmental niche: a comparative study of two co-occurring Eucalyptus species. Journal of Ecology 97:496–507.; 6.17 Adapted from Mokany, K., R.J. Raison, and A.S. Prokushkin. 2006. Critical analysis of root : shoot ratios in terrestrial biomes. Global Change Biology 12:84–96. Figure 5, pg. 92; 6.18b Givnish TJ. 1984. Leaf and canopy adaptations in tropical forests. In: Medina E, Mooney HA, Va´zquez-Ya´nes C, eds. Physiological ecology of plants of the wet tropics. The Hague, the Netherlands: Dr W Junk, 51–84; 6.19 Adapted from Mooney et al., "Photosynthetic capacity of Death Valley plants," Carnegie Institute Yearbook 75: 310–413 (1976).; 6.21 Republished with permission of the AMerican Society of Plant Biologist from R. W. Pearcy, "Acclimation of Photosynthetic and Respiratory Carbon Dioxide Exchange to Growth Temperature in Atriplex lentiformis (Torr.) Wats.," Plant Physiology 59 (5): 795–799 (May 1977), Copyright © 1977 American Society of Plant Biologists, Figure 1 (top half) p. 796. Permission conveyed through Copyright Clearance Center, Inc.; 6.23 a&b Adapted from Woodward and Smith 1994; 6.23c Adapted from Woodward and Smith 1994; 6.24 Data from Austin et al 1986.; 6.26a Republished with permission of the Ecological society of America from P. B. Reich et al., "Leaf Life-Span in Relation to Leaf, Plant, and Stand Characteristics among Diverse Ecosystems," Ecological Monographs 62 (3): 365–392, (1982). Copyright 1982 by Ecological Society of America. Permission conveyed through Copyright Clearance Center, Inc.; 6.30a Data from Ainsworth, E.A. and A. Rogers. 2007. The response of photosynthesis and stomatal conductance to rising CO_2: mechanisms and environmental interactions. Plant, Cell and the Environment 30:258–270. Fig. 1, pg. 260.; 6.30b Data from Ainsworth, E.A. and A. Rogers. 2007. The response of photosynthesis and stomatal conductance to rising CO_2: mechanisms and environmental interactions. Plant, Cell and the Environment 30:258–270. Fig. 3, pg. 263.; 6.31 Field, C.B., R.B. Jackson and H.A. Mooney. 1995. Stomatal response to increased CO_2: implications from plant to the global scale. Plant, Cell and the Environment 18:1214–1225 Fig. 2, pg. 1218.; 6.32 Adapted from H. Poorter and M. Pérez-Soba, "Plant Growth at Elevated CO_2," Vol. 2, The Earth system: biological and ecological dimensions of global environmental change, edited by H. A. Mooney and J. G. Canadell, from Encyclopedia Of Global Environmental Change. Copyright © John Wiley & Sons, Ltd, 2002; 6.33 Adapted from H. Poorter and M. Pérez-Soba, "Plant Growth at Elevated CO_2," Vol. 2, The Earth system: biological and ecological dimensions of global environmental change, edited by H. A. Mooney and J. G. Canadell, from Encyclopedia Of Global Environmental Change. Copyright © John Wiley & Sons, Ltd, 2002; 6. FS.03 Republished with permission of Springer Science and Bus Media B.V. from Kaoru Kitajima, "Relative importance of photosynthetic traits and allocation patterns as correlates of seedling shade tolerance of 13 tropical trees," Oecologia 98: 419–428, 1994, Fig. 3, Springer-Verlag © 1994; permission conveyed through Copyright Clearance Center, Inc.; 6.FS.04a Myers, J. A. and Kitajima, K. 2007. Carbohydrate storage enhances seedling shade and stress tolerance in a neotropical forest. Journal of Ecology 95: 383–395. Fig. 6a, pg. 391.; 6.12b Iio, A., H. Fukasawa, Y. Nose, and Y.Kakubari.2004. Stomatal closure induced by high vapor pressure deficit limited midday photosynthesis at the canopy top of Fagus crenata Blume on Naeba mountain in Japan. Trees 18:510–517. Figure 1, pg. 512; p. 94 John Updike, Couples, copyright © 1968 and renewed 1996 by John Updike (Alfred A. Knopf, 1968) p 115.; 6.26b Republished with permission of the Ecological society of America from P. B. Reich et al., "Leaf Life-Span in Relation to Leaf, Plant, and Stand Characteristics among Diverse Ecosystems," Ecological Monographs 62 (3): 365–392, (1982). Copyright 1982 by Ecological Society of America. Permission conveyed through Copyright Clearance Center, Inc.

Chapter 7 7.7 Bauwens, D., A.M. Castilla, and P. Mouton. 1999. Field body temperatures, activity levels and opportunities for thermoregulation in an extreme microhabitat specialist, the girdled lizard (Cordylus macropholis). Journal of Zoology 249:11–18.; 7.15 Stillman, J. and G.N. Somero. 2000. A comparative analysis of the upper thermal tolerance limits of eastern Pacific porcelain crabs, genus Petrolisthes: Influences of latitude, vertical zonation, acclimation, and phylogeny. Physiological and Biochemical Zoology. 73(2): 200–208. Fig. 2, pg. 203. Reprinted by permission of the University of Chicago Press.; 7.17 Republished with permission of the Ecological Society of America from Blouin-Demers, G. and P.J.Weatherhead. 2001. Thermal ecology of black rates snakes (Elapheobsoleta) in a thermally challenging environment. Ecology 82:3025–3043. Fig.4b, pg.3. Permission conveyed through Copyright Clearance Center, Inc.; 7.18 Glanville, E.J. and Seebacher, F.J. 2006. Compensation for environmental change by complementary shifts of thermal sensitivity and thermo regulatory behavior in an ectotherm. Journal of Experimental Biology 209:4869–4877 Fig. 2, pg. 4872.; 7.20 Data from Schmidt-Nielsen, K. Animal Physiology: Adaptation and Environment, 5th Edition 1997; 7.21 Adapted from Schmidt-Nielson 1979; 7.22 McKechnie, A.E. and B.G.Lovegrove. 2001. Thermoregulationand the energetic significance of clustering behavior in the White-backed Mousebird (Coliuscolius). Physiological and Biochemical Zoology 74:238–249. Fig. 5D, pg.718. Reprinted by permission of the University of Chicago Press.; 7.23d Adapted from Schmidt-Nielsen, K. Animal Physiology: Adaptation and Environment, 5th Edition 1997.; 7.25 Adapted from James 1971.; 7.27a Teplitsky, C., J.A. Mills, J.S. Alho, J.W. Yarrall, and J. Merila. 2008. Bergman's rule and climate change revisited" disentangling environmental and genetic responses in a wild bird population. Proceedings of the National Academy of

Sciences 105:13492–13496. Figures 1 and 2 a, pg. 13493. Copyright (2008) National Academy of Sciences, U.S.A.; 7.27b Teplitsky, C., J.A. Mills, J.S. Alho, J.W. Yarrall, and J. Merila. 2008. Bergman's rule and climate change revisited" disentangling environmental and genetic responses in a wild bird population. Proceedings of the National Academy of Sciences 105:13492–13496. Figures 1 and 2 a, pg. 13493. Copyright (2008) National Academy of Sciences, U.S.A.; 7.02 Republished with permission of the Ecological Society of America, from Martin Wikelski, "Energy Limits to Body Size in a Grazing Reptile, the Galapagos Marine Iguana," Ecology 78 (7): 2204–2217 (Oct. 1997). Copyright 1997 by Ecological Society of America; permission conveyed through Copyright Clearance Center, Inc.; 7.03 Adapted from From M. Wikelski and F. Trillmich, "Body Size and Sexual Size Dimorphism in Marine Iguanas Fluctuate as a Result of Opposing Natural and Sexual Selection: An Island Comparison," Evolution 51 (3): 922–936 (June 1997). Copyright © 1997 Society for the Study of Evolution; 7.04 Adapted from M. Wikelski et al., "Marine iguanas shrink to survive El Niño," Nature 403, p. 37, Fig. 2b, (Jan 6, 2000). Copyright © 2000.

Chapter 8 8.7 Krebs, Charles J., Ecology: The Experimental Analysis of Distribution and Abundance, 5th Ed., © 2001. Reprinted and Electronically reproduced by permission of Pearson Education, Inc., Upper Saddle River, New Jersey.; 8.8 Bleich, V.C., J. D. Wehausen and S. A. HollSource. 1990. Desert-Dwelling Mountain Sheep:Conservation Implications of a Naturally Fragmented Distribution.Conservation Biology 4:383–390 Fig. 1, pg. 387; 8.20a Valverde, T. and J. Silvertown. 1998. Variation in the Demography of a Woodland Understorey Herb (Primula vulgaris) along the Forest Regeneration Cycle: Projection Matrix. Journal of Ecology 86:545–562 Fig. 4, pg. 556.; 8.20b Valverde, T. and J. Silvertown. 1998. Variation in the Demography of a Woodland Understorey Herb (Primula vulgaris) along the Forest Regeneration Cycle: Projection Matrix. Journal of Ecology 86:545–562 Fig. 4, pg. 556.; 8.20 Table Valverde, T. and J. Silvertown. 1998. Variation in the Demography of a Woodland Understorey Herb (Primula vulgaris) along the Forest Regeneration Cycle: Projection Matrix. Journal of Ecology 86:545–562 Table 1pg. 548.; 8.21a Donald, P.F. 2007. Adult sex ratios in adult bird populations. Ibis 149:671–692. Fig. 1, pg. 674 and Fig. 2, pg. 676.; 8.21b Donald, P.F. 2007. Adult sex ratios in adult bird populations. Ibis 149:671–692. Fig. 1, pg. 674 and Fig. 2, pg. 676.; 8.22 Adapted from Engle (1960) "Yellow-poplar seedfall pattern" Central States Forest Experiment Station Note 143; 8.24 Krebs, Charles J., Ecology: The Experimental Analysis of Distribution and Abundance, 5th Ed., © 2001. Reprinted and Electronically reproduced by permission of Pearson Education, Inc., Upper Saddle River, New Jersey.; 8.24 http://www.fs.fed.us/foresthealth/technology/fdar_maps.shtml; 8.26a Reprinted by permission. http://www.eddmaps.org/distribution/; 2 Alberto F., L. Gouveia, S. Arnaud-Haond, J.L. Pérez-Lloréns, C.M. Duarte CM, and E.A. Serrão. 2005. Within-population spatial genetic structure, neighbourhood size and clonal subrange in the seagrass Cymodocea nodosa.Molecular Ecology14: 2669–2681.Fig. 2, pg. 2674; 3 Alberto F., L. Gouveia, S. Arnaud-Haond, J.L. Pérez-Lloréns, C.M. Duarte CM, and E.A. Serrão. 2005. Within-population spatial genetic structure, neighbourhood size and clonal subrange in the seagrass Cymodocea nodosa.Molecular Ecology14: 2669–2681.Fig. 2, pg. 2674; 1 Figure 1 - Sampling grid of Cymodocea nodosa in Cadiz Bay., Source: Alberto F., L. Gouveia, S. Arnaud-Haond, J.L. Pérez-Lloréns, C.M. Duarte CM, and E.A. Serrão. 2005. Within-population spatial genetic structure, neighbourhood size and clonal subrange in the seagrass Cymodocea nodosa. Molecular Ecology14: 2669–2681.Fig. 2, pg. 2674, Wiley-Blackwell Publishing Ltd.

Chapter 9 9.5 From V. B. Scheffer, "The Rise and Fall of a Reindeer Herd," The Scientific Monthly, 73: 356–362, (Dec. 1951). Reprinted with permission from AAAS.; 9.9 Republished with the permission of the Ecological Society of America from P. W. Sherman and M. L. Morton. 1984. Demography of Belding's Ground Squirrels. Ecology 65:1617–1628. Data from : Tale 1, pg. 1621, Adapter from: Figure 1a, pg. 1622. Permission conveyed through Copyright Clearance Center, Inc.; 9.11 E. S. Deevey, Jr. 1947. Life Tables for Natural Populations of Animals. The Quarterly Review of Biology 22:283–314. Data from: Table 1, pg. 289; D. W. Tinkle, J. D. Congdon and P. C. Rosen. 1981. Nesting Frequency and Success: Implications for the Demography of Painted Turtles. Ecology 62:1426–1432. Data from: Table 3, pg. 1431; S. G. Tilley. 1980. Life Histories and Comparative Demography of Two Salamander Populations. Copeia 4:806–821. Data from: Table 8, pg. 819; M. B. Vinegar. 1975. Demography of the Striped Plateau Lizard, Sceloporus virgatus. Ecology 56:172–182. Data from: Table 9, pg. 179.; 9.13 From V. B. Scheffer, "The Rise and Fall of a Reindeer Herd," The Scientific Monthly, 73: 356–362, (Dec. 1951). Reprinted with permission from AAAS. Reprinted with permission from AAAS.; 9.14 The INCN Red List of Threatened Species (www.iucnredlist.org); 9.15 J. P. M. Syvitski and A. Kettner. 2011. Sediment flux and the Anthropocene. Philosophical Transactions of the Royal Society A 369:957–975. Fig. 3, pg. 962. By permission of the Royal Society.; 9.16 N. M. Burkhead. 2012. Extinction Rates in North American Freshwater Fishes, 1900–2010. BioScience 62: 798–808. Data source–Map by USGS - http://fl.biology.usgs.gov/extinct_fishes/; 9.17 Based on W.R. Ostlie and J.L. Haferman. 1999. Ecoregional conservation in the Great Plains. Proceedings of the North American Prairie Conference 16:136–148. Fig. 5, pg. 142; 9.19 Data from U.S. Department of Interior 2012.; Table 9.2 Data from R. W. Campbell 1969.; Table 9.3 Data from Sharitz and McCormick 1973.; 9.19 Source: Data from U.S. Department of Interior 2012.; p. 173 Source: © Pearson Education.

Chapter 10 Table 10.1 © Pearson Education; 10.1 © Pearson Education; 10.6 From Primack, Richard B., Hall, Pamela, "cost of Reproduction in the Pink Lady's Slipper Orchid: A Four-Year Experimental Study," The American Naturalist, 136 (5): 638–656, (Nov., 1990). Copyright © 1990 The University of Chicago Press. Used by permission of the publisher; 10.5 From Primack, Richard B., Hall, Pamela, "cost of Reproduction in the Pink Lady's Slipper Orchid: A Four-Year Experimental Study," The American Naturalist, 136 (5): 638–656, (Nov., 1990). Copyright © 1990 The University of Chicago Press. Used by permission of the publisher; 10.7a Adapted from Eis 1965; 10.7b From A. C. Tsikliras, "A phenotypic trade-off between previous growth and present fecundity in round sardinella Sardinella aurita," Population Ecology 49: 221–227 (2007), © The Society of Population Ecology and Springer 2007.; 10.8a Data from Hines, A.H. (1991) "Fecundity and Reproductive Output in Nine Species of Cancer Crabs (Crustacea, Brachyura, Cancridae)" Canadian journal of Fisheries and Aquatic Sciences, 48: 267–275; 10.8b From L. Wauters and A. Dhondt, "Body weight, longevity, and reproductive success in red squirrels (Sciurus vulgaris)," Journal of Animal Ecology 58: 637–651 (1989). Copyright © 1989 British Ecological Society.; 10.9 © Pearson Education; 10.10 From David A. Reznick, et al, "Experimentally induced life-history evolution in a natural population," Nature 346: 357–359 (July 26, 1990). Copyright © 1990 Macmillan Publishers Ltd; 10.13a © Pearson Education; 10.12 Data from Dijkstra et al., "Brood size manipulations in the kestrel (Falco tinnunculus): Effects on offspring and parent survival" Journal of Animal Ecology Vol. 59, no. 1, February, 1990.; 10.14 © Pearson Education; 10.15a From P. Werner and W. Platt, "Ecological relationships of co-occurring goldenrods," The American Naturalist 110: 959–971 (1976). Copyright © 1976 The University of Chicago Press. Used by permission of the publisher.; 10.15b From T.A. Oksanen et al., "Hormonal Manipulation of Offspring Number: Maternal Effort and Reproductive Costs," Evolution 56 (7): 1530–1537 (July 2002). Copyright © 2002 The Society for the Study of Evolution. All rights reserved.; 10.17 © Pearson Education; 10.18 N. E. Tamburi and P.R. Martin, "Reaction norms of size and age at maturity of Pomacea canaliculata (Gastropoda: Ampullariidae) under a gradient of food deprivation," Journal of Molluscan Studies 75 (1): 19–26, (2009), Oxford University Press on behalf of The Malacological Society of London.; 10.20a From Marion Petrie, "Improved growth and survival of offspring of peacocks with more elaborate trains," Nature 371: 598–599, (Oct. 1994). Copyright © 1994. Macmillan Publishers Ltd.; 10.20b From Marion Petrie, "Improved growth and survival of offspring of peacocks with more elaborate trains," Nature 371: 598–599, (Oct. 1994). Copyright © 1994. Macmillan Publishers Ltd.; 10.23b Reid, Jane M., et al., Fitness Correlates of Song Repertoire Size in Free-Living Song Sparrows (Melospiza melodia). The American Naturalist, Vol. 165, No. 3 (March 2005), pp. 299–310. Reprinted by permission of the University of Chicago Press.; 10.24a Republished with permission of The Royal Society, from Temeles, E.J. and W. J. Kress. 2010. Mate choice and mate competition by a tropical hummingbird at a floral resource. Proceedings of the Royal Society B 277:160–1613. Fig.1 and Fig. 3, pg. 1610; permission conveyed through Copyright Clearance Center, Inc.; 10.24b Republished with permission of Royal Society from Temeles, E.J. and W. J. Kress. 2010. Mate choice and mate competition by a tropical hummingbird at a floral resource. Proceedings of the Royal Society B 277:1607–1613. Fig.1 and Fig. 3, pg. 1610: permission conveyed through Copyright Clearance Center, Inc.; 10.25 © Pearson Education; 10.27 © Pearson Education; 10.28 © Pearson Education; 10.29 Data from U.S. Center for Disease Control.; 10.30 Data from U.S. Center for Disease Control; 10FS2 Based on Basolo, A. L. 1990b. Female preference for male sword length in the green swordtail, Xiphophorus helleri. Animal Behaviour 40:332–338.; 10FS3 Based on Basolo, A. L., and G. Alcaraz. 2003. The turn of the sword: Length increases male swimming costs in swordtails. Proceedings of the Royal Society of London 270:1631–1636.; 10.1 © Pearson Education; 10.1 © Pearson Education; p. 205 Shaffer, M. L. 1981. "Minimum population sizes for species conservation." BioScience 31:131–134.; 10.1QE © Pearson Education.

Chapter 11 11.1 © Pearson Education; 11.2 © Pearson Education; 11.3 © Pearson Education; 11.4 © Pearson Education; 11.6 a, b, c, d Data from L. Wang, "Effects of intraspecific competition on growth and photosynthesis of Atriplex prostrata," Aquatic Botany 83 (3): 187–192, Nov. 2005.; 11.7 a, b, c Republished with permission of the Ecological Society of America from R. Relyea, "Competitor-Induced Plasticity in Tadpoles: Consequences, Cues, and Connections to Predator-Induced Plasticity," Ecological Monographs 72(4): 523–540, (Nov 2002). Copyright 2002 by Ecological Society of America; permission conveyed through Copyright Clearance Center, Inc.; 11. Flockhart, D.T.T., Martin, T.G., and Norris, D.R. 2012. Experimental Examination of Intraspecific Density-Dependent Competition during the Breeding Period in Monarch Butterflies (Danaus plexippus). PLoS ONE 7(9): e45080. doi:10.1371/journal.pone.0045080 Figure 2; 11.10 From R. N. Hughes and C. L. Griffiths, "Self-Thinning in Barnacles and Mussels: The Geometry of Packing," The American Naturalist 132 (4) (Oct. 1988). Copyright © 1988 The University of Chicago Press. Used by permission of the publisher.; 11.11 Republished with permission of the Ecological Society of America from T. M. Jenkins, et al., "Effects of Population Density on Individual Growth of Brown Trout in Streams," Ecology 80 (3): 941–956, (April 1999). Copyright 1999 by Ecological Society of America. Permission conveyed through Copyright Clearance Center, Inc.; 11.12a Republished with the permission of the Ecological Society of America from E. R. Keeley, "Demographic Responses to Food and Space Competition by Juvenile Steelhead

Trout," Ecology 82(5): 1247–1259, (May 2001). Copyright 2001 by Ecological Society of America. Permission conveyed through Copyright Clearance Center, Inc.; 11.12b Republished with the permission of the Ecological Society of America from E. R. Keeley, "Demographic Responses to Food and Space Competition by Juvenile Steelhead Trout," Ecology 82(5): 1247–1259, (May 2001). Copyright 2001 by Ecological Society of America. Permission conveyed through Copyright Clearance Center, Inc.; 11.13a Based on "Density-dependent processes and management strategy for the northwest Atlantic harp seal populations" by P. F. Lett, from Dynamics Of Large Mammal Populations, edited by C. W. Fowler and T. D. Smith.; 11.13b "Density-dependent processes and management strategy for the northwest Atlantic harp seal populations" by P. F. Lett, from Dynamics Of Large Mammal Populations, edited by C. W. Fowler and T. D. Smith.; 11.14a Adapted from A. M. Schueller et al., "Density dependence of walleye maturity and fecundity in Big Crooked Lake, Wisconsin, 1997–2003," North American Journal of Fisheries Management. Copyright © 2005. Reprinted by permission from American Fisheries Society.; 11.14b Adapted from A. M. Schueller et al., "Density dependence of walleye maturity and fecundity in Big Crooked Lake, Wisconsin, 1997–2003," North American Journal of Fisheries Management. Copyright © 2005. Reprinted by permission from American Fisheries Society; 11.14c Adapted from A. M. Schueller et al., "Density dependence of walleye maturity and fecundity in Big Crooked Lake, Wisconsin, 1997–2003," North American Journal of Fisheries Management. Copyright © 2005. Reprinted by permission from American Fisheries Society.; 11.15a Adapted from Fery and Janick 1971; 11.15b Based on Ball, Rosalind A., Purcell, Larry C., Vories, Earl D. 2000. Short-Season Soybean Yield Compensation in Response to Population and Water Regime. Crop Science 40: 1070–1078.; 11.17a Source: Republished with permission from the Ecological Society of America from J. Krebs, "Territory and breeding density in the great tit Parus major," Ecology 52: 2–22 (Jan. 1971). Copyright 1971 by Ecological Society of America. Permission conveyed through Copyright Clearance Center, Inc.; 11.17b Republished with permission from the Ecological Society of America from J. Krebs, "Territory and breeding density in the great tit Parus major," Ecology 52: 2–22 (Jan. 1971). Copyright 1971 by Ecological Society of America. Permission conveyed through Copyright Clearance Center, Inc.; 11.18a From Both, Christian, and Visser, Marcel E., "Breeding Territory Size Affects Fitness An Experimental Study on Competition at the Individual Level," Journal of Animal Ecology, Vol. 69, No. 6 (Nov., 2000). pp. 1021–1030. Table 1 pg. 1025.; 11.48b Data from Both, Christian, and Visser, Marcel E., "Breeding Territory Size Affects Fitness An Experimental Study on Competition at the Individual Level," Journal of Animal Ecology, Vol. 69, No. 6 (Nov., 2000). pp. 1021–1030. Table 1 pg. 1025.; 11.19 Data from Smith and Goodman 1986.; 11.20a From J. F. Cahill, Jr. and B. B. Casper, "Investigating the Relationship between Neighbor Root Biomass and Belowground Competition: Field Evidence for Symmetric Competition Below ground," Oikos, 90 (2): 311–320, (Aug. 2000). © 2000 Nordic Society Oikos.; 11.20b From J. F. Cahill, Jr. and B. B. Casper, "Investigating the Relationship between Neighbor Root Biomass and Belowground Competition: Field Evidence for Symmetric Competition Belowground," Oikos, 90 (2): 311–320, (Aug. 2000). © 2000 Nordic Society Oikos.; 11.20 © Pearson Education Inc.; 11.21 © Pearson Education; 11.22 Republished with permission of John Wiley and Sons Inc., from, E.E. Hackney and J.B. McGraw, "Experimental Demonstration of an Allee Effect in American Ginseng," Conservation Biology 15 (1): 129–136, (Feb. 2001). Permission conveyed through Copyright Clearance Center, Inc.; 11.23 Adapted from J. S. Brashares, J. R. Werner and A. R. E. Sinclair. 2010. Social 'meltdown' in the demise of an island endemic: Allee effects and the Vancouver Island marmot. Journal of Animal Ecology 79:965–973. Fig. 1, pg. 966.; 11.24 DeLoach, C. J. 1974. Rate of increase of populations of cabbage, green peach, and turnip aphids at constant temperatures. Annals of the Entomological Society of America 67:332–340.; 11.25a From L. D. Mech et al., "Relationship of Deer and Moose Populations to Previous Winters' Snow," Journal of Animal Ecology, 56 (2): 615–627, (June 1987). Copyright © 1987 British Ecological Society.; 11.25b From L. D. Mech et al., "Relationship of Deer and Moose Populations to Previous Winters' Snow," Journal of Animal Ecology, 56 (2): 615–627, (June 1987). Copyright © 1987 British Ecological Society.; 11.27 Krebs, Charles J., Ecology: The Experimental Analysis of Distribution and Abundance, 5th Ed., © 2001. Reprinted and Electronically reproduced by permission of Pearson Education, Inc., Upper Saddle River, New Jersey.; 11.28 Adapted from From Joel Berger, "Persistence of Different-Sized Populations: An Empirical Assessment of Rapid Extinctions in Bighorn Sheep," Conservation Biology 4(1): 91–98 (Mar, 1990). Copyright © 1990 Blackwell Publishing.; 11.2FS From T. S. Sillett et al., "Impacts of a Global Climate Cycle on Population Dynamics of a Migratory Songbird," Science 288 (June 16, 2000). Copyright © 2000 AAAS.; 11.3FS From T. S. Sillett et al., "Impacts of a Global Climate Cycle on Population Dynamics of a Migratory Songbird," Science 288 (June 16, 2000). Copyright © 2000 AAAS.; 11.4FS From Rodenhouse, N. L., T. S. Sillett, P. J. Doran, and R. T. Holmes. 2003. Multiple density- dependence mechanisms regulate a migratory bird population during the breeding season. Proceedings of the Royal Society of London 270:2105–2110. Figure 2, pg. 2106. By permission of the Royal Society.

Chapter 12 12.2a © Pearson Education; 12.2b © Pearson Education; 12.3a © Pearson Education; 12.3b © Pearson Education; 12.4 © Pearson Education; 12.5 © Pearson Education; 12.7 © Pearson Education; 12.8b Brodie Jr., E.D., B.J. Ridenhour and E.D.Brodie 2002. The evolutionary response of predators to dangerous prey: hotspots and cold spots in the geographic mosaic of co-evolution between garter snakes and newts. Evolution 56:2067–2082. Fig 3b, pg. 2074; 12.8c Brodie Jr., E.D., B.J. Ridenhour and E.D.Brodie 2002. The evolutionary response of predators to dangerous prey: hotspots and cold spots in the geographic mosaic of co-evolution between garter snakes and newts. Evolution 56:2067–2082. Fig 5, pg. 2076; 12.9 a&b © Pearson Education; 12.10 © Pearson Education; 12.11 a, b, c © Pearson Education; 12.12 a, b © Pearson Education; 12.13a From J. B. Grace and R.B. Wetzel, "Habitat Partitioning and Competitive Displacement in Cattails (Typha): Experimental Field Studies," The American Naturalist, Vol. 118, No. 4: 463–474, (Oct 1981). Copyright © 1981 The University of Chicago Press. Used by permission of the publisher.; 12.13b From J. B. Grace and R.B. Wetzel, "Habitat Partitioning and Competitive Displacement in Cattails (Typha): Experimental Field Studies," The American Naturalist, Vol. 118, No. 4: 463–474, (Oct 1981). Copyright © 1981 The University of Chicago Press. Used by permission of the publisher.; 12.16a Based on data from the U.S. Census Bureau; 12.16b Based on World Health Organization; 12.17 Shochat et al. 2010. Invasion, competition and biodiversity loss in urban ecosystems. BioScience 60:199–208. Fig. 3, pg. 203; 12.1 Table © Pearson Education. 12.13c From J. B. Grace and R.B. Wetzel, "Habitat Partitioning and Competitive Displacement in Cattails (Typha): Experimental Field Studies," The American Naturalist, Vol. 118, No. 4: 463–474, (Oct 1981). Copyright © 1981 The University of Chicago Press. Used by permission of the publisher.

Chapter 13 p. 263 Charles Darwin, On the Origin of the Species (John Murray 1859).; 13.1a © Pearson Education; 13.1b © Pearson Education; 13.2 a, b, c, d © Pearson Education; 13.14 D. Tilman et al., "Competition and Nutrient Kinetics Along a Temperature Gradient: An Experimental Test of a Mechanistic Approach to Niche Theory," Limnology and Oceanography, 26 (6): 1020–1033, (Nov. 1981). © 1981 by the Association for the Sciences of Limnology and Oceanography, Inc. Reprinted by permission of ASLO.; 13.4b D. Tilman et al., "Competition and Nutrient Kinetics Along a Temperature Gradient: An Experimental Test of a Mechanistic Approach to Niche Theory," Limnology and Oceanography, 26 (6): 1020–1033, (Nov. 1981). © 1981 by the Association for the Sciences of Limnology and Oceanography, Inc. Reprinted by permission of ASLO.; 13.4c D. Tilman et al., "Competition and Nutrient Kinetics Along a Temperature Gradient: An Experimental Test of a Mechanistic Approach to Niche Theory," Limnology and Oceanography, 26 (6): 1020–1033, (Nov. 1981). © 1981 by the Association for the Sciences of Limnology and Oceanography, Inc. Reprinted by permission of ASLO.; 13.5 From Vittor, A.Y., W. Pan, R. H. Gilman, J. Tielsch, G. Glass, T. Shields, W. Sa´nchez-Lozano, V. V. Pinedo, E. Salas-Cobos, S. Flores, and J. A. Patz. 2009. Linking Deforestation to Malaria in the Amazon: Characterization of the Breeding Habitat of the Principal Malaria Vector, Anopheles darlingi. American Journal of Tropical Medicine and Hygene 81:5–12. Fig. 3, pg. 10; 13.7 Adapted from Groves and Williams, 1975.; 13.8b From M. P. Austin et al., "Relative Growth of Six Thistle Species Along a Nutrient Gradient with Multispecies Competition," Journal of Ecology 73(2): 667–684, (July 1985). Copyright © 1985 British Ecological Society.; 13.8b From M. P. Austin et al., "Relative Growth of Six Thistle Species Along a Nutrient Gradient with Multispecies Competition," Journal of Ecology 73(2): 667–684, (July 1985). Copyright © 1985 British Ecological Society.; 13.9 From M. P. Austin et al., "Relative Growth of Six Thistle Species Along a Nutrient Gradient with Multispecies Competition," Journal of Ecology 73(2): 667–684, (July 1985). Copyright © 1985 British Ecological Society.; 13.10 From R. W. S. Fynn et al., "Plant strategies and trait trade-offs influence trends in competitive ability along gradients of soil fertility and disturbance," Journal of Ecology 93: 384–394 (2005). Copyright © 2005 British Ecological Society.; 13.11a Republished with permission of Ecological Society of America, from N. C. Emery et al., "Competition and Salt-Marsh Plant Zonation: Stress Tolerators May Be Dominant Competitors," Ecology 82 (9): 2471–2485, (Sep. 2001). Copyright 2001 by Ecological Society of America; permission conveyed through Copyright Clearance Center, Inc.; 13.11b Data from Bertness. M.D. 1991. Zonation of Spartina patens and Spartina alterniflora in New England Salt Marsh. Ecology 72: 138–148. Fig. 2, pg. 143.; 13.11c Data from Bertness. M.D. 1991. Zonation of Spartina patens and Spartina alterniflora in New England Salt Marsh. Ecology 72: 138–148. Fig. 2, pg. 143; 13.13a Republished with the permission of the Ecological Society of America, from Jessica Gurevitch, "Competition and the Local Distribution of the Grass Stipa Neomexicana," Ecology 67 (1): 46–57, (Feb. 1986). Copyright 1986 by Ecological Society of America, permission conveyed through Copyright Clearance Center, Inc.; 13.13b Republished with permission of Ecological Society of America, from Jessica Gurevitch, "Competition and the Local Distribution of the Grass Stipa Neomexicana," Ecology 67 (1): 46–57, (Feb. 1986). Copyright 1986: pemission conveyed through Copyright Clearance Center, Inc.; 13.13c Reprinted with permission of the Ecological Society of America, from Jessica Gurevitch, "Competition and the Local Distribution of the Grass Stipa Neomexicana," Ecology 67 (1): 46–57, (Feb. 1986). Copyright 1986 by Ecological Society of America; permission conveyed through Copyright Clearance Center, Inc.; 13.14 From D. I. Bolnick, et al., "Ecological release from inter specific competition leads to decoupled changes in population and individual niche width," Proceedings of The Royal Society B 277: 1789–1797 (17 Feb 2010). © The Royal Society. Reprinted by permission of The Royal Society.; 13.15 Reprinted with permission of the Ecological Society of America from N. K. Wieland, F. A. Bazzaz, "Physiological Ecology of Three Codominant Successional Annuals," Ecology: 56 (3):

681–688, (May 1975). Copyright 1975 by Ecological Society of America; permission conveyed through Copyright Clearance Center, Inc.; 13.16 From T. Dayan et al., "Feline Canines: Community-Wide Character Displacement Among the Small Cats of Israel," The American Naturalist 136 (1): 39–60, (July 1990). Copyright © 1990 The University of Chicago Press. Used by permission of the publisher.; 13.17a From Ecology And Evolution Of Darwin's Finches by Peter R. Grant. © 1986 Princeton University Press. Reprinted by permission of Princeton University Press.; 13.17b From Ecology And Evolution Of Darwin's Finches by Peter R. Grant. © 1986 Princeton University Press. Reprinted by permission of Princeton University Press.; 13.17c From Ecology And Evolution Of Darwin's Finches by Peter R. Grant. © 1986 Princeton University Press. Reprinted by permission of Princeton University Press.; 13.18 2013 International Mapping.; 13.19 Reprinted by permission of ubancoyoteresearch.com; 13.20 Berger, K.M and E.M.Gese. 2007. Does interference competition with wolves limit the distribution and abundance of coyotes? Journal of Animal Ecology 76:1075–1085. Fig. 4, pg. 1081; 13.2aQE © Pearson Education; 13.2bQE © Pearson Education; 13.2Fig2QE © Pearson Education.

Chapter 14 p. 286 Alfred, Lord Tennyson, In Memoriam A.H.H. (1849); p. 309 John Cabot, 1497; 14.1 © Pearson Education; 14.2a, b, c © Pearson Education; 14.3 a, b © Pearson Education; 14.6 a, b © Pearson Education; 14.7a B. W. Frost, "Effects of Size and Concentration of Food Particles on the Feeding Behavior of the Marine Planktonic Copepod Calanus Pacificus," Limnology and Oceanography, 17 (6): 805–814 (Nov 1972). © 1972 by the Association for the Sciences of Limnology and Oceanography, Inc. Reprinted by permission of ASLO.; 14.7b Reprinted by permission of the Ecological Society of America from O'Donoghue, M., S. Boutin, C.J. Krebs, G. Zuleta, D.L. Murray and E.J. Hofer. 1998. Functional responses of coyotes and lynx to the snowshoe hare cycle. Ecology 79:1193–1208 Fig. 5b, pg. 1200. © 1998 Ecological Society of American; permission conveyed through the Copyright Clearance Center, Inc.; 14.7c Reprinted with permission of the Ecological Society of America, from R.D. Seitz, et al., "Density-dependent Predation, Habitat Variation, and the Persistence of Marine Bivalve Prey," Ecology 82 (9): 2435–2451, (Sept. 2001). Copyright 2001 by Ecological Society of America; permission conveyed through Copyright Clearance Center, Inc.; 14.8 ab © Pearson Education; 14.9a © Pearson Education; 14.9b From R. N. Hughes and M. I. Croy, "An Experimental Analysis of Frequency-Dependent Predation (Switching) in the 15-Spined Stickleback, Spinachia spinachia," Journal of Animal Ecology, 62 (2): 341–352, (Apr. 1993). Copyright © 1993 British Ecological Society.; 14.11 a, b, c Reprinted with permission of the Ecological Society of America, from M. Salamolard, et al., "Responses of an Avian Predator to Variations in Prey Density at a Termperate Latitude," Ecology 81 (9): 2428–2441, (2000). Copyright 2000 by Ecological Society of America; permission conveyed through Copyright Clearance Center, Inc.; 14.12 Reprinted with permission of the Ecological Society of America from W. Jedrzejewski, et al., "Weasel Population Response, Home Range, and Predation on Rodents in a Deciduous Forest in Poland," Ecology 76 (1): 179–195, (Jan 1995). Copyright 1995 by Ecological Society of America; permission conveyed through Copyright Clearance Center, Inc.; 14.13a Republished with permission of Ecological Society of America from O'Donoghue, M., S.Boutin, C. J. Krebs, G. Zuleta, D. L. Murray and E. J. Hofer. 1998. Functional responses of coyotes and lynx to the snowshoe hare cycle. Ecology 79: 1193–1208 Fig. 1, pg. 1198, permission conveyed through Copyright Clearance Center, Inc.; 14.13b O'Donoghue, M., S. Boutin, C. J. Krebs and E. J. Hofer. 1997. Numerical responses of coyotes and lynx to the snowshoe hare cycle. Oikos 80:150–162. Fig. 6a, pg. 155.; 14.14 a, b From N. B. Davies, "Prey Selection and Social Behaviour in Wagtails (Aves: Motacillidae)," Journal of Animal Ecology 46 (1): 37–57, (Feb. 1977). Copyright © 1977 British Ecological Society.; 14.23 Reprinted with permission by the Ecological Society of America from K. S. Williams, et al., "Emergence of 13-Yr Periodical Cicadas (Cicadidae: Magicicada): Phenology, Mortality, and Predators Satiation," Ecology 74 (4): 1143–1152, (June 1993). Copyright 1993 by Ecological Society of America; permission conveyed through Copyright Clearance Center, Inc.; 14.29 From "Consequences of reduced activity in the mayfly Baetis tricaudatus" in Steven L. Lima, "Nonlethal Effects in the Ecology of Predator-Prey Interactions," Bioscience vol. 48, no. 1(January 1998), © 1998 by the American Institute of Biological Sciences. Published by the University of California Press.; 14.30 Reprinted with permission by the Ecological Society of America from E. H. Nelson et al., "Predators Reduce Prey Population Growth by Inducing Changes in Prey Behavior," Ecology, 85 (7): 1853–1858, (July 2004). Copyright 2004 by Ecological Society of America; permission conveyed through Copyright Clearance Center, Inc.; 14.31 © Pearson Education; 14.32 United Nations Food and Agriculture Organization (FAO), Fishstat-PC (FAO, Rome, 1997).; 14.33 © Pearson Education; FS4 Reprinted with permission of the Ecological Society of America from R. Relyea, "The Lasting Effects of Adaptive Plasticity: Predator Induced Tadpoles Become Long-Legged Frogs," Ecology 82 (7): 1947–1955 (Jul 2001). Copyright 2001 by Ecological Society of America; permission conveyed through Copyright Clearance Center, Inc.; 13.1QE © Pearson Education; p. 300 Carroll, Lewis: Through the Looking-Glass and What Alice Found There, Chapter 2.

Chapter 15 15.10b lichens.science.oregonstate.edu; 15.03b © Pearson Education; 15.4 © Pearson Education; 15.06 Based on Schall 1983.; 15.07 b, c Republished with the permission of the Ecological Society of America, from K. D. Lafferty and A. K. Morris, "Altered Behavior of Parasitized Killifish Increases Susceptibility to Predation by Bird Final Hosts," Ecology 77 (5): 1390–1397, (July 1996). Copyright 1996 by Ecological Society of America; permission conveyed through Copyright Clearance Center, Inc.; 15.08 © Pearson Education; 15.09b © Pearson Education; 15.11a © Pearson Education; 15.12b © Pearson Education; 15.16 © Pearson Education; 15.18 a, b, c Allan, B.F., F. Keesing and R.S. Ostefeld. 2003. Effect of Forest Fragmentation on Lyme Disease Risk. Conservation Biology 17:267–272. Fig.1, pg. 270.; 15.19 Vittor, A.Y., W. Pan, R. H. Gilman, J. Tielsch, G. Glass, T. Shields, W. Sa´nchez-Lozano, V. V. Pinedo, E. Salas-Cobos, S. Flores, and J. A. Patz. 2009. Linking Deforestation to Malaria in the Amazon: Characterization of the Breeding Habitat of the Principal Malaria Vector, Anopheles darlingi. American Journal of Tropical Medicine and Hygene 81:5–12. Fig. 3, pg. 10; 15.1QE © Pearson Education; 15.2QE Peter J. Morin, Community Ecology. Copyright © 1999 by Blackwell Science, Inc.; 15.2FS Republished with permission of the Ecological Society of America, from J. J. Stachowicz and M. E. Hay, "Mutualism and Coral Persistence: The Role of Herbivore Resistance to Algal Chemical Defense," Ecology 80 (6): 2085–2101, (Sep 1999). Copyright 1999 by Ecological Society of America; permission conveyed through Copyright Clearance Center, Inc.; 15.3FS "The crab Mithrax forceps within the branches of the coral Oculina obscura" in John J. Stachowicz, "Mutualism, Facilitation, and the Structure of Ecological Communities," BioScience vol. 51 no. 3 (March 2001), pp. 235–246. © 2001 by the American Institute of Biological Sciences. Published by the University of California Press.; p. 315 Margulis, Lynn and Dolan, Michael F., Early Life: Evolution on the PreCambiran Earth (Jones & Bartlett, 2001) pg. 157.

Chapter 16 16.1 © Pearson Education; 16.2 a, b Data from Phillip 1998 as presented in Magurran, A.E. 2004. Ecological diversity and its measurement. Princeton University Press, Princeton, NJ. Figure 2.2, pg. 20.; 16.03 aData from Williams 1963 as presented in Magurran, A.E. 2004. Measuring biological diversity. Blackwell Scientific, Oxford.; 16.03b Data from Edgell, R.A. and C.C. Long. 2009. Wabash River: Carroll, Cass, Huntington, Miami, Tippecanoe, and Wabash Counties 2008 Fish Management Report. Fisheries Section, Indiana Department of Natural Resources, Division of Fish and Wildlife. Indianapolis, IN. Data from Table 1, pg. 16; 16.04b Reprinted by permission of the Ecological Society of America from Riginos, C., J.B. Grace, D.J. Augustine, and T.P. Young. 2009. Local versus landscape scale effects of savanna trees on grasses. Journal of Ecology 97:1337–1345. Figure 2a, pg. 1341 © 2009 Ecological Society of America: permission conveyed through Copyright Clearance Center, Inc.; 16.04c Reprinted by permission of the Ecological Society of America from Riginos, C. and J.B. Grace. 2008. Savanna tree density, herbivores and herbaceous community: bottom-up vs top-down effects. Ecology 89: 2228–2238. Figure 4b, pg. 2234. © 2008 Ecological Society of America: permission conveyed through Copyright Clearance Center, Inc.; 16.05 a, b Data from Parker, D.M. 2008. The effects of elephants at low densities and after short occupation time on ecosystems of the Eastern Cape Province, South Africa. Ph.D.. Dissertation. Rhodes University South Africa. Figure 3.2b, pg. 64 and 3.4b, pg. 66.; 16.06 Estes, J.A., M. T. Tinker, T. M. Williams, and D. F. Doak. 1998. Killer whale predation on sea otters linking oceanic and nearshore ecosystems. Science 282:473–476.Fig. 1, pg. 474.; 16.07 © Pearson Education; 16.08 © Pearson Education; 16.09 © Pearson Education; 16.10 © Pearson Education; 16.11 a, b Rezende, E.L., E.M. Albert, M. A. Fortuna, and J. Bascompte. 2009. Compartments in a marine food web associated with phylogeny, body mass, and habitat structure. Ecology Letters 12:779–788. Figure 1, pg. 783.; 16.12 a, b © Pearson Education, Inc.; 16.13 Adapted from Anderson and Shugart 1974; 16.14 a, b Data from Schukat, A., M. Bode, H. Auel, R. Carballo, B. Martin, R. Koppelmann, and W. Hagen. 2013. Pelagic decapods in the northern Benguela upwelling system: Distribution, ecophysiology and contribution to active carbon flux. Deep-Sea Research I 75:146–156. Figure 3e and c, pg. 151.; 16.16 © Pearson Education; 16.17 © Pearson Education; 16.18 a, b Reprinted with permission by the Ecological Society of America from R. H. Whittaker, "Vegetation of the Great Smoky Mountains," Ecological Monographs, 26 (1): 1–80 (Jan. 1956). Copyright 1956 by Ecological Society of America; permission conveyed through Copyright Clearance Center, Inc.; 16.19 J. M. Dyer, "Revisiting the Deciduous Forests of Eastern North America," BioScience 56 (4): 341–352, (April 2006). Copyright © 2006 American Institute of Biological Sciences.; 16.20 a, b © Pearson Education; 16.1 © Pearson Education; 16.2 © Pearson Education.

Chapter 17 17.01a, b © Pearson Education; 17.03 © Pearson Education Inc.; 17.03b Data from Connell 1961; 17.04 a, b Reprinted from Thomas Smith and Michael Huston, "A Theory of the Spatial and Temporal Dynamics of Plant Communities," Vegetatio Vol. 83, No. 1/2, Progress in Theoretical Vegetation Science (Oct 1, 1989): pp, 49–69; Fig. 7, p. 58, © 1989 Kluwer Academic Publishers, with kind permission from Springer Science+Business Media B.V.; 17.05 © Pearson Education; 17.06 Republish with permission of Ecological Society of America, from Emery et al. "Competition and Salt-Marsh Plant Zonation: Stress Tolerators May Be Dominant Competitors," ECOLOGY, September 2001 Fig. 1, p 2473; permission conveyed through Copyright Clearance Center, Inc.; 17.07 a, b © Pearson Education; 17.08 Krebs, Charles J., Ecology: The Experimental Analysis of Distribution and Abundance, 5th Ed., © 2001. Reprinted and Electronically reproduced by permission of Pearson Education, Inc..; 17.09b Based on Payne, Robert T., "Food Web Complexity and Specie Diversity" The American Naturalist, 100 (910), Fig. 1 p. 67; 17.09c © Pearson Education; 17.10 a, b © Pearson Education'; 17.11 © Pearson Education; 17.13 © Pearson Education; 17.14

Reprinted by permission of the Ecological Society of America from Power, Mary E., Matthews, William J., and Stewart, Arthur J., Grazing Minnows, Piscivorous Bass, and Stream Algae: Dynamics of a Strong Interaction. Ecology, Vol. 66, No. 5 (Oct., 1985), po. 1448–1456. Figure 4, pg. 1452. © 1985 Ecological Society of America; permission conveyed through the Copyright Clearance Center Inc.; 17.15 a, b © Pearson Education; 17.16 a, b Reprinted by permission of the Ecological Society of America from G. P. Robertson, et al., "Spatial Variability in a Successional Plant Community: Patterns of Nitrogen Availability," Ecology 69 (5): 1517–1524 (Oct. 1988). Copyright 1988 by Ecological Society of America; permission conveyed through Copyright Clearance Center, Inc.; 17.17 Reprinted with permission of the Ecological Society of America from MacArthur and MacArthur, "On Bird Species Diversity," Ecology 42 (3): 594–598, (July 1961). Copyright 1961 by Ecological Society of America; permission conveyed through Copyright Clearance Center, Inc.; 17.18 Catherine Reed. 1993. Reconstruction of Pollinator Communities on Restored Prairies in Eastern Minnesota. Final Report to the Minnesota Department of Natural Resources Nongame Wildlife Program. © 1993 State of Minnesota, Department of Natural Resources. Reprinted with permission.; 17.19 Michael Huston, "Soil Nutrients and Tree Species Richness in Costa Rican Forests," Journal of Biogeography, 7 (2): 147–157 (Jun 1980). Copyright © 1980 Blackwell Publishing.; 17.20 Based on Brenchley, 1958; 17.22 Hillebrand, H. et al. 2007. Consumer versus resource control of producer diversity depends on ecosystem type and producer community structure. Proceedings of the National Academy of Sciences 104: 10904–10909. Fig. 1, pg. 10905. Copyright (2007) National Academy of Sciences, U.S.A.; 17.25 a, b Ripple, W.J. et al. 2013. Trophic cascades from wolves to grizzly bears in Yellowstone. Journal of Animal Ecology Figures 1 and 3.; 17.26 Data from Baril, L. M. 2009. Change in deciduous woody vegetation, implications of increased willow (Salix spp.) growth for bird species diversity, and willow species composition in and around Yellowstone National Park's northern range. Thesis. Montana State University, Bozeman, Montana, USA.; FS1 From M. D. Bertness and S. D. Hacker, "Physical Stress and Positive Associations Among Marsh Plants," The American Naturalist 144 (3): 363–372, (Sep 1994). Copyright © 1994 The University of Chicago Press. Used by permission of the publisher.; FS2 © Pearson Education; FS3 © Pearson Education.

Chapter 18 18.01 © Pearson Education; 18.02b Reprinted with permission by the Ecological Society of America from Wayne P. Sousa, "Experimental Investigations of Disturbance and Ecological Succession in a Rocky Intertidal Algal Community," Ecological Monographs, 49 (3): 227–254 (Sep 1979). Copyright 1979 by Ecological Society of America; permission conveyed through Copyright Clearance Center, Inc.; 18.03b Data from Reiners, William A., Twenty Years of Ecosystem Reorganization Following Experimental Deforestation and Regrowth Suppression. Ecological Monographs, Vol. 62, No. 4 (Dec., 1992), pp. 503–523.; 18.04b Republished with permission of the Ecological Society of America fromLichter, John, Primary Succession and forest Development on Coastal Lake Michigan Sand Dunes. Ecological Monographs, Vol. 68, No. 4 (Nov., 1998), pp. 487–510. Figure 3a, pg. 291; permission conveyed through Copyright Clearance Center, Inc.; 18.05a R. L. Crocker and J. Major, "Soil development in relation to vegetation and surface age at Glacier Bay, Alaska," Journal of Ecology 43:427–488, (July 1955). Copyright © 1955 British Ecological Society.; 18.05c Erik A. Hobbie, "(Controls on) Nitrogen cycling during primary succession in Glacier Bay, Alaska," University of Virginia, 1994 (Masters thesis). Used by permission of the author.; 18.07b Based on Duggins 1980.; 18.09a Republished with permission of the Ecological Society of America from Williams, Susan L., Experimental Studies of Caribbean Seagrass Bed Development. Ecological Monographs, Vol. 60, No. 4 (Dec., 1990), pp. 449–469; permission conveyed through Copyright Clearance Center, Inc.; 18.09b © Pearson Education; 18.10 a, b Data from Grime 1979.; 18.11 a, b Adapted from Tilman 1988.; 18.12 © Pearson Education; 18.14 R. L. Crocker and J. Major, "Soil development in relation to vegetation and surface age at Glacier Bay, Alaska," Journal of Ecology 43:427–488, (July 1955). Copyright © 1955 British Ecological Society.; 18.18 © Pearson Education; 18.19 a, b Reprinted with permission by the Ecological Society of America from Johnston, D.W. and E.P Odum. 1956. Breeding bird populations in relation to plant succession on the Piedmont of Georgia. Ecology 37:50–62. Table 1, pg. 54. © 1956 Ecological Society of America; permission conveyed through Copyright Clearance Center, Inc.; 18.20 Data from Huntley, N. and R. Inouye. (1987) "Small mammal populations of an old-field chronosequence: Successional patterns and associations with vegetation" Journal of Mammology 68:739–745. Fig. 1, pg. 741.; 18.21 a, b, c Barberaena-Arias, M. and T.M. Aide 2003. Species diversity and trophic composition of litter insects during plant secondary succession. Caribbean Journal of Science 29:161–169 (a) –Fig. 1, pg. 163 (b) - Fig. 3, pg. 164 (c) - Fig. 2, pg. 163; 18.22 Reprinted by permission of the Ecological Society of America from W. G. Crumpton and R.G. Wetzel, "Effects of Differential Growth and Mortality in the Seasonal Succession of Phytoplankton Populations in Lawrence Lake, Michigan," Ecology 63 (6): 1729–1739 (Dec. 1982). Copyright 1982 by Ecological Society of America; permission conveyed through Copyright Clearance Center, Inc.; 18.23 © Pearson Education; 18.24 © Pearson Education; 18.25 a, b, c, d M. B. Davis, "Quarternary history and the stability of forest communities" from Forest Succession: Concepts And Applications, edited by D. C. West et al., (1981), figs 10.3, 10.8, 10.9, 10.11, © 1981 by Springer-Verlag New York, Inc.; 18.26 Data from P. A. Delcourt, "Vegetation maps for Eastern North America: 40,000 yr BP to present," from GeobotanyII edited by Robert Romans, © 1981 Plenum Press; 18.27 Vegetatio by International Society For Vegetation Science; International Society for Plant Geography and Ecol; International Society of Vegetation Science Reproduced with permission of W. JUNK, in the format Book via Copyright Clearance Center.; 18.28b Reprinted by permission of the Ecological Society of America from R. H. Whittaker, "Vegetation of the Great Smoky Mountains," Ecological Monographs, 26 (1): 1–80 (Jan. 1956). Copyright 1956 by Ecological Society of America; permission conveyed through Copyright Clearance Center, Inc.; 18.28c Based on Whittaker, R.H. (1956), "Vegetation of the Great Smokey Mountains" Ecological monographs, 26(1) Fig 19; 18.29 a, b Based on data from U.S. Census Bureau; 18.31 C.H. Flather, T.H. Ricketts, C.H. Sieg, M.S. Knowles, J.P. Fay, and J. McNees. 2003. Criterion 1: Conservation of Biological Diversity. Indicator 6: The Number of Forest-Dependent Species. In Darr, D. (comp.), Technical Document Supporting the 2003 National Report on Sustainable Forests. Washington, DC: USDA ForestService. http://www.fs.fed.us/research/sustain/. Accessed November 2006.; 18.1 Adapted from Bazzaz 1979.

Chapter 19 19.07 Based on Biological Conservation, Vol. 17, N. R. Webb and L. E. Haskins, "An Ecological survey of the heath-lands in the Poole Basin, Dorset, England in 1978," pp. 281–296, copyright © 1980, Elsevier.; 19.08 © Pearson Education; 19.09 a, b © Pearson Education; 19.11 a, b, c, d © Pearson Education; 19.12 Adapted from Sisk et al. 1997.; 19.14 Kroodsma, R.L. 1982. Edge effect on breeding forest birds along a power-line corridor. Journal of Applied Ecology 19:361–370. Fig. 3d, pg. 366.; 19.15 a, b Winter, M., D.H. Johnson, and J. Faaborg. 2000. Evidence for edge effects on multiple levels in tallgrass prairie. Condor 102: 256–266. Fig. 1b pg. 261 and Fig. 2 pg. pg. 262.; 19.16 © Pearson Education; 19.17 a, b, c © Pearson Education; 19.18 a, b, c Based on C.S. Robbins et al., "Habitat Area Requirements of Breeding Forest Birds of the Middle Atlantic States," Wildlife Monographs, No. 103, Habitat Area Requirements of Breeding Forest Birds of the Middle Atlantic States: 3–34, (July 1989). Copyright © 1989 Allen Press.; 19.19 a, b Echeverría, C., A.C. Newton, A. Lara, J.M.R. Benayas, and D.A. Coomes. 2007. Impacts of forest fragmentation on species composition and forest structure in the temperate landscape of southern Chile. Global Ecology and Biogeography 16:426–439. Fig. 2, pg. 432.; 19.21 a, b Honnay, O., K. Verheyen, J. Butaye, H. Jacquemyn, B. Bossuyt and M. Hermy. 2002. Possible effects of habitat fragmentation and climate change on the range of forest plant species. Ecology Letters 5:525–530. Fig. 3, Fig. 4, pg. 528; 19.21c Jacquemyn, H., J. Butaye and M. 2001. Forest plant species richness in small, fragmented mixed deciduous forest patches: the role of area, time and dispersal limitation. Biogeography 28: 801–812. Fig. 5, pg. 810; 19.23 Based on F. W. Preston, "The Canonical Distribution of Commonness and Rarity: Part I," Ecology, 43 (2): 185–215, (Apr 1962). Copyright 1962 by Ecological Society of America.; 19.24 © Pearson Education; 19.25 a, b © Pearson Education; 19.26 © Pearson Education; 19.27 b, c, d Based on Kindvall and I. Ahlen, "Geometrical Factors and Metapopulation Dynamics of the Bush Cricket, Metrioptera bicolor Philippi (Orthoptera: Tettigoniidae)," Conservation Biology, 6 (4): 520–529, (Dec. 1992). © 1992 Blackwell Publishing.; 19.28 Based on O. Kindvall and I. Ahlen, "Geometrical Factors and Metapopulation Dynamics of the Bush Cricket, Metrioptera bicolor Philippi (Orthoptera: Tettigoniidae)," Conservation Biology, 6 (4): 520–529, (Dec. 1992). © 1992 Blackwell Publishing.; 19.29b Based on C. D. Thomas and T. M. Jones, "Partial recovery of a Skipper Butterfly (Hesperia comma) from Population Refuges: Lessons for Conservation in a Fragmented Landscape," Journal of Animal Ecology, 62 (3): 472–481, (July 1993). Copyright © 1993 British Ecological Society.; 19.30 © Pearson Education; 19.31 © Pearson Education; 19.32 © Pearson Education; 19.35 © Pearson Education; 19.FS1 © Pearson Education; 19.FS2 Source: Based on Haddad, N. 1999a. "Corridor and distance effects on interpatch movements: A landscape experiment with butterflies." Ecological Applications 9:612–622.; p. 411 © Pearson Education; p. 427 Daniel S. Simberloff, Equilibrium Theory of Island Biogeography and Ecology, Annual Reviews. 1974;

Chapter 20 20.1 © Pearson Education; 20.02 a, b Based on Helmut Lieth, "Primary Production: Terrestrial Ecosystems," Human Ecology 1(4): 303–332 (Sep 1973), Fig. 5A, p. 326 and Fig. 4A, p. 324, Copyright © 1973 Springer Netherlands.; 20.03 Based on Lieth and Whittaker, "Modeling primary productivity of the world," Primary Productivity of the Biosphere: Ecological Studies, Vol. 14 (1975), copyright © 1975 Springer-Verlag.; 20.04 Based on R.H. MacArthur and J.H. Connell, The Biology Of Populations (Wiley, © 1966).; 20.05 Based on NODIS, Numerical Terradynamic Simulation Group; 20.06 Based on John Pastor et al., "Aboveground Production and N and P Cycling Along a Nitrogen Mineralization Gradient on Blackhawk Island, Wisconsin," Ecology 65 (1): 256–268 (Feb 1984). Copyright 1984 by Ecological Society of America.; 20.07 Based on Peter Reich, et al., "Fire and Vegetation Effects on Productivity and Nitrogen Cycling across a Forest-Grassland Continuum,"Ecology 82 (6): 1703–1719 (June 2001). Copyright 2001 by Ecological Society of America.; 20.08 © Pearson Education; 20.09 a, b Based on John A. Downing, "Meta-Analysis of Marine Nutrient-Enrichment Experiments: Variation in the Magnitude of Nutrient Limitation," Ecology 80 (4): 1157–1167 (June 1999). Copyright 1999 by Ecological Society of America.; 20.10 Based on http://www.nasa.gov/centers/goddard/news/topstory/2003/0815oceancarbon.html; 20.11 Based on P. J. Dillon, "The Phosphorus-Chlorophyll Relationship in Lakes," Limnology and Oceanography, 19 (5): 767–773 (Sep 1974). © 1974 by the Association for the Sciences of Limnology and Oceanography, Inc.; 20.12 a, b From Rosemond, A.D., P.J. Mulholland, and S.H. Brawley. 2000. Seasonally shifting limitation of stream periphyton: response of algal populations and assemblage biomass and productivity to variation in light, nutrients, and herbivores. Canadian Journal of Fisheries and Aquatic Science 57:66–75. Fig. 1, pg. 70; 20.13 From Wetzel 1975.; 20.14 © Pearson Education; 20.15 Gough, C.M. (2012) Terrestrial Primary Production: Fuel for Life. Nature Education Knowledge 3(10):28 Fig. 3; 20.16 a, b Based on T. H. Sparks and J. M. Potts, "Late Summer Grass Production," Agriculture and Forestry (Indicators of Climate Change in the UK), Feb. 1999, updated Oct. 2003; 20.17 © Pearson Education; 20.18 Gough, C.M. (2012) Terrestrial Primary Production: Fuel for Life. Nature Education Knowledge 3(10):28 Fig. 1; 20.19 a, b Based on S. Gower, et al., "Aboveground net primary production decline with stand age: potential causes," Trends in Ecology & Evolution, 11(9): 378–382 (Sept. 1, 1996).; 20.20 a, b, c Based on S.J. McNaughton, et al., "Ecosystem-level patterns of primary productivity and herbivory in terrestrial habitats," Nature 341: 142–144, (Sep 1989). Copyright © 1989.; 20.21 Based on Brylinsky and Mann 1973, "An Analysis of Factors Governing Productivity in Lakes and Reservoirs," Limnology and Oceanography.; 20.22 Based on Copyright © 2008 Pearson Education, Inc., publishing as Pearson Benjamin Cummings.; 20.23 © Pearson Education; 20.24 a, b © Pearson Education; 20.25 a, b, c Based on Macmillan Publishers Ltd: H. Cyr and M. Pace, "Magnitude and patterns of her-bivory in aquatic and terrestrial ecosystems," Nature 361: 148–150 (Jan. 1993). Copyright © 1993.; 20.26 Pauly, D. and V. Christensen. 1995. Primary production required to sustain global fisheries. Nature 374:255–257. Fig. 2, pg. 257.; 20.27 © Pearson Education; 20.28 a, b Based on Macmillan Publishers Ltd: Marc L. Imhoff, et al., "Global patterns in human consumption of net primary production," Nature 429: 870–873, (June 2004). Copyright © 2004.; 20.FS.2 a, b Silliman, B. R., and M. D. Bertness. 2002. A trophic cascade regulates salt marsh primary production. Proceedings of the National Academy of Sciences USA 99:10500–10505. Fig. 2, pg. 10501.; 20.FS.3 © Pearson Education; 20.1QE Thomas C. Hart from Goetz, S. J. and S. D. Prince (1996). Remote sensing of net primary production in boreal forest stands. Agricultural and Forest Meteorology 78(3):149–179. Fig. 7, pg. 168; 20.2QE Fensholt, R., I. Sandholt, and M.S Rasmussen. 2004. Evaluation of MODIS LAI, fAPAR and the relation between fAPAR and NDVI in a semi-arid environment using in situ measurements. Remote Sensing of Environment 91: 490–507. Fig. 10 (bottom graph), pg. 502; p. 444 Adapted from Leith, H., and R. H. Whittaker, eds. 1975. Primary productivity in the biosphere. Ecological Studies Vol. 14. New York: Springer-Verlag.; p. 453 Data from Humphreys 1979.

Chapter 21 21.1 © Pearson Education; 21.04 © Pearson Education; 21.06 Based on Carri Leroy and Jane Marks, "Litter quality, stream characteristics and litter diversity influence decomposition rates and macroinvertebrates," Freshwater Biology 51: 601–607 (April 2006). Copyright © 2006, Blackwell Publishing Ltd.; 21.07 Based on Swift, Heal, and Anderson 1979.; 21.08 a, b Adapted from Klap et al. 1999.; 21.09 Based on Ivan Valiela, Marine Biology Processes, 2nd ed., p. 301. Copyright © 1984 Ivan Valiela.; 21.10 Based on Melillo et al. 1982, New Hampshire; Mudrick et al. 1994, West Virginia; Smith 2002, Virginia.; 21.11 Adapted from Whitkamp and Frank 1969, "Evolution of CO2 from litter, humus and subsoil of a pine stand," Pedobiologia.; 21.13 © Pearson Education; 21.14 a, b, c Based on Beare et al. 1992, "Microbial and Faunal Interactions and Effects on Litter Nitrogen and Decomposition in Agroecosystems," Ecological Monographs; 21.15 © Pearson Education; 21.16 Based on Staaf and Berg 1982, "Accumulation and release of plant nutrients in decomposing Scots pine needle litter. Long-term decomposition in a Scots pine forest II," Canadian Journal of Botany.; 21.17 a, b, c, d Data from Berg et al. 1982.; 21.18 © Pearson Education; 21.19 Based on Odum and Heywood 1978, "Decomposition of Intertidal Freshwater Marsh Plants," Freshwater Wetlands: Ecological Processes and Management Potential; 21.20 Based on Chapin 1980, "The Mineral Nutrition of Wild Plants," Annual Review of Ecology and Systematics.; 21.21 a, b Based on John Pastor et al., "Aboveground Production and N and P Cycling Along a Nitrogen Mineralization Gradient on Blackhawk Island, Wisconsin," Ecology 65 (1): 256–268 (Feb 1984). Copyright 1984 by Ecological Society of America.; 21.22 a, b © Pearson Education; 21.23 a, b © Pearson Education; 21.24 a, b, c © Pearson Education; 21.25 a, b Based on Newbold et al. 1982, "Nutrient Spiralling in Streams: Implications for Nutrient Limitation and Invertebrate Activity," The American Naturalist.; 21.26 Based on Correll 1978, "Estuarine Productivity," Bioscience; Fig. 2, p. 649; 21.28 © Pearson Education; 21.29 Data from Ewel et al. 1981, "Slash and Burn Impacts on a Costa Rican Wet Forest Site," Ecology.; 21.31 Data from USDA; 21.1QE © Pearson Education; 21.3.FS Based on Edward A.G. Schuur, "The Effect of Water on Decomposition Dynamics in Mesic to Wet Hawaiian Montane Forests," Ecosystems 4 (3): 259–273 (Apr. 2001), adapted from parts of Fig. 2, p. 265 and Fig. 6, p. 268, copyright © 2001 Springer New York.; 21.4.FS Based on Edward A. G. Schuur, "Productivity and Global Climate Revisited: The Sensitivity of Tropical Forest Growth to Precipitation," Ecology 84 (5): 1165–1170 (May 2003). Copyright 2003 by Ecological Society of America.; 21.2 Table © Pearson Education; 21.2 © Pearson Education.

Chapter 22 22.01 © Pearson Education; 22.02 © Pearson Education; 22.03 Based on Baumgartner 1968, Functioning of Terrestrial Ecosystems at the Primary Production Level. Proceedings of Copenhagen Symposium Natural Resources Research, UNESCO, Paris, pp. 367–374.; 22.04 Based on Pearman and Hyson 1981, "The Annual Variation of Atmospheric CO2 Concentration Observed in the Northern Hemisphere," Journal of Geophysical Research.; 22.05 Based on Edmonds 1992, "Why Understanding the Natural Sinks and Sources of CO2 is Important: A Policy Analysis Perspective," Water, Air, and Soil Pollution.; 22.06 © Pearson Education; 22.07 © Pearson Education; 22.08 Based on Schlesinger 2013.; 22.09 © Pearson Education; 22.10 Based on Schlesinger 2013, Biogeochemistry: An Analysis of Global Change.; 22.11 © Pearson Education; 22.12 Based on Schlesinger 2013, Biogeochemistry: An Analysis of Global Change.; 22.13 Adapted from Schlesinger 2013.; 22.14 National Atmospheric Deposition Program.; 22.15 Based on "Hypothesized response of temperate forest ecosystems to long-term nitrogen additions," in John Aber, et al., "Nitrogen Saturation in Temperate Forest Ecosystems," BioScience vol. 48, no. 11 (November 1998), pp. 921 – 934. © 1998 by the American Institute of Biological Sciences. Published by the University of California Press.; 22.16 Dise, N.B. and R.F. Wright. 1995. Nitrogen leaching from European forests in relation to nitrogen deposition. Forest Ecology and Management 71: 153–161. Fig. 3a, pg. 157; 22.17 McNulty, S.G., J. D. Aber, and S. D. Newman. 1996. Nitrogen saturation in a high elevation New England spruce-fir stand. Forest Ecology and Management 84:109–121. Fig. 6, pg. 117.

Chapter 23 23.01 D. M. Olson et al., "Terrestrial Ecoregions of the World: A New Map of Life on Earth," BioScience: 51 (11): 933–938, (Nov. 2001). Copyright © 2001 American Institute of Biological Sciences. Used by permission of Oxford University Press on behalf of the American Institute of Biological Sciences.; 23.02 Based on Whittaker 1970, Communities and Ecosystems; fig. 3.8, p. 65.; 23.05 Based on O. W. Archibold, Ecology Of World Vegetation, fig. 2.1 © 1995 Chapman and Hall, Springer Science+Business Media B.V.; 23.08 © Pearson Education; 23.12 Based on O. W. Archibold, Ecology Of World Vegetation, fig. 2.1 © 1995 Chapman and Hall, Springer Science+Business Media B.V.; 23.13 © Pearson Education; 23.14 Based on O. W. Archibold, Ecology Of World Vegetation, p. 60 © 1995 Chapman and Hall, Springer Science+Business Media B.V.; 23.15 Based on O. J. Reichman, Konza Prairie, A Tallgrass Natural History. Used by permission of University Press of Kansas.; 23.18 Based on Lauenroth 1979, "Grassland primary production: North American grasslands in perspective," from Perspectives in Grassland Ecology (Springer, NY); fig. 1.4 (p. 7).; 23.19 Based on O. W. Archibold, Ecology Of World Vegetation, fig. 4.1, p. 95 © 1995 Chapman and Hall, Springer Science+Business Media B.V; 23.23 Based on O. W. Archibold, Ecology Of World Vegetation, fig 5.1, pg. 131, © 1995 Chapman and Hall, Springer Science+Business Media B.V.; 23.27 Based on O. W. Archibold, Ecology Of World Vegetation, fig. 8.1, p. 238; fig. 9.1, p. 280, © 1995 Chapman and Hall, Springer Science+Business Media B.V.; 23.28 Based on "Large scale distribution of temperate forest communities in the eastern United States," in James M. Dyer, "Revisiting the Deciduous Forests of Eastern North America," BioScience vol. 56 no. 4 (April 2006), pp. 341 – 352. © 2006 by the American Institute of Biological Sciences. Published by the University of California Press.; 23.30 Based on "Large scale distribution of temperate forest communities in the eastern United States," in James M. Dyer, "Revisiting the Deciduous Forests of Eastern North America," BioScience vol. 56 no. 4 (April 2006), pp. 341 – 352. © 2006 by the American Institute of Biological Sciences. Published by the University of California Press.; 23.32 Based on S. Payette, "The Subarctic Forest-Tundra: the structure of a biome in a changing climate," BioScience 51 (9): 709–718, (Sept. 2001). Copyright © 2001 American Institute of Biological Sciences.; 23.34 Based on S. Payette, "The Subarctic Forest-Tundra: the structure of a biome in a changing climate," BioScience 51 (9): 709–718, (Sept. 2001). Copyright © 2001 American Institute of Biological Sciences.; 23.38 © Pearson Education; 23.39 Adapted from United Nations, FAO.; 23.41 Adapted from Oriens et al. 1986.; 23.42 Adapted from Oriens et al. 1983.; 23.43 Based on Likens and Borman 1995, Biogeochemistry of a Forested Ecosystem, 2e.; 23.1QE © Pearson Education

Chapter 24 24.01 © Pearson Education; 24.03 © Pearson Education; 24.04 © Pearson Education; 24.08 © Pearson Education; 24.11 © Pearson Education; 24.12 © Pearson Education; 24.13 © Pearson Education; 24.17 © Pearson Education; 24.20 © Pearson Education; 24.24 Based on Archibold, 1995, Ecology of World Vegetation; 24.25 Based on Nybakken 1997; 24.27 Based on map by Robert Simmon, data from Robert Diaz, Virginia Institute of Marine Science.; 24.28 Based on N.N. Rabalais, Louisiana Universities Marine Consortium, R.E. Turner, Louisiana State University. Funded by: Noaa, Center for Sponsored Coastal Ocean Research.; 24.29a http://water.usgs.gov/nawqa/sparrow/gulf_findings/delivery.htm; 24.29b http://water.usgs.gov/nawqa/sparrow/gulf_findings/primary_sources.html; 24.1QE © Pearson Education, Inc.; 24.2QE © Pearson Education, Inc.; 24.3QE © Pearson Education, Inc.

Chapter 25 25.01b © Pearson Education, Inc.; 25.02 © Pearson Education, Inc.; 25.12 © Pearson Education, Inc.; 25.13 © Pearson Education, Inc.; 25.18b © Pearson Education, Inc.; 25.19 © Pearson Education, Inc.; 25.22 Adapted from U.S. Department of Agriculture.

Chapter 26 26.01 Data from Wilson 1999, The Diversity of Life.; 26.02 © Pearson Education; 26.03 © Pearson Education; 26.04 Adpated from Niklas et al. 1983, "Patterns in vascular land plant diversification," NATURE; based on Fig. 1.; 26.05 © Pearson Education; 26.06a Based on Mutke, J. and Barthlott, W. 2005. Patterns of vascular plant diversity at continental to global scales. Biologiske Skrifter 55: 521–531.; 26.06b Based on Kaufman, D.M. and M.R. Willig. 1998. Latitudinal patterns of mammalian species richness in the New World: the effects of sampling method and faunal group. Journal of Biogeography 25:795–805.; 26.06c Based on Hawkins, B.A. et al. 2006. Post-Eocene climate change, niche conservatism, and the latitudinal diversity gradient of New World birds. Journal of Biogeography 33:770–780. Fig. 1, pg. 774; 26.07a Mutke, J. and Barthlott, W. 2005. Patterns of vascular plant diversity at continental to global scales. Biologiske Skrifter 55: 521–531. Fig. 6, pg. 529.; 26.07b Adapted from Kaufman, D.M. and M.R. Willig. 1998. Latitudinal patterns of mammalian species richness in the New World: the effects of sampling method and faunal group. Journal of Biogeography 25:795–805. Fig. 2a, pg. 801.; 26.07c Adapted from Blackburn, T.M. and K. J. Gaston. 1996. Spatial Patterns in the Species Richness of Birds in the New World. Ecography 19:369–376. Fig. 2, pg. 372; 26.08a Macpherson, E. and Duarte, M. C. 1994. Patterns in species richness, size, and latitudinal range of East Atlantic fishes. Ecography 17: 242–248.(a) - Fig. 1, pg. 243; 26.08b, c Macpherson, E. 2002. Large-Scale Species-Richness Gradients in the Atlantic Ocean. Proc. R. Soc. Lond. B. 269:1715–1720 (b) - Fig. 1, pg. 1716; (c) - Fig. 2, pg. 1718.; 26.09 Based on Mutke, J. &and Barthlott, W. 2005. Patterns of vascular plant diversity at continental to global scales. Biologiske Skrifter 55: 521–531. Fig. 6, pg. 529; 26.10 Based on Mutke, J. and Barthlott, W. 2005. Patterns of vascular plant diversity at continental to global scales. Biologiske Skrifter 55: 521–531. Fig. 8, pg. 530; 26.11 © Pearson Education; 26.12 © Pearson Education; 26.13 © Pearson Education; 26.14a, b David J. Currie, "Energy and Large-Scale Patterns of Animal- and Plant-Species Richness," The American Naturalist 137 (1): 27–49, (Jan. 1991). Copyright © 1991 The University of Chicago Press. 26.14c, d Currie, D.J. 1991. Energy and Large-Scale Patterns of Animal- and Plant-Species Richness. American Naturalist 137: 27–49. Fig. 6, pg. 37; 26.15a, b, c Adapted from Roy, K., D. Jablonski, J.W. Valentine, and G. Rosenberg. 1998. Marine latitude diversity gradients: Tests of causal hypotheses. Proceedings of the National Academy of Science 95:3699–3702. Fig. 1, pg. 3700; Fig. 5, pg. 3701, Fig. 6, pg. 3702; 26.16a, b, c, d Adapted from Astorga, A. et al. 2003. Two oceans, two taxa and one mode of development: latitudinal diversity patterns of South American crabs and test for possible causal processes. Ecology Letters 6:420–427. (a) and (b)–Fig. 1, pg. 422, (c) and (d)–Fig. 4, pg. 425; 26.17 Adapted from Mutke, J. &and Barthlott, W. 2005. Patterns of vascular plant diversity at continental to global scales. Biologiske Skrifter 55: 521–531. Fig. 7, pg. 530.; 26.18 Based on Currie and Paquin 1987.; 26.19 Adapted from Adams, J.M. and F.I. Woodward. 1989. Patterns in tree species richness as a test of the glacial extinction hypothesis. Nature 339: 699-701. Fig. 1a, pg. 700; 26.20a, b Adapted from Currie, D.J. 1991. Energy and Large-Scale Patterns of Animal- and Plant-Species Richness. American Naturalist 137: 27-49. (a) Fig. 4, pg. 35; (b) Fig. 5, pg. 35; 26.21a, b Adapted from Xiao, X. et al. 1997. Linking a Global Terrestrial Biogeochemical Model with a 2-Dimensional Climate Model: Implications for Global Carbon Budget. Tellus, Vol. 49B, pp. 18-37. Fig. 4a, pg. 25; 26.22a Based on Kikkawa and Williams 1971.; 26.22b Based on Hunter and Yonzon 1992.; 26.22c Based on Whittaker 1977; 26.33a, b Hurlbert, A.H. and J.P. Haskell. 2003. The Effect of Energy and Seasonality on Avian Species Richness and Community Composition. American Naturalist 161:83–97. Fig. 4, pg. 89; 26.24 Adapted from Gentry 1986.; 26.25 Adapted from Conservation International.orghttp://www.conservation.org/where/priority_areas/hotspots/Pages/hotspots_main.aspx; 26.1QE © Pearson Education

Chapter 27 27.01 Based on Hansen, J., R. Ruedy, Mki. Sato, and K. Lo, 2010: Global surface temperature change. Rev. Geophys., 48, RG4004, doi:10.1029/2010RG000345.; 27.02a, b, c, d From Intergovernmental Panel on Climate Change 2013; 27.03 Data from Dai, A. (2011), Drought under global warming: a review. WIREs Clim Change, 2: 45–65.; 27.04 Dillon, M.E. et al. 2010. Global metabolic impacts of recent climate warming. Nature 467:704-707. Fig. 1, pg. 705.; 27.05 Modified from Stillman and Somero 2000 as presented in Somero, G.N. 2010. The physiology of climate change: how potentials for acclimatization and genetic adaptation will determine 'winners' and 'losers'. The Journal of Experimental Biology 213: 912-920. Fig. 1, pg. 913; 27.06a, b, c Van Mantgem, P.J. et al. 2009. Widespread increase of tree mortality rates in the western United States. Science 323:521-524. Fig. 2, Pg. 522; 27.08a, b, c Charmantier, A. et al. 2008. Adaptive phenotypic plasticity in response to climate change in a wild bird population Science 320:800-803. Fig. 1, pg. 801; 27.09 Cotton, P.A. 2003. Avian migration phenology and global climate change. Proceedings of the National Academy of Sciences 100: 12219-12222. Fig. 2, pg. 12221.; 27.10a, b, c Adapted from Beaubien, E. and A. Hamann. 2011. Spring flowering response to climate change between 1936 and 2006 in Alberta, Canada. Bioscience 61:514-524, Fig. 3, pg. 518, Fig. 4, pg. 519.; 27.11 Parmesan, C. 2007. Influences of species, latitudes and methodologies on estimates of phenological response to global warming. Global Change Biology 13: 1860-1872. Fig. 2, pg. 1865; 27.12a Data from Crummett and Eernisse 2007 as modified by Somero, G.N. 2010. The physiology of climate change: how potentials for acclimatization and genetic adaptation will determine 'winners' and 'losers'. The Journal of Experimental Biology 213:912-920. Fig. 4, pg. 917.; 27.12b Data from Somero, G.N. 2010. The physiology of climate change: how potentials for acclimatization and genetic adaptation will determine 'winners' and 'losers'. The Journal of Experimental Biology 213:912-920. Fig. 4, pg. 917.; 27.14 Data from: UK – Thomas and Lennon 1999; Finland – Brommer 2004; North America/breeding – Hitch and Leberg 2007; North America/wintering – LaSorte and Thompson 2007; 27.15 Data from Kelly, A.E. and M.L. Goulden. 2008. Rapid shifts in plant distribution with recent climate change. Proceedings of the National Academy of Sciences 105: 11823-11826.; 27.16a, b, c Post, E. and M.C. Forchhammer. 2008. Climate change reduces reproductive success of an Arctic herbivore through trophic mismatch. Philosophical Transactions of the Royal Society B 363: 2369– 2375. Fig. 2, pg. 2372, Fig. 3, pg. 2373.; 27.18 Mitton, J.B. and S.M. Ferrenberg. 2012. Mountain pine beetle develops an unprecedented summer generation in response to climate warming. The American Naturalist 179: E163-E171. Fig. 1, pg. E164.; 27.19a Data from U.S. National Weather Service.; 27.19b Brusca, R.C. et al. 2013. Dramatic response to climate change in the Southwest: Robert Whittaker's 1963 Arizona Mountain plant transect revisited. Ecology and Evolution 3: 3307-3319.; 27.20a, b Based on Beaugrand, G. et al. 2002. Reorganization of North Atlantic Marine Copepod Biodiversity and Climate. Science 296:1692-1694. Fig. 1, pg. 693.; 27.21a, b, c, d, e Hofstede, R. et al. 2010. Regional warming changes fish species richness in the eastern North Atlantic Ocean. Marine Ecology Progress Series 414:1-9. Fig. 1, pg. 3, Fig. 2, pg. 5, Fig. 3, pg. 6.; 27.22 Adapted from Nemani, R.R., et al. 2003. Climate-Driven Increases in Global Terrestrial Net Primary Production from 1982 to 1999. Science 300:1560-1563. Fig. 2, pg. 1561, Fig. 3, pg. 1562.; 27.23 a, b sed on Zhao, M. and S.W. Running. 2010. Drought-Induced Reduction in Global Terrestrial Net Primary Production from 2000 Through 2009. Science 329:940-943. Fig. 2, pg. 941, Fig. 3, pg. 941.; 27.24a, b Based on Climate Change 2007: The Physical Science Basis. Working Group I Contribution to the Fourth Assessment Report of the Intergovernmental Panel on Climate Change, Figure 10.9; Figure 10.5; FAQ 5.1, Figure 1. Cambridge University Press. Used by permission of IPCC.; 27.25a © Pearson Education; 27.26a, b, c, d Adapted from Rustad, L.E. et al. 2001. A meta-analysis of the response of soil respiration, net nitrogen mineralization, and aboveground plant growth to experimental ecosystem warming. Oecologia 126:543–562. Fig. 1, pg. 545, Fig. 2, pg. 547, Fig. 3, pg. 550.; 27.27 © Pearson Education; 27.28a, b Iverson, L.R., et al. 2008. Estimating potential habitat for 134 eastern US tree species under six climate scenarios. Forest Ecology and Management 254:390–406. Fig. 4, pg. 401, Fig. 5, pg. 402.; 27.29 Iverson, L.R., et al. 2008. Estimating potential habitat for 134 eastern US tree species under six climate scenarios. Forest Ecology and Management 254:390–406. Fig. 6, pg. 403.; 27.30a, b, c McKenney, D.W., et al. 2011. Revisiting projected shifts in the climate envelopes of North American trees using updated general circulation models. Global Change Biology 17:2720–2730. Fig. 1, pg. 2724, Fig. 2, pg. 2725, Fig. 3, pg. 2727.; 27.31a, b Matthews, S.N., et al. 2011. Changes in potential habitat of 147 North American breeding bird species in response to redistribution of trees and climate following predicted climate change. Ecography 34: 933-945. Fig. 3, pg. 936.; 27.32a, b David Currie, "Projected Effects of Climate Change on Patterns of Vertebrate and Tree Species Richness in the Conterminous United States," Ecosystems (2001): 216-225, Fig. 5, p. 221 and Fig. 7, p. 222; © 2001 Springer-Verlag.; 27.33 Schaphoff, S., et al. 2006. Terrestrial biosphere carbon storage under alternative climate projections. Climatic Change 74: 97–122. Fig. 2, pg. 102.; 27.2FS © Pearson Education; 27.3 FS © Pearson Education

Photo Credits

Chapter 1 Chapter Opener Jonathan & Angela Scott/NHPA/Photoshot/Newscom; 1.1 NASA; 1.2d Glenn Bartley/All Canada Photos/Alamy; 1.2e John T. Fowler/Alamy; 1.2a Yuriy Poznukhov/Fotolia; 1.2 bottom Zentilia/Fotolia; 1.2 bottom center Cecelia Spitznas/Getty Images; 1.2 bottom right Vinicius Tupinamba/Shutterstock; 1.2b Joyce Photographics/Science Source; 1.2c John Burnley/Science Source; 1.3 third from top Jason Lindsey/Alamy; 1.3 middle Dennis Frates/Alamy; 1.3 third from bottom Jim Richardson/National Geographic/Getty Images; 1.3 bottom NASA; 1.3 top Bjorn Forsberg/Nature Picture Library; 1.3 second from top Motorolka/Shutterstock; 1.6 David G Tilman; 1.7a Harvard Forest Archives, Harvard Forest, Petersham, MA, U.S.A.. Photography by Moshe D. Roberts; 1.7b Harvard Forest Archives, Harvard Forest, Petersham, MA, U.S.A.. Photography by Moshe D. Roberts; 1.9 The American Heritage Center; 1.10 Bettmann/Corbis.

Chapter 2 Chapter Opener Dr. Paul A. Zahl/Science Source; 2.14 National Oceanic and Atmospheric Administration (NOAA); 2.22b Andre Jenny/Alamy; 2.22a Estivillml/Fotolia.

Chapter 3 Chapter Opener Beyond/SuperStock; 3.6 James R.D. Scott/Getty Images; 3.5 Felix Büscher/Panther Media/AGE Fotostock; 3.15a Hydromet/Shutterstock; 3.10a Hydromet/Shutterstock; 3.15b Jps/Shutterstock; 3.20 Steve Gschmeissner/Science Source.

Chapter 4 Chapter Opener Radius Images/Getty Images; 4.1a Wishfaery14/Fotolia; 4.8 Joyce Photographics/Science Source; 4.12a Loyal A. Quandt/USDA Forest Service Southern Region; 4.12b USDA Forest Service Southern Region; 4.12c USDA Forest Service Southern Region; 4.12d Loyal A. Quandt/USDA Forest Service Southern Region; 4.12e USDA Forest Service Southern Region; 4.12g USDA Forest Service Southern Region; 4.12i USDA Forest Service Southern Region; 4.12j Loyal A. Quandt/USDA Forest Service Southern Region; 4.12k USDA Forest Service Southern Region; 4.12l USDA Forest Service Southern Region; 4.14a Universal Images Group/SuperStock; 4.15c Library of Congress Prints and Photographs Division [LC-USF34-004072-E]; 4.15b FDR Library/National Archives [195691]; 4.16a Farming Today/Alamy; 4.16c AgStock Images, Inc./Alamy; 4.16b USDA Forest Service Southern Region.

Chapter 5 Chapter Opener Wendy Dennis/FLPA; Chapter Opener Thorsten Milse/ Robert Harding World Imagery; 5.2 The Print Collector/Alamy; 5.6a Sonia Sultan; 5.5a Whitman D.W. and A.A. Agrawal, Science Publishers; 5.5b Whitman D.W. and A.A. Agrawal, Science Publishers; p. 84 Figure 1 Dr. Hopi Hoekstra; p. 84 Figure 1 Dr. Hopi Hoekstra; p. 84 Figure 1 Dr. Hopi Hoekstra; p. 85 Figure 2 Dr. Hopi Hoekstra; p. 85 Figure 2 Dr. Hopi Hoekstra; p. 85 Figure 2 Dr. Hopi Hoekstra; p. 85 Figure 2 Dr. Hopi Hoekstra; 5.22c David Hosking /Alamy; 5.22a David Hosking/FLPA; 5.22b McGraw-Hill Michael Stubblefield/Alamy.

Chapter 6 Chapter Opener Ulrich Doering/Alamy; 6.7 Stokes, V.J. Morecrof, and J.I.L. Morison/Agricultural and Forest Meteorology; p. 102 Figure 1a Kaoru Kitajima; p. 102 Figure 1b Kaoru Kitajima; p. 102 Figure 2 Kaoru Kitajima; 6.29 Yavor Parashkevov/Duke University.

Chapter 7 Chapter Opener John Warburton Lee/SuperStock; 7.1d Daniel Heuclin/ AGE Fotostock; 7.1b Doug Perrine/Nature Picure Library; 7.1c Robert Hardholt/ Shutterstock; 7.1a Lebendkulturen.de/Shutterstock; 7.3c Comstock/Stockbyte/ Thinkstock; p. 130 Figure 1 Ian MacDonald; p. 131 Figure 2 Photo Researchers DNU-use (Science Source); p. 131 Figure 2 Photo Researchers DNU- use (Science Source); 7.4b Ian R. MacDonald; 7.10 Karl Ammann/Digital Vision/Getty Images; 7.11 Chris Mattison/AGE Fotostock; 7.26 William Leaman/Alamy.

Chapter 8 Chapter Opener Mazzzur/Shutterstock; 8.3a Ralf Hettich/Panther Media/AGE Fotostock; 8.3b Marty Snyderman/Spirit/Corbis; 8.6a Carl S. Keener; 8.6b Dean Walton/DCR Natural Heritage; 8.10b Kirk Anderson/Alamy; 8.10c DJ/ Fotolia LLC; 8.10a SusaZoom/Shutterstock; 8.12b Cotswolds Photo Library/Alamy; 8.12a Bryan Eastham/Shutterstock; 8.11b Thomas M. Smith; 8.15 National Wildlife Refuge System/U.S. Fish and Wildlife; 8.13 Department of Ecology and Evolutionary Biology, The University of Connecticut; 8.17a Zastolskiy Victor/Shutterstock; 8.25b KrystynaSzulecka/Alamy; 8.25c National Park Service; 8.26 Steve Nudson/Alamy; 8.25a Andrew J. Martinez/Science Source; 8.28a Blickwinkel/Hartl/Alamy; 8.28b Larry D. Hodge/Texas Parks and Wildlife Department/AP Images; 8.27b John Burnley/Science Source; 8.27a Mark Robinson /USDA Forest Service Southern Region.

Chapter 9 Chapter Opener Graeme Shannon/Shutterstock; 9.2 Roland Burke/Photo Researchers, Inc.; 9.6 Lynn M. Stone/Nature Picture Library; 9.20b Stavrida/Fotolia; 9.20a MODIS Rapid Response Team/Nasa.

Chapter 10 Chapter Opener Tony Campbell/Shutterstock; 10.2 Premaphotos/ Nature Picture Library; 10.3 Georgie Holland/AGE Fotostock; 10.4 Jerry & Marcy Monkman/AGE Fotostock; 10.10 Juniors Bildarchiv/AGE Fotostock; 10.11 McPhoto/ Blickwinkel/AGE Fotostock; 10.12 Regis Cavignaux/Biosphoto/Photo Researchers, Inc.; 10.16a Derrick Hamrick Image Broker/Newscom; 10.16b Dr. Paul Zahl/Science Source; 10.18 Stijn Ghesquiere; 10.19 Tony Heald/Nature Picture Library; 10.20a Eky Studio/ Shutterstock; 10.21 Craig K. Lorenz/Science Source; p. 210 Figure 1 Alexandra Basolo; p. 210 Figure 2 Alexandra Basolo; p. 211 Figure 3 Alexandra Basolo.

Chapter 11 Chapter Opener Georgie Holland/AGE Fotostock; 11.10 Joy Prescott/ Shutterstock; p. 230 Figure 1 G. Ronald Austing/Science Source; p. 230 Scott Sillett; p. 231 Figure 2 Scott Sillett; p. 231 Figure 3 Scott Sillett; 11.19 Tom Smith; 11.20 James Cahill; 11.22a John T. Fowler/Alamy; 11.28a Michael S. Quinton/National Geographic Image Collection/Alamy.

Chapter 12 Chapter Opener McPhoto/Blickwinkel/AGE Fotostock; 12.1 JJM Stock Photography/Alamy; 12.6 Rolf Nussbaumer Photography/Alamy; 12.8a Edmund D. Brodie, Jr./AP Images; 12.10h Kip Evans/Alamy; 12.10a Michael Tieck/Fotolia; 12.10d Gino Santa Maria/Fotolia; 12.10c Ronnie Howard/Fotolia; 12.10e Indigo/Fotolia; 12.10f Rafael Ben-Ari/Fotolia; 12.10g Dariusz Kopestynski/Fotolia; 12.10i Vladimir Voronin/ Fotolia; 12.10j Alessandrozocc/Fotolia; 12.10k Pavel Timofeev/Fotolia; 12.10b 5D2/ Shutterstock; 12.14a Pascal Goetgheluck/Ardea; 12.14b Pascal Goetgheluck/Ardea; 12.15b Rosa Ribas; 12.15a Aaron C./Flickr; 12.15c Moritz Muschick.

Chapter 13 Chapter Opener Nature Ingo Bartussek/Nature Picture Library; 13.3b M. I. Walker/Science Source; 13.3a Michael Abbey/Science Source; 134b Maurice Loir; 13.4a National Park Service; 13.9b Andrea Innocenti/CuboImages SRL/Alamy; 13.9c Richard Becker/Frank Lane Picture Agency; 13.9a Ene/Shutterstock; 13.13 Joseph Sohm/Visions of America, LLC/Alamy; 13.16c Christian Heinrich/imageBroker/Alamy; 13.16b Gea Strucks/Dreamstime; 13.16a EcoPrint/Shutterstock; 13.21 David Renwald.

Chapter 14 Chapter Opener Anup Shah/AGE Fotostock; 14.7a Dr. D. P. Wilson/ Science Source; 14.7c Millard H. Sharp/Science Source; 14.7b Critterbiz/Shutterstock; 14.9b Paul Kay/NHPA/Photoshot/Newscom; 14.11a Prasad Basavaraj/Alamy; 14.18 Robert Shantz/Alamy; 14.16 WaterFrame/Alamy; 14.15 Cirrus Digital Imaging; 14.17 S & D & K Maslowski/FLPA; 14.19a Sylvana Rega/Shutterstock; 14.19b Worldswildlifewonders/Shutterstock; 14.21b J.H. Robinson/Science Source; 14.20a Patrick K. Campbell/Shutterstock; 14.21a Frank Greenaway/DK Images; 14.21c Philip Kinsey/Fotolia; 14.20b Pe Willie Davis/Shutterstock; 14.22 Ben Cranke/Digital Vision/ Getty Images; 14.24 Scott Camazine/Alamy; 14.25b Bigemrg/Fotolia; 14.26b W. Perry Conway/Ramble/Corbis; 14.25a Kurt G/Shutterstock; 14.26a John Bora/Science Source; p. 304 Rick Relyea; p. 304 Figure 1 Rick Relyea; p. 305 Figure 2 Rick Relyea; p. 305 Figure 3 Rick Relyea; p. 305 Figure 4 Rick Relyea; 14.27 Neil Thompson/Alamy; 14.29a blickwinkel/Alamy; 14.29b Adrian Davies/Nature Picture Library; 14.30 Ken Gray/ Oregon State University Crop & Soil Science.

Chapter 15 Chapter Opener Inga Spence/Science Source; 15.2 Marvin Dembinsky Photo Associates/Alamy; 15.1a Horst Sollinger/Getty Images; 15.1b Biosphoto/ SuperStock; 15.3a center The Science Picture Company/Alamy; 15.3a top right Science Source; 15.5a Interfoto /Alamy; 15.5b Nigel Cattlin/Alamy; 15.5c Nic Hamilton/Alamy; 15.5d Bill Brooks/Alamy; 15.7 Todd D. Huspeni; 15.10b Lichens Science, Oregon State University; 15.10a Fedorov Oleksiy/Shutterstock; 15.9 Universal Images Group/ SuperStock; p. 324 Stachowicz, John J; p. 324 Figure 1 Stachowicz, John J; p. 325 Figure 2(a) Stachowicz, John J; p. 325 Figure 2(b) Stachowicz, John J; 15.11b Nigel Cattlin/ FLPA; 15.12a Thomas M. Smith; 15.13a Reinhard Dirscherl/Alamy; 15.13b JSPix/ Imagebroker/Alamy; 15.14 Benoit Guenard; 15.15 S & D & K Maslowski/FLPA; 15.17 Centers for Disease Control and Prevention.

Chapter 16 Chapter Opener AGE Fotostock; 16.4 Manuel Romaris/Moment/Getty Images; 16.21 National Marine Fisheries Services; 16.22a Courtesy of the University of Wisconsin Arboretum; 16.22b Courtesy of the University of Wisconsin Arboretum; 16.23 Courtesy of the University of Wisconsin Arboretum.

Chapter 17 Chapter Opener Greg Vaughn/Alamy; p. 364 Sally Hacker; p. 364 Figure 1 Sally Hacker; p. 365 Figure 2 Sally Hacker; p. 365 Figure 3 Sally Hacker; 17.9 Bob Pool/Photographer's Choice RF/Getty Images; 17.12 misc animals Mick Hoult/ Photoshot Holdings Ltd /Alamy; 17.12 top right Roberto Nistri/Alamy; 17.12 botom right M. I. Walker/Science Source; 17.12 bottom left Tan, Derek.

Chapter 18 Chapter Opener Larry Michael/Nature Picture Library; 18.1 Robert Leo Smith; 18.1 top right Robert Leo Smith; 18.2 Ron Niebrugge/Alamy; 8.4 Aerial Archives/Alamy; 18.5b Caleb Foster/Fotolia; 18.1.2 Robert Leo Smith; 18.6 top left David Stephens; 18.6 center right AGE Fotostock/SuperStock; 18.7 David Duggins; 18.8 Robert Leo Smith; 18.28a Melinda Fawver/Shutterstock.

Chapter 19 Chapter Opener Aerial Archives/Alamy; 19.3a Micha Pawlitzki/Getty Images; 19.3b Jupiterimages/Getty Images; 19.1 Robert Leo Smith; 19.4a B.A.E. Inc./ Alamy; 19.4b The Post Crescent, Dan Powers/AP Images; 19.6b Clearviewstock/Fotolia; 19.5b John Stennes/KRT/Newscom; 19.5a David Hosking/Science Source; 19.6a Calvin Larsen/Science Source; 19.5c Steve Schwettman/Shutterstock; 19.7 Derek Payne/Alamy; 19.9 left Ainars Aunins/Alamy; 19.9 right Vespasian/Alamy; 19.9 center right AGE Fotostock/Alamy; 19.9 center left John M. Coffman/Getty Images; 19.10b Saulius/AGE Fotostock; 19.10a Givaga/Shutterstock; 19.13 Birds and Dragons/Shutterstock; 19.14 Minik/Shutterstock; 19.22b Bill Bachman/Alamy; 19.22a Chrislofotos/Shutterstock; p. 424 Nick Haddad; p. 424 Figure 1 Nick Haddad; p. 425 Figure 2 Nick Haddad; p. 425 Figure 3 Nick Haddad; 19.27a Peter Kocarek; 19.27b George McCarthy/Nature Picture Library; 19.36 Drew Rush/Getty Images; 19.37 right John E Marriott/Alamy; 19.37 left Jackson Hole News & Guide/Bradly J. Boner/AP Images.

Chapter 20 Chapter Opener John Barger/Alamy; 20.13 top left Maxim Tupikov/ Shutterstock; 2.13 top right MIMOHE/Shutterstock; p. 454 Brian Silliman; p. 454 Figure 1 Brian Silliman; p. 455 Figure 2 Brian Silliman; p. 455 Figure 3 Brian Silliman.

Chapter 21 Chapter Opener Justus de Cuveland/Getty Images; 21.2b Center for Food Safety & Applied Nutrition; 21.2d left Bill Curtsinger/Getty Images; 20.2d right David A. Harvey/Getty Images; 20.2c-L Jack Goldfarb/Design Pics Inc./Alamy; 21.2c right Vinicius Tupinamba/Shutterstock; 21.2 top right David Scharf/Science Source; 21.2a right Martin Pateman/Shutterstock; 21.3 Robert Leo Smith; 21.5b left Mark Gessner; 21.5a right Steve McBeath/Getty Images; 21.5a left Stockstudios/Shutterstock; 21.5b right Wayne A.Wurtsbaugh; 21.12 Kristin Piljay/Pearson Education bottom; 21.12 Cecelia Spitznas/Getty Images top; 21.12 David Scharf/Science Source midright; 21.12a Meredith Carlson/Science Source; p. 474 Edward Schuur; p. 474 Figure 1 Edward Schuur; p. 475 Figure 2 Edward Schuur; p. 475 Figure 3 Edward Schuur; 21.18 top left Science Source; 21.18 top right Kristin Piljay/Pearson Education; 21.18 middle left Meredith Carlson/Science Source; 21.18 bottom right M. I. Walker/Science Source; 21.18 bottom left Sinclair Stammers/Science Source; 21.27a Aflo/Nature Picture Library; 21.27b Thomas M. Smith; 21.30b Dusan Kostic/123RF; 21.30a Library of Congress Prints and Photographs Division[LC-USF34-027920-D].

Chapter 22 Chapter Opener Martin Shields/Alamy; 22.4 left Hugh Rose/Danita Delimont/Alamy; 22.4 right Ron Niebrugge/Alamy.

Chapter 23 Chapter Opener Mayovskyy Andrew/Shutterstock; 23.3a Thomas M. Smith; 23.3b Thomas M. Smith; 23.3c Thomas M. Smith; 23.4a Thomas M. Smith; 23.4b Thomas M. Smith; 23.3d Thomas M. Smith; 23.6a Dr. Morley Read/Shutterstock;

23.6b Nature Picture Library/Alamy; 23.7b Donyanedomam/Fotolia; 23.7a Anup Shah/ Nature Picture Library; 23.9 Hugh Lansdown/Shutterstock; 23.11a Dorival Moreira Lucin/AGE Fotostock; 23.11b Blue Gum Pictures/Alamy; 23.10a Erin Eldermire; 23.10b Erin Eldermire; 23.13 middle Nigel Dennis/AGE Fotostock; 23.13 left Exactostock/ SuperStock; 23.13 right Thomas M. Smith; 23.16b Clint Farlinger/Alamy; 23.16c Bill Brooks/Alamy; 23.16a NHPA/Photo Researchers/Photoshot; 23.17a Jim Parkin/ Shutterstock; 23.17b Galina Dreyzina/Shutterstock; 23.20b Australia Lands Department; 23.20a NASA; 23.21a Robert Leo Smith; 23.21b Ray Ellis/Science Source; 23.22 Chris Mattison/AGE Fotostock; 23.24c Gary Monroe;' 23.24a Lynn Watson; 23.24b Denise Dethlefsen/Alamy; 23.25 Martin Harvey/Alamy; 23.26a Bob Gibbons/Alamy; 23.26b George Ostertag/SuperStock; 23.29b Robert Leo Smith; 23.29a Corbis; 23.33 Robert Leo Smith; 23.35a Alan Majchrowicz/Alamy; 23.35b Neil Nightingale/Nature Picture Library; 23.36b Steven J. Kazlowski/Alamy; 23.37 Tom Uhlman/Alamy; 23.36a Kristin Piljay/Wanderlust Photos; 23.40b Stockbyte/Getty Images; 23.40a Joe Sohm/Alamy; 23.44 NASA;

Chapter 24 24.31a John S. Sfondilias/Shutterstock; 24.2b Steve Morgan/Alamy; 24.2 Evgeny Dubinchuk/Fotolia; 24.2d Robert Leo Smith; 24.2c James Steinberg/Science Source; 24.5a NHPA/M. I. Walker/Photoshot; 24.5b Laguna Design/Science Source; 24.6b Robert Leo Smith; 24.6a Michael P. Gadomski/Science Source; 24.7 Jean-Paul Ferrero/Auscape/The Image Works; 24.9b Vkimages/Shutterstock; 24.9a Hydromet/ Shutterstock; 24.10 Darren K. Fisher/Shutterstock; p. 547 Figure 2a Michael Doolittle/ Alamy; 24.14 Thomas M. Smith; 24.16 Marevision/AGE Fotostock; 24.15 South Carolina Department of Natural Resources; 24.19 National Marine Fisheries Service; 24.18 Steve Gschmeissner/Science Source; 24.23 Darryl Leniuk/Digital Vision/Getty Images; 24.22 Richard A. Lutz; 24.21 Verena Tunnicliffe; 24.26 SIPA USA/SIPA/Newscom.

Chapter 25 Chapter Opener Steven Kaufman/Photolibrary/Getty Images; 25.1a Thomas M. Smith; 25.3 Maine Department of Environmental Protection; 25.4 Radius/ Superstock; 25.6 Florida Images/Alamy; 25.5 Michael P. Gadomski/Science Source; 25.7 Dr. Carleton Ray/Science Source; 25.9 Mark Conlin/Alamy; 25.10 Mark Conlin/Alamy; 25.8 Robert Leo Smith; 25.11b Sergio Pitamitz/Robert Harding Picture Library Ltd/Alamy; 25.11a NASA; 25.14 Nancy Carter/North Wind Picture Archives/Alamy; 15.15b Michael P. Gadomski/Science Source; 15.15a USDA Forest Service Southern Region; 25.18a James Dyas Davidson/Getty Images; 25.16 Robert Leo Smith; 25.17 Science Source; 25.20 Larry Mayer/Getty Images; 25.21a J-C.&D. Pratt/Photononstop/Glow Images.

Chapter 26 Chapter Opener Jean E. Roche/Nature Picture Library.

Chapter 27 Chapter Opener Edmund Lowe Photography/Shutterstock; 27.13 Roque141/ Fotolia; 27.15 M.Sobreira/Alamy; 27.25b Gregg Henry; p. 612 Bernie Tershy; p. 612 Figure 1 Bernie Tershy; p. 613 Figure 2 Bernie Tershy; p. 613 Figure 3 Bernie Tershy.

INDEX

Note: Boldface page numbers refer to material contained in tables, boxes, figures, and legends.

B

D

E

F

G

Q

R

S

T

U

V

W